OXFORD MEDICAL PUBLICATIONS

Oxford Textbook of Psychiatry

Oxford Textbook of Psychiatry

MICHAEL GELDER
Professor of Psychiatry, University of Oxford

DENNIS GATH
Clinical Reader in Psychiatry, University of Oxford

RICHARD MAYOU
Clinical Reader in Psychiatry, University of Oxford

PHILIP COWEN
Medical Research Council Clinical Scientist, University of Oxford

THIRD EDITION

Oxford   New York   Melbourne
OXFORD UNIVERSITY PRESS

Oxford University Press, Great Clarendon Street, Oxford OX2 6DP
Oxford New York
Athens Auckland Bangkok Bogota Bombay Buenos Aires
Calcutta Cape Town Dar es Salaam Delhi Florence Hong Kong
Istanbul Karachi Kuala Lumpur Madras Madrid Melbourne
Mexico City Nairobi Paris Singapore Taipei Tokyo Toronto
and associated companies in
Berlin Ibadan

Oxford is a trade mark of Oxford University Press

Published in the United States
by Oxford University Press Inc., New York

©Michael Gelder, Dennis Gath,
Richard Mayou, and Philip Cowen 1983, 1989, 1996

First edition 1983
Second edition 1989
Third edition 1996, reprinted 1996

A catalogue record for this book is available from the British Library

Library of Congress Cataloging in Publication Data
(Data available)

ISBN 0 19 262500 4 (Hbk)
ISBN 0 19 262501 2 (Pbk)

Printed in Great Britain by Butler & Tanner Ltd, Frome, Somerset

Preface to the third edition

This third edition has been revised extensively to take account of new knowledge and advances in practice reported since the publication of the second edition. It includes the latest revisions of the *International Classification of Diseases* (ICD10) and of the *Diagnostic and Statistical Manual for Mental Disorders* (DSMIV). Least change has been made in Chapters 1 and 2, on signs and symptoms and the clinical method respectively; most revision has been made in Chapter 6 which is now concerned with stress-related and adjustment disorders, and Chapter 19 in which the account of psychiatric services has been revised extensively. In other chapters, necessary changes have been made to bring them up to date while preserving the original clinical descriptions. In this revision the authors of the second edition have been joined by Dr Philip Cowen.

Oxford
1995

In preparing this third edition we have been greatly helped by Dr R. S. Abell, Dr Gwen Adshead, Professor David Barlow, Dr Jennifer Barraclough, Professor Sidney Bloch, Dr David Clark, Dr Robert Cloninger, Professor John Cooper, Dr Tim Crow, Dr Christopher Fairburn, Dr Ray Fitzpatrick, Dr Bill Fulford, Dr David Geaney, Dr Guy Goodwin, Dr Paul Harrison, Dr Keith Hawton, Dr Tony Hope, Dr Robin Jacoby, Dr David Julier, Professor Eve Johnstone, Dr Harold Koenig, Professor Alwyn Lishman, Professor Joseph LoPiccolo, Dr Marie-Anne Martin, Dr Max Marshall, Professor Gethin Morgan, Professor William Parry-Jones, Dr Phil Robson, Professor Gene Paykel, Professor John Rush, Dr David Schaffer, Dr Michael Sharpe, Dr Alan Stein, Dr Greg Stores, Dr Eric Taylor, and Professor Peter Tyrer.

M.G.
D.G.
R.M.
P.C.

Preface to the second edition

In this edition the main aims and general approach are the same as in the first edition, but the text has been extensively revised. This revision has been undertaken with three aims: to incorporate the latest systems of classification, namely the draft ICD10 and DSMIIIR, in the sections dealing with clinical syndromes; to introduce advances in knowledge and practice; and to correct errors.

In preparing this second edition we have been greatly helped by:

Dr D. H. Clark
Dr P. J. Cowen
Dr K. E. Hawton
Dr D. Jones

Dr I. B. Glass
Dr A. Hope
Mr H. C. Jones
Professor D. Shaffer

Professor P. McGuffin
Dr G. Stores

Professor Sir David Weatherall

We acknowledge with gratitude the permission of the World Health Organization to quote material from the 1988 Draft of Chapter V of ICD10, Categories F00–F99, Mental, Behavioural and Developmental Disorders; Clinical Descriptions and Diagnostic Guidelines, World Health Organization, Division of Mental Health, Geneva, 1988 (MNH/MEP/87.1, Rev.2). The copyrights of ICD and the trial drafts are held by WHO. The authors wish to make it clear that the trial drafts are provisional and subject to alteration in the final version.

Oxford
1988

M.G.
D.G.
R.M.

Preface to the first edition

This book is written primarily as an introductory textbook for trainee psychiatrists, and also as an advanced textbook for clinical medical students. We hope that the book will also be useful, for purposes of revision and reference, to psychiatrists who have completed their training and to general practitioners and other clinicians.

The subject matter of this book is the practice of clinical psychiatry. Recent years have seen the increasing development of sub-specialties such as child and adolescent psychiatry, forensic psychiatry, and the psychiatry of mental retardation. This book is mainly concerned with general psychiatry, but it also contains chapters on the sub-specialties. Throughout the whole book, our purpose has been to provide an introduction to each subject, rather than a fully documented account. It is assumed that the trainee psychiatrist will go on to consult more comprehensive works such as the *Handbook of psychiatry* (Shepherd 1983), the *Comprehensive textbook of psychiatry* (Kaplan *et al.* 1980), and specialized textbooks dealing with the sub-specialties. In some chapters references are made to basic sciences such as psychology, genetics, biochemistry, and pharmacology. Discussion of these subjects is based on the assumption that the reader already has a working knowledge of them from previous study.

The chapters dealing with psychiatric treatment fall into two groups. First, there are three chapters wholly devoted to treatment and concerned only with general issues. In this group, Chapter 17 deals mainly with drug treatment and electroconvulsive therapy; Chapter 18 deals with psychological treatments; and Chapter 19 discusses the organization of services for the rehabilitation and care of patients with chronic psychiatric disorders. Second, there are the various chapters on individual syndromes, which include sections on the treatments specific to those syndromes. In

these chapters, treatment is us[...] two parts. The first part exam[...] that a particular treatment is [...]ed in particular syndrome; the second [...]ce (under the heading of managem[...] issues in treatment, such as ways of u[...] treatments singly or in combination [...] stages of a patient's illness. The sepa[...] chapters on general issues from those or[...] issues means that the reader has to consu[...] than one chapter for complete informati[...] the treatment of any disorder. None the les[...] arrangement is preferred because a si[...] treatment may be used for several syndrom[...] For example, antipsychotic drugs are used to treat mania and schizophrenia, and supportive psychotherapy is part of the treatment of many disorders.

In this book there is no separate chapter on the history of psychiatry. Instead certain chapters on specific topics include brief accounts of their history. For example, the chapter on psychiatric services contains a short historical review of the care of the mentally ill; and the chapter on abnormal personality includes some information about the development of ideas about the subject. This arrangement reflects the authors' view that, at least in an introductory text, historical points are more useful when related to an account of modern ideas. The historical references in this book can be supplemented by reading a history of psychiatry such as that by Ackerknecht (1968) or Bynum (1983).

The use of references in this book also needs to be explained. As this is an introductory postgraduate text, we have not provided references for every statement that could be supported by evidence. Instead we have generally followed two principles: to give references for statements that may be controversial; and to give more references for issues judged to be of topical interest. In an introductory text it also seemed appropriate to

Preface to the first

...y to the Anglo–American
...se reasons the coverage of the
give referenc...m uneven; but—as explained
literature...is written in the expectation that
literat...
ab...

the trainee psychiatrist will progress to other
works for more detailed literature surveys.
Suggestions for further reading are given at the
end of each chapter.

M.G.
D.G.
R.M.

Preface to the first edition

This book is written primarily as an introductory textbook for trainee psychiatrists, and also as an advanced textbook for clinical medical students. We hope that the book will also be useful, for purposes of revision and reference, to psychiatrists who have completed their training and to general practitioners and other clinicians.

The subject matter of this book is the practice of clinical psychiatry. Recent years have seen the increasing development of sub-specialties such as child and adolescent psychiatry, forensic psychiatry, and the psychiatry of mental retardation. This book is mainly concerned with general psychiatry, but it also contains chapters on the sub-specialties. Throughout the whole book, our purpose has been to provide an introduction to each subject, rather than a fully documented account. It is assumed that the trainee psychiatrist will go on to consult more comprehensive works such as the *Handbook of psychiatry* (Shepherd 1983), the *Comprehensive textbook of psychiatry* (Kaplan *et al.* 1980), and specialized textbooks dealing with the sub-specialties. In some chapters references are made to basic sciences such as psychology, genetics, biochemistry, and pharmacology. Discussion of these subjects is based on the assumption that the reader already has a working knowledge of them from previous study.

The chapters dealing with psychiatric treatment fall into two groups. First, there are three chapters wholly devoted to treatment and concerned only with general issues. In this group, Chapter 17 deals mainly with drug treatment and electroconvulsive therapy; Chapter 18 deals with psychological treatments; and Chapter 19 discusses the organization of services for the rehabilitation and care of patients with chronic psychiatric disorders. Second, there are the various chapters on individual syndromes, which include sections on the treatments specific to those syndromes. In

these chapters, treatment is usually discussed in two parts. The first part examines the evidence that a particular treatment is effective for a particular syndrome; the second part discusses (under the heading of management) practical issues in treatment, such as ways of using various treatments singly or in combination at different stages of a patient's illness. The separation of chapters on general issues from those on specific issues means that the reader has to consult more than one chapter for complete information on the treatment of any disorder. None the less this arrangement is preferred because a single treatment may be used for several syndromes. For example, antipsychotic drugs are used to treat mania and schizophrenia, and supportive psychotherapy is part of the treatment of many disorders.

In this book there is no separate chapter on the history of psychiatry. Instead certain chapters on specific topics include brief accounts of their history. For example, the chapter on psychiatric services contains a short historical review of the care of the mentally ill; and the chapter on abnormal personality includes some information about the development of ideas about the subject. This arrangement reflects the authors' view that, at least in an introductory text, historical points are more useful when related to an account of modern ideas. The historical references in this book can be supplemented by reading a history of psychiatry such as that by Ackerknecht (1968) or Bynum (1983).

The use of references in this book also needs to be explained. As this is an introductory postgraduate text, we have not provided references for every statement that could be supported by evidence. Instead we have generally followed two principles: to give references for statements that may be controversial; and to give more references for issues judged to be of topical interest. In an introductory text it also seemed appropriate to give references mostly to the Anglo–American literature. For all these reasons the coverage of the literature may seem uneven; but—as explained above—the book is written in the expectation that the trainee psychiatrist will progress to other works for more detailed literature surveys. Suggestions for further reading are given at the end of each chapter.

Oxford
May 1983

M.G.
D.G.
R.M.

Acknowledgements*

In writing this book we have been greatly helped by advice and comments generously given by colleagues. We wish to thank:

Dr S. Abel, Dr J. Bancroft, Mr J. Beatson, Miss V. L. Bellairs, Dr S. Bloch, Dr L. Braddocks, Dr S. Crown, Professor J. E. Cooper, Dr J. Corbett, Dr P. Cowen, Dr N. Eastman, Professor Griffith Edwards, Dr G. Forrest, Dr K. W. M. Fulford, Professor D. P. B. Goldberg, Dr G. Goodwin, Professor D. G. Grahame-Smith, Professor J. C. Gunn, Dr J. Hamilton, Dr K. E. Hawton, Dr A. Hope, Dr T. Horder, Professor R. E. Kendell, Professor I. Kolvin, Professor M. H. Lader, Dr J. P. Leff, Dr R. Levy, Dr P. F. Liddle, Professor W. A. Lishman, Professor H. G. Morgan, Dr J. McWhinnie, Dr D. J. Nutt, Dr W. Ll. Parry-Jones, Professor E. S. Paykel, Dr J. S. Pippard, Miss S. Rowland-Jones, Dr G. Stores, Dr C. A. Storr, Dr T. G. Tennent, Dr C. P. Warlow, Dr G. K. Wilcock, and Dr H. H. O. Wolff.

We are also grateful to many other colleagues who have given advice and to our secretaries. We are particularly indebted to Mrs Susan Offen who has given invaluable help at all stages of the preparation of the typescript and the checking of references.

*This Acknowledgements section is reprinted from the first edition of this book.

While every effort has been made to check drug dosages in this book, it is still possible that errors have been missed. Furthermore, dosage schedules are being continually revised and new side-effects recognized. For these reasons the reader is strongly urged to consult the drug companies' printed instructions before administering any of the drugs recommended in this book.

Contents

1 | *Signs and symptoms of mental disorder*

Psychiatry can be practised only if the psychiatrist develops two distinct capacities. One is the capacity to collect clinical data objectively and accurately by history taking and examination of mental state, and to organize the data in a systematic and balanced way. The other is the capacity for intuitive understanding of each patient as an individual. When the psychiatrist exercises the first capacity, he draws on his clinical skills and knowledge of clinical phenomena; when he exercises the second capacity, he draws on his general understanding of human nature to gain insights into the feelings and behaviour of each individual patient, and into ways in which life experiences have affected that person's development.

Both capacities can be developed by accumulating experience of talking to patients, and by learning from the guidance and example of more experienced psychiatrists. From a textbook, however, it is inevitable that the reader can learn more about clinical skills than about intuitive understanding. In this book several chapters are concerned with aspects of clinical skills. This emphasis on clinical skills in no way implies that intuitive understanding is regarded as unimportant but simply that it cannot be learnt from reading a textbook.

The psychiatrist can acquire skill in examining patients only if he has a sound knowledge of how each symptom and sign is defined. Without such knowledge, he is liable to misclassify phenomena and make inaccurate diagnoses. For this reason, questions of definition are considered in this first chapter before history taking and the examination of the mental state are described in the second.

Once the psychiatrist has elicited a patient's symptoms and signs, he needs to decide how far these phenomena resemble or differ from those of other psychiatric patients. In other words, he must determine whether the clinical features form a syndrome, which is a group of symptoms and signs that identifies patients with common features. When he decides on the syndrome, the psychiatrist combines observations of the patient's present state with information about the history of the disorder. The purpose of identifying a syndrome is to be able to plan treatment and predict the likely outcome by reference to accumulated knowledge about the causes, treatment, and outcome of the same syndrome in other patients. The principles involved are discussed in Chapter 4, which is concerned with classification, and also in the chapters dealing with the different syndromes.

Since the present chapter consists mainly of definitions and descriptions of symptoms and signs, it may be less easy to read than those which follow. It is suggested that the reader should approach the chapter in two stages. The first reading can be applied to the introductory sections and to a general understanding of the more frequent abnormal phenomena. The second can focus on details of definition and the less common symptoms and signs.

Before individual phenomena are described, it is important to consider some general issues concerning the methods of studying symptoms and signs and the terms used to describe them.

Psychopathology

The study of abnormal states of mind is known as **psychopathology**, a term that denotes three distinct approaches.

The first approach, **phenomenological psychopathology** (or **phenomenology**), is concerned with the objective description of abnormal states of mind in a way that avoids, as far as possible, preconceived theories. It aims to elucidate the basic data of psychiatry by defining the essential qualities of morbid mental experiences and by understanding what the patient is experiencing. It is entirely concerned with conscious experiences and observable behaviour. According to Jaspers (1963), phenomenology is 'the preliminary work of representing, defining and classifying psychic phenomena as an independent activity'.

The second approach, **psychodynamic psychopathology**, originates in psychoanalytical investigations. Like phenomenological psychopathology, it starts with the patient's description of his mental experiences and the doctor's observations of his behaviour. However, unlike phenomenological psychopathology, it goes beyond description and seeks to explain the causes of abnormal mental events, particularly by postulating unconscious mental processes. These differences can be illustrated by the two approaches to persecutory delusions. Phenomenology describes them in detail and examines how they differ from normal beliefs and from other forms of abnormal thinking such as obsessions. On the other hand, the psychodynamic approach seeks to explain the occurrence of persecutory delusions in terms of unconscious mechanisms such as repression and projection. In other words, it views them as evidence in the conscious mind of more important disorders in the unconscious.

In the third approach, often called **experimental psychopathology**, relationships between abnormal phenomena are examined by inducing a change in one of the phenomena and observing associated changes in the others. Hypotheses are formulated to explain the observed changes, and then tested in further experiments. The general aim is to explain the abnormal phenomena of mental disorders in terms of psychological processes that have been shown to account for normal experiences in healthy people. (An example is given on p. 17 where mood disorders are discussed.)

It should be noted that the term experimental psychopathology is also used to cover a wider range of experimental work that might throw light on psychiatric disorder. This usage includes studies of animals as well as of humans, for example studies of animal learning and behavioural responses to frustration or punishment.

This chapter is concerned mainly with phenomenological psychopathology, although reference will also be made to relevant ideas from dynamic or experimental psychopathology.

The most important exponent of phenomenological psychopathology was the German psychiatrist philosopher, Karl Jaspers. His classical work, *Allgemeine Psychopathologie* (*General psychopathology*), first appeared in 1913 and was a landmark in the development of clinical psychiatry. It provides the most complete account of the subject and contains much of interest, particularly in its early chapters. The seventh (1959) edition is available in an English translation by Hoenig and Hamilton (Jaspers 1963). Alternatively, useful outlines of the principles of phenomenology have been given by Scharfetter (1980), and by Sims (1988).

The significance of individual symptoms

It is often mistaken to conclude that a person has a mental disorder on the evidence of an individual symptom. Even hallucinations, which are generally regarded as hallmarks of mental disorder, are sometimes experienced transiently by healthy people. Symptoms are more likely to indicate mental disorder when they are intense and persistent. None the less, even when intense and persistent, a single symptom does not necessarily indicate mental disorder. It is the characteristic grouping of symptoms into a syndrome that is important.

Primary and secondary symptoms

The terms primary and secondary are used in describing symptoms, but with more than one meaning. The first is temporal: primary meaning antecedent, and secondary meaning subsequent. The second is causal: primary meaning a direct expression of the pathological process, and secondary meaning a reaction to the primary symptoms. The two meanings are often related—the symptoms appearing first in time being direct expressions of the pathological process.

It is preferable to use the terms primary and secondary in the temporal sense because this sense does not require an inference about causal mechanisms. However, many patients cannot give a clear account of the chronological development of their symptoms. In these cases a distinction between primary and secondary symptoms in the temporal sense cannot be made with certainty. If this happens, it is only possible to conjecture whether one symptom could be a reaction to another, for example whether the fixed idea of being followed by persecutors could be a reaction to hearing voices.

The form and content of symptoms

When psychiatric symptoms are described, it is usual to distinguish between form and content, a distinction that can be best explained by an example. If a patient says that, when he is entirely alone, he hears voices calling him a homosexual, then the form of his experience is an auditory hallucination (i.e. a sensory perception in the absence of an external stimulus) while the content is the statement that he is homosexual. A second person might hear voices saying that he is about to be killed; the form is still an auditory hallucination but the content is different. A third might experience repeated intrusive thoughts that he is homosexual but realize that these are untrue. He has an experience with the same content as the first (concerning homosexuality) but the form is different—in this case an obsessional thought. (In the case of delusions, the decision about form requires a knowledge of content since it involves a judgement as to whether the belief is false. Nevertheless, the distinction between form and content of delusions is useful.)

Description of symptoms and signs

Introduction

In the following sections symptoms and signs are described in a different order from the one adopted when the mental state is examined. The order is changed because it is useful to begin with the most distinctive phenomena—hallucinations and delusions. This change should be borne in mind when reading Chapter 2 in which the description of the mental state examination begins with behaviour and talk rather than hallucinations and delusions.

Before we consider individual symptoms it is appropriate to remind the reader that it is important not only to study individual mental phenomena but also to consider the whole person. The doctor must try to understand how the patient fulfils social roles such as worker, spouse, parent, friend, or sibling. He should consider what effect the disorders of function have had upon the remaining healthy parts of the person. Above all he should try to understand what it is like for this person to be ill, for example to care for small children while profoundly depressed or to live with the symptoms and disabilities of schizophrenia. The doctor will gain such understanding only if he is prepared to spend time listening to patients and their families, and to interest himself in every aspect of their lives.

Disorders of perception

Perception and imagery

Perception is the process of becoming aware of what is presented through the sense organs. **Imagery** is an experience within the mind, usually without the sense of reality that is part of perception. **Eidetic imagery** is a visual image which is so intense and detailed that it has a 'photographic' quality. Unlike perception, imagery can be called up and terminated by voluntary effort. It is usually obliterated by seeing or hearing. Occasionally, imagery is so vivid that it persists when the person looks at a poorly structured background such as plain wallpaper. This condition is called **pareidolia**, a state in which real and unreal percepts exist side by side, with the latter being recognized as unreal. Pareidolia can occur in acute organic disorders caused by fever, and in a few people it can be induced deliberately.

Alterations in perception

Perceptions can alter in intensity and quality. They can seem more intense than usual; for example when two people experience the same auditory stimulus, such as the noise of a door shutting, the more anxious person may perceive it as louder. In mania perceptions often seem very intense. Conversely, in depression colours may seem less intense. Changes in the quality of sensations occur in schizophrenia; they sometimes appear distorted or unpleasant. For example a patient may complain that food tastes bitter or that a flower smells like burning flesh.

Illusions

Illusions are misperceptions of external stimuli. They are most likely to occur when the general level of sensory stimulation is reduced. Thus at dusk a common illusion is to misperceive the outline of a bush as that of a man. Illusions are also more likely to occur when the level of consciousness is reduced, for example in an acute organic syndrome. Thus a delirious patient may mistake inanimate objects for people when the level of illumination is normal, though he is more likely to do so if the room is badly lit. Illusions occur when attention is not focused on the sensory modality or when there is a strong affective state ('affect illusions'); for example in a dark lane a frightened person is more likely to misperceive the outline of a bush as that of an attacker. (The so-called **illusion of doubles** or **Capgras syndrome** is not an illusion but a form of delusional misinterpretation. It is considered under paranoid syndromes (see p. 304).)

Hallucinations

A hallucination is a percept experienced in the absence of an external stimulus to the sense organs and with a similar quality to a true percept. A hallucination is experienced as originating in the outside world (or within one's own body) like a percept, and not within the mind like imagery.

Hallucinations are not restricted to the mentally ill. A few normal people experience them, especially when tired. Hallucinations also occur in healthy people during the transition between sleep and waking; they are called **hypnagogic** if experienced while falling asleep and **hypnopompic** if experienced during awakening.

Pseudohallucinations

This term has been applied to abnormal phenomena that do not meet the above criteria for hallucinations and are of less certain diagnostic significance. Unfortunately the word has two meanings which are often confused. The first, originating in the work of Kadinsky, was adopted by Jaspers (1913) in his book *General psychopathology*. In this sense, pseudohallucinations are especially vivid mental images, i.e. they lack the quality of representing external reality and seem to be within the mind rather than in external space. However, unlike ordinary imagery, they cannot be changed substantially by an effort of will. The term is still used with this meaning (e.g. Scharfetter 1980).

The second meaning of pseudohallucination is the experience of perceiving something as in the external world, while recognizing that there is no external correlate to the experience. This is the sense in which the term is used by Hare (1973) and Taylor (1979).

Both definitions are difficult to apply because they depend on patients' ability to give precise answers to difficult questions about the nature of their experience. Not surprisingly, judgements based on patients' recognition of the reality of their experience are difficult to make reliably because patients are often uncertain themselves. Although the percepts must be experienced as either in the external world or within the mind, patients often find this distinction difficult to make.

Table 1.1   *Description of hallucinations*

---

1. According to complexity
    Elementary
    Complex

---

2. According to sensory modality
    Auditory
    Visual
    Olfactory and gustatory
    Somatic (tactile and deep)

---

3. According to special features
    (a) Auditory: second-person
                      third-person
                      Gedankenlautwerden
                      écho de la pensée
    (b) Visual: extracampine

---

4. Autoscopic hallucinations

Taylor (1981) has suggested that the two kinds of pseudohallucination should be distinguished by separate names: 'imaged' pseudohallucinations that are experienced within the mind, and 'perceived' pseudohallucinations that are experienced as located in external space but recognized as unreal. In everyday clinical work it seems better to abandon the term pseudohallucinations altogether, and simply to maintain the term hallucination as defined at the beginning of this section. If the clinical phenomena do not meet this definition, they should be described in detail rather than labelled with a technical term that provides no additional information useful for diagnosis. Readers requiring a more detailed account of these problems of definition are referred to Hare (1973), Taylor (1981), and Jaspers (1963, pp. 68–74). Further information about the phenomena themselves will be found in Sedman (1966). Neither type of pseudohallucination is of diagnostic value.

Types of hallucination

Hallucinations can be described in terms of their complexity and their sensory modality (Table 1.1). The term **elementary hallucination** is used for experiences such as bangs, whistles, and flashes of light; **complex hallucination** is used for experiences such as hearing voices or music, or seeing faces and scenes.

Hallucinations may be auditory, visual, gustatory, olfactory, tactile, or of deep sensation. **Auditory hallucinations** may be experienced as noises, music, or voices. (Hallucinatory 'voices' are occasionally called **phonemes**.) Voices may be heard clearly or indistinctly, they may seem to speak words, phrases, or sentences, and they may seem to address the patient directly (**second-person hallucinations**) or talk to one another referring to the patient as 'he' or 'she' (**third-person hallucinations**). Sometimes voices seem to anticipate what the patient thinks a few moments later, or speak his own thoughts as he thinks them, or repeat them immediately after he has thought of them. In the absence of concise English technical terms, the last two experiences are sometimes called *Gedankenlautwerden* and *écho de la pensée* respectively.

Visual hallucinations may also be elementary or complex. They may appear normal or abnormal in size; if the latter, they are more often smaller than the corresponding real percept. Visual hallucinations of dwarf figures are sometimes called lilliputian. **Extracampine visual hallucinations** are experienced as located outside the field of vision, i.e. behind the head. **Olfactory and gustatory hallucinations** are frequently experienced together, often as unpleasant smells or tastes.

Tactile hallucinations, sometimes called **haptic hallucinations**, may be experienced as sensations of being touched, pricked, or strangled. They may also be felt as movements just below the skin which the patient may attribute to insects, worms, or other small creatures burrowing through the tissues. **Hallucinations of deep sensation** may occur as feelings of the viscera being pulled upon or distended, or of sexual stimulation or electric shocks.

An **autoscopic hallucination** is the experience of seeing one's own body projected into external space, usually in front of oneself, for short periods.

This experience may convince the person that he has a double (*doppelgänger*), a theme occurring in several novels, including Dostoevsky's *The Double*. In clinical practice this is a rare phenomenon, mainly encountered in a small minority of patients with temporal lobe epilepsy or other organic brain disorders (see Lukianowicz (1958) and Lhermitte (1951) for detailed accounts).

Occasionally, a stimulus in one sensory modality results in a hallucination in another, for example the sound of music may provoke visual hallucinations. This experience, sometimes called **reflex hallucinations**, may occur after taking drugs such as LSD or, rarely, in schizophrenia.

As already mentioned, **hypnagogic** and **hypnopompic hallucinations** occur at the point of falling asleep and of waking respectively. When they occur in healthy people, they are brief and elementary—for example hearing a bell ring or a name called. Usually the person wakes suddenly and recognizes the nature of the experience. In narcolepsy, such hallucinations are common but may last longer and be more elaborate.

Diagnostic associations

Hallucinations may occur in severe affective disorders, schizophrenia, organic disorders and dissociative states, and at times among healthy people. Therefore the finding of hallucinations does not itself help much in diagnosis. However, certain kinds of hallucination do have important implications for diagnosis.

Both the form and content of **auditory hallucinations** can help in diagnosis. Of the various types—noises, music, and voices—the only ones of diagnostic significance are voices heard as speaking clearly to or about the patient. As explained already, voices which appear to be talking to each other, referring to the patient in the third person (e.g. 'he is a homosexual') are called **third-person hallucinations**. They are associated particularly with schizophrenia. Such voices may be experienced as commenting on the patient's intentions (e.g. 'he wants to make love to her') or actions (e.g. 'she is washing her face'). Of all types of hallucination, commentary voices are most suggestive of schizophrenia.

Second-person hallucinations appear to address the patient (e.g. 'you are going to die') or give commands (e.g. 'hit him'). In themselves they do not point to a particular diagnosis, but their content and especially the patient's reaction may do so. For example voices with derogatory content suggest severe depressive disorder, especially when the patient accepts them as justified (e.g. 'you are wicked'). In schizophrenia the patient more often resents such comments.

Voices which anticipate, echo, or repeat the patient's thoughts also suggest schizophrenia.

Visual hallucinations may occur in hysteria, severe affective disorders, and schizophrenia, but they should always raise the possibility of an organic disorder. The content of visual hallucinations is of little significance in diagnosis.

Hallucinations of taste and smell are infrequent. When they do occur they often have an unusual quality which patients have difficulty in describing. They may occur in schizophrenia or severe depressive disorders, but they should also suggest temporal lobe epilepsy or irritation of the olfactory bulb or pathways by a tumour.

Tactile and somatic hallucinations are not generally of diagnostic significance, although a few special kinds are weakly associated with particular disorders. Thus hallucinatory sensations of sexual intercourse suggest schizophrenia, especially if interpreted in an unusual way (e.g. as resulting from intercourse with a series of persecutors). The sensation of insects moving under the skin occurs in people who abuse cocaine and occasionally among schizophrenics.

Perception and meaning

A percept has a meaning for the person who experiences it. In some psychiatric disorders an abnormal meaning may be associated with a normal percept. When this happens we speak of **delusional perception**. In some neurological disorders percepts lose their meaning. This is called **agnosia**. These abnormalities are considered further on pp. 12 and 20 respectively.

Table 1.2 *Disorders of thinking*

---

1. Particular kinds of abnormal thoughts
   Delusions
   Obsessions (see p. 15)

---

2. Disorders of the stream of thought (speed and pressure)

---

3. Disorder of the form of thought (linking of thoughts together)

---

4. Abnormal beliefs about the possession of thoughts

Disorders of thinking

Disorders of thinking are usually recognized from speech and writing. They can also be inferred from inability to perform tasks; thus one psychological test of thought disorder requires the person to sort objects into categories.

The term disorder of thinking can be used in a wide sense to denote four separate groups of phenomena (Table 1.2). The first group comprises particular kinds of abnormal thinking—delusions and obsessional thoughts. The second group, disorders of the stream of thought, is concerned with abnormalities of the amount and the speed of the thought experienced. The third group, known as disorders of the form of thought, is concerned with abnormalities of the ways in which thoughts are linked together. The fourth group, abnormal beliefs about the possession of thoughts, comprises unusual disturbances of the normal awareness that one's thoughts are one's own.

The second and third groups are considered here, whilst the first and last will be discussed later in the chapter.

Disorders of the stream of thought

In disorders of the stream of thought both the amount and the speed of thoughts are changed. At one extreme there is **pressure of thought**, when ideas arise in unusual variety and abundance and pass through the mind rapidly. At the other extreme there is poverty of thought, when the patient has only a few thoughts, which lack variety and richness, and seem to move through the mind slowly. The experience of pressure occurs in mania; that of poverty occurs in depressive disorders. Either may be experienced in schizophrenia.

The stream of thought can also be interrupted suddenly—a phenomenon which the patient experiences as his mind going blank, and which an observer notices as a sudden interruption in the flow of conversation. Minor degrees of this experience are common, particularly in people who are tired or anxious. By contrast, **thought blocking**, a particularly abrupt and complete interruption, strongly suggests schizophrenia. Because thought blocking has this importance in diagnosis, it is essential that it should be identified only when there is no doubt about its presence. Inexperienced interviewers often wrongly identify a sudden interruption of conversation as thought blocking. There are several other reasons why the flow of speech may stop abruptly: the patient may be distracted by another thought or an extraneous sound, or he may be experiencing one of the momentary gaps in the stream of thought that are normal in people who are anxious or tired. Thought blocking should only be identified when interruptions in speech are sudden, striking, and repeated, and when the patient describes the experience as an abrupt and complete emptying of his mind. The diagnostic association with schizophrenia is strengthened if the patient also interprets the experience in an unusual way, for example as having had his thoughts taken away by a machine operated by a persecutor.

Disorders of the form of thought

Disorders of the form of thought (*formal thought disorder*) can be divided into three subgroups, flight of ideas, perseveration, and loosening of associations. Each is related to a particular form of mental disorder, so that it is important to distinguish them, but in none of the three is the relationship strong enough to be regarded as diagnostic.

In **flight of ideas** the patient's thoughts and conversation move quickly from one topic to another so that one train of thought is not completed before another appears. These rapidly changing topics are understandable because the links between them are normal, a point that differentiates them from loosening of associations (see below). In practice, the distinction is often difficult to make, especially when the patient is speaking rapidly. For this reason it may be helpful to tape record a sample of speech and listen to it several times. The characteristics of flight of ideas are preservation of the ordinary logical sequence of ideas, using two words with a similar sound (clang associations) or the same word with a second meaning (punning), rhyming, and responding to distracting cues in the immediate surroundings. Flight of ideas is characteristic of mania.

Perseveration is the persistent and inappropriate repetition of the same thoughts. The disorder is detected by examining the person's words or actions. Thus, in response to a series of simple questions, the person may give the correct answer to the first but continue to give the same answer inappropriately to subsequent questions. Perseveration occurs in dementia but is not confined to this condition.

Loosening of associations denotes a loss of the normal structure of thinking. To the interviewer this appears as muddled and illogical conversation that cannot be clarified by further enquiry. Several features of this muddled thinking have been described (see below), but in the end it is usually the general lack of clarity in the patient's conversation that makes the most striking impression. This muddled thinking differs from that of people who are anxious or of low intelligence. Anxious people give a more coherent account when they have been put at ease, while those with subnormal intelligence can express ideas more clearly if the interviewer simplifies his questions. When there is loosening of associations, the interviewer has the experience that the more he tries to clarify the patient's thoughts the less he understands them. Loosening of associations occurs most often in schizophrenia.

Loosening of associations can take several forms. **Knight's move** or **derailment** refers to a transition from one topic to another, either between sentences or in mid-sentence, with no logical relationship between the two topics and no evidence of the forms of association described under flight of ideas. When this abnormality is extreme it disrupts not only the connections between sentences and phrases but also the finer grammatical structure of speech. It is then called **word salad**. The term **verbigeration** refers to a kind of stereotypy in which sounds, words, or phrases are repeated in a senseless way.

One effect of loosened associations on the patient's conversation is sometimes called **talking past the point** (also known by the German term *vorbeireden*). In this condition the patient seems always about to get near to the matter in hand but never quite reaches it.

Several attempts have been made to devise psychological tests to detect loosening of associations, but the results have not been particularly useful to the clinician. Attempts to use the tests to diagnose schizophrenia have failed.

In addition to these disorders of links between ideas, thoughts may become illogical through **widening of concepts,** i.e. the grouping together of things that are not normally regarded as closely connected with one another.

Neologisms

Although not a disorder of the form of thought, neologism is conveniently described here. In this abnormality of speech the patient uses words or phrases, invented by himself, often to describe his morbid experiences. Neologisms must be distinguished from incorrect pronunciation, the wrong use of words by people with limited education, dialect words, obscure technical terms, and the 'private words' which some families invent to amuse themselves. The interviewer should always record examples of the patient's words and ask what he means by them. Neologisms occur most often in chronic schizophrenia.

Explanations of thought disorder

Many attempts have been made to explain thought disorder but so far none has been convincing. Each of the theories focuses on a particular aspect of thought disorder and proposes that it is the basic problem. For example, Goldstein (1944) proposed that the basic problem is an inability to form abstract concepts ('concrete thinking'), Cameron (1938) emphasized the inability to form clear boundaries between concepts ('loosening of associations'), Payne and Friedlander (1962) suggested that concepts are too broad ('over-inclusive'), and Bannister (1962) first proposed that the concepts of thought-disordered people are less consistent than those of normal subjects and then devised a way of measuring this inconsistency (Bannister and Fransella 1966).

More recently enquiries have been directed less to thought disorder (which is inferred from speech or actions) and more to studies of speech itself. While these studies have confirmed the complexity of the speech disorder (and presumably therefore of thinking) they have not found an explanation of the disorder. It seems likely that the complexities of the disorders of both speech and thought in schizophrenia have more than one cause (Frith 1992).

Particular kinds of abnormal thoughts

Delusions

A delusion is a belief that is firmly held on inadequate grounds, is not affected by rational argument or evidence to the contrary, and is not a conventional belief that the person might be expected to hold given his educational and cultural background. This definition is intended to separate delusions, which are indicators of mental disorder, from other kinds of strongly held belief found among healthy people. A delusion is usually a false belief, but is not invariably so.

The hallmark of the delusion is that it is firmly held on inadequate grounds, i.e. the belief is not arrived at through normal processes of logical thinking. It is held with such conviction that it cannot be altered by evidence to the contrary. For example a patient who holds the delusion that there are persecutors in the adjoining house will not be convinced by evidence that the house is empty; instead he will retain his belief by suggesting, for example, that the persecutors left the house before it was searched. It should be noted that non-delusional ideas of normal people can sometimes be equally impervious to reasoned argument, for example certain shared beliefs of people with a common religious or ethnic background. Thus a person who has been brought up to believe in spiritualism is unlikely to change his convictions when presented with contrary evidence that convinces a non-believer.

Although delusions are usually false beliefs, in exceptional circumstances they can be true or subsequently become true. A well-recognized example is pathological jealousy (p. 301). A man may develop a jealous delusion about his wife in the absence of any reasonable evidence of infidelity. Even if the wife is actually being unfaithful at the time, the belief is still delusional if there is no rational ground for holding it. The point to stress is that it is not falsity that determines whether the belief is delusional, but the nature of the mental processes that led up to the belief. Conversely, it is a well-known pitfall of clinical practice to assume that a belief is false because it is odd, instead of checking the facts or finding out how the belief was arrived at. For example improbable stories of persecution by neighbours, or of attempts at poisoning by a spouse, may turn out to be arrived at through normal processes of logical thinking, and, in fact, to be correct.

The definition of a delusion emphasizes that the belief must be firmly held. However, the belief may not be so firmly held before or after the delusion has been fully formed. Although some delusions arrive in the patient's mind fully formed and with total conviction, other delusions develop more gradually. Similarly, during recovery from his disorder, a patient may pass through a stage of increasing doubt about his belief before finally rejecting it as false. The term *partial delusion* is sometimes used to denote these phenomena. It is safest to use the term partial delusion only when it is known to have been preceded by a full delusion or (with hindsight) to have later developed into a full delusion. Partial delusions are sometimes found during the early stages of schizophrenia. When partial delusions are met they cannot be given much weight in themselves, but a careful search should be made for other phenomena of mental illness.

Although a patient may be wholly convinced that a delusional belief is true, this conviction does not necessarily influence all his feelings and actions. This separation of belief from feeling and action is known as *double orientation*. It occurs most often in chronic schizophrenics. For example such a patient may believe that he is a member of a Royal Family while living contentedly in a hostel for discharged psychiatric patients.

Delusions must be distinguished from *over-valued ideas*, which were first described by Wernicke (1900). An overvalued idea is an isolated preoccupying belief, neither delusional nor obsessional in nature, which comes to dominate a person's life for many years and may affect his actions. The preoccupying belief may be understandable when the person's background is known. For example a person whose mother and sister suffered from cancer one after the other may become preoccupied with the conviction that cancer is contagious. Although the distinction between delusions and overvalued ideas is not always easy to make, this difficulty seldom leads to practical problems because diagnosis of mental illness depends on more than the presence or absence of a single symptom. (For further information about overvalued ideas the reader is referred to McKenna (1984).)

Delusions are of many kinds, which will now be described. In the following section, the reader may find it helpful to refer to Table 1.3.

Primary, secondary, and shared delusions

A **primary** or **autochthonous** delusion is one that appears suddenly and with full conviction but without any mental events leading up to it. For example a schizophrenic patient may be suddenly and completely convinced that he is changing sex, without ever having thought of it before and without any preceding ideas or events which could have led in any understandable way to this conclusion. The belief arrives in the mind suddenly, fully formed, and in a totally convincing form. Presumably it is a direct expression of the pathological process causing the mental illness—a primary symptom. Not all primary delusional experiences start with an idea; a **delusional mood** (see p. 12) or a **delusional perception** (see p. 12) can also arrive suddenly and without any antecedents to account for it. Of course, patients do not find it easy to remember the exact sequence of such unusual and often distressing mental events, and for this reason it is difficult to be certain what is primary. Inexperienced interviewers usually diagnose primary delusional experiences too readily because they do not probe carefully enough into the antecedents. Primary delusions are given considerable weight in the diagnosis of schizophrenia, and it is important not to record them unless they are present for certain.

Secondary delusions can be understood as derived from some preceding morbid experience. The latter may be of several kinds, such as a hallucination (e.g. someone who hears voices may come to believe that he is being followed), a mood (e.g. a person who is profoundly depressed may believe that people think he is worthless), or an existing delusion (e.g. a person with the delusion that he has lost all his money may come to believe he will be put in prison for failing to pay debts). Some secondary delusions seem to have an integrative function, making the original experiences more comprehensible to the patient, as in the first example above. Others seem to do the opposite, increasing the sense of persecution or failure, as in the third example.

The accumulation of secondary delusions may result in a complicated **delusional system** in which each belief can be understood as following from the one before. When a complicated set of interrelated beliefs of this kind has developed the delusions are sometimes said to be **systematized**.

Shared delusions: as a rule, other people recognize delusions as false and argue with the patient in an attempt to correct them. Occasionally, a person who lives with a deluded patient comes to share his delusional beliefs. This condition is known as shared delusions or *folie à deux*. Although the second person's delusional conviction is as strong as the partner's while the couple remain together, it

Table 1.3 *Descriptions of delusions*

---

1. According to fixity
   Complete
   Partial

---

2. According to onset
   Primary
   Secondary

---

3. Other delusional experiences
   Delusional mood
   Delusional perception
   Delusional memory

---

4. According to theme
   Persecutory (paranoid)
   Delusions of reference
   Grandiose (expansive)
   Delusions of guilt and worthlessness
   Nihilistic
   Hypochondriacal
   Religious
   Jealous
   Sexual or amorous
   Delusions of control
   Delusions concerning possession of thought
      Thought insertion
      Thought withdrawal
      Thought broadcasting

---

5. According to other features
   Shared delusions

---

often recedes quickly when they are separated. The condition is described more fully on p. 305.

Delusional moods, perceptions, and memories

As a rule, when a patient first experiences a delusion he also has an emotional response and interprets his environment in a new way. For example a person who believes that a group of people intend to kill him is likely to feel afraid. At the same time he may interpret the sight of a car in his driving mirror as evidence that he is being followed. In most cases the delusion comes first and the other components follow.

Occasionally the order is reversed: the first experience is change of mood, often a feeling of anxiety with the foreboding that some sinister event is about to take place, and the delusion follows. For example after a sense of foreboding lasting for several hours the person concludes that there is a plot to assassinate him. In German this change of mood is called *Wahnstimmung*, a term usually translated as **delusional mood**. The latter term is unsatisfactory because there is really a mood from which a delusion arises. At other times, the first change may be attaching a new significance to a familiar percept without any reason. For example a new arrangement of objects on a colleague's desk may be interpreted as a sign that the patient has been chosen to do God's work. This is called **delusional perception**; this term is also unsatisfactory, since it is not the patient's perceptions that are abnormal but the false meaning that has been attached to a normal percept. Although both terms are less than satisfactory, there is no generally agreed alternative and they have to be used if the experience is to be labelled. However, it is usually better simply to describe what the patient has experienced and to record the order in which changes have occurred in beliefs, affect, and the interpretation of sense data.

In another disorder a patient sees a familiar person and believes him to have been replaced by an impostor who is the exact double of the original. This symptom is sometimes described by the French term *l'illusion de sosies* (illusion of doubles), but of course it is a delusion, not an illusion. The symptom may be so persistent that a syndrome, the **Capgras syndrome**, has been described in which it is the central feature (see p. 304). The opposite false interpretation of experience occurs when a patient recognizes a number of people as having different appearances but believes that they are a single persecutor in disguise. This abnormality is called the **Fregoli delusion**. It is described further on p. 304.

Finally, some delusions concern past rather than present events, and are known as **delusional memories**. For example if a patient believes that there is a plot to poison him, he may attribute new significance to the memory of an occasion when he vomited after eating a meal, long before his delusional system began. This experience has to be distinguished from the accurate recall of a delusional idea formed at the time. The term is unsatisfactory because it is not the memory that is delusional but the interpretation that has been applied to it.

Delusional themes

For the purposes of clinical work, delusions are grouped according to their main themes. This grouping is useful because there is some correspondence between themes and the major forms of mental illness. However, it is important to remember that there are many exceptions to the broad associations mentioned below.

Persecutory delusions are often called **paranoid**, a term which strictly speaking has a wider meaning. The term paranoid was used in ancient Greek writings in the modern sense of 'out of his mind', and Hippocrates used it to describe febrile delirium. Many later writers applied the term to grandiose, erotic, jealous, and religious, as well as persecutory, delusions. For this reason, it is preferable not to use the term paranoid to describe a persecutory delusion. However, the term paranoid applied in its wide sense to symptoms, syndromes, and personality types retains its usefulness (see Chapter 10).

Persecutory delusions are most commonly concerned with persons or organizations that are thought to be trying to inflict harm on the patient, damage his reputation, make him insane, or poison him. Such delusions are common but of little help in diagnosis, for they can occur in organic states, schizophrenia, and severe affective disorders. However, the patient's attitude to the delusion may point to the diagnosis: in a severe depressive disorder he characteristically accepts the supposed activities of the persecutors as justified by his own guilt and wickedness, but in schizophrenia he resents them, often angrily. In assessing such ideas, it is essential to remember that apparently improbable accounts of persecution are sometimes true, and that it is normal in certain cultures to believe in witchcraft and to ascribe misfortune to the malign activities of other people.

Delusions of reference are concerned with the idea that objects, events, or people have a personal significance for the patient; for example an article read in a newspaper or a remark heard on television is believed to be directed specifically to himself. Alternatively, a radio play about homosexuals is thought to have been broadcast in order to tell the patient that everyone knows that he is a homosexual. Delusions of reference may also relate to actions or gestures made by other people which are thought to convey something about the patient; for example people touching their hair may be thought to signify that the patient is turning into a woman. Although most delusions of reference have persecutory associations, they may also relate to grandiose or reassuring themes.

Grandiose or expansive delusions are beliefs of exaggerated self-importance. The patient may think himself wealthy, endowed with unusual abilities, or a special person. Such ideas occur in mania and in schizophrenia.

Delusions of guilt and worthlessness are found most often in depressive illness, and therefore are sometimes called depressive delusions. Typical themes are that a minor infringement of the law in the past will be discovered and bring shame upon the patient, or that his sinfulness will lead to divine retribution on his family.

Strictly speaking, **nihilistic delusions** are beliefs about the non-existence of some person or thing, but their meaning is extended to include pessimis-tic ideas that the patient's career is finished, that he is about to die, that he has no money, or that the world is doomed. Nihilistic delusions are associated with extreme degrees of depressed mood. Comparable ideas concerning failures of bodily function (e.g. that the bowels are blocked with putrefying matter) often accompany nihilistic delusions. The resulting clinical picture is called **Cotard's syndrome** after the French psychiatrist who described it (Cotard 1882). The condition is considered further in Chapter 10.

Hypochondriacal delusions are concerned with illness. The patient may believe wrongly, and in the face of all medical evidence to the contrary, that he is ill. Such delusions are more common in the elderly, reflecting the increasing concern with health among mentally normal people at this time of life. Other delusions may be concerned with cancer or venereal disease, or with the appearance of parts of the body, especially the nose. Patients with delusions of the last kind sometimes request plastic surgery (see p. 355).

Religious delusions: delusions with a religious content were much more frequent in the nineteenth century than they are today (Klaf and Hamilton 1961), presumably reflecting the greater part that religion played in the life of ordinary people in the past. When unusual and firmly held religious beliefs are encountered among members of minority religions, it is advisable to speak to another member of the group before deciding whether the ideas (e.g. apparently extreme ideas about divine punishment for minor sins) are abnormal or not.

Delusions of jealousy are more common among men. Not all jealous ideas are delusions; less intense jealous preoccupations are common, and some obsessional thoughts are concerned with doubts about the spouse's fidelity. However, when the beliefs are delusional they have particular importance because they may lead to dangerously aggressive behaviour towards the person thought to be unfaithful. Special care is needed if the patient follows the spouse to spy on her, examines her clothes for marks of semen, or searches her handbag for letters. A person with delusional jealousy will not be satisfied if he fails to find evidence supporting his beliefs; his search will continue. These important problems are discussed further in Chapter 10.

Sexual or amorous delusions: both sexual and amorous delusions are rare but when they occur, they are more frequent among women. Delusions concerning sexual intercourse are often secondary to somatic hallucinations felt in the genitalia. A woman with amorous delusions believes that she is loved by a man who is usually inaccessible, of higher social status, and to whom she has never even spoken. Erotic delusions are the most prominent feature of **De Clérambault's syndrome** which is discussed in Chapter 10.

Delusions of control: the patient who has a delusion of control believes that his actions, impulses, or thoughts are controlled by an outside agency. Because the symptom strongly suggests schizophrenia, it is important not to record it unless it is definitely present. A common error is to diagnose it when it is not present. Sometimes the symptom is confused with the experience of hearing hallucinatory voices giving commands that the patient obeys voluntarily. At other times it is misdiagnosed because the patient has mistaken the question for one about religious beliefs concerning the divine control of human actions. The patient with a delusion of control firmly believes that individual movements or actions have been brought about by an outside agency, for example that his arms are moved into the position of crucifixion not because he willed them to do so, but because an outside force brought it about.

Delusions concerning the possession of thoughts: healthy people take it for granted that their thoughts are their own. They also assume that thoughts are private experiences that can be known to other people only if spoken aloud, or revealed by facial expression, gesture, or action. Patients with delusions about the possession of thoughts may lose these convictions in several ways.

Delusions of thought insertion are a person's beliefs that some of his thoughts are not his own but have been implanted by an outside agency. Often there are associated explanatory delusions, for example that persecutors have used radio waves to insert the thoughts. This experience differs from that of the obsessional patient who may be distressed by unpleasant thoughts but

never doubts that they originate within his own mind. As Lewis (1957) said, obsessional thoughts are 'home made but disowned'. The patient with a delusion of thought insertion will not accept that the thoughts have originated in his own mind.

Delusions of thought withdrawal are beliefs that thoughts have been taken out of the mind. The delusion usually accompanies thought blocking, so that the patient experiences a break in the flow of thoughts through the mind and believes that the 'missing' thoughts have been taken away by some outside agency. Often there are associated explanatory delusions similar to those accompanying delusions of thought insertion.

Delusions of thought broadcasting are the patient's belief that unspoken thoughts are known to other people through radio, telepathy or in some other way. Some patients also believe that their thoughts can be heard by other people (a belief which also accompanies the experience of hearing one's own thoughts spoken (*Gedankenlautwerden*)). Explanatory delusions often occur with delusions of thought broadcasting.

All three of these symptoms occur much more commonly in schizophrenia than in any other disorder.

The causes of delusions

So little is known about the process by which normal beliefs are formed and tested against evidence that it is not surprising that we are ignorant about the cause of delusions. However, this lack of knowledge has not prevented the development of several theories, mainly concerned with persecutory delusions.

One theory was proposed by Freud (1958):

the study of a number of cases of delusions of persecution has led me as well as other investigators to the view that the relation between the patient and his persecutor can be reduced to a simple formula. It appears that the person to whom the delusion ascribes so much power and influence is either identical with someone who played an equally important part in the patient's emotional life before illness, or an easily recognizable substitute for him. The intensity of the emotion is projected in the shape of external power, whilst its quality is changed into the opposite. The person who is now hated and feared for being a persecutor was at one time loved and honoured. The main purpose of the persecution asserted by the patient's delusion is to justify the change in his emotional attitude.

Freud further summarized his view as follows: delusions of persecution are the result of the sequence 'I do not *love* him—I *hate* him, because he persecutes me', erotomania of the sequence 'I do not love *him*—I love *her*, because *she loves me*', and delusions of jealousy of the sequence 'It is not I who loved the man—*she* loves him (Freud 1958, pp. 63–4, emphases in the original).

This hypothesis suggests that patients who experience persecutory delusions have repressed homosexual impulses. Attempts to test this theory have not produced convincing evidence in its favour (Arthur 1964). The theory that persecutory ideas involve the defence mechanism of projection can be more convincing when applied to the persecutory ideas of people with abnormal personalities (p. 111) than when applied to persecutory delusions.

Several existential analyses of delusions have been made. These describe in detail the experience of the deluded patient and make the important point that the delusion affects the whole being—it is not just an isolated symptom. Conrad (1958), using the approach of Gestalt psychology, described the delusional experience as having four stages starting from a delusional mood which he called trema (fear and trembling) and leading via the delusional idea which he called apophenia (the appearance of the phenomenon) to the person's efforts to make sense of the experience by revising his whole view of the world. These efforts break down in the last stage (apocalypse) when thought disorder and behavioural symptoms appear. While a sequence of this kind can be observed in a few patients it is certainly not invariable.

Attempts have been made to explain delusions as some kind of general failure of logical reasoning (Helmsley and Garety 1986). These attempts are not convincing because deluded patients can reason logically about matters unconnected with their delusions.

Obsessional and compulsive symptoms

These symptoms are more common than delusions but generally of less serious significance. Obsessional and compulsive symptoms are best described separately, although they often occur together.

Obsessions are recurrent persistent thoughts, impulses, or images that enter the mind despite the person's efforts to exclude them. The characteristic feature is the subjective sense of a struggle—the patient resisting the obsession which nevertheless intrudes into his awareness. Obsessions are recognized by the person as his own and not implanted from elsewhere. They are often regarded by him as untrue or senseless—an important point of distinction from delusions. They are generally about matters which the patient finds distressing or otherwise unpleasant.

The presence of resistance is important because, together with the lack of conviction about the truth of the idea, it distinguishes obsessions from delusions. However, when obsessions have been present for a long time, the amount of resistance often becomes less. This seldom causes diagnostic difficulties because, by the time it happens, the nature of the symptom has usually been established.

Obsessions can occur in several forms (Table 1.4). Obsessional thoughts are repeated and intrusive words or phrases which are usually upsetting to the patient, for example repeated obscenities or blasphemous phrases coming into the awareness of a religious person. Obsessional **ruminations** are repeated worrying themes of a more complex kind, for example about the ending of the world. Obsessional **doubts** are repeated themes expressing uncertainty about previous actions, for example whether or not the person turned off an electrical appliance that might cause a fire. Whatever the nature of the doubt, the person realizes that the action has, in fact, been completed safely. Obsessional **impulses** are repeated urges to carry out actions, usually actions that are aggressive,

Table 1.4 *Obsessional and compulsive symptoms*

1. Obsessions: thoughts
   ruminations
   doubts
   impulses
   obsessional phobias

2. Compulsions (rituals)

3. Obsessional slowness

dangerous, or socially embarrassing. Examples are the urge to pick up a knife and stab another person, to jump in front of a train, or to shout obscenities in church. Whatever the urge, the person has no wish to carry it out, resists it strongly, and does not act on it.

Obsessional phobia is an unsatisfactory term that is used to denote obsessional symptoms associated with avoidance as well as anxiety, for example obsessional impulses to injure another person with a knife may lead to consequent avoidance of knives. Sometimes obsessional fears of illness are called phobias, but this usage is confusing (see below under phobias).

Although the themes of obsessions are various, most can be grouped into one or other of six categories: dirt and contamination, aggression, orderliness, illness, sex, and religion. Thoughts about **dirt** and **contamination** are usually associated with the idea of harming others through the spread of disease. **Aggressive** thoughts may be about striking another person or shouting angry or obscene remarks in public. Thoughts about **orderliness** may be about the way objects are to be arranged or work is to be organized. Thoughts about **illness** are usually of a fearful kind, for example a dread of cancer or venereal disease. This fearfulness has resulted in the name **illness phobia**, but this term should be avoided because the phenomena are not examples of anxiety arising in specific situations (which is the hallmark

of a phobia (see below)). Obsessional ideas about sex usually concern practices which the patient would find shameful, such as anal intercourse. Obsessions about **religion** often take the form of doubts about the fundamentals of belief (e.g. 'does God exist?') or repeated doubts about whether sins have been adequately confessed (**scruples**).

Compulsions are repetitive and seemingly purposeful behaviours, performed in a stereotyped way (hence the alternative name of **compulsive rituals**). They are accompanied by a subjective sense that they must be carried out and by an urge to resist. Like obsessions, compulsions are recognized as senseless. A compulsion is usually associated with an obsession as if it has the function of reducing the distress caused by the latter. For example, a handwashing compulsion often follows obsessional thoughts that the hands are contaminated with faecal matter. Occasionally, however, the only associated obsession is an urge to carry out the compulsive act.

Compulsive acts are of many kinds, but three are particularly common. Checking rituals are often concerned with safety, for example checking over and over again that a gas tap has been turned off. **Cleaning** rituals often take the form of repeated handwashing but may also involve household cleaning. **Counting** rituals may be spoken aloud or rehearsed silently. They often involve counting in a special way, for example in threes, and are frequently associated with doubting thoughts such that the count must be repeated to make sure that it was carried out adequately in the first place. In **dressing** rituals the person has to lay out his clothes in a particular way or put them on in a special order. Again, the ritual is often accompanied by doubting thoughts that lead to seemingly endless repetition. In severe cases patients may take several hours to put on their clothes in the morning.

Obsessional slowness is usually the result of compulsive rituals or repeated doubts, but it can occur occasionally without them (primary obsessional slowness).

The differential diagnosis of obsessional thoughts is from the ordinary preoccupations of healthy people, from the repeated concerns of anxious and depressed patients, from the recurring ideas and urges encountered in sexual deviations or drug dependency, and from delusions. Ordinary preoccupations do not have the same insistent quality and can be resisted by an effort of will. Many anxious or depressed patients experience intrusive thoughts (for example the anxious person may think that he is about to faint, or the depressed person that he has nothing to live for), but they do not find these ideas unreasonable and they do not resist them. Similarly, sexual deviants and drug-dependent people often experience insistent ideas and images concerned with their sexual practices or habits of drug taking, but these ideas are usually welcomed rather than resisted. Likewise, delusions are not resisted and are firmly held to be true.

Theories about the **aetiology** of obsessions are discussed on p. 181–4, where obsessional neuroses are considered.

Disorders of mood

In psychiatry two terms are used to refer to an emotional state—mood and affect. The words are often used interchangeably, but increasingly the word mood is preferred. Both ICD10 and DSMIV have adopted this usage. In mental disorder, mood may be abnormal in three ways: its nature may be altered, it may fluctuate more or less than usual, and it may be inconsistent with the patient's thoughts or actions or with current events.

Changes in the nature of mood can be towards anxiety, depression, elation, or anger. Changes in any of these emotions may be associated with an obvious cause in the person's life, or may arise without reason. Mood disorders usually include several components other than the mood change itself. Thus feelings of anxiety are usually accompanied by autonomic over-activity and increased muscle tension, and feelings of depression by gloomy preoccupations and psychomotor slowness. These other features are part of the syndromes of anxiety and depressive disorders, and as such are described in later chapters.

Abnormal fluctuation of mood may take the extreme form of total loss of emotion and inability to feel pleasure. The latter is sometimes called **apathy** (i.e. without feeling), a sense of the word that contrasts with the everyday usage of indolence or lack of initiative. When the normal variation of emotion is reduced rather than lost, affect is described as **blunted** or **flattened**. When emotions change in an excessively rapid and abrupt way, affect is said to be **labile**. When mood changes are very marked, the term **emotional incontinence** is sometimes used.

Normally, emotional expression seems appropriate to a person's circumstances (e.g. looking sad after a loss) and congruent with his thoughts and actions (when a person looks sad he is likely to be thinking gloomy thoughts). In psychiatric disorders, there may be **incongruity** of mood (or affect). For example a patient may laugh when describing the death of his mother. Such incongruity must be distinguished from laughter indicating that someone is ill at ease when talking about a distressing topic. It should be noted that failure to show emotion in distressing circumstances, although equally incongruous in the everyday sense, is called flattening of affect, not incongruity.

Disorders of mood are found in all kinds of psychiatric disorder. They are the central feature of the mood (affective) disorders (depression and elation) and of anxiety disorders. They are also common in other neuroses, organic disorders, and schizophrenia.

Experimental psychopathology of mood

Some of the most fruitful applications of experimental psychopathology have been to anxiety and depressive disorders. In anxiety disorders, research has focused on ways in which thinking about symptoms can increase and prolong anxiety. Thus anxious patients frequently think that physical symptoms such as palpitations signify an impending heart attack, that feelings of dizziness are a precursor of losing consciousness, or that increasing mental tension will lead to loss of control (Beck *et al.* 1974*a*; Hibbert 1984*a*). Changing these conditions through cognitive therapy (see p. 638) appears to improve the outcome of anxiety neuroses.

Experimental studies of the psychopathology of depression have been mainly concerned with the relationship between mood and memory. Depressive mood change, whether in normal states of sadness or in depressive disorders, is associated with greater accessibility of unhappy as compared with happy memories (Teasdale and Fogarty 1979; Clark and Teasdale 1982). Since thinking about unhappy events leads to depressive mood (an everyday observation that has been confirmed experimentally by Teasdale and Bancroft 1977) a circular process may ensue, leading to progressive deterioration of mood. (See Teasdale (1983) or Williams *et al.* (1988) for a further account of these ideas.)

Phobias

A phobia is a persistent irrational fear of and wish to avoid a specific object, activity, or situation. The fear is irrational in the sense that it is out of proportion to the real danger and is recognized as such by the person experiencing it. The person finds it difficult to control his fear and often tries to avoid the feared objects and situations if possible. The object that provokes the fear may be a living creature such as a dog, snake, or spider, or a natural phenomenon such as thunder. Fear-provoking situations include high places, crowds, and open spaces. Phobic patients feel anxious not only in the presence of the objects or situations but also when thinking about them (**anticipatory anxiety**).

Isolated phobic symptoms are common among normal people and have been described since the earliest medical writings (see Lewis (1976) or Errera (1962) for a historical account). The variety of feared objects and situations is great. In the past, Greek names were given to each one (Pitres and Régis (1902) labelled some seventy in this way), but there is nothing to be gained by this practice.

As pointed out earlier, obsessional thoughts leading to anxiety and avoidance are often called **obsessional phobias**; for example a recurrent thought about doing harm with knives is sometimes called a phobia of knives because the person is anxious in the presence of these objects and avoids them. Similarly, obsessional thoughts about illness are sometimes called illness phobias (e.g. 'I may have cancer'). Strictly speaking, neither of these symptoms is a phobia, and nor is dysmorphophobia, which is a disorder of bodily awareness (see p. 355).

Depersonalization and derealization

Depersonalization is a change of self-awareness such that the person feels unreal. Those who have this condition find it difficult to describe, often speaking of being detached from their own experience and unable to feel emotion. A similar change in relation to the environment is called **derealization**. In this condition, objects appear unreal and people appear as lifeless two-dimensional 'cardboard' figures. Despite the complaint of inability to feel emotion, both depersonalization and derealization are described as highly unpleasant experiences.

These central features are often accompanied by other morbid experiences. There is some disagreement as to whether these experiences are part of depersonalization and derealization or separate symptoms. These accompanying features include changes in the experience of time, changes in the body image such as a feeling that a limb has altered in size or shape, and occasionally a feeling of being outside one's own body and observing one's own actions, often from above. These features do not occur in every case (Ackner 1954*a*).

Because patients find it difficult to describe the feelings of depersonalization and derealization, they often resort to metaphor. Unless careful enquiry is made, this can lead to confusion between descriptions of depersonalization and of delusional ideas. For example, a patient's description of depersonalization may be 'as if part of my brain had stopped working', or of derealization 'as if the people I meet are lifeless creatures'—statements which must be explored carefully to distinguish them from delusional beliefs that the brain is no longer working or that people have really changed. At times, this distinction may be very difficult to make.

Depersonalization and derealization are experienced quite commonly as transient phenomena by healthy adults and children, especially when tired. The experience usually begins abruptly and in normal people seldom lasts more than a few minutes (Sedman 1970). The symptoms have been reported after sleep deprivation (Bliss *et al.* 1959), after sensory deprivation (Reed and Sedman 1964), and as an effect of hallucinogenic drugs (Guttman and Maclay 1936). The symptoms also occur in many psychiatric disorders when they may be persistent, sometimes lasting for years. They are particularly associated with generalized and phobic anxiety disorders, depressive disorders, and schizophrenia. Depersonalization has also been described in epilepsy, especially the kind arising in the temporal lobe. Some psychiatrists, notably Shorvon *et al.* (1946), have described a separate depersonalization syndrome (see p. 195). Because depersonalization and derealization occur in so many disorders, they do not help in diagnosis.

There are several **aetiological theories** about depersonalization. Mayer-Gross (1935) proposed that it is a 'preformed functional response of the brain' in the sense that an epileptic fit is a preformed response. Others have suggested that depersonalization is a response to alterations in consciousness (which is consistent with its appearance during fatigue and sleep deprivation in normal people).

A third suggestion is that depersonalization occurs when anxiety becomes excessive. Thus Lader and Wing (1966) described one anxious patient who developed depersonalization during an experiment in which skin conductance and heart rate were being measured. An accompanying fall in these measures suggested that depersonalization might have been an expression of some mechanism that reduced anxiety. However, depersonalization can occur when consciousness is normal and anxiety is absent so that, at best, these ideas can explain only a proportion of cases. Moreover, in states with undoubted changes in consciousness (acute organic psychosyndromes) depersonalization is found in only a minority of patients. The same argument can be applied to states of anxiety. Other writers have suggested that

depersonalization is the expression of a disorder of perception mechanisms, and some psychoanalytic authors regard it as a defence against emotion. These various theories, none of which is satisfactory, have been reviewed by Sedman (1970).

Motor symptoms and signs

Abnormalities of social behaviour, facial expression, and posture occur frequently in mental illness of all kinds. They are discussed in Chapter 3 where the examination of the patient is considered. There are also a number of specific motor symptoms. With the exception of tics, these symptoms are mainly observed among schizophrenic patients. They are described briefly here for reference, and their clinical associations are discussed in Chapter 9.

Tics are irregular repeated movements involving a group of muscles, for example sideways movement of the head or the raising of one shoulder. **Mannerisms** are repeated movements that appear to have some functional significance, for example saluting. **Stereotypies** are repeated movements that are regular (unlike tics) and without obvious significance (unlike mannerisms), for example rocking to and fro. **Posturing** is the adoption of unusual bodily postures continuously for a long time. The posture may appear to have a symbolic meaning, for example standing with both arms outstretched as if being crucified, or may have no apparent significance, for example standing on one leg. Patients are said to show **negativism** when they do the opposite of what is asked and actively resist efforts to persuade them to comply. **Echopraxia** is the imitation of the interviewer's movement automatically even when asked not to do so. Patients are said to exhibit **ambitendence** when they alternate between opposite movements, for example putting out the arm to shake hands, then withdrawing it, extending it again, and so on. **Waxy flexibility** is detected when a patient's limbs can be placed in a position in which they then remain for long periods whilst at the same time muscle tone is uniformly increased.

Disorders of the body image

The body image or body schema is a person's subjective representation against which the integrity of his body is judged and the movement and positioning of its parts assessed. To the earlier neurologists the body schema was a postural model (Head 1920). Schilder (1935), in his book *The image and appearance of the human body*, argued that this postural model is only the lowest level of organization of the body schema, and that there are also higher psychological levels founded on emotion, personality, and social interaction.

It is certainly true that, in clinical practice, abnormalities of body image are encountered that affect far more than the appreciation of posture and movement. These abnormalities arise in neurological as well as psychiatric disorders, and in many cases organic and psychological factors appear to be acting together. Unfortunately, in neither neurological nor psychiatric disorders are the causes of body image disturbances understood completely. In the account that follows, we follow broadly the scheme proposed by Lishman (1987) and we recommend the relevant sections of his book (pp. 59–66) to the reader who requires more detailed information about these disorders.

Phantom limb is a continuing awareness of a part of the body that has been lost. As such, it is perhaps the most convincing evidence for the concept of a body schema. It usually occurs after limb amputation, but has been reported after removal of breasts, genitalia, or eyes (Lishman 1987, p. 91). Phantom limbs may be experienced as painful. The phantom limb is usually present immediately after amputation and normally fades gradually, although a minority persist for years (see textbooks of neurology or the review by Frederiks (1969) for further information).

Unilateral unawareness and neglect is the most frequent neurologically determined disorder of body image. It usually affects the left limbs and arises most often from lesions of the supramarginal and angular gyri of the right parietal lobe, often following a stroke. When the disorder is marked, the patient may neglect to wash one side of his body or to shave one side of his face, or may put on only one shoe. In its mildest form it can be detected only by special testing using double stimulation (e.g. if both wrists are touched with cotton wool but the patient reports a touch from only one side, even though the sensation on the other side is present when tested on its own). Further information is given by Critchley (1953) who provides detailed information about syndromes arising from lesions in the parietal lobes.

Hemisomatognosis, which is also known as hemidepersonalization, is much less frequent than unilateral unawareness. The patient reports the feeling that one of his limbs is missing, usually on the left. The disorder can occur on its own, or together with hemiparesis. There is often a coincident unilateral spatial agnosia. The nature of the patient's awareness is variable; some patients know that the limb is present though it feels as if it is absent, whilst others believe wholly or partly that the limb is really absent.

Anosognosia is a lack of awareness of disease, and it too is more often manifest on the left side of the body. Most often it occurs briefly in the early days after acute hemiplegia but occasionally it persists. The patient does not complain of the disability on the paralysed side and denies it when pointed out to him. There may also be denial of dysphasia, blindness (**Anton's syndrome**), or amnesia (most marked in Korsakov's syndrome). **Pain asymbolia** is a disorder in which the patient perceives a normally painful stimulus but does not recognize it as painful. Although these disorders are clearly associated with cerebral lesions, it has been suggested that there is a psychogenic element whereby the awareness of unpleasant things is repressed (e.g. Weinstein and Kahn 1955). Although it is hardly possible that structural damage could act in the absence of psychological reactions, it seems unlikely that the latter can be the sole cause of a condition that is so much more frequent on the left side of the body.

Autotopagnosia is the inability to recognize, name, or point on command to parts of the body. The disorder may also apply to parts of the body of another person, but not to inanimate objects. It is a rare condition which arises from diffuse lesions, usually affecting both sides of the brain. Nearly all the cases can be explained by accompanying apraxia, dysphasia, or disorder of spatial perception (Lishman 1987, p. 63).

Distorted awareness of size and shape includes feelings that a limb is enlarging, becoming smaller, or otherwise being distorted. Unlike the phenomena described so far, these experiences are not related closely to lesions of specific areas of the brain. They may occur in healthy people, especially when falling asleep, or in the waking state when very tired. They are sometimes reported in the course of migraine, in acute brain syndromes, as part of the aura of epilepsy, or after taking LSD. Changes of shape and size of body parts are also described by some schizophrenic patients. The person is nearly always aware that the experience is unreal, except in some cases of schizophrenia.

Reduplication phenomenon is the experience that part or all of the body has doubled. Thus the person may feel that he has two left arms, or two heads, or that the whole body has been duplicated. These phenomena have been reported rarely in the course of migraine and temporal lobe epilepsy as well as in schizophrenia. In an extreme form the person has the experience of being aware of a copy of his whole body, a phenomenon already described under the heading of autoscopic hallucinations.

Coenestopathic states are localized distortions of body awareness, for example the nose feels as if it is made of cotton wool.

Disorders of memory

Failure of memory is called **amnesia**. Several kinds of memory failure are met in psychiatric disorders, and it might be expected that these would correspond broadly to the process of memory believed to exist in healthy people. Although psychologists do not agree completely about the structure of normal memory, the following general scheme has had wide acceptance. Human memory behaves as if organized in three kinds of 'stores'. **Sensory stores** have a limited capacity to receive information from the sense organs and to retain it for a brief period (about 0.5 s), pre-

sumably so that processing can be undertaken. The second store is called **primary memory** or **short-term memory**. This nomenclature does not correspond with the usual clinical usage of the adjective short-term to denote memory for events over a period of hours rather than seconds. The second store has a limited capacity—on average seven items of information. Information is held longer than in the sensory store, being lost in about 15–20 s. Information can be retained longer by repeated rehearsal (as in repeating an unfamiliar telephone number until it has been dialled fully). There may be two short-term stores, one for verbal and the other for visual information. The general system supporting retention over short periods is often called working memory. Short-term memory holds information while it is being used in some other cognitive process. The information is then either forgotten or transferred to long-term memory. How this transfer takes place is uncertain.

The third kind of store is secondary memory or **long-term memory** which receives information that has been selected for more permanent storage. Unlike short-term memory, this kind of store has a large capacity and holds information for a long time. Some information in this store is 'processed' and stored according to verbal characteristics such as the meaning or sound of words. Other information is stored as visual imagery. Two useful distinctions can be made about the working of long-term memory. The first is between memory for events (**episodic memory**) and memory for language and knowledge (**semantic memory**). The second is between the **recognition** of material presented to the person, and **recall** without a cue; the latter is more difficult. Recall without a cue (retrieval) may vary from one time to another with well-learned material (the experience of being unable to recall the name of a person but remembering it a few minutes later).

Memory is affected by several kinds of psychiatric disorder. Organic brain disorder generally affects all aspects of secondary memory, but some organic conditions give rise to an interesting partial effect known as amnestic disorder (p. 314) in which the person is unable to remember events occurring a few minutes before (impaired episodic memory), but can converse normally (intact semantic memory). Some patients with memory disorder recall more when given cues. This observation suggests that the disorder is concerned with retrieval as well as with the registration or retention of memory.

After a period of unconsciousness there is poor memory for the interval between the ending of complete unconsciousness and the restoration of full consciousness (**anterograde amnesia**). Some causes of unconsciousness (e.g. head injury and electroconvulsive therapy (ECT)) also lead to inability to recall events before the onset of unconsciousness (**retrograde amnesia**).

In some neurological and psychiatric disorders, patients have a peculiar disturbance of recall, either failing to recognize events that have been encountered before (*jamais vu*) or reporting the recognition of events that are in fact novel (*déjà vu*). Some patients with extreme difficulty in remembering may report, as memories, events that have not taken place at the time in question (or may never have involved the person at all)—a disorder known as **confabulation**. In depressive disorders memories of unhappy events are recalled more readily than other events, a process which adds to he patient's low mood.

For a review of psychological studies of memory see Baddeley (1990).

Disorders of consciousness

Consciousness is awareness of the self and the environment. The level of consciousness can vary between the extremes of alertness and coma. The quality of consciousness can also vary: sleep differs from unconsciousness, and stupor differs from both (see below).

Many terms have been used for states of impaired consciousness. **Coma** is the most extreme form. The patient shows no external evidence of mental activity and little motor activity other than breathing. He does not respond even to strong stimuli. Coma can be graded by the extent of the remaining reflex responses and by the type of EEG activity. **Sopor** is an infrequently used term for a state in which the person can be aroused only by strong stimulation. **Clouding of consciousness** refers to a state in which the patient is drowsy and reacts incompletely to stimuli. Attention, concentration, and memory are impaired and orientation is disturbed. Thinking seems slow and muddled, and events may be interpreted inaccurately.

Stupor refers to a condition in which the patient is immobile, mute, and unresponsive but appears to be fully conscious, usually because the eyes are open and follow external objects. If the eyes are closed, the patient resists attempts to open them. Reflexes are normal and resting posture is maintained, though it may be awkward. (Note that in neurology the term implies impaired consciousness.)

Confusion means inability to think clearly. It occurs characteristically in organic states, but in some functional disorders as well. In acute organic disorder confusion occurs together with partial impairment of consciousness, illusions, hallucinations, delusions, and a mood change of anxiety or apprehension. The resulting syndrome has been called a confusional state, but this term is not well defined and it is preferable to avoid it (see p. 310). Three variations of this syndrome may be mentioned. The first is an **oneiroid** (dream-like) state in which the patient, although not asleep, describes experiences of vivid imagery akin to that of a dream. When such a state is prolonged it is sometimes called a **twilight state** (see p. 310). **Torpor** is a state in which the patient appears drowsy, readily falls asleep, and shows evidence of slow thinking and narrowed range of perception.

Disorders of attention and concentration

Attention is the ability to focus on the matter in hand. Concentration is the ability to maintain that focus. The ability to focus on a selected part of the information reaching the brain is important in many everyday situations, for example when conversing in a noisy place. It is also important to be able to attend to more than one source of information at the same time, for example when conversing while driving a car. Attention is modified by previous experience: if a person has found that a stimulus is irrelevant to a certain situation, he will attend to it less. If the stimulus subsequently becomes relevant, the person will take longer to learn this than he would to learn the significance of a new stimulus. This phenomenon is known as **latent inhibition**. Schizophrenic patients are less influenced by previous experience (show less latent inhibition) than normal subjects. It has been suggested that lack of latent inhibition could explain the schizophrenic patient's experience of being overwhelmed by stimuli, and could even contribute to the formation of other symptoms such as hallucinations (Gray *et al.* 1990) (see also p. 92).

Attention and concentration may be impaired in a wide variety of psychiatric disorders including depressive disorders, mania, anxiety disorders, schizophrenia, and organic disorders. Therefore the finding of abnormalities of attention and concentration does not assist in diagnosis. Nevertheless these abnormalities are important in management; for example they affect patients' ability to give or receive information when interviewed and poor concentration can interfere with a patient's ability to work or pass his leisure time, for example in reading or watching television.

Insight

Insight may be defined as awareness of one's own medical condition. It is difficult to achieve, since it involves some knowledge of what constitutes a healthy mind, and yet doctors cannot agree among themselves about the meaning of terms such as mental health and mental illness. Moreover, insight is not simply present or absent, but rather a matter of degree. For this reason it is better to consider four separate questions. First, is the patient aware of phenomena that other people have observed (e.g. that he appears to be unusually active and elated)? Second, if so, does he recognize that these phenomena are abnormal (or does he, for example, maintain that his unusual activity and cheerfulness are merely normal high spirits)? Third, if he recognizes the phenomena as abnormal, does he consider that they are caused by mental illness, as opposed to, for example, a physical illness or the

results of poison administered to him by his enemies? Fourth, if he accepts that he is ill, does he think that he needs treatment?

The answers to these questions are much more informative—and much more likely to be reliable—than those of the single question: is insight present or not? Newcomers to psychiatry often ask this question because they have read that loss of insight distinguishes psychoses from neuroses. While it is generally true that neurotic patients retain insight and psychotic patients lose it, this is not invariable; nor is this in practice a reliable way of distinguishing between the two. Also, the concepts of neurosis and psychosis are themselves unsatisfactory (see p. 60). On the other hand, the four questions listed above can help the clinician decide whether the patient is likely to cooperate with treatment.

Standard criteria for rating symptoms

For many kinds of research it is important to detect and record symptoms reliably. In some epidemiological studies symptoms have to be detected by interviewers who are not trained psychiatrists. Standardized methods of interviewing have been developed in which standard questions are used and criteria are provided to decide whether a symptom is present or absent and, in some schemes, to decide how severe it is. Among the best known schemes are the Schedule of Clinical Assessment in Neuropsychiatry (SCAN), the Structured Clinical Interview for DSMIIIR (SCID) and the Composite International Diagnostic Interview (CIDI). These formal definitions of symptoms are considered in Chapter 2 where they are related to the standardized methods of interview used in these schemes.

Further reading

Jaspers, K. (1963). *General psychopathology* (trans. from the 7th German edition by J. Hoenig and M. W. Hamilton), Chapter I, Phenomenology. Manchester University Press.

Scharfetter, C. (1980). *General psychopathology: an introduction* (trans. H. Marshall). Cambridge University Press.

Schneider, K. (1949). The concept of delusion. Reprinted and translated in *Themes and variations in European psychiatry* (ed. S. R. Hirsch and M. Shepherd). John Wright, Bristol, 1974.

Sims, A. (1995). *Symptoms in the mind; an introduction to descriptive psychopathology* (2nd edn). Baillière Tindall, London.

Wing, J. K., Cooper, J. E., and Sartorius, N. (1974). *The measurement and classification of psychiatric symptoms*, Glossary of definitions, pp. 141–88. Cambridge University Press.

2 | Interviewing, clinical examination, and record keeping

In medicine generally interviews are used to gather information, to develop rapport with the patient, and to inform and motivate the patient. This three-function approach (Cohen-Cole 1991) also applies to psychiatry. Although in medicine the information-gathering aspect of the interview concerns both the history and the mental state, in psychiatry the examination of the mental state is much more detailed and more important in diagnosis. This chapter begins with some advice about the technique of interviewing before describing history taking and mental state examination. Whilst this account draws attention to important points of technique, it should be remembered that interviewing is a practical skill that the trainee can acquire only through carrying out interviews under supervision and watching experienced interviewers at work.

The following section outlines an approach to interviewing that is widely used and effective in practice. It is hoped that it will be of value to readers who have not yet received training in interviewing.

The diagnostic interview

Before the interview begins the interviewer should make arrangements that will help to put the patient at ease. The extent to which the following recommendations can be achieved will depend where the interview is taking place: arrangements in the patient's home may be less ideal than those in an out-patient department, and those in a general hospital ward, an emergency department, or a police station may be still less ideal. If possible, the interview should be carried out in a room that is reasonably soundproof and free from interruptions. Patients should not be seated directly opposite the interviewer, nor should their chairs be so much lower that they have to look upwards. In this way they will feel at ease rather than under constant scrutiny. For a diagnostic interview the interviewer should sit at a writing table in order to take notes (a psychotherapeutic interview may require less formal arrangements, with both patient and therapist in armchairs). The interviewer should not attempt to memorize the interview and write the notes afterwards, as this is time consuming and likely to be inaccurate. The least obtrusive way of taking notes is to place the patient at the side of the desk, and on the left side of a right-handed interviewer. This creates a suitably informal atmosphere and allows the interviewer to attend to the patient whilst writing.

The first encounter with the patient is important. The interviewer should welcome him by name, and give his own name. If the patient is accompanied, it is good practice for the interviewer to welcome the companions, and to explain how long they may expect to wait and whether they are to be interviewed. If the patient is seen at the request of a general practitioner, the interviewer should indicate that the latter has written, though it is not usually appropriate to reveal the contents of the letter in detail.

The interviewer should explain how the interview will proceed: for example 'First I should like to hear about your present problems. Only when I am sure that I have understood these shall I ask you how they began'. The interviewer then asks an open question such as 'Tell me about the problems' or 'Tell me what you think is wrong', and the patient is encouraged to talk freely for several minutes. During this time the interviewer makes two separate kinds of observations—how the patient is talking, and what he has to say. The first helps the doctor to decide *how* to interview the patient, whilst the second helps him to decide *what* to ask about.

Certain general skills have been shown to be important in interviewing, and capable of being learnt (Goldberg *et al.* 1980, 1991):

- making eye contact throughout the interview
- adopting a relaxed posture
- making facilitatory noises when listening
- not appearing hurried
- picking up verbal and non-verbal cues of distress
- dealing with over-talkativeness
- not offering information too soon.

Whilst deciding how to interview the individual patient, the interviewer observes whether the patient seems cooperative, reasonably at ease, and able to express ideas coherently. The most frequent difficulty is that the patient is over-anxious. The interviewer should consider whether such anxiety is part of the presenting disorder or fear of seeing a psychiatrist. If the latter, the interviewer should take time to discuss the patient's apprehension before proceeding with the interview. Reassurance and a calm unhurried approach will usually put the patient more at ease.

Sometimes the patient seems uncooperative and resentful. This may be because the interview is taking place against the person's wishes; for example the spouse or general practitioner may have persuaded the patient to attend, or the interview may be after admission to a general hospital for a drug overdosage. Faced with this kind of resentment, the interviewer should talk over the circumstances of the referral and try to persuade patients that the interview is intended to be in their own interest. Patients may appear resentful for other reasons. Some patients act in a hostile way when anxious, and some depressed or schizophrenic patients may seem uncooperative because they do not regard themselves as ill. At times it becomes apparent that a patient cannot respond adequately to the interview because of impaired consciousness. When this seems likely, orientation, concentration, and memory should be tested, and if impaired consciousness is confirmed, an informant should be seen before returning to the patient. Provided that there are no immediate problems of this kind, the interviewer should consider whether there are likely to be difficulties

in guiding the interview effectively. Some patients, such as successful businessmen, attempt to dominate the interview, especially if the interviewer is younger then themselves. Others adopt an unduly friendly attitude that threatens to convert the interview into a social conversation. In either case, the interviewer should explain why he needs to guide the patient to relevant issues.

As mentioned above, the interviewer, whilst listening to the patient's opening remarks, also begins to consider what questions should be asked. These usually begin with further enquiries about the *nature* of the patient's presenting symptoms. It is a common mistake to start asking about the timing of such symptoms before their nature is clearly established. For example patients sometimes say that they are depressed, but further enquiry shows that they are experiencing anxiety rather than low spirits. If there is any doubt, the patient should be asked to give examples of the experiences. The interviewer should clearly understand the nature of the symptoms before asking about their timing and the factors that make them better or worse.

When all the presenting complaints have been explored in this way, direct questions are used to ask about other relevant symptoms. For example a person who complains of feeling depressed should be asked about ideas concerning the future, sleep pattern, appetite, etc. The subsidiary questions required for each presenting symptom will be apparent from reading the chapters in this book on psychiatric syndromes.

Next, the mode of onset of the complaint is asked about and its course noted, including any exacerbations or periods of partial remission. Considerable persistence may be needed to date the onset accurately, and if necessary it should be related to events that the patient can remember accurately (Was it before or after your birthday? Had it already started before Christmas?).

Controlling the interview

As the interview continues, the doctor's task is to keep the patient to relevant topics by bringing him back to the point if he strays from it. In doing this the interviewer should use a minimum of leading or closed questions (a leading question suggests the answer; a closed question allows only the answers yes or no, thus preventing the person from volunteering information). Thus instead of the closed question 'Are you happily married?' the interviewer might ask 'How do you and your wife get on with one another?'. When there is no alternative to a closed question, the answer should be followed by a request for an example.

Taciturn patients can often be encouraged to speak more freely if the interviewer shows non-verbal expressions of concern (e.g. leaning forward a little in the chair with an expression of interest). It is less easy to curb the flow of an over-talkative patient. Sometimes this can be done only by waiting for a natural break in the flow of speech to explain that, because time is limited, the interviewer proposes to interrupt the patient when appropriate to help him focus on the issues that are important for planning treatment. Provided that such advice is offered tactfully, most garrulous patients are relieved to be given it.

Although it is essential to ask direct questions about specific items of information, it is equally important to give the patient an opportunity to talk spontaneously, as unexpected material may be revealed in this way. Spontaneous talk can be encouraged by prompting rather than by questioning, for example by repeating in an enquiring tone the patient's reply to previous questions or by using non-verbal prompts. Also, before ending the interview, it is useful to ask a general question such as 'Is there anything else that you think I should know?'

History taking

Whenever possible, the history from the patient should be supplemented by information from a close relative or another person who knows him well. This is much more important in psychiatry than in the rest of medicine, because psychiatric patients are not always aware of the extent of their symptoms. For example a manic patient may not realize how much embarrassment he has caused by his extravagant social behaviour, or a demented patient may not fully understand the extent to which his work is impaired. Alternatively, patients may know what their problems are, but not wish to reveal them; for example alcoholics often conceal the extent of their drinking. Also, when personality is being assessed, patients and relatives often give quite different accounts of characteristics such as irritability, obsessional traits, and jealousy.

The history should always be recorded systematically and in the same order to ensure that important themes are not forgotten by the interviewer and to make it easier for colleagues to refer to the notes. However, it is not always possible to gather information in the same order with every patient. Some flexibility must be allowed if the patient is not to feel unduly restricted by the interviewer.

In this section, a standard scheme of history taking is given in the form of a list of topics to be covered. This will serve as a check-list for the beginner, and a reminder for the more experienced interviewer, of the topics that make up a complete history. However, it is neither necessary nor possible to ask every question of every patient. Common sense must be used in judging how far each topic needs to be explored with a particular patient. The trainee must learn by experience how to adjust his questioning to problems that emerge as the interview proceeds. This is done by keeping in mind the decisions about diagnosis and treatment that will have to be made at the end of the interview.

The scheme given below is followed by notes explaining how to record the different items, and why they are important. After this the assessment of personality is discussed in more detail.

A scheme of history taking is given below. For ease of reference, this scheme is presented simply as a list of headings and items. However, it is essential that the trainee should understand how to record the different items and why they are important; these topics are outlined in the subsequent notes on history taking. It is recommended that the scheme and subsequent notes should be studied together. Throughout the interview it is important to gain an understanding of the patient's experience of the illness, how the illness has interfered with his life and his plans for the future, what he thinks is wrong, what he believes to be the nature of the illness, and what he fears may happen to him if the illness continues.

The scheme of history taking (Table 2.1)

Informant

Name, relation to patient, intimacy, and length of acquaintance. Interviewer's impression of informant's reliability.

Source of referral and reasons for referral

Present condition

Symptoms with duration and mode of onset of each. Description of the time relations between symptoms and physical disorders and psychological or social problems. Effects on work, social functioning, and relationships. Associated disturbance in sleep, appetite, and sexual drive. Any treatment given by other doctors.

Table 2.1 *Outline of the psychiatric history*

Name, age, and address of patients; name of any informants and their relationship to the patient
History of present condition
Family history
Personal history
Past illness
Personality
Drugs, alcohol, tobacco

Family history

Father: age now or at death (if dead give cause of death), health, occupation, personality, quality of relationship with patient. **Mother:** the same items. **Siblings:** names, ages, marital status, occupation, personality, psychiatric illness, and quality of relationship with patient. **Social position of family:** atmosphere in the home.

Family history of mental illness: psychiatric disorder, personality disorder, epilepsy, alcoholism. Other neurological or relevant medical disorders (e.g. Huntington's chorea).

Personal history (Table 2.2)

Early development: abnormalities during pregnancy and at birth; difficulties in habit training and delay in achieving milestones (walking, talking, sphincter control, etc.); separation from parents and reaction to it. **Health during childhood:** serious illness, especially any affecting the central nervous system, including febrile seizures. **'Nervous problems' in childhood:** fears, temper tantrums, shyness, stammering, blushing, food fads, sleepwalking, prolonged bed-wetting, frequent nightmares (though the significance of these behaviours is doubtful). **School:** age of starting and finishing each school; types of school; academic record; sporting and other achievements; relationships with teachers and pupils. **Higher education:** comparable enquiries. **Occupations:** chronological list of jobs, with reasons for changes; present financial circumstances, satisfaction in work. **Service or war experience:** promotion and awards; disciplinary problems; service overseas.

Menstrual history: age of menarche, attitude to periods, regularity and amount, dysmenorrhoea, premenstrual tension, age of menopause and any symptoms at the time, date of last menstrual period.

Marital history: age of patient at marriage; how long spouse known before marriage and length of engagement; previous relationships and engagements; present age, occupation, health, and personality of spouse; quality of the marital relationship.

Table 2.2 *The personal history*

Mother's pregnancy and the birth

Early development

Childhood
    Separations
    Emotional problems
    Illnesses

Schooling and higher education

Occupations

Menstrual history

Marriage

Sexual relationships

Children

Social circumstances

Past medical history

Past psychiatric history

Forensic history

Sexual history: attitude to sex; experience of sexual abuse; heterosexual and homosexual experience; current sexual practices, contraception.

Children: names, sex, and age of children; date of any abortions or stillbirths; temperament, emotional development, mental, and physical health of children.

Present social situation

Housing, composition of household, financial problems.

Previous medical history

Illness, operations, and accidents.

Previous psychiatric illness

Nature and duration of illness. Date, duration, and nature of any treatment. Name of hospital and of doctors. Outcome.

Forensic history

Arrests, convictions, imprisonment. Nature of the offences.

Personality before present illness (Table 2.3)

Relationships: friendships, few or many, superficial or close, with own or opposite sex; relations with workmates and superiors. **Use of leisure**: hobbies and interests; membership of societies and clubs. **Predominant mood**: anxious, worrying, cheerful, despondent, optimistic, pessimistic, self-depreciating, over-confident; stable or fluctuating; controlled or demonstrative. **Character**: sensitive, reserved, timid, shy; suspicious, jealous, resentful; quarrelsome, irritable, impulsive; selfish, self-centred; timid, reserved, shy, self-conscious, lacking in confidence; dependent; strict, fussy, rigid;

Table 2.3 *Assessment of personality*

Relationships

Leisure activities

Prevailing mood

Character

Attitudes and standards

Habits, including alcohol/drug use

---

meticulous, punctual, excessively tidy. **Attitudes and standards**: moral and religious; attitude towards health and the body. **Habits**: food, alcohol, tobacco, drugs.

Notes on history taking

The scheme just outlined lists the items to be considered when a full history is taken, but gives no indication as to why these items are important or what sort of difficulties may arise in eliciting them. These issues are discussed in this section, which is written in the form of notes referring to the headings used above.

The reason for referral

State in everyday language why the patient has been referred, for example 'Severe depression, failing to respond to drug treatment'.

The present illness

In an out-patient clinic, it is usually better to consider this item first because the patient probably wants to talk about it straight away. However, with in-patients the doctor may already have substantial information about the present illness, either from doctors dealing with the case before admission or from relatives. In these circumstances the interviewer may find it better to begin with the family and personal history.

Always record which complaints have been volunteered by the patient, and which revealed by questioning. Record the severity and duration of each symptom, how it began, and what course it has taken (increasing gradually, diminishing stepwise, staying the same, intermittent). Indicate which symptoms co-vary and which take an independent course (e.g. obsessional thoughts and rituals may have fluctuated together, whilst depressed mood may have been a recent addition). Any recent treatment should be noted, together with its apparent effects. When a drug has not been effective, note whether the patient took it in the required dosage.

Family history

Mental illness among parents or siblings suggests that the cause of illness may in part be hereditary. Because the family is the environment in which the patient grew up, the personality and attitudes of the parents are important, as are separations from the parents for any reason. Ask about the parents' relationship with one another, for example whether there were frequent quarrels. Enquire about separations, divorce, and remarriage. Rivalry between siblings may be important, as may favouritism towards one child by the parents. The occupation and social standing of the parents reflect the material circumstances of the patient's childhood.

Recent events in the family may have been stressful to the patient. Serious illness of either parent or divorce of a sibling are likely to be relevant problems in other family members. Finally, the family history may throw light on the patient's concerns about himself. For example, the death of an older brother from a brain tumour may partly explain a patient's extreme concern about headaches.

Personal history

Pregnancy and birth: events in pregnancy are occasionally relevant, especially when the patient is mentally handicapped. An unwanted pregnancy may be followed by a poor relationship between mother and child. Similarly, serious problems during delivery sometimes account for intellectual impairment.

Early development: few patients know whether they have passed through developmental stages normally. However, this information is more important if the patient is a child or adolescent, in which case the parents are likely to be interviewed routinely. This information may also be important in cases of mental handicap, when the parents or other relatives should be questioned and previous medical records should be obtained. (A summary of the main developmental milestones will be found in Chapter 20.)

Notes should also be made of any prolonged periods of separation from the mother, for example through illness. The effects of such separations vary considerably (see Chapter 20), and it is important to ask an appropriate informant whether the patient was emotionally upset at the time and, if so, for how long.

Health in childhood: there is little point in recording minor childhood ailments such as uncomplicated chickenpox, but it is appropriate to enquire about encephalitis or convulsions, any illness leading to prolonged admission to hospital, or prolonged disability.

Early neurotic traits: it is conventional to enquire about such symptoms as fears, sleep-walking, shyness, stammering, and food fads. However, there is no evidence that these behaviours in childhood are precursors of neurosis in adult life.

Schooling: the school record not only gives an indication of intelligence and scholastic achievements, but also reflects social development. The type of school and examination results should be noted. The interviewer should also ask whether the patient had friends and was popular, whether he played games and with what success, and how he got on with teachers. Similar questions are asked about **higher education**.

Occupational history: information about the present job helps the interviewer to understand the circumstances of the patient's life and to judge whether he is under stress at work. A list of previous jobs is mainly relevant to the assessment of personality. If the patient has had many jobs, it is important to ask why he left each one. Repeated dismissals may reflect an awkward, aggressive, or otherwise abnormal personality (though there are, of course, many other reasons for repeated sackings). When each job is inferior to the last, it is necessary to consider declining efficiency caused by chronic mental illness or by alcohol abuse. Information about relationships with colleagues, senior and junior, helps to assess personality.

When the patient has served in the armed forces or worked abroad, details should be obtained, and an enquiry should be made about tropical disease later in the history.

Menstrual history: it is usual to enquire about the age of menarche and how the patient first learnt about menstruation. These questions were more important in earlier times when ignorance about sexual matters was widespread and the unexpected onset of periods in an unprepared girl could give rise to lasting anxieties. This rarely happens in the United Kingdom today, except among some immigrant groups. When interviewing immigrants to this country or when working in other countries, the interviewer may find the answers more informative. Questions about current menstrual function should be asked in all relevant cases. Dysmenorrhoea, menorrhagia, and premenstrual tension should be identified, and amongst women in middle life the menopause should be noted. The date of the last period should also be noted.

Marital history: the interviewer should enquire about previous lasting relationships with the opposite sex and about the present marital relationship. Sexual relationships are considered in the next section; in this part of the history it is the personal aspects that are considered. Frequent broken relationships before marriage may reflect abnormalities of personality. A previous relationship may determine the patient's attitude to the present marriage; for example when a first marriage has ended in divorce because of the husband's infidelity, a woman may over-react to minor difficulties in her second marriage.

The spouse's occupation, personality, and state of health give information of obvious relevance to the patient's circumstances. Present difficulties can often be understood better by enquiring about each partner's original expectations of the marriage. It is also useful to ask about the sharing of decisions and responsibilities in the marriage. The dates of birth of the children, or of any miscarriages, may indicate whether marriage was forced by pregnancy.

Sexual history: in taking the sexual history, the interviewer should use common sense in deciding how much to ask the individual patient. For example a detailed account of masturbation and sexual techniques may be essential when the patient is seeking help for sexual impotence, but the interviewer is often more concerned to establish generally whether the patient's sexual life is satisfying or not. Only if there are problems need he enquire into all the details under this heading. Judgement must also be used about the optimal timing and amount of detail of questions about experience of sexual abuse.

Finally, the interviewer should ask about methods of contraception and, when relevant, a woman's wishes about bearing children.

Children: pregnancy, childbirth, miscarriages, and induced abortions are important events which are sometimes associated with adverse psychological reactions in the mother. Information about the patient's children is relevant to present worries and the pattern of family life. Since children may be affected by the parent's illness, it is important to know, for example, whether a seriously depressed woman has the care of a baby, or whether a violent alcoholic man has children in the home. If admission to hospital is being considered for a patient who has the care of children it is important to find out their needs and if necessary arrange for their care. This is obvious but is sometimes overlooked.

Previous illness: previous medical or surgical treatment should always be asked about, and particularly careful inquiries should be made about previous mental illness. Patients or relatives may be able to recall the presenting symptoms of illness and the main points about treatment. However, details of diagnosis and treatment can usually be obtained only from the doctors who treated the patient at the time. In psychiatry the nature of previous illness is an important guide to the present disorder, and it is nearly always appropriate to request information from other hospitals.

Present circumstances: questions about housing, finances, and the composition of the household help the interviewer to understand the patient's circumstances and to judge more clearly what aspects of his life are likely to be stressful and how illness may affect him. There can be no general rule about the amount of detail to elicit, and this must be left to common sense.

Assessment of personality

Aspects of a patient's personality can be judged by asking him for his own self-rating, by asking other people who know him well, and by observing his behaviour at interview. Mistakes can arise from paying too much attention to the patient's own assessment of his personality. Some people give an unduly favourable account of themselves; for example antisocial people may conceal the extent of their aggressive behaviour or dishonesty. Conversely, depressed patients often judge themselves too severely, for example as being ineffectual, selfish, or unreliable, an impression that is not confirmed by other people. Therefore it is essential to interview other informants whenever possible.

Good indications of personality can often be obtained by asking the patient or others how he has behaved in particular circumstances. For example if a patient says he is self-confident, it is useful to enquire how he behaves in particular situations when he has to convince other people or speak in public. Similarly, personality can often be assessed by asking about occasions when social roles are changing, such as leaving school, starting work, marrying, or becoming a parent.

When assessing a patient's personality from behaviour at interview, it is essential to allow for the possible effects of psychiatric illness. Thus, when depressed, a normally self-possessed and sociable person may appear abnormally shy and lacking in self-confidence.

Whatever the source of information, it is important to assess the strengths as well as the weaknesses in a patient's personality.

Enquiries about personality are most fruitful when they are systematic. The scheme outlined on p. 29 is widely used, and covers the most important areas of enquiry in clinical work. The points given below refer to patient interviews but can be adapted to informant interviews.

The assessment begins with enquiries about **relationships** with friends and people at work. Is the person shy or does he make friends easily? Are his friendships close and are they lasting? **Leisure activities** can throw light on personality, not only by reflecting a person's interests but also by indicating his preference for company or solitude, and his levels of energy and resourcefulness.

Mood is considered next. The interviewer tries to find out whether the patient is generally cheerful or gloomy and whether he has marked changes of mood, and if so, how quickly they appear, how long they last, and if they follow life-events. The interviewer should also find out whether the patient shows emotions or hides them.

Character: the interviewer will already have gathered some impression of this while taking the personal history. Further information about the patient's character should be sought, for example by asking whether he is reserved, timid, shy, or self-conscious, sensitive or suspicious, resentful or jealous, irritable, impulsive, or quarrelsome, selfish or self-centred, lacking in confidence, and strict, fussy, rigid, meticulous, punctual, or excessively tidy.

These are chiefly negative attributes of character but, as mentioned above, it is also important to ask about positive ones. It is not appropriate to go though a complete list with every patient; common sense will indicate what to enquire about as the picture of the patient gradually builds up. However, it is good practice to determine how resilient every patient is in the face of adversity.

Answers should not always be taken at face value; for example when readiness to anger is asked about, the interviewer should not simply accept the answer that the patient never feels angry. Instead he should persist in questioning, for example by remarking that everyone feels angry at times and asking what makes the patient angry. The interviewer should also find out whether the patient expresses anger or bottles it up, and if the former, whether through angry words or violent acts. If the patient contains anger, he should be asked how he feels when he does so.

Attitudes and standards: in this part of the interview it is usual to ask about attitudes to the body, health, and illness, as well as religious and moral standards. The personal history will usually have provided general indications about these matters so that extensive questioning is seldom necessary.

Habits: this final section deals with habits of taking tobacco, alcohol, or drugs.

Mental state examination

In the course of history taking, the interviewer will have noted the patient's symptoms up to the time of the consultation. The mental state examination is concerned with the symptoms and behaviour at the time of the interview. Hence there is a degree of overlap between the history and the mental state, mainly in observations about mood, delusions, and hallucinations. If the patient is already in hospital, there will also be some overlap between mental state examination and the observations made by nurses and occupational therapists of his behaviour outside the interview room. The psychiatrist should pay considerable attention to these accounts from other staff, which are at times more revealing than the small sample of behaviour observed at mental state examination. For example, a patient may deny hallucinations at interview, but the nurses may notice him repeatedly talking alone as if replying to voices. On the other hand, mental state examination may reveal information not disclosed at other times, for example suicidal intentions in a depressed patient.

The examination of the mental state is described in the following paragraphs. The symptoms and signs that are referred to here have been described in Chapter 1. These descriptions will be repeated only when there is a special reason to do so. Carrying out the mental state examination is a practical skill that can be learnt only by watching experienced interviewers and by practising repeat-

Table 2.4 *Summary of the mental state examination*

Behaviour
Speech
Mood
Depersonalization, derealization
Obsessional phenomena
Delusions
Hallucinations and illusions
Orientation
Attention and concentration
Memory
Insight

edly under supervision. More detailed accounts are provided by Leff and Isaacs (1978), and by Wing *et al.* (1974) in an account of the Present State Examination.

The mental state examination follows the headings in Table 2.4.

Appearance and behaviour

Although the mental state examination is largely concerned with what the patient says, much can also be learnt from observing appearance and behaviour.

The patient's **general appearance** and clothing repay careful observation. Self-neglect, as shown by a dirty unkempt look and crumpled clothing, suggest several possibilities, including alcoholism, drug addiction, depression, dementia, or schizo-phrenia. Manic patients may wear bright colours, adopt incongruous styles of dress, or appear poorly groomed. Occasionally an oddity of dress may provide the clue to diagnosis; for example a rainhood worn on a dry day may be the first evidence of a patient's belief that rays are being shone on her head by persecutors.

The interviewer should also note the patient's body build. An appearance suggesting recent weight loss should alert the observer to the possibility of physical illness, or of anorexia nervosa, depressive disorder, or chronic anxiety neurosis.

Facial appearance provides information about mood. Although judgements are usually made on a general impression of the person's appearance, it is useful to be aware of the specific changes that make up this overall appearance. In depression the most characteristic features are turning down of the corners of the mouth, vertical furrows on the brow, and a slight raising of the medial aspect of each brow. Anxious patients generally have horizontal creases on the forehead, raised eyebrows, widened palpebral fissures, and dilated pupils. Although depression and anxiety are especially important, the observer should look for evidence of the whole range of emotions, including elation, irritability, and anger, together with the unchanging 'wooden' expression of patients taking drugs with parkinsonian side-effects. The facial appearance may also suggest physical conditions such as thyrotoxicosis and myxoedema.

Posture and movement also reflect mood. A depressed patient characteristically sits leaning forwards, with shoulders hunched, the head inclined downwards, and gaze directed to the floor. An anxious patient usually sits upright with head erect, often on the edge of the chair and with hands gripping its sides. Anxious people and patients with agitated depression are often tremulous and restless, touching their jewellery, adjusting clothing, or picking at the fingernails. Manic patients are over-active and restless.

Social behaviour is important. Manic patients often break social conventions and are unduly familiar with people they do not know well. Demented patients sometimes respond inappropriately to the conventions of a medical interview, or continue with their private preoccupations as if the interview were not taking place. Schizophrenic patients may behave oddly when interviewed; some are over-active and socially disinhibited, some are withdrawn and preoccupied, and others are aggressive. Patients with antisocial personality disorders may also appear aggressive. In recording abnormal social behaviour, the psychiatrist should give a clear description of what the patient actually does. He should avoid general terms such as 'bizarre', which are uninformative. Instead he should describe what is unusual.

Finally, the interviewer should watch for certain uncommon **disorders of motor behaviour** encountered mainly in schizophrenia (see p. 19). These include stereotypes, posturing, negativism, echopraxia, ambitendence, and waxy flexibility. He should also look for tardive dyskinesia, a motor disorder seen chiefly in elderly patients, especially women, who have taken antipsychotic drugs for long periods (see p. 552). This disorder is characterized by chewing and sucking movements, grimacing, and choreo-athetoid movements affecting the face, limbs, and respiratory muscles.

Speech

How the patient speaks is recorded under this heading, whilst what he says is recorded later. The **rate and quantity** of speech are assessed first. Speech may be unusually fast, as in mania, or slow, as in depressive disorders. Depressed or demented patients may pause for a long time before replying to questions and may then give short answers, producing little spontaneous speech. The same may be observed among shy people or those of low intelligence. The amount of speech is increased in manic patients and in some anxious patients.

Next the interviewer should consider the **patient's utterances**, keeping in mind some unusual disorders found mainly in schizophrenia. He should note whether any of the words are neologisms, i.e. private words invented by the patient, often to describe morbid experiences. Before assuming that a word is a neologism it is essential to make sure that it is not merely mispronounced or a word from another language.

Disorders of the **flow of speech** are recorded next. Sudden interruptions may indicate thought blocking but are more often merely the effects of distraction. It is a common mistake to diagnose thought blocking when it is not present (see p. 7). Rapid shifts from one topic to another suggest flight of ideas, while a general diffuseness and lack of logical thread may indicate the kind of thought disorder characteristic of schizophrenia (see p. 8). It can be difficult to be certain about these abnormalities at interviews, and it is often helpful to record a sample of conversation for more detailed analysis.

Mood

The assessment of mood begins with the observations of behaviour described already, and continues with direct questions such as 'What is your mood like?' or 'How are you in your spirits?'

If **depression** is detected, further questions should be asked about a feeling of being about to cry (actual tearfulness is often denied), pessimistic thoughts about the present, hopelessness about the future, and guilt about the past. Suitable questions are 'What do you think will happen to you in the future?' or 'Have you been blaming yourself for anything?'

Trainees are often wary of asking about suicide in case they should suggest it to the patient, but there is no evidence to warrant this caution. Nevertheless, it is sensible to enquire about suicide in stages, starting with the question 'Have you thought life is not worth living?' and, if appropriate, going on to ask 'Have you wished you could die?' or 'Have you considered any way in which you might end your life?'

Anxiety is assessed further by asking about physical symptoms and thoughts that accompany the affect. These are discussed in detail in Chapter 12; here we need only note the main questions. The interviewer should start with a general question such as 'Have you noticed any changes in your body when you feel anxious?', and then go on to specific enquiries about palpitations, dry mouth, sweating, trembling, and the various other symptoms of autonomic activity and muscle tension. To detect anxious thoughts, one can ask

'What goes through your mind when you are feeling anxious?' Possible replies include thoughts of fainting, losing control, and going mad. Inevitably many of these questions overlap with inquiries about the history of the disorder.

Questions about **elation** correspond to those about depression; for example 'How are you in your spirits?, followed if necessary by direct questions such as 'Do you feel unusually cheerful?' Elated mood is often accompanied by ideas reflecting excessive self-confidence, inflated assessment of one's abilities, and extravagant plans.

As well as assessing the prevailing mood, the interviewer should find out **how mood varies** and whether it is appropriate. When mood varies excessively, it is said to be labile; for example the patient appears dejected at one point in the interview but quickly changes to a normal or unduly cheerful mood. Any persisting lack of affect, usually called blunting or flattening, should also be noted.

In a normal person, mood varies in parallel with the main themes discussed; he appears sad while talking of unhappy events, angry while describing things that have annoyed him, and so on. When the mood is not suited to the context, it is recorded as incongruent, for example if a patient giggles when describing the death of his mother. This symptom is often diagnosed without sufficient reason, and so it is important to record specific examples. Further knowledge of the patient may later provide another explanation for the behaviour; for example giggling when speaking of sad events may result from embarrassment.

Depersonalization and derealization

Patients who have experienced depersonalization and derealization usually find them difficult to describe; patients who have not experienced them frequently misunderstand the question and give misleading answers. Therefore it is particularly important to obtain specific examples of the patient's experiences. It is useful to begin by asking 'Do you ever feel that things around you are unreal?' and 'Do you ever feel unreal or have the experience that part of your body is unreal?' Patients with derealization often describe things in

the environment as seeming artificial and lifeless, whilst those with depersonalization may describe themselves as feeling detached from their surroundings, unable to feel emotion, or as if acting a part. Some patients use illustrations to describe their experience, for example 'as if I were a robot'; such descriptions should be distinguished carefully from delusions. If a patient has described these experiences, he should be asked to explain them. Most cannot suggest a reason, but a few give a delusional explanation, for example that the feelings are caused by a persecutor (this should be recorded later under the heading of delusions).

Obsessional phenomena

Obsessional thoughts are considered first. An appropriate question is 'Do any thoughts keep coming into your mind, even though you try hard not to have them?'. If the patient says 'yes', he should be asked for an example. Patients are often ashamed of obsessional thoughts, especially those about violence or sexual themes, and therefore persistent but sympathetic questioning may be required. Before recording thoughts as obsessional, the interviewer should be certain that the patient accepts them as his own (and not implanted by someone or something else).

Some **compulsive rituals** can be observed, but others are private events (such as counting silently) which are detected only because they interrupt the patient's conversation. Appropriate questions are 'Do you have to keep checking activities that you know you have really completed?'. 'Do you have to do things over and over again when most people would have done them only once?', and 'Do you have to repeat actions many times in exactly the same way?'. If the patient answers 'yes' to any of these questions, the interviewer should ask for specific examples.

Delusions

A delusion is the one symptom that cannot be asked about directly, because the patient does not recognize it as differing from other beliefs. The interviewer may be alerted to delusions by information from other people or by events in the history. In searching for delusional ideas it is useful to begin by asking for an explanation of other symptoms or unpleasant experiences that the patient has described. For example, if a patient says that life is no longer worth living, he may also believe that he is thoroughly evil and that his career is ruined, though there is no objective evidence. Many patients hide delusions skilfully, and the interviewer needs to be alert to evasions, changes of topic, or other hints of information being withheld. However, once the topic of the delusion has been uncovered, patients often elaborate on it without much prompting.

When ideas are revealed that may or may not be delusional, the interviewer must find out how strongly they are held. To do this without antagonizing the patient requires patience and tact. The patient should feel that he is having a fair hearing. If the interviewer expresses contrary opinions to test the strength of the patient's beliefs, his manner should be enquiring rather than argumentative. On the other hand, the interviewer should not agree with the patient's delusions.

The next step is to decide whether the beliefs are culturally determined convictions rather than delusions. This judgement may be difficult if the patient comes from another culture or is a member of an unusual religious group. In such cases any doubt can usually be resolved by finding a healthy informant from the same country or religion, and by asking him whether the patient's ideas would be shared by other people from that background.

Some **special forms of delusion** present particular problems of recognition. Delusions of thought broadcasting must be distinguished from the belief that other people can infer a person's thoughts from his expression or behaviour. In eliciting such delusions an appropriate question is 'Do you believe that other people know what you are thinking, even though you have not spoken your thoughts aloud?' If the patient says 'yes', the interviewer should ask how other people know this. (Many patients answer 'yes' when they mean that others can infer their thoughts from their facial expression.) A corresponding question about delusions of thought insertion is 'Have you ever felt that some of the thoughts in your mind were not your own but were put there from

outside?' A suitable question about delusions of thought withdrawal is 'Do you ever feel that ideas are being taken out of your head?' In each case, if the patient answers 'yes', detailed examples should be sought.

Delusions of control present similar difficulties to the interviewer. It is appropriate to ask 'Do you ever feel that some outside force is trying to take control of you?' or 'Do you ever feel that your actions are controlled by some person or thing outside you?' Since these experiences are far removed from the normal, some patients misunderstand the question and answer 'yes' when they mean that they have a religious or philosophical conviction that man is controlled by God or the devil. Others think that the questions refer to the experience of being 'out of control' during extreme anxiety; some schizophrenic patients say 'yes' when they have heard commanding voices. Therefore positive answers must be followed by further questions to eliminate these possibilities.

Finally, the reader is reminded of the various **categories of delusion** described in Chapter 1, namely persecutory, grandiose, nihilistic, hypochondriacal, religious, and amorous delusions together with delusions of reference, guilt, unworthiness, and jealousy. The interviewer should also distinguish between primary and secondary delusions, and should look out for the experiences of delusional perception and delusional mood that may precede or accompany the onset of delusions.

Illusions and hallucinations

When asked about hallucinations, some patients take offence because they think that the interviewer regards them as mad. Therefore enquiries should be made tactfully, and common-sense judgement used to decide when it is safe to omit them altogether. Questions can be introduced by saying 'Some people find that, when their nerves are upset, they have unusual experiences'. This can be followed by enquiries about hearing sounds or voices when no one else is within earshot. Whenever the history makes it relevant, corresponding questions should be asked about visual hallucinations, or those of taste, smell, touch, and deep bodily sensations.

If the patient describes hallucinations, certain further questions are required depending on the type of experience. The interviewer should find out whether the patient has heard a single voice or several and, if the latter, whether the voices appear to talk to each other about the patient in the third person. This experience must be distinguished from that of the patient who hears actual people talking in the distance and believes that they are discussing him (delusion of reference). If the patient says that the voices are speaking to him (second-person hallucinations), the interviewer should find out what they say and, if the words are experienced as commands, whether the patient feels that they must be obeyed. It is important to record examples of the words spoken by hallucinatory voices.

Visual hallucinations should be distinguished carefully from visual illusions. Unless the hallucination is experienced at the time of the interview, this distinction may be difficult because it depends on the presence or absence of a visual stimulus which has been misinterpreted.

The interviewer must also distinguish dissociative experiences from hallucinations. The former are described by the patient as the feeling of being in the presence of another person or a spirit with whom he can converse. Such experiences are reported by people with hysterical personality, though not confined to them; they are encouraged by some religious groups and have little importance in diagnosis.

Orientation

This is assessed by asking about the patient's awareness of time, place, and person. If the question of orientation is kept in mind throughout the interview, it may not be necessary to ask specific questions at this stage of the examination because the interviewer will already know the answers.

Specific questions begin with the day, month, year, and season. In assessing the replies, it is important to remember that many healthy people do not know the exact date and that, understandably, patients in hospital may be uncertain about the day of the week, particularly if the ward

has the same routine every day. When enquiring about orientation in place, the interviewer asks what sort of place the patient is in (such as a hospital ward or an old people's home). Questions are then asked about other people such as the spouse or the ward staff; for example who they are and what their relationship to the patient is. If the patient cannot answer these questions correctly, he should be asked about his own identity.

Attention and concentration

Attention is the ability to focus on the matter in hand. Concentration is the ability to sustain that focus. Whilst taking the history, the interviewer should look out for evidence of attention and concentration. In this way he will already have formed a judgement about these abilities before reaching the mental state examination. Formal tests add to this information and provide a semi-quantitative indication of changes as illness progresses. It is usual to begin with the **serial sevens test**. The patient is asked to subtract seven from 100 and then subtract seven from the remainder repeatedly until this is less than seven. The time taken is recorded, together with the number of errors. If poor performance seems to be due to lack of skill in arithmetic, the patient should be asked to do a simpler subtraction or to say the months of the year in reverse order. If mistakes are made with these, he can be asked to give the days of the week in reverse order.

Memory

Whilst taking the history, questions will have been asked about everyday difficulties in remembering. During the examination of mental state, tests are given of immediate, recent, and remote memory. None is wholly satisfactory, and the results should be assessed alongside other information about the patient's ability to remember and, if there is doubt, supplemented by standardized psychological tests.

Short-term memory in the psychologist's sense (see p. 21) is assessed by asking the patient to repeat sequences of digits that have been spoken slowly enough for him reasonably to be expected to register them. An easy short sequence is given first to make sure that the patient understands the task. Then five different digits are presented. If the patient can repeat five correctly, six are given and then seven; if he cannot repeat five digits, the test is repeated with a different sequence of five. A normal response from a person of average intelligence is to repeat seven digits correctly. The test also involves concentration, and so it cannot be used to assess memory if tests of concentration are definitely abnormal. Short-term memory in the clinician's sense of the memory over a few minutes is assessed by asking the patient to memorize a name and a simple address, to repeat it immediately (to make sure it has been registered correctly), and to retain it.

The interview continues on other topics for 5 minutes before recall is tested. A healthy person of average intelligence should make only minor errors.

Memory for **recent events** is assessed by asking about news items from the last day or two, or about events in the patient's life that are known to the interviewer (such as the ward menus on the previous day). Questions about news items should be adapted to the patient's interests, and should have been widely reported in the media.

Remote memory can be assessed by asking the patient to recall personal events or well-known public items from some years before, such as the birth dates of his children or grandchildren (provided of course that the latter are known to the interviewer), or the names of earlier political leaders. Awareness of the **sequence of events** is as important as the recall of individual items.

When a patient is in hospital, important information about memory is available from observations made by nurses and occupational therapists. These observations include how fast the patient learns the daily routine and the names of staff and other patients, and whether he forgets where he has put things, or where to find his bed, the sitting room, and so on.

For elderly patients the questions about memory in the clinical interview discriminate poorly between those who have a cerebral pathology and those who do not. For these patients it is more informative to use a standard set of questions that lead to a rating. One widely used rating, the Mini-Mental State Examination, is reproduced in the appendix to this chapter.

Standardized psychological tests of learning and memory can help in diagnosis and provide a quantitative assessment of the progression of memory disorder. One useful example is the Wechsler logical memory test (Wechsler 1945) in which the patient has to recall the contents of a short paragraph immediately and after 45 minutes. The score is based on the number of items recalled. Kopelman (1986) found this test to be a good discriminator between patients with organic brain disease on the one hand, and healthy controls and patients with depressive disorder on the other.

Insight

When insight is assessed, it is important to keep in mind the complexity of the concept (see Chapter 1). By the end of the mental state examination, the interviewer should have a provisional estimate of how far the patient is aware of the morbid nature of his experiences. Direct questions should then be asked to assess this awareness further. These questions are concerned with the patient's opinion about the nature of his individual symptoms, for example whether he believes that his extreme feelings of guilt are justified or not. The interviewer should also find out whether the patient believes himself to be ill (rather than, say, persecuted by his enemies) and, if so, whether he thinks that the illness is physical or mental, and whether he sees himself as needing treatment. The answers to these questions are important because they determine, in part, how far the patient is likely to collaborate with treatment. A note that merely records 'insight present' or 'no insight' is of little value.

Some difficulties in mental state examination

Apart from the obvious problem of examining patients who speak little or no English—a problem which requires the help of an interpreter—several difficulties commonly arise.

The unresponsive patient

The doctor will encounter occasional patients who are mute or stuporous (conscious but not speaking or responding in any way). He can then only make observations of behaviour, but this can be useful if done properly.

It is important to remember that some stuporous patients change rapidly from inactivity to over-activity and violence. Therefore it is wise to have help at hand when seeing such a patient. Before deciding that the patient is mute, the interviewer should allow adequate time for reply and should try a variety of topics. He should also find out whether the patient will communicate in writing. Apart from the observations of behaviour described earlier in this chapter, the examiner should note whether the patient's eyes are open or closed. If open, he should note whether they follow objects, move apparently without purpose, or are fixed; if closed, he should note whether the patient opens them on request, and, if not, whether he resists attempts at opening them.

A physical examination including neurological assessment is essential in all such cases. Also, certain signs found in catatonic schizophrenia should be sought, namely waxy flexibility of muscles and negativism (see Chapter 9).

In such cases it is essential to interview an informant who can give a history of the onset and course of the condition.

Over-active patients

Some patients are so active and restless that systematic interviewing is difficult. The interviewer may have to limit his questions to a few that seem particularly important, and to base his conclusions mainly on observations of the patient's behaviour and spontaneous utterances. However, if the patient is being seen for the first time during an emergency consultation, some of his over-activity may be a reaction to other people's attempts to restrain him. In such a case a quiet but confident approach by the interviewer often calms the patient enough to allow more adequate examination.

The patient who appears confused

When the patient gives a history in a muddled way, or appears perplexed or frightened, the interviewer should test cognitive functions early in the interview. If there is evidence of impaired consciousness, the interviewer should try to orientate the patient and to reassure him before starting the interview again in a simplified form. In such cases every effort should be made to interview another informant.

Interviewing people with mental handicap

The procedures for interviewing people with mental handicap are similar to those for people with normal intelligence but certain points should receive particular attention. Questions should be brief and worded in a simple way, avoiding structures such as subordinate clauses, passive verb constructions, and figures of speech. It may be difficult to avoid closed questions, but if they are used, the answers should be checked; for example if the question 'Are you sad?' is answered 'yes', the question 'Are you happy?' should not be answered in the same way. Some mentally handicapped people repeat the interviewer's last word, and such a response should not be accepted as agreement unless checked. (For example, Interviewer: Do you feel sad? Patient: Feel sad.)

Mentally handicapped people may have difficulty in timing the onset of symptoms or describing their sequence, and to obtain this and other information it is important to interview an informant. Nevertheless an attempt should always be made to obtain from patients an account of their symptoms and their concerns.

Special investigations

Special investigations vary according to the nature of the patient's symptoms and the differential diagnosis. No single set of routine investigations is essential for every case. Investigations most relevant to suspected specific psychiatric disorders are reviewed in the chapters on individual syndromes, particularly the section on organic psychiatry. If there is any reason to suspect physical ill-health, relevant investigations should be carried out.

Physical examination

When patients attend as day-patients or become in-patients, the psychiatrist becomes responsible for their physical as well as their mental health, and he should conduct a thorough physical examination. When out-patients are seen, they are usually referred by a general practitioner or another specialist who has often carried out the appropriate physical examination. Moreover, the care of such patients is usually shared between the psychiatrist and the other doctor. However, the psychiatrist should always determine what physical examination is relevant; he should then carry it out himself, or ensure that it has been done adequately by the referring doctor, or in certain cases arrange for another doctor to complete it. The latter course may be appropriate, for example, when the referral is made by a consultant physician who knows the patient well.

How extensive the physical examination should be must be judged in every case on the basis of diagnostic possibilities. However, the psychiatrist is most likely to be concerned with examination of the central nervous system (including its vascular supply) and the endocrine system. Of course, this does not imply that physical examination should be limited to these systems and, as indicated above, any patient admitted to the hospital should certainly have a full routine examination.

Additional neurological examination when an organic syndrome is suspected

When an organic syndrome is suspected a routine neurological examination should be performed. In this section it is assumed that the reader has some knowledge of clinical neurology; those without such knowledge are referred to a standard text-book of neurology, such as *Brain's diseases of the nervous system* (Walton 1985). Further information about tests of parietal lobe function will be found in the monograph of Critchley (1953). These tests do not help in the localization of disorders of the frontal or temporal lobes, which are diagnosed mainly from the history (see Chapter 11). The neurology of psychiatric disorders has been reviewed by Pincus and Tucker (1985).

Language abilities

Partial failure of language function is called **dysphasia** or **aphasia** (the terms are used interchangeably). Language may be affected in its expression or reception, or both, and in either its spoken or written form. Gross disorders of language function will have been noted when taking the history and mental state. Special tests will reveal less severe degrees of dysfunction. Before conducting them, tests for **dysarthria** should be done by giving difficult phrases such as 'West Register Street' or a tongue twister.

Receptive aspects of language ability are tested in several ways. The patient can be asked to read a passage of appropriate difficulty or, failing this, individual words or letters. If he can read the passage, he is asked to explain it. Comprehension of spoken language is tested by asking a patient to listen to speech. Thus he can be asked to explain what has been heard or to respond to simple commands, for example by pointing to named objects.

Expressive aspects of language are tested by asking the patient to speak and write. He can be asked to talk about his work or hobbies, and then to name objects (for example pen, key, watch, and component parts of these objects) and parts of the body. Next he can be asked to write a brief passage to dictation, and then to make up and write a passage (for example about the members of his family). If he cannot do these tests, he should be asked to copy a short passage.

Language disorders point to the left hemisphere in right-handed people. In left-handed patients localization is less certain, but in many it is still the left hemisphere. The type of language disorder gives some further guide to localization: expressive dysphasia suggests an anterior lesion, receptive dysphasia suggests a posterior lesion, mainly auditory aphasias suggest a lesion towards the temporal region, and mainly visual aphasias suggest a more posterior lesion.

Construction abilities

Apraxia is inability to perform a volitional act even though the motor system and sensorium are sufficiently intact for the person to do so. Apraxia can be tested in several ways. **Constructional** apraxia is tested by asking the patient to make simple figures with matchsticks (a square, triangle, cross) or to draw them. He can also be asked to draw a bicycle, house, or clock face. **Dressing** apraxia is tested by asking the person to put on his clothes. **Ideomotor** apraxia is tested by asking him to perform increasingly complicated tasks to command, ending for example with touching the right ear with the left middle finger while placing the right thumb on the left elbow.

Constructional apraxia, especially if the patient fails to complete the left side of figures, suggests a right-sided lesion in the posterior parietal region. It may be associated with other disorders related to this region, namely sensory inattention and anosognosia.

Agnosias

Agnosia is the inability to understand the significance of sensory stimuli even though the sensory pathways and sensorium are sufficiently intact for the patient to be able to do so. Agnosia cannot be diagnosed until there is good evidence that the sensory pathways are intact and consciousness is not impaired. Several kinds of agnosia are tested. **Astereognosia** is failure to identify three-dimensional form; it is tested by asking the patient to identify objects placed in his hand while his eyes are closed. Suitable items are keys, coins of different sizes, and paper clips.

Atopognosia is failure to know the position of an object on the skin. In **finger agnosia** the patient cannot identify which of his fingers has been touched when he has his eyes shut. Right–left confusion is tested by touching one hand or ear and asking the patient which side of the body has been touched. **Agraphognosia** is failure to identify letters or numbers 'written' on the skin. It is tested by tracing numbers on the palms with a closed fountain pen or similar object. **Anosognosia** is failure to identify functional deficits caused by disease. It is seen most often as unawareness of left-sided weakness and sensory inattention after a right parietal lesion.

Agnosias point to lesions of the association areas around the primary sensory receptive areas. Lesions of *either* parietal lobe can cause contralateral astereognosia, agraphognosia, and atopognosia. Sensory inattention and anosognosia are more common with right parietal lesions. Finger agnosia and right–left disorientation are said to be more common with lesions of the dominant parietal region.

Psychological assessment

In the past, clinical psychologists were largely concerned with the assessment of patients by standardized tests. Nowadays they are more concerned with treatment, and with assessment in the form of quantified observation of the patient's behaviour.

Many standardized tests are available. The most useful for the clinician are tests of intelligence and of higher neurological functions. Other tests are still in use but have less general value, for example those of personality, 'brain damage', and thought disorder. In this section a knowledge of the principles of psychological testing is assumed; no detailed account will be attempted here. When psychological testing is an important part of assessment, it will be mentioned in the chapters on clinical syndromes. At this stage a few general comments are appropriate.

In general adult psychiatry it is not necessary to have an accurate assessment of every patient's intelligence. If a patient seems to be of borderline subnormal intelligence, or if his psychological symptoms appear to be a reaction to work beyond his intellectual capacity, **intelligence tests** are essential. Such tests together with standard **tests of reading** ability are also essential in child and adolescent psychiatry (see Chapter 20) and in the assessment of mentally retarded patients (see Chapter 21).

In the past, much use was made of **'tests of brain damage'** in the diagnosis of possible organic syndromes. The recent advent of computerized axial tomography has reduced the need for such indirect ways of assessing diffuse cerebral pathology, although **specific neuropsychological tests** are still of some value as pointers to specific lesions of the frontal or parietal cortex. Such tests are also valuable in measuring the progression of deficits caused by disease. Further discussion of these issues will be found in Chapter 11.

Personality tests have some value in clinical research, but contribute little to everyday clinical practice because more can usually be learnt from the clinical assessment described earlier in this chapter. Projective tests such as the Rorschach test are not recommended because their validity has not been established.

Tests of thought disorder were developed to improve the accuracy of diagnosis of schizophrenia, but they proved unsuccessful. However, they are occasionally helpful in charting the progression of thought disorder.

Standardized **rating scales of behaviour** are among the most useful applications of psychometric principles in everyday clinical practice. When no ready-made rating scale is available, a clinical psychologist can often devise *ad hoc* ratings that are sufficiently reliable to chart the effects of treatment in the individual patient. For example, in measuring the progress of a depressed in-patient, a scale could be devised for the nurses to show how much of the time he was active and occupied. This could be a five-point scale, in which the criteria for each rating refer to behaviour (such as playing cards or talking to other people) appropriate to the individual patient.

Psychological principles are also used to make a **behavioural assessment**. This is a detailed account of the component elements of a patient's disorder (for example in a phobic state the elements of anticipatory anxiety, avoidance behaviour, and coping strategies) and their relationship to stimuli in the environment (for example heights), or more general circumstances (for example crowded places), or internal cues (for example awareness of heart action). A detailed description of this kind can aid diagnosis and provide a basis for behavioural treatment.

Special kinds of interview

Interviewing relatives

In psychiatry, interviews with one or more close relatives of the patient are highly important. Generally such interviews are used to obtain additional information about the patient's condition; sometimes they are used to involve the relative in the treatment plan, and sometimes to enlist his help in persuading the patient to comply with treatment.

A history from a relative or close friend is essential when the patient is suffering from a mental illness or personality disorder severe enough to impair his ability to give an unbiased and accurate account. In less severe disorders, a relative can still help by giving another view of the patient's illness and personality. For example a relative is sometimes more able than the patient to date the onset of illness accurately, especially if it was gradual. A relative can also give a useful indication of how disabling the illness is and how it affects other people. Finally, when it is important to know about the patient's childhood, an interview with a parent or older sibling is important.

With few exceptions, the patient's permission should be obtained before interviewing a relative. Exceptions occur when the patient is a child (the referral is usually initiated by the parents), and when adult patients present as emergencies and cannot give a history because they are mute, stuporous, confused, violent, or extremely retarded. In other cases, the doctor should explain to the patient that he wishes to interview a relative to obtain additional information needed for diagnosis and treatment. He should emphasize that confidential information given by the patient will not be passed to the relative. If any information needs to be given to a relative, for example about treatment, the patient's permission should be obtained. It is important to remember that relatives may misunderstand the purpose of the interview. Some assume that demands will be made on them; for example the married daughter of an elderly demented woman may think that she will be asked to take her mother into her own small home. Other relatives expect to be blamed for the patient's illness; for example the parents of a young schizophrenic may expect the doctor to imply that they have failed as parents. It is important for the interviewer to be sensitive to such ideas and, when appropriate, to discuss them in a reassuring way. He should always begin the interview by explaining its purpose.

Adequate time should be allowed for the interview; relatives are likely to be anxious, and time is needed to put them at ease, gather facts, and impart any necessary information.

The interview will enable the doctor to discover whether the relatives are having any problems as a result of the patient's illness. If they need help, the doctor should collaborate with the general practitioner to arrange it. However, he should not become involved in the relative's problem to an extent that conflicts with his primary duty to his patient.

After the interview the psychiatrist should not let the patient know what the relative has said unless the latter has given permission. It is important to seek permission if the relative has revealed something that should be discussed with the patient, for example an account of excessive drinking previously denied by the patient. However, if the relative is unwilling that information should be passed on, this must be respected by the doctor; for example a wife may fear violent retaliation from her husband. When the relative is unwilling in this way, the psychiatrist should try to find ways of enabling the patient to reveal the behaviour himself in a further interview. Since, in most places, patients are generally entitled to read their medical notes, interviews with relatives

should be recorded and kept in a way that respects their wishes about confidentiality.

Problems sometimes arise when someone other than the nearest relative telephones the psychiatrist about the patient. Information should not be given over the telephone, even if the doctor is certain of the caller's identity. Instead the patient should be consulted and, if he agrees, an interview arranged. The psychiatrist must never allow a conspiratorial atmosphere to develop in which he conceals conversations with family members or takes sides in their disputes.

Family interviews at home

It is sometimes appropriate to add to information about the patient's social circumstances by visiting the home, or arranging for a psychiatric nurse or a social worker to do so. Such a visit often throws new light on the patient's home life. It can sometimes lead to a more realistic evaluation of the relationship between family members than can be obtained from interviews in hospital. Before arranging a visit the psychiatrist should if possible talk to the general practitioner, who often has first-hand knowledge of the family and their circumstances from home visits over the years. If another member of the staff is going to make the visit, the psychiatrist should discuss the purpose of the visit with him.

Emergency consultations

When time is limited and an immediate decision is required about diagnosis and management, it may be possible to obtain an outline history. However short the time, it is essential to obtain a clear account of the presenting symptoms, including their onset, course, and severity. A knowledge of the major clinical syndromes will then guide the interviewer to enquiries about other relevant symptoms, including those which arise in organic brain syndromes. Recent stressful events should always be asked about, together with any previous physical or mental illness. An account of previous personality is important, though it may be difficult to obtain unless there are relatives or close friends present. Habits regarding alcohol and drugs are especially important.

The family and personal history will often have to be covered quickly by asking a few salient questions. Throughout the interview the psychiatrist should be thinking which questions need to be asked immediately and which can be deferred until later.

A brief but relevant physical examination should be carried out unless already performed by another doctor.

If the above points are borne in mind when conducting an emergency consultation, common sense coupled with a sound knowledge of the major clinical syndromes should prove a satisfactory guide.

Interviewing in primary care

Much of a general practitioner's work is concerned with the identification of minor psychiatric disorders among patients presenting with a combination of physical and mild psychological symptoms. A useful brief method of identifying psychiatric disorders has been described by Goldberg and Huxley (1980). It focuses on the emotional symptoms encountered most commonly in general practice, and takes account of faults in interviewing that were found by watching general practitioners at work.

Goldberg and Huxley (1980) make a useful distinction between bias and accuracy in interviewing. General practitioners with a positive bias towards mental disorder (who over-diagnose it compared with standard assessment) ask many questions about psychological distress, ask about home circumstances, and are sensitive to cues for distress in the interview. General practitioners who are accurate (who agree closely with standard assessments) are also sensitive to cues for distress but differ from the former group in having a better interview technique and a better knowledge of general medicine. The latter point may relate to the fact that accuracy involves identifying those who are normal as well as identifying those who are emotionally disordered. When patients are interviewed by doctors with poor interviewing skills, they reveal fewer cues of emotional distress than when interviewed by more skilful people. Thus the skilled interviewer has an extra advantage.

In these brief interviews the first few minutes are extremely important. However short the time, it is essential to give the patient an adequate opportunity to express his problem. Family doctors sometimes omit this because they assume that the patient has come back for further advice about a previous problem. Sometimes they start questioning too early, with the result that opening questions may be answered as if they were social pleasantries (for example 'How are you feeling now?'—'Fine thanks'). On the other hand the doctor should not sit in silence reading his previous notes, as the patient may then begin to feel ill at ease and become unable to reveal his real concerns. The interview can begin with an open question such as 'What seems to you to be wrong?', and then proceed mainly by prompts and clarifying questions. As in a longer interview, the doctor should be as alert to non-verbal behaviour as he is to the spoken word.

The next task is to understand clearly the nature of the symptoms. In general practice the presenting complaint is often physical even when the disorder is psychiatric. The patient should always be allowed adequate time to describe the complaint in his own words before questions are asked. Thus a complaint of headache should not be followed immediately by questions about the side of head on which it is felt. Instead the patient should be encouraged to describe the symptom in more detail. It may then become apparent that he has a tight feeling over the brows rather than a painful headache. Although this may seem obvious, it was found to be a common cause of error in Goldberg and Huxley's study of interviews in general practice.

For any complaint that may have psychological causes Goldberg and Huxley put forward a simple scheme of assessment with four components:

- general psychological adjustment
- the presence of anxiety and worries
- symptoms of depression
- the psychological context.

General psychological adjustment is assessed by asking about fatigue, irritability, poor concentration, and the feeling of being under stress. To enquire into *anxieties and worries* the interviewer asks about physical symptoms as well as tension, phobias, and persistent worrying thoughts. *Symptoms of depression* are covered next, including persistent depressive mood, tearfulness, crying, hopelessness, self-blame, thoughts that life is unbearable, ideas about suicide, early morning waking, diurnal variation of mood, weight loss, and loss of libido. Of these, Goldberg and Huxley found that general practitioners were most likely to overlook questions concerned with depressive thoughts.

Often the family doctor already knows the *psychological context* of the patient's problem since he is aware of his family and occupational circumstances. If so, the doctor can omit some of the questions required in an interview with a new patient. However, he should think systematically about the patient's work, leisure, marriage, and other relationships, and should ask any questions needed to bring his knowledge up to date.

An interview of this kind can be conducted within the short time available for first consultations in general practice. Usually a conclusion can be reached by the end of the interview. If not, a preliminary plan can be made and a later interview arranged for completion of a full psychiatric history and mental state examination.

Case notes

The importance of case notes

Good case records are important in every branch of medicine. In psychiatry they are even more vital because a large amount of information is collected from a variety of sources. Unless material is recorded clearly, with facts separated from opinions, it is difficult to think clearly about clinical problems and to make appropriate decisions about treatment. Equally, it is important to summarize the information in a way that allows essential points to be grasped readily by someone new to the case. Case notes are not just an *aide mémoire* for a doctor's own use, but an essential source of information for others who may see the patient in the future. Therefore they must be legible and well thought out.

It is important to remember the medicolegal importance of good case records. On the rare occasions when a psychiatrist is called upon to justify his actions in the coroner's court, at a trial, or after a complaint lodged by a patient, he will be greatly assisted by good case notes. The psychiatrist should remember that in certain circumstances case notes can be called upon by lawyers acting for patients (and in many countries may be read by the patient at any time).

The admission note

When a patient is admitted to hospital urgently, the doctor may have limited time for the interview. It is then particularly important to select the right topics to elicit and record. The admission note should contain at least: (i) a clear account of the reasons for admission, (ii) any information required for a decision about immediate treatment, and (iii) any relevant information that will not be available later including details of the mental state on admission and information from any informant whose presence at a later date cannot be relied upon. The account of the mental state should include well-chosen verbatim extracts to illustrate phenomena such as delusions or flight of ideas. If there is time, a systematic history should be added. However, it is a common mistake among trainees to spend too much time on details that are not essential to immediate decisions and can be taken next day, while failing to record details of mental state that may be transitory and yet of great importance to final diagnosis.

The admission note should end with a brief statement of a provisional plan of management. This plan should be agreed with the senior nurses caring for the patient at the time.

Progress notes

Progress notes should not be written in such general terms as to be of little value when the case is reviewed later. Instead of recording merely that the patient feels better or is behaving more normally, the note should state in what ways he feels better (for example less despondent or less preoccupied with thoughts of suicide) or is less disturbed in behaviour (for example no longer so restless as to be unable to sit at table throughout a meal).

Progress notes should also refer to treatment. Details of drug treatment often go unrecorded in the progress notes, presumably because they appear on the prescription sheet. However, when a patient's progress is reviewed, it is much more convenient to have the timing and dosage of medication recorded alongside the mental state and behaviour. Psychological and social treatment should also be noted. A verbatim account of a psychotherapy session is difficult to write and seldom of value in management. Instead, notes should be made of the main themes of therapeutic interviews, together with any relevant observations of the patient's response. An additional note summarizing progress made in the course of several sessions can be made at intervals.

A careful note should also be made of any information or advice given by the doctor to the patient or his relatives. This should enable anyone giving advice later to know whether or not it differs from what was said before, so that an appropriate explanation can be given.

Observations of progress are made not only by doctors, but also by nurses, occupational therapists, clinical psychologists, and social workers. As a rule, these other members of staff keep separate notes for their own use, but it is desirable that important items of information are also written in the medical record.

A careful note chould be kept of decisions reached at ward rounds and case conferences, and on any other occasions when the management of the patient is discussed with the consultant or his deputy. It is particularly important to set out clearly the plans made for the patient's further care on discharge from hospital.

The case summary

This section can best be understood by referring to the specimen case summary on pp. 47–8.

The case summary is usually written in two parts. The first is completed within a week of the patient's admission. It has two main purposes. First, after extensive history taking has been completed, it is a useful exercise to select the salient features of the case. Second, the Part I summary is valuable to any doctor called to see the patient when the usual psychiatrist is not available. The items included in the first summary are all those from 'Reason for referral' to 'On examination' in the specimen summary.

The Part II summary is usually prepared within a day or two of the patient's discharge; it complements the Part I summary by adding special investigations and all subsequent items in the specimen summary. The whole summary is important if the patient becomes ill again, especially if he is under the care of another psychiatrist.

Summaries should be brief but comprehensive. They should be written in telegraphic style and laid out in a standard form that makes it easy for other people to find particular items. It is sometimes appropriate to omit particularly confidential details, noting instead that relevant information will be found in the case history. The Part I summary seldom occupies more than one and a half sides of a typed page, while the Part II summary is about half a page in length. A longer summary often means that the case has not been understood clearly.

Some of the items in the summary call for comment. The reason for referral should be a brief statement avoiding technical terms; for example it might read: 'having been found wandering at night in an agitated state, shouting about God and the devil', rather than 'for treatment of schizophrenia'. The description of personality is often the most difficult section to complete briefly but informatively. However, with practice it is usually possible to list well-chosen words and phrases which bring the person to life. This part of the summary is important and repays considerable thought.

If no abnormality is found on physical examination, there is no need to make a separate entry for each system; it is usually sufficient to enter a single statement that routine physical examination showed no abnormality. However, when the mental state is recorded, a comment should be made under each heading whether or not any abnormality has been found.

When possible, the entry under diagnosis should use the categories of the current edition of the *International classification of disease* (or DSMIV in countries using this classification) (see p. 68). However it may be necessary to add some additional comments to convey the complexities of an unusual case. If the diagnosis is uncertain, alternatives should be given, with an indication of the likelihood of each.

The summary of treatment should indicate the main treatments used, including the dosage and duration of any medication. The prognosis should be stated briefly but as definitely as possible. Statements such as 'prognosis guarded' are of little help to anyone. Unless the doctor commits himself more firmly he will be unable to learn from comparing his predictions with the actual outcome. At the same time, it is appropriate to note how certain the writer is about the prognosis and why any uncertainties have arisen; for example 'The depressive symptoms are not likely to recur in the next year provided that the patient continues taking drugs. The subsequent course is uncertain because it depends on the course of her son's leukaemia'.

The plan for further treatment should specify not only what is to be done but also who is to do it. The roles of the hospital staff and of the family doctor should be made clear.

An example of a widely used method of recording the case summary is given below.

Example of a case summary

Consultant: Dr A   Admitted 27.6.91
Registrar: Dr B   Discharged 4.8.91

Mrs C. D. Date of birth 7.2.60

Reason for referral Increasingly low in mood and inactive despite out-patient treatment.

Family history *Father* 66, retired gardener, good physical health, mood swings, poor relationship with patient. *Mother* 57, housewife, healthy, convinced spiritualist, distant relationship. *Sibling* Joan, 35, divorced, healthy.

Home materially adequate, little affection. *Mental illness* father's brother in hospital four times: 'manic depression'.

Personal history *Birth and early development* normal. *Childhood health* good. *School* 6–16 uneventful; male friends. *Occupations* 16–22 shop assistant. *Marital* several boyfriends; married at 22, husband two years older, lorry driver. Unhappy in last year following husband's infidelity. *Children* Jane, 7, well; Paul, 4, epileptic. *Sexual* satisfactory until last year. *Menses* no abnormality. *Circumstances* council house, financial problems.

Previous illness Aged 20, appendicectomy. Aged 24 (post-natal) depressive illness lasting three weeks.

Previous personality Few friends, interests within the family, variable mood, worries easily, lacks self-confidence, jealous, no obsessional traits, no conventional religious beliefs but shares mother's interest in the supernatural. Drinks occasionally, non-smoker, denies drugs.

History of present illness For six weeks, since learning of husband's infidelity, increasingly low-spirited and tearful, waking early, inactive, neglecting children. Eating little. Low libido. Believes herself to be in contact with dead grandmother through telepathy. Progressive worsening despite amitriptyline 125 mg per day for three weeks.

On examination *Physical* n.a.d. *Mental* dishevelled and distraught. *Talk* slow, halting, normal form. Preoccupied with her unhappy state and its effect on her children. *Mood* depressed, with self-blame, hopelessness but no ideas of suicide. *Delusions* none. *Hallucinations* none. *Compulsive phenomena* none. *Orientation* normal. *Attention and concentration* poor. *Memory* not impaired. *Insight* thinks she is ill but believes she cannot recover.

Special investigations Haemoglobin and electrolytes n.a.d.

Treatment and progress Amitriptyline increased to 175 mg per day, graded activities, joint interviews with husband to improve marital relationship. Advice from social worker to husband about management of financial problems. Progressive improvement in hospital with three weekends at home before final discharge. Amitriptyline reduced to 100 mg per day at time of discharge.

Condition on discharge Not depressed but still uncertain about future of marriage.

Diagnosis Depressive disorder.

Prognosis Depends on further progress with marital problems. If these improve, the short-term prognosis is good. However, vulnerable to further depressive illness in the long term.

Further management (1) Continue amitriptyline 100 mg for six months (prescriptions from hospital); (2) out-patient attendance to continue marital interviews (first appointment 14.8.91); (3) review progress and return to GP's care in three months.

Formulation (Table 2.5)

A formulation is a concise assessment of the case. Unlike a summary, it is a discussion of alternative ideas about diagnosis, aetiology, treatment, and prognosis, and of the arguments for and against each alternative. A good formulation is based on the arguments for and against each alternative. A good formulation is based on the facts of the case and not on speculation, but it may contain verifiable hypotheses about matters that are uncertain at the time of writing. A formulation is concerned not only with disease concepts, but also with the understanding of how the patient's lifelong experiences have influenced his personality and his ways of reacting to adversity.

There is more than one way of setting out a formulation, and the following is an approach recommended by the authors. The formulation begins with a concise statement of the essential features of the case. This should seldom be more than two or three sentences; for example 'Mrs Jones is a 60-year-old divorced woman with depressed mood and sleep disturbance which started after an operation for cancer of the bowel and which have not responded to out-patient treatment'.

Table 2.5 *The formulation*

Statement of the problem
Differential diagnosis
Aetiology
Further investigations
Plan of treatment
Prognosis

The differential diagnosis is considered next. This should be a list of reasonable possibilities in the order of their probability. The writer should avoid listing every conceivable diagnosis however remote. A note is made of the evidence for and against each diagnosis, with an assessment of the balance. At the end, the writer's conclusion about the most probable diagnosis should be stated clearly.

Aetiology comes next. The first step is to identify predisposing, precipitating, and maintaining causes. The reasons for any predisposition are then considered, usually in chronological order to show how each factor may have added to those that went before. For example a family history of manic depressive psychosis suggests a genetic predisposition to similar illness; in a particular case this may have been added to by the death of the patient's mother when he was a child, and increased further by adverse influences in a children's home.

After aetiology the conclusions about diagnosis and aetiology should be summarized with a list of outstanding problems and any further investigations needed. Next a concise plan of treatment is outlined. This should mention social measures as well as psychological treatment and medication, together with the role of nurses and occupational therapists.

Finally, a statement is made about prognosis. This is often the most difficult part of the formulation. As with the summary, it is wrong to avoid commitment by writing down a vague statement. It is better to make a firm prediction, for example 'These depressive symptoms should recover quickly in hospital but are likely to recur if her husband begins to drink heavily again'. If his prediction is proved wrong, the doctor can learn by comparing it with the actual outcome, but nothing can be learnt from a non-committal statement.

Example of a formulation

(*Note*: This formulation refers to the same hypothetical case that was presented above in an example of a summary. By comparing the two, the reader can appreciate the difference between the material selected for each.)

Mrs C. D. is a 31-year-old married woman who for six weeks has been feeling increasingly low-spirited and unable to cope at home, despite out-patient treatment with antidepressant drugs.

Diagnosis

Depressive disorder As well as feeling low-spirited, Mrs C. D. has woken unusually early, felt worse in the morning, and lost her appetite. She has little energy or initiative. She blames herself for being a bad mother and believes that she cannot recover. The only feature apparently against this diagnosis—her belief that she is in contact with her dead grandmother—is discussed below.

Schizophrenia Mrs C. D.'s belief that she is in contact with her dead grandmother was present before she became ill. It relates clearly to her own and her mother's interest in spiritualism. It is an over-valued idea, not a delusion. She has no first-rank symptom of schizophrenia.

Personality disorder Although Mrs C. D. has mood variations from mild depression to an energetic cheerful state, these are not sufficiently intense to constitute a cyclothymic personality disorder.

Conclusions Depressive disorder.

Aetiology The symptoms appear to have been *precipitated* by news of her husband's infidelity. She was *predisposed* to react severely to this news by the insecure and jealous traits in her personality. She also appears to be predisposed to develop a depressive disorder in that (a) she became depressed after the birth of her first child, (b) she is subject to mood variations, and (c) her father suffers similar, but more extreme, mood variations and his brother has been admitted to hospital four times for treatment of a manic depressive disorder.

The depressive disorder may have been *maintained* in part by continuing quarrels with the husband and by worry about debts he has incurred. Her knowledge of her sister's divorce and subsequent unhappiness has added to her concerns about the future of her own marriage.

Treatment The pattern of symptoms of the depressive disorder suggests that it is likely to respond to amitriptyline given in adequate dosage. The few side-effects experienced with 125 mg per day may indicate a lower than average blood concentration. The dose should be increased to 175 mg per day. Joint interviews with the patient and her husband are needed to attempt to resolve the marital problems. (It appears that she has a genuine wish for a reconciliation.) The husband should be advised by the social worker about the steps that he can take to deal with his debts.

Prognosis If the marital problems improve, the immediate prognosis is good. However, the several predisposing factors noted above indicate that she may develop further depressive disorder, particularly at times when she encounters further stressful events.

Problem lists

A problem list is a useful addition to the formulation in cases with complicated social problems. Such a list makes it easier to identify clearly what can be done to help the patient, and to monitor progress in achieving agreed objectives of treatment.

The use of a problem list can be shown with two examples. The first (Table 2.6) is a list that might be compiled for a young married women who had taken a small overdose impulsively and had no psychiatric disorder or definite personality disorder.

As progress is made in dealing with problems in this list, new ones may be added or existing ones modified. For example after a few joint interviews it might appear that the patient's sexual difficulties are a cause of the marital problem rather than a result, and that counselling about sexual matters should be carried out. Item 4 would then be amended appropriately. Likewise, if the assessment of the child by the general practitioner were to confirm speech delay, an appointment for a specialist opinion might follow.

It is often appropriate to draw up the list with the patient so that he understands which problems can be changed and what he must do himself to bring this about.

Similar lists can be a valuable aid to the review of cases during ward rounds. For this purpose, an important component of therapy is likely to be the treatment of mental disorder, often by drugs. Although such treatment can be shown on the same sheet as other problems, it should be separated clearly from them, as in the list in Table 2.7 drawn up for a 45-year-old depressed woman.

The life chart

A life chart is a way of showing the time relations between episodes of physical and mental disorder and potentially stressful events in the patient's life.

Table 2.6

Problem	Action	Agent	Review
1. Frequent quarrels with husband	Joint interviews	Dr A	3 weeks
2. 3-year-old son retarded in speech	Assessment	General practitioner	1 week
3. Housing said to be damp and unsatisfactory	Visit housing department of local authority	Patient	2 weeks
4. Sexual dysfunction (?secondary to 1)	Defer		

It is often useful when the history is long and complicated. The chart has three columns—one for life events, and one each for physical and mental disorder. Its rows represent the years in the patient's life.

Completion of a life chart requires detailed enquiry into the timing of events, and this may clarify the relationships between stressors and the onset of illness, and also between physical and mental disorders. For example in the case of a recurrent illness previously thought to be provoked by stressful events, the chart may show that it has run a regular course and that comparable events have occurred at other times without consequent illness. In another case the chart may provide convincing evidence of a relationship between stressful events and illness.

Letters to general practitioners

When a letter is written to a general practitioner, whether after an out-patient assessment or on discharging a patient from hospital, the first step is to think what the general practitioner already knows about the patient, and what questions he asked on referral. If the family doctor's referral letter outlined the salient features of the case, there is no need to repeat them in reply. When the patient is less well known to the general practitioner, more detailed information should be given; it is then often appropriate to use subheadings (family history, personal history, etc.) so that information can be found readily if needed later.

Similarly, if the diagnosis given in the referral letter is correct, it is only necessary to confirm it; otherwise the reasons for the diagnosis should be outlined.

Treatment and prognosis are dealt with next. When discussing treatment, the dosage of drugs should always be stated. The psychiatrist should indicate whether he has issued a prescription, how long a period it covers, and whether he or the general practitioner is to issue any subsequent prescriptions. If psychotherapy, behaviour therapy, or social work are planned, the letter should name the therapist or agent concerned and indicate his profession (for example supportive psychotherapy from Mr Smith, hospital social worker). The date of the patient's next visit to hospital should be stated, so that the general

Table 2.7

Problem	Action	Agent	Review
1. Depressive disorder	Amitriptyline 150 mg per day	Dr A	3 weeks
2. Loneliness (children now grown up)	Seek paid or voluntary work	Patient and social worker	5 weeks
3. Shy and awkward in company	Social skills training group	Psychologist and nurse	4 weeks
4. Heavy irregular periods	Gynaecological opinion	Dr A	1 week

practitioner knows whether to see the patient himself in the meantime.

At the time of discharge from in-patient or day-patient treatment, it is often appropriate to telephone the general practitioner to discuss subsequent management before the discharge letter is written. This telephone discussion ensures that the division of responsibilities is acceptable to the family doctor. If this is not done, the plans formulated may be well-intentioned but inappropriate.

Standardized definitions of symptoms

For research, symptoms need to be recorded in a reliable way, and for some kinds of research (for example community surveys) interviews have to be carried out by people who have not trained in psychiatry. For these purposes, standard instruments have been developed, some designed primarily to rate symptoms (for example in a clinical trial) and others for making diagnoses. This account begins with examples of the standard definitions of symptoms in a commonly used instrument, the Present State Examination (PSE) (Wing *et al.* 1974).

Standardized interviewing

In most instruments, standard questions are provided to ensure that the interviews are thorough and reproducible. The following examples illustrate how the standard interview is carried out in PSE (Example 1) and in SCID (Example 2).

Example 1 Delusion of thought being read
This is usually an explanatory delusion. Often it goes with delusions of reference or misinterpretation which require some explanation of how other people know so much about the subject's future movements. It may be an elaboration of thought broadcast, thought insertion, auditory hallucinations, delusions of control, delusions of persecution, or delusions of influence. It can even occur with expansive delusions (the subject wishing to explain how Einstein, for example, stole his original ideas). Therefore the symptom is in no way diagnostic. It is most important that it should not be mistaken for diagnostically more important symptoms such as thought insertion or broadcast.

If the subject merely entertains the possibility that his thought might be read but is not certain about it, rate (1); rate delusional conviction (2). Exclude those who think that people can read their thoughts as a result of belonging to a group that practices 'thought reading'—this would be rated (1) or (2) on symptom no. 83.

The section on hallucinations in PSE continues with questions about third person hallucinations and non-verbal hallucinations.

Example 2 Questions about delusions (reproduced from Spitzer *et al.* 1990)

SCID-P (Version 1.0)
B. *Psychotic and Associated Symptoms*

This module is for coding psychotic and associated SXS that have been present at any point in the person's lifetime.

For all psychotic and associated symptoms coded '3', determine whether the symptom is 'not organic', or whether there is a possible or definite organic cause. The following questions may be useful if the overview has not already provided the information: When you were (PSYCHOTIC SXS), were you taking any drugs or medicine: Drinking a lot? Physically ill?

If has not acknowledged psychotic SXS: Now I am going to ask you about unusual experiences that people sometimes have.

If has acknowledged psychotic SXS: You have told me about (PSYCHOTIC EXPERIENCES). Now I am going to ask you more about those kinds of things.

Delusions
False personal belief(s) based on incorrect inference about external reality and firmly sustained in spite of what almost everyone else believes, and in spite of what constitutes incontrovertible and obvious proof or evidence to the contrary. Code overvalued ideas (unreasonable and sustained belief(s) that is/are maintained with less than delusional intensity) as '2'.

Note: A single delusion may be coded '3' on more than one of the following items.

Did it ever seem that people were talking about you or taking special notice of you?

Delusions of reference, i.e. personal significance is falsely attributed to objects or events in environment.

What about receiving special messages from the TV, radio, or newspaper, or from the way things were arranged around you? Describe:

$$? \quad 1 \quad 2 \quad 3$$

1, Poss/def organic; 3, Not organic.

Persecutory delusions, i.e. the individual (or his or her group) is being attacked, harassed, cheated, persecuted, or conspired against.

What about anyone going out of the way to give you a hard time, or trying to hurt you?

Describe:

$$? \quad 1 \quad 2 \quad 3$$

1, Poss/def organic; 3, Not organic.

Did you ever feel that you were especially important in some way, or that you had powers to do things that other people could not do?

Grandiose delusions, i.e. content involves exaggerated power, knowledge, or importance.

Describe:

$$? \quad 1 \quad 2 \quad 3$$

1, Poss/def organic; 3, Not organic.

? = inadequate information/ 1 = absent or false; 2 = subthreshold; 3 = threshold or true.

The section on delusions in SCID continues with questions about other kinds of delusions (somatic, nihilistic, etc.).

Instruments for measuring symptoms

In research, and at times in clinical practice, it is necessary not only to record whether symptoms are present or absent but also to measure their severity. Some instruments rate a single symptom or a narrow group of symptoms (for example anxiety or depression), whereas others rate a broad group of symptoms as an overall measure of the severity of a disorder. Most diagnostic instruments also rate the severity or certainty of symptoms rather than simply recording whether they are present or absent. Some instruments are completed by the patient, and others are filled in by an interviewer who may choose the interviewing method or be required to use a standard set of questions.

Ratings of single symptoms and narrow groups of symptoms

Anxiety symptoms

1. **Hamilton Anxiety Scale (HAS)** (Hamilton 1959) Although this scale is used widely to measure anxiety it should be employed solely with anxiety disorders and not for rating anxiety in patients with other disorders.

Thirteen items are rated by an interviewer on five-point scales, each on the basis of a brief description. The interviewing method is for the rater to decide. Some depressive symptoms are included so that the scale is in fact a measure of the severity of the anxiety syndrome and not of the symptom of anxiety.

2. **Clinical Anxiety Scale (CAS)** (Snaith *et al.* 1982) This scale for interviewers was developed from the HAS, to focus more clearly on the symptom of anxiety by leaving out the depressive and somatic symptoms included in the HAS. Its use is not restricted to patients with a diagnosis of anxiety disorder.

3. **The State–Trait Anxiety Inventory (STAI)** (Speilberger *et al.* 1970) is a self-rating scale with 20 statements which is completed in two ways: as the person feels when he completed the scale (trait) and how he feels generally (state).

Depressive symptoms

1. **Hamilton Rating Scale for Depression (HRSD)** (Hamilton 1967) This scale is filled in by an interviewer who uses an unstructured interview. It measures the severity of the depressive syndrome (p. 197) rather than the symptom of depression.

2. **Beck Depression Inventory (BDI)** (Beck *et al.* 1961) This 21 item inventory is usually completed by the patient: each item has four to six statements, one of which is chosen as best describing the symptom at the time.

3. **Montgomery–Asberg Depression Rating Scale (MADRS)** (Montgomery and Asberg 1979) This inventory has 10 items rated on a four-point scale by an interviewer using definitions for each point. Only psychological symptoms of depression are rated.

Obsessional symptoms
Yale–Brown Obsessive Compulsive Scale (YBOCS) (Goodman *et al.* 1989*a*) This instrument rates obsessive compulsive symptoms in patients diagnosed as having obsessive–compulsive disorder. This scale is rated by a clinician using a four-point

scale for each of 10 symptoms. Depressive and anxiety symptoms and obsessional personality traits are not rated.

Other ratings of narrow groups of symptoms

(a) Rating for negative symptoms of schizophrenia, for example the **Positive and Negative Syndrome Scale (PANSS)**, (Kay *et al.* 1987) which rates blunted affect, emotional withdrawal, poor rapport, social withdrawal, difficulty in abstract and stereotyped thinking, and lack of spontaneity. A 30 minute interview is required.

(b) Ratings for extrapyramidal symptoms such as the **Extrapyramidal Symptoms Rating Scale (ESRS)**, (Chouinard *et al.* 1980) on which the clinician makes quantitative ratings of the symptoms of parkinsonism, dystonia, and dyskinesia.

Ratings of broad groups of symptoms

1. **General Health Questionnaire (GHQ)** (Goldberg 1972) This instrument contains 60 items, but shorter versions have been developed with 30, 28, and 20 items. It is designed for use as a screening instrument in primary care, general medical practice, or community surveys. The full version can be completed within 10 minutes, and the shorter versions even more rapidly. The symptom ratings are added to a score which indicates overall severity which is expressed by the judgement of whether a psychiatrist would judge the patient to be a 'case' or a 'non-case'. There is also a version with symptom subscales for somatic symptoms, anxiety and insomnia, depression, and social dysfunction (Goldberg and Hillier 1979).

2. **PSE Index of Definition** The PSE (Wing *et al.* 1974) and its successor SCAN are primarily designed for diagnosis. However, the profile of symptoms obtained from the interview can be used to calculate an overall severity rating, the Index of Definition, which varies from 1 to 8.

A rating of 5 or more indicates that the patient probably has a psychiatric disorder.

3. **Brief Psychiatric Rating Scale (BPRS)** (Overall and Gorham 1962) This instrument has 16 items each scored on a seven-point scale. There are criteria to define the symptom items but not for the severity ratings. The time period is not defined and must be decided by the rater. It is suitable for rating severe psychiatric illness but not minor disorders.

Several instruments rate broad groups of symptoms for the purpose of making standard diagnoses. These instruments are described in the chapter on classification (p. 64). (For a comprehensive review of rating methods see Thompson (1989).)

Appendix

The Mini-Mental State Examination

(Add points for each correct response.)

Orientation		Score	Points
1. What is the	Year?	—	1
	Season?	—	1
	Date?	—	1
	Day?	—	1
	Month?	—	1
2. Where are we?	State	—	1
	County?	—	1
	Town or city?	—	1
	Hospital?	—	1
	Floor?	—	1

Registration

3. Name three objects, taking 1 second to say each. Then ask the patient all three after you have said them. Give one point for each correct answer. — 3
Repeat the answers until patient learns all three.

Attention and calculation

4. Serial sevens. Give one point for each correct answer. Stop after five answers. *Alternative:* Spell WORLD backwards. — 5

Recall

5. Ask for name of three objects learned in Q.3. Give one point for each correct answer. — 3

Language

6. Point to a pencil and a watch. Have the patient name them as you point. — 2
7. Have the patient repeat 'No ifs, ands, or buts'. — 1
8. Have the patient follow a three-stage command: 'Take a paper in your right hand. Fold the paper in half. Put the paper on the floor'. — 3
9. Have the patient read and obey the following: 'CLOSE YOUR EYES'. (Write it in large letters.) — 1
10. Have the patient write a sentence of his or her choice. (The sentence should contain a subject and an object, and should make sense. Ignore spelling errors when scoring.) — 1
11. Enlarge the design printed below to 1.5 cm per side and have the patient copy it. (Give one point if all sides and angles are preserved and if the intersecting sides form a quadrangle.) — 1

—— =Total 30

Reprinted with permission from J. C. Anthony, L. Le Resche, U. Niaz, M. R. Von Korf, and M. F. Folstein (1982). Limits of the 'Mini-Mental State' as a screening test for dementia and delirium among hospital patients. *Psychological Medicine*, **12**, 397–408.

Further reading

Cohen-Cole, S. A. (1991). *The medical interview: the three function approach.* Mosby Year Book, St Louis, M.

Goldberg, D. and Huxley, P. (1992). *Common mental disorders: a biosocial model,* pp. 37–46. Tavistock/Routledge, London.

Leff, J. P. and Isaacs, A. D. (1978). *Psychiatric examination in clinical practice.* Blackwell, Oxford.

Pincus, J. H. and Tucket, G. J. (1985). *Behavioural neurology* (3rd edn). Oxford University Press.

Strauss, G. D. (1995). The psychiatric interview, history and mental status examination. In *Comprehensive textbook of psychiatry* (6th edn) (ed. H. I. Kaplan and B. J. Saddock). Williams and Wilkins, Baltimore, MD.

Thompson, C. (1989). *The instruments of psychiatric research.* John Wiley, Chichester.

Trzepacz, P. T. and Baker, R. W. (1993). *The psychiatric mental status examination.* Oxford University Press, New York.

3 | *Classification in psychiatry*

In psychiatry, classification attempts to bring order into the great diversity of phenomena met in clinical practice. The purpose of classification is to identify groups of patients who share similar clinical features, so that suitable treatment can be planned and the likely outcome predicted. Most systems of classification are based on diagnostic categories such as schizophrenia or affective disorder. These systems are classifications of disorders (rather than patients). Since one patient may have more than one disorder, it is necessary to devise simple procedural rules for classification. It may also be convenient to add various other characteristics of a patient such as social functioning.

In general medicine, classification is fairly straightforward. Most physical conditions can be classified on the basis of aetiology (for example pneumococcal or viral pneumonia) and structural pathology (for example lobar or broncho-pneumonia). Some general medical conditions, such as migraine or trigeminal neuralgia, are not yet classifiable in this way; therefore they are classified solely on symptoms. Psychiatric disorders are mainly analogous to this second group. Although some psychiatric disorders have a well-understood physical aetiology (such as phenylketonuria, Down's syndrome, or Alzheimer's disease), most can be classified only on symptoms.

This chapter begins with a brief discussion of the concept of mental illness. Then an outline is given of the principles that underlie most systems of classification, and the system used in this book is summarized. Some contentious issues are reviewed, including objections that have been raised to psychiatric classification itself, and also the question of categorical versus non-categorical classification. Next, an account is given of methods for achieving greater diagnostic agreement between psychiatrists. This account leads to a description of individual systems of classification, including the main international systems. Finally, guidelines are given on classification in everyday clinical practice.

The concept of mental illness

In everyday speech the word 'illness' is used loosely. Similarly, in psychiatric practice the term 'mental illness' is used with little precision. A good definition of mental illness is difficult to achieve. In everyday clinical practice this difficulty is important mainly in relation to ethical and legal issues such as compulsory admission to hospital. In forensic psychiatry the definition of mental illness is particularly important in relation to the assessment of issues such as fitness to plead and criminal responsibility.

It is easy to understand why the concept of mental illness is not prominent in most psychiatric practice. The psychiatrist is not usually concerned with a concept of such generality. He is more concerned with organizing the wide-ranging phenomena encountered in psychiatry, so that he can plan treatment rationally and predict outcome. In practice the best approach is to start with the basic data (symptoms and signs) and to group them into syndromes, i.e. constellations of symptoms that occur together frequently and have implications for treatment and prognosis. The psychiatrist habitually works from the particular to the general, and not vice versa.

Although the major systems of classification refer specifically to mental disorders rather than to mental illnesses (World Health Organization 1992*b*; American Psychiatric Association 1994), the concept of mental illness is intellectually interesting and (as mentioned above) has ethical and legal implications. For these reasons an outline of the main arguments will be given here.

Definitions of mental illness

Many attempts have been made to define mental illness (Clare 1979; Caplan *et al.* 1981; Fulford 1989). A common approach is to examine the concept of illness in general medicine and to identify any analogies with mental illness. In general medicine there are three types of definition: absence of health, presence of suffering, and pathological process whether physical or psychological.

First, illness of any kind can be defined as the *absence of health*. This approach changes the emphasis of the problem but does not solve it, because health is even more difficult to define. The World Health Organization, for example, defined health as 'a state of complete physical, mental and social well-being, and not merely the absence of disease or infirmity'. As Lewis (1953b) rightly commented, 'a definition could hardly be more comprehensive than that, or more meaningless'. Many other definitions of health have been proposed, all equally unsatisfactory.

A second approach is to define illness in terms of the *presence of suffering*. This approach has some practical value because it defines a group of people likely to consult doctors. A disadvantage is that the term cannot be applied to everyone who would usually be regarded as ill in everyday terms. For example, patients with mania may feel unusually well and may not experience suffering, though most people would regard them as mentally ill.

A third approach is to define mental illness in terms of *pathological process*. Some extremists, such as Szasz (1960), take the view that illness can be defined only in terms of physical pathology. Since most mental disorders have no demonstrable physical pathology, on this view they are not illnesses. Szasz takes the further step of asserting that most mental disorders are therefore not the province of doctors. This kind of argument can be sustained only by taking an extremely narrow view of pathology. It is also incompatible with the available evidence; thus there are genetic and biochemical grounds for supposing that both schizophrenia and depressive disorders have a physical basis (see pp. 265 and 213), and it is possible that physical pathology processes will be found to underlie other psychiatric disorders.

Mental illness can be defined in terms of *psychopathology*. Such a view was taken by Lewis (1953b), who suggested that illness could be characterized by 'evident disturbance of part functions as well as general efficiency'. In psychiatry part functions refer to perception, memory, learning, emotion, and other such psychological functions. As an example, a disturbance of the part function of perception would be an illusion or hallucination. One difficulty with this approach is that there is no definition of 'evident disturbance'. Recent work has focused more on the notion of *incapacity* as a result of a disturbance of function (Fulford 1989). Several writers (e.g. Lewis 1953b; Fulford 1989) have warned strongly against defining mental illness solely in terms of socially deviant behaviour. The argument is often made that someone must have been mentally ill to commit a particularly cruel murder or grossly abnormal sexual act (the word 'sick' is often used in this context). Although such antisocial behaviour is highly unusual, there is no justification for equating it with mental illness. Moreover, if mental illness is inferred from socially deviant behaviour alone, political abuse may result. For example opponents of a political system may be confined to psychiatric hospitals simply because they do not agree with the authorities. A further reason for excluding social criteria from the definition of mental illness (and from diagnostic criteria) is that many behaviours are appraised differently in different countries.

The above examples show that mental illness is difficult to define. As already mentioned, the concept of mental illness need not be defined for most purposes, but the law in England and Wales requires psychiatrists to diagnose 'mental illness' in relation to compulsory admission to hospital and certain court procedures. Faced with this task, most psychiatrists begin by separating mental handicap and personality disorder from mental illness, as explained in the next section. Whether implicitly or explicitly, they usually invoke Lewis's concept of part functions to define mental illness; thus they diagnose mental illness if there are delusions, hallucinations, severe alterations of mood, or other major disturbances of psychological functions. In practice, most psychiatrists allocate psychiatric disorders to diagnostic categories such as schizophrenia, affective disorders, organic mental states, and others; by convention, they agree to group these diagnostic categories together under the rubric mental illness. Problems may arise with certain abnormalities of behaviour such as abnormalities of sexual preference or drug abuse.

The concept of mental illness is exceedingly complicated, and in a brief space only a few of the issues can be outlined. Readers seeking more information are referred to the papers by Lewis (1953*b*), Wootton (1959), Farrell (1979), and Häfner (1987*b*), and to the books by Clare (1979), Roth and Kroll (1987), and Fulford (1989).

Disease, illness, and sickness

In general medicine, a distinction is made between disease and illness. A further distinction can be made between illness and sickness (Susser 1990). These terms are generally used as follows:

disease refers to objective pathology;
illness is subjective awareness of distress;
sickness refers to a loss of capacity to fill normal social roles.

It is possible to describe patients in terms of one, two, or three of these terms. Most patients suffering from physical disorders can be said to suffer from disease, illness, and sickness. However, in many of the conditions dealt with by psychiatrists, only the terms illness and sickness are applicable, and in some disorders of personality and behaviour only the term sickness is relevant.

It is useful to describe the consequences of any disorder, either physical or mental, in terms of *impairment, disability,* and *handicap* (Susser 1990). These concepts, which are derived from medical sociology and social psychology, have been incorporated in the *International Classification of Impairment, Disease and Handicap ICIDH—80* by the World Health Organization (WHO) (see World Health Organization 1988). The terms are used in the following way:

impairment is analogous to disease, referring to a pathological defect;
disability is the stable persistent limitation of physical or psychological function which results from impairment and the individual psychological reaction to it;

handicap is analogous to sickness, referring to continuing social dysfunction, arising from inability to fill individual and social expectations.

Two instruments have been used in WHO epidemiological studies to assess impairment, disability, and handicap. They are the *Psychiatric Disability Assessment Schedule* (DAS) (World Health Organization 1988) and the *Psychological Impairments Ratings Schedule* (PIRS) (Biehl 1989).

The need for classification

In psychiatry, as in the rest of medicine, classification is needed for two main purposes. The first purpose is to enable clinicians to communicate with one another about their patients' symptoms, prognosis, and treatment. The second purpose is to ensure that research can be conducted with comparable groups of patients. In the past, the use of psychiatric classification has been criticized as inappropriate or even harmful (see below). Such criticisms have greatly diminished since syndromes have been shown to predict prognosis and to respond to specific treatments, and as familiarity with modern classifications has increased.

Among psychiatrists, the main critics of classification were psychotherapists whose work was concerned more with neurotic and personality disorders than with the whole range of psychiatric disorders. Psychotherapists tended to make two main criticisms. The first was that allocating patients to a diagnostic category distracts from the understanding of their unique personal difficulties. The second was that individual patients do not fit neatly into the available categories. Although these criticisms are important, they are arguments only against the improper use of classification. The use of classification can certainly be combined with consideration of a patient's unique qualities; indeed, it is important to combine the two because these qualities can modify prognosis and should be taken into account in treatment. The critics of standard classifications have themselves used idiosyncratic classifications with their own technical terms as a means of summarizing information. Also, whilst it is not feasible to classify a minority of disorders, this is not a reason for abandoning classification for the majority.

Some sociologists have suggested that to allocate a person to a diagnostic category is simply to label deviant behaviour as illness (Lemert 1951; Scheff 1963). They argue that such labelling serves only to increase the person's difficulties. There can be no doubt that terms such as epilepsy or schizophrenia attract social stigma, but this does not lessen the reality of disorders that cause suffering and require treatment. Disorders such as epilepsy and schizophrenia cannot be made to disappear simply by ceasing to give names to them.

The history of classification

The early Greek medical writings contained descriptions of different manifestations of mental disorder, for example excitement, depression, confusion, and memory loss. This simple classification of mental disorders was adopted by Roman medicine and developed by the Greek physician Galen, whose system of classification remained in use until the eighteenth century. Thus, in a widely read textbook of medicine published in 1583, Barrough divided mental disorders into *frenzy* (fever, madness, and disturbed sleep), *mania*, *melancholy*, *fatuities* (loss of both memory and reasoning—dementia in modern terminology), and *memory loss* (amnesia with intact reasoning—the amnesic syndrome in modern usage) (Hunter and MacAlpine 1963, pp. 24–8).

Interest in the classification of natural phenomena developed in the eighteenth century, partly as a result of the publication of a classification of plants by Linnaeus, a medically qualified professor of botany. Linnaeus also devised a less well-known classification of diseases in which one major class was mental disorders. These disorders were divided into three groups: (1) ideales—delirium, amentia, mania, vesania (or madness), and melancholia; (2) imaginarii—hypochondriasis, phobia, somnambulism, and vertigo; (3) pathetici—bulimia, polydipsia, satyriasis, and erotomania (Thompson 1814, pp. 188). Thus Lin-

naeus's classification of mental disorder was more comprehensive than that of Galen, and included not only the serious mental disorders but also those of less severity. Linnaeus's three categories have some resemblance to the present-day rubrics of psychosis, neurosis, and behaviour disorders.

Many other classifications have been proposed. A particularly well-known classification was published in 1772 by William Cullen, a Scottish physician. He grouped mental disorders together, though with one exception, delirium, which he classified with febrile conditions. In his scheme, mental disorders were part of a broad class of 'neuroses', a term he used to denote diseases affecting the nervous system. Cullen defined neuroses as 'preternatural affections of sense and motion which are without pyrexia as part of the primary disease and do not depend on a topical affection of the organs' (Hunter and McAlpine 1963, p. 495).

In Cullen's classification the neuroses were divided into four Orders: comata—apoplexy and paralysis; adynamiae—including hypochondriasis and syncopy; spasmi—including tetanus, chorea, epilepsy, hysteria, and palpitation; vesaniae—insanity. The last group, vesaniae, had three subgroups: amentia—which could be congenital, acquired, or senile; melancholia; oneirodynia—a term used to describe excessive imaginings during sleep. Therefore Cullen's classification contained an aetiological principle—that mental illnesses were disorders of the nervous system—as well as a descriptive principle for distinguishing individual clinical syndromes within the neuroses. In Cullen's usage, the term neurosis covered the whole range of mental disorders as well as many neurological conditions; the modern narrower usage developed later (see below).

In the early years of the nineteenth century, several French writers published influential classifications of major disorders. Phillipe Pinel's Treatise on insanity, which appeared in an English edition in 1806, divided mental disorders into mania with delirium, mania without delirium, melancholia, dementia, and idiocy. One of Pinel's compatriots, Esquirol, wrote another widely read textbook which was published in an English edition in 1845. Esquirol adopted a scheme similar to Pinel's, adding a new category of monomania which was characterized by 'partial insanity', in which there were fixed false ideas that could not be changed by logical reasoning. Esquirol divided the monomanias into subgroups including reasoning monomania, erotic monomania, incendiary monomania, homicidal monomania, and monomania resulting from drunkenness. Like other psychiatrists of the time, Pinel and Esquirol did not discuss neuroses (in the modern sense) or behaviour disorders in their textbooks because these conditions were generally treated by physicians.

These various schemes of classification were based on symptoms elicited on a single occasion. Two developments followed, both based on serial observations of the course of symptoms over time. The first derived from observations of cases of general paralysis of the insane, in which the patient could present in several ways as the disease progressed. This finding led to the idea of a unitary psychosis, i.e. a single pathology presenting with different manifestations. It was suggested that various serious mental disorders had a single common origin. The second development followed observations of conditions other than general paralysis of the insane; in these conditions distinct clinical courses could be discerned. In 1854 these findings led Falret to describe folie circulaire, which was broadly similar to the modern concept of bipolar disorder. Meanwhile, in Germany Kahlbaum formulated two requirements for research on classification: that the entire course of a mental illness was fundamental to the definition of that illness, and that the total clinical picture must be used in formulating definitions in a system of classification. These ideas were adopted by Kraepelin, who studied the course of mental diseases over many years and thereby made the important distinction between manic depressive psychosis and schizophrenia. The successive editions of Kraepelin's textbook led to further refinements in the classification of mental illness that are the basis of today's systems.

At the same time, developments in the emerging specialty of neurology led to decreasing medical interest in the 'nervous patient', a term used throughout the nineteenth century in Britain and North America (Bynum 1985). The writings of Freud and his contemporaries led to greater recognition of the psychological causes of nervous symptoms and 'neurotic' disorders, and to the modern concepts of hysteria and anxiety disorder.

The more recent development of national and international classifications is described in this chapter; in the remainder of this section two terms which have been widely used in most classifications are considered (see Pichot (1994) for a review of nosological models in psychiatry).

Neurosis and psychosis

In the past the concepts of psychosis and neurosis were included in most systems of classification. As is explained later in this chapter, neither of these terms is used as an organizing principle in ICD10 or DSMIV. In practice, however, these terms are still used widely; hence it is of practical importance to understand their history and usage.

It has been explained above that the term *neurosis* was introduced by Cullen to denote diseases of the nervous system. Cullen's concept of neurosis included neurological disorders such as apoplexy and epilepsy as well as the serious mental disorders (called vesaniae in his system). Gradually the category of neurosis narrowed, first as neurological disorders with a distinct neuropathology (such as epilepsy and stroke) were removed, and later as the vesaniae were transferred to a separate category of psychosis. The term *psychosis* was suggested by Feuchterleben, who published a book entitled *Principles of medical psychology* in 1845. This author proposed psychosis as a term for severe mental disorders. He also accepted the term neurosis for mental disorders as a whole; thus he wrote, 'every psychosis is at the same time a neurosis but not every neurosis is a psychosis' (Hunter and MacAlpine 1963, p. 950). As the concept of neurosis was narrowed, psychoses ceased to be subgroups of neuroses and were regarded as independent conditions. Many of the difficulties encountered today in defining the terms neurosis and psychosis are related to these origins.

In modern usage, the term **psychosis** refers broadly to severe forms of mental disorder such as organic mental disorders, schizophrenia, and affective disorders. Numerous criteria have been proposed to achieve a more precise definition. Greater severity of illness is a common criterion, but the conditions in this group can occur in mild or severe forms. Lack of insight is often suggested as a criterion for psychosis, but the term insight is itself difficult to define (see p. 23). A more straightforward criterion is the patient's inability to distinguish between subjective experience and reality, as evidenced by hallucinations and delusions. Since none of these three criteria is easy to apply, the term psychosis is unsatisfactory. However, it is not only difficulty of definition that makes the term psychosis unsatisfactory. There are two other reasons: first the conditions embraced by the term have little in common, and second it is less informative to classify a disorder as psychosis than it is to classify it as a particular disorder within the rubric of psychosis (for example schizophrenia). It is for these reasons that the distinction between neurosis and psychosis, which was a fundamental classificatory principle in ICD9, was abandoned in DSMIII and subsequently in ICD10.

Although the term psychosis has little value in a scheme for classifying mental disorders, it is still in everyday use as a convenient term for disorders that cannot be given a more precise diagnosis because insufficient evidence is available, for example when it is still uncertain whether a disorder is schizophrenia or mania. Similarly, it is useful to retain such terms as 'psychotic disorders not otherwise specified' (as in DSMIV) and 'acute or transient psychotic disorders' (as in ICD10). Finally, the adjectival form psychotic is in general use, for example in the terms psychotic symptom (generally meaning delusions, hallucinations, and excitement) and antipsychotic drug (meaning a drug that controls these symptoms).

The term **neurosis** refers to mental disorders that are generally less severe than the psychoses and characterized by symptoms closer to normal experience (for example anxiety). The history of the term is referred to on p. 59. Here the value of the term in classification is discussed. The objections to the term neurosis are similar to the objections to the term psychosis. First neurosis is difficult to define, second the conditions that it

embraces have little in common, and third more information can be conveyed by using a more specific diagnosis, such as anxiety disorder or obsessional disorder, than by calling the condition a neurosis. A final objection was put forward in the manual to DSMIII, namely that neurosis has been widely used with an aetiological meaning in psychodynamic writings. It is true that the term has been so used and that such usage is historically unjustified (see p. 60), but if this misuse were the only objection to the term neurosis, it would be appropriate to ensure correct usage rather than to abandon the term. In practice, it has been possible to organize the main groups of disorders in terms of common features without the need to group them as neuroses and psychoses in the classification. The term neurosis is not used in DSMIV, although it is retained in ICD10 in the heading of one group of disorders: 'neurotic, stress-related, and somatoform disorders'. Like psychosis, the term neurosis continues to be used in everyday clinical practice as a convenient term for disorders that cannot be assigned to a more precise diagnosis.

Types of classification

Categorical classification

Traditionally, psychiatric disorders have been classified by dividing them into **categories** which are supposed to represent discrete entities. Categories have been defined in terms of symptom patterns and of the course and outcome of the different disorders. Such categories have proved useful in both clinical work and research. However, three objections are often raised against them: (i) there is uncertainty about the validity of categories as representing distinct entities; (ii) many systems of classification do not provide adequate definitions and rules of application, and so categories cannot be used reliably; (iii) many psychiatric disorders do not fall neatly within the boundaries of a category but are intermediate between two categories (for example cases intermediate between schizophrenia and affective disorder). Recently, multivariate statistical techniques have been used in attempts to define categories more clearly. The results have been interesting, but so far not conclusive.

Categorical systems often include an implicit **hierarchy** of categories. If two or more diagnoses are made, it is often conventional (though not always made explicit) that one takes precedence. For example, organic mental disorders take precedence over schizophrenia. Not only is this convenient, but there is some clinical evidence for an in-built hierarchy of significance within the disorders themselves. For instance, affective symptoms occur commonly with schizophrenia and they often improve when the schizophrenia is treated. Similarly, it is well recognized that anxiety symptoms occur commonly with depressive disorders and are sometimes the presenting feature. If the anxiety is treated, there is little response, but if the depressive disorder is treated, there may be improvement in anxiety as well as in the depressive symptoms.

Dimensional classification

Dimensional classification rejects the use of separate categories. In the past it was advocated by Kretschmer and other psychiatrists. It has also been strongly promoted by the psychologist Eysenck, who argues that there is no evidence to support the traditional grouping into discrete entities. Instead, Eysenck (1970*b*) proposed a system of three dimensions: psychoticism, neuroticism, and introversion–extroversion. Patients are given scores which locate them on each of these three axes. For example in the case of a person with a disorder that would be assigned to dissociative disorder in a categorical system, in Eysenck's system the person would have high scores on the axes of neuroticism and extroversion, and a low score on the psychoticism axis. Subsequent research has not confirmed specific predictions of this kind, but the example brings out the principles.

Eysenck's three dimensions were established by various procedures of multivariate analysis. They are attractive in theory, but it should be remembered that they depend considerably on the initial

assumptions and the choice of methods. The dimension of 'psychoticism' bears little relation to the concept of psychosis as generally used. For example, artists and prisoners score particularly highly on this dimension. The dimensions of neuroticism and introversion–extroversion have been useful in research with groups of patients, but they are difficult to apply to the individual patient in clinical practice.

The multiaxial approach

In one sense, the term multiaxial can be applied to the three dimensions just described. However, the term is usually applied to schemes of classifications in which two or more separate sets of information (such as symptoms and aetiology) are coded. In 1947 Essen-Møller proposed that clinical syndrome and aetiology should be coded separately. It would then be possible to identify cases with a similar clinical picture on the one hand, and those with a similar aetiology on the other (Essen-Møller 1971). Such a scheme should avoid the unreliability of schemes in which clinical picture and aetiology can be combined in the definition of a single category, such as reactive depression. Multiaxial systems are attractive, but there is an obvious danger that they will be so comprehensive and complicated as to be difficult for everyday use (Williams 1985). Several multiaxial systems have been proposed. The introduction of multiaxial classification was a major innovation in DSMIII. The DSMIV and ICD10 systems are described on pp. 68 and 69.

The basic categories for classification in psychiatry

Several categorical systems of classification have been used in psychiatry, but they all contain the same basic categories (see Table 3.1). The first category is **mental retardation**, i.e. impairment of intellectual functioning present continuously from early life. The second category is **personality disorder**, i.e. dispositions to behave in certain abnormal ways present continuously since early adult life. The third category is **mental disorder,** i.e. abnormalities of behaviour or psychological experience with a recognizable onset after a period of normal functioning. To qualify for a diagnosis of mental disorder, the abnormalities of experience or behaviour have to reach a certain level of severity. Disorders that fail to meet this criterion and that occur in relation to stressful events or changed circumstances are called **stress-related** or **adjustment disorders.** A fifth category of 'other disorders' is required for conditions that do not fit into the first four groups, for example abnormalities of sexual preference and drug dependence. The last two categories are for disorders of childhood and adolescence—the sixth for disorders of learning and development, and the seventh for other kinds of disorder specific to this time of life.

Table 3.1 *The basic classification*

Mental retardation
Personality disorder
Mental disorder
Adjustment disorder (reaction to stress)
Other disorders
Developmental and learning disorders
Disorders with onset in childhood or adolescence

The process and reliability of diagnosis

Diagnosis is the process of identifying disease and allocating it to a category on the basis of symptoms and signs. It involves four main components (Cooper 1993): (1) the interviewing technique of the psychiatrist; (2) the perception of the patient's speech and behaviour; (3) a complicated series of processes by which the psychiatrist sorts out the available information and decides how to use it and what task to perform next; (4) a final stage in which the psychiatrist chooses one or more terms from a stated classification of psychiatric disorders.

Reliability

Systems of classification are of little value unless psychiatrists can agree with one another in attempting to make a diagnosis. In the past 40 years there has been increasing interest in improving the level of diagnostic agreement between psychiatrists (Kendell 1975). Early studies consistently showed poor diagnostic reliability. In

Philadelphia, Ward *et al.* (1962) concluded that overall disagreement was made up of the following elements: inconsistency in the patient, 5 per cent; inadequate interview technique, 33 per cent; inadequate use of diagnostic criteria, 62 per cent. The last two factors will be discussed in turn. (See Spitzer and Williams (1985) for a review of the process of diagnosis.)

Interviewing technique

Psychiatrists vary widely not only in the amount of information that they elicit at interview, but also in their interpretation of the information. Thus a psychiatrist may or may not elicit a phenomenon, and he may or may not regard it as a significant symptom or sign. Variations have been found between groups of psychiatrists trained in different countries, and between individual psychiatrists in the same country. When shown filmed interviews, American psychiatrists reported many more symptoms than did British psychiatrists (Sandifer *et al.* 1968). Presumably this reflected differences in training between the two countries.

Standardized interview schedules

Differences in eliciting and rating symptoms can be reduced when psychiatrists are trained to use standardized interview schedules, such as the Present State Examination (PSE) (Wing *et al.* 1974) which was designed for use by trained clinicians who make judgements about the presence and severity of symptoms. Another example is the Diagnostic Interview Schedule (DIS) which was introduced by the National Institutes of Mental Health for use by non-specialist interviewers who record patients' complaints without making a judgement as to whether these are symptoms. Both types of schedule specify sets of items that must be enquired about, and they also provide definitions and give instructions on rating severity. These and other instruments now available will be reviewed next.

Present State Examination (PSE)

The development of this instrument began in the late 1950s. Early versions were used solely in the authors' own research: the ninth edition (PSE 9) was the first to be published for use by others (Wing *et al.* 1974). It is available in at least 35 languages and has been widely used in many countries. The main principle is that the interview, although clearly structured, retains the features of a clinical examination. A trained interviewer seeks to identify abnormal phenomena that have been present during a defined period of time and to rate their severity. Each of the 140 items is defined in detail in a glossary. Computer programs generate a symptom score, a diagnosis (CATEGO), and a clinically derived measure of the severity of non-psychotic symptoms (the Index of Definition).

Schedules for Clinical Assessment in Neuropsychiatry (SCAN)

The tenth edition of the Present State Examination has been incorporated into this more extensive schedule which can be used to diagnose a broader range of disorders, including eating, somatoform, substance abuse, and cognitive disorders (World Health Organization 1992*a*). Although the prin-

ciples are similar to those of PSE 9, there are several differences including a change from three possible ratings to four, an improved system for rating episodes of disorder, and procedures for including information about history and aetiology. The system is compatible with PSE 9, and it allows ICD10, DSMIII-R, and DSMIV diagnoses (Janca *et al.* 1994). A computer-assisted version is available (Glover 1992).

International Personality Disorder Examination (IPDE)

This instrument assesses phenomenology and life experiences so as to enable psychiatric diagnosis according to ICD10 and DSMIIIR. It has 153 items and includes open-ended enquiries and questions to determine frequency, duration, and age of onset (Janca *et al.* 1994).

Diagnostic Interview Schedule (DIS)

This schedule was developed in the United States as part of the Epidemiological Catchment Point Area (ECPA) project (Robins *et al.* 1981). The fully structured interview schedule was developed for use by non-clinicians but employs diagnostic criteria used by clinicians. The DIS covers the most common adult diagnoses that can be evaluated by assessing the interview alone (for example it omits delirium and bulimia). Diagnoses are first made on a lifetime basis. Then the interviewer asks how recently the last symptom was experienced. On the basis of the answer, the disorder is recorded as occurring within the last two weeks, the last month, the last six months, or the last year. This procedure enables diagnosis of a disorder either within the previous year or at any time during the person's life. Reliability and validity, as determined by a second DIS given by a psychiatrist and by a clinical interview by a psychiatrist, are reasonably satisfactory.

There are problems in making lifetime diagnoses with the DIS (and with all other such instruments) because respondents may have forgotten or denied an illness, and because populations of varying ages have had different durations of time in which to develop a disorder.

Structured Clinical Interview for Diagnosis (SCID)

Soon after the publication of DSMIII, work began on a clinical diagnostic assessment procedure for making DSM diagnoses. The draft instrument was field-tested and then issued for use with DSMIIIR. It can be used by the clinician as part of a normal assessment procedure to confirm a particular diagnosis or in research or screening as a systematic evaluation of a whole range of medical states. The instrument covers all the criteria of the diagnoses included in the various modules and the interviewer makes a clinical judgement as to whether each criterion is met (Spitzer *et al.* 1990). It is available in a patient edition for use with subjects who have been identified as psychiatric patients and in a non-patient edition which is suitable for use in epidemiological studies. In addition the SCID-II is available for making 12 Axis II, i.e. personality disorder, diagnoses in DSMIIIR.

Composite International Diagnostic Interview (CIDI)

This interview was produced for the WHO and the US Alcohol Drug Abuse and Mental Health Administration. It is a comprehensive and standardized interview derived from the DIS. It is used for the assessment of mental disorders and to provide diagnoses according to ICD and DSMIV. It is available in 16 languages and is designed to be used by clinicians and non-clinicians in different cultures. The CIDI package includes a core interview in a researcher's version and an interviewer's version (the latter has a diagnostic index that allows linkage of specific CIDI questions to specific diagnostic criteria of ICD10 and DSMIII-R), additional modules (concerned, for example, with antisocial personality and post-traumatic stress disorders), as well as training manuals and computer programs. The interview includes ques-

tions about symptoms and problems experienced at any time in life, as well as questions about current state (World Health Organization 1989; Essau and Wittchen 1993; Janca *et al.* 1994).

Criteria for diagnosis

International studies have compared the diagnostic criteria used by different psychiatrists. In the US–UK Diagnostic Project, for example, American and British psychiatrists were shown the same video-taped clinical interviews and were asked to make diagnoses (Cooper *et al.* 1972). Compared with psychiatrists in London, psychiatrists in New York diagnosed schizophrenia twice as often and diagnosed mania and depression correspondingly less often. Further investigation suggested that New York was not typical of North America, and that diagnostic practice in other places in the United States and in Canada was closer to British practice.

A second study, the International Pilot Study of Schizophrenia (World Health Organization 1973) was carried out in nine countries: Colombia (Cali), Czechoslovakia (Prague), Denmark (Aarhus), England (London), India (Agra), Nigeria (Ibadan), Taiwan (Taipei), the USA (Washington), and the USSR (Moscow). The main purposes of this study were first to establish whether standardized interviews could be used in different languages in different cultures, and second to determine whether typical schizophrenic patients could be found in all the different cultures. As a side issue, the study examined some of the differences between the psychiatrists, even though the latter had trained together and had been asked to use the ICD9 criteria in a standard way. Psychiatrists in all these countries carried out lengthy interviews which included the PSE. The psychiatrists made their own diagnoses and these were compared with those of the PSE computer program CATE-GO. There was substantial agreement between seven of the centres, but Washington and Moscow differed from the rest. The findings in Washington confirmed the results of the US–UK project described above. The Moscow psychiatrists also appeared to have an unusually broad concept of schizophrenia; this apparently reflected a particular local emphasis on the course of the disorder as a diagnostic criterion.

Diagnostic unreliability can be reduced by providing a clear definition of each category in a diagnostic scheme. Each definition should specify discriminating symptoms rather than characteristic symptoms. **Discriminating symptoms** are those that may occur in the defined syndrome but seldom in other syndromes. Discriminating symptoms are important in diagnosis but may be of little concern to patients and relatively unimportant in treatment. An example is the delusion that thoughts are being inserted into the mind, a symptom that seldom occurs except in schizophrenia. **Characteristic symptoms** occur frequently in the defined syndrome but also occur in other syndromes. Such symptoms may be important to the patient and relevant in planning treatment, but do not help in diagnosis. An example is thoughts of suicide, which occur in depressive disorders but also in other conditions.

Diagnostic criteria can be descriptive statements, as in the ICD10 clinical criteria, or more precise operational criteria, as in DSMIV. Operational definitions were originally suggested by the philosopher Carl Hempel, and were incorporated in an important report to the WHO on ways of overcoming the problems of diverse national classifications (Stengel 1959). The term **operational definition** in this context means the specification of a category by a series of precise inclusion and exclusion statements.

The first published operational criteria were those devised for a longitudinal study of psychiatric hospital patients. The first detailed set of rules was drawn up by Feighner *et al.* (1972) in the United States, who provided specific inclusion and exclusion criteria. A similar approach was adopted in the Research Diagnostic Criteria (Spitzer *et al.* 1978) and in the American system DSMIII, which is described later in this chapter. When criteria of this kind are used, a substantial number of patients may not fit into any of the designated categories and may have to be allocated to an 'atypical' category. In some kinds of research this atypical group may not matter, but it can be a problem in everyday clinical practice.

Diagnosis by computer

Computer diagnosis ensures that the same rules will be applied to every case. Computer programs to generate diagnoses have been based either on a logical decision tree or on statistical models. *A decision-tree program* evaluates a sequence of yes/no answers, and so successively narrows the diagnosis. Thus it resembles differential diagnosis in clinical practice. Spitzer and Endicott (1968) first used this procedure to develop the program DIAGNO. Later, Wing *et al.* (1974) developed the program CATEGO for use with the PSE. CATEGO has proved valuable in epidemiological studies of major and minor psychiatric disorders, and comparison data are now available from a variety of patient groups and normal populations. Computer programs are available for the major international diagnostic instruments SCAN and CIDI (see pp. 64–5).

In the alternative *statistical approach*, data are collected from a sample of patients whose diagnoses are known. A system of classification is then devised from this database by statistical methods. Whereas the decision-tree method follows a sequence of arbitrary rules that underlie ordinary clinical practice, this second method estimates the probability that a given patient's symptoms match the symptoms of previously diagnosed patients.

The validity of schemes of classification

Whilst the unreliability of diagnosis can be reduced by the measures just described, a scheme of classification must also be valid. Even if different interviewers can be trained to reach high levels of agreement in making diagnoses, little has been achieved unless the diagnostic categories have some useful relationship to the disorders met in clinical practice. To be valid, a scheme of classification should have categories that fit well with clinical experience (*face validity*). The categories should also be able to predict the outcome of psychiatric disorders (*predictive validity*); ideally they should also point to associations between psychiatric disorders and independent variables such as biochemical measures (*construct validity*).

So far little progress has been made towards establishing the validity of existing schemes of classification (see Spitzer and Williams (1985) for a discussion of reliability and validity in diagnosis).

Other features of schemes of classification

Two other features contribute to the value of a system of classification: *coverage* and *ease of use*. Coverage refers to the extent to which a scheme has categories for all the disorders that are encountered in clinical practice.

Individual systems of classification

In European countries national systems of classification remain largely within Kraepelin's framework. The two main exceptions are Scandinavia and France. In Scandinavia much emphasis is placed on the concept of **psychogenic** or **reactive psychoses**, which are said to have paranoid, depressive, or confusional symptoms, or sometimes a mixture of all three (Strömgren 1985; Cooper 1986; Taylor 1994).

In French psychiatry, classification is based on a combination of psychopathology and elements of existential philosophy (Pichot 1984, 1994). Certain diagnostic categories in France differ from those in the rest of Europe and North America. They include two special categories: *bouffée délirante* and *délires chroniques*. **Bouffée délirante** is the sudden onset of a delusional state with trance-like feelings, of short duration and good prognosis. Although this condition may develop into schizophrenia, it is clearly separated from acute schizophrenia and acute manic depressive illness (see Chapter 9). This disorder has been included in ICD10 in the category of 'acute transient psychotic disorder', which also incorporates features of the Scandinavian concept of reactive psychosis. *Délires chroniques* are conditions which, in the ICD system, would be classified as 'persistent delusional disorders'; they are

separated from schizophrenia, a diagnosis which is used in France only when there is definite evidence of deterioration of personality. The *délires chroniques* are subdivided into the 'non-focused', in which several areas of mental activity are affected, and the 'focused' with a single delusional theme. The latter include several conditions such as erotomania (described on p. 303).

In the 1920s and 1930s American views on psychiatric classification diverged widely from those in Europe. Psychoanalysis and the teaching of Adolf Meyer directed American psychiatry towards a predominant concern with the uniqueness of individuals rather than with their common features. At the same time, diagnostic concepts were increasingly based on presumed psychodynamic mechanisms. Attitudes towards psychiatric classification in the United States have changed considerably. An important first step was the introduction of strict criteria for classification in research, as described above (Feighner *et al.* 1972). This step was followed by the thorough work that led to the new American scheme DSMIII (Wilson 1993). Attitudes towards classification in the United States changed considerably after the development and general acceptance of the American Psychiatric Association's classification DSMIII. For example, in the post-war period up to the early 1960s, the concept of schizophrenia was very broad. This has been succeeded by a particularly narrow concept of schizophrenia.

Another example of international variation is the Chinese national classification, which is derived from both ICD9 and DSMIII. This classification includes the category of neurasthenia, which is one of the most frequently used diagnoses in Chinese psychiatry (Kleinman 1982).

Classification in developing countries

Classifications developed in Europe and North America have not proved entirely satisfactory in developing countries where behavioural disturbances can be different. In developing countries acute psychotic symptoms may present particular difficulties of diagnosis; they are often atypical and raise doubt as to whether they represent separate entities or merely variations of syndromes seen in developed countries. Investigation of these issues is difficult for outsiders who may not appreciate important cultural factors or the varying use of language to describe emotions and behaviour. For further information about the cultural aspects of classification see Yap (1951), Murphy (1977), Leff (1981), Simons and Hughes (1985), and Fabrega (1987).

The International Classification of Diseases (ICD)

Mental disorders were not included in the ICD until its sixth edition produced by the WHO in 1948. This first scheme for mental disorders was widely criticized. As a preliminary to a major revision of the scheme, a survey of principles of classification in different countries was carried out (Stengel 1959) and wide variations were found. Stengel recommended a new approach based on operational definitions and supported by a glossary, but not linked to any theories of aetiology.

The eighth edition (ICD8) was published in 1968. It made some progress towards solving the earlier problems but was still unsatisfactory in several ways. It contained too many categories and allowed alternative codings for some syndromes. This probably reflected an endeavour to make the scheme widely acceptable. One major advance was the publication of a glossary, which was largely based on the British version produced by a Working Party chaired by Sir Aubrey Lewis (General Register Office 1968). ICD9 was very similar to ICD8 because the WHO believed that national governments would be unwilling to accept many changes. Although the mental health section had asked to be allowed to wait until the publication of ICD10 before making changes, in the meantime a series of seminars were held which led to a revised and improved glossary (World Health Organization 1978*a*).

Major changes in the mental health section of ICD10 have resulted from several initiatives, including the continuing WHO research programme on diagnosis and classification, collaborative studies, and innovations by national organizations,

especially the American Psychiatric Association. Public consultation has been achieved through a series of research projects and the work of many advisors. As a result of close collaboration with the American Psychiatric Association and an overlap of the membership of the ICD and DSM working parties, the two systems are now broadly similar.

The aims of the ICD10 working party were that the scheme should (1) be suitable for international communication about statistics of morbidity and mortality, (2) be a reference for national and other psychiatric classifications, (3) be acceptable and useful in research and clinical work, and (4) contribute to education. To achieve these aims, the classification has to be *acceptable* to a wide range of users in different cultures; it also has to be *practical* in that it is easy to understand and can be translated into many languages. It also has to be *versatile;* for this reason a policy of 'different versions for different purposes' was used from the start and led to the several versions shown in Table 3.2.

The clinical descriptions and diagnostic guidelines contain descriptions of each of the disorders in the classification; these allow some latitude for clinical judgement in making the diagnosis. The diagnostic criteria for research contain lists of specific criteria that have to be met before a diagnosis can be made. The format resembles that used in DSMIV for clinical purposes as well as for research. The provision of these two sets of criteria—the more flexible ones for clinical purposes and the more precise ones for research—is

Table 3.2   *ICD10 Chapter V*

1. Clinical descriptions and diagnostic guidelines
2. Diagnostic criteria for research
3. Primary care version
4. Multiaspect (axial) systems

one of the main differences between ICD10 and DSMIV.

A second difference between ICD10 and DSMIV is the inclusion in ICD10 of a simplified classification for use in primary care. This version of the classification contains only broad categories such as dementia, delirium, eating disorders, acute psychotic disorder, chronic psychotic disorder, depression, and bipolar disorder. These categories are not subclassified, and the clinical descriptions are simpler than those in the main classification and are adapted for use in primary care.

ICD10, like DSMIV, is a descriptive classification. However, aetiology is included in some general categories, namely organic, substance-use-related, and stress-related. Therefore the classification is a mixture of symptoms and aetiology.

Mental disorders are classified in Chapter F of the ICD. The chapter is divided into ten groups, shown in Table 3.3. A decimal system is used in which each group can be subdivided into ten, and each of these into a further ten. Categories are denoted by the letter F (for the mental disorders chapter) followed by a number for the main group (for example F2 schizophrenia) followed by a further number for the category within the group (for example F25 schizoaffective disorder). A fourth character is used when it is necessary to subdivide further (for example F25.1 schizoaffective disorder, depressive type). The traditional division of neurosis and psychosis is no longer an organizing principle.

After wide consultation about a draft of the classification, the WHO carried out an international field trial to evaluate both the clinical descriptions and guidelines and the diagnostic criteria for research which have been translated into all the widely spoken languages of the world.

Table 3.3 *The main categories in ICD10*

F0	Organic, including symptomatic, mental disorders
F1	Mental and behaviour disorders due to psychoactive substance use
F2	Schizophrenia, schizotypal, and delusional disorders
F3	Mood (affective) disorders
F4	Neurotic, stress-related, and somatoform disorders
F5	Behavioural syndromes associated with physiological disturbances and physical factors
F6	Disorders of adult personality and behaviour
F7	Mental retardation
F8	Disorders of psychological development
F9	Behavioural and emotional disorders with onset usually occurring in childhood or adolescence

This trial aimed to assess whether the classification fitted diagnostic practice in different countries, how easy it was to use, and whether psychiatrists could reach agreement about diagnoses after brief training. The trial was carried out at 112 clinical centres in 39 countries. It was found that the classification was generally easy to use and applicable to most common disorders. Reliability was less good for a few conditions, notably personality disorder (Sartorius *et al.* 1993). The field trials of ICD10 carried out in Canada and the United States showed comparable reliability to the findings world-wide and to DSM (Regier *et al.* 1994).

The Diagnostic and Statistical Manual (DSM)

In 1952 the American Psychiatric Association published the first edition of the Diagnostic and Statistical Manual (DSMI) as an alternative to ICD6 which, as mentioned above, had been widely criticized. DSMI was influenced by the views of Adolf Meyer and Karl Menninger, and its simple glossary reflected the prevailing acceptance of psychoanalytic ideas in the United States.

DSMII was published in 1968 as the American National Glossary to ICD8. It combined psychoanalytic ideas with those of Kraepelin. In 1974 the American Psychiatric Association set up a task force to produce a revised version of DSM to coincide with the publication of ICD9. The chairman of the task force was a member of the group at Washington University, St Louis, which had published the Feighner criteria for diagnosis, and he had subsequently developed the Research Diagnostic Criteria published in 1975 (Endicott and Spitzer 1978). These developments strongly influenced DSMIII which was published in 1980. The criteria were prepared with great care. Advisory committees prepared detailed drafts, opinions were obtained from 550 clinicians, and the results were subjected to field tests (American Psychiatric Association 1980). DSMIII was intended to provide a comprehensive classification with clear criteria for each diagnostic category. It contained five main innovations.

1. Precise operational criteria were provided for each diagnosis, with rules for inclusion and exclusion.
2. A multiaxial classification was adopted with five axes: I, clinical syndromes and 'conditions not attributable to mental disorder that are the focus of attention and treatment'; II, personality disorders; III, physical disorders and conditions; IV, severity of psychosocial stressors; V, highest level of adaptive functioning in the last year.
3. The nomenclature was revised and some syndromes were regrouped, for example the terms neurosis and hysteria were discarded, and all affective disorders were grouped together.
4. Classification relied less on psychodynamic concepts than it had done in previous versions.
5. For some conditions, duration of illness was introduced as one of the criteria for diagnosis.

(See Wilson (1993) for an account of the development of DSMIII and a discussion of the consequences for American psychiatry.)

The production of DSMIII was an important achievement even though it contained compromises to accommodate different theoretical and clinical viewpoints among American psychiatrists. Not only did it introduce advances in the method of classification and the preparation of diagnostic criteria, but it also marked a major change in attitude to classification in American psychiatry. This revision resulted in many differences between this American national classification and the international classification. The differences are attributable to the different aims of the two classifications, and to unresolved differences in particularly controversial areas of classification (for example somatoform disorders). In 1987 a further revision of the American classification (DSMIIIR) was produced as an interim scheme to remedy some of the defects of DSMIII pending the production of a fourth version (DSMIV) to coincide with the tenth edition of the ICD (American Psychiatric Association 1987).

In 1988 a meeting of the American Psychiatric Association considered how the United States could best fulfil its treaty obligation with the

WHO to maintain coding and technological consistency with ICD. The conclusion was that work on DSMIV and ICD10 should be closely coordinated. As a result, DSMIV is technically compatible with ICD, although there are a number of specific differences (Table 3.4). Thirteen work groups were established, each responsible for a section of the classification. Each work group carried out a three-stage procedure: (1) comprehensive reviews of published literature; (2) analysis of data already collected; (3) field trials to evaluate proposed changes (Kline 1993). Drafts of DSMIV were circulated for discussion and comments before publication of the final manual in 1994 (American Psychiatric Association 1994). An International Version (American Psychiatric Association 1995) is identical in content but uses ICD10 codes. The review material has been published in a series of DSMIV source books.

Comparison of ICD10 and DSMIV

ICD10 and DSMIV are closely similar because they are derived from a common base of knowledge and research, and because their authors collaborated closely. The two classifications are complementary rather than competing. Thus DSM has been designed for use in a single country—it is a national classification—whereas ICD has been designed for use in all countries with their varied cultures and needs. Table 3.4 summarizes the main differences between the two classifications.

Classification in this book

In this book both the DSMIV and the ICD10 classifications are discussed in the chapters dealing with clinical syndromes. As in other textbooks,

Table 3.4 *Differences between ICD10 and DSMIV*

	ICD10	DSMIV
Origin	International	American Psychiatric Association
Presentation	Different versions for clinical work, research, and use in primary care	A single document
Languages	Available in all widely spoken languages	English version only
Structure	Part of overall ICD framework	
	Single axis in Chapter V; separate multiaxial systems available	Multiaxial
Content	Guidelines and criteria do not include social consequences of the disorder	Diagnostic criteria usually include significant impairment in social, occupational, or other areas of functioning

disorders are grouped in chapters for convenience and ease of understanding. The headings of the chapters do not always correspond exactly to the terms used in DSMIV and ICD10; any difference means that the heading more appropriately summarizes the scope of the chapter.

In this book we have generally followed the usage in ICD10 and DSMIV by adopting the term mental disorder instead of the term mental illness. The former is defined as 'a clinically significant behaviour or psychological syndrome or pattern that occurs in a person and that is associated with present distress (a painful symptom) or disability (impairment of one or more important areas of functioning) or with a significantly increased risk of suffering death, pain, disability, or an important loss of freedom. In addition, this syndrome or pattern must not be merely an expectable response to a particular event, e.g. the death of a loved one.'

Classification in everyday practice

This subject was considered briefly in the section on formulation (p. 48). It is discussed in more detail here. It is important to reiterate the distinction made on p. 56 between classifying disorders and classifying patients. The clinical formulation of the problems of an individual patient include statements about each of the areas covered in a multiaxial system (see p. 70).

A diagnosis is made after the history and examination of mental state have been completed. The first step is to review the pattern of the symptoms occurring in the past month (as reported by the patient and any other informants) and the pattern of symptoms and signs elicited by mental state examination. An attempt is then made to match this pattern to one or more of the diagnostic categories in the system of classification used. Appropriate reference is made to the definitions and rules of application provided in the scheme. An important distinction needs to be made between characteristic symptoms and discriminating symptoms (see p. 66). In practice, only a few categories need be considered; the rest are obviously inapplicable.

In attempting to match a patient's pattern of symptoms and signs to a diagnostic category, problems may arise when most symptoms fit well but one or two are incongruous. For example, a patient may have depressed mood, morbid self-blame, early morning waking, and diurnal mood variation—all symptoms typical of a depressive disorder. In addition, he may have the delusion that people talk about him on television—a typical symptom of schizophrenia. When this kind of incongruity occurs, the clinician should review the case thoroughly and search for other evidence of the alternative syndrome. If only a single incongruous symptom is found among many that are congruous, generally the diagnostic category remains unchanged.

This kind of problem can sometimes be resolved by looking at the diagnostic category longitudinally as well as cross-sectionally. The process described so far is cross-sectional, i.e. allocation to a category is based on present mental state and the history of symptoms in the past few weeks. The longitudinal approach deals with the nature and course of a disorder since it first began. For example the present symptoms can be compared with those of any previous episodes of disorder. If it is found that the patient described above had had two previous episodes of clear-cut affective disorder and no episodes of schizophrenia, then the clinician will more readily discount the current atypical symptom (the delusion of being talked about on television). If there have been two episodes of definite schizophrenia, the opposite conclusion will be justified. The time course of previous illness is also informative; a history of intermittent episodes with complete recovery between them occurs more often with affective disorder than with schizophrenia. Of course, these principles can be applied to other differential diagnoses.

Further reading

American Psychiatric Association (1994). *Diagnostic and statistical manual of mental disorders* (4th edn). American Psychiatric Association, Washington, DC.

Kendell, R. E. (1975). *The role of classification in psychiatry*. Blackwell Scientific Publications, Oxford.

Lewis, A. J. (1953). Health as a social concept. *British Journal of Sociology*, 4, 109–24. Reprinted in Lewis, A. J. (1967). *The state of psychiatry*, pp. 179–94. Routledge and Kegan Paul, London.

World Health Organization (1992). *The ICD10 classification of mental and behavioural disorders: clinical descriptions and diagnostic guidelines*. World Health Organization, Geneva.

World Health Organization (1993). *The ICD10 classification of mental and behavioural disorders: diagnostic criteria for research*. World Health Organization, Geneva.

4 | *Aetiology*

Psychiatrists are concerned with aetiology in two ways. First, in everyday clinical work they try to discover the causes of the mental disorders presented by individual patients. Second, in seeking a wider understanding of psychiatry they are interested in aetiological evidence obtained from clinical studies, community surveys, or laboratory investigations. Correspondingly, the first part of this chapter deals with some general issues about aetiology in the assessment of the individual patient, whilst the second part deals with the various scientific disciplines that have been applied to the study of aetiology.

When the clinician assesses an individual patient, he draws on a common fund of aetiological knowledge that has been derived from the study of groups of similar patients, but he cannot understand the patient in these terms alone. He also has to use everyday insights into human nature. For example, in assessing a depressed patient, the psychiatrist should certainly know what has been discovered about the psychological and neurochemical changes accompanying depressive disorders, and what evidence there is about the aetiological role of stressful events and about genetic predisposition to depressive disorder. At the same time he will need intuitive understanding to recognize that this particular patient feels depressed because he has been informed that his wife has cancer.

Common-sense ideas of this kind are nearly always an important part of aetiological formulation in psychiatry, but they must be used carefully if superficial explanation is to be avoided. Aetiological formulation can be done properly only if certain conceptual problems are clearly understood. These problems can be illustrated by a case history.

For four weeks a 38-year-old married man became increasingly depressed. His symptoms started soon after his wife left him to live with another man.

In the past the patient's mother had received psychiatric treatment on two occasions, once for a severe depressive disorder and once for mania; on neither occasion was there any apparent environmental cause for the illness. When the patient was 14 years old, his mother went away to live with another man, leaving her children with their father. For several years afterwards the patient felt rejected and unhappy but eventually settled down. He married and had two children aged 13 and 10 at the time of his illness.

Two weeks after leaving home, the patient's wife returned, saying that she had made a mistake and really loved her husband. Despite her return the patient's symptoms persisted and worsened. He began to wake early, gave up his usual activities, and spoke at times of suicide.

In thinking about the causes of this man's symptoms, the clinician would first draw on knowledge of aetiology derived from scientific enquiries. Genetic investigations have shown that, if a parent suffers from mania as well as depressive disorder, a predisposition to depressive disorder is particularly likely to be transmitted to the children. Therefore it is possible that this patient received the predisposition from his mother. Clinical investigation has also provided some information about the effects of separating children from their mothers. In the present case the information is not helpful because it refers to people who were separated from their mothers at a younger age than the patient. On scientific grounds there is no particular reason to focus on the departure of the patient's mother, but intuitively it seems likely that this was an important event. From everyday experience it is understandable that a man should feel sad if his wife leaves him; it is also understandable that he is likely to feel even more distressed if this event recapitulates a similar distressing experience in his own childhood. Therefore, despite the lack of scientific evidence, the clinician would recognize intuitively that the patient's depression is likely to be a reaction to the wife's departure. The same sort of intuition might suggest that the patient would recover when his wife came back. In the event he did not recover. Although his symptoms seemed understandable when his wife was away, they no longer seem so after her return.

This simple case history illustrates the following aetiological issues: the complexity of causes in psychiatry, the classification of causes, the concept of stress, the concept of psychological reaction, and the roles that intuition and scientific knowledge should play in aetiology. These problems will be considered in turn.

The complexity of causes in psychiatry

In psychiatry the study of causation is complicated by two problems. Both these problems are met in other branches of medicine, but to a lesser degree.

The first problem is that causes are often *remote in time* from the effects that they produce. For example it is widely believed that childhood experiences partly determine the occurrence of neuroses in adult life. It is difficult to test this idea because the necessary information can only be gathered either by studying children and tracing them many years later, which is difficult, or by asking adults about their childhood experiences, which is unreliable.

The second problem is that a *single cause* may lead to *several effects*. For example deprivation of parental affection in childhood has been reported to predispose to antisocial behaviour, suicide, depressive disorder, and several other disorders. Conversely, a *single effect* may arise from *several causes*. The latter can be illustrated either by different causes in different individuals or by multiple causes in a single individual. For example mental handicap (single effect) may occur in several children, but the cause may be a different genetic abnormality in each child. On the other hand depressive disorder (single effect) may occur in one individual through a combination of causes, such as genetic factors, adverse childhood experiences, and stressful events in adult life.

The classification of causes

A single psychiatric disorder, as just explained, may result from several causes. For this reason a scheme for classifying causes is required. A useful approach is to divide causes chronologically into predisposing, precipitating, and perpetuating.

Predisposing factors

There are factors, many of them operating from early life, that determine a person's vulnerability to causes acting close to the time of the illness. They include genetic endowment and the environment *in utero*, as well as physical, psychological, and social factors in infancy and early childhood. The term **constitution** is often used to describe the mental and physical make-up of a person at any point in his life. This make-up changes as life goes on under the influence of further physical, psychological, and social influences. Some writers restrict the term constitution to the make-up at the beginning of life, whilst others also include characteristics acquired later (this second usage is adopted in this book). The concept of constitution includes the idea that a person may have a predisposition to develop a disorder (such as schizophrenia) even though the latter never manifests itself. From the standpoint of psychiatric aetiology, one of the important parts of the constitution is the personality.

When the aetiology of an individual case is formulated, the **personality** is always an essential element. For this reason the clinician should be prepared to spend considerable time in talking to the patient and to people who know him in order to build up a clear picture of his personality. This assessment often helps to explain why the patient responded to certain stressful events, and why he reacted in a particular way. The obvious importance of personality in the individual patient contrasts with the small amount of relevant scientific information so far available. Therefore in the evaluation of personality it is particularly important to acquire sound clinical skills through supervised practice.

Precipitating factors

These are events that occur shortly before the onset of a disorder and appear to have induced it. They may by physical, psychological, or social.

Whether they produce a disorder at all, and what kind of disorder, depends partly on constitutional factors in the patient (as mentioned above). Physical precipitants include cerebral tumours or drugs for example. Psychological and social precipitants include personal misfortunes such as the loss of a job, and changes in the routine of life such as moving home. Sometimes the same factor can act in more than one way; for example a head injury can induce psychological disorder either through physical changes in the brain or through its stressful implications to the patient.

Perpetuating factors

These factors prolong the course of a disorder after it has been provoked. When planning treatment, it is particularly important to pay attention to these factors. The original predisposing and precipitating factors may have ceased to act by the time that the patient is seen, but the perpetuating factors may well be treatable. For example, in their early stages many psychiatric disorders lead to secondary demoralization and withdrawal from social activities, which in turn help to prolong the original disorder. It is often appropriate to treat these secondary factors, whether or not any specific measures are carried out.

The concept of stress

Discussions about stress are often confusing because the term is used in two ways. First, it is applied to events or situations, such as working for an examination, which may have an adverse effect on someone. Second, it is applied to the adverse effects that are induced, which may be psychological or physiological change. In considering aetiology it is advisable to separate these components.

The first set of factors can usefully be called **stressors**. They include a large number of physical, psychological, and social factors that can produce adverse effects. The term is sometimes extended to include events that are not experienced as adverse at the time, but may still have adverse long-term effects. For example intense competition may produce an immediate feeling of pleasant tension, though it may sometimes lead to unfavourable long-term effects.

The effect on the person can usually be called the **stress reaction** to distinguish it from the provoking events. This reaction includes autonomic responses (such as a rise in blood pressure), endocrine changes (such as the secretion of adrenaline and noradrenaline), and psychological responses (such as a feeling of being keyed up).

The concept of a psychological reaction

As already mentioned, it is widely recognized that psychological distress can arise as a reaction to unpleasant events. Sometimes the association between event and distress is evident, for example when a man becomes depressed after the death of his wife. In other cases, it is far from clear whether the psychological disorder is really a reaction to an event or whether the two have coincided fortuitously, for example when a man becomes depressed after the death of a distant relative. Jaspers (1963, p. 392) suggested three criteria for deciding whether a psychological state is a reaction to a particular set of events. First, there must be events that seem adequate in severity and closely related in time to the onset of the psychological state. Second, there must be a clear connection between the nature of the events and the content of the psychological disorder (in the example just given, the man should be preoccupied with ideas concerning his distant relative). Third, the psychological state should begin to disappear when the events have ceased (unless, of course, it can be shown that perpetuating factors are acting to maintain it). These three criteria are quite useful in clinical practice, though they can be difficult to apply in many cases (particularly the second criterion).

Understanding and explanation

As already mentioned, aetiological statements about individual patients must combine knowledge derived from research on groups of patients with intuitions derived from everyday experience. Jaspers (1963, p. 302) has called these two ways of making sense of psychiatric disorders **Erklären** and **Verstehen** respectively. In everyday German, these terms mean 'explanation' and 'understanding' respectively, and they are usually translated as such in English translations of Jaspers' writing. However, Jaspers used them in a special sense. He used *Erklären* to refer to the sort of causative statement that is sought in the natural sciences. It is exemplified by the statement that a patient's aggressive behaviour has occurred because he has a brain tumour. He used *Verstehen* to refer to psychological understanding, or the intuitive grasp of a natural connection between events in a person's life and his psychological state. In colloquial English, this could be called 'putting oneself in another person's shoes'. It is exemplified by the statement, 'I can understand why the patient became angry when his wife was insulted by a neighbour'.

These distinctions are reasonably clear when we consider an individual patient. Confusion sometimes arises when attempts are made to generalize from insights obtained in a single case to widely applicable principles. Understanding may then be mistaken for explanation. Jaspers suggested that some psychoanalytic ideas are special kinds of intuitive understanding that are derived from the detailed study of individuals and then applied generally. They are not explanations that can be tested scientifically. They are more akin to insights into human nature that can be gained from reading great works of literature. Such insights are of great value in conducting human affairs. It would be wrong to neglect them in psychiatry, but equally wrong to confuse them with statements of a scientific kind.

The aetiology of a single case

How to make an aetiological formulation was discussed in Chapter 2 (p. 48). An example was given of a woman in her thirties who had become increasingly depressed. The formulation showed how aetiological factors could be grouped under headings of predisposing, precipitating, and per-petuating factors. It also showed how information from scientific investigations (in this case genetics) could be combined with an intuitive understanding of personality and the likely effects of family problems on the patient. The reader may find it helpful to re-read the formulation on p. 49 before continuing with this chapter.

Approaches to aetiology

Before considering the contribution that different scientific disciplines can make to psychiatric aetiology, attention needs to be given to the kinds of aetiological model that have been employed in psychiatry. A model is a device for ordering information. Like a theory, it seeks to explain certain phenomena, but it does so in a broad and comprehensive way that cannot readily be proved false.

Reductionist and non-reductionist models

Two broad categories of explanatory model can be recognized. Reductionist models seek to understand causation by tracing back to simpler and simpler early stages. Examples are the medical model described below, and the psychoanalytic model. This type of model can be exemplified by the statement that the cause of schizophrenia lies in a disordered neurotransmission in a specific area of the brain.

Non-reductionist models try to relate problems to wider rather than narrower issues. The explanatory models used in sociology are generally of this kind. In psychiatry, this type of model can be exemplified by the statement that the cause of a patient's schizophrenia lies in his family; the patient is the most conspicuous element in a disordered group of people.

It is unlikely that psychiatric aetiology can be understood by using either of these models exclusively. Different types of disorder are likely to require different kinds of explanation.

The 'medical model'

Several models are used in psychiatric aetiology, but the so-called medical model is the most prominent. It represents a general strategy of research that has proved useful in medicine, particularly in studying infectious diseases. A disease entity is identified in terms of a consistent pattern of symptoms, a characteristic clinical course, and specific post-mortem findings. When an entity has been identified in this way, a set of necessary and sufficient causes is sought. In the case of tuberculosis, for example, the tubercle bacillus is the necessary cause, but it is not by itself sufficient. The tubercle bacillus in conjunction with either poor nutrition or low resistance is sufficient cause.

This medical model has been useful in psychiatry, though not for all conditions. It is most relevant to organic syndromes, with the best example being general paralysis of the insane which is caused by syphilitic infection of the brain. It is least appropriate to the neuroses, which seem more like an exaggeration of normal psychological reactions to events. The medical model might be better named the organic model, because general medicine now adopts a broader aetiological framework including the idea that certain disorders such as hypertension are in some cases quantitative variations from the normal.

The behavioural model

Amongst the disorders that psychiatrists treat, some do not fit readily into the medical model. The latter include hysteria, sexual deviations, deliberate self-harm, the abuse of drugs and alcohol, and repeated acts of delinquency. The behavioural model is an alternative way of comprehending these disorders. In this model the disorders are explained in terms of factors that determine normal behaviour: drives, reinforcements, social and cultural influences, and internal psychological processes such as attitudes, beliefs, and expectations. The behavioural model predicts that there will not be a sharp distinction between the normal and the abnormal but a continuous gradation.

This model can be a useful way of considering many conditions seen by psychiatrists.

Although the behavioural model is mainly concerned with psychological and social causes, it does not exclude genetic, physiological, or biochemical causes. This is because normal patterns of behaviour are partly determined by genetic factors, and because psychological factors such as reinforcement have a basis in physiological and biochemical mechanisms. Also, the behavioural model employs both reductionist and non-reductionist explanations. For example, abnormalities of behaviour can be explained in terms of abnormal conditioning (a reductionist model), or in terms of a network of social influences (a non-reductionist mode). For a review of the various models used in psychiatric aetiology the reader is referred to McHugh and Slavney (1986).

Developmental models

Medical and behavioural models incorporate the idea of predisposing as well as precipitating causes, i.e. the idea that past events may determine whether or not a current cause gives rise to a disorder. Some models place even more emphasis on past events in the form of a sequence of experiences leading to the present disorder. This approach has been called the 'life story' approach to aetiology (McHugh and Slavney 1986). One example is Freud's psychoanalysis; another is Meyer's psychobiology. Freud's theories are considered later; Meyer's will be referred to next.

Adolf Meyer was born in Switzerland but spent most of his working life in the United States. Although he practised first as a neuropathologist, he made his most important contribution by emphasizing the role of psychological and social factors in the aetiology of psychiatric disorders. He applied the term **psychobiology** to this approach to aetiology, in which a wide range of previous experiences were considered and then common-sense judgement was used to decide which experiences might have led to the present disorder. Meyer recognized the importance of heredity and brain disorder, but he emphasized that these factors were modified by life experiences which could increase or decrease a person's basic vulnerability. The clinician was left to decide which experiences were relevant without relying on preconceived ideas or focusing exclusively on scientific findings.

Psychobiology is valuable as an approach to the aetiology of the individual patient rather than as a method of discovering general causes of mental disorder. For the latter purpose it is too general and does not lead to testable hypotheses.

Meyer's approach was influential for a time in the United States, where he was director of the Phipps Clinic at Johns Hopkins University Medical School for 32 years until his retirement in 1942. His ideas had a more lasting impact in Great Britain, where they were disseminated through the teaching of three of his pupils, Sir Aubrey Lewis, Sir David Henderson, and Desmond Curran (Gelder 1991). These three psychiatrists promulgated his ideas after the Second World War, at a time when American psychiatry was increasingly influenced by psychoanalytic teaching. Meyer's approach remains the basis of the evaluation of aetiology for the individual patient.

The historical development of ideas of aetiology

From the earliest times, theories of the causation of mental disorder have recognized both somatic and psychological influences. Greek medical literature referred to the causes of mental disorders, mainly in the Hippocratic writings (fourth century BC). Serious mental illness was ascribed mainly to physical causes, which were represented in the theory that health depended on a correct balance of the four body 'humours' (blood, phlegm, yellow bile, and black bile). Melancholia was ascribed to an excess of black bile. Most of the less severe psychiatric disorders were thought to have supernatural causes and to require religious healing. An exception was hysteria, which was thought to be physically caused by the displacement of the uterus from its normal position. Nowadays hysteria is attributed mainly to psychological causes.

Roman physicians generally accepted the causal theories of Greek medicine and developed them in some respects. Galen accepted that melancholia was caused by an excess of black bile, but suggested that this excess could result either from cooling of the blood or from overheating of yellow bile. Phrenitis, the name given to an acute febrile condition with delirium, was thought to result from an excess of yellow bile (Bynum 1983).

Throughout the Middle Ages these early ideas about the causes of mental illness were largely neglected, though maintained by scholars such as Bartholomeus Anglicus. The causes of mental illness were now formulated in theological terms of sin and evil, with the consequence that many mentally ill people were persecuted as witches. It was not until the middle of the sixteenth century that beliefs in the supernatural and witchcraft were strongly rejected as causes of mental disorder, notably by the Flemish writer Johan Weyer (1515–88) in his book *De praestigiis demonum*, published in 1563. Earlier, Paracelsus (1491–1541), the renowned physician, had emphasized the natural causes of mental illness.

In the seventeenth and eighteenth centuries a more scientific approach to the causation of mental illness developed as physicians became interested in mental disorders, mainly hysteria and melancholia. The English physician Thomas Willis attributed melancholia to 'passions of the heart', but considered that madness (illness with thought disorder, delusions, and hallucinations) was due to a 'fault of the brain'. Willis realized that this fault was not a recognizable gross structural lesion, but a functional abnormality. In the terminology of the time, he referred to a disorder of the 'vital spirits' that were thought to account for nervous action. Willis also pointed out that hysteria could not be caused by a displacement of the womb because the organ is firmly secured in the pelvis. (See Dewhurst (1980) and also Hunter and MacAlpine (1963, pp. 187–92).) Another seventeenth century English physician, Thomas Sydenham, rejected the alternative theory that hysteria was caused by a functional disorder of the womb ('uterine suffocation') because he had observed it in men. Despite this renewed medical interest in the causes of mental disorder, the most influential seventeenth century treatise was written by a clergyman, Robert Burton. This work, *The anatomy of melancholy* (1621), described in detail the psychological and social causes (such as poverty, fear, and solitude) that were associated with melancholia and seemed to cause it.

Aetiology depends on nosology. Unless it is clear how the various types of mental disorder relate to one another, little progress can be made in understanding causation. From his observations of patients with psychiatric disorders, the Italian physician Morgagni became convinced that there was not one single kind of madness but many (Morgagni 1769). Further attempts at classification followed. One of the best known was proposed by William Cullen, who included a category of neurosis for disorders not caused by localized disease of the nervous system (see p. 59).

The idea that individual mental disorders are caused by lesions of particular brain areas can be traced back to the theory of phrenology proposed by Gall (1758–1828) and his pupil Spurzheim (1776–1832). Gall proposed that the brain was the organ of the mind, that the mind was made up of specific faculties, and that these faculties originated in specific brain areas. He also proposed that the size of a brain area determined the strength of the faculty that resided in it, and that the size of brain areas was reflected in the contours of the overlying skull. Hence the shape of the head reflected a person's psychological make-up. Although the last steps in Gall's argument were false, the ideas of cerebral localization were to develop further. An increased interest in brain pathology led to theories that different forms of mental disorder were associated with lesions in different parts of the brain.

It had long been observed that serious mental illness ran in families, but in the nineteenth century this idea took a new form. In 1809 Morel, a French psychiatrist, put forward ideas that became known as the 'theory of degeneration'. He proposed not only that some mental illnesses were inherited, but also that environmental influences (such as poor living conditions and the abuse of alcohol) could lead to physical changes that could be transmitted to the next generation. Morel also proposed that, as a result of the successive effect of environmental agents in each generation, illnesses appeared in increasingly severe forms in successive generations. It was inherent in these ideas that mental disorders did not differ in kind but only in severity—neuroses, psychoses, and mental handicap were increasingly severe manifestations of the same inherited process. These ideas were consistent with the accepted theories of the inheritance of acquired characteristics, and they were accepted widely. They had the unfortunate effect of encouraging a pessimistic approach to treatment. They also supported the Eugenics Movement, which held that the mentally ill should be removed from society in order to prevent them from reproducing. These developments are an important reminder that aetiological theories may determine undesirable attitudes to the care of patients.

Mid-nineteenth century views of the causation of mental illness can be judged from the widely acclaimed textbooks of Esquirol, a French psychiatrist, and of Griesinger, a German psychiatrist. Esquirol (1845) focused on the causes of illness in the individual patient and was less concerned with general theories of aetiology. He recorded psychological and physical factors which he believed to be significant in individual cases, and he distinguished between predisposing and precipitating causes. He regarded heredity as the most important of the predisposing causes, but he also stressed that predisposition was acted on by psychological causes and by social (at that time called 'moral') causes such as domestic troubles, 'disappointed love', and reverses of fortune. Important physical causes of mental disorder included epilepsy, alcohol abuse, excessive masturbation, childbirth and lactation, and suppression of menstruation. Esquirol also observed that age influenced the type of illness; thus dementia was not observed among the young, but mania was uncommon in old age. He recognized that personality was often a predisposing factor.

In *Pathology and therapy of mental disorders*, which was first published in 1845, Wilhelm Griesinger maintained that mental illness was a physical disorder of the brain, and he considered at length the neuropathology of mental illness. He paid equal attention to other causes, including heredity, habitual drunkenness, 'domestic unquiet', disappointed love, and childbirth. He emphasized the multiplicity of causes when he wrote:

A closer examination of the aetiology of insanity soon shows that in the great majority of cases it was not a single specific cause under the influence of which the disease was finally established but a complication of several, sometimes numerous causes, both predisposing and exciting. Very often the germs of the disease are laid in those early periods of life from which the commencement of the formation of character dates. It grows by education and external influences... (Griesinger 1867, p. 130)

British views on aetiology in the late nineteenth century can be judged from *A manual of psychological medicine* by Bucknill and Tuke (1958), and from *The pathology of mind* by Henry Maudsley (1879). Maudsley described the causes of mental disorder in terms similar to those of Greisinger; thus causes were multiple, whilst predisposing causes (including heredity and early upbringing) were as important as the more obvious proximal causes. Maudsley held that mistakes in determining causes were often due to 'some single prominent event, which was perhaps one in a chain of events, being selected as fitted by itself to explain the catastrophe. The truth is that in the great majority of cases there has been a concurrence of steadily operating conditions within and without, not a single effective cause' (Maudsley 1879, p. 83).

Although these nineteenth-century writers and teachers of psychiatry emphasized the multiplicity of causes, many practitioners focused narrowly on the findings of genetic and pathological investigations, and adopted a pessimistic approach to treatment. A strong reaction to these attitudes was led by Adolf Meyer, a Swiss psychiatrist working in the United States, who emphasized the long sequence of events leading up to a mental illness. (Meyer's ideas are also referred to on p. 78.)

The nineteenth-century aetiological theories considered so far were mainly concerned with the more severe mental disorders. The conditions that came to be called neurosis, hysteria, and hypochondriasis, and the less severe states of depression, were treated mainly by physicians. Pierre Charcot, a French neurologist, carried out extensive studies of patients with hysteria and of their response to hypnosis. He believed that hysteria resulted from a functional disorder of the brain and could be treated by hypnosis. In the United States Weir Mitchell proposed that conditions akin to mild chronic depression were due to exhaustion of the nervous system—a condition he called neurasthenia. In Austria another neurologist, Sigmund Freud, tried to develop a more comprehensive explanation of nervous diseases, first of hysteria and then of other conditions. After an initial interest in physiological causes, Freud proposed that the causes were psychological, but hidden from the patient because they were in the unconscious part of the mind (Freud's ideas are considered further on p. 82). In France, Pierre Janet developed an alternative psychological explanation which was based on variations in the strength of nervous activity and on narrowing of the field of consciousness. Both systems were largely speculative rather than based on new observations.

Interest in psychological explanations of mental disorders grew as neuropathological and genetic studies failed to yield new insights. Freud and his followers attempted to extend their theory of the neuroses to explain the psychoses. Although the psychological theory was elaborated, no new objective data were obtained about the causes of severe mental illness. Nevertheless the theories provided explanations which some psychiatrists found more acceptable than an admission of ignorance. Psychoanalysis became increasingly influential, particularly in American psychiatry where it predominated until the 1970s. Renewed interest in genetic, biochemical, and neuropathological causes of mental disorder followed—an approach that became known as biological psychiatry.

From the history of ideas on the causation of mental disorder the most important lesson is that each generation bases its theories of aetiology on the scientific approaches most active and plausible at the time. Sometimes psychological ideas prevail, sometimes neuropathological, and sometimes genetic. Throughout the centuries, however, observant clinicians have been aware of the complexity of the causes of psychiatric disorders, and have recognized that neither aetiology nor treatment should focus narrowly on the scientific ideas of the day. Instead, the approach should be broader, encompassing whatever psychological, social, and biological factors seem most important in the individual case. Modern psychiatrists are working at a time of rapid development of the neurosciences, and they need to keep the same broad clinical perspective of aetiology whilst assimilating any real scientific advances.

Psychoanalysis

In psychoanalysis the method of investigation differs from the scientific methods reviewed later in this chapter in that it was developed specifically for the study of psychiatric disorders. It arose from clinical experience and not from work in the basic sciences. Psychoanalysis is characterized by a particularly elaborate and comprehensive theory of both normal and abnormal mental functioning. Compared with experimental psychology, it is much more concerned with the irrational parts of mental activity. Psychoanalytic theory provides a comprehensive range of explanations for clinical phenomena, and therefore has a wide appeal. However, the features that make it all-embracing also make it impervious to scientific testing.

Freud originated psychoanalytic theory, but many other workers contributed to it or developed alternative theories. This section refers only to Freud's theory and not to the other theories, some of which are mentioned elsewhere in the book. This section focuses on the basic ideas of psychoanalysis; hypotheses about particular syndromes are discussed in other chapters.

Psychoanalytic theories are mainly derived from data obtained in the course of psychoanalytic treatment. These data relate to the patient's thoughts, fantasies, and dreams, together with his memories of childhood experiences. By adopting a passive role, Freud tried to ensure that the material consisted of the patient's free associations and not of Freud's own preconceptions. However,

Freud also made interpretations of the patient's reports, and in some of Freud's writings it is difficult to distinguish clearly between the patient's statements and Freud's interpretations. It is recommended that the reader consult some of Freud's original writings, for example the *Introductory lectures on psychoanalysis* or the *New introductory lectures*, and the papers listed in the references as Freud (1924*a,b*). In this way Freud's method of working can be understood better. It is also valuable to study a critical evaluation of psychoanalytic theory (for example the shorter account by Farrell (1981) or the longer account by Dalbiez (1941)).

Farrell (1981) pointed out that psychoanalysis is an example of a broad theory of a kind found in other branches of knowledge. Such theories can be useful in science by providing a framework within which other ideas can be developed. These theories should not be judged solely on their ability to generate testable hypotheses; however, to be useful, such theories must be able to incorporate new observations as they arise. Darwin's theory of evolution is an example of a useful theory; it survives because it has proved compatible with later observations from genetics and from the fossil record. On the other hand, psychoanalytic theory has not proved compatible with advances in neurosciences in such a satisfactory way.

As pointed out earlier in this chapter, an important distinction between understanding and explanation can be made in psychiatry. In the sense of this distinction, psychoanalysis is a highly elaborate form of understanding which seeks to make both normal mental processes and psychiatric disorders more intelligible. Psychoanalysis does not lead to explanatory hypotheses that can be tested experimentally, although attempts have been made to test some of the low-level hypotheses (Fisher and Greenberg 1977). The value of psychological understanding has been discussed earlier (see p. 77), and is repeated here before psychoanalytic ideas are reviewed. These ideas can deepen our understanding of patients, but they are not the only way of doing so.

At this point a summary of the main features of Freud's theory will be presented. It is too short to do full justice to Freud's ideas, but it is long in relation to the space devoted to some other methods of scientific enquiry later in this chapter.

The structure of the healthy mind

Many of the ideas in the theory were current before Freud began his psychological studies, for example the idea of an unconscious part of the mind (Sulloway 1979). However, Freud developed and combined these ideas in an ingenious way. A central feature was his elaborate concept of the *unconscious mind*. He supposed that all mental processes originated there. Some of these processes were allowed to enter the conscious mind freely (for example sensations), some not at all (the unconscious proper), and some occasionally (most memories, which made up the 'preconscious'). According to Freud, the unconscious mind had three characteristics that were important in the genesis of neurosis: it was *divorced from reality*, it was *dynamic* in that it contained powerful forces, and it was *in conflict with the conscious mind*. These three characteristics will be discussed in turn.

The unconscious mind was held to be *divorced from reality* in several ways. It contained flagrant contradictions and paradoxes, and it tended to telescope situations and fantasies that were widely separated in time. In Freud's view, these features were well illustrated by dream analysis. Freud believed that the manifest content of a dream (what the dreamer remembered) could be traced back through analysis to a 'latent' content, which was an infantile wish. The sleeper was thought to perform 'dream work' to translate the latent to the manifest content. This translation was effected by a series of mechanisms, such as condensation (several images fused into one), displacement (of feelings from an essential feature to non-essential features of an object), and secondary elaboration (rearrangement of the assembled elements). Freud attached importance to this dream theory because he supposed the composition of neurotic symptoms to be like that of dreams, though with greater secondary elaboration.

Secondly, the unconscious mind was *dynamic*, i.e. it contained impulses that were kept in equilibrium by a series of checks and balances. In Freud's early writings, these impulses were regarded as entirely sexual. Later, he placed more emphasis on aggressive impulses. Sexual impulses were supposed to be active even in infancy, receding by about the age of four and then remaining latent until re-emergence at puberty. In Freud's view, *psychosexual development* not only began early but was long and complicated. The first stage of organization was *oral*, i.e. the sexual drive was activated by stimulation of the mouth by sucking and touching with the lips. The second stage was *anal*, i.e. the drive was activated by expelling or retaining faeces. Only in the third stage did the *genital organs* become the primary source of sexual energy. Sometimes these stages were not passed through smoothly. The *libido* (the energy of the sexual instincts) could become *fixated* (partially arrested) at one of the early stages. When this happened, the person would engage in infantile patterns of behaviour or regress to such patterns under stress. In this way the point of fixation determined the nature of any neurosis that developed later in life.

As libido developed, not only was it activated in these three successive ways but its object was supposed to change. Self-love came first, to be followed in both boys and girls by love of the mother. Next, still in infancy, boys focused their sexual wishes more intensely upon the mother while developing hostile feelings towards the father (the *Oedipus complex*). Girls developed the reverse attachments. These attachments came to an end through repression of sexual impulses. As a result the capacity to feel shame and disgust developed, and the child passed into the latency period. Finally, the sexual impulses emerged again at puberty and were directed into relationships with other adults.

The third aspect of the unconscious mind was its *struggle against the conscious mind*. This conflict was regarded as giving rise to anxiety that could persist throughout life and generate neurotic symptoms. One of Freud's lasting contributions was his idea that anxiety could be reduced by a variety of defence mechanisms, which could be discerned at times in the behaviour of healthy people. These mechanisms are considered on pp. 135–6.

Applications to psychiatric disorder

The application of psychoanalytic theory to psychiatric aetiology can be illustrated by summarizing the development of Freud's ideas about the aetiology of neurosis (for a longer account see Freud (1935)). These ideas originated in Freud's work with Breuer on the causes of hysteria. This work led Freud to conclude that hysteria was caused by a disorder of sexual function. Initially, writing in 1895, Freud postulated two kinds of disturbance that caused separate kinds of neurosis. First, he suggested that suppression of sexual function had direct toxic effects that caused anxiety neurosis and neurasthenia. He referred to these conditions as '*aktual*', a term that means 'contemporary' and refers to their direct and current causes. Suppressed sexual function was thought to have other indirect psychological effects, including the causation of hysteria, anxiety hysteria (agoraphobia), and obsessional neurosis. Before long, the idea of *aktual* neurosis was abandoned, and instead all neuroses were thought to be psychologically caused by suppressed memories of disturbing sexual events. Freud had difficulty in eliciting the supposed suppressed memories, and this problem led him to postulate an active process keeping the memories from consciousness. He named this process 'repression'. He developed a method of overcoming repression which led to frequent revelations by the patient of childhood sexual trauma. Freud later concluded that some of these accounts were not true memories but fantasies. Nevertheless, he maintained that these fantasies were important in aetiology. Thus he wrote, 'neurotic symptoms were not related directly to actual events but to phantasies embodying wishes and psychical reality was of more importance than material reality' (Freud 1935, p. 61).

In general there are three components in all later versions of Freud's theory of the aetiology of the neuroses. First, it is proposed that anxiety is the central symptom of all neuroses; other symptoms arise secondarily through mechanisms of defence (see p. 161) which act to reduce this anxiety. Second, anxiety arises when the ego fails to deal on the one hand with the mental energy reaching it from the id, and on the other hand with the demands of the superego. Third, the predisposition to develop neurosis in adult life originates in childhood from a failure to pass normally through one or other of three postulated stages of development—oral, anal, and genital. Readers who wish to know more about Freud's theories of neurosis should consult Fenichel (1945).

Freud's ideas have had considerable influence in certain countries. In Great Britain most psychiatrists take the view (which is shared by the authors of this book) that some of the basic ideas are useful in understanding patients, for example the ideas about defence mechanisms, but that the details of the theory are not generally helpful, either as an aetiological explanation of clinical syndromes or as a guide to practice. However, it is stressed again that it is impossible to do justice to Freud's theories in the space of this chapter, and readers are recommended to read some of his original papers in order to form their own judgements.

Useful summaries of psychoanalytic concepts and their application to psychiatric disorder are contained in Moore and Fine (1990).

The contribution of scientific disciplines to psychiatric aetiology

Among the disciplines that have contributed to the knowledge of psychiatric aetiology, the main groups are as follows: clinical studies and epidemiology; genetics, biochemistry, pharmacology, physiology, and neuropathology; experimental psychology, ethology, and psychoanalysis (see Table 4.1). In this section each group is discussed in turn, and the following questions are asked. What sort of problem in psychiatric aetiology can be answered by each discipline? How, in general, does each discipline attempt to answer the questions? Are any particular difficulties encountered in applying its methods to psychiatric disorders?

Table 4.1 *Scientific disciplines contributing to psychiatric aetiology*

Clinical descriptive studies	Genetics
Epidemiology	Biochemical studies
Social sciences	Pharmacology
Experimental and clinical psychology	Endocrinology
	Physiology
	Neuropathology

Clinical descriptive studies

Before reviewing more elaborate scientific approaches to aetiology, attention is drawn to the continuing value of simple clinical investigations. Psychiatry was built on such studies. For example the view that schizophrenia and the affective disorders are likely to have separate causes depends ultimately on the careful descriptive studies and follow-up enquiries carried out by earlier generations of psychiatrists.

Only two examples can be given here of the many clinical investigations that have contributed in important ways to knowledge. Both are from the British literature, but similar examples could have been chosen from the literature of continental Europe or America.

Anyone who doubts the value of clinical descriptive studies should read the paper by Aubrey Lewis on 'melancholia' (Lewis 1934). The paper describes a detailed investigation of the symptoms and signs of 61 cases of severe depressive disorder. It provided the most complete account in the English language and it remains unsurpassed. It is an invaluable source of information about the clinical features of depressive disorders untreated by modern methods. Lewis's

careful observations drew attention to unsolved problems, including the nature of retardation, the relation of depersonalization to affective changes, the presence of manic symptoms, and the validity of the classification of depressive disorders into reactive and endogenous groups. None of these problems has yet been solved completely, but the analysis by Lewis was important in focusing attention on them.

The second example is a clinical follow-up study by Roth (1955). Elderly psychiatric patients were classified on the basis of their symptoms into five diagnostic groups: affective disorder, late paraphrenia, acute or subacute delirious states, senile dementia, and arteriosclerotic dementia. These groups were found to differ in their course. Two years later, about two-thirds of the patients with affective psychoses had recovered, about four-fifths of those with senile dementia and almost as many with arteriosclerotic dementia had died, over half the patients with paraphrenia were alive but still in hospital, and of those with acute confusional states, half had recovered and half had died. These findings confirmed the value of the original diagnoses, and refuted the earlier belief that affective and paranoid disorders in old age were part of a single degenerative disorder that could also present as dementia. This investigation clearly illustrates how careful clinical follow-up can clarify issues of aetiology.

Although many opportunities for this kind of research have been taken already, it does not follow that clinical investigation is no longer worthwhile. For example a more recent clinical study describing the syndrome of bulimia nervosa had aetiological implications (Russell 1979). Well-conducted clinical enquiries are likely to retain an important place in psychiatric research for many years to come.

Epidemiology

Epidemiology is the study of the distribution of a disease in space and time within a population, and of the factors that influence this distribution. Its concern is with disease in groups of people, not in the individual person.

Aims of epidemiological enquiries

In psychiatry, epidemiology attempts to answer three main kinds of question. What is the prevalence of psychiatric disorder in a given population at risk? What are the clinical and social correlates of syndromes or forms of behaviour? What factors may be important in aetiology?

Prevalence can be estimated in community samples or among people attending general practitioners. It is important to use standardized methods of assessment since, if general practitioners are asked to report the frequency of emotional disorders among their patients, their estimates vary as much as ninefold (Shepherd *et al.* 1966). Objective assessments show that these differences are due to differences in the ability of doctors to detect and diagnose such disorders, and not to any substantial difference in frequency. In general, family doctors detect emotional disorder more readily among women, and among the middle-aged, separated, and widowed (Goldberg and Huxley 1980).

Studies of prevalence in different locations, social groups, or social classes can contribute to aetiology. Studies of associations between a disorder and personal and social variables can do the same and may be useful for clinical practice; for example epidemiological studies have shown that the risk of suicide is increased in elderly males with certain characteristics, such as living alone, abusing drugs or alcohol, suffering from physical or mental illness, and having a family history of suicide.

Epidemiological studies of **aetiology** have been concerned with predisposing and precipitating factors, and with the social correlates of mental illness. Amongst predisposing factors, the influence of heredity has been examined in studies of families, twins, and adopted people, as described in the later section on genetics (p. 93). Other examples are the influence of maternal age on the risk of Down's syndrome, the psychological development of premature babies in later life, and the psychological effects of parental loss during childhood. Studies of precipitating factors include life-events research, which is described in the following section on the social sciences.

There have been numerous studies of the **social correlates** of psychiatric disorder. For example Hollingshead and Redlich (1958) in the United States found that schizophrenia was 11 times more frequent in social class V than in social class I. In itself this finding throws no light on aetiology, but it suggests studies of other factors associated with social class (for example poor housing). It also raises questions about the interpretation of associations. For example do schizophrenics drift into the lower social classes when they become disabled, or were they in the lower social classes before the disorder began? Several epidemiological studies have focused on place of residence. Generally, high rates of schizophrenia have been found amongst people living in districts with poor housing and a large proportion of single-person households, whilst high rates of manic-depressive psychosis and neurosis have been found amongst those living in more prosperous districts (Hare 1956b).

Such results are difficult to interpret, since it is not clear whether they are due to the neighbourhood environment directly, or to patterns of life shared by people living in the district, such as methods of child rearing and patterns of marriage, or to the drift of mentally ill people into poor neighbourhoods after the start of the illness.

Concepts and methods of epidemiology

The basic concept of epidemiology is that of **rate**, or the ratio of the number of instances to the numbers of people in a defined population. Instances can be episodes of illness, or people who are or have been ill. Rates may be computed on a particular occasion (**point prevalence**) or over a defined interval (**period prevalence**). Other concepts include **inception rate**, which is based on the number of people who were healthy at the beginning of a defined period but became ill during it, and **lifetime expectation**, which is based on an estimate of the number of people who could be expected to develop a particular illness in the course of their whole life. In **cohort studies**, a group of people are followed for a defined period of time to determine the onset or change in some characteristic with or without previous exposure to a potentially important agent (for example lung cancer and smoking).

Three aspects of method are particularly important in epidemiology—defining the population at risk, defining a case, and finding cases. It is essential to **define the population at risk** accurately. Such a population can be all the people living in a limited area (for example a country, an island, or a catchment area), or a subgroup chosen by age, sex, or some other potentially important defining characteristic.

Defining a case is the central problem of psychiatric epidemiology. It is relatively easy to define a condition such as Down's syndrome, but until recently it has been difficult to define cases of affective disorder or schizophrenia. A major advance has been the development of standardized techniques for defining, identifying, rating, and classifying mental disorders. An example is the Present State Examination (Wing *et al.* 1974) and the associated computer programme CATEGO, now incorporated in the Schedule of Clinical Assessment in Neuropsychiatry. This standardized interview was used, for example, in the Interna-

tional Pilot Study of Schizophrenia (see p. 255). As well as showing some differences in diagnostic practices (referred to on p. 66), the study showed that disorders meeting strict criteria for schizophrenia occurred in a wide variety of cultures—a finding which must be explained by any aetiological theory of this condition.

Two methods are used for **case-finding**. The first is to enumerate all cases known to medical or other agencies (*declared cases*). Hospital admission rates may give a fair indication of rates of major mental illnesses, but not of alcoholism or phobias. Moreover, hospital admission rates are influenced by many extraneous variables, such as the geographical accessibility of hospitals, attitudes of doctors, admission policies, and the law relating to compulsory admissions. The second method is to search for both declared and undeclared cases in the community. In community surveys, the best technique is often to use two stages: preliminary screening to detect potential cases with a self-rated questionnaire such as the General Health Questionnaire (Goldberg 1972), followed by detailed clinical examination of potential cases with a standardized psychiatric interview.

Causes in the environment

Epidemiological approaches to aetiology can be illustrated by the results of studies of environmental causes of mental disorders. It is commonly supposed that **poor living conditions** can predispose to mental disorder, either directly or through their effects on family life. If this supposition is correct, people who move from poor to better housing should experience fewer disorders. Two well-known studies examined this possibility. Taylor and Chave (1964) investigated people moving from poor urban conditions to a new town; Hare and Shaw (1965) studied people moving from an old to a new housing estate in the same town. In neither investigation was the rate of mental disorder reduced after the move. A possible explanation is that the beneficial effect of better housing was cancelled by the adverse effect of greater social isolation in new surroundings.

(The relationship between mental health and living conditions has been reviewed in the book edited by Freeman (1984).)

Another suggested environmental cause of mental disorder is **noise**. The noise of aircraft has been studied most thoroughly. A causal relationship is suggested by the finding that, near to a large airport, people who complain most of noise tend to have more psychiatric symptoms than others. This finding might indicate that noise causes minor mental disorder, but it is equally possible that intolerance to noise is a symptom of mental disorder caused by something else. Several investigations have examined this issue (Meecham and Smith 1977; Tarnopolsky *et al.* 1980; Jenkins *et al.* 1981); whilst the evidence is not conclusive, it seems unlikely that noise is an important cause of mental disorder.

It has been suggested that some kinds of **working conditions** cause mental disorder. This possibility was studied extensively during the Second World War, when it was concluded that work requiring constant attention but little initiative or responsibility (such as repetitive machine work) can cause mental disorder (Fraser 1947). More recent studies have shown that men on paced assembly lines report more neurotic symptoms than do comparable men who have more control over their rate of work (Broadbent and Gath 1979; Broadbent 1981). Taken alone this finding could be due to the selective movement of healthy people away from more unpleasant types of work. In other circumstances, however, it has been shown that the same person has more neurotic symptoms when working in more stressful conditions. Thus student nurses have been studied in different kinds of wards; they reported more symptoms when working in conditions that they judged to be more stressful and less satisfying (Parkes 1982). It seems fair to conclude that stressful conditions of work can play a part in causing minor mental disorder.

Prolonged **unemployment** is associated with increased reporting of minor affective symptoms (Banks and Jackson 1982). The explanation for this finding may be that unemployment causes these symptoms, or that people prone to develop such symptoms are less likely to find work. Warr

and Jackson (1985) suggested that the second explanation was improbable because they found that the severity of symptoms soon after the loss of a job did not predict length of subsequent unemployment. If unemployment is a cause of minor affective disorder, the effect might be related to loss of self-esteem and social role, to financial problems, or to increased emotional conflicts within the family. (The relationship between employment and mental disorder has been reviewed by Smith (1985).)

The social sciences

Many of the **concepts** used by sociologists are relevant to psychiatry. It has been noted (p. 86) that the concepts of *social class* and subculture have been informative in epidemiological studies in which a higher prevalence of schizophrenia has been found in the lower social classes. The concepts of *stigma* and *labelling* have been useful in analysing the handicaps of people with chronic mental illness who are living in the community, and the effects of in-patient treatment have been understood better by considering hospitals as institutions that can affect the behaviour of those who stay in them. The concept of *social deviance* has been useful in the study of delinquent behaviour among adolescents. Finally, the concept of *illness behaviour* has been of value when examining the psychological consequences of physical illness.

Unfortunately, some of these potentially fruitful ideas have been used uncritically, for example in the suggestion that mental illness is no more than a label for socially deviant people—the 'myth of mental illness'. This development points to the obvious need for sociological theories to be tested in the same way as other theories by collecting appropriate data.

Some of the concepts of sociology overlap with those of social psychology, for example attribution theory (which deals with the way in which people interpret the causes of events in their lives, and ideas about self-esteem). An important part of research in sociology, the study of life events, uses epidemiological methods (see below).

Transcultural studies

Studies in different societies help in making an important causal distinction. Biologically determined features of mental disorder are likely to be similar in different cultures, whilst psychologically and socially determined features are likely to be dissimilar. Thus the 'core' symptoms of schizophrenia are present in patients from widely different societies (see p. 255), but the symptoms of the less severe forms of anxiety disorder and depressive disorders differ considerably. For example in India patients with anxiety disorders are especially likely to complain of bodily symptoms rather than emotional symptoms. Knowledge of these variations is also important in understanding the aetiology of mental disorder in individuals from immigrant groups.

The study of life events

Epidemiological methods have been used in social studies to examine associations between illness and certain kinds of events in a person's life. Wolff (1962) studied the morbidity of several hundred people over many years and found that episodes of illness clustered at times of change in the person's life. Rahe and his colleagues attempted to improve on the highly subjective measures used by Wolff (Rahe *et al.* 1967; Holmes and Rahe 1967). They used a list of 41 kinds of life change (e.g. work, residence, finance, and family relationships) and weighted each according to its apparent severity, for example 100 for the death of a spouse and 13 for a spell of leave for a serviceman.

As these last two examples show, the changes could be desirable or undesirable, and within or outside the person's direct control. In a study of men serving in the United States Navy, Rahe *et al.* (1970) found that those with the highest scores on the list of life changes developed more illnesses of all kinds. This finding suggested that the risk of illness was greater at a period of life change that at an uneventful time.

In subsequent studies the term life events has been commonly used. Research workers have modified the earlier methods in five ways:

(1) in order to reduce memory distortion, limits are set to the period over which events are to be recalled;

(2) efforts are made to date the onset of the illness accurately;

(3) attempts are made to exclude events that are not clearly independent of the illness, for example losing a job because of poor performance;

(4) events are characterized in terms of their nature (for example losses or threats) as well as their severity;

(5) data are collected with a semistructured interview, and improved rating methods are used.

Although significant, life events may be less important than at first appears. For example in one study (Paykel *et al.* 1969) events involving the loss or departure of a person from the immediate social field of the respondent ('exit events') were reported in 25 per cent of patients with depressive disorders but in only 5 per cent of controls. This difference was significant at the 1 per cent level and appears impressive, but Paykel (1978) has questioned its real significance and carried out the following calculation.

The incidence of depressive disorder is not accurately known, but if it is taken to be 2 per cent for new cases over a six-month period, then a hypothetical population of 10 000 people would yield 200 new cases.

Paykel's study showed that exit events occurred to 5 per cent of people who did not become cases of depressive disorder; therefore, in the hypothetical population, exit events would occur to 490 of the 9800 people who were not new cases. Amongst the 200 new cases, exit events would occur to 25 per cent, i.e. 50 people. Thus the total number of people experiencing exit events would be 490 plus 50, or 540, of whom only 50 (less than 1 in 10) would develop depressive disorders. Hence the greater part of the variance in determining depressive disorder must be attributed to something else. Subsequent studies have given further reasons for caution. For example, in a community study, Henderson *et al.* (1982) found that antecedent life stress accounted for only about

4 per cent of the variance of scores on the General Health Questionnaire.

Vulnerability and protective factors

People may differ in their response to life events for three reasons. First, the same event may have different meanings for different people, according to their previous experience. For example a family separation may be more stressful to an adult who has suffered separation in childhood. The other reasons are that certain contemporary factors may increase vulnerability to life events or protect against them. Ideas about these last two factors derive largely from the work of Brown and Harris (1978) who have found evidence that, among women, **vulnerability factors** include having the care of small children and can be decreased by having a confidant who can share problems. The idea of **protective factors** has been used to explain the observation that some people do not become ill even when exposed to severe adversities—a finding that is particularly evident in studies of the effects of adverse family factors on children (Rutter 1985b).

There are two major difficulties about the ideas of vulnerability factors and protective factors. First, if the findings are accepted there could be wholly different explanations for them, such as genetic or psychological differences between individuals. Second, the findings can be disputed on two kinds of grounds—technical and conceptual. The technical grounds are concerned with the validity of measures and the appropriateness of the statistical procedures. The conceptual grounds are concerned with the uncertainties of making a sharp distinction between protective and vulnerability factors on the one hand, and stressors on the other. Thus a confiding marriage has been treated as a protective factor, but divorce as a stressor. At present the notions of protective and vulnerability factors are attractive but still controversial (e.g. Tennant and Bebbington 1978; Henderson *et al.* 1982; Paykel 1983).

Causes in the family

It has been suggested that some mental disorders are an expression of emotional disorder within a whole family, not just a disorder in the person seeking treatment (the 'identified patient'). Although family problems are common among neurotic patients, their general importance is almost certainly overstated in this formulation since emotional difficulties in other family members may be the result of the patient's neurosis rather than its cause. A study by Kreitman *et al.* (1970) illustrates this point. Compared with wives of controls, wives of neurotic men were found to have higher neuroticism scores and more neurotic symptoms, and these symptoms were more frequent in longer marriages, suggesting that they resulted at least in part from living with a neurotic husband. Such an interaction may be increased by the tendency of neurotic men to spend more time with their wives and less in outside social activities (Kreitman *et al.* 1970; Henderson *et al.* 1978).

Migration and psychiatric disorder

Moving to another country, or even to an unfamiliar part of the same country, is a life change that has been suggested as a cause of mental disorder. Immigrants have been shown to have higher rates of mental disorder than similar people who remained in their own county. For example in a well-known study Ødegaard (1932) found higher rates of schizophrenia among Norwegian-born immigrants who lived in the United States than among the population of Norway (this study is also considered on p. 280). This finding may indicate that migration is a cause of mental disorder, but it can be explained in another way. Thus, amongst people who migrate, personal characteristics may make them unsettled in their country of origin and may also predispose to mental disorder. While Ødegaard's study compared migrants with people remaining in the country of origin, other studies have compared migrants with the native-born population of the new country. The latter studies are even more difficult to interpret because higher rates of mental disorder among the migrants might reflect generally higher rates in their country of origin (i.e. amongst those staying at home as well as those moving away). Alternatively, a higher rate among migrants might not reflect migration in itself, but a fall in their social class due to difficulty in finding work for example. As explained above, rates of certain psychiatric disorders are higher among people of lower social class. A further complexity is that different groups migrate for different reasons. For example Europeans may move to the United States for economic reasons, whilst the Vietnamese migrated to escape from war. For all these reasons, it is not surprising that there is no simple relationship between migration and mental disorder (see Leff (1981) for a review).

Experimental and clinical psychology

A characteristic feature of the psychological approach to psychiatric aetiology is the idea of a continuity between the normal and abnormal. This idea leads to investigations that attempt to explain psychiatric abnormalities in terms of processes determining normal behaviour. An example is research into learning mechanisms as causes of anxiety disorders (see p. 167).

A second characteristic of the psychological approach is its concern with the interaction between the person and his environment. The psychological approach differs from the social approach in being concerned less with environmental variables and more with the person's ways of processing information coming from the external environment and from his own body. Some of these ideas will become clearer when coping mechanisms are discussed later in this section.

A third characteristic of psychological research into mental disorder is an emphasis on factors maintaining abnormal behaviour. Psychologists are less likely to regard behaviour disorders as resulting from internal disease processes, and more likely to assume that persisting behaviour is maintained by reinforcement. This has led, for example, to research findings suggesting that some abnormal behaviour of chronic schizophrenic patients is maintained by social reinforcement, and that some anxiety neuroses are maintained by avoidance of situations that provoke anxiety.

Experimental psychology, more than neurochemistry or neurophysiology, makes use of broad theoretical schemes. Familiar examples are operant and Pavlovian conditioning. Such schemes can be used to provide a framework for experimental work and to construct explanations of mental disorders.

Learning theories propose mechanisms by which experiences in childhood and later life give rise to neurosis. The theories are of two kinds. The first kind, exemplified by the writings of Mowrer (1950) and Dollard and Miller (1950), accepts some of the aetiological mechanisms proposed by Freud and attempts to account for them in terms of learning mechanisms. For example repression is equated with avoidance learning, emotional conflict, and approach–avoidance conflict, and displacement is equated with association learning. Although these parallels are interesting, the approach has not led to a major advance in the understanding of neurosis.

The second kind of theory rejects Freudian ideas and attempts to explain neurosis directly in terms of concepts derived from experimental psychology. In this approach anxiety is regarded as a drive state, while the other symptoms are regarded as learned behaviour reinforced by their effects in reducing this drive. This formulation has to overcome the objections that learned behaviour extinguishes quickly unless reinforced, whilst neurotic behaviour can persist for years without obvious reinforcement. Mowrer (1950) tried to resolve this 'neurotic paradox' by proposing a **two-stage theory**: first neutral stimuli become sources of anxiety through classical conditioning, and second avoidance responses reduce this anxiety. This secondary reduction of anxiety is thought to reinforce and thereby perpetuate neurotic behaviour. Eysenck (1976) suggested a related explanation, the **'incubation effect'**. This idea is based on the observation that conditioned stimuli that do not produce a drive are subject to extinction (as in Pavlov's bell–salivation experiments), whilst conditioned stimuli that produce a drive are not extinguished by repetition but are enhanced. This enhancement is called incubation. Eysenck proposed that in neuroses the relevant conditioned stimuli produced anxiety which acts as a drive, causing incubation and prolonging the disorder.

Eysenck linked the learning theory of neurosis with personality variables. He proposed that the personality variable of neuroticism reflects autonomic reactivity, i.e. a readiness to respond to stressors by developing anxiety. He suggested that a second personality variable, introversion–extroversion, reflects the ease with which inhibition builds up during learning. People with little tendency to inhibition (introverts) are supposed to be more responsive to social conditioning in childhood and more likely to develop anxiety, phobic, and obsessional disorders later in life. People with much tendency to inhibition (extroverts) are supposed to be less responsive to conditioning, and more likely to develop hysteria or antisocial behaviour in adult life. Although intellectually satisfying, this theory is not well supported by the results of investigations of patients (see Gossop (1981) for a review).

Principles of conditioning help to explain the onset and persistence of phobic symptoms (see p. 175). Principles of operant learning help to explain the persistence of certain problems, such as disturbed patterns of behaviour in children, which are reinforced by the extra attention provided by parents when the child is behaving abnormally (see p. 697). These approaches have led to useful forms of treatment, which are known collectively as behaviour therapy (see p. 627), but their scope is limited because they undervalue thinking in the genesis and persistence of human behaviour. This limitation in earlier work on learning led to an increasing interest in cognitive psychology, especially in the approach known as information theory.

The **information theory** approach to psychology proposes that the brain can be regarded as an information channel, which receives, filters, processes, and stores information from sense organs, and retrieves information from memory stores. This approach, which compares the brain to a computer, suggests useful ways of thinking about some of the abnormalities in psychiatric disorders. Attention is viewed as an active process of selecting, from the mass of sensory input, the elements that are relevant to the processing that is being carried out at the time. One method of limiting incoming information to a manageable level is *latent inhibition*, which excludes information that past experience has shown to be irrelevant to the task in hand. Latent inhibition may be defective in schizophrenia, and this defect could contribute to the disruption of other psychological processes in this condition. Latent inhibition can be demonstrated in animals, and has been localized to specific brain areas; hence the finding of an abnormality in patients could indicate which brain areas are abnormal in schizophrenia. At a later stage in processing, information from memory stores is merged with new information from sense organs; normally the two can be identified separately, and it is possible that hallucinations are caused by some breakdown of this process with the result that stored sensory information is misidentified as current sensory information. In this case the information-processing analogy is less useful than in the previous example because it does not suggest any immediate experimental test.

The information-processing model has been applied fruitfully to the study of *memory*. It suggests that there are different kinds of memory store: sensory stores in which sensory information is held for short periods while awaiting further processing, a short-term store in which information is held for only 20 seconds unless it is continually rehearsed, and a long-term store in which information is retained for long periods. There is a mechanism for retrieving information from this long-term store when required, and this mechanism could break down while memory traces are intact. This model has led to useful experiments. For example patients with the amnesic syndrome (see p. 314) score better on memory tests requiring recognition of previously encountered material than on tasks requiring unprompted recall; this finding suggests a breakdown of information retrieval rather than of information storage.

The model also predicts that responses to information, including the emotional response, are determined by *attitudes and expectations*. This part of the model has been used to develop ideas for cognitive therapy. For example experimental work has shown that patients with panic disorder (p. 179) have inaccurate expectations that sensory information about rapid heart action predicts an imminent heart attack. This expectation results in anxiety when the information is received, with the result that the heart rate accelerates further and a vicious circle of mounting anxiety is set up. Changing these expectations can alleviate panic attacks (see also p. 179).

Another application of the 'psychological' model to psychiatry is concerned with **coping mechanisms**. This term has been applied to certain ways in which people deal with changes in their environment. It is used in a narrow sense and a wide sense. Thus some psychologists limit the word to those responses to a stressor that reduce any stress reaction that might otherwise ensue, whereas others apply it more widely to any response whether or not it reduces the stress reaction.

Coping mechanisms have two components: internal events and observable behaviour. A person's coping mechanisms after bereavement might be first a return to former religious beliefs (an internal mechanism) and second joining a social club to combat loneliness (an observable behaviour). Research into coping mechanisms is much concerned with the ways in which meaning is attached to events. The same event, for example a change of job, can be seen as a threat by one person and a challenge by another. It is presumed that the meaning attached to an event by a person is an important determinant of his response to it.

To date, in the study of psychiatric problems, psychology has been more successful in the use of experimental methods than in the application of theoretical constructs. This kind of experimental approach to patients, sometimes known as experimental psychopathology, has been referred to already (p. 2).

Many psychological studies involve quantitative observations of behaviour. In some of these investigations use is made of methods developed originally in the related discipline of **ethology**. Complex behaviour is divided into simpler components and counted systematically. Regular sequences are noted as well as interactions between individuals, for example between a mother and her infant. Such methods have been used, for example, to study the effects of separating infant primates from their mothers, and to compare this primate behaviour with that of human infants separated in the same way (e.g. Blurton-Jones 1972). Similarities in the responses of primate and human infants suggests a biological basis for this aspect of the behaviour. Similar methods have been used to identify possible delays in the social development of human infants whose mothers are depressed by comparison with infants of healthy mothers (Stein *et al.* 1991). (For an account of aspects of ethology relevant to psychiatry see Hinde (1985).)

Genetics

Genetic investigations are concerned with three issues: the relative contributions of genetic and environmental factors to aetiology, the mode of inheritance of disorders that have a hereditary basis, and the mechanisms of inheritance. In psychiatry, important advances have been made with the first two issues, but so far little progress has been made with the third. Research methods in genetics are of three broad kinds: population and family studies, cytogenetics, and molecular genetics. Population and family studies are mainly concerned with estimating the contribution of genetic factors and the mode of inheritance, whilst cytogenetics and molecular genetics provide information about mechanisms of inheritance. To date, genetic research in psychiatry has relied mainly on methods of population genetics. For a detailed account of the topics in this section, see McGuffin (1984) and Pardes *et al.* (1989).

The contribution of genetic factors

Methods of population genetics are used to assess risk in three groups of people: families, twins, and people who have been adopted. In **family risk studies** the investigator determines the risk of a psychiatric condition among the relatives of affected persons and compares it with the expected risk in the general population. (The affected persons are usually referred to as index cases or **probands**.) Such studies require a sample selected in a strictly defined way. Moreover, it is not sufficient to ascertain the current prevalence of a psychiatric condition among the relatives because some of the population may go on to develop the condition later in life. For this reason, investigators use corrected figures known as **expectancy rates** (or morbid risks).

Family risk studies have been used extensively. Examples will be found in the chapters on mood disorders and schizophrenia (see pp. 213 and 265). Since these studies cannot distinguish between inheritance and the effects of family environment, they are the least satisfactory way of determining the genetic contribution. They are useful chiefly in pointing to the need for other kinds of investigation.

In **twin studies** the investigator seeks to separate genetic and environmental influences by comparing concordance rates in uniovular (monozygous (MZ)) and binovular (dizygous (DZ)) twins. Such studies depend crucially on the accurate determination of zygosity. If concordance for a psychiatric disorder is substantially higher in MZ twins than in DZ twins, a major genetic component is presumed. More precise estimates of the relative importance of heredity and environment can be made by comparisons of MZ twins reared together and MZ twins reared apart from early infancy. A high concordance between MZ twins reared apart is strong evidence of a genetic aetiology. An example of such studies will be found in Chapter 9 on schizophrenia (see p. 267).

Adoption studies provide another useful method of separating genetic and environmental influences. These studies are concerned with children who, since early infancy, have been reared by nonrelated adoptive parents. Two main comparisons can be made. First the frequency of the disorder can be compared between two groups of adopted people: those whose biological parents had the illness, and those whose biological parents did not have it. If there is a genetic cause, the rate will be greater in the former. Secondly, in the case of adopted people who have a psychiatric disorder, the frequency of the disorder can be compared between the biological parents and the adoptive parents. If there is a genetic cause, the rate will be greater in the former. Such studies may be affected by a number of biases, such as the reasons why the child was adopted, non-random assignment of the

children on socio-economic status, and the effects on adoptive parents of raising a difficult child. An example is provided by the studies of schizophrenia reviewed on p. 268.

The mode of inheritance

This is assessed by using special statistical methods to test the fit of pedigree or family data with alternative models of inheritance. Usually four models are considered: the single major locus model, which may be dominant, recessive, or sex-linked, and a mixed model of major genes operating together. This approach has been successful in studies of certain dementias such as Alzheimer's disease and Huntington's chorea. Attempts to study other psychiatric disorders in this way have generally led to equivocal results despite considerable research, particularly on schizophrenia and affective disorder.

Linkage studies

Linkage studies seek to identify the locus of a gene on the chromosomes by studying the extent to which it co-segregates with a 'marker' gene. Genetic markers are readily identifiable characters with known single modes of inheritance and two or more common alleles (alternative genes). They include blood groups, human leucocyte antigens (HLAs), and certain physiological abnormalities (e.g. colour blindness). Large family pedigrees are studied to determine to what extent two genes 'stick together', departing from Mendel's law of independent assortment. In this way, by using appropriate mathematical techniques, it can be estimated how closely the gene loci are likely to be linked on a chromosome. Many such studies have been carried out with psychiatric disorders, but so far no linkage has been found with a marker of this kind.

Linkage studies work best when there is an established mode of inheritance for the disorder and high penetrance, and diagnosis is reliable and supported by pathological findings. Neither schizophrenia nor mood disorders are of this kind, and the value of linkage studies thus far has been limited. Among disorders of interest to psychiatrists, this method has so far been applied successfully only to Huntington's chorea (see p. 324), although recent studies of Alzheimer's disease have yielded important linkage findings (see p. 523). As mentioned below, recent advances in molecular genetics are linkely to increase greatly the scope for linkage analysis in other psychiatric disorders.

Cytogenetic studies

These studies are concerned with identifying structural abnormalities in chromosomes and associating them with disease. The most important example in psychiatry concerns Down's syndrome (mongolism). In this condition two kinds of abnormality have been detected: in the first kind there is an additional chromosome (trisomy); in the second kind the chromosome number is normal but one chromosome is unusually large because a segment of another chromosome is attached to it (translocation) (see p. 738). Other examples involve the X and Y chromosomes. In Turner's syndrome there is only one sex chromosome (XO), while in three other syndromes there is an extra one—XXY (Klinefelter's syndrome), XXX, and XYY.

Molecular genetics

Advances in cell and molecular biology are starting to contribute to knowledge of psychiatric aetiology. These advances stem from technical innovations. The first is the discovery of bacterial enzymes called **restriction endonucleases** which cut DNA at sites with particular base sequences. The second advance is in techniques for reproducing fragments of human DNA by inserting them into bacterial plasmids (small circular pieces of DNA in bacteria) and encouraging the bacteria to reproduce. Techniques of this kind have been used to develop **genomic libraries**, i.e. bacterial cultures which contain the entire genome of the individuals from which they were made. It is possible to make libraries of complementary DNA (cDNA); this represents only the DNA that is transcribed into

RNA in the cells from which the messenger RNA (mRNA) was isolated. The third advance is the construction of **gene probes**, i.e. short sequences of DNA, either from the genome or copied from mRNA. Copies are made by employing enzymes from tumour viruses that reverse the normal sequence, i.e. in which RNA is made from DNA. It is possible to incorporate radioactive bases into the short DNA sequences, thereby allowing a sequence to be identified; because DNA sequences stick to sequences that are similar, the radioactive fragment can be used to search for and label specific base sequence in mixtures of DNA that have been separated by electrophoresis. This procedure is called **gene mapping**. A further important advance has been the development of the **polymerase chain reaction (PCR)** which allows short sequences of DNA to be reproduced in significant quantities over a few hours. The technique involves the use of DNA primers and heat-stable DNA polymerase to produce repeated cycles of DNA synthesis at elevated temperatures. The use of PCR has greatly aided the sequencing of human genes and the identification of single-point mutations.

Molecular genetic linkage studies Human DNA shows many harmless variations in its base sequences. These variations may provide new sites for restriction endonucleases, or may remove existing ones. By changing the sites at which the enzymes cut the DNA, the variations lead to alterations in the length of the cut fragments. For this reason these harmless variations in base sequence are called **restriction fragment length polymorphisms**. Because they are scattered through the human genome, they can be used to carry out a systematic search for genetic linkage even if knowledge of the pathological process is lacking. Once linkage has been established, the disease gene can be located and its mutation identified. This approach has proved useful in Huntington's chorea where the disease gene and its mutations have been characterized (Ross *et al.* 1993), though the nature of the resulting pathophysiological process is still obscure. However, linkage analyses in major psychoses have proved disappointing, and initial reports of linkage to chromosome 11 in bipolar disorder and to chromosome 5 in schizophrenia seem incorrect (Owen 1992).

There are several reasons why linkage analysis in psychiatric disorders, particularly the psychoses, may be problematic. Linkage studies are of most value when there is a single major gene abnormality, as in Huntington's chorea. Linkage studies are of less value when a disorder may result from the effects of several different genes (polygenic inheritance) unless a few of them produce proportionately major effects in the disease process. Another problem is that many psychiatric conditions may be genetically heterogeneous. In the absence of good knowledge of pathophysiology, the classification of a psychiatric disorder such as schizophrenia is based on identification of a clinical syndrome. However, there may be genetic variants of schizophrenia that cannot be distinguished clinically at present. Other cases may be phenocopies lacking a significant genetic element. Lumping these cases together in the same linkage analysis might obscure a major gene effect even if one were present.

While we lack definite evidence concerning the pathophysiology of most psychiatric disorders, current hypotheses can be used to explore a **candidate gene** approach. Thus linkage studies can be carried out in affected pedigrees by using DNA probes for particular neuropharmacological mechanisms that have been postulated to be involved in the disease process. For example the demonstration of abnormalities in noradrenaline receptor sensitivity in panic disorder prompted a linkage study between various adrenoceptor gene variants and panic disorder in several multiply affected families (Wang *et al*. 1992). In fact, the findings of the study make it unlikely that panic disorder is associated with a mutation in any of the genes studied. Similarly, the candidate gene approach can be used in **association studies** in which frequencies of different alleles for a particular gene are examined in patients and a control population. For example apolipoprotein E (apoE) is a polymorphic protein found in the plaques associated with Alzheimer's disease. Therefore it is of great interest that individuals with a particular genetic variant of apoE (apoE4) have a high risk of developing late-onset Alzheimer's disease. This suggests that apoE may be an important susceptibility factor in the development of this disorder (Poirier *et al*. 1990).

Gene expression in psychiatric disorders A defective gene may manifest itself at one or more of the points that lead to the synthesis of a particular protein. In general, the end result is either a structurally abnormal protein or a protein product that is present in reduced amounts or is entirely absent. Therefore, to understand the phenotypic presentation of psychiatric disease, it is necessary to study gene expression and protein synthesis in the human brain. The molecular genetic techniques described above can be readily adapted for this purpose. Thus cDNA libraries made from neuronal tissue can be used to determine whether the complementary mRNAs are expressed differently in patients with psychiatric disorders and in controls. The technique of *in situ* hybridization employs labelled DNA or RNA probes to hybridize to complementary RNA coding for a particular protein. Using autoradiographic methods it is possible to study gene expression in individual neurones in a quantitative manner. Of course, these techniques can be used to study acquired changes in gene expression caused by environmental changes, brain injury, drugs, or hormones. Most information on the nature and localization of a neuronal abnormality is likely to come from studies that combine molecular genetic methods with established neurochemical techniques (Harrison and Pearson 1989). For an excellent review of this detailed and complex developing area of research, the reader should consult Weatherall (1991).

Biochemical studies

These studies can be directed either to the causes of diseases or to the mechanisms by which disease produces its effects. The methods of biochemical investigation are too numerous to consider here, and it is assumed that the reader has some knowledge of them. The main aim here is to consider some of the problems of using biochemical methods to investigate psychiatric disorder.

It will be clear from the above account that the scope for molecular genetic studies is greatly enhanced by the presence of a biochemical abnormality that reliably distinguishes patients with a particular psychiatric disorder. The value of such an abnormality would be greater still if the biochemical abnormality concerned played a significant role in the cause of the illness or its pathophysiology. However, the nature of the biochemical changes associated with most psychiatric disorders remains unknown. This is due both to our lack of knowledge about the biochemical complexities of the normal brain and to the difficulty of investigating the biochemistry of the living human brain directly. Moreover, because most psychiatric disorders do not lead to death (other than by suicide), post-mortem material is not widely available except among the elderly.

Because of these problems, workers have adopted a variety of indirect methods involving sampling of peripheral tissues and fluids such as cerebrospinal fluid, blood cells, and urine. These studies, while more feasible to carry out, are not always easy to interpret. For example concentrations of neurotransmitters and their metabolites in lumbar cerebrospinal fluid have an uncertain relationship to the corresponding functionally active neurotransmitter in the brain. Equally, neurotransmitter receptors and their second messengers in blood platelets and lymphocytes often appear to be regulated in a different way to their brain counterparts. Finally, measures in plasma and urine are very susceptible to confounding dietary and behavioural changes (see below).

The reader will find accounts of the results of biochemical research in subsequent chapters, especially those on affective disorders and schizophrenia. At this point a few examples will be given of the different kinds of investigation.

Post-mortem studies

Post-mortem studies of the brain provide the most direct evidence of chemical changes within it. Unfortunately, interpretation of the findings is difficult because it must be established that any changes in the concentrations of neurotransmitters or enzymes did not occur after death. Moreover, because psychiatric disorders do not lead directly

to death, the ultimate cause of death is another condition (often bronchopneumonia or the effects of a drug overdose) that could have caused the observed changes in the brain. Even if this possibility can be ruled out, it is still possible that the chemical findings are the results of treatment rather than of disease. For example the density of dopamine receptors has been found to be increased in the nucleus accumbens and caudate nucleus in schizophrenic patients (Owen *et al.* 1978). This finding might be interpreted as supporting the hypothesis that schizophrenia is caused by changes in dopamine function in these areas of the brain. The finding could equally be the result of long-term treatment with antipsychotic drugs which block dopamine receptors and might lead to a compensatory increase of receptors. Owen *et al.* provided some evidence against this alternative when they demonstrated similar changes in two patients who had apparently never received antipsychotic drugs, but the point is still unsettled. In any case, even if it is eventually possible to rule out the effects of treatment, it must still be shown that the observed changes in receptors are not merely a compensatory mechanism to balance a primary disorder in another neurotransmitter system.

As mentioned above, molecular genetic techniques can be used to complement biochemical investigations in post-mortem brain. For example, *in situ* hybridization provides information about the gene expression of neurotransmitter receptors of interest. Using this technique it was shown that the mRNA for glutamate receptors is decreased in the hippocampus of patients with schizophrenia, a finding which complements ligand-binding studies of the glutamate receptors in this area of the brain (Harrison *et al.* 1991). An important development in post-mortem studies is the combined use of gene expression, neurochemical, and neuropathological techniques to investigate abnormalities in neurotransmitter function in carefully defined brain regions.

Brain biochemistry and brain imaging

Novel methods of studying biochemical events in the living brain have recently become available and have been used in some studies of psychiatric disorders. These methods include magnetic resonance imaging (MRI), single-photon emission tomography (SPET), and positron emission tomography (PET). The use of these techniques to measure cerebral structure and blood flow is discussed below under the relevant headings. However, brain imaging can also be employed to measure aspects of brain biochemistry. For example it is possible to carry out *in vivo* receptor binding in different groups of psychiatric patients using positron-labelled ligands and PET imaging.

Using this technique, Wong *et al.* (1986) reported that unmedicated patients with schizophrenia had increased binding of ^{11}C-labelled methylspiperone to dopamine D_2 receptors in the caudate nucleus. This finding is of great interest because it supports the hypothesis that schizophrenia may be associated with increased levels of D_2 receptors in the brain (see above). However, using ^{11}C-raclopride as a D_2 ligand, Farde *et al.* (1990) were unable to confirm the increase in D_2 receptors. The possible reasons for these discrepant reports are unclear but seem likely to depend on differences in the ligand used and details of the *in vivo* binding methodology. Nevertheless, the use of PET, and more recently SPET, to label specific neurotransmitter receptors in the human brain has greatly increased the scope of research in this area. For reasons of cost, studies employing PET are likely to remain restricted to a small number of specialist research centres. However, SPET imaging is more widely available and increasing numbers of specific receptor ligands suitable for SPET studies are being developed.

MRI has the advantage over SPET and PET that subjects are not exposed to radiation. While MRI has proved an excellent tool for structural brain imaging and more recently for the examination of cerebral blood flow, its application to the study of brain biochemistry (magnetic resonance spectroscopy (MRS)) has been somewhat limited by lack of sensitivity. However, proton MRS has recently been used to detect regional changes in brain γ-aminobutyric acid (GABA). MRS can also be used to identify the spectrum of phosphorus-containing compounds and thereby can provide information about energy metabolism and intracellular pH. A

number of psychotropic drugs possess fluorine atoms which can be imaged by MRS; this provides a means of imaging the distribution of such drugs at their specific receptor sites in the brain. MRS has also been used to image lithium in the human brain (Lock *et al.* 1990).

Peripheral measures

There have been long-standing doubts as to whether changes in the composition of neurotransmitters in the cerebrospinal fluid (CSF) reflect functionally significant changes in the brain. However, recent studies suggest that, in some circumstances, measurement of neurotransmitter metabolites can provide an index of turnover in certain brain regions, particularly following administration of psychotropic drugs. In fact, it may be more useful in CSF studies to examine changes in the ratios of different monoamine metabolites. This approach appears to yield more consistent abnormalities in psychiatric disorders than does examination of a single metabolite in isolation, and may provide important data on the functional interactions between neurotransmitters in the central nervous system (Potter and Manji 1993). The major limitation of CSF studies is that it is often ethically and practically difficult to obtain CSF samples from psychiatric patients. In addition, it is not feasible to monitor time-dependent changes in neurotransmitter metabolism through repeated sampling.

Ingenious attempts have been made to infer biochemical changes in the brain from measurements of substances in the blood. For example it is known that the rate of synthesis of 5-hydroxytryptamine (5-HT) depends on the concentration of the 5-HT precursor tryptophan in the brain. Several studies have shown that plasma tryptophan is decreased in patients with major depression, a finding which supports the hypothesis that brain 5-HT function may be impaired in depressive disorders. However, it cannot be assumed that a modest reduction in concentrations of plasma tryptophan will necessarily be associated with impaired brain 5-HT neurotransmission. Furthermore, the same reduction in plasma tryptophan concentrations is found when healthy people lose weight through dieting. Therefore it is quite possible that the decrease in plasma tryptophan found in depressed patients is simply a consequence of concomitant weight loss (Anderson *et al.* 1990). In general, investigations of biochemical abnormalities in blood and urine have not proved particularly fruitful in understanding the aetiology of psychiatric disorders. The real advances from such studies are in the field of mental handicap, where measurement of metabolites in blood and urine have sometimes provided a useful picture of the abnormalities present in the brain as well as valuable diagnostic tests. A good example is phenylketonuria (see p. 734).

Peripheral blood cells such as platelets and lymphocytes possess receptors for neurotransmitters that often resemble the analogous receptor binding sites in the brain. There have been many studies of monoamine receptors in platelets of depressed patients, but the findings tend to be inconsistent and easily confounded by factors such as drug treatment. In addition, it is far from clear that abnormalities found in these peripheral binding sites will necessarily also be present in the brain. Similar comments apply to the use of blood cells to investigate neurotransmitter-linked second messengers and ion flux processes such as calcium entry.

Pharmacology

The study of effective treatment of disease can often throw light on aetiology. In psychiatry, because of the great problems of studying the brain directly, research workers have examined the actions of effective psychotropic drugs in the hope that the latter might indicate the biochemical abnormalities in disease. Of course, such an approach must be used cautiously. If an effective drug blocks a particular transmitter system, it cannot be concluded that the disease is caused by an excess of that transmitter. The example of parkinsonism makes this clear; anticholinergic drugs modify the symptoms, but the disease is due to a deficiency in dopaminergic transmission and not an excess of cholinergic transmission.

It is assumed here that the general methods of neuropharmacology are familiar to the reader, and attention is focused on the particular difficulties of using these methods in psychiatry. There are two main problems. First, most psychotropic drugs have more than one action and it is often difficult to decide which is relevant to the therapeutic effects. For example, although lithium carbonate has a large number of known pharmacological effects, it has so far been impossible to explain its remarkable effect of stabilizing the mood of manic depressive patients. The second difficulty arises because the therapeutic effects of many psychotropic drugs are slow to develop, while most pharmacological effects identified in the laboratory are quick to appear. For example it has been suggested that the beneficial effect of antidepressant drugs depends on alterations in the re-uptake of transmitter at presynaptic neurones. However, changes in re-uptake occur quickly, while the therapeutic effects are usually delayed for about two weeks.

Recent studies in animals have concentrated on changes that occur in brain neurotransmitter receptors during long-term psychotropic drug treatment. These changes are interesting because the time course is similar to that of the development of therapeutic effects. Also, antidepressant treatments that have different pharmacological effects when first given may, after repeated administration, produce similar effects on neurotransmitter receptors. Thus it appears that the late effects of both antidepressant drugs and electroconvulsive shock are to produce a common change in postsynaptic noradrenergic receptors, a change which may be important in mediating the antidepressant effect (see Green *et al.* (1986) for a review).

Other pharmacological studies in animals and human subjects have identified subtypes of receptor for many of the classical neurotransmitters, and these subtypes seem to have distinct functional roles. It is possible that drugs could be developed which would affect the particular receptor subtypes, thereby reducing side-effects (Cowen 1991).

The introduction of new drugs with different pharmacological actions from conventional compounds can often be used to generate hypotheses about the mode of action of beneficial treatments and the pathophysiology of the disorder concerned. For example, with the introduction of selective serotonin re-uptake inhibitors (SSRIs) it has become clear that only drugs with potent 5-HT re-uptake inhibitor properties are effective in the pharmacological treatment of obsessive–compulsive disorder. Conventional tricyclic antidepressants (with the exception of clomipramine) are not useful (Insel 1991). This suggests that the pathophysiology of obsessive compulsive disorder differs from that of major depression, for which both classes of compounds are equally effective. The SSRIs may be effective for obsessive–compulsive disorder because they produce larger overall increases in 5-HT neurotransmission than conventional tricyclic antidepressants, or because they activate particular 5-HT receptor subtypes which are not affected by tricyclics. Specific experiments can be designed to test these hypotheses.

Another drug which has stimulated research in this way is clozapine, an antipsychotic drug which is effective in a significant proportion of patients who are unresponsive to traditional antipsychotic agents. Most antipsychotic drugs are believed to produce their therapeutic effects through blockade of dopamine D_2 receptors, but clozapine has a weak affinity for this binding site. In fact clozapine binds most potently to the recently described dopamine D_4 receptor. This finding has led to the hypotheses that selective D_4 receptor antagonism may be a particularly effective means of producing an antipsychotic effect, and that changes in D_4 receptors may be implicated in the pathophysiology of schizophrenia (Seeman *et al.* 1993).

Endocrinology

Changes in circulating concentrations of hormones can have profound effects on mood and behaviour, whilst abnormalities in endocrine function are responsible for a number of well-defined clinical syndromes, some of which have characteristic neuropsychiatric presentations. Measurement of plasma hormone levels in psychiatric disorders has not, in general, shown consistent abnormalities in psychiatric patients or thrown much light on aetiology. The exception is major depression, in which a significant proportion of patients hypersecrete cortisol. However, the pathophysiological role of cortisol in depression has remained uncertain, as has the specificity of the association (see Braddock (1986) for a review).

Recently, knowledge of how hormones may alter brain function has increased, and so have implications for the pathophysiology of certain psychiatric disorders. For example corticosteroids act on the cell nucleus to alter the expression of receptors for various neurotransmitters. In animal experimental studies the density of 5-HT_{1A} receptors is modulated by circulating corticosterone levels, and it has been proposed that excessive cortisol secretion may predispose to a depressive disorder through an attenuation of 5-HT_{1A} receptor function in limbic brain regions (Deakin 1991*a*). Animal studies have also indicated that corticosteroid administration can cause cell loss in the hippocampus. This finding has led to the hypothesis that the cognitive impairment seen in elderly depressed patients may be a consequence of neuronal damage produced by excessive cortisol secretion (Rubinow *et al.* 1984; Axelson *et al.* 1993).

Another use of plasma hormone measurement is to monitor the functional activity of brain neurotransmitters. The secretion of pituitary hormones is controlled by a variety of neurotransmitters. Under certain circumstances changes in the concentration of a plasma hormone can be used to assess the function of the neurotransmitters involved in its release. For example stimulating brain 5-HT function with a specific drug gives rise to an increase in plasma prolactin levels; accordingly, the rise in prolactin concentration that accompanies administration of a standard dose of the drug gives a measure of the functional state of brain 5-HT pathways. These **neuroendocrine challenge tests** provide dynamic functional measures of brain neurotransmitter pathways, and in certain psychiatric disorders they have yielded consistent evidence of impairments in neurotransmitter function. For example in depressed patients there is good evidence that the prolactin response to 5-HT stimulation is blunted but that this impairment remits following effective treatment and clinical recovery. This suggests that depressive disorders are associated with a reversible deficit in brain 5-HT neurotransmission. However, as with other biological measures, great care must be taken to control for possible confounding effects such as weight loss and impaired sleep. In fact, weight loss does alter brain 5-HT function but causes the opposite change in 5-HT-mediated prolactin release as is seen in depressed patients. Thus in 5-HT neuroendocrine studies of depressed patients it is important to assess and control for concomitant weight loss, otherwise the impairment in 5-HT-mediated prolactin release may be obscured (Cowen and Anderson 1991).

Neuroendocrine challenge tests can also be used to assess the effect of psychotropic drugs on brain neurotransmitter function. For example the cortisol response to the 5-HT receptor agonist *m*-chlorophenylpiperazine is blocked in patients receiving treatment with the atypical antipsychotic drug clozapine, but not in patients receiving a conventional antipsychotic agent such as fluphenazine (Owen *et al.* 1993). This suggests that clozapine treatment attenuates neurotransmission at a specific subpopulation of 5-HT receptors, and this action may relate to its unusual therapeutic efficacy or perhaps to aspects of its side-effect profile such as excessive weight gain.

Physiology

Physiological methods can be used to investigate the cerebral and peripheral disorders associated with disease states. Several methods have been used: studies of cerebral blood flow, particularly in chronic organic syndromes, electroencephalographic (EEG) studies, and a variety of psychophysiological methods including measurements of pulse rate, blood pressure, blood flow, skin conductance, and muscle activity. These psychophysiological measures can be interpreted in at least two ways. The first interpretation is straightforward. The data are used as information about the activity of peripheral organs in disease, for example to determine whether electromyographic (EMG) activity is increased in the scalp

muscles of patients who complain of tension headaches. The second interpretation depends on the assumption that peripheral measurements can be used to infer changes in the state of arousal of the central nervous system. Thus increases in skin conductance, pulse rate, and blood pressure are taken to indicate greater arousal.

Measurement of cerebral blood flow and metabolism

Advances in brain imaging methods have led to increasing sophistication in the measurement of cerebral blood flow in psychiatric disorders. Studies using PET and SPET have largely replaced older techniques using xenon inhalation because the addition of tomographic techniques allows a three-dimensional measurement of regional cerebral blood flow to be achieved. Another important recent development is the demonstration that MRI techniques using the water proton signal are sufficiently sensitive to define regional increases in cerebral blood flow following neuronal activation (McCarthy *et al.* 1993). This technique promises the development of highly sensitive non-invasive methods for studying associations between psychological state and altered regional cerebral blood flow.

PET imaging can be used to measure either cerebral metabolism or cerebral blood flow. Usually the two measures are closely correlated. In the adult brain, functional activity is almost entirely dependent on oxidative metabolism which requires glucose and oxygen as substrates. Hence rates of metabolism can be determined by measuring the utilization of oxygen or accumulation of deoxyglucose. Measurement of regional cerebral blood flow can be made by assessing the accumulation of radioactivity in the brain during inhalation of suitably labelled CO_2 (Bench *et al.* 1990).

Measurement of blood flow with SPET employs lipophilic radiotracers such as technetium-labelled hexamethyl propyleneamine oxime ($^{99}Tc^m$-HMPAO). Following intravenous administration, these compounds are retained in the brain in a stable form for several hours. This enables high resolution images to be obtained with the use of a conventional detector such as a rotating gamma camera. The uptake of ^{99}Tc^m-HMPAO is linearly related to cerebral blood flow. However, unlike PET, SPET cannot provide an absolute measure of regional cerebral blood flow; therefore the results of SPET studies are often expressed by comparing the radioactive counts in each brain region of interest with a reference area, usually either whole brain or cerebellum (Geaney and Abou-Saleh 1990).

There have been many studies of basal blood flow in various psychiatric disorders, but the results of different investigations have often been contradictory. To a large extent the conflicting data may result from the considerable methodological difficulties in standardizing the imaging conditions and the patient population. It is noteworthy that more recent carefully controlled investigations in rigorously assessed drug-free patients are reaching a greater level of consensus. For example, both PET and SPET studies of patients with obsessive compulsive disorder have revealed increased metabolic activity and blood flow in the frontal cortex, notably in orbitofrontal regions (Insel 1992). The information provided by functional brain imaging can be enhanced by correlating regional cerebral blood flow with the psychopathology of the patients at the time of scanning. Using this technique it has been shown, for example, that auditory hallucinations in schizophrenia are associated with increased blood flow in Broca's area (McGuire *et al.* 1993).

Psychological activation paradigms have been widely used in studies of healthy volunteers to map the brain regions and distributed neuronal circuits involved in fundamental processes such as memory and language. Similar techniques can be applied to studies in psychiatric patients. For example when normal control subjects undertake the Wisconsin Card Sort Test, there is an increase in blood flow in the dorsolateral prefrontal cortex. Patients with schizophrenia perform less well than controls on this test, and produce a much smaller change in blood flow in the corresponding cortical area. This suggests that some patients with schizophrenia may have a dysfunction of the dorsolateral prefrontal cortex which is associated

with poor performance on tasks that depend on increased neuronal activity in this brain region (Berman *et al.* 1993).

Electroencephalography

Much use has been made of EEG techniques in psychiatric research. Routine EEG recordings have been most helpful in studying the relationships between epilepsy and psychiatric disorders but otherwise have not been particularly informative about aetiology. Sleep EEG shows fairly consistent abnormalities in depressed patients, notably a decrease in the latency to the onset of rapid eye movement (REM) sleep. Some of these abnormalities may persist into clinical remission. There are also reports that first-degree relatives of patients with major depression, who have not themselves been depressed, may show similar abnormalities in sleep EEG. While the relationship of changes in sleep EEG to the pathophysiology of depression is unclear, it is possible that a shortened REM latency may prove to be a marker of vulnerability to depressive disorders (Giles *et al.* 1993).

EEG provides a measure of cortical neuronal activity through detection of potential differences across the scalp. Magnetoencephalography (MEG) uses changes in extracranial magnetic fields to detect ion fluxes in cortical neurones (Pizella and Romani 1990). Like EEG, MEG has the ability to detect changes in physiological signals over millisecond time intervals. In some circumstances MEG can provide better localization of signals than EEG, but the most useful information may come from using the techniques in combination. Neither MEG nor EEG is generally helpful in identifying changes in subcortical neuronal activity. In some studies of psychiatric patients MEG has been used to measure evoked potentials (see below).

EEG techniques can also be used to detect changes in brain electrical activity in response to environmental stimuli. These evoked (or event-related) potentials can be detected by computerized averaging methods, and can be identified as waveforms occurring at particular times after the stimulus. For example, the P300 response is a positive deflection that occurs 300 milliseconds after a subject identifies a target stimulus embedded in a series of irrelevant stimuli. The P300 wave probably corresponds to the cognitive processes required for the recognition, retrieval from memory, and evaluation of a specific stimulus. In patients with schizophrenia, the amplitude of the P300 wave is reduced. It is notable that the same abnormality can be found in first-degree relatives of schizophrenic patients and in patients with borderline and schizotypal personalities. In these subjects the change in the P300 response is likely to stem from an abnormality in information processing, and may represent a vulnerability trait marker factor for the development of schizophrenia. It also suggests that certain kinds of personality disorder found in association with schizophrenia may stem in part from similar pathophysiological processes (Blackwood and Muir 1990).

Neuropathology

Neuropathological studies attempt to answer the question as to whether a structural change in the brain (localized or diffuse) accompanies a particular kind of mental disorder. Such studies have an obvious application to the aetiology of dementia and other psychiatric disorders in which organic lesions can readily be found. In the past, many post-mortem studies were carried out on the brains of patients who had suffered from schizophrenia and mood disorders. Consistent changes were not identified and therefore it was assumed that these psychiatric conditions were disorders of function rather than of structure (hence the name functional psychoses was sometimes used as a collective name for these conditions).

Improved methods of structural brain imaging have played an important role in the resurgence of interest in the neuropathology of psychosis. For example, Johnstone *et al.* (1976) showed that computerized tomography (CT) scanning could be used to demonstrate enlargement of the lateral ventricles in schizophrenia. More recently, structural MRI has allowed the examination of cortical and subcortical structures with a high degree of

resolution. These studies have shown that medial temporal structures are often reduced in volume in patients with schizophrenia. Post-mortem investigations have confirmed several of the abnormalities detected by structural imaging, whilst detailed pathological studies have identified neuronal loss and architectural disarray in the temporal lobe and other cortical regions. These findings have stimulated post-mortem neurochemical studies on temporal lobe structures and the brain regions with which they are closely connected (Pilowsky 1990).

The recent discovery of consistent brain neuropathological changes in patients with schizophrenia is a useful reminder that methods of investigation available at a particular time may fail to detect relevant biological abnormalities even when the latter are undoubtedly present. In addition, as neuropathological investigations embrace the molecular level, drawing distinctions between 'functional' and 'structural' disorders becomes somewhat arbitrary. Finally, the recent neurobiological studies in schizophrenia emphasize that progress in determining aetiology is most likely to be made through the integration of different kinds of pathological and biochemical investigation, so that the various approaches can be used to inform and guide each other.

Relationship of this chapter to the psychiatric syndromes

This chapter has reviewed several diverse approaches to aetiology. It may be easier for the reader to put these approaches into perspective when reading the sections on aetiology in the chapters on the different psychiatric syndromes, especially those on depression on pp. 213–27 and schizophrenia on pp. 264–81.

Further reading

Bloom, F. E. (ed.) (1995). *Psychopharmacology: the fourth generation of progress.* Raven Press, New York.

Freud, S. (1916–17). *Introductory lectures on psychoanalysis.*

Reprinted in Penguin Freud Library, Vol. 1. Penguin, Harmondsworth.

Jaspers, K. (1963). *General psychopathology* (trans, J. Hoenig and M. W. Hamilton), pp. 301–11, 355–64, 383–99. Manchester University Press.

Sackett, D. L., Hayes, R. B., Guyatt, G. H., and Tugwell, P. (1985). *Clinical epidemiology* (2nd edn). Little Brown, Boston, Mass.

Weatherall, D. J. (1991). *The new genetics and clinical practice* (3rd edn). Oxford University Press.

5 | *Personality disorder*

The term personality refers to enduring qualities of an individual that are shown in his ways of behaving in a wide variety of circumstances. All doctors should be able to assess personality so that they can predict how patients are likely to behave when ill. The psychiatrist shares this general concern about the personality of his patients. He also has a wider interest because personality can prepare the ground for illness and is sometimes mistaken for illness.

Features of personality can make some people more vulnerable to emotional disorders when experiencing stressful events. Thus difficult circumstances are more likely to induce an anxiety disorder in a person who has always worried about minor problems than in a person who has been less prone to worry. In worry-prone people, abnormal behaviour occurs only in response to stressful events. In people with more abnormal personalities, unusual behaviour occurs even in the absence of stressful events. In such people anomalies of behaviour may be so great that it is difficult to decide, solely on the patient's state at the time, whether they are due to personality or to mental disorder.

The treatment of a psychiatric disorder is more difficult when the patient has a personality disorder. For example, the management of schizophrenia or severe affective disorder is particularly difficult in patients with antisocial personality disorder because they are more likely to be aggressive or uncooperative. In countries such as the United Kingdom general practitioners treat mostly uncomplicated depressive disorders, whilst psychiatrists treat mainly depressive disorders that are complicated by abnormal personality traits or disorders.

The distinction between personality disorder and mental disorder is valuable in everyday clinical practice, but is not always easy to make. Central to the concept is the duration of unusual behaviour. If the person has previously behaved normally and then begins to behave abnormally, he is said to have a mental disorder. If the person has always behaved abnormally, he is said to have a personality disorder. The distinction is usually easy when behaviour changes quickly (as in an acute manic disorder), but difficult when behaviour changes slowly (as in some cases of schizophrenia).

Some German psychiatrists (e.g. Jaspers 1963) maintained that mental illness arises from causes within the person and is not a reaction to circumstances. This proposal led, in turn, to the idea that mental conditions that are clearly provoked by stressful events should not be regarded as illnesses but as reactions of the personality. Although this idea has some merit, it can no longer be sustained because research has shown that stressful events may occur before the onset of some conditions (such as schizophrenia) which Jaspers and others regarded as illnesses rather than reactions.

The assessment of personality

The assessment of personality has been discussed in Chapter 2, but two points need to be mentioned again. The first point is that a distinction must be made between assessments in everyday life and those in clinical practice. In everyday life, if we meet a new colleague at work, we are likely to judge his personality largely from his behaviour in the first few weeks. We assume that this behaviour represents his habitual way of behaving. Occasionally we are wrong; for example the new colleague may have been more guarded than usual. Generally, however, everyday assessment of this kind is accurate and useful.

The personalities of patients should not be judged in the same way. Thus too much weight should not be given to the pattern of behaviour observed in the ward or the out-patient clinic, where the patient's behaviour is likely to reflect a combination of personality and mental disorder.

Personality can be judged only from reliable accounts of past behaviour.

The second point concerns **psychological tests of personality**. Such tests are valuable for purposes such as vocational assessment. It is tempting to suppose that they give better information about personality than the clinician can obtain from interviews with the patient and informants. In fact personality tests do not give better information, because they are affected by the presence of mental disorder and because they measure traits that are seldom important in clinical practice.

Standardized assessment of personality

Standardized interview schedules are systematic and reproducible ways of gathering information about personality. Several schedules have been developed of which the following are examples.

The *Personality Assessment Schedule* (Tyrer and Alexander 1979) combines information from the patient and from an informant with observations of the subject to yield ratings of 24 personality variables. The *Standard Assessment of Personality* (Mann *et al.* 1981) is a brief interview with an informant. The *Structured Interview for DSM Personality Disorders—SCID II* (Spitzer *et al.* 1990) contains a self-rated questionnaire and a standardized interview. (For a review of methods of assessment of personality see Skodol and Oldham (1991).)

The concept of abnormal personality

Some personalities are obviously abnormal; for example people with paranoid personalities are suspicious, sensitive, and constantly vigilant for attempts by others to deceive them. However, it is impossible to draw a sharp dividing line between the normal and the abnormal. Indeed, it is even difficult to decide what criterion should be used to make this distinction. Two kinds of criteria have been suggested, the first statistical and the second social. On the statistical criterion, abnormal personalities are quantitative variations from the normal and the dividing line is decided by a cut-off score. In principle, this scheme is attractive as it

parallels the approach used successfully in defining abnormalities of intelligence. It has obvious value in research where tests are required to measure personality in groups of patients. However, it is of limited value in clinical work with individual patients.

When social criteria are used, an arbitrary dividing line is drawn between normal and abnormal personalities. The criterion is that the individual suffers from his personality, or that other people suffer from it. Thus a person with an abnormally sensitive and gloomy personality suffers himself, whilst an emotionally cold and aggressive person makes other people suffer. Although such criteria are subjective and lack the precision of the first approach, they correspond well to the realities of clinical practice and they have been adopted widely.

Given the conceptual problems, it is hardly surprising that it is difficult to frame a satisfactory definition of abnormal personality. The definition in ICD9 is not without difficulties but is widely accepted. It refers to:

deeply ingrained maladaptive patterns of behaviour recognizable by the time of adolescence or earlier and continuing through most of adult life although often becoming less obvious in middle or old age. The personality is abnormal either in the balance of its components, their quality and expression or in its total aspect. Because of this . . . the patient suffers or others have to suffer and there is an adverse effect on the individual or on society.

(The wording in ICD10 is less concise but conveys the same meaning.)

It is important to recognize that people with abnormal personalities may have both favourable and unfavourable traits. No matter how abnormal the personality, the clinician should always make enquiries about positive as well as unfavourable features. Such enquiries are particularly important in planning treatment.

Personality change

In some circumstances, during adult life there may be a profound and enduring change in a personality that has previously been established and stable. The circumstances leading to enduring change of personality are as follows:

(i) injury or organic disease of the brain;
(ii) severe mental disorder, especially schizophrenia;
(iii) exceptionally severe stressful experiences, for example those experienced by hostages or prisoners undergoing torture.

These changes are listed in ICD10 as personality disorder due to brain disease, damage, and dysfunction (F07), and enduring personality changes not attributable to brain damage and disease (F62).

How ideas about abnormal personality developed

In psychiatry the concept of abnormal personality can be traced back to the beginning of the nineteenth century, when the French psychiatrist Pinel described '*manie sans délire*'. Pinel applied this term to patients who were prone to outbursts of rage and violence but were not deluded (at that time delusions were regarded as the hallmark of mental illness, and *délire* is the French term for delusion). Presumably this group of patients included not only those who would now be regarded as having an antisocial personality, but also those who were mentally ill but not deluded, for example some with mania. (See Kavka (1949) for a translation of the relevant section of the second edition of Pinel's book, first published in 1801.)

Although other writers, such as the American Benjamin Rush, were interested in similar clinical problems, it was an English physician who took the next important step forward. In 1835, J. C. Prichard, senior physician to the Bristol Infirmary, published his *Treatise on insanity and other disorders of the mind*. After referring to Pinel's *manie sans délire*, he suggested a new term, **moral insanity**, which he defined as a 'morbid perversion of the natural feelings, affections, inclinations, temper, habits, moral dispositions and natural impulses without any remarkable disorder or defect of the intellect or knowing or reasoning faculties and in particular without any insane delusion or hallucination' (Prichard 1835, p. 6). Although this description included the violent patients described by Pinel, Prichard clearly had a wider group in mind, since he added: 'a propensity to theft is sometimes a feature of moral insanity and sometimes it is its leading if not sole characteristic' (p. 27). Prichard's category of moral insanity, like Pinel's *manie sans délire*, may have included affective disorders, for he wrote: 'a considerable proportion among the most striking instances of moral insanity are those in which a tendency to gloom or sorrow is the predominant feature'. (p. 18). He added: 'a state of gloom and melancholy depression occasionally gives way... to the opposite condition of preternatural excitement' (p. 19).

Under moral insanity Prichard also included patients whose behaviour would now be regarded as characteristic of antisocial personality disorder. Thus he wrote 'eccentricity of conduct, singular and absurd habits, a propensity to perform the common actions of life in a different way from that usually practised, is a feature of many cases of moral insanity but can hardly be said to contribute sufficient evidence of its existence' (p. 23). Prichard did not confine the term moral insanity to people who had always behaved in these ways: 'When however such phenomena are observed in connection with a wayward and intractable temper with a decay of social affections, an aversion to the nearest relatives and friends formerly beloved—in short, with a change in the moral character of the individual, the case becomes tolerably well marked'. In this passage the reference to change in character indicates that Pritchard had in mind not only patients who would now be classified as mentally ill, but also those who would now be classified as having personality disorder.

Later in the nineteenth century, it was recognized that mental illness could occur without delusions, and that affective disorders and schizophrenia were separate disorders. Nevertheless, the concept of moral insanity continued, although with a more restricted meaning. Thus Henry Maudsley applied the term to someone whom he described as having 'no capacity for true moral feeling—all his impulses and desires, to which he yields without check, are egoistic, his conduct appears to be governed by immoral motives, which are cherished and obeyed without any evident desire to resist them' (Maudsley 1885, p. 171). Maudsley commented on the current dissatisfaction with the term moral insanity, which he referred to as 'a form of mental alienation which has so much the look of vice or crime that many people regard it is an unfounded medical invention' (p. 170).

The next step towards modern ideas was the introduction by Koch (1891) of the term **psychopathic inferiority** to denote this same group of people who have marked abnormalities of behaviour in the absence of mental illness or intellectual impairment. Later, the word inferiority was replaced by personality to avoid judgemental overtones. At first Kraepelin shared the general doubt about the best way to classify these people, and it was not until the eighth edition of his textbook that he finally adopted the term **psychopathic personality** and devoted a long chapter to it. He described not only the antisocial type but also six others: excitable, unstable, eccentric, liars, swindlers, and quarrelsome.

A further step towards broadening the concept of personality disorder was taken by another German psychiatrist, Schneider. Whereas Kraepelin's seven types of psychopathic personality were applied only to people causing inconvenience, annoyance, or suffering to other people, Schneider extended the concept of psychopathic personality to people causing suffering to themselves and not necessarily to others. For example, he included people with markedly depressive or insecure characters. Thus, in Schneider's usage, psychopathic personality covered the whole range of abnormal personality, not just antisocial personality. In this way the term came to have two meanings: the wider meaning of abnormal personality of all kinds, and the narrower meaning of antisocial personality.

Confusion about the term psychopathic personality does not end with Schneider's broader definition. Two other usages call for attention. The first originates in the work of the Scottish psychiatrist, Sir David Henderson, who in 1939 published the influential book *Psychopathic states*. Henderson began by defining psychopaths as people who, although not mentally subnormal, 'throughout their lives or from a comparatively early age, have exhibited disorders of conduct of an antisocial or asocial nature, usually of a recurrent or episodic type which in many instances have proved difficult to influence by methods of social, penal and medical care or for whom we have no adequate provision of a preventative or curative nature'. So far this corresponds to the familiar narrow definition of psychopathic personality. However, Henderson extended his definition by referring to three groups of psychopaths.

The first group, the predominantly aggressive group, included not only those who are repeatedly aggressive, but also those prone to suicide, drug addiction, and alcohol abuse. The second group, those with passive and inadequate personalities, included unstable, hypochondriacal and sensitive people, pathological liars, and those with a schizoid nature. The third group, creative psychopaths, was so wide as to be of little value; thus the examples given by Henderson included T. E. Lawrence and Joan of Arc, who were creative in different ways but had little in common. In retrospect, Henderson's main contribution was to draw attention to the group of inadequate personalities.

Yet another variation in the meaning of the term psychopathic was introduced in the 1959 Mental Health Act for England and Wales. In this Act, psychopathic disorder was defined in Section 4(4) as 'a persistent disorder or disability of mind (whether or not including subnormality of intelligence) which results in abnormally aggressive or seriously irresponsible conduct on the part of the patient, and requires or is susceptible to medical treatment'. This definition was a return to the central idea of aggressive or irresponsible acts that cause suffering to other people. The definition is unsatisfactory because it includes the requirement for or response to treatment—criteria that may be administratively convenient but cannot be justified logically. Not surprisingly, many difficulties have attended the use of this definition. (More recent mental health legislation is discussed in the Appendix.)

The two meanings of psychopathic personality—the wider meaning of all abnormal personality, and the narrower meaning of antisocial personality—persist to the present day in, for example, the practice of forensic psychiatry (Dolan and Coid 1993). The term psychopathic personality remains ambiguous and difficult to define, in the words of Sir Aubrey Lewis 'a most elusive category' (Lewis 1974). Because of this ambiguity, this textbook avoids the term and uses instead the terms personality disorder and antisocial personality to denote the wide and narrow senses respectively.

The classification of abnormal personalities

Before considering ways of classifying abnormal personalities, it is important to realize that any category in any classification scheme represents an ideal type which few patients fit exactly. To quote Schneider (1950), 'Any clinician would be greatly embarrassed if asked to classify into appropriate types the psychopaths (that is abnormal personalities) encountered in any one year'. There are only a few cases in which one of the characteristic types of description or combinations can be applied without further qualification. Human beings resist precise measurement and, unlike the phenomena of disease, abnormal individuals cannot be classified neatly in the manner of clinical diagnosis. The consequence is that many abnormal personalities fulfil criteria for more than one of the personality disorders listed in the systems of classification (see Fyer *et al.* 1988).

Two kinds of term are used for abnormal personalities. The first is descriptive, and includes terms such as anxious or dependent. The second is aetiological, and includes terms that relate abnormal personalities to a syndrome of mental disorder to which they bear some resemblance. For example personalities characterized by eccentricity and emotional coldness are called schizoid because these features resemble some of those found in schizophrenic patients, and because in the past it was thought that the personality and the mental disorder shared a common cause. (At present schizotypal personality disorder is thought to be more closely related to schizophrenia. In both ICD10 and DSMIV, the terms are mainly descriptive.

In Table 5.1, the classification of personality disorders in ICD10 is compared with that in DSMIV. The two schemes are broadly similar, with the few differences being of two kinds, the use of different names for similar types of personality disorder and the inclusion in each scheme of a small number of types that do not appear in the other. DSMIV includes three types that do not appear in ICD10, and ICD10 includes one that does not appear in DSMIV. A more fundamental difference is that in DSMIV personality is recorded on a separate axis (axis II) from disorders which are recorded on axis I. In ICD10 personality and mental disorder are coded on the same axis.

Different names used to describe similar personality disorders

1. The use in ICD10 of the term *dissocial* to describe the personality disorder referred to as *antisocial* in DSMIV (the term antisocial is used in this book).
2. The use in ICD10 of *anankastic* as the preferred term for the personality disorder called *obsessive–compulsive* in DSMIV.
3. The use in ICD10 of *anxious* as the preferred term for the personality disorder called *avoidant* in DSMIV.

Categories of abnormal personality found in one system but not the other

The clinical features are described on the pages given in parentheses.

(a) Present in ICD10 but not in DSMIV:
   (i) emotionally unstable impulsive type (p. 113).
   (ii) *enduring* personality change, not attributable to brain damage or disease.

Table 5.1 *Classification of personality disorders*

ICD10	DSMIV
Paranoid	Paranoid
Schizoid (schizotypal, see text)	Schizoid Schizotypal
Dissocial	Antisocial
Emotionally unstable Impulsive type Borderline type	Borderline
Histrionic	Histrionic Narcissistic
Anankastic (obsessive–compulsive)	Obsessive–compulsive
Anxious (avoidant)	Avoidant
Dependent	Dependent
Other	Passive–aggressive

(b) Present in DSMIV but not in ICD10:
    (i) narcissistic (p. 115);
    (ii) passive aggressive (p. 117).

Schizotypal personality disorder (p. 112) appears in ICD10 and DSMIV, but in ICD10 it is classified with schizophrenia as schizotypal disorder (rather than under personality disorder).

Neither classification has categories of **depressive** or **cyclothymic** personality disorder because these conditions are classified with affective disorders (under the names of dysthymia and cyclothymia) (see p. 201). **Multiple personality** is classified with dissociative disorders (and considered in this book on p. 190).

A final difference between the two schemes is that personality disorders in DSMIV are grouped together as 'clusters'.

Cluster A: paranoid, schizoid, schizotypal
Cluster B: antisocial, borderline, histrionic, narcissistic
Cluster C: avoidant, dependent, obsessive–compulsive.
This grouping is not employed in ICD10.

Classification versus description

Although it is necessary to classify personality disorders for the purpose of collecting statistics, in everyday clinical work it is often better to give a brief description of the main features of the personality. Examples of such descriptions are sensitive, lacking in self-confidence, and prone to worry unreasonably, or abnormally aggressive with little evidence of feelings for other people or of remorse. Such descriptions are clinically useful.

They also help the clinician to avoid the error which Jaspers (1963) called 'pseudo-insight through terminology', i.e. the error of thinking that, because a personality disorder has been assigned to an ICD category, more is known about the patient.

Sjöbring's classification

Before leaving the topic of classification, a brief mention should be made of a scheme devised by Sjöbring. This scheme has been used mainly in Scandinavia, and it may be encountered in some of the important Scandinavian publications on psychiatry. It uses three dimensions to characterize personality (there is a fourth for intelligence, which is called **capacity**). The first dimension of personality is **stability**, which resembles introversion–extroversion. A superstable person is cold, introverted, and interested in ideas, whilst a substable person is warm, sociable, and active. The second dimension is **solidity**. A supersolid person is dependable, deliberate, and self-possessed, whilst a subsolid person is inconstant, quick, and subjective in judgements. The third dimension is **validity**. A supervalid person is venturesome, expansive, and self-confident, while a subvalid person is retiring, cautious, and easily worried. The interested reader will find an account of this scheme in the paper by Sjöbring (1973).

Clinical features of abnormal personalities

This section begins with an account of the abnormal personalities that are included in the ICD. A brief review of the additional or alternate classes used in DSMIV is then given. The account follows the broad scheme of the ICD and DSM. The disorders are considered in the order used in ICD10 (see Table 5.1).

Paranoid personality disorder

This term is used in both ICD10 and DSMIV. The central features of this kind of abnormal personality are suspiciousness and sensitivity (Table 5.2).

Table 5.2 *Features of a paranoid personality disorder*

Suspicious
Sensitive
Mistrustful
Argumentative
Stubborn
Self-important

As already mentioned, minor obsessional and histrionic traits can add socially desirable qualities to a normal personality. There is no such positive side to paranoid traits. Even when they form only a small part of the personality, paranoid traits add a distrust that goes beyond ordinary caution together with a sensitivity to rebuff that is handicapping to social relationships. In people with paranoid personality disorder, suspiciousness can be shown in several ways. The person may be constantly on the look out for attempts by others to get the better of him, to deceive him, or to play tricks on him. He may doubt the loyalty of other people and may be unable to put his trust in them. As a result, he appears touchy and suspicious. He does not make friendships easily and may avoid involvement in groups. He may be perceived by other people as secretive, devious, and self-sufficient to a fault. He seems to have little sense of humour or capacity for enjoyment. Such personality traits are fertile grounds for jealousy. (See also Chapter 10.)

People with paranoid personalities appear argumentative and stubborn. Presented with a new proposal, they are overcautious and look for ways in which it might be designed to harm their own interests. Some engage in litigation that is prolonged long after any non-paranoid person would have abandoned it.

Table 5.3 *Features of schizoid personality disorder*

Emotionally cold

Detached

Aloof

Humourless

Introspective

An important feature of the paranoid personality is a strong sense of self-importance. The paranoid person often has a powerful inner conviction that he is unusually talented and capable of great achievements. This idea is maintained, despite only modest accomplishments, by paranoid beliefs that other people have prevented him from fulfilling his real potential, that he has been let down, tricked, swindled, or deceived. Sometimes these self-important ideas are crystallized round a central overvalued idea that persists for many years.

Sensitivity is another important aspect of the paranoid personality. People of this kind readily feel shame and humiliation. They take offence easily and see rebuffs where none are intended. As a result, other people find them difficult, prickly, and unreasonable. Both Schneider (1950) and Kretschmer (1927) used the term **sensitive** to describe such a person. Kretschmer also described how such people, when faced with a deeply humiliating experience, may develop suspicious ideas that can easily be mistaken for persecutory delusions. These 'sensitive ideas of reference' are considered further in Chapter 10.

Schizoid personality disorder (Table 5.3)

This term is used in both ICD10 and DSMIV. A person with this disorder is introspective and prone to engage in fantasy rather than to take action. He is emotionally cold, self-sufficient, and detached from other people. The name schizoid was suggested by Kretschmer (1936), who held that there is an aetiological relationship between this kind of personality and schizophrenia (see Chapter 9). However, the two are not closely associated, and the term should be used descriptively without implying any causal relationship with schizophrenia.

The most striking feature is a lack of emotional warmth and rapport. People with this disorder appear detached, aloof, and humourless, and seem incapable of expressing affection or tenderness. As a result, they do not make intimate friendships and they often remain unmarried. They show little concern for the opinions of other people and they pursue a lonely course through life. Their hobbies and interests are solitary and are more often intellectual than practical.

Schizoid people tend to be introspective. Their inner world of fantasy is often extensive but lacks emotional content. They are more likely to be concerned with intellectual problems than with ideas about other people. If the disorder is extreme, the individual is cold, callous, seclusive, ill at ease in company, and without friends. Lesser degrees of the same traits, appearing as part of a normal personality, may confer advantages in some ways of life. For example some forms of academic work may be carried out more effectively by a person who can detach himself from social activities for long periods, and can concentrate in a detached and unemotional way on intellectual problems.

Schizotypal personality disorder

The term schizotypal is used to denote a personality disorder characterized by social anxiety, inability to make close friendships, eccentric behaviour, oddities of speech (for example speech that is vague and excessively abstract), inappropriate affect, suspiciousness, ideas (but not delusions) of reference, other odd ideas (for example, ideas about telepathy and clairvoyance that are not normal in the culture), and unusual perceptual experiences (such as the sensing of the presence of

Table 5.4 *Features of antisocial personality disorder*

Failure to sustain relationships
  Disregard of the feelings of others

Impulsive actions
  Low tolerance of frustration
  Tendency to violence

Lack of guilt

Failure to learn from experience

---

a dead person by a person who is not recently bereaved). This personality disorder is said to be more closely related than the schizoid type to schizophrenia.

Antisocial (dissocial) personality disorder

In ICD10 this personality disorder is called dissocial. The preference of this book is for the term antisocial, which appears in DSMIV. People with this disorder show a bewildering variety of abnormal features. Several attempts have been made to identify an essential core to the disorder. The most useful core features are failure to make loving relationships, impulsive actions, lack of guilt, and failure to learn from adverse experiences (Table 5.4).

Failure to make loving relationships is accompanied by self-centredness and heartlessness. In its extreme form there is a degree of callousness that allows the person to inflict cruel, painful, or degrading acts on others. This lack of feeling is often in striking contrast to a superficial charm that enables the person to make shallow and passing relationships. Sexual activity is without evidence of tender feelings. Marriage is often marked by a lack of concern for the partner, and sometimes by physical violence. Many marriages end in separation or divorce.

The characteristic impulsive behaviour is often reflected in an unstable work record marked by frequent dismissals. It is also shown in the whole pattern of life, which seems to lack any plan or persistent striving towards a goal.

This impulsive behaviour, coupled with a lack of guilt or remorse, is often associated with repeated offences against the law. Such offences begin in adolescence with petty acts of delinquency, lying, and vandalism; many of them show a striking indifference to the feelings of other people, and some include acts of violence or callous neglect. Often the behaviour is made more extreme by the effects of alcohol or drugs.

People with antisocial personality disorder make seriously inadequate parents, and may neglect or abuse their children. Some have difficulty in managing their finances or in organizing family life in other ways.

Vivid descriptions of severe forms of antisocial personality disorder are contained in Cleckley's book *The mask of sanity* first published in 1941 (see Cleckley 1964).

Impulsive personality disorder

People with impulsive personality disorder cannot control their emotions adequately and are subject to sudden unrestrained outpourings of anger which they regret subsequently. These outbursts are not always confined to words, but may include physical violence, leading at times to serious injury. Unlike people with antisocial personality disorder, who also exhibit explosions of anger, the impulsive group does not have other difficulties in relationships. In ICD10 this disorder is recognized as a subtype of emotionally labile personality disorder. It is not included in DSMIV.

Borderline personality disorder

The term borderline has been used in psychiatry in two ways. The first usage was in adoption studies of schizophrenia, in which the term was applied to people who showed behaviours which were thought to be related to schizophrenia. The term schizotypal personality disorder is now applied to

Table 5.5 *Features of borderline personality disorder*

Unstable relationships

Impulsive behaviour

Variable moods

Lack of control of anger

Recurrent suicidal threats or behaviour

Uncertainty about personal identity

Chronic feelings of 'emptiness'

Efforts to avoid abandonment

Transient stress-related paranoid or dissociative symptoms

such people (see p. 112). In the second usage, the term was applied to people who showed 'instability'. Originally, this instability was usually described in psychodynamic terms (Kernberg 1975; Gunderson and Kolb 1978) and was said to be characterized by (a) ego weakness, for example poor ability to control impulses, (b) a propensity for 'primary process' (i.e. irrational) thinking despite intact reality testing, (c) use of less 'mature' defence mechanisms such as projection and denial, and (d) diffuse personal identity. Kernberg (1975, Chapter 1) further described the clinical picture and stressed the following features: diffuse anxiety, the presence of multiple neurotic symptoms, 'perverse sexual trends', addiction, and 'prepsychotic personality structures' by which he meant paranoid, schizoid, or hypomanic personality features. Spitzer *et al.* (1979) suggested more objective criteria for the diagnosis of borderline personality disorder, and similar criteria are now

incorporated in DSMIV and ICD10 (the latter scheme uses the term emotionally unstable personality, borderline type).

Borderline personality disorder is now characterized by nine features (Table 5.5), of which five are required to make a DSMIV diagnosis. These nine features are unstable relationships, impulsive behaviour that is harmful to the person (for example reckless spending, binge eating, or sexual behaviour), variable moods, lack of control of angry feelings, recurrent suicidal threats or behaviour, uncertainty about personal identity, chronic feelings of emptiness, efforts to avoid real or imagined abandonment, and transient stress-related paranoid ideas or severe dissociative symptoms. The category is broad, encompassing several abnormal aspects of personality that can be classified in other ways. Indeed, many people who meet the criteria for borderline personality disorder also meet the criteria for histrionic, narcissistic, and antisocial personality disorder (Pope *et al.* 1983; Oldham *et al.* 1992).

Histrionic personality disorder

This term is used in both ICD10 and DSM. The important features of this kind of personality are self-dramatization, a craving for novelty and excitement, and a self-centred approach to personal relationships (Table 5.6).

In a normal personality, minor histrionic traits can be socially advantageous. People with such traits make lively engaging company and are popular guests; they do well in amateur dramatics and are entertaining public speakers. They tend to wear their emotions on their sleeves and are easily moved to joy or tears, but their feelings soon pass.

When these qualities are exaggerated in histrionic personality disorder, they become less acceptable. These people dramatize themselves as larger than life characters; they seem to be playing a part, incapable of being themselves. They often seem unaware that other people can see through their defences. Instead of the enjoyment of novelty characteristic of a person with histrionic personality traits, in histrionic personality disorder there is a relentless search for new experiences, coupled

Table 5.6 *Features of histrionic personality*

Less intense	More intense
Outwardly confident	Vain, self-centred
Lively	Short-lived enthusiasms
Sociable	Acts a part, self-deceiving
Emotionally responsive	Unrestrained emotional display

with short-lived enthusiasms, readiness to boredom, and craving for novelty. The tendency to be self-centred may be greatly exaggerated in histrionic personality disorder. People with this disorder lack consideration for others, and think only of their own interests and enjoyment. They appear vain, inconsiderate, and demanding, and may go to extreme lengths to force other people to fall in with their wishes. Emotional 'blackmail', angry scenes, and demonstrative suicide attempts are all part of the stock-in-trade. Histrionic people display emotions readily, exhausting others with tantrums of rage or dramatic expressions of despair. They seem to feel little of the emotions that they express, recover quickly, and often seem surprised that other people are not prepared to forget the scenes as quickly as they are themselves.

With these qualities is combined a capacity for self-deception that can at times reach astounding proportions. Such people go on believing themselves to be in the right when all the facts show that they are not. They are able to maintain elaborate lies long after people have seen through them. This pattern of behaviour is observed in its most extreme form in 'pathological liars' and swindlers.

Some of these qualities are normal in children, particularly the transient enthusiasms, the easy change from laughter to tears, the enjoyment of make-believe, and the egocentricity. This has led some psychiatrists to apply the term immature to this type of personality. However, the term is imprecise and is best avoided (see p. 118).

In histrionic personality disorder, sexual life is also affected. There is often sexual provocation combined with frigidity, especially in women. Histrionic women engage in displays of affection and are flirtatious, but they are often incapable of deep feelings and may fail to reach orgasm. The features of histrionic personality disorder have been described by Chodoff and Lyons (1958) and by Slavney and McHugh (1974).

Narcissistic personality disorder

People with this disorder are characterized by a grandiose sense of self-importance and by a preoccupation with fantasies of unlimited success, power, and intellectual brilliance. They crave attention from other people but show few warm feelings in return. They exploit others and seek favours that they do not return. Most people of this kind could be classified as having histrionic personality disorders, and others seem to fit into the group of antisocial or borderline personalities. Intermediate forms are inevitable in any scheme of classification of personality, and there seems to be no strong reason at present for assigning an additional category to people with these characteristics. (See Akhtar and Thomson (1982) for a review.)

Table 5.7   *Features of obsessional personality*

Less intense	More intense
Dependable	Obstinate
Persistent	Inflexible
Cautious	Indecisive
Careful	Lost in detail
Stable mood	Lacking emotion or humour
High standards	Bigoted
Law-abiding	Judgemental

Obsessive–compulsive personality disorder (Table 5.7)

The term obsessive–compulsive personality disorder is used in DSMIV. Following the usage of Kahn (1928), the preferred term in ICD10 is anankastic. The only advantage of this term is that it avoids the erroneous implication of an inevitable link between this type of personality and obsessional disorders. (People with this kind of personality are also liable to develop anxiety and depressive disorders.)

Before describing obsessive–compulsive personality disorder, it is useful to review the expression of obsessional traits in someone with a normal personality. A person with obsessional traits is dependable, precise, and punctual. He sets high standards and keeps to social rules. He is determined and persistent in tasks, despite difficulties. His moods are reliably the same from day to day. However, even within a normal personality, these qualities have another side; thus, at times determination may give way to obstinacy, precision to preoccupation with unimportant detail, and high moral standards to bigotry. More-over, the qualities that make for stable moods can be expressed as a humourless approach to life.

In obsessive–compulsive personality disorder, these features are more extreme. One of the most striking features is a lack of adaptability to new situations. The person is rigid in his views and inflexible in his approach to problems. He is upset by change, and prefers a safe and familiar routine. Such a person lacks imagination and fails to take advantage of opportunities. In obsessive–compulsive personality disorder there is an inhibiting perfectionism that makes ordinary work a burden and leaves the person immersed in trivial detail. High moral standards become painful guilty preoccupation with wrongdoing, which stifles enjoyment. People with this disorder seem without humour, ill at ease when others are enjoying themselves, moralistic in their opinions, and judgemental in their attitudes. They are often mean to the point of being miserly and do not enjoy giving or receiving gifts.

Such people are often indecisive. They find it hard to weigh up the advantages and disadvantages of new situations; they delay decisions, and often ask for more and more advice. They fear making mistakes, and after coming to a decision they worry lest the choice was wrong.

Sensitivity to criticism is a related feature of this personality. There is an undue concern about other people's opinions and an expectation of being judged harshly.

Obsessive–compulsive people often show little emotion. However, they are given to smouldering and unexpressed feelings of anger and resentment, often directed to people who have interfered with their routine of life. Such angry feelings may be accompanied by obsessional thoughts and images of an aggressive kind, even in those who do not develop the full syndrome of an obsessional disorder.

Anxious (avoidant) personality disorder

People with this disorder are persistently anxious. They are ill at ease in company, fearing disapproval, criticism, or rejection, and worrying that they will be embarrassed or ridiculed. They are cautious about new experiences and meeting unfamiliar people, and they are timid in the face of everyday hazards. They lack self-esteem, believing themselves inferior to others, unappealing, and socially inept. As a result they have few close friends and they avoid social demands such as taking on new responsibilities at work. They differ from people with schizoid personalities in not being emotionally cold; indeed, they crave social relationships that they cannot attain. In DSMIV the term avoidant is used to denote this kind of personality disorder, whilst in ICD10 the term anxious personality is preferred, with 'avoidant' as an accepted alternative.

Dependent personality disorders

People with this disorder appear weak-willed and unduly compliant with the wishes of others. They lack vigour and show little capacity for enjoyment. They avoid responsibility and lack self-reliance. Some dependent people are more determined, but achieve their aims by persuading other people to assist them whilst protesting their own helplessness.

If married, such people may be protected from the full effects of their personality by support from a more energetic and determined spouse who is willing to make decisions and arrange activities. Left to themselves, some drift down the social scale and others are found among the long-term unemployed and the homeless.

Passive–aggressive personality disorder

This term is applied to a person who, when demands are made upon him for adequate performance, responds with some form of passive resistance, such as procrastination, dawdling, stubbornness, deliberate inefficiency, pretended forgetfulness, and unreasonable criticism of people in authority.

Affective personality disorders

Some people have lifelong disorders of mood regulation. They may be persistently gloomy (depressive personality disorder) or habitually in a state of inappropriate elation (hyperthymic personality disorder). A third group alternates between these two extremes (cycloid or cyclothymic personality disorder). These types of personality disorder have been described for many years and are readily recognized in clinical practice. However, they do not appear in either the ICD10 or DSMIV systems of classification. The reason is that in both systems these disorders are classified under disorders of mood and not under disorders of personality. Thus they are classified under 'persistent mood (affective) states' (cyclothymia or dysthymia) in ICD10, and under cyclothymia or dysthymia in DSMIV. People with **depressive personality disorder** always seem to be in low spirits. They take a persistently gloomy view of life, anticipating the worst outcome of every event. They brood about their misfortunes and worry unduly. They often have a strong sense of duty. They show little capacity for enjoyment and they express dissatisfaction with their lives. Some are irritable and bad-tempered.

People with **hyperthymic personality** disorder are habitually cheerful and optimistic, and show a striking zest for living. If they have these traits to a moderate degree, they are often effective and successful. If they have these traits to an extreme degree (as rarely happens), they show poor judgement and may be uncritical and hasty in coming to conclusions. Their habitual cheerfulness is often interrupted by periods of irritability, especially when their aims are frustrated. In the past, exceptionally contentious people in this group were called pseudoquerulant.

People with **cycloid personality disorder** alternate between the extremes of depressive and hyperthymic states described above. This instability of mood is much more disruptive than either of the persisting conditions. People with this disorder are periodically extremely cheerful, active, and productive. At such times they take on additional commitments in their work and social lives. Eventually their mood changes. Instead of confident optimism, they have a gloomy defeatist approach to life. Their energy is reduced. Whereas in the elated phase they took up activities with much relish, now they find them a burden. They make different but equally unwise decisions, and they refuse opportunities that could be managed.

Eventually they return to a normal mood or to further elation.

Terms to avoid

It has been explained above that the term psychopathic personality is unsatisfactory (see p. 108). Two other commonly used terms are also unsatisfactory and should be avoided. Both tend to be used when the doctor has not thought clearly enough about the precise nature of his patient's difficulties. The first, **inadequate personality**, is often used pejoratively. In place of this term, it is better to specify precisely the ways in which the person is inadequate to the demands of life. Such a specification will lead to more constructive ideas about helping the person to cope better.

The second term, **immature personality**, is often used vaguely to denote a non-specific discrepancy between the patient's behaviour and his chronological age, such that the behaviour is more appropriate to a younger person than to a person of the patient's age. To avoid the vagueness of the term immature, it is better to specify the exact nature of the problem, whether it is in social relationships, the control of emotions, willingness to take responsibility, or elsewhere. Such specification of the patient's problems is more likely to lead to a constructive approach than is the mere labelling of the personality as immature. It also avoids the implication of an unsubstantiated cause, namely a failure of maturation.

Epidemiology

Until recently there was little information about the frequency of personality disorders in the general population. In epidemiological surveys using DSMIII criteria, estimates of the prevalence of overall personality disorders vary from 10 to 13 per cent (Weissman 1993). When a structured interview was used in the UK, disorders of personality were found in 13 per cent of adults in an urban population (Casey and Tyrer 1986). A common finding in epidemiological studies is that rates of personality disorder are higher in men and decrease with age.

By comparison with the general population, higher rates of personality disorder are reported amongst patients with conspicuous psychiatric morbidity attending a general practitioner (30 per cent) (Casey *et al.* 1984) and amongst patients consulting a psychiatrist (nearly 50 per cent) (Cutting *et al.* 1986).

These estimates of the frequency of personality disorder are likely to be less reliable than those of mental disorder because it is difficult to identify personality disorder reliably in community surveys in which interviewers may not have access to information from informants. Information on **antisocial personality disorder** has recently come from a large community survey in three sites in the USA (Baltimore, Newhaven, and St Louis) (Robins *et al.* 1984; Regier *et al.* 1988). At each site over 3000 people were interviewed using DSMIII criteria. Overall the lifetime prevalence of antisocial personality disorder was 1.5–3.2 per cent.

Estimates of the prevalence of other personality disorders are shown in Table 5.8. The substantial range of some of these estimates reflects, among other sources of variation, the use of different assessment instruments. Estimates based on the use of one instrument (the Personality Diagnosis Questionnaire of Hyler *et al.* (1983)) show less variation. Most of these studies have been in the USA with a few in Europe. A study of antisocial personality disorder in Taiwan gave a lifetime prevalence of 0.0–0.1 per cent, a much lower figure than that reported in the USA studies (Weissman 1993). For a review of the epidemiology of personality disorder see de Girolamo and Reich (1993).

Aetiology

Since little is known about the factors accounting for normal variations in personality, it is not surprising that knowledge about the causes of personality disorder should be incomplete. Research is made difficult by the long interval between potentially relevant events in early life

Table 5.8 *Prevalence of personality disorder using DSMIII criteria in epidemiological survey or in relatives*

Personality cluster	Personality disorder	Rate (%)	Rate (PDQ* measures only) (%)
Cluster A	Paranoid	0.4–1.8	0.4–0.8
	Schizoid	0.5–0.9	0.8–0.9
	Schizotypal	0.6–5.6	5.1–5.6
Cluster B	Antisocial	1.5–3.2	0.4–0.9
	Borderline	1.1–4.6	1.3–4.6
	Histrionic	1.4–3.0	2.1–2.7
	Narcissistic	0.0–0.4	0.4
Cluster C	Avoidant	0.0–1.3	0.0–0.4
	Dependent	1.6–6.7	5.1–6.7
	Compulsive	1.7–6.4	4.0–6.4
	Passive–aggressive	0.0–3.0	0.0–0.4

*Personality Diagnosis Questionnaire (Hyler *et al.* 1987)
Data from sources reviewed by Weissman (1993).

and the time when disorder comes to attention in the adult. It might be expected that the more extreme the disorder, the more obvious its causes would be. In keeping with this expectation, there is more information about antisocial personality disorder than about the other disorders.

The origins of normal personality

Personality types and personality traits

In understanding normal personality, the first step is to identify basic common features of individual personalities. This was done originally by intuition; certain personality types seemed to be generally recognizable, for example sociable and outgoing types as against solitary and self-conscious types. Psychologists have tried to identify finer divisions of personality, known as traits, in a more objective way by applying statistical methods to responses to personality tests. For example Cattell (1963) identified five traits, which he referred to as 'factors'. (There were six factors in all, with the sixth for intelligence.) Other investigators have used statistical methods to identify a smaller number of superordinate dimensions which might be more easily related to differences in brain organization. Some psychologists have grouped these traits together into 'dimensions'. For example Eysenck (1970c) proposed a scheme which originally had two dimensions, extroversion–introversion and neuroticism. Later, a third dimension of 'psychoticism' was added (Eysenck and Eysenck 1976). In this scheme the term psychoticism denotes the characteristics of being cold, aggressive, cruel, and given to antisocial behaviour, rather than features related more directly to the clinical concept of psychosis. Eysenck developed a test, the Eysenck Personality Inventory (EPI), to measure these dimensions.

The inheritance of normal personality

Everyday observation suggests that children often resemble their parents in personality. Such similarities could be inherited, or acquired through social learning. Indirect evidence for inheritance was obtained many years ago in studies of the association between personality and body build; the only obvious link between the two is that they could both be genetically determined.

Kretschmer (1936) attempted to define psychological types and to link them to recognizable types of body build. He described three types of body build: **pyknic** (stocky and rounded), **athletic** (with strong development of muscles and bones), and **asthenic** (lean and narrow). He suggested that the pyknic body build was linked to the cyclothymic personality type, characterized by varying moods and sociability, while the asthenic build was related to the schizotypal personality type, which is cold, aloof, and self-sufficient.

Kretschmer's ideas were based on subjective judgements of body build. Sheldon *et al.* (1940) repeated the earlier studies using more quantitative methods for assessing physique. Instead of assigning each person to a type, these workers rated body build on three dimensions: **endomorphy** ('predominance of softness and rounded'), **mesomorphy** ('predominance of muscle, bone and connective tissue'), and **ectomorphy** ('predominance of linearity and fragility'). Each person was given a score indicating his position on the three dimensions; for example 711 indicated extreme endomorphy. Sheldon also attempted to rate personality objectively but chose dimensions which are no longer in general use, namely **viscerotonia**, which indicated enjoyment of comfort and relaxation, **somatotonia**, which indicated assertiveness and energy, and **cerbrotonia**, which indicated strong inhibitory controls and a tendency to think rather than to act (Sheldon *et al.* 1942). These quantitative methods did not reveal any simple relationship between body build and personality type to support genetic origins of personality.

More convincing evidence of a genetic basis of normal personality was obtained by studying the degree of similarity of scores on personality tests of identical twins reared together or apart. These studies have produced some evidence that normal personality is partly inherited (Loehlin *et al.* 1988). Thus the scores of pairs of twins reared apart are as similar as those of pairs of twins reared together (Shields 1962; Pedersen *et al.* 1988). The general conclusion from these and other studies is that there is a hereditability of 35–50 per cent for traits such as extroversion and neuroticism (McGuffin and Thapar 1992).

The neurophysiological basis of personality

Several attempts have been made to link differences in personality to differences in brain functions. Eysenck (1970*c*, 1976) suggested that differences in neuroticism are related to differences in the activity of the visceral brain, while differences in introversion–extroversion are related to differences in arousal systems. These ideas have stimulated research but the physiological basis of personality remains uncertain.

Cloninger (1986*b*) developed another model for brain function which he applied first to basic behavioural dispositions, and thence to personality. He proposed that three brain functions are relevant to personality: *behavioural activation, behavioural inhibition,* and *behavioural maintenance*. According to Cloninger, behavioural activation is associated with a disposition to seek novelty, avoid monotony, and avoid punishment or non-reward; the personality characteristics associated with this disposition should include curiosity, readiness to boredom, enthusiasm, and unconventional behaviour. Cloninger suggests that behavioural inhibition is associated with harm avoidance, whilst behavioural maintenance is associated with reward dependence; he proposes personality traits that could be the expression of these basic dispositions. Cloninger also proposes that certain other features of personality such as cooperativeness and 'self-directedness' are related to learning and to the conceptual organization of the brain.

Although going beyond the present evidence, Cloninger's scheme is noteworthy for its attempt to account for both inherited differences in brain function and the effects of experience. Much more research is needed before these ideas can be evaluated.

Childhood temperament and adult personality

Even in young infants, marked differences can be seen in patterns of sleeping and waking, approach and withdrawal from new situations, the intensity of emotional responses, and span of attention. Although these differences have been shown to persist into the childhood years, they do not seem to be closely related to adult personality traits (Berger 1985). Personality features of shyness, ill-temper, and dependency may be more persistent.

Childhood experience and personality development

Everyday experience suggests that experiences in childhood play a part in shaping personality. However, it is not easy to produce objective evidence to test this impression. Experiences that seem relevant are difficult to record reliably and still more difficult to quantify, whilst the long intervals between early events and the final emergence of adult personality make it extremely hard to arrange prospective studies. Retrospective investigations are easier to arrange, but the recall by adults of their experiences in childhood is unreliable. These difficulties in producing objective data have left the field open for subjective accounts based on selective observations and unconfirmed theories. Of these accounts, the best known are those developed many years ago by Freud and others. A brief account of the main theories will be given next. For more information the reader is referred to Hall and Lindzey (1980).

Psychodynamic theories of personality development

Freud's theory

In Freud's scheme, emphasis is placed on events in the first five years of life. It is proposed that crucial stages of development (oral, anal, and genital) must be passed through successfully if personality development is to proceed normally. Certain predictions are made about the effects of failure at particular stages; for example serious difficulties at the anal stage result in an obsessional personality. The scheme allows for some modification of personality at later stages of development through identification with people other than the parents, but this influence is thought to be less important than those in earlier years.

The scheme is comprehensive and flexible enough to enable clinicians to construct a retrospective explanation of many of the personality types and personality disorders that they encounter among their patients. However, the scheme is unsatisfactory as a scientific account of abnormal personality because this same flexibility makes it impossible to generate crucial tests of the hypotheses.

Jung's theory

Jung's theory of personality development resembles Freud's theory in placing the greatest emphasis on internal psychic events rather than on social influences. Unlike Freud, Jung thought of personality development as a lifelong process. He referred to events in the first part of life as merely 'fulfilling one's obligations', and he applied this term to events such as severing ties with parents, finding a spouse, and starting a family. Jung was more concerned with adjustments that occur later in life and reach completion only when a person is ready to face death. These ideas are sometimes useful when treating elderly patients.

Other theories

In his individual psychology, **Adler** rejected Freud's ideas of libido development; instead he proposed that personality develops through efforts to compensate for basic feelings of inferiority. The **neo-Freudians** (Fromm, Horney, and Sullivan) emphasized social factors in development rather than the biologically determined stages of Freud's scheme, though the three differed amongst themselves in their opinions about the details of this social development.

Erikson's scheme was essentially similar to Freud's though the nomenclature was different. Thus Erikson referred to the oral stage as the stage of trust versus mistrust, and emphasized that this was the period in which feelings of security were developing. He referred to the anal stage as the stage of autonomy versus doubt because it was the period in which the child learned self-control, social rules, and self-confidence. Erikson referred to the genital stage as the stage of initiative versus guilt, this being the stage at which the child developed an image of himself as a person. If this stage went well, the child emerged with confidence and initiative; if it did not, the child emerged with inhibitions and guilt that impaired further development. The latency period was referred to by Erikson as the period of industry versus inferiority, in which the child learnt the value of achievements in work and at school, and in social relationships outside the family. Thus Erikson placed emphasis on events in adolescence to which Freud gave little importance. His ideas have been influential in the psychiatry of adolescence.

The origins of personality disorder

Genetic causes of personality disorder

Most genetic studies of personality disorder concern the antisocial type; these studies are considered on p. 124. Anxious–avoidant and obsessional personality disorders have not been studied in a way that separates them clearly from anxiety disorders and obsessional disorders respectively. The few studies of histrionic personality disorder have had inconsistent findings (McGuffin and Thapar 1992).

Some investigators have reported that **paranoid personality disorder** is more frequent among first-degree relatives of probands with schizophrenia

than among the general population (Kendler *et al.* 1984) but others have not confirmed this finding (Coryell and Zimmerman 1989). There is no evidence that **schizoid personality** is associated with schizophrenia (Fulton and Winokur 1993).

Some studies have found increased rates of **schizotypal personality disorder** amongst relatives of probands with this disorder compared with controls, both among co-twins (Torgersen 1984) and among other family members (Baron *et al.* 1985), findings that suggest a genetic aetiology for the disorder. It has been suggested that schizotypal personality disorder is related genetically to schizophrenia since it has been reported to be more frequent among biological relatives of probands with schizophrenia than among adopted relatives or controls (Kendler *et al.* 1981). When investigations have begun with probands with schizotypal personality disorder, rates of schizophrenia have not been found to be increased in the same way (Soloff and Millward 1983; Schultz *et al.* 1986), but this finding may reflect groups that were too small to reveal an effect. Questionnaire studies of 'schizotypy' in normal subjects have shown an inherited component (Claridge and Hewitt 1987), but it is not certain how far the scores of normal subjects on this questionnaire are related to schizotypal personality disorder.

Antisocial personality disorder is discussed on p. 124.

Conflicting results have come from studies of the prevalence of **borderline personality disorder** among the relatives of probands with this disorder (Dahl 1993). Taken together, these results do not suggest that there is an important genetic aetiology for this kind of personality disorder.

The genetics of **histrionic personality disorder** has not been studied with standardized methods of assessment. **Obsessional personality** appears to have a substantial genetic aetiology (Murray and Reveley 1981).

Relation of personality disorder to mental disorder

Kretschmer suggested an association between personality disorder and mental disorder. On this view, some disorders of personality result from the same process—most probably genetic—that causes mental disorder. **Schizoid personalities** are considered to be partial expressions of schizophrenia, and **cycloid personalities** to be partial expressions of manic depressive psychosis. Although this theory is without convincing support, it lingers on in the names cycloid and schizoid. More recently attention has shifted to **schizotypal personality** disorder as being possibly related to schizophrenia (see p. 259). There could be some less specific genetic connection between mental disorder and personality disorder, as suggested by reports of an increased frequency of various kinds of personality abnormalities among relatives of schizophrenic patients (see p. 277) and among relatives of patients with manic depressive disorder (see p. 215).

Personality disorder and upbringing

Considerable attention has been given to disturbances in parent–child relationships, particularly maternal deprivation, as factors influencing personality development. Although maternal deprivation has been proposed as a cause of antisocial personality (see p. 126), there is no convincing evidence that it leads to other kinds of personality disorder. Despite the lack of objective evidence for specific associations between early experience and personality disorder, it is good clinical practice to assess the patient's childhood carefully and to consider whether there have been experiences (such as persistent rejection by parents) that seem understandably related to the features of the personality disorder (such as low self-esteem).

Psychological causes have been suggested for most types of personality disorder, but there is no scientific evidence on which to judge their importance. In the absence of other explanations psychoanalytic ideas were accepted for a time but are now much less influential. It has to be accepted that very little is known about the psychological causes of abnormalities of personality.

Paranoid personality disorder

Freud suggested that paranoid thinking results from the projection of unacceptable feelings and impulses on to others, either directly ('I hate him' becomes 'he hates me') or after a preliminary stage of reaction formation ('I love him' first becomes 'I hate him' and then 'he hates me') (see Freud 1911). Cameron (1963) suggested that a central feature of the disorder is a basic absence of trust which results from a lack of consistent parental affection and mistreatment in childhood.

Schizoid personality disorder

Psychoanalytic ideas focus on the inability to give or receive love. This inability is thought to be a defence that developed early in life in response to inadequate mother–child relationships. Klein (1952) suggested that all infants pass through a 'schizoid position' stage of development in which oral and sadistic impulses are experienced as dangerous and are projected on to the parent. Most children pass through this stage, but people with a schizoid personality have retained some of the projective defences.

Antisocial personality disorder

See p. 126.

Borderline personality disorder

Essentially, psychoanalytic explanations propose that widespread problems in personal relationships result from severe disturbance of early relationships with the parents. People with the borderline personality are more likely than controls to report physical and sexual abuse in childhood (Berelowicz and Tarnopolsky 1993). Such retrospective accounts do not establish causality, and the necessary prospective studies have not been reported. Beck and Freeman (1990) proposed a cognitive model in which people with borderline personality disorder have three basic inappropriate beliefs—that the world is dangerous, that the person is vulnerable and powerless, and that the person is 'inherently unacceptable'. Beck does not attempt to establish how these ways of thinking originated, but focuses on how they can be changed.

Histrionic personality disorder

Psychoanalytic explanations relate this disorder to failure to resolve either Oedipal conflicts (Fenichel 1945) or oral conflicts, and to excessive use of repression as a mechanism of defence. No coherent cognitive theory has been proposed.

Obsessional personality disorder

Psychoanalytic theory suggests that obsessional personality disorder originates in the same disturbances of early development (those at the anal stage, see p. 83) as those that cause obsessional symptoms (see p. 184). To explain the features of the disorder, a set of defence mechanisms is proposed which includes regression, reaction formation, and isolation.

Anxious (avoidant) personality disorder

This term was introduced recently and there are no well-established psychological theories to account for the disorder. However, it seems that this disorder fits better than the others with a cognitive model (Beck and Freeman 1990) which proposes that the central features are fear of rejection, self-criticism, and inaccurate evaluations of the reactions of other people. Although this model applies to the present condition of the person, it does not explain how the features developed.

Dependent personality disorder

Psychoanalytic ideas attempt to explain the person's current dependency as a fixation at the oral stage of development, or as a regression to that stage because of failure to resolve Oedipal conflicts.

It should be noted again that all the psychoanalytic explanations listed above are speculative. They are considered as part of the development of thought about personality disorder, not as secure contributions to knowledge.

Causes of antisocial personality

Genetic causes

There are no satisfactory twin studies concerned directly with the inheritance of antisocial personality. However, there have been several studies in which repeated convictions for criminal offences were used as a surrogate for the diagnosis of antisocial personality on the grounds that many, though not all, of those who repeat criminal offences have antisocial personality. A well-known early enquiry of this kind was carried out by Lange (1931) who studied 13 pairs of monozygotic (MZ) twins, in which one of each pair had committed a criminal offence. Of the 13 co-twins, as many as 10 had offended. Moreover, in one of the three discordant pairs, the proband had committed his offence after a head injury. On the other hand, among 17 pairs of dizygotic (DZ) twins of the same sex, in which the proband was a criminal, only two co-twins had offended. Lange also reported that the personalities of the MZ twins were usually similar, for example both explosive and excitable, or both weak-willed and shy. The personalities of the DZ twins were less similar.

Lange's findings must be viewed cautiously because his study had methodological shortcomings. The numbers were small and the selection of cases may have been biased. Moreover, criminal behaviour is associated with alcoholism, so that inheritability of the former may simply reflect the inheritability of the latter (see p. 757). However, similar conclusions were drawn from part of a larger study carried out in America by Rosanoff *et al.* (1934). In a study of 340 twin pairs, these investigators identified the pairs in which at least one twin had offended in adult life. Among the 33 MZ pairs, 22 had co-twins who had offended. Among the 23 DZ pairs, only three had a co-twin who had offended.

Another source of evidence is the study of people separated from antisocial parents by adoption at birth. Conflicting results have been obtained from such investigations, perhaps because the criterion of antisocial behaviour (usually criminal convictions) was subject to many influences other than those of personality.

Cadoret (1978*b*) attempted to study 190 adoptees who had been separated at birth from parents who showed persistent antisocial behaviour. He examined those adoptees who had been reared in a permanent home and compared them with a control group of adoptees whose parents were not antisocial. The findings must be accepted cautiously because about 30 per cent of the subjects refused to be interviewed. Of the adult descendants of antisocial parents, 22 per cent had been diagnosed as having antisocial personality while none of the descendants of controls had been so diagnosed. This finding held irrespective of whether it was the biological father or mother who had shown antisocial behaviour. Among the offspring, however, antisocial behaviour disorder was diagnosed more often in men than women. (Cadoret also found an increased rate of hysteria among the women, and suggested that this condition might be an alternative expression of the same genetic endowment.) These findings did not appear to be related to differences in the families of adoption. However, in the sample as a whole the number of antisocial symptoms in the offspring was related to psychiatric problems in the adoptive parents. Two other investigations have supported this evidence that antisocial behaviour is increased amongst the adopted children of antisocial biological parents (Crowe 1974; Cadoret *et al.* 1975).

Other studies have examined the parents of adoptees who have shown antisocial behaviour. These studies found an excess of antisocial behaviour in the biological parents of the adoptees compared with the biological parents of children who were not antisocial (Schulsinger 1982). The numbers of subjects in these studies have not been large. Thus, in Schulsinger's study of 57 biological parents of antisocial adoptees, only four showed antisocial personality disorders. Even when a wide definition of antisocial personality disorder was adopted to include criminality, alcoholism, and hysterical personality traits, the figure was only 14.

Chromosomal abnormalities

It has been suggested that these abnormalities are an occasional cause of abnormally aggressive behaviour. This suggestion followed the discovery that about 3 per cent of patients in a maximum security hospital had the XYY karyotype (Jacobs *et al.* 1965). Since this finding was first reported, it has become known that the incidence of the XYY karyotype in the general population is higher than was thought at the time. The number found in the maximum security hospital, though high, is much less remarkable than originally supposed.

Cerebral pathology and cerebral maturation

People with antisocial personality seem so different from normal people, and so similar in their behaviour to some patients with brain injuries, that organic causes have been suggested for the personality disorder. There is no convincing evidence linking antisocial personality in adult life with brain injury in childhood. However, it has been suggested that antisocial behaviour in childhood can be caused by minor degrees of damage to the brain ('minimal brain dysfunction'). There is evidence linking antisocial behaviour in childhood with antisocial personality in adult life. Taken together, these observations may be indirect evidence for an association between damage to the brain in childhood and antisocial personality in adult life. While there is fairly strong evidence for a continuity between behaviour disorders in childhood and antisocial personality in adult life (see p. 668), there is only weak evidence that minimal brain dysfunction is a cause of behaviour disorder in childhood (see p. 670).

A related view is that antisocial personality disorder may result from delay in the maturation and development of the brain. EEG abnormalities consistent with maturational delays have been reported in people with antisocial personalities. For example Hill (1952) carried out an uncontrolled study of 194 antisocial and aggressive people, none of whom had epilepsy. Three patterns of EEG abnormality were found, all of which could have arisen from maturational defects. The most frequent were bilateral excess of slow waves (theta activity) and foci of 3–5 Hz activity in the posterior temporal regions. Both kinds of abnormality were usually bilateral but, if not, were more often on the right. The abnormalities were less frequent among the older subjects. Williams (1969) confirmed these findings in a study of 333 men convicted of violent offences; of these men, 206 had been habitually aggressive whilst 127 were known to have had only a single aggressive outburst, usually in response to provocation. After exclusion of subjects with mental subnormality, epilepsy, or a previous head injury, 57 per cent of the habitually aggressive group had abnormal EEGs as against only 12 per cent of the single-outburst group. Abnormalities were found most often in the anterior temporal region. Williams speculated that these abnormalities might indicate a primary disorder in the reticular activating system or limbic mechanisms. He concluded that disturbed cerebral physiology was an important predisposing cause of the propensity to seriously aggressive behaviour, though single aggressive outbursts were usually provoked by environmental factors.

5-Hydroxytryptamine and aggression

Recent biological studies have focused on possible abnormalities in brain 5-hydroxytryptamine (5-HT) neurotransmission in patients with impulsive and aggressive behaviours. Low levels of the 5-HT metabolite, 5-hydroxyindole acetic acid (5-HIAA), have been found in the cerebrospinal fluid of subjects who have committed acts of unpremeditated violence. Such subjects often have diagnoses of borderline or sociopathic personality (Linnoila and Virkkunen 1992). Similarly, 5-HT neuroendocrine studies have shown lower 5-HT-mediated prolactin release in subjects with histories of impulsive aggressiveness (Coccaro *et al.* 1989). The data suggest that low brain 5-HT neurotransmission is not confined to certain categories of personality disorder but instead may correlate with the propensity to engage in irritable and aggressive acts. Therefore it is interesting that a study of free-ranging rhesus monkeys also found a negative correlation between levels of cerebrospinal fluid 5-HIAA and aggressive behaviour (Dee Higley *et al.* 1992).

The effects of upbringing

Two departures from the normal pattern of upbringing have been thought to contribute to the development of antisocial personality disorders: separation from parents, and disordered behaviour in the parents.

From an uncontrolled retrospective study of 44 young delinquents, Bowlby (1944) suggested that separation of a young child from its mother leads to a personality characterized by antisocial behaviour and failure to form close relationships. Later, Bowlby expanded these ideas in an influential book, *Forty-four juvenile thieves* (Bowlby 1946). This work stimulated much research into the immediate and long-term effects of separating children from their mothers. The general conclusions are referred to on p. 671–2). In relation to the aetiology of antisocial personality disorder, it is now known that the effects of separation from the mother are much more varied than Bowlby originally suggested, and that not all children are affected adversely. Moreover, the original ideas proposed a unitary process—maternal deprivation—but in reality the effects of separation depend on many factors, for example the child's age, his previous relationship with his mother (and father), and the reasons for the separation. These last two points lead to the second kind of departure from the normal pattern of upbringing—disordered behaviour in the parents.

Two kinds of evidence point indirectly to the importance of the parents' behaviour and of enduring family relationships as causes of antisocial personality disorder. First there is evidence that parental separation usually follows a long period of tension and arguments that could themselves affect the child's development. The second line of evidence is in two parts. The first part (reviewed in an earlier paragraph) links behaviour disorder in childhood with antisocial personality disorder in later life. The second part indicates that the behaviour of parents is an important cause of childhood behaviour disorders. For example Rutter (1972) showed that the association between separation and antisocial disorder in sons is determined by disharmony in the marriage. Taken together, these items of indirect evidence point quite strongly to the importance of upbringing as a cause of antisocial personality disorder.

Social learning and personality disorder

Several authors have suggested that antisocial personality results from a failure of social learning. For example Eysenck (1970a) suggested that antisocial personality disorder is more likely to develop in people who condition slowly and so fail to learn normal social behaviour. Although attractively simple, this explanation takes no account of the complexities of social learning and does not explain why people with antisocial personality learn other behaviour patterns normally. Scott (1960) proposed a broad scheme which is based not on experimental evidence but on common-sense considerations of how children learn. This scheme can be useful to the clinician. Scott suggested four ways in which repeated antisocial behaviour could develop. First people may acquire socially unacceptable behaviour because they grow up in antisocial families. Second they may have had no opportunity to learn because they were not presented with consistent rules of behaviour in the family. Third they may have learnt antisocial behaviour as a way of overcoming some emotional problem; for example a young man who feels inferior with women may adopt aggressive behaviour to hide this inferiority. Fourth they may have a learning difficulty such as poor ability to sustain attention.

Childhood behaviour problems and antisocial personality

If constitution and experience of childhood are causes of antisocial personality, an association would be expected between behaviour problems in childhood and antisocial personality in adult life. Such an association has been reported from a study in the USA of 524 people who, as children 30 years before, had attended a child guidance clinic (Robins 1966). Among those whose behaviour in childhood had been seriously antisocial, a substantial minority had persistent antisocial behaviour in adult life. Most of the adults with antisocial personality disorder had shown this behaviour as children. (In contrast, the prognosis for neurotic children was generally good see p. 668).) The outcome was particularly poor if, in childhood, several different antisocial behaviours coexisted and antisocial acts were repeated. Stealing among boys and sexual delinquency among girls had a poor prognosis.

The prognosis of personality disorder

Just as normal personalities may show small changes with increasing age, so abnormal personalities may become less abnormal. There is little factual information about the outcome of personality disorders, and almost all of it concerns borderline and antisocial disorders. In the American study by Robins mentioned above, information was collected about people with persistent antisocial behaviour in early adult life. At later follow-up, about a third of these people had improved, as judged by the number of arrests and contacts with social agencies, but they still had problems in relationships, as shown by hostility to wives and neighbours. They also had an increased rate of death by suicide. Research in England showed that, amongst offenders with **antisocial personalities**, those whose first offences were aggressive did not subsequently commit offences that were predominantly aggressive (Gibbens *et al.* 1959). This finding accords with a general impression among clinicians that antisocial people over the age of 45 present fewer problems of aggressive behaviour but continue to have problems of personal relationships.

Studies of the outcome of **borderline personality disorder** (Stone *et al.* 1987) showed that only about one in four of people with this diagnosis in their twenties still met criteria for the same diagnosis in middle age, although most met the criteria for another personality disorder (including histrionic, avoidant, and obsessive). Those who continued to have the original personality disorder more often had associated substance abuse or a criminal record. Other studies have shown more agreement (about 60 per cent) between initial diagnosis of borderline personality disorder and diagnosis at follow-up (Berelowicz and Tarnopolsky 1993, p. 99). It is uncertain whether the suicide rate is increased in these people; the rate was high in some studies (for example 8.5 per cent in the study by Stone *et al.* (1987)) but not in others; for example, Mehlum *et al.* (1991) reported a three to five year follow-up in which schizotypal personality disorder had a worse outcome than borderline personality disorder, whilst 'cluster C' personality disorders (see p. 110) had the best outcome (see Stone (1993) for a review).

The management of personality disorders

It has been said that people cannot change their natures, but can only change their situations. There has been some progress in finding ways of effecting small changes in disorders of personality, but management still consists largely of helping the person to find a way of life that conflicts less with his character. It is recognized that some changes in abnormal personalities may occur slowly over many years (see prognosis above). The psychiatrist's role is often to help patients to avoid adding to their problems (for example by abusing drugs or alcohol, or by entering into unsatisfactory relationships) until natural changes occur. Whatever treatment is used, aims should be modest and considerable time should be allowed to achieve them.

Assessment

Thorough assessment is the first step in the management of personality disorder. Information from independent informants should be sought in every psychiatric assessment, but it is particularly important when personality disorder is being considered.

In assessing someone with a personality disorder, it is more useful to describe the main features of the disorder than to attach a diagnostic label. The description should refer to both strengths and weaknesses, because treatment attempts to build on favourable features as well as to modify unfavourable features. The patient's circumstances should be examined with care, with particular attention to any that regularly provoke undesirable behaviour. This last step is often overlooked to the patient's detriment. Aggressive people are not aggressive in all circumstances, nor are shy and self-conscious people ill at ease in every social encounter. To discover what provokes undesirable behaviour, there must be detailed observation over several weeks to identify any recurring patterns. This method is often useful for people with antisocial personalities because such people appreciate a practical approach.

Enquiries may show that specific factors are making abnormal behaviour worse. For example a man with an anankastic personality may need encouragement to move to a job with less responsibility for people who have lower standards than his own. A man with antisocial personality may be provoked into anger when he feels rejected by women. Sometimes the enquiries suggest that attention should be given to a problem which was not apparent at first. For example the man just described may provoke rejection by his own clumsy approach to women. He might be helped by counselling and social skills training directed to this clumsiness.

Psychological treatment in general

Counselling

This kind of treatment is most likely to help young people who lack confidence, have difficulty in making relationships, and are uncertain about the direction that their lives should take. It is important that they be highly motivated to work at solving their problems by examining their attitudes and emotions. People with antisocial personality disorders are least likely to be helped by psychological treatment, but some are helped by special forms of large-group treatment in a therapeutic community. (This form of treatment is considered on p. 130.)

Dynamic psychotherapy is not indicated for most patients with personality disorder, but psychological support is often beneficial. Such support can be given by a doctor, though for many patients it can be given equally well by an experienced social worker or psychiatric nurse. For some personality disorders modest but useful readjustments can be achieved over a period of months, but several years may be needed for antisocial personality disorders. Some antisocial people are put on probation after breaking the law; in this way useful external control can be provided when motivation for treatment is poor at the start.

Whatever the nature of the disorder, the treatment plan usually includes attempts to bring about limited changes whereby the patient has less contact with situations that provoke difficulties and more opportunity to develop assets in the personality. It is essential to attempt to build a trusting relationship so that patients can talk openly and learn from their mistakes. Patients will certainly experience setbacks, and at these times the therapist should avoid any suggestion of failure. Often progress can be made only by a series of small steps whereby the patient gradually moves nearer to a satisfactory adjustment. These steps can often be taken most effectively when setbacks occur, since it is then most likely that the patient will be willing to face the real problems. The therapist should also encourage patients to develop more satisfying relationships, for example by taking part in leisure interests, pursuing further education, or joining clubs.

When the personality disorder includes traits of dependency, the patient should not be seen at short intervals. Some patients become excessively dependent if they are seen too often or for too long, with a consequent increase in their difficulties. Even if little progress is made, a supportive and watchful relationship can often prevent the accumulation of additional problems until some fortuitous change in the patient's life brings about some improvement.

Although the therapist should try hard and long to help these patients, it should be recognized that, however skilful the therapist, some patients will not benefit. The therapist should not be discouraged by this lack of response.

Dynamic psychotherapy

Treatment by dynamic psychotherapy is much the same for personality disorders as for neuroses. It can be carried out individually or in groups (see Chapter 18).

The individual treatment of personality disorders differs in emphasis from the treatment of neuroses. With personality disorders there is less emphasis on the reconstruction of past events and more on the analysis of current behaviour. In so-called character analysis there is detailed examination of the person's ways of relating to other people, coping with external difficulties, and dealing with his own feelings. The approach is more directive than in the classical methods of analysis for neurotic symptoms, although the analysis of transference remains an important element. To emphasize any discrepancies between the patient's habitual ways of relating to others and his real life situation, the therapist has to reveal more of himself than is usual in classical analysis. At the same time the analysis of counter-transference (the therapist's emotional attitudes to the patient) can be an important guide to the likely reactions of other people to the patient. For a review of dynamic psychotherapy for severe personality disorder see Kernberg (1993).

Cognitive therapy

Attempts have recently been made to adapt cognitive therapy methods for the management of personality disorder. These methods have not yet been evaluated adequately. They focus on the modes of thinking that characterize the personality disorder and attempt to change them with basic cognitive therapy techniques (see p. 641). Thus, for obsessive–compulsive personality disorder the treatment focuses on characteristic beliefs, for example that all mistakes must be avoided or that the person must be in control of the immediate environment (Beck and Freeman 1990).

Psychological treatment for specific personality disorders

Paranoid personality disorder

Patients with this disorder do not engage well in psychological treatment because they are touchy and suspicious.

Schizoid personality disorder

The patient with a schizoid personality avoids close personal contact, and so makes any kind of psychotherapy difficult. He often drops out after a few sessions; if he does stay in treatment, he tends to intellectualize his problems and to question the scientific status of the treatment. The therapist should try to penetrate these intellectual defences gradually, and to help the patient to recognize his emotional problems. Only then can the therapist begin to explore ways of dealing with the problems. At best the process is slow and often is not successful.

Antisocial (dissocial) personality disorder

Several forms of treatment have been proposed for antisocial personality disorder, but none has been shown to be effective in a controlled evaluation (for a review see Quality Assurance Project 1991). Without such evaluation the following paragraphs are based on clinical opinions.

Individual psychotherapy

Most psychiatrists agree that individual psychotherapy seldom helps antisocial personality disorder, and that the conduct of interviews is often made difficult by the patient's behaviour. Schmideberg (1947) was exceptional in reporting good results from a psychotherapy in which patients were confronted repeatedly and directly with evidence of their own abnormal behaviour. If such treatment is effective at all, it is only with therapists who have a particularly forceful and robust personality.

Small-group therapy

If one person with an antisocial personality disorder joins a conventional therapeutic group, he seldom benefits and often disrupts the treatment of the others. On the other hand, groups composed entirely of antisocial patients can sometimes be more constructive. Running such groups requires skill and experience, together with determination to set limited goals and to encourage group members to share responsibility and help one another. This kind of treatment should not be undertaken without special training (Whiteley 1975).

The therapeutic community

The principles of the therapeutic community are outlined in Chapter 18. They have been used to treat antisocial personalities since the work of Jones (1952) in the Social Rehabilitation Unit at Belmont Hospital, later called The Henderson Hospital. In such a unit antisocial patients live and work together, and meet several times a day for group discussions in which each person's behaviour and feelings are examined by the other group members. Frank discussion is encouraged, and patients are required to consider their own behaviour and its effects on other people. Such discussion often provokes much outpouring of emotion including anger. The therapists hope that patients, by repeatedly facing these issues, will gradually learn to control their own antisocial behaviour and to adopt more acceptable ways of dealing with their feelings and relationships. Rapoport (1960) has described four aspects of treatment that may be important in bringing about change: permission to act on feelings without the usual social restraints, sharing of tasks and responsibilities, group decision-making to involve patients in making rules as well as breaking them, and confrontation of each person with the effects of his actions on others. No controlled evaluation of this treatment has been carried out, and opinions of its value are divergent. One- to two-year follow-up studies have reported improvement rates of 40–60 per cent, depending on whether the criterion for improvement was general social functioning, employment, or reconviction (Taylor 1966).

Other group regimes

In contrast with communities which allow patients the freedom to learn from their mistakes, Craft (1965) has advocated a more authoritarian regime for people who have antisocial personality disorder combined with subnormality of intelligence, and who become more disturbed when treated in a therapeutic community.

Stürup (1968) described the application of therapeutic community principles to offenders who had committed violent crimes or serious sexual offences, and who were detained on indefinite sentences in the Herstedvester Detention Centre in Denmark. At Grendon Prison in England similar principles have been applied to the care of prisoners who have committed less serious offences and who are not serving indefinite sentences. In neither case has any controlled enquiry been possible. Hence it is difficult to evaluate Stürup's report (Stürup 1968, p. ix) that 90 per cent of offenders passing through his unit committed no further offences.

Borderline personality disorder

Exploratory psychotherapy

The management of borderline personality disorder is similar to that of histrionic personality disorder, as described below. People with borderline personality disorder do not respond well to exploratory psychotherapy, and in response to unskilled treatment they may show reduced emotional control and increased impulsiveness. Although 10–15 year outcome has been reported in two-thirds of borderline personality disorders treated with expressive psychotherapy (Stone *et al.* 1987), there was no randomized control group in the study and improvement may have been unrelated to treatment. Nevertheless some authors consider that psychoanalysis-based methods are appropriate for these patients (Higgitt and Fonagy 1993, p. 236; Kernberg 1993, Chapter 10),

although they accept that intense and sometimes hostile transference reactions are frequent.

Group treatment has the advantages that transference relationships are spread over the group instead of focused on the therapist, and that the group's comments on acting out behaviour may be accepted more readily than those of the therapist. The therapist needs considerable skill if the borderline patient is to be helped without disrupting the treatment of the other group members. It is seldom practicable to treat more than one borderline patient in the same group. It can be useful to employ *problem-solving counselling* (see p. 607) for borderline patients and to focus on practical goals in dealing with everyday problems.

For a review of psychotherapy for borderline personality disorder see Quality Assurance Project (1991) and Higgitt and Fonagy (1993); for a review of group treatment see Clarkin *et al.* (1991).

Histrionic personality disorder

Murphy and Guze (1960) have given an interesting account of the difficulties that can arise in the treatment of histrionic personalities. These authors describe direct and indirect demands that patients can place on the doctor. Direct demands include reasonable requests for medication, repeated seeking for assurances of continuing help, telephoning at unreasonable times, and attempts to impose impractical conditions on treatment. Indirect demands include seductive behaviour, threats of dangerous actions such as drug overdoses, and repeated unfavourable comparisons of the present treatment with any received in the past. The doctor has to be alert for the first signs of such demands, and should clearly set limits by indicating how much of the patient's behaviour he is prepared to tolerate. This must be done before the patient's demands become too great.

Obsessional personality disorder

People with obsessional personality disorder often express great eagerness to please the therapist. Usually, however, this kind of personality disorder does not respond well to psychotherapy. Unskilled treatment can lead to excessive morbid introspection which leaves the person worse rather than better.

Anxious (avoidant) personality disorders

These patients have low self-esteem and they fear disapproval and criticism. They can be helped by a therapeutic relationship in which they feel valued and able to review their perception of themselves. They may break off treatment, fearing criticism or rejection; such feelings should be detected early so that they can be interpreted appropriately to the patient.

Dependent personality disorder

For dependent patients, dependence may be increased by dynamic psychotherapy. Such patients are usually helped more by problem-solving counselling in which they are encouraged to take responsibility for themselves.

Drug treatment of personality disorder

Until recently it was generally agreed that drugs had little part to play in the management of personality disorder. Recently, specific uses for drug treatment have been proposed mainly to control unstable and aggressive aspects of personality disorder.

Anti-psychotic drugs have short-term beneficial effects for patients with *borderline personality disorder*. Controlled trials have shown that these drugs, when given in doses smaller than those generally used for psychosis, have beneficial effects for periods of increased disturbance including deliberate self-harm (Goldberg *et al.* 1986; Soloff *et al.* 1986, 1993; Cowdrey and Gardner 1988).

Anxiolytics should generally be *avoided* because they may lead to dependence and may produce disinhibition in people prone to act violently. These drugs are sometimes useful for short-term relief of symptoms in anxious or obsessional personalities at times of increased stress.

Tricyclic antidepressants are not generally useful in uncomplicated personality disorders (as opposed to chronic dysthymic disorders) but may help when there is a secondary depressive disorder. Tricyclic antidepressants sometimes appear to worsen borderline personality disorders (Soloff *et al.* 1986), but there is some evidence that the SSRI fluoxetine may be helpful even when there is no associated depressive disorder (Marcovitz and Schultz 1993).

Monoamine oxidase inhibitors have been found in one small controlled study to benefit patients with borderline personality disorder (Cowdrey and Gardner 1988). The trial was conducted with tranylcypromine which has stimulant as well as antidepressant effects (see p. 569); until other MAOIs have been tested it cannot be certain which effect was beneficial.

Lithium carbonate appears to reduce mood variation in some patients with *cyclothymic personality disorder*, but there is no convincing evidence from clinical trials to confirm this clinical impression. If lithium carbonate is used, it should be only after a long period of observation to make sure that the mood changes are not responses to life events that could be modified by psychological treatment or social measures. It has been reported that aggressive behaviour in people with various kinds of personality disorder can be reduced by lithium carbonate in doses comparable to those used to prevent the recurrence of affective disorder. Aggressive behaviour is usually reduced within two weeks (see Nilsson (1993) for a review). However, there is a practical difficulty that people with aggressive behaviour may not comply with the strict regime for the safe use of lithium.

Anti-epileptic drugs have been used to treat people who have sudden episodes of violent or otherwise disruptive behaviour ('behavioural dyscontrol', see p. 759). One small controlled study showed carbamazepine to have beneficial short-term effects (Gardner and Cowdrey 1986) but there is insufficient evidence to recommend this treatment at present. (For a review of the drug treatment of personality disorder see Stein (1992).)

Further reading

Dolan, B. and Cord, J. (1993). *Psychopathic and antisocial personality disorders; treatment and research issues.* Gaskell, London.

Gunderson, J. G. and Phillips, K. A. (1995). Personality disorders. In *Comprehensive textbook of psychiatry* (6th edn) (ed. H. I. Kaplan and B. J. Sadock), Chapter 25. Williams and Wilkins, Baltimore, Maryland.

Schneider, K. (1950). *Psychopathic personalities* (trans. M. W. Hamilton). Cassell, London.

Tyrer, P. and Stein, G. L. (ed.) (1993). *Personality disorder reviewed.* Gaskell, London.

6 | *Reactions to stressful experiences (and minor affective disorders)*

Stressful events frequently provoke psychiatric disorders. Such events can also provoke emotional reactions that are distressing but not severe or prolonged enough to meet the diagnostic criteria for anxiety disorders or depressive disorders. These less severe reactions are the subject of this chapter.

The chapter begins with a description of the components of responses to stressful events. An account is then given of coping strategies and mechanisms of defence. The classification of reactions to stressful experience is discussed. Various syndromes are described, including acute stress reactions, post-traumatic stress disorder, special forms of response to severe stress, and adjustment disorders. Other special forms of reaction, such as adjustment to serious physical illness or trauma, are also described.

At the end of the chapter an account is given of the conditions known as **minor affective disorders** in which neither anxiety symptoms nor depressive symptoms are severe enough to meet criteria for anxiety disorder or depressive disorder. Although not severe, these disorders are an important cause of family tension and absence from work.

Components of responses to stressful events

Responses to stressful events have three components: an emotional response, a somatic response, and a psychological response that reduces the impact of the experience. *Emotional and somatic responses* may be of two kinds. The first kind is anxiety, with autonomic arousal that leads to tachycardia, increased muscle tension, and dry mouth; the second kind is depression with reduced physical activity. Anxiety responses are generally associated with events that pose a threat, whilst depression is usually associated with events that involve separation or loss. The third component of the response to a stressful event comprises psychological mechanisms that normally reduce the impact of the event, thus limiting the emotional and somatic reactions. Two groups of psychological mechanisms have been described: *coping strategies* and *mechanisms of defence*.

Coping strategies and mechanisms of defence

Coping strategies and mechanisms of defence can reduce the effects of the strong emotions generated by stressors; in this way normal performance can be maintained. The term **coping strategy** is derived from research in social psychology; it is applied to activity of which the person is aware, for example deliberately avoiding stressors (Lazarus 1966). The term **mechanism of defence** is derived from psychoanalytic studies; it refers to unconscious mental processes such as denial. A third term, **adaptation**, is sometimes applied to the psychological processes involved in adjustment to stressful events associated with chronic illness.

Coping strategies may be adaptive or maladaptive. **Adaptive** coping strategies are of two kinds: problem-solving strategies, which can be used to make adverse circumstances less stressful, and emotion-reducing strategies, which can enhance adaptation to stressful circumstances.

Problem-solving strategies include:

- seeking help from another person, or obtaining information or advice that would help to solve the problem;
- **problem solving**—making and implementing plans to deal with the problem;

- **confrontation**—defending one's own rights, or persuading another person to change his behaviour, when either action would help the patient.

Emotion-reducing strategies include:

- **ventilation of emotion**—talking to another person and expressing emotion;
- **avoidance**—refusing to think about the problem, avoiding people who are causing problems, or avoiding reminders of the problem;
- **positive reappraisal**—recognizing that the problem has led to some good, for example self-betterment;
- **accepting or rejecting responsibility**—recognizing that one is wholly or partly responsible for the problem and can deal with it, or that one is not responsible and need not react.

These coping strategies are generally useful in reducing the problem or in lessening the emotional reaction. However, they are not always adaptive. For example avoidance may not be adaptive in the early stages of physical illness because it can lead to delay in seeking appropriate treatment. Hence a person needs not only the ability to use coping strategies but also the ability to judge which strategy should be used in particular circumstances. This point is expressed in the encouragement given to members of Alcoholics Anonymous, who are advised to seek 'the courage to change what can be changed, the serenity to accept what cannot be changed, and the wisdom to know the difference'.

Maladaptive coping strategies reduce a person's emotional response to stressful circumstances in the short term, but lead to greater difficulties in the long term. **Maladaptive coping strategies** include the following.

- **Use of alcohol or unprescribed drugs** to reduce the emotional response or to reduce awareness of stressful circumstances.
- **Deliberate self-harm** either by drug overdose or self-injury. Sometimes the skin is cut with a sharp instrument to induce pain and draw blood as a way of reducing tension.

- **Histrionic behaviour:** an unrestrained display of emotion may reduce tension, and in some societies such behaviour is sanctioned in particular circumstances, for example grieving. Such behaviour can be adaptive, but it can also damage relationships that would otherwise have been supportive.
- **Aggressive behaviour:** overt aggression may release pent-up feelings of anger and provide immediate release. In the longer term, however, it damages relationships and increases the person's difficulties.

When particular coping mechanisms are used repeatedly by a person in different situations they are said to constitute a *coping style*. Some people change their coping strategies according to the circumstances; for example they use problem-solving strategies in one kind of situation but employ avoidance in another. Some people habitually use maladaptive strategies; for example they repeatedly abuse alcohol or take overdoses of drugs when under stress. For a review of coping strategies see Lazarus (1993).

Mechanisms of defence (Table 6.1) were originally described by Sigmund Freud and later elaborated by his daughter Anna Freud (1936). These authors described numerous mechanisms. As responses to stressful circumstances, the most frequent mechanisms are repression, denial, displacement, projection, and regression. Defence mechanisms are *unconscious* processes, i.e. people do not use them deliberately and are unaware of their own real motives, although they may become aware later through introspection or through another person's comments. Freud identified defence mechanisms in his study of the 'psychopathology of everyday life', a term that he applied to slips of the tongue and lapses of memory. The concept of defence mechanisms has proved useful in understanding many aspects of the day-to-day behaviour of people under stress, notably those with physical or psychiatric illness. Freud also used the concept of mechanisms of defence to explain the aetiology of mental disorders, but this extension of his original observations has not proved useful.

Table 6.1 *Mechanisms of defence*

Repression	Reaction formation
Denial	Rationalization
Displacement	Sublimation
Projection	Identification
Regression	

The main mechanisms of defence can be illustrated by the following examples.

- **Repression** is the exclusion from consciousness of impulses, emotions, or memories that would otherwise cause distress. For example, the memory of distressing events such as sexual abuse in childhood may be kept out of awareness for many years.
- **Denial** is a related concept: it is inferred when a person behaves as if unaware of something that he may reasonably be expected to know. For example on learning that he is dying of cancer, a patient may ward off depression by denial and may continue to live normally as if unaware of the diagnosis. In this example, denial is adaptive. However, in the early stage of illness denial may delay seeking help and may lead to refusal of necessary investigations or treatment. In this second example, denial is maladaptive.
- **Displacement** is the transfer of emotion from a person, object, or situation with which it is properly associated, to a less potent source of distress. For example, after the recent death of his wife, a man may blame the doctor for failure to give adequate care, and may thus avoid blaming himself for putting his work before her needs in the last months of her life.

- **Projection** is the attribution to another person of thoughts or feelings similar to one's own, thereby rendering one's own thoughts or feelings more acceptable. For example a person who dislikes a colleague may attribute reciprocal feelings of dislike to him; it is then easier to justify his own feelings of dislike for the colleague.
- **Regression** is the adoption of behaviour appropriate to an earlier stage of development, for example dependence on others. Regression often occurs among physically ill people. In the acute stages of illness it can be adaptive, enabling the person to acquiesce to the requirements of passively accepting intensive medical and nursing care. If regression persists into the stage of recovery and rehabilitation, it can be maladaptive because it reduces the patient's ability to take responsibility for himself.
- **Reaction formation** is the unconscious adoption of behaviour opposite to behaviour that would reflect true feelings and intentions. For example excessively prudish attitudes to sex are sometimes (but not always) a reaction to the person's own sexual urges that he cannot accept.
- **Rationalization** is the unconscious provision of a false but acceptable explanation for behaviour that has a less acceptable origin. For example a husband may leave his wife at home because he does not enjoy her company, but he may reassure himself falsely that she is shy and would not enjoy going out.
- **Sublimation** is the unconscious diversion of unacceptable impulses into more acceptable outlets, for example turning the need to dominate others into the organization of good works for charity.
- **Identification** is the unconscious adoption of the characteristics or activities of another person, often to reduce the pain of separation or loss. For example a widow may undertake the same voluntary work that her husband used to do.

Previous experience and response to stressful events

Studies of life events provide another useful way of considering response to stressful events (see p. 89). Brown and Harris (1978) have pointed out that response to stressful life events may be modified by other circumstances of life at the time and by past experience of life events. Some current circumstances make a person vulnerable to stressful life events, for example the lack of a confidant with whom to share problems. Such circumstances are called *vulnerability factors*. Previous experience also increases vulnerability; for example the experience of losing a parent in childhood may make a person more vulnerable in adult life to stressful events involving loss. It is more difficult to examine these more remote associations scientifically. (See p. 90 for further discussion of life events and vulnerability factors.)

Classification of reactions to stressful experience

Although included within the classifications of diseases, not all reactions to stressful events are abnormal. Grief is a normal reaction to the stressful experience of bereavement, and only a minority of people have a very severe or abnormally prolonged reaction. There is also a normal pattern of reaction to a dangerous or traumatic event such as a car accident. Thus most people have an immediate feeling of great anxiety, are dazed and restless for a few hours afterwards, and then recover; a few people have more severe and prolonged symptoms—an abnormal reaction. It is difficult to decide where to make a separation between normal and abnormal reactions to stressful events in terms of severity or of duration, and in practice the division is arbitrary. Similarly, amongst patients in hospital for medical or surgical treatment, most are anxious but a few are severely anxious and show extreme denial or other defence mechanisms that impair cooperation with treatment.

ICD10 and DSMIV reactions to stressful experiences are classified into three groups (Table 6.2).

1. *Acute reactions* are immediate and brief responses to sudden intense stressors in a person who does not have another psychiatric disorder at the time. The ICD10 definition sets a low threshold, requiring only that the response should last for 'a few hours' to no more than 'about three days'. This definition includes most reactions that would generally be considered normal responses to highly stressful events. DSMIV uses a different convention: the term used is *acute stress disorder*, which indicates an abnormal state defined as lasting for at least two days and up to four weeks. This definition excludes some conditions that would be included as acute stress reactions in ICD10.
2. *Post-traumatic stress disorder* is a prolonged and abnormal response to exceptionally intense stressful circumstances such as a natural disaster or a sexual or other physical assault.
3. *Adjustment disorder* is a more gradual and prolonged response to stressful changes in a person's life. In both ICD10 and DSMIV adjustment disorders are subdivided, according to the predominant symptoms, into depressive, mixed anxiety and depressive, with disturbance of conduct, and with mixed disturbance of emotions and conduct. DSMIV has an additional category of adjustment disorder with anxiety. ICD10 has an additional category of 'predominant disturbance of other emotion' which includes not only adjustment disorder with anxiety but also adjustment disorder with anger.

In ICD10 the three types of reaction to stressful experience are classified together under 'reactions to stress and adjustment disorders', which is a subdivision of section F4, 'neurotic, stress related and somatoform disorders'. The defining characteristics of this group of reactions to stress and of adjustment disorders are (i) they arise as a direct

Table 6.2 *Classification of reactions to stressful experience*

ICD10	DSMIV
Acute stress reaction	Acute stress disorder
Post-traumatic stress disorder	Post-traumatic stress disorder

Adjustment disorders
  Depressive reaction
  Mixed anxiety and depressive reaction
  Predominant disturbance of other emotions
  Predominant disturbance of conduct
  Mixed disturbance of emotions and conduct
  Other specified symptoms
	Adjustment disorder*
	  With depressed mood
	  With mixed anxiety and depressed mood
	  With anxiety
	
	  With disturbance of conduct
	  With mixed disturbance of emotions and conduct
	  Unspecified

*The order of the subgroup has been changed to show similarities and differences between the two systems

consequence of either acute stress or continued unpleasant circumstances, and (ii) it is judged that the disorder would not have arisen without these factors. A different organizing principle is used in DSMIV: acute stress disorder and post-traumatic stress disorder are classified as anxiety disorders, whilst adjustment disorders have their own place in the classification, separate from the anxiety disorders.

In ICD10, if any of these reactions is accompanied by an act of deliberate self-harm, another code can be added to record this fact (one of the codes is X60–X82 which refers to 23 different methods of self-harm). It is also possible to specify certain kinds of stressful event by adding a code from Chapter Z; examples are Z58 problems related to employment and unemployment, or Z63 related to family circumstances.

Although post-traumatic stress disorder occurs only after exceptionally stressful events, not every response to such events is a post-traumatic stress disorder. Studies of combat veterans have shown high rates of *depression, somatization disorder,* and *alcohol and drug abuse* (see Rundell *et al.* 1989). Post-traumatic stress disorder occurs after road accidents, but *generalized* and *phobic anxiety disorders* are more frequent. (Mayou 1992). Survivors of concentration camps may develop post-traumatic stress disorder, and also persistent *irritability* and poor memory (Eitinger 1960), and survivors of disasters may develop *marital problems* and *family violence* (Raphael 1986). Put in another way, many people who meet the criteria for post-traumatic stress disorder also meet the criteria for another disorder. For example 90 per cent of Vietnam war veterans met diagnostic criteria for post-traumatic stress disorder and 43 per cent had at least one other diagnosis, the most frequent being atypical depression, alcohol dependence, anxiety disorder, substance abuse, and somatization disorder (MacFarlane 1985).

Acute stress reaction or disorder

Clinical picture

As explained above, this condition is referred to as *acute stress reaction* in ICD10 and as *acute stress disorder* in DSMIV. Acute stress reaction is a brief response (lasting from several hours to about 3 days) to severely stressful events, whilst acute stress disorder is a more prolonged response lasting from at least two days to at most four weeks.

The core symptoms of an acute reaction to stress are anxiety or depression. **Anxiety** is the response to threatening experiences; **depression** is the response to loss. Anxiety and depression often occur together, partly because stressful events often combine danger and loss; an extreme example is a road accident in which a companion is killed. **Other symptoms** include *feelings of being numb or dazed, insomnia, restlessness, poor concentration*, and *physical symptoms of auto-nomic arousal* especially sweating, palpitations, and tremor. *Anger* or *histrionic behaviour* may be part of the response. Occasionally there is a *flight* reaction, for example when a driver runs away from the scene of a road accident.

Coping strategies and defence mechanisms are also part of an acute stress reaction. **Avoidance** is the most frequent coping strategy; the person avoids talking or thinking about the stressful events and also avoids reminders of them. The most frequent defence mechanism is **denial,** which is experienced as a belief that the events have not really happened, or as inability to remember them. Usually avoidance and denial recede as anxiety diminishes; memories of the events return and the person can think or talk about them with less distress. This process allows *working through* and coming to terms with the stressful experience, though there may be continuing difficulty in recalling details of the most stressful events.

Not all acute stress reactions follow this orderly sequence, in which coping strategies and defences are maintained long enough to allow the person to function until anxiety and depression subside and working through can occur. *Coping strategies* may be maladaptive, for example

excessive use of alcohol or drugs to reduce distress. *Defence mechanisms* may be of the less adaptive types such as regression, displacement, or projection. Sometimes defence mechanisms persist longer than usual; for example denial may persist so long that 'working through' is delayed. Sometimes vivid memories of the stress-ful events intrude into awareness as images ('flashbacks') or disturbing dreams. When this state persists the condition is called a post-traumatic stress disorder (see p. 140).

Diagnostic conventions

The different requirements for the duration of symptoms in the two systems of classification has been explained above—at least a few hours in ICD10, and at least two days in DSMIV. In ICD10 there must be a clear and immediate temporal connection between the impact of the stressor and the onset of symptoms; in DSMIV the onset may be delayed up to four weeks after the event. In either scheme, a reaction that lasts longer than the arbitrary limit is classified as either a post-traumatic stress disorder (p. 140) or an adjustment disorder (p. 145)

Both systems of classification describe typical symptoms of the disorder. In DSMIV the diagnosis of acute stress disorder requires fear, helplessness, or horror, together with three from a list of five 'dissociative' symptoms, namely (i) a sense of numbing or detachment, (ii) absence of emotional response and reduced awareness ('being in a daze'), (iii) derealization, (iv) depersonalization, and (v) dissociative amnesia. There must be avoidance of stimuli that arouse recollections of the trauma, and significant distress or impaired social functioning. In ICD10 it is stressed that the clinical picture is usually mixed and changing, including anxiety, depression, anger, and 'daze'.

The terms acute stress reaction and acute stress disorder are used only when the person was free from psychiatric disorder at the time of the impact of the stressful event; otherwise the response is classified as an exacerbation of pre-existing psychiatric disorder.

Aetiology

Many kinds of event can provoke an acute reaction to stress. Examples are involvement in a significant but brief event such as a motor accident or a fire, an event that involves actual or threatened injury such as a physical assault or rape, or the sudden discovery of serious illness. Some of these stressful events involve life changes to which further adjustment is required, for example the serious injury of a close friend involved in the same accident. In the past acute reactions to the stress of war have been referred to as shell shock or battle fatigue by military staff; these terms included what is now called post-traumatic stress disorder (see below). Not all people exposed to the same stressful situation develop the same degree of reaction; this variation suggests that differences in constitution, previous experience, and coping styles may play a part in aetiology.

Treatment

The treatment of acute reactions to stressful events has three elements: to *reduce the emotional response*, to *encourage recall*, and to help with more *effective coping* with *residual problems*. By definition acute stress reactions are brief and most are dealt with by general practitioners, physicians, or surgeons caring for the patient (for example after a road accident) or by counsellors attached to medical teams. Psychiatrists seldom treat these acute and transitory reactions, though they may be called upon to advise others about care.

1. *Reducing the emotional response* A reduction can often be achieved if the affected person talks to sympathetic relatives or friends, or to members of the professional staff dealing with the consequences of an experience such as a road accident or a medical emergency. More structured counselling may be required if there is no confidant, if the stressful circumstances cannot easily be discussed with a relative or friend (for example in some cases of rape), or if the response is severe. When anxiety is severe an *anxiolytic drug* may be prescribed for a few days, and when sleep is severely disrupted a hypnotic drug may be given for a few nights.

2. *Encouraging recall* Discussion of the stressful events helps to prevent the persistence of avoidance or denial, which may prolong or intensify the problem and lead to the development of phobias or post-traumatic stress disorders. Sometimes gradual repeated questioning is needed to help the person to remember and express the associated emotion.

3. *Learning effective coping skills* If there has been an acute crisis for which coping has been maladaptive (for example excess use of alcohol or a drug overdose), there may be an opportunity for the person to learn how to adopt effective coping skills in future. Counselling for this purpose (crisis intervention) is described on p. 607.

Post-traumatic stress disorder

This term denotes an intense, prolonged, and sometimes delayed reaction to an intensely stressful event. The essential features of a post-traumatic stress reaction are *hyperarousal*, *re-experiencing* of images of the stressful events, and *avoidance* of reminders. Examples of extreme stressors that may cause this disorder are natural disasters such as floods and earthquakes, man-made calamities such as major fires, serious transport accidents, or the circumstances of war, and rape or serious physical assault on the person. The original concept of post-traumatic stress disorder was of a reaction to such an extreme stressor that any person would be affected. Epidemiological studies have shown that not everyone exposed to the same extreme stressor develops post-traumatic stress disorder; hence personal predisposition plays a part. In many disasters the victims suffer not only psychological distress but also physical injury, which may increase the likelihood of a post-traumatic stress disorder. Other predisposing factors are reviewed below under aetiology.

The condition now known as post-traumatic stress disorder has been recognized for many years, though under many names. The concept of post-traumatic stress disorder originated in part from the study of United States servicemen returning from the Vietnam War. The diagnosis meant that affected servicemen could be given medical and social help without being diagnosed as suffering from another psychiatric disorder. Similar psychological effects have been reported (under other names) among servicemen in both World Wars, and amongst survivors of a disastrous fire at the Coconut Grove nightclub in America (Adler 1943). (For a historical review see Gersons and Carlier (1992).)

Clinical picture (Table 6.3)

The clinical features of post-traumatic stress disorder can be divided into three groups. The first is **hyperarousal**: persistent anxiety, irritability, insomnia, and poor concentration. The second group is **intrusions**: difficulty in recalling stressful events, intense intrusive imagery (flashbacks), and recurrent distressing dreams; depressive symptoms also occur. The third group is **avoidance**: avoidance of reminders of the events, detachment, inability to feel emotion ('numbing'), and diminished interest in activities. Guilt is often present among survivors of disasters. There may be **maladaptive coping responses** of persistent aggressive behaviour, the excessive use of alcohol or drugs, and deliberate self-harm (Davidson *et al.* 1991) and suicide (Hendin and Haas 1991). After some traumatic events, survivors feel forced into a painful reconsideration of their beliefs about the meaning and purpose of life (Jannoff-Bulman, 1985). It has been suggested that dissociative symptoms and depersonalization are important symptoms of the disorders (Foa *et al.* 1995).

Post-traumatic stress disorder may begin very soon after the stressful event or after an interval usually of days, but occasionally of months, though rarely more than six months (McFarlane 1988). If the person experiences a new traumatic event, symptoms may return even if the second event is less severe than the original. Most cases

Table 6.3 *Post-traumatic stress disorder*

Hyperarousal
  Persistent anxiety
  Irritability
  Insomnia
  Poor concentration

Intrusions
  Difficulty in recalling stressful
    events at will
  Intense intrusive imagery
    ('flashbacks')
  Recurring distressing dreams

Avoidance
  Avoidance of reminders of the
    events
  Detachment
  Inability to feel emotion
    ('numbness')
  Diminished interest in activities

resolve within about three months but some may persist for years (Blank 1992).

By convention, post-traumatic stress disorder can be diagnosed in people who have a history of mental disorder before the stressful events (contrast acute stress disorder (p. 139–40)).

Epidemiology

Estimates of the prevalence of post-traumatic stress disorder in the general population vary from 1.0 to 2.6 per cent (Helzer *et al.* 1987; Shore *et al.* 1989; Davidson *et al.* 1991). Studies of groups subjected to unusual stress yield higher rates: it is common immediately after the stress but declines within a few weeks, for example to 3.6 per cent in a population affected by a volcanic eruption (Shore *et al.* 1989), about 30 per cent among volunteer firefighters (McFarlane 1989) and victims of torture (Ramsay *et al.* 1993), and

45 per cent among battered women (Housekamp and Foy 1991). In several studies it has been reported that there were some patients who had symptoms similar to those of post-traumatic stress disorder but who did not meet the criteria for this condition.

Aetiology

The stressor

The necessary cause of post-traumatic stress disorder is an exceptionally stressful event. It is not necessary that the person should have been harmed physically or threatened personally; those involved in other ways may develop the disorder, for example the drivers of a train that someone has used for suicide (Farmer *et al.* 1992*b*) and bystanders at such events. Intensity of personal involvement seems unimportant. Thus in a study of people affected by a volcanic eruption, the highest rate of post-traumatic stress disorder was amongst those who experienced the greatest exposure to the stressful events (Shore *et al.* 1989). Even so, not all of those most affected by the stressor developed post-traumatic stress disorder, a finding that suggests that some form of personal vulnerability plays a part. Such vulnerability might be genetic or acquired after birth.

Genetic factors

Studies of twins suggest that differences in susceptibility are in part genetic. True *et al.* (1993) studied 2224 monozygotic and 1818 dizygotic male twin pairs who had served in the US armed forces during the Vietnam War. After allowance had been made for the amount of exposure to combat, genetic variation accounted for about one-third of the variance in susceptibility to self-reported post-traumatic stress disorder. Self-reported childhood and adolescent environment did not contribute substantially to this variance.

Other predisposing factors

Vulnerability to developing post-traumatic stress disorder appears to be related to temperament (McFarlane 1989), and in particular to neuroticism. Other factors determining vulnerability are age (children and old people are more vulnerable) and a history of psychiatric disorder (Andreasen 1985; Smith *et al.* 1990).

Neurophysiological factors

Attempts to find physiological causes of post-traumatic stress disorder have drawn on information from experiments in which animals were exposed to severe aversive stimuli. Some patients with post-traumatic stress disorder experience vivid memories of the traumatic events in response to smell and sounds related to the stressful situation. This finding suggests that classical conditioning may be involved—a form of learning that involves subcortical mechanisms. Animals subjected to uncontrollable stress show evidence of increased turnover of norepinephrine in limbic and cortical areas, and increased dopamine release in the nucleus accumbens and frontal cortex. It has been suggested that such changes could account for the symptoms of patients with post-traumatic stress disorder, but there is no evidence to support this notion. The main abnormal findings in such patients are hyperarousal and increased noradrenergic activity (Southwick *et al.* 1993). Abnormalities in the hypothalamic–pituitary–adrenal axis have been reported in patients with post-traumatic stress disorder, but the evidence is contradictory (Charney *et al.* 1993).

Psychological explanations

Psychological explanations of post-traumatic stress disorder include psychodynamic, and cognitive, theories. **Psychodynamic** theories emphasize the role of previous experience in determining individual variations in the response to severely stressful events (Horowitz 1986). The **cognitive** theory suggests that post-traumatic stress disorder arises when the normal processing of emotionally charged information is overwhelmed to the extent that intrusive memories persist, and that high arousal leads to an increased liability to interpret ambiguous information as threatening (Foa *et al.*

1991). Although this formulation succeeds in attributing the phenomena of post-traumatic stress disorder to abnormalities in well-recognized psychological mechanisms, it has not been subjected to a critical test. (For a review of information processing in post-traumatic stress disorder see Litz and Keane (1989).)

Prognosis

The prognosis of post-traumatic stress disorder is uncertain. Probably the less severe cases recover, especially when there is an opportunity to talk about the traumatic events and to express emotion soon after the events have been experienced. Many persistent cases have been reported among Vietnam veterans (McFarlane 1989), but there have been no satisfactory long-term studies of representative groups of adequate size.

Assessment

This should include enquiries about the nature and duration of **symptoms**, previous **personality**, and **psychiatric history**. When the traumatic events have included head injury (for example in an assault or transport accident), a neurological examination should be carried out.

Treatment

Treatment is in two parts: care at the time of the stressful events, and treatment of an established post-traumatic stress disorder. The first steps in care are often taken by ambulancemen, firemen, or police present at the disaster, and these emergency workers may need help themselves at the time or later.

Immediate measures

The initial treatment of post-traumatic stress disorder is the same as that described already for acute reactions to stress, namely encouragement to recall the stressful experiences and to express associated emotions to an understanding person (see p. 140). A few doses of a benzodiazepine drug may be needed to reduce anxiety, and a hypnotic drug may be required for a few nights to restore sleep. This simple early care is best carried out as soon as possible and therefore close to the scene of the trauma. Such care can usually be best provided by personnel treating any physical injury resulting from the traumatic events, or else by other emergency workers (see below). Sometimes stressful events have to be talked about many times ('worked through') before the symptoms begin to subside. Although widely used, there is no good evidence that early conselling is effective.

Planning for disaster Planning is needed to ensure an immediate and appropriate response to the psychological effects of a major disaster. Such a response can be achieved by enrolling and training helpers who can counsel victims and are willing to be called on at short notice; and by agreeing procedures for contacting these helpers promptly. At the time of the disaster, priorities have to be decided between the needs of the victims of the disaster, those of relatives (including children), and those of members of the emergency services who may be severely affected by their experiences.

Later treatment

Counselling The treatment of established post-traumatic stress disorder is difficult. The general approach is to provide emotional support, to encourage recall of the traumatic events, and to facilitate working through the associated emotions. Treatment may also need to deal with the person's feelings of guilt about his or her perceived shortcomings in responding during the events, grief and guilt about surviving when others have died, and existential concerns about the meaning and purpose of life and death (Horowitz 1986). (Victims of personal assault or rape have additional concerns (see p. 145).)

Behavioural techniques have been used in attempts to desensitize patients to intrusive memories by encouraging them to recall these memories either while relaxing, or while becoming more anxious (flooding). These methods reduce the impact of intrusive images in some patients, but flooding may increase the impact in others (Solomon *et al.* 1992).

Psychodynamic psychotherapy has been advocated for patients with post-traumatic stress disorder (Marmar 1991). As well as helping the patients to work through traumatic events, this treatment also helps them to understand how distressing past experiences have increased vulnerability to the recent stressful events (Horowitz 1976).

Drug treatment. The results of psychological measures are often unsatisfactory, and therefore drug treatment has been tried. *Anxiolytic drugs* should be avoided for established post-traumatic stress disorder because prolonged use may lead to dependence. *Monoamine oxidase inhibitors* (Shestatzky *et al.* 1988) have been recommended for post-traumatic symptoms, but clinical trials have given contradictory results (Davidson 1992). Clinical trials of *tricyclics* have shown only a modest effect even with high doses (Davidson *et al.* 1990). Fluoxetine, a *specific serotonin uptake inhibitor*, has been found effective by Nagy *et al.* (1993) in an open trial. Without clearer guidance from clinical trials, the clinician can try the effect of a sedative antidepressant or possibly a specific serotonin uptake inhibitor if drug treatment seems an appropriate part of management.

For a review of the treatment of post-traumatic stress disorder see Solomon *et al.* (1992) and Foa *et al.* (1995).

Special forms of response to severe stress

Shell shock, battle fatigue, or war neurosis

During the First World War, psychological reactions to battle in British and American servicemen were called shell shock or battle fatigue. Some of these reactions were probably post-traumatic stress disorder, whilst others may have been anxiety or depressive disorders. Some soldiers developed a combination of anxiety attacks and concern about heart disease known as Da Costa's syndrome or disorderly action of the heart, a condition that is now diagnosed as panic disorder (see p. 176). Army psychiatrists were few in number and were unable to deal with the many cases. Moreover, their experience of mental hospital work with severely ill patients did not equip them to treat

reactions to battle. At that time some patients with shell shock were treated by neurologists, while others were cared for by psychologists with knowledge of the so-called medical psychology that had been developing before the war. W. H. Rivers, William Brown, and William McDougall were British psychologists who treated shell shock during the war, and this wartime experience led them to write influential books on medical psychology in the years after the war (Rivers 1920; McDougall 1926; Brown 1934).

At first, shell shock was treated with the methods in use at the time for neurasthenia (i.e. mild chronic anxiety and mild depressive disorders), namely rest, isolation, massage, and diet, but these methods had a low success rate. Hypnosis achieved some dramatic cures but was not generally effective. Medical psychologists tried psychotherapeutic methods advocated by Freud, including the recall of stressful events to remove repression and the expression of associated emotion. There was an increasing emphasis on early treatment, and it became evident that psychotherapy had to be combined with military drill to maintain general fitness and morale. This combined treatment led to improved results.

The general principles of early treatment, abreaction, and maintenance of fitness and morale were adopted in the Second World War and in subsequent conflicts. Abreaction with anxiolytic drugs was used widely in the Second World War (Sargant and Slater 1940). In recent conflicts it has been reported that, with immediate counselling (but not drug-induced abreaction), about 70 per cent of 'battle shock' personnel can be returned to their units within five days (Brandon 1991).

A longer account of the treatment of shell shock in the British Army in the First World War is given by Stone (1985).

Rape and physical assault

After rape or physical assault victims experience various acute reactions to stress, including anxiety and depressive disorders, psychosexual dysfunction, and post-traumatic stress disorder. The most frequent of these reactions is post-traumatic stress disorder. In one study of women victims of rape, 94 per cent had post-traumatic stress disorder immediately after the assault, and 47 per cent had the symptoms three months later (Rothbaum *et al.* 1992). In another study of rape, two thirds of women reported reduced sexual activity, whilst 40 per cent gave up intercourse or had impaired orgasm for six months (Burgess and Holmstrom 1979*a*).

In addition to experiencing symptoms of post-traumatic stress disorder, victims of rape and assault feel humiliated, ashamed, and vulnerable to further attack. They lose self-esteem, question why they were chosen as victims, and blame themselves for putting themselves in unnecessary danger (Janoff-Bulman and Frieze 1983). To these problems are added issues of betrayal and secrecy when the rapist is a family member or a friend (Nadelson 1989).

It might be expected that the psychological outcome would be related to the amount of violence involved in the rape, and to the number and kinds of sexual act; however, systematic studies have not confirmed this expectation (Calhoun and Atkeson 1991, pp. 29–34). Earlier psychiatric difficulties and inability to cope with other life problems may be related to a poor outcome of rape (Nadelson 1989).

Treatment is similar to that of other kinds of post-traumatic stress disorder; thus it includes prolonged exposure by reliving the events in imagination (Foa *et al.* 1991), but it also emphasises the overcoming of feelings of vulnerability and self-blame. As mentioned above, half the victims of rape no longer have symptoms three months after the assault; this finding indicates that early treatment should be focused on those who are most distressed. Social support is important in providing opportunities for the victim to talk over the problem and to regain self-esteem (Burgess and Holmstrom 1979*b*).

Victims of torture

Victims of torture often experience post-traumatic stress disorder as well as the physical consequences of the experience. The therapist should take special care to establish a trusting relationship and to help the victim to cope with feelings of personal humiliation and of remorse for the suffering of others. Some victims experience problems related to refugee status, separation, and bereavement. It is important to consider these other issues and not to focus narrowly on the post-traumatic stress disorder. For a review of the treatment of victims of torture see Allodi (1991).

Adjustment disorders

This term refers to the psychological reactions involved in adapting to new circumstances. Adjustment disorders are commonly provoked by life changes, such as divorce and separation, a major change of work and abode such as transition from school to university or migration, and the birth of a handicapped child. Bereavement, the onset of a terminal illness, and sexual abuse involve special kinds of adjustment that are discussed below.

Clinical features

The symptoms of an adjustment disorder include anxiety, worry, poor concentration, depression, and irritability, together with physical symptoms of autonomic arousal such as palpitations and tremor. There may be outbursts of dramatic or aggressive behaviour, single or repeated episodes of deliberate self-harm, or the abuse of alcohol or drugs. The onset is more gradual than the onset of an acute reaction to stress, and the course is more prolonged. Usually social functioning is impaired.

Stressful life events may precipitate depressive, anxiety, schizophrenic, and other psychiatric disorders; for this reason the diagnosis of adjustment disorder is not made when diagnostic criteria for another psychiatric disorder are met. Therefore, in practice the diagnosis is mostly made by excluding an anxiety or depressive disorder. A further requirement for diagnosis is that the disorder starts soon after the change of circumstances. Both ICD10 and DSMIV require that the disorder starts within three months, and ICD10 indicates that it usually starts within one month.

An essential point is that the reaction is understandably related to and in proportion to the stressful experience when account is taken of the patient's previous experiences and personality.

As explained on p. 137, in ICD10 adjustment disorders are divided into depressive reactions, mixed anxiety and depressive reactions, reactions with disturbance of other emotions, and reactions with disturbed conduct with or without emotional disturbance.

Aetiology

Stressful circumstances are the necessary cause of an adjustment disorder, but individual vulnerability must be an important cause because not all people exposed to the same stressful circumstances develop an adjustment disorder. The nature of this vulnerability is not known; it seems to vary from person to person, and probably relates in part to previous life experiences.

Prognosis

There have been no systematic studies of prognosis, but clinical experience suggests that most adjustment disorders last for several months and a few persist for years.

Treatment

Treatment is designed to help a resolution of the stressful problems if possible, and to aid the natural processes of adjustment by reducing denial and avoidance of the stressful events, by encouraging problem-solving, and by discouraging maladaptive coping responses. *Anxiety* can usually be reduced by encouraging the patient to talk about the problems and to express his feelings. Occasionally an anxiolytic or hypnotic drug is needed for a few days.

Problem-solving counselling (p. 607) encourages the patient to seek solutions to stressful problems, and to consider the advantages and disadvantages of various kinds of action. The patient is then helped to select and implement a course of action to solve the problem. If this action succeeds, another problem is considered. If the first attempt fails another approach to the original problem is tried.

Special forms of adjustment reaction

Adjustment to serious physical illness

Physical illness as a stressor

When people become physically ill, they may feel anxious, depressed, or angry. Usually this emotional reaction is transient, subsiding as the patient comes to terms with the new situation. As in other adjustment reactions, denial or minimization can protect the patient against overwhelming anxiety when the diagnosis is first known. Although helpful in this way, denial may be maladaptive in other ways; thus if denial begins before a diagnosis has been made it may lead to delay in seeking help, or if it persists long after diagnosis it may lead to poor compliance with treatment.

Physical illness as a direct cause of psychiatric symptoms

As well as acting as a stressor and causing an adjustment reaction, physical illness may induce psychiatric symptoms directly. Anxiety, depression, fatigue, weakness, weight loss, or abnormal behaviour may all be caused directly by physical disorders; common examples of these are listed in Table 6.4. Sexual function may be impaired by physical illness or its treatment (see p. 491–2). Any of these symptoms may be the reason for referral, and psychiatrists should always be alert to the possibility of undetected physical illness in their patients.

Treatment of physical illness as a cause of psychiatric symptoms

Some *drugs* used in the treatment of physical illness may affect mood, behaviour, and consciousness; those most likely to have these effects are listed in Table 6.5.

Table 6.4 *Some organic causes of common psychiatric symptoms*

Depression	Carcinoma, infections, thyroid disorder, neurological disorders including dementias, diabetes, Addison's disease, systemic lupus erythematosus
Anxiety	Hyperthyroidism, hyperventilation, hypoglycaemia, neurological disorders, drug withdrawal, phaeochromocytoma
Fatigue	Anaemia, sleep disorders, chronic infection, diabetes, hypothyroidism, Addison's disease, carcinoma, Cushing's syndrome, radiotherapy
Weakness	Myasthenia gravis and other muscle disorder, peripheral neuropathy, other neurological disorders
Episodes of disturbed behaviour	Epilepsy, hypoglycaemia, porphyria, early dementia, toxic states, transient global amnesia, phaeochromocytoma
Headache	Migraine, giant-cell arteritis, space-occupying lesions
Loss of weight	Carcinoma, diabetes, tuberculosis, hyperthyroidism, malabsorption, chronic infections including tuberculosis and HIV

The sick role and illness behaviour

The term **sick role** was suggested by Parsons (1951) to denote the social role that other people bestow on a person who is ill. According to Parsons, the sick role has four components: *exemption* from certain social responsibilities, the *right to expect help and care* from others, the *obligation* to seek and cooperate with treatment, and the expectation that the sick person will have a *desire to recover*. During physical illness, the sick role is adaptive. If it continues when disease is no longer present, recovery may be delayed; patients may continue to avoid responsibilities and may depend on others when they should be striving to help themselves. The concept of sick role is useful, but the ideas are stated in general terms and therefore may be difficult to apply specifically to an individual case.

Mechanic (1978) suggested the term **illness behaviour** for behaviour associated with physical or mental disorder, whether adaptive or not. Illness behaviour includes *consulting doctors, taking medicines, seeking help* from relatives and friends, and *giving up* inappropriate activities. These behaviours are adaptive in the early stages of illness, but may become maladaptive if they persist into the stage of convalescence when the patient should be becoming independent of others. Illness behaviour results from the person's appraisal of his illness rather than the objective presence of disease, and it may develop when no physical disease is present. Illness behaviour without disease is an important problem in general

Table 6.5 *Drugs that may cause psychiatric symptoms*

Delirium	Central nervous system depressants (hypnotics, sedatives, alcohol, antidepressants, neuroleptics, anticonvulsants, antihistamines), anticholinergic drugs, beta-blockers, digoxin, cimetidine
Psychotic symptoms	Hallucinogenic drugs, appetite suppressants, sympathomimetic drugs, beta-blockers, corticosteroids, L-dopa, indomethacin
Depression	Antihypertensive drugs, oral contraceptives, neuroleptics, anticonvulsants, corticosteroids, L-dopa
Elation	Antidepressants, corticosteroids, anticholinergic drugs, isoniazid
Behavioural disturbance	Benzodiazepines, neuroleptics

practice, and once firmly established it is difficult to treat.

Quality of life

Quality of life is a general term applied to the totality of physical, psychological, and social functioning. It is determined by physical impairment, emotional reaction, personality, illness behaviour, and sick role. Illness usually impairs quality of life but occasionally enhances it, for example as a result of changing excessive commitments to work or excessive use of alcohol. Quality of life is difficult to measure since it covers so many aspects of functioning that rating scales must be either very lengthy or, if brief, very general in their questioning. For a review of measures of quality of life see Fitzpatrick *et al.* (1992) and Fletcher *et al.* (1992).

Treatment

Most people adjust well to physical illness, but when adjustment is slow and incomplete psychological treatment may be needed. This treatment need not be complicated and can usually be provided effectively by the general practitioner or hospital doctors or nurses dealing with the physical illness. Generally the psychiatrist has a role in treating only the most severe problems or in supporting the medical and nursing staff. *Counselling* requires a trusting relationship with the patient. To establish this relationship there should be adequate time for interviews. Counselling begins with an explanation of the nature of the illness and its treatment; the patient is then helped to accept the implication of the diagnosis, to adjust to illness, and to give up any maladaptive behaviours such as excessive dependence on others or denial of the need for treatment.

If the reaction to physical illness is a depressive or anxiety disorder, treatment appropriate to the disorder should be given (see pp. 228 and 168).

Adjustment to terminal illness

Amongst patients dying in hospital, about half have emotional symptoms of anxiety, depression, anger, or guilt. **Anxiety** may be provoked by the prospect of severe pain, disfigurement, or incontinence, by fear of death, and by concerns about the future of the family. Families and carers sometimes try to spare the patient anxiety by concealing the truth about the condition. Since most patients become aware of the diagnosis, attempts at concealment only increase their fear of possible consequences of the disease such as pain or incontinence. **Depression** may be provoked by the prospect of separation from family and friends and the loss of valued activities. Changes in physical appearance may be caused by the illness, the effects of surgery, and the debilitating effects of radiotherapy, and are important causes of low mood. Depressive disorders develop in 5–15 per cent of patients (Cody 1990). Some patients experience **guilt** because they believe that excessive demands are being placed on relatives or friends, whilst patients with religious beliefs may believe that illness is a punishment for wrongdoing. **Anger** may be felt about the unjustness of impending death; it may be displaced onto doctors, nurses, and relatives, making care more difficult. Understandably, among dying patients these emotional reactions are more common in the young than in the elderly, and less common among those who believe in an afterlife. Determinants of emotional reactions include the patient's personality, and the amount and quality of support from family, friends, and carers.

Kubler-Ross (1969) was one of the first investigators to study in detail the problems and needs of the dying. She described five phases of psychological adjustment to death. The phases do not necessarily occur in the same sequence, and some may not occur at all, but they are a useful guide. The phases are (a) *denial* and isolation, (b) *anger*, (c) *'bargaining for time'* (partial acceptance), (d) *depression*, and (e) *acceptance*.

The experience of terminal illness may provoke the defence mechanisms of denial, dependency, and displacement. **Denial** is usually the first reaction to the news of fatal illness. It may be experienced as a feeling of disbelief and may lead to an initial period of calm. Denial diminishes as the patient becomes reconciled to the illness. Denial may return as the disease progresses, and the patient may again behave as if unaware of the nature of the illness. Some **dependency** is common among ill people and is adaptive in the early stages of medical treatment when the patient needs to comply passively with treatment. Excessive or prolonged dependency makes subsequent stages of treatment more difficult, and increases the burden on the family. There may be **displacement** of anger onto staff and relatives, who may not understand this reaction, may find it difficult to tolerate, and may then be less inclined to spend time with the patient, thereby increasing his feelings of despair. Denial, dependency, and displacement are usually followed by **acceptance**. The doctor's aim should be to help the patient to reach this acceptance before the final stage of the illness. This aim is more likely to be achieved when there is good communication between patients, the staff caring for them, and relatives.

Adjustment to terminal illness is made more difficult by psychological symptoms induced by the disease or its treatment. The associations between disease and psychological symptoms are summarized in Table 6.4. There is a particularly strong association between dyspnoea and anxiety. Acute organic syndromes are common. The associations between drug treatment and psychological symptoms are summarized in Table 6.5.

Treatment

According to Hackett and Weissman (1962) the aim of treating the dying patient should be to achieve an 'appropriate death'. This term means that 'the person should be relatively free from pain, should operate on as effective a level as possible, should recognize and resolve remaining conflicts, should satisfy as far as possible remaining wishes, and should be able to yield control to others in whom he has confidence' (Hackett and Weissman 1962). Usually medical and nursing staff help the patient to achieve these aims, and psychiatrists are called upon only when there are special problems (see below).

The first step in treatment is to reduce symptoms as much as possible. Adequate *control of pain and breathlessness* and the *reduction of confusion* are particularly important. Anxiety and depression may diminish as pain and breathlessness are controlled. The causes of confusion are listed on p. 312; among dying patients important remediable causes are dehydration, the side-effects of drugs, secondary infection, cardiac or respiratory failure, and hypercalcaemia (Stedeford and Regnard 1991).

The next step is to establish a good relationship with the patient and to enable him to talk about his problems and to ask questions. A straightforward *explanation* is given of the nature of the illness. Sometimes doctors are apprehensive that such an explanation will increase the patient's distress. Although excessive detail given unsympathetically can have this effect, it is seldom difficult to decide how much to say about diagnosis and prognosis provided that patients are allowed to lead the discussion, express their worries, and say what they want to know. If patients ask about the prognosis, they should be told the truth; evasive answers undermine trust in the carers. If patients do not communicate a desire to know the full extent of their problems, it is usually better to withhold this information until later. At an appropriate stage patients should be told what can be done to make their remaining time as comfortable as possible. While the whole account should be truthful, the amount disclosed on a single occasion should be judged by patients' reactions and by their questions. If necessary, the doctor should be prepared to return for further discussion when patients are ready to continue. It is important to bear in mind that most dying patients become aware of their prognosis whether or not they are told directly, because they infer the truth from the behaviour of those who are caring for them. They notice when answers to questions are evasive and when people avoid talking to them. Patients who are anxious, angry, or despairing need to be able to express these feelings and to discuss the ideas that induce them.

The information given to the patient should be known to all the staff and to the relatives, otherwise conflicting advice and opinions may be given. If all those involved know what has been said, they will feel more at ease in talking to the patient. Otherwise they will draw back from the patient, isolating him and increasing his difficulty in adjusting.

Relatives often need as much help as the patient because of their own anxiety and depression. They may respond with guilt, anger, or denial which may make it difficult for them to communicate helpfully with the patient or the staff. They need information and opportunities to talk about their feelings and to prepare for the impending bereavement. Without these steps an increasing alienation may develop between patients and their families.

In many hospitals there are specialized nurses who work with the family doctor and with the hospital staff caring for dying patients. These nurses are skilled in the psychological as well as the physical care of the dying (Corr and Corr 1983). Sometimes care is provided in special hospices where it is possible to provide close attention to the details that can improve quality of life for the dying person. These hospices care for the patient when home care is impractical, and also provide periods of respite care for the patient in order to relieve those who are caring for the patient at home.

Referral to a psychiatrist is appropriate when psychiatric symptoms or behaviour disturbance are severe (the cause is often an acute organic syndrome). Studies have found that about 10–15 per cent of patients in terminal care units were referred to a psychiatrist (Stedeford and Bloch 1979; Ramsay 1992). The referrals were for the following main types of problem: (i) assessment of low mood to determine whether it was part of a depressive disorder requiring treatment; (ii) difficulty of the patient in talking to the staff or the relatives about the illness; (iii) difficulties in accepting social restrictions, making appropriate plans, and taking decisions about the illness and everyday life; (iv) difficulties in accepting the effects of the illness on the pattern of family life; (v) long-term personality and family problems that were increased by physical illness. Anxiety and delirium are common, and are often dealt with appropriately by medical staff rather than referred to a psychiatrist. The exception is *delirium with paranoid symptoms*, for which specialist help is usually requested (Ramsay 1992).

A **depressive disorder** may be caused by pain, breathlessness, or confusion, all of which should be treated appropriately. Any drugs that can cause depression (see Table 6.5) should be reviewed and if possible given in lower dose or replaced. Some symptoms of depressive disorder are difficult to evaluate in patients with advanced cancer; thus weight loss, anorexia, insomnia, loss of interest,

Table 6.6 *Normal grief reaction*

Stage I	Hours to days Denial, disbelief 'Numbness'
Stage II	Weeks to 6 months Sadness, weeping, waves of grief Somatic symptoms of anxiety Restlessness Poor sleep Diminished appetite Guilt, blame of others Experience of presence Illusions, vivid imagery Hallucinations of the dead person's voice Preoccupation with memories of deceased Social withdrawal
Stage III	Weeks to months Symptoms resolve Social activities resumed Memories of good times (symptoms may recur at anniversaries)

and fatigue may be caused by the physical illness. Early morning wakening, extreme hopelessness, and self-blame are surer guides to diagnosis. Suicidal ideation should be assessed carefully. If counselling and improved medical management do not improve the low mood, antidepressant drugs should be prescribed with careful supervision. Among physically ill patients, tricyclic antidepressants may induce side-effects including delirium, nausea, and urinary retention; hence the starting dose should be small, and medication should be changed if necessary to find a compound that is well tolerated (see Haig (1992) for a review).

Liaison with medical staff is important. Often medical staff can provide treatment when the psychiatrist has formulated a plan. For reviews of the care of the dying see Cassidy (1986) and Patterson (1977).

Grief: the adjustment to bereavement

Normal grief is a continuous process, but it can be described conveniently as having three stages (Table 6.6). The **first stage** lasts from a few hours to several days. There is denial, which is manifested as a lack of emotional response ('numbness'), often with a feeling of unreality, and incomplete acceptance that the death has taken place. The bereaved person may be restless, as if searching for the dead person.

The **second stage** usually lasts from a few weeks to about six months but may be much longer. There may be extreme sadness, weeping, loneliness, and often overwhelming waves of yearning for the dead person. Anxiety is common; the bereaved person is anxious and restless, sleeps poorly, lacks appetite, and may experience panic attacks. Many bereaved people feel *guilt* that they failed to do enough for the deceased. Some feel anger and project their feelings of guilt, blaming doctors or others for failing to provide optimal care for the dead person. Many bereaved people have a vivid experience of being in the presence of the dead person, and about one in ten experience brief hallucinations (Clayton 1979). The bereaved person is preoccupied with memories of the dead person, sometimes in the form of intrusive images. Withdrawal from social relationships is frequent. Complaints of physical symptoms are common (Parkes and Brown 1972), and widows seek medical care more often than comparable people who are not bereaved (Stein and Susser 1969).

In the **third stage** these symptoms subside and everyday activities are resumed. The bereaved person gradually comes to terms with the loss and recalls the good times shared with the deceased in the past. Often there is a temporary return of symptoms on the anniversary of the death.

Abnormal or pathological grief

The term abnormal grief reaction is applied if the symptoms are more *intense* than usual and meet the criteria for a depressive disorder, if they are *prolonged* beyond six months, or if they are *delayed* in onset.

Abnormally intense grief Up to 35 per cent of bereaved people meet the criteria for a **depressive disorder** at some time during grieving (Clayton *et al.* 1974; Zisook and Schuchter 1993). Most of these depressive disorders resolve within six months but about 20 per cent persist for longer (Jacobs 1993). People who meet the criteria for a depressive disorder are more likely to have poor social adjustment, to visit doctors frequently (Zisook and Shuchter 1993), and to use alcohol (Clayton and Darvish 1979). Therefore, it is of practical value to record the additional diagnosis of a depressive disorder when the criteria for this disorder are met. When there is doubt whether depressive disorder should be recorded, particular attention should be paid to symptoms of retardation and global loss of self-esteem (clearly greater

than regret about omissions of care during the terminal illness) because these features are seldom present in uncomplicated grief (Clayton *et al.* 1974; Jacobs *et al.* 1989).

Suicidal thoughts may occur when grief is intense. The rate of suicide is increased most in the year after bereavement, but continues to be high for five years after the death of a spouse or parent. Young widows and elderly widowers are at higher risk than other bereaved people (Mac-Mahon and Pugh 1965; Bunch 1972). The presence of suicidal ideas should prompt appropriate assessment of suicide risk (see p. 421).

Prolonged grief This is often defined as grief lasting for more than 6 months. However, it is difficult to set such a definite limit to normal grief, and complete resolution may take much longer. One study found that only a minority of widows had completely ceased to grieve a year after the death (Parkes 1971). Instead of the normal progression, symptoms of the first and second stages persist. Such prolongation may be associated with a depressive disorder but can occur without such a disorder.

Delayed grief Parkes (1965) described delayed grief as a form of pathological guilt. By convention it is said to occur when the first stage of grief does not appear until more than *two weeks* after the death. It is said to be more frequent after sudden, traumatic, or unexpected deaths (Jacobs 1993, p. 175).

Inhibited and distorted grief. Deutsch (1937) suggested that *absence of grief* is a pathological variant of grieving. The term *inhibited grief* refers to a reaction which lacks some normal features. *Distorted grief* refers to features (other than depressive symptoms) that are either unusual in degree, for example marked hostility, overactivity, and extreme social withdrawal, or else unusual in kind, for example physical symptoms that were part of the last illness of the deceased. These distorted presentations were described by Lindemann (1944) in a study of survivors of a fire in a nightclub.

In all these forms of grief, persistent avoidance of situations and of other reminders of death are common.

The mortality of bereavement

Several studies (reviewed by Stroebe and Stroebe (1993)) have shown an increased rate of mortality among bereaved spouses and other close relatives, with the greatest increase being in the first six months after bereavement. Most studies report increased rates of death from heart disease, and some have reported increased rates of death from cancer, liver cirrhosis, suicide, and accidents. The reasons for these associations are uncertain, and are likely to be different for different conditions.

Causes of pathological grief

Abnormal grief reactions are more likely in the following circumstances:

- when the death was sudden and unexpected;
- when the bereaved person had a very close, or dependent, or ambivalent relationship with the deceased (the role of relationships was emphasized by Freud (1917) in a much quoted paper on 'Mourning and melancholia');
- when the survivor is insecure, or has difficulty in expressing feelings, or has suffered a previous psychiatric disorder;
- when the survivor has to care for dependent children and so cannot show grief easily.

It might be expected that lack of social support would be a cause of abnormal grief, but the available research evidence does not support this idea (Jacobs 1993, Chapter 6) even though social support certainly assists people who are bereaved (see Parkes (1985) or Stroebe and Stroebe (1993) for a review).

Management

Although bereaved people have some problems in common, they also have problems that are individual. Thus a young widow with small children has many difficulties that are not shared by an elderly widow whose adult children can support her. A mother grieving for a stillborn child will have special problems (discussed below). In planning management it is important to take into account the individual circumstances of the patient as well as the general guidelines outlined below.

Counselling for the bereaved follows the general rules that apply to other kinds of adjustment reaction. The bereaved person needs to talk about the loss, to express feelings of sadness, guilt, or anger, and to understand the normal course of grieving. It is helpful to forewarn a bereaved person about unusual experiences such as feeling as if the dead person were present, illusions, and hallucinations, otherwise these experiences may be alarming. Help may be needed:

- to accept that the loss is real
- to work through the stages of grief
- to adjust to life without the deceased.

The bereaved person may need help to progress from the first stage of denial of loss to the acceptance of reality. Viewing the dead body and putting away the dead person's belongings help this transition, and a bereaved person should be encouraged to perform these actions. Practical problems may need to be discussed, including funeral arrangements and financial difficulties. A young widow may need help in maintaining and caring for young children, and in supporting them without inhibiting her own grief excessively. As time passes, the bereaved person should be encouraged to resume social contacts, to talk to other people about the loss, to remember happy and fulfilling experiences that were shared with the deceased, and to consider positive activities that the latter would have wanted survivors to undertake. (For further information about grief counselling see Worden (1991).)

Parents **grieving for a stillborn child** need special help. They should be encouraged to name the dead baby and to view the body. If they do not feel able to take these steps, it is often helpful to obtain a photograph of the body which the parents can see later if they wish. If these steps are combined with counselling, the mothers of stillborn children experience less distress than those not given this help (Forrest and Standish 1984).

Drug treatment cannot remove the distress of normal grief, but it can relieve severe anxiety or depression that is distressing and may be interfering with social adjustment. In the first stage of grief, a *hypnotic* or *anxiolytic* drug may be needed for a few days to restore sleep or to relieve any severe anxiety. In the second stage *antidepressant* drugs may be beneficial if the criteria for depressive disorder are met, though such usage has not been evaluated in this special group.

Support groups have been developed to help recently bereaved people, particularly young widows. One such organization in the United Kingdom is known as CRUSE. By sharing their experience with others who have dealt successfully with bereavement, recently bereaved people can share grief, obtain practical advice, and discuss ways of coping. In one study a support group was found to be as effective as brief psychotherapy, although more people dropped out of the support group (Marmar *et al.* 1988); in another study of support groups the benefit was small (Barrett 1978).

Psychotherapy: it is not practical, nor is there evidence that it is helpful, to provide psychotherapy for all bereaved persons. Several clinical trials suggest that some kind of psychotherapy may be helpful when the person is at high risk of an abnormal grief reaction. Thus Raphael (1977) found that *crisis intervention* benefited a group of people selected for their high risk of abnormal grief, but Polak *et al.* (1975) did not find this treatment beneficial for unselected grieving people.

When Marmar *et al.* (1988) studied *brief dynamic psychotherapy*, they found it no more effective than a mutual support group. Similarly, Lieberman and Yalom (1992) found no significant difference in outcome between bereaved spouses treated with group psychotherapy and a control group who were not treated.

Avoidant behaviour can be reduced by *guided mourning*, a behavioural treatment in which the bereaved person is helped to confront memories of the dead person and to enter situations that provoke these memories (a form of exposure treatment, see p. 631). In a controlled evaluation, such an approach produced modest benefit (Mawson *et al.* 1981). When exposure treatment was compared with a form of anxiety management ('stress inoculation training') the former produced better long-term results, while the latter led to more immediate relief of symptoms (Foa *et al.* 1991).

For a fuller description of the psychological treatment of grief, see Parkes and Weiss (1983) or Jacobs (1993).

Long-term adjustment to sexual abuse in childhood

When sexually abused, children may experience anxiety, depression, and post-traumatic stress disorder. These effects usually subside during childhood, but people who have been abused in childhood appear to be more vulnerable than others to psychiatric disorder in adult life. Also, sexual abuse in childhood may be followed by persistent low self-esteem and psychosexual difficulties whether or not a psychiatric disorder develops. Among some people who were previously unaware of sexual abuse, there may be an experience of a sudden, vivid, and disturbing recall of such abuse. Sometimes this recall occurs during counselling given for another purpose, and sometimes it is provoked by an event that is a reminder of the relevant period of childhood. In recent years an increasing number of adult patients have reported childhood sexual abuse to doctors and others. Some of this increase is likely to result from publicity that has reduced unwillingness to report the problem. It has been suggested that some of the reports are not of actual events but of childhood fantasies remembered vividly.

Epidemiology

There have been no satisfactory prospective studies of people who were sexually abused as children, but people who report sexual abuse may have higher rates of psychiatric disorder in adult life (Beitchman *et al.* 1992). Most of the retrospective information is about the effects on women. Female psychiatric patients are more likely than healthy controls to report sexual abuse in childhood, and such reporting is particularly frequent in those with eating disorders (Palmer *et al.* 1992), somatization disorder (Morrison 1989), borderline personality disorder (Bryer *et al.* 1987), multiple personality disorder (Putnam *et al.* 1986), and sexual dysfunction (Briere 1988). It is not clear what proportion of women who were sexually abused in childhood develop these disorders in adult life, but some make a good adjustment.

Aetiology

There could be three explanations for an association between the reporting of childhood sexual abuse and the symptoms of psychiatric disorder in adult life. First, people with psychiatric disorder may be more likely than controls to report childhood sexual abuse, perhaps because they have been asked questions about their childhood in the course of psychiatric assessment. Second, childhood sexual abuse may be a direct cause of vulnerability to adult psychiatric disorder. Third, sexual abuse may be a marker of some other factor, such as disturbed relationships within the family, which is the real cause of the excess psychiatric disorder in adult life.

It seems unlikely that the association can be explained solely by the greater recall of sexual abuse by women who have psychiatric disorder because community studies have also found an association between the reporting of childhood sexual abuse and the reporting of psychiatric symptoms (Bushnell *et al.* 1993). There is some evidence in favour of the second explanation (a direct causal relationship). Thus, in another community survey, Mullen *et al.* (1993) found higher rates of psychiatric disorder among women reporting severe abuse involving penetration than among women reporting less severe abuse, a finding that suggests a causal relationship between abuse and subsequent disorder.

It appears that, whilst extreme forms of abuse can increase vulnerability, in other cases an association between childhood abuse and adult psychiatric disorder may be explained in part by disturbed relationships in the family of the abused child. Clinical observations suggest that effects are more severe when the abusing person is a parent.

From the study by Mullen *et al.* (1993) there is evidence that the disturbed social relationships in incestuous families are themselves causes of subsequent disorder. These investigators used statistical methods to correct for the influence of family factors on the association between child abuse and adult disorder. They found that, with less severe forms of abuse, the relationship could be accounted for by the family factors alone; it was only when abuse was severe that it increased vulnerability. Also, people who have been abused as children are more likely to report to others that their parents were uncaring or emotionally distant (Alexander and Lupfer 1987).

Treatment

Several forms of psychological treatment have been used to treat the late effects of childhood sexual abuse, including counselling about current problems, dynamic psychotherapy, and cognitive therapy. Group treatment has been used as well as individual therapy.

The various methods have several common features. The first is the development of a therapeutic relationship in which the patient can trust the therapist and feel understood and respected. Second, the abused woman should be allowed to set the pace at which she talks about the experience of being abused. Otherwise, the present adjustment may be overwhelmed by an extreme emotional response to the memories of abuse, and the patient may withdraw from treatment. Although this procedure is probably necessary, particularly when the patient is experiencing vivid flashbacks or dreams of the experience, before long the focus of treatment should move to present problems of adjustment and to ways in which they can be improved.

In helping with current difficulties, particular attention is given to eliciting any avoidance of problems in relationships and to encouraging the patient to confront such issues directly. Attempts are made to increase self-esteem and, when appropriate, to help the patient to find constructive ways of expressing feelings of anger. Some patients need help with psychosexual problems. From time to time the patient may need to be reminded that, although the problems originated in earlier experience, the task is to improve adjustment to present life circumstances. The main difference between dynamic treatment and cognitive behavioural treatment is the greater emphasis given in the former to understanding the effects of the trauma on self-esteem and emotional expression, and the greater emphasis given in the latter to more precise specification of ways in which current patterns of thinking affect present behaviour. When treatment is in a group, it is important to emphasize these constructive aims, while encouraging appropriate personal revelation and support from other group members. For a review of psychological treatment of the long-term effects of sexual abuse see Hobbs (1994) or Paddison (1993).

Minor affective disorders

Anxiety and depressive symptoms often occur together. Indeed, earlier writers, including Mapother (1926) and Lewis (1956), considered that anxiety and depressive disorders could not be separated clearly in patients admitted to hospital with severe disorders. Although most psychiatrists now accept that the distinction can usually be made among the more severe forms presenting in psychiatric practice, the distinction is less easy to make in the milder forms presenting in primary care.

Classification

As psychiatrists work increasingly with general practitioners, the importance of minor anxiety–depressive disorders has been recognized but without any agreement about classification.

ICD10 includes a category of 'mixed anxiety and depressive disorder' which can be applied when neither anxiety symptoms nor depressive symptoms are severe enough to meet criteria for an anxiety disorder or a depressive disorder, and when the symptoms do not have the close association with stressful events or significant life changes required for a diagnosis of acute stress reaction or

adjustment disorder. According to ICD10, patients with this presentation are seen frequently in primary care and there are many others in the general population who are not seen by doctors. In ICD10 this diagnosis appears amongst anxiety disorders, although some psychiatrists consider that the condition is more closely related to the mood disorders, a view that is reflected in the alternative term minor affective disorder.

In DSMIV no comparable diagnosis appears in the classification. The appendix to the classification contains two provisional diagnoses that might be used for these cases: mixed anxiety and depressive disorder, and minor depressive disorder. It is stated that there is insufficient factual information to justify the inclusion of either in the classification. Although little is known about these conditions or about their relationship to other disorders, patients present to doctors with this group of symptoms. A suitable category is needed even if it is not possible to write strict criteria for diagnosis. In this book the term minor affective disorder has been chosen as a purely descriptive term that does not imply a relationship to anxiety disorders (as is implied in ICD10 by placing the term anxiety–depressive disorder under the broad heading of anxiety disorder) or to depressive disorders (as is implied by the term minor depressive disorder). For the same reason these conditions are discussed in this chapter (following reactions to stressful experiences), rather than in the chapter on anxiety disorders or the chapter on mood disorders. The title of the chapter separates them from reactions to stress because, as explained above, cases which are closely related to stressful events or significant life changes are excluded from this group.

Clinical picture

One of the best descriptions has been given by Goldberg *et al.* (1976) who studied 88 patients from general practice in Philadelphia. The most frequent symptoms are listed in Table 6.7. They include fatigue, anxiety, depression, irritability, poor concentration, and insomnia. *Somatic symptoms* were present in about half the cases and concern with bodily functions in about a quarter.

Table 6.7 *Relative frequency of 12 common symptoms in 88 patients diagnosed as having mental disorder in general practice*

Anxiety and worry	82
Despondency, sadness	71
Fatigue	71†
Somatic symptoms*	52
Sleep disturbance	50
Irritability	38
Excessive concern with bodily function	27
Depressive thoughts, inability to concentrate	21
Obsessive and compulsions	19
Phobias	11
Depersonalization	6

*Only those precipitated, exacerbated, or maintained by psychological factors.
†Not all the same patients who complain of despondency and sadness.
Source: Goldberg *et al.* 1976.

The reason is uncertain; some symptoms are autonomic features of anxiety, and it is possible that patients expect somatic complaints to be received more sympathetically than emotional problems. Whatever the reasons, the observation is not new (see for example the lengthy descriptions of physical symptoms of neurosis in Déjerine and Gauckler (1913).

In the patients studied by Goldberg *et al.* (1976), complaints of sleep disturbance were also common, particularly difficulty in falling asleep and restlessness during the night. (Complaints of early waking suggest that the condition may be the early stage of a depressive disorder requiring antidepressant medication.) About a fifth of the patients complained of obsessional thoughts and mild compulsions. Definite phobic symptoms were less common than either of these, although mild phobias are of course very common among normal people. Complaints of fatigue and irritability were also frequent (Table 6.3), and were often accompanied by poor concentration and lack of enjoyment. As mentioned earlier, these symptoms were grouped together as neurasthenia in the past. Presentation with somatic complaints is especially common in both Western and non-Western cultures, and this is one reason why family doctors fail to detect some psychiatric disorders (Bridges and Goldberg 1985).

The more frequent somatic complaints of patients with minor neurotic disorders are as follows. Complaints related to the digestive system include feelings of abdominal discomfort or distension, and preoccupations with the effects of certain foods in producing indigestion or flatulence. There may also be complaints of poor appetite, nausea, epigastric pain, weight loss, or difficulty in swallowing, and discomfort in the left iliac fossa. Complaints related to the cardiovascular system include palpitations, precordial discomfort, and worries about heart disease. Other complaints include aching in the neck, shoulders, and back. Headaches are commonly described as tightness and pressure, or a dull constant ache, or throbbing. Pain may have other regional localization. The pattern of symptoms differs in different cultures and in the same culture at different times. See Ndetei and Muhangi (1979) for a description of physical symptoms among Africans with minor affective disorder, and Déjerine and Gauckler (1913) for a historical account.

All these physical symptoms require a thorough investigation for physical causes because they they may originate in physical disease while the anxiety and depression is secondary to this state. Thorough physical investigation is specially important when the complaint is of difficulty in swallowing or of weight loss, because these symptoms are particularly likely to have a physical cause. For these reasons the diagnosis of minor affective disorder requires a wide range of clinical skills.

Epidemiology

Estimates of the frequency of minor affective disorders show wide variations because the different studies have not defined cases in the same way. Minor affective disorders and other neurotic disorders together form about two-thirds of psychiatric cases seen in general practice (Shepherd *et al.* 1966). Estimates of prevalence vary from 18 per 1000 for men and 27 per 1000 for women (in a Danish survey by Fremming (1951)) to 79 per 1000 for men and 165 per 1000 for women (in a Swedish investigation by Hagnell (1966)). One-year prevalence rates vary even more (Carey *et al.* 1980).

Prognosis

Among people aged 20–50 identified in community surveys, about half recover in three months (Hagnell 1970; Tennant *et al.* 1981*a*). Goldberg and Huxley (1980, p. 104) calculated from the data of Harvey Smith and Cooper (1970) that recent onset cases seen in general practice have a turnover of 70 per cent per year and chronic cases have a turnover of 3 per cent per year. In one study about half improved in three months and about a further quarter within six months (Catalan *et al.* 1984). The rest either persist as mixed anxiety–depressive disorders or continue to have social disability (Ormel *et al.* 1993). Among those with depressive symptoms, this disability is comparable to that of patients with chronic medical disorder (Wells *et al.* 1989).

Assessment

When a patient with minor symptoms of both anxiety and depression is assessed it is essential to make sure that symptoms of a more severe disorder have not been missed. These other disorders include anxiety disorder, depressive disorder, dementia, the early stages of schizophrenia, and undeclared abuse of alcohol or drugs. Physical disease should also be considered carefully as a primary cause of the symptoms. The relative probability of these disorders varies according to the age of the patient. Even if there is no evidence of a primary cause when the patient is first examined, it should always be considered again if there is no improvement in the symptoms after adequate treatment. In weighing up the likelihood of physical disease, it should always be remembered that stressful events are commonplace and their presence does not exclude the possibility of primary organic disease. This is particularly important in middle-aged patients who have not had anxiety–depressive symptoms before.

The search for organic disease requires a thorough history, an appropriate physical examination, and investigations. What is appropriate varies with the age of the patient, the nature of the symptoms, and any clues from the history. If, after appropriate investigation, a degree of uncertainty remains, it is important to make a note of it; the diagnosis should be recorded as provisional and subject to review after a suitable interval.

Treatment

For patients with less severe symptoms, brief counselling is as effective as drugs; 15 minutes per week for a few weeks is usually sufficient (Catalan *et al.* 1984). When somatic symptoms are present, a clear explanation should be given of their nature and causation, and the doctor should find out whether the patient has any special reason to be worried about their significance. For example a patient with palpitations may fear heart disease because other members of the family have had this condition.

Whenever possible, **problems** should be solved by the patient and not by other people, though relatives should be encouraged to play a part when appropriate. Problem-solving counselling (see p. 607) has been shown to be effective in these cases. However, when problems are overwhelming or prolonged, the help of a doctor, nurse, or social worker may be needed. Thus Shepherd *et al.* (1979) found that social work was effective in two-thirds of chronic neuroses seen in general practice. Whatever the approach, patients should be encouraged to take a part in identifying the relevant problems, considering what can be done about each one, and deciding the order in which they should be tackled. In this way they will be better equipped to help themselves in the future. When problems cannot be resolved the patient should be helped to come to terms with them.

Although many patients who develop minor affective disorders have only temporary problems, others have more prolonged **social difficulties**. Some lack friends in whom they can confide, or have few enjoyable activities. Such people should be encouraged to join a club or, in the case of a housewife who is lonely at home, take part-time paid or voluntary work. For most patients, such normal pursuits are better than those involving other people who are ill.

Anxiolytic drugs should be reserved for patients with severe anxiety symptoms, and prescribed for no more than six weeks (because of the risk of dependency, see p. 544). *Antidepressant drugs* should be kept for patients with more severe or persistent symptoms.

Further reading

Calhoun, K. S. and Atkeson, B. M. (1991). *Treatment of rape victims. Facilitating psychosocial adjustment.* Pergamon Press, New York.

Charney, D. S., Deutsch, A. Y., Krystal, J. H., Southwick, S. M., and Davis, M. (1993). Psychobiologic mechanisms of posttraumatic stress disorder. *Archives of General Psychiatry*, 50, 294.

Davidson, J. R. T. and Foa, E. A. (1992). *Posttraumatic stress disorder: DSMIV and beyond.* American Psychiatric Press, Washington, DC.

Shepherd, M., Cooper, B., Brown, A. C., and Kalton, G. (1981). *Psychiatric illness in general practice* (2nd edn with new material by M. Shepherd and A. Clare). Oxford University Press.

Stroebe, M. S., Stroebe, W., and Hansson, R.O. (1993). *Handbook of bereavement.* Cambridge University Press.

7 | *Anxiety, obsessive–compulsive, and dissociative disorders*

Anxiety, obsessional, and dissociative disorders are sometimes referred to as neurotic disorders, a term that denotes conditions which are not caused by organic disease of the brain and, however severe, do not involve hallucinations and delusions. The term neurosis (or neurotic disorder) has fallen out of favour recently because it seems better to consider the individual features of these three disorders rather than their common factors. The term neurosis is not used as an organizing principle in modern systems of classification, though it appears in ICD10 as part of a wider grouping of 'Neurotic, stress-related, and somatoform disorders'.

Table 7.1 shows how anxiety, obsessional, and dissociative disorders are classified in DSMIV and ICD10. The two classifications are broadly similar. The first important difference is that in DSMIV obsessive–compulsive disorders are regarded as a type of anxiety disorder, while in ICD10 they are classified separately. The second difference is that in DSMIV depersonalization disorder is regarded as a type of dissociative disorder, while in ICD10 it is classified separately. In DSMIV acute stress disorder and post-traumatic stress disorder are classified as anxiety disorder, but in this book they are considered in Chapter 6. These and other differences between the two classifications will be explained further when each group of disorders is considered later in this chapter.

Anxiety disorders

Anxiety disorders are abnormal states in which the most striking features are mental and physical symptoms of anxiety (these symptoms are described on p. 163) which are not caused by organic brain disease or another psychiatric disorder. Anxiety disorders are divided as follows:

(i) generalized anxiety disorders in which anxiety is unvarying and persistent;
(ii) phobic anxiety disorders in which anxiety is intermittent and arises in particular circumstances;
(iii) panic disorder in which anxiety is intermittent and unrelated to particular circumstances.

The development of ideas about anxiety disorders

Anxiety has long been recognized as a prominent symptom of many psychiatric disorders. It has been particularly associated with the symptom of depression, and until the last part of the nineteenth century anxiety disorders were not classified separately from other disorders of mood. For many years the less severe cases were referred to as neurasthenia, a term suggested by the American physician Beard in 1880. It was Freud (1895*b*) who first suggested that cases with mainly anxiety symptoms should be separated from neurasthenia under the name of anxiety neurosis.

Freud's original anxiety neurosis included patients with phobias and panic attacks, but subsequently he divided it into two groups. The first, which retained the name anxiety neurosis, was a group of cases with mainly mental symptoms; the second group, which he called anxiety hysteria, was for cases with physical symptoms of anxiety and phobias including those we now call agoraphobia, a term originally proposed by Westphal (see below). Freud originally proposed that the causes of anxiety neurosis and anxiety hysteria were related to sexual conflicts, though he later broadened his view of the causes. Many psychiatrists, including those who

Table 7.1

DSMIV*	ICD10
Anxiety disorders	
Phobic anxiety disorders	F40 Phobic anxiety disorder
Panic disorder	
Generalized anxiety disorder	}F41 Other anxiety disorder
Obsessive–compulsive disorder	F42 Obsessive–compulsive disorder
Acute stress disorder	
Post-traumatic stress disorder	}F43 Reactions to severe stress and adjustment disorder
Dissociative disorders	F44 Dissociative (conversion) disorders
Dissociative amnesias	
Dissociative fugue	
Dissociative identity disorder	
Depersonalization disorder	F48 Other neurotic disorders (includes depersonalization disorder)

*The order has been changed to facilitate comparison of the two systems.

followed the teaching of Adolf Meyer (see p. 78), considered that a very wide range of stressful problems could cause anxiety neurosis (for example, Henderson and Gillespie 1930, pp. 416–17).

Phobic disorders have been recognized since antiquity, but the first systematic medical study of these conditions was probably that of Le Camus in the eighteenth century (Errera 1962). In the early nineteenth century the psychiatric classifications included phobias with the monomanias, which were disorders of thinking rather than emotion. Later, Westphal (1872) studied phobias of open spaces and stressed the role of anxiety in the condition that he named agoraphobia.

In 1895 Freud proposed that two groups of phobias could be recognized: common phobias, with an exaggerated fear of things that are commonly feared such as darkness and solitude, and specific phobias, i.e. the fear of situations not feared by healthy people such as open spaces (Freud 1895a, pp. 135–6).

These views on phobias were accepted widely until the 1960s when the use of behaviour therapy to treat the condition led to an increased interest in classification. The different responses of various phobias to behavioural methods suggested a grouping into simple phobias, social phobia, and agoraphobia, and these groups were also found to differ in their age of onset. Simple phobias generally begin in childhood, social phobia in late adolescence, and agoraphobia in early adult life (Marks and Gelder 1966). It was also observed that phobias accompanied by marked panic attacks responded poorly to behaviour therapy and better to treatment with imipramine (Klein 1964). It was proposed that these cases should be classified separately under the rubric of panic disorder. This led to the development of the present scheme of classification into generalized anxiety disorder, phobic anxiety disorder (simple, social, and agoraphobic), and panic disorder.

Table 7.2   *Classification of anxiety disorders*

ICD10 F4 Anxiety disorders	DSMIV Anxiety disorders*
F40 Phobic anxiety disorders	

Agoraphobia
        without panic disorder
        with panic disorder
    Social phobia
    Specific phobia
	Agoraphobia without a history of panic
	    disorder
	Panic disorder with agoraphobia
	Social phobia
	Specific phobia

F41 Other anxiety disorders Panic disorder	Panic disorder without agoraphobia

Generalized anxiety disorder
    Mixed anxiety and depressive
        disorder
	Generalized anxiety disorder

*The order of presentation has been altered slightly to facilitate comparison of the schemes.

The relationship between obsessive–compulsive disorders and the phobic and anxiety disorders has been and remains uncertain. Freud thought at first that phobias and obsessions were closely related (see Freud 1895a). He proposed later that anxiety is the central problem in both conditions and that their characteristic symptoms—phobias and obsessions—resulted from different kinds of defence mechanisms against anxiety. Other workers, less convinced by the proposed aetiology, considered obsessional disorders as a separate group of neuroses of uncertain aetiology. This division of opinion is reflected today in the two major classification systems. In DSMIV, obsessive–compulsive disorders are classified as a subgroup of the anxiety disorders; in ICD10, anxiety disorders and obsessive–compulsive disorders have separate places in the classification.

The classification of anxiety disorders

DSMIV and ICD10 adopt the same general scheme (Table 7.2). The main differences are that in ICD10 the disorders are divided into (a) phobic anxiety disorder (F40) and (b) other kinds of anxiety disorder (F41), which include panic disorder and generalized anxiety disorder.

Panic disorder is classified rather differently in the two schemes for reasons that are explained on p. 176, and (as noted above) DSMIV classifies obsessive–compulsive disorder as a type of anxiety disorder while ICD10 separates the two conditions. An important difference is that ICD10 contains a category of mixed anxiety–depressive disorder which is not present in DSMIV. This category is discussed on p. 180.

Generalized anxiety disorders

Clinical picture

The symptoms of generalized anxiety disorder (Table 7.3) are persistent and are not restricted to or strongly predominating in any particular set of

Table 7.3 *Symptoms of generalized anxiety disorder*

1. Psychological	Fearful anticipation
	Irritability
	Sensitivity to noise
	Restlessness
	Poor concentration
	Worrying thoughts

2. Physical	
Gastrointestinal	Dry mouth
	Difficulty in swallowing
	Epigastric discomfort
	Excessive wind
	Frequent or loose motions
Respiratory	Constriction in the chest
	Difficulty inhaling
	Overbreathing
Cardiovascular	Palpitations
	Discomfort in chest
	Awareness of missed beats
Genitourinary	Frequent or urgent micturition
	Failure of erection
	Menstrual discomfort
	Amenorrhoea
Neuromuscular system	Tremor
	Prickling sensations
	Tinnitus
	Dizziness
	Headache
	Aching muscles

3. Sleep disturbance	Insomnia
	Night terror

4. Other symptoms	Depression
	Obsessions
	Depersonalization

circumstances (this second criterion distinguishes generalized from phobic anxiety disorders (see p. 169)). There are three characteristic features.

● *Worry* and apprehension, which are difficult to control and more prolonged than the ordinary worries and concerns of healthy people. The worries are widespread and not focused on a specific issue such as the possibility of having a panic attack (as in panic disorder) or of being embarrassed (as in social phobia) or contaminated (as in obsessive–compulsive disorder).

- *Motor tension,* which may be experienced as restlessness, trembling, inability to relax, and headache (which is usually bilateral and often frontal or occipital).
- *Autonomic hyperactivity,* which may be experienced as sweating, palpitations, dry mouth, epigastric discomfort, and dizziness.

Other psychological symptoms of generalized anxiety disorder are *irritability, poor concentration,* and *sensitivity to noise.* Some patients complain of poor memory when they are experiencing the effects of failure to concentrate; if a true memory impairment is found, a careful search should be made for a primary organic mental disorder.

Other motor symptoms include aching and stiffness in muscles, especially of the back and shoulders.

Autonomic symptoms can be grouped according to systems of the body as follows.

- *Gastro-intestinal:* dry mouth, difficulty in swallowing, epigastric discomfort, excessive wind, borborygmi, and frequent loose motions.
- *Respiratory:* feeling of constriction in the chest, difficulty in inhaling (which contrasts with the difficulty in exhaling in asthma), and the consequences of hyperventilation.
- *Cardiovascular:* feeling of discomfort over the heart, palpitations, awareness of missed beats, and throbbing in the neck.
- *Genitourinary:* frequency and urgency of micturition, failure of erection, and lack of libido, and among women menstrual discomfort and at times amenorrhoea.
- *Nervous system:* tinnitus, feeling of blurring of vision, dizziness which is not rotational, prickling sensations.

Patients may ask for help with any of these symptoms without mentioning spontaneously the psychological symptoms of anxiety (see differential diagnosis on p. 165).

Sleep disturbance: patients may have difficulty in falling asleep and may lie awake worrying. Sleep is often intermittent and accompanied by unpleasant dreams. Occasionally patients experience *night terrors* in which they wake suddenly feeling intensely fearful. Sometimes they recall a nightmare; sometimes they are uncertain why they are anxious. Patients with generalized anxiety disorder may wake unrefreshed. Early morning waking is not a feature of this disorder, and if present should suggest the possibility that any associated anxiety symptoms are part of a depressive disorder.

The appearance of a person with generalized anxiety disorder is characteristic. His face looks strained, with a furrowed brow, his posture is tense, and he is restless and often tremulous. The skin looks pale, and sweating is common, especially from the hands, feet, and axillae. Readiness to tears, which may at first suggest depression, reflects a generally apprehensive state.

Other symptoms of generalized anxiety disorder include *tiredness, depressive* symptoms, *obsessional* symptoms, and *depersonalization.* However, these symptoms are not the most prominent feature of the syndrome. (If they are prominent, other diagnoses should be considered; see differential diagnosis below.) Some patients hyperventilate at times, thereby adding further symptoms to the clinical picture, especially tingling in the extremities and dizziness. (Because hyperventilation is more frequent in panic disorder than generalized anxiety disorder, it is described on p. 177.)

Diagnostic conventions

DSMIV and ICD10 require the presence of similar symptoms for diagnosis of generalized anxiety disorder but they require different durations for these symptoms. DSMIV requires that symptoms have been present for six months, and this criterion is used in the research version of ICD10. However, the clinical version of ICD10 has the less stringent requirement that symptoms should have been present for 'most days for at least several weeks at a time, and usually several months'.

Comorbidity: when anxiety and depressive symptoms coexist, the diagnostic criteria may be met for both depressive disorder and generalized anxiety disorder. When there are intermittent exacerbations of continuous anxiety, two anxiety disorder diagnoses could be made. The guidance in DSMIV and ICD10 about the circumstances in which both diagnoses should be made are as follows. In ICD10 it is stated that to diagnose generalized anxiety disorder as the main diagnosis the patient 'must not meet the full criteria for a depressive episode (F32), phobic anxiety disorder (F40), panic disorder (F41) or obsessive compulsive disorder (F42)'. In DSMIV the advice is that the anxiety should not 'occur exclusively during a mood disorder, psychotic disorder or pervasive developmental disorder'. When this rule is adopted most cases of generalized anxiety disorder receive a second diagnosis, most often social phobia, dysthymia, or major depressive disorder, but also panic disorder and simple phobia (Sanderson and Barlow 1990). The diagnosis of cases with mixed anxiety and depressive symptoms is considered further on p. 180.

Differential diagnosis

Generalized anxiety disorders have to be distinguished from certain other psychiatric and physical disorders. Anxiety symptoms may occur in any psychiatric disorder, but there are some in which particular diagnostic difficulties arise. The most frequent problem is the distinction between anxiety and depressive disorder. Anxiety is a common symptom in **depressive disorder**, and generalized anxiety disorder often includes some depressive symptoms. It is conventional to make the diagnosis on the basis of the severity of two kinds of symptom and by the order in which they appeared. Information on these two points should be obtained from a relative or other informant as well as from the patient. Whichever type of symptoms appeared first and is more severe is considered primary. An important diagnostic error is to misdiagnose the agitated type of severe depressive disorder for generalized anxiety disorder. This mistake will seldom be made if anxious patients are asked routinely about symptoms of a depressive disorder including depressive thinking and, when appropriate, suicidal ideas.

Schizophrenic patients sometimes complain of anxiety before other symptoms are recognized. The chance of this misdiagnosis can be reduced by asking anxious patients routinely what they think caused their symptoms. In reply to this question, a schizophrenic patient may give an unusual reason which leads to the uncovering of previously unexpressed delusional ideas. **Presenile** or **senile dementia** occasionally comes to notice when the person complains of anxiety. When this happens, the clinician may not detect an associated impairment of memory or may dismiss this as the result of poor concentration. Therefore memory should be assessed appropriately in middle-aged or older patients presenting with anxiety. Occasionally a patient dependent on **drugs or alcohol** reports that he is taking these substances to relieve anxiety, either because he wishes to deceive the doctor or because he genuinely mistakes symptoms of drug withdrawal for anxiety. If there is a report of anxiety that is particularly severe on waking in the morning, this should suggest the possibility of alcohol dependence because withdrawal symptoms are most likely at this time. (It should be remembered also that anxiety symptoms occurring as part of a depressive disorder are characteristically worse in the morning.)

Some **physical illnesses** may present with anxiety symptoms. This possibility should be considered especially when no obvious psychological cause can be found and when the personality is normal. An important differential diagnosis is from *thyrotoxicosis,* in which the patient may be irritable and restless with tremor and tachycardia. Physical examination may reveal characteristic signs of thyrotoxicosis, such as enlarged thyroid, atrial fibrillation, and exophthalmos. If there is any doubt, thyroid function tests should be arranged. When symptoms are episodic but without any obvious situational causes, *phaeochromocytoma* and *hypoglycaemia* should be considered and appropriate physical examination carried out. Sometimes anxiety is the presenting symptom of other physical illness because the patient fears that the early symptoms portend a fatal illness such as cancer. This is particularly likely when the patient has a special reason to fear serious illness, for example if a relative or friend died after developing similar symptoms. It is good practice to ask anxious patients whose symptoms are unexplained whether they know anyone who has had similar symptoms.

The opposite diagnostic error can also be made. Thus *generalized anxiety disorder* with prominent physical symptoms may be *mistaken for physical disease*. When this happens, patients may then undergo unnecessary investigations that may increase anxiety. Correct diagnosis is more likely if the doctor remembers the diversity of the anxiety symptoms: palpitations, headache, frequency of micturition, and abdominal discomfort can all be the primary complaint of an anxious patient. Correct diagnosis is also more likely when the doctor enquires systematically about other symptoms of generalized anxiety disorder and about the order in which the various symptoms appeared when the illness developed.

Epidemiology

Because methods for identifying generalized anxiety disorders in community surveys have been developed only in the last few years, epidemiological data are scarce. In a large population study in the United States, the one-year prevalence rates ranged from 2.5 to 6.4 per cent (Weissman and Merikangas 1986); in another study the one-year rate was 2 per cent in men and 4.3 per cent in women (Kessler *et al.* 1994).

Aetiology

Generalized anxiety disorder appears to be caused by stressors acting on a personality predisposed by a combination of genetic factors and environmental influences in childhood.

Stressful events

Clinical observations indicate that generalized anxiety disorders often begin in relation to stressful events, and some become chronic when stressful problems persist. Stressful events involving threat are particularly related to anxiety disorder (loss events are associated more with depression (Finlay-Jones and Brown 1981)). However, there have been no satisfactory studies directed to the aetiological role of 'life events' in cases with a specific diagnosis of generalized anxiety disorder.

Genetic causes

Most population genetic studies have not distinguished between generalized anxiety disorder and other kinds of anxiety disorder. Anxiety disorders of all kinds are more frequent (about 15 per cent) among the relatives of patients with anxiety disorders than in the general population (about 3 per cent) (Brown 1942; Noyes *et al.* 1978). Evidence that this familial association may be due in part to genetic causes was obtained by Slater and Shields (1969) in a study of 17 monozygous and 28 dizygous twin pairs, each containing a proband with an anxiety disorder. Forty-one per cent of monozygous co-twins had an anxiety disorder, compared with only 4 per cent of the dizygous co-twins.

A study of a population sample of 1033 female twins confirmed that the tendency of anxiety disorders to run in families applies to cases diagnosed specifically as generalized anxiety disorder, and showed that this is largely due to genetic factors rather than shared family environment. Hereditability was estimated to be about 30 per cent, and this rate could not be accounted for by the giving of an additional diagnosis of generalized anxiety disorder to cases of depressive or panic disorder (Kendler *et al.* 1992b). It should be noted that, in order to increase the sample size, these authors required a duration of only one month instead of the six months required by DSMIV.

Personality

Some generalized anxiety disorders occur in people with personalities characterized by a persistent tendency to anxiety (see p. 116). However, the condition may occur with other kinds of personality disorder, notably the obsessional and asthenic types, and in people with normal personality.

Psychoanalytic theories

Psychoanalytic theories suggest that generalized anxiety disorder is caused not by a specific type of personality but by a particular way of dealing with painful emotions. In psychoanalytic theory anxiety is the central symptom of all kinds of neurotic disorder; other symptoms such as obsessions and phobias are the result of defences used to limit the experience of anxiety. In these terms, the principal causes of generalized anxiety neurosis are intrapsychic conflicts which generate anxiety when the ego is overwhelmed by excitation from any of three sources: (i) the outside world (*realistic anxiety*); (ii) the instinctual levels of the id, including love, anger, and sex (*neurotic anxiety*); (iii) the superego (*moral anxiety*). In generalized anxiety disorder, anxiety is experienced directly unmodified by the defence mechanisms that in other people cause phobias to develop.

The ego is more likely to be overwhelmed when it has been weakened by development failure in childhood. Several kinds of experience can produce this weakening, but separation and loss are thought to be particularly important (Bowlby 1969). In the early stages of childhood, anxiety is linked particularly to separation from the mother, and children need the security of loving parents if they are to become capable as adults of experiencing separation without undue anxiety. At a later stage of childhood, anxiety is linked to rivalry with the father—in psychoanalytic terminology the Oedipal conflict leading to castration anxiety (see p. 83). Failure to surmount this stage of development successfully in a loving family is thought to leave the person unusually vulnerable to anxiety in adult life.

Other studies of early experience

Accounts given by anxious patients of their experience in childhood suggest that early adverse experiences are important. Brown and Harris (1993) studied the relation between such experience and anxiety disorder in adult life in 404 working-class women living in an inner city. Adverse early experience was assessed from patients' accounts of parental indifference and of physical or sexual abuse.

Women reporting early adversity had increased rates of generalized anxiety disorder (and also agoraphobia, but not mild agoraphobia or simple phobia). Rates of depressive disorder were also increased in these women.

Psychological responses to stressful events

As well as the psychoanalytic theories outlined above, two other theories have been proposed to explain why some people respond to stressful events by developing a generalized anxiety disorder.

Conditioning theories propose that generalized anxiety disorders arise when there is an inherited predisposition to excessive lability of the autonomic nervous system, and when the initial fear responses become generalized by conditioning to previously neutral stimuli. Although this is a plausible explanation of the spread and generalization of anxiety in these cases, it has not been shown convincingly that patients with generalized anxiety disorder differ from controls on measures of conditioning. (For a review of these theories and the possible neural substrates for anxiety responses see Gray (1982).)

Cognitive theories propose that generalized anxiety disorders arise as the result of a tendency to worry unproductively about problems and to focus attention on potentially threatening circumstances. These theories provide a plausible explanation of the role of thinking in generalized anxiety disorder and complement the conditioning theories. They are supported directly by studies of thinking in anxious patients and controls, and indirectly by the efficacy of cognitive behavioural treatments (see p. 168). (For a review of research on the relationship between anxiety and worry, see Matthews (1990).)

Prognosis

Most follow-up studies of prognosis were carried out before modern criteria for the diagnosis of generalized anxiety disorder were decided, and many included patients who would now be diagnosed as having phobic or panic disorder. One of the DSMIV criteria for generalized anxiety disorder is that the symptoms should have been present for six months before the diagnosis can be made. ICD10 requires 'several weeks at a time and usually several months'. Many anxiety disorders improve within six months; of those lasting longer, about 80 per cent persist for three years (Kedward and Cooper 1966). Poorer prognosis is associated with severe symptoms and with syncopal episodes, agitation, derealization, hysterical features, and suicidal ideas (Kerr *et al.* 1974). When patients complain mainly of physical symptoms of anxiety and attribute these symptoms to physical causes, they generally seem more difficult to help. In a study of medical patients with anxiety disorder, two-thirds improved substantially or recovered within six years, and the remaining third had persistent symptoms (Noyes and Clancy 1976).

On follow-up, episodes of major depression occur frequently among many patients with anxiety disorders (Clancy *et al.* 1978). (The relationship between anxiety and depression is considered further on p. 180). The rates of schizophrenia and manic depressive disorder found in patients with anxiety disorders are no greater than in the general population (Greer 1969; Kerr *et al.* 1974).

Treatment

Counselling

In the early stages of a generalized anxiety disorder (before symptoms have been established for the six months necessary for a DSMIV diagnosis), simple methods of counselling are often effective. This counselling has two components. First the nature of the disorder is explained and reassurance is given about any fears that physical symptoms of anxiety are caused by physical disease. Since anxious people often concentrate poorly on new information, it is useful to provide an information leaflet that repeats the same points.

Second the patient is helped to deal with or adjust to any relevant social problems. Anxiety is prolonged by uncertainty, and therefore it is important to set out a clear plan of treatment. Patients with generalized anxiety disorder are more sensitive than normal subjects to the anxiogenic effects of caffeine (Bruce *et al.* 1992).

Most patients discover this for themselves and reduce their caffeine intake; if they have not done so, reduction of excessive caffeine intake may be helpful. Patients with recent onset anxiety need no more than counselling, but the more severe and persistent cases usually require additional cognitive or behavioural or drug treatment.

Cognitive–behavioural treatments

For all but the severe cases of generalized anxiety disorder *relaxation training* may be helpful. If practised regularly, the effects of relaxation can equal those of anxiolytic drugs. However, many patients fail to persevere with practice. Training in a group may improve motivation, and some patients do better when relaxation is part of a programme such as yoga exercises which contain more to engage their interest.

Anxiety management training combines relaxation with cognitive procedures designed to help patients to control worrying thoughts (see p. 637). A clinical trial by Butler *et al.* (1987) indicated that this treatment produces worthwhile improvements in generalized anxiety disorder. However, anxiety management training has not been compared directly with drug treatment for generalized anxiety disorder.

Treatment with drugs

The use of drugs for generalized anxiety disorders should be selective. Drugs can be used to bring symptoms under control quickly, while the effects of psychological treatment are awaited. Drugs are also helpful in the minority of patients who do not improve with psychological measures. However, there is a general tendency to prescribe drugs too often and for too long.

One of the longer acting **benzodiazepines** is appropriate for the short-term treatment of generalized anxiety disorders, for example diazepam in a dose from 5 mg twice daily in mild cases to 10 mg three times a day in the most severe. Anxiolytic drugs should seldom be prescribed for more than three weeks because of the risk of dependence when given for longer. In the past, barbiturates were used to treat generalized anxiety disorders, but the risk of dependence is particularly great and these drugs should no longer be used for this purpose. (Anxiolytic drugs are discussed further on p. 547.)

Buspirone, an azaspirone, is as effective as the benzodiazepines in the short-term management of generalized anxiety disorder and is much less likely to cause dependency (Cowen 1992).

Beta-adrenergic antagonists have a limited use for patients with generalized anxiety disorders, and for controlling severe palpitations that have not responded to short-term treatment with an anxiolytic. Care should be taken to observe the contraindications to the use of these drugs and to follow the advice given on p. 545, where beta-adrenergic antagonists are discussed further.

Tricyclic antidepressants and related compounds have anxiolytic as well as antidepressant effects. They act more slowly than benzodiazepines but their effect is equivalent or greater (Kahn *et al.* 1986; Rickels *et al.* 1993). They are much less likely to cause dependence than benzodiazepines.

One of the more sedative antidepressants such as amitriptyline or trazodone can be used to treat generalized anxiety disorder, though imipramine (which is less sedating) appears to have a comparable effect (Rickels *et al.* 1993). If imipramine is used, initial dosage should be low and should be increased gradually as described on p. 179.

Monoamine oxidase inhibitors (see p. 568) have been used to treat generalized anxiety disorders. This usage was described many years ago (Sargant and Dally 1962) but has not been widespread because the drugs interact with some drugs and foodstuffs (see p. 570).

Phobic anxiety disorders

Phobic anxiety disorders have the same core symptoms as generalized anxiety disorders, but these symptoms occur only in particular circumstances. In some phobic disorders these circumstances are few and the patient is free from anxiety for most of the time; in other phobic disorders many circumstances provoke anxiety with the result that anxiety is more frequent, but even so there are situations in which no anxiety is experienced. Two other features characterize phobic disorders: the person *avoids* circumstances that provoke anxiety, and he experiences *anticipatory anxiety* when there is the prospect of encountering these circumstances. The circumstances provoking anxiety include situations (for example crowded places), 'objects' (for example spiders), and natural phenomena (for example thunder). For clinical purposes three principal phobic syndromes are recognized: simple phobia, social phobia, and agoraphobia. These syndromes will be described next, and some less common phobic syndromes will be mentioned later in this section.

Phobic disorders are classified in slightly different ways in DSMIV and ICD10. In both systems phobic disorders are divided into simple phobia, social phobia, and agoraphobia. In DSMIV agoraphobic patients who experience more than four panic attacks in four weeks, or one attack followed by a month of persistent fear of having another attack, are classified as having a type of panic disorder. The reasons why panic attacks are given this importance in DSMIV are explained on p. 176.

Simple phobia

Clinical picture

In this disorder a person is inappropriately anxious in the presence of one or more particular objects or situations. The whole range of anxiety symptoms (described under generalized anxiety disorder (p. 163)) may be experienced in the presence of the object or in the situation. Simple phobias may be characterized by adding the name of the stimulus; for example spider phobia. In the past it was common practice to use terms such as arachnophobia (instead of spider phobia) or acrophobia (instead of phobia of heights), but this adds nothing useful.

Three types of specific phobia merit separate consideration.

Phobia of dental treatment About 5 per cent of adults have fears of the dentist's chair; these fears can become so severe that all dental treatment is avoided and serious caries develops (Gale and Ayer 1969; Kleinknecht *et al.* 1973).

Phobia of flying Anxiety during aeroplane travel is common. A few people have such intense fear that they are unable to travel in an aeroplane and they seek treatment. This fear occurs occasionally among pilots who have had an accident while flying.

Phobia of illness Patients experience repeated fearful thoughts that they might have cancer, venereal disease, or some other serious illness. The thoughts are recognized as irrational and are not accompanied by the resistance that characterizes obsessional thoughts. Such fears may be associated with avoidance of hospitals, but are not otherwise specific to situations. If the patient is also convinced that he has the disease, the condition is classified as hypochondriasis (see p. 354); if the thoughts are recognized as irrational and are accompanied by resistance, the condition is classified as obsessional–compulsive disorder.

Prevalence

Among adults the lifetime prevalence of simple phobias has been estimated, using DSMIIIR criteria, as 4 per cent in men and 13 per cent in women (Kessler *et al.* 1994).

Aetiology

Most simple phobias of adult life are a continuation of childhood phobias. Simple phobias are common in childhood (see p. 702). By early teenage years most of these childhood fears have been lost, but a few persist into adult life. Why the few persist is not certain, except that the most severe phobias are likely to last the longest. Genetic predisposition may play a part. In a study of over 2000 female twins the results from those with simple phobia fitted an aetiological model in which a modest genetic vulnerability combined with phobia-specific stressful events (Kendler *et al.* 1992*a*). The psychoanalytic explanation is that phobias persist because they are not related to the obvious stimulus but to a hidden source of anxiety. In the terms used in this theory, the source of anxiety is excluded from consciousness by repression and is attached to the manifest object by displacement. A minority of simple phobias begin in adult life, in relation to a highly stressful experience; for example a phobia of horses may follow a dangerous encounter with a bolting horse.

Differential diagnosis

Diagnosis is seldom difficult. The possibility of an underlying depressive disorder should always be kept in mind, since some patients seek help for long-standing simple phobias when a depressive disorder makes them less able to tolerate their phobic symptoms. Obsessional disorders sometimes present with fear and avoidance of specific objects (for example knives). A systematic history and mental state examination will reveal the associated obsessional thoughts (for example harming a person with a knife).

Prognosis

The prognosis of simple phobia in adult life has not been studied systematically. Clinical experience suggests that simple phobias that originate in childhood continue for many years, whilst those starting in adult life after stressful events have a better prognosis.

Treatment

The main treatment is the exposure form of behaviour therapy (see p. 631). Usually the phobia can be reduced considerably in intensity and with it the social disability. However, it is unusual for the phobia to be lost completely.

Social phobia

Clinical picture

In this disorder inappropriate anxiety is experienced in situations in which the person is observed and could be criticized. Socially phobic people tend to avoid such situations and not engage in them fully; for example they avoid making conversation, or they sit in a place where they are least conspicuous. Anxiety is also felt in anticipation of entering situations. The situations include restaurants, canteens, dinner parties, seminars, board meetings, and other places where the person feels observed by other people. Patients may experience any of the symptoms of an anxiety disorder (see p. 163), but complaints of blushing and trembling are particularly frequent. Socially phobic people are often preoccupied with the idea of being observed critically, though aware that the idea is groundless (Amies *et al.* 1983). Some patients take alcohol to relieve the symptoms of anxiety, and alcohol abuse is more common in social phobia than in other phobias.

Some patients become anxious in a wide range of social situations (generalized social phobia), whilst others are anxious only in specific situations such as public speaking. The tendency to avoid the stimulus is strong; in most cases the stimulus is avoided, and even the prospect of encountering the object or situation causes anticipatory anxiety. Some patients experience anxiety in a more limited set of circumstances, for example writing in front of others, speaking in public, or playing a musical instrument in public (Clark and Agras 1991). These limited or discrete social phobias are classified separately in DSMIV (but not in ICD10).

Two discrete social phobias require separate consideration: phobias of excretion and phobias of vomiting.

Phobias of excretion Patients with these phobias either become anxious and unable to pass urine in public lavatories, or have frequent urges to pass urine and an associated dread of incontinence. Such patients often arrange their lives so as never to be far from a lavatory. A few have comparable symptoms centred around defecation.

Phobias of vomiting Some patients fear that they may vomit in a public place, often a bus or train; in these surroundings they feel anxious and nauseated. A smaller group have repeated fears that other people will vomit in such places.

Epidemiology

Social phobias are about equally frequent in men and women. The one-year prevalence of social phobia has been estimated as 7 per cent for men and 9 per cent for women (Davidson *et al.* 1993; Kessler *et al.* 1994).

Prognosis

The condition usually begins between the ages of 17 and 30. The first episode occurs in some public place, usually without any apparent reason. Subsequent anxiety occurs in similar places. The episodes gradually become more severe and avoidance increases. There have been no systematic follow-up studies but clinical experience suggests that, without treatment, the condition often lasts for many years. Although there have been reports of an increased rate of deliberate self-harm in these patients, it seems that this behaviour occurs only when there is a second condition such as depressive disorder and alcohol abuse (Schneier *et al.* 1992).

Differential diagnosis

The symptom of social phobia can occur in *agoraphobia* and *panic disorder*. When this occurs both diagnoses can be made, but it is more useful for the clinician to decide which symptoms occurred first, and which symptom is more severe and should be given priority.

Social phobia has to be distinguished from *generalized anxiety disorder* (by establishing the situations in which anxiety occurs), *depressive disorder* (by examining the mental state), and schizophrenia. Patients with *schizophrenia* may avoid social situations because of persecutory delusions when anxiety has subsided; patients with social phobia know that their insistent ideas of being observed are untrue. Social phobia has to be distinguished from *avoidant personality disorder* characterized by life-long shyness and lack of self-confidence. In principle the phobia has a recognizable onset and a shorter history, but in practice the distinction may be difficult since social phobia may often begin in the teenage years and the onset may be difficult to recall. Many cases meet criteria for both diagnoses (Schneier *et al.* 1991). Finally, a distinction needs to be made between social phobia and **social inadequacy**. The latter is a primary lack of social skills with secondary anxiety; it is not a phobic disorder but a type of behaviour that occurs in personality disorders and schizophrenia, and among people of low intelligence. Its features include hesitant, dull, and inaudible diction, inappropriate use of facial expression and gesture, and failure to look at other people in conversation (see Bryant *et al.* (1976) for a more detailed account).

Aetiology

The causes of social phobia are not well understood. *Genetic factors* are suggested by the finding that social phobias (but not other anxiety disorders) are more common among the relatives of social phobics than in the population (Fyer *et al.* 1993). In a population-based sample of over 2000 female twins the results from probands with social phobia fitted a model in which moderately strong genetic influences interact with *non-specific environmental factors* (Kendler *et al.* 1992a). Most social phobias begin with a sudden episode of anxiety in circumstances similar to those which become the stimulus for the phobia, and it is possible that the subsequent development of phobic symptoms is through conditioning and cognitive learning.

The principal *cognitive factor* in the aetiology of social phobia is an undue concern that other people will be critical (often referred to as a 'fear of negative evaluation'). Whether this cognition precedes the disorder or develops with it is unknown, but in either case it is likely to increase and prolong the phobic anxiety. Social phobia usually begins in late adolescence, when young people are expand-

ing their social contacts and are particularly concerned about the impression that they are making on other people. It is possible that social phobias occur particularly among people in whom these concerns are pronounced; however, there is no evidence on which to decide the matter. See Amies *et al.* (1983) for a more detailed discussion.

Treatment

Psychological treatment Psychological treatments used for social phobia include social skills training (see p. 632), relaxation training (see p. 630), and exposure (see p. 631). Some patients with social phobia lack specific social skills, such as those involved in initiating conversation; for these patients social skills training can be a useful component of treatment, directed also to the reduction of phobic anxiety.

The psychological treatment of choice for social phobia is *cognitive behavioural therapy*, in which exposure to feared situations is combined with anxiety management. (This treatment is described on p. 637.) The relapse rate is lower after this combined treatment than after exposure alone (Butler *et al.* 1984; Mattick and Peters 1988). For a review of studies evaluating psychological treatment for social phobia see Heimburg (1989).

Drug treatment Benzodiazepines reduce symptoms of social phobia more than does placebo (Gelernter *et al.* 1991) but there is a risk of dependency if their use is prolonged. Their main use is to help patients cope with social encounters until other treatment has led to improvement. **Beta-adrenergic inhibitors** such as atenolol help to control tremor and palpitations, which are often the most distressing symptoms especially in the specific forms of social phobia (Liebowitz *et al.* 1992). Among the **monoamine oxidase inhibitors**, phenelzine has been shown to be more effective in the treatment of social phobia than placebo (Liebowitz *et al.* 1988) and atenolol (Liebowitz *et al.* 1992). Moclobemide, the reversible inhibitor of monoamine oxidase, appears to be equally effective (Versiani *et al.* 1992) and should be less prone to the hazards of interaction with food stuffs or drugs (see p. 570). The effects of cognitive behaviour therapy and phenelzine appear comparable in the short term (Gelernter *et al.* 1991). This therapeutic effect may not be confined to monoamine oxidase inhibitors. The **specific serotonin re-uptake inhibitor** fluoxetine has also been reported to be an effective treatment for social phobia, but controlled comparisons with placebo and other drugs are needed. **Dynamic psychotherapy** may help some patients, particularly those whose social phobia is associated with pre-existing problems in personal relationships. However, there are no clinical trials for this form of treatment.

For a general review of social phobia see Liebowitz *et al.* (1985).

Agoraphobia

Clinical features

Agoraphobic patients are anxious when they are away from home, in crowds, or in situations that they cannot leave easily. In these circumstances the symptoms are similar to those of other phobic disorders (see p. 169), but other symptoms such as depression, depersonalization, and obsessional thoughts are more frequent in agoraphobia than in other phobic disorders.

Two groups of **anxiety symptoms** are more marked in agoraphobia than in other kinds of phobic disorder. First *panic attacks* are more frequent, whether in response to environmental stimuli or arising spontaneously. (In DSMIV, cases with more than four panic attacks in four weeks are classified as panic disorder with secondary agoraphobic symptoms; this convention is discussed on p. 176.) Second *anxious cognitions* about fainting and loss of control are frequent among agoraphobic patients.

Many **situations** provoke anxiety and avoidance. They seem at first to have little in common but, as explained above, there are three common themes of distance from home, overcrowding, and confinement. The situations include buses and trains, shops and supermarkets, and places that cannot be left suddenly without attracting attention, such as the hairdresser's chair or a seat in the middle row of a place of entertainment. As the condition progresses, patients avoid more and more of these situations until in severe cases they may be more or less confined to their homes (sometimes called the 'housebound housewife syndrome', though not all of these patients are housewives). Apparent variations in this pattern are usually due to additional factors that reduce symptoms for a short time. For example most patients are less anxious when accompanied by a trusted companion and some are helped even by the presence of a child or pet dog. The variability in anxiety produced in this way may suggest erroneously that the patient is exaggerating symptoms when they are said to be severe, rather than that the symptoms have been alleviated at times when less severe.

Anticipatory anxiety is a common symptom. In severe cases this anxiety appears hours before the person enters the feared situation, adding to the patient's distress and sometimes misleading doctors into thinking that the anxiety is generalized rather than phobic.

Other symptoms include depressive symptoms, depersonalization, and obsessional thoughts. Depressive symptoms are common, and often seem to be consequent upon the limitations to normal life caused by anxiety and avoidance. Depersonalization was at one time thought to signify a subgroup of agoraphobia with a special cause. Thus Roth (1959) described the *phobic anxiety depersonalization syndrome* and suggested that it might result from a disorder of the temporal lobes. Subsequent work has not supported this view.

An association has been reported between agoraphobia and *prolapse of the mitral valves*. For example Kantor *et al.* (1980) reported that 44 per cent of agoraphobic women had prolapse of the mitral valves. Such findings have not been confirmed by subsequent studies.

The **onset and course** of agoraphobia differ in several ways from those of other phobic disorders. Most cases begin in the early or middle twenties, though there is a further period of high onset in the mid-thirties. Both these ages are later than the average ages of onset of simple phobias (childhood) and social phobias (mostly late teenage years or early twenties) (Marks and Gelder 1966).

Typically, the first episode occurs while the person, more often a woman (see below), is waiting for public transport or shopping in a crowded store. Suddenly she becomes extremely anxious without knowing why, feels faint, and experiences palpitations. She rushes away from the place and goes home or to hospital, where she recovers rapidly. When she enters the same or similar surroundings, she becomes anxious again and makes another hurried escape. This sequence recurs over the next weeks and months; the panic attacks are experienced in more and more places, and a habit of avoidance follows. It is unusual to discover any serious immediate stress that could account for the first panic attack, though some patients describe a background of serious problems (e.g. worry about a sick child); in a few cases the symptoms begin soon after a physical illness or childbirth.

While most patients associate the onset of agoraphobic symptoms with a panic attack, some describe an onset without such an attack. In one study, two-thirds of 260 new cases of agoraphobia reported an onset without a panic attack (Eaton and Keyl 1990). This finding is relevant to the theory that agoraphobia develops as a consequence of panic disorder. It has been suggested that the methods of interviewing in community surveys underestimate the frequency of initial panic attacks; the use of more detailed interviews seems to support this opinion (Horwath *et al.* 1993).

As the condition progresses, agoraphobic patients become increasingly dependent on the spouse or other relatives for help with activities, such as shopping, that provoke anxiety. The consequent demands on the spouse often lead to arguments, but serious marital problems are no more common among agoraphobics than among other people of similar social background (Buglass *et al.* 1977).

Differential diagnosis

Agoraphobia has to be distinguished from social phobic disorder, generalized anxiety disorder, panic disorder, depressive disorder, and paranoid disorder. When DSMIV is used, the cases with the agoraphobic pattern of anxiety and avoidance are divided into those meeting the criteria for *panic disorder* (panic disorder with agoraphobia) and those without a history of panic disorder (see p. 176). Agoraphobia may be confused with *social phobia* because many patients with agoraphobia feel anxious in social situations, whilst some social phobics avoid crowded buses and shops where they feel under scrutiny. Detailed enquiry into the present pattern of avoidance and into the order in which the two sets of symptoms developed will usually settle the point.

Some patients with *generalized anxiety disorder* experience an increase of anxiety in public places, but they are also anxious in situations not characteristic of agoraphobia and they do not show the avoidance pattern characteristic of agoraphobia. When the agoraphobia is severe, involving many situations, the distinction may be difficult to make on the basis of current psychiatric state, but the history of development of the disorder will usually point to the correct diagnosis. Agoraphobic symptoms can occur in a *depressive disorder,* and agoraphobic patients may have depressive symptoms; usually careful history taking and mental state examination will show which set of symptoms developed first. Sometimes a recent depressive disorder is superimposed on a long-standing agoraphobic disorder; it is important to identify these cases and treat them appropriately (see below). Occasionally a patient with *paranoid delusions* (as part of early schizophrenia or a delusional disorder) avoids going out and meeting people in shops and other places. In such a patient there may be an initial resemblance to agoraphobia, but this is dispelled when examination of the mental state reveals delusions of persecution or of reference that account for the avoidance.

Epidemiology

In a study using DSMIIIR criteria, the one-year prevalence of agoraphobia without panic disorder was estimated as 1.7 per cent in men and 3.8 per cent in women (Kessler *et al.* 1994) and the lifetime prevalence about 6 to 10 per cent (Weissman and Merikangas 1986).

Aetiology

Theories of the aetiology of agoraphobia have to explain why the initial anxiety attacks occur, and why attacks spread and recur persistently. The two problems will be considered in turn.

There are three explanations for the initial anxiety attacks. The *cognitive* hypothesis proposes that the anxiety attacks develop in people who are unreasonably afraid of minor physical symptoms (for example palpitations which are misinterpreted as evidence of serious heart disease). Although such fears are found in patients with established agoraphobia, it is not known whether they predate the disorder or are a consequence of it. Hence the cognitive hypothesis does not provide a convincing explanation of the onset of the disorder. The *biological* hypothesis proposes that the initial anxiety attack results from environmental stimuli acting on a person with a biological predisposition to respond with anxiety, possibly because normal inhibitory mechanisms are defective. There is some evidence for a genetic component in agoraphobia (Kendler *et al.* 1992*a*) and this could act through a biological mechanism. The issues are considered under panic disorder on p. 178. The *psycho-analytic* hypothesis proposes unconscious mental conflicts related to unacceptable sexual or aggressive impulses. However, the only evidence for this hypothesis derives from psychoanalytic interviews with selected patients. (For a more detailed account see Mathews *et al.* (1981).)

The spread and persistence of anxiety responses can also be explained in more than one way. It is reasonable to suggest that learning mechanisms are important; conditioning could account for the association of anxiety with more and more situations, whilst avoidance learning could account for the subsequent tendency to avoid. Although this explanation seems logical, there is no evidence that agoraphobic patients learn more readily than patients who have panic attacks without becoming agoraphobic. *Personality* may be important; agoraphobic patients are often described as dependent, and prone to avoid rather than confront problems. This dependency could have arisen from over-protection in childhood, which is reported more often by agoraphobics than by controls. However, despite such retrospective reports it is not certain that the dependency was present before the onset of the agoraphobia (see Matthews *et al.* (1981) for a review). Furthermore, Buglass *et al.* (1977) found no difference between agoraphobics and controls in history of separation anxiety or of other indices of dependency. It has been suggested that, once started, agoraphobia could be maintained by family problems. However, in a well-controlled study Buglass *et al.* (1977) found no evidence that agoraphobics had more family problems than controls. Clinical observation suggests that symptoms are sometimes prolonged by over-protective attitudes of other family members, but this feature is not found in all cases.

Prognosis

Although brief cases may be seen in general practice, agoraphobia lasting for one year changes little in the next five years (Marks 1969). Brief episodes of depressive symptoms often occur in the course of chronic agoraphobia, and clinical experience suggests that patients are more likely to seek help during these episodes.

Treatment

Psychological treatment In early cases, patients should be strongly encouraged to return to the situations which they are avoiding, since avoidance seems to prolong the disorder. The treatment of choice for established cases is a form of *behaviour therapy* combining exposure to phobic situations with training in coping with panic attacks (see p. 637). Compared with exposure alone, this combination gives better long-term results, including substantial and lasting changes in avoidance behaviour, and a reduction in phobic anxiety and panic attacks (Cohen *et al.* 1984). However, most patients continue to experience mild anxiety in the situations in which symptoms were originally most severe (Mathews *et al.* 1981). The prognosis for this kind of treatment may be better in patients with good marital relationships before treatment (Monteiro *et al.* 1985) and worse in those experiencing chronic life stress (Wade *et al.* 1993).

Drug treatment **Anxiolytic drugs** may be used for a specific purpose such as helping the patient to undertake an important engagement before other treatment has taken effect. Anxiolytic drugs should not be used regularly in this way for more than a few weeks because of the risk of dependence (see p. 544). An exception to this general rule may be the use of alprazolam to treat agoraphobia with frequent panic attacks (in DSMIV terms, panic disorder with agoraphobia). This treatment is discussed further under panic disorder (p. 179), since most studies have included patients with panic disorder as well as patients with agoraphobia with panic. An exception is the trial by Marks *et al.* (1993*a*,*b*) in which all patients had agoraphobia; these authors found that the short-term benefit with alprazolam was about half that obtained with exposure treatment, and that during follow-up after treatment relapse rates were greater with alprazolam.

Antidepressant drugs may be used to treat a concurrent depressive disorder; they also have a therapeutic effect in agoraphobic patients who are not depressed but have frequent panic attacks (Zitrin *et al.* 1983). Imipramine has been tested most thoroughly, but the therapeutic effects may not be confined to this compound since similar short-term effects have been reported with clomipramine (Modigh *et al.* 1992, Gentil *et al.* 1993). The dose of imipramine used in the treatment of agoraphobia with panic is high; if no response is obtained at lower levels, doses of up to 225 mg per day may be used but only if patients are free from cardiac or other physical disease (see p. 179). A high rate of relapse has been reported when imipramine is stopped (Zitrin *et al.* 1983).

Monoamine oxidase inhibitors (p. 568) have also been reported to reduce agoraphobic symptoms (Sargant and Dally 1962), but they are used less often than imipramine because they interact with some drugs and foodstuffs (see p. 570). As with imipramine, the relapse rate is high when monoamine oxidase inhibitors are stopped (Tyrer and Steinberg 1975), even after many months of treatment.

Patients who have relapsed after drug treatment can be offered behaviour therapy though no controlled evaluation has been carried out specifically with such patients. The use of imipramine for this purpose is discussed further under panic disorder (see p. 179).

Panic disorder

Although the diagnosis of panic disorder was not used until it was introduced in DSMIII in 1980, cases fitting this rubric have been described under a variety of names for more than a century. The central feature is the occurrence of panic attacks, i.e. sudden attacks of anxiety in which physical symptoms predominate and are accompanied by fear of a serious consequence such as a heart attack. In the past these symptoms have been variously referred to as irritable heart, Da Costa's syndrome, neurocirculatory asthenia, disorderly action of the heart, and effort syndrome. These early terms assumed that patients were correct in fearing a disorder of cardiac function. Some later authors suggested psychological causes, but it was not until the Second World War (when interest in this condition revived) that Wood (1941) convincingly showed the condition to be a form of anxiety disorder. From then until 1980 patients with panic attacks were classified as having either generalized or phobic anxiety disorders. In 1980 the authors of DSMIII introduced a new diagnostic category, panic disorder, which included patients whose panic attacks occurred with or without generalized anxiety, but excluded those whose panic attacks appeared in the course of agoraphobia. In DSMIV, all patients with frequent panic attacks are classified as having panic disorder whether or not they have agoraphobia. (Agoraphobia without panic attacks has a separate rubric (see p. 162).) The category panic disorder did not appear in ICD9; it appears in ICD10, but it is not applied to patients who have marked agoraphobic anxiety and avoidance (a point of difference from DSMIV).

Clinical features

The symptoms of a panic attack are listed in Table 7.4. Not every patient has all these symptoms. For the diagnosis of panic disorder, DSMIV requires at least four of the symptoms in at least one attack of panic. Important features of panic attacks are that anxiety builds up quickly, the response is severe, and there is fear of a catastrophic outcome. In DSMIV the diagnosis is made when panic attacks occur unexpectedly (i.e. not in response to a known phobic stimulus), and when more than four attacks have occurred in four weeks or one attack has been followed by four weeks of persistent fear of another attack. Some patients with panic disorder hyperventilate, with the effect that further symptoms are added.

Hyperventilation is breathing in a rapid and shallow way with a resultant fall in the concentration of carbon dioxide in the blood. The resulting hypnocapnoea may cause dizziness, tinnitus, headache, a feeling of weakness, faintness, numbness and tingling in the hands, feet, and face, carpopedal spasms, and precordial discomfort. There is also a feeling of breathlessness, which may prolong the condition if the patient concludes from this feeling that he should breathe even more vigorously. When a patient has unexplained bodily symptoms, the possibility of persistent overbreathing should always be borne in mind. The diagnosis can usually be made by watching the pattern of breathing. If there is doubt, blood gas analysis should decide the matter in acute cases though the findings may be normal in chronic cases (Hibbert 1984*b*).

Differential diagnosis

Panic attacks occur in generalized anxiety disorders, phobic anxiety disorders (most often agoraphobia), depressive disorders, and acute organic disorder. In DSMIV panic disorder can be diagnosed when these disorders are present, but in the United Kingdom it is the custom not to diagnose panic disorder in the presence of these other disorders.

Epidemiology

Most epidemiological studies have used the criteria of DSMIIIR and include cases with repeated panic attacks whether or not accompanied by marked agoraphobia. Using this criterion the one-year prevalence of panic disorder is about 13 per 1000 in men

Table 7.4 *Symptoms of a panic attack (from DSMIV)*

Shortness of breath and smothering sensations
Choking
Palpitations and accelerated heart rate
Chest discomfort or pain
Sweating
Dizziness, unsteady feelings or faintness
Nausea or abdominal distress
Depersonalization or derealization
Numbness or tingling sensations
Flushes or chills
Trembling or shaking
Fear of dying
Fears of going crazy or doing something uncontrolled

and 32 per 1000 in women (Kessler *et al.* 1994). In most studies, the prevalence in women is about twice that in men. For panic attacks that are too mild or too infrequent to meet criteria for panic disorder, the six-month prevalence has been estimated as 30 per 1000 (Von Korff *et al.* 1985) and the lifetime prevalence as about 56 per 1000 (Katerndahl 1993). No sharp cut-off was found for panic attacks meeting or not meeting the criteria for panic disorders; instead there seemed to be a continuous variation. (For a review, see Weissman and Merikangas (1986).)

Aetiology

There are three main hypotheses about the origin of panic disorder. The first proposes a biochemical abnormality, the second hyperventilation, and the third a cognitive abnormality. Each will be briefly considered in turn.

The *biochemical hypothesis* is reflected in the term 'endogenous anxiety' which has been proposed for these cases. The hypothesis is based on three sets of observations. First, chemical agents such as sodium lactate (Pitts and McClure 1967) and yohimbine (Charney *et al.* 1984) can induce panic attacks more readily in patients with panic disorder than in healthy people. Second, panic attacks are reduced by certain drugs. Third, there is some evidence for a genetic basis for the disorder. The multitude of *chemical agents that provoke panic attacks* in panic disorder patients make it difficult to identify a single common mechanism. They include the benzodiazepine receptor antagonist flumazenil, cholecystokinin, and the 5-hydroxtryptamine (5-HT) receptor agonist MCCP (Bradwejn *et al.* 1991; Nutt and Lawson 1992). Suggestions include abnormalities in the presynaptic alpha-adrenoceptors that normally restrain the activity of presynaptic neurones in brain areas concerned with the control of anxiety, and an abnormality of benzodiazepine or 5-HT receptor function (see Nutt and Lawson (1992) for a review).

The effects of drugs suggest that 5-HT mechanisms are important in panic disorder. Imipramine affects both 5-HT and noradrenergic systems. Clomipramine and fluvoxamine (which mainly affect 5-HT transmission) are effective antipanic drugs, but maprotiline, a selective noradrenergic uptake blocker, is not (Den Boer and Westerberg 1988). Clomipramine appears to be more potent than imipramine as an antipanic agent. It is possible that very high doses of imipramine are needed to suppress panic attacks because this compound is a relatively weak 5-HT uptake blocker.

There is some evidence that panic disorder occurs more often among relatives, suggesting a genetic and therefore possibly a biochemical basis for the disorder (Crowe *et al.* 1983). Twin studies support this idea, but the numbers of pairs studied are too small for a definitive conclusion.

The *hyperventilation hypothesis* is based on the observation that in some people voluntary overbreathing produces symptoms like those of a panic attack (Hibbert 1984*b*). The hypothesis is that 'spontaneous' panic attacks result from involuntary hyperventilation. However, although some panics appear to originate or to be exacerbated in this way, hyperventilation does not appear to be a general cause of panic disorder (Hibbert and Pilsbury 1988). Panic is also provoked by the *inhalation of carbon dioxide* more readily in panic disorder patients than in controls. This observation is one of the bases of hypotheses that panic disorder is caused by hypersensitivity of a biological system that responds to feelings of suffocation by anxiety and an urge to escape. It is suggested that this response is set off in panic patients by minimal cues for suffocation (Klein 1993).

The *cognitive hypothesis* is based on the observation that fears about serious physical or mental illness are more frequent among patients with panic attacks than among anxious patients without panic attacks (Hibbert 1984*a*). It has been proposed that there is a spiral of anxiety in panic disorder as the physical symptoms of anxiety activate fears of illness and thereby generate more anxiety (Clark 1986). These observations have led to a cognitive treatment for panic disorder (see pp. 179 and 638).

Course and prognosis

Recent follow-up studies have generally included patients with panic attacks and agoraphobia as well as patients with panic disorder alone. Earlier studies used categories such as effort syndrome. One study of patients with effort syndrome found that 90 per cent still had symptoms 20 years later, though most had a good social outcome (Wheeler *et al.* 1950). In more recent studies of patients with panic disorder the course was characterized by fluctuating anxiety and depression (Noyes *et al.* 1990). Mortality rates from unnatural causes and, among men, from cardiovascular disorders have been found to be higher than average (Coryell *et al.* 1982).

Treatment

Apart from supportive measures and attention to any causative personal or social problems, treatment is with mainly either drugs or cognitive therapy.

Benzodiazepines control panic attacks when given in high doses. Alprazolam, a high potency benzodiazepine, can be given in such doses without marked sedation, although it is probably no more effective in reducing panic attacks than an equivalent dose of diazepam (Dunner *et al.* 1986). Benzodiazepines should be withdrawn very gradually to avoid withdrawal symptoms (see p. 544).

The antidepressant drug **imipramine** also controls panic attacks (Klein 1964). The first effect of the drug is often to produce an unpleasant feeling of apprehension, sleeplessness, and palpitations. For this reason the initial dose should be small, for example 10 mg per day for three days, increasing by 10 mg every three days to 50 mg per day, and then by 25 mg per week to 150 mg per day. If symptoms are not controlled at this dose, further increments of 25 mg may be given to physically fit patients up to a maximum of 175–225 mg per day. Before high doses are given, an ECG should be obtained if there is any doubt about cardiac function. Full dosage is continued for three to six months. A relapse rate of up to 30 per cent has been reported after stopping imipramine (Zitrin *et al.* 1978), but the rate may be less if imipramine is continued in reduced dosage for a further few months (Mavissakalian and Perel 1992). There is some evidence that **clomipramine** is at least as effective as imipramine (Gentil *et al.* 1993). The specific serotonin uptake inhibitor **fluvoxamine** has been reported to have a therapeutic effect comparable to that with imipramine (Black *et al.* 1993).

Patients who hyperventilate can be helped in two ways. An immediate treatment is to rebreathe expired air from a bag in order to increase the concentration of carbon dioxide in the alveolar air. This **rebreathing** is also an effective way of demonstrating the connection between symptoms and hyperventilation. When the connection has been made clear, patients can practise controlled breathing first under supervision and then on their own.

Recently, a technique of **cognitive therapy** has been used to reduce fears of the physical effects of anxiety, on the assumption that fears prolong the disorder. Common fears are that palpitations indicate an impending heart attack, or that dizziness indicates impending loss of consciousness. The relevant symptoms are induced by voluntary means—usually hyperventilation but occasionally in other ways such as exercise—and it is explained that the symptoms of a panic attack have an equally benign origin. This demonstration is followed by further explanation of the origin of the feared symptoms and questioning of the patient's beliefs about them. Substantial improvement in symptoms has been reported in uncontrolled investigations of this treatment (Clark *et al.* 1985), and controlled studies have confirmed that cognitive therapy is at least as effective as imipramine given in high doses of 225 mg (Clark *et al.* 1994).

Transcultural variations of anxiety disorder

In several cultures the presenting symptoms of anxiety disorder are more often somatic than mental. Leff (1981) has pointed out that this difference in symptomatology parallels the different vocabulary that is available for describing anxiety in the corresponding languages. Thus there is no word for anxiety in a number of African, Oriental, and American Indian languages; instead, a phrase denoting bodily experience is used. For example in Yoruba, an African language, the phrase is 'the heart is not at rest'. Several conditions have been described that may be transcultural variants of anxiety disorders, though their exact relation to these disorders is uncertain.

Koro may be an extreme variant of anxiety disorder. It occurs amongst men in Southwest Asia, more commonly among the Chinese; the Cantonese people call it *suk-yeong*, which means shrinking of the penis. There are episodes of acute anxiety, lasting from 30 minutes to a day or two, in which the person complains of palpitations, sweating, pericardial discomfort, and trembling. At the same time he is convinced that the penis will retract into the abdomen and that when this process is complete he will die. Most episodes occur at night, sometimes after sexual activity. To prevent the feared outcome patients may tie the penis to an object, or ask another person to hold the organ. This belief resembles the conviction held by patients during a panic attack that the heart is damaged and they will die. Epidemics of koro have been described among people made anxious by social stressors and superstitious ideas (Tseng *et al.* 1988). (See Yap (1965) for a more detailed account.)

Variants of social phobia have been described in the east, originally among Japanese people where it is known as *taijin-kyofu-sho* or phobia of interpersonal relations. There is an intense anxiety in social situations and an intense conviction bordering on the delusional that the person is being thought of unfavourably by others. Other symptoms include fears of producing body odours, dysmorphophobia, and aversion to eye contact (Tseng *et al.* 1992).

Mixed anxiety and depressive disorder

As explained on p. 165, anxiety and depressive symptoms often occur together. The overlap is greatest when the symptoms are mild (52 per cent in the study of psychiatric patients by Hiller *et al.* (1989)) and least when they are severe enough for a diagnosis of psychiatric disorder (29 per cent in the study by Hiller *et al.*). Similar findings were obtained in a community epidemiological study of people meeting diagnostic criteria for anxiety disorder; 28 per cent also met criteria for minor depression and a further 21 per cent met criteria for major depression (Angst and Dobler-Mikola 1985). Similar rates of comorbidity have also been reported with panic disorder, agoraphobia, and major depression (Clayton 1990).

When the anxiety and depressive symptoms are not severe enough for a diagnosis, the condition may be referred to as a minor affective disorder (these conditions are considered in Chapter 6). In the ICD10 classification these minor affective disorders can be classified as mixed anxiety and depressive disorders. This category is for cases in which anxiety and depressive symptoms are both present, but neither set of symptoms, considered separately, is severe enough to make a diagnosis of depressive disorder or anxiety disorder. When minor anxiety and depressive symptoms are related to a change in life circumstances, adjustment disorder is diagnosed. The remaining group of persistent mixed disorder is seen commonly in general practice. The category of mixed anxiety–depressive disorders is not included in the main classification of DSMIV, though it is included among the list of categories provided for further study.

There are several reasons why anxiety and depression may occur together. First, the antecedent causes may be similar. Brown *et al.* (1993) found that childhood adversity is associated with both anxiety and depressive disorders in adult life. A second reason is that many stressful events combine elements of loss (which is known to be associated with depression) and danger (which is associated with anxiety). A possible third reason for the association is that persistent anxiety can lead to secondary depression of mood. Follow-up studies support this idea because onsets of depression among people with persistent anxiety are more common than onsets of anxiety among people with persistent depression.

The *prognosis* of these mixed cases is not clearly established. *Treatment* is generally with a tricyclic antidepressant or monoamine oxidase inhibitor, both of which have anxiolytic as well as antidepressant effects (Rickels *et al.* 1974; Davidson *et al.* 1980; Johnstone *et al.* 1980).

Obsessive–compulsive disorder

A concise description of obsessive–compulsive disorder was contained in ICD9, where the disorder was characterized as a state in which:

the outstanding symptom is a feeling of subjective compulsion—which must be resisted—to carry out some action, to dwell on an idea, to recall an experience, or ruminate on an abstract topic. Unwanted thoughts, which include the insistency of words or ideas, ruminations or trains of thought, are perceived by the patient to be inappropriate or nonsensical. The obsessional urge or idea is recognized as alien to the personality but as coming from within the self. Obsessional actions may be quasi ritual performances designed to relieve anxiety, e.g. washing the hands to deal with contamination. Attempts to dispel the unwelcome thoughts or urges may lead to a severe inner struggle, with intense anxiety.

Although ICD9 is no longer in use, this description fits well with current concepts of the disorder.

Clinical picture

Obsessive–compulsive disorders are characterized by obsessional thinking, compulsive behaviour, and varying degrees of anxiety, depression, and depersonalization. Obsessional and compulsive symptoms are described on pp. 15–17, but the reader may find it helpful if the main features are repeated here.

Obsessional thoughts are words, ideas, and beliefs, recognized by the patient as his own, that intrude forcibly into his mind. They are usually unpleasant, and attempts are made to exclude them. It is the combination of an inner sense of compulsion and of efforts at resistance that characterize obsessional symptoms, but the effort of resistance is the more variable of the two. Obsessional thoughts may take the form of single words, phrases, or rhymes; they are usually unpleasant or shocking to the patient, and may be obscene or blasphemous. **Obsessional images** are vividly imagined scenes, often of a violent or disgusting kind, involving abnormal sexual practices for example.

Obsessional ruminations are internal debates in which arguments for and against even the simplest everyday actions are reviewed endlessly. Some **obsessional doubts** concern actions that may not have been completed adequately, such as turning off a gas tap or securing a door; other doubts concern actions that might have harmed other people, for example that driving a car past a cyclist might have caused him to fall off his bicycle. Sometimes doubts are related to religious convictions or observances ('scruples')—a phenomenon well known to those who hear confession.

Obsessional impulses are urges to perform acts, usually of a violent or embarrassing kind, for example leaping in front of a car, injuring a child, or shouting blasphemies in church.

Obsessional rituals include both mental activities, such as counting repeatedly in a special way or repeating a certain form of words, and repeated but senseless behaviours, such as washing the hands 20 or more times a day. Some of these have an understandable connection with obsessional thoughts that precede them, for example repeated handwashing and thoughts of contamination. Other rituals have no such connection, for example routines concerned with laying out clothes in a complicated way before dressing. Some patients feel compelled to repeat such actions a certain number of times; if this cannot be achieved, they have to start the whole sequence again. Patients are invariably aware that their rituals are illogical and usually try to hide them. Some fear that their symptoms are a sign of incipient madness and are greatly helped by reassurance that this is not so.

Both obsessional thoughts and rituals inevitably lead to slow performance of everyday activities. However, a minority of obsessional patients are afflicted by extreme **obsessional slowness** that is out of proportion to other symptoms (Rachman 1974).

Obsessional thoughts and compulsive rituals may worsen in certain situations; for example obsessional thoughts about harming other people often increase in a kitchen or other place where knives are kept. Because patients often avoid such situations, there may be a superficial resemblance to the characteristic pattern of avoidance found in a phobic anxiety disorder. It is partly for this reason that obsessional thoughts with fearful content (such as thoughts about knives) have been called **obsessional phobias**.

Anxiety is an important component of obsessive–compulsive disorders. Some rituals are followed by a diminution of anxiety, whilst others are followed by increased anxiety (Walker and Beech 1969). It is because anxiety is such a prominent symptom that these conditions are classified as anxiety disorders in DSMIV.

Obsessional patients are often **depressed**. In some this seems to be an understandable reaction to the obsessional symptoms, but others have recurring depressive mood swings that arise independently. A proportion of obsessional patients also report **depersonalization**.

The **obsessional personality** is described in Chapter 5. It is important to realize that obsessional personality and obsessive–compulsive disorders do not have a simple one-to-one relationship. Obsessional personality is over-represented among patients who develop obsessive–compulsive disorder, but about a third of obsessional patients have other types of personality (Lewis 1936*b*). Moreover, people with obsessional personality are more likely to develop depressive disorders than obsessive–compulsive disorders (Pollitt 1960).

Differential diagnosis

Obsessive–compulsive disorders must be distinguished from other disorders in which obsessional symptoms occur. The distinction from **generalized anxiety disorder, panic disorder, or phobic disorder** should seldom be difficult provided that a careful history is taken and the mental state is examined thoroughly. The course of obsessive–compulsive disorder is often punctuated by periods of depression in which the obsessional symptoms increase; when this happens the depressive disorder may be overlooked. **Depressive disorders** may also present with obsessional symptoms; it is particularly important to bear these conditions in mind because they usually respond well to antidepressant treatment. Obsessive–compulsive disorder is occasionally difficult to distinguish from **schizophrenia**, especially when the degree of resistance is doubtful, the obsessional thoughts are peculiar in content (for example mingling sexual and blasphemous themes), or the rituals are exceptionally odd. In such cases it is important to search for schizophrenic symptoms and to question relatives carefully about other aspects of the patient's behaviour. Obsessional symptoms are found occasionally in **organic cerebral disorders** and were especially common after the encephalitis lethargica epidemic in the 1920s.

Epidemiology

Obsessive–compulsive disorders are less frequent than anxiety neuroses. Estimates of one-year prevalence vary from 1.1 to 1.8 per cent when DSMIIIR criteria are used, and lifetime prevalence varies from 1.9 to 2.5 per cent. Closely similar estimates have been obtained in the United States, Germany, Puerto Rico, Taiwan, and New Zealand. The ratio of lifetime prevalence of women to men varies from 1.2 (in Puerto Rico) to 3.8 (in New Zealand). In clinic populations the ratio is usually about 1.0. In community samples, from 20 to 60 per cent of people reported obsessions only. This is in contrast with those referred to psychiatric clinics among whom 70–94 per cent report both obsessions and compulsions (Weissman *et al.* 1994).

Aetiology

Healthy people experience occasional intrusive thoughts, some of which are concerned with sexual, aggressive, and other themes similar to those of obsessional patients (Rachman and Hodgson 1980). It is the frequency, intensity, and above all the persistence of obsessional phenomena that have to be explained.

Genetics

Obsessive–compulsive disorders have been found in about 5–7 per cent of the parents of patients with these disorders (Brown 1942; Rüdin 1953); although low, this rate is higher than in the general population. Of course, these findings could reflect environmental as well as genetic causes. Twin studies would help to identify the genetic component, but too few cases have been reported to allow firm conclusions.

Evidence of a brain disorder

Two kinds of evidence suggest a disorder of brain function in obsessive–compulsive disorder: associations between the condition and disease that has established effects on brain function, and evidence from brain scanning of various kinds.

Associations with other disorders Obsessional symptoms were recorded frequently among patients affected by *encephalitis lethargica* after the epidemic of the 1920s. Gilles de la Tourette included obsessional symptoms in his original description of the disorder that now bears his name (Gilles de la Tourette 1885), and recent studies have confirmed this observation (Cummings and Frankel 1985; Robertson *et al.* 1988).

Brain imaging studies *Magnetic resonance imaging* has not revealed any consistent structural brain abnormality specific to patients with obsessive–compulsive disorder (Garber *et al.* 1989; Kellner *et al.* 1991). Studies with *single-photon emission tomography* (SPET) have shown increased uptake in the frontal lobe (Machlin *et al.* 1991; Rubin *et al.* 1992), which is reduced after treatment (Hoehn-Saric *et al.* 1991). Several studies using *positron emission tomography* (PET) have shown increased metabolic activity in the orbitofrontal cortex and possibly also in the caudate nucleus and anterior cingulate (Insel 1992; Rauch *et al.* 1994). Although not wholly consistent, the findings taken together suggest abnormalities in the orbitofrontal cortex and perhaps in the caudate nucleus (Baxter *et al.* 1992; Swedo *et al.* 1992). These findings have led to the suggestion that there may be abnormal activity in a neurological circuit involving the orbitofrontal cortex, cingulate gyrus, and caudate nucleus (Insel 1992).

Abnormal serotonergic function

The finding that drugs affecting 5-HT function reduce obsessive–compulsive symptoms suggests that 5-HT function might be abnormal in obsessive–compulsive disorder. 5-HT uptake inhibitors have complex effects on 5-HT function. Their immediate effect is to increase the availability of 5-HT at the synapse, but this effect is followed by down-regulation of 5-HT_2 receptors. The effect of 5-HT uptake inhibitors on obsessive–compulsive symptoms takes several weeks and therefore is likely to be related to the later effects of the drugs, although it is not certain what these are. In animals chronic treatment with 5-HT uptake inhibitors results in an increase of 5-HT neuro-

transmitters but it is not certain whether the same occurs in humans. Even if it were certain whether 5-HT function was increased or decreased by 5-HT uptake blockers, this knowledge would not necessarily show that abnormal 5-HT function causes obsessive–compulsive disorder. Anticholinergic drugs control parkinsonian symptoms by acting on the normal cholinergic systems of patients whose disorder is due to abnormal dopaminergic function.

The uncertainty as to whether 5-HT function is abnormal in obsessive–compulsive disorder is increased by the results of neuroendocrine tests of 5-HT function in humans (see p. 101). These tests have not shown any consistent abnormality in patients with obsessive–compulsive disorder (Barr *et al.* 1992; Hollander *et al.* 1992).

Early experience

It is uncertain whether early experience plays a part in the aetiology of obsessive–compulsive disorder. Mothers with the disorder might be expected to transmit symptoms to their children by imitative learning. However, although the children of patients with obsessive–compulsive disorder have an increased risk of non-specific neurotic symptoms, they do not have more obsessional symptoms (Cowie 1961).

Psychoanalytic theories

Freud (1895*a*) originally suggested that obsessional symptoms result from unconscious impulses of an aggressive or sexual nature. These impulses could potentially cause extreme anxiety, but anxiety is reduced by the action of the defence mechanisms of repression and reaction formation. This idea fits with the aggressive and sexual fantasies of many obsessional patients, and with their restraints on their own aggressive and sexual impulses. Freud also proposed that obsessional symptoms occur when there is a regression to the anal stage of development as a way of avoiding impulses related to the subsequent genital and Oedipal stages. This idea reflects the obsessional patient's frequent concerns over excretory functions and dirt. Although Freud's ideas draw attention to aspects of the disorder other than the obvious symptoms, as an explanation of obsessive–compulsive disorder it is convincing only within the framework of psychoanalytic theory.

Learning theory

Learning theory attempts to explain obsessive–compulsive disorder in terms of abnormalities in normal mechanisms of learning. It has been suggested that obsessional rituals are the equivalent of avoidance responses, but as a general explanation this idea cannot be sustained because anxiety increases rather than decreases after some rituals (Walker and Beech 1969). A useful review of this and other aspects of aetiology was given by Rachman and Hodgson (1980). Subsequent research has not added support to the original ideas.

Prognosis

About two-thirds of cases improve by the end of a year. Cases lasting for more than a year usually run a fluctuating course, with periods of partial or complete remission lasting from a few months to several years (Pollitt 1957). Prognosis is worse when the personality is obsessional and symptoms are severe (Kringlen 1965), and when there are continuing stressful events in the patient's life. Severe cases may be exceedingly persistent; for example in a study of obsessional patients admitted to hospital, Kringlen (1965) found that three-quarters remained unchanged 13–20 years later.

Treatment

In treatment, it is important to remember that obsessive–compulsive disorder often runs a fluctuating course with long periods of remission. Depressive disorder often accompanies obsessive–compulsive disorder, and in such cases effective treatment of the depressive disorder often leads to improvement in the obsessional symptoms. For this reason a thorough search for depressive disorder should be made in every patient presenting with obsessive–compulsive disorder.

Counselling

Treatment should begin with an explanation of the symptoms, and if necessary with reassurance that these symptoms are not an early sign of madness (a common concern of obsessional patients). Obsessional patients often involve other family members in their rituals; hence in planning treatment it is important to interview relatives and encourage them to adopt a firm but sympathetic attitude to the patient.

Drugs

Anxiolytic drugs give some short-term symptomatic relief but should not be prescribed for more than a few weeks at a time. If anxiolytic treatment is needed for more than a month or two, small doses of a tricyclic antidepressant or an antipsychotic may be used. If there is a coincident depressive disorder this should be treated.

Clomipramine is a tricyclic antidepressant with potent 5-HT uptake blocking effects. When given in doses of 200–250 mg per day it is more effective than placebo in reducing the obsessional symptoms of patients with obsessional compulsive disorder (Clomipramine Collaborative Study Group 1991). Most patients tolerate the treatment well, but at these high doses anticholinergic side-effects are common and a few patients develop seizures. A clinically useful effect may not be reached until about six weeks after starting treatment; further improvement may occur over the next six weeks. Many patients relapse in the first few weeks after the drug is stopped. Tricyclic drugs such as imipramine are less potent 5-HT uptake blockers, and do not have this therapeutic effect in obsessive–compulsive disorder (Foa *et al.* 1987).

Specific serotonin uptake inhibitors (SSRIs), such as fluoxetine and fluvoxamine, are also effective in reducing obsessional symptoms (Goodman *et al.* 1989b; Jenike *et al.* 1990). They may be as effective as clomipramine and produce fewer side-effects (Piggot *et al.* 1990). Not all patients with obsessional–compulsive disorder improve with either drug; with SSRI about half improve substantially. The reasons for this failure of response are not known, but attempts have been made to increased the response rate by adding a second drug to the SSRI. Beneficial effects have been reported from the addition of a neuroleptic (McDougle *et al.* 1990), but at the time of writing this report requires confirmation. As with clomipramine, relapse is common in the few weeks after the drug has been stopped abruptly. It is possible that longer treatment might result in fewer relapses.

When there are prominent rituals, drug treatment should be combined with response prevention and exposure (see Jenike and Rauch 1994).

Behaviour therapy

Obsessional rituals usually improve with a combination of response prevention and exposure to any environmental cues that increase them (see p. 632 for a description of this treatment). About two-thirds of patients with moderately severe rituals can be expected to improve substantially but not completely (Rachman and Hodgson 1980). When rituals are reduced by this treatment, the accompanying obsessional thoughts usually improve as well. Patients who do not consistently recognize that their beliefs are untrue appear to respond to behaviour therapy as well as those with more typical obsessional symptoms (Lelliot *et al.* 1988).

Behavioural treatment is considerably less effective for obsessional thoughts occurring without rituals. The technique of thought stopping has been used for many years, but there is no good evidence that it has a specific effect. Indeed, Stern *et al.* (1973) found an effect which did not differ from that of thought-stopping directed to irrelevant thoughts. Drug treatment is indicated for these patients.

Psychotherapy

We have noted that obsessive–compulsive disorder runs a fluctuating course and may improve

eventually whatever treatment is given. Until recovery, supportive interviews can benefit patients by providing continuing hope. Joint interviews with the spouse are indicated where marital problems seem to be aggravating the symptoms. However, exploratory and interpretive psychotherapy seldom help. Indeed, some obsessional patients are made worse because these procedures encourage painful and unproductive rumination about the subjects discussed during treatment.

Psychosurgery

The immediate results of psychosurgery for severe obsessive–compulsive disorder are often striking, with a marked reduction in tension and distress. However, it has not been proved that the long-term prognosis is improved, since no prospective controlled trial has been carried out although the treatment has been in use for at least 25 years. Several types of operation have been used. Tan *et al.* (1971) studied 23 patients who had undergone *bimedial leucotomy* operations. Compared with retrospectively chosen controls, the operated patients improved more in obsessional symptoms and social handicap over the next five years. For other forms of operation only uncontrolled assessments were available, and some have follow-up periods too short to guide practice (Göktepe *et al.* 1975; Mitchell-Heggs *et al.* 1976). Jenike *et al.* (1991) studied 33 obsessional patients treated with *cingulotomy* at periods of up to 25 years after the operation. Although a generally good outcome was reported, 23 patients had more than one operation and six patients died by suicide. Two-thirds of the survivors were assessed, and only a third had improved.

Hay *et al.* (1993) reported a ten-year follow up of 26 obsessive–compulsive patients treated with orbitomedial or cingulate lesions or both; of the 18 patients interviewed eight had a second operation, two died by suicide, and about a third of the survivors improved.

The frequency of second operations and the low improvement rates indicate the limitations of this treatment. Although intractable cases were chosen for surgery, most were declared intractable at a time when modern drugs and behavioural treatment were not available. If neurosurgery is considered for these patients, it should be only for the most chronic cases that have resisted inpatient or day-patient treatment, including drug and behavioural methods, for at least a year. Using these criteria the authors have not referred patients for surgical treatment.

Conversion and dissociative disorders

The terms conversion and dissociative disorder refer to disorders that until recently were known as hysteria, i.e. conditions in which physical symptoms and certain mental symptoms occur without the physical pathology with which they are usually associated and with psychological causes. The change in terminology has been adopted because the word hysteria is used in everyday speech to denote extravagant behaviour, and it is confusing to use the same word for the different phenomena of the syndrome considered here. In DSMIV two terms, conversion and dissociative, are used to distinguish conditions with physical and mental symptoms respectively. In ICD10 the two terms are interchangeable. Before considering these conditions further it is appropriate to describe briefly the history of ideas about hysteria.

Conversion and dissociative symptoms and syndromes

The concept of conversion or dissociative symptom is that symptoms of physical illness or certain kinds of mental illness have occurred in the absence of physical pathology with which they are normally associated, and that the symptoms have been produced unconsciously, not deliberately. There are two obvious difficulties with this concept. First, physical pathology can seldom be excluded with complete certainty when a patient is first seen, and even after extensive investigation such pathology may not be excluded. The second difficulty is that there may be no certainty that the

symptoms are produced by unconscious mechanisms. Often, uncertainty about diagnosis can be overcome only by waiting for follow-up information; until this information is available the diagnosis of conversion or dissociative disorder has to be provisional, and has to be reviewed as new evidence becomes available. Uncertainty as to how far symptoms arise from unconscious mechanisms is even more difficult to resolve because there often seems to be a mixture of conscious and unconscious mechanisms.

Conversion and dissociative symptoms can occur as the major feature of a conversion or dissociative disorder, or as a feature of another psychiatric disorder, for example an anxiety, depressive, or organic mental disorder. It is important to recognize this possibility and to search carefully for other symptoms of these primary disorders before concluding that conversion symptoms result from a conversion disorder or that dissociative symptoms result from a dissociative disorder.

Although conversion and dissociative symptoms are not produced deliberately, they represent the patient's ideas about illness. Sometimes the symptoms imitate those of a relative or friend who has been ill. Sometimes they originate in the patient's experience of illness, for example dissociative memory loss in a person with a previous head injury. Conversion or dissociative symptoms will be more similar to those of physical illness in a person who is well informed, such as someone working in a hospital, and less similar in someone who is ill informed, such as a child or a mentally retarded adult. Usually there are obvious discrepancies between signs and symptoms of conversion and dissociative disorder and those of organic disease, for example a pattern of anaesthesia that does not correspond to the anatomical innervation of the part. Therefore thorough physical example is essential in every case.

The symptoms of conversion and dissociative disorder usually confer some advantage on the patient. For this reason, following Freud's terminology, these disorders have been said to produce a **secondary gain** (Freud referred also to primary gain, which was the exclusion from consciousness of anxiety due to psychological conflict). For example a woman would be said to have secondary gain if she developed a psychogenic paralysis of the arm and was consequently spared the care of an elderly relative. Although characteristic of conversion and dissociative disorders, secondary gain is not confined to them; people with physical illness may also gain advantage from secondary gain. For example a woman with an organic paralysis of the arm could be spared the unwanted task of nursing an elderly relative. Thus secondary gain is an important feature of conversion and dissociative disorder; the diagnosis should be reconsidered if it is absent, but the diagnosis is not established if it is present.

Patients with conversion and dissociative symptoms often show less distress than would be expected of someone with their symptoms. This state is sometimes called *belle indifference*, a term used first by French writers of the nineteenth century. It is unwise to base the diagnosis on this state of indifference because patients may adopt a stoical attitude to the symptoms of physical disease. A point of difference is that patients with conversion disorder may be unconcerned by the symptoms, but often show exaggerated emotional reactions in other ways (and may have a high level of autonomic arousal (Lader and Sartorius 1968)). These distinctions depend on subjective judgements and are not a reliable basis for the diagnosis of conversion disorder.

The development of ideas about the causes of hysteria

Descriptions of hysteria were included in ancient Greek medical texts. At that time the disorder was thought to result from movement of the uterus from its normal position (hence the name of the condition). In the second century AD, Galen rejected the idea that the uterus had been displaced and suggested instead that the condition was caused by an undue retention of uterine secretions. One or other form of uterine pathology was generally thought to cause hysteria until the seventeenth century, when the English physician Thomas Willis (1621–75) suggested that hysteria was caused by a disorder of the brain (Dewhurst 1980). Gradually this idea became accepted; by the

nineteenth century, the importance of predisposing constitutional and organic causes of this brain disorder were recognized, and it was accepted that strong emotion was the usual provoking cause.

The studies of hysteria by Charcot, a French neurologist, were particularly influential in the late nineteenth century. He believed that hysteria was caused by a functional disorder of the brain which caused symptoms, and which also rendered patients susceptible to hypnosis so that new symptoms could be produced by suggestion. Charcot's ideas were developed further by his pupil, Pierre Janet, who proposed that the disorder in hysteria was a tendency to dissociate, i.e. to lose the normal integration between various parts of mental functioning, together with a restriction of personal awareness so that the person became unaware of certain aspects of psychological functioning which would otherwise be within his awareness. (For a review of Janet's ideas of dissociation see Van der Hart and Horst (1989).)

Another advance in thinking about hysteria was made by Freud. He visited Charcot in the winter of 1895–6 and was impressed by demonstrations of the susceptibility of patients to hypnosis, and of the power of suggestion to hypnotized patients (Sulloway 1979). On his return to Vienna, Freud and his colleague Breuer studied patients with hysteria and reported their findings in a paper 'On the psychical mechanisms of hysterical phenomena' (Freud 1893). In a subsequent monograph *Studies in hysteria* (1895), Breuer and Freud suggested that hysteria was caused by emotionally charged ideas which had become lodged in the unconscious of the patient at some previous time, and which were excluded from conscious awareness by a process which the authors called repression. They summarized this idea in the phrase 'hysterics suffer mainly from reminiscences' (Freud 1895–7, p. 7). Freud suggested that the repressed ideas were usually sexual. Other clinicians followed Charcot's pupil, Janet, and took a wider view of the emotionally charged ideas that could cause hysteria. This wider view was supported by experience in the First World War, when hysteria was observed in response to the stressful experience of battle in people who had no apparent antecedent sexual problems.

Kretschmer (1961) proposed an interpretation of the observation of hysterical disorders in wartime. He suggested that the symptoms resulted from an innate biological mechanism that counteracts highly stressful experiences. He believed that hysterical symptoms could develop in psychologically stable people as a result of this 'reflex' mechanism, and could subside quickly when the stressors receded. Kretschmer suggested that a minority of cases become chronic for two reasons. First, the symptoms may be deliberately cultivated by someone who wishes to take advantage of them. Second, if hysterical behaviour is repeated many times, it becomes a habit ('slips into a groove' to use Kretschmer's phrase). Although these ideas have not been substantiated, they have the merit of drawing attention to the apparent mixture of voluntary and involuntary causes in many cases of hysteria.

For a more complete account of the history of hysteria see Veith (1965) and Ellenberger (1970). The interested reader should also consult Breuer and Freud's studies on hysteria which are available in Volume 3 of the Pelican Freud Library.

Classification

As explained above, the two major systems of classification have adopted somewhat different ways of classifying the conditions formerly known as hysteria (see Table 7.5). In DSMIV two terms, conversion and dissociative disorder, are used. Conversion denotes cases in which physical symptoms are the principal manifestations of the disorder; whilst dissociative disorder denotes cases in which psychological symptoms are the principal manifestations. In ICD10 both kinds of manifestation are called dissociative disorders (with the term conversion used as an alternative means of indicating the same meaning). In this chapter the DSMIV convention is adopted. The word hysteria is used occasionally when discussing results or theories which originated before the new terminology was adopted and which cannot be related readily to the newer terms.

In DSMIV, conversion disorders (those with physical symptoms) are classified under the rubric 'Somatoform disorder' (a rubric for mental disorders with mainly physical symptoms), while dissociative disorders have a rubric of their own. In ICD10 all these disorders are classified together under the rubric 'Dissociative disorder'. In this book the DSMIV convention has been adopted, and conversion disorders are described on p. 352 in the chapter on psychiatry and medicine.

Dissociative disorders

Dissociative amnesia

Dissociative amnesia starts suddenly. Patients are unable to recall long periods of their lives and sometimes deny any knowledge of their previous life or personal identity. Amongst patients who present in this way, some have concurrent organic disease, especially epilepsy, multiple sclerosis, or the effects of head injury (Kennedy and Neville 1957). These patients with organic disorders have similar symptoms to the psychogenic cases, and are also likely to start suddenly. Moreover, patients with organic disease may be as suggestible as those without it, and may recover their memory just as well.

Dissociative pseudodementia

Pseudodementia is a disorder with extensive abnormalities of memory and behaviour that suggest generalized intellectual impairment. Simple tests of memory are answered wrongly, but often in a way that strongly suggests that the correct answer is in the patient's mind. It is sometimes difficult to be certain how much of the behaviour is outside the patient's control and how much is deliberately produced. The diagnosis may be difficult, since a similar clinical picture is sometimes associated with organic brain disease, epilepsy, or schizophrenia. (The term pseudodementia is also applied in a different sense to the apparent dementia of depressed elderly patients (see p. 527).)

Table 7.5 *Classification of dissociative or conversion disorders*

ICD10 F4 Dissociative (conversion) disorders*	DSMIV Dissociative disorders
Dissociative amnesia Dissociative fugue Dissociative stupor	Dissociative amnesia Dissociative fugue
Other dissociative disorder Ganser's syndrome	
Multiple personality disorder	Dissociative identity disorder (multiple personality disorder) Depersonalization disorder
Trance and possession disorder	Dissociative disorder not otherwise specified
Dissociative motor disorder Dissociative convulsions Dissociative anaesthesia	Conversion disorder (not further divided)

*The order of presentation has been changed slightly to aid comparison of the two systems

Dissociative fugue

In a dissociative fugue patients not only lose their memory but also wander away from their usual surroundings. When found they usually deny all memory of their whereabouts during the period of wandering, and may also deny knowledge of personal identity. Fugues also occur in epilepsy, severe depressive disorders, and alcoholism. They may be associated with suicide attempts. Many patients who present in fugue give a history of seriously disturbed relationships with their parents in childhood, and others are habitual liars (Stengel 1941).

Dissociative stupor

In dissociative stupor, patients show the characteristic features of stupor. They are motionless and mute, and they do not respond to stimulation, but they are aware of their surroundings. Dissociative stupor is rare. It is essential to exclude other possible conditions, namely schizophrenia (p. 249), depressive disorder (p. 200) mania (p. 203), and organic brain disorder.

Ganser's syndrome

Ganser's syndrome is a rare condition with four features: giving 'approximate answers' to questions designed to test intellectual functions, psychogenic physical symptoms, hallucinations, and apparent clouding of consciousness. The syndrome was first described among prisoners (Ganser 1898) but is not confined to them. The term 'approximate answers' denotes answers (to simple questions) that are plainly wrong but are clearly related to the correct answer in a way that suggests that the latter is known. For example, when asked to add two and two a patient might answer five, when asked to add two and three he might answer six, and when asked to add five and four he might answer ten; each answer is one greater than the correct answer. When present, hallucinations are usually visual and may be elaborate. The obvious advantage to be gained from illness, coupled with the approximate answers, often suggests a crude form of malingering. However, the condition is maintained so consistently that unconscious mental mechanisms are generally thought to play a part. Some authors have suggested that the syndrome is an unusual form of psychosis (Whitlock 1961). It is important to exclude an organic or functional psychosis; the former should be considered particularly carefully when muddled thinking and visual hallucinations are part of the clinical picture.

Dissociative identity disorder

In this disorder, also known as *multiple personality disorder*, there are sudden alternations between two patterns of behaviour, each of which is forgotten by the patient when the other is present. Each pattern of behaviour is a complex and integrated scheme of emotional responses, attitudes, memories, and social behaviour, and the behaviour usually contrasts strikingly with the patient's normal state. In some cases there is more than one additional behaviour pattern or 'personality'. The condition is rare, though it may be observed more frequently at some times than at others, possibly because it is fostered by the increased interest of doctors at that time. In the past, striking examples of this disorder were described by Morton Prince (1908) and in the book *The three faces of Eve* (Thigpen *et al.* 1957). Interest in the condition has recently been revived in the United States after increased reporting of cases. Like psychogenic memory disorder, dissociative identity disorder may occasionally have a basis in organic disease of the central nervous system (Lewis 1953*a*). For a review of the history

and present status of the condition see North *et al.* (1993).

The *causes* of this disorder are uncertain. Such cases often receive wide publicity, which may provoke the behaviour in some suggestible people. Many of these patients report physical or sexual abuse during their childhood; thus Ross *et al.* (1990) used a standardized interview and reported a history of abuse in 95 per cent of 102 patients with this condition. It has been suggested that this abuse might be causal (Kluft 1985), but there is seldom any corroborating evidence to decide whether patients' recollections are accurate or the response of a suggestible person to direct enquiries (Frankel 1993).

Patients who develop this pattern of behaviour often meet the criteria for other diagnoses, especially antisocial personality disorder and alcohol or drug abuse; they also have symptoms of anxiety and depression (Bliss 1984). The relationship between multiple personality disorder and these other conditions would be illuminated by follow-up studies over many years, but no systematic studies of this kind have been reported.

Dissociative trance

This condition is referred to as trance and possession disorder in ICD10. Trance and possession states are characterized by a temporary loss of the sense of personal identity and of full awareness of the person's surroundings. Some cases resemble multiple personality disorder, with the person acting as if taken over by another personality for a brief period. When the condition is induced by religious ritual, the person may feel taken over by a deity or spirit. The focus of attention is narrowed to a few aspects of the immediate environment, for example to the priest carrying out a religious ceremony. The affected person may repeatedly perform the same movements, or adopt postures, or repeat utterances.

Such states are induced temporarily in willing participants in religious or other ceremonies. In these circumstances these states are not recorded as disorders. The terms dissociative trance disorder, and trance and possession disorder, are used for states that are unwanted and occur outside religious or other ceremonies, or having begun in these ceremonies are exceptionally prolonged. The causes and treatment of such disorders are not understood.

Some cases of latah, amok, and similar conditions (see p. 192) may be classified under this heading.

Conversion disorder

Conversion disorders (dissociative motor disorder, convulsions, and sensory disorder) are considered on p. 352.

Related syndromes

Epidemic hysteria

Occasionally, dissociative (or conversion) disorder spreads within a group of people as an 'epidemic'. This spread happens most often in closed groups of young women, for example in a girls' school, a nurses' home, or a convent. Usually anxiety has been heightened among the members of the group by some threat to the community, such as the possibility of being involved in an epidemic of actual and serious physical disease present in the neighbourhood. Typically, the epidemic starts in one person who is highly suggestible, histrionic, and a focus of attention in the group. This first case may result from a general apprehension about the threat of physical illness, or from a specific concern about an acquaintance who has contracted the illness. Gradually other cases appear, first in the most suggestible and then, as anxiety mounts, among those with less predisposition. The symptoms are variable, but fainting and dizziness are common. Outbreaks among schoolchildren have been reported by Benaim *et al.* (1973) and Moss and McEvedy (1966). Some writers believe that the 'dancing manias' of the Middle Ages may have been hysterical epidemics in people aroused by religious fervour.

Transcultural variations in dissociative disorders

Certain patterns of unusual behaviour are found in particular cultures. They may be variants of dissociative disorders, though they may have more than one cause.

Latah, which is found among women in Malaya (Yap 1951), is characterized by echolalia, echopraxia, and other kinds of abnormally compliant behaviour. The condition usually begins after a frightening experience.

Amok has been described among men in Indonesia and Malaya (Van Loon 1927). It begins with a period of brooding, which is followed by violent behaviour and sometimes dangerous use of weapons. Amnesia is usually reported afterwards. It is unlikely that all patients with this pattern of behaviour are suffering from a dissociative disorder; the others may be manic, schizophrenic, or postepileptic.

Arctic hysteria is seen among the Eskimo, more often in the women. The affected person tears off her clothing, screams and cries, runs about wildly, and may endanger her life by exposure to cold. Sometimes the behaviour is violent. The relationship of this syndrome to dissociative disorder is not firmly established, and there may be more than one cause.

For further information about cultural variations of hysteria see Kiev (1972), Leff (1981), and Simons and Hughes (1985).

Differential diagnosis

There are three ways in which physical disease may be wrongly diagnosed as a dissociative (or conversion) disorder. First, the symptoms may be those of physical disease that has not yet been detected, for example an undiagnosed brain tumour causing amnesia. Second, undiscovered brain disease may, in some unknown way, 'release' hysterical symptoms; for example dissociative symptoms may be the first evidence of a small tumour in the frontal or parietal lobe, or an early dementia. Third, the anxiety caused by the awareness of the early symptoms of physical disease may act as a non-specific stimulus which provokes additional dissociative symptoms such as fugue in a minority of people.

Differential diagnosis from organic disease may be difficult. The greatest problem arises with **organic diseases of the central nervous system.** The first step is to determine the exact form of the symptoms and signs, and to compare them carefully with those arising from known neurological disease. Dissociative disorders may be difficult to distinguish from **partial complex seizures** (temporal lobe epilepsy) in which unusual disorders of behaviour can occur (see Chapter 11). These points should be considered afresh each time such behaviour occurs in a patient who has more than one episode of symptoms.

Mistakes in diagnosis arise because dissociative and conversion disorders are confused with the extravagant behaviour of a **histrionic (or hysterical) personality** (p. 114). When distressed, people with this kind of personality display emotions readily and tend to react in a demonstrative way that attracts attention. They respond in the same way to physical illness as to other events in their life—by exaggeration. Such over-reaction to organic disease may be mistaken for the wholly psychological dissociative or conversion disorders. (Exaggeration of physical symptoms is sometimes called 'hysterical overlay' or 'functional overlay'.) Similarly, the histrionic personality can put its stamp on psychiatric disorder, and histrionic behaviour can occur in depressive disorders, anxiety disorders, and many other conditions.

The distinction between dissociative and conversion disorders and **malingering** should be considered particularly among prisoners, military servicemen, or others who may have reasons to feign illness to avoid some unpleasant duty or punishment or to gain compensation. As explained above, the distinction may be difficult to make because some patients add conscious embellishments to the core of unconsciously produced dissociative or conversion symptoms. This embellishment happens more often when the patient believes that his doctor is sceptical about his original complaint. Unlike dissociative

symptoms, the complaints of malingerers can rarely be sustained continuously; for this reason, discreet and prolonged observation will usually provide valuable information. Malingering and factitious disorders are considered further on p. 356–7.

Diagnostic errors will be minimized if four other points are taken into account. First, dissociative and conversion disorders seldom appear for the first time after the **age of 40**, presumably because most predisposed patients have already encountered problems stressful enough to provoke the reaction at an earlier age. Second, dissociative and conversion disorders are **provoked by stressors**. If no stressor can be found, the diagnosis is in serious doubt. It is important to question other informants, since the patient may not reveal stressful circumstances of which he feels ashamed. On the other hand, it is essential to remember that finding stressors does not prove the diagnosis of conversion or dissociative disorder because they often precede physical illness. The third point concerns **secondary gain**, which is usually present in dissociative disorder. As already noted, secondary gain does not prove the diagnosis because patients may gain advantage from physical illness as well as from dissociative disorder. The fourth point, also noted above, is that **hysterical indifference** can seldom be judged reliably, and should be given little weight in diagnosis.

Epidemiology

The lifetime prevalence of dissociative disorder in the general population is difficult to determine. The combined prevalence of dissociative and conversion disorders is probably between 3 and 6 per 1000 for women, and substantially less for men (Carey *et al.* 1980). Clinical experience suggests that most cases begin before the age of 35 and few new cases appear after the age of 40, although dissociative and conversion *symptoms* commonly occur as part of some other disorder well beyond this age.

Aetiology

Most information about aetiology applies equally to dissociative and conversion disorders. To avoid repetition the aetiology of both disorders is considered here and not repeated in Chapter 12.

Genetics

The few genetic studies have been inconclusive. Ljungberg (1957) studied first-degree relatives of 281 patients, of whom almost half had conversion disturbances of gait and a further fifth had dissociative seizures. He found rates of any dissociative or conversion disorder among the relatives (2.4 per cent among males and 6.4 per cent among females) that were probably higher than in the general population. A small twin study by Slater (1961) did not support a genetic aetiology since no concordance was found between identical twins, one of whom had a dissociative or conversion disorder.

Organic disease

As already noted, dissociative and conversion disorders are sometimes associated with organic disease of the nervous system. However, they can undoubtedly occur in the absence of such pathology, and when the pathology is present the intervening mechanisms are not known.

Psychological causes

It is generally accepted that the immediate cause of dissociative and conversion disorders is psychological, although the mechanisms may not be those proposed in Freudian theory (see p. 188). The essential feature seems to be the capacity to dissociate (as proposed by Janet), i.e. to disconnect one aspect of psychological function from the rest when the person is subjected to severely stressful events. The mechanisms underlying dissociation are not known. Also, it is not known whether or how far the tendency to dissociate is innate or is developed through learning in childhood.

Prognosis

Most patients with dissociative and conversion disorders of recent onset seen in general practice or hospital emergency departments recover quickly. However, those cases that last longer than a year are likely to persist for many years. For example, Ljungberg (1957) found that, among patients who still had symptoms after a year, half still had them after 10 years.

It has already been noted that organic disease is often missed among these patients. In an important study, Slater and Glithero (1965) followed-up a series of patients who had been referred to a specialist neurological hospital and diagnosed at that time as having hysteria (those with 'hysterical overlay' in known physical disease were excluded). About a third of the patients developed a definite organic illness within 7–11 years of the original diagnosis, and a further third developed depression or schizophrenia. Although this study teaches an important lesson, it must be remembered that the patients were unrepresentative in being referred to a neurological hospital.

Treatment

For acute dissociative and conversion disorders seen in general practice or hospital casualty departments, treatment by reassurance and suggestion is usually appropriate, together with immediate efforts to resolve any stressful circumstances that provoked the reaction. More active treatment is required for cases that have lasted more than a few weeks. The general approach is to focus on the elimination of factors that are reinforcing the symptoms and on the encouragement of normal behaviour. It should be explained to the patient that he has a disability (as in remembering, or moving his arm) which is not caused by physical disease but by psychological factors. It is often helpful to explain the disorder as due to a blocking of the psychological process between, for example, patients' intentions to remember or move their arms and the nervous mechanisms that bring about recall or movement. Patients should then be told that if they try hard to regain control, they will succeed. If necessary, they can be offered help in doing this, for example in the form of physiotherapy to the arm. Attention is then directed away from the symptoms and towards problems that have provoked the disorder. The hospital staff should show concern to help the patient, and this is best done by encouraging self-help. It is important not to make undue concessions to patients' disability; for example a patient who cannot walk should not be provided with a wheelchair, and a patient who has collapsed on the floor should be encouraged to get up unaided. To achieve these ends, there must be a clear plan so that all members of staff adopt a consistent approach to the patient.

Abreaction

Abreaction was used in the past to treat dissociative and conversion disorders, especially those arising in battle. It is of much less value in civilian practice, where more gradual methods allow patients to take responsibility for overcoming their symptoms and for finding solutions to the problems that provoked them. If abreaction is used, it can be brought about by hypnosis or by intravenous injection of small amounts of amylobarbitone. In the resulting state, the patient is encouraged to relive the stressful events that provoked the disorder and to express the accompanying emotions.

Dynamic psychotherapy

Patients with dissociative and conversion disorders often appear to respond well in the early stages of exploratory psychotherapy concerned with their past life, and may produce striking memories of childhood sexual behaviour and other problems apparently of significance to a psychodynamic view of the case. However, it is seldom fruitful to explore these ideas at length because such exploration is likely to deflect attention from the patient's current difficulties, and in some cases may lead to transference reactions that are difficult to manage.

Other treatments

Medication has no part to play in the treatment of these dissociative and conversion disorders, although treatment of a depressive or anxiety disorder usually leads to an improvement of any secondary dissociative or conversion symptoms. Specific methods of **behaviour therapy** are also of little value. In the past operant conditioning methods were tried, for example in the treatment of psychogenic blindness (Parry-Jones *et al.* 1970), but there is no evidence that these or other techniques are more effective than suggestion.

Subsequent care

Most patients with dissociative and conversion disorders improve with simple treatment unless there is clear advantage in remaining ill, as in some compensation cases. Those who do not improve should be reviewed thoroughly for undiscovered physical illness. All patients, whether improved or not, should be followed carefully for long enough to exclude any organic disease that might not have been detected. Six months to a year will usually be needed, but if a condition such as multiple sclerosis has to be excluded, a much longer follow-up may be required. This must be done discreetly and tactfully in order to identify any symptoms suggestive of organic disease without perpetuating the psychological problems, The general practitioner is often best placed to undertake this follow-up.

Depersonalization disorder

Depersonalization disorder is characterized by an unpleasant state of disturbed perception in which external objects or parts of the body are experienced as changed in their quality, unreal, remote, or automatized. The patient is aware of the subjective nature of this experience. The symptom of depersonalization is quite common as a minor feature of other syndromes, but depersonalization disorder is quite uncommon.

Depersonalization disorder is classified as a dissociative disorder in DSMIV but has a separate place in ICD10. There is insufficient evidence about the aetiology of depersonalization disorder to be certain how closely it is related to the dissociative disorders.

Clinical picture

Patients describe feelings of being unreal and experiencing an unreal quality to perceptions. They say that their emotions are dulled and that their actions feel mechanical. Paradoxically they complain that this lack of feeling is extremely unpleasant. Insight is retained into the subjective nature of their experiences. These symptoms may be intense, and accompanied by *déjà vu* and by changes in the experience of passage of time. Some patients complain of sensory distortions affecting a single part of the body (usually the head, the nose, or a limb), which may be described as feeling as if made of cotton wool.

Two-thirds of the patients are women. The onset is often in adolescence or early adult life, with the condition starting before the age of 30 in about half the cases (Shorvon *et al.* 1946). The symptoms usually begin suddenly, sometimes when the person feels aroused but sometimes in the course of relaxation after intense physical exercise (Shorvon *et al.* 1946). Once established, the disorder often persists for years, though with periods of partial or complete remission.

Differential diagnosis

Before diagnosing depersonalization disorder, any primary disorder must be excluded, especially temporal lobe epilepsy, schizophrenia, depressive disorder, obsessional disorder, conversion or dissociative disorder, and generalized and phobic anxiety disorders. Severe and persistent depersonalization is experienced by some people with schizoid personality disorder. Most patients who present with depersonalization will be found to have one of these other disorders. The primary syndrome is rare. Ackner (1954*a,b*) found that all cases could be allocated to organic, depressive,

anxiety, or hysterical syndromes, or to schizoid personality disorder.

Aetiology

The causes of a primary depersonalization disorder are not known. Apart from the possible association with schizoid personality disorder, no definite constitutional factors have been identified. Lader (1969) suggested that the symptoms represent a restriction of sensory input that serves to reduce intolerably high levels of anxiety. He reported a striking example in a patient who was undergoing physiological recordings at the time. Since many cases begin when the patient is relaxed or tired, this mechanism, if it is important, cannot be invariable.

Prognosis

Secondary cases have the prognosis of the primary condition. The rare primary depersonalization disorder has not been followed systematically; clinical experience indicates that cases lasting longer than a year have a poor long-term outcome.

Treatment

When depersonalization is secondary to another disorder, treatment should be directed to the primary condition. In the rare primary depersonalization disorder, anxiolytic drugs may give short-term relief but they should not be prescribed for long periods because of the risk of dependency. Other drugs are generally ineffective, as are cognitive behavioural treatments and dynamic psychotherapy. Supportive interviews may help the patient to function normally despite the symptoms, and any stressors should be reduced if possible. As in other untreatable conditions, it is important to resist the temptation to give ineffective treatments in order to appear to be helping the patient; it is better to give adequate time for supportive care that enables the patient to live as normal a life as possible.

Further reading

Coryell, W. and Winokur, G. (1991). *The clinical management of anxiety disorders*. Oxford University Press, New York.

Goldberg, D. and Huxley, P. (1992). *Common mental disorders: a biosocial model*. Routledge, London.

Marks, I. M. (1987). *Fears, phobias, and rituals*. Oxford University Press, New York.

Mersky, H. (1979). *The analysis of hysteria*. Baillière Tindall, London.

8 | *Mood disorders*

The mood disorders are so called because one of their main features is abnormality of mood. Nowadays the term is usually restricted to disorders in which this mood is **depression** or **elation**, but in the past some authors have included states of anxiety as well (Lewis 1956). In this book, anxiety disorders are described in Chapter 7.

It is part of normal experience to feel unhappy at times of adversity. The *symptom* of depressed mood is a component of many psychiatric syndromes and is also commonly found in certain physical diseases, for example in infections such as hepatitis and some neurological disorders. In this chapter we are concerned neither with normal feelings of unhappiness nor with depressed mood as a symptom of other disorders, but with the *syndromes* known as **depressive disorders**. The central features of these syndromes are depressed mood, pessimistic thinking, lack of enjoyment, reduced energy, and slowness. Of these, depressed mood is usually, but not invariably, the most prominent symptom. The other elements are variable enough to suggest that there is not one disorder but several.

Similar considerations apply to states of elation. A degree of elated mood is part of normal experience at times of good fortune. Elation can also occur as a *symptom* in several psychiatric syndromes, though it is less widely encountered than depressed mood. In this chapter we are concerned with a *syndrome* in which the central features are over-activity, mood change, and self-important ideas. The mood change may be towards elation or towards irritability. This syndrome is called **mania**. Some diagnostic classifications distinguish a less severe form of mania—**hypomania** (see below). As the dividing line between mania and hypomania is arbitrary, this book uses only the term mania whilst indicating the severity of the disorder by adding mild, moderate, or severe.

Clinical features

Depressive syndromes

The clinical presentations of depressive syndromes are so varied that they cannot be described fully in a short space. In the following account disorders are grouped by their severity. The account begins with a description of the clinical features of depressive disorders of moderate severity, and continues with a description of severe disorders. Certain important variants of these moderate and severe disorders are then described. Finally the special features of the least severe depressive disorders are outlined.

In depressive disorders of moderate severity, the central features are low mood, lack of enjoyment, pessimistic thinking, and reduced energy, all of which lead to decreased functioning. The patient's **appearance** is characteristic. Dress and grooming may be neglected. The facial features are characterized by a turning downwards of the corners of the mouth and by vertical furrowing of the centre of the brow. The rate of blinking may be reduced. The shoulders are bent and the head is inclined forwards so that the direction of gaze is downwards. Gestural movements are reduced. It is important to note that some patients maintain a smiling exterior despite deep feelings of depression.

The **mood** of the patient is one of misery. This mood does not improve substantially in circumstances where ordinary feelings of sadness would be alleviated, for example in pleasant company or after hearing good news. Moreover the mood may be experienced as different from ordinary sadness. Patients sometimes speak of a black cloud pervading all mental activities. Some patients can conceal this mood change from other people, at least for short periods. Some try to hide their low mood during clinical interviews, making it more difficult for the doctor to detect.

Pessimistic thoughts ('depressive cognitions') are important symptoms which can be divided into three groups. The first group is concerned with the *present*. The patient sees the unhappy side of every event; he thinks that he is failing in everything that he does and that other people see him as a failure; he no longer feels confident, and discounts any success as a chance happening for which he can take no credit.

The second group of thoughts is concerned with the *future*. The patient expects the worst. He foresees failure in his work, the ruin of his finances, misfortune for his family, and an inevitable deterioration in his health. These ideas of hopelessness are often accompanied by the thought that life is no longer worth living and that death would come as a welcome release. These gloomy preoccupations may progress to thoughts of, and plans for, **suicide**. It is important to ask about these ideas in every case. (The assessment of suicidal risk is considered further in Chapter 13.)

The third group of thoughts is concerned with the *past*. They often take the form of unreasonable **guilt** and **self-blame** about minor matters; for example a patient may feel guilty about past trivial acts of dishonesty or letting someone down. Usually these events have not been in the patient's thoughts for years but, when he becomes depressed, they flood back into his memory accompanied by intense feelings. Preoccupations of this kind strongly suggest depressive disorder. Some patients have similar feelings of guilt but do not attach them to any particular event. Other memories are focused on unhappy events; the patient remembers occasions when he was sad, when he failed, or when his fortunes were at a low ebb. These gloomy memories become increasingly frequent as the depression deepens.

Lack of interest and enjoyment is frequent, though not always complained of spontaneously. The patient shows no enthusiasm for activities and hobbies that he would normally enjoy. He feels no zest for living and no pleasure in everyday things. He often withdraws from social encounters. **Reduced energy** is characteristic (though sometimes associated with a degree of physical restlessness that can mislead the observer). The patient feels lethargic, finds everything an effort, and leaves tasks unfinished. For example, a normally houseproud woman may leave the beds unmade and dirty plates on the table. Understandably, many patients attribute this lack of energy to physical illness.

Psychomotor retardation is frequent (though, as described later, some patients are agitated rather than slowed up). The retarded patient walks and acts slowly. Slowing of thought is reflected in the patient's speech; there is a long delay before questions are answered, and pauses in conversation may be so long that they would be intolerable to a non-depressed person.

Anxiety is frequent though not invariable in moderately severe depressive disorder. (As described later, it is common in some less severe depressive disorders.) Another common symptom is **irritability**, which is the tendency to respond with undue annoyance to minor demands and frustrations. **Agitation** is a state of restlessness which is experienced by the patient as inability to relax and is seen by an observer as restless activity. When it is mild, the patient is seen to be plucking at his fingers and making restless movements of his legs; when it is severe, he cannot sit for long but paces up and down.

A group of symptoms often called biological is important. These **biological symptoms** include sleep disturbance, diurnal variation of mood, loss of appetite, loss of weight, constipation, loss of libido, and, among women, amenorrhoea. These symptoms are frequent but not invariable in depressive disorders of moderate degree. (They are less usual in mild depressive disorders, but particulary common in the severe disorders.) Some of these symptoms require further comment.

Sleep disturbance in depressive disorders is of several kinds. Most characteristic is early morning waking, but delay in falling asleep and waking during the night also occur. Early morning waking occurs two or three hours before the patient's usual time; he does not fall asleep again, but lies awake feeling unrefreshed and often restless and agitated. He thinks about the coming day with pessimism, broods about past failures, and ponders gloomily about the future. It is this combination of early waking with depressive thinking that is important in diagnosis. It should be noted that some depressed patients sleep excessively rather than wake early, but they still report waking unrefreshed.

Weight loss in depressive disorders often seems greater than can be accounted for merely by the patient's **lack of appetite**. In some patients the disturbances of eating and weight are towards excess—they eat more and gain weight. Usually it seems that eating brings temporary relief to their distressing feelings.

Complaints about **physical symptoms** are common in depressive disorders. They take many forms, but complaints of constipation, fatigue, and aching discomfort anywhere in the body are particularly common. Complaints about any pre-existing physical disorder usually increase and hypochondriacal preoccupations are common.

Several **other psychiatric symptoms** may occur as part of a depressive disorder, and occasionally one of them dominates the clinical picture. They include depersonalization, obsessional symptoms, phobias, and disassociative symptoms such as fugue or loss of function of a limb. Complaints of **poor memory** are also common; depressed patients commonly show deficits on a wide range of neuropsychological tasks, but impairments in the retrieval and recognition of recently learned material may be particularly prominent (Austin *et al.* 1992; Robbins *et al.* 1992). Sometimes the impairment of memory in a depressed patient is so severe that the clinical presentation resembles that of dementia. This presentation, which is particularly common in the elderly, is sometimes called **depressive pseudodementia** (see p. 527).

Severe depressive disorder

As depressive disorders become more severe, all the features described above occur with greater intensity. In addition, certain distinctive features may occur in the form of delusions and hallu-cinations; the disorder is then sometimes called **psychotic depression.**

The **delusions** of severe depressive disorders are concerned with the same themes as the non-delusional thinking of moderate depressive disorders. Therefore they are termed **mood congruent.** These themes are worthlessness, guilt, ill-health, and, more rarely, poverty. Such

delusions have been described in Chapter 1, but a few examples may be helpful at this point. The patient with a *delusion of guilt* may believe that some dishonest act, such as a minor concealment in making a tax return, will be discovered and that he will be punished severely and humiliated. He is likely to believe that such punishment is deserved. A patient with *hypochondriacal delusions* may be convinced that he has cancer or venereal disease. A patient with a *delusion of impoverishment* may wrongly believe that he has lost all his money in a business venture. *Persecutory delusions* also occur. The patient may believe that other people are discussing him in a derogatory way or are about to take revenge on him. When persecutory delusions are part of a depressive syndrome, typically the patient accepts the supposed persecution as something that he has brought upon himself. In his view, he is ultimately to blame. Some depressed patients experience delusions and hallucinations that are not clearly related to themes of depression or grandiosity ('**mood-incongruent**'). Their presence appears to worsen the prognosis of the illness (Kendler 1991).

Particularly severe depressive delusions are found in **Cotard's syndrome**, which was described by a French psychiatrist (Cotard 1882). The characteristic feature is an extreme kind of *nihilistic delusion* (sometimes called by the French name *délire de négation*—*délire* meaning delusion in this context). The nihilism is extreme. For example some patients may complain that their bowels have been destroyed so that they will never pass faeces again. Others may assert that they are penniless and without any prospect of having money again. Still others may be convinced that their whole family has ceased to exist. Although the extreme nature of these symptoms is striking, such cases do not appear to differ in important ways from other severe depressive disorders.

Perceptual disturbances may also be found in severe depressive disorders. Sometimes these fall short of true hallucinations ('pseudohallucina-tions'; see Chapter 1). In a minority of cases definite hallucinations occur; they are usually auditory, and take the form of voices addressing repetitive words and phrases to the patient. The voices seem to confirm his ideas of worthlessness (for example, 'you are an evil man; you should die'), or make derisive comments, or urge the patient to take his own life. A few patients experience visual hallucinations, sometimes in the form of scenes of death and destruction.

Agitated depression

This term is applied to depressive disorders in which agitation is prominent. As already noted, agitation occurs in many severe depressive disorders, but in agitated depression it is particularly severe. Agitated depression is seen more commonly among the middle-aged and elderly than among younger patients. However, there is, no reason to suppose that agitated depression differs in other important ways from the other depressive disorders.

Retarded depression

This name is sometimes applied to depressive disorders in which psychomotor retardation is especially prominent. There is no evidence that they represent a separate syndrome, though retardation does predict a good response to electroconvulsive therapy (ECT). If the term is used, it should be in a purely descriptive sense. In its most severe form, retarded depression shades into depressive stupor.

Depressive stupor

In severe depressive disorder, slowing of movement and poverty of speech may become so extreme that the patient is motionless and mute. Such depressive stupor is rarely seen now that active treatment is available. The description by Kraepelin (1921, p. 80) is of particular interest: 'The patients lie mute in bed, give no answer of any sort, at most withdraw themselves timidly from approaches, but often do not defend themselves from pinpricks . . . They sit helpless before their food, perhaps, however, they let themselves be spoon-fed without making any difficulty . . .'. Kraepelin commented that recall

of the events taking place during stupor was sometimes impaired when the patient recovered. Nowadays, the general view is that on recovery patients are able to recall nearly all the events taking place during the period of stupor (Lishman 1978). It is possible that in some of Kraepelin's cases there was clouding of consciousness (possibly related to inadequate fluid intake, which is common in these patients). Patients in states of depressive stupor may exhibit **catatonic** motor disturbances (see p. 249).

Masked depression

The term 'masked depression' is sometimes used for cases where depressive mood is not conspicuous. Although there is no reason to think that these cases form a separate syndrome, the term is useful in drawing attention to a mode of presentation that is easily missed. Masked depression is discussed in Chapter 12 on psychiatry and medicine. Here it should be noted that diagnosis depends on a careful search for the other features of depressive disorder, especially sleep disturbance, diurnal mood variation, and depressive cognitions. It is also important to detect loss of pleasure and poor concentration. Masking is most likely to occur with mild or moderate disorders, but it occasionally occurs with severe disorders.

Atypical depression

The term atypical depression has been applied to several different clinical syndromes; it has included features such as variably depressed mood, pronounced phobic anxiety, and reversal of the usual biological symptoms such that patients tend to overeat and oversleep. In recent years the Columbia group in the United States has proposed that the essential feature of atypical depression is *mood reactivity*, as shown by the capacity to be significantly cheered by positive interpersonal events. In addition, patients with atypical depression may show other associated symptoms such as *overeating* and *oversleeping*, together with a feeling of fatigue amounting to a subjective sensation of *leaden paralysis* (extreme heaviness) in the limbs. In addition, many of these patients have a lifelong tendency to react in an exaggerated way to perceived rejection *(rejection sensitivity)*. The importance of recognizing atypical depression is that, because of their interpersonal sensitivity, many of these patients, are hard to manage and are not infrequently regarded as having 'difficult' personalities rather than depression. Also atypical depression seems to be associated with a poor response to tricyclic antidepressant treatment but a better response to monoamine oxidase inhibitors (MAOIs) and perhaps to selective serotonin re-uptake inhibitors (Quitkin *et al.* 1989; Davidson 1992).

Brief recurrent depression

Some individuals experience recurrent depressive episodes of short duration, typically two to seven days. These episodes recur with some frequency, about once a month on average. There is no apparent link with the menstrual cycle in female sufferers. While the depressive episodes are short, they are as severe as the more enduring depressive disorders. Hence they are associated with much personal distress and social and occupational impairment. The risk of a manic episode is low. Therefore it seems unlikely that this disorder is a manifestation of the rapid cycling form of bipolar disorder (Angst 1992; Merikangas *et al.* 1994) (see below).

Mild depressive disorder

It might be expected that mild depressive disorders would present with symptoms similar to those of the depressive disorders described already, but with less intensity. Sometimes this is so, with the patient complaining of low mood, lack of energy and enjoyment, and poor sleep. However, in mild depressive disorder there are frequently other symptoms that are less prominent in severe disorders. These symptoms have been characterized as 'neurotic', and they include anxiety, phobias, obsessional symptoms, and, less often, dissociative symptoms. Although anxiety may be a symptom in all degrees of depressive disorder, it can be just as severe in the mild disorders as in the severe ones. This finding has suggested to many people that these mild depressive disorders are not just a minor variant of the moderate and severe cases but a separate syndrome. Because of the nature of the additional symptoms, this syndrome has been called 'neurotic depression', though in current diagnostic schemes the terms **minor** or **mild** depression are used (see p. 156).

Apart from the 'neurotic' symptoms found in some cases, mild depressive disorders are characterized by the expected symptoms of low mood, lack of energy and interest, and irritability. There is sleep disturbance, but not the early morning waking that is so characteristic of more severe depressive disorders. Instead, there is more often difficulty in falling asleep and periods of waking during the night, usually followed by a period of sleep at the end of the night. 'Biological' features (poor appetite, weight loss, and low libido) are not usually found. Although mood may vary during the day, it is usually worse in the evening than in the morning. The patient may not be obviously dejected in his appearance or slowed in his movement. Delusions and hallucinations are not present.

In their mildest forms these cases shade into the minor affective disorders considered in Chapter 6. As described later, they pose considerable problems of classification. Many of these mild depressive disorders are brief, starting at a time of personal misfortune and subsiding when fortunes have changed or a new adjustment has been achieved. However, some cases persist for months or years, causing considerable suffering even though the symptoms do not increase. These chronic depressive states are often called **dysthymia**. The term **cyclothymia** refers to a persistent instability mood in which there are numerous periods of mild elation or mild depression. It is not unusual for episodes of more severe mood disorder to supervene in patients who experience dysthymia or cyclothymia.

Mania

As already mentioned, the central features of the syndrome of mania are elevation of mood, increased activity, and self-important ideas. When the mood is elevated, the patient seems cheerful and optimistic, and he may have a quality described by earlier writers as infectious gaiety. However, other patients are irritable rather than euphoric, and this irritability can easily turn to anger. The mood often varies during the day, though not with the regular rhythm characteristic of many severe depressive disorders. In patients who are elated, not uncommonly high spirits are interrupted by brief episodes of depression.

The patient's appearance often reflects his prevailing mood. Clothes may be brightly coloured and ill-assorted. When the condition is more severe, the patient's appearance is often untidy and dishevelled. Manic patients are **over-active**. Sometimes their persistent over-activity leads to physical exhaustion. Manic patients start many activities but leave them unfinished as new ones catch their fancy. Their **speech** is often rapid and copious as thoughts crowd into their minds in quick succession. When the disorder is more severe, there is **flight of ideas** (see p. 8) with such rapid changes that it is difficult to follow the train of thought. **Sleep** is often reduced. Patients wake early, feeling lively and energetic; often they get up and busy themselves noisily, to the surprise of other people. **Appetite** is increased and food may be eaten greedily with little attention to conventional manners. **Sexual desires** are increased and behaviour may be uninhibited. Women may neglect precautions against pregnancy, a point calling for particular attention when a patient is of childbearing age.

Expansive ideas are common. Patients believe that their ideas are original, their opinions important, and their work of outstanding quality. Many patients become extravagant, spending more than they can afford on expensive cars or jewellery. Others make reckless decisions to give up good jobs, or embark on plans for hare-brained and risky business ventures.

Sometimes these expansive themes are accompanied by **grandiose delusions**. Some patients may believe that they are religious prophets or destined to advise statesmen about great issues. At times there are delusions of persecution, with patients believing that people are conspiring against them because of their special importance. Delusions of reference and passivity feelings also occur. Schneiderian first-rank symptoms (see Chapter 9, Table 9.3) have been reported in about 10–20 per cent of manic patients (Carpenter *et al.* 1973). Neither the delusions nor the first-rank symptoms last long—most disappear or change in content within days.

Hallucinations also occur. They are usually consistent with the mood, taking the form of voices speaking to the patient about his special powers or, occasionally, of visions with a religious content.

Insight is invariably impaired. Patients see no reason why their grandiose plans should be restrained or their extravagant expenditure curtailed. They seldom think of themselves as ill or in need of treatment.

Most patients can exert some **control** over their symptoms for a short time, and many do so when the question of treatment is being assessed. For this reason it is important to obtain a history from an informant whenever possible. Henry Maudsley expressed the problem well: 'Just as it is with a person who is not too far gone in intoxication, so it is with a person who is not too far gone in acute mania; he may on occasion pull his scattered ideas together by an effort of will, stop his irrational doings and for a short time talk with an appearance of calmness and reasonableness that may well raise false hopes in inexperienced people' (Maudsley 1879, p. 398).

Carlson and Goodwin (1973) have described three stages of mania which, while not in any way distinctly separated from one another, may help the reader to judge the pattern of symptoms in mild, moderate, and severe cases. In **mild** cases there is increased physical activity and speech, mood is labile, being mainly euphoric but giving way to irritability at times, ideas are expansive and the patient often spends more than he can afford, and sexual drive increases. In **moderate** cases, there is marked over-activity and pressure of speech which seems disorganized, the euphoric mood is increasingly interrupted by periods of irritability, hostility, and depression, and grandiose and other preoccupations may pass into delusions. In **severe** cases, there is frenzied over-activity, thinking is incoherent, delusions become increasingly bizarre, and hallucinations are experienced. It should be emphasized that this description is merely a guide, and there is no invariable sequence.

Mixed affective states

Depressive and manic symptoms sometimes occur at the same time. Patients who are over-active and over-talkative may be having profoundly depressive thoughts. In other patients mania and depression follow each other in a sequence of rapid changes; for example a manic patient may become intensely depressed for a few hours and then return quickly to his manic state. These changes were mentioned in early descriptions of mania by Griesinger (1867), and have been re-emphasized in recent years, for example by Kotin and Goodwin (1972).

Manic stupor

In this unusual disorder, patients are mute and immobile. Their facial expression suggests elation and on recovery they describe having experienced a rapid succession of thoughts typical of mania. The condition is rarely seen now that active treatment is available for mania. Hence an earlier description by Kraepelin (1921, p. 106) is of interest: 'The patients are usually quite inaccessible, do not trouble themselves about their surroundings, give no answer, or at most speak in a low voice . . . smile without recognizable cause, lie perfectly quiet in bed or tidy about at their clothes and bedclothes, decorate themselves in an extraordinary way, all this without any sign of outward excitement'. On recovery, patients can remember the events that occurred during their period of stupor. The condition may begin from a state of manic excitement, but at times it is a stage in the transition between depressive stupor and mania.

Periodic psychoses and rapid cycling disorders

Some bipolar disorders recur regularly with intervals of only weeks or months between episodes. In the nineteenth century these regularly recurring disorders were designated *folie circulaire* (circular insanity) by the French psychiatrist Falret (1854). They have also been referred to as **periodic psychoses**. The latter term is not satisfactory because some writers use it in a different way to include other disorders such as recurrent schizophrenic illnesses, unusual syndromes like periodic catatonia (see Chapter 9), and the recurring atypical psychoses to which Leonhard (1957) gave the name cycloid psychoses (see p. 263).

At present, the frequent recurrence of mood disturbance in bipolar patients is usually called **rapid cycling** disorder. These recurrent episodes may be depressive, manic, or mixed. The main features are that recurrence is frequent (conventionally, at least four distinct episodes a year) and that episodes are separated by a period of remission or a switch to an episode of opposite polarity. Rapid cycling disorders are much more common in females, and may predict a poor response to lithium treatment. An issue of practical importance is that rapid cycling can sometimes be triggered by antidepressant medication; in these circumstances withdrawal of the antidepressant can lead to resolution of the frequent mood swings (Wehr *et al.* 1988; Bauer and Whybrow 1991).

Classification of depressive disorders

There is no general agreement about the best method of classifying depressive disorders. Three broad approaches have been tried. The first attempts to base classification on *aetiology*, the second on *symptoms*, and the third on the *course* of the disorder. One of the more useful results of this work has been the delineation of a subtype of depression formerly called *endogenous depression* but now known as *melancholic* or *somatic depression*. This classification of depression has been derived from both aetiological and symptomatic approaches and its development will be described first. Other classifications based on aetiology and course will then be described more briefly.

Classifications based on aetiology and symptoms

Reactive and endogenous depression

This scheme was based on the view that depressive disorders could be classified on the basis of aetiology into two groups—**endogenous** and **reactive** (less commonly called **exogenous**). In endogenous disorders, symptoms were caused by factors within the individual person, and were independent of outside factors. In reactive disorders, symptoms were a response to external stressors. This distinction between endogenous and reactive causes is unsatisfactory because it produces categories that are not mutually exclusive but overlapping. For this reason, the endogenous–reactive distinction was judged to be of little use in classification by many influential psychiatrists, notably Mapother (1926), Lewis (1934, 1936a, 1938), and Curran (1937). For example, Lewis (1934) wrote, 'every illness is a product of two factors—of the environment working on the organism—whether the constitutional factor is the determining influence or the environmental one, is never a question of kind, never a question to be dealt with as either/or'. Nowadays most psychiatrists agree that it is pointless to try to allocate depressive syndromes exclusively to endogenous or reactive categories in terms of aetiology; in seeking to understand the development of depression in individual cases, the relative contributions of both endogenous and reactive factors must be carefully weighed. Neither ICD10 nor DSMIV contains categories of reactive or endogenous depression.

The reactive–endogenous classification of depressive disorders carried a further complication. Many supporters of this classification maintained that the two aetiological categories were associated with characteristic *patterns of symptoms*. Thus endogenous disorders were said to be characterized by *loss of appetite, weight loss, constipation, reduced libido, amenorrhoea, and early morning waking* (the 'biological' symptoms mentioned on p. 198). Reactive disorders were said to be characterized by a pattern of *anxiety, irritability, and phobias*. The latter three symptoms are also used in another system of classification (described below) to distinguish **neurotic** from **psychotic** depressive disorders. In this way, confusion arose between two systems of classification, the reactive-endogenous classification (based on aetiology but incorporating symptoms as well) and the neurotic-psychotic classification (based only on symptoms). Some authors made no sharp distinction between the two systems (Kiloh *et al.* 1972).

Neurotic and psychotic depressions

As already explained, certain *symptoms* are frequently more intense in the mild depressive disorders than in the severe disorders. This difference in symptom intensity led to the suggestion that there were two distinct forms of depressive disorder, *neurotic* and *psychotic*. Over the last thirty years this hypothesis has been pursued by statistical means; standardized information has been gathered either from case notes or by interviewing, and then examined by some form of multivariate analysis. The results have been contradictory. In a series of papers, Roth and his colleagues in Newcastle held that two separate syndromes could be distinguished (Kiloh and Garside 1963; Carney *et al.* 1965). However, Kendell (1968) did not confirm this distinction, but found evidence for a unimodal distribution of cases.

The problems surrounding these issues are made more difficult by the imprecise use of the term psychotic. In one sense, this term means a disorder in which there is evidence of loss of contact with reality, usually in the form of hallucinations or delusions. However, in the literature on depressive disorders, the term has also been applied to cases with so-called biological symptoms, namely early morning waking, weight loss, poor appetite, impaired libido, and diurnal variation.

Melancholic and somatic depression

Current views are that the term neurotic depression is probably not useful because it

Table 8.1 *Clinical features of melancholic and somatic depression*

  Melancholic features (DSMIV)
    Loss of interest or pleasure in usual activities*
    Lack of reactivity to pleasurable stimuli*
  plus at last three of the following
    Distinct quality of mood (unlike normal sadness)
    Morning worsening of mood*
    Early morning waking*
    Psychomotor agitation or retardation*
    Significant anorexia or weight loss*
    Excessive guilt
  Marked loss of libido* (ICD10 only)

---

*Somatic symptoms of depression in ICD10 (at least four required for diagnosis)

covers several different disorders including, for example, mild depressive episodes, atypical depressions, and dysthymia. However, the clinical syndrome of endogenous depression (*major depression with melancholia* in DSMIV, or *depressive episode with somatic symptoms* in ICD10) (Table 8.1) as defined by symptom profile appears to have a number of validating characteristics that distinguish it from depressions lacking these features:

- better response to certain somatic treatments (for example ECT)
- poorer response to placebo drug treatment
- more evidence of neurobiological abnormalities (decreased latency to rapid eye movement sleep, impaired growth hormone response to clonidine).

Of course, it is possible that these characteristics simply reflect a greater severity of somatic or melancholic endogenous depressions; nevertheless the predictions for treatment response can be clinically useful (for a review of these issues see Grove and Andreasen (1992)).

Primary and secondary depression

This scheme, which is based solely on aetiology, was introduced mainly for research purposes.

The aim was to exclude cases of depression that might be caused by another disorder. This exclusion was attempted by applying the term 'secondary' to all cases with a history of previous non-affective psychiatric illness (such as schizophrenia or anxiety neurosis) or of alcoholism, medical illnesses, or the taking of certain drugs (such as steroids). At first it was suggested (Guze *et al.* 1971) that primary and secondary depressive disorders might differ in prognosis and response to treatment. No such difference has been found, nor is there any convincing evidence for a difference between the two groups in the pattern of symptoms (Weissman *et al.* 1977). Therefore, although this classification may have some value for research, it has little value for the clinician.

Classification by course and time of life

Unipolar and bipolar disorders

Kraepelin was guided by the course of illness when he brought mania and depression together as a single entity. He found that the course was essentially the same whether the disorder was manic or depressive, and so he put the two together in a single category of manic depressive psychosis. This view was widely accepted until 1962 when Leonhard *et al.* suggested a division into three groups: patients who had had a depressive disorder only (**unipolar depression**), those who had had mania only (**unipolar mania**); and those who had had both depressive disorder and mania (**bipolar**). Nowadays it is the usual practice not to use the term unipolar mania, but to include all cases of mania in the bipolar group on the grounds that nearly all patients who have mania eventually experience a depressive disorder.

In support of the distinction between unipolar and bipolar disorders, Leonhard *et al.* (1962) described differences in heredity and personality between the groups. However, it is generally agreed that the two groups do not differ in their symptoms when depressed. There must be some overlap between the two groups, because a patient classified as having unipolar depression at one time may have a manic disorder later. In other words, the unipolar group inevitably contains some bipolar cases that have not yet declared themselves. Despite this limitation, the division into unipolar and bipolar cases is a useful classification because it has some implications for treatment (see p. 229).

Seasonal affective disorder

Some patients repeatedly develop a depressive disorder at the same time of year. In some cases the timing reflects extra demands placed on the person at a particular season of the year, either in his work or in other aspects of his life. In other cases there is no such cause and it has been suggested (Rosenthal *et al.* 1984) that they are related in some way to the changes in the seasons, for example to the length of daylight. Although these seasonal affective disorders are characterized mainly by the time at which they occur, some symptoms are said to occur more often than in other affective disorders. These symptoms are hypersomnia and increased appetite with craving for carbohydrate.

The most common pattern is onset in autumn or winter, and recovery in spring or summer. Some patients show evidence of hypomania or mania in the summer suggesting that they have a seasonal bipolar illness. This pattern has led to the suggestion that shortening of daylight is important, and to attempts at treatment by exposure to bright artificial light during hours of darkness. It has been reported that symptoms are reduced after two to five days of this treatment (in some patients longer is needed), though relapse generally occurs soon after the bright light is stopped (Rosenthal *et al.* 1984). This effect seems to be due to the extra light rather than the accompanying reduction in sleep (sleep deprivation can reduce depressive symptoms in some patients (see p. 233)). Such changes might be due to placebo effect, but this explanation is made less likely by the observation that the effect of bright light is greater than that of dim illumination (Kripke *et al.* 1983; Rosenthal *et al.* 1984, 1985). However, it has also been reported that extra light given in daytime hours is as effective as extra light given in hours of darkness (Wehr *et al.* 1986). This finding casts doubt on the idea that these disorders are caused by the shortage of daylight hours and corrected by adding to them. Speculations that the improvements are related to the known effect of light in suppressing the nocturnal secretion of **melatonin** are not supported by strong evidence (Checkley *et al.* 1993).

Involutional depression

In the past, depressive disorders starting in middle life were thought to be a separate group characterized by agitation and hypochondriacal symptoms. It was suggested that they might have a distinct aetiology, such as involution of the sex glands or some kind of relationship with schizophrenia. Family studies do not support the idea of a separate group. Thus, among the relatives of patients with so-called involutional depression, the frequency of affective disorders is increased and there is no excess of involutional disorders (early onset disorders are just as frequent). Similarly, the rate of schizophrenia among relatives is not increased (Stenstedt 1952; Slater and Cowie 1971, p. 86).

Senile depression

In the past, elderly patients with depressive disorders were also regarded as a separate group. There has been debate as to whether elderly

Table 8.2 *Classification of bipolar disorder*

ICD10	DSMIV
Manic episode Hypomania Mania Mania with psychosis	*Hypomanic episode* *Manic episode* Mild Moderate Severe Severe with psychosis
Bipolar affective disorder Currently hypomanic Currently manic Currently depressed Currently mixed In remission	*Bipolar I and bipolar II disorder* Current (or most recent episode) Hypomanic Manic* Depressed Mixed*
	Cyclothymia

*Excludes bipolar II

depressives have a worse prognosis than younger depressives. Recent studies suggest that the outcome is fairly similar in both groups (Cole 1990). However, elderly patients with a late onset of depressive disorder have larger cerebral ventricles than age-matched patients with an early onset, suggesting that structural brain disease may contribute to aetiology in the former group (Alexopoulos *et al.* 1992). Whether this disease is associated with a worse prognosis is unclear.

Classification in ICD and DSM

The main categories in the sections on mood disorders in DSMIV and the ICD10 are shown in Tables 8.2 and 8.3. Broad similarities are evident, together with some differences. The first similarity is that both systems contain categories for single episodes of mood disorder as well as categories for recurrent episodes. The second is that both recognize **mild** but **persistent** mood disturbances in which there is either a repeated alternation of high and low mood (**cyclothymia**) or a sustained

depression of mood (**dysthymia**). In neither case are the mood disturbances sufficiently severe to meet criteria for a hypomanic or depressive episode. In DSMIV, mood disorders judged to be secondary to a medical condition are included as a subcategory of mood disorders, while in ICD10 these conditions are classified as mood disorders under 'Organic mental disorders'.

Bipolar disorder

Both classifications delineate **hypomania** from **mania** on grounds of severity which include both symptomatic features and degree of social incapacity. A manic episode can be further subdivided according to **severity** and whether or not **psychotic symptoms** are present. In DSMIV the presence of a single episode of hypomania or mania is sufficient to meet criteria for bipolar affective disorder. In ICD10, however, at least two episodes of mood disturbance are needed for this diagnosis. DSMIV also separates bipolar disorder into **bipolar I** (in which mania has occurred on

Table 8.3   *Classification of depressive disorders*

ICD10	DSMIV
Depressive episode 　Mild 　Moderate 　Severe 　Severe with psychosis	*Major depressive episode* 　Mild 　Moderate 　Severe 　Severe with psychosis
Other depressive episodes 　Atypical depression	
Recurrent depressive disorders 　Currently mild 　Currently moderate 　Currently severe 　Currently severe with psychosis 　In remission	*Major depressive disorder recurrent*
Persistent mood disorders 　Cyclothymia 　Dysthymia	*Dysthymic disorder*
Other mood disorders 　Recurrent brief depression	*Depressive disorders not otherwise specified* 　Recurrent brief depression

least one occasion) and **bipolar II** (where hypomania has occurred but mania has not). The diagnosis of bipolar II disorder is intended to indicate the importance of detecting mild hypomanic episodes in patients who might otherwise be diagnosed as having recurrent major depression. The presence of such episodes may have implications for treatment response. In DSMIV, if a hypomanic or manic episode appears to have been precipitated by a somatic treatment, for example drug therapy, it is not counted towards the diagnosis of bipolar I or bipolar II disorder.

Depressive disorders

Both ICD10 and DSMIV classify depressive episodes on the basis of **severity** and whether or not **psychotic** features are present. It is also possible to specify whether the depressive episode has **melancholic** (DSMIV) or **somatic** (ICD10) features. In DSMIV, an episode of major depression with appropriate clinical symptomatology (see above) can be specified as **atypical depression**. In ICD10, **atypical depression** is classified separately under 'Other depressive episodes'. Both ICD10 and DSMIV allow the diagnosis of **recurrent brief depression**, but under slightly different headings (Table 8.3).

Classification and description in everyday practice

Although neither DSMIV nor ICD10 is entirely satisfactory, it seems unlikely that further

Table 8.4 *A systematic scheme for the clinical description of mood disorders*

The episode	
Severity	Mild, moderate, or severe
Type	Depressive, manic, mixed
Special features	With melancholic symptoms
	With neurotic symptoms
	With psychotic symptoms
	With agitation
	With retardation or stupor
The course	Unipolar or bipolar
Aetiology	Predominantly reactive
	Predominantly endogenous

rearrangement of descriptive categories will be better. The solution will come only when we have a better understanding of aetiology. Meanwhile, either ICD or DSMIV should be used for statistical returns. For most clinical purposes, it is better to *describe* disorders systematically than to classify them. This can be done for every case by referring to the severity, the type of episode, and the course of the disorder, together with an evaluation of the relative importance of endogenous and reactive causation.

This scheme is shown in Table 8.4. The **severity** of the episode is described as mild, moderate, or severe. The **type** of episode is described as depressive, manic, or mixed. Any **special features** are noted, namely melancholic symptoms, neurotic symptoms, psychotic symptoms, agitation, retardation, or stupor.

The presence of **recurrent episodes** of mood disorder is noted, and the **course** of the disorder is characterized as unipolar or bipolar. If the term bipolar is used descriptively, it is logical to restrict it to cases that have had both manic and depressive episodes. However, it has become conventional to record all cases with a manic episode as bipolar even if there has been no depressive disorder, on the grounds that (a) most manic patients develop a depressive disorder eventually and (b) in several important ways

manic patients resemble patients who have had both types of episode. This convention is followed in this textbook.

Finally the predominant **aetiology** is noted, remembering that all cases have both endogenous and reactive causes.

Differential diagnosis

Depressive disorders have to be distinguished from normal sadness and from other psychiatric disorders, namely neuroses, schizophrenia, and organic brain syndromes. As already explained on p. 197, the distinction from normal sadness is made on the presence of other symptoms of the syndrome of depressive disorder.

Mild depressive disorders are sometimes difficult to distinguish from **anxiety neuroses**. This point is also discussed in Chapter 6; here it need only be observed that accurate diagnosis depends on assessment of the relative severity of anxiety and depressive symptoms, and on the order in which they appeared. Similar problems arise when there are prominent **phobic** or **obsessional** symptoms, or when there are hysterical conversion symptoms with or without histrionic behaviour. In each case, the clinician may fail to identify the depressive symptoms and so prescribe the wrong treatment.

As cases with mixed anxiety and depressive symptoms are common, it can be asked whether anxiety and depressive disorders can really be distinguished from one another. In a follow-up study of 66 patients with anxiety disorders and 45 patients with depressive disorders, differences were found in the course of the two conditions over an average period of nearly four years (Kerr *et al.* 1972; Schapira *et al.* 1972). Of the 66 patients originally diagnosed as having anxiety disorders, 24 (40 per cent) relapsed and all but one developed a further anxiety disorder. Of those originally diagnosed as having depressive disorders, 12 (26 per cent) relapsed and all but two developed a further depressive disorder. Moreover, features predicting relapse in the two groups were different, a finding that was strongly against the hypothesis that all the cases were really manifestations of the same disorder.

The differential diagnosis from **schizophrenia** depends on a careful search for the characteristic features of this condition (see Chapter 9). Difficult diagnostic problems arise when the patient has persecutory delusions, but here again the distinction can usually be made on careful examination of the mental state and on the order in which symptoms appeared. Particular difficulties also arise when symptoms characteristic of depressive disorder and of schizophrenia are found in equal measure in the same patient; these so-called **schizoaffective** disorders are discussed on p. 261.

In middle and late life, depressive disorders are sometimes difficult to distinguish from **dementia** because some patients with depressive symptoms complain of considerable difficulty in remembering. In depressive disorders, difficulty in remembering occurs mainly because poor concentration leads to inadequate registration. The distinction between the two conditions can often by made by careful memory testing (if necessary by a clinical psychologist), though it can be extremely difficult. If memory disorder does not improve with recovery of normal mood, an organic brain syndrome is probable (see also p. 527).

Manic disorders have to be distinguished from schizophrenia, organic brain disease involving the frontal lobes (including brain tumour and general paralysis of the insane), and states of brief excitement induced by amphetamines. The diagnosis from **schizophrenia** can be most difficult. Auditory hallucinations and delusions, including some that are characteristic of schizophrenia such as delusions of reference, can occur in manic disorders. However, these symptoms usually change quickly in content, and seldom outlast the phase of over-activity. When there is a more or less equal mixture of features of the two syndromes, the term schizoaffective (sometimes schizomanic) is often used. This term is discussed further in Chapter 9.

An **organic brain lesion** should always be considered, especially in middle-aged or older patients with expansive behaviour and no past history of affective disorder. In the absence of gross mood disorder, extreme social disinhibition (for example urinating in public) strongly suggests frontal lobe pathology. In such cases appropriate neurological investigation is essential.

The distinction between mania and excited behaviour due to **drug abuse** depends on the history and an examination of the urine for drugs before treatment with psychotropic drugs is started. Drug-induced states usually subside quickly once the patient is in hospital (see Chapter 17).

The epidemiology of mood disorders

It is difficult to determine the prevalence of depressive disorder, partly because different investigators have used different diagnostic definitions. Many studies in the United States have been concerned not with the syndrome of depression, but with depressive symptoms arising in any circumstances (Weissman and Klerman 1978). Such data are of little value because of the failure to distinguish between depressive symptoms as part of another syndrome (for example schizophrenia), and depressive symptoms as part of a depressive disorder. Recently, important advances in both the United States and the United Kingdom have stemmed from the use of structured diagnostic interviews linked to standardized diagnostic criteria. In the United Kingdom, the principal structured interview has been the **Present State Examination** (PSE) (Wing *et al.* 1974), which is used to produce diagnoses related to the ICD classification. In the United States, the **Schedule for Affective Disorders and Schizophrenia** (Endicott and Spitzer 1978) can be used to make diagnoses based on the Research Diagnostic Criteria and the DSM system. These diagnostic criteria can also be generated by the **Diagnostic Interview Schedule** (DIS) (Robins *et al.* 1982) which is a highly structured interview designed to be used by non-clinical investigators.

Depressive symptoms

Depressive symptoms are common, as shown by reported point prevalences of between 13 and 20 per cent of the population. They are more frequent among **women**, the **lower socioeconomic groups**, and the **divorced** or **separated** (Boyd and Weissman 1982).

Depressive syndromes

The information about depressive syndromes concerns **bipolar** cases (in which mania has occurred at some time) and **unipolar** cases (also classified as **major depression**). Recent surveys have also investigated the epidemiology of **dysthymia**.

Bipolar disorder

Bipolar cases are probably identified most reliably. More recent community surveys in industrialized countries have suggested that the lifetime risk for bipolar disorder ranges from about 0.5 to 1.5 per cent. In the **Epidemiological Catchment Area** (ECA) study (Regier *et al.* 1988), the one-month prevalence rate for bipolar disorder was 0.4 per cent. Rates of bipolar disorder are significantly higher in urban areas. The prevalence in men and women is about the same. Some studies, but not all, have suggested that the prevalence of bipolar illness is increased in the higher social classes. The rate of divorce is increased in patients with bipolar disorder; this is likely to be a consequence of the

disruptive effects of the illness on the marital relationship.

There is some evidence that the mean age of onset of bipolar illness (now about 21 years) is lower than previous estimates (25–30 years); there is also the possibility of a **cohort effect** with those born since 1935 having a higher apparent lifetime prevalence of bipolar illness. Similar trends are apparent in the epidemiology of major depression (Smith and Weissman 1992) (Table 8.5).

Major depression

Prevalence figures for major depression vary substantially between surveys (Smith and Weissman 1992). The one-month prevalence in the ECA survey was 1.6 cases per 100 males and 2.9 cases per 100 females (Regier *et al.* 1988) with an overall lifetime prevalence of about 5 per cent. These rates are low compared with other studies. In Europe particularly, the rates are higher. For example, from a ten-year prospective study in Zurich, Angst (1992) estimated the lifetime prevalence of major depression as about 16 per cent with a six-month prevalence of about 6%. In general, when the PSE has been used, rates of unipolar depression have been higher, probably because milder cases are included. Nevertheless, the recently published National Comorbidity Survey in the United States (Kessler *et al.* 1994) found a lifetime prevalence rate of 17 per cent for DSMIII major depression, which is very similar to that found in the Zurich study.

Rates of depressive disorder seem to be higher in industrialized countries. They are consistently increased in **women** across different cultures. The reasons for increased rates among women are uncertain. The increase could be due in part to a greater readiness in women to admit depressive symptoms, but such selective reporting is unlikely to be the whole explanation. It is possible that some depressed men abuse alcohol and are diagnosed as alcoholic rather than depressed, with the consequence that the true number of depressive disorders is underestimated. Misdiagnosis of this kind is unlikely to account for the whole of the difference. Depression is more common among the **unemployed**, but is not

Table 8.5 *Epidemiology of bipolar and unipolar disorder (major depression)*

	Bipolar disorder	Unipolar disorder
Lifetime risk	About 1%	5%–10%
Sex ratio (M:F)	1:1	2:1
First-degree relatives: lifetime risk for bipolar disorder	About 10%	About 2%
First-degree relatives: lifetime risk for unipolar disorder	10%–15%	10%–15%
Average age of onset*	21 years	27 years

*ECA study.

clearly related to other socio-economic indices. Rates of depression are higher in the **divorced** but not in people who have never married (Table 8.5).

While it used to be thought that the risk of depressive disorders increased with age, the ECA survey found that major depression was most prevalent in the 18–44 age group. There is some agreement that people born since 1945 in industrialized countries have both a higher lifetime risk of major depression and an earlier age of onset. These phenomena do not appear to be artefacts of data collection or of a tendency for older people to forget earlier depressive episodes. It seems likely that the increased risk of depression in younger people is environmentally mediated, but the factors involved are not known (Klerman and Weissman 1989).

Dysthymia and recurrent brief depression

In the ECA survey the lifetime prevalence of dysthymia was about 3 per cent. The one month prevalence rate was 2.2 per 100 cases in males and 4.2 per 100 cases in females (Regier *et al.* 1988). Rates of dysthymia are increased in **women** and the **divorced**. There is less epidemiological information about recurrent brief depression. In the Zurich prospective study the six-month prevalence for recurrent brief depression was about 5 per cent, a little less than for major depression (Angst 1992).

Transcultural factors

Bipolar disorder appears to be present in all cultures. In some non-Western countries, for example Nigeria, patients in hospital usually have features of mania and seldom of depression. It is not known whether Nigerian patients with manic depressive disorders experience fewer depressive episodes, or whether depressed patients are better tolerated in their local communities and therefore do not present to hospital. Symptoms of mania seem to be similar in Nigerian and European patients, although reports of auditory hallucinations are more frequent in Nigerians (Mankanjuola 1982).

The **presentation** of less severe depressive disorders seems to be influenced by culture. In developing countries, the presenting complaints are somatic in most patients (Harding *et al.* 1980). In China, many patients with mixed somatic and psychological symptoms receive a diagnosis of neurasthenia. However, in a detailed assessment over 80 per cent of these patients were given diagnoses of major depression (Kleinman 1986). It is important to note that many cultures do not have words that convey the Western concepts of depression or depressive illness. Instead, emotional distress is often expressed in terms of somatic symptoms, for example the feeling that 'the heart is uncomfortable' (Good 1977).

The aetiology of mood disorders

There have been many different approaches to the aetiology of mood disorders. In this section, consideration is first given to the role of genetic factors and of childhood experience in laying down a predisposition to mood disorders in adult life. Next, an account is given of stressors that may provoke mood disorders (Table 8.6). This is followed by a review of psychological and biochemical factors through which predisposing factors and stressors might lead to episodes of mood disorder. In all these topics, investigators have paid more attention to depressive disorders than to mania. More space is given to aetiology in this chapter than in most others in this book; the purpose is to show how several different kinds of enquiry can be used to throw light on the same clinical problem.

Genetic causes

Genetic causes have been studied mainly in moderate to severe cases of mood disorder, rather than in milder cases (those called 'neurotic depression' by some investigators). Most *family studies* have shown that parents, siblings, and children of severely depressed patients have a morbid risk of about 20 per cent for mood disorders, as against about 7 per cent in the relatives of controls (Nurnberger and Gershon 1992). It is also generally agreed that there is no excess of schizophrenia among the relatives of depressed probands

Twin studies suggest strongly that these high rates within families are largely due to genetic factors. Thus, from a review of seven of the larger twin studies, Rush *et al.* (1991) concluded that the

Table 8.6 *Some aetiological factors in major depression*

Genetic
  Family history of depression

Early development
  Parental discord in childhood
  Childhood abuse

Personality
  Neuroticism

Environmental factors
  Recent stressful life events
  Lack of social support

---

concordance rates for manic depressive disorder were 69 per cent for monozygotic twins reared together, 67 per cent for monozygotic twins reared apart (12 pairs), and 13 per cent for dizygotic twins.

Studies of adoptees also point to a genetic aetiology. Cadoret (1978*a*) studied eight children, each born to a parent with a mood disorder and then adopted by a healthy couple. Three of the eight developed a mood disorder, as against only eight of 118 adoptees whose biological parents either suffered from a different psychiatric disorder or were healthy. In a study of 29 adoptees who had suffered from a bipolar affective disorder, Mendlewicz and Rainer (1977) found psychiatric disorder in 31 per cent of their biological parents (mainly but not exclusively mood disorders) compared with only 12 per cent of adoptive parents. In a study of 71 Danish adoptees previously treated for a major affective disorder Wender *et al.* (1986) found a significantly increased frequency of similar disorders among the biological relatives but not among the adoptive relatives (comparing each group of relatives with the corresponding relative of healthy adoptees).

Bipolar disorder and unipolar disorder

So far no distinction has been made between cases with depression only (unipolar disorders) and those with a history of mania (bipolar disorders). Leonhard *et al.* (1962) were the first to present evidence that bipolar disorders were more frequent among the families of bipolar probands than among the families of unipolar probands. Several subsequent investigations (reviewed by Nurnberger and Gershon (1992)) have confirmed this finding. However, these studies have also shown that unipolar cases are frequent in the families of both unipolar and bipolar probands (Table 8.5).

Bipolar patients appear to have a greater genetic loading for mood disorder than do unipolar patients. For example, Bertelsen *et al.* (1977) reported higher pairwise concordance rates among bipolar (74 per cent) than among unipolar (43 per cent) monozygotic twins. There is also evidence that the risk of mood disorder in the first-degree relatives of bipolar patients (about 30 per cent) is higher than that in the relatives of unipolar probands (about 15 per cent). Unipolar patients with early onset of illness appear to have a greater familial morbidity for depression. At present there is little evidence that dysthymia or minor depression are found with increased frequency in first-degree relatives of patients with bipolar or unipolar depression. Therefore these disorders do not seem to be genetically related to bipolar disorder or major depression (Nurnberger and Gershon 1992).

There are conflicting theories about the **mode of inheritance** of major mood disorders because no simple genetic model fits the frequencies of cases observed among family members of different degrees of relationship to the proband. Most family studies of depressive disorders have shown more affected women than men, suggesting sex-linked inheritance perhaps with a dominant but incompletely penetrant gene. However, the many reports of transmission from father to son (Gershon *et al.* 1975) argue against this mode of transmission. This is because sons must inherit the X chromosome from the mother, since only the father can provide the Y chromosome.

Attempts to find **genetic markers** for mood disorder have not been successful. Linkages have been reported between mood disorder and colour blindness, Xg blood group, and certain HLA antigens, but none has been consistently confirmed (Nurnberger and Gershon 1992). If bipolar disorder is heterogenous, some of these linkages may occur in occasional pedigrees. Recently, techniques of molecular genetics have been used to look for linkage between identifiable genes and manic depressive disorder in the members of large families. A study from North America of Old Order Amish kinship suggested linkage with two markers on the short arm of chromosome 11, namely the insulin gene and the cellular oncogene *Ha-ras-1* (Egeland *et al.* 1987). However, this linkage with the insulin and *Ha-ras-1* markers was not confirmed in an Icelandic kinship studied by Hodgkinson *et al.* (1987), or in three families in North America studies by Detera-Wadleigh *et al.* (1987). Moreover, in an extension of the Amish study, the original finding could not be confirmed (Kelsoe *et al.* 1989). These studies illustrate the difficulties of linkage studies in psychiatric disorders, even when there is a large genetic contribution to aetiology (see p. 96).

Physique and personality

Kretschmer (1936) proposed that patients of **pyknic body build** (thickset and rounded) were particularly prone to mood disorders. Subsequent investigations using objective measurements have not shown any strong association of this kind (von Zerssen 1976).

Kraepelin (1921) suggested that people with **cyclothymic personality** (i.e. those with repeated and sustained mood swings) were more prone to develop manic depressive disorder. Subsequently Leonhard *et al.* (1962) reported that this association was stronger among patients with bipolar disorders than among those with unipolar disorders. However, when personality was assessed without knowledge of the type of disorder, bipolar patients were not found to have mainly cyclothymic personality traits (Tellenbach 1975).

Unfortunately most reported investigations of personality in depressed patients are of little value because measurements were made when the patients were depressed. Assessments of patients who are currently depressed may not accurately reflect the *premorbid personality*. Clinical experience suggests that the most relevant personality features are obsessional traits and readiness to develop anxiety. Presumably these features are important because they influence the way in which people respond to stressful life events. *Neuroticism* as assessed by the Eysenck Personality Questionnaire may predispose to major depression, perhaps because neuroticism is associated with lower levels of social support (Kendler *et al.* 1993*a*).

Early environment

Parental deprivation

Psychoanalysts have suggested that childhood *deprivation of maternal affection* through separation or loss predisposes to depressive disorders in adult life. Epidemiologists have attempted to discover what proportion of adults with depressive disorder have experienced parental separation or loss in childhood. Many of the studies have methodological defects but overall it does not appear that loss of a parent by death in childhood does predispose to depressive disorder in adult life. In contrast, there is more support for the proposal that depressive disorder in later life is associated with *parental separation*; the main factor here appears to be *parental discord*. For example, in the Second World War children separated from their parents by evacuation did not experience increased affective morbidity in adult life, whereas children separated from their parents as a result of marital problems and divorce do subsequently have increased rates of depression (Tennant 1988). It is possible that some of the effects of parental discord on the child's liability to develop depression are mediated through deficiencies in parental caring styles (see below).

Relationships with parents

When assessing a depressed patient, it is difficult to determine retrospectively what kind of relationship he had with his parents in childhood. The patient's recollection of the relationship may be distorted by many factors, including the depressive disorder itself. These problems make it difficult to decide the aetiological importance of reports that, compared with normal controls or patients with severe depressive disorders, patients with mild depressive disorders (neurotic depression) remember their parents as having been less caring and more over-protective (Parker 1979). However, after recovery from depression, patients make very similar assessments of their parents' caring style. The impact of the latter on the subsequent development of depression is likely to be somewhat indirect, perhaps mediated through impaired self-esteem or increased neuroticism (for a review see Parker (1992)).

Precipitating factors

Recent life events

It is an everyday clinical observation that **depressive disorders** often follow stressful events. However, several other possibilities must be discounted before it can be concluded that stressful events cause the depressive disorders that succeed them. First, the association might be coincidental. Second, the association might be non-specific; there might be as many stressful events in the weeks preceding other kinds of illness. Third, it might be spurious; the patient might have regarded the events as stressful only in retrospect when seeking an explanation for his illness, or he might have experienced them as stressful only because he was already depressed at the time.

In recent years, research workers have tried to overcome each of these *methodological difficulties*. The first two problems—whether the events are coincidental or whether any association is non-specific—require the use of control groups suitably chosen from the general population and from people with other illnesses. The third problem—whether the association is spurious—requires two other approaches. The first approach (Brown *et al.* 1973*b*) is to separate events that are undoubtedly independent of illness (such as losing a job because a whole factory closes) from events that may have been secondary to the illness (such as losing a job when no one else is dismissed). The second approach (Holmes and Rahe 1967) is to assign a rating to each event according to the consensus view of healthy people about its stressful qualities.

These methods have shown that there is an excess of life events in the months before the onset of depressive disorder (Paykel *et al.* 1969; Brown and Harris 1978). However, an excess of similar events has also been shown to precede suicide attempts, and the onset of neurosis and schizophrenia. To estimate the relative importance of life events in each condition, Paykel (1978) applied a modified form of the epidemiological measure of *relative risk*. He found that the risk of developing depression increased sixfold in the six months after experiencing markedly threatening life events. The comparable increase for schizophrenia was two- to fourfold, and for attempted suicide it was sevenfold. These conclusions were similar to those of Brown *et al.* (1973a) who used another estimate, 'the brought forward time'.

Are any *particular kinds* of events more likely to provoke a depressive disorder? Because depressive symptoms are part of the normal response to bereavement, it has been suggested that loss by separation or death might be particularly important. However, research shows that not all people with depressive symptoms report losses. For example, Paykel and Cooper (1992) reviewed 21 studies that reported specifically on recent separations. In 11 of these studies depressives reported more separation than the control groups, suggesting some specificity; however, discord in key relationships also seemed to be associated with the onset of depressive disorders. Overall, the strongest relationship between life events and depression is with events that can be categorized in a general way as *threatening or undesirable* (Paykel and Cooper 1992).

It is less certain whether **mania** is provoked by life events. In the past, mania was thought to arise entirely from endogenous causes. However, clinical experience suggests that a proportion of cases are precipitated, sometimes by events that might have been expected to induce depression,

for example bereavement. The findings of life-event studies are contradictory. It is possible that the impact of life events may be more important early in the course of a recurrent manic depressive illness because once the illness is established environmental precipitants become less important. It also appears that life events may have more effect on patients with a later age of onset, perhaps because their genetic loading is relatively less. (Paykel and Cooper 1992; McPherson *et al.* 1993).

Predisposing life events and life difficulties

It is a common clinical impression that the events immediately preceding a depressive disorder act as a 'last straw' for a person who has been subjected to a long period of adverse circumstances such as an unhappy marriage, problems at work, or unsatisfactory housing. Brown and Harris (1978) divided predisposing events into two kinds. The first are prolonged stressful circumstances which can themselves cause depression as well as adding to the effects of short-term life events. Brown and Harris gave the name *long-term difficulties* to these circumstances. The second kind of predisposing circumstance does not itself cause depression problems; instead, it acts only by increasing the effects of short-term life events. This kind of circumstance is known as a *vulnerability factor*. In practice, the distinction between the two kinds of circumstance is not clear-cut. Thus long continued marital problems (a long-term difficulty) are likely to be associated with a lack of a confiding relationship, and the latter has been identified by Brown as a vulnerability factor.

Brown and Harris studied working-class women living in Camberwell, London, and found three circumstances acting as vulnerability factors: having the care of young children, not working outside the home, and having no one to confide in. They also found that events in the past increased vulnerability, namely loss of the mother by death or separation before the age of 11. Brown's four factors have not been supported consistently by subsequent enquiries. In a study in a rural community in the Outer Hebrides, Brown was able to confirm only one of his four factors at a significant level, namely having three children aged under 14 years (Brown and Prudo 1981). One other study confirmed this last finding (Campbell *et al.* 1983) but three others did not (Costello 1982; Solomon and Bromet 1982; Bebbington *et al.* 1984). Another vulnerability factor—the lack of someone to confide in (lack of 'intimacy')—has received more support; Brown and Harris (1986) cite eight studies that confirm it, and two that do not. In a recent community study Brown and Harris (1993) found that depression in adult life was associated with childhood adversity (particularly physical and sexual abuse) and major losses in adulthood, especially when levels of social support were low.

Overall there is good evidence that *poor social support*, measured as lack of intimacy or social integration, is associated with an increased risk of depression (Paykel and Cooper 1992). The mechanism of this association is not clear and is open to different interpretations. First, it may be that a lack of opportunities to confide makes people more vulnerable. Second, it may indicate that depressed people have a distorted perception of the degree of intimacy that they achieved before becoming depressed. Third, some other factor, presumably an abnormality in personality, may result in both difficulty confiding in others and vulnerability to depression.

Recently, attention has turned from these external factors to an intrapsychic factor, low self-esteem. Brown has suggested that vulnerability factors might act in part by lowering self-esteem and, intuitively, this seems likely to be important. However, self-esteem is difficult to measure and its role as a predisposing factor is not yet established by research findings. (For reviews of the evidence for and against the vulnerability model see Brown and Harris (1986) and Tennant (1985).)

The effects of physical illness

All medical illnesses and their treatment can act as non-specific stressors which may lead to mood disorders in predisposed subjects. Sometimes, however, medical conditions are believed to play a more direct role in causing the mood disorder;

examples of such medical conditions are brain disease, certain infections, and endocrine disorders. The resulting mood disorders are known as organic mood disorders; they are discussed fully in Chapter 11.

Inevitably, the above distinction is arbitrary. For example major depression occurs in about half of patients with Cushing's disease (see p. 404); since not all patients with Cushing's disease suffer from depressive disorder, it follows that variables other than raised plasma cortisol levels are involved. However, organic mood disorders, can give clues to aetiology. For example depressive disorders in Cushing's disease remit after cortisol levels are restored to normal; this finding had led to the proposal that increased cortisol secretion may play a role in the pathophysiology of major depression (see p. 223). It is also worth noting here that the *puerperium* (although not an illness) is associated with an increased risk of mood disorders (see p. 397).

Psychological theories of aetiology

These theories are concerned with the psychological mechanisms by which recent and remote life experiences can lead to depressive disorders. Most of the literature on this subject fails to distinguish adequately between the symptom of depression and the syndrome of depressive disorder. The three principal approaches to the problem are derived from the ideas of psychoanalysis, learned helplessness, and cognitive theories.

Psychoanalytic theory

The psychoanalytic theory of depression began with a paper by **Abraham** in 1911, and was developed by **Freud** in 1917 in a paper called 'Mourning and melancholia'. Freud drew attention to the resemblance between the phenomena of mourning and symptoms of depressive disorders, and suggested that their causes might be similar. It is important to note that Freud did not suppose that all severe depressive disorders necessarily had the same cause. Thus he commented that some disorders 'suggest somatic rather than psychogenic affections' and indicated that his ideas were to be applied only to those 'whose psychogenic nature was indisputable' (Freud 1917, p.243). Freud suggested that, just as mourning results from loss by death, so melancholia results from *loss* of other kinds. Since it was apparent that not every depressed patient had suffered an actual loss, it was necessary to postulate a loss of 'some abstraction' or internal representation, or in Freud's terms the loss of an 'object'.

Freud pointed out that depressed patients often appear critical of themselves, and he proposed that this self-accusation was really a disguised accusation of someone else for whom the patient 'felt affection'. In other words, depression was thought to occur when feelings of love and hostility were present at the same time (ambivalence). When a loved 'object' is lost the patient feels despair; at the same time any hostile feelings attached to this 'object' are redirected against the patient himself as self-reproach.

Freud also put forward *predisposing factors*. He proposed that the depressed patient regresses to an earlier stage of development, the oral stage, at which sadistic feelings are powerful, and that problems at this stage somehow predispose to depression in later life. Klein (1934) developed this idea by suggesting that at this early stage of development, the infant gradually acquires confidence that, when his mother leaves him, she will return even when he has been angry. This proposed stage of learning was called the 'depressive position'. Klein suggested that, if this stage is not passed through successfully, the child will be more likely to develop depression when faced with loss in adult life.

Further important modifications of Freud's theory were made by Bibring (1953) and Jacobson (1953). They suggested that loss of self-esteem is of central importance in depressive disorders. They also proposed that self-esteem depends not only on experiences at the oral stage, but also on failures at later stages of development. However, although it is part of a depressive disorder, there is no clear evidence that it is more common among people who subsequently develop depressive disorders than among those who do not.

Psychoanalytic theory explains mania as a defence against depression; this is not a convincing explanation of most cases.

For a review of the psychoanalytic literature on depression, see Mendelson (1992).

Learned helplessness

This explanation of depressive disorders is based on experimental work with animals. Seligman (1975) suggested a similarity between depression and the behaviour of animals subjected to punishment that is not clearly contingent on their actions. When animals are exposed to experimental situations in which they cannot control punishing stimuli, they develop a behavioural syndrome known as 'learned helplessness'; this syndrome bears some resemblance to depressive disorders in humans, notably reduced voluntary activity and reduced intake of food. Seligman's original hypothesis was later broadened by stating that depression results when 'highly desired outcomes are believed improbable or highly aversive outcomes are believed probable and the individual expects that no response (of his) will change their likelihood' (Abrahamson *et al.* 1978, p.68). At the time this work attracted considerable attention, perhaps as a result of the name 'learned helplessness' rather than as a result of scientific strength (the latter has not been increased by subsequent experimental studies).

Experimenters continue to use stressors in animal studies to model depressive states of human beings. For example when rats are exposed to mild but unpredictable environmental stressors over several weeks, they show a decreased response to natural rewards such as a sweet sucrose solution. It has been proposed that this effect resembles the clinical syndrome of anhedonia (Willner *et al.* 1992).

Cognitive theories

Depressed patients characteristically have pessimistic thoughts. Beck (1967) has proposed that these *depressive cognitions* consist of *automatic thoughts* that reveal negative views of the *self*, the *world*, and the *future* (the depressed patient usually reviews the past in a similar vein). These automatic thoughts appear to be sustained by cognitive distortions, which include: 'arbitrary inference' (drawing a conclusion when there is no evidence for it and even some against it), 'selective abstraction' (focusing on a detail and ignoring more important features of a situation), 'overgeneralization' (drawing a general conclusion on the basis of a single incident), and 'personalization' (relating external events to oneself in an unwarranted way).

While most psychiatrists regard these cognitions as secondary to a primary disturbance of mood, Beck regards them as manifestations of long-standing *dysfunctional beliefs* that are activated by stress and adversity. An example of a dysfunctional belief is, 'If I am not perfectly successful then I am a nobody'. A person with this belief may be expected to develop depressive cognitions and low mood after a failure at work (Haaga and Beck 1992). Therefore, in this model latent dysfunctional beliefs lead to both depressive cognitions and depressed mood.

Beck gives a detailed description of the cognitions found in depressive disorders; these cognitions are found more often in depressed patients than in patients from other diagnostic groups. Other studies have demonstrated that depressed patients show selective retrieval of unpleasant memories (Clark and Teasdale 1982), suggesting that depressive cognitions are associated with negative biases in information processing.

It is not known whether dysfunctional attitudes and cognitive distortions are predisposing factors present before the depressive disorder, or whether they are features of the disorder. If the former, they might be expected to persist after the disorder recovers, but they do not seem to do so (Peselow *et al.* 1990). Beck has suggested that dysfunctional attitudes may be important in predisposing only to some forms of depression, particularly non-melancholic kinds. He has also proposed that, if dysfunctional attitudes are latent beliefs, they may not be readily elicited until activated by an appropriate stressor (Haaga and Beck 1992)—a hypothesis that is very difficult to test.

Biochemical theories

Whatever their aetiology, the clinical manifestations of depressive disorders must ultimately be mediated through changes in brain neurochemistry. Biochemical investigations in depressed patients have focused on the **monoamine neurotransmitters** because monoamine pathways appear to play an important role in the actions of effective antidepressant drugs (see below). Of course, the actions of antidepressant drugs may not reverse the cause of depression but may merely change the expression of symptoms. However, another reason for studying monoamines, is that they play an important role in mediating adaptive responses to stress, and depressive disorders could be viewed as a failure of these adaptive responses.

A major problem in assessing neurobiological studies in depression is the lack of reproducibility of many of the abnormal findings reported by different laboratories. To some extent this may lack reflect clinical and biochemical heterogeneity in depressive disorders. However, another important point is that previous *antidepressant drug treatment* almost invariably alters the biochemical processes that have been investigated in depressed patients. Hence it is difficult to be certain that observed effects are due to the disorder rather than to the residual effects of previous drug treatment.

The **monoamine hypothesis** suggests that depressive disorder is due to an abnormality in a monoamine neurotransmitter system at one or more sites in the brain. In its early form, the hypothesis suggested a changed provision of the monoamine; more recent elaborations postulate alterations in receptors as well as in the concentrations or the turnover of the amines (Delgado *et al.* 1992). Three monoamine transmitters have been implicated: 5-hydroxytryptamine (5-HT), noradrenaline, and dopamine. The hypothesis has been tested by observing three kinds of phenomenon: the metabolism of neurotransmitters in patients with mood disorders, the effects of selective drugs on measurable indices of the function of monoamine

Table 8.7 *Abnormalities in monoamine neurotransmitters in depression*

5-HT
   Decreased plasma tryptophan concentration
   Blunted 5-HT-mediated prolactin release
   Increased cortical 5-HT$_2$ receptor binding (suicide victims)
   Decreased 5-HIAA level in CSF (suicide attempters)

Noradrenaline
   Blunted noradrenaline-mediated growth hormone release

Dopamine
   Decreased HVA level in CSF

---

systems (usually neuroendocrine indices), and the pharmacological properties shared by anti-depressant drugs. The biochemical effects of antidepressant drugs are considered in Chapter 17. The present chapter will consider the evidence for abnormalities in monoamine neurotransmitters in untreated depressed patients (Table 8.7). Much less work of this kind has been carried out in mania, and there is no consistent evidence that monoamine neurotransmitter function is altered in manic patients. Therefore the present discussion will be largely restricted to depressive disorders, although for completeness the dopamine hypothesis of mania will be outlined.

5-HT function

Plasma trypytophan The synthesis of 5-HT in the brain depends on the availability of its precursor amino acid L-tryptophan. Recent studies have shown that plasma tryptophan levels are decreased in untreated depressed patients, particularly in those with melancholic depression. Studies in healthy subjects have shown that weight loss through dieting can lower plasma tryptophan, and this factor appears to explain some, but not all, of the reduction in plasma tryptophan seen in depression. Decreases in plasma tryptophan may contribute to the impairments seen in brain 5-HT function in depressed patients but are probably not an important causal factor (Anderson *et al.* 1990).

Studies of cerebrospinal fluid Indirect evidence about 5-HT function in the brains of depressed patients has been sought by examining cerebrospinal fluid (CSF). Numerous studies have been carried out, but overall the data do not suggest that drug-free patients with major depression have a consistent reduction in CSF concentrations of *5-hydroxyindolacetic acid (5-HIAA)*, the main metabolite of 5-HT formed in the brain (Koslow *et al.* 1983). However, there is more consistent evidence that depressed patients who have made *suicide attempts* have low CSF 5-HIAA levels. This finding is not restricted to patients with depression. It has also been reported in, for example, patients with schizophrenia and personality disorder who have histories of aggressive behaviour directed towards themselves or other people (Roy *et al.* 1990). It has been proposed that low levels of CSF 5-HIAA, while not related specifically to depression, may be associated with a tendency of individuals to respond in an impulsive and hostile way to life difficulties (Brown and Linnoila 1990).

Studies of post-mortem brain Measurements of 5-HT and 5-HIAA have been made in the brains of depressed patients who have died, usually by suicide. Although this is a more direct test of the monoamine hypothesis, the results are difficult to interpret for two reasons. First, the observed changes may have taken place after death.

Second, the changes may have been caused before death but by factors other than the depressive disorder, for example by anoxia or by drugs used in treatment or taken to commit suicide. These limitations may explain why some investigators have reported lowered concentrations of 5-HT in the brainstem of depressed patients (Lloyd *et al.* 1974), whilst others have not (Cochran *et al.* 1976). Overall there is little consistent evidence that depressed patients dying from natural causes or suicide have lowered brain concentrations of 5-HT or 5-HIAA (Horton 1992). The observation that patients dying by suicide do not have consistently lowered brain levels of 5-HIAA is at variance with the observation that CSF 5-HIAA levels are lower in patients who have made suicide attempts. This may be because levels of 5-HIAA in the brainstem and CSF do not correlate reliably with each other. Another possibility is that decreased 5-HIAA levels may be found, particularly in patients whose suicidal behaviour is of an impulsive or violent nature (Brown and Linnoila 1990). Hence, if all kinds of suicidal behaviour are classed together, an abnormality in brain 5-HIAA in a subgroup could be obscured.

Recently, multiple 5-HT receptors have been described in mamalian brain. There have been a number of reports that the density of *brain 5-HT$_2$ receptors* in certain cortical regions is increased in patients dying by suicide, particularly if violent means were used (Arora and Meltzer 1989). However, there are some negative studies (Horton 1992). An increase in 5-HT$_2$ receptors has also been reported in some depressed patients dying of natural causes (Yates *et al.* 1990). It has been proposed that the increase in brain 5-HT$_2$ receptors may be an adaptive change that compensates for a decline in 5-HT release from presynaptic terminals. This decline, if it occurs, does not seem to result from an obvious decrease in brain 5-HT metabolism (see above).

Neuroendocrine tests The functional activity of 5-HT systems in the brain has been assessed by giving a substance that stimulates 5-HT function and by measuring a neuroendocrine response that is controlled by 5-HT pathways—usually the release of prolactin. Neuroendocrine challenge tests have the advantage that they measure an aspect of brain *5-HT function*. However, the 5-HT synapses involved presumably reside in the hypothalamus, which means that important changes in 5-HT pathways in other brain regions could be missed. A number of drugs have been used to increase brain 5-HT function for the purposes of neuroendocrine challenge. These include intravenous infusions of L-tryptophan, a precursor of 5-HT, oral doses of fenfluramine, which releases 5-HT and blocks its re-uptake; and intravenous clomipramine, a potent inhibitor of 5-HT re-uptake.

The prolactin response to *L-tryptophan* and *clomipramine* is consistently reduced in depressed patients, while the response to *fenfluramine* is lowered in most studies (Cowen and Anderson 1991). In contrast, the prolactin responses to the dopamine antagonist metoclopramide and to the pituitary lactotroph stimulant thyrotropin-releasing hormone are not decreased in depression, suggesting that the decrement in 5-HT-mediated release is not due to a generalized disturbance in the regulation of prolactin secretion. Another finding of interest is that, following clinical recovery and withdrawal of antidepressant treatment, the prolactin responses to L-tryptophan return to normal. The abnormality in L-tryptophan-induced prolactin release may be specific to depression, as it is not found in patients with panic disorder or obsessive–compulsive disorder. This work has been reviewed by Power and Cowen (1992). Taken together, these findings suggest that depressed patients have an impairment in brain 5-HT function which remits with clinical recovery. The mechanism underlying this impairment is not known; from the studies reviewed earlier it is unlikely to be caused by an abnormality in 5-HT metabolism. One interesting suggestion is that impaired brain 5-HT function in depression may be a consequence of cortisol hypersecretion (Deakin 1991*a*, *b*).

Noradrenaline function

Metabolism and receptors There is no consistent evidence that brain or CSF concentrations of noradrenaline or its major metabolite 3-methoxy-4-hydroxy-phenylethylene glycol (MHPG) are altered in depressed patients (Delgado *et al.* 1992). As with 5-HT receptors, noradrenaline receptors in the brain can be divided into a number of subclasses. There is some evidence that depressed patients dying of natural causes and those who commit suicide have lowered α_1-adrenoceptor binding in some brain regions, though different studies have implicated different brain regions. It is also possible that α_2-adrenoceptor concentrations are increased in patients dying of suicide, but similar changes have not been found in depressed patients who died of natural causes. Some studies have found increases in β-adrenoceptor binding in post-mortem studies of depressed suicide victims, but again there are a number of contradictory reports (Horton 1992).

Neuroendocrine tests Increasing brain noradrenaline function elevates plasma concentrations of ACTH, cortisol, and growth hormone. There is fairly consistent evidence that the growth hormone response to both the noradrenaline re-uptake inhibitor *desipramine* and the noradrenaline receptor agonist *clonidine* is blunted in patients with melancholic depression (Checkley 1992). Clonidine acts directly on postsynaptic α_2-adrenoceptors in the hypothalamus to increase plasma growth hormone, and therefore the blunted response in depressed patients suggests a decreased responsivity of postsynaptic α_2-adrenoceptors or of mechanisms linked to them. Other biochemical and behavioural responses to clonidine, including changes in plasma MHPG and sedation, are not altered in depressed patients, suggesting that any deficit in α_2-adrenoceptor function is fairly localized, probably to the hypothalmus. Impaired growth hormone responses to clonidine appear to persist in recovered depressed patients (Checkley 1992). However, the fact that tricyclic antidepressants can produce long-standing impairments in clonidine-induced growth hormone release suggests that the latter finding should be received with some caution, because recovered depressed patients have usually undergone treatment with these drugs (Schittecatte *et al.* 1989).

Dopamine function

The function of dopamine in depression has been less studied than that of 5-HT or noradrenaline. CSF levels of the dopamine metabolite homovanillic acid (HVA) have been reported as low in several studies of drug-free depressed patients (Brown and Gershon 1993). However, no consistent changes have been found in post-mortem studies of brain dopamine or HVA levels. Overall, dopamine-mediated neuroendocrine responses seem to be unchanged in depressed patients (Delgado *et al.* 1992).

An influential hypothesis links *excessive* dopamine activity to the pathophysiology of schizophrenia (see p. 274). It has also been proposed that **manic states** may be attributable to dopamine overactivity (Silverstone and Cookson 1982). Studies of dopamine metabolism and function have provided little direct evidence for this suggestion, but one study has suggested that puerperal mania may be associated with an increase in the growth hormone response to the dopamine agonist apomorphine (Wieck *et al.* 1991). Mania can be provoked by other dopamine agonists such as bromocriptine, and the euphoriant and arousing effects of psycho-stimulants such as amphetamine and cocaine are well known (Jacobs and Silverstone 1986). The latter drugs are potent dopamine-releasing agents. In animal studies, increases in brain dopamine function can lead to increased locomotor behaviour and rapid switching of behaviour, both of which show some superficial resemblances to the clinical presentation of mania. Finally, dopamine receptor antagonist drugs such as haloperidol are useful in the treatment of mania, though it has been suggested that such agents largely reduce psychotic symptoms and over-activity rather than elevated mood (Johnstone *et al.* 1988) (see below).

Monoamines and depression

Studies in depressed patients have shown changes in a number of aspects of monoamine neurotransmission. The most reliable abnormalities are found in neuroendocrine challenge tests which provide dynamic measures of monoamine function, but not necessarily in areas of the brain concerned with mood regulation. These tests reveal that depressed patients have impairments in some aspects of 5-HT and noradrenaline neurotransmission. The changes are modest and do not clearly separate individual patients from controls. In addition, they are not found in all patients, even those with similar clinical presentations. An important consideration is that noradrenaline and 5-HT pathways are very widely distributed throughout the brain. These neuro-transmitters probably act as neuromodulators, influencing the underlying activity of diverse brain regions and circuits. This may explain why monoamines play a prominent role in mediating responses to stress in animal experimental studies. The abnormalities in 5-HT and noradrenaline function found in depressed patients could be isolated examples of an overall dysregulation in a highly organized adaptive system (Delgado *et al.* 1992). Quite how far changes in monoamine function are involved in the underlying patho-physiology of the depressed state is unclear. It is possible, for example, that they could relate to the expression of individual symptoms. However, it does appear, that the clinical depressive syndrome can be relieved by facilitation of 5-HT and noradrenaline neurotransmission (see Chapter 17).

Water and electrolytes

There have been several reports of changes in water and electrolytes in depressive disorders and mania. Thus 'residual sodium' (more or less equivalent to intracellular sodium) has been reported to be increased in both conditions (Coppen and Shaw 1963; Coppen *et al.* 1976). There have also been reports of changes in erythrocyte membrane sodium potassium ATPase, such that active transport of sodium and potassium increases on recovery from mania and depressive disorders (Naylor *et al.* 1973, 1976). Such findings are of some interest because they could reflect an abnormality of the mechanisms subserving nerve conduction.

The activity of the sodium pump can be studied *in vivo* using the technique of rubidium loading. Such a study confirmed that the erythrocytes of

drug-free **manic** patients have increased sodium pump activity of their erythrocytes (Wood *et al.* 1989). Other work *in vitro* has suggested that the genetic regulation of the sodium pump is abnormal in euthymic bipolar patients both on and off lithium (Wood *et al.* 1991). While these studies cannot establish whether a similar abnormality occurs in sodium pump activity in the brain, increasing knowledge of the genetic control of sodium pump expression makes this an attractive marker for candidate gene studies in bipolar disorder (see p. 96).

Endocrine abnormalities

Abnormalities in endocrine function may be important in aetiology for three reasons. First, some disorders of endocrine function are followed by depressive disorders more often than would be expected by chance, suggesting a causative relationship. Second, endocrine abnormalities found in depressive disorder indicate that there may be a disorder of the hypothalamic centres controlling the endocrine system. Third, hormones modulate the activity of monoamine neuro-transmitters and could play a part in producing some of the changes in monoamine function found in depressed patients.

Endocrine pathology and depression

About half of patients with **Cushing's syndrome** suffer from major depression, which usually remits when the cortisol hypersecretion is corrected (Checkley 1992). Depression also occurs in **Addison's disease, hypothyroidism,** and **hyperpara-thyroidism.** Endocrine changes may account for depressive disorders occurring premenstrually, during the menopause, and after childbirth. As noted above, there is a report that puerperal manic illness may be associated with a hyper-dopaminergic state, perhaps induced by the abrupt decline in oestradiol levels that follows delivery. These clinical associations are discussed further in Chapter 12.

Hypothalamic–pituitary–adrenal axis Much research effort has been concerned with abnormalities in the control of cortisol in depressive disorders. In about half of patients whose depressive disorder is at least moderately severe, plasma cortisol is increased throughout the 24 hour cycle. Despite this, patients do not show clinical features of excess production of cortisol, possibly because the number of glucocorticoid receptor sites is reduced (Whalley *et al.* 1986). In any case, excess cortisol production is not specific to depressed patients, for similar changes occur in drug-free manic and schizophrenic patients (Christie *et al.* 1986).

In studying depressed patients, much use has been made of the **dexamethasone suppression test** which suppresses cortisol levels via inhibition of ACTH release at pituitary level. About 50 per cent of depressed in-patients do not show the normal suppression of cortisol secretion induced by giving a 1 mg dose of the synthetic corticosteroid dexamethasone. However, not all cortisol hypersecretors are dexamethasone resistant. Dexamethsone non-suppression is more common in depressed patients with melancholic features, but it has not been reliably linked with any specific psychopathological feature (Holsboer 1992). Even in normal subjects, however, severe weight loss can induce dexamethasone resistance (Mullen *et al.* 1986). Abnormalities in the dexamethasone suppression test are not confined to mood disorders; they have also been reported in mania, chronic schizophrenia, and dementia (Braddock 1986). There were early hopes that dexamethasone non-suppression could be used as a diagnostic marker of melancholic depression, but these have not been borne out. Nevertheless, dexamethasone non-suppression does predict a poorer response to *placebo treatment* in drug trials and a high risk of *relapse* in apparently recovered depressed patients (Ribeiro *et al.* 1993).

The cause of cortisol hypersecretion in depressed patients is not clearly established. There seem to be abnormalities at various points in the hypothalamic–pituitary–adrenal axis including an increased cortisol response to ACTH, an increased number of ACTH secretory episodes, and a blunted ACTH response to **corticotropin-releasing hormone (CRH)**. It has been suggested that, taken together, abnormalities in cortisol regulation in depression can best be explained by hypersecretion of CRH in the hypothalamus with a resultant increase in ACTH and cortisol release. Also, ACTH is a trophic hormone which can increase the responsiveness of the adrenal gland; as a consequence more cortisol will be secreted when the adrenal is stimulated (Holsoboer 1992).

In addition to its effects on cortisol secretion, CRH may play a more direct role in the aetiology of depression. It is well established that CRH has a neurotransmitter role in limbic regions of the brain where it is involved in regulating biochemical and behavioural responses to stress. Administration of CRH to animals produces changes in neuro-endocrine regulation, sleep, and appetite that parallel those found in depressed patients. Furthermore, CRF levels may be increased in the CSF of depressed patients. Therefore it is possible that hypersecretion of CRH could be involved in the pathophysiology of the depressed state (Heilig *et al.* 1994).

There is also increasing interest in how cortisol hypersecretion may be involved in the aetiology of depression. Exogenous corticosteroids have pronounced effects on mood in some people; both depressive and manic reactions have been reported. Recent work has shown that corticosteroids regulate the *genomic expression* and function of a number of *monoamine receptors* in the brain. For example, it has been shown that corticosteroids can decrease the expression of postsynaptic 5-HT receptors in the hippocampus; this finding has led to the suggestion that excessive cortisol secretion may precipitate depressive states through decreasing 5-HT neurotransmission (Deakin 1991*a,b*). Experimental studies of animals have linked excessive cortisol secretion in depression to damage to neurones in the hippocampus (Sapolsky *et al.* 1990). Subsequently, it has been suggested that chronic cortisol hypersecretion could be associated with the cognitive impairment, which may be a particular feature in elderly depressed patients with a poor prognosis (Sapolsky 1992).

Thyroid function Circulating plasma levels of free thyroxine appear to be normal in depressed patients, but levels of free triiodothyronine may be decreased. About a third of depressed in-patients have a blunted thyrotropin-stimulating hormone (TSH) response to intravenous thyrotropin-releasing hormone (TRH); this abnormality is not specific to depression, as it is also found in alcoholism and panic disorder (Checkley 1992). It is notable that TRH levels may be increased in the CSF of depressed patients (Banki *et al.* 1988). Like CRH, TRH has a role in brain neurotransmission and is found in brain neurones colocalized with classical monoamine neurotransmitters such as 5-HT. Therefore it is possible that the abnormalities in thyroid function found in depressed patients may be associated with changes in TRH in monoamine neurones.

Melatonin The nocturnal secretion of the pineal hormone melatonin in depression has been studied with two main aims: first as a measure of *β-adrenergic function*, and second as a marker of *circadian rhythm*. Although there are some studies suggesting that melatonin secretion is decreased in drug-free depressed patients, one well-controlled investigation found normal levels and so this finding remains in doubt (Thompson *et al.* 1988). If confirmed, it would suggest that depressed patients have decreased activity at *β*-adrenergic synapses before treatment.

There is no evidence for a phase shift of melatonin secretion in depressed patients. In general, despite the obvious circadian changes in mood in depressed patients and the disruption of the sleep–wake cycle, there is little evidence from endocrine studies that depressed patients have a disturbance in the regulation of circadian rhythm.

Sleep changes in depression

Disturbed sleep is characteristic of depression. Recordings of the sleep EEG (polysomnogram) have shown a number of abnormalities in sleep architecture in patients with major depression:

- impaired sleep continuity and duration
- decreased deep sleep (stages 3 and 4)
- decreased latency to the onset of rapid eye movement (REM) sleep
- an increase in the proportion of REM sleep in the early part of the night (see Rush *et al.* 1991).

Decreased *REM sleep latency* is of interest in relation to aetiology because there is some evidence that it may persist in recovered depressed patients and indicate a vulnerability to relapse (Giles *et al.* 1988). A further link between REM sleep and depression is that many effective antidepressant drugs decrease REM sleep time and the latency to its onset (Nicholson and Pascoe 1991). In addition, both total sleep deprivation and selective REM sleep deprivation can produce a temporary alleviation of mood in depressed patients (for reviews see Wu and Bunney (1990) and Leibenluft and Wehr (1992)). The neuro-chemical mechanism that links changes in REM sleep and mood is not known. The abnormalities in REM sleep in depressed patients could be attributable to excessive sensitivity of muscarinic cholinergic receptors (Mendelson 1991).

Brain imaging in depression

Structural brain imaging Computerized tomography (CT) and magnetic resonance imaging (MRI) have found *enlarged lateral ventricles* in patients with mood disorders, but the data are inconsistent (Robbins *et al.* 1992). The abnormality may be more frequent in elderly depressives with a late age of onset (Alexopoulos *et al.* 1992). There may also be a correlation between enlarged ventricles and cognitive impairment (Robbins *et al.* 1992). There is no evidence of generalized cortical atrophy in depressed subjects, but individual MRI studies have found loss of volume in temporal and frontal lobes (Coffey *et al.* 1993). MRI investigations have also revealed changes in white matter in some patient groups. In elderly depressives *subcortical white matter hyperintensities* may represent arteriosclerotic changes, but similar abnormalities have been reported in young bipolar patients who perform poorly on tests of verbal fluency and learning (Dupont *et al.* 1990). The relationship between the white matter changes and mood disorder is unclear.

Cerebral blood flow and metabolism Cerebral blood flow can be measured with single-photon emission tomography (SPET) and xenon inhalation or positron emission tomography (PET); the latter can also measure cerebral metabolism. Cerebral blood flow and metabolism are normally highly correlated. Numerous studies have examined these two factors in groups of depressed patients. The findings have sometimes been contradictory, and there are many methodological factors such as patient selection, drug status, and imaging techniques that may account for the discrepant findings. Nevertheless, there is some consensus that depressed patients have decreased blood flow in the *dorsolateral prefrontal cortex, the cingulate cortex,* and perhaps some regions of the *basal ganglia* (Baxter *et al.* 1991; Bench *et al.* 1990). It is thought that the prefrontal cortex is involved in allocating attention to particular tasks and that impaired functioning in this region of the brain could be associated with some of the cognitive deficits shown by depressed patients, particularly poor concentration and difficulty in planning (Robbins *et al.* 1992).

Conclusions

The **predisposition** to develop mania and severe depressive disorders has important genetic determinants. There is no convincing evidence that this inherited predisposition is modified in important ways by specific childhood experiences of the kind postulated by psychoanalysts. Nevertheless, adverse early experience such as parental discord may play a part in shaping features of personality which in turn determine whether, in adult life, certain events are experienced as stressful. If such a predisposition exists, it is not expressed as a single personality type that is invariably associated with mood disorder, but in several different kinds of personality. An effect on the characteristic level of self-esteem may also be important.

The **precipitating** causes are stressful life events and certain kinds of physical illness. Some progress has been made in discovering the types of event that provoke depression and in quantifying their stressful qualities. Such studies show that loss can be an important precipitant, but not the only one. The effects of particular

events may be modified by a number of background factors that may make a person more vulnerable, for example caring for several small children without help and being socially isolated. As noted in the preceding paragraph the impact of potentially stressful events also depends on personality factors.

Two kinds of **mechanism** have been proposed to explain how precipitating events lead to the phenomena observed in depressive disorders. The first mechanism is psychological and the second is biochemical. The two sets of mechanism are not mutually exclusive, for they may represent different levels of organization of the same pathological process. The psychological studies are at an early stage. Abnormalities have been shown in the thinking of depressed patients. and they may play a part in perpetuating depressive disorder. However, there is no convincing evidence that they induce it. Various **biochemical** theories can be used to explain how life stress and difficulty could be translated into the neuro-chemical changes that characterize depressive disorders. **Monoamine neurotransmitters** are involved in regulating responses to stress and in modifying behaviours known to be altered in mood disorders. It seems likely that both external events and a genetic predisposition bring about the changes in brain monoamine function that are seen in depressed patients. How far these changes are of direct pathophysiological importance to the development of the depressive syndrome is unclear. Measurements of **cerebral blood flow** in depressed patients show alterations in brain regions involved in attention and planning. At present these abnormalities are more readily associated with the cognitive deficits seen in depressed patients than with depressed mood and pessimistic styles of thinking.

Recent studies have used sophisticated statistical techniques in an attempt to derive quantitative estimates of the roles of different risk factors in the development of depressive disorders. For example, from a prospective study of 680 female twin pairs, Kendler *et al.* (1993*a*) calculated that about half the liability to major depression was attributable to four main factors (in order of relative importance):

- recent stressful life events
- genetic factors
- previous history of major depression
- neuroticism (as measured on the Eysenck scale).

These factors interact with each other. For example while about 60 per cent of the effect of genetic factors was directly on the risk of developing depression, the remaining 40 per cent appeared to be expressed through increased liability to experience stressful life events and to have a high neuroticism score. In contrast, ratings of parental warmth had no direct effect on the liability to major depression, but influenced other factors of aetiological importance such as level of recent difficulties and social support.

Therefore major depression is a disorder with important genetic, environmental, and interpersonal determinants. These factors do not interact in a simple additive manner but modify each other in direct and indirect ways. While this formulation precludes the use of simple models to explain the aetiology of major depression, it does correspond more closely to clinical experience. In addition, it suggests that different kinds of intervention could be useful in decreasing the liability of individuals to develop major depression.

Course and prognosis

In considering course and prognosis it is convenient to deal with biopolar and unipolar disorders separately because more information is available about the bipolar cases.

Bipolar disorders

As explained earlier, in these disorders there has been at least one episode of mania irrespective of whether or not there has been a depressive disorder. In recent epidemiological studies the mean age of **onset** of bipolar disorder is about 21 years, which is earlier than the mean age of about 30 years found in previous retrospective studies (Smith and Weissman 1992). There is a wide variation in age onset, with a significant proportion of cases starting in the teens and a few in late life.

Angst *et al.* (1973) reported that almost 90 per cent of cases began before the age of 50.

The natural **course** of manic episodes can be judged from reports written before the introduction of modern treatment. According to Kraepelin (1921, p.73) 'while occasionally attacks run their course within a few weeks or even a few days, the great majority extend over many months. Attacks of two or three years duration are very frequent; isolated cases may last longer, for ten years or more'. In 1945, before phenothiazines had been introduced, Lundquist reported an average duration of 13 months. These estimates contrast with the average duration of under three months reported by Angst *et al.* in 1973. Among patients who have repeated manic attacks, the length of each episode does not seem to alter systematically in the later attacks (Angst *et al.* 1973).

It is generally agreed that nearly all manic patients **recover** eventually, though patients with **mixed affective states** seem to have a poorer short-term outcome. Before modern treatment was available, about 5 per cent of manic disorders persisted for several years (Lundquist 1945; Stenstedt 1952). Also, manic disorder had a very high acute mortality at that time. For example Derby (1933) reported that between 1912 and 1932 the death rate of manic patients in hospital was over 20 per cent; of these patients, 40 per cent died from 'exhaustion'.

Manic disorders often **recur**, and subsequent depressive disorders are frequent. The proportion of patients having a single episode of mania is uncertain, with estimates varying from 50 per cent (Kraepelin 1921; Lundquist 1945) to 1 per cent (Angst *et al.* 1973). Nowadays it is often difficult to decide whether a patient has a succession of illnesses or a single illness interrupted by periods of partially successful treatment. Whilst the length of episodes remains fairly constant, the period of remission tends to become shorter as the number of episodes increases (Roy-Byrne *et al.* 1985). Manic and depressive episodes are about equally frequent in men, but depressive episodes predominate in women (Roy-Byrne *et al.* 1985). Patients with bipolar disorder have more episodes of illness than unipolar patients, but they are less likely to have chronic disorders. A family history of mania increases the risk of multiple episodes in bipolar patients (Winokur *et al.* 1993).

Unipolar depressive disorders

The age of onset of unipolar disorders varies widely; it is generally agreed to be later than for bipolar cases. The course of the episode is equally variable. Kraepelin (1921, p. 97) wrote that 'the duration of the attack is usually longer than in mania; but it may likewise fluctuate between a few days and more than a decade. The remission of the morbid phenomena invariably takes place with many fluctuations . . .'. Studies that have included both unipolar and bipolar disorders have not revealed consistent differences between them in rates of recovery (Tsuang *et al.* 1979). Overall, about 10 per cent of patients remain persistently depressed after an index episode (Piccinelli and Wilkinson 1994).

It is estimated that about 25 per cent of patients with unipolar depression have a recurrence within a year, and 75 per cent have a recurrence in the following 10 years (Piccinelli and Wilkinson 1994). About 10 per cent of depressed patients will eventually have a manic illness. Recurrence is associated with female sex and an early age of onset. The long-term outcome of major depression is very variable, but may be less good than was previously thought. For example in an 18-year follow-up of 89 depressed in-patients at the Maudsley Hospital, London, Lee and Murray (1988) found that nine had died from unnatural causes while a further 16 had severe chronic residual symptoms with significant social handicap. Other long-term follow-up studies have also found that many depressed patients have a poor outcome (Kiloh *et al.* 1988*a*; Surtees and Barkley 1994).

Mood disorders and suicide

Between 11 and 17 per cent of people who have suffered a severe depressive disorder at any time will eventually commit suicide (Fremming 1951; Helgason 1964; Pitts and Winokur 1964; Black *et al.* 1987). It has been disputed whether the relative risk of suicide is greater in unipolar or bipolar disorder, but overall the two disorders appear to carry a fairly similar risk (Coryell and Winokur 1992).

Acute treatment of depression

This section is concerned with the effectiveness of various forms of treatment in the acute management of depression. *Details of treatment with drugs and ECT are given in Chapter 17,* which should be consulted before reading this section. Advice on the selection of treatments and the day-to-day care of patients is given in the section on management.

Antidepressant drugs

Tricyclic antidepressants

In severe depression Tricyclic antidepressants have been extensively compared with placebo in both in-patients and out-patients with major depression. In all but the most severely depressed patients, tricyclic antideprssants are clearly more effective than placebo (Ball and Kiloh 1959; Morris and Beck 1974; Paykel 1989). There is little evidence that tricyclic antidepressant drugs differ from one another in clinical efficacy, but they do differ in side-effect profile (see Chapter 17). Among the other classes of antidepressant drug none is more effective than the tricyclics, although individual patients may show a preferential response to other compounds (see below).

Response to tricyclic antidepressants alone is not good in patients with the most severe depressive disorders, and particularly those with depressive psychosis (Kocsis *et al.* 1990). However, combination of a tricyclic antidepressant with an antipsychotic drug often proves effective in the latter group (Spiker *et al.* 1985).

In less severe depression The value of tricyclic antidepressants in less severe depressive disorders has been debated (Rogers and May 1975). In a placebo-controlled trial in general practice, Paykel *et al.* (1988) found that amitriptyline produced significant benefits in patients with mild depression, as shown by an average score of a little under 15 on the Hamilton Depression Rating Scale. Only patients with the least symptom severity (Hamilton Depression scores of less than 13) showed no amitriptyline–placebo differences. Overall, the data suggest that tricyclic antidepressants are effective in a broad range of depressive states, although their therapeutic activity is most pronounced in the mid-range of symptom severity (Paykel 1989).

Monoamine oxidase inhibitors (MAOIs)

The efficacy of monoamine oxidase inhibitors (MAOIs) in the treatment of major depression (particularly with melancholic features) has been a matter of controversy. However, placebo-controlled trials have shown that MAOIs are effective antidepressants and equal in therapeutic activity to tricyclic antidepressants for moderate to severe depressive disorders (Davidson *et al.* 1988; Davidson 1992). Like tricyclics, MAOIs are also effective for milder depressive disorders (Rowan *et al.* 1982).

MAOIs are liable to produce dangerous reactions with other drugs and some foods; therefore they are not recommended as first line antidepressant drugs. However, it has been reported that MAOIs appear more effective than imipramine for some depressive disorders such as *atypical depression* (Quitkin *et al.* 1989) and for *anergic bipolar depression,* which is characterized by fatigue, retardation, increased appetite, and sleep (Himmelhoch *et al.* 1991). It appears that MAOIs can sometimes be effective in severe depressive states that have failed to respond to other antidepressant drugs and to ECT (Nolen *et al.* 1988).

The reversible type-A MAOI *moclobemide* has the advantage of not requiring adherence to a low-tyramine diet. However, it can still produce hazardous interactions with other drugs (see p. 573). While controlled trials have shown that moclobemide is as effective as tricyclic anti-depressants in the treatment of uncomplicated major depression (Angst *et al.* 1993), it is not established whether this compound is as effective

as conventional MAOIs for atypical and tricyclic-resistant depression.

Combination of MAOIs and tricyclic antidepressants It has been reported that, in cases resistant to treatment with a tricyclic or MAOI, a combination of the two drugs is more effective than either drug given alone in corresponding dosage. This claim has not been proved; indeed, two clinical trials have failed to confirm it. (Razani *et al.* 1983). However, it can be argued that neither trial was concerned with patients known to be resistant to single drugs (the patient group to whom combined treatment is most often given).

Other antidepressant drugs

In addition to MAOIs and tricyclic anti-depressants, other antidepressant drugs are now available. They are described in detail in Chapter 17. They include the selective serotonin re-uptake inhibitors (SSRIs), venlafaxine, trazodone, refazodone and mianserin. These drugs differ from the tricyclic antidepressants in that they are not *anticholinergic* and are much *safer in overdose.* There is good evidence that they are as effective as tricyclics in the broad range of depressed patients (Song *et al.* 1993), although there is less compelling evidence for the efficacy of some of them in severely depressed in-patients (Rudorfer and Potter 1989). For example, two studies have found clomipramine superior to SSRIs in this patient group (Danish University Antidepressant Group 1990).

Lithium

Lithium as a sole treatment

This section is concerned only with lithium as a treatment for depressive disorders. The use of lithium in prevention is considered later. Six placebo-controlled studies have shown that lithium has antidepressant effects in depressed patients. In depressed in-patients, the efficacy of lithium is equivalent to that of tricyclic anti-depressants, although the onset of action is probably slower. Placebo-controlled trials indicate that the response rate to lithium is higher in depressed bipolar patients (almost 80 per cent) than in unipolar depressives (about 35 per cent) (for a review of the antidepressant efficacy of lithium see Goodwin and Jamison (1990), pp. 639–42).

Lithium in combination with other treatments

It has been reported that the therapeutic effects of tricyclic antidepressants can be increased if lithium is added (Lingjaerde *et al*. 1974). An important finding was that, in patients who did not respond to tricyclic antidepressants alone, improvement occurred when *lithium was added* (de Montigny *et al*. 1983; Heninger *et al*. 1983*b*). The evidence from controlled trials suggests that about 50 per cent of patients who have not responded to tricyclic antidepressants show significant clinical benefit when lithium is added to their tricyclic medication. In one randomized trial, lithium augmentation and ECT were about equally effective in the treatment of non-psychotically depressed patients who had not responded to full-dose tricyclic antidepressant treatment (Dinan and Barry 1989).

While some studies have reported a rapid amelioration of the depressed state within as little as 48 hours after the addition of lithium, the more usual pattern of response is a gradual resolution of symptoms over two to three weeks. The latter effect leaves open the question of whether this clinical response may be due to the antidepressant properties of lithium alone rather than a synergistic action of lithium–tricyclic combination. However, unipolar depressed patients seem to respond as well as bipolar patients (Price 1989). So far there are no reliable clinical or biochemical predictors that identify depressed patients likely to respond to lithium augmentation.

Non-responsive patients can also benefit when lithium is added to other antidepressant drugs such as mianserin and SSRIs, though caution is needed in the latter case because of the possibility of *5-HT neurotoxicity* (see p. 571). Lithium can produce useful clinical remissions when added to MAOI treatment, and in one study a combination of lithium and tranylcypromine was effective in patients who had not responded to combined lithium and tricyclic treatment (Price *et al*. 1985).

Anticonvulsants

Carbamazepine

This drug was introduced as an anticonvulsant. It has since been reported to have antimanic effects and to have prophylactic effects in bipolar disorder (see p. 583). Carbamazepine is also reported to have antidepressant properties in some severely depressed patients (Post *et al*. 1986), but there is not yet enough evidence from controlled trials to assess the size and clinical importance of this effect. It has been reported that, in one series of severely depressed patients who were unresponsive to carbamazepine, the addition of lithium produced a rapid clinical improvement (Kramlinger and Post 1989*a*).

Sodium valproate

Like carbamazepine, sodium valproate was introduced as an anticonvulsant. Controlled trials have established that it is effective for acute mania, and there is growing clinical evidence that it can be useful in the management of bipolar disorder. The value of this compound in the treatment of depressive states is not established (McElroy *et al*. 1992).

Electroconvulsive therapy

This treatment is described in Chapter 17, where its unwanted effects are also considered. The present section is concerned with evidence about the therapeutic effects of ECT for depressive disorders.

Comparison with simulated ECT

Six double-blind controlled trials have compared the efficacy of ECT and simulated ECT (anaesthesia with electrode application but no passage of current) in patients with major depression. Five of these studies found ECT to be more effective than the simulation (Freeman *et al*. 1978; Johnstone *et al*. 1980; West 1981;

Brandon *et al.* 1984; Gregory *et al.* 1985). In the study which did not find the full procedure more effective (Lambourn and Gill 1978) unilateral low-dose ECT was used, a procedure which is thought on other grounds to be relatively ineffective (Sackheim *et al.* 1993) (see p. 591).

Overall, the studies show that the difference in effects between the full procedure and simulated ECT is greatest about four weeks after treatment and disappears by about 12 months. When Janicak *et al.* (1985) combined data from several studies, they found a response rate of about 70 per cent for ECT and 40 per cent for the simulated treatment.

Comparison with other treatments

Several studies have compared depressed in-patients receiving ECT with those receiving antidepressant drugs. In nine comparisons with tricyclic antidepressants, ECT was therapeutically more effective in six studies and equally effective in the remaining three. In five comparisons with MAOIs, ECT was superior in each trial. These data suggest that in severely depressed in-patients ECT is probably superior to antidepressant drug treatment, at least in the short-term (Paykel 1989).

Indications for ECT

Clinicians generally agree that the therapeutic effects of ECT are greatest in severe depressive disorders, especially those with marked **weight loss, early morning waking, retardation**, and **delusions.** Studies by Hobson (1953), Carney *et al.* (1965), and Mendels (1965) confirmed this view and identified certain features predicting an unfavourable response to ECT, namely poor premorbid adjustment, neurotic traits, a fluctuating course, and hypochondriacal and histrionic features. However, these predictors are not specific to ECT; they are similar to those suggesting response to tricyclic antidepressants (Kiloh *et al.* 1962). From the trials comparing full ECT with simulated ECT, it appears that **delusions** and (less strongly) **retardation** are the features that distinguish patients who respond to full ECT from

those who respond to placebo (Brandon *et al.* 1984; Clinical Research Centre 1984).

Other studies have established that patients with depressive psychosis respond better to ECT than to tricyclic antidepressants or antipsychotic drugs given alone (Perry *et al.* 1982; Kocsis *et al.* 1990). However, *combined* treatment with tricyclic antidepressants and antipsychotic drugs may be about as effective as ECT (Spiker *et al.* 1985), although no direct comparisons have been made. Another point of practical importance is that ECT may often prove effective in depressed patients who have *not responded* to full trials of medication, whether or not psychotic features are present (Sackheim *et al.* 1990).

Psychotherapy

The psychological treatments mainly used for depressive disorders can be divided as follows:

- clinical management
- supportive psychotherapy
- dynamic psychotherapy
- marital therapy
- interpersonal psychotherapy
- cognitive therapy.

Clinical management

All depressed patients, whatever other treatment they may be receiving, need appropriate clinical management. Such management should provide education, reassurance, and encouragement. These measures can provide some symptomatic relief and can also ensure that pessimistic patients comply with specific treatments. Education and reassurance should also be given to the patient's spouse, other close family members, and other people involved in care.

Supportive psychotherapy

Supportive psychotherapy goes beyond clinical management in focusing on the identification and resolution of current life difficulties, and in using the patient's strengths and available coping resources. A development of this approach is

problem-solving, in which the therapist and patient identify the main problems of concern and devise feasible step-by-step ways of tackling them (Hawton and Kirk 1989).

Dynamic psychotherapy

Dynamic psychotherapy has a different aim, which is to resolve underlying personal conflicts and attendant life difficulties that are believed to cause or maintain the depressive disorder. Opinions differ about its value. Arieti (1977) suggested that it has a role in treating most cases of depression, including severe cases. Other clinicians restrict the use of dynamic psychotherapy to the less severe cases. There have been a few attempts to evaluate treatment, and they have mainly been concerned with brief dynamic psychotherapy in *groups*. The results for the treatment of depression suggest that dynamic psychotherapy may be somewhat less effective than more structured psychotherapies such as cognitive therapy (Depression Guideline Panel 1993). However, these results may not apply to dynamic psychotherapy given individually.

Marital therapy

Marital therapy can be given to depressed patients for whom marital discord appears to have contributed to causing or maintaining the depressive disorder. Friedman (1975) compared weekly marital therapy with amitriptyline, placebo, and minimal contact over a period of 12 weeks. He found that amitriptyline led to the greatest change in symptoms, but marital therapy led to more beneficial changes in marital relationships. More recent studies have shown that both marital therapy and cognitive therapy are effective in reducing depressive symptoms in women with marital discord. However, marital therapy is more effective than other psychotherapies or waiting-list control in improving marital satisfaction (O'Leary and Beach 1990; Depression Guideline Panel 1993). Overall, marital therapy is worth considering for depressed patients with significant marital discord, perhaps as an adjunct to drug treatment.

Interpersonal psychotherapy

This is a systematic and standardized treatment approach to personal relationships and life problems. Weissman *et al.* (1979) reported that the effects of interpersonal therapy on the symptoms of moderately severe depressive disorders were equal to those of amitriptyline and greater than those of minimal treatment. Drug treatment had more effect on sleep disturbance, appetite, and somatic complaints; while interpersonal therapy had more effect on depressed mood, guilt, suicidal ideas, interests, and work. The drug effects began in the first week, whilst the effects of psychotherapy began only after four to eight weeks (DiMascio *et al.* 1979). 'Endogenous' depression was less likely to respond to psychotherapy alone, while 'situational' depression responded equally well to drugs or psychotherapy (Prusoff *et al.* 1980).

The National Institute of Mental Health (NIMH) organized an important multicentre psychotherapy study in which 240 out-patients with major depression were randomly allocated to one or other of four treatments given over 16 weeks (Elkin *et al.* 1989):

- interpersonal therapy
- cognitive therapy
- imipramine with clinical management (see above)
- placebo with clinical management.

Most patients had improved after 16 weeks. In the patient group as a whole there were few significant differences between the treatments; imipramine was somewhat more effective, as was interpersonal therapy but to a lesser extent. When patients with more *severe depression* (Hamilton Depression scores of 20 or greater) were considered separately, *imipramine* was consistently better than placebo and clinical management, while *interpersonal therapy* was almost as effective as imipramine. In this study cognitive therapy did not appear superior to clinical management with placebo. The results of this investigation suggest that, in patients with comparatively mild depressive symptoms, *clinical management* (with a placebo pill) is useful. The results suggest that psychotherapy studies using 'no treatment' or 'waiting-list treatment' controls may overestimate the size of the specific treatment effect.

Cognitive therapy

Overall, the results of cognitive therapy have been comparable with those of interpersonal therapy. For depressive disorder, the essential aim of cognitive treatment is to help patients to modify their ways of thinking about life situations and depressive symptoms (see p. 639 for further information). For unipolar depressive disorders of moderate severity, clinical trials indicate that the results of cognitive therapy are about equal to those of antidepressant drug treatment (Rush *et al.* 1977; Blackburn *et al.* 1981; Murphy *et al.* 1984). However, in the NIMH trial discussed above (Elkin *et al.* 1989), cognitive therapy was less effective than imipramine or interpersonal therapy in the more severely depressed patients. When cognitive therapy is combined with antidepressant drugs, the outcome appears to be no better than with either treatment alone (Murphy *et al.* 1984; Hollon *et al.* 1992). If cognitive therapy were shown to be effective for drug-resistant patients, its value would be great, but so far this effect has not been demonstrated.

Cognitive therapy may have a role in preventing *relapse* of depression. Several follow-up studies have indicated that patients receiving acute treatment with cognitive therapy had lower relapse rates and sought less treatment subsequently than patients who were treated with tricyclic antidepressants (Clark 1990; Evans *et al.* 1992). It is important to discover whether cognitive therapy can reduce relapse rates in depression, and prospective studies are needed. At present, interpretation of the findings has been hindered by the definition of relapse and by naturalistic methods of follow-up; for example, if patients seek less help after cognitive therapy, it does not necessarily mean that they are free of depression.

Other treatments

Sleep deprivation

Several studies suggest that, in some depressive disorders, rapid short-term changes in mood can be brought about by keeping patients awake overnight. Specific *REM sleep deprivation* can also alleviate depressed mood, but more slowly (Wu and Bunney 1990). The alleviation of depressed mood after total sleep deprivation is nearly always temporary; it disappears after the next night's sleep or even a during a day-time nap after the night of sleep deprivation. While the antidepressant effect of sleep deprivation is of great theoretical interest, its brevity makes it unpractical. However, there are reports that sleep deprivation can be used to quicken the onset of effect of antidepressant drugs. (See Leibenluft and Wehr (1992) for a review.)

Bright light treatment

Over 50 per cent of patients with **recurrent winter depression** (p. 206) respond to bright light treatment (10 000 lux). As noted above, the timing of the light treatment does not seem critical; for example light administered in the morning or the evening or at midday can be effective. The duration of exposure usually needs to be one to two hours. Designing placebo-controlled trials of bright light for seasonal affective disorder presents problems, because most patients are aware before treatment that bright light is believed to be the important therapeutic ingredient. Within this limitation, most studies have found that dim light is less effective than bright light. The usual onset of the antidepressant effect of bright light is within two to five days, but longer periods of treatment seem to be needed in some patients. Patients with 'atypical' depressive features such as *over-eating* and *over-sleeping* appear to respond best. To avoid relapse, light treatment usually needs to be maintained until the usual time of natural remission, in the early spring (for a review of bright light treatment see Blehar and Rosenthal (1989)).

Acute treatment of mania

Antipsychotic drugs and lithium

Antipsychotic drugs, particularly chlorpromazine and haloperidol, are widely used to treat mania, but their role is currently being re-assessed in the light of increasing concern about toxic side-effects, particularly at the high doses that have sometimes been used to treat manic patients (see p. 555). Four placebo-controlled trials have shown that **lithium** is effective in the acute treatment of mania, and further controlled studies have compared the effects of lithium and of antipsychotic drugs, and also a combination of the two (for a review of these studies see Chou (1991)).

From these studies it appears that antipsychotic drugs produce an earlier symptomatic improvement than lithium, with their main effect being to reduce *psychotic symptoms* and *over-activity*. In contrast, the main effect of lithium is to decrease *elevated mood* (Johnstone *et al.* 1988). There is little evidence from the controlled studies that patients receiving combined treatment with lithium and antipsychotic drugs have a better short-term outcome than patients receiving either drug alone. Prominent depressive symptoms, marked over-activity, or prominent delusion predicts a poorer response to lithium alone (Chou 1991). In patients who respond, the controlled studies indicate that lithium produces a greater overall improvement than antipsychotic drugs.

As in the treatment of schizophrenia, the current trend in the management of mania is to use lower doses of antipsychotic drugs, such as 5–20 mg of haloperidol daily. If additional sedation is needed, adjunctive treatment with a benzodiazepine, for example lorazepam, can be given (Chou 1991) (see p. 556). Chlorpromazine is often used for the treatment of mania because it provides greater sedation than haloperidol; however, the dose required for this purpose may well be higher than the dose for an antipsychotic effect.

Carbamazepine and sodium valproate

Controlled investigations have shown that **carbamazepine** is effective in the acute treatment of mania (Chou 1991; Small *et al.* 1991). The overall efficacy of this compound seems to be similar to that of lithium and antipsychotic drugs; however, it may have an earlier onset of action than lithium. Carbamazepine may be useful for

patients who have not been helped by lithium, including those with dysphoric symptoms and rapid-cycling disorders. In some patients, the *combination* of lithium and carbamazepine may be effective when neither drug alone has proved beneficial (Kramlinger and Post 1989*b*).

Five controlled trials have shown that **sodium valproate** is beneficial in the acute management of mania, with the overall response rate being about 65 per cent (McElroy *et al.* 1992). Although some studies have suggested that valproate may be effective in patients with dysphoric mania and rapid-cycling bipolar disorder, the evidence is inconsistent (McElroy *et al.* 1992).

Electroconvulsive therapy

ECT has been widely used to treat mania, although there are only two recent prospective controlled trials. In one, bilateral ECT was found to be superior to lithium (Small *et al.* 1988). In another study, unilateral and bilateral ECT were compared with a combination of lithium and haloperidol in patients who had not responded to antipsychotic drugs alone. The response rate of these patients to ECT (13 of 22) was greater than that to combined drug treatment (none of five). The efficacy of unilateral and bilateral ECT did not differ (Mukherjee *et al.* 1988).

Retrospective investigations have shown ECT to be effective in acute mania, with the overall response rate being about 80 per cent. Many of the patients had been unresponsive to medication (Chalmers 1990; Mukherjee *et al.* 1994). In clinical practice, there is a tendency to give ECT to patients *unresponsive* to drug treatment, or to patients with a manic illness of life-threatening proportions due to extreme *over-activity* and *physical exhaustion*. ECT is often given at shorter intervals in the treatment of mania than in the treatment of depression, but there is no evidence that this regime is necessary or speeds treatment response (Mukherjee *et al.* 1994). It is unclear whether bilateral electrode placement is better than unilateral placement in manic patients (Milstein *et al.* 1987; Mukherjee *et al.* 1994).

Longer-term treatment of mood disorders

Follow-up studies have shown that mood disorders often recur and have a poor long-term prognosis. For this reason there is now increasing emphasis on long-term management.

Prevention of relapse and recurrence

Strictly, the word *relapse* refers to the worsening of symptoms after an initial improvement during the treatment of a single episode of affective disorder, whilst *recurrence* refers to a new episode after a period of complete recovery. Treatment to prevent relapse should be called *continuation treatment*, and treatment to prevent recurrence should be called *preventative* (or prophylactic) treatment. In practice, however, it is not always easy to maintain the distinction between these two kinds of treatment because a therapy may be given at first to prevent relapse and then may be used to prevent recurrence.

Drug treatment

Unipolar depression For unipolar depressive disorders, *continuation treatment* for six months with tricyclic antidepressants has been shown to reduce the relapse rate (Mindham *et al.* 1973; Paykel *et al.* 1975*b*). Another study has shown that relapse occurs frequently unless antidepressant drugs are continued for four to five months after the patient has been free from even mild depressive symptoms (Prien and Kupfer 1986). Continuation treatment with antidepressant drugs is also important after *ECT* (see p. 547).

If antidepressant drugs are continued for longer periods of time as *prophylactic* treatment, they reduce the risk of new episodes of depression (Medical Research Council 1981; Glen *et al.* 1984). There is good evidence for the efficacy of *imipramine*, *SSRIs*, and *phenelzine* used in this way (Montgomery and Montgomery 1992). In patients with recurrent depression the effects of treatment can be substantial. In a three-year study of 128 patients, Frank *et al.* (1990) found a relapse rate of 22 per cent in patients taking imipramine compared with 78 per cent in patients treated with placebo. In this latter study, patients were maintained on the same dose of imipramine as they had received during their acute illness. During longer-term treatment it is common practice to lower the dose of tricyclic antidepressant to a 'maintenance' level, but this may lessen prophylactic efficacy (Frank *et al.* 1993). *Lithium carbonate* is also effective in the prevention of recurrent unipolar depression, but its efficacy in relation to tricyclic antidepressants is unclear. Clinical experience suggests that some patients who do not show a prophylactic response to antidepressant drugs may be helped by lithium given alone or in combination with the antidepressant. Lithium may be the preferred treatment for patients who have shown any sign of elevated mood because (unlike antidepressant drugs) it also prevents mania (Prien 1992).

Bipolar disorder There is substantial evidence for the efficacy of *lithium* in the prevention of recurrent mood disturbances in patients with bipolar disorders. About 50 per cent of such patients respond well to lithium, whilst the rest show either a partial response or no response (Goodwin and Jamison 1990, pp. 684–700). In bipolar patients, lithium seems to be equally effective in preventing recurrences of depression and recurrences of mania. In bipolar patients, long-term treatment with lithium has been found as effective as imipramine in preventing further depressive disorder, and more effective in preventing manic recurrences (Prien *et al.* 1972; Prien and Kupfer 1986). From a review of the literature Abou-Saleh (1992) concluded that a poor prophylactic response to lithium is associated with the following:

- rapid cycling disorders or chronic depression
- mixed affective states
- alcohol and drug misuse
- family history of schizophrenia or alcoholism
- mood-incongruent psychotic features
- marked subjective side-effects to lithium.

Although the number of patients studied in randomized controlled studies is relatively few, *carbamazepine* appears to be about as effective as lithium in the prophylaxis of bipolar disorder (Post 1991). Therefore carbamazepine can be considered in patients who are *intolerant* of lithium. Some patients who *respond poorly* to lithium, particularly those with rapid-cycling disorders, may benefit from carbamazepine given either alone or in *combination* with lithium. On the results of both controlled and uncontrolled trials, Post (1991) estimated that about 65 per cent of bipolar patients show a good response to carbamazepine, or to lithium and carbamazepine given in combination. It is not clear from the literature what proportion of these patients had not responded to lithium before receiving carbamazepine. Also, in studies of treatment-resistant patients carbamazepine is often added to lithium; it is then difficult to determine whether the therapeutic effect is caused by carbamazepine alone, or by lithium and carbamazepine together.

While these results with carbamazepine are promising, it is worth noting that a group of 24 treatment-resistant patients initially showed a good response to carbamazepine; but about half reported the re-emergence of mood swings after two years of continuous treatment (Post *et al.* 1990). This finding suggests that in some patients *tolerance* to the mood-stabilizing properties of carbamazepine may attenuate the therapeutic effect.

There are also reports that *sodium valproate* may have useful prophylactic effects in patients with refractory bipolar illness, even when there has been a poor response to lithium and carbamazepine (McElroy *et al.* 1992; Post 1991). Valproate may prevent episodes of mania more effectively than episodes of depression (Calabrese *et al.* 1992). In treatment-resistant patients, valproate has often been combined with lithium or carbamazepine.

Psychotherapy

As noted above, there is a little evidence that *cognitive therapy* given during an acute phase of depression decreases the risk of subsequent relapse and recurrence. There are no randomized trials of *continuation* cognitive therapy after symptomatic remission; however, some open studies suggest

that continuing cognitive therapy may help to sustain the acute treatment response (Blackburn *et al.* 1986).

In patients with recurrent depressive episodes, *interpersonal therapy* given once monthly delayed, but did not prevent, depressive relapse compared with standard clinical management (see above). In the same study interpersonal therapy did not increase the prophylactic effect of imipramine (Frank *et al.* 1990).

Social factors

Various social and interpersonal problems may also contribute to depressive relapse. *Unhappy marriage* seems to be important (Kerr *et al.* 1974), and a particularly relevant feature of the marriage is criticism of the patient by the spouse. Thus it has been shown that relapse was predicted by a high frequency of critical comments during an interview at the time of the original illness (Vaughn and Leff 1976; Hooley *et al.* 1986). It is possible that intervention to reduce the relative's critical attitude may lower the relapse rate, as in schizophrenia (see p. 289), but this effect has not yet been demonstrated for depressive disorders.

Assessment of depressive disorders

The steps in assessment are (a) to decide whether the diagnosis is depressive disorder, (b) to judge the severity of the disorder, including the risk of suicide, (c) to form an opinion about the causes, (d) to assess the patient's social resources, and (e) to gauge the effect of the disorder on other people.

Diagnosis depends on thorough **history taking** and examination of the **physical** and **mental state**. It has been discussed earlier in this chapter. Particular care should be taken not to overlook a depressive disorder in the patient who does not complain spontaneously of being depressed ('masked depression'). It is equally important not to diagnose a depressive disorder simply on the grounds of prominent depressive symptoms; the latter could be part of another disorder such as an organic syndrome caused by a cerebral neoplasm. It should also be remembered that certain drugs can induce depression (see p. 361).

The **history** of previous mood disturbance is important in assessment. Some patients will have had recurrent episodes of mood disorder. A history of these episodes often provides clues to the probable course of the current disorder and its response to treatment. It is particularly important to ask about possible previous episodes of **mania**, even if mild and short-lived. If there is a history of mania, the mood disorder is bipolar. Interviews with relatives and close friends may help to establish whether such episodes have occurred.

The **severity** of the disorder is judged from the symptoms. Considerable severity is indicated by 'biological' symptoms, **hallucinations**, and **delusions**, particularly the latter two. It is also important to assess how the depressive disorder has reduced the patient's capacity to *work* or to engage in *family life* and *social activities*. In this assessment the duration and course of the condition should be taken into account as well as the severity of the present symptoms. Not only does the length of history affect prognosis, but it also gives an indication of the patient's capacity to tolerate further distress. A long-continued disorder, even if not severe, can bring the patient to the point of desperation. The risk of **suicide** must be judged in every case (the methods of assessment are described on pp. 421).

Aetiology is assessed next, with reference to precipitating, predisposing, maintaining, and pathoplastic factors. No attempt need be made to allocate the syndrome to an exclusively endogenous or reactive category; instead, the importance of external and internal causes should be evaluated in every case.

Provoking causes may be psychological and social (the *'life events'* discussed earlier in this chapter) or they may be *physical illness* and its treatment. In assessing such cases, it is good practice to enquire routinely into the patient's work, finances, family life, social activities, general living conditions, and physical health. Problems in these areas may be recent and acute, or may take the form of chronic background difficulties such as prolonged marital tension, problems with children, and financial hardship.

The patient's **social resources** are considered next. Enquires should cover family, friends, and work. A loving family can help to support a patient through a period of depressive disorder by providing company, encouraging him when he has lost confidence, and guiding him into suitable activities. For some patients **work** is a valuable social resource, providing distraction and comradeship. For others it is a source of stress. A careful assessment is needed in each case.

The **effects** of the disorder **on other people** must be considered carefully. The most obvious problems arise when a severely depressed patient is the mother of young children who depend on her. This clinical observation has been confirmed in objective studies which have demonstrated that depressive disorder in either parent is associated with emotional disorder in the children (Keller *et al.* 1986). It is important to consider whether the patient could endanger other people by remaining at work, for example as a bus driver. When there are depressive delusions, it is necessary to consider what would happen if the patient were to act on them. For example severely depressed mothers may occasionally kill their children because they believe them doomed to suffer if they remain alive.

The management of depressive disorders

This section starts with the management of a patient with a depressive disorder of moderate or greater severity. The first question is whether the patient requires in-patient or day-patient care. The answer depends on the severity of the disorder and the quality of the patient's social resources. In judging severity, particular attention should be paid to the risk of suicide (or any risk to the life or welfare of family members, particularly dependent children) and to any failure to eat or drink that might endanger the patient's life. Provided that these risks are absent, most patients with a supportive family can be treated at home, even when severely depressed. Patients who live alone, or whose families cannot care for them during the day, may need in-patient or day-patient care.

If the patient is to remain out of hospital, the next question is whether he should continue to work. If the disorder is mild, work can provide a valuable distraction from depressive thoughts and a source of companionship. When the disorder is more severe, retardation, poor concentration, and lack of drive are likely to impair performance at work, and such failure may add to the patient's feelings of hopelessness. In severe disorders there may be dangers to other people if the patient remains at his job.

The need for **antidepressant drug treatment** should be considered next. This treatment is indicated for most patients with a depressive syndrome of at least moderate severity, and particularly those with *melancholic* symptoms. Other indications include a *family history or personal history* of depression, particularly if there has been a clear response to drug treatment (Potter *et al.* 1991).

Choice of antidepressant drug

Several kinds of antidepressant drug treatment are available, and the choice should be made according to the needs of the individual patient, with particular consideration of likely side-effects. These are fully described in Chapter 17. The following points may be used to guide selection.

Tricyclic antidepressants may be the first choice for patients with *severe depression*, and particularly for depressed in-patients (Potter *et al.* 1991). Tricyclics should also be used when patients have previously shown both a therapeutic response to them and adequate tolerance. Secondary tricyclics such as desipramine have fewer anticholinergic side-effects and cause less sedation than tertiary tricyclics such as amitriptyline. Tertiary compounds may be helpful where there is severe *sleep disturbance* and when rapid sedation is needed. Tricyclic antidepressants will be contraindicated in some patients because of their *anticholinergic* and *cardiovascular* side-effects. In addition, they should not be used where there is a significant risk of deliberate *overdose* (although the newer compound *lofepramine* is relatively safe).

SSRIs may be considered when patients are unable to tolerate a therapeutic dose of a tricyclic antidepressant, when tricyclics are contra-indicated because of their anticholinergic side-effects, or because of pre-existing cardiac disease. SSRIs should also be used when *sedation* is *undesirable*, for example in patients striving to carry on with normal work and social activities. SSRIs may be appropriate for patients in whom tricyclic-induced *weight gain* would be a problem, and should also be used when the risk of overdose cannot be minimized. Finally, SSRIs (or clomipramine) should be used when a depressive disorder occurs in the context of an obsessional disorder (see p. 185).

Non-sedating compounds that are alternatives to SSRIs include venlafaxine and refazodone (see p. 575).

Trazodone or **mianserin** may be considered if patients need sedation but anticholinergic effects are contraindicated. Both drugs are safer than tricyclics for patients with *cardiac disease*, and mianserin is the safer of the two in this respect. Both compounds are less toxic than tricyclics in *overdose*.

The dosage of these drugs, the precautions to be observed in using them, and the instructions to be given to patients are described in Chapter 17. Here it is necessary only to emphasize again the importance of explaining to the patient that, although **side-effects** will appear quickly, the **therapeutic effect** is likely to be delayed for two weeks, with maximum improvement occurring over 6–12 weeks. During this time patients should be seen regularly to provide suitable clinical management (see above); those with more severe disorders may need to be seen every two or three days, and the others once a week. During this time it is important to make sure that the drugs are taken in the prescribed dose. Patients should be warned about the effects of taking alcohol. They should be advised about driving, particularly that they should not drive while experiencing sedative side-effects or any other effects that might impair their performance in an emergency.

The use of ECT

ECT will seldom be part of the first-line treatment of depression and will usually be considered only for patients already admitted to hospital. The only indication for ECT as a first measure is the need to bring about improvement as rapidly as possible. In practice this applies to two main groups of patients: those who **refuse to drink** enough fluid to maintain an adequate output of urine (including the rare cases of depressive stupor), and those who present a **highly dangerous suicidal** risk.

Occasionally, ECT is indicated for a patient who is suffering such extreme distress that the most rapid form of treatment is deemed justifiable. Such cases are rare. It should be remembered that, with the exception of patients who are *unresponsive* to antidepressant drugs, the effects of ECT differ from those of antidepressant drugs in greater rapidity of action rather than in the final therapeutic result. In patients with **depressive psychosis**, ECT is considerably more effective than a tricyclic antidepressant given alone, but probably about the same therapeutic effect can be achieved, albeit more slowly, if a combination of a tricyclic antidepressant and an antipsychotic drug is used.

Activity

The need for suitable **activity** should be considered for every patient. Depressed patients give up activities and withdraw from other people. In this way they become deprived of social stimulation and rewarding experiences, and their original feelings of depression are increased. It is important to make sure that the patient is occupied adequately, though he should not be pushed into activities where he is likely to fail because of slowness or poor concentration. Hence there is a fairly narrow range of activity that is appropriate for the individual depressed patient, and the range changes as the illness runs its course. If the patient remains at home, it is important to discuss with relatives how much he should be encouraged to do each day. If he is in hospital the question should be decided in collaboration with nurses and occupational therapists. Relatives may

also need to be helped to accept the disorder as an illness and to avoid criticizing the patient.

Psychotherapy

The need for **psychological treatment** should also be decided in every case. All depressed patients require support, encouragement, and a thorough explanation that they are suffering from illness and not moral failure. Similar counselling of spouses and other family members is often useful.

The use of one of the more specific psychotherapies discussed earlier should also be considered. These treatments can be used as the sole therapy for patients with *mild to moderate* depression without melancholic features, particularly when the patient prefers not to take drug therapy (Depression Guideline Panel 1993). The kind of psychotherapy used depends largely on the availability of a suitably trained therapist and the preference of the patient, even though the more structured therapies (such as interpersonal therapy and cognitive therapy) have greater efficacy in the treatment of depression. The therapeutic response to antidepressant drugs is usually quicker than that to psychotherapy.

In general, there is little evidence that combining psychotherapy with medication adds much to the effect of the latter in resolving an acute depressive episode. In individual patients, however, it often seems worthwhile tackling psychosocial problems that seem to make an important contribution to the depression. For example marital therapy may be given to patients with marital problems.

If the depressive disorder is *severe*, too much discussion of problems at an early stage is likely to increase the patient's feeling of hopelessness. Therapy directed to self-examination is particularly likely to make the disorder worse. During intervals between acute episodes, such therapy may be given to patients who have recurrent depressive disorders largely caused by their ways of reacting to life events.

Resistant depression

If a depressive disorder does not respond within a reasonable time to a chosen combination of

Table 8.8 *Some pharmacological treatments for resistant depression*

Increase antidepressant to maximum tolerated dose; measure tricyclic antidepressant levels; if patient has depressive psychosis add an antipsychotic drug; try different class of antidepressant drug
Add lithium to antidepressant drug treatment
Add triiodothyronine to tricyclic antidepressant treatment
MAOIs (can be usefully combined with lithium)
ECT

*Pharmacokinetic interactions may occur.

antidepressant drugs, graded activity, and psychological treatment, the plan should be reviewed. The first step is to check again that the patient has been taking his medication as prescribed. If not, the reasons should be sought. The patient may be convinced that no treatment can help, or may find the side-effects unpleasant. The diagnosis should also be reviewed carefully, and a check made that important stressful life events or continuing difficulties have not been overlooked.

If this enquiry reveals nothing, antidepressants should be continued in full dosage. With certain tricyclic antidepressants, particularly in doses greater than 150 mg daily, it may be worth checking plasma concentrations (see p. 562). If it becomes clear that the patient is not improving, a number of further steps can be taken (Table 8.8).

Change in antidepressant drug treatment

If a patient cannot *tolerate* a full dose of an antidepressant of one class, for example a tricyclic antidepressant, it is worthwhile trying another kind of drug that may be better tolerated such as an SSRI or trazodone. Some patients do not respond to one class of antidepressant despite an adequate trial, but improve if switched to an antidepressant of a different class. For example about half of depressed patients who did not respond to tricyclic antidepressant treatment improved when fluoxetine was substituted (Beasley *et al.* 1990). Similarly, in a group of patients who did not respond to tricyclics or SSRIs, about half responded to the MAOI tranylcypromine (Nolen *et al.* 1988). In view of the food and drug interactions with MAOIs, it is unusual to use them early in drug-resistant depression unless a patient has clear-cut features of **atypical depression** (see above) or has responded well to them in the past.

Augmentation of antidepressant drug treatment

In changing antidepressant preparations, a problem is that withdrawal of the first compound may not be straightforward. Patients may have gained some small benefit from the treatment, for example improved sleep or reduced tension, and this benefit may be lost. Also, if the first medication is stopped quickly, *withdrawal symptoms* may result (see p. 563); however, if the dose is reduced gradually, the changeover in medication may be protracted and may not be easily tolerated by a depressed and despairing patient. For this reason, in patients unresponsive to first-line medication it may be more appropriate to add a second compound to the antidepressant—so-called *'augmentation therapy'*. A disadvantage of augmentation therapy is the increased risk of adverse effects through *drug interaction*.

Lithium augmentation

Provided that there are no contraindications (see p. 582), the addition of **lithium** to antidepressant drug treatment is usually safe and well tolerated. About half of depressed patients will show a useful response over one to three weeks. Lithium can be added to tricyclic antidepressants, mianserin or SSRIs with good effect, although the last combination should be used with caution because of the risk of *5-HT neurotoxicity* (see p. 571). In augmentation treatment the aim should be to obtain the same lithium level as that used for prophylactic purposes (0.5–0.8 mmol/l) and the dosage schedule is the same.

Combined SSRIs and tricyclics

A second procedure is to add an *SSRI* to the tricyclic antidepressant. Conversely, in patients taking SSRIs it is possible to add a tricyclic antidepressant such as desipramine or nortriptyline (Weilburg *et al.* 1989, 1991). There are no published controlled studies of these combination treatments, but open trials have obtained response rates at least equal to that of lithium augmentation. The combination of SSRIs and tricyclics carries a significant risk of pharmacokinetic interaction because SSRIs can markedly *increase plasma tricyclic* levels, and adverse reactions including seizures have been reported. If combined treatment with SSRIs and tricyclics is used, it is recommended that small doses (25–50 mg) of tricyclic antidepressant be used initially, with plasma monitoring of tricyclic levels. Recent studies suggest that sertraline may cause less elevation in plasma tricyclic levels than other SSRIs (Preskorn *et al.* 1994). In general, it may be preferable to use a single agent that potently inhibits the re-uptake of both 5-HT and noradrenaline, such as clomipramine or venlafaxine.

Tri-iodothyronine

Among patients who have not responded to tricyclic antidepressants, some improve promptly when tri-iodothyronine (20–50 µg) is added (Chalmers and Cowen 1990). This improvement is unlikely to be a pharmacokinetic effect because it is not accompanied by a change in plasma concentrations of the antidepressant drug. There seems to be no relationship between thyroid status and the efficacy of this combination. In a recent study of depressed out-patients who had not responded to imipramine, the addition of tri-iodothyronine (37.5 µg) was as effective as the addition of lithium and both were superior to the addition of placebo (Joffe *et al.* 1993). In general, the combination appears safe, but it should be used with caution in patients with cardiovascular disease.

MAOI combinations

In patients who do not respond to the augmentation strategies described above, the use of an **MAOI** may be considered. At this stage patients often receive lithium as part of their treatment; if the lithium is well tolerated, it should be continued with the MAOI because evidence from open studies indicates that this combined treatment can be effective in refractory depressed states (Price *et al.* 1985). *MAOIs must not be given with clomipramine or SSRIs*, but can be added cautiously to some tricyclic antidepressants (*the method of addition is explained on p. 571*). It is not established whether MAOIs and tricyclic antidepressants can potentiate one another in resistant depression (see above). However, this combination can help to prevent MAOI-induced sleep disturbance, although there is an increased risk of weight gain and postural hypotension (Chalmers and Cowen 1990). Of the MAOIs available, *tranylcypromine* is often used for the more severely depressed patients, but it has been implicated more often than *phenelzine* and *isocarboxazid* in food and drug interactions. *Moclobemide* is probably the safest of the MAOIs from the point of view of food and drug interaction (Dingemanse 1993), but it has not been studied systematically in resistant depression.

The use of ECT

If severe depression persists, ECT should be considered. ECT is often beneficial to patients who have not responded to antidepressant drugs, and it is probably more effective that these drugs in severe depression, particularly when *psychotic features* are present. It is important to note that, among patients who have not responded to full-dose antidepressant medication, the relapse rate after ECT is high if the same previously ineffective antidepressant is continued (Sackheim *et al.* 1990). In these circumstances it seems reasonable to use an antidepressant of a different class or lithium for continuation treatment after ECT, but there is currently no evidence that this procedure lowers the relapse rate.

Other measures

Continuing support is required for patients who do not respond to treatment. Lack of improvement increases the pessimism experienced by the depressed patient. Hence it is important to give *reassurance* that depressive disorders almost invariably recover eventually, whether or not treatment speeds recovery. Meanwhile, if the patient is not too depressed, any problems that have contributed to his depressed state should be discussed further. Techniques derived from cognitive therapy can be used to help the patient to limit distorted thinking and judgement. In resistant cases it is particularly important to watch carefully for **suicidal intentions**.

Prevention of relapse and recurrence

Drug treatment

After recovery, the patient should be followed-up for several months by the psychiatrist or the family doctor. If the recovery appears to have been brought about by an antidepressant drug, the drug should usually be continued for about six months and then gradually withdrawn over several weeks. At follow-up interviews, a careful watch should be kept for signs of relapse. Relatives should be warned that relapse may occur, and they should be asked to report any signs of it. Major depression is often recurrent, and long-term prophylactic treatment will need to be considered. Several factors should be taken into account (Prien 1992):

- the likelihood of relatively early recurrence
- the impact of a recurrence on the patient's life
- the previous response to drug treatment
- the patient's view of long-term drug treatment.

The crucial factor is the likelihood of early recurrence, which is difficult to predict. It is estimated that, among patients who have had three episodes of major depression, the chance of another episode is 90 per cent (National Institute of Mental Health 1985). The usual recommendation is that maintenance drug treatment should be considered if a patient has had two episodes of depression in a five-year period, particularly if there is a family history of recurrent major depression, bipolar disorder, or onset at an early age (Prien 1992).

If the patient has had both **manic** and **depressive** disorders in the past, and if recurrences have occurred at short intervals (see above), **lithium prophylaxis** should be considered carefully and possible advantages should be weighted against the risks (see Chapter 17). Among patients poorly responsive to lithium, **carbamazepine** or **sodium valproate** can be considered either alone or in combination with lithium.

Other measures

If a depressive disorder was related to self-imposed stressors, such as overwork or complicated social relationships, the patient should be encouraged to change to a life-style that is less likely to lead to further illness. These readjustments may be helped by **psychotherapy**, which may be individual, marital, or group therapy (for a review of psychotherapy in the treatment of bipolar disorder, see Jamison (1992)).

The assessment of mania

In the assessment of mania, the steps are the same as those for depressive disorders, as outlined above: (a) decide the diagnosis, (b) assess the severity of the disorder, (c) form an opinion about the causes, (d) assess the patient's social resources, and (e) judge the effects on other people.

Diagnosis depends on a careful history and examination. Whenever possible, the history should be taken from **relatives** as well as from the patient because the latter may not recognize the extent of his abnormal behaviour. Differential diagnosis has been discussed earlier in this chapter. It is always important to remember that mildly disinhibited behaviour can result from frontal lobe lesions (such as tumours) as well as from mania.

Severity is judged next. For this purpose it is essential to interview an informant as well as the patient. Manic patients may exert some self-control during an interview with a doctor, and may then behave in a disinhibited and grandiose way immediately afterwards (see p. 202). At an early stage of mania, the doctor can easily be misled and may lose the opportunity to persuade the patient to enter hospital before causing himself long-term difficulties, for example ill-judged decisions or unjustified extravagance.

Usually, the **causes** of a manic disorder are mainly endogenous, but it is important to identify any life events that may have provoked the onset. Some manic episodes follow physical illness, treatment by drugs (especially steroids), or operations.

The patient's **resources** and the **effect of the illness** on other people are assessed in the ways described above for depressive disorders. Even for the most supportive family, it is extremely difficult to care for a manic patient at home for more than a few days unless the disorder is exceptionally mild. The patient's responsibilities in the care of dependent children or at work should always be considered carefully.

The management of manic patients

The first decision is whether to **admit** the patient to hospital. In all but the mildest cases admission is nearly always advisable to protect the patient from the consequences of his own behaviour. If the disorder is not too severe, the patient will usually agree to enter hospital after some persuasion. When the disorder is more severe, compulsory admission is likely to be needed.

The immediate treatment is usually with an **antipsychotic drug**. Haloperidol is a suitable choice, whilst chlorpromazine is a more sedating alternative. The addition of benzodiazepine sedation can permit the use of lower doses of haloperidol. Benzodiazepines can also be used to ensure adequate sleep, which is an important aim in the treatment of mania. In emergencies, the first doses of medication may need to be given by parenteral injection (see p. 556 for further advice about the use of antipsychotic drugs in urgent cases).

The subsequent dosage should be adjusted frequently to take account of the degree of overactivity. The doctor should visit the ward repeatedly until the patient's condition has been brought under control. During this early phase, it is important for all staff to avoid the angry confrontations which often arise because the patient makes unreasonable demands that cannot be met. It is often possible to avoid an argument by taking advantage of the manic patient's easy distractibility; instead of refusing his demands, it is better to delay until his attention turns to another topic that he can be encouraged to pursue.

As soon as the patient is well enough to cooperate, tests of renal and thyroid function are carried out, so that the results are available if **lithium** treatment should be required later. Practices vary about the use of lithium. Some psychiatrists continue to use haloperidol or chlorpromazine throughout the manic episode, and reserve lithium for **prophylactic treatment**. Others start lithium as soon as the acute symptoms are under control, though taking precautions not to introduce it when the dose of haloperidol is high (because occasional interactions have been reported) (see p. 582). The use of lithium rather than haloperidol at an early stage sometimes has the advantage of leaving the patient more alert; possibly it also has the advantage of making a depressive disorder less likely to follow (Goodwin and Jamison 1990, pp. 613–17). In addition, overall improvement in manic symptoms may be greater with lithium.

For these reasons there is a growing trend in the United States to manage mania predominantly with **mood-stabilizing agents** such as lithium and carbamazepine together with adjunctive benzodiazepines, and to use antipsychotic drug treatment as little as possible (Gerner and Stanton 1992). Recently, however, Goodwin (1994) has pointed out that discontinuation of prophylactic lithium in patients with a bipolar disorder may be associated with a significantly *greater relapse rate* than that attributable to natural recurrence. From this viewpoint, starting lithium for short-term management is not advisable; if significant benefit is to be gained from lithium, at least two years continuous treatment should be planned. However, it is not clear what the rate of lithium discontinuation relapse is in patients who have received relatively short-term treatment for acute mania. Also, the relapse rates after withdrawal of other medications, such as carbamazepine or antipsychotic drugs, are not known. Further studies of the most appropriate pharmacological management for acute mania are needed.

Progress of a manic episode can be judged not only by the mental state and general behaviour, but also by the pattern of sleep and the regaining of any weight lost during the illness. As progress continues, antipsychotic drug treatment is reduced gradually. It is important not to discontinue the drug too soon; otherwise there may be a relapse and a return of all the original problems of management. Since the untreated duration of a manic episode can be several months (see above), some form of *continuation treatment*, as in the management of depression, is advisable.

As explained already (p. 234), **ECT** was used to treat mania before antipsychotic drugs were introduced, but evidence about its effectiveness is still limited. It is appropriate to consider the treatment for the unusual patients who do not respond to drug treatment even when given at maximum doses. In such cases clinical experience suggests that ECT may often be followed by a significant reduction in symptoms sufficient to allow treatment to be continued with drugs. ECT may also be helpful for mixed affective states in which depressive symptoms are prominent.

Whatever treatment is adopted, a careful watch should be kept for symptoms of **depressive disorder.** It should be remembered that transient but profound depressive mood change, accompanied by depressive ideas, is common among manic patients. The clinical picture may also change rapidly to a sustained depressive disorder. If either change happens, the patient may develop **suicidal ideas.** A sustained change to a depressive syndrome is likely to require antidepressant drug treatment unless the disorder is mild.

The practical aspects of using lithium to prevent further episodes are discussed on p. 582. There are two aspects of prophylaxis that require emphasis. First, patients should be seen and plasma lithium measured at regular intervals. Second, some patients stop lithium of their own accord because they fear that the drug may have harmful long-term effects or because it makes them feel 'flat'. The risk of stopping lithium is that it may lead to relapse. As pointed out above, some investigators believe that the relapse rate shortly after lithium discontinuation is greater than that of patients not receiving prophylactic treatment. It does appear that slow withdrawal of lithium decreases the risk of relapse after discontinuation (Faedda et al. 1993), and therefore patients should be warned not to stop lithium suddenly.

With carefully supervised follow-up, the chances of relapse can be reduced substantially by prophylactic treatment, though lesser degrees of mood change often continue. These mood changes can often be treated successfully with adjunctive antidepressant or antipsychotic drug treatment (Peselow et al. 1994). Some patients with bipolar disorder respond to antidepressant drug treatment by developing a **rapid-cycling mood disorder.** In this case, the best plan is to try to withdraw the antidepressant drug (for a review of the epidemiology and treatment of rapid-cycling mood disorder see Bauer and Whybrow (1991)). There is some evidence that, among patients with mood disorders, rates of relapse and rates of suicide are lower in those who receive carefully supervised lithium treatment in a specialized mood disorder clinic (Coppen et al. 1991; Muller-Oerlinghausen et al. 1992).

Further reading

Goodwin, F. K. and Jamison, K. R. (1990). *Manic depressive illness*. Oxford University Press.

Kraepelin, E. (1921). *Manic-depressive insanity and paranoia* (transl. R. M. Barclay), pp. 1–164. Churchill Livingstone, Edinburgh. [Reprinted in 1976 by Arno Press, New York.]

Lewis, A. J. (1930). Melancholia: a clinical survey of depressive states. *Journal of Mental Science*, 80, 277–378. [Reprinted in Lewis, A. J. (1967). *Inquiries in psychiatry*, pp. 30–117. Routledge and Kegan Paul, London.]

Lewis, A. J. (1938). States of depression: their clinical and aetiological differentiation. *British Medical Journal*, 2, 875–8. [Reprinted in Lewis, A. J. (1967). *Inquiries in psychiatry*, pp. 133–40. Routledge and Kegan Paul, London.]

Paykel, E. S. (ed.) (1992). *Handbook of affective disorders*. Churchill Livingstone, Edinburgh.

9 | *Schizophrenia and schizophrenia-like disorders*

Of all the major psychiatric syndromes, schizophrenia is much the most difficult to define and describe. The main reason for this difficulty is that, over the past 100 years, many widely divergent concepts of schizophrenia have been held in different countries and by different psychiatrists. Radical differences of opinion persist to the present day. If these conflicting ideas are to be made intelligible, it is useful to start with a simple comparison between two basic concepts—**acute schizophrenia** and **chronic schizophrenia**. Such a comparison can pave the way for a description of the many varieties of clinical picture encountered in clinical practice, and for discussion of the main theories and arguments about schizophrenia. Accordingly, after a brief section on epidemiology, the features of 'typical' acute and chronic syndromes are described. The reader should bear in mind that these will be idealized descriptions, but it is useful to over-simplify at first before introducing controversial issues.

Essentially, the predominant clinical features in acute schizophrenia are *delusions, hallucinations,* and *interference with thinking.* Features of this kind are often called **'positive'** symptoms. Some patients recover from the acute illness, whilst others progress to the chronic syndrome. In contrast, the main features of chronic schizophrenia are *apathy, lack of drive, slowness,* and *social withdrawal.* These features are often called **'negative'** symptoms. Once the chronic syndrome is established, few patients recover completely.

Most of the disagreements about the diagnosis of schizophrenia are concerned with the acute syndrome. The criteria for diagnosis are concerned with both the pattern of symptoms and the course of the disorder. There are disagreements about both the range of symptoms that are required and the length of time that these symptoms should have been present in order to make the diagnosis. This point is illustrated later in the chapter when the criteria for diagnosis in DSMIV and ICD10 are discussed.

Epidemiology

Estimates of the incidence and prevalence of schizophrenia depend on the criteria for diagnosis and the population surveyed (problems of diagnosis are discussed on p. 252). The annual **incidence** is probably between 0.1 and 0.5 per 1000 population. Thus Wing and Fryers (1976) found a rate of between 0.11 and 0.14 per 1000 for first contact with services in Camberwell, London, whilst Häfner and Reimann (1970) reported a rate of 0.54 per 1000 in Mannheim. A World Health Organization study of ten countries found a rather consistent annual inception rate for schizophrenia ranging from 0.07 per 1000 in Aarhus, Denmark, to 0.14 per 1000 in Nottingham, United Kingdom (Jablensky et al. 1992). The onset of schizophrenia characteristically occurs between the ages of 15 and 45. While schizophrenia occurs equally in men and women, the mean age of onset is about five years earlier in men (Häfner et al. 1989).

The **lifetime risk** of developing schizophrenia is probably between 7.0 and 9.0 per 1000 (Jablensky 1986). In island communities, for example, cohort studies gave rates of 9.0 per 1000 on a Danish island (Fremming 1951) and 7.0 per 1000 in Iceland (Helgason 1964). Two recent multicentre studies in the United States have suggested that the lifetime risk for schizophrenia lies between 7.0 and

13.0 per 1000 (Regier *et al.* 1988; Kessler *et al.* 1994).

The **point prevalence** of schizophrenia in European countries is probably between 2.5 and 5.3 per 1000 (Jablensky 1986). Collaborative studies by the World Health Organization have shown that the prevalence of schizophrenia, when assessed in comparable ways, is similar in different countries (Jablensky *et al.* 1992). The similarities are greatest when Schneider's first-rank symptoms (see p.253) are used as diagnostic criteria (Jablensky *et al.* 1986). There are exceptions to this general uniformity of rates. Böök (1953) reported a high annual prevalence (11 per 1000) in the extreme north of Sweden. High rates have also been reported from Slovenia and western Ireland, and among Catholics in Canada and the Tamils of southern India (Cooper 1978). Conversely, a low prevalence of 1.1 per 1000 has been reported among the Hutterites and the Anabaptist sect in the United States (Eaton and Weil 1955). These reported differences in prevalence may have more than one cause. First, they may reflect differences in *diagnostic criteria*. Second, they may be affected by differences in *migration*. For example people predisposed to schizophrenia may stay in the remote north of Sweden because they are more able to tolerate extreme isolation, whilst those predisposed to schizophrenia may leave a Hutterite community because they are less able to tolerate a close-knit society. The third reason, which is related to the second, is that prevalence rates may reflect differences in methods of *case-finding*. It is possible that the findings reported by Eaton and Weil may be partly explained by such differences of method, since a study in Canada found no differences between Hutterite and other areas in admission rates for schizophrenia (Murphy 1968). Also, a further study in western Ireland (NiNuallain *et al.* 1987) did not confirm a high incidence.

Some recent studies have suggested that the incidence of schizophrenia, as measured by the diagnosis rate in first admissions to hospital, may be *falling* in industrialized countries. For example Eagles and Whalley (1985) reported a 40 per cent fall in first admission rates for schizophrenia in Scotland between 1969 and 1978. This observation has been supported by some, but not all, subsequent investigations (Jablensky 1993). Whether the observed decline represents a true fall in the inception rate for schizophrenia is doubtful; for example Kendell *et al.* (1993) found that in Scotland, at least, the apparent decline may be attributable to changing diagnostic criteria and inaccurate recording of true first admissions.

Epidemiological studies of the demographic and social correlates of schizophrenia are considered later under aetiology.

Clinical features

The acute syndrome

Some of the main clinical features are illustrated by a short description of a patient. A previously healthy 20-year-old male student had been behaving in an increasingly odd way. At times he appeared angry and told his friends that he was being persecuted; at other times he was seen to be laughing to himself for no apparent reason. For several months he had seemed increasingly preoccupied with his own thoughts. His academic work had deteriorated. When interviewed, he was restless and awkward. He described hearing voices commenting on his actions and abusing him. He said that he believed that the police had conspired with his university teachers to harm his brain with poisonous gases and take away his thoughts. He also believed that other people could read his thoughts.

This case history illustrates the following common features of acute schizophrenia: prominent persecutory ideas with accompanying hallucinations, gradual social withdrawal and impaired performance at work, and the odd idea that other people can read one's thoughts. In describing the features of schizophrenia here, it is assumed that the reader has read the descriptions of symptoms and signs in Chapter 1. Reference will be made to that chapter in the following account.

In **appearance and behaviour** some patients with acute schizophrenia are entirely normal. Others seem changed and different, although not always in a way that would immediately point to psychosis. They may be preoccupied with their health, their appearance, religion, or other intense interests. Social withdrawal may occur. Some patients smile or laugh without obvious reason. Some appear to be constantly perplexed. Some are restless and noisy, or show sudden and unexpected changes of behaviour. Others retire from company, spending a long time in their rooms, perhaps lying immobile on the bed apparently preoccupied in thought.

The **speech** often reflects an underlying **thought disorder**. In the early stages, there is vagueness in the patient's talk that makes it difficult to grasp his meaning. Some patients have difficulty in dealing with abstract ideas (a phenomenon called concrete thinking). Other patients become preoccupied with vague pseudoscientific or mystical ideas.

When the disturbance is more severe, two characteristic kinds of abnormality may occur. **Disorders of the stream of thought** include pressure of thought, poverty of thought, and thought blocking, which are described on p. 7. Thought withdrawal (the conviction that one's thoughts have been taken away) is sometimes classified as a disorder of the stream of thought, but it is more usefully considered as a form of delusion (see p. 14).

Loosening of association denotes a lack of connection between ideas. This may be detected in illogical thinking ('knight's move') or talking past the point (*Vorbeireden*). In the severest form of loosening the structure and coherence of thinking is lost, so that utterances are jumbled (word salad or verbigeration). Some patients use ordinary words in unusual ways (paraphrases or metonyms), and a few coin new words (neologisms).

Abnormalities of mood are common and are of three main kinds. First, there may be sustained abnormalities of mood such as anxiety, depression, irritability, or euphoria. Second, there may be *blunting of affect*, sometimes known as flattening of affect. Essentially this is sustained emotional indifference or diminution of emotional response. Third, there is *incongruity of affect*. Here the expressed mood is not in keeping with the situation or with the patient's own feelings. For example, a patient may laugh when told about a bereavement. Incongruity of affect is usually observed as a sign, and it can be difficult to distinguish from a social gaucherie. Occasionally patients are able to describe this experience, which can resolve diagnostic doubt. Without such a description, although the abnormality of incongruity of affect is said to be highly characteristic of schizophrenia, different interviewers tend to disagree about its presence.

Auditory hallucinations are among the most frequent symptoms. They may take the form of noises, music, single words, brief phrases, or whole conversations. They may be unobtrusive or so severe as to cause great distress. Some voices seem to give commands to the patient. Some patients hear their own thoughts apparently spoken out loud either as they think them (*Gedankenlautwerden*) or immediately afterwards (*écho de la pensée*). Some voices seem to discuss the patient in the third person. Others comment on his actions. As described later, these last three symptoms can have particular diagnostic value.

Visual hallucinations are less frequent and usually occur with other kinds of hallucination. **Tactile, olfactory, gustatory,** and **somatic** hallucinations are reported by some patients. They are often interpreted in a delusional way; for example hallucinatory sensations in the lower abdomen are attributed to unwanted sexual interference by a persecutor.

Delusions are characteristic. Primary delusions (see p. 10) are infrequent and are difficult to identify with certainty. Delusions may originate against a background of so-called primary delusional mood—*Wahnstimmung* (see p. 12). Persecutory delusions are common, but are not specific to schizophrenia. Less common but of greater diagnostic value are delusions of reference and of control, and delusions about the possession of thought. The latter are delusions that thoughts are being inserted into or withdrawn from one's mind, or 'broadcast' to other people (see p. 14).

In acute schizophrenia **orientation** is normal. Impairment of **attention** and **concentration** is common, and may result in memory impairment. So-called delusional memory occurs in a few patients (see p. 12).

Insight is usually impaired. Most patients do not accept that their experiences result from illness, but usually ascribe them to the malevolent actions of other people. This lack of insight is often accompanied by unwillingness to accept treatment.

Schizophrenic patients do not necessarily experience all these symptoms. The clinical picture is variable, as described later in this chapter. At this stage the reader is referred to Table 9.1, which lists the most frequent symptoms found in one large survey.

The chronic syndrome

In contrast with the 'positive' symptoms of the acute syndrome, the **chronic syndrome** is characterized by thought disorder and the 'negative' symptoms of under-activity, lack of drive, social withdrawal, and emotional apathy. The syndrome can be illustrated by a brief example. A middle-aged man lives in a group home and attends a sheltered workshop. He spends most of his time alone. He is usually dishevelled and unshaven, and cares for himself only when encouraged to do so by others. His social behaviour seems odd and stilted. His speech is slow, and its content is vague and incoherent. He shows few signs of emotion. For several years this clinical picture has changed little except for brief periods of acute symptoms which are usually related to upsets in the ordered life of the hostel.

This description illustrates several of the negative features of what is sometimes called schizophrenic defect state. The most striking feature is diminished **volition**, that is a lack of drive and initiative. Left to himself, the patient may be inactive for long periods, or may engage in aimless and repeated activity. He withdraws from social encounters and his behaviour may deteriorate in ways that embarrass other people. A few patients neglect themselves to the point of incontinence.

A variety of **motor disturbances** occur, but most are uncommon. They are outlined here because they are the only symptoms of schizophrenia not described in Chapter 1.

Table 9.1 *The most frequent symptoms of acute schizophrenia*

Symptom	Frequency (%)
Lack of insight	97
Auditory hallucinations	74
Ideas of reference	70
Suspiciousness	66
Flatness of affect	66
Voices speaking to the patient	65
Delusional mood	64
Delusions of persecution	64
Thought alienation	52
Thoughts spoken aloud	50

Source: World Health Organization 1973.

Disorders of motor activity are often called **catatonic**. In the past a separate syndrome of catatonia was recognized, but nowadays the symptoms more often occur individually than as a distinct syndrome. **Stupor** and **excitement** are the most striking catatonic symptoms. A patient in stupor is immobile, mute, and unresponsive, although fully conscious. Stupor may change (sometimes quickly) to a state of uncontrolled motor activity and excitement.

Occasionally, schizophrenic patients show a disorder of muscle tone called **waxy flexibility** (or *flexibilitas cerea*). The patient allows himself to be placed in an awkward posture which he then maintains apparently without distress for much

Table 9.2 *Behavioural characteristics of chronic schizophrenic patients in rank order of frequency*

Characteristic	Frequency (%)
Social withdrawal	74
Under-activity	56
Lack of conversation	54
Few leisure interests	50
Slowness	48
Over-activity	41
Odd ideas	34
Depression	34
Odd behaviour	34
Neglect of appearance	30
Odd postures and movements	25
Threats or violence	23
Poor mealtime behaviour	13
Socially embarrassing behaviour	8
Sexually unusual behaviour	8
Suicidal attempts	4
Incontinence	4

Source: Creer and Wing 1975.

longer than most people could achieve without severe discomfort. This phenomenon is also called **catalepsy** (a term that is also used to describe similar phenomena in patients who have been hypnotized). Some patients themselves take up odd and uncomfortable postures and maintain them for long periods. At times these postures have obvious symbolic significance (for example crucifixion). Occasionally a patient lies for a long period with his head raised a little above the pillow (the so-called psychological pillow). Healthy people would experience extreme discomfort if they tried to do the same.

Various disorders of **movement** occur in schizophrenia (Manschreck *et al.* 1982). A **stereotypy** is a repeated movement that does not appear to be goal directed. It is more complex than a tic. The movement may be repeated in a regular sequence, for example rocking forwards and backwards or rotating the trunk.

A **mannerism** is a normal goal-directed movement that appears to have social significance but is odd in appearance, stilted, and out of context, for example a repeated hand movement resembling a military salute. It is often difficult to decide whether an abnormal movement is a stereotypy or a mannerism, but the distinction is of no diagnostic importance.

Ambitendence is a special form of ambivalence in which a patient begins to make a movement but, before completing it, starts the opposite movement, for example putting the hand back and forth to an object but without reaching it. Some patients, on the point of entering a room, repeatedly walk backwards and forwards across the threshold. *Mitgehen* is moving a limb in response to slight pressure on it, despite being told to resist the pressure (the last point is important). *Mitgehen* is often associated with forced grasping, which is repeated grasping (despite instructions to the contrary) at the interviewer's outstretched hand. In **automatic obedience** the patient obeys every command, though he has first been told not to do so. These disorders are described in more detail by Hamilton (1984).

Social behaviour may deteriorate. Self-care may be poor and, particularly in women, the style of dress and presentation may be careful but somewhat inappropriate. Some patients collect and hoard objects, so that their surroundings become cluttered and dirty. Others break social conventions by talking intimately to strangers or shouting obscenities in public.

Speech is often abnormal, showing evidence of **thought disorder** of the kinds found in the acute syndrome described above. **Affect** is generally blunted; when emotion is shown, it is often incongruous. **Hallucinations** are common, again in any of the forms occurring in the acute syndrome.

Delusions are often systematized. In chronic schizophrenia, delusions may be held with little emotional response. For example, patients may be convinced that they are being persecuted but show neither fear nor anger. Delusions may also be 'encapsulated' from the rest of the patient's beliefs. Thus, although a patient may be convinced that his private sexual fantasies and practices are widely discussed by strangers, his remaining beliefs may be normal and his working and social life well preserved.

The symptoms and signs are combined in many ways so that the clinical picture is variable. The reader is referred to Table 9.2 which shows the range and frequency of the symptoms and behavioural abnormalities found in one survey of chronic schizophrenic patients.

Variations of the clinical picture

As anticipated in the introduction to this chapter, so far an account has been given of the typical features of acute and chronic syndromes. Such an account makes description easier, but it is an over-simplification. Two points need to be stressed. First, different features may predominate within a syndrome; for example, in the acute syndrome one patient may have predominantly paranoid delusions, and another mainly thought disorder. Second, some patients have features of both the acute and the chronic syndromes. Clinicians have attempted to identify various clinical subtypes, and these will be described later in the chapter.

Depressive symptoms in schizophrenia

It has been recognized since the time of Kraepelin that **depressive symptoms** commonly occur in schizophrenia, in both the acute and chronic stages. More recently these clinical observations have been confirmed using standardized methods of assessment (World Health Organization 1973). It is not uncommon for patients to experience depressive symptoms some time before more florid symptoms of schizophrenia become apparent. In addition, following resolution of psychotic symptoms about a quarter of patients exhibit persistent and significant depression which may respond to antidepressant drug treatment (Siris *et al.* 1987).

There are several reasons why depressive symptoms may be associated with schizophrenia. First, they may be a side-effect of antipsychotic medication. This is not the full explanation since depressive symptoms can occur in the absence of antipsychotic drug therapy, for example before the diagnosis of schizophrenia has been made (see above). In addition, where patients are receiving antipsychotic medication, motor side-effects may be interpreted wrongly as depressive retardation. Second, in the post-psychotic phase, depressive symptoms may be a response to recovery of insight into the nature of the illness and the problems to be faced. Again, this may happen at times, but it does not provide a convincing general explanation. Third, depression may be an integral part of schizophrenia. In a study of patients after the acute onset of schizophrenia, Knights and Hirsch (1981) found depression to be most common in the acute phase, decreasing during the following three months. They concluded that depression is a symptom of schizophrenia, although it may be recognized only after more striking symptoms have improved (see Hirsch (1986a) and Johnson (1986) for reviews.)

Water intoxication in schizophrenia

A few chronic schizophrenic patients drink excessive amounts of water, thus developing a state of 'water intoxication' characterized by polyuria and hyponatraemia (de Leen *et al.* 1993). When this condition is severe, it may give rise to seizures, coma, visceral and cerebral oedema, and sometimes death. The reasons for this behaviour cannot be explained by the patients and has not been uncovered by investigation. The possible mechanisms have been discussed by Ferrier (1985) and include a response to a delusional belief, changes in the secretion of antidiuretic hormone, or abnormalities in the hypothalamic centre that regulates thirst and fluid intake. Treatment involves fluid restriction, but in more severe cases diuretics or urea may be needed.

Factors modifying the clinical features

The amount of **social stimulation** has a considerable effect on the clinical picture. Under-stimulation increases 'negative' symptoms such as poverty of speech, social withdrawal, apathy, and lack of drive. Over-stimulation precipitates 'positive' symptoms such as hallucinations, delusions, and restlessness. Modern treatment is designed to avoid under-stimulation, and as a result 'negative' features are less frequent than in the past.

The **social background** of the patient may affect the content of some symptoms. For example, religious delusions are less common now than a century ago (Klaf and Hamilton 1961). **Age** also seems to modify the clinical features of schizophrenia. In adolescents and young adults the clinical features often include thought disorder, mood disturbance, passivity phenomena, and thought insertion, and withdrawal. With increasing age paranoid symptomatology is more common, with more organized delusions (Häfner *et al.* 1993).

Low intelligence also affects the clinical features. Patients of subnormal intelligence usually present with a less complex clinical picture, sometimes referred to as *pfropfschizophrenie* (described in Chapter 21, p 728).

Diagnostic problems

The development of ideas about schizophrenia

Some of the diagnostic problems encountered today can be understood better with some knowledge of the historical developments of ideas about schizophrenia.

In the nineteenth century, one view was that all mental disorders were expressions of a single entity which Griesinger called *Einheitpsychose* (unitary psychosis). The alternative view, which was put forward by Morel in France, was that mental disorders could be separated and classified. Morel searched for specific entities, and argued for a classification based on cause, symptoms, and outcome (Morel 1860). In 1852 he gave the name *démence précoce* to a disorder which he described as starting in adolescence and leading first to withdrawal, odd mannerisms, and self-neglect, and eventually to intellectual deterioration. Not long after, Kahlbaum (1863) described the syndrome of catatonia, and Hecker (1871) wrote an account of a condition he called hebephrenia (an English translation has been published by Sedler (1985)). **Emil Kraepelin (1855–1926)** derived his ideas from study of the course of the disorder as well as the symptoms. His observations led him to argue against the idea of a single psychosis, and to propose a division into dementia praecox and manic depressive psychosis. This grouping brought together as subclasses of dementia praecox the previously separate entities of hebephrenia and catatonia. Kraepelin's description of dementia praecox appeared for the first time in 1893, in the fourth edition of his textbook, and the account was expanded in subsequent editions (see Kraepelin (1897) for a translation of the description in the fifth edition). He described the illness as occurring in clear consciousness and consisting of 'a series of states, the common characteristic of which is a peculiar destruction of the internal connections of the psychic personality. The effects of this injury predominate in the emotional and volitional spheres of mental life' (Kraepelin 1919, p. 3). He originally divided the disorder into three subtypes (catatonic, hebephrenic, and paranoid) and later added a fourth (simple). Kraepelin separated paraphrenia from dementia praecox on the grounds that it started in middle life and seemed to be free from the changes in emotion and volition found in dementia praecox.

It is commonly held that Kraepelin regarded dementia praecox as invariably progressing to chronic deterioration. However, he reported that, in his series of cases, 13 per cent recovered completely (though some relapsed later) and 17 per cent were ultimately able to live and work without difficulty.

Eugen Bleuler (1857–1959) was the Director of Burghlözli Clinic and Professor of Psychiatry in Zurich. He based his work on that of Kraepelin, and in his own book wrote 'the whole idea of dementia praecox originates with Kraepelin' (Bleuler 1911, p. 1). He also acknowledged the help of his younger colleague, Carl Gustav Jung, in trying to apply some of Freud's ideas to dementia praecox. Compared with Kraepelin, Bleuler was concerned less with prognosis and more with the mechanisms of symptom formation. It was Bleuler who proposed the name **schizophrenia** to denote a 'splitting' of psychic functions, which he thought to be of central importance.

Bleuler believed in a distinction between **fundamental** and **accessory symptoms**. Fundamental symptoms included disturbances of *associations*, changes in *emotional reactions*, and *autism* (withdrawal from reality into an inner world of fantasy). It is interesting that, in Bleuler's view, some of the most frequent and striking symptoms were accessory (secondary), for example hallucinations, delusions, catatonia, and abnormal behaviours. Bleuler was interested in the psychological study of his cases, but did not deny the possibility of a neuropathological cause for schizophrenia. Compared with Kraepelin, Bleuler took a more optimistic view of the outcome, but still held that one should not 'speak of cure but of far reaching improvement'. He also wrote: 'as yet I have never released a schizophrenic in whom I could not still see distinct signs of the disease, indeed there are very few in whom one would have to search for such signs' (Bleuler 1911, pp. 256, 258). Since Bleuler was preoccupied more with psychopathological mechanisms than with symptoms themselves, his approach to diagnosis was less precise than that of Kraepelin.

Kurt Schneider (1887—1967) tried to make the diagnosis more reliable by identifying a group of symptoms characteristic of schizophrenia, but

Table 9.3 *Schneider's symptoms of the first rank*

Hearing thoughts spoken aloud
Third-person hallucinations
Hallucinations in the form of a commentary
Somatic hallucinations
Thought withdrawal or insertion
Thought broadcasting
Delusional perception
Feelings or actions experienced as made or influenced by external agents

rarely found in other disorders. Unlike Bleuler's fundamental symptoms, Schneider's symptoms were not supposed to have any central psychopathological role. Thus Schneider (1959) wrote:

Among the many abnormal modes of experience that occur in schizophrenia, there are some which we put in the first rank of importance, not because we think of them as basic disturbances, but because they have this special value in helping us to determine the diagnosis of schizophrenia. When any one of these modes of experience is undeniably present and no basic somatic illness can be found, we may make the diagnosis of schizophrenia ... Symptoms of first rank importance do not always have to be present for a diagnosis to be made.

The last point is important. Schneider's first-rank symptoms are considered more fully by Mellor (1982) and are listed in Table 9.3. Some of these symptoms are included in the diagnostic criteria for schizophrenia used in DSMIV and ICD10 (see pp. 257–60).

Several German psychiatrists tried to define subgroupings within schizophrenia. **Karl Kleist,** a pupil of the neurologist Wernicke, looked for associations between brain pathology and different subtypes of psychotic illness. He accepted Kraepelin's main diagnostic framework but used careful clinical observation in an attempt to distinguish various subdivisions within schizophrenia and other atypical disorders. He then attempted to match these subtypes to specific kinds of brain pathology (Kleist 1928, 1930). This attempt was ingenious but not successful.

Leonhard continued this approach of careful clinical observation, but did not pursue Kleist's interest in cerebral pathology. He published a complicated classification which distinguishes schizophrenia from the 'cycloid' psychoses, which are a group of non-affective psychoses of good outcome (Leonhard 1957). Cycloid psychoses are described later in this chapter. Leonhard also divided schizophrenia into two groups. The first group is characterized by a progressive course, and is divided into catatonias, hebephrenias, and paraphrenias. Leonhard gave this group a name which is often translated as 'systematic'. The second group, called **non-systematic,** is divided into affect-laden paraphrenia, schizophasia, and periodic catatonia. **Affect-laden paraphrenia** is characterized by paranoid delusions and the expression of strong emotion about their content. In **schizophasia** speech is grossly disordered and difficult to understand. **Periodic catatonia** is a condition with regular remissions; during an episode akinetic symptoms are sometimes interrupted by hyperkinetic symptoms.

Originally Leonhard's views had little impact outside his own country of East Germany, but recently an increasing concern with the heterogeneity of schizophrenia has led to a wider interest in his system of classification and particularly the concept of cycloid psychoses. Further information can be found in the accounts by Hamilton (1984) and Ban (1982).

Scandinavian psychiatrists have been influenced by Jaspers' distinction between process schizophrenia and reactive psychoses. In the late 1930s **Langfeldt,** using follow-up data on patients in Oslo, proposed a distinction between **true schizophrenia,** which had a poor prognosis, and **schizophreniform states,** which had a good prognosis (Langfeldt 1961). True schizophrenia was defined narrowly and was essentially similar to Kraepelin's dementia praecox. It was characterized by emotional blunting, lack of initiative, paranoid symptoms, and primary delusions. Schizophreniform states were described as often precipitated by stress and frequently accompanied by confusional and affective symptoms. Langfeldt's proposed distinction between cases with good and bad prognosis has been influential but, as explained later, other psychiatrists have not found that his criteria predict prognosis accurately (see p. 282). According to modern diagnostic criteria, most of Langfeldt's cases of schizophreniform psychosis would, in fact, be classified as mood disorders (Bergen *et al.* 1990).

In Denmark and Norway cases arising after stressful events have received much attention. The terms **reactive psychosis** or **psychogenic psychosis** are commonly applied to conditions which appear to be precipitated by stress, are to some extent understandable in their symptoms, and have a good prognosis. They were first described by Wimmer (1916); see also Strömgren (1968, 1986), Faergeman (1963), and Cooper (1986).

International differences in diagnostic practice

By the 1960s there were wide divergences in the criteria for the diagnosis of schizophrenia. In the United Kingdom and continental Europe, psychiatrists generally employed Schneider's approach, using *first rank symptoms* to identify a narrowly delineated group of cases. In the United States, however, interest in psychodynamic processes led to diagnosis on the basis of mental mechanisms and to the inclusion of a much wider group of cases.

First admission rates for schizophrenia were much higher in the United States than in the United Kingdom. This discrepancy prompted two major cross-national studies of diagnostic practice. The **US–UK Diagnostic Project** (Cooper *et al.* 1972) showed that the diagnostic concept of schizophrenia was much wider in New York than in the United Kingdom. In New York the concept included cases that were diagnosed as depressive illness, mania, or personality disorder in the United Kingdom. The **International Pilot Study of Schizophrenia** (IPSS) was concerned with the diagnosis of schizophrenia in nine countries (World Health Organization 1973). The main finding was that similar criteria were adopted in seven of the nine countries: Colombia, Czechoslovakia, Denmark, India, Nigeria, Taiwan, and the United Kingdom. Broader criteria were used in the United States and the former USSR. Despite these differences it was found, by using standard diagnostic techniques, that a core of cases with similar features could be identified in all the countries. In France, schizophrenia is not defined in Kraepelian terms. It is regarded as a chronic disorder with an onset before the age of 30. The diagnosis is made largely on symptoms, especially those proposed by Bleuler, which are referred to as 'dissociation' and 'discordance'. The term *bouffée délirante* is used for a sudden-onset syndrome of good prognosis (Pichot 1982, 1984).

Reasons for diagnostic inconsistencies

To agree about the diagnosis of schizophrenia, clinicians must first agree about the symptoms that make up the syndrome, about the criteria for deciding whether these symptoms are present, and about the length of time that these features must last before the diagnosis can be made. The criteria for diagnosing the various symptoms of schizophrenia were discussed in Chapter 1. Here we consider the other two issues.

The *narrower* the range of symptoms that are accepted as diagnostic of schizophrenia, the more *reliable* is the diagnosis. However, a narrow definition may exclude cases that are aetiologically related; this problem cannot be settled until we know more about the causes of schizophrenia. An example of a well-known narrow definition is that requiring the presence of Schneider's first-rank symptoms (Table 9.3). The use of this criterion leads to high reliability in diagnosis (though not to effective prediction of outcome (Brockington 1986)). In addition, while genetic factors are known to play a part in the aetiology of schizophrenia, Schneider's first rank symptoms delineate a syndrome with no apparent heritability (McGuffin *et al.* 1984). It is also now clear that first-rank symptoms can occur in cases that otherwise meet generally agreed diagnostic criteria for mania. Both ICD10 and DSMIV use wider definitions of schizophrenia, including other symptoms as well as Schneider's first-rank symptoms.

There is disagreement about the duration of symptoms required before schizophrenia can be diagnosed. ICD10 requires one month, and DSMIV requires six months. Both periods are arbitrary. The longer time is more likely to identify patients with a poor prognosis, but the importance of prognosis in defining schizophrenia is not yet clear. (For reviews of this topic, the reader is referred to Farmer *et al.* (1992*a*) and Andreasen and Carpenter (1993).)

'Standardized diagnoses'

An important example of a standardized diagnosis is CATEGO, a computer program designed to process data from a standard interview known as the **Present State Examination** (Wing *et al.* 1974). The program incorporates diagnostic rules which give a series of standard diagnoses. The narrowest syndrome (S+) is diagnosed mainly on the symptoms of thought intrusion, broadcast, or withdrawal, delusions of control, and voices discussing the patient in the third person or commenting on his actions.

The **Feighner criteria** (Feighner *et al.* 1972) were developed in St Louis to identify patients with a poor prognosis. They include symptomatic criteria that are less precise than those in CATEGO, and a criterion of six months' continuous illness. They also require the exclusion of cases meeting diagnostic criteria for mood disorder, drug abuse, or alcoholism. The criteria are reliable but restrictive, leaving many cases without a diagnosis. Patients with a poor prognosis are identified well (probably because of the criterion of six months' continuous illness). These criteria have been widely used in research.

Table 9.4 *Classification of schizophrenia and schizophrenia-like disorders in ICD10 and DSMIV**

ICD10	DSMIV
Schizophrenia	*Schizophrenia*
Paranoid	Paranoid
Hebephrenic	Disorganized
Catatonic	Catatonic
Undifferentiated	Undifferentiated
Residual	Residual
Simple schizophrenia	
Post-schizophrenia depression	
Other schizophrenia	
Unspecified schizophrenia	
Schizotypal disorder	
Schizoaffective disorder	*Schizoaffective disorder*
Persistent delusional disorders	*Delusional disorder*
Delusional disorder	
Other persistent delusional disorders	
Acute and transient psychotic disorders	*Brief psychotic disorder*
Acute polymorphic psychotic disorder	
Schizophrenia-like psychotic disorder	*Schizophreniform disorders*
Other acute psychotic disorder	
Induced delusional disorder	*Shared psychotic disorder*
Other non-organic psychotic disorders	
Unspecified non-organic psychosis	*Psychotic disorder not otherwise specified*

*The order of categories in the two systems has been changed slightly to show their comparable features more clearly.

The **Research Diagnostic Criteria** (RDC) were developed from those of Feighner by Spitzer *et al.* (1978). The main difference is in the length of history required before the diagnosis of schizophrenia can be made: two weeks, instead of six months in the Feighner criteria. A structured interview, the Schedule of Affective Disorders and Schizophrenia (SADS), has been developed for use with RDC. Both the Feighner criteria and the RDC influenced the development of DSMIII and DSMIV.

Schizophrenia secondary to other medical conditions

If the symptoms of schizophrenia occur as a result of some identifiable disease of the central nervous system, the condition is generally referred to as a *psychotic disorder secondary to a general medical condition* (DSMIV) or an *organic delusional disorder* (ICD10). These conditions are sometimes known as *'symptomatic schizophrenia'*. Certain disorders that can produce a secondary schizophrenic syndrome are discussed elsewhere in this book. They include **temporal lobe epilepsy** (complex partial seizures) (p. 340), **encephalitis** (p. 329), **amphetamine abuse** (p. 475), and **alcohol abuse** (p. 451). In addition, schizophrenia-like syndromes can occur **post partum** (see p. 396) and in the **postoperative** period (p. 367).

Classification in DSMIV and ICD10

The classification of schizophrenia and schizophrenia-like disorders is summarized in Table 9.4. Both classifications distinguish schizophrenic illness, though in DSMIV more emphasis is laid on course and functional impairment, while in ICD10 Schneider's first rank signs are given more weight. Both classifications separate out disorders with especially prominent mood disturbance (*schizoaffective disorders*), and also identify illnesses which meet symptomatic criteria for schizophrenia but have a brief duration of symptomatology (*brief psychotic disorder* and *schizophreniform disorder* in DSMIV and *acute and transient psychotic disturbance* in ICD10). In addition, both DSMIV and ICD10 distinguish illnesses that centre around relatively enduring non-bizarre delusions without other features of schizophrenia (*delusional disorder*).

DSMIV

In this classification **schizophrenia** is defined in terms of the symptoms in the acute phase and also of the course, for which the requirement is *continuous signs of disturbance for at least six months*. The acute symptoms can include delusions, hallucinations, disorganized speech or behaviour, and negative symptoms. At least two of these symptoms must have been present for a one-month period (unless successful treatment has occurred). Where subjects have major disturbances of mood occurring concurrently with the acute phase symptoms a diagnosis of a **schizoaffective disorder** or **mood disorder with psychotic features** should be made (Table 9.5) (see below).

A further criterion for the diagnosis of schizophrenia in DSMIV is that the patient must have exhibited deficiencies in their expected level of occupational or social functioning since the onset of the disorder. As noted above, for the diagnosis to be made there must have been at least six months of continuous disturbance; this can include prodromal and residual periods when acute phase symptoms, as described above, are not evident. Patients whose symptom duration does not meet this criterion six-month period will be classified as suffering from a **schizophreniform disorder** or a **brief psychotic disorder** (see below).

In DSMIV, schizophrenic disorders are divided into a number of subtypes which are defined by the predominant symptomatology at the time of evaluation (Table 9.4) (see below). In addition, after at least one year has elapsed since the onset of active phase symptoms, DSMIV allows the schizophrenic disorder to be classified by *longitudinal course*. This classification takes into account the pattern of illness so that remissions and their quality and the presence of relapse can be coded.

In DSMIV schizophrenia is distinguished from two other major categories: **'delusional disorder'** (discussed in the next chapter) and the mixed group of **'psychotic disorders not elsewhere classified'** .

ICD10

The ICD10 definition of **schizophrenia** places more reliance than DSMIV on first-rank symptoms, and

Table 9.5 *Criteria for schizophrenia in DSMIV*

A *Characteristic symptoms of the active phase*
Two (or more) of the following, each present for a significant portion of time during a
one-month period (or less if successfully treated)
  (1) Delusions
  (2) Hallucinations
  (3) Disorganized speech (e.g. frequent derailment or incoherence)
  (4) Grossly disorganized or catatonic behaviour
  (5) Negative symptoms, i.e. affective flattening, alogia, or avolition

B *Social/occupational dysfunction*
For a significant portion of the time since the onset of the disturbance, one or more major
areas of functioning such as work, interpersonal relations, or self-care are markedly
below the level achieved prior to the onset (or when the onset is in childhood or
adolescence, failure to achieve expected level of interpersonal, academic, or occupational
achievement)

C *Duration*
Continuous signs of the disturbance persist for at least six months. This six-month period
must include at least one month of symptoms (or less if successfully treated) that meet
criterion A (i.e. active-phase symptoms) and may include periods of prodromal or
residual symptoms. During these prodromal or residual periods, the signs of the
disturbance may be manifested by only negative symptoms or by two or more symptoms
listed in criterion A present in an attenuated form (e.g. odd beliefs, unusual perceptual
experiences)

D *Schizoaffective and mood disorder exclusion*
Schizoaffective disorder and mood disorder with psychotic features have been ruled out
because either (1) no major depressive, manic, or mixed episodes have occurred
concurrently with the active-phase symptoms, or (2) if mood episodes have occurred
during active-phase symptoms, their total duration has been brief relative to the duration
of the active and residual periods

E *Substance/general medical condition exclusion*
The disturbance is not due to the direct physiological effects of a substance (e.g. a drug of
abuse, a medication) or a general medical condition

F *Relationship to a pervasive developmental disorder*
If there is a history of autistic disorder or another pervasive development disorder, the
additional diagnosis of schizophrenia is made only if prominent delusions or
hallucinations are also present for at least one month (or less if successfully treated)

requires only one month's duration of illness with prodromal symptoms being excluded (Table 9.6) There are several clinical subtypes of schizophrenia; these differ most obviously from DSMIV in the inclusion of categories of *simple schizophrenia* and *post-schizophrenic depression*. The ICD10 category of *hebephrenic schizophrenia* corresponds to the DSMIV category of *schizophrenia, disorganized type*. Like DSMIV, cases in which symptoms of schizophrenia are accompanied by prominent mood disturbances are classified as schizoaffective disorders.

ICD10 includes **schizotypal disorder** among schizophrenic disorders, whereas in DSMIV, this condition is classified as a personality disorder. Family studies show that schizotypal disorder is likely to be part of the phenotypic expression of a schizophrenic disorder (see p. 267). Like *schizoid personality disorder*, **schizotypal disorder** is associated with social isolation and restriction of affect, but in addition there are cognitive or perceptual distortions with disorders of thinking and speech and striking eccentricity or oddness of behaviour.

Similarly to DSMIV, ICD10 classifies **delusional disorders** separately from schizophrenia (see next chapter). In addition, a group of **acute and transient psychotic disorders** are recognized that have an acute onset and complete recovery within two to three months (see below).

Subtypes of schizophrenia

The variety of the symptoms and course of schizophrenia has led to several attempts to define subgroups. This section is concerned only with the traditional subgroups of hebephrenic, catatonic, paranoid, and simple schizophrenia.

Patients with **hebephrenic** schizophrenia (disorganized type in DSMIV) often appear silly and childish in their behaviour. Affective symptoms and thought disorder are prominent. Delusions are common and not highly organized. Hallucinations also are common, and are not elaborate.

Catatonic schizophrenia is characterized by motor symptoms of the kind described on p. 19 and by changes in activity varying between excitement and stupor. Hallucinations,

delusions, and affective symptoms occur but are usually less obvious. In **paranoid schizophrenia** the clinical picture is dominated by well-organized paranoid delusions. Thought processes and mood are relatively spared, and the patient may appear normal until his abnormal beliefs are uncovered.

Simple schizophrenia is characterized by the insidious development of odd behaviour, social withdrawal, and declining performance at work. Since clear schizophrenic symptoms are absent, simple schizophrenia is difficult to identify reliably.

With the possible exception of paranoid schizophrenia, these 'subgroups' are of uncertain validity. Some patients present symptoms of one group at one time, and then symptoms of another group later. There is some genetic evidence for separating cases with the paranoid picture (McGuffin *et al.* 1987; Onstad *et al.* 1991) but not enough to recommend doing so in everyday clinical work. Follow-up studies suggest that paranoid schizophrenia has a later and more acute onset than other subtypes of schizophrenia and tends to run a more a remittent course (Fenton and McGlashan 1991).

Catatonic symptoms are much less common now than 50 years ago, perhaps because of improvements in the social environment in which patients are treated or because of the use of more effective physical treatments. Other explanations are possible; for example organic syndromes may have been included in some earlier series (Mahendra 1981).

Schizophrenia-like disorders

Whatever definition of schizophrenia is adopted, there will be cases that resemble schizophrenia in some respects and yet do not meet the criteria for diagnosis. Schizophrenia-like disorders can be divided into four groups: (a) delusional or paranoid disorders; (b) brief disorders;(c) disorders accompanied by prominent affective symptoms; (d) disorders without all the required symptoms for schizophrenia. The latter three groups will be discussed below: **delusional or paranoid disorders** are discussed in the next chapter.

Table 9.6 *Symptomatic criteria for schizophrenia in ICD10*

The normal requirement for a diagnosis of schizophrenia is that a minimum of one very clear symptom (and usually two or more if less clear-cut) belonging to any one of the groups listed as (a)–(d) below, or symptoms from at least two of the groups referred to as (e)–(h), should have been clearly present for most of the time *during a period of one month or more.*

(a) Thought echo, thought insertion or withdrawal, and thought broadcasting

---

(b) Delusions of control, influence, or passivity, clearly referred to body or limb movements or specific thoughts, actions, or sensations; delusional perception

---

(c) Hallucinatory voices giving a running commentary on the patient's behaviour, or discussing the patient among themselves, or other types of hallucinatory voices coming from some part of the body

---

(d) Persistent delusions of other kinds that are culturally inappropriate and completely impossible

---

(e) Persistent hallucinations in any modality, when accompanied either by fleeting or half-formed delusions without clear affective content, or by persistent over-valued ideas, or when occurring every day for weeks or months on end

---

(f) Breaks or interpolations in the train of thought, resulting in incoherence or irrelevant speech, or neologisms

---

(g) Catatonic behaviour, such as excitement, posturing, or wavy flexibility, negativism, mutism, and stupor

---

(h) 'Negative' symptoms such as marked apathy, paucity of speech, and blunting or incongruity of emotional responses, usually resulting in social withdrawal and lowering of social performance; it must be clear that these are not due to depression or to neuroleptic medication

---

(i) A significant and consistent change in the overall quality of some aspects of personal behaviour, manifest as loss of interest, aimlessness, idleness, a self-absorbed attitude, and social withdrawal

---

Brief disorders

DSMIV uses the term 'brief psychotic disorder' for a syndrome characterized by at least one of the acute phase positive symptoms shown in Table 9.4. The disorder lasts for at least one day but not more than one month by which time full recovery has occurred. The disorder may or may not follow a stressor, but psychoses induced by the direct physiological effects of drugs or medical illness are excluded. **Schizophreniform psychosis** is a syndrome similar to schizophrenia (meeting criterion

A) which has lasted more than one month (and so cannot be classified as brief psychotic disorder) but less than the six months required for a diagnosis of schizophrenia to be made. Social and occupational dysfunction are not needed to make the diagnosis (although they may occur).

In ICD10 more short-lived psychotic illnesses are grouped under the term **acute and transient psychotic disorder.** These disorders are of acute onset, and complete recovery within two to three months is the rule. The disorder may or may not be precipitated by a stressful life event. In the first two subtypes *(acute polymorphic psychotic disorders)* hallucinations, delusions, and perceptual disturbance are obvious but change rapidly in nature and extent. There are often accompanying changes in mood and motor behaviour. The two subtypes are distinguished from each other by whether or not typical schizophrenic symptoms are also present, although if they are, the duration of the disorder must not have exceeded one month or schizophrenia will be diagnosed. The terms *bouffée délirante* and *cycloid psychosis* are given as synonyms for these categories. A third subtype, *schizophrenia-like episode*, is a non-committal term for cases meeting the symptom criteria for schizophrenia but lasting for less than a month. Such cases could turn out to be schizophrenia or acute delusional episodes depending on the subsequent course. The fourth subtype, for cases that do not fit the other three, is called *other acute psychotic episodes*.

Disorders with prominent affective symptoms

Some patients have a more or less equal mixture of schizophrenic and affective symptoms. As mentioned earlier, such patients are classified under **schizoaffective disorder** in both DSMIV and ICD10.

The term **schizoaffective disorder** has been used in several distinct ways (Lapensée 1992). It was first applied by Kasanin (1933) to a small group of young patients with severe mental disorders 'characterized by a very sudden onset in a setting of marked emotional turmoil with a distortion of the outside world. The psychosis lasts a few weeks and is followed by recovery.' The definitions of schizoaffective disorder in DSMIV and ICD10 differ substantially from this description. For example DSMIV requires that 'there should have been an uninterrupted period of illness during which, at some time there is either a major depressive episode, a manic episode or a mixed episode concurrent with symptoms that meet criterion A for schizophrenia'. During this continuous episode of illness, the acute-phase psychotic symptoms must have been present for at least two weeks in the absence of prominent mood symptoms (or the diagnosis would be a mood disorder with psychotic features). However, the episode of mood disturbance must have been present for a substantial part of the illness.

The definition of schizoaffective disorder in ICD10 is similar. It specifies that the diagnosis 'should only be made when *both* definite schizophrenic and definite affective symptoms are equally prominent and present *simultaneously,* or within a few days of each other'. ICD10 classifies schizoaffective disorder by whether the mood disturbance is *depressive, manic,* or *mixed.* In DSMIV schizoaffective disorder is specified either as *depressive* type or *bipolar.*

Family studies have shown that the more recent diagnostic concepts of schizoaffective disorder delineate a syndrome in which first-degree relatives have an increased risk of both mood disorders and schizophrenia (see below). In addition, the *outcome* of schizoaffective disorder may be generally rather better than that for schizophrenia, particularly where *manic* symptoms have been prominent (Maj and Perris 1990).

As noted above, it is not uncommon for patients with schizophrenia to develop depression as the symptoms of acute psychosis subside. This is recognized in ICD10 as **post-schizophrenic depression** where prominent depressive symptoms have been present for at least two weeks while some symptoms of schizophrenia (either positive or negative) still remain. It is important to distinguish depressive symptoms in these circumstances from the side-effects of antipsychotic drug treatment or the impaired volition and affective flattening that may occur in the context of a schizophrenic disorder.

Disorders without all the required symptoms for schizophrenia

A difficult problem is presented by cases with symptoms which resemble those of schizophrenia but fail strict criteria for diagnosis and which endure for years. There are three groups. The first group consists of people who from an early age have behaved oddly and shown features seen in schizophrenia, for example ideas of reference, persecutory beliefs, and unusual types of thinking. When long-standing, these disorders can be classified as personality disorders as in DSMIV (**schizotypal personality disorder** (see p. 112)), or with schizophrenia as in ICD10 (**schizotypal disorder**). Because of a suggested close relationship to schizophrenia, these disorders have also been called **latent schizophrenia**. They are more frequent in families of schizophrenic patients than in other families, suggesting a possible genetic relationship (see p. 267). The second group is of people who develop symptoms after a period of more normal development. There is a gradual development of social withdrawal, lack of initiative, odd behaviour, and blunting of emotion. These 'negative' symptoms are not accompanied by any of the 'positive' symptoms of schizophrenia such as hallucinations and delusions. Such cases are classified as **simple schizophrenia** in ICD10; in DSMIV they are classified as schizotypal personality disorder. The third group is of patients who have shown the full clinical picture of schizophrenia in the past but no longer have all the symptoms required to make the diagnosis. These cases are classified as **residual schizophrenia** in both DSMIV and ICD10.

Borderline disorders

In the United States there has been much interest in states intermediate between schizophrenia and the neuroses and personality disorders. The term **'borderline states'** has been applied to them, but it is used in several different ways (Dahl 1985; Tarnopolsky and Berelowitz 1987). There have been three main usages.

1. A borderline state is regarded as an independent entity which is wholly distinct from all other diagnostic categories. The main features have not been defined precisely, but include anger, difficulty in relationships, and lack of self-identity.
2. A borderline state is a mild expression of schizophrenia. This usage resembles Bleuler's concept of latent schizophrenia; it also covers the 'schizophrenia spectrum' referred to in some genetic studies (see p. 266) and the concept of pseudoneurotic schizophrenia (Hoch and Polantin 1949).
3. The term 'borderline' is applied to a form of personality abnormality characterized by impulsive unstable relationships, inappropriate anger, identity disturbance, unstable mood, and chronic boredom.

The most common usage at present is to apply the term borderline to certain kinds of personality disorder, for example *borderline personality* in DSMIV and *emotionally unstable personality, borderline type* in ICD10. The characteristics of this personality disorder are described on p. 113. Although patients with borderline personality disorder may experience transient psychotic symptoms, family studies do not suggest that this disorder is related to schizophrenia (Kendler *et al.* 1993*a,b*; Parnas *et al.* 1993).

Some other notable terms

Schizophreniform states

As mentioned earlier, this term was applied by Langfeldt (1961) to good prognosis cases as distinct from 'true' schizophrenia. The main features of schizophreniform states were the presence of a precipitating factor, acute onset, clouding of consciousness, and depressive and hysterical features. Although Welner and Strömgren (1958) confirmed the better outcome of schizophreniform cases, recent work has cast considerable doubt on the diagnostic and predictive value of Langfeldt's criteria (Bergen *et al.* 1990). It is unfortunate that DSMIV uses the term schizophreniform quite differently to describe a condition identical to schizophrenia but with a course of less than six months.

Cycloid psychoses

Kleist introduced the term cycloid marginal psychoses to denote functional psychoses which were neither typically schizophrenic nor manic depressive. Leonhard (1957) developed these ideas by describing three forms of cycloid psychosis which are distinguished by their predominant symptoms, these conditions are all bipolar and are described as having a good prognosis and leaving no chronic defect state. The first is **anxiety elation psychosis**, in which the prominent symptom is a mood change. At one 'pole' anxiety is associated with ideas of reference and sometimes with hallucinations. At the other 'pole' the mood is elated, often with an ecstatic quality. The second is **confusion psychosis**, in which thought disorder is the prominent symptom and the clinical picture varies between excitement and a state of underactivity with poverty of speech. The third is **motility psychosis**, in which the striking changes are in psychomotor activity. These syndromes are described briefly by Hamilton (1984) and in detail by Leonhard (1957). A recent study, using the statistical technique of latent class analysis, suggested that a group of variously described disorders including cycloid psychosis, *bouffée délirante*, and Kasanin's schizoaffective disorder may form a cluster of atypical psychoses which can be distinguished on clinical grounds from ICD10 and DSMIV concepts of schizophrenia (McGorry *et al.* 1992).

Type I and Type II schizophrenia

For many years a distinction has been made between **positive** and **negative** symptoms of schizophrenia. Recently there has been renewed interest in defining negative symptoms, and using them as an alternative approach to subtyping (Crow 1980, 1985; Andreasen 1982). There is no complete consensus on the specification of negative symptoms but it is generally agreed that they include *poverty of speech*, *affective blunting*, *lack of volition*, and *social withdrawal* (Malmberg and David 1993).

Crow and his colleagues have described two syndromes. **Type I** is said to have an *acute onset*, mainly *positive symptoms*, and *good social functioning* during remissions. It has a *good response to antipsychotic drugs* with biochemical evidence of *dopamine over-activity*. In contrast, **Type 2** is said to have an *insidious onset*, mainly *negative symptoms*, and *poor outcome*. It has a *poor response to antipsychotic drugs* without evidence of dopamine over-activity. In Type 2 there is evidence of structural change in the brain (especially ventricular enlargement). (Cognitive, biochemical, and neurological aspects of schizophrenia are considered later in this chapter.) In practice, although some patients can be recognized to have Type I syndrome, those with Type II are much less common and most patients show a mixture of Type I and II symptoms. For example, Farmer *et al.* (1987) found no 'pure' Type II patients in a series of 35 chosen to represent typical cases of chronic schizophrenia.

In a reappraisal of the positive and negative symptom distinction, Andreasen *et al.* (1990*a*) found that patients with a predominantly negative symptom syndrome had evidence of *poor premorbid adjustment*, an *earlier age of onset*, *poor performance on cognitive tests*, and a *poor response to treatment*. These subjects were mainly *male and unemployed*. Therefore in these respects they did have features of the Type II syndrome described by Crow (1980, 1985). However, they did not have significantly larger ventricular–brain ratios than patients with predominantly positive symptoms or mixed symptomatology, suggesting that the Type I and Type II syndromes cannot be distinguished on grounds of structural brain pathology (at least, that which is related to changes in ventricular–brain ratio).

Recent studies have proposed more complex delineations of psychopathology than that offered by the positive and negative dichotomy. For example Liddle (1987), studying patients with chronic schizophrenia, described three overlapping clinical syndromes each linked to a particular symptom cluster. Thus negative symptoms are associated with a syndrome of *psychomotor poverty*, hallucinations and delusions with a syndrome of *reality distortion,* and thought disorder and inappropriate affect with a syndrome of *disorganization*. Attempts have been made to validate these clinical syndromes by studying their association with patterns of neuropsychological deficit and regional cerebral blood flow (Liddle and Morris 1991; Liddle *et al.* 1992). At present the most reproducible finding seems to be the link between psychomotor poverty (negative symptoms), impaired performance on frontal lobe tasks, and decreased frontal blood flow (see below).

Gjessing's syndrome

Gjessing (1947) described a rare disorder in which catatonic symptoms recurred in phases. He also found changes in nitrogen balance, which were not always in phase with the symptoms. Gjessing believed that there were underlying changes in thyroid function and that the disorder could be treated successfully with thyroid hormone. The condition, if it exists, is exceedingly rare.

Differential diagnosis

Schizophrenia has to be distinguished mainly from **organic syndromes, mood disorder,** and **personality disorder.** Among younger patients, the most relevant organic diagnoses are *drug-induced states* and *temporal lobe epilepsy*. Among older patients, various brain diseases must be excluded. For example an *acute brain syndrome* could be mistaken for schizophrenia, and the same holds for *dementia*, particularly with prominent persecutory delusions. Some diffuse brain diseases can present a schizophrenia-like picture in the absence of neurological disorder; the most important example is *general paralysis of the insane* (see Chapter 11). In seeking to exclude such organic disorders, it is important to obtain a thorough history, mental state examination, and physical examination with particular reference to neurological abnormalities. There must be careful observation for *clouding of consciousness,* obvious *disorientation*, and other symptoms and signs which are not characteristic of schizophrenia.

The distinction between **mood disorder** and schizophrenia depends on the degree and persistence of the mood disorder, the relation of any hallucinations or delusions to the prevailing mood, and the nature of the symptoms in any previous episodes. The distinction from **mania** in young people can be particularly difficult, and sometimes the diagnosis can be clarified only by longer-term follow-up. A family history of mood disorder may be a useful pointer. Differential diagnosis from **personality disorder** can also be difficult when insidious changes are reported in a young person. Prolonged lengthy observation for acute symptoms of psychosis may be required.

Aetiology

Before reviewing evidence on the causation of schizophrenia, it may be helpful to outline the main areas of inquiry. Of the **predisposing causes,** *genetic factors* are most strongly supported by the evidence, but the fact that some patients with schizophrenia have no family history of the disorder suggests that *environmental factors* are likely to play a part as well. The nature of these environmental factors is uncertain. Neurological damage around the time of birth has been suggested, and so have interpersonal and social influences. The evidence for all these factors is indirect and incomplete. While *psychodynamic theories* of aetiology are now largely of historical interest, there is growing evidence that schizophrenic disorders presenting in adulthood have distinguishable *childhood antecedents* characterized by cognitive and social impairments. Research on **precipitating causes** has been concerned mainly with *life events*. Although such precipitants seem to have a definite effect, the size of the effect is uncertain. Among **perpetuating** factors, *social and family* influences seem important; however, they are not considered in this section but in the later section on course and prognosis.

The neurodevelopmental hypothesis

In recent years there has been accumulating evidence that patients with schizophrenia have abnormalities in *brain structure* and *morphology*. In addition, appropriate neuropsychological testing provides evidence of deficits in aspects of *psychological performance*, often in association with altered patterns of *cerebral blood flow*. Some of these abnormalities have been shown to be present when patients with schizophrenia first present for treatment and they do not seem to worsen as the disease progresses. In addition, while neuropathological studies provide evidence of disordered *cellular architecture* in some cortical and subcortical brain regions, gliosis is not consistently present (Bruton *et al.* 1990), suggesting that the pathological changes are neurodevelopmental in origin, or that if an injury of some kind occurred to the brain, it happened before birth.

These findings have given rise to the **neurodevelopmental hypothesis** of schizophrenia in which the pathological changes of the disorder are laid down early in life, presumably through genetic influences, and then modified by maturational and environmental factors (Murray and Lewis 1987; Weinberger 1987). The neurodevelopmental theory provides a useful framework in which to consider the aetiology of schizophrenia, particularly the various neurobiological changes that have been linked to the clinical disorder. A number of these changes will be discussed below, followed by a consideration of how far psychological, social, and epidemiological findings in schizophrenia are consistent with the neurodevelopmental approach to aetiology (Table 9.7).

Genetics

The genetic study of schizophrenia has been directed towards three questions:

(i) Is there a genetic basis?
(ii) Is there a relationship between clinical form and inheritance?
(iii) What is the mode of inheritance?

Table 9.7 *Some findings compatible with the neurodevelopment hypothesis of schizophrenia*

Non-progressive structural brain lesions

Non-progressive cognitive impairment

Cytoarchitectural disturbances without gliosis

Cognitive and social impairments in childhood

'Soft' neurological signs

Excess of winter births

These three questions will be considered in turn. (Readers requiring a more detailed account are referred to Kendler (1986) and McGuffin (1988).)

Is there a genetic basis?

Family studies The first systematic family study of dementia praecox was carried out in Kraepelin's department by Ernst Rüdin, who showed that the rate of dementia praecox was higher among the *siblings* of probands than in the general population (Rüdin 1916). Another comprehensive study of families was conducted by Kallman (1938), whose sample of schizophrenic probands numbered more than a thousand. He found increased rates not only among the probands' siblings but also among their *children*. Table 9.8 shows approximate risks for various degrees of kinship to a schizophrenic patient.

More recent research has used better criteria for diagnosis in probands and relatives, and better ways of selecting the probands (Weissman *et al.* 1986). These improved methods yield estimates of an average lifetime risk of about 5 per cent among first-degree relatives of schizophrenics, compared with 0.2–0.6 per cent among first-degree relatives of controls (Kendler 1986). To some extent, the risk depends on the definition of the phenotype. For example in a prospective study of over 200 children of mothers with schizophrenia, Parnas *et al.* (1993) found not only a significant excess of *schizophrenia* (16.2 per cent versus 1.9 per cent in controls) but also an excess of *schizotypal, paranoid,* and *schizoid* personality disorders (21.3 per cent versus 5 per cent). Similar findings were reported by Kendler *et al.* (1993b,c) in a study of first-degree relatives of patients with schizophrenia.

Taken together, the family studies provide clear evidence of a familial aetiology but do not distinguish between the effects of genetic factors and those of the family environment. To make this distinction twin and adoption studies are required.

An additional value of family studies is to determine whether the liability to schizophrenia and mood disorders is transmitted *independently,* which should be the observed pattern if the two

disorders are separate syndromes with differing aetiology. However, the findings from the various studies have been contradictory, with some workers arguing for separate familial transmission of mood disorders and schizophrenia, while others suggest that what is transmitted is a general vulnerability to psychotic illness (Crow 1994a,b). The following conclusions can be drawn from the more recent studies (Kendler *et al.* 1993b,c,d; Maier *et al.* 1993; Parnas *et al.* 1993) (Table 9.9).

- The risk of schizophrenia, schizoaffective disorder, and schizotypal personality is increased in first-degree relatives of patients with schizophrenia.
- The risk of both schizophrenia and mood disorder is increased in first-degree relatives of patients with schizoaffective disorder (as defined in DSMIIIR).

Table 9.8 *Approximate lifetime expectancy of developing schizophrenia for relatives of schizophrenics*

Relationship	Risk (%)	
	Definite cases only	Definite and probable cases
Parents	4.4	5.5
All siblings	8.5	10.2
Siblings (one parent schizophrenic)	13.8	17.2
Children	12.3	13.9
Children (both parents schizophrenic)	36.6	46.3
Half siblings	3.2	3.5
Nephews and nieces	2.2	2.6

Adapted from Shields (1980).

Table 9.9 *Lifetime risk of psychiatric disorder in first-degree relatives of probands with DSMIIIR schizophrenia, schizoaffective disorder, and controls*

	Diagnosis of proband		
Disorder	Schizophrenia	Schizoaffective disorder	Control
Schizophrenia	6.5%	6.7%	0.5%
Schizotypal disorder	6.9%	2.8%	1.4%
Schizoaffective disorder	2.3%	1.8%	0.7%
All affective illness	24.9%	49.7%	22.8%
Psychotic affective illness	33.0%	25.0%	11.5%
Bipolar disorder	1.2%	4.8%	1.4%

Data from Kendler *et al.* (1993*b,c,d*).

● The risk of bipolar illness is not increased in first-degree relatives of patients with schizophrenia.

Possible familial links between schizophrenia and mood disorders remain controversial (Kendler 1993*d*). For example Maier *et al.* (1993) reported an excess of *unipolar depression* in first-degree relatives of patients with schizophrenia. While this observation is of interest in view of the frequent coexistence of depression and schizophrenia, it has not been reliably confirmed (Kendler *et al.* 1993*d*; Parnas *et al.* 1993). Another interesting finding comes from the Roscommon study carried out by Kendler *et al.* (1993*b,c,d*) who investigated the families of patients with schizophrenia in County Roscommon, Ireland. While the overall incidence of mood disorders was not increased in first-degree relatives of schizophrenic probands, if the relatives suffered a mood disorder, it was more likely to be of a *psychotic type* (Table 9.9). Again, other workers have not identified such an effect (Parnas *et al.* 1993).

At present, the findings from family studies suggest some familial specificity for schizophrenia and schizotypal disorder and for bipolar illness. However, schizoaffective illness, even quite narrowly defined, appears to predispose to both schizophrenia and mood disorder. Whether, as the data of Kendler *et al.* (1993*d*) suggest, a vulnerability to psychosis might be transmitted in families together with a vulnerability to schizophrenia remains to be determined.

Twin studies These studies compare the concordance rates for schizophrenia in monozygotic (MZ) and dizygotic (DZ) twins. The methodological problems of such studies have been considered already (see p. 94).

The first substantial twin study was carried out in Munich by Luxenberger (1928), who found concordance in 11 of his 19 MZ pairs and in none of his 13 DZ pairs. Although this finding for DZ pairs casts doubt on the selection of the sample, other workers (Kallmann 1946; Slater 1953) confirmed that concordance among MZ pairs was higher than that among DZ pairs. Subsequent investigations using improved methods have led to similar results. In these studies concordance rates in MZ pairs have varied considerably, but they have always been higher than the concordance rates in DZ pairs (Kringlen 1967; Tienari 1968; Gottesman and Shields 1972; Fischer 1973). Representative figures for concordance are about 50 per cent for MZ pairs and about 10 per cent for DZ pairs (McGuffin 1988). It might be expected that studies of MZ twins discordant for schizophrenia would reveal some environmental factors relevant to aetiology (see below). However, it is worth noting that the risk of schizophrenia in the *offspring* of an *unaffected twin* is the same as that of an affected twin. This means that an unaffected twin has the same genetic susceptibility to developing schizophrenia as the affected twin but for some reason the susceptibility is not expressed. This lack of expression may not necessarily be linked to environmental factors; for example even identical twins may be subject to differential gene expression (McGuffin *et al.* 1994).

Adoption studies Heston (1966) studied 47 adults who had been born to schizophrenic mothers and separated from them within three days of birth. As children they had been brought up in a variety of circumstances, though not by the mother's family. At the time of the study their mean age was 36. Heston compared them with controls who were matched for circumstances of upbringing, but whose mothers had not been schizophrenic. Amongst the offspring of schizophrenic mothers, five were diagnosed as schizophrenic compared with none of the controls. There was also an excess of antisocial personality and neurotic disorders among the children of schizophrenic mothers. The age-correlated rate for schizophrenia among the index cases was comparable with that among non-adopted children with one schizophrenic parent. In this investigation, no account was taken of the fathers.

Further evidence has come from a series of studies started in 1965 by a group of Danish and American investigators. The work has been carried out in Denmark, which has national registers of psychiatric cases and adoptions. In a study of children separated from schizophrenic mothers at an average age of six months, the findings confirmed those of Heston described above (Rosenthal *et al.* 1971). The major project (Kety *et al.* 1975) employed a different design. Two groups of adoptees were identified: 33 who had schizophrenia, and a matched group who were free from schizophrenia. Rates of disorder were compared in the biological and adoptive families of the two groups of adoptees. The rate for schizophrenia was greater among the *biological relatives* of the schizophrenic adoptees than among the relatives of controls, a finding which supports the genetic hypothesis. Furthermore, the rate for schizophrenia was not increased amongst couples who adopted the schizophrenic adoptees, suggesting that environmental factors were not of substantial importance. The reverse situation was studied by Wender *et al.* (1974), who found no increase in schizophrenia amongst adoptees who had normal biological parents and a schizophrenic adoptive parent (Kety 1983). Recent follow-up studies using a national sample of adoptees in Denmark has confirmed that biological first-degree relatives of patients with schizophrenia have an approximately tenfold increased risk of suffering from schizophrenia or a related disorder (Kety *et al.* 1994) (see below).

The Danish investigators viewed schizophrenia as a spectrum of illnesses with four divisions: (i) process schizophrenia, (ii) reactive schizophrenia, (iii) borderline schizophrenia, and (iv) schizoid states. The last three are sometimes referred to collectively as schizophrenia spectrum disease. The adoption study findings reported above were for process schizophrenia, but the investigators also reported an excess of schizophrenia spectrum disease in biological relatives. These data have been reanalysed using DSMIII criteria for diagnosis, and an excess of *schizophrenia, schizoaffective psychosis* (mainly schizophrenic type), and *schizotypal personality disorder* has been confirmed among the biological relatives of schizophrenic probands (Kendler and Gruenberg 1984; Kendler *et al.* 1994*a*).

Adoption studies cannot rule out environmental causes in the adoptive family, but they indicate that, if there are such environmental causes, they act only on genetically predisposed children.

Is there a relationship between clinical form and inheritance?

Family and twin studies of hebephrenic and paranoid subtypes have shown that there is some tendency for these subtypes to 'breed true' in families, although this is by no means clear cut (McGuffin 1988). There is some evidence that, compared with the hebephrenic subtype, probands with *paranoid schizophrenia* have a lower incidence of affected first-degree relatives and also that in paranoid schizophrenia the concordance ratio between MZ and DZ twins is less than that in other subtypes (Onstad *et al.* 1991). This may mean that paranoid schizophrenia is a less genetic form of schizophrenia. Another way of viewing this distinction is by means of a threshold liability model in which hebephrenic and catatonic subtypes of schizophrenia are more severe forms of the illness and carry a greater genetic loading (McGuffin 1988).

What is the mode of inheritance?

The genetic evidence does not permit definite conclusions about the mode of inheritance. There are three main theories (Kendler 1986; McGuffin 1988).

Monogenic theory (single-gene models) As the ratios of the frequencies of schizophrenia among people with different degrees of relationship with the proband do not fit any simple Mendelian pattern, it is necessary to propose modifying factors. Slater (1958) suggested a dominant gene of variable penetrance. *Penetrance* depends on the definition of the phenotype, and as mentioned above this can be expanded to include various other disorders including schizotypal disorder. In addition, some unaffected relatives of patients with schizophrenia may show neurophysiological

dysfunctions, such as disordered eye tracking, which are present in patients with schizophrenia (see below). Such abnormalities can be included in descriptions of the phenotype and can be used in attempts to identify a single major gene with varying clinical expressivity (Holzman *et al.* 1988; Keefe *et al.* 1991). Thus far, however, there is no evidence that schizophrenia, broadly or narrowly defined, is linked to a single major gene.

Polygenic theory This theory proposes a cumulative effect of several genes (Gottesman and Shields 1967). It is proposed that the liability to schizophrenia lies on a continuum in the population and is expressed when a certain *threshold* of susceptibility is exceeded. In the polygenic model this liability is composed of predominantly additive effects of genes at different loci, together with additional environmental effects (McGuffin 1988). This model is less precise than the monogenic theory and more difficult to test, particularly with molecular genetic linkage techniques.

Genetic heterogeneity theory Monogenic and polygenic theories assume that schizophrenia is a single disease. Heterogeneity theories explain the observed patterns of inheritance by proposing that schizophrenia is a group of disorders of different genetic make-up or perhaps with genetic and non-genetic forms. Several attempts have been made to test these hypotheses, but the results have been ambiguous and will remain so until a definite biological or genetic marker is identified.

Molecular genetic studies

As described in Chapter 4, two main approaches have been used to investigate the molecular genetic basis of schizophrenia. First, *linkage analysis* has been applied to multiply affected families. There was an initial report of an association between schizophrenia and a locus on chromosome 5 in Icelandic pedigrees, but this finding could not be confirmed and is now believed to have been a false positive. No other linkage markers have been clearly identified, but there is interest in the possibility of linkage to the pseudo-autosomal locus of the sex chromosomes (Collinge *et al.* 1991; Asherson *et al.* 1992; d'Amato *et al.* 1992) and a site on chromosome 6 (Wang *et al.* 1995).

In addition, linkage studies have been applied using the *candidate gene* approach where genes for biological mechanisms that may be involved in schizophrenia have been localized and cloned. By application of this technique a number of genes coding for dopamine receptors, including the D_2, D_3, and D_4 receptors as well as the 5-HT_{1A} and 5-HT_2 receptors, have been excluded from linkage to schizophrenia in various pedigrees (Sharma and Murray 1993).

In view of the neurodevelopmental hypothesis of schizophrenia (see below), genes coding for *brain development* are also candidates (Jones and Murray 1991). For example, Crow (1994*a*) has hypothesized that the genes predisposing to psychosis may be those that code for brain size and lateralization, both of which may be disturbed in schizophrenia (see below). Increasing knowledge of the genetic control of neurodevelopment will facilitate investigation of these interesting ideas.

Changes in brain structure in schizophrenia

Neuropathological studies

In the past, many investigators searched for gross pathological changes in the brains of schizophrenic patients but, with a few exceptions, found none. The exceptions were the reports by Southard (1910) of 'internal hydrocephalus' (ventricular enlargement), and by Alzheimer (1897) of cell loss and gliosis in the cerebral cortex.

Recent post-mortem studies have shown that, in comparison with psychiatric and healthy controls, the brains of patients with schizophrenia are *lighter* and somewhat *smaller*. There is consistent enlargement of the *lateral ventricles*, particularly in the anterior and temporal horns. This is associated with a reduction in the volume of *medial temporal structures* such as the hippocampus and parahippocampal gyrus. These

changes tend to be more evident on the left side of the brain. (Brown *et al.* 1986; Crow *et al.* 1989).

More detailed histopathological investigations have shown evidence of *cytoarchitectural disturbances* in the hippocampus, frontal cortex, cingulate gyrus, and entorhinal cortex. These subtle abnormalities, which are not consistently associated with gliosis (Bruton *et al.* 1990), are likely to be due to disturbances in the late migration or in the final differentiation of neurones during brain development. The consequence could be the misplacement of neurones from their intended sites and a disruption of the normal pattern of cortical connections. The cause of this disturbance in neuronal migration is not known, but genetic factors and viral infections could play a part (for reviews see Falkai and Bogerts (1993) and Bloom (1993)).

As noted above, there is a tendency for the pathological changes in schizophrenia to be more apparent on the *left side* of the brain. The maturation of the left hemisphere occurs later than the right hemisphere, and accordingly at certain time points the left hemisphere may be more vulnerable to disruption by a pathological process. Another possibility is that the process of *lateralization* itself is disturbed in schizophrenia. This could account for evidence of disrupted cellular migration of cortical neurones (Falkai and Bogerts 1993).

Structural brain imaging

Lateral ventricular enlargement in schizophrenia was first reported from studies using air-encephalography (Haug 1962). Using computerized tomographic (CT) scanning, which provides a non-invasive alternative, Johnstone *et al.* (1976) found significantly *larger ventricles* in 17 elderly institutionalized schizophrenics than in eight normal controls. Much work has followed with both CT scanning and MRI, which allows better detection of localized brain changes.

The subsequent studies have confirmed that patients with schizophrenia have enlarged cerebral ventricles, but suggest that mean increase in size is somewhat modest with only about 30 per cent of patients having an increase one standard deviation greater than the healthy control mean (Andreasen *et al.* 1990*a*). The ventricular size of patients with schizophrenia shows a unimodal distribution, indicating that the abnormality is not confined to a subgroup of subjects. Repeated scanning through the course of the illness does not show a consistent increase in ventricular size, suggesting that the abnormality is not caused by a progressive pathological process.

Enlarged lateral ventricles have been correlated with a number of clinical and epidemiological variables. The most consistent associations are with *male sex, early age of onset, neuropsychological impairment*, and *poor response to treatment*. In a study of MZ twins, discordant for schizophrenia, the affected twins had larger cerebral ventricles and smaller hippocampi bilaterally, suggesting that these structural changes in the brain are associated with the development of clinical schizophrenia and are not an invariable product of genetic inheritance (Suddath *et al.* 1990).

MRI studies have confirmed reductions in the volume of *temporal lobe* in schizophrenia. However, abnormalities have also been found in the frontal lobes, basal ganglia, and thalamus, suggesting that the disturbance in cerebral morphology in schizophrenia is fairly widespread (Ziparsky *et al.* 1992; Chua and McKenna 1995). As with previous imaging studies, structural abnormalities with MRI have been associated with poor response to treatment and neuropsychological impairment (Lieberman and Sobel 1993; Liddle 1994).

Clinical neurological aspects of schizophrenia

The increasing evidence of abnormal brain morphology has restimulated interest into whether patients with schizophrenia have objective evidence of clinical **neurological signs**. Clinicians have often detected signs of minor neurological abnormalities in schizophrenic patients. Although it is possible that some of these signs resulted from coincidental neurological disease, they might reflect the neuropathological changes described above.

'**Soft signs**' (neurological signs without localizing significance) have been reported in many studies. Rochford *et al.* (1970) found them in about 65 patients examined before starting drug treatment, a finding recently confirmed by Sanders *et al.* (1994) who studied patients during a first episode of illness. Pollin and Stabenau (1968) found at least one sign in nearly three-quarters of a series of schizophrenics. It is generally found that the commonest abnormalities are in *stereognosis, graphaesthesia, balance*, and *proprioception*. It has been suggested that these abnormalities reflect defects in the integration of proprioceptive and other sensory information (Quitkin *et al.* 1976; Cox and Ludwig 1979).

Movement disorders are also common in patients in schizophrenia and have often been attributed to antipsychotic drugs. McCreadie and Ohaeri (1994) concluded that in younger patients drug treatment makes a major contribution to such abnormal movements. However, in older patients movement disorders such as tardive dyskinesia can be found even in those who have never apparently received antipsychotic drug treatment (Owens and Johnstone 1982). This suggests that ageing can interact with a schizophrenic illness to increase the risk of movement disorders.

In view of the changes found in the temporal lobe in schizophrenia, it is of interest to note that patients with chronic *temporal lobe epilepsy* have an increased risk of developing schizophrenic symptoms (see p. 340), as do patients with Huntington's chorea (see p. 324). An association of schizophrenia with brain injury was described by Achté *et al.* (1969), but was not apparent in patients described by Lishman (1968). From a review of published work, Davison and Bagley (1969) concluded that neurological conditions associated with schizophrenia have the common feature of affecting the *temporal lobe*. The fact that the abnormalities in the temporal lobe in schizophrenia may be more evident on the left side is consistent with the observation by Flor-Henry (1969) that epileptics with a schizophrenia-like psychosis are more likely to have a focus in the left temporal lobe.

Neurophysiological changes in schizophrenia

The EEGs of schizophrenic patients generally show increased amounts of theta activity, fast activity, and paroxysmal activity compared with those of healthy controls. The significance of these findings is unknown. Studies of evoked potentials have also shown more abnormalities among schizophrenics than among normal subjects, but no single abnormality is present in all patients (Connolly *et al.* 1983; Morstyn *et al.* 1983).

There has been recent interest in the *P300 response*, an evoked potential which occurs 300 milliseconds after a subject identifies a target stimulus embedded in a series of irrelevant stimuli. Therefore this response provides a measure of auditory information processing. In patients with schizophrenia, the amplitude of the P300 wave is reduced and the same abnormality can be found in a proportion of first-degree relatives of schizophrenic patients (Blackwood and Muir 1990). However, similar abnormalities can be found in other patient groups, such as those with Alzheimer's disease, making the specificity of this abnormality doubtful. In general, studies of evoked potentials offer support for the idea that *attentional processes* are defective in schizophrenia (Corcoran and Frith 1993), a proposal which receives some support from neuropsychological investigations (see below).

There is much evidence that patients with schizophrenia have defective performance in tests of *eye tracking*, but the specificity of the association is not clearly established (Friedman *et al.* 1992; Muir *et al.* 1992). As with the P300, the fact that the eye tracking abnormality has been detected in a significant proportion of non-psychotic first degree relatives has suggested to some that it could be a biological marker for the phenotype (Holzman *et al.* 1988).

Functional brain imaging in schizophrenia

As described in Chapter 4, a number of techniques, particularly PET and SPET, have been used to assess patterns of cerebral blood flow in patients with schizophrenia. Probably the most consistent abnormality found is a decrease in blood flow in *frontal and prefrontal cortex* (**hypofrontality**). This abnormality is most readily demonstrated during psychological performance tasks, such as the Wisconsin Card Sorting Test, which require frontal lobe activation (Weinberger and Kleinman 1986; Liddle 1994). However, the performance of patients with schizophrenia on such tasks is often impaired, raising the possibility that the decrease in blood flow could be secondary to lessened task performance, perhaps as a consequence of *poor motivation*.

The abnormality in frontal blood flow is apparent in patients who have not received antipsychotic drug treatment (Buchsbaum *et al.* 1992) and may correlate with the presence of *negative symptoms* (Andreasen *et al.* 1992). Liddle *et al.* (1992) found in a group of medicated patients with chronic schizophrenia that reduced blood flow in the left dorsolateral prefrontal cortex correlated with the presence of the 'psychomotor poverty syndrome', which reflects the presence of negative symptoms. The other syndromes described by Liddle (1987) (see p. 264) also correlated with characteristic patterns of blood flow in this subject group. For example the syndrome of 'reality distortion' was associated with increased blood flow in the left parahippocampal gyrus, while the syndrome of 'disorganization' appeared to correlate with increased blood flow in the right anterior cingulate. While these associations need confirmation, the recent studies of functional imaging indicate that patterns of cerebral blood flow in schizophrenia need to be understood in the context of both clinical symptomatology and neuropsychological function.

The clinical importance of the changes in blood flow in prefrontal cortex were demonstrated by Berman *et al.* (1992) who studied a group of MZ twins, discordant for schizophrenia. These workers found that all affected twins had lower prefrontal blood flow than their unaffected co-twin during the Wisconsin Card Sort Test. Moreover, the affected twin almost invariably had *reduced hippocampal volume* which correlated with the decrement in prefrontal blood flow. This study suggests that the expression of clinical schizophrenia is associated with abnormal function of temporal and frontal lobes (Weinberger *et al.* 1992).

Neuropsychological abnormalities in schizophrenia

Careful neuropsychological testing reveals many *cognitive deficits* in patients with schizophrenia (Johnstone *et al.* 1978; Taylor and Abrams 1984; Tamlyn *et al.* 1992), although these abnormalities have to take into account factors such as poor motivation and the effects of antipsychotic drug treatment. Neuropsychological approaches to the aetiology of schizophrenia have two main aims: first to identify the brain regions and networks that function abnormally, and second to clarify the neuropsychological mechanisms that may give rise to symptoms of the schizophrenic syndrome such as delusions, hallucinations, and negative symptoms.

Neuropsychological performance

In a recent study of drug-free patients suffering a first episode of schizophrenia, Saykin *et al.* (1994) found generalized cognitive impairments in verbal memory and learning, attention and vigilance, and visuo-motor processing. The most striking difference between patients and controls was in *verbal learning and memory*, and controlling for this impairment attenuated the other differences considerably. A number of other studies have concluded that patients with schizophrenia perform poorly on *memory tasks* and that this impairment may correlate with the presence of *negative symptoms* (Corcoran and Frith 1993). As noted above, cognitive impairment can be detected in patients presenting with their first episode of schizophrenia, but subsequently there is little evidence for progressive decline. This is consistent with the view that the neuropsychological impairments in schizophrenia are part of abnormal neurodevelopment rather than a progressive dementing illness (Hyde *et al.* 1994).

The hypofrontality seen in functional imaging studies of patients with schizophrenia has led to much assessment with tests sensitive to frontal lobe impairment. There is general agreement that schizophrenic patients perform poorly on such tests, and Shallice *et al.* (1991) concluded that the evidence favoured a defect in a *supervisory attentional system* whose role is to allocate attentional resources to specific cognitive tasks. Gold and Weinberger (1991) have argued that the clinical features of *frontal lobe lesions,* for example apathy, attentional impairment, and flat affect, resemble the negative symptoms of schizophrenia.

Neuropsychological mechanisms and symptom production

Attention deficits are common in patients with schizophrenia and probably account for many of the impairments seen in psychological performance tests (see above). It has been proposed that these attention deficits stem from an inability of schizophrenic patients to inhibit irrelevant information from gaining attentional resources and thus entering conscious awareness (Frith 1979).

Disorders of *attention* and *conscious awareness* have been proposed to account for some of the clinical features of schizophrenia, particularly the positive symptoms. For example, Hemsley (1993) has argued that patients with schizophrenia *over-attend* to irrelevant stimuli with the result that they are unable to apply past experience to organize current perceptions. The result is an ambiguous unstructured sensory input, which predisposes to formation of delusions. Using a somewhat different model, Frith (1992) has argued that in schizophrenia there is a breakdown in the *internal representation* of mental events. For example a failure to monitor and identify one's own willed intentions might give rise to the idea that thoughts and actions arise from external sources, resulting in delusions of control. Such formulations are difficult to test clinically, and presumably might apply to the pathophysiology of psychotic symptoms in general rather than schizophrenic illness specifically.

Biochemical abnormalities in schizophrenia

Many findings of early biochemical investigations turned out to be the result of unusual diet or medication rather than of schizophrenia itself. Recent studies have usually attempted to control these extraneous variables. Several hypotheses have been suggested, but the most dominant proposal concerns the role of **dopamine.**

The dopamine hypothesis

Two lines of research have converged on the neurotransmitter dopamine. The first concerns *amphetamine* which, among other actions, releases dopamine at central synapses. Amphetamine also induces a disorder indistinguishable from acute schizophrenia in some normal people, and worsens schizophrenic symptoms. The second approach starts from the finding that the various *antipsychotic drugs* share *dopamine receptor blocking* effects. Carlsson and Lindquist (1963) showed that such drugs increase dopamine turnover. This effect was interpreted as a feedback response of the presynaptic neurone to blockade of postsynaptic dopamine receptors. There is now much additional evidence that antipsychotic drugs block postsynaptic dopamine receptors. They also antagonize *dopamine-sensitive adenyl cyclase*, and the extent to which these effects are produced *in vitro* by the different antipsychotic drugs correlates closely with their clinical potency (Miller *et al.* 1974). Further, it has been shown that α-flupenthixol, an effective dopamine antagonist, has significant antipsychotic activity, whilst the D-isomer, which lacks receptor-blocking properties, is therapeutically inert (Johnstone *et al.* 1978).

Recent studies have demonstrated that *dopamine receptors* in the brain exist in multiple subtypes which differ in their pharmacological characteristics and neuroanatomical localization. Virtually all antipsychotic drugs bind to *dopamine* D_2 *receptors* and it is this property which correlate

best with their clinical potency (see Chapter 17, Table 17.4). In addition, drugs that are highly selective D_2 receptor antagonists, such as *sulpiride*, are effective antipsychotic agents. The exception to this rule is *clozapine* which during usual clinical dosing occupies fewer D_2 receptors than other antipsychotic drugs (see Chapter 17, Table 17.4). However, clozapine produces a relatively more effective blockade of the related *dopamine D_4 receptor*, and it has been hypothesized that this may account for its unusual therapeutic efficacy (for a review of dopamine receptor subtypes and their interactions with antipsychotic drugs, see Seeman and Van Tol (1993)).

Although the evidence that dopamine is central to the action of antipsychotic drugs is strong, evidence for the corollary—that dopamine neurotransmission is abnormal in schizophrenia—is weak. The antipsychotic drugs do not have effects specific to schizophrenia; they are equally effective in mania. Also, it is important to recall the analogy of Parkinson's disease (mentioned on p. 99). In this condition, anticholinergic drugs have therapeutic effects even though the biochemical lesion is not an excess of acetylcholine but a deficiency in dopaminergic neurones due to selective degeneration.

Most direct evidence comes from biochemical studies of post-mortem brains from schizophrenic patients. There are reports of increased dopamine receptor density in the caudate nucleus, putamen and nucleus accumbens (Owen *et al.* 1978), of increased concentrations of dopamine in the amygdala of the left hemisphere, with smaller increases in the caudate nucleus (Reynolds 1983), and of increases of the peptides cholecystokinin, somatostatin, and vasoactive polypeptide in the limbic regions (these peptides are usually associated with dopaminergic neurones (Ferrier *et al.* 1983).). However, particularly in the case of the increase in D_2 receptors, these changes could be a consequence of previous treatment with *antipsychotic drugs* which produce reliable increase in brain D_2 receptor binding when administered to experimental animals (Kornhuber *et al.* 1989). More recently, Seeman *et al.* (1993) reported a sixfold increase in D_4 receptor binding in the striatum of patients with schizophrenia. A similar increase was not apparent in patients with Alzheimer's disease or Huntington's chorea, who had also received treatment during life with antipsychotic drugs. However, Reynolds and Mason (1994) could not confirm an increase in D_4 receptors in the putamen of patients with schizophrenia.

PET studies provide a way of investigating dopamine receptor binding in the brain of living patients by using appropriately labelled dopamine receptor ligands (Sedvall *et al.* 1986). There is a report that D_2 receptor densities in the caudate nuclei of both sides, as measured by [^{11}C]methylpiperone, were greater in schizophrenic patients than normals, and this abnormality was present in small groups of patients who had never received neuroleptic drugs (Wong *et al.* 1986). However, this finding was not confirmed by Farde *et al.* (1987) or Hietala *et al.* (1994), both of whom used [^{11}C]raclopride to define D_2 receptors. Seeman *et al.* (1993) have suggested that these discrepant findings might be attributable to the different ligands used, in that [^{11}C]methylpiperone, whilst labelling D_2 receptors would also label the D_4 receptor subtype, while [^{11}C]raclopride would label D_2 receptors alone. Thus the binding of [^{11}C]raclopride would not reveal a selective increase in D_4 receptor number. More work will be needed before definite conclusions can be reached about dopamine receptors in untreated schizophrenic patients.

Amino acids in schizophrenia

There is increasing interest in the possible role of the amino acid neurotransmitter *glutamate* in the pathogenesis of schizophrenia. Glutamate is the major excitatory neurotransmitter in the cortex and has extensive interactions with dopamine pathways (Carlsson and Carlsson 1990). Postmortem studies in patients with schizophrenia have revealed a variety of abnormalities in glutamatergic neurotransmission, which vary in nature depending on the brain region examined. For example in the orbitofrontal cortex there is evidence of an increased number of presynaptic

uptake sites for neuronal glutamate, which has been attributed to an abnormal increase in the density of glutamatergic innervation. There is also evidence for an increase in the density of certain postsynaptic glutamate receptors in frontal brain regions. In contrast, in the hippocampus, particularly on the left side, there appears to a reduction in the density of some glutamate receptor subtypes and a decrease in the expression of the corresponding messenger RNA (Kerwin *et al.* 1988; Harrison *et al.* 1991).

These complex changes are of interest in the light of the clinical effects of phencyclidine and ketamine, both of which are glutamate receptor antagonists. Like amphetamine, these drugs can produce hallucinations and delusions but, in addition, they can cause negative-like symptoms, such as blunted affect and emotional withdrawal (Krystal *et al.* 1994). While the studies of glutamate function in patients with schizophrenia are of interest, the possible effects of antipsychotic drug treatment could be a confounding factor. If disturbances in glutamate function are indeed associated with schizophrenia, it seems probable that the abnormalities are bound up with the morphological changes in cytoarchitecture found in temporal and frontal cortex (for a review of these issues see Royston and Simpson (1991)).

Developmental aspects of aetiology

Perinatal factors

While genetic influences are of major importance in the aetiology of schizophrenia, it seems likely that *environmental factors* also play a role (see above). A number of workers have suggested that factors acting before or around the time of birth may contribute to the neuropathological abnormalities found in adult patients with schizophrenia. In support of this idea, there is indirect evidence from studies of *birth complications* and *season of birth*.

As evidence of birth complications, retrospective studies of schizophrenic patients report more obstetric complications than do studies of normal controls (Woerner *et al.* 1973; O'Callaghan *et al.* 1992). Also, in studies of identical twins discordant for schizophrenia, it was found that the affected twin more often had a complicated birth (Pollin and Stabenau 1968), although a further investigation did not confirm this (Onstad *et al.* 1992). The more recent investigations have not found a consistent relation between obstetric complications and increased risk of schizophrenia (Done *et al.* 1991). It has been suggested that the association between obstetric complications and schizophrenia may be confined to *males*, or that birth injury may correlate with other aspects of the disorder such as *early age of onset*. However, in children with a genetic predisposition to schizophrenia obstetric complications may increase the risk of a clinical schizophrenic disorder (Fish *et al.* 1992) and of lateral ventricular enlargement in brain imaging studies (Cannon *et al.* 1993). It is worth noting that *pre-existing brain dysfunction* may predispose to the development of obstetric complications.

Schizophrenia is more frequent among people born in the winter than among those born in the summer, with about an 8 per cent excess of schizophrenic births taking place in the first three months of the year (Hare 1975; Bradbury and Miller 1985). Interestingly, winter birth may be more common in patients without a family history of schizophrenia (O'Callaghan *et al.* 1991). Kendell *et al.* (1993) found that the incidence of schizophrenia in those born between February and May correlated inversely with climate temperature the previous autumn, i.e. the colder the temperature around the third month of gestation, the higher is the risk of a subsequent schizophrenic illness. It has been argued that these findings suggest a seasonal environmental influence of aetiological importance, perhaps viral in origin (see below). Some studies have linked increases in the incidence of schizophrenia with prenatal exposure to *influenza* (Sham *et al.* 1992), but the data do not suggest that such an effect has an important role in aetiology (Crow *et al.* 1994c).

Childhood development and antecedents

In 1968 Mednick and Schulsinger reported a study of 207 children of schizophrenic mothers who were interviewed and assessed for cognitive disturbance in 1962, when they were aged 8–12 years. Their galvanic skin responses were measured, their school reports were examined, and their parents were questioned. The subjects were reassessed 18 years later in 1980 (Parnas *et al.* 1982), when they were aged 26–30 years. Thirteen had developed schizophrenia (CATEGO diagnoses), 29 had 'borderline schizophrenia', and five had died by suicide. Of the measures made on the first occasion, those predicting schizophrenia on the second were poor rapport at interview, social isolation or disciplinary problems mentioned in school reports, reports by the parents that the person had been passive as a baby and had shown a short attention span as a child, and fast recovery of galvanic skin responses. More recently, Done *et al.* (1994) found that in a cohort of more than 16 000 children prospectively studied over a 16-year period, those who developed schizophrenia could be distinguished at age 11 by *greater hostility* towards adults and *speech and reading difficulties*.

Fish *et al.* (1992) reviewed the prevalence of *pan-dysmaturation* in children with a schizophrenic parent in a series of prospective studies of at risk individuals. Pan-dysmaturation is proposed to mark the inherited neurological defect associated with schizophrenic genotype and consists of transient retardation of motor and visual development, an abnormal pattern of test scores on cross-sectional developmental examination, and retardation of skeletal growth. The findings of the various studies suggest that pan-dysmaturation is common in the offspring of patients with schizophrenia and predicts poor motor and cognitive development at 10 years of age. Fish *et al.* (1992) propose that pan-dysmaturation is associated with the development of schizotypal personality disorder and in some cases schizophrenic psychosis.

While it is clear that children of patients with schizophrenia, particularly boys, show some evidence of cognitive dysfunction and poor social competence compared with controls (Marcus *et al.* 1993), it is not clear how specific these abnormalities are for the children of schizophrenics or how they relate to the development of a schizophrenic illness. The work of Fish *et al.* (1992) suggests that neurodevelopmental impairment may be associated with the subsequent development of schizotypal personality, but further prospective studies are needed.

Personality factors

Several early writers, including Bleuler (1911), commented on the frequency of abnormalities of personality preceding the onset of schizophrenia. Kretschmer (1936) proposed that both personality type and schizophrenia were related to the *asthenic* type of body build. He suggested a continuous variation between normal personality, schizoid personality (see p. 112), and schizophrenia. He regarded schizoid personality as a partial expression of the psychological abnormalities that manifest in their full form in schizophrenia. Such ideas must be treated with caution, since it is difficult to distinguish between premorbid personality and the prodromal phase of slowly developing illness. However, current findings from family studies suggest that the schizophrenic illness may share a common diathesis with schizotypal personality and some related personality disorders (Kendler *et al.* 1993*b*, 1994; Parnas *et al.* 1993).

Taken together, the findings suggest that abnormal personality is common among people who later develop schizophrenia and among their first-degree relatives. However, many schizophrenics have no obvious disorder of personality before the onset of the illness (Cutting *et al.* 1986), and only a minority of people with schizoptypal or schizoid personalities develop schizophrenia.

Sex and age of onset

As noted above, although the incidence of schizophrenia in men and women is about equal, the age of onset of schizophrenia is *earlier in men than in women*. Also, many of the neurodevelopmental abnormalities described above are more common in males. For example men with schizophrenia are more likely than females to have poor premorbid adjustment, with more frequent and severe relapses and more evidence of structural brain abnormalities. In contrast, women diagnosed as having schizophrenia more frequently experience later-onset psychoses with obvious affective symptoms and a better prognosis. Castle and Murray (1991) suggested that there may be two forms of clinical schizophrenia, the first an early-onset neurodevelopmental disorder occurring predominantly in males, and the second a later-onset illness, related to mood disorders, occurring mainly in females. However, Eaton *et al.* (1992) found that, if the age of onset was controlled for, the sex of the patient did not predict outcome. This suggests that the key variable determining severity of the illness is, in fact, age of onset.

The neurodevelopmental view of schizophrenia is that abnormalities in brain structure and function are present for many years before the onset of a schizophrenic illness. While the studies reviewed above provide some evidence that cognitive and social abnormalities may be present before the onset of psychosis, these changes are usually modest and many patients appear to have functioned normally or even at a high level. This raises the question as to why, in most cases, the onset of schizophrenia becomes apparent only in early adulthood. Various ideas have been proposed. For example the *frontal cortex* matures relatively late in development, and in non-human primates the full behavioural effects of early frontal lesions are not apparent until maturity, with the attendant development of cerebral specialization (Gold and Weinberger 1991). Also, following puberty, predisposed individuals will be subject to effects of *late myelination* in frontal and limbic cortex, changing *sex hormone environments*, and the *psychosocial stressors* consequent upon development into adulthood (Falkai and Bogerts 1993). At the moment there is no firm evidence to favour one proposal rather than another.

Dynamic and interpersonal factors in aetiology

Psychodynamic theories

From the point of view of aetiology, psychodynamic theories of schizophrenia are now largely of historical interest. However, they do focus attention on interpersonal aspects of schizophrenia which are important in view of the evidence described above suggesting that patients with schizophrenia have abnormal social development during childhood. Freud's theory of schizophrenia was stated most clearly in his 1911 analysis of the Schreber case (see p. 295) and in his 1914 paper 'On narcissism: an introduction'. According to Freud, in the first stage libido was withdrawn from external objects and attached to the ego. The result was exaggerated self-importance. Since the withdrawal of libido made the external world meaningless, the patient attempted to restore meaning by developing abnormal beliefs. Because of libidinal withdrawal, the patient could not form a transference and therefore could not be treated by psychoanalysis. Although Freud developed his general ideas considerably after 1914, he elaborated his original theory of schizophrenia but did not replace it.

Melanie Klein believed that the origins of schizophrenia were in infancy. In the 'paranoid–schizoid position' the infant was thought to deal with innate aggressive impulses by splitting both his own ego and his representation of his mother into two incompatible parts, one wholly bad and the other wholly good. Only later did the child realize that the same person could be good at one time and bad at another. Failure to pass through this stage adequately was the basis for the later development of schizophrenia.

Hartmann (1964) and other writers developed Freudian ideas about schizophrenia in another way. They took the view that defects in the ego result in problems in the defence and 'neutralization' of libido and aggression. Psychodynamic views on aetiology and treatment have been discussed by Arieti (1974) and Jackson and Cawley (1992).

The family as a cause of schizophrenia

Two kinds of theory have been proposed about the family as a cause of the onset of schizophrenia: *deviant role relationships* and *disordered communication* (Liem 1980). The different role of the family in determining the *course* of established schizophrenia is discussed later (p. 284).

Deviant role relationships The concept of the 'schizophrenogenic' mother was suggested by the analyst Fromm-Reichmann in 1948. In a comparison of the mothers of schizophrenia patients, neurotic patients, and normal controls, Alanen (1958, 1972) found that mothers of schizophrenics showed an excess of psychological abnormalities. He suggested that these abnormalities might be an important cause of the child's schizophrenia. Lidz and his colleagues (Lidz and Lidz 1949; Lidz *et al.* 1965) used intensive psychoanalytic methods to study the families of 17 schizophrenic patients, of whom 14 were in social classes I or II. There was no control group. Two types of abnormal family pattern were reported: (i) 'marital skew', in which one parent yielded to the other's (usually the mother's) eccentricities, which dominated the family; (ii) 'marital schism', in which the parents maintained contrary views so that the child had divided loyalties. It was suggested that these abnormalities were the cause rather than the result of the schizophrenia. Investigations by other clinicians have not confirmed these findings (Ferreira and Winter 1965; Sharan 1965). Even if they were confirmed, the abnormalities in the parents could be an expression of genetic causes or secondary to the disorder in the patient. These and other speculations about the causative role of family relationships have had the unfortunate consequence of inducing unjustified guilt in parents.

Disordered family communication Research on disordered communication in families originated from the idea of the **double bind** (Bateson *et al.* 1956). A double bind is said to occur when an instruction is given overtly, but is contradicted by a second more covert instruction. For example a mother may overtly tell her child to come to her, whilst conveying by manner and tone of voice that she rejects him. According to Bateson, double binds leave the child able to make only ambiguous or meaningless responses, and schizophrenia develops when this process persists. The theory is ingenious but not supported by evidence (see Leff (1978) for a more detailed discussion).

Wynne and his colleagues suggested that different patterns of *disordered communication* occurred among the parents of schizophrenics (Wynne *et al.* 1958). These investigators first gave projective tests to such parents, and identified 'amorphous communications' ('vague, indefinite, and loose') and 'fragmented communications' ('easily disrupted, poorly integrated, and lacking closure'). In a further study using blind interpretation of these tests, the investigators found more of these disordered communications in parents of schizophrenics than in parents of neurotics (Singer and Wynne 1965). In an independent replication, Hirsch and Leff (1975) found a similar but smaller difference between parents of schizophrenic patients and controls.

Subsequent attempts to test Wynne's hypothesis have used more elaborate methods such as observing family communication during the performance of a task (Liem 1980; Wynne 1981). So far the hypothesis must be judged as not proven. However, it is worth noting that an association between abnormal social communication in parents and schizophrenic illness in children may, in fact, be a consequence of a shared genetic inheritance. This is because the incidence of schizotypal disorder (characterized by abnormalities of thought and speech) is raised in first-degree relatives of patients with schizophrenia (Table 9.9). Interestingly, parents of children with schizophrenia are more likely to suffer from schizotypal disorder than schizophrenia, presumably because the former impairs reproductive fitness somewhat less than the latter (Kendler 1993*b,c*).

Social factors in aetiology

Culture

If cultural factors are important in the aetiology of schizophrenia, differences in the incidence of the disorder might be expected in countries with contrasting cultures. As explained already (p. 247), the incidence rates of schizophrenia are remarkably similar in widely different places (Jablensky 1993), and the possible exceptional rates are in areas (northern Sweden and Slovenia) with cultures that do not differ much from others in the Western world.

Occupation and social class

Several studies have shown that schizophrenia is over-represented among people of *lower social class*. In Chicago, for example, Hollingshead and Redlich (1958) found both the incidence and the prevalence of schizophrenia to be highest in the lowest socio-economic groups. At first, these findings were thought to be of aetiological significance, but more recent evidence suggests that they could be a consequence of schizophrenia. For instance Goldberg and Morrison (1963) found that schizophrenics were of lower social status than their fathers and that the change had usually occurred after the illness began.

Place of residence

Faris and Dunham (1939) studied the place of residence of mentally ill people in Chicago, and found that schizophrenics were over-represented in the *disadvantaged inner-city areas*. This distribution has been confirmed in other cities including Bristol (Hare 1956*a*) and Mannheim (Häfner and Reimann 1970). Faris and Dunham suggested that unsatisfactory living conditions caused schizophrenia. However, the findings can be explained equally plausibly by the occupational and social decline described above, or by a search for social isolation by people about to develop schizophrenia. A search for isolation would be consistent with the finding that schizophrenics in disadvantaged areas usually live alone, and not with their families (Hare 1956*b*). However, Castle *et al.* (1993) found that, compared with controls, patients with schizophrenia were more likely to have been *born into socially deprived households*. The authors proposed that some environmental factor of aetiological importance was more likely to affect those of lower economic status living in inner cities. (For a review see Freeman (1994).)

Migration

High rates of schizophrenia have been reported among *migrants* (Malzberg and Lee 1956). In a study of Norwegians who had migrated to Minnesota, Ødegaard (1932) found that the inception rate for schizophrenia was twice that of Norwegians in Norway. The reasons for these high rates are not clear, but they are probably due mainly to a disproportionate migration of people who are unsettled because they are becoming mentally ill. The effects of a new environment may also play a part in provoking illness in predisposed people. Thus 'social selection' and 'social causation' may both contribute to an excess of schizophrenia among migrants. (See Murphy (1977) for a review.)

Recent studies in the United Kingdom have suggested a strikingly increased incidence of schizophrenia in *Afro-Carribeans* (Harrison *et al.* 1988; Wessely *et al.* 1991), particularly in the 'second generation' born in the United Kingdom. Jablensky (1993) proposed that these apparently high figures be received with caution because of the lack of accurate data on the size and age structure of the Afro-Caribbean population in the areas concerned. The familial nature of schizophrenia in Afro-Carribeans living in the United Kingdom seems similar to that of the remainder of the population (Sugarman and Crawford 1994). Therefore if the incidence of schizophrenia is indeed increased, some environmental factor is likely to be involved. Various possibilities have been put forward including an increased risk of *viral infection in utero*, high prevalence of *cannabis use*, and increased *social adversity* through racial discrimination (Jablensky 1993; Sugarman and Crawford 1994).

Social isolation

Schizophrenics often live alone, unmarried, and with few friends (Hare 1956b). A retrospective study comparing schizophrenics with controls (Clausen and Kohn 1959) suggested that the pattern of isolation began before the illness, sometimes in early childhood. Schizophrenics who were not isolated in early life were not isolated as adults.

Psychosocial stresses

Life stresses have often been put forward as precipitants of schizophrenia, but few satisfactory studies have been carried out. In one of the most convincing, Brown and Birley (1968) used a standardized procedure to collect information from 50 patients newly admitted with a precisely datable first onset or relapse of schizophrenia. By comparison with a control group, the rate of 'independent' events in the schizophrenics was increased in the three weeks before the onset of the acute symptoms. (Independent events are those that could not be the result of illness (see p. 89).) When the events (which included moving house, starting or losing a job, and domestic crises) were compared with events preceding depression, neurosis, and suicide attempts, they were found to be non-specific. As a rough guide to the size of the effect, Paykel (1978) calculated that experiencing a life event doubles the risk of developing schizophrenia over the subsequent six months. Other workers have confirmed these findings for both first episodes and relapse of schizophrenia (Jacobs *et al.* 1974; Jacobs and Myers 1976; Bebbington *et al.* 1993). In a review of recent studies, Norman and Malla (1993) concluded that in patients with chronic schizophrenia, the level of symptoms over time correlated with life events; however, there was little evidence that patients with schizophrenia suffered more life events than the general population.

Conclusions

There is strong evidence that schizophrenia has important *genetic causes*, although the mode of inheritance is not known. There is increasing evidence that many cases are of *neurodevelopmental* origin, but whether neurodevelopmental abnormalities are present in all patients to a greater or lesser extent is not clear. Most believe that schizophrenia results from an *interaction* of *genetic predisposition* and *environmental factors*, although some have concluded that genetic explanations will eventually prove sufficient (Crow 1994*a*; McGuffin *et al.* 1994). Environmental factors proposed as important include *infection in utero*, *obstetric injury*, and *social adversity*. There is good reason to think that *stressful life events* often provoke the disorder; the events appear to be non-specific and are similar to those which precede affective disorders.

Recent studies suggest that schizophrenia may be preceded by *cognitive and social impairment* in childhood, and that the presence of these impairments and certain kinds of personality disorder are forms of expression of a schizophrenia genotype. Patients with a schizophrenic illness have changes in cerebral structural and function, particularly involving the *temporal and frontal* lobes. These deficits are associated with *non-progressive neuropsychological impairment*. Although neurochemical disorders are likely to accompany these changes in cerebral morphology, no clear-cut reproducible biochemical abnormalities has been identified. Dopamine receptors are blocked by drugs that control schizophrenic symptoms, but there is no compelling evidence at present that over-activity of dopaminergic systems is the central disorder in schizophrenia.

Course and prognosis

Although it is generally agreed that the outcome of schizophrenia is worse than that of most psychiatric disorders, there have been surprisingly few long-term follow-up studies of schizophrenic patients. Fewer still have included satisfactory criteria for diagnosis, samples of adequate size, and outcome measures that distinguish between symptoms and social adjustment. It is generally accepted that there are wide variations in outcome. This variation can be explained in three ways: first, schizophrenia may be a single condition with a course that is modified by extraneous factors; second, schizophrenia may consist of separate subtypes with different prognoses; third, the good prognosis cases may not be schizophrenia but some other condition. The second and third explanations have already been discussed; the rest of this section is concerned with the first explanation.

It is essential to distinguish between data from studies of first admissions to hospital, and data from investigations that study second or subsequent admissions or give no information on this point (see Harding *et al.* (1987) for review of outcome). When successive reports are studied, it appears that prognosis may have improved since the beginning of the century. Kraepelin (1919) concluded that only 17 per cent of his patients in Heidelberg were socially well-adjusted many years later. In 1932, Mayer-Gross, from the same clinic, reported social recovery in about 30 per cent of patients after 16 years. By 1966, Brown *et al.* reported social recovery in 56 per cent of patients after five years. Against this, in a study of patients identified as schizophrenic in a single centre between the early years of the century and 1962, Ciompi (1980) found little change in the proportion with a good or fair social outcome. Schizophrenics have a mortality substantially higher than that of the general population. The excess is contributed to by a variety of natural causes and by suicide (Allebeck and Wisledt 1986). All studies with prolonged follow-up report that up to *10 per cent of schizophrenics die by suicide* (Roy 1982).

An important long-term study was carried out by Manfred Bleuler (1972, 1974) who personally followed-up 208 patients who had been admitted to hospital in Switzerland between 1942 and 1943. Twenty years after admission, 20 per cent had a complete remission of symptoms and 24 per cent were severely disturbed. Bleuler considered that these proportions had changed little since the introduction of modern treatments, although advances in drug and social treatments had substantially benefited patients whose illnesses had a fluctuating course. When social adjustment was examined, a good outcome was found in about 30 per cent of the whole group, and in 40 per cent of those who had originally been first admissions. When full recovery had occurred, it was usually in the first two years and seldom after five years of continuous illness. Bleuler's diagnostic criteria were narrow, and his findings suggest that the traditional view of schizophrenia as a generally progressive and disabling condition must be reconsidered. Nevertheless, 10 per cent of his patients suffered an illness of such severity that they required long-term sheltered care. When the illness was recurrent, each subsequent episode usually resembled the first in its clinical features.

Bleuler's conclusions are broadly supported by Ciompi's larger but less detailed study of long-term outcome in Lausanne (Ciompi 1980). The study was based on the well-kept records of 1642 patients diagnosed as schizophrenic from the beginning of the century to 1962. The average follow-up was 37 years. A third of the patients were found to have a good or fair social outcome. Symptoms often became less severe in the later years of life. Huber *et al.* (1975) reported similar findings from a 22-year follow-up study of 502 patients in Bonn.

More recent studies are in general agreement with these findings. For example, in a 3–13 year follow-up study of schizophrenic patients discharged between 1975 and 1985, Johnstone (1991) found that almost half had a good social outcome. In a 15-year follow-up of 330 Chinese patients with first admission schizophrenia, almost one-third recovered but about 17 per cent remained unable to function outside the hospital (Tsoi and Wong 1991). In the United States, however, Carone *et al.* (1991) found that only about 15 per cent of patients meeting DSMIII criteria for schizophrenia recovered after five years, perhaps consistent with the proposal that DSM criteria identify a group of subjects with a poorer prognosis (see below).

Predictors of outcome

There have been several attempts to find satisfactory predictors of the outcome of schizophrenia (Stephens 1978). Langfeldt (1961) identified a set of criteria, and reported them to be successful. However, in the International Pilot Study of Schizophrenia (World Health Organization 1979) tests were made of the predictive value of several sets of criteria based on symptoms, including Langfeldt's criteria, Feighner's diagnostic criteria, and others. All these symptom criteria proved to be largely unsuccessful at predicting outcome at two years (Strauss and Carpenter 1974) or five years (Strauss and Carpenter 1977). The best predictors of poor outcome appear to be the criteria used for diagnosis in DSMIII, in part though not entirely because they stipulate that the syndrome should have been present for six months before the diagnosis can be made (Helzer *et al.* 1983).

More recent studies suggest that poor outcome in schizophrenia is associated with *younger age of onset, male sex*, and *poor premorbid functioning* (Lieberman and Sobel 1993). Interestingly, Loebel *et al.* (1992) found that the *duration of symptoms prior to antipsychotic drug treatment* correlated significantly with time to remission and level of remission. While reluctance to seek treatment could be associated with other predictors of poor outcome, it is also possible that early antipsychotic drug treatment may improve the prognosis for some patients. *Negative symptoms* that persist after resolution of positive symptoms also may also predict a poorer outcome (Lieberman and Sobel 1993).

Some of the predictors that have emerged from the various studies are listed in Table 9.10 and may be taken as a moderately useful guide. However, it is wise for clinicians to be cautious when asked to predict the outcome of individual cases.

So far this discussion has been concerned with factors operating before or at the onset of schizophrenia. An account will now be given of factors acting after the illness has been established.

Social environment and course

Cultural background

Recent international studies suggest that the incidence of schizophrenia is similar in different countries, but that the course and outcome are not similar. In a 12-year follow-up study of 90 patients in Mauritius, Murphy and Raman (1971) observed a better prognosis than that reported in the United Kingdom by Brown *et al.* (1966). More Mauritian patients were able to leave hospital and return to a normal way of life. Nearly two-thirds were classified as socially independent and symptom-free at follow-up, compared with only half of the English sample.

Comparable differences were reported at two-year follow-up in the International Pilot Study of Schizophrenia (World Health Organization 1979). Outcome was better in India, Colombia, and Nigeria than in the other centres. This finding could not be explained by any recorded differences in the initial characteristics of the patients. The possibility of selection bias remains; for example in these three countries patients with acute illness may be more likely to be taken to hospital than patients with illness of insidious onset (Stevens 1987). However, more recent studies designed to overcome these objections also found a more favourable course of illness in less developed countries (Jablensky *et al.* 1986, 1992). The major difference in outcome was that patients in developing countries were more likely to achieve a complete remission than those in developed countries, who tended to be impaired by continual residual symptoms.

Table 9.10 *Factors predicting the outcome of schizophrenia*

Good prognosis	*Poor prognosis*
Sudden onset	Insidious onset
Short episode	Long episode
No previous psychiatric history	Previous psychiatric history
Prominent affective symptoms	Negative symptoms
Paranoid type of illness	Enlarged lateral ventricles Male gender
Older age at onset	Younger age at onset
Married	Single, separated, widowed, divorced
Good psychosexual adjustment	Poor psychosexual adjustment
Good previous personality	Abnormal previous personality
Good work record	Poor work record
Good social relationships	Social isolation
Good compliance	Poor compliance

Life events

As explained above (p. 280), some patients experience an excess of life events in the weeks before the onset of acute symptoms of schizophrenia. This applies not only to first illnesses but also to relapses (Brown and Birley 1968; Bebbington *et al.* 1993). There is evidence that patients with increased numbers of life events experience a more symptomatic course (Norman and Malla 1993).

Social stimulation

In the 1940s and 1950s clinicians recognized that among schizophrenics living in institutions many clinical features were associated with an *unstimulating environment*. Brown and coworkers (Brown *et al.* 1966; Wing and Brown 1970) investigated patients at three mental hospitals. One was a traditional institution, another had an active rehabilitation programme, and the third had a reputation for progressive policies and short admissions. The research team devised a measure of 'poverty of the social milieu' which took into account little contact with the outside world, few personal possessions, lack of constructive occupation, and pessimistic expectations on the part of ward staff. Poverty of social milieu was found to be closely related to three aspects of the patients' clinical condition: *social withdrawal*, *blunting of affect*, and *poverty of speech*. The causal significance of these social conditions was strongly supported by a further survey of the same hospitals four years later. Improvements had taken place in the environment of the hospitals, and these changes were accompanied by corresponding improvements in the three aspects of the patients' clinical state.

While an understimulating hospital environment is associated with worsening of the so-called poverty syndrome, an *over-stimulating* environment can precipitate florid symptoms and lead to *relapse*. Since factors in a hospital environment play an important part in determining prognosis, it seems likely that similar factors are important to patients living in the community (Hirsch and Bristow 1993).

Family life

Brown *et al.* (1958) found that, on discharge from hospital, schizophrenics returning to their families generally had a worse prognosis than those entering hostels. Brown *et al.* (1962) found that relapse rates were greater in families where relatives showed 'high expressed emotion' by making critical comments, expressing hostility, and showing signs of emotional over-involvement. In such families the risk of relapse was greater if the patients were in contact with their close relatives for more than 35 hours a week. The work was confirmed and extended when Leff and Vaughn (1981) investigated the interaction between 'expressed emotion' in relatives and life events in the three months before relapse. The onset of illness was associated either with a high level of expressed emotion or with an independent life event.

Vaughn and Leff (1976) suggested an association between expressed emotion in relatives and the patient's response to antipsychotic medication. Among patients who were spending more than 35 hours a week in contact with relatives showing high emotional expression, the relapse rate was 92 per cent for those not taking antipsychotic medication and only 53 per cent for those taking antipsychotic medication. Among patients taking antipsychotic drugs and spending less than 35 hours in contact with relatives showing high emotional expression, the relapse rate was as low as 15 per cent. In this study patients had not been allocated randomly to the treatment conditions. However, a further study (Leff *et al.* 1982, 1985*a*) strongly suggests that high emotional expression has a causal role. Twenty-four families were selected in which a schizophrenic patient had extended contact with relatives showing high emotional expression. All patients were on maintenance neuroleptic drugs. Half the families were randomly assigned to routine out-patient care. The other half took part in a programme including education about schizophrenia, relatives' groups, and family sessions for relatives and patients. The relapse rate was significantly lower in this group than in the controls at nine-month and two-year follow-up. Apart from providing further evidence of the importance of relatives' expressed emotion in relapse, this study showed the effectiveness of combined social intervention and drug treatment.

Expressed emotion has now been widely investigated as a predictor of schizophrenic relapse. While not all studies concur, overall there seems little doubt that patients living in families with high levels of expressed emotion have a *two- to threefold increased risk of psychotic relapse*. Similar findings have been found across cultures, although high levels of expressed emotion are more common in Western countries (Kavanagh 1992). Further evidence has accumulated that family intervention aimed at improving the knowledge and skills of relatives can lower expressed emotion levels and decrease relapse rate (Lam 1991; Kavanagh 1992). However, in a meta-analysis of the published studies, De Jesus Mari and Streiner (1994) found that inclusion of treatment withdrawals and drop-outs (an *intention to treat analysis*) reduced the apparent effectiveness of family intervention to prevent relapse. This suggests that the results of family intervention may be less useful in clinical practice than would be expected from the controlled trials.

Levels of expressed emotion probably depend on interactions between the patient, their symptomatology, and their relatives. For example, certain types of symptoms may be more likely to provoke high expressed emotion. However, no clear relationships have emerged between the form and severity of the schizophrenic disorder and levels of expressed emotion (Kavanagh 1992), although relatives with low levels of expressed emotion are more likely than those with high levels to regard the schizophrenic disorder and associated behaviour as an illness beyond the control of the patient. Interestingly, such causal attributions, like expressed emotion, are significant predictors of relapse (Barrowclough *et al.* 1994).

Conclusion

Social factors play a role in the *onset* of schizophrenia and the *level of symptomatology* in established cases. The *emotional involvement* of

relatives can affect the risk of relapse, and it is possible that the lower degree of expressed emotion typically found in families in developing countries could play a part in the improved prognosis of schizophrenia in these cultures. In terms of the general environment in which patients live, too much stimulation appears to precipitate relapse into positive symptoms, while under-stimulation leads to worsening of negative symptoms.

Effects of schizophrenia on the family

With the increasing care of patients in the community rather than in hospital, difficulties have arisen for some families. Relatives of schizophrenics describe two main groups of problems (Creer 1978). The first group relates to *social withdrawal*: schizophrenic patients do not interact with other family members; they seem slow, lack conversation, have few interests, and neglect themselves. The second group relates to more obviously *disturbed and socially embarrassing behaviour* such as restlessness, odd or uninhibited social behaviour, and threats of violence. These problems are likely to play a part in the development of high levels of expressed emotion in relatives, which may in turn worsen the behaviours.

Creer found that relatives often felt anxious, depressed, guilty, or bewildered. Many were uncertain how to deal with difficult and odd behaviour. Further difficulties arose from differences in opinion between family members, and more commonly from a lack of understanding and sympathy among neighbours and friends. The effects on the lives of such relatives were often serious. Unfortunately, in the United Kingdom and other countries community services for chronic schizophrenic patients and their relatives are often less than adequate (Johnstone *et al.* 1984; Hirsch and Bristow 1993).

Treatment

The treatment of schizophrenia is concerned with both the acute illness and chronic disability. In general, the best results are obtained by combining *drug* and *social treatments*, while methods aimed at providing psychodynamic insight are unhelpful.

This section is concerned with evidence from clinical trials about the efficacy of various forms of treatment. A later section on management deals with the use of these treatments in everyday clinical practice. The pharmacological aspects of antipsychotic drugs and their use are discussed in Chapter 17. The organization of community care and care planning is discussed in Chapter 19.

Antipsychotic drugs

Treatment of acute schizophrenia

The effectiveness of antipsychotic medication in the treatment of acute schizophrenia has been established by several well-controlled double-blind studies. For example the National Institute of Mental Health collaborative project (Cole *et al.* 1964) compared chlorpromazine, fluphenazine, and thioridazine with placebo. Three-quarters of the patients receiving antipsychotic treatment for six weeks improved, whatever the drug, whilst half of those receiving placebo worsened. Drug treatment has most effect on the positive symptoms of schizophrenia, such as hallucinations and delusions, and least effect on the negative symptoms. The sedative and calming action is immediate, but the antipsychotic effect develops more slowly, sometimes taking up to several weeks to be clearly apparent.

The various antipsychotic drugs do not differ in therapeutic effectiveness, with the exception of *clozapine* and possibly *risperidone*. However, their side-effects vary (see p. 552). Recent studies suggest that, for most patients, modest doses of antipsychotic drugs (for example 5–15 mg of haloperidol) are as effective as higher doses and have less risk of side-effects (Kane and Marder 1993) (a table of dose equivalents is given on p. 555). If *rapid sedation* is required, it is usually preferable to combine the antipsychotic drug with a benzodiazepine (Pilowsky *et al.* 1992). For the acute illness there is no proven way of distinguishing patients who require medication if they are to improve from those who would improve without medication (Davis *et al.* 1980).

Clozapine is a dibenzodiazepine with relatively weak dopamine D_2 receptor blocking properties. Initial studies suggested that clozapine was an effective antipsychotic drug which caused little in the way of extrapyramidal movement disorders. However, because of the risk of *white blood cell dyscrasia* (about 2 per cent annually) the drug was withdrawn from use in many countries. Despite this, clozapine continued to be used to some extent because some patients *unresponsive to other antipsychotic drugs* appeared to derive particular benefit from it (McKenna and Bailey 1993). Subsequently, Kane *et al.* (1988), in a double-blind multicentre six-week trial, found that clozapine (up to 900 mg daily) was therapeutically superior to chlorpromazine (up to 1800 mg daily) in patients who met stringent criteria for treatment resistance. When a 20 per cent decrease in the Brief Psychiatric Rating Scale (BPRS) was used to define a clinically meaningful response, it was found that 30 per cent of clozapine-treated patients responded but only 4 per cent of the chlorpromazine-treated patients met this criterion. In the responders to clozapine, *negative symptoms* of schizophrenia improved as well as *positive symptoms*.

Subsequent studies have confirmed the value of clozapine in treatment-resistant patients (Kane 1992; Meltzer 1992; Clozapine Study Group 1993). The length of a treatment trial is obviously an important issue, given the expense of clozapine and its liability to cause blood dyscrasia. Meltzer (1992) found that although about 50 per cent of patients who responded to clozapine did so by about six weeks, others showed more delayed improvement over six to nine months. Few clinical predictors of therapeutic response to clozapine have emerged, although preliminary data suggest that patients with prominent paranoid ideation and thought disorder may benefit the most (Kane 1992; Meltzer 1992).

The risk of *granulocytopenia* with clozapine is reported to be about 2.8 per cent and that of *agranulocytosis* is 0.6 per cent (Lieberman and Alvir 1992), with most cases occurring in the first 18 weeks of treatment (Krupp and Barnes 1992). At present it is recommended that blood monitoring be carried out weekly for the first 18 weeks of treatment and at two-weekly intervals thereafter.

Because of the risk of agranulocytosis, the prescription of clozapine is now restricted to patients who are *unresponsive* to adequate doses of other antipsychotic drugs or who suffer *intolerable extrapyramidal side-effects* with the latter agents.

Risperidone is a benzisoxazole derivative designed to produce blockade of both dopamine D_2 receptors and 5-HT_2 receptors. In two multicentre double-blind placebo-controlled trials risperidone (6 mg) was at least as effective as haloperidol (20 mg) in attenuating positive symptoms, but unlike haloperidol also benefited negative symptoms (Chouinard *et al.* 1993; Marder and Meibach 1994). These studies have been criticized on the grounds that a dose of 20 mg of haloperidol could, through toxicity, induce negative-like symptoms. At a dose of 6 mg, risperidone did not produce more extrapyramidal side-effects than a placebo control. The studies with risperidone reported here have not included patients who were selected for treatment resistance, and so it is not known whether risperidone will benefit this group of subjects.

Treatment after the acute phase

Since the original demonstration by Pasmanick *et al.* (1964), many controlled trials have shown the effectiveness of *continued oral and depot therapy* in preventing relapse (Leff and Wing 1971; Hirsch *et al.* 1973). It has also become clear that some chronic schizophrenics do not respond even to long-term medication, and that other patients remain well without drugs. Unfortunately, there has been no success in predicting which patients benefit from such treatment. Since long-continued antispychotic medication may lead to irreversible dyskinesias (see p. 552), it is important to know how long such treatment needs to be given. There is still no clear answer to this question, but Hogarty and Ulrich (1977) reported that, over a three-year period, maintenance antipsychotic medication was two and a half to three times better than placebo in preventing relapse.

There is a widespread clinical impression that depot injections are more successful than continued oral medication in preventing relapses of schizophrenia. However, Schooler *et al.* (1980) found that depot injections offered no such advantage. According to Davis *et al.* (1980), in the long-term management of schizophrenia there is no difference in the usefulness of the various antipsychotic drugs available. Studies of *intermittent treatment*, where antipsychotic drugs are given at the first sign of relapse, have shown it to be less effective than continuous prophylaxis (Jolley *et al.* 1990).

Interaction of maintenance treatment and social treatment

Since both *medication* and *social casework* appear effective in the management of schizophrenia, it is reasonable to enquire whether the two kinds of treatment interact. Hogarty *et al.* (1974) studied the use of 'major role therapy' (i.e. social casework) with and without drugs. Given alone, social casework had only a small effect in reducing relapse rate; combined with medication, it had a larger effect. This difference may have occurred partly because patients took their drugs more regularly when seeing social workers, but such an effect seems unlikely to be the whole explanation. However, as noted above, there is good evidence that medication and family work designed to lower levels of expressed emotion can have additive effects in the prevention of relapse (Kavanagh 1992). In a study of the effect of adding day hospital treatment to continued medication, Linn *et al.* (1979) found that day care conferred extra benefit on patients when it was of low intensity and based on occupational therapy, but not when it included more active treatments such as group therapy.

Antidepressants and mood stabilizers

As already explained, symptoms of *depression* occur commonly in the syndrome of schizophrenia (p. 251). Since it is not easy to distinguish between depressive symptoms and apathy, it is difficult to assess the effects of antidepressant medication in chronic schizophrenia. As yet, there have been few satisfactory clinical trials. Siris *et al.* (1987) carried out a placebo-controlled trial of *imipramine* in schizophrenic patients receiving maintenance fluphenazine in whom a depressive episode became apparent following resolution of the psychotic symptoms. Imipramine was significantly better than placebo in these patients, improving depressed mood but not altering psychotic symptoms. A more recent study by the same group confirmed the beneficial effect of imipramine in post-psychotic depression and also suggested some benefit on negative symptoms (Siris *et al.* 1991). In patients where depressive symptoms coexist with an *active psychosis*, antidepressant drugs do not seem to be helpful and in fact there is some evidence that they may worsen psychotic symptoms (Plasky 1991).

The value of *lithium* in treating schizophrenia is uncertain. Occasional beneficial effects could be due to the treatment of schizoaffective cases. There is some evidence that lithium has a therapeutic action in this diagnostic group. In two small trials, Brockington *et al.* (1978) found that chlorpromazine was more effective than lithium for schizodepressive patients (i.e. those satisfying criteria for both depressive disorder and schizophrenia). However, lithium and chlorpromazine were equally effective for patients with 'schizomania'. In the Northwick Park Functional Psychosis study lithium decreased elevated mood, whatever the diagnosis of the patient, but had no significant effect on positive or negative symptoms (Johnstone *et al.* 1988). Prophylactic studies of other mood stabilizers such as carbamazepine and sodium valproate often include patients with diagnoses of schizoaffective disorder. The current evidence suggests that patients with more prominent manic symptoms are likely to be helped by these drugs (see Chapter 17).

Electroconvulsive therapy

In the treatment of schizophrenia, the traditional indications for ECT are *catatonic stupor* and *severe depressive symptoms* accompanying schizophrenia. The effects of ECT are often rapid and striking in both these conditions. Nowadays, ECT is seldom used for other presentations of schizophrenia, although there is some evidence that it is rapidly effective in acute episodes (Taylor and Fleminger 1980). ECT is occasionally used in patients whose positive symptoms have not responded to adequate antipsychotic drug treatment, but its effectiveness for this indication has not been formally evaluated.

Psychotherapy

In the past, individual psychotherapy was used quite commonly for schizophrenia, though much more in the United States than in the United Kingdom. Evidence from clinical trials is scanty, but it does not support the use of psychotherapy. An investigation by May (1968) found that psychotherapy had little benefit, but the treatment was short and provided by relatively inexperienced psychiatrists. Apart from the lack of convincing evidence that intensive individual psychotherapy is effective in schizophrenia, there may be some danger that the treatment will cause over-stimulation and consequent relapse (Mosher and Keith 1980). However, some authorities believe that psychoanalytic psychotherapy by skilled therapists can produce significant benefit in schizophrenia (Benedetti 1987). The characteristics of patients who are said to do well include good motivation and productivity, features unlikely to apply to the majority of patients with a chronic schizophrenic illness.

Many kinds of group therapy have been used to treat schizophrenia. When the results have been compared with routine hospital treatment, the general finding in the better-controlled evaluations has been that group therapy is of little benefit in the acute stage of the disorder (Mosher and Keith 1980). There may be a place for supportive groups for patients who are resettling after the resolution of an acute illness.

Work with relatives

There have been few controlled studies of intensive family therapy in the treatment of schizophrenia. They were all concerned with short interventions for acute illness, and all found slight benefits from the treatment (Mosher and Keith 1980). The work on *emotional expression* reviewed above (see p. 284) suggests that specific interventions to reduce high levels of expressed emotion in relatives are beneficial in preventing relapse. Education of relatives about schizophrenia does not, by itself, produce a useful effect on relapse rate, though it may help them come to terms with illness and facilitate subsequent family interventions (Lam 1991; Leff 1994).

Behavioural treatment

The results of behavioural treatment for schizophrenia have not been fully evaluated. Most of the reported results could be due to receiving increased attention. **Individual methods** include social skills training (Wallace *et al.* 1980; Liberman *et al.* 1986).

Token economies also use positive and negative reinforcement to alter behaviour, but they are applied to all the patients in a ward rather than to a single patient. Rewards may be praise and interest, but it is usual to give tokens that can be used to purchase goods or privileges (hence the name of the treatment). While some short-term benefits may occur, many patients relapse when they move from a token economy to a new environment in which there is not the same system of rewards. These methods are now little used.

Cognitive behaviour therapy

The use of cognitive behaviour therapy in schizophrenia is based on the rationale that positive psychotic symptoms are amenable to structured reasoning and behavioural modification (Kingdon *et al.* 1994). With delusional beliefs, for example, individual ideas are traced back their origin and alternative explanations are explored. However, direct confrontation is avoided. Similarly, it may be possible to modify a patient's beliefs about the omnipotence, identity, and purpose of auditory hallucinations, with a resulting decrease in the distress that accompanies the experience and, perhaps, in its frequency (Chadwick and

Birchwood 1994). A number of open studies have shown that such techniques may benefit patients who have not responded well to antipsychotic drugs (Tarrier *et al.* 1993). However, the techniques involved are time consuming, and it is not clear how far any improvement reflects a specific effect of the cognitive approach. In addition, it is not known how far the gains made are sustained.

Assessment

Assessment begins with *differential diagnosis*, which is particularly concerned with the exclusion of *organic disorder* (especially a drug-induced state), *mood disorder*, and *personality disorder*. In practice, the main difficulty is often to elicit all the symptoms from a withdrawn or suspicious patient. This procedure may require several psychiatric interviews as well as careful observations by the nursing staff. The differential diagnosis from mood disorder can be particularly difficult and may require prolonged observation for discriminating symptoms of schizophrenia. Information from relatives or close friends will always be helpful.

While the psychiatric diagnosis is being confirmed, a *social assessment* should be carried out. This includes assessment of the patient's previous personality, work record, accommodation, and leisure pursuits, and especially the attitudes to the patient of relatives and any close friends. The doctor or social worker can proceed with this enquiry, while an evaluation of the patient's social functioning in the ward is made by nurses and occupational therapists.

In the assessment of schizophrenia, formal *psychological testing* seldom adds much to clinical observation. In the past, projective tests such as the Rorschach Test and the Thematic Apperception Test were used to examine thought processes, but they are unreliable and lack validity. Standardized tests of personality are less useful than a thorough history from an informant who knows the patient well. While there is little place for psychometric assessment in diagnosing schizophrenia, the clinical psychologist has a valuable role in making quantitative assessments of *specific abnormalities of behaviour* as a basis for planning and evaluating social rehabilitation. Both social and psychological assessments are important parts of assessment for *community care* and *discharge planning* (see Chapter 19).

Management

Success in management depends on establishing a good relationship with the patient so that his cooperation is enlisted. It is often difficult to establish a working relationship with chronic patients who are paranoid or emotionally unresponsive, but with skill and patience progress can usually be made. It is important to make plans that are realistic, especially for the more handicapped patient. Over-enthusiastic schemes of rehabilitation may increase the patient's symptoms and, if he lives in the community, place unacceptable burdens on relatives. *Minimizing the side-effects of medication* is also important if compliance is to be facilitated.

The acute illness

Treatment in hospital is usually needed for both first episodes of schizophrenia and acute relapses, although with adequate resources home treatment is possible and may carry some advantages (Dedman 1993). However, hospital admission allows a thorough assessment and provides a secure environment for the patient. It also gives a period of relief to the family, who have often experienced considerable distress during the period of prodromal symptoms of the illness (Johnstone *et al.* 1986).

There are important advantages in a few days of observation without drugs, although some acutely disturbed patients may require immediate treatment. A drug-free period allows thorough assessment of the patient's mental state and behaviour as described above. It also shows whether mental abnormalities and disturbed behaviour are likely to improve simply with change of environment. If they do not improve, an *antipsychotic drug* should be prescribed, with the dose depending on the severity of the symptoms. There is a wide choice of drugs (see

p. 549), but the clinician should become thoroughly familiar with a few. For acutely disturbed patients, the sedating effects of chlorpromazine are useful. An alternative approach is to use a modest dose of a high potency agent with additional *benzodiazepine treatment* for a short period (Schatzberg and Cole 1991*a,b*). This approach permits lower doses of antipsychotic drugs to be used, which is important in lessening toxicity and improving patient compliance. Potential complications such as behavioural disinhibition and benzodiazepine dependence seem to be rare if benzodiazepine treatment is co-prescribed with an *antipsychotic drug* and given *as needed* for a limited period of time.

The timing and dosage of antipsychotic drug treatment should be reviewed frequently with the ward staff and adjusted to changes in the patient's condition. Oral medication is usually given at this stage, although occasional intramuscular doses may be needed for patients who exhibit acutely disturbed behaviour and are unwilling to comply with oral treatment. If there are doubts whether the patient is swallowing tablets, the drug can be given as syrup. An alternative is *zuclopenthixol decanoate* given as a short-acting depot preparation, but this approach must be used with caution because it is not possible to titrate the dose with precision. It has been recommended that zuclopenthixol decanoate should not be given to patients whose tolerance of antipsychotic drugs is undetermined (Royal College of Psychiatrists 1993).

After the first few days, medication is continued at a constant daily amount for several weeks, with a gradual transfer to twice daily dosage or a single dose at night. *Antiparkinsonian drugs* should be prescribed if parkinsonian side-effects are troublesome, but they need not be given routinely. Symptoms of excitement, restlessness, irritability, and insomnia can be expected to improve within days. Symptoms of mood disturbance, delusions, and hallucinations respond more slowly, often persisting for six to eight weeks. Lack of improvement at this stage suggests inadequate dosage or failure to take the drugs prescribed, but a few cases resist all efforts at treatment. Once there is undoubted evidence of sustained improvement, dosage can be reduced cautiously while careful watch is kept for any return of symptoms. This reduced dose is continued for a further period (see below).

During the early days of treatment, the clinician will have taken histories from the patient, relatives, and other informants, so as to build up a picture of the *patient's previous personality*, *premorbid adjustment*, and *social circumstances*, and any *precipitants* of illness. By the time that symptomatic improvement has taken place, the clinical team should have formulated a provisional plan for continuing care. Although it is difficult to predict the long-term prognosis at this stage, a judgement has to be made about the likely immediate outcome. The judgement is based on the *degree* and *speed of response* to treatment, the *history of previous episodes*, and on the factors listed in Table 9.10. The aim is to decide how much after-care patients will require, and to make realistic plans accordingly.

After-care of 'good prognosis' patients

After a first episode of schizophrenia, patients judged to have a good immediate prognosis have two principal needs for treatment following discharge from hospital. The first is to take medication in reducing dosage for three to six months. The second is to be given advice about avoiding obviously stressful events. The patient should be seen regularly as an out-patient or at home by a community nurse until a few months after medication has been stopped and symptoms have ceased. Thereafter, a cautiously optimistic prognosis can be given, but the patient and his family should be warned to seek medical help quickly if there is any suggestion of the condition returning.

The after-care of the 'poor prognosis' patients without major social handicaps

When it is judged that further relapse is likely, continuing care will be required and this will probably include *prophylactic medication*. It is often better to give such medication by injection (for example as fluphenazine or flupenthixol decanoate) since some patients fail to take oral medication regularly over long periods. The dosage, which should be the minimum required to suppress symptoms, can be determined by cautiously varying its size and frequency while observing the patient's clinical state. Some patients show little response to continued antipsychotic medication even in high doses. For such patients it is worthwhile considering treatment with *clozapine* (see above*). Risperidone* may also merit a trial in treatment-resistant patients, particularly if negative or depressive symptoms are prominent.

Where patients with a poorer prognosis have responded to drug therapy, it is uncertain how long prophylactic treatment with neuroleptics should be continued. The treatment should be reviewed at least once a year, taking into account side-effects as well as symptoms. Medication may need to be withdrawn if *tardive dyskinesia* occurs (see p. 552); otherwise a balance has to be struck between benefits and adverse effects.

Whether maintenance drug therapy is given or not, the patient should be seen regularly to review his mental state, social adjustment, and occupation. When necessary he should be advised to avoid stressful situations and helped to reduce the amount of time spent with the family if it is emotionally arousing. Specific interventions may be needed to help the family understand the patient's illness, to have realistic expectations of what the patient can accomplish, and to decrease their emotional involvement where appropriate. Community psychiatric nurses can undertake most treatment of this kind, and social workers can help with those measures not involving the administration of medication. The most difficult problem in continuing management is likely to be the patient's tendency to withdraw from treatment.

Patients with chronic handicap

When patients have poor social adjustment and behavioural defects characteristic of chronic schizophrenia, they require more elaborate after-care. They should be identified as early as possible so that long-term plans can be made for both rehabilitation in hospital and resettlement outside hospital. Maintenance drug therapy plays an important part, but the main emphasis is on a programme of rehabilitation tailored to the needs of the individual patient.

It can be expected that the least handicapped patients will live more or less independently. Sheltered work and accommodation are likely to be needed for the remainder. The essential requirements are a management plan that focuses on one or two aspects of behaviour disorder at any one time, and a consistent approach between the members of staff carrying it out.

Despite the present emphasis on treatment outside hospital, early rehabilitation in hospital has significant advantages because it allows greater consistency of approach over the whole of the patient's day. Wards organized mainly for the treatment of acute illness are often too stimulating for chronic schizophrenics in need of rehabilitation. Hence it is appropriate to set aside a special area for rehabilitation. Treatment can be based either on the principles of a therapeutic community or on those of behavioural management. These two approaches produce similar results, and are compatible with one another (Hall 1983).

Most handicapped patients are able to live outside hospital, albeit in sheltered provisions. A minority require long-term care in hospital. The components of a community service for these patients are described in Chapter 19. Success depends less on physical provisions than on well-trained staff who have tolerant attitudes and the capacity to obtain satisfaction from work that produces small improvements over long periods.

When a patient has persisting abnormalities of behaviour, particular attention needs to be given to the problems of his family, as noted above. Relatives may be helped by joining a voluntary group and meeting others who have learnt to deal with similar problems. They also need to know that professional help will be provided whenever problems become too great. Advice is needed about the best ways of responding to abnormal behaviour, and about the expectations that they should have of the patient. Such advice is often given best by community nurses who have experience of treating chronic schizophrenic patients in hospital as well as in the community. Social workers also have a part to play in advising and helping relatives.

Care plans

There is increasing emphasis on the use of *care plans* in the management of psychiatric disorders. While this approach is applicable to all patients, it is particularly appropriate for those with schizophrenic illness who will often require long-term support in the community. The essential elements of a care plan are *systematic assessment* of health and social needs, a *treatment plan* agreed by the relevant staff, the patient, and the relatives, and the allocation of a *key worker* whose task is to maintain contact with patient, monitor progress, and ensure that the treatment programme is being delivered. *Regular review* of the patient's progress and needs is also important. (For further discussion of care planning and the role of psychiatric services, see Chapter 19.)

The violent patient

Over-activity and disturbances of behaviour are common in schizophrenia, and although major violence towards others is not common, it certainly does occur. For example Humphreys *et al.* (1992) found that, in a group of 253 patients with a first schizophrenic episode, 52 behaved in way threatening to the lives of others. In about half the patients, the behaviour was directly attributable to psychotic symptoms, usually delusions. While homicide is rare, self-mutilation is more frequent, and about one schizophrenic patient in ten dies by suicide. These self-harmful behaviours may sometimes be associated with positive symptoms such as delusions of control or auditory hallucinations (see p. 247). Drake *et al.* (1984) found that those at risk tended to be male, with a previous high educational attainment, and in a relatively non-psychotic stage of their illness. These patients had depressive symptoms in the context of high premorbid expectations of themselves together with a negative appraisal of the way that their illness had damaged their prospects for the future.

General management for the potentially violent patient is the same as that for any other schizophrenic, although a compulsory order is more likely to be needed. While medication is often needed to bring disturbed behaviour under immediate control, much can be done by providing a calm, reassuring, and consistent environment in which provocation is avoided. A special ward area with an *adequate number of experienced staff* is much better than the use of heavy medication. Threats of violence should be taken seriously, especially if there is a history of such behaviour in the past, whether or not the patient was ill at the time. The danger usually resolves as acute symptoms are brought under control, but a few patients pose a continuing

threat. The management of violence is considered further in Chapter 22.

Further reading

Bleuler, E. (1911). *Dementia praecox or the group of schizophrenias*. (English edition 1950). International University Press, New York.

Frith, C. D. (1992). *The cognitive neurospsychology of schizophrenia*. Erlbaum, Hove.

Hamilton, M. (1984). *Fish's schizophrenia*. Wright, Bristol.

Kavanagh, D. J. (1992). *Schizophrenia: an overview and practical handbook*. Chapman & Hall, London.

Kraepelin, E. (1919). *Dementia praecox and paraphrenia*. Churchill Livingstone, Edinburgh.

Kraepelin, E. (1986). Dementia praecox. *Psychiatrie* (5th edn.), pp. 426–41. Translated in *The clinical roots of schizophrenia concept* (ed. J. Cutting and M. Shepherd). Cambridge University Press.

McGuffin, P. and Murray, R. (ed.) (1991). *The new genetics of schizophrenia*. Butterworth Heinemann, Oxford.

McKenna, P. J. (1994). *Schizophrenia and related syndromes*. Oxford University Press.

10 | *Paranoid symptoms and paranoid syndromes*

Introduction

The term *paranoid* can be applied to symptoms, syndromes, or personality types. Paranoid symptoms are delusional beliefs which are most commonly persecutory but not always so. Paranoid syndromes are those syndromes in which paranoid symptoms form part of a characteristic constellation of symptoms, such as pathological jealousy or erotomania (described later). Paranoid personalities are those personalities in which there is excessive self-reference and undue sensitiveness to real or imaginary humiliations and rebuffs, often combined with self-importance, combativeness, and aggressiveness. The term paranoid is descriptive and not diagnostic. If we recognize a symptom or syndrome as paranoid, this is not making a diagnosis, but it is a preliminary to doing so. In this respect it is like recognizing stupor or depersonalization.

Paranoid syndromes present considerable problems of classification and diagnosis. The reasons for this can be understood by dividing them into two groups. In the first group, paranoid features occur in association with a *primary mental illness* such as schizophrenia, mood disorder, or an organic mental disorder. In the second group, paranoid features occur but no other primary disorder can be detected—the paranoid features appear to have arisen independently. In this book, following the DSMIV and ICD10 classifications, the term delusional disorders is applied to this second group. It is this second group that has given rise to difficulties and confusion over classification and diagnosis. For example there has been much argument as to whether these conditions are an alternative form of schizophrenia, or a stage in the evolution of schizophrenia, or a quite separate entity. It is because these problems arise frequently in clinical practice that a whole chapter is devoted to them.

This chapter begins with definitions of the commonest paranoid symptoms and then reviews the causes of such symptoms. Next comes a short account of paranoid personality. This is followed by discussion of primary psychiatric disorders such as organic mental states, mood disorders, and schizophrenia, with which paranoid features are frequently associated. These primary disorders are dealt with elsewhere in the book, but the focus here is on differentiating them from delusional disorders. These disorders are then reviewed, with particular reference to paranoia and paraphrenia. These latter terms are considered against their historical background. Next, an account is given of a number of distinctive paranoid symptoms and syndromes, some of which are fairly common and some exceedingly rare. This chapter finishes with a description of the assessment and treatment of patients with paranoid features.

Paranoid symptoms

In the introduction it was pointed out that the commonest paranoid delusions are *persecutory*. The term paranoid is also applied to the less common delusions of *grandeur* and *jealousy*, and sometimes to delusions concerning *love*, *litigation*, or *religion*. It may seem puzzling that such varied delusions should be grouped together. The reason is that the central abnormality implied by the term paranoid is a morbid distortion of beliefs or attitudes concerning relationships between oneself and other people. If someone believes falsely or on inadequate grounds that he is being victimized, or exalted, or deceived, or loved by a famous person, then in each case he is construing the relationship between himself and other people in a morbidly distorted way.

The varieties of paranoid symptom are discussed in Chapter 1, but the main ones are outlined here for convenience. The following definitions are derived from those in the glossary to the Present State Examination (PSE) (Wing *et al.* 1974).

Ideas of reference are held by people who are unduly self-conscious. The subject cannot help feeling that people take notice of him in buses, restaurants, or other public places, and that they observe things about him that he would prefer not to be seen. He realizes that this feeling originates within himself and that he is no more noticed than other people, but all the same he cannot help the feeling, quite out of proportion to any possible cause.

Delusions of reference consist of a further elaboration of simple ideas of self-reference, and the person does not recognize that the ideas are false. The whole neighbourhood may seem to be gossiping about the subject, far beyond the bounds of possibility, or he may see references to himself on the television or in newspapers. The subject may hear someone on the radio say something connected with some topic that he has just been thinking about, or he may seem to be followed, his movements observed, and what he says tape-recorded.

Delusions of persecution: the subject believes that someone, or some organization, or some force or power is trying to harm him in some way, to damage his reputation, to cause him bodily injury, to drive him mad, or to bring about his death. The symptom may take many forms, ranging from the direct belief that people are hunting him down to complex and bizarre plots with every kind of science fiction elaboration.

Delusions of grandeur: the glossary of the PSE proposes a division into delusions of grandiose ability, and delusions of grandiose identity.

The subject with delusions of *grandiose ability* thinks that he is chosen by some power, or by destiny, for a special mission or purpose because of his unusual talents. He thinks that he is able to read people's thoughts or that he is particularly good at helping them, that he is much cleverer than anyone else, or that he has invented machines, composed music, or solved mathematical problems beyond most people's comprehension.

The subject with delusions of *grandiose identity* believes that he is famous, rich, titled, or related to prominent people. He may believe that he is a changeling and that his real parents are royalty.

The causes of paranoid symptoms

When paranoid symptoms occur in association with a primary *organic illness, a mood disorder, or schizophrenic illness*, the main aetiological factors are those determining this primary illness. The question still arises as to why some people develop paranoid symptoms, whilst others do not. It has usually been answered in terms of premorbid personality and of factors causing social isolation.

Many writers including Kraepelin have held that paranoid symptoms are most likely to occur in patients with *premorbid personalities of a paranoid type* (see next section). Kretschmer (1927) also believed that paranoid disorders were more likely in people with predisposed or *sensitive personalities*. In such people a precipitating event could induce what Kretschmer called sensitive delusions of reference (*sensitive Beziehungswahn*), occurring as an understandable psychological reaction.

Modern studies of so-called late-onset paraphrenia have supported these views (see Chapter 16, p. 530). Thus Kay and Roth (1961) found paranoid or hypersensitive personalities in over half of their group of 99 late-onset paraphrenics.

Freud (1911) proposed that, in predisposed people, paranoid symptoms could arise through the *defence mechanisms* of denial and projection. He held that a person does not consciously admit his own inadequacy and self-distrust, but projects them on to the outside world. Clinical experience generally confirms this idea. If one examines paranoid patients, one often finds an inner dissatisfaction associated with a sense of inferiority and with self-esteem and ambition which are inconsistent with achievement.

Freud also held that paranoid symptoms could arise when denial and projection were being used as defences against unconscious homosexual tendencies. These ideas were derived from his study of Daniel Schreber, the presiding judge of the Dresden appeal court (Freud 1911). Freud never met Schreber, but read the latter's autobiographical account of his paranoid illness (now generally accepted to be paranoid schizophrenia), together with a report by Weber, the physician in charge. Freud held that Schreber could not consciously admit his homosexuality, and so the idea 'I love him' was dealt with by denial and changed by a reaction formation to 'I hate him'; this was further changed by projection into 'it is not I who hate him, but he who hates me', and this in turn became transformed to 'I am persecuted by him'. Freud believed that all paranoid delusions could be represented as contradictions of the idea 'I (a man) love him (a man)'. He went so far as to argue that delusions of jealousy could be explained in terms of unconscious homosexuality; the jealous husband was unconsciously attracted to the man whom he accused his wife of loving. In this case the formulation was 'it is not I who love him; it is she who loves him'. At one time these ideas were widely taken up, but nowadays they gain little acceptance. They are certainly not supported by clinical experience.

Apart from psychological factors within the patient, *social isolation* may also lead to the emergence of paranoid symptoms. As mentioned later in this chapter, prisoners in solitary confinement, refugees, and migrants may be prone to paranoid developments, although the evidence on this is conflicting.

Social isolation can also be produced by *deafness*. In 1915, Kraepelin pointed out that chronic deafness could lead to paranoid attitudes. Houston and Royse (1954) found an association between deafness and paranoid schizophrenia, whilst Kay and Roth (1961) found hearing impairment in 40 per cent of late-onset paraphrenics. However, it should be remembered that the great majority of deaf people do not become paranoid. (See Corbin and Eastwood (1986) for a review of the association of deafness and paranoid disorders in the elderly.)

Family studies of paranoid disorders

Paranoid personality disorder, delusional disorder, and schizophrenia have been the subject of family studies designed to clarify the possible relationships between these disorders. The findings of the various investigations have been somewhat contradictory, but recent studies have found that the incidence of *paranoid personality disorder* is increased in first-degree relatives of patients with schizophrenia, although not to the same extent as *schizotypal personality* (Kendler *et al.* 1993*b,c*; Parnas *et al.* 1993). First-degree relatives of patients with delusional disorder also have an increased incidence of *paranoid personality disorder*, perhaps more so than first-degree relatives of patients with schizophrenia (Kendler *et al.* 1985*a,b*).

The familial relationship of delusional disorder to schizophrenia is less clear. At present it appears that, while the risk of *delusional disorder* is increased in first-degree relatives of patients with schizophrenia, relatives of patients with delusional disorder do not have an increased risk of schizophrenia or schizotypal personality (Kendler *et al.* 1985*a,b*). This familial association pattern has been called *asymmetric co-aggregation* and may be due to (i) differences in the incidence rates of the two disorders in the general population, (ii) differences in diagnostic error rate between probands and relatives (probands are usually subject to more intensive assessment), and (iii) a higher genetic loading for severe illness in those who come to medical attention (and therefore are assessed as probands) (Kendler *et al.* 1993*b*). Taken together the findings suggest a modest, but somewhat complex, familial relationship between schizophrenia, delusional disorder, and paranoid personality.

Paranoid personality disorder

The concept of personality disorder was discussed in Chapter 5, and paranoid personality disorder was briefly described there. It is characterized by extensive sensitivity to setbacks and rebuffs, suspiciousness and a tendency to misconstrue the actions of others as hostile or contemptuous, and a combative and inappropriate sense of personal rights. It is implied in the DSMIV and ICD10 definitions that paranoid personality embraces a wide range of types. At one extreme is the excessively sensitive youth who shrinks from social encounters and thinks that everyone disapproves of him. At the other is the assertive and challenging man who flares up at the least provocation. Many grades lie between these two extremes.

Because of the implications for treatment, it is important to distinguish these *paranoid personalities* from the *paranoid syndromes* to be described later. The distinction can be very difficult to make. Sometimes one shades into the other in the course of a single life history, as exemplified by the life of the philosopher Jean Jaques Rousseau. The basis for making the distinction is that in paranoid personalities there are no delusions but only dominant ideas, and no hallucinations. Separating paranoid ideas from delusions calls for considerable skill. The criteria for doing so are given in Chapter 1 (p. 9).

Primary psychiatric disorders with paranoid features

It was mentioned in the introduction to this chapter that paranoid features occur in association with primary mental disorders. This association occurs commonly in clinical practice. As the primary disorders are described at length in other chapters, they are mentioned only briefly here.

Cognitive mental disorders

Paranoid symptoms are common in *delirium*. Impaired grasp of what is going on around the patient may give rise to apprehension and mis-interpretation, and so to suspicion. Delusions may then emerge which are usually transient and disorganized; they may lead to disturbed behaviour, such as querulousness or aggression. Examples are drug-induced states. Similarly, paranoid delusions may occur in dementia arising from any cause, including trauma, degenerations, infections, meta-bolic, and endocrine disorders (see Chapter 11).

In clinical practice it is important to remember that in elderly patients with dementia, paranoid delusions may appear before any intellectual deterioration is detectable.

Mood disorders

Paranoid delusions not uncommonly occur in patients with severe *depressive disorders*. The latter are often characterized by *guilt* and *retardation*, and by *biological features* such as loss of appetite and weight, sleep disturbance, and reduced sex drive. These disorders are more common in middle or later life. In depressive disorder, the patient typically accepts the supposed activities of the persecutors as justified by his own guilt or wickedness, but in schizophrenia he often resents them bitterly. It is sometimes difficult to determine whether the paranoid features are secondary to depressive disorder, or whether depressed mood is secondary to paranoid symptoms arising from another cause. Primary depressive disorder is likely if the mood changes have occurred earlier and are of greater intensity than the paranoid features. Previous psychiatric history and family history may also be useful pointers.

The distinction may have some importance from the point of view of *prognosis,* because patients with a primary mood disorder will usually have a better long-term outcome. Furthermore, if depressive disorder is primary, patients will probably need antidepressant medication or ECT, in addition to treatment with antipsychotic drugs (see Chapter 17).

Paranoid delusions also occur in *manic* patients. Often the delusions are grandiose rather than persecutory—the patient claims to be extremely wealthy or of exalted rank or importance.

Paranoid schizophrenia

Paranoid schizophrenia has been described in Chapter 9. In contrast with the hebephrenic and catatonic forms of schizophrenia, the paranoid form usually begins later in life—in the thirties rather than in the twenties. The dominant feature of paranoid schizophrenia is *delusions* that are relatively stable over time. The delusions are frequently of *persecution*, but may also be of *jealousy, exalted birth, Messianic mission,* or *bodily change.* They may be accompanied by *hallucinatory voices* which sometimes, but not invariably, have a persecutory or grandiose content.

It is important to consider the differential diagnosis of paranoid schizophrenia from other paranoid conditions. The criteria for the diagnosis of schizophrenia in DSMIV and ICD10 were described on p. 257. In cases of doubt, the diagnosis of schizophrenia rather than delusional disorder is suggested if the paranoid delusions are particularly odd in content (often referred to by psychiatrists as bizarre delusions). If the delusions are grotesque, then there may be no room for doubt. For example a middle-aged woman became convinced that a Cabinet Minister was taking a special interest in her and was promoting her well-being. She believed that he was the pilot of an aeroplane that flew over her house shortly after noon each day. Therefore she waited in her garden each day and threw a large red beach-ball into the sky when the plane flew over. She maintained that the pilot always acknowledged this action by 'waggling the wings' of the plane. When the delusions are less extreme than this, a judgement as to how bizarre they are must be arbitrary.

There are two other diagnostic points. First, in schizophrenia delusions are most likely to be fragmented and multiple, rather than systematized and unitary. Second, paranoid schizophrenics often have auditory hallucinations that seem to be totally unrelated to their delusions, whereas patients with paranoid conditions other than schizophrenia frequently have no auditory hallucinations, or else fleeting hallucinations that are related in content to their delusional ideas.

Schizophrenia-like syndromes

Paranoid syndromes are also common in several schizophrenia-like syndromes discussed in Chapter 9 (and listed in Table 9.4). These include the DSMIV categories of **brief psychotic disorders** and **schizophreniform disorder**, and the ICD10 categories grouped under the heading **acute and transient psychotic disorders**.

Paranoid psychoses (delusional disorders)

DSMIV uses **delusional disorder** for a disorder with persistent, non-bizarre delusion that is not due to any other mental disorder. ICD10 has a rather similar category of **persistent delusional disorders**.

Historical background: paranoia and paraphrenia

These two terms have played a prominent part in psychiatric thought. Much can be learnt from reviewing the conceptual difficulties associated with them. For this reason, their history will be traced in some detail, starting with paranoia.

The term **paranoia**, from which the modern adjective paranoid is derived, has a long and chequered history (Bynum 1983). It has probably given rise to more controversy and confusion of thought than any other term used in psychiatry. A comprehensive review of the large literature, which is mostly German, has been provided by Lewis (1970).

The term paranoia came into special prominence in the last quarter of the nineteenth century, but its origins are much older. The word paranoia is derived from the Greek *para* (beside) and *nous* (mind). It was used in ancient Greek literature to mean 'out of mind', i.e. of unsound mind or insane. This broad usage was revived in the eighteenth century. However, in the mid-nineteenth century German psychiatrists became interested in conditions that were particularly characterized by delusions of persecution and grandeur. The German term *verrücktheit* was often applied to these conditions, but eventually was superseded by paranoia. There were many different conceptions of these disorders. The main issues can be summarized as follows.

1. Did these conditions constitute a primary disorder, or were they secondary to a mood disorder or other disorder?
2. Did they persist unchanged for many years, or were they a stage in an illness which later manifested deterioration of intellect and personality?
3. Did they sometimes occur in the absence of hallucinations, or were hallucinations an invariable accompaniment?
4. Were there forms with good prognosis?

Kahlbaum raised these issues as early as 1863, when he classified paranoia as an independent or primary delusional condition which would remain unchanged over the years.

Kraepelin had a strong influence on the conceptual history of paranoia, although he was never comfortable with the term, and his views changed strikingly over the years (Kendler and Tsuang 1981). In 1896 he used the term only for incurable, chronic, and systematized delusions without severe personality disorder. In the sixth edition of his textbook he wrote:

The delusions in dementia praecox are extremely fantastic, changing beyond all reason, with an absence of system and a failure to harmonize them with events of their past life; while in paranoia the delusions are largely confined to morbid interpretations of real events, are woven together into a coherent whole, gradually becoming extended to include even events of recent date, and contradictions and objects are apprehended and explained (Kraepelin 1904, p. 199).

In later descriptions Kraepelin (1912, 1919) used the distinction made by Jaspers (1913) between personality development and disease process. He proposed paranoia as an example of the former, in contrast to the disease process of dementia praecox. In his final account, Kraepelin (1919) developed these ideas by distinguishing between dementia praecox, paranoia, and a third paranoid psychosis, paraphrenia. Dementia praecox had an early onset and a poor outcome ending in mental deterioration, and was fundamentally a disturbance of affect and volition. Paranoia was restricted to patients with the late onset of completely systematized delusions and a prolonged course usually without recovery but not inevitably deteriorating. An important point was that the patients did not have hallucinations. Kraepelin regarded paraphrenia as lying between dementia praecox and paranoia; in paraphrenia, the patient had unremitting systematized delusions but did not progress to dementia. The main difference from paranoia was that the patient with paraphrenia had hallucinations.

Bleuler's concept of the paranoid form of dementia praecox (which he later called paranoid schizophrenia) was broader than that of Kraepelin (Bleuler 1906, 1911). Thus Bleuler did not regard paraphrenia as a separate condition, but as part of dementia praecox. However, he accepted Kraepelin's view of paranoia as a separate entity, although he differed from Kraepelin in maintaining that hallucinations could occur in many cases. Bleuler was particularly interested in the psychological development of paranoia; at the same time he left open the question of whether paranoia had a somatic pathology.

From this time, two main themes were prominent in the history of paranoia. The first theme was that paranoia was distinct from schizophrenia and mainly psychogenic in origin. The second theme was that paranoia was part of schizophrenia.

Two celebrated studies of individual cases supported the first theme—the psychogenic origins of paranoid delusions. Gaupp (1914) made an intensive study of the diaries and mental state of the mass murderer Wagner who murdered his wife, four children, and eight other people as part of a careful plan to revenge himself on his supposed enemies. Gaupp concluded that Wagner suffered from paranoia in the sense described by Kraepelin, as discussed above. At the same time, he believed that Wagner's first recognizable delusions developed as a psychogenic reaction. The second study, mentioned earlier in the chapter, was Freud's analysis of the memoirs of Schreber. Freud called this a case of paranoia, although Schreber's illness conformed much more to the clinical picture of schizophrenia than to any of the prevailing notions of paranoia.

The most detailed argument for psychogenesis was put forward by Kretschmer (1927) in his monograph *Der sensitive Beziehungswahn*. Kretschmer believed that paranoia should not be regarded as a disease, but as a psychogenic reaction occurring in people with particularly sensitive personalities. Many of Kretschmer's cases would nowadays be classified as suffering from schizophrenia.

In 1931, Kolle put forward evidence for the second theme, that paranoia is part of schizophrenia. He analysed a series of 66 patients with so-called paranoia, including those diagnosed by Kraepelin in his Munich clinic. For several reasons, both symptomatic and genetic, Kolle came to the conclusion that so-called paranoia was really a mild form of schizophrenia.

Considerably less has been written about **paraphrenia**. However, it is interesting that Mayer (1921), following up Kraepelin's series of 78 paraphrenic patients, found that 50 of them had become schizophrenic. He found no difference in original clinical presentation between those who became schizophrenic and those who did not. Since then paraphrenia has increasingly been regarded as late-onset schizophrenia or schizophrenia-like disorder of good prognosis. Kay and Roth (1961) used the term late paraphrenia to denote paranoid conditions in the elderly which were not due to primary organic or affective illnesses. These authors found that a large majority of their 99 patients had the characteristic features of schizophrenia (see Chapter 9). Modern classifications do not use separate categories for early- and late-onset schizophrenia.

Modern usage: DSMIV and ICD10

In DSMIV, **delusional disorder** replaces the traditional category of paranoia. The criteria require *non-bizarre delusions* of at least one month's duration and that auditory or visual hallucinations, if present, are not prominent. However, olfactory and tactile hallucinations may be present as part of the delusional disorder (for example a patient may experience itching associated with a delusion of infestation).

In DSMIV, there are five specific subtypes of delusional disorder: persecutory, jealous, erotomanic, somatic, and grandiose. The somatic form covers the disorder sometimes referred to as *monosymptomatic hypochondriacal psychosis* (Munro 1980). Delusional disorder appears to be rare, to occur predominantly in mid-life, and to have a prolonged course (Kendler and Tsuang 1981; Kendler 1982).

ICD10 gives a similar definition for the principal category of **persistent delusion disorders**, of which *delusional disorder* is the main diagnosis. However, the symptoms must have been present for at least three months rather than the one month required in DSMIV and the subtypes are not specified.

The essence of the modern concept of delusional disorder is that of a permanent and unshakeable delusional system developing insidiously in a person in middle or late life. This delusional system is encapsulated, and there is no impairment of other mental functions. The patient can often go on working, and his social life may sometimes be maintained fairly well. In clinical practice, cases conforming strictly to the definitions are rare. The term *paraphrenia* does not appear in DSMIV or ICD10. The term is no longer used in modern psychiatric practice.

Special paranoid conditions

Certain paranoid conditions are recognizable by their distinctive features. They can be divided into two groups—those with special symptoms and those occurring in special situations. The special symptoms include jealous, erotic, and querulant delusions, and also the delusions associated with the names of Capgras and Fregoli. The special situations include intimate relationships (*folie à deux*), also known as *induced delusional disorder* (ICD10) or *shared psychotic disorder* (DSMIV), as well as migration and imprisonment. Many of these symptoms have been of particular interest to French psychiatrists (Pichot 1982, 1984). None, apart from 'shared psychotic disorder' or 'induced delusional disorder', is recognized as a separate category in DSMIV or ICD10.

Among the conditions with special symptoms, pathological jealousy will be described first and in the greatest detail because of its importance in clinical practice. It is probably the most common, and is often dangerous.

Pathological jealousy

In *pathological (or morbid) jealousy*, the essential feature is an abnormal belief that the marital partner is being unfaithful. The condition is called pathological because the belief, which may be a *delusion* or an *over-valued idea*, is held on inadequate grounds and is unaffected by rational argument. Pathological jealousy has been reviewed by Shepherd (1961) and by Mullen and Maack (1985).

The belief is often accompanied by strong emotions and characteristic behaviour, but these do not in themselves constitute pathological jealousy. A man who finds his wife in bed with a lover may experience extreme jealousy and may behave in an uncontrolled way, but this should not be called pathological jealousy. The term should be used only when the jealousy is based on unsound evidence and reasoning.

Pathological jealousy has often been described in the literature, generally in reports of one or two cases. Various names have been given to it, including sexual jealousy, erotic jealousy, morbid jealousy, psychotic jealousy, and the Othello syndrome. The main sources of information are surveys of patients with pathological jealousy carried out by Shepherd (1961), Langfeldt (1961), Vaukhonen (1968), and Mullen and Maack (1985). Shepherd examined the hospital case notes of 81 patients in London and Langfeldt did the same for 66 patients in Norway, Vaukhonen made an interview study of 55 patients in Finland; and Mullen and Maack examined the hospital notes of 138 patients.

The frequency of pathological jealousy in the general population is unknown, although jealous feelings are ubiquitous (Mullen and Martin 1994). Pathological jealousy is not uncommon in psychiatric practice, and most full-time clinicians probably see one or two cases a year. They merit careful attention, not only because of the great distress that they cause within marriages and families, but also because they may be highly *dangerous*.

All the evidence suggests that pathological jealousy is more common in men than in women. In the three surveys mentioned above, the male-to-female ratios were 3.76:1 (Shepherd), 1.46:1 (Langfeldt), and 2.05:1 (Vaukhonen). However, the precise sex ratio, may depend on the particular group studied. For example, in a retrospective survey of psychiatric in-patients, Soyka *et al.* (1991) found that amongst patients with paranoid schizophrenia more females than males had delusions of jealousy, whilst amongst patients with alcoholic psychosis more men developed delusional jealousy (even allowing for the fact that more men than women were affected with alcoholic psychosis).

Clinical features

As indicated above, the main feature is an *abnormal belief* in the partner's *infidelity*. This may be accompanied by other abnormal beliefs, for example that the spouse is plotting against the patient, trying to poison him, taking away his sexual capacities, or infecting him with venereal disease.

The *mood* of the pathologically jealous patient may vary with the underlying disorder, but often it is a mixture of misery, apprehension, irritability, and anger.

The *behaviour* of the patient is often characteristic. Commonly there is intensive seeking for evidence of the partner's infidelity, for example by searching in diaries and correspondence, and by examining bed-linen and underwear for signs of sexual secretions. The patient may follow the spouse about, or engage a private detective to spy on her. Typically the jealous person cross-questions the spouse incessantly. This may lead to violent quarrelling and paroxysms of rage in the patient. Sometimes the partner becomes exasperated and worn out, and is finally goaded into making a false confession. If this happens, the jealousy is inflamed rather than assuaged.

An interesting feature is that the jealous person often has no idea as to who the supposed lover may be, or what kind of person he may be. Moreover, he may avoid taking steps that could produce unequivocal proof one way or the other.

The behaviour of patients with pathological jealousy may be strikingly abnormal. A successful City businessman carried a briefcase that contained not only his financial documents but also a machete for use against any lover who might be detected. A carpenter installed an elaborate system of mirrors in his house so that he could watch his wife from another room. A third patient avoided waiting alongside another car at traffic lights, in case his wife in the passenger seat might surreptitiously make an assignation with the other driver.

Aetiology

In the surveys described above, pathological jealousy was found to be associated with a range of primary disorders. The frequencies varied, depending on the population studied and the diagnostic scheme used. For example *paranoid schizophrenia* (or paranoia or paraphrenia) was reported in 17–44 per cent of patients, *depressive disorder* in 3–16 per cent, *neurosis and personality disorder* in 38–57 per cent, *alcoholism* in 5–7 per cent, and *organic disorders* in 6–20 per cent. Primary organic causes include exogenous substances such as amphetamine and cocaine, but more commonly a wide range of brain disorders including infections, neoplasms, metabolic and endocrine disorders, and degenerative conditions.

The role of *personality* in the genesis of pathological jealousy should be stressed. It is often found that the patient has a pervasive sense of his own inadequacy, together with *low self-esteem*. There is a discrepancy between his ambitions and his attainments. Such a personality is particularly vulnerable to anything that may threaten this sense of inadequacy, such as loss of status or advancing age. In the face of such threats the person may project the blame onto others, and this may take the form of jealous accusations of infidelity. As mentioned earlier, Freud believed that unconscious homosexual urges played a part in all jealousy, particularly the delusional kind. He held that this could occur when such urges were dealt with by repression, denial, and reaction formation. However, none of the three surveys mentioned above found any association between homosexuality and pathological jealousy.

Many writers have held that pathological jealousy may be induced by the onset of erectile difficulties in men or sexual dysfunction in women. In their surveys, Langfeldt and Shepherd found little or no evidence of such associations. However, Vaukhonen reported sexual difficulties in over half the men and women in his series, but his sample was drawn partly from a marriage guidance clinic.

The prognosis depends on a number of factors, including the nature of any underlying psychiatric disorder and the patient's premorbid personality. There is little statistical evidence on prognosis. Langfeldt followed-up 27 of his patients after 17 years, and found that over half of them still had persistent or recurrent jealousy. This confirms a general clinical impression that the prognosis is often poor.

Risk of violence

Although there is no direct statistical evidence of the risks of violence, there is no doubt that pathological jealousy can be highly dangerous. Mowat (1966) made a survey of homicidal patients admitted to Broadmoor Hospital over several years, and found pathological jealousy in 12 per cent of men and 15 per cent of women. In Shepherd's series of 81 patients with pathological jealousy, three had shown homicidal tendencies. In addition to homicide, the risk of physical injury inflicted by jealous patients is undoubtedly considerable. In Mullen and Maack's (1985) series few of the 138 patients had received criminal convictions, but a quarter had threatened to kill or injure their partner, and 56 per cent of men and 43 per cent of women had been violent to or threatened the supposed rival (see Chapter 22). There may also be a risk of suicide, particularly when an accused spouse finally decides to end the relationship.

Assessment

The assessment of a patient with pathological jealousy should be painstaking and thorough. Full psychiatric assessment of the patient is essential and the spouse should be seen alone at first, and with the patient afterwards.

The spouse may give a much more detailed account of the patient's morbid beliefs and actions than can be elicited from the patient. The doctor should try to find out tactfully how firmly the patient believes in the partner's infidelity, how much resentment he feels, and whether he has contemplated any vengeful action. What factors provoke outbursts of resentment, accusation, and cross-questioning? How does the partner respond to such outbursts by the patient? How does the patient respond in turn to the partner's behaviour? Has there been any violence so far? If so, how was it inflicted? Has there been any serious injury?

In addition to these enquiries, the doctor should take a detailed marital and sexual history from both partners. It is also important to diagnose any underlying psychiatric disorder, as this will have implications for treatment.

Treatment

The treatment of pathological jealousy is often difficult, because the jealous person may regard it as obtrusive and may show little compliance. Adequate treatment of any underlying disorder such as schizophrenia or a mood disorder is a first requisite. In other cases the pathological jealousy may be the symptom of a delusional disorder, or an overvalued idea in a patient with low self-esteem and personality difficulties. When the jealousy seems to be delusional in nature, when a careful trial of an antipsychotic drug is worthwhile (Munro 1984) though results are often disappointing. There are also some anecdotal accounts of successful treatment with selective serotonin re-uptake inhibitors in patients whose jealous thoughts are overvalued ideas (Lane 1990); however, no randomized trials have yet been reported. As noted above, even when depressive disorder is not the primary diagnosis, it frequently complicates pathological jealousy and may worsen it. Treatment with an antidepressant drug may help in these circumstances.

Psychotherapy may be given to patients with neurotic or personality disorders. The aims may be to reduce tensions by allowing the patient (and spouse) to ventilate feelings. Behavioural methods have also been advocated (Cobb and Marks 1979). These include encouraging the partner to produce behaviour that reduces jealousy, for example by counter-aggression or by refusal to argue, depending on the individual case. An open study of cognitive therapy, in which the patient was encouraged to identify and challenge the assumptions underlying jealous thoughts, also gave encouraging results (Bishay *et al.* 1989).

If there is no response to out-patient treatment or if the risk of violence is high, in-patient care may be necessary. Not uncommonly, however, the patient appears to improve as an in-patient, only to relapse on discharge.

If there appears to be a risk of violence, the doctor should warn the spouse. In some cases the safest procedure is to advise separation. It is not uncommon for feelings of pathological jealousy to wane once a relationship has ended. Sometimes, however, the problem re-emerges if the patient enters a new relationship.

Erotic delusions (De Clérambault's syndrome)

De Clérambault (1921, see also 1987) proposed that a distinction should be made between *paranoid* delusions and delusions of *passion*. The latter differed in their pathogenesis and in being accompanied by excitement. They also had a sense of purpose: 'patients in this category whether they display erotomania, litigious behaviour, or morbid jealousy all have a precise aim in view from the onset of the illness, which brings the will into play from the beginning. This constitutes a distinguishing feature of the illness.' This distinction is of historical interest only, as it is not made nowadays. However, the syndrome of *erotomania* is still known as *De Clérambault's syndrome*. It is exceedingly rare (see Enoch and Trethowan (1979) for further information). Although erotomania is usually a disorder of women, Taylor *et al.* (1983) have reported four cases in a series of 112 men charged with violent offences.

In erotomania, the subject, usually a single woman, believes that an exalted person is in love with her. The supposed lover is usually inaccessible, as he is already married, or of much higher social status, or famous as an entertainer or public figure. According to De Clérambault, the infatuated woman believes that it is the supposed lover who first fell in love with her, and that he is more in love than she. She derives satisfaction and pride from this belief. She is convinced that the 'object' cannot be a happy or complete person without her.

The patient often believes that the supposed lover is unable to reveal his love for various reasons that he has withheld from her, and that he has difficulties in approaching her, has indirect conversations with her, and has to behave in a paradoxical and contradictory way. The woman may be a considerable nuisance to the supposed lover, who may complain to the police and the courts. Sometimes the patient's delusion remains unshakeable, and she invents explanations for the other person's paradoxical behaviour. She may be extremely tenacious and impervious to reality. Other patients turn from a delusion of love to a delusion of persecution. They become abusive and make public complaints about the 'object'. This was described by De Clérambault as two phases—hope followed by resentment.

Many patients with erotic delusions are suffering from paranoid schizophrenia. Sometimes there is insufficient evidence to make a final diagnosis at the time, and they can be classified as delusional disorder erotomanic type in DSMIV.

Querulant delusions and reformist delusions

Querulant delusions were the subject of a special study by Krafft-Ebing (1888). Patients with this kind of delusion indulge in a series of complaints and claims lodged against the authorities. Closely related to querulant patients are *paranoid litigants* who undertake a succession of lawsuits; they become involved in numerous court hearings, in which they may become passionately angry and make threats against the magistrates. Baruk (1959) described '*reformist delusions*' which are centred on religious, philosophical, or political themes. People with these delusions constantly criticize society and sometimes embark on elaborate courses of action. Their behaviour may be violent, particularly when the delusions are political. Some political assassins fall within this group. It is extremely important that this diagnosis is made on clear psychiatric grounds rather than political grounds (Bloch and Chodoff 1981).

Capgras delusion

Although there had been previous case reports, the condition now known as *Capgras syndrome* was well described by Capgras and Reboul-Lachauz in 1923 (see Serieux and Capgras 1987). They called it *l'illusion des sosies* (illusion of doubles). Strictly speaking, it is not a syndrome but a symptom, and it is better termed the *delusion* (rather than illusion) of doubles. The patient believes that a person closely related to him has been replaced by a double. He accepts that the misidentified person has a great resemblance to the familiar person, but still believes that they are different people. It is an extremely rare condition. It is more common in women than in men and is usually associated with schizophrenia or mood disorder. A history of depersonalization, derealization, or *déjà vu* is common. The misidentified person is usually the patient's spouse or another relative. It has been said that in most cases there is strong evidence of an organic component as shown by the clinical features, psychological testing, and radiological studies of the brain (Christodoulou 1977). However, a review of 133 published cases concluded that more than half were schizophrenic; 31 cases had a proven physical disorder (Berson 1983). Similar findings were reported by Förstl *et al.* (1991). Some recent neuropsychological case studies have suggested that patients with Capgras syndrome have impairments in facial recognition tasks, in terms of both recognizing familiar faces and labelling facial expressions (Young *et al.* 1993). It is worth noting that some patients with Capgras syndrome may behave *dangerously* by attacking the presumed doubles.

Fregoli delusion

This is usually referred to as the Fregoli syndrome, and derives its name from an actor called Fregoli who had remarkable skill in changing his facial appearance. The condition is even rarer than the Capgras delusion. It was originally described by Courbon and Fail in 1927. The patient identifies a familiar person (usually someone whom he believes to be his persecutor) in various other people he encounters. He maintains that, although there is no physical resemblance between the familiar person and the others, nevertheless they are psychologically identical. This symptom is usually associated with schizophrenia. Here again the clinical features, psychological testing, and radiological examination of the brain have been said to suggest an organic component in the aetiology (Christodoulou 1976).

Paranoid conditions occurring in special situations

An account will now be given of these conditions, beginning with induced psychosis.

Induced delusional disorder (*folie à deux*)

An induced delusional disorder is a paranoid delusional system which appears to have developed in a person as a result of a close relationship with another person who already has an established and similar delusional system. The delusions are nearly always persecutory. These cases are classified as *shared psychotic disorder* in DSMIV, and as *induced delusional disorder* (*folie à deux*) in ICD10. The frequency of induced psychosis is not known, but it is low. Sometimes more than two people are involved, but this is exceedingly rare.

The condition has occasionally been described in two people who are not family relations, but 90 per cent or more of reported cases are members of the same family. Usually there is a dominant partner with fixed delusions who appears to induce similar delusions in a dependent or suggestible partner, sometimes after initial resistance. Generally the two have lived together for a long time in close intimacy, often in isolation from the outside world. Once established, the condition runs a chronic course.

Induced delusional disorder is more common in women than in men. Gralnick (1942) studied a series of patients with *folie à deux* and found the following combinations in order of frequency: two sisters, 40; husband and wife, 26; mother and child, 24; two brothers, 11; brother and sister, 6; father and child, 2; not related, 9.

The principles of treatment are the same as for other paranoid conditions (p. 307). It is usually necessary to advise separation of the affected people. This separation sometimes leads to disappearance of the delusional state, but not invariably; improvement is more likely in the recipient than in the inducer. Induced delusional disorder is comprehensively reviewed by Enoch and Trethowan (1979).

Migration psychoses

It might be expected that people migrating to foreign countries would be likely to develop paranoid symptoms because their appearance, speech, and behaviour attract attention. Ødegaard (1932) found that rates for schizophrenia (including paranoid schizophrenia) were twice as high amongst Norwegian-born immigrants to the United States as amongst the general population of Norway. However, the explanation of the finding appeared to be not so much that emigration was a pathogenic experience as that pre-psychotic Norwegians were more likely than others to emigrate. Astrup and Ødegaard (1960) later found that hospital first admission rates for psychotic illness in general were significantly lower amongst people who had migrated inside their own country than amongst those who stayed where they were born. The authors suggested that migration within one's own country might be a natural step for enterprising young people, whilst migrating abroad was likely to be a much more stressful experience. To this extent they favoured the environmental hypothesis.

Studies of immigrants are difficult to interpret. If one controls for factors such as age, social class, occupational status, and ethnic group, it becomes doubtful whether there is a significant association between migration and rates of mental illness (Murphy 1977). The highest rates of mental disorder have been reported in refugees whose migration was enforced (Eitinger 1960). However, such people may have been exposed to persecution, in addition to the experiences of losing their homeland and readjusting to another country. (The apparent increase in the rate of schizophrenia amongst Afro-Carribean immigrants to the United Kingdom is discussed on p. 280.)

Prison psychosis

The evidence about imprisonment is conflicting. The work of Birnbaum (1908) suggests that isolation in prison, and especially solitary confinement, may lead to paranoid disorders that clear up when the prisoners are allowed to mix with others. Eitinger (1960) reported that paranoid states were not uncommon in prisoners of war. However, Faergeman (1963) concluded that such developments were rare even amongst the inmates of concentration camps (see Chapter 22 for a review of psychiatric disorders in prisons).

Cultural psychoses

It appears that in some developing countries there is a high incidence of transient and acute psychotic states, in which paranoid features commonly occur. Some of these acute states may be due to organic causes such as tropical infections. Because of the conditions of observation, information about these disorders is incomplete.

Paranoid symptoms: assessment and diagnosis

In the assessment of paranoid symptoms there are two stages, the recognition of the symptoms themselves and the diagnosis of the underlying condition.

Sometimes it is obvious to everyone that the patient has persecutory ideas or delusions. At other times recognition of paranoid symptoms may be exceedingly difficult. The patient may be suspicious or angry. He may offer little speech, simply staring silently at the interviewer, or he may talk fluently and convincingly about other things, whilst steering away from delusional ideas or beliefs, or denying them completely. Considerable skill may be needed to elicit the false beliefs. The psychiatrist should be tolerant and impartial. He should present himself as a detached but interested listener who wants to understand the patient's point of view. He should show compassion and ask how he can help, but without colluding in the delusions and without giving promises that cannot be fulfilled. Tact is required to avoid any argument which may cause the patient to take offence. Despite skill and tact, experienced psychiatrists may interview a patient for a long time without detecting the morbid thoughts.

If apparently false beliefs are disclosed, before concluding that they are delusions it may be necessary to check the patient's statements against those of an informant and to ensure that the patient has had an opportunity to recognize the falsity of his beliefs. As with all apparent delusions, they must be judged against the cultural background, since the patient may hold a false belief which is generally held by his own group.

If paranoid delusions are detected, the next step is to diagnose the underlying psychiatric disorder. This means looking for the diagnostic features of organic mental states, schizophrenia, and mood disorders, which are described in other chapters.

It is important to determine whether any persecutory or jealous delusions are likely to make the patient behave *dangerously* by trying to kill or injure his supposed persecutor. This calls for close study of the patient's personality and the characteristics of his delusions and any associated hallucinations. Hints or threats of homicide should be taken seriously, in the same way as for suicide. The doctor should be prepared to ask tactfully about possible homicidal plans and preparations to enact them. In many ways the method of enquiry resembles the assessment of suicide risk: 'Have you ever thought of doing anything about it?' 'Have you made any plans?' 'What might prompt you to do it?'

Sometimes a patient with persecutory delusions does not know the identity of a supposed persecutor, but may still be dangerous. For example, an overseas visitor in his early twenties was seen in a psychiatric emergency clinic. Careful enquiry revealed that he believed that unidentified conspirators were trying to kill him, and that his life was in imminent danger. When asked if he had taken any steps to protect himself, he said that he had made a brief trip to Brussels to buy a pistol, which he was now carrying. When asked what he might do with the gun, he said that he was waiting until the voices told him to shoot someone.

The assessment of dangerousness is discussed further in Chapter 22. The most reliable guideline is that the risk of violence is greatest in patients with a history of previous violence.

The treatment of patients with paranoid symptoms

In the management of patients with paranoid symptoms, both psychological and physical measures should be considered.

Psychological management is frequently difficult. The patient may be suspicious and distrustful, and may believe that psychiatric treatment is intended to harm him. Even if he is not suspicious, he is likely to regard his delusional beliefs as justified, and to see no need for treatment. Considerable tact and skill are needed to persuade patients with paranoid symptoms to accept treatment. Sometimes this can be done by offering to help non-specific symptoms such as anxiety or insomnia. Thus a patient who believes that he is surrounded by persecutors may agree that his nerves are being strained as a result, and that this nervous strain needs treatment.

It is usually necessary at an early stage to decide whether to admit the patient for in-patient care. This may be indicated if the delusions are causing aggressive behaviour or social difficulties. In assessing such factors, it is usually best to consult other informants and to obtain a history of the patient's behaviour in the past. If voluntary admission is refused, compulsory admission is often justified to protect the patient or other people, although this is likely to add to the patient's resentment.

During treatment the psychiatrist should strive to maintain a good relationship. He should be dependable and should avoid provoking resentment by letting the patient down. He should show compassionate interest in the patient's beliefs, but without condemning them or colluding in them.

Patients with paranoid delusions may be helped by psychological support, encouragement, and assurance. Interpretative psychotherapy and group psychotherapy are unsuitable because suspiciousness and hypersensitivity may easily lead the patient to misinterpret what is being said.

Treatment by medication may be indicated for a primary psychiatric disorder, such as schizophrenia, mood disorder, or an organic mental state. In paranoid states with no detectable primary disorder, symptoms are sometimes relieved by *antipsychotic medication* such as trifluoperazine, chlorpromazine, or haloperidol. The choice of drug and dosage depends on the patient's age, physical condition, degree of agitation, and response to previous medication. Several case reports support the use of pimozide for various forms of delusional disorder, particularly monosymptomatic hypochondriacal psychosis (delusional disorder, somatic type in DSMIV) and pathological jealousy (Munro 1984); however, it is doubtful whether pimozide is more effective than other nonsedating antipsychotic drugs.

A common reason for failure of drug treatment is that patients do not take their medication because they suspect that it will harm them. It may then be necessary to prescribe a long-acting preparation such as fluphenazine decanoate. In some patients the dosage can be reduced or stopped later without ill effects, whilst in others it must be maintained for long periods of time. This issue can be judged only by trial and error.

Further reading

Hirsch, S. R. and Shepherd, M. (ed.) (1974). *Themes and variations in European psychiatry*. John Wright, Bristol. See following sections: E. Strömgren, Psychogenic psychoses; R. Gaupp, The scientific significance of the case of Ernst Wagner; and The illness and death of the paranoid mass murderer schoolmaster Wagner: a case history; E. Kretschmer, The sensitive delusion of reference; H. Baruk, Delusions of passion; H. Ey, P. Barnard, and C. Brisset, Acute delusional psychoses (*Bouffées délirantes*).

Lewis, A. (1970). Paranoia and paranoid: a historical perspective. *Psychological Medicine*, 1, 2–12.

McKenna, P. J. (1994). *Schizophrenia and related syndromes*. Oxford University Press, Oxford.

11 | *Delirium, dementia, and other cognitive disorders*

The term 'organic psychiatry' has been applied to a group of topics that are related only loosely one to another. The term is used most often to denote psychiatric disorders that arise from demonstrable structural disease of the brain, such as brain tumours, injuries, or degenerations. It is also applied to psychiatric disorders arising from brain dysfunction that is clearly caused by disease outside the brain, such as myxoedema. By convention, the term also includes epilepsy, which is sometimes, but not always, associated with psychiatric disorder; however, it excludes mental retardation, even though the latter is sometimes associated with demonstrable brain disease.

The term 'organic disorders' is increasingly thought to be unsatisfactory because many so-called 'functional' disorders have a neuro-biological basis. Although the term organic disorder is retained in ICD10, this book mainly uses the terminology of DSMIV, namely *delirium, dementia, amnestic and other cognitive disorders*. Although this new term is cumbersome and not without problems, it has a more precise meaning than the term organic disorder.

The syndromes considered in this chapter can be subdivided on the basis of three criteria. The first criterion is whether the impairment of psychological functioning is *generalized or specific*. Generalized impairment affects cognition, mood, and behaviour globally. Specific impairment affects just one or two functions, such as memory, thinking, perception, or mood. The second criterion is whether the syndrome is *acute or chronic*. The third criterion is whether the underlying *dysfunction of the brain is generalized or focal*. Generalized dysfunction of the brain may result, for example, from raised intracranial pressure; focal dysfunction may arise from a localized tumour, for example in the temporal lobe (though such a lesion may also cause generalized dysfunction).

In this chapter the following division is employed: acute generalized impairment; chronic generalized impairment; specific impairment. (These groups are based on the first two criteria above.)

Acute generalized psychological impairment is usually referred to as *delirium*. The most important clinical feature is impairment of consciousness. The underlying brain dysfunction is generalized, and the primary cause is often outside the brain, for example, anoxia due to respiratory failure.

Chronic generalized psychological impairment is referred to as *dementia*. The main clinical feature is generalized intellectual impairment, but there are also changes in mood and behaviour. The underlying brain dysfunction is generalized. The primary cause is usually within the brain, and is often a degenerative condition such as Alzheimer's disease.

Specific psychological impairment may take the form of a specific impairment of memory (the *amnesic syndrome*), thinking, perception, or mood. It may also take the form of personality change or a schizophrenia-like syndrome. In some of these conditions, but not all, focal lesions in the brain can be demonstrated.

These three syndromes are described in the first part of this chapter. Some primary neurological conditions that may cause organic mental states are described in the second part. General medical conditions (such as HIV infection and endocrine disorders) that are sometimes included in an account of organic psychiatric disorders because they may cause secondary or symptomatic cognitive impairment are described in Chapter 12 on psychiatry and medicine.

Table 11.1 *Classification of organic mental disorders*

DSMIV	ICD10
Delirium, dementia, amnestic and other cognitive disorders	**Organic, including symptomatic mental disorders**
Dementia	
Dementia of the Alzheimer's type with early onset	Dementia in Alzheimer's disease with early onset
Dementia of the Alzheimer's type with late onset	Dementia in Alzheimer's disease with late onset
Vascular dementia	*Vascular dementia*
	Multi-infarct dementia
	Subcortical vascular dementia
	Dementia in other diseases, classified elsewhere
Dementia due to HIV disease	HIV disease
Dementia due to head trauma	
Dementia due to Parkinson's disease	Parkinson's disease
Dementia due to Huntington's disease	Huntington's disease
Dementia due to Pick's disease	Pick's disease
Dementia due to Creutzfeldt–Jakob disease	Creutzfeldt–Jakob disease
Amnesic disorders	*Organic amnesic syndrome, not induced by*
Amnesic disorder due to general medical condition	*alcohol or other psychoactive substance*
Substance induced persisting amnesic disorder	
Delirium	*Delirium, not induced by alcohol and other*
Delirium due to a general medical condition	*psychoactive drugs*
Substance-induced delirium	*Other mental disorders due to brain damage*
Substance-withdrawal delirium	*and dysfunction and to physical disease*
Delirium due to multiple aetiologies	Organic hallucinosis
	Organic delusional (schizophrenia-like) disorder
	Organic mood (affective) disorders
	Organic anxiety disorder
	Personality and behavioural disorders due to brain disease, damage, and dysfunction
	Organic personality disorder
	Postencephalitic syndrome
	Postconcussional syndrome

Classification of 'organic' psychiatric syndromes

Although, as explained above, the general term for these syndromes differs in the two classifications, they are classified similarly. In ICD10 the term is *organic, including symptomatic mental disorders*, and in DSMIV it is *delirium, dementia, amnesic and other cognitive disorders*. The two classifications are compared in Table 11.1. The main differences are as follows.

1. The decision by the authors of DSMIV to omit the word organic from the section title has led to rearrangement of the classification of some conditions formerly grouped under the heading organic. Thus major depression, of whatever aetiology, is classified under mood disorders, where those with organic aetiology can be classified as secondary to a general medical condition, or substance induced. As a result of these changes, DSMIV avoids the problems within ICD10 of the definition of the terms organic, symptomatic and secondary (Spitzer *et al.*, 1992; Tucker *et al.* 1994).

2. In both classifications the specific medical conditions causing cognitive disorder can be coded in addition to the latter order. In DSMIV, this additional code is recorded on Axis III. In ICD10, the section on organic disorders includes a category for mental disorders due to brain damage and dysfunction and to physical disease, and another for personality and behavioural factors due to brain disease, damage, and dysfunction. In DSMIV these conditions are not classified under delirium, dementia, and amnesic disorders but under the relevant psychiatric disorder with the addition of a code to indicate that the disorder is secondary to a medical condition.

3. DSMIV includes under delirium, dementia and amnesic disorders, a category for *substance-induced delirium, substance-induced dementia, and substance-induced amnestic syndrome*. In ICD10 these conditions are recorded in the section on *mental and behavioural disorders due to psychoactive substance abuse*. (In this book, these disorders are described in Chapter 14.)

Delirium

Delirium is characterized by impairment of consciousness. It is a common accompaniment of physical illness, occurring in about 5–15 per cent of patients in general medical or surgical wards, and in about 20–30 per cent of patients in surgical intensive care units (Lipowski 1990). Most cases recover quickly; therefore, despite their frequency, few are seen by psychiatrists.

Historical background

In the past, the word *delirium* was used in two ways. Until the early nineteenth century it was generally used to denote a disorder of thinking. Later, especially in the English literature, it was used to denote an organic brain disorder associated with impaired consciousness (Berrios 1981*b*; Lipowski 1990). In French psychiatry the term *confusion mentale* was used to denote the core features of inability to think coherently, impaired perception, and defective memory.

In 1909 Karl Bonhoeffer, professor of psychiatry in Berlin, defined delirium as the stereotyped manifestation of acute brain failure. He proposed several different 'exogenous reactions', i.e. distinct psychiatric syndromes resulting from the effects of agents or disorders outside the brain. These proposed reactions included delirium, hallucinosis, epileptic excitement, twilight state, and amentia (which in Bonhoeffer's scheme meant a syndrome of incoherent thinking). Since the early part of the twentieth century, the last three syndromes have been discarded; the first two have been retained, but are now applied to the effects of cerebral as well as extracerebral disorders (Cooper 1986).

The clinical features of delirium were well described by Wolff and Curran (1935), and more detailed accounts of the clinical and psychological features and of associated EEG abnormalities were provided by Engel and Romano (1959). Until DSMIII the word *delirium* was used in a variety of ways, but in both DSMIV and ICD10 it is used as a general term for the acute cognitive impairment syndrome (see Liptzin *et al.* (1994) for a review of this classification in DSMIV).

The term *confusional state* has also been applied to acute organic psychiatric syndromes. This term is unsatisfactory because the word confusion properly refers to muddled thinking. The latter is an important symptom of acute organic disorders, but is not confined to them.

Clinical features

The most important feature is impairment of consciousness (defined in DSMIV as reduced clarity of awareness of the environment), though it is not always the most obvious. It often varies in intensity through the day and is usually worse at night. It is recognized by slowness, poor concentration, and uncertainty about the time of day. Lishman (1987) has summarized the main features as 'slight impairment of thinking, attending, perceiving, and remembering, in other words as mild global impairment of cognitive processes in association with reduced awareness of the environment'. Occasionally, it is difficult to establish whether consciousness is impaired. After recovery, memory is poor for the period of impaired consciousness, a point which may allow retrospective diagnosis in previously doubtful cases.

Apart from impaired consciousness, the other features vary widely between different patients, and in the same patient at different times. The features are often influenced by the patient's personality; for example ideas of persecution are more likely in a person who is habitually suspicious and touchy. Lipowski (1980) distinguished two patterns of presentation: in the first, the patient is restless and oversensitive to stimuli, and has psychotic symptoms; in the second, he is lethargic and quiet, and has few psychotic symptoms.

The patient's **behaviour**, as Lipowski's distinction implies, may take the form of over-activity, irritability, and noisiness, or else of inactivity, slowness, reduced speech, and perseveration. In either case, repetitive purpose-less movements are common.

Thinking is slow and muddled, but is often rich in content. Ideas of reference and delusions (often persecutory) are common, but are usually transient and poorly elaborated.

Visual perception may be distorted. Illusions, misinterpretations, and visual hallucinations are frequent, and may have a fantastic content. Tactile and auditory hallucinations also occur.

Changes in **mood** such as anxiety, depression, or lability are common. Some patients are frightened and agitated, whilst others are perplexed. Experiences of depersonalization and derealization are also reported by some patients. **Disorientation** in time and place is an invariable and important feature. Disturbance of **memory** affects registration, retention, and recall, and new learning is impaired. As mentioned above, on recovery there is usually amnesia for most of the illness. **Insight** is impaired. (See Cutting (1987) and Liptzin *et al.* (1994) for a description of the clinical features of delirium.)

Aetiology

The main causes of delirium are shown in Table 11.2. The condition appears particularly in association with increasing age, anxiety, sensory under- or over-stimulation, drug dependence, and brain damage of any kind. Wolff and Curran (1935) showed that the clinical picture, as noted above, is much influenced by previous experience and personality. Engel and Romano (1959) showed that the severity of the clinical state was related to the degree of impairment of brain function as reflected by slowed rhythms in the EEG. (See Trzepacz (1994) for a review of recent research on neuropathogenesis.)

The assessment, diagnosis, and treatment of acute organic psychiatric syndromes are described later in this chapter.

Dementia

Dementia is a generalized impairment of intellect, memory, and personality, with no impairment of consciousness. It is an acquired disorder, as distinct from amentia which is present from birth. Although most cases of dementia are irreversible, a small but important group are remediable. (See Berrios (1994) for an account of

Table 11.2 *Some causes of delirium*

---

1. Drug intoxication
   Anticholinergics, anxiolytic-hypnotics, anticonvulsants, digitalis, opiates, L-dopa,
   also some industrial poisons

---

2. Withdrawal of alcohol and anxiolytic sedatives

---

3. Metabolic failures
   Uraemia, liver failure, respiratory failure, cardiac failure, disorders of electrolyte
   balance

---

4. Endocrine causes: hypoglycaemia

---

5. Systemic infection
   Exanthemata, septicaemia, pneumonia

---

6. Intracranial infection
   Encephalitis, meningitis, HIV infection, cerebral malaria

---

7. Other intracranial causes
   Space-occupying lesions, raised intracranial pressure

---

8. Head injury

---

9. Nutritional and vitamin deficiency
   Thiamine, nicotinic acid, vitamin B_{12}

---

10. Epileptic
    Epileptic status, post-ictal states

---

the history of the concept of dementia, Rebok and Folstein (1994) for a review of classification in DSMIV, and the book edited by Burns and Levy (1994) for reviews of all aspects of dementia.)

Clinical features

Dementia usually presents with impairment of memory. Other features include change in personality, mood disorder, hallucinations, and delusions. Though dementia generally develops gradually, if often comes to notice after an exacerbation caused by either a change in social circumstances or an intercurrent illness.

Again, the clinical picture is much determined by the patient's premorbid personality. For example neurotic traits become exaggerated in some patients. People with good social skills may maintain a social facade despite severe intellectual deterioration, whilst those who are socially isolated or deaf are less likely to compensate for failing intellectual abilities.

Behaviour is often disorganized, inappropriate, distractible, and restless. There are few signs of interest or initiative. Changes in personality may manifest as antisocial behaviour, which sometimes includes sexual disinhibition or shoplifting. In middle-aged or elderly people, any social lapse that is out of character should always suggest dementia.

Goldstein (1975) described the ways in which behaviour can be affected by the cognitive defects. Typically, there is a reduction of interests ('shrinkage of the milieu'), rigid and stereotyped routines ('organic orderliness'), and, when the person is taxed beyond restricted abilities, a sudden explosion of anger or other emotion ('catastrophic reaction').

As dementia worsens patients care for themselves less well and neglect social conventions. Behaviour becomes aimless, and stereotypes and mannerisms may appear. Eventually, the patient becomes disorientated, incoherent, and incontinent of urine and faeces.

Thinking slows and becomes impoverished in content. There may be concrete thinking, reduced flexibility, and perseveration. Judgement is impaired. False ideas, often of a persecutory kind, gain ground easily. In the later stages, thinking becomes grossly fragmented and incoherent. Disturbed thinking is reflected in the quality of **speech**, in which syntactical errors and nominal dysphasia are common. Eventually, the patient may utter only meaningless noises or become mute.

In the early stages, change of **mood** may include anxiety, irritability, and depression. As dementia progresses, emotions and responses to events become blunted, and sudden mood changes may occur without apparent cause.

Disorders of **cognitive function** are salient features. Forgetfulness is usually early and prominent, but may sometimes be difficult to detect in the early stages. Difficulty in new learning is generally the most conspicuous sign. Memory loss is more obvious for recent than for remote events. Patients often make excuses to hide these memory defects, and some confabulate. Other cognitive defects include impaired attention and concentration. Disorientation for time, and at a later stage for place and person, is almost invariable once dementia is well established.

Insight into the degree and nature of the disorder is lacking.

Subcortical dementia

In 1974, Albert and his colleagues introduced the term subcortical dementia to denote intellectual deterioration seen in progressive supranuclear palsy. The meaning of the term has been extended to cover a syndrome of slowing of cognition, difficulty with complex intellectual tasks, and affective disturbance without impairment of language, calculation, or learning. Possible causes of subcortical dementia include Huntington's chorea, Parkinson's disease, progressive supranuclear palsy, hydrocephalus, and AIDS dementia complex. In contrast, Alzheimer's disease is usually regarded as an example of cortical dementia. So far, a clear distinction between the two forms of dementia has not been convincingly established (Dunne 1993).

Aetiology

Dementia has many causes, of which the most important are listed in Table 11.3. The aetiology of dementia in the elderly is discussed separately in Chapter 16. Among elderly patients, degenerative and vascular causes predominate, but no subgroups predominate at other ages. Therefore the clinician should keep in mind the whole range of causes when assessing a patient, and should take care not to miss any cause that might be partly or wholly arrested by treatment, such as an operable cerebral neoplasm, cerebral syphilis, or normal pressure hydrocephalus.

Several of the conditions listed in Table 11.3 are described in detail later in this chapter. Alzheimer's disease is described in the chapter on psychiatry of the elderly (p. 52). The assessment, diagnosis, and treatment of dementia are also discussed later in this chapter. So far, two main generalized organic psychiatric syndromes (acute and chronic) have been described. An account will now be given of the third group of syndromes, in which there is impairment of specific psychological functions.

Table 11.3   *Some causes of dementia*

Degenerative	Alzheimer's disease, Pick's disease and other frontal dementias, Huntington's chorea, Parkinson's disease, normal pressure hydrocephalus, multiple sclerosis*
Intracranial space-occupying lesions	Tumour, subdural haematoma
Traumatic	Severe single head injuries, repeated head injury in boxers and others
Infections and related conditions	Encephalitis of any cause, neurosyphilis, cerebral sarcoidosis, HIV, prion diseases
Vascular	Vascular dementia, occlusion of the carotid artery, cranial arteritis
Metabolic	Sustained uraemia, liver failure, remote effects of carcinoma or lymphoma, renal dialysis
Toxic	Alcohol, poisoning with heavy metal (lead, arsenic, thallium)
Anoxia	Anaemia, post-anaesthesia, carbon monoxide, cardiac arrest, chronic respiratory failure
Vitamin lack	Sustained lack of vitamin B_{12}, folic acid, thiamine

*Cause uncertain, see p. 333.

Cognitive impairment syndromes with specific psychological dysfunctions

In this group, as explained earlier, psychological impairment is partial rather than general, i.e. a limited number of specific functions are affected, such as memory, thinking, perception, or mood. Personality disorder is another highly important complication (described later in this chapter).

In some of these conditions, but not all, focal lesions in the brain are demonstrable. The amnesic syndrome and other focal brain damage are discussed in this section, followed by the symptomatic and secondary syndromes.

Amnestic disorder

Amnestic disorder (DSMIV) or amnesic syndrome (ICD10) is characterized by a prominent disorder of recent memory and by disordered time sense, in the absence of generalized intellectual impairment. The psychological disorder has been reviewed by Lishman (1987). The condition usually results from lesions in the posterior hypothalamus and nearby midline structures, but occasionally it is due to bilateral hippocampal lesions.

The Russian neuropsychiatrist Korsakov described a chronic syndrome in which memory deficit was accompanied by confabulation and irritability (Korsakov 1889). His patients also suffered from peripheral neuropathy. They either abused alcohol or developed the syndrome in association with puerperal sepsis or an infection causing persistent vomiting. Therefore it is likely that they were suffering from thiamine deficiency. Nowadays, peripheral neuropathy is not regarded as an essential feature of the amnesic syndrome, and vitamin deficiency is not regarded as the only cause.

The term Korsakov's syndrome has been used in more than one way, sometimes to denote a combination of symptoms, and at other times to denote pathology as well as symptoms. Now the term usually implies impairment of memory and learning out of proportion to impairment of other cognitive functions. Confabulation may be present but is not an essential feature.

The term Wernicke–Korsakov syndrome is also used, as suggested by Victor *et al.* (1971), because the chronic amnesic syndrome often follows an acute neurological syndrome described by Wernicke in 1881. The main features of this acute syndrome are impairment of consciousness, memory defect, disorientation, ataxia, and ophthalmoplegia. Neuropsychiatric studies have reported prevalences of 1–2 per cent (Harper *et al.* 1989). At post-mortem examination Wernicke found haemorrhagic lesions in the grey matter around the third and fourth ventricles and the aqueduct. More recent investigation has shown that lesions occur in these same anatomical sites in both the acute Wernicke syndrome and the chronic Korsakov syndrome.

In DSMIV, amnestic disorder is defined by inability to learn new information and the inability to recall previously learned knowledge or past events, not occurring during the course of delirium or dementia. The definition also requires 'significant impairment in social or occupational functioning', and evidence of a general medical condition 'judged to be aetiologically related to the memory impairment' (Caine 1994). The definition of amnesic syndrome in ICD10 is similar, but does not include social criteria (see Kopelman 1995 for a review of the syndrome).

Clinical features

The central feature of amnestic disorder is a profound impairment of recent memory. The patient can recall events immediately after they occur, but cannot do so a few minutes or hours afterwards. Thus on a test of digit span, recall is good in the first few seconds but is impaired 10 minutes later. New learning is grossly defective, but remote memory is relatively preserved. There is some evidence that the disorder may not be entirely an inability to lay down memories, but may also be a failure to recall established memories—possibly because of interference from irrelevant memories (Warrington and Weiskrantz 1970). One consequence of this profound disorder of memory is an associated disorientation in time.

Gaps in memory are often filled by confabulating. The patient may give a vivid and detailed account of recent activities, all of which, on checking, turn out to be inaccurate. It is as though he cannot distinguish between true memories and the products of his imagination or the recollection of events from times other than those that he is trying to recall. Such a patient is often suggestible; in response to a few cues from the interviewer, he may give an elaborate account of taking part in events that never happened. Confabulation is not a feature of the amnesic syndrome associated with bilateral hippocampal lesions.

Other cognitive functions are relatively well preserved. The patient seems alert and able to reason or hold an ordinary conversation, so that the interviewer is often surprised when the extent of the memory disorder is revealed. However, the disorder is not limited entirely to memory; some emotional blunting and lack of volition are often observed as well.

Aetiology and pathology

Alcohol abuse, the most frequent cause, seems to act by causing a deficiency of thiamine. Several other causes also seem to act through thiamine deficiency, for example gastric carcinoma and severe dietary deficiency. As mentioned above, Korsakov described cases due to persistent vomiting in puerperal sepsis and typhoid fever, but these diseases are rarely seen today. At post-mortem, such cases generally have haemorrhagic lesions in the mamilliary bodies, the region of the third ventricle, the periaqueductal grey matter, and parts of certain thalamic nuclei or cortical damage. The mamilliary bodies or the medial dorsal nucleus of the thalamus (Victor 1964) are the structures most often involved.

Other causes involve the brain directly and not through thiamine deficiency. The brain areas listed above may be damaged by vascular lesions, carbon monoxide poisoning, or encephalitis; and by tumours in the third ventricle. Another cause is bilateral hippocampal damage due to surgery. When the syndrome is due to causes other than thiamine deficiency, patients seem less likely to show confabulation and more likely to retain insight into the memory disorder.

Course and prognosis

Victor *et al.* (1971) studied 245 patients who had developed an acute Wernicke–Korsakov syndrome, most of whom had histories of many years of alcohol abuse. There was a 17 per cent death rate in the acute stage. All except 4 per cent of cases presented with Wernicke's encephalopathy. Of those who were followed-up, 84 per cent developed a typical amnesic syndrome. Once established there was no improvement in a half, complete recovery in a quarter, and partial recovery in the rest. The best predictors of a better prognosis are a short history before diagnosis, and little delay between diagnosis and the start of a replacement treatment in cases due to thiamine deficiency.

Rarely, an improvement occurs in cases due to causes other than alcoholism, for example carbon monoxide poisoning or thiamine deficiency due to simple malnutrition. Sometimes the amnesia is progressive, as in cases with slowly expanding brain tumours.

Other psychiatric syndromes due to focal brain damage

An account will now be given of other 'focal' syndromes that are relevant to psychiatry. The many forms of dysphasia, agnosia, and dyspraxia will not be described, as they are part of neurology; they are to be found in the textbook on organic psychiatry by Lishman (1987) or one of the standard textbooks of neurology. (See Tucker (1994) for a discussion of classification in DSMIV.)

Persistent denial

Persistent denial of a deficit or function (such as hemiplegia) or of the handicaps resulting from such a deficit is not uncommon (House and Hodges 1988; Hodges 1994). Such denial may have a specific neurological basis (anosognosia) or may be a psychological unwillingness to recognize disability and its consequences. Management requires a systematic assessment of possible causes of denial, followed by a combination of physical rehabilitation and psychological treatment to overcome unwillingness to accept disability (Langer and Padrone 1992).

Frontal lobe syndrome

Frontal lobe damage has distinctive effects on temperament and behaviour which are generally referred to as personality change. In **behaviour** the patient is disinhibited, over-familiar, tactless, and over-talkative. He makes jokes and engages in pranks (a feature sometimes referred to in the literature by the German word *Witzelsucht*). He may make errors of judgement, commit sexual indiscretions, and disregard the feelings of others. The **mood** is generally one of fatuous euphoria. **Concentration** and **attention** are reduced. Measures of formal intelligence are generally unimpaired, but special testing may show deficits in abstract reasoning. **Insight** is impaired.

Encroachment of a frontal lobe lesion on the motor cortex or deep projections may result in contralateral spastic paresis or dysphasia. Other possible signs are optic atrophy on the same side as the frontal lobe lesion, anosmia, a grasp reflex, and, if the lesion is bilateral, incontinence of urine (see Ron (1989) for further information about the psychiatric manifestations of frontal lobe tumours).

Parietal lobe

Compared with lesions of the frontal or temporal lobe, lesions of the parietal lobe are less likely to induce psychiatric changes (Lishman 1987), but they do cause various neuropsychological disturbances which are easily mistaken for hysteria. Lesions of the non-dominant parietal lobe cause visuospatial difficulties. Lesions of the dominant lobe are associated with dysphasia, motor and dressing apraxias, right–left disorientation, finger agnosia, agraphia, and body image disorders (see p. 20). These clinical features present in various combinations, some of which are designated as syndromes (see a textbook of neurology). If these conditions are not to be misdiagnosed, thorough neurological assessment is required. Important signs may include cortical sensory loss, sensory inattention, and agraphaesthesia. There may also be evidence of a mild contralateral hemiparesis.

Temporal lobe

Although some temporal lobe lesions are asymptomatic, there is usually impairment of intellectual function, especially with a lesion on the dominant side. There may be personality change resembling that of frontal lobe lesions, though more often accompanied by intellectual deficits and neurological signs. With chronic temporal lobe lesions another kind of personality change is characterized by emotional instability and aggressive behaviour.

Temporal lobe lesions may cause epilepsy, and also an increased risk of a schizophrenia-like psychosis (see p. 341). Unilateral temporal lobe lesions produce specific learning impairments (in right-handed people verbal on the left, non-verbal on the right). Rare bilateral lesions of medial temporal lobe structures can produce an amnesic syndrome. An important neurological sign of a deep temporal lobe lesion is a contralateral homonymous upper quandrantic visual field defect due to interference with the visual radiation. Sometimes a deep lesion causes a mild contralateral hemiparesis. Dominant lesions may produce language difficulties.

Occipital lobe

Occipital lobe lesions may cause complex disturbances of visual recognition which are easily misdiagnosed as hysterical. Complex visual hallucinations can also occur and may be mistaken for signs of non-organic mental illness. The visual fields should be examined thoroughly and tests carried out for visual agnosias.

The corpus callosum

Corpus callosum lesions typically extend laterally into both hemispheres. They then produce a picture of severe and rapid intellectual deterioration, with localized neurological signs varying with the degree of extension into the frontal or occipital lobes or the diencephalon.

Diencephalon and brainstem

With lesions of midline structures, the most characteristic features are the amnesic syndrome, hypersomnia, and the syndrome of 'akinetic mutism'. There may also be progressive intellectual deterioration, emotional lability with euphoria and abrupt outbursts of temper, excessive eating, and endocrine signs of pituitary disorder.

The assessment of suspected cognitive impairment disorder

Any suspicion of a cognitive impairment disorder should lead to detailed questioning about intellectual function and neurological symptoms. It is particularly important to interview other informants. The mode of onset and progression of symptoms should be determined in detail. An appropriate physical examination is essential.

Special investigations

With every patient the psychiatrist should use his judgement about the extent of the special investigations required. The aim should be to perform the minimum of investigations that will allow accurate diagnosis. A common basic routine is haemoglobin, erythrocyte sedimentation rate, and automated blood biochemistry. Experience is needed to interpret the results. If, on clinical grounds, there is the least suspicion of physical disorder or if any of the screening tests is abnormal, the clinician should judge what further investigations are required. They are likely to include mid-stream urine (MSU) microscopy, urine analysis for drugs and, in some cases, liver function tests, serum calcium and phosphate, thyroid function (serum T4, T3, TSH), serum B12, and red cell folate. HIV testing is not normally routine for ethical reasons, but in the presence of specific risk factors or other indications it should be considered. Serology for syphilis should be arranged in certain cases. Chest and lateral skull radiography should also be considered. Further skull views may be valuable. Possible findings include abnormalities of the vault, intracranial calcification, changes in the sella turcica, the shift of a calcified pineal, and unsuspected fractures.

Further investigations

A review of the clinical findings and the results of this first round of investigations will usually indicate whether cognitive impairment disorder can be excluded or special investigations are needed. The latter may include CT scan, MRI scan, SPECT scan, and further laboratory investigations. Psychological tests may be required at this stage. Some of these investigations will now be considered in more detail.

Neuro-imaging

Computerized tomography ('CT scanning') has a key role in the diagnosis of both focal and diffuse cerebral pathology. CT scans are requested if there is any suspicion of organic brain disease in patients up to late middle age, or if there is any suggestion of a focal brain lesion in the elderly (radiological investigation of the elderly is discussed further in Chapter 16). Serial medial temporal views may be useful in the diagnosis of Alzheimer's disease. Burns and Pearlson (1994) have reviewed the value of CT scanning in dementia.

Magnetic resonance imaging (MRI) has specific applications to the study of cognitive impairment disorders, for instance possible posterior fossa lesions. *Single-photon-emission computed tomography* (SPECT) enables three-dimensional images of regional cerebral blood flow to be obtained. It may be useful in the diagnosis of dementia and in the differential diagnosis of other disorders (Geaney 1994).

Electroencephalography

EEG has an important role, limited to the diagnosis of the epilepsies. Although it is sometimes suggested as being useful in the diagnosis of psychotic mental states, this is true only when there is a history of epilepsy.

Further physical investigation

Psychiatrists who have the appropriate neurological skills may go on to perform a lumbar puncture or order further special imaging procedures. However, it is usually more appropriate to seek the opinion of a neurologist or general physician before performing more specialized investigations.

Psychological testing

Psychometric tests depend on the patient's cooperation but can be valuable when given by an experienced tester. They may help in localizing lesions in certain sites, for example the parietal lobe. Even when interpreted skilfully, they discriminate poorly between organic and functional disorders. They are of more value in monitoring changes in psychological functioning over time, and in assessing patterns of disability as a basis for planning rehabilitation. Some of the most frequently used tests will now be considered briefly. (For further information see Lishman (1987) and Hodges (1994)).

1. *Wechsler Adult Intelligence Scale (WAIS)* This is a well-standardized text providing a profile of verbal and non-verbal abilities. Analysis of

subscores can provide useful information for diagnosis. It is often said that organic impairment is indicated by a discrepancy between performance IQ (as an estimate of current capacity) and verbal IQ (as an estimate of previous capacity), but there is no strong evidence to support this view. Usually, the more specific tests mentioned below are more helpful in diagnosis, but the WAIS is useful for screening.

2. *Perceptual functions, especially spatial relationships* This kind of test is exemplified by the Benton Revised Visual Retention Test, which requires the patient to study and reproduce ten designs. Parallel versions of the test are available, thus allowing serial testing.

3. *New learning as a test of memory* There are many new word learning tasks, for example the Walton–Black Modified Word Learning Test and the Paired Associated Learning Test, both of which give a useful quantitative estimate of memory impairment.

4. *Specific tests* Examples of specific tests are the Wisconsin Card Sorting Test for frontal lobe damage, and the Token Test for receptive language disturbance.

5. *Standardized mental state schedules* There are numerous mental state schedules, ranging in complexity from the ten-item Hodkinson Mental Test to lengthy research instruments (see also p. 53). Two of the most commonly used brief schedules are the Mini-Mental State Examination (MMSE) and the Hodkinson Mental Test. An example of a longer research procedure is the Cambridge Cognitive Examination (CAMCOG) which is part of a standardized assessment schedule (CAMDEX) designed specifically for the use with elderly people with the diagnosis of dementia.

Aspects of differential diagnosis

Symptomatic or secondary syndromes

A variety of mental disorders, as well as changes in personality, perception, and mood, may be attributable to brain dysfunction due to either cerebral disease or systemic disease secondarily affecting the brain. Examples include organic delusional syndromes (Cornelius *et al.* 1991), including the schizophrenia-like delusional syndrome reported to be associated with epilepsy (see p. 341). There has been particular interest in organic mood disorder, and Winokur *et al.* (1988) have proposed the term 'secondary depression' to describe depressive disorders secondary to medical conditions (or to other psychiatric disorder).

Awareness of the co-morbidity between psychiatric and physical disorders is reflected in the systems of classification (Spitzer *et al.* 1992). These conditions were first brought together in DSMIII, and the criteria for diagnosis have been developed further in DSMIV and ICD10. They are classified differently in the two systems. In ICD10 they are included in the *organic mental disorders* section, whereas in DSMIV they are classified under the relevant phenomenological grouping (for example mood disorder) and the organic aetiology is indicated by the use of an extra digit (see p. 310).

Since these syndromes do not have characteristic features enabling them to be identified as having a cerebral aetiology, the diagnosis depends on other features. In ICD10 the clinical guidelines are as follows.

(a) *evidence of cerebral disease, damage or dysfunction, or of systematic physical disease*, known to be associated with one of the listed syndromes;

(b) *temporal relationship* (weeks or a few months) between the development of the underlying disease and the onset of the mental syndrome;

(c) *recovery* from the mental disorder following removal or improvement of the underlying presumed cause;

(d) *absence of* evidence suggesting an *alternative cause* of the mental syndrome (such as a strong family history or precipitated and stress).

Although there can be no doubt that symptomatic or secondary psychiatric disorders occur in relation to brain and systemic diseases, the aetiological mechanisms are uncertain. Even where there is convincing evidence of an association with such diseases, psychological and social factors usually play a part in the aetiology of psychiatric disorder. (See Popkin and Tucker (1994) for a review of classification in DSMIV.)

Cognitive impairment or functional disorder?

Usually, there is little difficulty in distinguishing between cognitive impairment and functional disorders, but occasionally one may be mistaken for the other. Thus a cognitive disorder may sometimes be misdiagnosed as a functional disorder if the patient has abnormalities of personality that modify the clinical presentation, for example by adding prominent depressive or paranoid features. Conversely, a functional disorder may be misdiagnosed as organic if there is apparent cognitive impairment, for example a patient with a depressive disorder may complain of poor memory and 'confusion'. Of the two kinds of misdiagnosis, missing a cognitive disorder is of course more serious. Therefore, it is vital that the psychiatrist should be constantly vigilant to the possibility of a cognitive disorder. The first requirement is to take a full history and make a thorough examination of the physical and mental state. Certain features call for alertness. For example complaints of physical symptoms should always be taken seriously, and suitable questions should be asked about their nature and time of onset. If psychiatric symptoms cannot be understood psychologically, enquiry should always be directed to a possible primary cognitive disorder. The first requirement is to take a full history and make a thorough examination of the physical and mental state.

In making the distinction between cognitive and functional disorders, several principles should be borne in mind. The first concerns three modes of presentation—as a conversion or dissociative disorder, as episodic disturbed behaviour, and as depression. Conversion or dissociative disorder should not be diagnosed unless there is an adequate psychological explanation, and unless every symptom has been adequately investigated. This principle holds even for patients with a previous history of one of these disorders. Brain disease may present with symptoms that resemble those of a conversion or dissociative disorder, for example with the parietal lobe lesions described above. Also, conversion or dissociative symptoms (such as global amnesia) may be 'released' by organic brain disease (see p. 193).

Unexplained episodes of disturbed behaviour in a patient with a previously stable personality suggest brain disease. Possible causes include epilepsy, early dementia, transient global amnesia, and extracerebral conditions such as hypoglycaemia, porphyria, or other metabolic disorders.

Finally, it should always be borne in mind that depressed mood may be the first manifestation of organic brain disease. Certain symptoms need to be analysed with particular care. For example, muscular weakness must be differentiated from the psychological experience of 'feeling weak', a distinction that is important in the diagnosis of myasthenia gravis. Another principle is that certain symptoms should arouse suspicion of a cognitive lesion, for example visual hallucinations, or complaints of 'confusion', or any complaint that would be unusual in a functional disorder, such as ataxia and incontinence.

As mentioned above, a functional disorder may be misdiagnosed as organic if there is apparent cognitive impairment. The term 'pseudo-dementia' is applied to patients who have a functional disorder and show intellectual impairment resembling that of cognitive disease. Pseudo-dementia is most common in elderly depressed patients (see p. 527). In differentiating pseudo-dementia from true dementia, it is important to know which symptoms developed first, since in functional disorders other psychological symptoms precede the apparent intellectual defects. Hence it is important to interview other informants to determine the precise mode of onset. (See Kopelman (1987) for a review of this topic.)

Acute or chronic?

In making the differential diagnosis, it is sometimes difficult to distinguish between an acute and a chronic cognitive psychiatric syndrome. This difficulty usually arises because a clear history is lacking. It should be remembered that an acute syndrome may be superimposed on a long-standing dementia; such an event may obscure the diagnosis, or alternatively may draw attention to the underlying chronic disorder. The characteristic features of the acute and chronic syndromes have been described above. In distinguishing the acute syndrome, the most helpful features are impairment of consciousness, perceptual abnormalities, disturbed attention, poor sleep, and thinking that is disorganized but rich in content.

Differential diagnosis of stupor

This condition, which is discussed on p.22, requires specific mention. The main psychiatric causes are severe depression, schizophrenia, and rarely hysteria and mania. Organic causes are relatively uncommon in cases seen in psychiatric practice (Johnson 1984). They include focal lesions in the posterior diencephalon or upper midbrain (for example tumours, especially craniopharyngiomas, infarction, meningitis, and epilepsy) and a number of extracerebral causes (for example uraemia, hypoglycaemia, electrolyte and fluid disturbance, endocrine disorder, and alcohol and drug intoxication). Diagnosis can usually be made on the history and examination. An EEG and CT scan can be helpful in distinguishing between organic and psychogenic causes. (See Lishman (1987) for a review of the differential diagnosis of stupor and Berrios (1981a) for a history of the concept of stupor.)

Diagnosis of the cause

A final aspect of differential diagnosis is to identify the cause of a cognitive psychiatric syndrome. If the cause is not readily apparent, the history and findings on physical and mental examination should be reviewed. Careful enquiry should be made about any history of head injuries, fits, alcohol or drug abuse, and recent physical illness. Dietary deficiency should be considered if the patient is elderly or of low intelligence. It is important to enquire about the symptoms of raised intracranial pressure (headaches, vomiting, and visual disturbance), as well as those suggesting a focal lesion in the brain. Physical examination should be directed towards signs of disease in other systems as well as the nervous system. Any appropriate investigations should then be arranged.

The management of delirium

The fundamental treatment is directed to the physical cause. General measures are necessary to relieve distress, and to prevent behaviour that might lead to accidents or other difficulties affecting the patient or other people. Amongst these general measures, the most important are to reduce disorientation, and to avoid too much or too little sensory stimulation.

Apart from good nursing care, the patient should be given repeated explanations of his condition. Disorientation and misinterpretation of the environment can be reduced by a calm and consistent approach, and by avoiding too many changes in the staff caring for the patient. If possible, relatives and friends should visit the patient frequently; it is good practice to explain the patient's condition to them, and to advise them how to reassure and orientate him. There are many advantages in nursing the patient in a quiet single room. At night, the room should have enough light to enable him to know where he is; however, it is desirable to avoid the high levels of illumination found in some intensive care units because it is important to ensure adequate sleep. (See Task Force of the American Psychiatric Association (1989) for a review.)

Drug treatment

In general, it is important to give as few drugs as possible and to avoid any that may increase impairment of consciousness. Nevertheless, medication often has an important role. Over-active, frightened, and disturbed patients may require medication to control distress and prevent accidents. There are two main requirements. First, during the daytime it may be necessary to calm the patient without inducing drowsiness. Second, at night it may be necessary to help him to sleep. For the first purpose (calming by day), the drug of choice is an antipsychotic drug such as haloperidol, which calms without causing drowsiness, hypotension, or cardiac side-effects (Chapter 17). The effective daily dose usually varies between 3 and 20 mg. If necessary, the first dose of 2–5 mg can be given intramuscularly. It is essential to start with low dosages in those who may have brain damage. Chlorpromazine and other phenothiazines are also widely used, but their usefulness is limited by their side-effects such as hypotension and sedation. Chlorpromazine should be avoided when there is liver disease, or when the organic state may be due to withdrawal from alcohol (since chlorpromazine can increase the risk of seizures). Although short-acting benzodiazepines may be appropriate at night to promote sleep, they should be avoided in the daytime because their sedative effects may make the patient more disoriented. However, they may be used by day if there is liver failure, since they are unlikely to provoke hepatic coma. Chlormethiazole is often used to treat alcohol withdrawal states; this should be done only under close supervision in hospital since, if further alcohol is taken, a dangerous interaction can occur (see p. 459).

The management of dementia

If possible, the cause should be treated. Otherwise management begins with an assessment of the degree of disability and the social circumstances of the patient. The plan of treatment should seek to improve functional ability as far as possible, relieve distressing symptoms, make practical provisions for the patient, and support the family.

Plans for long-term care should make clear the part to be played by the doctors, nurses, and social workers. This applies whether the patient is living in hospital or outside. For certain conditions there are some benefits from carrying out the first stage of rehabilitation in special units, for example dementia caused by head injury or by a stroke. It

should be remembered that personality change and restlessness at night cause particular difficulties for the family. It may be necessary to make legal arrangements for the administration of the patient's financial affairs.

Drug treatment

There is no established specific drug treatment for dementia. Medication is mainly used to alleviate certain symptoms, such as anxiety and agitation. For example, anxiety may be treated with a low dosage of haloperidol, a benzodiazepine, or a phenothiazine such as chlorpromazine or thioridazine. At night, a benzodiazepine or a sedating phenothiazine may be useful. Patients with cognitive impairment may be unusually sensitive to antipsychotic drugs, and so the first doses should be small. In particular, patients with Lewy body disease (p. 524) are said to be especially liable to develop extrapyramidal side-effects. If the patient is over-active, deluded, or hallucinated, a phenothiazine may be appropriate, but care is needed to find the optimal dose. If the patient has depressive symptoms, a trial of antidepressant medication is worthwhile even in the presence of dementia.

Behavioural methods

Much rehabilitation is based on the analysis of problems and the setting of goals. Behavioural methods share these principles, but add specific procedures for assessment and treatment. Behavioural procedures include positive reinforcement, shaping, desensitization and prompts and other practical methods for coping with memory deficit (Wilson 1989; Fisher and Carstensen 1990).

SPECIFIC PHYSICAL CONDITIONS GIVING RISE TO MENTAL DISORDERS

Primary dementia

Among the important causes of dementia are intrinsic degenerative diseases of the central nervous system presenting in middle or late life. The dementias arising from these diseases are sometimes called primary dementias. The most common examples are those occurring mainly in old age, namely Alzheimer's disease, multi-infarct dementia, and Lewy body dementia, which are discussed in Chapter 16.

The category of **presenile dementia** was introduced in 1894 by Binswanger and later included in Kraepelin's classification. Nowadays the term presenile is usually taken to mean younger than 65 years of age. It includes Alzheimer's disease of presenile onset, Pick's disease, and Huntington's chorea.

Twenty years ago, most cases of primary dementia were labelled as Alzheimer's disease. It is now clear that there are several types of primary dementia. Alzheimer's disease may not be a single entity, and its characteristic histopathology may be a final common pathway of different diseases.

Follow-up studies have shown that it is not easy to make an accurate diagnosis of presenile dementia. For example, Ron *et al.* (1979) studied 51 patients 5–15 years after a confident diagnosis of presenile dementia had been made. Follow-up information led to rejection of the original diagnosis in almost a third of the cases. Accurate diagnosis depends on careful history taking together with mental and physical examination, and increasingly on sophisticated neurological investigations. In doubtful cases, admission for observation is often informative. A 'functional' diagnosis is suggested by a history of affective disturbance, ability to learn in everyday activities or in psychological tests, and marked inconsistencies in performance. However, all these features may also occur in organic disorders.

Alzheimer's disease, Lewy body dementia, and vascular dementia

For clinical convenience these conditions are described in Chapter 16 on psychiatry of the elderly.

Pick's disease

This disease, which was described by Pick in 1892, is much less common than Alzheimer's disease. It appears to be inherited as an autosomal dominant (Sjögren *et al.* 1952). The gross pathology is circumscribed asymmetrical atrophy of the frontal or temporal lobes accompanied by a lesser degree of general atrophy. The gyri are said to have a characteristic brownish 'knife blade' atrophy. There is severe neuronal loss in the outer layers of the atrophic cortex with argentophilic inclusions (Pick bodies) and swollen chromatophilic neurones (Pick cells).

Although the onset may be at any adult age, most cases start between 50 and 60 years. Women are affected twice as often as men. There are no specific clinical features to separate Pick's disease from Alzheimer's disease, and the distinction is generally made at autopsy and not in life. None the less, it is sometimes said that the presenting symptoms of Pick's disease are changes in character and social behaviour more often than memory disturbance. There is loss of inhibition (sometimes affecting sexual behaviour), deterioration of conventional manners, and sometimes marked loss of drive.

These features correspond to the predominant pathology in the frontal lobes, and to a lesser extent the temporal lobes. Parietal lobe features and extrapyramidal symptoms are less common. A small proportion of cases are familial, but the aetiology of other cases is obscure. The condition is progressive, with death after 2–10 years. (See Brown (1992) for a review of Pick's disease.)

Dementia of frontal lobe type

Frontal lobe dementia of non-Alzheimer type (FLD) is the second most common form of primary dementia in the pre-senium (Neary *et al.* 1988; Gustafson *et al.* 1992), although it is still uncertain whether it is a separate disorder or a variant of Pick's disease. The cortical degeneration is mainly in the frontal lobe and is non-specific, with neural loss, slight gliosis, and spongiosis. The clinical picture is characterized by personality change and social disinhibition, together with progressive dementia in which memory loss and spatial impairment are usually relatively spared (Mann *et al.* 1993). Onset is usually in the mid-fifties and the normal duration is 7–10 years. The aetiology is unknown, but a family history of similar dementia is found in 50 per cent of cases verified at post-mortem. Diagnosis is made by clinical and neuropsychological examination, together with brain imaging.

Huntington's chorea

This was described in 1872 by George Huntington, a New England physician (see Critchley 1984). Since then, epidemiological studies have shown that the condition occurs in many countries; the estimated prevalence varies widely, with the average being about 4–7 per 100 000 (Oliver 1970). Men and women are affected in equal numbers. The pathological changes mainly affect the frontal lobes and the caudate nucleus. Neuronal loss, which is most marked in the frontal lobes, is accompanied by gliosis. The basal ganglia are strikingly atrophied.

Clinical features

Huntington's chorea usually begins at an age between 25 and 50, with the mean being in the forties (Minski and Guttman 1938). Earlier onset in successive generations (anticipation) has been reported. A rare juvenile form has been reported. The onset of neurological and psychiatric symptoms may be several years apart. The early neurological signs are choreiform movements of the face, hands, and shoulders. These movements are sudden, unexpected, aimless, and forceful. They are associated with dysarthria and changes in gait. Patients often attempt to disguise an involuntary movement by following it with a voluntary movement in the same direction. Gradually, abnormal movements become increasingly obvious with gross writhing contortions and ataxia. Patients begin to drop objects and later to fall over. Extrapyramidal rigidity and epilepsy also occur, especially in younger patients. Eventually walking, eating, and even sitting become difficult or impossible.

Dementia is usual in the later stages, after the development of chorea, but its severity and progress vary widely. There appear to be some clinical differences from Alzheimer and vascular dementias in which cortical destruction is more extensive. Memory is less affected than other aspects of cognitive function, and insight is often retained at a late stage. Initially, distractibility is characteristic, with reduced ability to regulate attention and psychomotor speed, both seen to be important determinators of everyday functioning (Rothlind *et al.* 1993). In the later stages, apathy is prominent.

Although psychiatric symptoms of all types occur at an early stage, depressive symptoms are particularly frequent. A high rate of affective disorder, including bipolar illnesses, has been reported (Peyser and Folstein 1993). The clinical features of depression are similar to those of major depression. Biological factors seem important in the aetiology.

Paranoid symptoms are common and schizophrenia may occur more frequently than in the general population. These symptoms may occur before the onset of chorea. It would seem that the incidence of schizophrenia is not raised amongst relatives unaffected by Huntington's chorea.

Quality of life is severely affected for the patient and family members. Carers experience a prolonged and heavy burden of care, and many complain of lack of help for their special needs. The effects seem to be particularly severe on the young children of patients with early-onset chorea.

Aetiology

Huntington's chorea is normally inherited as an autosomal dominant, although sporadic cases have been described. The Huntington's chorea gene was localized to a region on chromosome 4 in 1983 (Gusella *et al.* 1983). Ten years later the gene itself was identified, and the disease was found to be caused by a trinucleotide repeat mutation. (Huntington's Disease Collaborative Research Group 1993). Trinucleotide repeat and the age of disease onset are inversely correlated (Harding 1993; Ross *et al.* 1993; Gusella and MacDonald 1994). All cases can be accounted for by this mutation, i.e. Huntington's chorea is genetically aetiologically homogeneous. The pathogenic mechanism by which the mutation causes Huntington's chorea is unknown. Changes mainly in neurotransmitters have been described. For example Perry *et al.* (1973) were the first to report decreased concentrations of gamma-aminobutyric acid (GABA), a neuroinhibitory transmitter, in the caudate nucleus. Later work has shown decreased GABA biosynthesis and increased dopamine concentrations in parts of the basal ganglia.

Management

In general, the treatment is similar to that of other dementing disorders. Phenothiazines and butyrophenones have been reported as effective for the specific control of choreiform movements. It is uncertain whether they act as non-specific tranquillizers or through their specific effect on the dopamine systems. Pallidectomy and thalamotomy have also been used for the involuntary movements; some success has been reported in younger patients, but there is a risk of worsening the dementia or causing neurological side-effects. Antidepressants are useful for major depressive symptoms.

Predictive testing

The introduction of linked DNA markers has enabled predictive testing for a number of years. The recent availability of direct molecular tests now enables a certain diagnosis in most cases as well as prenatal diagnosis. In some circumstances, it may be possible to predict the age of onset. The psychological and ethical aspects of testing are considerable. Wiggins *et al.* (1992) have described a psychological follow-up of 135 participants in a Canadian programme of genetic testing. They found that predictive testing appeared to have benefits for psychological health for people whose results indicated either an increase or a decrease in their risk of inheriting the genes for the disease.

Ethical issues include the possible consequences for employment, insurance, and personal relationships, the use of the test in those with psychological symptoms which may be due to another psychiatric disorder, and the implications of the test for other relatives who may not themselves wish to be tested (Harper 1993; Rubinsztein *et al.* 1994). It is essential that testing should be undertaken only by those familiar with the issues, that counselling be provided before and after testing, and that written consent be obtained.

Prion diseases

The **spongiform encephalopathies**, or transmissible dementias, are a group of neurodegenerative disorders. The first of these names derives from the characteristic vacuolar spongy appearance of the brain seen post mortem. Early research focused on the disease *scrapie* which occurs in sheep and goats. By 1938 the experimental transfer of scrapie between animals began to suggest an infectious aetiology.

Four forms of prion disease were recognized in humans, of which Creutzfeldt–Jakob is the best known, and at least six identical diseases with different names are now recognized in a number of species. Despite their apparent rarity, the spongiform encephalopathies have been the subject of great research interest because each of the human forms has been shown to be transmissible from person to person by surgical and therapeutic procedures, and to other species by inoculation. There is considerable concern that the bovine form (bovine spongiform encephalopathy (BSE)) may be transmissible to human beings, as it is to mice, through the consumption of contaminated beef products.

It has been possible to characterize these conditions on the basis of mutations in the PrP gene or by the demonstration of abnormal PrP in the brain. All the conditions are characterized by deposits of an abnormal proteinaceous infectious particles (hence prion); in other ways the neuropathology is heterogenous (Palmer and Collinge 1992). Aetiology is by transmission of the prion protein. There is a wide variety of clinical presentations, but all are associated with dementia (see Prusiner (1993) and Prusiner and Hsiao (1994) for reviews).

Although rare, **Creutzfeldt–Jakob** disease is the most common of the prion diseases; it is of particular importance because it can be transmitted in blood or tissues between human beings (for example during surgery using contaminated instruments). The disorder was described by Creutzfeldt in 1920 and independently by Jakob in 1921. It is a rapidly progressive degenerative disease of the nervous system, characterized by intellectual deterioration and various neurological deficits including cerebellar ataxia, spasticity, and extrapyramidal signs. Evidence for an infective agent was preceded by the discovery that the rare neurological disease kuru, which is pathologically similar to Creutzfeldt–Jakob disease, could be transmitted. (**Kuru** was identified in a group of native people in Papua New Guinea, with a particularly high incidence among women. It appears to have been transmitted during ritual cannibalism.) Gibbs *et al.* (1968) were able to transmit Creutzfeldt–Jakob disease by inoculation of brain biopsy homogenate to a chimpanzee, an observation that has been frequently repeated. Precautions should be taken to avoid contamination with blood from these patients. Essentially, they involve extra care in handling samples of blood or tissue from these patients.

Normal pressure hydrocephalus

In this variety of hydrocephalus, there is no block within the ventricular system (Hakim and Adams 1965). Instead, there is an obstruction in the subarachnoid space such that cerebral spinal fluid can escape from the ventricles but is prevented from flowing up over the surface of the hemispheres. There is marked hydrocephalus with a generally normal or even low ventricular pressure (though sometimes with episodes of high pressure).

The characteristic features are progressive memory impairment, slowness, marked unsteadiness of gait, and later urinary incontinence. The condition is more common in the elderly, but sometimes occurs in middle life. Often, no cause for the obstruction can be discovered although there may be a history of subarachnoid haemorrhage, head injury, or meningitis. It is most important to be alert to the possibility of this syndrome and to differentiate the condition from the primary dementias, or possibly from depressive disorder with mental slowness. Treatment is a shunt operation to improve the circulation of cerebrospinal fluid. The results are difficult to predict. The dementia may improve but generally it does not (Lishman 1987).

Head injury

Head injuries are common, with the majority being closed injuries occurring in road traffic accidents. The peak incidence is between the ages of 15 and 24 years. Although most head injuries do not have serious medical consequences, improved care of severe injuries is resulting in larger numbers of young and otherwise physically fit people surviving with severe neuropsychiatric consequences.

The psychiatrist is likely to see two main kinds of patients who have suffered a head injury. First, there are a small number of patients with serious and lasting cognitive sequelae, such as persistent defect of memory. Second, there is a larger group with emotional symptoms and personality change; these symptoms may be less conspicuous but they often cause persistent disability (see Silver *et al.* (1994) for a review).

Acute psychological effects

Impairment of consciousness occurs after all but the mildest closed injuries, but is less common after penetrating injuries. The cause is uncertain but is probably related to rotational stresses within the brain. On recovery of consciousness, defects of memory are usually apparent. The period of **post-traumatic amnesia** is the time between the injury and the resumption of normal continuous memory. The duration of post-traumatic amnesia is closely correlated with first neurological complications such as motor disorder and dysphasia, and persistent deficits in memory and calculation; second with psychiatric disability and generalized intellectual impairment; and third with change of personality after head injury. The period of **retrograde amnesia** is the time between the injury and the last clearly recalled memory *before* the injury. It is not a good predictor of outcome.

After severe injury there is often a prolonged phase of **delirium**, sometimes disordered behaviour, anxiety and mood disturbance, hallucinations, delusions, and disorientation.

Chronic psychological effects

Damage to the brain is of central importance in determining the effects of head injury on subsequent cognitive performance and on personality disorder. However, other factors are important, including premorbid personality and environmental factors such as type of job, amount of social support available, and whether there is a compensation claim.

Lasting cognitive impairment

When head injuries are followed by post-traumatic amnesia of more than 24 hours, they are likely to give rise to persisting cognitive impairment proportional to the amount of damage to the brain. After a closed injury, the impairment is usually global, and varies in severity from obvious dementia to slight defects that only become apparent during intellectually demanding activities. After a penetrating or other localized injury, there may be only focal cognitive defects but some evidence of general impairment is usually found.

Lishman (1968) found that the site and extent of brain damage after penetrating injuries were related to the mental state one to five years later. The amount of tissue destruction was related both to intellectual impairment and to 'organic' psychological symptoms such as apathy, euphoria, poor judgement, and disinhibition. Lishman's study suggested that cognitive disorder was particularly

associated with parietal and temporal damage (especially on the left side).

Less severe head injury, followed by only transient loss of consciousness, can cause diffuse brain damage and may be followed by cognitive impairment.

Personality change

Personality change is common after severe head injury, and is particularly likely after frontal lobe damage. There may be irritability, loss of spontaneity and drive, some coarsening of behaviour, and occasionally reduced control of aggressive impulses. These changes may improve gradually, but as long as they are present may cause serious difficulties for the patient and the family (Prigatano 1992). Management may be difficult and may include much social support and rehabilitation over a prolonged period.

Emotional disorder

Depression and anxiety are very common after brain damage. Whilst there is usually a gradual improvement, persistent depression (Fedoroff *et al.* 1992; Jorge *et al.* 1993) and anxiety occur in about a quarter of the patients, a frequency similar to that reported in other serious physical disorders. Antidepressant medication is often useful. Mania has been reported, and may sometimes have an organic cause (Jorge *et al.* 1993). Post-traumatic stress disorder can occur, but is much less common than in those who have intact memories of trauma (Mayou *et al.* 1993). It is reported that the risk of suicide is increased after head injury, though the reason for this increase is not clear.

Schizophrenia-like syndromes

Delusions are not uncommon during recovery from prolonged unconsciousness. It is difficult to draw firm conclusions about the incidence of schizophrenia-like syndromes after head injury. Davison and Bagley (1969) reviewed the literature and concluded that the rate of schizophrenia-like syndromes was well above expectation and could not be explained by chance.

Post-concussional syndrome

After head injury some patients describe a group of symptoms known as the post-concussional syndrome. The main features are anxiety, depression, and irritability, accompanied by headache, dizziness, fatigue, poor concentration, and insomnia. Since this syndrome often occurs after mild head injury, clinicians have often suggested that the aetiology is largely psychological. Lewis (1942) examined prolonged emotional reactions amongst soldiers with head injuries, and concluded they occurred in 'much the same person as develops a psychiatric syndrome, anyway'. In a study of patients who were making claims for compensation, Miller (1961) found no relation between the severity of head injury and the extent of neurotic symptoms. In his study of penetrating injuries, Lishman (1968) found no demonstrable relationship between the extent of brain damage and the main symptoms of a post-traumatic syndrome.

It is now evident that minor head injuries can cause brain damage (Levin *et al.* 1989) and it is probable that both physical and psychological factors contribute to the aetiology of post-traumatic syndrome (Lishman 1988; Bohnen and Jolles 1992; Fenton *et al.* 1993; Jacobson 1995). Lishman (1988) concludes that 'there is certainly an interplay of factors in the genesis of post-concussional syndrome with an intertwining of organic and non-organic contributions. But this interplay is time-dependent. Over many weeks and months there is a shifting balance as the patient's innate proclivities, and his individual problems and conflicts, affect the initial symptoms of cerebral dysfunction'. (See Jacobsen (1995) for a review.)

Social consequences of head injury

The physical and psychosocial consequences of head injury for the patient often place a heavy burden on relatives. Many relatives experience severe distress and have to make substantial changes in their way of life. Family life is particularly affected if the patient shows personality change, and there may be consequences for children (Urbach and Culbert 1991). Rehabilitation after head injury should take account of this burden on families and should include help for them (Brooks 1991).

Treatment

A plan for long-term treatment should be made as early as possible after head injury. Planning begins with a careful assessment of three aspects of the problem. The first is the degree of physical disability. It has been shown by prospective study that early assessment of the extent of neurological signs provides a useful guide to the likely pattern of long-term physical disability. Second, any neuropsychiatric problems should be assessed and their future course anticipated. Third, a social assessment should be made.

Treatment includes physical rehabilitation, to which the clinical psychologist can sometimes contribute behavioural and cognitive techniques (Benedict 1989). Practical and social support is needed for the family. Any problems of compensation and litigation should be settled as quickly as possible. Ideally, continuing help should be provided by a special team. (See Brooks (1984), and Silver *et al.* (1994) for reviews of the treatment of head injury.)

Boxing and head injury

For many years there has been dispute about the significance of 'punch drunk' states after repeated minor head injury in the boxing ring. There have been numerous published reports of a characteristic syndrome (punch drunk syndrome, dementia pugilistica), but most of the findings could be due to biased selection of cases with coincidental neurological disease. However, in an important study of a random sample of 224 retired professional boxers, Roberts (1969) found that 37 had a characteristic syndrome related to the extent of exposure to head injury during boxing. The principal early features are affective symptoms and mild incoordination, followed by dysphasia, apraxia, agnosia, apathy, and neurological signs. Later, there is global cognitive decline and parkinsonism. When the syndrome is fully developed, there are cerebellar, pyramidal, and extrapyramidal signs as well as intellectual disorientation. The condition usually progresses until retirement from the ring, and may continue afterwards. Cerebral atrophy has been shown radiologically. Corsellis *et al.* (1973) examined the post-mortem brains of ex-boxers and found excessive loss of cortical neurones and neurofibrillary degeneration. The psychiatrist should bear this condition in mind as an occasional cause of dementia (Roberts *et al.* 1990*a*; British Medical Association 1993).

Intracranial infections

Many intracranial infections, including cerebral malaria and other endemic tropical diseases, may cause cognitive impairment disorders. This section briefly considers neurological conditions of past or present clinical interest. *HIV infection is considered in Chapter 11.*

Neurosyphilis

Neurosyphilis is now rare in psychiatric and neurological practice in Western countries. It is still a significant clinical problem in some developing countries.

Of every twelve patients with neurosyphilis, approximately five have general paresis, four have meningovascular syphilis, and three have tabes dorsalis. Of these groups, general paresis (general paralysis of the insane (GPI)) is the most important to psychiatrists. The discovery of its cause was an important landmark in the history of psychiatry because it stimulated a search for organic causes of other psychiatric syndromes. A further important discovery was that this one cause could give rise to many different kinds of clinical picture (see Hare (1959) for a historical review).

General paresis usually presents in middle age after an average of 10–15 years from infection. It may present with minor emotional symptoms, evidence of personality change, or striking lapses

in social conduct. In the past, a grandiose presentation was frequently described, but dementia and depression are now more common. Many other combinations of psychiatric symptoms can occur. As the disease progresses, there is increasing dementia, paralysis, ataxia and seizures. In untreated cases, death usually occurs within four to five years. If treatment is given early, the condition usually remits; if treatment is given in established cases, the progression of the disease can generally be halted.

Encephalitis

Encephalitis may be due to either a primary viral disease of the brain or a complication of bacterial meningitis, septicaemia, or a brain abscess. Many viral causes have been identified, of which herpes simplex is the most common in the United Kingdom. Encephalitis sometimes occurs after influenza, measles rubella, and other infectious diseases, and also after vaccination.

In the acute stage, headache, vomiting, and impaired consciousness are usual, and seizures are common. There may be an acute organic psychiatric syndrome. Rarely, encephalitis presents with predominant psychiatric symptoms. For example herpes simplex encephalitis may result in damage to the temporal lobes which may lead to amnesic and other psychiatric syndrome. However, the psychiatrist is more likely to see the complications that follow the acute episode; these may include prolonged anxiety and depression, dementia, personality change, or epilepsy. In childhood, encephalitis may be followed by behaviour disorders (Lishman 1987).

Encephalitis lethargica (epidemic encephalitis)

A small outbreak of encephalitis lethargica was first reported in 1917 by Von Economo at the Vienna Psychiatric Clinic (Von Economo 1929). The condition increased in the 1920s. By the 1930s, it had largely disappeared, although rare sporadic cases may still occur. The acute stage was usually characterized by somnolence and ophthalmoplegia. The chronic sequelae were of most interest to the psychiatrist. Parkinsonism was a disabling complication, and oculogyric crises were particularly striking. Another disabling sequel was personality change towards more anti-social behaviour. Some patients developed a clinical state resembling schizophrenia. Davison and Bagley (1969) analysed 40 of these schizophrenia-like cases from the literature, and found that they had fewer schizophrenic family members than did schizophrenic probands.

Mental symptoms were commonly associated with parkinsonism. Some parkinsonian patients had marked slowing and apathy. Sacks (1973) has given a vivid description of such cases, and the striking but temporary improvements brought about in some by L-dopa.

Cerebral abscess

A cerebral abscess may present rapidly and obviously with headache, epileptic seizures, papilloedema, and focal signs. However, a cerebral abscess may develop insidiously, and may then be mistaken for a psychiatric disorder. It is vital that the psychiatrist should always be alert to such a possibility. For example the diagnosis of cerebral abscess should always be considered when depressive symptoms are accompanied by mild confusion and fever, especially when the patient seems generally ill. In such cases, papilloedema often appears late and there may be few other neurological signs. Radiological studies using CT scan are likely to be important. The primary focus of infection is usually outside the brain and may be difficult to detect; common sites include the mastoid, middle ear, and nasal sinuses, and chronic suppurative lung disease. Penetrating head injury is another cause.

Tuberculous meningitis

Nowadays, tuberculous meningitis is uncommon and notoriously difficult to diagnose. The psychiatrist occasionally encounters the condition when it presents with apathy, irritability, and 'personality change'. Pyrexia, neck stiffness, and clouding of consciousness are often late to appear, and therefore should be looked for repeatedly.

Cerebrovascular disease

Cerebrovascular accident

Amongst people who survive a cerebrovascular accident, just over half return to a fully independent life. The rest suffer some loss of independence because of disabilities that may be psychological as well as physical. The psychological changes are often the more significant, and many patients do not return to normal life even after physical disability has ceased to be a serious obstacle. Rehabilitation can improve psychological well-being and reduce social problems (Wade 1992), but there are doubts about the best way to design, use, and evaluate programmes of rehabilitation (Pollock *et al.* 1993).

Cognitive defects

A single stroke can cause dementia and other deficits of higher cortical function such as dysphasia and dyspraxia, which may handicap the patient to a degree that is often underestimated by doctors. After a first stroke repeated small strokes may lead to progressive dementia.

The separate condition of vascular dementia is described in Chapter 16 on psychiatry of the elderly (p. 525).

Personality change

Irritability, apathy, lability of mood, and occasionally aggressiveness may occur. Inflexibility in coping with problems is common and may be seen in extreme form as a 'catastrophic reaction'. Such changes are probably due more to associated widespread arteriosclerotic vascular disease than to a single stroke; they may continue to worsen even though the focal signs of a stroke are improving. When a person has had a stroke, the family is likely to complain mainly of the subsequent personality change.

Mood disturbance

Depressed mood is a common reaction to the handicap caused by a stroke (House *et al.* 1991; Starkstein and Robinson 1993). It may contribute to the apparent intellectual impairment and is often an important obstacle to rehabilitation. It is not certain whether depressed mood is more common after stroke than after other disabling illnesses in elderly people. Depression may be associated with the size of the brain lesion (Sharpe 1994), and it has been argued that depression is associated with lesions in the anterior part of the left hemisphere (Robinson *et al.* 1984; Starkstein and Robinson 1993). However, there is as yet no conclusive proof and it seems equally likely that depression after stroke is caused by social and psychological factors, especially in the long-term (Sharpe *et al.* 1994)

Treatment of the mood disorder depends mainly on active rehabilitation and help for the family. Severe and persistent depressive symptoms often respond to treatment with an antidepressant, but the drug should be given cautiously because side-effects are frequent in these patients (Starkstein and Robinson 1993). (See Johnson (1991) for a review of mood disorders after cerebrovascular accident.)

Anxiety is also common after a stroke (Castillo *et al.* 1993; Burrill *et al.* 1995), particularly in relation to falling and the risk of another stroke. It is uncertain whether biological factors contribute to aetiology.

Abnormal emotionalism (emotional lability) is frequent after a stroke (House *et al.* 1989) and appears to be unrelated to mood disturbance. It may be particularly frequent in patients with left-frontal temporal lesions. Emotionalism is excessive and patients find it distressing and socially disabling, but it is not inappropriate to the general mood at the time. Treatment is not definitely established, but antidepressants may be helpful (Robinson *et al.* 1993; van Gijn 1993).

Other problems

Psychological problems include denial, excessive invalidism, poor motivation, and disturbed behaviour. Ability to work, enjoy leisure, and take part in other social activities may be reduced. Many *relatives* are anxious or otherwise distressed, and the degree of distress is not directly related to the extent of the patient's physical disability, mood, and personality change (Wade *et al.* 1986; Carnwath and Johnson 1987).

Subarachnoid haemorrhage

A high incidence of mental disorder has been reported after subarachnoid haemorrhage. However, prospective studies suggest that most survivors without neurological deficits return to premorbid cognitive function. Adverse affects on quality of life are frequent in the first year, and may persist much longer. Patients with neurological deficits are likely to suffer persistent cognitive impairment and social disability (McKenna *et al.* 1989).

Subdural haematoma

The psychiatrist should remember that subdural haematoma is not uncommon after falls associated with alcoholism. The symptoms may then be easily overlooked or misdiagnosed. Acute haematomas may cause coma or fluctuating impairment of consciousness, and are often associated with hemiparesis and oculomotor signs. The psychiatrist is more likely to see the chronic syndromes, in which patients present with headache, poor concentration, vague physical complaints, and fluctuating consciousness, but often few localizing neurological signs. If there is any suspicion of a subdural haematoma, radiological investigation is required. Treatment is by surgical evacuation, which may reverse the symptoms in some chronic cases but leaves many others with continuing deficits.

Other neurological conditions

Cerebral tumours

Many cerebral tumours cause psychological symptoms at some stage, and a significant minority first present with such symptoms. Psychiatrists are most concerned with slow-growing tumours in brain areas that produce psychological effects but few neurological signs, for example frontal meningiomas. The nature of the psychological symptoms depends not only on the site of the tumour but also on the presence or absence of raised intracranial pressure. The rate of tumour growth is also important; fast-growing tumours with raised intracranial pressure can present as an acute organic syndrome, whilst less rapidly growing tumours are more likely to cause cognitive defects. The nature of the psychological symptoms is also much affected by the patient's personality. It should be remembered that focal lesions can give rise to one of the specific syndromes discussed already (p. 316); these may take the form of personality change or may be mistaken for neurotic symptoms.

Cerebral tumours are easily overlooked in psychiatric practice unless the psychiatrist is constantly alert. Unexplained 'changes in personality' are particularly suspicious. (See Ron (1989) for a review.)

Transient global amnesia

This syndrome is important in the differential diagnosis of episodes of unusual behaviour. It occurs in middle or late life. It is characterized by abrupt episodes, lasting several hours, in which there is a global loss of recent memory. The patient apparently remains alert and responsive, but usually appears bewildered by his inability to understand his experience. There is impairment of new learning but not of other cognitive functions, and the patient may be able to continue a set task or find his way around. There is complete recovery, but with amnesia for the episode. The aetiology is obscure, although there may be a link with migraine. The prognosis is good.

Patients with this condition often present as emergencies to general practitioners and casualty officers. Doctors who are not familiar with the syndrome may misdiagnose it as a dissociative fugue.

Multiple sclerosis

Multiple sclerosis is one of the most common neurological diseases. It is often difficult to diagnose in its early stages and the physical symptoms may be erroneously diagnosed as having a psychological cause, thereby adding to the patient's distress (Skegg *et al.* 1988). Psychological symptoms may also occur early in the disorder, but are rarely the presenting feature (Skegg 1993). Transient mood changes in anxiety probably occur in all patients from time to time. In the course of a year such symptoms occur in two-thirds of patients, and they are severe enough to meet the criteria for major depression in a third (Ron and Feinstein 1992). Euphoria occurs in around 10 per cent, usually in those with the most severe cognitive impairment. Fatigue is also a prominent symptom; it may result either from physical causes or from emotional disturbance.

Depression is more common in multiple sclerosis than in other disabling illnesses (Ron and Feinstein 1992); it does not appear to be clearly related to the severity of the clinical symptoms or the site of lesions. The occurrence of depression is related to social variables, but high prevalence and good premorbid personality suggest that biological factors are also important. The lifetime risk of depression in this condition is probably about 50 per cent (Ron and Logsdail 1989). Standard antidepressant medication is usually effective (Schiffer and Wineman 1990).

Cognitive impairment may be one of the early signs of the illness, and occasionally a rapidly progressive dementia occurs early in the course. In most cases, however, intellectual deterioration is less severe and progresses slowly. Dementia is common in the late stage of the disease, and is present in about 50–60 per cent of clinic attenders (Ron *et al.* 1991). Psychometric testing shows that, in the early stages, well-practised verbal skills are often preserved despite deficits in problem-solving, dealing with abstract concepts, memorizing and learning (see Rao *et al.* 1992; Ron *et al.* 1991). The extent of cognitive impairment is generally in line with the severity of MRI abnormalities, but cognitive deficits do not relate simply to the pathology.

Inevitably there are considerable social consequences of this chronic fluctuating disorder affecting mobility. Sexual problems (especially impotence) are frequent, and the condition places heavy burdens on patients' families.

Parkinson's disease

Parkinson's disease causes progressive limitation of movement which may greatly impair quality of life. The main psychological consequences are cognitive impairment and depression.

Cognitive impairment

Estimates of the prevalence of dementia in Parkinson's disease have varied widely. These differences in estimates are likely to be due to differences in the definitions of dementia, and differences in the populations studied. Thus Ebmeier *et al.* (1991) found a prevalence of 23 per cent in an elderly population cohort, but recent estimates using DSMIIIR criteria have been 11–15 per cent (Biggins *et al.* 1992). In a controlled longitudinal study the incidence was 19 per cent in 54 months, a higher rate than in the general population (Biggins *et al.* 1992).

The nature of the dementia is unclear; some subjects have symptoms of the Alzheimer type, whilst others show a pattern of cognitive impairment that is consistent with pathology in the structures known to be affected by Parkinson's disease. It remains unclear how far the dementia is of a subcortical type (Gibb 1989; Ring 1993). It is possible that Lewy body disease (see p. 524) accounts for much of the dementia (Perry *et al.* 1991).

Depression

The association of Parkinson's disease with depression is well established, but again it is not well understood. Estimates of prevalence vary, partly because of differences in the definition of depression, and partly because of difficulties in distinguishing between the features of depression and those of Parkinson's disease. It seems likely that depression occurs in approximately 40 per cent of those with Parkinson's disease, a higher proportion than in other disabling conditions (Cummings 1992; Rao *et al.* 1992; Starkstein and Robinson 1993). The mechanism of depression is unknown, but one suggestion is that it may be related to frontal lobe abnormalities or to disturbance of serotonergic mechanisms.

Some of the mental symptoms of parkinsonism may be induced by the drugs used to treat the movement disorder. Anticholinergic drugs may cause excitement, agitation, illusions, and hallucinations. L-Dopa is associated with an acute organic syndrome as well as with depressive symptoms. Stereotactic surgery for the treatment of tremor is often followed by transient deficits in cognitive function and rarely by lasting cognitive impairment. Antidepressant medication has a important role in treatment together with individually planned rehabilitation (Montgomery *et al.* 1994).

Wilson's disease

In his original account of the syndrome in 1912, Wilson recognized psychiatric symptoms. In a recent series of 195 cases, Dening (1989) found evidence of psychopathology in 51 per cent. The most common features were abnormalities in behaviour and personality, whilst depression and cognitive impairment were also frequent. Psychiatric symptoms seemed to be related to the severity of the neurological rather than the hepatic symptoms.

Epilepsy

The psychiatrist is likely to meet four kinds of problem in relation to epilepsy: differential diagnosis of episodic disturbance of behaviour

Table 11.4 *Classification of seizures*

1. *Partial seizures or seizures beginning focally*
   Simple motor or sensory (without impaired consciousness)
   Complex partial (with impaired consciousness)
   Partial seizures with secondary generalization

2. *Generalized seizures without focal onset*
   Tonic–clonic convulsion
   Myoclonic, atonic
   Absences

3. *Unclassified*

during the day or night (particularly of atypical attacks, aggressive behaviour, and sleep problems), the treatment of the psychiatric and social complications of epilepsy, the treatment of epilepsy itself in patients who consult him, and the psychological side-effects of anticonvulsant drugs. Reviews of these problems and of other aspects of epilepsy can be found in the texts by Laidlaw *et al.* (1993).

Types of epilepsy

To understand the psychiatric aspects of epilepsy, it is necessary to know how seizures and the epilepsies are classified and what the clinical features are of their common forms. It is useful to remember that the term **ictus** refers to seizure itself and is characterized by abnormal electrical activity. The **aura** is nothing more than a partial seizure, and therefore may be a complete seizure or the first stage in which consciousness is preserved. It should be distinguished from **prodromata**, symptoms which sometimes precede the seizure.

The International League Against Epilepsy drew up the original version of the classification of seizures (Gastaut 1969), which is now in general use in a revised version (Dreifuss *et al.* 1981). Traditional terms such as **petit mal** and **grand mal** are not used because of their ambiguity. The

scheme is elaborate, and the outline shown in Table 11.4 is much simplified. The principal distinction is between partial seizures which start focally, and generalized seizures which are generalized from the beginning. Since focal seizures often become generalized, a description of the initial stages of the attack is of the greatest importance in the use of this diagnostic scheme. It is important to distinguish between classification of types of epilepsy (Shorvon 1990) and types of seizure (Gram 1990).

Simple partial seizures

This group includes Jacksonian motor seizures and a variety of sensory seizures in which the phenomena are relatively unformed. Consciousness is not impaired. These seizures may secondarily become generalized into atonic–clonic seizures.

Complex partial seizures

This category replaces the earlier categories of 'psychomotor' seizures and 'temporal lobe epilepsy', and is characterized by impaired consciousness. These seizures arise most commonly in the temporal lobe but may have other focal origins. Those originating in the frontal lobe are particularly likely to be misdiagnosed as psychiatric disorder (Williamson and Spencer

Table 11.5 *Clinical features of complex partial seizures*

Consciousness	Impaired
Autonomic and visceral	'Epigastric aura', dizziness, flushing, tachycardia, and other bodily sensations
Perceptual	Distorted perceptions, *déjà vu*, visual, auditory, olfactory, and somatic hallucinations
Cognitive	Disturbances of speech, thought, and memory
Affective	Fear and anxiety
Psychomotor	Automatisms, grimacing and other bodily movements; repetitive or more complex stereotyped behaviour

1986). Complex partial seizures originating in the temporal lobe are often preceded by an aura which lasts for a few seconds and may take the form of hallucinations of smell, taste, vision, hearing, or bodily sensation. The patient may also experience intense disturbances of thinking, perception, or emotion.

The clinical features of complex partial seizures are summarized in Table 11.5. An important point is that in the individual patient the sequence of events in the seizure tends to be the same on each occasion. A particularly common feature is the 'epigastric aura', a sensation of churning felt in the stomach and spreading towards the neck. Patients often have great difficulty in describing these phenomena.

The whole ictal phase usually lasts up to one or two minutes. During this phase and the post-ictal phase, the subject appears to be out of touch with his surroundings and may show automatisms. After recovery, only the aura may be recalled.

Status epilepticus of two kinds may occur—a prolonged single seizure or a rapid succession of brief seizures. In such cases a prolonged period of automatic behaviour may be mistaken for a dissociative fugue or other psychiatric disorder.

Generalized tonic–clonic seizure

This is the familiar epileptic seizure with a sudden onset, tonic and clonic phases, and a final period of several minutes or longer in which the patient is unrousable, sleeping, or confused. Many generalized tonic–clonic seizures are initiated by a partial seizure whose features may be overlooked. This is clinically important, as primary and secondary generalized tonic–clonic seizures are due to different forms of epilepsy.

Myoclonic, atonic

There are several types of generalized epilepsy with predominantly motor symptoms, such as widespread myoclonic jerks or drop attacks with loss of postural tone. They are unlikely to present problems to the psychiatrist.

Absences

There are several clinical types (all of which have impaired consciousness as the cardinal features) as part of a number of different epilepsy syndromes. The attack starts suddenly without an aura, lasts for seconds, and ends abruptly. There are no post-ictal abnormalities, and simple automatisms are often present. For a number of purposes, including treatment, it is important to distinguish between absence seizures and the less florid forms of complex partial seizures. The latter often begin with an aura, last longer, and are followed by slow return to recovery. An EEG may be required to make the distinction with certainty.

Epidemiology

Population surveys show that epilepsy is usually of short duration and only becomes a chronic and potentially handicapping condition in about a fifth of subjects. This means that regular attenders at specialist clinics are a minority of those with epilepsy, and are especially likely to suffer from medical and social complications related to epilepsy. In the United Kingdom, surveys in general practice have shown the prevalence of epilepsy to be at least 4–6 per 1000. The inception rate is highest in early childhood, and there are further peaks at adolescence and over the age of 65. In a few cases starting in childhood, epilepsy is associated with mental handicap (Shorvon 1990; Hauser and Annegers 1993).

Aetiology

Many causes of epilepsy are known and their frequency varies with age. In the newborn, birth injury, congenital malformations, metabolic disorders, and infections are the most common. In the elderly, the most common causes are cerebrovascular disease, head injury, and degenerative cerebral disorder. Although advances in neuroimaging are leading to increased identification of structural causes, no specific cause is found in around three-quarters of patients with epilepsy. In such patients, genetic factors appear to be of greater significance than in patients with demonstrable pathology. It is likely that epilepsy in children in the absence of a specific cause is of genetic aetiology. The prognosis for remission by early adult life is usually good in these cases.

Seizures can occur as a result of drug therapy, but many neurologists would not diagnose them as epilepsy on the grounds that the latter must have a primary cause in the brain. Amongst the psychotropic drugs, chlorpromazine is implicated most often, and amitriptyline and imipramine less often. Sudden withdrawal of substantial doses of any drug with antiepileptic properties may be followed by seizures. The withdrawal of large doses of diazepam and alcohol are the most common examples among psychiatric patients.

The diagnosis of epilepsy

Epilepsy is essentially a clinical diagnosis which depends upon detailed accounts of the attacks given by witnesses as well as by the patient. The rest of the history, the physical examination, and special investigations are concerned with aetiology. The extent of investigation is guided by the initial findings, the type of attack, and the patient's age. Only an outline can be given here; for a full account the reader is referred to a textbook such as that by Laidlaw *et al.* (1993), and the review by Chadwick (1990).

An EEG can confirm but cannot exclude the diagnosis of epilepsy. It is more useful in determining the type of epilepsy and site of origin. The standard recording may be supplemented by sleep recording, ambulatory monitoring, and split-screen video techniques.

Since epilepsy is often erroneously diagnosed in children and adults (Jeavons 1983), it is important to keep in mind the differential diagnosis (Table 11.6), including dissociative seizures ('pseudoseizures') which may occur with or without epilepsy. Features that suggest dissociative disorder are an unusual or variable pattern of attacks, occurrence only in public, and absence of autonomic signs or changes in reflexes. Factors that point strongly to epilepsy are tongue-biting, incontinence, definite loss of consciousness, and sustaining injury during the attack. Epilepsy is often considered in the diagnosis of aggressive outbursts, but is a rare cause (Treiman and Delgado-Escueta 1983). Some forms of frontal lobe epilepsy are particularly likely to be

Table 11.6 *Differential diagnosis of epilepsy*

Organic
    Syncope
    Hypoglycaemia
    Transient ischaemic attacks
    Migraine
    Sleep disorders

Non-organic
    Temper tantrums
    Breath-holding
    Hyperventilation
    Dissociative disorder
    Panic attacks
    Schizophrenia
    Aggressive outburst in unstable personality
    Night terrors

misdiagnosed as psychogenic or as having other non-epileptic causes in children (Stores *et al.* 1991) and in adults (Fusco *et al.* 1990). If the diagnosis remains uncertain, closer observation in hospital, including video recording and EEG telemetry and ambulatory monitoring should be considered (though seizures often stop when the patient is in hospital). If an epileptic patient has an abnormal personality and shows aggressive outbursts, it is sometimes difficult to decide whether the latter are due to epileptic automatism or simply an expression of personality disorder (see pp. 113 and 339).

A particular difficulty may arise in distinguishing complex forms of epilepsy from dissociative disorder causing non-epileptic seizures ('pseudoseizures') (Betts 1990). The difficulty is the greater because epilepsy and pseudoseizures may occur in the same patient.

A careful description of the seizures and detailed history are important. Ambulatory EEG can be helpful, but only if a seizure occurs while the recording is being made and is well described or videotaped. (See Fenton (1986) and Lowman and Richardson (1987) for reviews of the differential diagnosis between epilepsy and psychiatric disorder.)

Social aspects of epilepsy

Although people with epilepsy are often at a social disadvantage, their quality of life is not greatly affected in recently diagnosed epilepsy (Chaplin *et al.* 1992) or in chronic well-controlled epilepsy (Jacoby 1992). Effects are greater for patients with severe symptoms and in those with associated brain pathology. Many patients suffer more from other people's misconceptions and prejudices of about epilepsy than from the condition itself. Problems arise in school, at work, and in the course of family life. Marriage prospects may be affected. In the care of people with epilepsy, it is important to attempt to reduce these misunderstandings and to support the patient and his family (Levin *et al.* 1988; Griffin and Wyles 1991). Problems are greater in less developed countries where treatment is less satisfactory and there is less understanding of the nature of the problems (Kleinman *et al.* 1995). Restrictions on driving

Table 11.7 *Associations between epilepsy and psychological disturbance*

---

1. Psychiatric disorder associated with the underlying cause

---

2. Behavioural disturbance associated with the seizure
   Pre-ictal: prodromal states and mood disturbance
   Ictal: complex partial seizures
                absence status
                complex partial status
   Post-ictal: automatisms
                impaired consciousness

---

3. Inter-ictal disorders
   Cognitive
   Personality
   Sexual behaviour
   Depression and emotional disorder
   Suicide and deliberate self-harm
   Crime
   Psychoses

---

may increase the person's difficulties. To obtain a British driving licence, the patient must have had at least two years with no seizures whilst awake, whether or not he is still taking antiepileptic drugs. Those who suffer seizures only whilst asleep may have a licence if this pattern has been present for at least two years.

Psychiatric consequences of epilepsy

As mentioned already, it used to be thought that people with epilepsy suffered an inevitable deterioration of personality. This belief has been repeatedly disproved, but continues as a popular misconception that causes much unnecessary distress. However, there are several important ways in which epilepsy predisposes to psychiatric disturbance (Table 11.7).

Prevalence

In a survey of people with epilepsy in general practice, Pond and Bidwell (1960) found that nearly 30 per cent had conspicuous psychological difficulties, 7 per cent had had in-patient psychiatric care, and 10 per cent were educationally subnormal. More recent studies have generally found a raised prevalence of most forms of psychiatric disorder (Levin *et al.* 1988), especially in people with severe chronic epilepsy, but the prevalence seems to be little raised when the epilepsy is well controlled (Jacoby 1992) and in those with idiopathic epilepsy (Fiordelli *et al.* 1993).

Similar conclusions apply in childhood. In their Isle of Wight survey of schoolchildren aged between 5 and 14, Graham and Rutter (1968) diagnosed psychiatric disorder in 7 per cent of non-epileptic children, 30 per cent of those with uncomplicated epilepsy, and nearly 60 per cent of those with epilepsy complicated by evidence of brain damage. Prevalence is particularly high amongst attenders of specialist epilepsy clinics (Hoare and Kerley 1991).

Clinical features

A classification of the psychiatric consequences of epilepsy is given in Table 11.7.

Psychiatric disorder associated with the underlying cause

The underlying cause of epilepsy may contribute to intellectual impairment or personality problems, especially if there is extensive brain damage. Epilepsy is common in the mentally retarded (see Chapter 21).

Behavioural disturbance associated with the seizure

Increasing tension, irritability, and depression are sometimes apparent as prodromata for several days before a seizure. Transient confusional states and automatisms may occur during seizures (especially complex partial seizures) and after seizures (usually those involving generalized convulsions, and complex partial seizures).

Less commonly, non-convulsive seizures may continue for days or even weeks (absence status and complex partial status). An abnormal mental state may be the only sign of this condition (Stores 1986) and the diagnosis is easily overlooked (see *Lancet* 1987).

Inter-ictal disorders

There is no convincing evidence of a direct relationship between epilepsy and psychiatric disturbance. However, there are several indirect associations.

Cognitive function

In the nineteenth century it was widely believed that epilepsy was associated with an inevitable decline in intellectual functioning. Subsequent investigations tended to support this belief, but the findings were misleading because they were based on residents in institutions. Early research was also unsatisfactory because it was retrospective, and therefore unable to distinguish between dementia that progressed over time and lifelong intellectual retardation. Nowadays, it is established that relatively few people with epilepsy show cognitive changes.

When intellectual changes do occur, the significant aetiological factors are likely to be brain damage, poor concentration and memory during periods of abnormal electrical activity, and the adverse effects of antiepileptic drugs given in high doses or even in doses optimal for the control of seizures (particularly barbiturates in the past).

A few epileptic patients show a progressive decline in cognitive function. In such cases, careful investigation is required to exclude the progress of an underlying neurological disorder, toxic drug levels, and repeated non-convulsive epileptic status.

Learning problems are more common in children with epilepsy than in non-epileptic children (Stores 1981). Apart from the factors listed above, possible causes include poor school attendance and the general social difficulties of having epilepsy. When intellectual deterioration occurs in children, it may be due to medication, or to the underlying pathological process causing the seizures, or to the seizures themselves (Ellenberg *et al.* 1986).

Personality

As already explained, nineteenth-century writers suggested that seizures cause personality deterioration. Early in the twentieth century, it was held that epilepsy and personality changes resulted from some common underlying abnormality. The 'epileptic personality' was said to be characterized by egocentricity, irritability, religiosity, quarrelsomeness, and 'sticky' thought processes. It is now recognized that these ideas arose from observations of severely affected people with brain disease living in institutions. Surveys of people with epilepsy living in the community have shown that only a minority have serious personality difficulties. Even when they occur, such personality problems show no distinctive pattern (Tizard 1962; Fenton 1983). For example, there is little evidence to support an association between epilepsy and aggressiveness (Treiman and Delgado-Escueta 1983; Levin *et al.* 1988; Treiman 1993). It has been suggested that abnormalities of personality are mainly associated with temporal lobe lesions (Pond and Bidwell 1960; Guerrant *et al.* 1962).

When personality disorder does occur, social factors probably play an important part in aetiology. These factors include the social limitations imposed on people with epilepsy, their own embarrassment, and the reactions of other people. It is also possible that brain damage sometimes contributes to the development of personality disorder.

Sexual dysfunction

Lack of libido and sexual dysfunction are probably more common in epileptic than in non-epileptic people. This is thought to apply particularly to patients with temporal lobe foci. Possible causes include the general social maladjustment of some epileptics and the effects of antiepileptic medication (Toone 1985; Griffin and Wyles 1991).

Depression and other emotional disorder

Depressed mood has consistently been found to be more common in people with epilepsy than in the general population. It is usually of moderate severity. There is no evidence of a raised incidence of bipolar affective disorder. There is an increased risk of suicide (see below). Depression is most common in those with adverse social factors (Ring and Trimble 1993). It is not certain whether depression is more frequent in patients with brain damage. Emotional disorders other than depression also seem to be more common among people with epilepsy than in the general population (Levin *et al.* 1988; Jacoby 1992).

Suicide and deliberate self-harm

Suicide is four times more frequent amongst people with epilepsy than amongst the general population (Sainsbury, 1986), and deliberate self-harm is six times as frequent (Hawton *et al.* 1980). (See Barraclough (1987) for a review.)

Epilepsy and crime

Nineteenth-century writers such as Lombroso thought that crime was far more common among epileptic than among non-epileptic people. It is now well established that there is no such close association between epilepsy and crime.

In a survey of the prison population of England and Wales, Gunn (1977*a*) found the proportion of people with epilepsy to be 7–8 per 1000 which is probably greater than that in the general population. Prisoners with epilepsy were no more aggressive than other prisoners, but they had a greater rate of psychiatric disturbance. There was no relationship between epilepsy and the type of crime. The explanation for the apparently disproportionate number of epileptics in prison is not known; it may be that the social difficulties of epileptics lead them into more conflict with the law.

Gunn and Fenton (1971) carried out a survey of special hospitals and concluded that crimes committed during epileptic automatisms are extremely rare. This conclusion, which has important medicolegal implications, is supported by evidence from other countries (see *Lancet* (1981) and p. 759). (For reviews of the relationship between epilepsy and crime see Treiman and Delgado-Escueta (1983) and Treiman (1993).)

Inter-ictal psychoses

The nature of inter-ictal psychoses is controversial. Some writers have suggested that psychotic disorder is less common in people who suffer from epilepsy than in the general population (the antagonism hypothesis); others have argued the opposite view, that such illnesses are more common in epilepsy (the affinity hypothesis). It is still not possible to reach definite conclusions but most recent research has concentrated on possible associations between certain forms of epilepsy and schizophrenia-like disorders.

Inter-ictal disorders resembling schizophrenia Some patients with temporal lobe epilepsy develop 'chronic paranoid hallucinatory psychosis'

(Hill 1953; Pond 1957). The clinical picture closely resembles schizophrenia, except that affective responses are preserved. Slater *et al.* (1963) studied 69 patients with unequivocal epilepsy who developed an illness diagnosed as schizophrenia. Almost all these patients suffered from temporal lobe epilepsy and, although there was no control group, the authors argued that this association was unlikely to have been due to chance. Like Hill and Pond, these authors reported that normal affective responses were usually preserved. They also found that some cases progressed towards a more 'organic' clinical picture. A family history of schizophrenia was usually lacking. The psychosis usually began many years after the onset of epilepsy, in patients with a normal premorbid personality.

There has been continuing argument as to whether there is a specific association between temporal lobe epilepsy and schizophrenia, an association which might seem understandable in that both conditions may have a neuro-developmental abnormality in the medical temporal lobe. It has been difficult to obtain epidemiological evidence that is free from selection bias and is based on precise diagnostic criteria. Nevertheless, it is likely that there is such an association and that the schizophrenia-like psychosis is more common among patients with foci in the temporal lobes than in those with primary generalized seizures (Roberts *et al.* 1990*b*; Trimble 1992; Mendez *et al.* 1993). Other possible risk factors include the onset of central nervous system disorder in adolescents, left-sided lesions, abnormal neurological signs, and left-handedness.

Treatment

The drug treatment of epilepsy is summarized in Chapter 17. For a more detailed account of the care of epileptic patients, a textbook of epilepsy should be consulted (Laidlaw *et al.* 1993). Here it is necessary only to emphasize the importance of distinguishing between peri-ictal and inter-ictal psychiatric disorders. For peri-ictal psychiatric disorders, treatment is aimed at control of the seizures. For inter-ictal psychiatric disorders, the treatment is the same as it would be for a non-epileptic patient, though it should be remembered that many antidepressants and other psychotropic drugs may increase seizure frequency. Emotional arousal can do the same. In some cases behavioural and cognitive interventions aimed to reduce stress may be helpful in reducing the frequency of seizures (Goldstein 1990).

Further reading

Adams, R. D. and Victor, M. (ed.) (1993). *Principles of neurology* (5th edn). McGraw-Hill, New York.

Yudofsky, S. C. and Hales, R. E. (ed.) (1992) *Textbook of neuropsychiatry* (2nd edn). American Psychiatric Press, Washington, DC.

Laidlaw, J., Richens, A., and Chadwick D. (ed.) (1993). *A textbook of epilepsy* (4th edn). Churchill Livingstone, Edinburgh.

Lishman, W. A. (1987). *Organic psychiatry* (2nd edn). Blackwell, Oxford.

Swash, M. and Oxbury, J. (ed.) (1991). *Clinical neurology*. Churchill Livingstone, Edinburgh.

Introduction

This chapter is concerned with the co-occurrence of psychiatric disorder and physical illness. Psychiatric and physical illness are more often seen together in patients treated in primary care or in general hospital practice than in the practice of psychiatry. When treatment is needed for the psychiatric disorder, it can often be provided by the practitioner who is treating the physical illness. However, the more severe disorders are likely to require treatment by psychiatrists who should have a good knowledge of the problems.

The chapter is in two parts: general considerations, and a review of specific syndromes.

GENERAL CONSIDERATIONS

This first section begins with an account of the epidemiology, recognition, clinical significance, and nature of the association between psychiatric and physical illness. This account is followed by a consideration of four types of association between physical and psychiatric disorder and general treatment:

- psychological factors as causes of physical illness
- psychiatric disorder presenting with physical symptoms
- psychiatric consequences of physical illness
- psychiatric and physical disorders occurring together by chance.

This section ends with an account of the management of psychiatric emergencies, and of psychiatric aspects of medical procedures.

Epidemiology

Physical symptoms and psychiatric disorders commonly occur together in the general population, in patients who consult primary care practitioners, and in those referred to psychiatrists, physicians, surgeons, and other specialists. Although this chapter is principally concerned with psychiatric disorder, it is important to be aware that psychological symptoms of severity insufficient to satisfy diagnostic criteria for a psychiatric disorder may cause considerable morbidity and lead to an increased use of medical services (Johnson *et al.* 1992).

In the *general population* physical illness and physical symptoms are associated with an increase of psychiatric disorder as shown for example in the large American Epidemiological Catchment Area Survey (Wells *et al.* 1988; Kroenke and Price 1993). In *primary care*, surveys have consistently found that patients identified as suffering from psychiatric disorders have high rates of physical morbidity (Shepherd *et al.* 1966). Conversely, as many as 20 per cent of new consultations in primary care are for somatic symptoms for which no physical cause can be found (Bridges and Goldberg 1985).

Numerous studies have found associations between physical and psychiatric disorders among *general hospital in-patients and out-patients*. For example surveys in medical wards, have shown that over a quarter of in-patients have a psychiatric disorder (Feldman *et al.* 1984). The frequency and nature of these disorders depend on the age and sex of the patients and the nature of the specialty of the ward. For example affective and adjustment disorders are more common in the elderly, and drinking problems in younger men (see Chick 1994). Psychological problems are frequent among patients attending emergency clinics, as well as gynaecological and medical out-patient clinics (Van Hemert *et al.* 1993). Organic mental disorders are frequent in geriatric wards, and drinking problems in liver units. (See Mayou and Hawton (1986) for a review of the prevalence of psychiatric disorder in general hospitals.)

Conversely, physical illness is frequent among psychiatric in-patients and out-patients. At least half of this physical illness is undetected even when it may be influencing the psychopathology (Koranyi and Potoczny 1992).

Recognition

Primary-care practitioners often fail to recognize psychiatric disorder in patients with physical illness (Goldberg 1990). Similarly, psychiatric disorder in patients in medical and surgical wards often goes undetected. In a survey of medical wards in an English hospital, Maguire *et al.* (1974) found that half the psychiatric morbidity had not been recognized by physicians or nurses. Subsequent studies have confirmed that many affective and organic disorders and most drinking problems are not detected among patients in general hospitals.

Clinical significance

It is important for several reasons that psychiatric disorder should not be missed when it accompanies physical illness.

(1) Severe conditions are likely to need psychiatric treatment, and may carry a risk of suicide. Moderately severe disorders may also require treatment, and if persistent may delay recovery from the physical illness. As explained above, even mild disorders may cause suffering which could be alleviated.

(2) Psychiatric disorder in patients with chronic medical conditions is associated with markedly worse quality of life (Stewart *et al.* 1989).

(3) Psychiatric disorder may be a cause for poor compliance with medical treatment and of excessive or inappropriate use of medical services.

The possibility that psychiatric disorder may interfere with recovery from physical illness was suggested by the findings of Querido (1959) who studied 1630 patients in a general hospital. He found that medical outcome seven months later was significantly worse in patients who had more psychiatric symptoms at the time of the original illness. Of course it is possible that the original physical illness had been more serious in the patients who originally had the most psychiatric symptoms. However, more recent evidence indicates that mood may affect physical outcome. For example depression in hospital has been reported to predict mortality following acute myocardial infarction (Frasure-Smith *et al.* 1993) and psychosocial interventions to reduce mood disturbance have been claimed to prolong the life of patients with metastatic breast cancer (Spiegel *et al.* 1989).

Classification in DSMIV and ICD10

Psychiatric patients who present to physicians or surgeons rather than to psychiatrists fall into several diagnostic groups. The most relevant of the categories in ICD10 and DSMIV are listed in Table 12.1. Some of these conditions are considered in this chapter or in Chapter 11; the remainder are described in other chapters.

DSMIV has a rubric for **psychological factors affecting physical conditions** in one of four ways: influencing the *course* of the medical condition, interfering with its *treatment*, constituting additional *health risks*, and precipitating or *exacerbating* the *physical condition*. When using this rubric, one or more of six factors may be specified as affecting the medical disorder: a mental disorder; psychological symptoms; personality traits or coping style, maladaptive health behaviours; stress-related physiological responses, or other unspecified factors. The description is so broad that it could be applied to most medical disorders.

DSMIV includes a code for **non-compliance with medical treatment** (there is no ICD equivalent). ICD10 has a category of somatoform disorder named **psychogenic autonomic dysfunction**. The definition states 'the symptoms

Table 12.1 *Classification of psychiatric disorder most often encountered in medical practice**

DSMIV	ICD10
Disorders diagnosed first in infancy, childhood, or adolescence	Behavioural and emotional disorders with onset usually occurring in childhood or adolescence
	Eating disorder (other than pica)
Feeding and eating disorders of infancy or early childhood	
Tic disorders	Tic disorders
Delirium, dementia, amnesic, and other cognitive disorders	Organic, including symptomatic, mental disorder
Mood disorders	Mood (affective) disorders
Anxiety disorders	Neurotic, stress-related, and somatoform disorders
	Reactions to stress and adjustment disorders
Adjustment disorders	Post-traumatic stress disorder
Post-traumatic stress disorder	Dissociative disorders
Dissociative disorders	Somatoform disorders
Somatoform disorder	Intentional production or feigning of symptoms or disabilities, either physical or psychological (factitious disorder)
Factitious disorders	
	Behavioural syndrome associated with psychological disturbance and physical factors
Eating disorders	Eating disorders
Sleep disorders	Non-organic sleep disorder
	Mental and behavioural disorder associated with puerperium, not elsewhere classified
Psychological factors affecting medical condition	Psychological and behavioural factors associated with disorder or disease classified elsewhere
	Disorders of adult personality or behaviour
Non-compliance with medical treatment	Elaboration of physical symptoms for psychological reasons
Malingering	Intentional production or feigning of symptoms

*The order of categories has been changed somewhat to aid comparison of the two systems of classification are presented by the patient as if they were due to physical disorder in a system or organ which is largely or completely under autonomic control, i.e. the cardiovascular, gastrointestinal, and respiratory systems'. Examples include hyperventilation, cardiac neurosis, and gastric neurosis.

The nature of the association between psychiatric and physical disorders

In discussing the association between physical and psychiatric disorder, it is useful to recognize the following five kinds of relationship.

1. Psychological factors as causes of physical illness.
2. Psychiatric disorders presenting with physical symptoms.
3. Psychiatric consequences of physical illness:
   (a) dementia, delirium, and other cognitive disorders (see also Chapter 11)
   (b) stress-related, anxiety, and mood disorders (see also Chapters 6, 7, and 8).
4. Psychiatric and physical disorder occurring together by chance.
5. Psychiatric problems with physical complications:
   (a) deliberate self-harm;
   (b) alcohol and other substance abuse;
   (c) eating disorders.

The first three types of association are discussed in this chapter. The fourth type is clinically important but needs no discussion. Among associations of the fifth type, eating disorders are discussed in this chapter, whilst deliberate self-harm and substance abuse are reviewed in Chapters 13 and 14 respectively.

Psychological factors as causes of physical illness

In the nineteenth century it was widely accepted that psychological factors can play a part in the aetiology of physical illness (Tuke 1872). In the twentieth century psychoanalysts, including Ferenczi, Groddeck, and Adler, suggested that Freud's theory of the conversion of mental disorder into physical symptoms (the theory to explain hysteria) could be applied to physical illness, and they presented case studies to support this idea. From these beginnings there emerged a body of theory and practice known as psychosomatic medicine, in which the term *psychosomatic* was used in a narrower aetiological sense than its original use in the early nineteenth century. (The term was first used by Heinroth to mean 'belonging to the body and the mind' (Bynum 1983).) Other contributions to ideas on psychosomatic medicine came from Pavlov's research on conditioning, Cannon's work on the visceral response of animals to rage and fear, and Wolf and Wolff's studies of human psychological responses to emotion. (See Weiner (1977) for a review of early literature on psychosomatic medicine.)

Psychosomatic theory held that emotional changes in human beings were accompanied by physiological changes and that when emotional changes were persistent or frequent, pathological physical changes could follow. Once physical pathology was established, psychological factors could help to maintain or aggravate it, or to trigger relapse. It was assumed that physical conditions induced in this way would improve if the psychological disturbance improved, either spontaneously or as a result of psychological treatment.

These ideas were taken up enthusiastically in some places, notably in the United States, and two further theories were developed. In the first theory it was proposed that *specific* types of emotional conflict or personality structure could cause specific physical pathology; in the second it was proposed that *non-specific* emotional conflict or stressors could contribute to pathology in constitutionally predisposed people. Two prominent exponents of the first theory were Franz Alexander and Flanders Dunbar. Alexander, a German-born psychoanalyst who migrated to the United States, proposed that seven diseases were psychosomatic: bronchial asthma, rheumatoid arthritis, ulcerative colitis, essential hypertension, neurodermatitis, thyrotoxicosis, and peptic ulcer

(Alexander 1950). Dunbar, who founded the American Psychosomatic Society, published a substantial survey of her own and other clinicians' observations in the book *Emotions and bodily changes* (Dunbar 1954). She described specific personality types which she believed were associated with specific disorders. (This view persists today in relation to ischaemic heart disease (see p. 381). In support of the second theory (that non-specific psychological stressors lead to physical illness in constitutionally predisposed people) several different intervening processes were proposed, including psychophysiological mechanisms (Wolf and Wolff 1947), behavioural processes (Mahl 1953), and hormonal factors (Selye 1950). (These theories and the relevant research have been reviewed by Weiner (1977).)

Subsequent developments

These ideas about psychosomatic medicine were most influential in the years between 1930 and the late 1950s. Subsequently a more general approach to the psychological causation of physical illness developed. It became apparent that many findings reported in the earlier psychosomatic literature were based on faulty research. For example observations were often made on biased sample of patients, notably those selected by physicians for referral to psychiatrists, often there were no control groups, and methods of psychological assessment were subjective and unstandardized. Above all, most of the work was retrospective in that the clinician saw patients with established physical diseases and tried to determine whether certain emotional conflicts, personality characteristics, or stressful events had preceded the onset of those diseases.

Later theories have been more modest and have accepted that physical illness has multiple causes. For instance, Engel and his colleagues proposed a *biopsychosocial* approach to illness. They maintained that a particular kind of emotional disturbance can induce physical pathology, but they applied their idea to a wider group of physical diseases than the seven psychosomatic

diseases proposed by Alexander (Engel 1962). They believed that people are particularly likely to become physically ill if they develop a 'giving-up, given-up complex' in response to actual or threatened 'object loss'. This 'complex' is a combination of depressed mood, 'helplessness', and 'hopelessness'. Clinicians recognize the state of demoralization described by Engel as a common and important reaction to physical illness which may hinder psychological adjustment. However, there is no convincing evidence that it is of causal significance in either the onset or course of physical disease. Nevertheless, Engel's emphasis on the 'bio-psychosocial' nature of illness has been influential, especially as he was himself a physician, not a psychiatrist.

Research has been concerned with a number of psychological and physical processes which may mediate the psychosocial effects on the onset or course of physical disease.

Personality and behavioural characteristics

Research has continued into psychological factors as causes of ischaemic heart disease. The type A behaviour pattern, a list of characteristics derived from the concept of the coronary personality, has not been shown to be a useful concept. However, there is some evidence that more specific behaviours, especially hostility and difficulty in time management, may be of significance (see p. 381).

Alexithymia

Nemiah and Sifneos (1970) suggested that a cause of psychosomatic illness may be *alexithymia*, i.e. inability to recognize and describe feelings, difficulty in discriminating between emotional states and bodily sensations, and inability to fantasize. Although these characteristics are recognizable in clinical practice, there is no agreed way of defining them for research. They have not been shown to constitute a valid syndrome, to be specifically associated with physical disorder, or to have significance in treatment (see Salminen *et al.* (1995) for a review).

Life events

Research has focused on the objective study of stressful *life events* using the Schedule of Recent Experience (Holmes and Rahe 1967) or a modification of it. This method, which has limitations, has been replaced by improved methods of assessment. Despite these advances, the role of life events is still uncertain. If they have a role in the aetiology of physical illness, it is likely to be as precipitants rather than as 'formative' causes (Rosengren *et al.* 1993).

Physiological mechanisms

Other work has dealt with the *mechanisms* through which emotions could induce physical changes. In a pioneering study, Wolf and Wolff (1947) studied a patient with a gastric fistula and found that emotional changes were accompanied by characteristic changes in the colour, motility, and secretory activity of the stomach mucosa. Subsequent research by others has focused on *neuroendocrine mechanisms* and *immune processes* (Herbert and Cohen 1993). These investigations have shown that emotions can be accompanied by physiological changes, but they have not yet shown whether or how such changes can result in pathological conditions.

The effects of treatment

Franz Alexander believed that the psychosomatic disease would clear up or improve in response to psychological treatment. There have been many reports of individual patients with asthma, peptic ulcer, or ulcerative colitis who appeared to improve physically after psychotherapy. These reports do not prove that psychological treatment affected the outcome, and are not supported by the few clinical trials that have been completed (Karasu 1979).

Conclusion

The traditional psychosomatic theories have been disproved, and there is no reason for separating a group of 'psychosomatic disorders'. It is doubtful whether psychological factors by themselves can lead to the onset of physical disease, though they may contribute to its relapse or aggravation. The nature of any mediating immune, metabolic, and other mechanisms is still uncertain. It is now clear that certain behaviours predispose to physical disease, for example smoking, overeating, and excessive drinking of alcohol. Psychological techniques may be valuable in modifying these behaviours and thereby reducing the risk of certain diseases. (See Steptoe and Wardle (1994) for a collection of major papers.)

Psychiatric disorder presenting with physical symptoms

General considerations

Bodily symptoms with no medical explanation occur frequently in the general population (Kroenke and Price 1993), in people attending primary care (Goldberg and Huxley 1980), and in the patients of general hospitals. Most such symptoms are transient and are not associated with psychiatric disorder, and most improve with reassurance. The minority that persist are often associated with psychiatric disorder. Psychiatrists see a small and atypical proportion of this last group. There is some evidence that psychological and behavioural treatments can be effective in these conditions. (See Mayou *et al.* (1995) for reviews.)

The terminology which has been used to describe non-organic physical symptoms is unsatisfactory. In the past the word **hypochondriasis** has been used in a general sense to denote a syndrome with by prominent somatic symptoms. The contemporary usage is more restrictive (see p. 354). The term **somatization** is preferred by some (Lipowski 1988), but unfortunately it too is used with more than one meaning—to denote a psychological mechanism underlying the formation of somatic symptoms, and (as in DSMIV and ICD10) as a syndrome within the category of somatoform disorder. An alternative term, used in this chapter, is **functional somatic symptoms** (Kellner 1994).

Epidemiology

At some time, almost everyone experiences physical symptoms which do not result from significant organic disease. Almost half of people questioned over a six-month period, describe symptoms severe enough to limit their ability to carry out their usual activities (Von Korff *et al.* 1988). Only a minority of these symptoms result in a consultation with a doctor; nevertheless, as many as one in five new consultations in primary care are for somatic symptoms for which no specific cause can be found (Bridges and Goldberg 1985). Whilst many of these symptoms are transient, around 20 per cent are distressing and persistent and take up considerable resources for healthcare (Kroenke and Mangels-dorff 1989). They are also frequent in children (Campo and Futsch 1994).

Non-specific physical complaints are amongst the most common reasons for referral to specialist out-patient clinics and they are frequent among patients in emergency departments. Van Hemert *et al.* (1993) diagnosed a psychiatric disorder in 38 per cent of those who presented with unexplained physical complaints at a Dutch general medical clinic. Functional complaints are especially frequent in certain specialist clinics, in particular pain clinics (see p. 370), neurological clinics (Ron 1994), gastroenterology clinics, and cardiac clinics.

Although the proportion of in-patients with functional somatic symptoms is small, they consume a considerable amount of medical resources. Fink (1992) used Danish case registers to document the use of medical and surgical treatment resources by patients who, during an eight year period, were admitted at least ten times to general hospitals for physical symptoms for which no organic cause was found. Severe psychiatric illness was common (Fink 1995). It was estimated that 3 per cent of the general hospital budget was used for these patients (excluding any costs of psychiatric care). These patients also consume psychiatric resources, because the assessment of unexplained physical symptoms is among the commonest reasons for referral to psychiatric consultation–liaison services.

Aetiology

The aetiological mechanisms underlying functional somatic symptoms are poorly understood. Single physical or psychiatric explanations have been proposed, but it is more probable that most persistent physical symptoms have a complex aetiology involving the misinterpretation of normal bodily sensations or the autonomic symptoms of anxiety. Social and psychological factors seem to predispose to functional somatic symptoms and to reinforce and maintain them; examples of such factors include the experience of previous illness, knowledge of the illness of friends and relatives, overconcern by members of the patient's family, and iatrogenic factors such as inadequate explanation or inconsistent or ambiguous information (Table 12.2). The media may have a role in promoting some symptoms (David and Wessely 1995).

Functional symptoms are a feature of many *psychiatric disorders* (see Table 12.3), most frequently associated with adjustment, mood anxiety disorders, and somatoform.

Unexplained physical symptoms do not fit easily into the diagnostic categories developed for specialist psychiatric practice, and multiple diagnoses are common (for example mood disorder and somatoform disorder). There have been particular problems in the classification of disorders in which there are few psychological symptoms; these are now grouped together as somatoform disorders. There are also cultural differences in the ways that doctors interpret symptoms and make diagnoses. For example Chinese psychiatrists are more likely to diagnose neurasthenia for symptoms which Western psychiatrists would diagnose as major depressive disorder (Kleinman 1982; Lee 1995). As well as recording a psychiatric diagnosis, it is appropriate to record the number or type of symptoms, their effects on behaviour, and whether there is any identifiable pathophysiological disturbance (see Table 12.3). (See Mayou *et al.* 1995 for a review.)

Management

The treatment of functional somatic symptoms presents two general problems. The first is to ensure that all those involved in care have a consistent approach. The second is to make sure

Table 12.2 *Some aetiological factors for functional somatic symptoms*

Predisposing factors
  Personality
  Beliefs about illness
  Previous physical illness

Precipitating factors
  Physiological and pathological factors:
    minor physical pathology
    muscle tension
    hyperventilation
    autonomic arousal
    side-effects of medication
    physiological consequences of diet, alcohol
  Psychiatric disorder
  Stressful events

Perpetuating factors
  Psychiatric disorder (primary or secondary)
  Continuing physical and psychological factors:
    lack of confidants
    attitudes of relations and friends
    iatrogenic factors
    litigation and financial gain
    alternative therapy

---

that patients understand that, although their symptoms are not due to physical illness, their complaints are accepted as real and are taken seriously. To these ends physicians should explain clearly the purpose and results of all investigations, and the likely value of psychological assessment and referral. The psychiatrist should be aware of the results of physical investigations, and the explanation and advice given to the patient by other clinicians.

Assessment

Many patients are reluctant to accept that their somatic symptoms may have psychological causes and that referral to a psychiatrist may be appropriate. Hence the clinician should adopt a tactful and sensitive approach. As noted above, it is important to discover the patient's views about the causes of his symptoms and to discuss them seriously. The patient should be made aware that the doctor believes his symptoms to be real. Physicians and psychiatrists should work together to achieve a consistent approach. The usual principles of history taking and assessment should be applied, though the interview may need to be adapted to suit the patient. Attention should be directed to any of the patient's thoughts and behaviours that accompany the physical symptoms, and to the relatives' reactions. It is important to obtain information from other informants as well as from the patient.

A point about diagnosis needs emphasis. When a patient has unexplained physical symptoms, a

Table 12.3 *Classification of psychiatric disorders which may present with somatic symptoms**

DSMIV	ICD10
Adjustment disorder (Chapter 6) 　Adjustment disorder with physical 　　complaints	Reaction to severe stress and adjustment 　disorders 　　Acute stress reaction 　　Post-traumatic stress disorder 　　Adjustment disorder
Mood (affective) disorders (Chapter 8)	Mood (affective) disorders
Anxiety disorders (Chapter 7) 　Panic disorder 　Obsessive–compulsive disorder 　Generalized anxiety disorder 　Post-traumatic stress disorder	Anxiety disorders
Somatoform disorders 　Somatization disorder 　Conversion disorder (or hysterical 　　neurosis, conversion type) 　Pain disorder 　Hypochondriasis (or hypochondriacal 　　neurosis) 　Body dysmorphic disorder 　Undifferentiated somatoform disorder 　Somatoform disorder not otherwise 　　specified	Somatoform disorders 　Multiple somatoform disorder 　Undifferentiated multiple somatoform 　　disorder 　Hypochondriacal syndrome 　　(hypochondriasis, hypochondriacal 　　neurosis) 　Psychogenic autonomic dysfunction 　Pain syndrome without organic cause 　Other psychogenic disorders of sensation, 　　function and behaviour
Dissociative disorders (Chapter 7)	Other neurotic disorders 　Neurasthenia
Schizophrenia and related disorders 　Schizophrenic disorders (Chapter 9) 　Delusional disorder (Chapter 10)	Schizophrenia, schizotypal states, and 　delusional disorders
Substance-related disorders (Chapter 14) 　Alcohol use 　(Other substances)	Mental and behavioural disorders due to 　psychoactive substance use
Factitious disorder 　With physical symptoms 　With psychological symptoms	
Malingering (V code)	

*Some disorders are considered in other chapters. For these disorders, the chapter number is given in brackets in the column showing the DSMIV classification

Table 12.4 *Somatoform disorders in DSMIV and ICD10*

DSMIV	ICD10
Somatization disorder	Multiple somatization disorder
Conversion disorder	(Dissociative disorder*)
Hypochondriasis	Hypochondriasis
Body dysmorphic disorder	
Pain disorder	Pain syndromes without specific organic cause
Undifferentiated somatoform disorder	Undifferentiated somatoform disorder
Somatoform disorder not otherwise specified	Other psychogenic disorders of sensation, function, and behaviour
	Psychogenic autonomic dysfunction

*In ICD10, dissociative disorders with physical symptoms are classified under Neurotic, stress-related disorders and somatoform, and not under somatoform disorders.

psychiatric diagnosis should be made only on positive grounds. It should not be assumed that such symptoms are of psychological origin merely because they occur in relation to stressful events. Such events are common and may coincide with physical disease which, while not yet identifiable, is far enough advanced to produce symptoms. For the diagnosis of psychiatric disorder, the same strict criteria should be used for the physically ill as for the physically healthy.

Treatment

Many patients with somatic complaints seek repeated medical investigation and reassurance. When all necessary investigations have been carried out, the patient should be told clearly that no further investigation is required. This information should be given with a combination of authority and willingness to discuss the investigations and

their results. After this explanation, the aim should be to combine psychological management with treatment of any associated physical disorder.

It is important to avoid arguments about the causes of the symptoms. Among patients who do not fully accept psychological causes for their symptoms, many are willing to accept that psychological factors may influence their perception of these symptoms. Such patients may then accept guidance on learning to live more positively with their symptoms. Explanation and reassurance are usually effective in cases of recent onset. In chronic cases, however, reassurance is seldom helpful; sometimes repeated reassurance may even encourage increased complacency and seeking of more and more frequent reassurance (Salkovskis and Warwick 1986).

Specific treatments should be based on a formulation of the individual's difficulties. They include antidepressant medication for an underlying mood disorder and specific behavioural techniques such as anxiety management, graded practice, and cognitive therapy (see Sharpe *et al.* 1992) to modify beliefs about the origin and significance of symptoms (see Mayou *et al.* (1995) for reviews).

Somatoform disorders

This category was introduced into DSMIII with the defining feature of 'physical symptoms suggesting a physical disorder for which there are no demonstrable organic findings or known physiological mechanism, and for which there is a strong evidence, or a strong presumption, that the symptoms are linked to psychological factors or conflicts'. Unfortunately, despite the close collaboration in the preparation of ICD10 and DSMIV, differences have arisen between these two classifications (Table 12.4). There are difficulties about the whole concept of somatoform disorder.

1. The lack of any clear operational definitions for the overall category and the unsatisfactory description of some of the subcategories.
2. Some types of somatoform disorder (especially hypochondriasis and somatization disorder) are so enduring that they could be classified as personality disorders rather than as mental disorders (Bass and Mayou 1995).
3. Most patients have clinical features which fit the criteria for more than one diagnostic category.

Somatization disorder

The essential feature of somatization disorder is multiple somatic complaints of long duration, beginning before the age of 30. A group of psychiatrists in St Louis (Perley and Guze 1962) originally described a syndrome of chronic multiple somatic complaints with no identified organic cause which they regarded as a form of hysteria and named Briquet's syndrome after the nineteenth-century French physician who wrote an important monograph on hysteria. In fact, Briquet did not describe the syndrome to which his name was given and the symptoms do not satisfy the usual criteria for hysteria. A similar syndrome was introduced in DSMIII and named somatization disorder. In DSMIIIR diagnostic criteria were very restrictive, so that many patients with chronic multiple physical complaints were excluded and had to be allocated to the residual category of *undifferentiated somatoform disorder* (Escobar *et al.* 1989). In DSMIV the criteria for diagnosis have been made somewhat less restrictive, but it is unlikely that this change will fully resolve the diagnostic difficulties.

Although the prevalence in the general population of the syndrome as defined in DSMIIIR was 0.38 per cent in the ECA epidemiological study (Swartz *et al.* 1986), the prevalence of a less narrowly defined syndrome requiring fewer types of symptoms was much greater at 4.4 per cent (Escobar *et al.* 1989).

The St Louis group proposed a familial association between somatization disorder in females and sociopathy and alcoholism in their male relatives. The group also proposed that follow-up studies and family studies show somatization disorder to be a single stable syndrome (Guze *et al.* 1986). However, there must be doubts about these views because some patients diagnosed as having somatization disorder fulfil criteria for other DSM diagnoses, especially depression. The chronicity essential for the diagnosis of somatization disorder has led to the suggestion that it may best be conceived as a personality disorder.

Somatization disorder is difficult to treat. Continuing care by one doctor using only the minimum of essential investigations can reduce the patients' use of health services and may improve their functional state (Smith *et al.* 1986; Smith 1995). Psychiatric assessment can help to clarify a complicated history, to negotiate a simplified pattern of care, and to agree the aims of treatment with the patient, the family, and the responsible physician. The aim of treatment is often to limit further progression rather than to cure.

Conversion disorder

In DSMIV this category denotes symptoms or deficits involving voluntary motor or sensory functions. Dissociative amnesia, fugue, and identity disorder are classed together in a separate category of dissociative disorders. In ICD10 all these conditions are classified as dissociative disorder (see p. 186). In the following account the term conversion disorder is used as a convenient term for a particular group of medically unexplained physical symptoms.

Conversion and dissociative disorders were previously known as hysteria, a condition which has a controversial history (Micale 1990; Mace 1992; Porter 1995). The term conversion is derived from Freud's hypothesis that the somatic symptom is a symbolic solution of an unconscious conflict which enables reduction in anxiety and keeps the conflict out of conscious awareness (a process called primary gain). Current theories stress the importance of social factors, and regard conversion symptoms as maladaptive variants of normal processes which in some circumstances are culturally accepted or even encouraged (for example possession states). Current diagnostic criteria do not refer to conversion or other hypothetical psychological mechanisms but require that psychological factors are associated with the onset and course of the disorder. *Belle indifference*, an apparent lack of concern about the symptoms, was stressed in past accounts of the condition, but it is not an invariable feature.

Conversion symptoms are common among people attending doctors. Conversion disorder, as defined in DSMIV, is much less common, with a reported prevalence of between 1 and 3 per cent among patients referred to psychiatrists. However, short-lived conversion syndromes, such as difficulty in walking or sensory complaints, are seen regularly in emergency departments.

The pattern of symptoms reflects the medical knowledge and sophistication of the patient and his immediate environment. Symptoms are often more dramatic in developing than in developed countries. The influence on social factors on the symptoms makes it difficult to assess claims that the incidence of hysteria is decreasing in developed countries. It is more probable that the symptoms are changing, and that functional somatic symptoms are now more common than dramatic motor or sensory disturbances.

In DSMIV conversion disorder is divided as follows:

1. **With motor symptoms or deficits** When a limb with psychogenic *paralysis* is examined, the lack of movement is often seen to result from simultaneous action of flexors and extensors. If no muscle activity follows a request to move the part, other tests usually show that the muscles are capable of reacting when the patient's attention is directed elsewhere. The pattern of paralysis does not conform to the innervation of the part. Appropriate changes in reflexes are not present; in particular the plantar response always remains flexor. Wasting is absent except in chronic cases, when disuse atrophy may occasionally be seen. Psychogenic *disorders of gait* are usually of a striking kind that draws attention to the patient and are worse when is observed. The pattern does not resemble any described in known neurological conditions. Although dramatic unsteadiness may appear when balance is tested, it often disappears when the patient's attention is directed elsewhere. Typically, psychogenic *tremor* is coarse and involves the whole limb. It worsens when attention is drawn to it, but so do many tremors with neurological causes. Choreoathetoid movements with organic cause are easily mistaken for psychogenic symptoms. Disease of the nervous system should always be considered carefully before diagnosing any abnormal movement as psychogenic. Psychogenic *dysphonia* and *mutism* are not accompanied by any disorder of the lips, tongue, palate, or vocal cords, and the patient is able to cough normally (House and Andrews 1988). They are usually more extreme than corresponding conditions caused by organic lesions. *Globus hystericus*, a feeling of a lump in the throat, is described on p. 386.

2. **With sensory symptoms or deficits** Dissociative sensory symptoms include anaesthesia, paraesthesia, hyperaesthesia, and pain, as well as deafness and blindness. In general, the sensory changes are distinguished from those in organic disease by a distribution that does not confirm to the known innervation of the part, by the varying intensity, and by their response to suggestion. The last point must be used cautiously in diagnosis because, among suggestible patients, sensory

symptoms with an organic cause may also respond to suggestion. *Hyperaesthesiae* are usually felt in the head or abdomen, and may be described as painful or burning.

Psychogenic *blindness* may taken the form of a concentric diminution of the visual field ('tunnel vision') but other patterns of field defect occur as well. The blindness is not accompanied by changes in pupillary reflexes, and there may be indirect evidence that the person can see, for example avoidance of bumping into furniture. The findings of perimetry are variable, whilst visual evoked responses are normal. Similar considerations apply to psychogenic *deafness*.

3. **With seizures and convulsions** Psychogenic *convulsions* can usually be distinguished from epilepsy in three ways. The patient does not become unconscious, though he may be inaccessible, the pattern of movements does not show a regular and stereotyped form of seizure, there is no incontinence, cyanosis, or injury, and the tongue is not bitten. Also, EEG findings are normal (see also p. 337).

Assessment of conversion disorder depends on history and examination to elicit the characteristic features, and on very considerable caution in excluding neurological and other physical causes. Many medical conditions may cause symptoms that are initially mistaken for conversation, for example multiple sclerosis and myasthenia gravis. Apart from physical disorders, differential diagnosis includes other somatoform conditions, disassociated disorders, factitious disorders, and malingering.

The course of the condition is very variable. It is probable that most episodes are transient but a minority have a chronic course. Reoccurrence is common.

Treatment follows the general principles outlined on pp. 348 and 351. It is essential that the doctor is sympathetic and offers the patient an acceptable explanation and prognosis which will encourage rapid recovery. Offers of physiotherapy or other symptom relief may have a role, but it is necessary to avoid measures which in any way reinforce the value of the symptom to the patient or put him in a position in which recovery would be seen as an admission of psychological weakness.

Hypochondriasis

The term hypochondriasis is one of the oldest medical terms, which was originally used to describe disorders believed to be due to disease of the organs situated in the hypochondrium. Since then, the term has been used in many ways. There have been disputes as to the extent to which it should refer to excessive concern about symptoms resulting from a physical disorder (Shorter and Parker 1995) as well as to symptoms for which no disorder can be found. There is also disagreement as to whether hypochondriasis is a primary psychiatric disorder or is always secondary to another condition.

Hypochondriasis is defined by DSMIV and ICD10 in terms of *disease conviction* and *disease phobia*. DSMIV requires a 'preoccupation with a fear or belief of having a serious disease based on the individual's interpretation of physical signs of sensations as evidence of physical illness. Appropriate physical evaluation does not support the diagnosis of any physical disorder than can account for the physical signs or sensations or for the individual's unrealistic interpretation of them. The fear of having, or belief that one has a disease, persists despite medical reassurance.' The definition goes on to exclude patients with panic disorder or delusions, and requires that symptoms have been present for at least six months. Not all hypochondriasis fits these criteria. For example Barsky *et al.* (1992) have described transient forms of the disorder. They have suggested also that primary and secondary types should be recognized.

The true **prevalence** of hypochondriasis is uncertain, but it is common amongst patients attending general medical clinics (Barsky *et al.* 1991; Van Hemert *et al.* 1993). Co-morbidity with depression and other disorders is frequent. The **cause** is unknown; recent formulations have emphasized the similarity of hypochondriasis to panic disorder in which there are also misinterpretations of the significance of minor physical symptoms (autonomic symptoms in the case of panic disorder, Salkovskis and Clark (1993)). The **course** is often persistent (Noyes *et al.* 1994).

Treatment of hypochondriasis follows the general principles outlined on p. 351. Since the disorder is often chronic or recurrent, management is difficult. Repeated reassurance is unhelpful but measures to control investigations, to correct misattributions, and to encourage constructive ways of coping with symptoms can be helpful. A controlled trial has shown that benefit can result from intensive cognitive behavioural treatment (Warwick *et al.* 1995).

Dysmorphophobia

Body dysmorphophobic disorder is the DSM term for a subgroup of the broader but ill-defined clinical syndrome of **dysmorphophobia** which was first described by Morselli (1886) as 'a subjective description of ugliness and physical defect which the patient feels is noticeable to others'. For convenience, this wider syndrome is considered in this section.

The typical patient with dysmorphophobia is convinced that some part of his body is too large, too small, or misshapen. To other people the appearance is normal, or there is a trivial abnormality. In the latter case, it may be difficult to decide whether the preoccupation is disproportionate. The common complaints are about the nose, ears, mouth, breasts, buttocks, and penis, but any part of the body may be involved. The patient may be constantly preoccupied with and tormented by his mistaken belief. It seems to him that other people notice and talk about his supposed deformity. He may blame all his other difficulties on it: if only his nose were a better shape, he would be more successful in his work, social life, and sexual relationships. The severe cases described in the psychiatric literature are infrequent, but less severe forms of dysmorphophobia are more common, especially in plastic surgery and dermatology clinics (see Phillips (1991) and Hollander *et al.* (1992) for reviews).

Some patients with this syndrome meet diagnostic criteria for other disorders. Thus Hay (1970*b*) studied twelve men and five women with the condition, and found that eleven had severe personality disorder, five had schizophrenia, and one had a depressive illness. In these patients the preoccupation is usually delusional.

The term **body dysmorphic disorder** was introduced in DSMIII to denote dysmorphophobia that is not secondary to another psychiatric disorder. It denotes 'a preoccupation with some imagined defect in appearance' which 'is not of delusional intensity' and which is usually an overvalued idea (McKenna 1984). The separate validity of this syndrome has not yet been established, and it is not included in ICD10.

Treatment When dysmorphophobia is secondary to a psychiatric disorder such as schizophrenia or major depression, the latter should be treated in the usual way. The treatment of body dysmorphic disorder is often difficult. It should be explained tactfully that there is no real deformity and that some people develop mistaken beliefs about their appearance, for example through misinterpreting overheard remarks. Some patients are helped by such reassurance and by continued support, but many are not. Counselling should be provided for any occupational, social, or sexual difficulties that accompany the condition.

Cosmetic surgery is usually successful for patients who have clear reasons for requesting an operation, whether there is a definite deformity or the reasons are connected with fashion or vanity. It is often said that surgery is contraindicated for those who have body dysmorphic disorder, but there is evidence that many such patients are helped by surgery which is followed by improvement of self-esteem and confidence (Goodacre and Mayou 1995). However, a small proportion of such patients remain very dissatisfied after operations. Selection for surgery requires careful assessment of the patient's expectations; in general, those who do not have realistic views of the possible benefits of surgery have a poor prognosis. This assessment is difficult, and collaboration between plastic surgeons, psychiatrists, and psychologists is valuable (Bradbury 1994).

There is some evidence of beneficial effects from *antidepressant medication*, especially in patients with prominent depressive symptoms (Phillips *et al.* 1993). Whatever specific treatment is offered, it is essential that patients feel that their views have been listened to sympathetically. Unintended rebuffs by surgeons or psychiatrists can exacerbate the difficulties of management. (See Goodacre and Mayou (1995) for a review of treatment of dysmorphophobia.)

Somatoform disorder not otherwise specified

This residual diagnostic category is used for a wide range of somatoform symptoms that do not meet the criteria for the specific somatoform categories discussed so far or for adjustment disorder with physical complaints. It includes the specific syndrome of pseudocyesis as well as non-psychotic hypochondriacal symptoms of less than six months' duration.

Pain disorder

This term denotes patients with chronic pain which is not caused by any physical or specific psychiatric disorder. DSMIV states that the essential feature of this disorder is pain that is the predominant focus of the clinical presentation and is of sufficient severity to cause distress or impairment of functioning, and that either no organic pathology or pathophysiological mechanism has been found to account for the pain, or, when there is related organic pathology, the pain or resulting social or occupational impairment is grossly in excess of what would be expected from the physical findings.

Factitious disorder

This term was introduced in DSMIII to refer to self-inflicted signs and symptoms. The term Munchausen syndrome is sometimes used to denote a small but striking subgroup of the condition which is seen mainly in emergency departments and has a poor prognosis (see below).

As defined in DSMIV factitious disorder denotes the 'intentional production or feigning of physical or psychological symptoms which can be attributed to a need to assume the sick role'. The category is divided further into cases with psychological symptoms only, those with physical symptoms only, and those with both. The classification in ICD10 is similar. Factitious disorder differs from malingering in that it does not bring any external reward such as financial compensation. The prevalence of factitious disorder is not known with certainty, but it accounts for about 1 per cent of referrals to consultation liaison services.

Reich and Gottfried (1983) described 41 cases, of whom 30 were women. Most had worked in occupations related to medicine. There were four principal clinical groups: self-induced infections, simulated specific illnesses with no actual disorder, chronic wounds, and self-medication. Many were willing to accept psychological assessment and treatment. Common presentations of factitious disorder include dermatitis artefacta (Sneddon 1983), pyrexia of unknown origin, bruising disorders, brittle diabetes (Schade *et al.* 1985), and diarrhoea of unknown cause. Psychological presentations include feigned psychosis (Hay *et al.* 1993) and feigned grief over fictitious bereavement. Diagnosis depends upon careful observation for discrepancies among the clinical findings. (See Sutherland and Rodin (1990), Wallach (1994), and Nordmeyer (1994) for reviews of factitious disorder.)

Treatment is difficult. The suspicion of a factitious cause for physical symptoms should lead to a careful review of the available information, including the history given by informants as well as that provided by the patient. A psychiatrist may be able to assist in this assessment, and in cases of doubt further specialist physical investigation may be needed (Wallach 1994). Additional evidence may be obtained by careful observation of the patient, but the ethical and legal aspects of any decision to make covert observations should be considered. Once the diagnosis seems likely the senior doctor should explain to the patient the findings and their implications. This should be done in a way that conveys an understanding of the patient's distress and enables a full discussion of its possible causes. Although some patients admit their behaviour, others persistently deny that the symptoms are self-inflicted. In the latter cases management should be directed to any associated psychological and social difficulties, and other appropriate care provided with the patient clearly aware of the doctors' views about aetiology.

Munchausen syndrome

Asher (1951) suggested the term Munchausen syndrome for the patient who is 'admitted to hospital with apparent acute illness supported by a plausible or dramatic history. Usually his story is largely made up of falsehoods; he is found to have attended and deceived an astounding number of hospitals; and he nearly always discharged himself against advice after quarrelling violently with both doctors and nurses. A large number of scars is particularly characteristic of this condition.' This type of factitious disorder occurs mainly in early adult life and among men, and is usually encountered in dramatic manner in emergency departments. The presenting symptoms may be of any kind, including psychiatric symptoms; initially the impression of a medical emergency may be convincing. There is gross lying (pseudologia fantastica), which includes the giving of false names and invented medical histories. There may be self-inflicted wounds or infections, and patients frequently demand powerful analgesics. They may obstruct efforts to obtain information about themselves and may interfere with diagnostic investigations. They invariably discharge themselves prematurely. When further information is obtained, it often reveals many recurrent previous simulated illnesses.

Such patients suffer from profound disorder of personality, and often describe major deprivation and disturbance in early life. Treatment is almost invariably refused. The prognosis is very poor with no published evidence of recovery.

Munchausen syndrome by proxy

In 1977 Meadow described a form of child abuse in which parents (or other carers) give false accounts of symptoms in their children and may fake signs of illness (Meadow 1985; Sigal *et al.* 1989). They seek repeated medical investigations and needless treatment for the children. Although mentioned here because its popular name resembles that of one kind of factitious disorder, it is not a factitious disorder as defined in DSMIV.

The signs reported most commonly are neurological signs, bleeding, and rashes. Some children collude in the production of the symptoms and signs. The perpetrators usually have severe personality disorders and may themselves suffer from factitious disorder. Hazards for the children include disruption of education and social development. The prognosis is probably poor for both children and perpetrators, and there is a significant mortality. Some children may progress to the adult factitious disorder (Bools *et al.* 1993). Occasional cases of murder of children by professional carers have been described as an extreme form of this disorder.

Malingering

Malingering is the fraudulent simulation or exaggeration of symptoms. In DSMIV it is said to differ from factitious disorder in that the production of symptoms is motivated by external incentives, whereas in factitious disorder there are no external incentives but a psychological need for the sick role. This distinction as well as the distinction from somatoform disorder can be difficult.

Malingering occurs most often among prisoners, the military, and people seeking compensation for accidents. Before malingering is diagnosed, there should always be a full medical examination. DSMIV suggests that malingering should be strongly suspected if the following are observed: marked discrepancy between the person's claimed stress or disability and the objective findings, lack of cooperation during the diagnostic evaluation and treatment, the presence of antisocial personality disorder, and the medicolegal context. When the diagnosis is certain, the patient should be informed tactfully. He should be encouraged to deal more appropriately with any problems that led to the symptoms, and in appropriate cases offered a face-saving way to give up the symptoms.

Psychiatric consequences of physical illness

Many serious physical illnesses or surgical procedures are accompanied by delirium or dementia, and these disorders are more common among the elderly. *Delirium*, *dementia*, and *cognitive disorders* associated with specific medical conditions are discussed in Chapter 11. This section is concerned with *emotional disorders* consequent upon physical illness.

The most conspicuous feature of the emotional response of most people to serious physical illness is their resilience. Although distress is common, only in a minority is it severe enough to be classified as a psychiatric disorder. The general features of emotional reactions to physical illness have been described on p. 146, together with accounts of psychological defence mechanisms, coping responses, and the concepts of illness behaviour and sick role.

Although this chapter is concerned principally with psychiatric disorders, those who care for physically ill people should be aware of other psychological consequences of such illness for patients and for families. These consequences are:

- adjustment disorders
- impact on social function
- effects on quality of everyday life
- effects on the family
- effects on compliance and on the use of medical resources.

The usual emotional response to acute illness is anxiety, but as illness progresses depressive symptoms become more prominent. The defence mechanisms of denial or minimization (see p. 136) are frequent, and may result in an unwillingness to accept the diagnosis or its implications. Such denial is usually short lived but, if prolonged, may lead to poor compliance with treatment.

There are difficulties in the use of standard criteria for the diagnosis of psychiatric disorder in the physically ill (Cooper 1990; Von Korff 1992). Physical illness may cause symptoms such as fatigue and poor sleep, which are among the criteria of psychiatric disorder. Although modifications to the criteria have been suggested to make them more appropriate for patients with physical as well as psychiatric disorder, none is wholly satisfactory. Psychiatric disorder is present in up to 30 per cent of patients with serious acute, recurrent, or progressive physical illness. The most frequent psychiatric diagnosis among the physically ill is adjustment disorder. Apart from 'organic' diagnoses, and adjustment disorders, the most frequent psychiatric disorders are, major depression, and anxiety disorder (Fifer *et al.* 1994). Other conditions which are less frequent include somatoform disorder and blood injury phobia (Page 1994). (See Mayou and Sharpe (1995) for a review of the epidemiology of psychiatric disorder in association with physical illness.)

Determinants of psychological and social responses to illness

The prevalence and nature of psychiatric disorder is determined by factors relating to the illness and its treatment on one hand, and to the patient's personality and circumstances on the other (Table 12.5).

The physical illness

Certain types of physical illness are particularly likely to provoke serious psychiatric consequences. These include life-threatening acute illnesses and recurrent progressive conditions. Psychiatric disorder is more common in *chronic illness* when it is particularly associated with *distressing symptoms* such as severe pain, persistent vomiting, and severe breathlessness, with *unpleasant treatment*, with demanding self-care, and with disability. Physical illness is likely to have more serious psychological consequences when its effects are particularly significant to the patient's life, for example

Table 12.5 *Determinants of the occurrence of psychiatric disorder among physically ill patients*

The physical disease as a cause of:
Symptomatic psychiatric disorder
Threat to normal life
Disability
Pain

Nature of the treatment
Side-effects
Mutilation
Demands for self-care

Factors in the patient
Psychological vulnerability
Social circumstances
Other life stresses

Reactions of others
Family
Employers
Doctors

---

arthritis of the hands of a pianist. Physical illnesses particularly associated with psychiatric symptoms include infections, endocrine disorders, neurological disorders, and malignant conditions.

Psychological symptoms induced directly by physical illness

The many psychological symptoms that can be caused directly by physical illness are shown in Table 12.6. Because all the symptoms listed in this table are commonly encountered in psychiatric practice, the psychiatrist should be on the look-out for undetected physical illness in any patient presenting with such symptoms. Occasionally they are presenting features before physical symptoms have become conspicuous (Fava *et al.* 1994). Some of the physical conditions listed in Table 12.6 are

considered further in the second part of this chapter dealing with individual syndromes.

Physical illnesses may induce affective disorders, anxiety disorders, and paranoid disorders. In the latter suspicion and resentment may be directed to staff and relatives. There is no clouding of consciousness. Response to antipsychotic medication is good.

Treatments causing physical symptoms

Table 12.7 lists some commonly used therapeutic **drugs** that may produce psychiatric symptoms as side-effects. Whenever psychological symptoms are found in a medical or surgical patient, the possibility should be considered that they have been induced by medication. **Other treatments** associated with psychiatric disorder include radiotherapy, renal dialysis, cancer chemotherapy, and mutilating operations such as mastectomy.

Individual psychological vulnerability

Patients who have histories of previous psychological problems in relation to stress, and those who have suffered other adverse life events or are in difficult social circumstances, are at greater risk of both acute and persistent psychiatric distress associated with physical illness.

The reaction of others

The reactions of family, friends, employers, and doctors may affect the psychological impact of physical illness. They may reduce its effect by their support, reassurance, and other help, or they may increase it by their excessive caution, contradictory advice, or lack of sympathy.

Quality of life

Physical illness and disability may have widespread effects on social adjustment. Patients vary in their capacity to adjust; most manage well but a few develop social handicaps out of proportion to the severity of the illness. A few patients find advantage in physical illness, for example as an excuse to avoid responsibilities, or as an opportu-

Table 12.6 *Some organic causes of common psychiatric symptoms*

Depression	Carcinoma, infections, neurological disorders including dementias, diabetes, thyroid disorder, Addison's disease, systemic lupus erythematosus
Anxiety	Hyperthyroidism, hyperventilation, phaeochromocytoma, hypoglycaemia, neurological disorders, drug withdrawal
Fatigue	Anaemia, sleep disorders, chronic infection, diabetes, hypothyroidism, Addison's disease, carcinoma, Cushing's syndrome, radiotherapy
Weakness	Myasthenia gravis, McArdle's disease and primary muscle disorder, peripheral neuropathy, other neurological disorders
Episodes of epilepsy	Hypoglycaemia, phaeochromocytoma, disturbed behaviour porphyria, early dementia, toxic states, transient global amnesia
Headache	Migraine, giant-cell arteritis, space-occupying lesions
Loss of weight	Carcinoma, diabetes, tuberculosis, hyperthyroidism, malabsorption

nity to reconsider their way of life and improve its quality. Sexual function is often affected by physical illnesses and treatments, of which the most important are listed in Table 12.8. This topic is discussed further on p. 491.

Effects on the family

Physical illness may place a considerable burden on the family (Covinsky *et al.* 1994). Although family ties may be strengthened when a member becomes ill, others may experience considerable distress or develop psychiatric disorder. These effects may result in increased use of medical services by other family members (Patrick *et al.* 1992). The consequences of illness in a patient are significant for children, especially when they have to assist in caring for a parent (Armistead *et al.* 1995; Jenkins and Wingate 1994).

Prevention

Good care for the physical illness is the most important factor in minimizing the psychological and psychiatric consequences. Additional help should be focused on patients suffering illnesses or undergoing treatment which are known to be associated with psychiatric disorder, and on patients who are psychologically vulnerable. Advice on ways of coping with the effects of treatment are often helpful. Written booklets and audio or video tapes are useful in providing information about the effects of the disorder, but patients and families also need to be able to ask questions and discuss their individual concerns.

Management of psychiatric disorder associated with physical illness

Assessment is similar to that for psychiatric disorder in other circumstances, except that the

Table 12.7 *Some drugs with psychiatric side-effects*

Delirium	Central nervous system depressants (hypnotic withdrawal, sedatives, alcohol, antidepressants, neuroleptics, anticonvulsants, antihistamines), anticholinergic drugs, beta-blockers, digoxin, cimetidine
Psychotic symptoms	Appetite suppressants, sympathomimetic drugs, beta-blockers, corticosteroids, L-dopa, indomethacin
Mood disorder, depression	Antihypertensive drugs, oral contraceptives, neuroleptics, anticonvulsants, corticosteroids, L-dopa, calcium-channel blockers
Elation	Antidepressants, corticosteroids, anticholinergic drugs, isoniazid
Behavioural disturbance	Benzodiazepines, neuroleptics

doctor needs to be well informed about the nature and prognosis of the physical illness. Doctors should be aware that certain symptoms, such as tiredness and malaise, may be features of both physical and psychiatric disorders. They should be able to distinguish anxiety and depressive disorders from normal emotional responses to physical illness and its treatment. This distinction is based partly on clinical experience of the reactions of other patients with similar illness, and partly on eliciting symptoms that seldom occur in normal distress (such as hopelessness, guilt, loss of interest, and severe insomnia). Although emotional distress is inevitable in physical illness, it can be reduced by emotional support, appropriate reassurance, advice, information, and practical help. When physical illness is accompanied by a psychiatric disorder, it can be treated with the methods appropriate for the same disorder occurring in a physically healthy person.

Drug therapy Hypnotic and anxiolytic drugs are valuable for short periods when distress is severe, for example during treatment in hospital. The indications for antidepressants are probably the same as those for patients who are not physically ill, but there have been too few satisfactory trials to be certain about the precise indications for the drug treatment of depression in medically ill patients. The side-effects and interactions of a drug should be considered before prescribing for patients who are physically ill.

Counselling Counselling should include information about the physical illness, and a discussion of practical issues concerning its likely effects. Counselling can be given to a group, but this is not always acceptable to the patients most in need of help.

Cognitive and behavioural methods Behavioural interventions (Epstein 1992) have been widely used and much medical care is behavioural in a general sense. Specific interventions have varied aims, for example to promote primary prevention, prevent side-effects of treatment, increase medication compliance, reduce disproportionate disability, and delay progression of chronic desease (e.g. cancer, angina). Cognitive behavioural methods are increasingly popular, but most clinical trials have

Table 12.8 *Some medical conditions associated with impaired sexual function*

Endocrine disorder
    Diabetes
    Hypogonadism
    Hypopituitarism

Cardiovascular disorders
    Myocardial infarction
    Vascular disease

Repiratory failure

Chronic renal failure

Neurological disorders
    Spinal cord damage
    Damage to higher centres

Pelvic surgery

Disabling arthritis

Medication
    Anticholinergic drugs
    Hormones
    Psychotropic drugs (phenothiazines,
        antidepressants)
    Antihypertensive drugs
    Diuretics
    L-Dopa
    Indomethacin

---

been concerned with the treatment of chronic pain (Steptoe and Wardle 1994).

Psychiatric and physical disorders occurring together by chance

Psychiatric and physical disorders often arise independently of one another and then interact.

Psychiatric disorder may affect the patient's response to physical symptoms and increase the problems of medical management. For example an eating disorder may greatly complicate the treatment of diabetes. Also, a coincident psychiatric disorder may be overlooked because the symptoms of that disorder have been ascribed, wrongly, to the accompanying physical illness.

The organization of psychiatric care for physically ill patients

Most psychological care of the physically ill is provided by doctors other than psychiatrists either in primary care or in the general medical services. Therefore it is important that medical, nursing, and other professional staff have the knowledge and skills to provide appropriate care and to identify patients who need specialist treatment. *Screening instruments* for psychological problems may have a role provided that staff have the skill to interpret and use findings (Goldberg 1990; Zimmerman *et al.* 1994).

The role of general hospital staff

General hospital medical teams assess and treat many of the psychological and social problems of the patients under their care. Although all the members of the team need to be able to manage these problems, units where psychological problems are frequent, such as oncology units, may have some staff with special training in counselling. Psychiatrists and psychologists may provide training and supervision for medical and other members of staff, and care for the patients who require specialist treatment.

Consultation and liaison psychiatry

Psychiatric services for general hospitals are widely referred to as consultation–liaison services. The two components of this term refer to two separate ways of conducting psychiatric work in a general hospital. In *consultation* work, the psychiatrist is available to give opinions on patients referred by physicians and surgeons. In *liaison* work the psychiatrist is a member of a medical or surgical team, and offers advice about any patient to whose care he feels able to contribute. The liaison psychiatrist also assists other staff to deal with day-to-day psychological problems encountered in their work, including the problems of patients whom he does not interview himself. One of the aims of liaison work is to increase the skills of other staff in assessment and management of psychological problems; in contrast, the consultation approach implicitly assumes that the other staff possess these skills. In practice, most psychiatrists work in a way that combines elements of the two approaches. Either approach requires close personal professional contact between the psychiatrist and the physician.

Consultation and liaison units vary in their size and organization. Some are staffed entirely by psychiatrists, and others by a team of psychiatrists, nurses, social workers, and clinical psychologists. In some countries clinical psychologists provide a separate 'behavioural medicine' service. Some liaison services have in-patient beds for patients who are both medically ill and psychiatrically disturbed. In a few North American general hospitals with large consultation–liaison services, up to 5 per cent of all admissions are referred to psychiatrists. In the United Kingdom and many other countries, a smaller proportion is referred and most referrals are for the assessment of emergencies including deliberate self-harm (see p. 422).

(For further information about consultation and liaison psychiatry see Stoudemire and Fogel (1993).)

The psychiatric consultation

The consultation has two parts: assessment of the patient and communication with the doctor making the referral. Assessment is not essentially different from that of any other patient referred for a psychiatric opinion; it is necessary to take into account the patient's physical state and willingness to see a psychiatrist. On receiving the request for a consultation the psychiatrist should make sure that the referring doctor has discussed the psychiatric referral with the patient. Before interviewing the patient the psychiatrist reads the relevant medical notes and asks the nursing staff about the patient's mental state and behaviour. The psychiatrist finds out what treatment the patient is receiving, and if necessary consults a reference book about the side-effects of any drugs.

At the start of the interview the psychiatrist makes clear to the patient the purpose of the consultation. It may be necessary to discuss the patient's anxieties about seeing a psychiatrist and to explain how the interview may contribute to the treatment plan. Next, an appropriately detailed history is obtained and the mental state examined. Usually the physical state is already recorded in the notes, but occasionally it will be necessary to extend the examination of the nervous system. It is essential for the psychiatrist to have a full understanding of the patient's physical condition. At this stage it may be necessary to ask further questions of the ward staff or social worker, to interview relatives, or to telephone the family doctor and enquire about the patient's social background and any previous psychiatric disorder.

The psychiatrist should keep separate full notes of the examination of the patient and of interviews with informants. His entry in the medical notes should differ from conventional psychiatric case notes. The entry should be brief and free of jargon, and should contain only essential background information. It should omit confidential information as far as possible and should concentrate on practical issues, including answering the questions raised by the referring doctor. When an opinion is entered in the medical notes, the principles are similar to those adopted in writing to the general practitioner (see p. 50). It is important to make clear the nature of any immediate treatment that is recommended, and who is to carry it out. If the assessment is provisional until other informants have been interviewed, the psychiatrist should state when the final opinion will be given. It is often appropriate to discuss the proposed plan of management with the consultant, ward doctor, or nurse in charge before writing a final opinion.

In this way the psychiatrist can make sure that recommendations are feasible and acceptable, and that answers have been given to the relevant questions about the patient. The note should be signed legibly, and should tell the ward staff where the psychiatrist or a deputy can be found should further help be required.

Recommendations about treatment are similar to those for a similar psychiatric disorder in a physically well patient. When psychiatric drugs are prescribed, attention should be paid to the possible effects of the patient's physical state on their metabolism and excretion, and to any possible interactions with other drugs prescribed for the physical illness. A realistic assessment should be made of the amount of supervision available on a medical or surgical ward, for example for a depressed patient with suicidal ideas. No undue demands should be made, but with support from a psychiatrist the nursing staff can manage most brief psychiatric disorders that arise in a general hospital.

Psychiatric emergencies in a general hospital

The successful management of a psychiatric emergency depends strongly on the initial clinical interview. The aims are to establish a good relationship with the patient, elicit information from the patient and other informants, and observe the patient's behaviour and mental state.

Although the pressures on the doctor in an emergency often make it difficult to follow this systematic approach, time can be saved and mistakes avoided if the assessment is as complete as the circumstances permit, and a calm and deliberate approach is adopted. If the patient is actually or potentially violent, it is essential to arrange for adequate but unobtrusive help to be available. If restraint cannot be avoided, it should be accomplished quickly by an adequate number of people using the minimum of force. Staff should always avoid attempting single-handed restraint. Physical contact (including physical examination) should not be attempted unless the purpose has been clearly understood by and agreed with the patient. Extreme caution is of course, required with a patient thought to possess any kind of offensive weapon.

Emergency drug treatment of disturbed or violent patients

Diazepam (5–10 mg) may be useful for a patient who is frightened. For a more disturbed patient, rapid calming is best achieved with 2–5 mg of haloperidol injected intramuscularly and repeated, if necessary, up to a usual maximum of 30 mg in 24 hours depending upon the patient's body size and physical condition. When distress and agitation are particularly severe, it may be helpful to combine haloperidol with a benzodiazepine. Alternatively, chlorpromazine (75–150 mg intramuscularly) is a more sedating alternative to haloperidol, but is more likely to cause hypotension. When the patient is calm, haloperidol may be continued in smaller doses, usually three to four times a day and preferably by mouth, using a syrup if the patient will not swallow tablets. The dosage depends on the patient's weight and on the initial response to the drug. Careful observations by nurses of the physical state and behaviour are necessary during this treatment. Extrapyramidal side-effects may require treatment with an anti-parkinsonian drug (see p. 558).

Patients who refuse to accept advice about treatment

Psychiatrists are sometimes asked to give urgent advice about patients who are refusing to accept medical treatment. Patients may be unwilling to accept their doctors' advice for many reasons. Commonly it is because they are frightened or angry, or do not understand what is happening; occasionally the cause is a mental illness that interferes with the patient's ability to make an informed decision. It has to be accepted that some patients will refuse treatment even after a full and rational discussion of the reasons for carrying it out, and, of course, it is a right of a conscious mentally competent adult to do so. However, in many countries (including the United Kingdom), it is accepted that the doctor in charge of the patient does have the right to give immediate treatment in life-threatening emergencies when he cannot obtain the patient's consent. If this has to be done, opinions should be obtained from medical and nursing colleagues, and, if possible, from the patient's relatives. Detailed records should be kept of the reasons for the decision. It is essential for all doctors to know the law about these matters in the country in which they are practising.

If a patient has a mental disorder that impairs the ability to give informed consent, it may be appropriate to use legal powers of compulsory assessment and treatment. The powers for compulsory treatment of a mental disorder do *not* give the doctor a right to treat concurrent physical illness against the patient's wishes, but successful compulsory treatment of the psychiatric disorder may result in the patient giving informed consent for the treatment of the physical illness.

Psychiatric aspects of medical procedures

Genetic counselling

The prevalence of genetic disease is between 4 and 5 per cent of the population. Major single-gene diseases include cystic fibrosis and Duchenne muscular dystrophy, as well as sickle cell anaemia which is common in some ethnic groups. Conditions with polygenic multifactorial inheritance are more common; they include coronary artery disease, hypertension, and diabetes.

At present genetic counselling is mainly concerned with the first group of conditions, although increasing knowledge of the genetics of polygenic conditions may increase the scope of counselling in the future. Genetic counselling is based on the results of two distinct approaches to detection: *genetic testing* for those who are known to be at high risk, such as close relatives of those who have inherited disorders, and *genetic screening* of populations. Genetic counselling about the risks of hereditary disease is mainly given to couples contemplating marriage or planning or expecting a child. The scope of counselling is growing rapidly because of increas-

ing understanding of genetic mechanisms, improved methods for identifying carriers and for making prenatal diagnoses, and the prospects of gene therapy. It includes providing information about risks, helping family members to cope with worries caused by the diagnosis, and enabling them to take informed decisions about family planning (Kuller and Laufer 1994).

Genetic counselling is usually provided by the staff of genetic clinics, but there is also a need for counselling by primary-care doctors and specialists other than geneticists. Genetic counselling needs to be guided by a clear understanding of ethical issues which affect both individuals and wider society (Nuffield Council on Bioethics 1993). These include confidentiality, consent to be screened, the implications for employment and insurance, likelihood of stigma, and the storage and use of genetic information for legal purposes. It is often difficult to weigh the benefits of genetic testing for the individual and for other family members. As genetic testing increases, more guidance will be required about the best ways of dealing with the psychosocial impact of the results (see Herman and Croyle (1994) for the example of hereditary breast cancer).

Counselling involves giving information about the results of tests and dealing with problems arising from these results. Several types of test may be carried out: chromosome tests (cytogenetics), direct genetic tests for disorders involving a single gene, indirect (biochemical) tests, and ultrasound imaging for the detection of genetic abnormalities in the fetus. Depending upon the mode of inheritance and tests available it may be possible to detect the following:

(1) persons who already have the disorder, for example phenylketonuria in the newborn;
(2) persons who may develop the disease in the future, for example Huntington's disease, Duchenne muscular dystrophy (Bradley *et al.* 1993), or hereditary forms of breast cancer (Herman and Croyle 1994);
(3) persons who are themselves unaffected but who carry a gene which increases the risk of having an affected child, for example the gene for cystic fibrosis.

Counselling should be non-directive, allowing those counselled to make decisions for themselves after receiving complete and up-to-date information. Counselling may be concerned with anxiety while awaiting the results of prenatal tests, distress before and after termination of pregnancy for medical reasons, distress about being a disease carrier, and worry about children at risk of an inherited disorder.

The nature of the information that should be given varies with the type of genetic risk. Difficult issues are raised by the newer methods which make it possible to identify carriers for Huntington's chorea before they have developed symptoms (see p.325). Usually, counselling is an opportunity to allay anxiety and to reassure, but occasionally it is necessary to help a couple to confront distressing facts and make difficult decisions. Successful counselling requires an awareness of the couple's circumstances, knowledge of the likely psychological and personal consequences of information about genetic risks, and, in appropriate cases, an understanding of the ethical and other aspects of termination of pregnancy. Receiving information about risk can be very distressing, especially for parents who have experienced a previous abnormal pregnancy.

After giving information about the risk of an inherited disorder, the counsellor should discuss alternative actions with the couple. These actions include effective contraception (which may include sterilization) and, where this is possible, prenatal diagnosis with the opportunity for termination. Unfortunately, counselling is more effective in imparting knowledge than in changing behaviour, and many couples ignore warnings that future children will be at high risk.

Screening

Presymptomatic screening for disease is becoming more widely established. Examples include prenatal screening for fetal abnormalities, and screening for breast and cervical cancers and for hypertension. The success of any programme depends in part upon psychological factors which determine whether people attend for screening and whether those who attend are distressed by the

experience (see Walker *et al.* 1994). Participation in many screening programmes is low, and there have been few studies of ways of increasing uptake. Screening can relieve long-standing worries about health, but a few of those screened suffer marked distress.

When the result of screening is negative, patients should be reassured. However, negative findings do not always allay anxiety, and screening may increase fears about health. These problems are greater in those recalled for further assessment. Although a positive finding on screening is distressing, so is the discovery of disease when symptoms or signs have developed.

Psychiatrists can contribute to screening programmes by assisting in staff training, by helping in the production of information for patients, and by assessing and treating the most distressed patients (see Marteau (1994) for a review).

Psychiatric aspects of surgical treatment

Pre-operative mental state and outcome

It is a matter of everyday observation that patients about to undergo surgery are often anxious. Many investigators have looked for relationships between psychological state before surgery and postoperative psychological state or that during recovery. One influential study (Janis 1958) reported that patients who were anxious before surgery were likely to be excessively anxious afterwards, those who were moderately anxious were least anxious afterwards, and those who were least anxious before were likely to be inappropriately angry and resentful afterwards. Other investigations have not confirmed these findings, but have found a linear relationship between anxiety before and after surgery.

Cohen and Lazarus (1973) compared two groups of patients with different attitudes to forthcoming surgery: a 'vigilant' group who sought information about the operation, and an 'avoidant' group who preferred not to know. Contrary to the authors' expectation, the vigilant group fared worse post-operatively as judged by the number of days that they stayed in hospital and by the reporting of minor complications. Most other research shows, as would be expected, that those who show more general ability to cope with stress suffer fewer post-operative problems (see Johnston 1986).

Many, but not all, studies of psychological preparation for surgery have shown that intervention can reduce post-operative distress and problems, especially if it includes cognitive coping techniques rather than mere information (Ridgeway and Mathews 1982).

Psychiatric problems in the post-operative period

Delirium is common after major surgery (Lipowski 1990), especially in the elderly (Cryns *et al.* 1990). The development of post-operative delirium also depends on the type of surgery, the type of anaesthetic, post-operative physical complications, and medication. Delirium is associated with increased mortality and longer stay in hospital.

Psychiatrists are sometimes asked to advise on the management of patients with unusually severe post-operative pain. There is evidence that when patients are given greater control over the timing of analgesia, they experience less pain without excessive drug use.

Long-term psychological problems after surgery

Long-term psychological problems may follow surgery. Adjustment problems are particularly common after mastectomy and laryngectomy, and after surgery that has not led to the expected benefit. A psychiatrist may be able to help with these problems of adjustment, especially when the surgery is part of the management of a relapsing, chronic, or progressive disorder.

The effects of plastic surgery and limb amputation are described below. The effects of certain other surgical procedures are described elsewhere in this chapter: organ transplantation (p. 368), cancer surgery (p. 382), gynaecological surgery (p. 399), and gastrointestinal surgery (p. 387). (The general psychological aspects of surgery have been reviewed by Riether and McDaniel (1993).)

Plastic surgery

People with physical deformities often suffer teasing, embarrassment, and distress, which may markedly restrict the lives of both adults (Harris 1982) and children (Hill-Beuff and Porter 1984). When patients are psychiatrically healthy, reconstructive plastic surgery usually gives good results. Even when there is no major objective defect, cosmetic surgery to the nose, face, breast, or other parts of the body is usually successful (Hay 1970*a*; Hay and Heather 1973; Khoo 1982). Nevertheless, it is appropriate to carry out psychological assessment before plastic surgery because the outcome is likely to be poor in patients who have unrealistic expectations, a history of dissatisfaction with previous surgery, or delusions about their appearance. Patients with dysmorphophobia are likely to be left with a greater sense of grievance after cosmetic surgery (see p. 355).

Limb amputation

Limb amputation has different psychological consequences for young and for elderly people (Frank *et al.* 1984). Young adult amputees, such as those losing a leg in military action or a road accident, characteristically show denial at first, and later experience depression and phantom limb pains which resolve slowly. The outcome for children and adolescents seems similar (Tyc 1992). Older subjects usually undergo amputation after prolonged medical and surgical problems associated with vascular disease. Such patients do not commonly report severe distress immediately after the operation (Parkes 1978), but they often develop phantom limb pain (Harwood 1992). Some have difficulty with the prosthesis and show a degree of functional incapacity disproportionate to the physical state (Sherman *et al.* 1987). (The psychiatric aspects of amputation have been reviewed by Lundberg and Guggenheim (1986) and Riether and McDaniel (1993).)

Drug side-effects as causes of psychiatric symptoms

Table 12.9 lists the drugs that are most likely to give rise to psychiatric side-effects. In addition to

Table 12.9 *Drugs with psychological side-effects*

Anti-parkinsonian agents	
Anticholinergic drugs (benzhexol, benztropine, procyclidine)	Disorientation, agitation, confusion, visual hallucinations
L-Dopa	Acute organic syndrome, depression, psychotic symptoms
Anti-hypertensive drugs	
Reserpine	Depression
Methyldopa	Tiredness, weakness, depression
Calcium-channel blockers	
Clonidine	
Sympathetic blockers	Impotence, mild depression
Digitalis	Disorientation, confusion, and mood disturbance
Diuretics	Weakness, apathy, and depression (due to electrolyte depletion)
Analgesics	
Salicyclamide	Confusion, agitation, amnesia
Phenacetin	Dementia with chronic abuse
Antituberculous therapy	
Isoniazid	Acute organic syndrome and mania
Cycloserine	Confusion, schizophrenia-like syndrome

this list, it is important to remember alcohol, drugs of addiction, psychotropic medication, and steroid therapy (all discussed elsewhere in this book). The psychological complications of chemotherapy are discussed on p. 392. The rare syndromes associated with heavy metals, such as lead, arsenic, and mercury, are reviewed by Lishman (1987) and in the larger medical textbooks.

Organ transplantation

Technical advances in surgery and intensive care, and particularly the introduction of the immuno-suppressant drug cyclosporin, have led to a rapid increase in availability of organ transplantation. As procedures become routine, medical and psychiatric complications have diminished and are in many ways comparable to reactions to other major medical procedures. However, the nature of the surgery and the need for intensive continuing medical assessment and treatment are associated with considerable neurological (Patchell 1994), emotional, and psychiatric consequences.

Selection for transplantation is stressful, particularly for seriously ill patients awaiting vital organ transplants compared with kidney transplantation in which continuing dialysis is available as an alternative treatment. Whilst psychological assessment is often said to have a role in selection for surgery, there are probably few psychological contraindications, most of which relate to inability to cope with demanding long-term medical care.

Transplantation is associated with the same psychiatric and emotional consequences as described for other major treatments, but the nature of the surgery, the immunosuppressive drug regime, and the social and family consequences may all cause anxiety and depression. In addition to the general consequences of organ transplantation, there are specific issues in each type of organ transplant. Liver transplantation is discussed on p. 407 and heart transplantation on p. 382.

The frequency and seriousness of psychiatric aspects is such that specific psychiatric liaison should be available both to train and support staff and to advise on management of individual patients. (See Craven and Rodin (1992) for a review of transplantation.)

A REVIEW OF SYNDROMES

Pain

Pain is widely reported in surveys of the general population. Most people report that the pain is transient, but a minority describe persistent or recurrent pain leading to disability (Von Korff *et al.* 1988; Goodman and McGrath 1991). Pain is the most common symptom among people who consult doctors. **Acute pain** usually has an organic cause such as trauma surgery or cancer. In general practice, pain is a common presenting symptom of emotional disorder (Bridges and Goldberg 1985). In psychiatric practice, it has been reported that pain is experienced by about one-fifth of in-patients and over a half of out-patients (Merskey and Spear 1967). Psychological factors affect the subjective response to pain. Pain is particularly associated with depression, anxiety, panic, and somatoform disorders. Conversely, patients with multiple pain are especially likely to suffer from psychiatric disorder (Dworkin 1990*b*).

The **assessment** of a patient complaining of pain of unknown cause should include appropriate examination for and a thorough investigation of possible physical causes. When the results of this investigation are negative, it should be remembered that pain may be the first symptoms of a physical illness that cannot be detected at an early stage. The psychiatric assessment of these cases should include a full description of the pain and the circumstances in which it occurs, and of any symptoms suggestive of a depressive or other psychiatric disorder.

The **treatment** of pain associated with a psychiatric disorder is the treatment of the primary condition. Skill is required to maintain a working relationship with patients unwilling to accept a psychological basis for their pain. Any associated physical disorder should be treated and adequate analgesics provided. (See Wall and Melzach (1994) for a review.)

Chronic pain occurs in many conditions, including neurological or musculoskeletal disorders, and in the specific pain syndrome (described below). Chronic pain often has both physical and psychological causes (Benjamin and Main 1995). Psychiatric disorder is common. Some patients (with or without physical pathology) have a depressive disorder but it is often difficult to make DSM or ICD diagnoses. In the past, it was suggested that a 'pain-prone disorder' existed which was a variant of depressive disorder (Blumer and Heilbronn 1982). There is little evidence to support this idea, and it is more likely that in these cases the pain arises from personal and social factors, and that beliefs about pain are important in maintaining it (Jensen 1994). Chronic pain may impose great burdens on the patient's family (Benjamin *et al.* 1992); conversely, the attitude of family members and other care givers can influence the perception of pain, its course, and the response to treatment (Benjamin and Main 1995).

The *management* of chronic pain should be individually planned, comprehensive, and involve the patient's family. Any physical cause must be treated. Psychological care is directed to two issues.

1. Whether there is an associated mental disorder. This assessment should be made on positive findings and not solely because no specific organic cause has been identified.
2. Whether the pain or any associated behaviours can be modified by using psychological techniques.

If depressive illness is present it should be treated vigorously. Antidepressant medication may also be effective in patients with chronic pain in the absence of evidence of a depressive disorder (Sullivan *et al.* 1992). Behavioural treatment is useful. However, many patients with chronic pain lack the motivation needed to make full use of these methods. In some cases such treatment aims to reduce social reinforcement of maladaptive behaviour, and to encourage the patient to seek ways to overcome disability (Benjamin and Main 1995).

Pharmacological and behavioural therapies may be combined for some patients. Multidisciplinary pain clinics have been set up to provide expertise in and resources for a range of treatments (Main and Benjamin 1995).

Specific pain syndromes

Complaints of specific kinds of pain are common in the population (Von Korff *et al.* 1988). This section is concerned with headache, facial pain, back pain, and pelvic pain. Certain other specific pain syndromes are discussed later in this chapter, namely non-cardiac chest pain (see p. 383), abdominal pain (see p. 386), and phantom limb pain (see p. 367).

Headache

Patients with **chronic or recurrent** headache may be referred to psychiatrists. There are many physical causes of headache, notably migraine which affects about one in ten of the population at some time of their life. Many patients attending neurological clinics have headaches for which no physical cause can be found. The commonest is 'tension' headache, which is usually described as a dull generalized feeling of pressure or tightness extending around the head. It is frequently of short duration and is relieved by analgesia or a good night's sleep, but may occasionally be constant and unremitting. Some patients describe depressive symptoms and others describe anxiety in relation to obvious life stresses. Psychological factors seem to contribute to aetiology, but there is no evidence that the headaches result from increased muscle tension, and vascular mechanisms are more likely to be involved (Martin *et al.* 1994).

Most patients with headaches for which no physical cause can be found can be reassured by an explanation of the results of investigations (Fitzpatrick and Hopkins 1981). Some patients with persisting headaches, whether or not they have psychiatric symptoms, respond to antidepressants, or cognitive behavioural methods, or psychotherapy. (See Hopkins (1992) for a general review and Blanchard (1992) for a review of psychological treatment.)

Facial pain

Facial pain has many physical causes, but two types may have psychological causes (Dworkin 1990*a*). The more common is *temporomandibular dysfunction* (Costen's syndrome or facial arthralgia). There is a dull ache around the temporomandibular joint, and the condition usually presents to dentists. *'Atypical' facial pain* is a deeper aching or throbbing pain which is more likely to present to neurologists. Patients with either of these symptoms are often reluctant to see a psychiatrist, but several trials suggest that antidepressants can relieve symptoms even when there is no evidence of a depressive disorder. In some cases, cognitive-behavioural methods are effective (Feinmann and Harris 1984).

Back pain

Back pain is the second leading cause for visits to primary-care doctors and a major cause of disability. Most acute pain is transient, but in about a fifth of patients it persists for more than six months. Psychological and behavioural problems at the start predict a poor outcome (Phillips *et al.* 1991). Treatment includes the provision of accurate information about the cause and outcome of the condition, limited use of analgesics, and a graded increase in activity. Von Korff (1994) showed that, in primary care, systematic advice about self-care led to a functional outcome as good as that with analgesia and bed rest, that cost less, and was more satisfactory to patients (see Main and Benjamin (1995) for a review).

Pelvic pain

Pelvic pain is one of the most common symptoms reported by women attending gynaecology clinics. The pain often persists despite negative investigations, and psychological factors appear to be significant causes of the pain and the associated disability. Cognitive-behavioural interventions may be effective in some cases (Slocumb 1989; Peters *et al.* 1991; Glover and Pearce 1995).

Chronic fatigue

Many terms, including postviral fatigue syndrome and neurasthenia, have been used to describe a disabling syndrome of chronic fatigue. Since the cause of the syndrome is uncertain, the descriptive term *chronic fatigue syndrome* is preferred. The main complaints are excessive and disabling fatigue at rest and prolonged exhaustion after minor physical and mental exertion, though other symptoms such as muscle pain and poor concentration are typically present. Operational criteria for diagnosis are essential (Fukuda *et al.* 1994; Sharpe 1994) and require that the illness has lasted for least six months and that other causes of fatigue have been excluded.

Surveys of the general population indicate that persistent fatigue is experienced by about a quarter of the population at any one time (Cope *et al.* 1994). It is a common complaint amongst people attending in primary-care and out-patient clinics. Only a small proportion of people who complain of excessive fatigue meet the criteria for chronic fatigue syndrome.

Despite the very considerable recent publicity, chronic fatigue syndrome has a long history (Wessely 1991). In the nineteenth century the symptoms were diagnosed as neurasthenia. More

recently the syndrome has been thought to be caused by infections such as chronic brucellosis or chronic Epstein–Barr virus infection. In ICD10 the syndrome can still be coded under neurasthenia (a diagnosis that is used widely in several countries including China (Lee 1995)). The prognosis is believed to be poor, but it is likely that most patients improve with time even without specialist treatment. A proportion (especially those with severe personality or social problems) remain chronically disabled.

Aetiology

The cause of chronic fatigue is controversial, and discussion of the problem has been made more difficult by the activities of pressure groups convinced that the cause is entirely physical without any psychological component. Whilst it is essential to exclude the many physical causes of fatigue in each case, there is no convincing evidence for a general physical cause of the chronic fatigue syndrome. Many have been suggested including chronic infection, immune dysfunction, a muscle disorder, and an ill-defined neurological disorder. It seems more likely that psychological factors are important, although the condition may begin with a physical cause for fatigue whose seriousness is misinterpreted by the patient. Indeed, in a general practice investigation the report of chronic severe fatigue six months after doctors diagnosed an acute viral illness was related to patients' symptom-attribual style and to doctors' behaviour, rather than to features of the viral illness (Cope *et al.* 1994). The condition seems to be maintained by inactivity and the resulting loss of physical fitness. The problems are often exacerbated by episodes of overactivity resulting in musculoskeletal symptoms which reinforce patients' beliefs about the dangers of activity.

Depression and anxiety may be conspicuous, or evident after systematic questioning. Up to half of the patients seen in specialist clinics are depressed and may benefit from antidepressant medication, but many patients with chronic fatigue do not have a depressive disorder.

Management

It is essential that the assessment excludes any treatable organic or psychiatric cause of chronic fatigue. There should be a detailed description of the course of the symptoms and their consequences for the patient. It is important to enquire carefully about depressive symptoms, especially as patients may not at first reveal them. Although extensive physical investigations are unlikely to be rewarding, the psychiatrist can usefully collaborate with a physician when assessing these patients. There have been encouraging reports of treatment with antidepressant drugs and psychological treatment (Sharpe 1994).

When there is a definite depressive disorder antidepressant drugs should be prescribed in usual doses. Low doses of antidepressant drugs are useful in reducing anxiety, improving sleep, and reducing pain. Any other associated personal or social difficulties should be discussed using a problem-solving approach.

Psychological (cognitive-behavioural) treatment includes education about the condition, correcting misconceptions about cause and treatment. It is appropriate to explain that there is no evidence that the condition is due to chronic viral infection even though the syndrome may be precipitated by an acute infection or other minor physical disorder. The role of psychological factors in perpetuating the disorder should be explained, and the doctor should emphasize that the syndrome is a real and disabling problem which is deserving of medical care. Arguments about the role of physical factors should be avoided, and instead the graded activities in reducing the fatigue should be explained together with the importance of avoiding bouts of strenuous exertion. Specific treatment includes cognitive strategies to correct misconceptions about the nature of the condition and excessive concern about activity, together with behavioural strategies to encourage a gradual increase in activity. In a recent controlled trial, cognitive-behavioural treatment was superior to standard care (Surawy *et al.* 1995).

The prognosis for full recovery from chronic fatigue is generally good in patients seen in general practice, but is less good in those referred to hospital clinics. Nevertheless, many patients referred to hospital improve with psychological treatment. Success is greater when the psychiatrist or psychologist collaborates with a physician who has a special interest in the disorder (see also p. 351).

Disorders of eating

Psychogenic vomiting

Psychogenic vomiting is chronic and episodic vomiting without an organic cause which commonly occurs after meals and in the absence of nausea. It should be distinguished from the more common syndrome of bulimia nervosa, in which self-induced vomiting follows episodes of uncontrolled overeating. Psychogenic vomiting appears to be more common in women than in men and usually presents in early or middle adult life. It is reported that psychotherapeutic and behaviour treatments can be helpful (Moraska *et al.* 1990; Lancet 1992*a*).

Anorexia nervosa

Anorexia nervosa was described by Marcé in 1859 and named in 1868 by the English physician William Gull, who emphasized the psychological causes of the condition, the need to restore weight, and the role of the family. The DSMIV criteria for diagnosis are listed in Table 12.10. The main features are a body weight 15 per cent below the standard weight (a body mass index below 17.5), an intense wish to be thin, and, in women, amenorrhoea. Most patients are young women (see epidemiology below). The condition usually begins in adolescence, most often between the ages of 16 and 17. It generally begins with ordinary efforts at dieting which then get out of control. The central psychological features are a set of characteristic overvalued ideas about body shape and weight, a fear of being fat, and a relentless pursuit of a low body weight. The patient may have a distorted image of her body, believing herself to be too fat even when severely under-

Table 12.10 *DSMIV criteria for anorexia nervosa*

(A) Refusal to maintain body weight over a minimally normal weight for age and height (e.g. weight loss leading to maintenance of body weight less than 85 per cent of that expected, or failure to make expected weight gain during period of growth leading to body weight less than 85 per cent of that expected)

(B) Intense fear of gaining weight and becoming fat, even though underweight

---

(C) Disturbance in the way in which one's body weight, size, or shape is experienced, undue influence of body weight or shape on self-evaluation, or denial of the seriousness of the current low body weight

---

(D) In postmenarcheal females, amenorrhoea, i.e. the absence of at least three consecutive menstrual cycles (a woman is considered to have amenorrhoea if her periods occur only following hormone (e.g. oestrogen) administration)

Two types can be specified:
(1) Restricting type
(2) Binge eating–purging type

---

weight. This belief explains why many patients do not want to be helped to gain weight.

The pursuit of thinness may take several forms. Patients generally eat little and set themselves daily calorie limits (often between 600 and 1000 calories). Some try to achieve weight loss by induced vomiting, excessive exercise, and misusing laxatives. Patients are often preoccupied with thoughts of food, and sometimes enjoy cooking elaborate meals for other people. Ten to twenty per cent of patients with anorexia nervosa admit to stealing food, either by shoplifting or in other ways. In a number of series up to half the patients had episodes of uncontrollable overeating, sometimes called binge eating or bulimia. This behaviour becomes more frequent with increasing age. During binges the patients typically eat foods that are usually avoided. After overeating they feel bloated and may induce vomiting. Binges are followed by remorse and intensified efforts to lose weight. If other people encourage them to eat, patients are often resent-

ful; they may hide food or vomit secretly as soon as the meal is over. In DSMIV, anorexia nervosa with binge eating and purging (self-induced vomiting or the misuse of laxatives or diuretics) is recognized as a distinct type, differing from the restricting type (Garner 1993).

Amenorrhoea is one of several physical abnormalities that have traditionally been incorporated in diagnostic criteria. It occurs early in the development of the condition, and in about a fifth of cases it precedes obvious weight loss. Some cases first come to medical attention with amenorrhoea rather than disordered eating.

Depressive symptoms, lability of mood, and social withdrawal are all common. In women and men lack of sexual interest is usual.

Physical consequences

A number of important symptoms and signs of anorexia nervosa are secondary to starvation, including sensitivity to cold, delayed gastric emptying, constipation, low blood pressure, bradycardia, and hypothermia. In most cases amenorrhoea is probably secondary to weight loss, but as mentioned above amenorrhoea precedes obvious weight loss in some cases. Investigations may show leucopenia and abnormalities of water regulation. Vomiting and abuse of laxatives may lead to a variety of electrolyte disturbances, the most serious being hypokalaemia. Rarely, these abnormalities may, cause epileptic seizures or death from cardiac arrhythmia. Hormonal abnormalities also occur: growth hormone levels are raised, plasma cortisol is increased and its normal diurnal variation lost, and levels of gonadotrophin are reduced. Thyroxine and thyrotropin-stimulating hormone are usually normal, but triiodothyronine (T_3) may be reduced (Sharp and Freeman 1993).

Epidemiology

Estimates of *incidence* based on case registers in the United Kingdom and the United States range from 0.37 to 4.06 per 100 000 population per year. These are likely to be underestimates. Reported incidence rates have increased recently, but it is likely that the changes reflect a greater awareness of the condition as well as some real increase of incidence (Lucas *et al.* 1991).

It is difficult to determine the true *prevalence* of anorexia nervosa because many people with the condition deny their symptoms. Surveys have suggested prevalence rates of up to 1 per cent among schoolgirls and female university students. Many more young women have amenorrhoea and weight loss less than that required for the diagnosis of anorexia nervosa. Amongst anorexic patients seen in clinical practice only 5–10 per cent are male. The onset of anorexia nervosa in females is usually between the ages of 16 and 17, and seldom after the age of 30. The condition is more common in the upper than the lower social classes, and is reported to be rare in non-Western countries or in the non-white population of Western countries. A survey of young Chinese women in Hong Kong found that, although they shared Caucasian views of the desirability of slimness, they rarely engaged in consistent efforts to lose weight (Lee 1993).

Aetiology

Anorexia nervosa appears to result from a combination of individual predisposition and social factors that encourage dieting (Cooper 1995). Once the disorder has started, the response of the family may help to perpetuate it.

Genetics Among the female siblings of patients with established anorexia nervosa, 6–10 per cent suffer from the condition (Strober 1995) compared with the 1–2 per cent found in the general population of the same age. This increase might be due to family environment or to genetic influences. Holland *et al.* (1984) found a much greater concordance in monozygotic than in dizygotic twins suggesting genetic influences. Family genetic studies also show a link between eating disorders and affective disorders (Strober 1995).

Hypothalamic dysfunction In anorexia nervosa there is profound disturbance of weight regulation. In some cases amenorrhoea begins before significant weight loss. This combination suggests a primary disorder of hypothalamic function, since it can occur with structural lesions of the hypothalamus. However, post-mortem studies have not revealed any regular occurrence of hypothalamic lesions in anorexia nervosa. The balance of evidence now suggests that endocrine and metabolic abnormalities are secondary to low weight and disturbed eating habits. (See Sharp and Freeman (1993) for a review of the endocrine abnormalities in anorexia nervosa.)

Social factors Surveys show that most schoolgirls and female college students diet at one time or another. Concern about body weight is more frequent, and anorexia nervosa more prevalent, in the middle and upper social classes. There is also a high prevalence of anorexia nervosa in occupational groups who are particularly concerned with weight, such as ballet students.

Individual psychological causes Bruch (1974) was one of the first writers to suggest that a disturbance of body image is of central importance in anorexia nervosa. She supposed that patients are engaged in 'a struggle for control, for a sense of identity and effectiveness with the relentless pursuit of thinness as a final step in this effort'. She also suggested three predisposing factors: dietary problems in early life, parents who are preoccupied with food, and family relationships that leave the child without a sense of identity. Crisp (1977) proposed that, while anorexia nervosa is at one level a 'weight phobia', the consequent changes in body shape and menstruation can be regarded as a regression to childhood and an escape from the emotional problems of adolescence. It is often said that psychosexual immaturity is characteristic of patients with anorexia nervosa. These findings are difficult to evaluate in the absence of a control group, but psychosexual problems did not appear to be characteristic of the group as a whole. However, clinical experience suggests that traits of low self-esteem, perfectionism, and undue compliance commonly precede the disorder (Vitousek and Manke 1994).

Causes within the family Disturbed relationships are often found in the families of patients with anorexia nervosa, and some authors have suggested that they have an important causal role. Minuchin *et al.* (1978) held that a specific pattern of relationships, could be identified consisting of 'enmeshment, over protectiveness, rigidity and lack of conflict resolution', could be identified. They also suggested that the development of anorexia nervosa in the patient served to prevent dissent within the family. From a study of 56 families in which one member had anorexia nervosa, Kalucy *et al.* (1977) concluded that the other family members had an unusual interest in food and physical appearance, and that the families were unusually close knit to an extent that might impede the patient's adolescent development. Neither these studies nor others in the literature have shown convincingly that such patterns of behaviour precede the illness or differ significantly from the patterns in families of normal adolescents.

Course and prognosis

In its early stages, anorexia nervosa often runs a fluctuating course with exacerbations and periods of partial remission. Full recovery is not uncommon in cases with a short history; the long-term prognosis is difficult to judge because most published series are based on selected cases or are incomplete in their follow-up. Outcome is very variable. Long-term outcome studies (Ratnasuriya *et al.* 1991) show that, although the disorder may run a chronic course, recovery can occur even after many years. Reported mortality rates from long-term follow-up studies are high at between 15 and 20 per cent; this represents a sixfold increase in the standardized mortality rate. It is believed that the mortality rate is falling with improved methods of treatment. About a fifth of patients make a full recovery, and another fifth remain severely ill; the remainder show some degree of chronic or fluctuating disturbance. Although weight and menstrual functioning usually improve, eating habits often remain abnormal and some patients become overweight while others develop bulimia nervosa. It does not evolve into other forms of psychiatric disorder.

The main factors predictive of outcome are the length of illness at presentation and age of onset; disorders with a short history and starting at a younger age are associated with a better prognosis. (For a review of outcome studies and of predictive studies, see Steinhausen *et al.* (1991).)

Assessment

Most patients with anorexia nervosa are reluctant to see a psychiatrist, and so it is important to try to establish a good relationship. A thorough history should be taken of the development of the disorder, the present pattern of eating and weight control, and the patient's ideas about body weight. In the mental state examination, particular attention should be given to depressive symptoms. More than one interview may be needed to obtain this information and gain the patient's confidence. The parents or other informants should be interviewed whenever possible. It is essential to perform a physical examination, with particular attention to the degree of emaciation, the state of the peripheral circulation, and signs of vitamin deficiency. Other wasting disorders, such as malabsorption, endocrine disorder, or cancer, should be excluded. Electrolytes should be measured if there is any possibility that the patient has been inducing vomiting or abusing purgatives.

Management

Starting treatment Success largely depends on establishing a good relationship with the patient. It should be made clear that achieving an adequate weight is an essential priority in order to reverse the physical and psychological effects of starvation. It is important to agree a definite dietary plan but not to become involved in wrangles about it. At the same time, it should be emphasized that weight control is only one aspect of the problem, and help should be offered with psychological problems.

Educating the patient and family about the disorder and its treatment is important. Admission to hospital is often needed if the patient's weight is dangerously low (for example less than 65 per cent of standard weight), if weight loss is rapid, or if there is severe depression. It is also indicated if out-patient care has failed. Less serious cases may be treated as out-patients. (See Hsu (1990) for a review.)

Restoring weight The patient's admission should be on the understanding that she will stay in hospital until her agreed target weight has been reached and maintained. The target usually has to be a compromise between a healthy weight (a body mass index above 20) and the patient's idea of what her weight should be. A balanced diet of at least 3000 calories is provided as three or four meals a day. Successful treatment depends on good nursing care, with clear aims and firmness, and understanding. Eating must be supervised by a nurse, who has two important roles: to reassure the patient that she can eat without the risk of losing control over her weight and to be firm about agreed targets, and to ensure that the patient does not induce vomiting or take purgatives. In the early stages it may be best for the patient to remain in bed in a single room while nurses maintain close observation. It is reasonable to aim for a weight gain of between a half and one kilogram each week. Weight restoration usually lasts between 8 and 12

weeks. Some patients demand to leave hospital before their treatment is finished, but with patience the staff can usually persuade them to stay.

Behavioural principles are sometimes used. The usual approach is to remove privileges when the patient enters hospital and to restore them gradually as rewards for weight gain. Suitable privileges include having visitors, newspapers, books, radio, or television for agreed periods of time. It is essential that the patient should agree freely to the programme before it starts and should feel free to withdraw whenever she wishes. These methods have been described more fully by Garfinkel and Garner (1982). They have not been shown convincingly to give better results than a general programme of the kind described above (which leads to satisfactory weight gain in hospital in about four patients out of five).

Rarely, the patient's weight loss is so severe as to pose an immediate threat to life. If such a patient cannot be persuaded to enter hospital, compulsory admission is necessary.

The role of psychotherapy Many forms of psychotherapy have been tried. It is generally agreed that intensive psychoanalytic methods are not helpful. Clinical experience suggests that there is some value in simple supportive measures directed to improving personal relationships and increasing the patient's sense of personal effectiveness.

In recent years, *family therapy* has been advocated (Minuchin *et al.* 1978). The results of a controlled evaluation suggest that there may be benefits with younger patients but do not support the general use of family therapy (Russell *et al.* 1987). If this treatment is used, it should be for selected cases in which family problems seem particularly relevant and the family members are willing to participate. *Cognitive therapy* has also been used with the aim of modifying abnormal cognitions about shape, weight, and eating, but it has to be formally evaluated. Many patients and families find *self-help groups* valuable. (For reviews of anorexia nervosa see Hsu (1990) and Brownell and Fairburn (1995).)

Table 12.11 *DSMIIIR criteria for bulimia nervosa*

(A) Recurrent episodes of binge eating. An episode of binge eating is characterized by both of the following:
 (i) eating in a discrete period of time (e.g. within any two-hour period) an amount of food that is definitely larger than most people would eat during a similar period of time and under similar circumstances
 (ii) a sense of lack of control over eating during the episode (e.g. a feeling that one cannot stop eating or control what or how much one is eating)

(B) Recurrent inappropriate compensating behaviour in order to prevent weight gain, such as self-induced vomiting, misuse of laxatives, diuretics, enemas, or other medications, fasting, or excessive exercise

---

(C) The binge eating and inappropriate compensating behaviours both occur on average at least twice a week for 3 months

---

(D) Self-evaluation is unduly influenced by body shape and weight

---

(E) The disturbance does not occur exclusively during episodes of anorexia nervosa

---

Two types may be specified:
 (1) purging type
 (2) non-purging type

Bulimia nervosa

Bulimia refers to episodes of uncontrolled excessive eating, sometimes called 'binges'. As mentioned above, the symptom of bulimia occurs in some cases of anorexia nervosa. Although the syndrome of bulimia nervosa was described by Russell (1979) as an 'ominous variant' of anorexia nervosa, at least two-thirds of cases occur without preceding anorexia nervosa. The central features are an irresistible urge to overeat, extreme measures to control body weight, and overvalued ideas concerning shape and weight of the type seen in anorexia nervosa. Two subtypes are recognized in DSMIV, the *purging type* characterized by the use of self-induced vomiting, laxatives, and diuretics to prevent weight gain, and the *non-purging type* in which 'purging' symptoms have not occurred

regularly but the person has used other behaviours to avoid weight gain, such as fasting and excessive exercise. Patients with bulimia nervosa are usually of normal weight; patients who are underweight usually qualify for a diagnosis of anorexia nervosa, which takes precedence. Most patients are female and they often have normal menses (see Table 12.11 for the DSMIII criteria).

Patients have a profound loss of control over eating. Episodes of bulimia may be precipitated by stress or by the breaking of self-imposed dietary rules, or may occasionally be planned. In the episodes enormous amounts of food are consumed; for example a loaf of bread, a whole pot of jam, a cake, and biscuits. This voracious eating usually takes place alone. At first it brings relief from tension, but relief is soon followed by guilt and disgust. The patient may then induce vomiting or take laxatives. There may be many episodes of bulimia and purging each day.

Depressive symptoms are more common than in anorexia nervosa, and are probably secondary to the eating disorder. A few patients appear to suffer from a depressive disorder requiring anti-depressant drugs. (See Brownell and Fairburn (1995) for a review of bulimia nervosa.)

Physical consequences

Repeated vomiting leads to several complications. Potassium depletion is particularly serious, resulting in weakness, cardiac arrythmia, and renal damage. Rarely, urinary infections, tetany, and epileptic fits may occur. The teeth become pitted by the acid gastric contents in a way that dentists can recognize as characteristic.

Epidemiology

The prevalence of bulimia nervosa is between 1 and 2 per cent among women aged between 16 and 40 years (Fairburn and Beglin 1990). It is uncommon among men. Bulimia nervosa has only been identified in developed countries.

Prognosis

This is uncertain since there have been few long-term studies. It is probable that abnormal eating habits persist for many years, but that they vary in severity. The overall outcome is very variable. The evidence to date suggests that cases of clinical severity tend to run a chronic course. Premorbid low self-esteem predicts a worse outcome. It is also probable that severe bulima nervosa in mothers of young children may have considerable consequences for maternal care (Stein and Fairburn 1989). It is also apparent that feeding disorders among children are specifically associated with disturbed eating habits and attitudes among mothers (Stein *et al.* 1995).

Management

The management of bulimia nervosa is similar in principle to that described for anorexia nervosa, but is easier because agreement about aims and a good working relationship can often be established and there is no need for weight restoration. It is necessary to assess the patient's physical state and to identify the few cases that might benefit from treatment with antidepressant drugs. Out-patient treatment is usually possible, and admission to hospital is indicated only if there are severe depressive symptoms or physical complications, or if out-patient treatment has failed.

The most extensively studied psychological treatment is a *cognitive behaviour therapy* which helps patients regain control over their eating (Fairburn *et al.* 1993). This out-patient treatment uses procedures which aim to normalize eating habits and cognitive techniques designed to modify the excessive concerns about shape and weight. There is good evidence that up to two-thirds of patients treated in this way achieve substantial and lasting change. Patients attend as out-patients several times a week, keep records of their food intake and episodes of vomiting, and attempt to identify and modify any environment stimuli, thoughts or emotional changes that regularly precede the urge to overeat. (See Fairburn *et al.* (1993) for a review of cognitive behavioural methods.) There is some evidence that *interpersonal psychotherapy*, which is a short-term focal psychotherapy, appears to be equally effective (Fairburn *et al.* 1993).

Treatment with SSRIs has been reported to be effective but more trials are needed. It is probable that a self-help treatment with minimal specialist supervision may be effective in milder cases (Fairburn 1995). (See Brownell and Fairburn (1995) for a general review of treatment.)

Eating disorder not otherwise specified (atypical eating disorders)

The DSMIV and ICD categories are for disorders of eating that do not meet the criteria for anorexia nervosa or bulimia nervosa. This disorder is frequent, and is diagnosed in a third of referrals. A *binge eating disorder* is described in DSMIV as requiring further study. It is characterized by recurrent bulimic episodes in the absence of the other diagnostic features of bulima nervosa

(Yanovski 1993). Treatment is similar in principle to that of cases that meet the criteria for diagnosis of anorexia nervosa or bulimia nervosa.

Obesity

By convention, obesity is diagnosed when the body mass index exceeds 25 per cent. Approximately a quarter of the adults in the United States are overweight (Williamson 1993). Obesity is associated with an increased mortality, and severe obesity (body mass index greater than 40 per cent) is associated with a 12-fold increase of risk in those aged 25–35 years (Kuczmaski 1994).

Most obesity is attributable to genetic factors exacerbated by social factors that encourage overeating. Psychological causes do not seem to be of great importance in most cases, but psychiatrists are sometimes asked to see obese people whose excessive eating seems to be determined by emotional factors. Many of these patients have binge eating disorder (Marcus 1995); they often have low self-esteem and lack social confidence, although overweight people in general show no more psychological disturbance than the general population. However, a few have serious psychiatric problems. Little is known about psychological aspects of obesity in infancy or childhood (Gortmaker *et al.* 1993).

Treatment

Mildly overweight people need nothing more than advice about diet and exercise. Although there has been concern that dieting may be associated with adverse physiological and psychological effects, it would seem that these are not of clinical significance for moderate changes in eating and exercise (French and Jeffery 1994). It is important to be aware that many obese people do not eat more than other people, and that aiming at an 'ideal' weight is unrealistic and even inappropriate.

The long-term results of all kinds of treatment are disappointing, whether supervised by a doctor or not (Wilson 1993). Group therapy and self-help groups produce short-term benefit but do not improve long-term results and self-esteem may become lower if there is failure (Wooley and Garner 1994). The same is true for appetite-suppressing drugs and behavioural methods. The results with children are somewhat more encouraging (Epstein *et al.* 1994), although there is concern that treatment may encourage the later development of bulimia nervosa (Wilson 1994).

Surgical treatment is probably indicated for very severe obesity. The most common form is gastric restriction. As the risks of this form of surgery are significant in severely overweight patients, it should be offered only when other treatments have failed. Jaw wiring is sometimes recommended as a means of losing substantial amounts of weight, but weight may be regained when the wires are removed. (See Stunkard and Wadden (1993) and Brownell and Fairburn (1995) for reviews of the epidemiology, causes, and treatment of obesity.)

Diabetes mellitus

Diabetes is a chronic condition requiring prolonged medical supervision and informed self-care, and many physicians emphasize the psychological aspect of treatment (Tattersall 1981).

Psychological factors and diabetic control

Psychological factors are highly important in the course of established diabetes (Helz and Templeton 1990) because they influence its control, and it is now generally accepted that good control of blood glucose is the single most important factor in preventing long-term complications. Psychological factors can impair control in two ways. First, stressful experience can lead directly to endocrine changes (Kemmer *et al.* 1986). Second, many diabetics show poor self-care and compliance with medical advice, especially at times of stress, and this is an important cause of 'brittle' diabetes (Tattersall *et al.* 1991). Psychological factors also limit the use of new methods of diabetic control, such as insulin pumps, which require well-informed and conscientious self-care.

Problems of being diabetic

For the diabetic person, psychological and social problems may be caused by restrictions of diet and activity, the need for self-care, and the possibility of serious physical complications such as vascular disease and impaired vision. Although most diabetic patients adapt well to the limitations of their illness, an important minority of those with insulin-dependent and non-insulin-dependent diabetes have difficulties; this is often particularly prominent in adolescence (Jacobson *et al.* 1994). Compliance with blood testing, diet, and insulin use is frequently unsatisfactory, and as a result glycaemic control is often less than optimal. Psychiatric problems include depression and anxiety (see Lustman *et al.* 1986; Stewart *et al.* 1989; Mayou *et al.* 1991). Both *psychiatric disorder* and less severe psychological distress are causes of poor compliance and have direct effects on the quality of life. Although the prevalence of *eating disorders* amongst adolescent and young adult diabetic women is not increased, when the two are combined treatment is more difficult (Fairburn *et al.* 1991). Insulin misuse to promote weight loss is frequent among young women with diabetes (Fairburn *et al.* 1991).

Psychological and social problems are more common in diabetics with severe *medical complications* such as loss of sight, renal failure, and vascular disease (Lloyd *et al.* 1992). *Pregnancy* is a difficult time for diabetic women, since there may be problems in the control of diabetes and increased risks of miscarriage and fetal malformations.

Sexual problems are believed to be common among diabetics. Two kinds of impotence occur in the men. The first is psychogenic impotence of the kind found in other chronic debilitating diseases. The second kind, which is more common in diabetes, may predate other features of the disease. It is thought to be associated with pelvic autonomic neuropathy, although vascular and endocrine factors may also contribute (McCulloch *et al.* 1984).

Organic psychiatric syndromes in diabetic patients

An **acute organic syndrome** is a prodromal sign of diabetic (hyperglycaemic) coma. It may present as an episode of disturbed behaviour, which may begin either abruptly or insidiously. Other prodromal physical symptoms include thirst, headaches, abdominal pain, nausea, and vomiting. The pulse is rapid and blood pressure is low. Dehydration is marked and acetone may be smelt on the breath. A second cause of an acute organic syndrome is hypoglycaemia (see below).

Mild **cognitive impairment** is not uncommon among chronic diabetics (Bale 1973; Perlmutter *et al.* 1984). It may be caused by recurrent attacks of hypoglycaemia or by cerebral arteriosclerosis. Dementia may develop in patients with associated cerebrovascular disease.

Psychiatric management

Medical treatment can be usefully supplemented with certain forms of specialist psychological intervention (Cox and Gonder-Frederick 1992; Bradley 1994). The latter includes the treatment of depressive disorder, blood glucose awareness training to improve the ability to recognize and act on fluctuations in blood glucose concentrations (Cox *et al.* 1994), behavioural methods to improve self-care and relieve associated psychological and social problems (Bradley *et al.* 1994), weight management programmes (Wing *et al.* 1994), and psychological treatment of sexual dysfunction. Tricyclic antidepressants may be helpful in relieving the pain of neuropathy (Young and Clarke 1985). (See Bradley *et al.* (1994) for a review of the treatment of adult diabetics; for the management of diabetic children see Lindsay (1985).)

Hypoglycaemia

Hypoglycaemia is usually induced by therapeutic insulin (when the patient has too much insulin or not enough food). Other causes are an insulin secreting tumour of the pancreas, alcoholism, or liver disease. Common psychological features of acute hypoglycaemia include anxiety and other abnormalities of mood, restlessness, irritability, aggressiveness, and behaving as if drunk. Physical symptoms include hunger and palpitations. Common physical signs are flushing, sweating, tremor, tachycardia, and ataxic gait. Occasionally other neurological signs occur. In severe episodes hypoglycaemic coma may develop. Hypoglycaemia is important in the differential diagnosis of episodic psychiatric symptoms and it is easily missed.

Cardiovascular disorders

Ischaemic heart disease

For many years, it has been assumed that emotional disorder predisposes to ischaemic heart disease (Osler 1910). Dunbar (1954) described a 'coronary personality'. Such ideas are difficult to test because only prospective studies can separate psychological factors present before the heart disease from the psychological effects of being ill. Recent research has concentrated on several groups of possible risk factors including chronic emotional disturbance, social and economic disadvantage, overwork or other chronic stress, and the type A behaviour pattern. The most intensively investigated of these factors is the type A behaviour pattern, which is defined as hostility, excessive competitive drive, ambitiousness, a chronic sense of urgency, and a preoccupation with deadlines (Friedman and Rosenman 1959). In the Western Collaborative Group study, Rosenman and his colleagues (1975) assessed over 3000 men aged 29–59 working in ten California companies, and followed them up for eight to nine years (Rosenman *et al.* 1975). The rate of ischaemic heart disease proved to be twice as high in type A subjects as in other subjects.

Although type A behaviour has been widely accepted to be an independent risk factor for ischaemic heart disease, recent evidence has cast doubt on this conclusion. Several recent prospective studies have failed to confirm an association between type A behaviour and coronary heart disease. Other studies have failed to confirm a clear association between type A behaviour and evidence of coronary artery disease on angiography. Some of the discrepancies may be due to the lack of standard criteria for defining type A behaviour and to the breadth of the concept, but it is clear that type A behaviour is not a major risk factor for coronary heart disease. Most current research has preferred to concentrate on more precisely defined psychological factors, such as hostility. Their role remains uncertain. (See Johnstone (1993) and Littman (1993) for reviews.)

Non-pharmacological trials of primary and secondary prevention have largely concentrated on changing lifestyle risk factors such as smoking, diet, and lack of physical activity (Lancet 1982). Attempts have also been made to alter type A behaviour. Thus the Stanford Heart Disease Prevention programme attempted to alter two behavioural characteristics: 'hostility' and 'time urgency'. The study was based on male volunteers who had experienced a heart attack. An experimental group of 600 patients received monthly group therapy, and a control group of 600 were simply seen by a cardiologist. In the experimental group, the frequency of components of type A behaviour was reduced, and the reinfarction rate was 7 per cent compared with 14 per cent among the controls (Friedman *et al.* 1986). The findings need replication and there are unanswered questions about the mechanisms and interpretation of the findings.

Angina

Angina is often precipitated by emotions such as anxiety, anger, and excitement. It can be a frightening symptom, and some patients become overcautious despite reassurance and encouragement to resume normal activities. Angina may be accompanied by atypical chest pain and breathlessness caused by anxiety or hyperventilation. There is often little relationship between objectively measured exercise tolerance and the patient's complaints of chest pain and limitation of activity. Patients with *silent ischaemia* on exercise testing appear to be less sensitive to pain and other bodily sensations than those with angina (Freedland *et al.* 1991).

Surgical and medical treatment together with regular and appropriate exercise can be highly effective. Individually planned programmes of information and training in self-help skills can produce increased confidence, a reduced frequency of symptoms, and less disability (Lewin *et al.* 1992). It has been claimed that an intensive programme designed to change lifestyle can result in regression of coronary atherosclerosis (Ornish *et al.* 1990), but this finding requires confirmation.

Myocardial infarction

Patients often respond to the early symptoms of myocardial infarction with denial, and consequently delay seeking treatment. In the first few days in hospital acute organic mental disorders and anxiety symptoms are common (Cay 1984). Emotional distress may be an important cause of arrhythmias and sudden death (Kamarck and Jennings 1991; Willich *et al.* 1994). Mood disorder may be associated with an increase of subsequent mortality (Frasure-Smith *et al.* 1993, 1995).

Survivors of cardiac arrest may suffer cognitive impairment. When such impairment is mild, it often manifests later as personality change or behavioural symptoms which may be attributed wrongly to an emotional reaction to the illness (Reich *et al.* 1983).

When patients return home from hospital, they commonly report non-specific symptoms such as fatigue, insomnia, and poor concentration, as well as excessive concern about somatic symptoms and an unnecessarily cautious attitude to exertion. Most patients overcome these problems and return to a fully active life. A few continue with emotional distress and social disability out of proportion to their physical state, often accompanied by atypical somatic symptoms. Such problems are more common in patients with long-standing psychiatric or social problems, overprotective families, and myocardial infarction with a complicated course (Littman 1993).

Attempts have been made to reduce these psychological problems by using various forms of rehabilitation, in which the most important component seems to be early mobilization. Other components include exercise training, education, and group therapy. Exercise training has been used widely but does not seem to be particularly effective in reducing psychological problems (Oldridge *et al.* 1991). A self-help programme about coping with stress has been reported to lead to an improved return to full activities (Lewin *et al.* 1992). It is important to treat the small minority of patients with persistent depression or with other emotional or social problems. Major depression should be treated with an antidepressant drug that is not cardiotoxic, such as a specific serotonin re-uptake inhibitor (Glassman *et al.* 1993; Roose and Glassman 1994).

Cardiac surgery

Coronary artery surgery for the relief of angina is one of the most common forms of major surgery. It is highly successful in relieving angina, but after the operation up to a quarter of the patients report persistent anxiety or depression and limitations of their everyday activities. The outcome is least satisfactory amongst patients who had severe emotional distress before the operation or responded overcautiously to angina. Neuropsychiatric symptoms are common in the early post-operative period but are usually shortlived. *Surgery for congenital heart disease* may be followed by long-term psychosocial sequelae even in those without residual cardiac problems.

Heart transplantation usually results in a much improved quality of life. However, psychosocial problems are frequent during and after the period of convalescence; they include the side-effects of medication, 'organic' psychiatric disorder, mood disorder, sexual problems, and marital and family problems (Craven and Rodin 1992; Dew *et al.* 1994). The psychological consequences of cardiac surgery in children have been described by Wray *et al.* (1994).

See Rosenthal (1993), Blumenthal and Mark (1994), and Jenkins (1994) for reviews of cardiac surgery.

Essential hypertension

Brief changes in blood pressure occur in the course of temporary emotional states. It has been suggested that prolonged emotional changes can lead to sustained hypertension (Alexander 1950) but the evidence is still unconvincing (Weiner 1977). There is some indirect supporting evidence from animal experiments; for example, Henry *et al.* (1967) found that rats became hypertensive if kept in crowded, and therefore stressful, conditions. In humans, there have been studies of people working in stressful occupations. Cobb and Rose (1973) found hypertension to be more common among air traffic controllers than in the general population. Theorell and Lind (1973) studied middle-aged men and reported that those with more responsible jobs had higher blood pressure. Attempts have also been made to relate hypertension to neurotic conflicts or personality type, but the findings have not been convincing. (See Mann (1986) for a review.)

Complaints of headache, dizziness, and fatigue are common among hypertensive patients who know that they are hypertensive but not among those who do not know (Kidson 1973). However, awareness does not necessarily lead to such consequences. In a screening programme, Mann (1977) found that patients who were told the diagnosis and also given psychological support did not develop adverse effects.

There is still uncertainty about the best treatment for the many people with mild hypertension. Such people comply poorly with drug treatment. Some hypotensive drugs notably reserpine, cause depression, and the side-effects of beta-blockers and clonidine have adverse effects on quality of life. These problems have led to the use of psychological methods to replace or supplement drugs. These methods have mainly involved some form of relaxation based on stress management. Despite some reports suggesting that these procedures reduce blood pressure significantly compared with controls, other controlled studies of stress management (Johnston 1993) and of cognitive behavioural techniques (Eisenberg *et al.* 1993) have had negative results. Neither approach is recommended for routine practice.

Low blood pressure

Low blood pressure has traditionally been seen in some countries as a cause of psychological symptoms, fatigue, and dizziness. In other countries hypotension has not been generally accepted as a cause of symptoms. Recent epidemiological studies (Wessely *et al.* 1990) have found an association between lower levels of blood pressure and complaints of tiredness and feeling faint. The evidence is not yet convincing and further evidence is necessary. Beliefs about the significance of low blood pressure may be more important than hypotension itself (Mann 1992).

Atypical cardiac symptoms

During the American Civil War, Da Costa (1871) described a condition which he called 'irritable heart'. This syndrome consisted of a conviction that the heart was diseased, together with palpitations, breathlessness, fatigue, and inframammary pain. This combination has also been named 'disorderly action of the heart', 'effort syndrome', and 'neurocirculatory asthenia'. The symptoms were originally thought to indicate a functional disorder of the heart. More recently they have been attributed to mitral valve prolapse, but such an association seems unlikely.

Atypical chest pain, in the absence of heart disease and often associated with complaints of breathlessness and palpitations, is very common among patients in primary care and in cardiac out-patient clinics. Most patients with the symptoms are reassured by a thorough assessment, but a significant minority continue to complain of physical and psychological symptoms and to limit their everyday activities. Follow-up studies of patients with chest pain and normal coronary angiograms have consistently found subsequent mortality and cardiac morbidity to be little greater than expectation, but persistent disability to be common.

Many causes have been suggested for atypical cardiac symptoms including pain originating in the chest wall, oesophageal reflex and spasm, microvascular angina, mitral valve prolapse, and psychiatric disorder. Many patients have chest pain which is due not to psychiatric disorder but to minor non-cardiac physical causes or to hyperventilation, which is then misconstrued as heart disease. The most common psychiatric cause of atypical chest pain with associated anxiety is panic disorder; less common causes are depressive disorder and hypochondriasis. Management should follow the general principles described on p. 351 with a particular emphasis on the treatment of hyperventilation, graded increase in activity, and discussion of beliefs about the course of the pain. Cognitive behavioural treatments are effective in the management of anxiety and hyperventilation (Klimes *et al.* 1990). Depressive disorder should be treated with antidepressant medication. (See Bass and Mayou (1995) for a review.)

Respiratory disorders

Breathlessness arising from respiratory and cardiac disorders may be exacerbated by psychological factors (Kellner 1992). Breathlessness can also be entirely psychological in origin, for example hyperventilation associated with anxiety disorders (see p. 163).

Asthma

Alexander (1950) suggested that asthma is caused by unresolved conflicts about dependency, but there is no satisfactory evidence for this idea. Suggestions that asthmatic attacks can be conditioned are equally unsupported (Steptoe 1984). There is more convincing evidence that emotions such as anger, fear, and excitement can provoke and exacerbate individual attacks in patients with established asthma.

The prevalence of psychiatric morbidity among asthmatic children is little greater than in the general population of children (Graham and Rutter 1970). However, when psychological problems occur in children they can add to management problems, and it has been reported that psychological and family problems are more common in severely asthmatic children who die of asthma than in other children with severe asthma (Strunk *et al.* 1985). In adults there is an association between asthma and agoraphobia and panic disorder (Shavitt *et al.* 1992). Educational programmes which teach patients to comply better with treatment lead to improved social adjustment (Wilson *et al.* 1993). Relaxation training and family therapy also appear to be helpful for some patients. (See Lehrer *et al.* (1992) for a review.)

Chronic bronchitis

Chronic obstructive airways disease impairs the quality of life and is often associated with anxiety and depression (McSweeney *et al.* 1982); also, it may lead to cognitive impairment by causing hypoxaemia (Grant *et al.* 1982; Prigatano *et al.* 1984). Breathing exercises, general physical exercise, and social support all appear to improve morale and reduce disability, and programmes of rehabilitation combining exercise training with behavioural methods are available. Some patients complain of breathlessness out of proportion to their physical disorder, and this problem may respond to psychological treatment (Burns and Howell 1969).

Cystic fibrosis

In the past the prognosis for life of cystic fibrosis was very poor, and consequently the psychological effects on the child and family were severe. Recently, the prognosis has much improved and children with cystic fibrosis experience emotional disturbance and behavioural problems with no greater frequency than children with other chronic physical illnesses. Adults with cystic fibrosis have to cope with chronic physical disability and substantially impaired fertility. Since any child born to an adult with cystic fibrosis has a one in forty chance of having the disease, genetic counselling is important (see p. 365).

Renal disorders

Uraemia

Uraemia causes an acute organic psychiatric syndrome characterized by drowsiness and fluctuating consciousness, there may be episodes of disturbed behaviour in up to a third of cases, and intellect may be impaired (Osberg *et al.* 1982). These symptoms are not closely related to the urea concentration in the plasma, probably because electrolyte disturbances and failure to excrete drugs also contribute to the aetiology.

Chronic renal failure

The treatment of end-stage renal failure is stressful to patients, families, and hospital staff. Symptoms of anxiety or depression occur in about half the patients. There is probably an increased risk of suicide. Impaired capacity to work, reduced physical activity, marital problems, and sexual dysfunction are common (Levenson and Glocheski 1991; Levy 1994). These psychosocial difficulties add to the problems of management and affect the course of the medical illness (Hussebye *et al.* 1987). Comparisons of dialysis patients with transplantation patients suggest that the latter have a better quality of life with fewer handicaps. Among dialysis patients, those treated at home report fewer problems than those treated in hospital. Continuous ambulatory peritoneal dialysis is less restricting than haemodialysis but is not effective for many patients with renal failure.

In childhood, end-stage renal failure impairs considerably the quality of life of some patients and their families (Winterborn 1987), but others have few emotional difficulties or family problems (Fielding *et al.* 1985).

In many renal units a liaison psychiatrist works directly with staff and patients and supports the work of the staff. Psychiatric help is mainly sought for mood disorders, difficulties in coming to terms with renal failure or its treatment, and disruptive or uncooperative behaviour. (See Levy (1994) or Brown and Brown (1995) for reviews.)

Haemodialysis

Patients with haemodialysis may develop cognitive deficits due to uraemia, anaemia, and other physical complications of the renal failure, and to drug toxicity. An acute organic psychiatric syndrome may occur when there is dialysis disequilibrium. The rare occurrence of dialysis dementia is probably due to the aluminium content of the perfusion fluid. Depression and anxiety are common. Symptoms such as lethargy, insomnia, and poor concentration may be due to either physical or psychological causes. Impotence is common, and often psychologically determined.

Psychological and social factors can cause difficulties in management, particularly when dialysis is carried out in the patient's home. The patient's capacity to cope may be influenced by his emotional state, personality, and understanding of the treatment. The willingness and ability of the family to help with treatment is important, and so is the quality of family relationships as well as their social and financial circumstances.

Renal transplantation

After receiving a kidney transplant, most patients experience a striking improvement in their sense of physical and mental well-being, their sexual functioning, and their ability to work. Despite these immediate improvements, psychological problems are common. Problems of transplant rejection or threatened rejection are frequently associated with depression and anger. Psychological symptoms may also occur as side-effects of immunosuppressive drugs, steroids in high dosage, and antihypertensive drugs. (See Levy 1994) for reviews of renal transplantation.)

Gastrointestinal disorders

Gastrointestinal symptoms are often an expression of psychiatric disorder. Complaints of poor appetite, abdominal pain, and constipation can have a psychiatric cause, especially a depressive disorder or an anxiety disorder. Medically unexplained gastrointestinal symptoms are common in the general population and are significantly associated with psychiatric disorder (Walker *et al.* 1992). In addition to the conditions reviewed in this section, the following conditions are discussed elsewhere: eating disorders (p. 372), and carcinoma of the pancreas (p. 392). (For a general review see Folks and Kinney (1992*b*) and for a review of behavioural treatment see Whitehead (1992).)

Oesophageal symptoms

In the past the condition known as **globus hystericus** (difficulty in swallowing or a persistent feeling of a lump at the level of the upper oesophageal stricture) was thought to be a form of hysteria. Mild forms of complaint are not infrequent in the general population, but persistent forms account for only about 4 per cent of laryngological referrals. Assessment requires thorough ear, nose, and throat investigation, since there may be an organic cause, together with psychiatric investigation. The symptom is not easy to treat, but antidepressants, the avoidance of unnecessary physical investigation, and behavioural methods may be helpful (Wilson *et al.* 1988; Deary and Wilson 1994).

Food and environmental allergy syndromes

In recent years, the medical profession has recognized food allergy as a disorder. However, many patients erroneously attribute psychological or somatic symptoms to this cause, and ask for help from their doctors. The minority of patients referred to psychiatrists often reject any psychological explanation of their symptoms or any offer of psychiatric treatment (Rix *et al.* 1984).

A few patients complain of *multiple chemical sensitivity* (also called environmental illness or total allergy syndrome) for which no consistent physical or immunological basis can be found. These complaints may be sporadic or epidemic, and may be associated with chemicals that have been discussed prominently in the media. There is often an associated psychiatric disorder and

there may be a wide range of functional somatic symptoms. Since many patients reject psychological treatment, management usually has to be directed towards correcting misunderstandings and promoting a return a more active life by reducing anxiety and social withdrawal (Howard and Wessely 1993; Simon *et al.* 1993).

Abdominal pain

Acute abdominal pain is frequent among patients seen as medical and surgical emergencies. Such pain may be psychologically determined. For example, amongst patients undergoing appendicectomy, psychiatric symptoms have been reported as more frequent in those with normal than in those with abnormal appendixes (Creed 1981; Beaurepaire *et al.* 1992).

Non-ulcer dyspepsia, i.e. dyspepsia for which no organic cause can be found, is frequent in the general population and is one of the most common reasons for consultation in primary care and medical out-patient clinics. Some of these patients are reassured by negative results of investigations, but others continue to complain of dyspepsia (Talley *et al.* 1987; C. Morris 1991). Abdominal pain in children is discussed on p. 704.

Peptic ulcer

It has long been held that mental activity can affect the stomach (see Weiner (1977) for a review.) Peptic ulcers have been produced in animals by electrical stimulation of the hypothalamus. If rats are allowed varying degrees of control over electric shock, the less control they have, the more likely they are to develop ulcers (Ader 1976). The direct effect of emotion of the gastric mucosa has been demonstrated in patients with gastric fistulae, such as the patient Tom described by Wolf and Wolff (1947).

Alexander (1950) suggested that 'the repressed longing for love is the unconscious psychological stimulus directly connected with the psychological processes leading finally to ulceration'. Although there is no support for this theory, some observations suggest that psychological factors may play a part in the aetiology of peptic ulcer. For example it has been reported that gastric and duodenal ulcers are more common at times of environmental stress, such as wartime bombing, and recent research indicates that patients with duodenal ulcers report more chronic background difficulties than do control subjects (see Lewin and Lewis (1995) for a review).

There is no evidence that specific psychological treatment is beneficial in the treatment of peptic ulcer.

Ulcerative colitis

Because no physical cause has been found for ulcerative colitis, some authors have suggested that psychological factors can lead to its onset. Alexander (1950) included the disease among his group of psychosomatic disorders. Clinical experience suggests that psychological stressors can provoke relapses of the established ulcerative colitis in some patients. However, there is no scientific evidence for the role of psychological or social factors in initiating the illness or provoking relapses.

Although many psychological and social problems have been described in patients with ulcerative colitis (Magni *et al.* 1991) most seem to adapt well to the disease (Hendriksen and Binder 1980). Many patients benefit from supportive counselling, but it is doubtful whether a more elaborate psychological treatment is helpful in most cases (Schwartz and Blanchard 1991). Karush *et al.* (1977) carried out a controlled trial comparing combined psychotherapy and medical treatment with medical treatment alone. They reported a better outcome after the combined treatment, but there are doubts about their diagnostic criteria, matching of controls, and assessment of outcome.

Crohn's disease

There is no convincing evidence that psychological factors contribute to onset or relapse in Crohn's disease. However, the unpredictable, fluctuating, and chronic course of the disease means that it often impairs the quality of life (Meyers *et al.* 1980). Morbidity is especially high in juvenile onset disease. Psychiatric symptoms are common (Helzer *et al.* 1984). Most patients report an improvement in quality of life and psychological symptoms after surgery.

Colostomy and ileostomy

Several bowel diseases require surgical treatment that results in a temporary or permanent stoma. Psychiatric morbidity seems to be greater after this treatment than after other bowel surgery, and is probably more frequent in patients with cancer than in those with other conditions. Amongst patients with a stoma, about half report post-operative depression or other psychiatric symptoms which mostly improve over a few months. Other common problems (which may be persistent) are social and leisure activities, embarrassment about possible leakage of bowel contents, and sexual difficulties (Thomas *et al.* 1987). Patients with stomas are often helped by practical advice from other stoma patients who have made a good adjustment. (See Bekkers *et al.* 1995.)

Irritable bowel syndrome

The irritable bowel syndrome is abdominal pain or discomfort, with or without an alteration of bowel habits, persisting for longer than three months in the absence of any demonstrable organic disease. The condition is common in gastroenterology clinics, and also amongst people who have not consulted a doctor (Heaton 1992). It is uncertain whether the condition is related to an intestinal motility disorder, dietary fibre deficiency, or other physical factors. Research has proved little evidence of an association with life events, social factors, or psychiatric symptoms. However, psychological and social factors make it more likely that patients with these symptoms will seek treatment. These factors are associated with greater disability and may make management more difficult. Psychiatric disorder is common among clinic attenders, particularly those who fail to respond to treatment (Drossman 1994; Heaton 1994).

Patients with mild symptoms usually respond to education, reassurance about the disorder, dietary modification, and, when required, antimotility agents. More severe and chronic symptoms may be helped by psychological treatment. There is evidence that psychotherapy (Guthrie *et al.* 1991), behavioural methods (Greene and Blanchard 1994), and antidepressants are effective with some patients (see Drossman (1994), and Creed (1995) for reviews).

Sensory disorders

Deafness

Deafness may develop before speech is learnt (prelingual deafness) or afterwards. Profound early deafness interferes with speech and language development, and with emotional development (Thomas 1981). When patients with this condition leave school at sixteen, they are on average eight years behind children with normal hearing. Prelingually deaf adults often keep together in their own social groups and communicate by sign language. They appear to develop behaviour problems and social maladjustment more often than emotional disorder. For the management of such problems, special knowledge of the practical problems of deafness is required (Denmark 1985; Steinberg 1991).

Deafness of later onset has less severe effects than those just described; Eriksson-Mangold and Carlsson 1991). However, the acute onset of profound deafness can be extremely distressing, whilst milder restriction of hearing may cause depression and considerable social disability.

Kraepelin suggested that deafness was an important factor in the development of persecutory delusions. This idea was supported by Kay *et al.* (1976), who carried out a large survey of elderly patients with chronic paranoid hallucinatory illnesses and found a high prevalence of deafness among them (see p. 531). It is not known whether there is any association between

deafness and paranoid disorders in younger patients.

Tinnitus

Tinnitus is very common, but few patients seek treatment and most are able to live a normal life. Persistent tinnitus may be associated with low mood. Some patients are helped by devices that mask tinnitus with a more acceptable sound. Antidepressant medication may improve mood and reduce the intensity of the tinnitus (Sullivan *et al.* 1993). Cognitive and behavioural methods may enable people to accept their tinnitus and to minimize their social handicaps.

Blindness

Although it imposes many difficulties, blindness in early life need not lead to abnormal psychological development in childhood (Ammerman *et al.* 1986; Graham and Rutter 1968) or to unsuccessful later development. In previously sighted people the later onset of blindness causes considerable distress. Initial denial and subsequent depression are common, as are prolonged difficulties in adjustment. For accounts of blindness as a cause of psychiatric disorder see Cooper (1984) and Corbin and Eastwood (1986).

Non-organic visual dysfunction is an uncommon problem which requires collaboration between psychiatrists and ophthalmologists to establish the diagnosis and management (Newman 1993).

Skin disorders

Alexander (1950) included 'neurodermatitis' among the psychosomatic disorders. Psychological causes have been suggested for many skin conditions including urticaria, lichen simplex, atopic dermatitis, psoriasis, alopecia areata, and pruritus. The evidence for psychological causation is not strong. However, psychiatric disorders are common among people with established skin disease, and Wessely and Lewis (1989) reported a prevalence of 40 per cent amongst new attenders at a dermatology clinic. Patients with conspicuous skin disorders such as acne, psoriasis, and eczema often describe the effects of embarrassment and lack of confidence on their social lives (Ramsay and O'Reagan 1988). A small minority fail to come to terms with their disability and suffer considerable emotional distress (Folks and Kinney 1992*a*). Antidepressants and other psychotropic drugs are useful (Folks and Kinney 1992*a*; Koo and Pham 1992). Behavioural treatments are sometimes effective in reducing scratching and the secondary exacerbation of the skin disorder which it may cause (Ehlers *et al.* 1994).

Primary psychiatric disorder may present to the dermatologist as dysmorphophobia, illness fears, pruritus, delusions of parasite infestation, and factitious disorder. For example patients with *dysmorphophobia* (see p. 355) may complain of supposed hair loss or abnormality of the appearance of the skin, where little or no abnormality is apparent to the doctor.

Factitious skin disorder

Factitious skin disorders include dermatitis arte-facta which is the name for self-inflicted skin lesions, usually areas of superficial necrosis. Most patients with this condition are young women, many of whom have abnormal personalities though there is no single personality type. The condition persists for many years in about a third of cases. Recovery is usually associated with a change in life circumstances rather than with treatment (Gupta *et al.* 1987). Another factitious skin disorder is self-induced purpura.

Some patients, particularly middle-aged to elderly women, present to dermatologists with *delusions* about parasites or other objects in the skin (Ekbom's syndrome). These delusions are usually encapsulated; they are sometimes associated with a depressive disorder, but more often constitute a monosymptomatic paranoid disorder (Berrios 1985). A few respond to antipsychotic drugs such as pimozide, but generally the prognosis is poor (Munro 1980).

Trichotillomania

Trichotillomania is an irresistible urge to pull hairs from the scalp and other parts of the body. The hair-pulling may be denied by the patient. Most cases start in early adolescence, though children are occasionally affected. Among adults, females are more commonly affected than males. The prevalence of trichotillomania is uncertain because many sufferers attempt to conceal the disorder, but in surveys it is reported by about 1 per cent of college students.

Some cases start at a time of stress and last for only a few months, others continue for years. Usually the scalp hair is pulled out, but eyelashes, eyebrows, axillary, and pubic hair may be removed. Hairs may be pulled out in tufts or one by one. Some patients save the hair and eat it, a practice that can lead to hair ball in the stomach or to intestinal obstruction. Trichotillomania may be associated with a variety of psychiatric disorders. It may also be an isolated symptom, which sometimes responds to behavioural treatment. It is usually a chronic disorder except in early childhood.

Treatment begins with an assessment of the circumstances in which the hair pulling occurs. Cognitive behavioural treatment is helpful in some cases. (See Rothbaum and Ninan (1994) and Minichiello and O'Sullivan (1994) for reviews.)

Connective tissue disorders

Rheumatoid arthritis

Rheumatoid arthritis was one of Alexander's psychosomatic disorders, but there is no convincing evidence that psychological factors are important in its aetiology (Weiner 1977). As with other physical illnesses, psychological abnormalities have often been described, but they are most likely to be the result rather than the cause of the illness, or simply coincidental. Attempts to describe a characteristic premorbid personality have been equally unconvincing, as have reports that psychological stressors can precipitate onset or relapse.

It is not surprising that this painful chronic disorder is associated with anxiety and depressive symptoms, and with limitation of work, leisure, family life, and sexual function. The prevalence of depression is similar to that in other chronic medical disorders. It is related to the severity of the physical illness, disability, social factors, and previous psychological state (Creed and Ash 1992). The capacity to cope with pain and disability is a major determinant of severity of depression and the quality of life (Smith and Walston 1992). Emotional problems are particularly severe in juvenile patients.

Physicians treating patients with rheumatoid arthritis spend much time in attending to psychological reactions to the illness, attempting to improve compliance with treatment, and reducing handicap. Psychological treatment can help to improve compliance, and patients cope with pain, alleviate depression, and reduce psychosocial handicaps (Skevington 1986). The methods employed include anxiety management, counselling, cognitive therapy, and self-help groups.

Systemic lupus erythematosus

Psychiatric disorders are common in the course of systemic lupus erythematosus, although they are seldom the first manifestation. They include acute and chronic organic syndromes, other psychoses, and emotional disorders. Most of these disturbances last for less than six weeks but some recur.

Cognitive impairment has been reported to be common in patients with active or inactive systemic lupus erythematosus, but this impairment may in part be secondary to another psychiatric disorder (Hay *et al.* 1994). In some patients, psychiatric symptoms result from prolonged treatment with high doses of steroids. The course of any psychiatric symptoms usually follows that of other features of the physical disease. The treatment of the psychiatric symptoms is mainly that of the primary condition. (See Calabrese and Stern 1995 for a review.)

Infections

Viral infections may cause encephalitis which may be accompanied by psychiatric symptoms. In addition, some infectious diseases, for example hepatitis A, influenza, and brucellosis, are frequently followed by periods of depression.

Psychological factors may affect the course of recovery from an acute infection. In one early study, psychological tests were completed by 600 people who subsequently developed Asian influenza. Delayed recovery from the influenza was no more common among people whose initial illness had been severe, but it was more frequent among those who had obtained more abnormal scores on the psychological tests before the illness (Imboden *et al.* 1961). Recent findings of research on viral illness in general practice (Cope *et al.* 1994) and infectious mononucleosis (White *et al.* 1995), have reported similar conclusions.

In the last few years there has been a considerable increase in the numbers of people presenting to doctors with chronic fatigue and emotional symptoms which they believe are consequences of viral infections (see p. 371). Numerous viruses and abnormalities of immune processes have been incriminated, but the significance of organic causes remain uncertain. It is probable that in many instances the aetiology is psychological, and that psychological factors contribute to the clinical features in other cases with organic causes (see Hotopf and Wessely 1994 for a review).

Sexually transmitted disease

Amongst patients attending clinics for sexually transmitted disease, 20–30 per cent have a psychiatric disorder, whether or not they have a venereal disease. A similar proportion report sexual dysfunction for which most would like further help. Apart from HIV infection, genital herpes is probably the most distressing sexually transmitted disease because it often causes intense discomfort and threatens the danger of transmission in pregnancy (Levenson *et al.* 1987; Van der Plate and Aral 1987). Some patients occasionally present with severe fears of venereal disease which persist despite reassurance; these fears are often related to depressive disorder.

HIV infection

The nature of the physical symptoms, their relentless progressive course, and the reactions of other people all explain why emotional distress is common in patients with HIV infection. A further reason is that groups at high risk for HIV (homosexuals, haemophiliacs, and drug abusers) may have other psychological problems. Neuro-psychiatric disorders also occur in people with HIV infection. Even so, many patients with AIDS manage to lead relatively normal lives for substantial periods. Men with AIDS and haemophilia do not appear to have psychological problems greater than those of other AIDS patients.

Fears of infection and reaction to testing: although surveys suggest that worry about having AIDS is not uncommon in the general population, severe concern is infrequent. Although HIV antibody testing is worrying for most of those who undergo it, the distress is usually short-lived whatever the outcome of the test. People who have persistent and unjustified worries about having AIDS require psychiatric help of the kind appropriate for other hypochondriacal concerns (see p. 351).

Psychiatric problems include adjustment disorder, depressive disorder, and anxiety disorder. These disorders may occur at any stage of the disease, but are particularly frequent at the time of diagnosis. People with previous psychological problems, long-standing social difficulties, or lack of social support are especially vulnerable.

Neuropsychiatric disorders are common, both secondary to the complications of immune suppression and as direct effects of HIV on the brain. Dementia has been described as a rare complication, sometimes associated with depression and seizures. Minor cognitive disorders are much more common (Maj 1990; Janssen *et al.* 1992). HIV infection can also result in neurological symptoms and dementia in those who do not have AIDS (Navia and Price 1987). Several acute and subacute organic syndromes have been described, of which the most frequent is subacute encephalitis. Delirium may occur when there is an opportunistic infection or cerebral malignancy. (See Simpson and Tagliatti (1994) for a review.)

Suicide and deliberate self-harm may occur in people who are concerned about the possibility of HIV infection as well as in people with proven disease. Among the latter, the risk is greater in those with advanced symptoms. However, it is not certain how much greater is the risk of suicide and deliberate self-harm in AIDS patients than in the general population (Marzuk *et al.* 1988).

Effects on the family may be considerable, as they may be with any serious medical disorder. These effects are particularly significant where the partner and especially the children also suffer from the infection.

In general, women have a similar psychiatric morbidity to men but they are especially concerned about the effects on **childbearing**.

Problems in relation to **illicit drug use** are considered on p. 464. The disorganized way of life of some drug users and their personal and social problems make the treatment of HIV difficult. Psychiatrists should be involved in planning services for AIDS patients; they may provide counselling and symptomatic treatment for neuro-psychiatric complications. It is not yet known what scale of facilities may be required for AIDS patients who develop dementia. See King (1993), Kalichman and Sikkema (1994), and Catalan *et al.* (1995), for reviews.

Cancer

Psychological factors in aetiology and prognosis

It is not surprising that cancer patients have emotional reactions to the disease. Some writers have suggested the opposite relationship, namely that psychological factors including depression, personality traits, the suppression of anger, and stressful life events may play a part in the aetiology of cancer. Overall, the evidence for this idea is not convincing, partly because the research on which it is based has severe limitations of methodology including reliance on retrospective accounts and on subjective or non-standardized methods of assessment (Levenson and Bernis 1991). Instead of studying the role of psychological factors in the onset of cancer, other workers have examined the influence of these factors on the course and outcome of cancer. Greer *et al.* (1979) reported that the prognosis of breast cancer was better in patients who reacted to their illness by denial or who had 'a fighting spirit' than in patients who reacted in other ways, but the considerable body of evidence is contradictory. Evidence from subsequent studies is equally conflicting, and it remains uncertain whether psychosocial factors significantly affect the course of the condition. Studies of group therapy (Spiegel 1990, 1994) have led to the claim that this treatment improves prognosis. Research with animals has indicated that the rate of tumour growth may be increased in animals exposed to stressful situations that are only partially under their control. This finding suggested the possibility that endocrine or immunological mechanisms may mediate the effects of emotion on the prognosis of malignancy (see Lewis *et al.* (1994) for a review of psychoimmunology).

It has been suggested that depressive symptoms may be a precursor of cancer in various sites. The issue remains unproved, and it is unlikely that, with one exception, any association is of practical significance (McGee 1994). The exception concerns patients with carcinoma of the pancreas, Fras *et al.* (1967) found that 76 per cent had psychiatric symptoms, mainly depression; in almost half these patients, depression had preceded the onset of physical symptoms and signs such as abdominal pain, weakness, jaundice, and weight loss. Holland *et al.* (1986) compared patients with pancreatic and gastric cancer and concluded that the former experienced significantly greater psychological disturbance. (See Green and Austin (1993) for a review.)

Psychological consequences of cancer

The psychological consequences of cancer are similar to those of any serious physical illness.

Some patients delay seeking medical help because they fear or deny symptoms (Facione 1993). Knowledge of the diagnosis of cancer may cause shock, anger, and disbelief, as well as anxiety and depression. The most common associated psychiatric disorder is adjustment disorder (Derogatis *et al.* 1985). The risk of suicide is increased in the early stages (Harris and Barraclough 1995). Depressed mood is particularly likely at the time of diagnosis and following relapse but is usually transient. Major depression occurs throughout the course of cancer affecting 10–20 per cent of patients (Noyes and Kathol 1986) and appears to be more frequent in those suffering pain (Spiegel 1994; McDaniel *et al.* 1995). However, patients with cancer are no more depressed than other physically ill patients, and the majority do not experience long-term distress unless the disorder progresses or unless they are particularly vulnerable to stress (Harrison and Maguire 1994). Both the progression and the recurrence of cancer are often associated with increased psychiatric disturbance, which may result from a worsening of physical symptoms such as pain and nausea, from fear of dying, or from the development of an organic psychiatric syndrome (Holland and Rowland 1989).

Organic mental disorder may arise from brain metastases which originate most often from carcinoma of the lung, but also from tumours of the breast and alimentary tract, and from melanomas (Clouston *et al.* 1992). Occasionally, brain metastases produce psychiatric symptoms before the primary lesion is discovered. Organic mental disorder is sometimes induced by certain kinds of cancer in the absence of metastases, notably by carcinoma of the lung, ovary, or stomach. The mechanism is unknown (Lishman 1987).

Several kinds of treatment for cancer may cause psychological disorder. Emotional distress is particularly common after mastectomy (see p. 393) and mutilating surgery. Radiotherapy causes nausea, fatigue, and emotional distress. Chemotherapy often causes malaise and nausea, and anxiety about chemotherapy may cause anticipatory nausea before the treatment. The latter may be helped by behavioural treatments used for anticipatory anxiety (see Andrykowski and Jacobsen 1993).

Family and other close relatives of cancer patients may experience psychological problems, which may persist even if the cancer is cured (Bearison and Mulhern 1994). Nevertheless many patients and relatives make a good adjustment to cancer. The extent of their adjustment depends partly on the information they receive. In the past doctors have been reluctant to tell patients that the diagnosis is cancer, but most patients prefer to know the diagnosis and how it will affect their lives. The quality of communication is often unsatisfactory and when this happens there may be consequences for psychological adjustment. The problem is particularly difficult when the patient is a child; even then it is generally better to explain the diagnosis in terms appropriate to the child's stage of development.

Psychological treatments

Various psychiatric treatments can be helpful for patients with cancer, including counselling and social support groups (Breitbart and Holland 1993). Recent studies of support groups (Spiegel 1990; Fawzy 1993) and cognitive behavioural treatments (Moorey and Greer 1989) have shown benefits to survival time and immune function.

Although the psychological problems of patients with cancer can often be alleviated, many are undetected. One possibility is to provide educational programmes, counselling, or group therapy for all cancer patients whether or not they report problems. However, it seems more appropriate to select suitable patients, particularly as there is some evidence that counselling may increase distress in some vulnerable patients who have denied their anxieties. The patients most likely to need psychological treatment include those with a history of previous psychiatric disorder or poor adjustment to other problems and those who lack a supportive family. Examples of programmes based on early identification and selective treatment have been described by Maguire (1992).

See Andersen (1992), Trijsburg *et al.* (1992), and Fawzy *et al.* (1995), for reviews of the efficacy of psychological intervention, and Watson (1991), Breitbart and Holland (1993), and Spiegel (1994) for reviews of treatment.

Breast cancer

About a quarter of patients undergoing mastectomy or other treatment for breast cancer develop depression or anxiety of clinical severity within eighteen months. Affective symptoms are especially common after a recurrence, and during radiotherapy and chemotherapy. Other responses to mastectomy are low self-esteem, embarrassment about disfigurement, and marital and sexual problems (Irvine *et al.* 1991). There is no evidence that psychiatric morbidity is any less after conservative treatment by lumpectomy and radiotherapy than after mastectomy or when a choice is offered (Fallowfield *et al.* 1994). Postmastectomy breast reconstruction appears to be associated with a substantial improvement in well-being (Rowland *et al.* 1993). Careful follow-up to detect and treat patients with psychiatric complications is probably more useful than counselling given routinely. Antidepressants and cognitive-behavioural treatment are useful in selected cases.

Childhood cancer

Childhood cancer presents special problems. The child often reacts to the illness and its treatment with behaviour problems. Many parents react at first with shock and disbelief, taking months to accept the full implications of the diagnosis. About one mother in five develops an anxiety or depressive disorder during the first two years of treatment of childhood leukaemia, and other family members may be affected. In the early stages of the illness parents are usually helped by advice about practical matters, and later by discussions of their feelings, which often include guilt (Bearison and Mulhern 1994). Adult survivors of cancer in childhood or adolescence appear to be at risk of social difficulties (Lancet 1992*b*).

Other types of cancer

Cancers of other primary sites, such as bowel, lung, head, and neck, may result in psychiatric morbidity. Several kinds of cancer are referred to elsewhere in this book: carcinoma of the pancreas (p. 392), and brain tumours (p. 332). The psychiatric aspects of these other types of cancer have been reviewed by Holland and Rowland (1989) Watson (1991), and Lesko *et al.* (1992).

Psychiatric aspects of obstetrics and gynaecology

Pregnancy

Psychiatric disorder is more common in the first and third trimesters of pregnancy than in the second. In the first trimester unwanted pregnancies are associated with anxiety and depression. In the third trimester there may be fears about the impending delivery or doubts about the normality of the fetus. Psychiatric symptoms in pregnancy are more common in women with a history of previous psychiatric disorder and probably also in those with serious medical problems affecting the course of pregnancy, such as diabetes. Although minor affective symptoms are common in pregnancy, serious psychiatric disorders are probably less common than in non-pregnant women of the same age (Pugh *et al.* 1963).

Amongst women who have chronic psychological problems when not pregnant, some report improvement in these problems during pregnancy whilst others require extra psychiatric care. The latter are often late or poor attenders at antenatal care, thus increasing the risk of obstetric and psychiatric problems. *Abuse of alcohol, opiates, and other substances* should be strongly discouraged in pregnancy, especially in the first trimester when the risk to the fetus is greatest (see p. 448). In a Canadian study 8.2 per cent of women reported drinking more than seven units of alcohol per week in the second half of pregnancy. Other adverse risk factors were also common in these women (Stewart and Streiner 1994). Eating disorders do not appear to be precipitated by pregnancy, and bulimic symptoms often improve (Fahy 1991; Davies and Wardle 1994). Great care must be taken in the *use of psychotropic drugs* during pregnancy because of the risk of fetal malformations, impaired growth, and prenatal problems (p. 539). Pharmacokinetics may be altered. Benzodiazepines should be avoided throughout the pregnancy and during breast-feeding because of the danger of depressed respiration and withdrawal symptoms in the neonate. Lithium should be stopped throughout the first trimester of pregnancy but can be restarted later if there are pressing reasons; it should be stopped again at the onset of labour. Mothers taking lithium should not breast-feed. It is preferable to avoid using tricyclic antidepressants or neuroleptics during pregnancy unless there are compelling clinical indications.

Hyperemesis gravidarum

About half of all pregnant women experience nausea and vomiting in the first trimester. Some authors have suggested that these symptoms, as well as the severe condition of **hyperemesis gravidarum,** are primarily of psychological aetiology. However, there is no reason to doubt that psychological factors are of primary importance, although psychological factors may substantially influence the severity and course of the symptoms (Katon *et al.* 1980).

Pseudocyesis

Pseudocyesis is a rare condition in which a woman believes that she is pregnant when she is not, and develops amenorrhoea, abdominal distension, and other changes similar to those of early pregnancy. The condition is more common in younger women. Pseudocyesis usually resolves quickly once diagnosed, but some patients persist in believing that they are pregnant. Recurrence is common (Small 1986; Drife 1987).

Couvade syndrome

In this syndrome, the husband of the pregnant woman reports that he is himself experiencing some of the symptoms of pregnancy. This condition may occur in the early months of the woman's pregnancy, when the man complains usually of nausea and morning sickness and often of tooth-ache. These complaints generally resolve after a few weeks (Bogren 1983).

Unwanted pregnancy

Until 1967, psychiatrists in the United Kingdom were often asked to see pregnant women who were seeking a therapeutic abortion on the grounds of mental illness. Since 1967 the law has allowed therapeutic abortion on the grounds of likely damage to the health of the mother and also of her children, and the law is similar in many other countries. The provisions now make it generally more appropriate for decisions to be made by the family doctor and the gynaecologist, without involving a psychiatrist. However, psychiatric opinions are still sought at times, not only about the grounds for termination of pregnancy but also for an assessment of the likely psychological effects of termination in a particular patient. Most of the evidence suggests that the psychological consequences of abortion are usually mild and transient, but that they are greater for mothers who have cultural or religious beliefs against abortion (Clare and Tyrrell 1994). (See also Gilchrist *et al.* 1995.)

Spontaneous abortion

Friedman and Gath (1989) interviewed 67 women four weeks after spontaneous abortion, and 32 (48 per cent) of them were PSE cases as defined by the Present State Examination (a rate four times higher than in the general population of women). All the women were diagnosed as suffering from depressive disorder. Many women showed features typical of grief. Depressive symptoms were most frequent in women with a history of previous spontaneous abortion.

Therapeutic abortion

Iles and Gath (1993) compared a group of women who had a termination of pregnancy for medical reasons with a group who suffered a spontaneous abortion. In both groups psychiatric morbidity was high at one month. At follow-up there were

significant improvements in psychiatric symptoms, guilt, and interpersonal and sexual adjustments. Adverse psychiatric and social consequences were rare.

Antenatal death

Antenatal death causes an acute bereavement reaction, long-term psychiatric problems, and concern about future pregnancy. Parents need to be helped to mourn and should be encouraged to see and hold the baby, to name it, and to have a proper funeral. The next pregnancy may be a particularly worrying time.

Caesarian section

Caesarian section is extremely frequent and is becoming more frequent in Western countries; it has been said to have adverse psychological consequences for parents and infants. Much of the research has failed to separate the effects of surgery from other adverse material factors. However, it would seem important to pay particular attention to parental support and to initial bonding (Mutryn 1993).

Post-partum mental disorders

These disorders can be divided into maternity blues, puerperal psychosis, and chronic depressive disorders of moderate severity.

Maternity blues

Amongst women delivered of a normal child, between a half and two-thirds experience brief episodes of irritability, lability of mood, and episodes of crying. Lability of mood is particularly characteristic taking the form of rapid alternations between euphoria and misery. The symptoms reach their peak on the third or fourth day post-partum. Patients often speak of being 'confused', but tests of cognitive function are normal. Although frequently tearful, patients may not be feeling depressed at the time but tense and irritable.

'Maternity blues' is more frequent among primigravida. The condition is not related to complications at delivery or to the use of anaesthesia. 'Blues' patients have often experienced anxiety and depressive symptoms in the last trimester of pregnancy; they are also more likely to give a history of premenstrual tension, fears of labour, and poor social adjustment (Kennerley and Gath 1989).

Both the frequency of the emotional changes and their timing suggest that maternity blues may be related to readjustment in hormones after delivery. Oestrogens and progesterone both increase greatly during late pregnancy and fall precipitously after childbirth. Changes also occur in adrenal steroids, but they are complicated by associated changes in corticosteroid-binding globulin. However, there is no convincing evidence that these changes lead to the emotional symptoms, and at present the cause of maternity blues is unknown.

No treatment is required because the condition resolves spontaneously in a few days.

Puerperal psychosis

In the nineteenth century, puerperal and lactational psychoses were thought to be specific entities distinct from other mental illnesses (Esquirol 1845; Marcé 1858). Later psychiatrists such as Bleuler and Kraepelin regarded the puerperal psychoses as no different from other mental illnesses. This latter view is widely held today on the grounds that puerperal psychoses generally resemble other psychoses in their clinical picture (see below).

The **incidence** of puerperal psychoses has been estimated in terms of admission rates to psychiatric hospital (Pugh *et al.* 1963; Kendell *et al.* 1987). The reported rates vary, but a representative figure is one admission per 500 births. This incidence is substantially above the expected rate for non-puerperal women of the same age. Puerperal psychoses are more frequent in primiparous women, those who have suffered previous major psychiatric illness, those with a family history of mental illness, and probably in unmarried mothers. There is no clear relationship between psychosis and obstetric factors (Kendell 1985). The onset of puerperal psychosis is usually within the first one to two weeks after delivery, but rarely in the first two days. Puerperal illnesses are especially common in develop-ing countries and the excess may be cases with an organic aetiology.

The early onset of puerperal psychoses has led to speculation that they might be caused by hormonal changes such as those discussed above in relation to the blues syndrome. There is no evidence that hormonal changes in women with puerperal psychoses differ from those in other women in the early puerperium. Hence if endocrine factors do play a part, they probably act only as precipitating factors in predisposed women.

Recent work has implicated a supersensitivity of dopamine receptors in the pathophysiology of puerperal psychoses, perhaps precipitated by oestrogen withdrawal (Wieck *et al.* 1991). In depressive disorders stress appears a less important factor than in depression at other times.

Three types of **clinical picture** are observed: acute organic, affective, and schizophrenic. Organic syndromes were common in the past, but are now much less frequent since the incidence of puerperal sepsis was reduced by antibiotics. Nowadays affective syndromes predominate. Dean and Kendell found that 80 per cent of cases were affective, and that the proportion of manic disorders was unusually high. Though less common than affective disorders, schizophrenic illnesses (using the then current diagnostic criteria which did not require a 6 month duration) were much more frequent that the expected rate. As mentioned above, the clinical features of these syndromes are generally regarded as being much the same as those of corresponding non-puerperal syndromes. The exceptions are that affective features are probably more common in puerperal schizophrenic disorders, and that disorientation and other organic features are more common in both the schizophrenic and the affective disorders.

In the **assessment** of patients with puerperal psychosis, it is necessary to ascertain their ideas concerning the baby. Severely depressed patients may have delusional ideas that the child is malformed or otherwise imperfect. These false ideas may lead to attempts to kill the child to spare it from future suffering. Schizophrenic patients may also have delusional beliefs about the child; for example they may be convinced that the child is abnormal or evil. Again, such beliefs may point to the risk of an attempt to kill the child. Depressed or schizophrenic patients may also make suicide attempts.

Treatment is given according to the clinical syndrome, as described in other chapters. For in-patient care it has been argued that there should be special mother and baby units. However, the benefits of such units have not been established (see Kumar *et al.* 1995). The layout of the mother and baby unit should have facilities for a separate nursery where the child can be nursed at times when the mother is too ill to care for him. The layout of the ward should enable staff to observe the mother closely while she is with the baby since the mental state may change quickly in puerperal psychosis. The nursing staff should have experience in the care of small babies as well as in psychiatry. These arrangements are demanding for the staff, but the risk to the child does not usually last long if treatment is vigorous.

Electroconvulsive therapy (ECT) is usually the best treatment for patients with depressive or manic disorders of marked or moderate severity, because it is rapidly effective and enables the mother to resume the care of her baby quickly. For less urgent depressive disorders, antidepressant medication may be tried first. If the patient has predominantly schizophrenia-like symptoms, an antipsychotic drug may be prescribed; if definite improvement does not occur within a short period, ECT should be considered, especially if the onset was acute.

Most patients recover fully from a puerperal psychosis, but some of those with a schizophrenic disorder, remain chronically ill, (Protheroe 1969). After subsequent childbirth the recurrence rate for depressive illness in the puerperium is approximately 20–30 per cent. According to Protheroe (1969), at least half of women who have suffered a puerperal depressive illness, will later suffer a depressive illness that is not puerperal.

Puerperal depression of mild or moderate severity

Less severe depressive disorders are much more common than the puerperal psychoses. Estimated rates vary, but are mainly within the range 10–15 per cent (Kendell 1985). These depressive disorders usually begin after the first two weeks of the puerperium. Tiredness, irritability, and anxiety are often more prominent than depressive mood change, and there may be prominent phobic symptoms. Most patients recover after a few months.

Clinical observation suggests that these disorders are often precipitated in vulnerable mothers by the psychological adjustment required after childbirth, as well as by the loss of sleep and hard work involved in the care of the baby. Previous psychiatric history and recent stressful events appear to be important aetiological factors. Paykel *et al.* (1980) assessed a series of women with mild clinical depression about six weeks post-partum, and found that the strongest associated factor was recent stressful life events. Previous history of psychiatric disorder, younger age, early post-partum blues, and a group of variables affecting poor marital relationship and absence of social support were also notable.

Cooper *et al.* (1988) examined 483 pregnant women six weeks before the expected date of delivery, and re-examined them three, six, and twelve months after childbirth. At all stages of assessment, the point prevalence of psychiatric disorder was no higher than in a matched sample of women from the general population. There was no evidence that the postnatal psychiatric disorder differed either diagnostically or in duration from psychiatric disorders arising at other times (see Cooper and Murray (1995) and Whiffen (1992) for reviews).

In treatment, psychological and social measures are usually as important as antidepressant drugs. Despite the medical and other care given to women after childbirth, many post-partum depressions are undetected or, if detected, untreated. There is evidence that postnatal depression may adversely affect the mother–infant relationship and the psychological development of the infant (Stein *et al.* 1991). The fathers may also experience distress (Ballard *et al.* 1994). A liaison obstetric service can help in the management of these patients (Appleby *et al.* 1989).

Menstrual disorders

Premenstrual syndrome

This term denotes a group of distressing psychological and physical symptoms starting a few days before and ending shortly after the onset of a menstrual period. The psychological symptoms include anxiety, irritability, and depression; the physical symptoms include breast tenderness, abdominal discomfort, and a feeling of distension. Premenstrual syndrome is not included in current classifications of psychiatric disorder, although premenstrual dysphoric disorder, is listed as a condition for future study in DSMIV. The syndrome should be distinguished from the much more frequent occurrence of similar symptoms that are not strictly premenstrual in timing.

The estimated frequency of the premenstrual syndrome in the general population varies widely from 30 to 80 per cent of women of reproductive age. There are several reasons for this wide variation in reported rates. First, there is a problem of definition. Mild and brief symptoms are frequent premenstrually, and it is difficult to decide when they should be classified as premenstrual syndrome. Second, information about symptoms is often collected retrospectively by asking women to recall earlier menstrual periods, and this is an unreliable way of establishing the time relations. Third, the description of premenstrual symptoms may vary according to whether or not the woman knows that the enquiry is concerned specifically with the premenstrual syndrome.

The aetiology is uncertain. Physical explanations have been based on ovarian hormones (excess oestrogen, lack of progesterone), pituitary hormones, and disturbed fluid and electrolyte balance. None of these theories has been proved. Dalton (1964) has argued that the premenstrual syndrome is caused by oestrogen–progesterone imbalance, but the evidence on this point remains inconclusive. Various psychological explanations have been based on possible associations of the syndrome with neuroticism or with individual or public attitudes towards menstruation.

The syndrome has been widely treated with progesterone, and also with oral contraceptives, bromocriptine, diuretics, and psychotropic drugs. There is no convincing evidence that any of these is effective, and treatment trials suggest a high placebo response (up to 65 per cent). Psychological support and encouragement may be as helpful as medication. There have been encouraging reports of the effectiveness of SSRI antidepressants and of a cognitive behavioural treatment. For reviews see Blake *et al.* (1995).

The menopause

In addition to the physical symptoms of flushing, sweating, and vaginal dryness, menopausal women often complain of headache, dizziness, and depression. It is not certain whether depressive symptoms are more common in menopausal women than in non-menopausal. Weissman and Klerman (1978) concluded that there is no such increase in symptoms at the menopause. Nevertheless, amongst patients who consult general practitioners because of emotional symptoms, a disproportionately large number of women are in the middle-age-group that spans the menopausal years (Shepherd *et al.* 1966).

Depressive and anxiety-related symptoms at the time of the menopause could have several causes. Hormonal changes have often been suggested, notably deficiency of oestrogen. In some countries, notably the United States, oestrogen has been used to treat emotional symptoms in women of menopausal age, but the results of trials of treatment with oestrogens have been disappointing. (See Pearce *et al.* 1995.) Psychiatric symptoms at this time of life could equally well reflect changes in the woman's role as her children leave home, her relationship with her husband alters, and her own parents become ill or die. In a community survey of over 500 women aged 35–59 it was found that both psychiatric symptoms and the personality dimension of neuroticism were associated with vasomotor symptoms (flushes and sweats) but not with the cessation of menstruation (Gath *et al.* 1987). It seems best to treat depressed menopausal women with methods that have been shown to be effective for depressive disorder at any other time of life. See Ballinger (1990) and Gath and Iles (1990) for reviews.

Hysterectomy

Several retrospective studies have indicated an increased frequency of depressive disorder after hysterectomy (Barker 1968). A prospective investigation using standardized methods showed that patients who are free from psychiatric symptoms before hysterectomy seldom develop them afterwards. Some patients with psychiatric symptoms before hysterectomy lose them afterwards, but others continue to have symptoms after the operation (Gath *et al.* 1982*a,b*). It is likely that these latter persisting cases (those with symptoms before and after surgery) are identified in the retrospective studies, and lead to the erroneous conclusion that hysterectomy causes depressive disorder. This finding provides a general warning about inferring the effects of treatment from the results of retrospective investigations.

Sterilization

Considerations similar to those for hysterectomy apply to these procedures. Retrospective studies have suggested that sterilization leads to psychiatric disorder, sexual dysfunction, and frequent regrets after the operation. However, prospective enquiry has shown that the operation does not lead to significant psychiatric disorder; sexual relationships are more likely to improve than worsen, and definite regrets are reported by fewer than one patient in twenty (Cooper *et al.* 1982).

Blood disorders

Leukaemia

Leukaemia in children and adults, especially in the acute form, causes great distress to patients and their families (Lesko *et al.* 1992). Chemotherapy and its side-effects are often extremely unpleasant. However, behavioural treatments may help to reduce anticipatory nausea and anxiety experienced during chemotherapy. Neuropsychological sequelae of treatment have been reported in 42 per cent of children reassessed six or more years after successful treatment of acute lymphoblastic leu-

kaemia (Wheeler *et al.* 1988). The long-term outcome of therapy of acute leukaemia following chemotherapy and bone marrow transplantation is generally good (Lesko *et al.* 1992).

Lymphomas

Although they have a better prognosis than other cancers, Hodgkin's disease and non-Hodgkin's lymphoma can cause considerable psychiatric and social morbidity, especially at the time when the diagnosis is learnt. These effects can be lessened by full discussion of the diagnosis and the plan of treatment with patients and relatives.

Haemophilia

Surveys amongst patients with haemophilia, surveys have shown that emotional reactions and chronic handicaps are common and disturbing to family life (Lineberger 1981; Klein and Nimorwicz 1982). New methods of treatment have greatly reduced the significance of bleeding episodes but have led to HIV infection in many sufferers (Bussing and Johnson 1992). Surprisingly, the prevalence of psychiatric disorder in men with haemophilia and HIV is little different from that in those who are not infected. See Shakin and Thompson (1993) for a review of the psychiatric aspects of haemophilia.

Muscle disease and dystonias

Muscular dystrophy

Muscular dystrophy is the most common progressive muscle disease in childhood and is inherited as a sex-linked recessive gene. Diagnosis is often delayed. Parents are often severely distressed by this delay and by the way in which information is given (Firth 1983). They need information about the illness and its course, and about the genetic implications.

There is a tendency for boys with this condition to be of less than average intelligence. The progressive physical handicaps lead to increasing restrictions and social isolation, with consequent boredom, depression, and anger. The children require considerable help and support from their families and others. The illness is extremely distressing for parents (Witte 1985).

Myasthenia gravis

Myasthenia gravis presents with weakness and fatigue. It is not uncommon for the diagnosis to be delayed because the symptoms suggest psychiatric disorder, especially if there is a history of previous psychological problems. For some patients it is difficult to adjust to a regular schedule of medication. Clinical neurologists have reported that the physical symptoms may be precipitated and aggravated by emotional influences (see Lishman 1987).

Dystonias

Focal dystonias are uncontrolled muscle spasms leading to involuntary movements of the eyelids, face, neck, jaw, shoulders, larynx, hands, and, rarely, other part of the body. They are uncommon but disabling. The aetiology is uncertain. In the past dystonias were regarded as hysterical phenomena; now there is evidence that they are idiopathic or drug-induced neurological disorders, and that psychogenic cases are rare. Clinical types include blepharospasm, torticollis, writer's cramp, and laryngeal dystonia. The most effective treatment is the injection of botulinum toxin directly into the affected muscles. Psychiatric symptoms secondary to the physical disorder can usually be treated with antidepressants or behavioural therapy.

Muscular problems are common amongst musicians and may threaten to end their careers. There are many causes, including overuse injury, pressure on peripheral nerves, and focal dystonias. These problems should be assessed by a physician with experience in this special field. Performance anxiety is also frequent, and may impair or prevent performance. Beta-blockers alleviate this symptom and are used by many musicians, sometimes without medical supervi-

sion. Anxiety management is effective in some cases (see Lockwood (1989) for a review).

Tics

Tics are purposeless, stereotyped, and repetitive jerking movements occurring most commonly in the face and neck. They are much more common in childhood than in adult life, though a few cases begin at ages of up to 40 years. The peak of onset is about seven years, and the onset is often at a time of emotional disturbance. They are especially common in boys. Most sufferers have just one kind of abnormal movement, but a few people have more than one (multiple tics). Like almost all involuntary movements, tics are worsened by anxiety. They can be controlled briefly by voluntary effort, but this results in an increasing unpleasant feeling of tension. Many tics occurring in childhood last only a few weeks; others last longer but 80–90 per cent of cases improve within five years. A few cases become chronic. (The subject of tics has been reviewed by Corbett and Turpin (1985).)

Gilles de la Tourette syndrome

This condition was described first by Itard in 1825 and subsequently by Gilles de la Tourette in 1885. The main clinical features are multiple tics beginning before the age of 16, together with vocal tics (grunting, snarling, and similar ejaculations). About a third of the people affected show coprolalia (involuntary uttering of obscenities), but few of these are children. Between 10 and 40 per cent show echolalia or echopraxia (Robertson 1994). There may be stereotyped movements such as jumping and dancing. The tics usually precede the other features, with the exception of abnormal sensations which may occur even before the tics. Associated features include over-activity, difficulties in learning, emotional disturbances, and social problems.

The reported prevalence of the condition varies according to the criteria for diagnosis and method of enquiry. A generally accepted figure is about 0.5 per 1000 population. The disorder is three to four times more common in males than in females and about ten times more prevalent in children and adolescents than in adults (see Robertson 1994).

Obsessive–compulsive symptoms occur frequently in patients with the Gilles de la Tourette syndrome, and more often among the families of these patients than in the general population (Eapen *et al*. 1993). Attention-deficit hyperactivity disorder has also been reported to be more frequent in these patients than in the general population.

The **aetiology** of the Gilles de la Tourette syndrome is unknown. A neurochemical disorder has been suggested, possibly of dopamine function since dopamine blocking drugs produce improvements. Post-mortem studies suggest decreased 5-hydroxytryptamine and glutamate in several areas of the basic ganglia (Anderson *et al*. 1992). Family genetic studies, including segregation analysis studies, strongly suggest a genetic basis for the disorder. The concordance rate in monozygotic twins is reported to be 53 per cent compared with 8 per cent for dizygotic twins (Pauls and Leckman 1988).

Multiple tics without vocal tics are more frequent in the families of probands with Gilles de la Tourette syndrome than in the general population, suggesting that the two conditions are related.

Many **treatments** have been tried. Haloperidol appears to be the most effective, but the side-effects (see p. 552) can be troublesome. Specific serotonin re-uptake inhibitors have been reported to control the obsessional symptoms of the patients as they do in obsessive–compulsive disorder (see p. 185). There is not enough follow-up information to indicate the **prognosis**, but clinical experience suggests that the outcome is generally poor. Coprolalia disappears in one-third of patients, but the tics and obsessive compulsive symptoms may be lifelong (Coffey *et al*. 1994). (For a review of the syndrome see Robertson (1994).)

Repetitive strain injury

The term repetitive strain injury (RSI) was introduced in Australia in the early 1980s and has been used increasingly in other countries. The condition is characterized by pain and fatigue in the arms. These symptoms are ascribed by the patient to repetitive movements, such as those involved in word-processing. The cause is not well understood and doctors may be involved in claims for compensation. Physical factors may play a part in aetiology, but it is likely that psychological factors also contribute to both the onset and the course of the disorder (Tyrer 1994; Reilly 1995).

Fibromyalgia

The term fibromyalgia refers to a syndrome of generalized muscle aching, tenderness, stiffness, and fatigue, often accompanied by poor sleep. Women are affected more than men, and the condition is more common in middle age. The most important sign is the presence of multiple tender points. The aetiology is uncertain. There may be physical causes, but the common association of the condition with functional somatic symptoms suggests that psychological factors may play a part. Controlled trials have shown that behavioural intervention is effective in preventing acute musculoskeletal pain from becoming chronic (Linton *et al*. 1993). There is no specific physical treatment, but the patient may be helped by the general psychological measures described on p. 351 for the management of functional somatic symptoms (Wolfe 1990; Magni 1993).

Accidents

Personality and psychiatric illness are among the important causes of accidents in the home, at work, and on the roads. In childhood, psychiatric reasons for 'accident proneness' include over-activity and conduct disorders. Among young adults, alcohol (Hingson and Hawland 1993), drug abuse, and mood disorders (Malt 1987) are important. In the elderly, organic mental disorders are important causes. Emotional symptoms are common, following accidents as are cognitive disorders if there is a head injury (see p. 327). The severity of affective symptoms including post-traumatic stress disorder are not closely related to the severity of the accident or of the physical injuries. In addition to the conditions discussed below, reactions to disasters are considered on p. 140 and criminal assault on p. 145. The effects of major railway accidents on train drivers have been reviewed by Karlehagen *et al.* (1994).

Road traffic accidents

Road traffic accidents are the leading cause of death in people aged under 40 and are a major cause of physical morbidity. As well as the abuse of alcohol and drugs, psychiatric factors in the causation of road accidents include severe psychiatric disorder, suicidal and risk-taking behaviour, and the side-effects of prescribed psychotropic drugs.

Psychiatric problems following a road accident include those consequent upon injury to the brain, acute stress reactions, anxiety and depression, post-traumatic stress disorder, and phobias of travel. Though some of these conditions are transient, many persist and give rise to considerable disability. Most of those affected do not seem to be have been psychologically vulnerable before the accident (Mayou 1992; Mayou *et al.* 1993; Blanchard *et al.* 1994).

Whiplash neck injury has often been attributed to psychological aetiology, but recent evidence suggests that physical trauma is the principal cause although psychological factors are important determinants of emotional and social outcome (Mayou and Radanou 1966).

Occupational injury

The psychiatric consequences of occupational injury resemble those of other accidents (Malt 1987). It is often alleged that hopes of compensation or other benefits are important in maintaining the symptoms and disability (see the discussion of compensation neurosis below). Although people who suffer occupational injury seem to have more time off work than those suffering similar injuries unrelated to work, deliberate exaggeration or simulation are rare.

Spinal cord injury

A survey found that about a quarter of patients admitted to a spinal injury unit suffer from psychiatric problems requiring treatment (Judd and Brown 1992). Depression is common in the period immediately after a spinal cord injury, but recent follow-up studies have generally found that most patients are not psychologically disturbed a year after the injury (Ditunno and Formal 1994). Nevertheless, suicide appears to be more common among these patients than in the general population. Depressive disorder is related to social isolation and unemployment rather than to the degree of medical impairment (Fuhrer *et al.* 1993).

Burns

Psychological and social problems may contribute to the causation of burns in both children and adults. In children, burns are associated with overactivity and mental retardation, and also with child abuse and neglect. In adults, burns are associated with alcohol and drug abuse, deliberate self-harm, and dementia. Severe burns and their protracted treatment may cause severe psychological problems. Hamburg *et al.* (1953) described three stages. In the first, lasting days or weeks, denial is common. The most frequent psychiatric disorders are organic syndromes. At this stage the relatives often need considerable help. The intermediate stage is prolonged and painful; here denial recedes and emotional disorders are more common. Patients need to be helped to withstand pain, to express their feelings, and gradually accept disfigurement. In the final stage the patient leaves hospital and has to make further adjustments to deformity or physical disability and the reaction of other people to his appearance. Recent reports have shown that post-traumatic disorder is a common complication of severe burn injury.

There are conflicting reports about the numbers of patients who have persistent emotional difficulties in adjusting after burns. Andreasen and Norris (1972) found persistent difficulties in about a third, but more recent research has usually reported higher figures. It is generally agreed that the outcome is worse in patients with burns affecting the appearance of the face. Such patients are likely to withdraw permanently from social activities. These patients need considerable support from the staff of the burns unit, but only a minority require referral to a psychiatrist. (See Patterson (1993) for a general review, and Tarnowski *et al.* (1991) for a review of psychosocial consequences of burns in childhood.)

Compensation neurosis

The term compensation neurosis (or accident neurosis) refers to psychologically determined physical or mental symptoms occurring when there is an unsettled claim for compensation. From his experience as a neurologist, Miller (1961) drew attention to the frequency of a psychological basis for persistent physical disability after occupational injuries and road accidents. He emphasized the role of the compensation claim in prolonging symptoms and suggested that settlement was followed by recovery. More recent evidence has failed to substantiate this extreme view, though it remains prevalent in medicolegal practice.

In fact few accident victims claim compensation, and even fewer become involved in prolonged litigation. For example amongst patients with mild head injuries, several studies found no association between prolonged psychological consequences and court proceedings or hope of compensation (see p. 328). However, it does appear that time off work and disability are affected by social factors, including the type of accident and the prospect of compensation, social security, or other benefits. It has usually been assumed that settlement of a compensation claim is followed by improvement. This assumption is not supported by follow-up studies (Mendelson 1995). (See Hoffman (1986) for guidance on writing a psychiatric report for compensation proceedings.)

Endocrine disorders

Hyperthyroidism

Alexander (1950) considered that thyrotoxicosis was a psychosomatic disorder (in the sense of a physical disorder induced by psychological factors), but this idea is not supported by evidence (Weiner 1977). However, there are reports suggesting that stressful life events may precipitate the onset of the disease (Rosch 1993). There are always some psychological symptoms in hyperthyroidism, including restlessness, irritability, and distractibility, which may be so marked as to resemble anxiety disorder. In the past acute organic psychiatric syndromes were observed as part of a 'thyroid crisis', but with modern treatment they are rare. However, mild degrees of memory impairment can often be demonstrated. Occasionally, delirium occurs soon after the start of treatment with anti-thyroid drugs. Using standard criteria, anxiety and depression are common (Kathol and Delahunt 1986), but the prevalences are probably somewhat inflated by scoring as anxiety the somatic symptoms of thyrotoxicosis which they closely resemble. The severity of emotional symptoms are not related to thyroid hormone levels (Trzepacz *et al.* 1989), but they usually improve following anti-thyroid therapy.

The differential diagnosis between thyrotoxicosis and anxiety disorder depends on a history of distinctive symptoms and on physical examination. The discriminating symptoms of thyrotoxicosis are preference for cold weather and weight loss despite increased appetite. The most discriminating signs of thyrotoxicosis are a palpable thyroid, a sleeping pulse above 90 beats per minute, atrial fibrillation, and tremor. T_4 and T_3 should be measured.

Hypothyroidism

Lack of thyroid hormones invariably produces mental effects. In early life it leads to retardation of mental development. When thyroid deficiency begins in adult life, it leads to mental slowness, apathy, and complaints of poor memory. These effects are important to psychiatrists because they easily lead to a mistaken diagnosis of dementia or a depressive disorder.

The symptoms of hypothyroidism are less distinctive than those of thyrotoxicosis. They include poor appetite and constipation, generalized aches and pains, and sometimes angina.

Occasionally, the psychiatric symptoms are the first evidence of myxoedema. On psychiatric examination, actions and speech are found to be slow, and thinking may be slow and muddled. Since these features are non-specific, myxoedema must be differentiated from dementia on the basis of its physical signs: distinctive facial appearance with non-pitting oedematous swelling and receding hair line, deep coarse voice, dry rough skin and lank hair, slow pulse, and delayed tendon reflexes.

In determining the cause of hypothyroidism, it is important to remember that lithium therapy may be a cause (see p. 579). Measurements of thyrotropin-stimulating hormone (TSH) help to distinguish primary thyroid disease (where TSH is high) from pituitary causes (where TSH is low).

Asher (1949) coined the phrase 'myxoedematous madness' to denote serious mental disorders associated with thyroid deficiency in adult life. There is no single form of psychiatric disorder specific to hypothyroidism. The most common are depression and an acute or subacute organic syndrome. Other patients develop a slowly progressive dementia or, more rarely, there may be a serious depressive disorder or schizophrenia. Paranoid features are said to be common in all the conditions.

Replacement therapy usually reverses the organic and psychiatric features provided that the diagnosis has not been long delayed. Depressed hypothyroid patients do not usually respond to antidepressant drugs (Denicoff *et al.* 1990).

Addison's disease (hypoadrenalism)

Psychological symptoms of withdrawal, apathy, fatigue, and mood disturbance are frequent and appear early. Hence Addison's disease may be misdiagnosed as dementia. When he first described the disease in 1868, Thomas Addison commented that memory disorder was common. Subsequent observations confirm this view; for example, Michael and Gibbons (1963) reported memory disorder in three-quarters of a series of patients. Addisonian crises are accompanied by the features of an acute organic psychiatric syndrome. The diag-nosis is usually apparent because the patient is obviously unwell, cold, and dehydrated, with low blood pressure and signs of failing circulation. Occasionally, a depressive or schizophrenic picture coincides with Addison's disease, but less commonly than in Cushing's syndrome (Lishman 1987).

Cushing's syndrome (hyperadrenalism)

Emotional disorder is common in this condition, as Cushing noted in his original description, and Michael and Gibbons (1963) reported emotional disorder in about half their cases. Cushing's disease usually comes to attention because of physical symptoms and signs, and any psychiatric disorders are usually encountered as complications in known cases. The physical signs include moon-face, 'buffalo hump', purple striae of the thighs and abdomen, hirsutism, and hypertension. Women are usually amenorrhoeic and men are often impotent.

Depressive symptoms are the most frequent psychiatric manifestations of Cushing's syndrome. Paranoid symptoms are less common and appear mainly in patients with severe physical illness (Cohen 1980; Kelly *et al.* 1985). The severity of the depressive symptoms is not related closely to plasma cortisol concentrations, and premorbid personality and stressful life events appear to predispose to the development of the affective disorder. Nevertheless, psychological symptoms usually improve quickly when the medical condition has been controlled. A few patients develop a severe depressive disorder with retardation, delusions, and hallucinations. Even these severe disorders generally improve when the endocrine disorder is brought under control. A textbook of medicine should be consulted for information about endocrine treatment.

Steroid therapy

Corticosteroids

The psychiatric symptoms induced by corticosteroid treatment might be expected to be identical to those of Cushing's syndrome, but they are not entirely the same. When the symptoms are not severe, euphoria or a mild manic syndrome is more common than depressive symptoms. When they are severe, they take the form of depressive disorder, as in Cushing's syndrome (Kershner and Wang-Cheng 1989).

Sometimes corticosteroid treatment induces an acute organic syndrome in which paranoid symptoms may be prominent. The severity of the mental disorder is not associated with the dosage. It appears that patients with a history of previous mental disorder are not specially prone to developing the psychological complications of corticosteroid treatment.

Less severe symptoms usually improve when the dose is reduced. A severe depressive disorder may require treatment with antidepressant medication, or a manic disorder may need antipsychotic drugs. Lithium prophylaxis should be considered for patients who need to continue steroid treatment after an affective disorder has been brought under control.

Rapid *withdrawal of corticosteroids* may cause lethargy, weakness, and joint pain. Delirium may follow the withdrawal of long-standing treatment for systemic lupus erythematosus and rheumatoid arthritis. A minority of patients become psychologically dependent on corticosteroids, and strongly resist withdrawal.

Anabolic steroids

Anabolic–androgenic steroids are very widely used by athletes. Mood disturbances and increased aggression have been reported among people using these drugs, but the frequency of the association is uncertain.

Phaeochromocytoma

Phaeochromocytomas are a rare and easily overlooked cause of episodic attacks of anxiety. They are tumours, usually benign, arising from the chromaffin cells of the adrenal medulla or ectopically in relation to the sympathetic ganglia. They secrete adrenaline and noradrenaline either continuously or paroxysmally, causing attacks characterized by palpitations, blushing, sweating, tremulousness, and violent headaches, together with hypertension and tachycardia. Intense anxiety is usual in the attacks. Occasionally, there is an episode of confusion. Between attacks blood pressure is usually continuously raised. The attacks may be precipitated by physical exertion or occasionally by emotion.

Diagnosis depends on the demonstration of increased concentrations of catecholamines in the blood or urine, or of their metabolites in the urine. For further information about the syndrome and its treatment, the reader should consult a textbook of medicine.

Acromegaly

In acromegaly, apathy and lack of initiative are common, but other psychiatric symptoms are uncommon. Depression sometimes occurs, but it may be a psychological reaction to the physical symptoms rather than a direct effect of the hormonal disturbance.

Hypopituitarism

Psychological symptoms are usual (Vance 1994). The main symptoms are depression, apathy, lack of initiative, and somnolence. Sometimes, cognitive impairment is severe enough for hypopituitarism to be misdiagnosed as dementia. Another possible misdiagnosis is mild depressive disorder. In the differential diagnosis from anorexia nervosa, hypopituitarism is distinguished by loss of body hair and by the absence of weight phobia. Psychological symptoms usually respond well and quality of life improves when hypopituitarism is treated by replacement therapy (McGauley 1989).

Hyperparathyroidism

Psychological symptoms are common and apparently related to the raised blood level of calcium (Petersen 1968). Depression, anergia, and irritability are the most frequent symptoms. Cognitive impairment also occurs. An acute organic psychiatric syndrome may develop as part of a 'parathyroid crisis'. A few patients first present with psychiatric symptoms, whilst many patients report, in retrospect, that they experienced mild anergia and low spirits for years before definite symptoms appeared (de Alarcón and Franchesini 1984).

Hyperparathyroidism should be considered when prolonged neurotic or minor intellectual symptoms are accompanied by thirst and polyuria. Mental symptoms usually recover after removal of a parathyroid adenoma (Solomon *et al.* 1992), but there may be episodes of hypocalcaemia after the operation giving rise to anxiety and sometimes tetany.

Hypoparathyroidism

Hypoparathyroidism is usually due to removal of or damage to the parathyroid glands at thyroidectomy, but a few cases are idiopathic. The main symptoms are tetany, ocular cataracts, and epilepsy. Denko and Kaelbling (1962) reviewed the literature on hypoparathyroidism and concluded that at least half the cases attributable to surgery had psychiatric symptoms, usually in the form of acute organic psychiatric syndromes. Chronic psychiatric syndromes are more common in idiopathic cases of hypoparathyroidism. Less frequent complications are depression, irritability, and nervousness ('pseudoneurosis'). Bipolar affective disorder and schizophrenic disorder are rare (Lishman 1987) and may be coincidental. The diagnosis is made on the characteristic physical symptoms and measurement of serum calcium.

Insulinomas

These usually present between the ages of 20 and 50. There is generally a long history of transient but recurrent attacks in which the patients behaves out of character, often in an aggressive and uninhibited way. At times the clinical features may resemble those of almost any psychiatric syndrome. The important diagnostic clue is the recurrence of attacks. Usually the patient cannot remember what happened during an attack.

Diagnosis depends on demonstrating a low blood glucose concentration during or immediately after an attack. The advice of a physician should be obtained in doubtful cases. (See Marks and Rose (1965) and Lishman (1987) for further information).)

Metabolic disorders

Liver disease

Psychiatric disorder among patients with liver disease is very frequent but has received rather little attention in the psychiatric literature. The recent development of liver transplantation has led to the establishment of a number of specialized psychiatric liaison services. (See Collis and Lloyd (1992) for a review of the psychiatric aspects of liver disease.)

Hepatic encephalopathy

The clinical picture of liver failure is an acute organic psychiatric syndrome (Summerskill *et al.* 1956), together with flapping tremor of the outstretched hands, facial grimacing, and fetor hepaticus. As in other acute organic syndromes, there may be hallucinations and confabulation. The condition may progress to coma, and there is a substantial mortality. Sedative drugs increase the severity of the encephalopathy and should be avoided.

Viral hepatitis

Whilst fatigue and malaise are very frequent in early convalescence, this is usually of limited duration (Hotopf and Wessely 1994).

Hepatic cirrhosis

Patients with cirrhosis may be cognitively impaired even in the absence of overt encephalopathy. Symptoms of lethargy, fatigue, and anorexia, which are common in depression, are frequent but may also have physical causes and differential diagnosis can be difficult.

Alcoholic liver disease

Alcohol problems are very common amongst those with liver disease, with the proportion varying with the population studied (see p. 447). Psychiatric morbidity is especially high amongst those whose liver disease is due to alcohol.

Liver transplantation

Liver transplantation is increasingly used for cirrhotic liver disease with good results. Psychiatric assessment is generally thought to be useful in screening and especially in the management of psychiatric problems post-operatively. Since many of those undergoing transplantation have abused alcohol (see below), this may be a cause of further problems. Overall the liver transplantation appears to have very considerable benefits for mental state (Collis *et al.* 1995) and quality of life of patients and for their families (see Surman (1994) for a review).

There has often been a tendency to reject those with alcohol problems for liver transplants, but current evidence suggests this may be inappropriate and increasing numbers of such patients are now receiving the operation. These patients appear to have a relatively good prognosis, but there is a need for better prospective studies on predictors of outcome of this group of patients (see Howard *et al.* (1994) for a review).

See Collis and LLoyd (1992) for a review of the psychiatric aspects of liver disease.

Acute porphyria

Acute porphyria is rare, but may resemble an acute psychiatric disorder. A much debated historical example is the madness of George III which was ascribed to acute porphyria by McAlpine and Hunter (1966); although their arguments were scholarly and ingenious, they are open to substantial doubt.

Psychiatric symptoms occur during the attack in a quarter to three-quarters of cases, and at times dominate the clinical picture (Ackner *et al.* 1962). They include depression, restlessness, and disturbed behaviour. Emotions are often labile. There may be an acute organic syndrome with impaired consciousness or eventually coma. Delusions and hallucinations often occur. Attacks may be precipitated by acute infection, alcohol, anaesthesia, and certain drugs, notably barbiturates, the contraceptive pill, dichloralphenazone, and methyldopa.

Cerebral anoxia

Cerebral anoxia can be divided into four categories: **anoxic** (respiratory failure, asphyxia, the effects of high altitude), **stagnant** (cerebral vascular disease, peripheral circulatory failure, cardiac failure, cardiac arrest and arrhythmias), and **metabolic** (hypoglycaemia, cyanide poisoning). The clinical picture depends substantially on the cause, but most forms of anoxia are temporary and present with impairment of consciousness which may be accompanied by muscular twitching or tremor and epileptic fits. Afterwards, there is a dense amnesic gap but usually no permanent consequences. In a small proportion of patients who have had severe anoxia, there may be permanent memory deficits and neurological symptoms.

Carbon monoxide poisoning

In the past, carbon monoxide poisoning was usually the result of deliberate self-harm with domestic gas supplies. Household gas no longer contains substantial amounts of carbon monoxide, but car exhaust fumes do and are sometimes used for self-poisoning.

After carbon monoxide poisoning, the course is variable. Milder cases recover over days or weeks. Recovery of consciousness is often followed by an organic psychiatric syndrome; this clears up, leaving an amnestic syndrome which in turn gradually improves. Extrapyramidal and other neurological signs occur at an early stage and then resolve. In more severe cases there is a characteristic period of partial recovery, followed by relapse with a return of an acute organic syndrome and extrapyramidal symptoms. Occasionally, death occurs at this stage. Some patients are left with permanent extrapyramidal symptoms or become demented.

The frequency of these complications is uncertain. Shillito *et al.* (1936) surveyed 21 000 cases of carbon monoxide poisoning in New York City and found few lasting problems. In contrast, Smith and Brandon (1973) made a detailed study of 206 cases from a defined area. They followed-up 74 patients for an average of three years: eight patients had sustained gross neurological damage; eight patients had died; of those alive at follow-up, eight had improved, 21 had shown personality deterioration, and 27 reported memory impairment.

Vitamin deficiency

Severe chronic malnutrition is accompanied by psychological changes such as apathy, emotional instability, cognitive impairment, and occasional delusions or hallucinations. These symptoms are well known among prisoners of war (Helweg-Larsen *et al.* 1952). In peace time, severe malnutrition with deficiency of several vitamins as well as protein and calories is common in some parts of the world. The psychological changes are usually reversed in adults when a normal diet is resumed. However, children may have permanent changes. Even in developed countries there are groups of people who lack balanced diets and are at special risk of deficiency; they include the aged, the chronically mentally ill, the mentally handicapped, alcoholics, and patients with chronic gastrointestinal diseases.

Vitamin B deficiency

Thiamine deficiency

Chronic depletion of thiamine leads first to fatigue, weakness, and emotional disturbance. Eventually it causes beri-beri, which is characterized by peripheral neuropathy, cardiac failure, and peripheral oedema. More acute and severe depletion of thiamine may lead to Wernicke's encephalopathy and thence to the amnestic syndrome (see p. 314).

Nicotinic acid deficiency

In established pellagra, disorientation and confusion may progress to outbursts of excitement and violence. Depression is often conspicuous and a paranoid hallucinatory state is sometimes seen. In these cases, response to treatment with nicotinic acid is often dramatic. More acute and severe nicotinic acid depletion leads to an acute organic psychiatric syndrome.

Vitamin B_{12} deficiency

Severe pernicious anaemia due to deficiency of gastric intrinsic factor causes the clinical picture of subacute combined degeneration of the cord accompanied by anaemia (macrocytic and megaloblastic) and a progressive dementia. In less advanced cases there is depression and lethargy. There may also be impairment of memory which improves after treatment with B_{12} (Shulman 1967).

It has been suggested that dementia and other psychological symptoms may occur before the characteristic physical features. Surveys have often shown that low serum B_{12} levels are common among psychiatric patients. However, it is highly likely that such findings can be explained by a poor diet consequent upon psychiatric disorder, rather than by B_{12} deficiency as a causal factor. Clinical experience indicates that it is unusual to diagnose B_{12} deficiency for the first time in a patient with early dementia. When B_{12} deficiency is found, replacement therapy rarely leads to improvement in dementia. It is reasonable to measure serum B_{12} in any unexplained acute or chronic organic psychiatric syndrome, but there is no justification for its routine estimation in all psychiatric patients.

Folic acid deficiency

It is common to find folic acid deficiency of dietary origin among the elderly and among psychiatric patients of all ages but it is difficult to assess its causal significance, if any. Low serum concentrations of folate and low red-cell levels are unusually frequent in epileptic patients, probably as a result of anticonvulsant medication. It has been suggested that these deficiencies may account for some of the psychological symptoms of epileptic patients, but at present the evidence is not convincing.

Overall, there is little evidence that folate deficiency is an important cause of psychiatric disorder. Routine screening is not justified, although measurement of red-cell folate may occasionally be appropriate when investigating an unexplained organic psychiatric disorder. If the folate is low, replacement therapy can be tried, though without great expectation of success.

Electrolyte and body fluid disorders

Various electrolyte and fluid disorders can cause mental symptoms, usually but not invariably in the form of an acute organic syndrome. (Other physical features of the various disorders will be found in a textbook of medicine.) The role of hypomagnesaemia is at present uncertain. Calcium metabolism was mentioned earlier in relation to parathyroid conditions (pp. 405–6); it is of particular interest because there seems to be a close relationship between the concentration of serum calcium and the extent of the mental changes (Petersen 1968; de Alarcón and Franchesini 1984).

Sleep disorders

Psychiatrists may be asked to see patients whose main problem is either difficulty in sleeping or, less often, excessive sleep. Such disorders are important as causes of psychological symptoms, as features of mental illness, and because they may be mistaken for psychological disorder (Table 12.12).

Many patients who sleep badly complain of tiredness during the day and mood disturbance. Although prolonged sleep deprivation leads to some impairment of intellectual performance and disturbance of mood, loss of sleep on occasional nights is usually of little significance. However, it is of importance in those whose responsibilities

or activities require maximum alertness. The daytime symptoms of people who sleep badly are probably related more to the cause of their insomnia (often a depressive disorder or anxiety disorder) than to insomnia itself. Table 12.13 shows the DSMIV classification of sleep disorders, which is compatible with the more elaborate International Classification of sleep disorders but is itself more complex than the classification in ICD10 (Buysse *et al.* 1994; Kupfer *et al.* 1994).

Sleep disorders are frequent in the general population. There is a wide range of variation in estimates of the prevalence of insomnia depending on the definition and on the population studied. Up to 30 per cent of adults complain of insomnia, a third as a chronic problem. Excessive sleepiness occurs in 5 per cent of adults, and possibly 15 per cent of adolescents and 14 per cent of the adult population suffer some form of chronic sleep–wake disorder. In the Epidemiologic Catchment Area Study (Ford and Kamerow 1989) 10.2 per cent of the community sample reported insomnia and 3.2 per cent reported hypersomnia. Forty per cent of those who suffered insomnia and 46.5 per cent of those with hypersomnia had a psychiatric disorder compared with 16.4 per cent of those with no sleep complaints. Groups at special risk of persistent sleep problems include young children, adolescents, the physically ill, those with learning disability, and those with dementia. The social and economic consequences are very considerable (Dement and Mitler 1993). See Shapiro and Dement (1993) for a review of the epidemiology of sleep disorders.

Insomnia

Transient insomnia occurs at times of stress or as 'jet lag'; short-term insomnia is often associated with personal problems, for example illness or bereavement. Insomnia in clinical practice is usually *secondary* to other disorders, notably painful physical conditions, depressive disorders, and anxiety disorders, and is often clinically overlooked (Berlin *et al.* 1984). Insomnia also

Table 12.12 *Sleep disorders mistaken for psychological disorder or behaviour problems*

Excessive daytime sleepiness
    Laziness and disinterest
    Misbehaviour
    Opting out

Automatic behaviour
    Misbehaviour
    Laziness
    Dissociative states

Parasomnias
    Panic attacks
    Sleep disorders causing violence in
      sleep

---

occurs with excessive use of alcohol or caffeine, and in dementia. Sleep may be disturbed for several weeks after stopping heavy drinking. Sleep problems are also common in association with any medical illness that results in significant pain or discomfort or is associated with metabolic disturbances. They may also be provoked by prescribed drugs.

In about 15 per cent of cases of insomnia, no cause can be found *(primary insomnia)*. People vary in the amount of sleep that they require, and some of those who complain of insomnia may be having enough sleep without realizing it.

Usually the diagnosis of insomnia can be based on the account given by the patient. EEG recordings made in a sleep laboratory or at home are occasionally helpful when there is continuing doubt about the extent and nature of the insomnia. These observations often show that, despite the patient's complaint, sleeping time is within the normal range.

If insomnia is secondary to another condition, the latter should be treated together with general measures to promote good sleep (see below), avoidance of the disruptions to sleep while the person is an in-patient, and short-term hypnotics.

(Shapiro and Dement 1993; Moran and Stoudemire 1992).

In primary insomnia it is useful to encourage regular habits and exercise, and discourage over-indulgence in tobacco, caffeine, and alcohol (Morin *et al.* 1994). Training in relaxation (see p. 630) helps some patients. Although it may sometimes be justifiable to give a hypnotic for a few nights, demands for prolonged medication should be resisted. This is because withdrawal of hypnotics may lead to insomnia that is as distressing as the original sleep disturbance. Continuation of hypnotics may be associated with impaired performance during the day, tolerance to the sedative effects, and dependency. The use of hypnotic drugs is described further on p. 547. See Lacks and Morin (1992) Morin *et al.* (1994) and Murtagh and Greenwood (1995) for reviews of the treatment of insomnia.

Excessive daytime sleepiness

Excessive daytime sleepiness is common and underdiagnosed. Many cases are secondary to loss of night-time sleep. Table 12.14 shows the principal causes.

Primary hypersomnia

This is the most prevalent of the primary hypersomnias. Patients complain that they are unable to wake completely until several hours after getting up. During this time they feel confused and possibly disorientated ('sleep drunkenness'). They usually report prolonged and deep night-time sleep. Almost half have periods of daytime automatic behaviour, the aetiology of which is obscure. Most patients respond well to small doses of central nervous system stimulant drugs. (See Parkes (1993) for a review of day time sleepiness.)

Narcolepsy

Narcolepsy usually begins between the ages of 10 and 20 years, though it may start earlier. Onset is rare after middle age. Narcolepsy is more frequent among males. Cataplexy (sudden temporary epi-

Table 12.13 *Classification of primary sleep disorders in DSMIV*

Dyssomnias
Primary insomnia
Primary hypersomnia
Narcolepsy
Breathing-related sleep disorder
Circadian rhythm sleep disorder
Dys-somnia not otherwise specified

Parasomnias
Nightmare disorder (dream anxiety disorder)
Sleep terror disorder
Sleep-walking disorder
Parasomnia not otherwise specified

Sleep disorder related to another mental disorder
Insomnia
Hypersomnia

Other sleep disorders
Secondary sleep disorder due to a general medical condition
Substance-induced sleep disorder

sodes of paralysis with loss of muscle tone) occurs in most cases, but sleep paralysis and hypnagogic hallucinations occur in only a quarter of patients. There is a family history of narcolepsy in about a third of patients, and in occasional families the disorder appears to be transmitted as an auto-somal dominant. Almost all cases of narcolepsy have the HLA type DR2 compared with about a quarter of the general population; the significance of this association is not understood, though it points to a genetic origin and links it with chromosome 6. Many aetiological theories have been advanced but none is convincing.

Psychiatric aspects of narcolepsy Strong emotions sometimes precipitate cataplexy but apparently not narcolepsy. Patients with narcolepsy often have secondary emotional and social difficulties, and their difficulties are increased by other people's lack of understanding. Schizophrenia-like mental disorders have been reported to occur more frequently in patients with narcolepsy than in the general population (Davison 1983).

Patients should be encouraged to follow a regular routine with planned short periods of sleep during the day. If stressful events seem to provoke attacks, efforts should be made to avoid them. Regular dosage with amphetamine or methylphenidate has some effect in reducing narcoleptic attack but little effect on cataplexy. These drugs have to be given in high doses that lead to side-effects and problems of dependency. Tricyclic antidepressants do not affect the sleep disorder but may reduce the frequency of cata-plexy. Some authors suggest the combined use of tricyclics and amphetamines, but this combina-tion is better avoided if possible because of the risk of hypertensive effects.

Table 12.14   *Causes of excessive daytime sleepiness*

Insufficient night-time sleep
Unsatisfactory irregular sleep routines or circumstances
Circadian rhythm sleep disorders
Frequent parasomnias
Chronic physical illness
Psychiatric disorders

Pathological sleep
Obstructive sleep apnoea
Narcolepsy
Other central nervous system disease
Drug effects
Kleine–Levin syndrome
Depressive illness

Breathing-related sleep disorder

This syndrome consists of daytime drowsiness together with periodic respiration and excessive snoring at night. It is usually associated with upper airways obstruction. The typical patient is a middle-aged overweight man who snores loudly. Treatment consists of relieving the cause of the respiratory obstruction or obesity. Continuous positive pressure ventilation using a face mask is often effective (Douglas 1993).

Circadian rhythm sleep disorder (sleep–wake schedule disorders)

There are several forms of circadian sleep disorder of which *jet lag* is the most familiar. *Shift-work type* is a common and increasing problem whose consequences are widely underestimated. Fatigue and transient difficulties in sleeping accompany regular changes of shift, or the irregular alternation of night work and days off may lead to chronic problems of poor sleep, fatigue, impaired concentration, and an increased liability to accidents as well as adverse effects on family life.

The Kleine–Levin syndrome

This rare secondary sleep disorder consists of episodes of somnolence and increased appetite, often lasting for days or weeks and with long intervals of normality between them. Patients can always be roused from the daytime sleep, but are irritable and occasionally aggressive on waking; some are muddled and experience depression, hallucinations, and disorientation. Although the combination of appetite disorder and sleep disturbance suggests a hypothalamic disorder, there is no convincing evidence about the aetiology.

Parasomnias (dream anxiety disorder)

Nightmares

A nightmare is an awakening from REM sleep to full consciousness with detailed dream recall. Children experience nightmares with a peak frequency around the ages of five or six years.

Nightmares may be stimulated by frightening experiences during the day, and frequent nightmares usually occur during a period of anxiety. Other causes include post-traumatic stress disorder, fever, psychotropic drugs, and alcohol detoxification.

Night terror disorder

Night terrors are much less common than nightmares. They are sometimes familial. The condition begins in childhood and usually ends there, but occasionally persists into adult life. A few hours after going to sleep, the child, whilst in stage 3–4 non-REM sleep, sits up and appears terrified. He may scream and usually appears confused. There are marked increases in heart and respiratory rates. After a few minutes the child slowly settles and returns to normal calm sleep. There is little or no dream recall. A regular bedtime routine and improved sleep hygiene have been shown to be helpful. Benzodiazepines and imipramine have been shown to be effective in preventing night terrors (Kales *et al.* 1987), but their prolonged use should be avoided.

Sleep-walking disorder

Sleep-walking is an automatism occurring during deep non-REM sleep, usually in the early part of the night. It is most common between the ages of 5 and 12 years, and 15 per cent of children in this age group walk in their sleep at least once. Occasionally, the disorder persists into adult life. Sleep-walking may be familial.

Most children do not actually walk, but sit up and make repetitive movements. Some walk around, usually with their eyes open, in a mechanical manner but avoiding familiar objects. They do not respond to questions and are very difficult to wake. They can usually be led back to bed. Most episodes last for a few seconds or minutes, but rarely as long as an hour.

As sleep-walkers can occasionally harm themselves, they need to be protected from injury. Doors and windows should be locked and dangerous objects removed. Adults with severe problems should be given advice about safety, avoidance of sleep deprivation, and any other circumstances that might make them excessively sleepy (for example drinking alcohol before going to bed). See Driver (1993) for a review of parasomnias.

Further reading

Mayou, R. A., Bass, C. B., and Sharpe, M. (1995). *Treatment of functional organic symptoms.* Oxford University Press.

Stoudemire, A. and Fogel, B.S. (1993). *Psychiatric care of the medical patient.* Oxford University Press, New York.

Stoudemmire, A. (ed.) (1995). *Psychological factors affecting medical conditions.* American Psychiatric Press, Washington, DC.

13 | *Suicide and deliberate self-harm*

In recent years a large number of the patients admitted to medical wards have deliberately taken drug overdoses or harmed themselves in other ways. It has become clear that only a small minority of these patients intend to take their lives; the rest have other motives for their actions. Equally, only a minority have a psychiatric disorder; the rest are facing difficult social problems. Psychiatrists are often called upon to identify and treat the minority with suicidal intent or psychiatric disorder, and to provide appropriate help for the rest.

In order to assess such patients properly, the psychiatrist should understand the differences between people who commit suicide (completed suicide), and those who survive after taking an overdose or harming themselves (deliberate self-harm). At this early stage in the chapter, it may be helpful to give a brief outline of the differences between the two.

In general, compared with people who harm themselves and survive, those who commit suicide are more often male and are usually suffering from a psychiatric disorder. They often plan their suicidal acts carefully, take precautions against discovery, and use dangerous methods. By contrast, amongst those who harm themselves and survive, a large proportion carry out their acts impulsively in a way that invites discovery and is unlikely to be dangerous. The two groups are not distinctly separate; they overlap in important ways. This point should be borne in mind throughout this chapter.

The chapter begins with an account of those who die by suicide. We then describe people whose drug overdoses or self-injury do not result in death. Each section starts with a description of the behaviour, its epidemiology, and its causes. We then consider assessment, management, and prevention.

Suicide

The act of suicide

People who take their lives do so in several ways. In England, drug overdoses account for about a half of suicides among women and about a third of those among men (Morgan 1979). The drugs used most often are analgesics and antidepressants. In the United Kingdom, in the period 1948–50 poisoning by domestic coal gas accounted for about 40 per cent of reported suicides among men and 60 per cent among women. Following the introduction of non-toxic North Sea gas, it became a much less frequent cause of death. Nowadays, deliberate carbon monoxide poisoning arises mainly from car exhaust fumes, and is the most frequent method among men, accounting for half of all suicide deaths. The remaining deaths are by a variety of physical means: hanging, shooting, wounding, drowning, jumping from high places, and falling in front of moving vehicles or trains. Violent methods, especially shooting, are more common in the United States than in the United Kingdom.

Most completed suicides have been planned. Some patients save drugs obtained from a series of prescriptions; others use drugs that can be bought without a prescription, such as aspirin. Precautions against discovery are often taken, for example choosing a lonely place or a time when no one is expected.

In most cases a warning is given before committing suicide. In a survey in the United States, interviews were held with relatives and friends of people who had committed suicide. It was found that suicidal ideas had been expressed by more than two-thirds of the deceased, and clear suicidal intent by rather more than a third. Often the warning had been given to more than one person (Robins *et al.* 1959). In a similar study of people who had committed suicide, Barraclough *et al.* (1974) found that two-thirds had consulted their general practitioner in the previous month, and 40 per cent had done so in the previous week. A quarter had been psychiatric out-patients at the time, of whom half had seen a psychiatrist in the week before their suicide. It is not certain whether these percentages still apply today; it is possible that they are lower among younger people who have committed suicide.

About one in six people committing suicide, leaves a note (Barraclough *et al.* 1974). The content of the notes varies: some ask for forgiveness, whilst others are accusing or vindictive, drawing attention to failings in relatives or friends. Such vindictive notes are more often left by younger people. Older people often express concern for those who remain alive.

The epidemiology of suicide

Accurate statistics about suicide are difficult to obtain. In England and Wales, official figures depend on the verdicts reached in coroners' courts, and comparable procedures are used in other countries. Such figures are affected by several sources of error and may only represent a half to two-thirds of all suicides. Occasionally it is uncertain whether a death is caused by suicide or murder. Much more often it is difficult to decide whether death was by suicide or accident. In many cases of uncertainty, the verdict will depend on legal criteria. In England and Wales there is a strict rule that suicide must be proved by evidence; if there is doubt, an open or accident verdict must be returned. In some other countries less stringent criteria are used (Robins and Kulbok 1988; Kreitman 1993).

For these reasons it is not surprising that official statistics appear to underestimate the true rates of suicide. In Dublin, psychiatrists ascertained four times as many suicides as the coroners did (McCarthy and Walsh 1975), and similar discrepancies have been reported from other places. Amongst people whose deaths are recorded as

accidental, many have recently been depressed or dependent on drugs or alcohol, thus resembling people who commit suicide (Barraclough 1973). For this reason, some investigators try to estimate suicide rates by combining official figures for suicide, accidental poisoning, and undetermined causes. For the purpose of comparing suicide rates between different countries, this procedure makes little difference since the rank ordering is not affected significantly by using it (Barraclough 1973).

It is difficult to make international comparisons of suicide rates, since different meanings and attitudes are associated with suicide in different cultures, and procedures for recording death vary considerably (Platt 1987; Charlton *et al.* 1993). Table 13.1 shows reported rates of suicide from eight countries for people aged 25–44 years. The official suicide rate in the United Kingdom is in the lower range of those reported in Western countries, but still accounts for 1 per cent of all deaths. Generally, higher rates are reported in Eastern and Northern European countries, and lower rates in Mediterranean countries. Indirect evidence that there are real differences between rates between nations and suicide rates was presented by Sainsbury and Barraclough (1968). They showed that, within the United States, the rank order of suicide rates among immigrants for eleven different nations was similar to the rank order of national rates within the eleven countries of origin. Trends within countries are likely to reflect changes which are genuinely occurring.

Changes in suicide rates

Over the years since 1900, suicide rates in the United Kingdom have changed substantially at different times (Charlton *et al.* 1992, 1993). Recorded rates for men and women fell during both World Wars. There were also two periods when rates were unusually high. The first, 1932–3, was a time of economic depression and high unemployment; the second, between the late 1950s and the early 1960s, was not. Another unusual period was 1963–74, when rates declined in England and Wales but not in other European

Table 13.1 *Reported suicide rates per 100 000 population in Western Europe and the United States (ages 25–44).*

Country	Sex and M : F ratio	Averages		Percentage change
		1969–71	1986–88	
France	Males	21	36	67
	Females	8	11	48
	M : F ratio	3	3	—
Greece	Males	5	6	21
	Females	2	2	−29
	M : F ratio	3	3	—
Italy	Males	7	10	35
	Females	3	3	−3
	M : F ratio	2	3	—
Netherlands	Males	8	15	83
	Females	6	9	45
	M : F ratio	1	2	—
Spain	Males	5	6	63
	Females	2	2	44
	M : F ratio	3	3	—
West Germany	Males	31	25	−19
	Females	13	9	−30
	M : F ratio	2	3	—
UK	Males	11	16	50
	Females	6	5	−24
	M : F ratio	2	3	—
USA	Males	20	24	18
	Females	10	6	−36
	M : F ratio	2	4	—
Finland	Males	49	59	20
	Females	12	13	9
	M : F ratio	4	5	—
Sweden	Males	35	31	−12
	Females	16	13	−16
	M : F ratio	2	2	—

Source: Charlton *et al.* 1993.

countries (except Greece) or in North America. Since 1975, the rates in England and Wales have risen among younger men but decreased among women, and there has been a particular increase in suicide by vehicle exhaust gas, hanging, and suffocating (McClure 1984). Similar changes occurred in cases where there is doubt and an open or accident verdict has been given. It has been suggested that these variations might reflect different rates in cohorts of people born at different times, rather than factors acting on all age groups at the same time. Recent figures for England and Wales suggest that both factors are present (Charlton *et al.* 1992).

Variations with the seasons

In England and Wales, suicide rates have been *highest in spring and summer* for every decade since 1921–30. A similar pattern has been found in other countries in the northern hemisphere. In the southern hemisphere a similar rise occurs during the spring and early summer, even though these seasons are in different months of the year. The reason for these fluctuations is not known.

Variations according to personal characteristics

Suicide is three times as common in men as in women but there is little association with age. Suicide among the elderly is particularly associated with depressive disorder, physical illness, and social isolation (Blazer 1986). Suicide rates are lowest among the married, and increase progressively with the never married, widowers and widows, and the divorced. Rates are higher in social class V (unskilled workers) and social class I (professional) than in the remaining social classes (Charlton *et al.* 1993).

Rates are particularly high in certain professions. Veterinary surgeons have three times the expected rate, and pharmacists, farmers, and medical practitioners have double the expected rate (Charlton *et al.* 1993). Easy access to drugs among these professions may be a factor. There is also an association between suicide rates and levels of unemployment, but its nature is complex (Crombie 1989).

Variations according to place of residence

Suicide rates in cities used to be higher than those in the country. In recent years this difference has grown less, and for men it has reversed. Within large cities, rates vary among different kinds of residential area. The highest rates are reported from areas in which there are many inhabitants of boarding houses, and many immigrants and divorced people. The common factor appears to be social isolation (Sainsbury 1955).

Some suicides occur during psychiatric in-patient care and soon after discharge from such care (Goldacre *et al.* 1993). High rates occur in prisoners, especially amongst those on remand (Charlton *et al.* 1993).

The causes of suicide

Social causes

In 1897, Durkheim published an important book in which he proposed a relationship between suicide and social conditions (see Durkheim 1951). He divided suicides into three main categories. **Egotistic suicide** occurred in individuals who had lost their sense of integration with their social group, so that they no longer felt subject to its social, family, and religious controls. **Anomic suicide** occurred in individuals who lived in a society that lacked 'collective order' because it was in the midst of major social change or political crisis. **Altruistic suicide** occurred in individuals who sacrificed their lives for the good of the social groups, thus reflecting the influence of the group's identity. Durkheim's views have been influential, even though it now seems that he overemphasized social factors at the expense of individual causes. Much of the research cited in the previous section and notably Sainsbury's work on social isolation, referred to above, follows directly from Durkheim's ideas.

Sometimes the means of suicide and its timing seems to be influenced by another suicide that has attracted attention in a community or received wide publicity in newspapers or on television (Eisenberg 1986; Gould *et al.* 1990). Suicide and attempted suicide rates also increase after the showing of fictional television programmes and films depicting suicide (Shaffer *et al.* 1988; Schmidtke and Häfner 1988).

Medical causes

Mental disorders are a most important cause of suicide. In several studies interviews have been held with doctors, relatives, and friends of people who have committed suicide, and detailed histories of the deceased have been compiled. These studies indicate that, amongst those who die from suicide, about nine in every ten have some form of mental disorder at the time of death (Robins *et al.* 1959; Barraclough *et al.* 1974). The most frequent of these mental disorders are depressive disorders (Isometsa *et al.* 1994) and alcoholism (Lesage *et al.* 1994). Suicide is particularly likely soon after discharge from psychiatric inpatient care (Goldacre *et al.* 1993).

The importance of **depressive disorders** is confirmed by findings that rates of suicide are increased among depressed patients (Fremming 1951; Pokorny 1964). Depressed patients who die by suicide cannot be distinguished from other severely depressed patients by their symptoms (Fawcett *et al.* 1987). However, they differ in having made more previous suicide attempts (Barraclough *et al.* 1974) and more often being single, separated, or widowed (Pitts and Winokur 1964) and older (Robins *et al.* 1959). The risk is also greater among men than women (Pitts and Winokur 1964).

There is a particular risk of suicide in the weeks following psychiatric contact which may be due to an improvement to psychomotor retardation on beginning treatment or to relapse following an initial improvement. Most of those with major depression have not been taking antidepressant medication immediately before death (Isacsson *et al.* 1994).

Alcohol abuse is the second most frequent psychiatric disorder amongst those who die from suicide, being present in 25 per cent of cases

(Robins *et al.* 1959; Barraclough *et al.* 1974; Murphy and Wetzel 1990). The lifetime risk is about 2 per cent amongst those with an untreated alcohol problem, 2.2 per cent amongst out-patients, and 3.4 per cent amongst those who have had been admitted for treatment (Murphy and Wetzel 1990). Follow-up studies of alcoholics show a continuing risk of suicide. Thus, among alcoholics who had received psychiatric treatment in hospital, the incidence of suicide over a five-year follow-up was about 80 times that of the general population (Kessel and Grossman 1965). The risk is greatest among older men with a long history of drinking, a definite depressive illness, and previous suicidal attempts. It is also increased among those whose drinking has caused physical complications, marital problems, difficulties at work, or arrests for drunkenness offences (Miles 1977).

Drug-dependent patients have an increased suicide risk (James 1967). Thus among 133 young people who committed suicide in California, 55 per cent had substance abuse as the principal psychiatric diagnosis. Usually, the abuse was long-standing and involved more than one substance (Fowler *et al.* 1986).

Personality disorder is detected in a third to a half of people who commit suicide (Seager and Flood 1965; Ovenstone and Kreitman 1974). This group tends to be younger, to come from broken homes, and to live in a subculture in which violence and alcohol or drug abuse are common. Personality disorder probably combines with other causes to increase the risk of suicide. In their study, Barraclough *et al.* (1974) found personality disorder in about half of the alcoholic and a fifth of the depressed patients who had died by suicide.

Schizophrenia accounts for only about 3 per cent of suicides, but the risk should be borne in mind when treating a schizophrenic patient. In schizophrenia, suicide is more likely among young men early in the course of the disorder, particularly when there have been relapses, when there are depressive symptoms, and when previous academic success has been turned to failure by the illness (Drake and Cotton 1986).

The suicide rate is higher after **deliberate self-harm** (see p. 430). **Chronic painful physical illness** is associated with suicide (Robins *et al.* 1959), especially among the elderly (Sainsbury 1962). The risk of suicide amongst people with **epilepsy** is about four times that in the general population (Sainsbury 1986; Barraclough 1987). The risk is known to be raised in a number of other **medical conditions** (MacKenzie and Popkin 1987) such as multiple sclerosis (Allebeck *et al.* 1989; Stenager *et al.* 1992), and HIV infection (Marzuk *et al.* 1988). Rates are low in pregnancy and the puerperium (Kendell 1991). [See Harris and Barraclough (1995) for a review.]

Conclusion

The associations between suicide and the various factors mentioned above do not, of course, establish causation. Nevertheless they point to the importance of two sets of interacting influences: amongst social factors, social isolation stands out; amongst medical factors, depressive disorders, alcoholism, and abnormal personality are particularly prominent.

'Rational' suicide

Despite the findings reviewed above, there can be no doubt that suicide is occasionally the rational act of a mentally healthy person. Moreover, mass suicides have been described among groups of people, and it is unlikely that they were all suffering from mental disorder. An example was the religious community at Jonestown, in which a large number of people died together by taking poison (Rosen 1981). Nevertheless, in the clinical assessment of someone who is talking of suicide, it is a good rule to assume that his suicidal inclinations are influenced by an abnormal state of mind.

If this assumption is correct—as it usually will be—the patient's urge to suicide is likely to diminish with recovery from the abnormal mental state. Even if the assumption is wrong (i.e. if the patient is one of the few who have reached a rational decision to die), the doctor should still try to protect him from harming himself. Give more time for reflection, most people with suicidal intent change their intentions. For example they may discover that

death from a cancer need not be as painful as they believed. Hence they may change a decision that was made rationally but on false premises (see Emanuel (1994) for a review of issues relating to euthanasia).

Special groups
Children and young adolescents

Accurate estimation of suicide rates is even more difficult for children than for adults. However, suicide is known to be rare in children. In 1989 the suicide rate for children aged 5–14 years was 0.7 per 100 000 in the United States and 0.8 per 100 000 in the United Kingdom. Among 15–19-year-olds the rate was 13.2 per 100 000 in the United States and 7.6 per 100 000 in the United Kingdom (Shaffer and Piacentini 1994). In several countries there is evidence of a recent marked increase in suicide by older adolescents (Charlton *et al.* 1992; Marttunen *et al.* 1993); in some countries, including the United States, there has also been an increase among younger children. In England and Wales the only recent change in rates has been an increase in males aged 15–19 years (McClure 1994). These increases may be due to many factors, including an increase in the number of broken homes, changing attitudes to suicide and to deliberate self-harm, and media publicity.

In the 12–14-year-old group, more boys than girls commit suicide. Boys are more likely to use violent methods such as hanging or shooting, and girls to take drug overdoses or jump from a high place. There are international differences in methods, with use of firearms being particularly common in the United States. Among children seen by child psychiatrists, most who threaten suicide do not carry it out. Nevertheless, of the children who died by suicide in Shaffer's (1974) series, almost half had previously talked about, threatened, or attempted suicide.

Little is known about factors leading to suicide in childhood. Shaffer (1974) reported that suicidal behaviour and depressive disorders were common among the parents and siblings, and that children who died by suicide had usually shown antisocial behaviour. Shaffer distinguished two groups of children. The first comprised children of superior intelligence who seemed to be isolated from less educated parents. Many of their mothers were mentally ill. Before death, the children had seemed depressed and withdrawn, and some had stayed away from school. The second group consisted of children who were impetuous, prone to violence, and resentful of criticism.

Little is known of the consequences of child or adolescent suicide on surviving family members, but it is likely to be profound. (See Hawton (1986) for a review of suicide in adolescence.)

Older adolescents and young adults

In the last 20 years suicide rates among 15–24-year-olds have increased (Platt 1987; Charlton *et al.* 1992; Marttunen *et al.* 1993). In most European countries the increase has been greater amongst males than amongst females. The specific risk factors are uncertain, but may include alcohol abuse, unemployment, and more family disruption due to rising rates in divorce (Shaffer *et al.* 1988).

A review of psychological autopsy studies shows that the majority of adolescents who kill themselves have severe psychosocial problems: approximately two-thirds have severe psychosocial problems, approximately two-thirds have expressed suicidal intent, and half made a previous suicide attempt (Marttunen *et al.* 1993).

University students

It is widely believed that suicide rates are high among university students. Accurate data are difficult to obtain and interpret (Whitaker and Slimak 1990). Schwartz (1993) reviewed all published reports and some unpublished data on the incidence of suicide among students at colleges and universities in the United States. Official data underestimate incidence by about 30 per cent, but even so the incidence of student suicide was significantly less than that for non-students. Hawton *et al.* (1995*a*,*b*) examined figures for Oxford University over a 14-year period. The observed number of suicides was greater than the number expected for people of the same age in England and Wales. However, when deaths due to undetermined cause were included in the comparison, the difference between the observed and expected numbers was much reduced. There were indications of a greater use of violent methods in students' suicides. There was no close association between suicide and examinations, although two-thirds of the students had been worried about academic achievement. Up to a half appeared to have had a psychiatric disorder. The authors suggested that media attention to a suicide may have contributed to the method chosen and possibly to the occurrence of some of the suicides.

Doctors

The suicide rate among doctors is greater than that in the general population, but similar to several other occupations. There are no consistent findings of differences between medical specialties (see Roy 1985; Arnetz *et al.* 1987). Many reasons have been suggested, such as the ready availability of drugs, increased rates of addiction to alcohol and drugs, the extra stresses of work, reluctance to seek treatment for depressive disorders, and the selection into the medical profession of predisposed personalities. Whatever the true reasons, it is clear that the profession could do useful preventive work within its own ranks.

Suicide pacts

In suicide pacts, two people agree that at the same time each will take his or her own life. Completed suicide pacts are uncommon. Cohen (1961) reported that completed pacts account for one in 180 of all completed suicides, and Parry-Jones (1973) estimated that there are about twice as many completed suicide pacts as uncompleted ones. Suicide pacts have to be distinguished from cases where murder is followed by suicide (especially when the first person dies but the second is revived), or where one person aids another person's suicide without intending to die himself.

The psychological causes for these pacts are not known. It seems paradoxical that suicide, an act that is so often associated with social isolation, should be carried out with another person. Usually there is a particularly close relationship between the two members of the pact. In Parry-Jones's series, half the pacts were between lovers; however, in this study at least one person from each pair survived and had been prosecuted, and so the finding may not be representative. According to Rosenbaum (1983), the initiator is usually a mentally ill man who influences a woman who is not mentally ill. There appear to be considerable cultural variations in the nature of suicide pacts (Fishbain and Aldrich 1985). The subject has been reviewed by Rosen (1981).

The assessment of suicidal risk

General issues

Every doctor should be able to assess the risk of suicide. The first requirement is a willingness to make tactful but direct enquiries about a patient's intentions. Asking a patient about suicidal inclinations does not make suicidal behaviour more likely. On the contrary, if the patient has already thought of suicide, he will feel better understood when the doctor raises the issue and this feeling may reduce the risk. If a person has not thought of suicide before, tactful questioning will not make him behave suicidally. For a review of general issues see Hawton and Catalan (1987).

The second requirement is to be alert to the general factors signifying an increased risk. Even so, prediction has a low sensitivity and specificity, and it is even more difficult to distinguish between long-term risk and risk at a particular time. For example Goldstein *et al.* (1991) tried to develop a statistical model to predict the occurrence of 46 suicides from amongst a group of high-risk hospital patients, but failed to identify a single patient who later committed suicide.

Assessing risk

The most obvious warning sign is a **direct statement of intent**. It is now well recognized, but cannot be repeated too often, that there is no truth in the idea that people who talk of suicide do not enact it. On the contrary, two-thirds of those who die by suicide have told someone of their intentions. The greatest difficulty arises with people who talk repeatedly of suicide. In time their statements may no longer be taken seriously but may be discounted as threats intended to influence other people. However, some repeated threateners do kill themselves in the end. Just before the act, there may be a subtle change in their way of talking about dying, sometimes in the form of oblique hints that need to be taken more seriously than the original open statements.

Risk is also assessed by considering the factors that surveys have shown to be associated with suicide (see p. 417). Older patients are more at risk, as are the lonely and those suffering from chronic painful illness. Those who have previously attempted suicide are especially at risk (Hawton and Fagg 1988) and 30–40 per cent of those who die by suicide have made a previous attempt. **Depressive disorders** are highly important, especially when there is severe mood change with insomnia, anorexia, and weight loss (Barraclough *et al.* 1974). It is important to remember that suicide may occur during recovery from a depressive disorder in patients who previously, when more severely depressed, had thought of the act but had lacked the energy and initiative to carry it out. **Hopelessness** is a predictor of subsequent as well as immediate suicide. In a 10-year follow-up study of patients with suicidal ideas admitted to hospital, hopelessness was found to be an important determinant of suicide at some time in the previous 10 years (Beck *et al.* 1985).

Other associations

As noted earlier, there is an increased risk of suicide with alcohol dependence, especially when associated with physical complications or severe social damage, drug dependence, epilepsy, and abnormal personality. In schizophrenia, suicide is particularly likely in young men with recurrent severe illness and intellectual deterioration (Drake and Cotton 1986).

Completing the history

When these general risk factors have been assessed, the rest of the history should be evaluated. The interview should be conducted in an unhurried and sympathetic way that allows the patient to admit any despair or self-destructive intentions. It is usually appropriate to start by asking about current problems and the patient's reaction to them. Enquiries should cover losses, both personal (such as bereavement or divorce) and financial, as well as loss of status. Information about conflict with other people and social isolation should also be elicited. Physical illness should always be asked about, particularly any painful condition in the elderly. (Some depressed suicides have unwarranted fears of some physical illness as a feature of the psychiatric disorder.)

In assessing previous personality, it should be borne in mind that the patient's self-description may be coloured by depression. Whenever possible, another informant should be interviewed. The important points include mood swings, impulsive or aggressive tendencies, and attitudes towards religion and death.

Mental state examination

The assessment of mood should be particularly thorough, and cognitive function must not be overlooked. The interviewer should then assess suicidal intent. It is usually appropriate to begin by asking whether patients think that life is too much for them, or whether they no longer want to go on. This question can lead to more direct questions about thoughts of suicide, specific plans, and actions such as saving tablets.

It is important always to remember that severely depressed patients occasionally have *homicidal ideas*; they may believe that it would be an act of mercy to kill other people, often the spouse or a child, to spare them intolerable suffering. Such homicidal ideas must not be missed, and should always be taken extremely seriously.

The management of suicidal patients

General issues

Having assessed the suicidal risk, the clinician should make a treatment plan and try to persuade the patient to accept it. The first step is to decide whether the patient should be admitted to hospital or treated as an out-patient or day patient. This decision depends on the intensity of the suicidal intention, the severity of any associated psychiatric illness, and the availability of social support outside hospital. If out-patient treatment is chosen, patients should be given a telephone number with which they can, at all times, obtain help if feeling worse. Frustrated attempts to find a doctor can be the last straw for a patient with suicidal inclinations.

If the suicidal risk is judged to be significant, in-patient care is nearly always required. An occasional exception may be made when the patient lives with reliable relatives, but only if those relatives wish to care for the patient themselves, understand their responsibilities, and are able to fulfil them. Such a decision requires an exceptionally thorough knowledge of the patient and his problems. If hospital treatment is essential but the patient refuses it, admission under a compulsory order will be necessary. The method of arranging this in England and Wales is outlined in the Appendix.

Management in the community

The current emphasis on community rather than hospital care means that there are an increased number of people in the community who are at risk of suicide. Good care depends upon recognition and continuing assessment of the suicidal risk, together with clear plans to provide appropriate treatment and social support. Table 13.2 summarizes the key points.

Management in hospital

The obvious first requirement is to prevent patients from harming themselves. Table 13.3 summarizes the principles of care. The arrangements require adequate staffing, vigilance, and good communication. Special nursing arrangements may be needed at times so that the patient is never alone. It is essential that a clear policy should be stated for each patient and understood by all the staff. For example, it

Table 13.2 *Care of the suicidal patient in the community*

Full assessment of patient and key relatives, including review of the suicidal risk
Organization of adequate social support
Regular review

Full dosage of safe psychiatric treatments
    Choose less toxic drugs
    Small prescriptions
    Involve relatives in care of tablets

Arrange immediate access to extra help for patient and relatives

should be clearly specified whether the patient is to be kept in pyjamas, how closely a member of staff is to remain at hand, and whether potentially dangerous objects such as scissors are to be removed. The policy should be drawn up as soon as the patient is admitted, and agreed between the doctor and the nursing staff. It should be reconsidered carefully at frequent intervals until the danger passes. It is particularly important that any changes in policy should be made clear when staff change between shifts.

When intensive supervision is needed for more than a few days, increasing difficulties may arise. Patients under constant observation may become irritated and resentful, and may evade supervision. Staff should be aware of such problems, and treatment of any associated mental illness should not be delayed. If the patient has severe depressive disorder, the rapid action of electroconvulsive therapy (ECT) may be required (see Chapter 8).

Appropriate physical treatment should be accompanied by simple psychotherapy. However determined the patient is to die, there is usually some small remaining wish to go on living. If doctors and nurses adopt a caring and hopeful attitude, these positive feelings can be encouraged and patients can be helped towards a more realistic and balanced view of their future. At the same time, they can be helped to see how an apparently overwhelming accumulation of problems can be dealt with one by one.

However carefully patients are managed, occasionally a patient will die by suicide despite all the efforts of the staff. The doctor then has an important role in supporting other staff, particularly any nurses who have come to know the patient well through taking part in constant observation. Although it is essential to review every suicide carefully to determine whether any useful lessons can be learnt, this review should never become a search for a culprit.

The relatives

When a patient has died by suicide, the relatives require not only the support that is appropriate for any bereaved person, but also help with particular difficulties. In a study by Barraclough and Shepherd (1976) the relatives usually reported that the police conducted their enquiries in a considerate way, but nearly all found the public inquest distressing. The subsequent newspaper publicity caused further grief, reactivating the events surrounding the death and increasing any feelings of stigma. Sympathetic counselling is likely to help relatives with these difficulties. See Wertheimer (1992) for a review of consequences of suicide for relatives.

Table 13.3   *Care of the suicidal patient in hospital*

Safe ward environment

Adequate well-trained staff with good working relationship

Clear policies for assessment, review, and observation

On admission
  Assess risk
  Agree level of observation
  Remove objects which might be used as means of suicide
  Discuss plans with patient
  Agree policy for visitors (number, duration, information)

During admission
  Regular review of risk and plans
  Clear plans for leave
  If patient leaves ward without notice, take immediate action

---

Discharge
  Plan and agree in advance
  Prescribe adequate but non-dangerous amount of drugs
  Early follow-up

Suicide prevention

As reported above, many people who commit suicide contact their doctors shortly beforehand. Most have a psychiatric illness, and four-fifths are being treated with psychotropic drugs, though not always with the most appropriate drug or the optimal dosage. These findings suggest a need to improve the skills of doctors in identifying high-risk patients and in planning their treatment (World Health Organization 1993). On the other hand, many depressed patients are treated successfully by general practitioners, and many of the problems associated with suicide are beyond the responsibility of doctors. Therefore it would be wrong therefore to suggest that the solution to the problem of suicide prevention lies wholly in better training of doctors, even though it might make a contribution. See Gunnell and Frankel (1994) for a review of issues and evidence.

Table 13.4 lists the main ways in which prevention has been approached.

Primary prevention

Improvements in psychiatric services might be expected to lead to earlier recognition and better treatment of the psychiatric disorders and problems associated with suicide. However, a review of published studies has found generally disappointing effects of psychosocial and psychiatric treatments of mental disorder on suicide rates (Diekstra 1994). However, therapy for depression seems to be associated with a reduction in suicide rate (Frank *et al.* 1990). Suicide prevention was the aim of an educational programme on the Swedish island of Gotland in 1983 and 1984 designed to increase general practitioners' understanding of the diagnosis and treatment of

Table 13.4 *Suicide prevention*

Primary
    Better and more available
        psychiatric services
    Restricting the means of suicide
    Educational programmes
    Restricting opportunities for
        imitation

Secondary
    Better and more available
        psychiatric care
    Crisis centres and 'hot lines'

---

affective disorder. After the programme was introduced the suicide rate dropped to an extent which was significantly different from both the long-term suicide trend on Gotland and that for Sweden as a whole (Rutz *et al.* 1989). A further study found that by 1988, three years after the project had ended, the suicide rate had returned almost to baseline values (Rutz *et al.* 1992). The researchers concluded that the programme had been effective but that it would need to be repeated every two years to have long-term benefits. These findings require replication.

Another approach is to attempt to restrict access to methods commonly used to commit suicide. The British experience with domestic gas (see p. 414) suggests that removal of means may be followed by an overall fall in the suicide rate. The evidence related to firearms is less clear, but there would appear to be good arguments for restricting their availability, and that of toxic drugs, while encouraging the use of the least toxic anti-depressant drugs, taking measures to reduce toxicity (such as modifying preparation of paracetamol), and restricting access to dangerous places, such as high bridges, that have achieved a reputation of places for suicide.

The evidence on the importance of imitation as a factor in precipitating suicide suggests the need to persuade the media to take a responsible view of the reporting and portrayal of suicide.

A different approach has been the growth of school-based programmes to increase awareness of problems and their solutions. Some programmes are directed solely at teachers, and others at students. There is little evidence of their effectiveness in suicide prevention (Shaffer *et al.* 1988).

Secondary prevention

Secondary prevention aims to improve the availability and quality of help for those who may be contemplating suicide. Apart from a general increase of awareness of diagnosis and treatment, as attempted in the Gotland project, numerous crisis or suicide prevention centres have been established. These centres attempt to provide crisis counselling in person or usually by a telephone service, or 'hot line'. Such services are anonymous, confidential, and offer 24-hour availability. It has been claimed that suicide rates, especially among young people, have decreased in areas with suicide prevention centres, but overall the evidence is contradictory and, whatever the benefits of such crisis centres, it seems unlikely that they are effective in preventing suicide.

A particularly well-known crisis counselling service is the Samaritan organization founded in London in 1953 by the Reverend Chad Varah. People in despair are encouraged to contact a widely publicized telephone number. The help offered ('befriending') is provided by non-professional volunteers, all trained to listen sympathetically without attempting to take on tasks that are in the province of a doctor or social worker. There is some evidence that, amongst people who telephone the Samaritans, the suicide rate in the ensuing year is higher than in the general population (Barraclough and Shea 1970). This finding suggests that the organization has attracted an appropriate group of people, but it also raises the question of the efficacy of the help offered. Comparisons of matched towns with and without services suggest little difference in suicide rates (Jennings *et al.* 1978). Even so, whether or not they prevent suicides, the Samaritans appear to perform a useful role providing for the needs of many lonely and despairing people.

A final approach to prevention is better treatment of suicide attempters who are a particularly high risk for suicide. This subject is considered further in the remainder of this chapter.

Deliberate self-harm

Introduction

Before the 1950s little distinction was made between people who killed themselves and those who survived after an apparent suicidal act. Stengel (1952) identified epidemiological differences between the two groups, and proposed the terms 'suicide' and 'attempted suicide' to distinguish the two forms of behaviour. He supposed that a degree of suicidal intent was essential in both groups; in other words, those who survived were failed suicides. These ideas were developed in an important monograph (Stengel and Cook 1958).

In the 1960s it was proposed that suicidal intent should no longer be regarded as essential because it was recognized that most 'attempted suicides' had 'performed their acts in the belief that they were comparatively safe; aware, even in the heat of the moment, that they would survive their overdosage, and be able to disclose what they had done in good time to ensure rescue' (Kessel and Grossman 1965). For this reason, Kessel proposed that 'attempted suicide' should be replaced by 'deliberate self-poisoning' and 'deliberate self-injury'. These terms were chosen to imply that the behaviour was clearly not accidental, without any assumption as to whether the desire for death was present. By the end of the 1960s, these ideas were widely accepted.

Kreitman and his colleagues introduced the term 'parasuicide' to refer to a 'non-fatal act—in which an individual deliberately causes self injury or ingests a substance in excess of any prescribed or generally recognized therapeutic dose' (Kreitman 1977, p.3). Thus the term parasuicide excludes the question of whether death was a desired outcome. Although 'parasuicide' has been used quite widely, 'self-poisoning' and 'self-injury' are retained by some workers. Morgan (1979) suggested the term 'deliberate self-harm' (sometimes abbreviated to

DSH) to provide a single term covering deliberate self-poisoning and deliberate self-injury. It has been objected that the term deliberate self-harm is sometimes a misnomer because the act is not invariably harmful (even though done in the knowledge that it might cause harm). In fact, no single term is wholly satisfactory. In this chapter, the term deliberate self-harm will be used rather than parasuicide.

The distinction between suicide and deliberate self-harm is not absolute. There is an important overlap. Some people who have no intention of dying succumb to the effects of an overdose. Others who intended to die are revived. Moreover, many patients were ambivalent at the time, uncertain whether they wished to die or live.

It should be remembered that, among patients who have been involved in deliberate self-harm, the suicide rate in the subsequent 12 months is about a hundred times greater than in the general population. For this reason, and for other reasons to be given later, deliberate self-harm should not be regarded lightly.

The act of deliberate self-harm

The drugs used in deliberate self-poisoning

In the United Kingdom, about 90 per cent of the cases of deliberate self-harm referred to general hospitals involve a drug overdose, and most of them presented no serious threat to life. The most commonly used drugs are the **non-opiate analgesics**, such as paracetamol and salicylates (Hawton and Fagg 1992*a,b*). In recent years paracetamol has been used increasingly; it is particularly dangerous because it damages the liver and may lead to the delayed death of patients who had not intended to die. It is particularly worrying that this drug is often taken by younger patients who are usually unaware of the serious risks (Gazzard *et al.* 1976). The type of drug used varies somewhat with age, local prescription practices, and the availability of drugs.

Amongst all cases of deliberate self-harm, about 40 per cent of people have taken **alcohol** in the six hours before the act, and in one study 15 per cent of males and 4 per cent of females were diagnosed as suffering from drinking problems (Hawton *et al.* 1989).

Methods of deliberate self-injury

Deliberate self-injury accounts for 5–15 per cent of all deliberate self-harm presenting to general hospitals in Britain (Hawton and Catalan 1987). The commonest method of self-injury is laceration, usually of the forearm or wrists; it accounts for about four-fifths of the self-injuries referred to a general hospital (Hawton and Catalan 1987). (Self-laceration is discussed further below.) Other forms of self-injury are jumping from heights or in front of a train or motor vehicle, shooting, and drowning. These violent acts occur mainly among older people who intended to die (Morgan *et al.* 1975). They are more common in North America than in the United Kingdom.

Deliberate self-laceration

There are three forms of deliberate self-laceration: deep and dangerous wounds inflicted with serious suicidal intent, more often by men, self-mutilation by schizophrenic patients (often in response to hallucinatory voices) or by transsexuals, and superficial wounds that do not endanger life. Only the last group will be described here.

The patients are mostly young. Generally they have severe personality problems characterized by low self-esteem, impulsive or aggressive behaviour, unstable moods, difficulty in interpersonal relationships, and a tendency to abuse alcohol and drugs. They are often said to have borderline personalities and are frequently forensic patients (Hillbrand 1994). Sexual identity problems have been reported in association with self-laceration (Hawton 1990).

Usually, increasing tension and irritability precede self-laceration, and are then relieved by it. Some patients say that the lacerations were inflicted during a state in which they felt detached from their surrounding and experienced little or no pain. The lacerations are usually multiple, and are made with glass or a razor blade on the forearms or wrists. Generally, some blood is drawn and the sight of this is often important to the patient. Some patients cause other injuries as well, for example by burning with cigarettes or inflicting bruises. After the act, the patient often feels shame and disgust. Some psychiatric in-patients, especially adolescents, lacerate themselves in imitation of others on the same ward (Walsh and Rosen 1985). For reviews see Hawton (1990) and Favazza and Rosenthal (1993).

The epidemiology of deliberate self-harm

During the 1960s and early 1970s there was a substantial increase in cases of deliberate self-harm admitted to general hospitals. Among women deliberate self-harm is now the most frequent single reason for admission to a medical ward, and among men it is second only to ischaemic heart disease.

Accuracy of statistics

The official statistics for the incidence of deliberate self-harm are likely to be less than the true rate because not all cases are referred to hospital. For example a survey in Edinburgh suggested that hospital referral rates underestimated the frequency of deliberate self-harm by at least 30 per cent (Kennedy and Kreitman 1973). Other reasons for inaccuracy in the rates are variations in the definition and the identification of deliberate self-harm.

Trends in the last three decades

In the early 1960s a substantial increase in deliberate self-harm began in most Western countries (Weissman 1974; Wexler *et al.* 1978). In the United Kingdom, the rates of admission to general hospitals increased about fourfold in the ten years up to 1973 (Kreitman 1977; Bancroft *et al.* 1975). The rates continued to increase more slowly in the mid-1970s, but then fell in the late 1970s and early 1980s for women. Rates are now rising again for both men and women (Hawton and Fagg 1992*a,b*). The reasons for the earlier decline are unknown.

Variations according to personal characteristics

Deliberate self-harm is more common among younger people, with the rates declining sharply in middle age. In all but the very old, the rates in the United Kingdom are 1.3–1.5 times higher for **females**, and particularly high rates are found among those aged 15–30 years. The peak age for men is older than that for women. For both sexes rates are very low under the age of 12 years. Deliberate self-harm is more prevalent in the lower **social classes** (Platt *et al*. 1988). There are also differences related to **marital status**. The highest rates for both men and women are among the divorced, and high rates are also found among teenage wives and younger single men and women (Bancroft *et al*. 1975; Holding *et al*. 1977).

There has been little recent change in the rates of deliberate self-harm in the **elderly**. The characteristics of such people seem to be rather more similar to those who kill themselves than do those of younger deliberate self-harm subjects (Hawton and Fagg 1990; Frierson 1991).

Variations according to place of residence

High rates are found in areas characterized by high unemployment, overcrowding, many children in care, and substantial social mobility (Holding *et al*. 1977; Buglass and Duffy 1978).

Causes of deliberate self-harm

Precipitating factors

Compared with the general population, people who deliberately harm themselves experience four times as many stressful life problems in the six months before the act (Paykel *et al*. 1975*a*). The events are various, but a recent quarrel with a spouse, girlfriend, or boyfriend is particularly common (Bancroft *et al*. 1977). Other events include separation from or rejection by a sexual partner, the illness of a family member, recent personal physical illness, and a court appearance.

Predisposing factors

The *precipitating events* often occur against a background of long-term problems concerning marriage, children, work, and health. About two-thirds of the patients in one study (Bancroft *et al*. 1977) had some kind of marital problem; half the men had been involved in an extramarital relationship, and a further quarter said that their wives had been unfaithful. Among the unmarried, a similar proportion have difficulties in their relationships with sexual partners.

Amongst men who deliberately harm themselves, the proportion who are unemployed has been increasing in recent years, and the rate of deliberate self-harm increases with length of *unemployment*. However, unemployment is related to other social factors associated with deliberate self-harm, such as financial difficulties, and it is difficult to determine whether unemployment is a direct cause (Hawton and Rose 1986). The association between female unemployment and attempted suicide has attracted less attention. The rates of attempted suicide are considerably higher amongst unemployed than amongst employed women, and are particularly high in women unemployed for more than a year (Hawton *et al*. 1988). As with men, it is unclear whether unemployment and its consequences are a direct cause of attempted suicide, or whether women already predisposed in other ways to attempt suicides are more likely to become unemployed.

A background of *poor physical health* is common (Bancroft *et al*. 1975). This applies particularly to people with epilepsy, who are found in the deliberate self-harm population about six times more frequently than would be expected (Hawton *et al*. 1980).

There is some evidence that early *parental loss* through bereavement, or a history of parental neglect or abuse, is more frequent among cases of deliberate self-harm (Hawton and Catalan 1987). There is also evidence of poor skills in solving interpersonal problems and in planning for the future.

Psychiatric disorder

Many of the patients who deliberately harm themselves have some affective symptoms, but few have full psychiatric disorder (Newson-Smith and Hirsch 1979a; Urwin and Gibbons 1979). This is in marked contrast to cases of completed suicide when *depressive disorders* are frequent (see p. 418). Personality disorder is more common, and has been reported in about a third to a half of self-harm patients (Kreitman 1977).

Dependence on alcohol is common (as it is in completed suicide), but estimates of the frequencies in different series vary between 15 and 50 per cent among men, and between 5 and 15 per cent among women (Hawton *et al.* 1989).

In a large series of people who deliberately harmed themselves, about half had consulted a general practitioner, psychiatrist, social worker, or another helping agency in the previous week (Bancroft *et al.* 1977).

Motivation and deliberate self-harm

The motives for deliberate self-harm are usually mixed and difficult to identify for certain. Even if patients know their own motives, they may try to hide them from other people. For example people who have taken an overdose in frustration and anger may feel ashamed and say instead that they wished to die. In the study, only about half of the patients who said that they intended to die were judged by psychiatrists to have had true suicidal intentions (Bancroft *et al.* 1979). Conversely, people who truly intended to kill themselves may deny it. For this reason, more emphasis should be placed on a common-sense evaluation of the patients' actions leading up to self-harm than on their subsequent accounts of their own motives.

Despite this limitation, useful information has been obtained by questioning groups of patients about their motives. Only a few say that the act was premeditated. About a quarter say that they wished to die. Some say that they are uncertain whether they wanted to die or not, others that they

were leaving it to 'fate' to decide, and others that they were seeking unconsciousness as a temporary escape from their problems. Another group admit that they were trying to influence someone; for example that they were seeking to make a relative feel guilty for having failed them in some way (Bancroft *et al.* 1979). This motive of influencing other people was first emphasized by Stengel and Cook (1958), who described the act of attempted suicide (as it was then called) as 'calling forth action from the human environment'. This behaviour has since been referred to as 'a cry for help'. Although some acts of deliberate self-harm result in increased help for the patient, others may arouse resentment, particularly if they are repeated (Hawton and Catalan 1987).

The outcome of deliberate self-harm

This section deals both with the risk that the act of self-harm will be repeated and with the risk that the patient will die by suicide at some later occasion.

The risk of repetition

Repetition rates are based on groups of patients, some of whom have received psychiatric treatment after the act. Reported rates vary between about 15 and 25 per cent in the year after the act (Kreitman 1977). There are three broad patterns: some patients repeat only once, some repeat several times, but only during a limited period of continuing problems, and a small group repeat many times over a long period as a habitual response to stressful events.

Several studies agree that the following factors distinguish patients who repeat self-harm from those who do not: previous deliberate self-harm, previous psychiatric treatment, personality disorder of the antisocial type, a criminal record, and alcohol or drug abuse. Lower social class and unemployment are also predictors (Kreitman 1977; Appleby 1993). These factors are summarized in Table 13.7 (p. 433).

The risk of completed suicide

Among people who have intentionally harmed themselves, the risk of later suicide is much increased. For example in the first year afterwards the risk of suicide is about 1–2 per cent, which is 100 times that of the general population (Kreitman 1977). An eight-year follow-up showed that, amongst patients who were previously admitted with deliberate self-harm, almost 3 per cent eventually take their own lives and about twice the expected number die from natural causes (Hawton and Fagg 1988). Looked at in another way, there is a history of previous deliberate self-harm in a third to a half of completed suicides (Kreitman 1977).

Among people who deliberately harm themselves, the risk of eventual suicide is greater in those with other risk factors for suicide. Thus the risk is greater among older patients who are male, depressed, or alcoholic (Kreitman 1977; Hawton *et al.* 1993). A non-dangerous method of self-harm does not necessarily indicate a low risk of subsequent suicide, partly because patients have little knowledge of the dangerousness of many methods. However, the risk is certainly high when violence or highly dangerous drug overdoses have been used.

In the weeks after deliberate self-harm, many patients report changes for the better. People with psychiatric symptoms often report a decrease in their intensity (Newson-Smith and Hirsch 1979*a*). Improvements may result from help provided by psychiatrists and other professionals, or from improvements in relationships, attitudes, and behaviour. However, some patients, fare much worse and repeatedly harm themselves within months of the first act, and some relatives are unsympathetic and even hostile.

There have been numerous attempts to construct scales to predict repetition with an accuracy that could be clinically useful. Kreitman and Foster (1991) have proposed a scale which allows allocation of patients to low, moderate, and high risk groups. However, all instruments are of relatively low specificity, identifying no more than half of those at high risk and an even lower proportion of the numerically largest group of repeaters at moderate risk.

The assessment of patients after deliberate self-harm

General aims

Assessment is concerned with three main issues: the immediate risk of suicide, the subsequent risks of further deliberate self-harm or suicide, and any current medical or social problems. The assessment should be carried out in a way that encourages the patient to undertake a constructive review of his problems and of the ways in which he can deal with them himself. This encouragement of self-help is important, because many patients are unwilling to be seen again by a psychiatrist.

Usually the assessment has to be carried out in an accident and emergency department or a ward of a general hospital, in which there may be little privacy. Whenever possible, the interview should be in a side room so that it will not be overheard or interrupted. If the patient has taken an overdose, the interviewer should first make sure that the patient has recovered sufficiently to be able to give a satisfactory history. If consciousness is still impaired, the interview should be delayed. Information should also be obtained from relatives or friends, the family doctor, and any other person (such as a social worker) already attempting to help the patient. Wide enquiry is important because sometimes information from other sources may differ substantially from the account given by

Table 13.5 *Circumstances suggesting high suicidal intent*

Planning in advance
Precautions to avoid discovery
No attempts to obtain help afterwards
Dangerous method
'Final acts'

the patient. See Hawton and Catalan (1987) for a review.

Specific enquiries

The interview is directed to five questions.

(1) What were the patient's intentions when he harmed himself?
(2) Does he now intend to die?
(3) What are the patient's current problems?
(4) Is there a psychiatric disorder?
(5) What helpful resources are available to this patient? Each question will be considered in turn.

1. *What were the patient's intentions when he harmed himself?* As mentioned already, patients sometimes misrepresent their intentions. For this reason the interviewer should reconstruct, as fully as possible, the events that led up to the act of self-harm. He will need to find the answers to five subsidiary questions (Table 13.5).

(a) Was the act **planned** or carried out on impulse? The longer and more carefully the plans have been made, the greater is the risk of a fatal repetition.

(b) Were **precautions** taken against being found? The more thorough the precaution, the greater is the risk of a further fatal overdose. Of course, events do not always take place as the patient expected; for example a spouse may arrive home earlier than usual so that the patient is discovered alive. In such circumstances it is the patient's reasonable expectations that count in predicting the future risk.

(c) **Did the patient seek help?** Serious intent can be inferred if there were no attempts to obtain help after the act.

(d) **Was the method dangerous?** If drugs were used, what were they and what amount was taken? Did the patient take all the drugs available? If self-injury was used, what form did it take? (As noted above, the greater the suicidal intent the greater is the risk of

a further suicide attempt.) Not only should the actual dangerousness be assessed, but also that anticipated by the patient, which may be inaccurate. For example some people wrongly believe that paracetamol overdoses are harmless or that benzodiazepines are dangerous.

(e) **Was there a 'final act'** such as writing a suicide note or making a will? If so, the risk of a further fatal attempt is greater.

By reviewing the answers to these questions, the interviewer makes a judgement of the patient's intentions at the time of the act. A similar approach has been formalized in Beck's suicide intent scale (Beck *et al.* 1974*b*) which gives a score for the degree of intent.

2. *Does the patient now intend to die?* The interviewer should ask directly whether the patient is pleased to have recovered or wishes that he had died. If the act suggested serious suicidal intent and if the patient now denies such intent, the interviewer should try to find out by tactful questioning whether there has been a genuine change of resolve.

3. *What are the current problems?* Many patients will have experienced a mounting series of difficulties in the weeks or months leading up to the act. Some of these difficulties may have been resolved by the time that he is interviewed; for example a husband who planned to leave his wife may now have agreed to stay. The more serious the problems that remain, the greater is the risk of a fatal repetition. This risk is particularly strong if there are problems of loneliness or ill-health. The review of problems should be systematic and should cover the following: intimate relationships with the spouse or another person; relations with children and other relatives; employment, finance, and housing; legal problems; social isolation, bereavement, and other losses. Drug and alcohol problems can be considered at this stage or when the psychiatric state is reviewed.

4. *Is there psychiatric disorder?* It should be possible to answer this question from the history

and from a brief but systematic examination of the mental state. Particular attention should be directed to depressive disorder, alcoholism, and personality disorder. Schizophrenia and dementia should also be considered, though they will found less often.

5. *What are the patient's resources?* These include his capacity to solve his own problems, his material resources, and the help that others may provide. The best guide to patients' ability to solve future problems is their record of dealing with difficulties in the past, for example the loss of a job or a broken relationship. The availability of help should be assessed by asking about the patients' friends and confidants, and about any support they may be receiving from their general practitioners, social workers, or voluntary agencies.

Is there a continuing risk of suicide?

The interviewer now has the information required to answer this important question. The answers to the first four questions outlined above are reviewed: (1) Did the patient originally intend to die? (2) Does he intend it now? (3) Are the problems which provoked the act still present? (4) Is he suffering from a mental disorder? The interviewer also decides what help other people are likely to provide after the patient leaves hospital (question (5) above). Having reviewed the individual factors in this way, the interviewer compares the patient's characteristics with those found in groups of people who have died by suicide. These characteristics are summarized in Table 13.6.

Is there a risk of further non-fatal self-harm?

The predictive factors, which have been outlined already (see p. 429), are summarized in Table 13.7. The interviewer should consider all the points in turn before making a judgement about the risk. Using their own six-item scale (slightly different from Table 13.7), Buglass and Horton (1974) gave a score according to the number of

Table 13.6 *Factors predicting suicide after deliberate self-poisoning*

Evidence of serious intent*
Depressive disorder
Alcoholism or drug abuse
Antisocial personality disorder
Previous suicide attempt(s)
Social isolation
Unemployment
Older age group (women only)
Male sex

*See Table 13.1.

items that were present in each case. Patients with a score of zero had a 5 per cent chance of repeating the act in the next year, while those with a score of five or more had an almost 50 per cent chance.

Is treatment required and will the patient agree to it?

If the patient is actively suicidal, the procedures are those outlined in the first part of this chapter (see p. 422). About 5–10 per cent of deliberate self-harm patients require admission to a psychiatric unit for further management; most need treatment for depressive disorders or alcoholism, but a few require only a brief respite from overwhelming domestic stress.

The best methods of treating the remaining patients are less certain. A quarter to a third are probably best referred to general practitioners, social workers, or others who may already be involved in their case. Many patients (up to half)

Table 13.7 *Factors predicting the repetition of non-fatal deliberate self-poisoning* *

Previously deliberate self-harm
Previous psychiatric treatment
Antisocial personality disorder
Alcohol or drug abuse
Criminal record
Low social class
Unemployment

* See Kreitman and Dyer 1980.

may benefit from out-patient care, usually problem-oriented counselling for personal problems rather than treatment for psychiatric disorder. Many patients refuse the offer of out-patient help; their care should be discussed with the general practitioner before they are allowed to return home. It is useful to provide an emergency telephone number enabling patients to obtain immediate advice or an urgent appointment in any further crisis.

Special problems

Mothers of young children

Mothers of young children require special consideration because of the known association between deliberate self-harm and child abuse (Roberts and Hawton 1980). It is important to ask about the mother's feeling towards her children, and to enquire about their welfare. In the United Kingdom information about the children can usually be obtained from the general practitioner, who may ask his health visitor to investigate the case.

Children and adolescents

Despite problems of case definition and identification, it appears that there has been a striking increase in the frequency of deliberate self-harm amongst children and adolescents in many parts of the developed world. Deliberate self-harm is rare but not unknown among pre-school children (Rosenthal and Rosenthal 1984); it becomes increasingly common after the age of 12. It is more common amongst girls except at younger ages. The most common method is drug overdosage which is usually not dangerous, though occasionally life-threatening. The more dangerous methods of self-injury are more frequent amongst boys. Epidemics of deliberate self-harm occasionally occur amongst adolescents in psychiatric hospitals and other institutions.

It is difficult to determine the motivation of self-harm in young children, especially as a clear concept of death is not usually developed until around the age of 12. It is probable that only a few of the younger children have any serious suicidal intent. Possibly their motivation is more often to communicate distress, to escape from stress, or to manipulate other people.

Deliberate self-harm in children and adolescents is associated with histories of broken homes, family psychiatric disorder, and child abuse. It is often precipitated by social problems such as difficulties with parents, boyfriends, or school work (Hawton and Catalan 1987; Pfeffer *et al.* 1994). It is also associated with mood disorder and personality disorder (Pfeffer *et al.* 1991). Hawton (1986) described three main groups: those with acute distress about problems of less than a month's duration, but without behavioural disturbance, those with chronic psychological and social problems, but without behavioural disturbance, and those with chronic psychological and social problems, and also behavioural disturbance such as stealing, truancy, drug taking, or delinquency.

For most children and adolescents, the outcome of deliberate self-harm is relatively good, but an important minority continue to have social and psychiatric problems, and to repeat acts of deliberate self-harm. A poor outcome is associated with poor psychosocial adjustment, a history of previous deliberate self-harm, and severe family problems. There is a significant risk of suicide amongst adolescents, especially boys (Hawton 1986).

When children harm themselves, it is better for them to be assessed by child psychiatrists rather than members of the adult services for deliberate self-harm. Treatment is usually directed towards the family. In the case of adolescents, treatment largely follows the general principles of management described in this chapter.

University students

Deliberate self-harm in university students has received less attention than completed suicide. Hawton *et al.* (1995*a,b*) found that, in Oxford University, the rates of deliberate self-harm during university time were two-thirds of those of other young people of similar age in the city of Oxford. The difference was especially marked in women. The lower rate for men partly reflect their generally higher socio-economic status. Very few of the attempted suicides appear to be failed suicides.

Who should assess?

In 1968 a British Government report recommended that all cases of deliberate self-harm should be assessed by a psychiatrist (Central Health Services Council 1968). The intention was to make sure that patients with depressive and other psychiatric disorders were identified and treated, and that appropriate assistance was given for other psychological and social problems. It is likely that a large proportion of the patients at that time were suffering from psychiatric disorder. The increase in cases since then has consisted mainly of younger patients in whom serious psychiatric disorders are less frequent. Such patients usually require assessment and counselling for social problems, rather than diagnosis and treatment of psychiatric disorders.

In England and Wales, a subsequent government report (Department of Health and Social Security 1984) recognized that such assessments could be carried out equally well by trained staff other than psychiatrists, as long as psychiatrists provide training and supervision, and interview any patients who may have a psychiatric disorder. It is essential that all emergency departments responsible for the care of deliberate self-harm patients have their own code of practice. It has been shown that, if they receive appropriate additional training, junior medical staff (Gardner *et al.* 1977), psychiatric nurses (Hawton *et al.* 1979), and social workers (Newson-Smith and Hirsch 1979*b*) can all assess these patients as well as psychiatrists. It is emphasized that when a nurse or social worker makes the assessment a psychiatrist reviews the question of psychiatric disorder.

For patients admitted to medical beds after deliberate self-harm, the best policy seems to be that the local consultant physicians and psychiatrists should agree who is to make the assessments and who will take the final responsibility for decisions about management. In this way each hospital can adopt the policy that makes the most effective use of the available medical, nursing, and social work staff.

Management

As indicated above, the assessment procedure divides patients into three groups. About 5–10 per cent need immediate in-patient treatment in a psychiatric unit, and about a quarter require no special treatment because their self-harm was a response to temporary difficulties and carried little risk of repetition. This section is concerned with the remaining two-thirds for whom some out-patient treatment may be appropriate.

The main aim of such treatment is to enable the patient first to resolve the difficulties that led up to the act of self-harm, and second to deal with any future crisis without resorting to further self-harm. The main problem is that, once they have left hospital, many patients are disinclined to take part in any treatment.

Treatment is psychological and social. Drugs are seldom required, but a small minority of patients require antidepressant medication. It is more often necessary to withdraw drugs for which there is no clear indication. The starting point of treatment is the list of problems compiled during the assessment procedure. Patients are encouraged to consider what steps that they could take to resolve each of these problems, and to formulate a practical plan for tackling them one at a time. Throughout this discussion, the therapist tries to persuade patients to do as much as possible for themselves.

Many cases are associated with interpersonal problems. It is often helpful to interview the other person involved, first alone and then in a few joint interviews with the patient. This procedure may help to resolve problems that the couple have been unable to discuss on their own.

When deliberate self-harm follows a bereavement or other kind of loss, a different approach is needed. The first step should be sympathetic listening while patients express their feelings of loss. Then patients are encouraged to seek ways of gradually rebuilding their lives without the lost person. Appropriate measures will depend on the nature of the loss—whether it was through death, or the break-up of a marriage, or the end of another relationship. Again, the emphasis should be on self-help.

Some special problems of management

Patients refusing assessment After deliberate self-harm, some patients refuse to be interviewed, and others seek to discharge themselves before the assessment is complete. In such cases it is essential to gather as much information as possible from other sources in order to exclude serious suicidal risk of psychiatric disorder before letting the patient leave hospital. Occasionally, detention under a compulsory order is appropriate.

Frequent repeaters Some patients take overdoses repeatedly at times of stress. Often the behaviour seems intended to reduce tension or gain attention. However, when overdoses are taken repeatedly, relatives often become unsympathetic or even overtly hostile, and staff of hospital emergency departments become angry and bewildered. These patients usually have a personality disorder and many insoluble social problems, but neither counselling nor intensive psychotherapy is

usually effective. It is helpful if all those involved in management agree a clear plan whereby the patient is rewarded for constructive behaviour. An opportunity for continuing support by one person should be arranged. However, whatever help is arranged, the risk of eventual death by suicide is high.

Self laceration The management of self-laceration presents many problems (see Hawton (1990) for a review). These patients often have difficulty in expressing their feelings in words, and so formal psychotherapy is seldom helpful. Simple efforts to gain the patients' confidence and increase their self-esteem are more likely to succeed.

Assessment should include a behavioural analysis of the sequences of events which lead to self-cutting. This may help to formulate ways in which treatment could either interrupt the chain of events which leads to self-cutting, or replace the cutting by an alternative method of relieving methods. There are four types of management which can be helpful:

(1) enabling the patient to gain control of the cutting by encouragement of alternative ways relieving tension, such as relaxation, expression of emotion, and physical exercise, and avoiding aggravating factors such as alcohol;
(2) dealing with underlying personal and social difficulties;
(3) medication and general measures within the in-patient setting;
(4) increased support from other people.

Medication appears to have a limited role, although neuroleptics such as chlorpromazine or haloperidol may be valuable as a short-term measure to reduce tension. Many people who cut themselves have severe personality difficulties (see Chapter 5). Treatment should be directed towards these problems, although it may be difficult and prolonged. Admission to a psychiatric unit is occasionally necessary, and it is essential that a clear and detailed policy is agreed by those involved since self-cutting is very often difficult to manage in an in-patient unit, and indeed may be imitated by other patients.

The results of treatment

As already indicated, the treatment of deliberate self-harm has two aims. The first is to help the patient to deal with the social and emotional problems that led up to the act. The second is to prevent further acts of self-harm. The process of assessment may well be therapeutic, and Hawton and Fagg (1992*a*) found a lower rate of repetition amongst those who had been assessed by psychiatrists than amongst those who had not been assessed.

The are considerable methodological problems in assessing the effectiveness of different treatments for the problems of deliberate self-harm patients (Hawton 1988). The most obvious and widely used measure of outcome has been repetition, but measures of adjustment and quality of life are equally important.

Prospective studies have compared different treatments for deliberate self-harm patients, but no study has included a no-treatment group. In a study of patients who had harmed themselves at least twice, Chowdhury *et al.* (1973) compared the usual follow-up service with an augmented service which included intensive follow-up home visits for defaulters and provided a continuously accessible on-call team. At six-month follow-up, there were no differences between the two groups in rates of repeated deliberate self-harm. The experimental group showed a slightly greater improvement in psychological symptoms, and a significantly greater improvement in handling problems of finance, housing, and employment. Hawton *et al.* (1981) found no difference in outcome between patients treated at home and those treated in an out-patient clinic. In another study, Hawton *et al.* (1987) found no overall difference in outcome between patients treated at an out-patient clinic and those receiving routine care from their general practitioner. However, women with interpersonal difficulties improved more with clinic treatment. Gibbons *et al.* (1978) compared task-centred social case work with routine treatment; they found no difference in rates of repetition of deliberate self-harm, even though the groups receiving social work improved more in social adjustment.

Two recent studies have examined the use of psychological treatment for selected patients at particular risk of repetition. Salkovskis *et al.* (1990) found that brief problem-solving treatment, supplemented by between-session homework, had significant benefits for mood and appeared to delay repetition, although it did not affect the overall rate. Linehan *et al.* (1994) found that cognitive behavioural treatment appeared to be effective for subjects who had had at least two incidents of deliberate self-harm in the previous five years and who met DSMIII criteria for borderline personality disorder. There is also evidence that brief cognitive behavioural treatment may be an effective intervention for adolescent suicide attempters and their families (Rotheram-Borus 1994).

In summary, it seems that most additional treatments of the kinds described do not reduce the overall risk of repeated self-harm. It may still be useful to deal with social problems, since many of them would have required attention even if the patients had not harmed themselves. Recent findings and clinical experience suggest that more psychological methods may be helpful with selected subjects.

Primary prevention

If prevention after the first episode of self-harm is so difficult, can primary prevention be achieved? Three main strategies have been suggested: reducing the availability of means of self-harm, encouraging the work of agencies that try to help people with social and emotional problems, and improving health education. (These issues are discussed further in relation to the prevention of suicide on p. 424.)

Reducing the means

It has been suggested that psychotropic drugs should be prescribed more cautiously, especially for patients whose affective symptoms are a reaction to life problems. Drugs may not help these patients but may merely provide the opportunity to take an overdose when the problems increase. Two factors should be weighed against this argument: up to a third of people who take deliberate drug overdoses use drugs originally prescribed for a person other than themselves, and another quarter use drugs that can be bought without prescription, notably analgesics, some of which are particularly dangerous in overdosage.

This last consideration has led to suggestions that, if any non-prescribed drug is dangerous in overdose, it should be sold in strip or blister packets to prevent people taking large amounts on impulse, or a small quantity of emetic should be added. It has also been suggested that paracetamol, one of the most dangerous analgesics, should be available only on prescription or in combined preparations that are safe.

Encouraging helping agencies

This at first appears sensible but, on closer enquiry, may be unlikely to help. In one study it was found that about three-quarters of patients taking deliberate overdoses already knew about the Samaritans and many were also aware of the social services provided in their area (Bancroft *et al.* 1977). Their overdoses were taken impulsively without thought about ways of obtaining help.

Education

Education about the dangers of drug overdoses and discussions of common emotional problems might be provided for teenagers. However, in the absence of any evidence that such measures reduce deliberate self-harm, there is an understandable reluctance to introduce them in schools (Shaffer *et al.* 1988). There may be a role for more general education in problem-solving and in coping with difficult relationships.

Further reading

Blumenthal, S. K. and Kupfer, D. (ed.) (1990). *Suicide over the life cycle: risk factors, assessment and treatment of suicidal patients.* American Psychiatric Press, Washington, DC.

Hawton, K. and Catalan, J. (1987). *Attempted suicide: a practical guide to its nature and management* (2nd edn). Oxford University Press.

14

The abuse of alcohol and drugs

The phrases **substance use disorder** (DSMIV) or **disorders due to psychoactive drug use** (ICD10) are used to refer to conditions arising from the abuse of alcohol, psychoactive drugs, and other chemicals such as volatile solvents. In this chapter, problems related to alcohol will be discussed first under the general heading of **alcohol use disorders**; problems related to drugs and other chemicals will be discussed second under the general heading of **other substance use disorders**.

Classification of substance use disorders

The two classification systems, DSMIV and ICD10, use similar categories of substance use disorders but group them in different ways. Both schemes recognize the following disorders: intoxication, abuse (or harmful use), dependence, withdrawal states, psychotic disorders, and amnestic syndromes. These and some additional categories are shown in Table 14.1.

In both diagnostic systems the first step in classification is to specify the *substance or class of substance* that is involved (Table 14.2): this provides the primary diagnostic category. Although many drug users take more than one kind of drug, the diagnosis of the disorder is made on the basis of the most important substance used. Where this judgement is difficult or where use is chaotic and indiscriminate, the categories *polysubstance-related disorder* (DSMIV) or *disorder due to multiple drug use* (ICD10) may be employed. Then the relevant disorder listed in Table 14.1 is added to the substance abused. In this system any kind of disorder can, in principle, be attached to any drug, though in practice certain disorders do not develop with individual drugs. ICD10 also has a specific category *residual and late-onset psychotic disorder*

which describes physiological or psychological changes that occur when a drug is taken but then persist beyond the period during which a direct effect of the substance would reasonably be expected to be operating. Such categories might include hallucinogen-induced flashbacks or alcohol-related dementia.

Definitions in DSMIV and ICD10

Both DSMIV and ICD10 provide definitions of **intoxication**. In both systems, intoxication is seen as a transient syndrome due to recent substance ingestion that produces clinically significant psychological and physical impairment. These changes disappear when the substance is eliminated from the body. The nature of the psychological changes varies with the person as well as with the drug; for example some people intoxicated with alcohol become aggressive, but others become maudlin.

The term **abuse** in DSMIV and **harmful use** in ICD10 refer to maladaptive patterns of substance use that impair health in a broad sense (Table 14.3). The definition in ICD10 is likely to depend more on the experience and judgement of the clinician and therefore may be more inclusive (Peachey and Loh 1994). Some individuals show definite evidence of substance abuse but do not meet criteria for **substance dependence** (Table 14.4). However, if they do, the diagnosis of dependence should be made and not that of abuse or harmful use.

The term **dependence** refers to certain *physiological* and *psychological* phenomena induced by the repeated taking of a substance; the criteria for diagnosing dependence are similar in DSMIV and ICD10, and include a strong desire to take the substance, progressive neglect of alternative sources of satisfaction, the development of tolerance (see below), and a physical withdrawal state (see Table 14.4). *Tolerance* is a state in which, after repeated

Table 14.1 *Substance-related disorders*

DSMIV	ICD10
Intoxication	Intoxication
Abuse	Harmful use
Dependence	Dependence syndrome
Withdrawal	Withdrawal state
Withdrawal delirium	Withdrawal state with delirium
Psychotic disorders	Psychotic disorder
Dementia	
Amnestic disorder	Amnestic syndrome
Mood disorders	Residual and late-onset psychotic disorder
Anxiety disorders	Other mental and behavioural disorders
Sexual dysfunctions	
Sleep disorders	

Table 14.2 *Classes of substances*

DSMIV	ICD10*
Alcohol	Alcohol
Amphetamines	Other stimulants, including caffeine
Caffeine	
Cannabis	Cannabinoids
Cocaine	Cocaine
Hallucinogens	Hallucinogens
Inhalents	Volatile solvents
Nicotine	Tobacco
Opioids	Opioids
Phencyclidine	
Sedatives, hypnotics, or anxiolytics	Sedatives or hypnotics
Polysubstance	Multiple drug use
Other	

* The order of entries in the classification has been amended to show parallels with DSMIV

administration, a drug produces a decreased effect, or increasing doses are required to produce the same effect. A *withdrawal state* is a group of symptoms and signs occurring when a drug is reduced in amount or withdrawn, which last for a limited time. The nature of the withdrawal state is related to the class of substance used. For a review of the implications of the DSMIV and ICD10 classifications for the identification of substance use disorders, see Peachey and Loh (1994).

Table 14.3 *Criteria for substance abuse (DSMIV) and harmful use (ICD10)*

DSMIV	ICD10
(A) A maladaptive pattern of substance use leading to clinically significant impairment or distress, as manifested by one (or more) of the following occurring within a 12-month period	(A) A pattern of psychoactive substance use that is causing damage to health; the damage may be to physical or mental health
(1) Recurrent substance use resulting in a failure to fulfil major role obligations at work, school, or home	
(2) Recurrent substance abuse in situations that are physically hazardous	
(3) Recurrent substance-abuse-related legal problems	
(4) Continued substance abuse despite having persistent or recurrent social or interpersonal problems caused or exacerbated by the effects of the substance	
(B) Has never met the criteria for substance dependence for this class of substance	

ALCOHOL-RELATED DISORDERS

Terminology

In the past, the term *alcoholism* was generally used in medical writing. Although the word is still widely used in everyday language, it is unsatisfactory as a technical term because it has more than one meaning. It can be applied to habitual alcohol consumption that is deemed excessive in amount according to some arbitrary criterion. Alcoholism may also refer to damage, whether mental, physical, or social, resulting from such excessive consumption. In a more specialized sense, alcoholism may imply a specific disease entity that is supposed to require medical treatment. However, to speak of an alcoholic often has a pejorative meaning, suggesting behaviour that is morally bad.

For most purposes it is better to use three terms that relate to the classifications outlined above: excessive consumption of alcohol, alcohol abuse, and alcohol dependence. Excessive consumption of alcohol refers to a daily or weekly intake of alcohol exceeding a specified amount. *Alcohol abuse* refers to any mental, physical, or social harm resulting from excessive consumption. The term *alcohol dependence* can be used when the additional criteria for a dependence syndrome listed in Table 14.4 are met.

A fourth term which is often used is *problem drinking*, which is applied to those in whom drinking has caused an alcohol-related disorder or disability. Problem drinkers may or may not be dependent on alcohol. The term alcoholism, if it is used at all, should be regarded as a shorthand way of referring to some combination of these four conditions. However, since these specific terms have been introduced fairly recently, the term alcoholism

Table 14.4 *Criteria for dependence in DSMIV and ICD10*

DSMIV	ICD10
(A) Diagnosis of dependence should be made if three (or more) of the following have been experienced or exhibited at any time in the same 12-month period	**(A) Diagnosis of dependence should be made if three or more of the following have been experienced or exhibited at some time during the last year**
(1) Tolerance defined by either need for markedly increased amounts of substance to achieve intoxication or desired effect, or markedly diminished effect with continued use of the same amount of the substance	(1) A strong desire or sense of compulsion to take the substance
(2) Withdrawal as evidenced by either of the following: the characteristic withdrawal syndrome for the substance or the same (or closely related) substance is taken to relieve or avoid withdrawal symptoms	(2) Difficulties in controlling substance-taking behaviour in terms of its onset, termination, or levels of use
(3) The substance is often taken in larger amounts over a longer period of time than was intended	(3) Physiological withdrawal state when substance use has ceased or been reduced, as evidenced by either of the following: the characteristic withdrawal syndrome for the substance or use of the same (or closely related) substance with the intention of relieving or avoiding withdrawal symptoms
(4) Persistent desire or repeated unsuccessful efforts to cut down or control substance use	(4) Evidence of tolerance, such that increased doses of the psychoactive substance are required in order to achieve effects originally produced by lower doses
(5) A great deal of time is spent in activities necessary to obtain the substance, use the substance, or recover from its effects	(5) Progressive neglect of alternative pleasures or interests because of psychoactive substance use and increased amount of time necessary to obtain or take the substance or to recover from its effects
(6) Important social, occupational, or recreational activities given up or reduced because of substance use	(6) Persisting with substance use despite clear evidence of overtly harmful consequences (physical or mental)
(7) Continued substance use despite knowledge of having had a persistent or recurrent physical or psychological problem that was likely to have been caused or exacerbated by the substance	

Source: Peachey and Loh 1994.

has to be used in this chapter when referring to much of the literature. Before explaining these specific terms further, it is appropriate to examine the moral and medical models of alcohol abuse.

The moral and medical models

According to the **moral model**, if someone drinks too much, he does so of his own free will, and if his drinking causes harm to himself or his family, his actions are morally bad. The corollary of this attitude is that public drunkenness should be punished. In many countries this is the official practice; public drunks are fined and if they cannot pay the fine, they go to prison. Many people now believe that this approach is too harsh and unsympathetic. Whatever the humanitarian arguments, there is little practical justification for punishment, since there is little evidence that it influences the behaviour of excessive drinkers.

According to the **medical model**, a person who abuses alcohol is sick rather than wicked. Although it had been proposed earlier, this idea was not strongly advocated until 1960 when Jellinek published an influential book *The disease concept of alcoholism*. The disease concept embodies three basic ideas. The first is that some people have a specific vulnerability to alcohol abuse. The second is that excessive drinking progresses through well-defined stages, at one of which the person can no longer control his drinking. The third is that excessive drinking may lead to physical and mental disease of several kinds. While the latter is true, it seems illogical to say that the abuse of alcohol is itself a disease because it can lead to recognized diseases.

One of the main consequences of the disease model is that attitudes towards excessive drinking become more humane. Instead of blame and punishment, medical treatment is provided. The disease model also has certain disadvantages. By implying that only certain people are at risk, it diverts attention from two important facts. First, *anyone* who drinks a great deal for a long time may become dependent on alcohol. Second, the best way to curtail the abuse of alcohol may be to limit consumption in the whole population, and not just among a predisposed minority.

Excessive alcohol consumption

In many societies the use of alcohol is sanctioned and even encouraged by sophisticated marketing techniques. Therefore the level of drinking at which an individual is considered to demonstrate excessive alcohol consumption is a somewhat arbitrary concept, usually defined in terms of the level of use associated with significant risk of alcohol-related health and social problems.

Table 14.5 *Alcohol consumption in men and women and risk of social and health problems*

Alcohol intake (units/week)	Risk of problems
Men 0–21 Women 0–14	Low
Men 22–50 Women 15–35	Increasing, particularly in smokers
Men >50 Women >35	High, particularly in smokers

Source: Austoker 1994

Table 14.6 *Alcohol content of some beverages*

Beverage	Approximate alcohol content (%)	Grams alcohol per conventional measure	Units of alcohol per conventional measure (approximate)
Beer and cider			
Ordinary beer	3	16 per pint	2 per pint
		12 per can	1.5 per can
Strong beer	5.5	32 per pint	4 per pint
		24 per can	3 per can
Extra-strong beer	7	40 per pint	5 per pint
		32 per can	4 per can
Cider	4	24 per pint	3 per pint
Strong cider	6	32 per pint	4 per pint
Wine			
Table wine	8–10	8 per glass	1 per glass
		56 per bottle	7 per bottle
Fortified wines (sherry, port, vermouth)	13–16	8 per measure	1 per measure
		120 per bottle	15 per bottle
Spirits (whisky, gin, brandy, vodka)	32	8–12 per single measure*	1–1.5 per measure*
		240 per bottle	30 per bottle

*Somewhat larger measures are used in Scotland and Northern Ireland (12 grams).
Adapted from the Royal College of Physicians (1987, p. 6).

It is usually expressed in units of alcohol consumed per week (Austoker 1994) (Table 14.5). While there is reason to suppose that anyone may become dependent on alcohol if he or she drinks a sufficiently large amount for long enough, because of substantial individual variation no exact threshold can be specified. However, women are more sensitive than men to the harm-inducing effects of alcohol.

If the concept of excessive alcohol consumption is to be understood and accepted, it is necessary to explain the units in which it is assessed. In everyday life, this is done by referring to conventional measures such as pints of beer or glasses of wine. These measures have the advantage of being widely understood, but they are imprecise because both beers and wines vary in strength (Table 14.6). Alternatively, consumption can be measured as the amount of alcohol (expressed in grams). This measure is precise and useful for scientific work, but is difficult for many people to relate to everyday measures. For this reason, the concept of a unit of alcohol has been introduced for use in health education. A unit can be related to everyday measures for it corresponds to half a pint of beer, one glass of table wine, one conventional glass of sherry or port, and one single bar measure of spirits. It can also be related to amounts of alcohol (see Table 14.6); thus on this measure a can of beer (450 ml) contains nearly 1.5 units, a bottle of table wine contains about 7 units, a bottle of spirits about 30 units, and 1 unit is about 8 g of alcohol.

Epidemiological aspects of excessive drinking and alcohol abuse

Epidemiological methods can be applied to the following questions concerning excessive drinking and alcohol abuse.

1. What is the annual per capita consumption of alcohol for a nation as a whole, and how does this vary over the years and between nations?
2. What is the pattern of alcohol use of different groups of people within a defined population?
3. How many people in a defined population abuse alcohol?
4. How does alcohol abuse vary with such characteristics as sex, age, occupation, social class, and marital status?

Unfortunately, we lack reliable answers to these questions, partly because different investigators have used different methods of defining and identifying alcohol abuse and 'alcoholism', and partly because excessive drinkers tend to be evasive about the amounts that they drink and the symptoms that they experience.

Consumption of alcohol in different countries

In the United Kingdom, the average annual consumption of alcohol per adult (calculated as absolute ethanol consumption) has shown a *small increase* over the last 10 years from 7.1 litres in 1980 to 7.6 litres in 1989. Several other Western European countries and the United States have registered small decreases in consumption since 1984, though in most cases the average level remains higher than that of the United Kingdom (Table 14.7). The traditional low level of consumption in Japan has shown a more pronounced increase from 5.4 litres in 1984 to 6.7 litres in 1989 (Ritson *et al.* 1993).

Current changes in the United Kingdom can be usefully considered in a historical perspective. For example, in Great Britain between 1860 and 1900 the consumption of alcohol was about 10 litres of

Table 14.7   *Annual consumption of pure alcohol (litres/adult)*

	1980	1989
United Kingdom	7.1	7.6
France	14.8	13.4
Austria	11.0	10.3
Italy	13.0	9.5
Portugal	11.0	10.4
United States	8.7	7.5
Australia	9.8	8.5
Japan	5.4	6.7

Source: Ritson *et al.* 1993.

absolute alcohol per head of population over 15 years old. Consumption then fell until the early 1930s (reaching about 4 litres per person over 15 years per annum). Consumption then increased slowly until the 1950s when it began to rise more rapidly.

These changes have been accompanied by alterations in the kinds of alcoholic beverages consumed. In Britain in 1900 beer and spirits accounted for most of the alcohol drunk; in 1980 the consumption of wine had risen about four times and accounted for almost as much of the consumption of alcohol as did spirits, though most alcohol was still consumed as beer.

Drinking habits in different groups

Surveys of drinking behaviour generally depend on self-reports, a method that is open to obvious errors. Enquiries of this kind have been conducted in several countries including the United Kingdom (Cochrane and Bal 1990; McKeigue and Karmi 1993) and the United States (Hilton 1988; Tracy *et al.* 1992).

Such studies show that the highest consumption of alcohol is generally amongst *young men* who are *unmarried, separated,* or *divorced*. However, over the last 15 years drinking by *women* has increased (Ritson *et al.* 1993). In 1992, men in the United Kingdom drank on average 15.9 units of alcohol a week (about 8 pints of beer), while women drank about a third of this amount. The alcohol consumption of 27 per cent of men and 11 per cent of women exceeded recommended limits; therefore these subjects could be considered at risk of developing an alcohol-related disorder, including alcohol dependence. Six per cent of men and 2 per cent of women drank more than 50 units a week (Austoker 1994).

The prevalence of alcohol abuse

This can be estimated in three ways: from hospital admission rates, from deaths from alcoholic cirrhosis, and by surveys in the general population.

Hospital admission rates

These give an inadequate measure of prevalence because a large proportion of excessive drinkers do not enter hospital. In the United Kingdom admissions for problems related to alcohol abuse account for 10 per cent of all *psychiatric admissions*. In France, Germany, and Eire the figure is almost 30 per cent. In the United Kingdom there has been a 30-fold increase in psychiatric hospital admission rates for alcohol-related disorders in the past 30 years, but this may reflect changes in the provision of services rather than in prevalence. More recent trends to treat patients with alcohol problems outside hospital are likely to result in a decline in these figures. Alcohol abuse also figures prominently in *admissions to general hospitals*, where screening questionnaires identify alcohol abuse in about 20–30 per cent of male admissions and 5–10 per cent of female admissions (Chick 1987).

Deaths from alcoholic cirrhosis

About 10–20 per cent of people who drink alcohol excessively develop *cirrhosis of the liver*, and there are correlations in a population between rates of liver cirrhosis and mean alcohol consumption. Therefore deaths from cirrhosis can be used as a means of estimating rates of alcohol abuse (Ritson *et al.* 1993). Rates of mortality from liver cirrhosis are showing a decline in a number of developed countries. For example, in Ontario, Mann and Smart (1990) found a 25 per cent reduction in mortality rates from cirrhosis between 1974 and 1984, during which time the mean alcohol consumption declined by about 3 per cent. The decrease in mortality from liver cirrhosis was related to levels of psychiatric intervention, suggesting that improved treatment may have played a part in delaying the development of cirrhosis in heavy drinkers.

General population surveys

One method of ascertaining the rate of problem drinking in a population is by seeking information from general practitioners, social workers, probation officers, health visitors, and other agents who are likely to come in contact with heavy drinkers. Another approach is the *community survey* in which samples of people are asked about the amount they drink and whether they experience symptoms. Two recent epidemiological investigations in the United States (the Epidemiological Catchment Area Programme and the National Comorbidity Survey) suggested a combined one-year prevalence rate for alcohol abuse and dependence of 7–10 per cent with a corresponding lifetime risk of about 14–20 per cent (Regier *et al.* 1994; Kessler *et al.* 1994).

In a cross-national study of ten different cultural regions, lifetime prevalence rates for alcohol abuse and dependence varied from about 0.5 per cent in Shanghai to 22 per cent in Korea (Helzer and Canino 1992). While there are undoubtedly wide variations between countries in the real prevalence of alcohol-related disorders, some of the apparent differences may stem from contrasting cultural perspectives on what constitutes alcohol abuse and the extent to which people chose to reveal their drinking habits to investigators.

Alcohol abuse and population characteristics

Sex

Rates of alcohol abuse and dependence are consistently higher in men than in women but the ratio of affected men to women varies markedly across cultures. In Western countries, about three times as many men as women suffer from alcohol abuse and dependence, but in Asian and Hispanic cultures over ten times as many men are affected (Helzer and Canino 1992). In the National Comorbidity Survey in the United States the one-year prevalence of alcohol abuse and dependence in men (14.1 per cent) was almost three times that in women (5.3 per cent) (Kessler *et al.* 1994).

Age

We have seen that the heaviest drinkers are *men* in their *late teens or early twenties*. In most cultures the prevalence of alcohol abuse and dependence is lower in those aged over 45 years (Helzer and Canino 1992). Recent studies suggest that the *homeless* young are at considerably greater risk of abusing alcohol and other substances (Medina-Mora 1992).

Ethnicity and culture

The followers of certain religions which proscribe alcohol, for example Islam, Hinduism, and the Baptist Church, are less likely than the general population to abuse alcohol. It is also worth noting that Afro-Caribbeans in the United Kingdom and blacks in the United States are less likely to drink excessively than the white population and therefore have a lower rate of alcohol-related disorders (McKeigue and Karmi 1993; Kessler *et al.* 1994).

In some instances the low consumption of alcohol in a particular ethnic group may be due to a biologically determined lack of tolerance to alcohol. For example Asians and Orientals with a particular variant of the isoenzyme of *aldehyde dehydrogenase* experience flushing, nausea, and tachycardia due to accumulation of acetaldehyde when they drink alcohol. Such subjects are likely to be at reduced risk of excess drinking and the consequent development of alcohol-related disorders (Wall *et al.* 1992). Thus, although the aldehyde dehydrogenase variant that causes the flushing reaction was present in 35 per cent of

the general Japanese population, it was found in only 7 per cent of Japanese patients with alcoholic liver disease (Shibuya and Yoshida 1988).

Occupation

The risk of problem drinking is much increased among several occupational groups: chefs, kitchen porters, barmen, and brewery workers, who have easy access to alcohol, executives and salesmen who entertain on expense accounts, actors and entertainers, seamen, and journalists and printers. Doctors are another important group with an increased risk of problem drinking, and they are often particularly difficult to help (Chick 1992).

The syndromes of alcohol dependence and alcohol withdrawal

Patients are described as *alcohol dependent* when they meet the criteria for substance dependence described in Table 14.3. A substantial minority of subjects who meet criteria for alcohol dependence do not experience any withdrawal phenomena when their intake of alcohol diminishes or stops. However, about 5 per cent of dependent subjects may experience severe withdrawal symptomatology including *delirium* and *grand mal seizures*.

Course of alcohol dependence

Schuckit *et al.* (1993) reviewed the course of over 600 men with alcohol dependence who received in-patient treatment at a single facility in the United States between 1985 and 1991. These subjects showed a general pattern of escalation of heavy drinking in their late twenties followed by evidence of serious difficulties in work and social life by their early thirties. In their middle to late thirties, following the perception that they could not control their drinking, subjects experienced increasing social and work problems together with a significant deterioration in physical health.

The alcohol withdrawal syndrome

Withdrawal symptoms occur in people who have been drinking heavily for years and who maintain a high intake of alcohol for weeks at a time. The symptoms follow a drop in blood concentration. They characteristically appear on waking, after the fall in concentration during sleep. Since they can stave off withdrawal symptoms only by further drinking, dependent drinkers often take a drink on waking. In most cultures, *early-morning drinking* is diagnostic of dependency.

With increasing need to stave off withdrawal symptoms during the day, the drinker typically becomes *secretive* about the amount consumed, hides bottles, or carries them in a pocket. Rough cider and cheap wines may be drunk regularly to obtain the most alcohol for the least money.

The earliest and commonest feature of alcohol withdrawal is acute *tremulousness* affecting the hands, legs, and trunk ('the shakes'). The sufferer may be unable to sit still, hold a cup steady, or do up buttons. He is also agitated and easily startled, and often dreads facing people or crossing the road. *Nausea*, *retching*, and *sweating* are frequent. *Insomnia* is also common. If alcohol is taken, these symptoms may be relieved quickly; if not, they may last for several days. As withdrawal progresses, *misperceptions* and *hallucinations* may occur, usually only briefly. Objects appear distorted in shape, or shadows seem to move; disorganized voices, shouting, or snatches of music may be heard. Later there may be *epileptic seizures*, and finally after about 48 hours *delirium tremens* may develop (Victor and Adams 1953) (see below).

Other alcohol-related disorders

The different types of damage—physical, psychological, and social—that can result from excessive drinking are described in this section. A person who suffers from these disabilities may or may not be suffering from alcohol dependence.

Physical damage

Excessive consumption of alcohol may lead to physical damage in several ways. First it can have a *direct toxic* effect on certain tissues, notably the brain and liver. Second it is often accompanied by *poor diet* which may lead to deficiency of protein and B vitamins. Third it increases the *risk of accidents*, particularly head injury. Fourth it is accompanied by *general neglect* which can lead to increased susceptibility to infection.

Physical complications of excessive drinking occur in several systems of the body. **Alimentary disorders** are common, notably liver damage, gastritis, peptic ulcer, oesophageal varices, and acute and chronic pancreatitis. Damage to the liver, including fatty infiltration, hepatitis, cirrhosis, and hepatoma, is particularly important.

For a person who is dependent on alcohol, the risk of dying from liver cirrhosis is almost ten times greater than the average. However, only about 10–20 per cent of alcohol-dependent people develop cirrhosis. Recent work suggests that vulnerability to alcohol-induced liver disease may be influenced by genetic differences in the enzymes that metabolize alcohol. For example the incidence of liver disease was higher in those excess alcohol users with a particular polymorphism for alcohol dehydrogenase (Sherman *et al.* 1993). There is also a suggestion that patterns of drinking may influence the risk of certain forms of liver disease.

Excess alcohol consumption also damages the **nervous system**. Neuropsychiatric complications are described later; other neurological conditions include *peripheral neuropathy*, *epilepsy*, and *cerebellar degeneration*. The last of these is characterized by unsteadiness of stance and gait, with less effect on arm movements or speech. Rare complications are *optic atrophy*, *central pontine myelinolysis*, and *Marchiafava–Bignami syndrome*. The last of these results from widespread demyelination of the corpus callosum, optic tracts, and cerebellar peduncles. Its main features are dysarthria, ataxia, epilepsy, and marked impairment of consciousness; in the more prolonged forms dementia and limb paralysis occur (see Lishman (1987) for a review). *Head injury* is also common in alcohol-dependent people.

Excess alcohol use is associated with **hypertension** and increased risk of stroke. Paradoxically, men who drink moderate amounts of alcohol (up to about 25 units a week) appear less likely than non-drinkers to die from coronary artery disease (Ritson and Patience 1994). Alcohol consumption has also been linked to the development of certain *cancers*, notably of the mouth, pharynx, oesophagus, and liver.

Other physical complications of excessive drinking are too numerous to detail here. Examples include anaemia, myopathy, episodic hypoglycaemia, haemochromatosis, cardio-myopathy, vitamin deficiencies, and tuberculosis. They are described in textbooks of medicine, for example the *Oxford textbook of medicine* (Weatherall *et al.* 1995).

Not surprisingly, the **mortality rate** is increased in excessive drinkers. Follow-up investigations have studied mainly middle-aged men in whom overall mortality is at least twice the expected rate (Shaper 1990; Marshall *et al.* 1994). In a study of 99 married men who had attended a specialist alcohol problems clinic in London, Marshall *et al.* (1994) found that 44 subjects had died during 20-year follow-up. The mortality rate in the moderately alcohol-dependent subjects was almost three times the expected value, whilst in those classified as severely dependent it was increased almost 4.5 times. The major causes of death were *diseases of the circulatory system* (13 deaths) and *cancer* (12 deaths). Six of the remaining deaths were caused by *injury or poisoning*, of which at least three were due to *suicide*.

Even allowing for the fact that heavy drinkers also tend to be heavy smokers, alcohol itself is almost certainly responsible for a substantial part of this increased mortality. In the United Kingdom it is estimated that excess alcohol consumption leads to about *28 000 deaths a year*, mainly from cardiovascular disorders, cirrhosis, accidents, and cancer (Austoker 1994).

Damage to the fetus

There is evidence that a **fetal alcohol syndrome** occurs in some children born to mothers who drink excessively. In France, Lemoine *et al.* (1968) described a syndrome of facial abnormality, small stature, low birth-weight, low intelligence, and psychological over-activity. In Seattle, USA, a series of reports confirmed this general clinical

picture (Jones and Smith 1973; Hanson *et al.* 1976). It is a limitation of this research that both studies were retrospective.

In a large prospective study in France (Kaminski *et al.* 1976), women who drank about 400 ml of wine (or equivalent of other drinks) daily were not found to have babies with higher rates of congenital malformation or neonatal mortality. However, the women had more than the expected number of *stillbirths*, and their babies' *birth-weights* were lower. Compared with other mothers, those who drank excessively were older, more often unmarried, of lower social status, and of greater parity, and they smoked more. They also had more bleeding in early pregnancy. When allowance was made for these factors, the authors still found that alcohol independently affected birth-weight, placental weight, and stillbirths.

It seems reasonable to conclude that, among the offspring of some mothers who drink excessively, there is a syndrome of the kind mentioned above, but that it occurs infrequently and only when the mother has been drinking very heavily indeed during pregnancy. When mothers drink excessively, though less heavily than this, their infants appear to have lower birth-weights and smaller stature than others, but the differences are not great.

Follow-up studies of infants of mothers who abuse alcohol have shown that development may be impaired (Jacobsen *et al.* 1993). A problem in assessing the role of alcohol in these effects is that these children may be subject to other factors that could impair development such as malnutrition and social deprivation (Ritson and Patience 1994). In a ten-year follow-up of children with fetal alcohol syndrome, Spohr *et al.*(1993) found that, while the characteristic craniofacial malformations decreased with time, many subjects had evidence of persisting mental retardation.

Psychiatric disorders

Alcohol-related psychiatric disabilities fall into four groups: intoxication phenomena, withdrawal phenomena, chronic or nutritional disorders, and associated psychiatric disorders.

Intoxication phenomena

The severity of the symptoms of alcohol intoxication correlate approximately with the blood concentration. As noted above, there is much individual variation in the psychological effects of alcohol, but certain reactions such as *lability of mood* and *belligerence* are more likely to cause social difficulties. At high doses, alcohol intoxication can result in serious adverse effects such as *falls, respiratory depression, inhalation of vomit, and hypothermia.*

The molecular mechanisms that underlie the acute effects of alcohol are not clear. An influential view has been that alcohol interacts with neuronal membranes to increase their fluidity, an action also ascribed to certain anaesthetic agents. Other studies have indicated that alcohol may enhance the actions of brain gamma-aminobutyric acid (*GABA*) pathways and perhaps decrease activity of *glutamatergic* neurotransmission.

The term **idiosyncratic alcohol intoxication** has been applied to marked maladaptive changes in behaviour, such as aggression, occurring within minutes of taking an amount of alcohol insufficient to induce intoxication in most people (with the behaviour being uncharacteristic of the person). In the past, these sudden changes in behaviour were called *pathological drunkenness*, or *manie à potu*, and the descriptions emphasized the explosive nature of the outbursts of aggression. There is doubt whether behaviour of this kind really is induced by small amounts of alcohol. Maletzky (1976) gave intravenous infusions of alcohol to 23 men who had a history of pathological drunkenness. Fifteen developed aggressive behaviour but they did so only when the blood alcohol level was substantially raised. The term idiosyncratic alcohol intoxication does not appear in DSMIV or ICD10.

Memory blackouts or short-term amnesia are frequently reported after heavy drinking. At first the events of the night before are forgotten, even though consciousness was maintained at the time. Such memory losses can occur after a single episode of heavy drinking in people who are not dependent on alcohol; if they recur regularly, they indicate habitual heavy drinking. With sustained excessive drinking, memory losses may become more severe, affecting parts of the daytime or even whole days.

Withdrawal phenomena

The *general withdrawal syndrome* has been described earlier under the heading of alcohol dependence. Here we are concerned with the more serious psychiatric syndrome of delirium tremens.

Delirium tremens This occurs in people whose history of excessive drinking extends over several years. There is a dramatic and rapidly changing picture of *disordered* mental activity, with clouding of consciousness, disorientation in time and place, and impairment of recent memory. Perceptual disturbances include misinterpretations of sensory stimuli and vivid hallucinations which are usually visual, but sometimes occur in other modalities. There is severe agitation, with restlessness, shouting, and evident fear. *Insomnia* is prolonged. The hands are grossly *tremulous* and sometimes pick up imaginary objects, and *truncal ataxia* occurs. *Autonomic disturbances* include sweating, fever, tachycardia, raised blood pressure, and dilatation of pupils. *Dehydration* and *electrolyte disturbance* are characteristic. Blood testing shows *leucocytosis* and *impaired liver function*.

The condition lasts three or four days, with the symptoms characteristically being worse at night. It often ends with deep prolonged sleep from which the patient awakens with no symptoms and little or no memory of the period of delirium.

Toxic or nutritional conditions

These include **Korsakov's psychosis** and **Wernicke's encephalopathy**, which are described in Chapter 11, and **alcoholic dementia**, which is described next.

Alcoholic dementia In the past there has been disagreement as to whether excessive alcohol intake can cause dementia. This doubt may have arisen because patients with general intellectual defects have been wrongly diagnosed as having Korsakov's psychosis. However, it is now generally agreed that chronic alcohol abuse can cause dementia.

Recently, attention has shifted to the related question of whether chronic alcohol abuse can cause *brain atrophy*. This question was studied by air encephalography and, more recently, by structural imaging techniques. Both CT scanning and magnetic resonance imaging (MRI) have shown that excess alcohol consumption is associated with *enlarged lateral ventricles* and *subarachnoid spaces*. Furthermore, MRI scans have shown focal deficits with *loss of grey matter* in both *cortical* and *subcortical areas*. Functional brain imaging *reveals decreased blood flow and glucose metabolism in cortical regions*; however, these changes may be secondary to loss of cerebral tissue. Many of the changes noted above occur in patients without obvious neurological disturbance, though psychological testing usually reveals *deficits in cognitive function*. The changes in brain structure and cognitive impairment seen in excessive alcohol users remit to some extent with cessation of alcohol use; however, many abnormalities can still be detected after long periods of abstinence. (For a review of the effect of alcohol on structural and functional brain imaging see Besson (1993).)

Neuropathological studies can provide more direct evidence of cerebral atrophy. Few have been reported. In one study, comparing 25 alcoholics with 44 controls, measurements of the 'pericerebral space' (which represents the space between brain and skull) were greater in the alcoholics. This was true of those without pathological signs of Wernicke's encephalopathy as well as those with such signs (Harper and Kril 1985). Atrophy of the vermis of the cerebellum has been found at autopsy in about a third of patients with chronic alcoholism (Harper and Kril 1985). The corresponding clinical evidence of cerebellar dysfunction was referred to on p. 447.

Studies of this kind suggest that alcoholic dementia is more common than was previously supposed, and it should be searched for carefully in every problem drinker. Older patients appear to be more at risk than younger ones with a similar length of heavy drinking, and those who have been drinking without respite seem to be more at risk than people who have periods in which they reduce their drinking. Women appear to be more vulnerable than men to alcohol-induced cognitive impairment (Mann *et al.* 1992; Ritson *et al.* 1993).

Associated psychiatric disorders

Personality deterioration As the patient becomes more and more concerned with the need to obtain alcohol, there is increasing self-centredness, a lack of consideration for others, and a decline in standards of conduct. Responsibilities at home and work are evaded, and behaviour may become dishonest and deceitful.

Mood disorder The relationship between alcohol consumption and mood is complex. On the one hand some depressed patients drink excessively in an attempt to improve their mood; on the other hand excess drinking may induce persistent depression or anxiety. In hospital populations *major depression* frequently coexists with alcohol abuse, but in community samples the relationship is weaker, particularly for men (Berglund and Nordstrom 1989). Davidson and Ritson (1993) concluded that the most common reason for depressive symptomatology in alcohol-dependent subjects was *adjustment disorder* or *alcohol-induced mood disorder*.

Suicidal behaviour Suicide rates amongst alcoholics are higher than among non-alcoholics of the same age. Kessel and Grossman (1965) found that 8 per cent of alcoholics admitted for treatment killed themselves within a few years of discharge. Reports from a number of countries suggest that 6–20 per cent of alcoholics end their lives by suicide (Ritson 1977). In a study of 50 alcohol abusers who had committed suicide, Murphy *et al.* (1992) identified a number of *risk factors for suicidal behaviour*, including continued drinking, co-morbid major depression, serious medical illness, unemployment, and poor social support. Suicide among alcoholics is discussed further on p. 418. Here it is worth noting that suicide in young men is associated with a high rate of substance abuse, including alcohol abuse (Marttunen *et al.* 1991).

Impaired psychosexual function *Erectile dysfunction* and *delayed ejaculation* are common. These difficulties may be worsened when drinking leads to marital estrangement, or if the wife develops a revulsion for intercourse with an inebriated partner.

Pathological jealousy Excessive drinkers may develop the delusion that the partner is being unfaithful. This syndrome of pathological jealousy is described on p. 300. Although it is a striking complication, delusional jealousy is less common than a non-delusional suspicious attitude to the spouse.

Alcoholic hallucinosis This is characterized by auditory hallucinations, usually voices uttering insults or threats, occurring in clear consciousness. The patient is usually distressed by these experiences, and appears anxious and restless.

There has been considerable controversy about the aetiology of the condition. Some follow Kraepelin and Bonhöffer in regarding it as organically determined; others follow Bleuler in supposing that it is related to schizophrenia.

Benedetti (1952) made a retrospective survey of 113 cases of alcoholic hallucinosis and divided them into 90 cases of less than six months' duration (acute cases) and the remaining chronic group. Among the former he found no evidence of a link with schizophrenia as judged by family history. However, he did find evidence of an organic cause, in that about half had experienced memory disorders. Despite this, the condition cleared up without residual defect. Among the cases that had lasted six months, nearly all went on for much longer, despite abstinence. Half developed the typical picture of schizophrenia and half developed amnesic syndromes or dementia. In their family histories, these chronic patients were intermediate between the acute cases and typical schizophrenic patients. Of course, this study cannot tell us whether these patients would have developed schizophrenia if they had never taken alcohol.

Some authors have held that alcoholic hallucinosis is not essentially different from the hallucinations occurring in excessive drinkers in the 24–48 hours after alcohol withdrawal (Knott and Beard 1971). It is true that auditory hallucinations may occur in simple withdrawal states (Hershon 1977) and may accompany the visual hallucinations of delirium tremens (Gross *et al.* 1971). However, they are fleeting and disorganized, in contrast with the persistent organized voices experienced in alcoholic hallucinosis.

Cutting (1978) has concluded that there is a small group of patients who have a true alcoholic hallucinosis, but that many of those who receive this diagnosis have depressive symptoms or first-rank symptoms of schizophrenia. Clinical experience also points to this conclusion. In both DSMIV and ICD10, alcoholic hallucinosis is subsumed under the heading of *substance-induced psychotic disorder*.

Social damage

Excessive drinking is liable to cause profound social disruption particularly in the family. *Marital and family tension* is virtually inevitable. The divorce rate amongst heavy drinkers is high, and the wives of such men are likely to become anxious, depressed, and socially isolated; the husbands of 'battered' wives frequently drink excessively, and some women admitted to hospital because of self-poisoning blame their husband's drinking. The home atmosphere is often detrimental to the children because of quarrelling and violence, and a drunken parent provides a poor role model. Children of heavy drinkers are at risk of developing emotional or behaviour disorders, and of performing badly at school.

At **work,** the heavy drinker often progresses through declining efficiency, lower-grade jobs, and repeated dismissals to lasting unemployment. There is also a strong association between **road accidents** and alcohol abuse. In the United States in 1990, 44 529 people were killed in traffic accidents, with alcohol being involved in 41 per cent of these fatalities (Zobeck *et al.* 1994). Similar findings have been obtained in the United Kingdom and Sweden (Naranjo and Bremner 1993). In one series a third of drivers arrested for driving under the influence of alcohol had raised gamma-glutamyl-transpeptidase (GGT) activity, suggesting chronic alcohol abuse

(Dunbar *et al.* 1985). This is consistent with the finding of excess mortality rates in drivers charged with drunk driving on more than one occasion. The causes of the increased mortality were very similar to those found in alcohol abusers (Mann *et al.* 1993).

Excessive drinking is also associated with **crime**, mainly petty offences such as larceny, but also with fraud, sexual offences, and crimes of violence including murder. Studies of recidivist prisoners in England and Wales have shown that many of them had serious drinking problems before imprisonment. It is not easy to know how far alcohol causes the criminal behaviour and how far it is just part of the life-style of the criminal. In addition, there is a link between certain forms of alcohol abuse and antisocial personality disorder (see below and p. 113).

The causes of excessive drinking and alcohol abuse

Despite much research, surprisingly little is known about the cause of excessive drinking and alcohol abuse. At one time it was believed that certain people were particularly predisposed, either through personality or an innate biochemical anomaly. Nowadays this simple notion of specific predisposition is no longer held. Instead, alcohol abuse is thought to result from a variety of interacting factors which can be divided into individual factors and those in society.

Individual factors

Genetic factors

Alcohol use Twin studies provide an opportunity to investigate the role of genetic and familial factors in patterns of alcohol use. A study employing this approach has suggested that, while genetic mechanisms play a part in determining levels of alcohol use, perhaps accounting for about one-third of the variance in alcohol consumption, most of the differences between subjects are due to *non-genetic familial factors* (Clifford *et al.* 1984).

Alcohol abuse and dependence Most genetic studies of alcoholism have investigated subjects with evidence of physiological alcohol dependence. If less severe diagnostic criteria are involved, for example fairly broadly defined alcohol abuse, the relative genetic contribution may appear somewhat less (Pickens *et al.* 1991; Kendler *et al.* 1994*b*).

It is well established that alcohol dependence aggregates in families. If this is partly the result of genetic factors (rather than social influences in the family), rates of dependence should be higher in monozygotic (MZ) than dizygotic (DZ) twins. While the results of MZ–DZ comparisons have been conflicting (Murray and McGuffin 1993), two recent large investigations have suggested a higher MZ concordance for alcohol dependence in male and female twins (Pickens *et al.* 1991; Kendler *et al.* 1994*b*). These studies have suggested that about 25–50 per cent of the liability to develop alcohol dependence may result from genetic factors.

Support for a genetic explanation also comes from investigations of adoptees. In an adoption study in Denmark, alcohol abuse and dependence was nearly four times more likely in the adopted-away sons of alcohol-dependent biological parents than in the adopted-away sons of non-alcohol-dependent biological parents (Goodwin *et al.* 1973). Similar findings have been reported from America (Cadoret and Gath 1978) and Sweden (Bohman 1978). Such studies suggest a genetic mechanism but do not indicate its nature.

Further analysis of the Swedish adoption data led to the suggestion that there were two separate kinds of alcohol dependence, one with a major genetic component where severe alcohol dependence passed from fathers to sons, and another with a smaller genetic component where less severe drinking problems affected both men and women in the family (Bohman *et al.* 1981; Cloninger *et al.* 1981). While this classification has not been entirely borne out, recent studies have confirmed that a useful distinction can be made between *familial* and *non-familial* cases of alcohol dependence with 'familial' alcohol dependence more likely to have a younger age at onset, be more severe in nature, and be associated with an antisocial personality (Marshall and Murray 1991).

If a genetic component to aetiology were confirmed, it would still be necessary to discover the mechanism. The latter might be biochemical, involving the metabolism of alcohol or its central effects, or psychological, involving personality. In addition, it is important to note that a *predisposition* to develop alcohol abuse and dependence will be expressed only if a person consumes excessive amounts of alcohol. Here non-genetic familial factors are likely to play a major role (Marshall and Murray 1991).

Biological factors

Several possible biochemical factors have been suggested to predispose to alcohol abuse and dependence, including abnormalities in alcohol dehydrogenase or in neurotransmitter mechanisms. As mentioned above, a significant proportion of Oriental subjects, who possess a particular allele of the isoenzyme *aldehyde dehydrogenase*, develop unpleasant reactions to alcohol and therefore are much less likely to abuse it.

The *sons of alcoholics* are at increased risk of developing alcohol dependence, and a number of studies have attempted to find biological abnormalities that may antedate and predict the development of alcohol dependence in these subjects. A variety of impairments have been described, including abnormal performance on *cognitive tasks* and on the *P300 visual evoked response*, which is a measure of visual information processing (Berman and Noble 1993). There is also reasonably consistent evidence that sons of alcoholics are less sensitive to the acute intoxicating effects of alcohol. A 10-year prospective follow-up of such subjects found that they had an increased risk of alcohol dependence. It is not clear which of these biological abnormalities in at-risk subjects is attributable to genetic factors, nor how the impairments that have been identified may relate to the development of alcohol dependence. Presumably, if subjects experience less subjective response to alcohol, they may tend to drink more, thus putting themselves at risk of developing alcohol dependence. While plausible, this hypothesis is not supported by direct evidence.

Experimental studies in animals have suggested that the reinforcing effects of drugs of abuse,

including alcohol, are mediated by increased *dopamine release* in subcortical regions, principally the nucleus accumbens (Koob 1992). There have been a number of reports that a particular allele of the *dopamine D_2* receptor gene is found with increased frequency in alcohol abusers. While this finding is of interest, several investigations failed to confirm it. One reason for these disparate findings may be the wide variation in the frequency of the allele concerned in different populations. Thus it is important that patient groups are carefully matched with a control population of similar ethnic composition. (See Goldman (1993) and Pato *et al.* (1993) for reviews of this issue.)

Learning factors

Alcohol use It has been reported that children tend to follow their parents' drinking patterns (Hawker 1978), and that from an early age boys tend to be encouraged to drink more than girls (Jahoda and Cramond 1972). Non-genetic familial factors appear to be important in determining levels of alcohol use (Clifford *et al.* 1984). Nevertheless, it is not uncommon to meet people who are abstainers although their parents drank heavily. More recently it has been proposed that an *expectation* of positive effects of alcohol in childhood correlates with the degree of subsequent alcohol use (Berman and Noble 1993).

Alcohol dependence It has also been suggested that learning processes may contribute in a more specific way to the development of alcohol dependence through the repeated experience of *withdrawal symptoms*. On this view, relief of withdrawal symptoms by alcohol may act as a reinforcer for further drinking.

Personality factors

Little progress has been made in identifying personality factors that contribute to alcohol abuse and dependence. In clinical practice it is common to find that excessive alcohol consumption is associated with *chronic anxiety*, a pervading sense of inferiority, or self-indulgent tendencies. However, many people with personality problems of this kind do not resort to excessive drinking or become alcohol dependent. More recent surveys have emphasized the role of personality traits that lead to *risk taking* and *novelty seeking* (Berman and Noble 1993). It seems likely that these characteristics apply to those with *antisocial personality disorder* who are known to be at increased risk of abusing alcohol and developing alcohol dependence (Lewis and Bucholz 1991). However, the majority of alcohol-dependent subjects do not have an antisocial personality disorder.

Psychiatric disorder

Although not a common cause of alcohol abuse, psychiatric disorder should always be borne in mind as it may be treatable. Some patients with *depressive disorders* take to alcohol in the mistaken hope that it will alleviate low mood. Those with *anxiety disorders*, including *social phobias*, are also at risk. Alcohol abuse is occasionally seen in patients with *organic brain disease* or *schizophrenia*.

Alcohol consumption in society

In recent years there has been increasing interest in the idea that rates of alcohol dependence and alcohol-related disorders are correlated with the general level of alcohol consumption in a society. Previously it had been supposed that levels of intake amongst excessive drinkers were independent of the amounts taken by moderate drinkers. The French demographer Ledermann (1956) challenged this idea, proposing instead that the distribution of consumption within a homogeneous population follows a logarithmic normal curve. If this is the case, an increase in the average consumption must inevitably be accompanied by an increase in the number of people who drink an amount that is harmful.

The mathematical details of Ledermann's work have been heavily criticized (Miller and Agnew 1974; Duffy 1977). None the less there are striking correlations between average annual consumption in a society and several indices of alcohol-related damage among its members (Smith 1981). For this reason, despite the criticisms of Ledermann's work, it is now widely accepted that the proportion

of a population drinking excessively is largely determined by the average consumption of that population.

What then determines the average level of drinking within a nation? Economic, formal, and informal controls must be considered. The **economic control** is the price of alcohol. There is now ample evidence from the United Kingdom and other countries that the real price of alcohol (i.e. the price relative to average income) profoundly influences a nation's drinking. Also, heavy drinkers as well as moderate drinkers reduce their consumption when the tax on alcohol is increased (Kendell *et al.* 1983).

The main *formal controls* are the licensing laws. It is difficult to be sure how these affect drinking behaviour because results in different countries have been conflicting. For example in Finland a new law in 1969 led to greatly increased availability of alcohol in restaurants, cafés, and shops. This was followed by a 47 per cent increase in consumption. A recent relaxation of licensing hours in Scotland did not apparently lead to a large increase in consumption, but increasing the availability of alcoholic drinks in shops seems to be associated with greater consumption (Royal College of Psychiatrists 1986, p. 117).

Informal controls are the customs and moral beliefs in a society that determine who should drink, in what circumstances, at what time of day, and to what extent. Some communities seem to protect their members from alcohol abuse despite general availability of alcohol; for example alcohol-related problems are uncommon among Jews even in countries with high rates in the rest of the community.

Recognition of alcohol abuse

Only a small proportion of alcohol abusers in the community are known to specialized agencies, and many opportunities for them are missed. When special efforts are made to screen patients in medical and surgical wards, between 10 and 30 per cent are found to have serious drinking problems, with the rates being highest in accident and emergency wards (Chick 1987).

Alcohol abuse often goes undetected because subjects conceal the extent of their drinking. However, doctors and other professionals often do not ask the right questions. It should be a standard practice to ask all patients—medical, surgical, and psychiatric—about their alcohol consumption. It is useful to ask four questions. Have you ever felt you ought to *cut down* on your drinking? Have people *annoyed* you by criticizing your drinking? Have you ever felt *guilty* about your drinking? Have you ever had a drink first thing in the morning (an '*eye-opener*') to steady your nerves or get rid of a hangover? These questions are known as CAGE, from the initial letters of the words cut, annoyed, guilty, and eye-opener. Two or more positive replies are said to identify alcohol abuse. Some patients will give false answers, but others find that these questions provide an opportunity to reveal their problems.

In a comparative study of alcohol-dependent subjects and non-dependent controls, the CAGE questionnaire was more effective at detecting the presence of alcohol dependence than routine laboratory blood tests such as the GGT level and the mean corpuscular volume (MCV) (Girela *et al.* 1994).

The next requirement is for the doctor to be suspicious about 'at-risk' factors. In general practice, alcohol abuse may come to light as a result of problems in the marriage and family, at work, with finances, or with the law. The wife may complain of the husband's boastfulness, lack of consideration, sexual dysfunction, or aggressiveness towards herself and the children. The alcohol abuser is likely to have many more days off work than the moderate drinker, and repeated absences on Monday are highly suggestive. The at-risk occupations (see p. 446) should also be remembered.

In hospital practice, the alcohol-dependent subject may be noticed if he develops *withdrawal symptoms* after admission. Florid delirium tremens is obvious, but milder forms may be mistaken for an acute organic syndrome, for example in pneumonia or post-operatively.

In both general and hospital practice, at-risk factors include physical disorders that may be alcohol related. Common examples are *gastritis*, *peptic ulcer*, and *liver disease*, but others such as *neuropathy* and *seizures* should be borne in mind. Repeated accidents should also arouse suspicion. Psychiatric at-risk factors include *anxiety, depression, erratic moods, impaired concentration, memory lapses*, and *sexual dysfunction*. Alcohol abuse should be considered in all cases of *deliberate self-harm*.

If at-risk factors raise suspicion, or if the patient hints at a drink problem, the next step is to ask tactful but persistent questions to confirm the diagnosis. The doctor should find out how much the patient drinks on a typical 'drinking day', starting with the amount that he takes in the second half of the day and working back to the earlier part of the day. The patient should be asked how he feels if he goes without drink for a day or two and how he feels on waking. Gradually a picture can be built up of what and how much a patient drinks throughout a typical day. Similarly, tactful enquiries should be made about the social effects of drinking, such as declining efficiency at work, missed promotion, accidents, lateness, absences, and extended meal breaks. The patient should be asked about any difficulties in relationships with the spouse and children. In this way, the patient may be led step by step to recognize and accept that he has a drinking problem which he has previously denied. Once this stage is reached, the doctor should be in a position to enquire about the typical features of dependency, and the full range of physical, psychological, and social disabilities described in preceding sections.

Laboratory tests

Several laboratory tests can be used to detect heavy drinkers, though none gives an unequivocal answer. This is because the more sensitive tests can give 'false positives' when there is disease of the liver, heart, kidneys, or blood, or if enzyme-inducing drugs, such as anticonvulsants, steroids, or barbiturates, have been taken. However, abnormal values point to the possibility of alcohol abuse. Only the three most useful tests are considered here.

Gamma-glutamyl-transpeptidase

Estimations of GGT in blood provide a useful screening test. The level is raised in about 70 per cent of alcohol abusers, both men and women, whether or not there is demonstrable liver damage. The heavier the drinking, the greater is the rise in GGT.

Mean corpuscular volume

MCV is raised above the normal value in about 60 per cent of alcohol abusers, and more commonly in women than in men. If other causes are excluded, a raised MCV is a strong pointer to excessive drinking. Moreover, it takes several weeks to return to normal after abstinence.

Blood alcohol concentration

A high concentration does not distinguish between an isolated episode of heavy drinking and chronic abuse. However, if a person is not intoxicated when the blood alcohol concentration is well above the legal limit of driving, he is likely to be unusually tolerant of alcohol. This tolerance suggests persistent heavy drinking. Alcohol is eliminated rather slowly from the blood and can be detected in appreciable amounts for 24 hours after an episode of heavy drinking.

The treatment of excessive consumption of alcohol and alcohol abuse

Early detection

Early detection of excessive consumption of alcohol and alcohol abuse is important because treatment of established cases is difficult, particularly when dependence is present. Many cases can be detected early by general practitioners, physicians, and surgeons when patients seek treatment for another problem. If counselling is given to such patients during their stay in a medical ward, their alcohol consumption is found to be reduced a year later (Chick *et al.* 1985).

Brief intervention in primary care

General practitioners are well placed to provide early treatment of alcohol problems, and they are likely to know the patient and his family well. It is often effective if the general practitioner gives simple advice in a frank matter-of-fact way, but with tact and understanding.

Brief intervention studies in general practice have shown that five to ten minutes of *simple advice* by a general practitioner plus an *educational leaflet* can lead to a mean reduction of about 25 per cent in alcohol intake over the next year and a corresponding reduction of about 45 per cent in the proportion of excessive drinkers (Chick 1993; Austoker 1994). While such results may not be achieved in more severely affected patients, nevertheless they are important because excess alcohol consumption without obvious dependence is more common and contributes more to social and economic cost (Chick 1993).

Motivational interviewing

Patients with problems of substance abuse, particularly those detected by screening methods, may be unsure whether or not to engage in treatment programmes. An appropriate interviewing style, particularly during the first assessment, can help to persuade the patient to engage in a useful review of their current pattern of substance use.

Confrontation is avoided in motivational interviewing, and a less directive approach is taken during which patients are helped to assess the balance of the effects of substance use on their lives. The clinician can help in this exercise by providing feedback to the patient on the personal risks that alcohol poses both to them and their family, together with a number of options for change. The aim of motivational interviewing is to persuade the patients to argue their own case for changing their pattern of substance use. (For a review of motivational interviewing and its role in the treatment of substance abuse see Miller and Rollnick (1991).)

Treatment plans for alcohol abuse and dependence

Where patients have significant alcohol-related disorders, particularly in the presence of alcohol dependence, treatment may need to be more intensive. Any intervention should be preceded by a full *assessment* and should include a *drinking history* and an appraisal of current *medical, psychological*, and *social* problems. An intensive and searching enquiry often helps the patient gain a new recognition and understanding of his problem, and this is the basis of treatment. It is usually desirable to involve the *husband or wife* in the assessment, both to obtain additional information and to give the spouse a chance to unburden feelings.

An explicit treatment plan should be worked out with the patient (and spouse if appropriate). There should be specific *goals* and the patient should be required to take responsibility for realizing them. These goals should deal not only with the drinking problem, but also with any accompanying problems in health, marriage, job, and social adjustment. In the early stages they should be short term and achievable, for example, complete abstinence for two weeks. In this way the patient can be rewarded by early achievement.

Longer-term goals can be set as treatment progresses. These will be concerned with trying to change factors that precipitate or maintain excessive drinking, such as tensions in the family. In drawing up this treatment plan, an important decision is whether to aim at total abstinence or at limited consumption of alcohol (controlled drinking).

Total abstinence versus controlled drinking

The disease model of alcoholism proposes that an alcohol-dependent person must become totally abstinent and remain so, since a single drink would lead to relapse. Alcoholics Anonymous have made this a tenet of their approach to treatment. In 1962 Davies reported that seven alcoholics who had failed to abstain when asked to do so had nevertheless succeeded in drinking in a controlled and moderate way. However, a 29–34-year follow-up of Davies' original cases showed that all but two had relapsed into uncontrolled drinking (Edwards 1985).

The issue of abstinence versus controlled drinking remains unresolved. A prevalent view is that controlled drinking may be a feasible goal for people whose alcohol abuse has been detected early and who are not heavily dependent or damaged, whilst abstinence is the better goal for those who are heavily dependent and have incurred physical damage, and who have attempted controlled drinking unsuccessfully. While there are few controlled studies of this issue, recent investigations have supported the idea that controlled drinking can be a realistic goal in patients with a *low severity of alcohol dependence* and *lesser levels of alcohol-related disorders* (Mattick and Heather 1993). If controlled drinking is to be attempted, then the doctor should advise the patient clearly about safe levels (see p. 442).

Withdrawal from alcohol

For patients with the dependence syndrome, withdrawal from alcohol is an important first stage in treatment which should be carried out carefully. In the less severe cases, withdrawal may be at home provided that there is someone to look after the patient. The general practitioner or health visitor should visit daily to check the patient's physical state and supervise medication. However, any patient likely to have severe withdrawal symptoms should be admitted to hospital.

Sedative drugs are generally prescribed to reduce withdrawal symptoms; chlordiazepoxide or chlormethiazole are often used. According to the severity of symptoms, chlordiazepoxide may be given in doses of 50 mg or 100 mg by intramuscular injection, repeated if necessary in two to four hours. Oral dosage may be 50–150 mg daily in divided doses. If convulsions occur, larger doses of chlordiazepoxide may be used.

Chlormethiazole may be prescribed in either of two ways: flexibly according to the patient's symptoms, or on a fixed six-hourly regime of gradually decreasing dosage over six to nine days.

It should not be prescribed for more than this because it can itself become a drug of dependence. Preparations and dosages of chlormethiazole are subject to revision, and before prescribing it the clinician should consult the current edition of a work of reference such as the *British National Formulary*. Chlormethiazole should not be given to patients who may continue to take alcohol, since the combination may cause fatal respiratory depression. *Vitamin supplements* are often given, and in some countries anticonvulsants, glucose, and magnesium infusions are added. During the first five days, there should be a daily check on the patient's temperature, pulse, blood pressure, hydration, level of consciousness, and orientation.

Withdrawal from alcohol is the main purpose of so-called **detoxification units**. In some places these units are used mainly for chronic drunkenness offenders who have little prospect of progressing to a treatment programme. In other places patients come mainly from general practitioners, and a substantial proportion move on to further treatment such as group or behavioural therapy (see below).

Psychological treatment

As noted above, provision of information and advice about the effects of excessive drinking is an important first stage in treatment. The information given should relate to the specific problems of the individual patient, both those that have occurred already and those likely to develop if drinking continues. The technique of *motivational interviewing* can be useful (see p. 456).

Group therapy

This has probably been the most widely used treatment for problem drinkers. Regular meetings are attended by about 10 patients and one or more members of staff. The aim is to enable patients to observe their own problems mirrored in other problem drinkers and to work out better ways of coping with their problems. They gain confidence, whilst members of the group jointly strive to reorganize their lives without alcohol. Until recently the most common plan of treatment in specialist alcohol units was an in-patient programme of group therapy lasting about eight weeks; this treatment could be preceded by detoxification if required. However, because of the lack of evidence that this kind of intensive treatment approach is particularly beneficial (see below), most clinics now offer a broader range of care including out-patient and day-patient programmes that utilize a variety of psychotherapeutic approaches including marital and family therapy and cognitive–behavioural methods of treatment (see below). Group therapy is discussed further on pp. 615–20.

Cognitive–behavioural therapy

Recently there has been increasing interest in cognitive–behavioural methods of treatment which tackle the drinking behaviour itself rather than underlying psychological problems. Such approaches stress the role of *education* and the improvement of *social and interpersonal skills* as these relate to alcohol abuse. For example it may be helpful to identify situational or interpersonal triggers that cause an individual to drink excessively, and then to plan and rehearse new methods of coping with these situations. This is called *relapse prevention*. Many patients who abuse alcohol have general deficiencies in problem-solving skills, and appropriate training may help reduce relapse rates, particularly if combined with marital therapy (McCrady *et al.* 1991; Mattick and Heather 1993).

Medication

Apart from the management of withdrawal discussed above, drug treatment plays only a small part in the management of excessive drinking. None the less drug treatment is described at some length here because it carries certain risks and may cause unpleasant side-effects.

Disulfiram (Antabuse) is sometimes prescribed as a deterrent to impulsive drinking. It acts by blocking the oxidation of alcohol so that acetaldehyde accumulates. If the patient drinks alcohol, he experiences unpleasant flushing of the face, headache, choking sensations, rapid pulse, and feelings of anxiety. The drug is not without risk, occasionally causing *cardiac irregularities* and *rarely cardiovascular collapse*. It also has unpleasant side-effects in the absence of alcohol: a metallic taste in the mouth, gastrointestinal symptoms, dermatitis, peripheral neuropathy, urinary frequency, impotence, and toxic confusional states. Treatment with disulfiram should not be started until at least 12 hours after the last ingestion of alcohol. On the first day the patient is warned carefully about the dangers of drinking alcohol while taking the drug, and then is given four tablets each of 200 mg and told not to take any alcohol whatever. The dosage is then reduced by one tablet a day over three days, with the maintenance dose being half to one tablet a day.

Citrated calcium carbimide is used in the same way. Compared with disulfiram, it is more rapidly absorbed and excreted, induces a milder reaction with alcohol, and has fewer side-effects. Details of the dosage can be found in standard works of reference.

The role of these drugs in the treatment of alcohol-related disorders is controversial. *Poor compliance* is common but can be improved if drug administration is supervised, usually by the spouse. In a double-blind controlled trial of 126 alcohol-dependent subjects, Chick *et al.* (1992) found that disulfiram (200 mg daily) increased total abstinent days by about one-third. This was reflected by a fall in GGT levels in the disulfiram-treated group, while a small increase was noted in patients receiving placebo. Patients in this study received a range of other treatments, including counselling and marital and group therapies.

It is worth noting, however, that therapy with disulfiram or citrated calcium carbimide therapy is most suited to patients who are compliant in treatment, attend regularly, and have high expectations of improvement. Such patients tend to have a better prognosis in any case.

Other agencies concerned with drinking problems

Alcoholics Anonymous (AA)

This is a *self-help* organization founded in the United States by two alcoholic men, a surgeon and stockbroker. It has since spread to most countries of the world. Members attend group meetings, usually twice weekly on a long-term basis. In crisis they can obtain immediate help from other members by telephone. The organization works on the firm belief that abstinence must be complete. At present there are about 1200 groups in the United Kingdom.

Alcoholics Anonymous does not appeal to all problem drinkers because the meetings involve an emotional confession of problems. However, the organization is of great value to some problem drinkers, and anyone with a drink problem should be encouraged to try it. The activities of Alcoholics Anonymous are described by Robinson (1979).

Al-Anon

This is a parallel organization providing support for the spouses of excessive drinkers. *Al-Ateen* does the same for their teenage children.

Councils on alcoholism

These are voluntary bodies that coordinate available services in an area and train counsellors. They advise problem drinkers and their families where to obtain help, and provide social activities for those who have recovered.

Hostels

These are intended mainly for homeless problem drinkers. They provide rehabilitation and counselling. Usually abstinence is a condition of residence.

Results of treatment

A number of investigations have combined results from different treatment centres. The Rand Report (Armor *et al.* 1976) describes a prospective study of 45 treatment centres in the United States, of which eight were followed for 18 months. Only a quarter of the patients remained abstinent for six months, and fewer than 10 per cent for 18 months. However, at 18 months 70 per cent of patients had reduced their consumption of alcohol. Patients with a better outcome had received more intensive treatment, but the form of treatment made no difference.

In a controlled trial with 100 male alcoholics, Edwards *et al.* (1977) compared simple advice with intensive treatment that included intro-ductions to Alcoholics Anonymous, medication, repeated interviews, counselling for their wives, and, where appropriate, in-patient treatment as well. The advice group received a three-hour assessment together with a single session of counselling with the spouse present. The two groups were well matched. After 12 months there was no significant difference between them in drinking behaviour, subjective ratings, or social adjustment (Orford and Edwards 1977).

Similar findings with regard to drinking behaviour were reported by Chick *et al.* (1988) who found that at a two-year follow-up, there was no difference in stable abstinence rates between patients who received a minimal treatment intervention, consisting mainly of advice, and those offered a broad range of therapies, including in-patient care and group therapy. However, the group offered intensive treatment suffered less alcohol-related harm, particularly in relation to *family life.*

Some have argued that measures of abstinence rates by themselves do not provide a useful measure of treatment outcome. For example, while subjects may still be drinking, the amount consumed may decrease. This can be associated with reductions in aspects of alcohol-related harm as shown by the study by Chick *et al.* (1988). From this viewpoint *harm reduction*, even where people continue to drink, is a worthwhile achievement (Heather 1993). The principles of the harm-reduction approach have been applied to the misuse of other substances (see below).

There is some evidence that the cognitive–behavioural approaches outlined above may benefit some problem drinkers, particularly those whose problems are milder in nature. Involvement

of the spouse may improve the efficacy of these interventions (Mattick and Heather 1993). Probably outcome depends as much upon factors in the patient as upon the particular treatment. There is some disagreement as to what these factors are, but the following generally predict a better prognosis whatever treatment is used: *good insight* into the nature of the problems; *social stability* in the form of a fixed abode, *family support*, and ability to *keep a job*; ability to control *impulsiveness*, to *defer gratification*, and to form deep *emotional relationships.*

Prevention of alcohol abuse and dependence

In seeking to prevent excessive drinking and alcohol-related disorders, two approaches are possible. The first is to improve the help and guidance available to the individual, as already described. The second is to introduce social changes likely to affect drinking patterns in the population as a whole. It is with this second group that we are concerned here. Consumption in a population might be reduced by four methods.

The pricing of alcoholic beverages

Putting up the price of alcohol would probably reduce the consumption (see above).

Advertising

Controlling or abolishing the advertising of alcoholic drinks might be another preventive measure. It is unclear how far advertising en-courages use of particular brands of alcohol rather than overall consumption; the evidence on this point is somewhat conflicting (Smart and Cutler 1976; McGuiness 1980).

Controls on sale

Another preventive measure might be to control sales of alcohol by limiting hours or banning sales in supermarkets. It is known that relaxation of restrictions led to increased sales in Finland and some other countries, but it does not follow that increased restrictions would reduce established rates of drinking. It is also possible that restricted licensing hours may produce more damaging patterns of drinking.

Health education

It is not known whether education about alcohol abuse is effective. Little is known as to how attitudes are formed or changed. Although education about alcohol seems desirable, it cannot be assumed that classroom lectures or mass media propaganda would alter attitudes. Indeed Plant *et al.* (1985) concluded, from a follow-up study of teenagers, that education had no significant effect on their drinking habits. More recently, MacKinnon *et al.* (1993) found that labelling containers of alcohol with a health warning made high school students more aware of the hazards of alcohol but did not significantly change their drinking habits.

OTHER SUBSTANCE USE DISORDERS

Under this heading we shall consider the use and abuse of substances other than alcohol. Although these substances include agents such as solvents (inhalants), the general term **drug** will be employed because it is more familiar.

Epidemiology

Illicit drug use

The *National household survey on drug abuse* (NHSDA) in the United States in 1991 found that 37 per cent of the whole population had used at least one illicit drug in their lifetime, while 13 per cent had used illicit substances in the past year and 6 per cent in the previous month (National Institute on Drug Abuse 1991; Kaplan *et al.* 1994). Use was highest in unemployed people in the age range 18–25. In this age group 15.4 per cent had used an illicit drug at some time in the previous month.

The most commonly used illicit drug was *cannabis* (Hughes 1992).

In a survey of London schoolchildren, Swadi (1988) found that 13 per cent of 11-year-olds and 26 per cent of 16-year-olds had used illicit drugs on at least one occasion. However, only 2 per cent and 16 per cent respectively admitted to using such drugs on a regular basis. In all countries, studies suggest that illicit substance use appears to be particularly prevalent among the *homeless young* (Medina-Mora 1992).

There are differing national temporal trends in illicit drug use. In the United Kingdom, for example, the prevalence of illicit drug use among people under 20 years of age began to rise steeply in the middle and late 1960s. In a survey in Wolverhampton, Wright and Pearl (1990) found that the proportion of secondary schoolchildren who knew someone taking drugs more than doubled between 1969 and 1989. There is some evidence that, in the United States, drug use by young adults has shown a gradual decline between 1975 and 1990 (Johnston *et al.* 1991). However, substance use, particularly involving *cocaine*, appears to be growing in developing countries.

Drug abuse and dependence

Little is known for certain about the *prevalence* of different types of drug abuse and drug dependence. In the United Kingdom, information comes from several sources: *criminal statistics*, mainly based on offences involving the misuse of drugs and thefts from chemists shops, *hospital admissions*, *Home Office statistics*, and *special surveys*. Unfortunately, none of these sources is satisfactory since much drug abuse goes undetected.

Reported national prevalence rates of drug abuse and drug dependence vary widely, partly because different methods of ascertainment have been used. The National Comorbidity Survey in the United States (Kessler *et al.* 1994) found that the one-year prevalence for drug abuse and drug dependence (excluding alcohol use) was 3.6 per cent, while the lifetime prevalence was 11.9 per cent. The one-year prevalence of drug abuse and dependence in men (5.1 per cent) was more than twice that in women (2.2 per cent).

Rates of drug abuse and dependence are high in *disadvantaged areas* of large cities. It has been estimated that 1–2 per cent of the adult population of several British cities inject illicit drugs (Frischer 1992). *Adolescents* are at risk, particularly around school-leaving age. A high proportion of attenders at drug-dependence clinics in large cities are *unemployed* with *few stable relationships* and leading *disorganized lives*. However, many young drug abusers remain in employment and their drug-taking is a passing phase (Plant 1975).

Causes of drug abuse

There is no single cause of drug abuse. It is generally argued that three factors are important, namely *availability of drugs*, a *vulnerable personality*, and *social environment*. Once regular drug-taking is established, *pharmacological factors* also play a role in determining abuse and dependence. Studies of the aetiology of abuse of substances other than alcohol are still in an early stage. For example it is unclear whether similar risk factors predict abuse of a range of substances, including alcohol, or whether there is some specificity in the mechanisms which leads certain individuals to abuse particular substances (Berman and Noble 1993).

Availability of drugs

People can abuse and become dependent on drugs by three routes. The first is by taking drugs *prescribed by doctors*. In the first part of the twentieth century much of the known dependence on *opioids and barbiturates* in Western countries was of this kind; nowadays *benzodiazepine* dependence is often acquired in this way. The second route is by taking drugs that can be *bought legally* without prescription: *nicotine* is an obvious contemporary example, and in the nineteenth century much dependence on *opioids* arose from taking freely available remedies containing morphia. (Alcohol dependence is also acquired in this way, but we are not concerned here with that problem.) The third route is by taking drugs that can be obtained only from *illicit sources* ('street drugs').

Personality factors

Many drug abusers, particularly younger people taking non-prescribed drugs, appear to have some degree of **personality vulnerability** before taking drugs. They often seem to be without resources to cope with the challenges of day-to-day life, inconsistent in their feelings, and critical of society and authority as shown by a *poor school record*, *truancy*, or *delinquency*. Traits such as *sensation-seeking* and *impulsivity* are also common. Many of those who abuse drugs report depression and anxiety, but it is seldom clear whether these are the causes or the consequences of drug abuse and dependence. Some give a history of mental illness or personality disorder in the family. Hospitalized groups of young drug abusers may report severe patterns of *family dysfunction* with high levels of *sexual abuse* and *parental drug use* (Maltzman and Schweiger 1991). Such subjects have an increased prevalence of *antisocial personality disorder*, which is a risk factor for all forms of substance abuse (Stabenau 1992). However, many people who abuse drugs show none of these features.

Social environment

The risk of drug abuse is greater in societies which condone drug-taking of one sort or another. Within the immediate group, there may be *social pressures* for a young person to take drugs to achieve status. There is evidence that drug use by individuals can be predicted by the substance use of their *peers* (Swadi 1992b). There are also links between drug abuse and indices of *social deprivation* such as unemployment and homelessness (Hammer 1992; Medina-Mora 1992).

Neurobiology of drug use, abuse, and dependence

Many subjects use drugs without abusing them, and not all drug abusers become drug dependent. Therefore it is useful to study the biological mechanisms underlying drug use, abuse, and dependence separately. Drugs are used and abused because they have the ability to serve as *positive reinforcers*, i.e. they increase the frequency of behaviours that lead to their use (Stolerman 1992). Drugs act as positive reinforcers because they cause *positive subjective experiences* such as euphoria or reduction in anxiety.

An important neurological substrate that may mediate such effects is the *midbrain dopamine system* whose cell bodies originate in the ventral tegmental area and innervate the forebrain, particularly the *nucleus accumbens*. It has been proposed that these dopamine pathways form part of a physiological *reward system* which has the property of increasing the frequency of behaviours that activate it. Therefore it is of interest that administration of different kinds of drugs of abuse, including alcohol, nicotine, and opiates, to animals *increases dopamine release in the nucleus accumbens* (Koob 1992). This suggests that activation of midbrain dopamine pathways may be a common property of drugs that have a propensity to be abused. While this hypothesis may explain in part the *social use* of particular drugs, it does not account for the *abuse* of drugs in some circumstances. Presumably this is a consequence of interactions between the *pharmacological properties* of the drug, the *biological disposition* and *personality* of the user, and the *social environment*.

Dependence on drugs has traditionally been described as either psychological or physiological. In DSMIV, dependence is separated into whether or not it is *physiological* in nature. **Physiological dependence** is diagnosed when a substance user demonstrates either *tolerance* to the pharmacological effects of the drug or a *characteristic withdrawal syndrome* when use of the drug is diminished. A cardinal feature of both physiological and non-physiological dependence, is desire for the drug and drug-seeking behaviour. These features probably result from the continued need to obtain the reinforcing properties of the drug, as described above. *Learning* and *conditioning* factors are likely to be important here.

It is important to note that, even in non-physiological dependence, phenomena such as craving and dysphoria are likely to be associated with altered brain function. For example in experimental studies the discontinuation of cocaine after a period of administration results in a sharp decline in dopamine release in the nucleus accumbens (Woolverton and Johnson 1992). It is possible that such an effect could correspond clinically to symptoms such as anhedonia and a desire to obtain more supplies of the drug.

The phenomena of *tolerance* and *withdrawal* are believed to be a result of *neuroadaptive changes* in the brain. These are part of a *homeostatic process* which counteracts the acute pharmacological effects that occur when a drug is administered. For example many drugs of abuse with anxiolytic and hypnotic properties, such as barbiturates, benzodiazepines, and alcohol, have, among their acute pharmacological effects, the ability to enhance brain *GABA* function. During continued treatment with these agents, adaptive changes occur in GABA and benzodiazepine receptor sensitivity that tend to offset the effect of the drugs to facilitate GABA neurotransmission. Such an effect could account for the phenomenon of *tolerance*, with the result that an individual needs to take more of the drug to produce the same pharmacological effect.

If the drug is now abruptly discontinued, persistence of the adaptive changes in receptor function could lead to a sudden *decline* in *GABA* activity. In fact, many of the clinical features of withdrawal from anxiolytic drugs, such as anxiety, insomnia, and seizures, can be explained on the basis of diminished brain GABA function (Cowen and Nutt 1982). Such an effect can also explain the well-known phenomenon of *cross-tolerance* between anxiolytics and hypnotics and alcohol, which makes it possible, for example, to treat alcohol withdrawal with a benzodiazepine.

Similar kinds of *adaptive changes* have been proposed to account for the tolerance and withdrawal phenomena seen with other drugs of abuse. For example, while acute administration of opiates decreases the firing of noradrenaline cell bodies in the brain stem, tolerance to this effect occurs during repeated treatment, probably because of adaptive changes in the sensitivity of opiate receptors (Jaffe 1989). If opiates are now suddenly withdrawn, there is a sudden *increase* in the firing of noradrenaline neurones and in the release of noradrenaline in terminal regions. Increased noradrenergic activity may account for several of the clinical features of acute opiate withdrawal, including sweating, tachycardia, hypertension, and anxiety. These studies have led to the use of the noradrenaline autoreceptor agonist *clonidine* in the management of opiate withdrawal (see p. 470).

While the *positive reinforcing* actions of drugs are seen as the major factor in promoting continued drug use, *withdrawal effects* are also likely to play a part because they are invariably unpleasant and individuals are likely to try to prevent them by taking more drug. It is worth noting that for many months following the cessation of a clear-cut withdrawal syndrome, dependent subjects may experience a sudden intense desire to consume the drug. Often, particular psychological and social stimuli previously associated with drug use may trigger intense *craving* associated with symptoms resembling a withdrawal state (Sell *et al.* 1995). It has been proposed that a single exposure to the drug during this period may rapidly lead to a full relapse, the so-called *reinstatement effect*. An analogous effect has been shown in previously drug-dependent animals where a single *priming* dose of the drug concerned can lead to a full recovery of drug seeking behaviours that had previously been extinguished. Whether the priming effect occurs to a great extent in dependent humans is not clear, and psychological and social factors are likely to be at least as important in sudden relapses (Jaffe 1989; Ritson *et al.* 1993).

Consequences of intravenous drug-taking

Some drug abusers administer drugs intravenously in order to obtain a marked and rapid effect. The practice is particularly common with opioid abuse, but barbiturates, benzodiazepines, amphetamines, and other drugs may be taken in this way. Intravenous drug use has important consequences, some local and others general.

Local effects include thrombosis of veins, infections at the injection site, and inadvertent damage to arteries. *General* consequences are due to transmission of infection, especially when needles are shared. Examples include *bacterial endocarditis, hepatitis,* and, most seriously, *HIV infection.* Intravenous drug users are the second-largest risk group for HIV infection, and also play an important role in the transmission of HIV to non-drug abusers through *sexual intercourse* and *perinatal transmission* (Cremers and Matot 1994).

Social consequences of drug abuse

There are three reasons why drug abuse has undesirable social effects. First, chronic intoxication may affect behaviour adversely, leading to unemployment, motoring offences, accidents, and family problems including neglect of children. Second, because illicit drugs are generally expensive, the abuser may cheat or steal to obtain money. Third, drug abusers often keep company with one another, and those with previously stable social behaviour may be under pressure to conform to a group ethos of antisocial or criminal activity.

Drug abuse in pregnancy and the puerperium

When a *pregnant woman* abuses drugs, the *fetus* may be affected. When drugs are taken in early pregnancy, there is a risk of increased rates of *fetal abnormality.* When drugs are taken in late pregnancy, the *fetus* may become *dependent* on them. The risk of fetal dependence is great with heroin and related drugs, and after delivery the neonate may develop serious withdrawal effects requiring skilled care. If the mother continues to take drugs after delivery, the infant may be neglected. Intravenous drug use may lead to infection of the mother with HIV or other conditions that can affect the fetus.

Diagnosis of drug abuse

It is important to diagnose drug abuse early, at a stage when dependence may be less established and behaviour patterns less fixed, and the complications of intravenous use may not have developed. Before describing the clinical presentations of the different types of drugs, some general principles will be given. The psychiatrist who is not used to treating drug-dependent people should remember that he may be in the unusual position of trying to help a patient who is attempting to deceive him. Patients dependent on heroin may overstate the daily dose to obtain extra supplies for their own use or for sale to others. Also, many patients take *more than one drug* but may not say so. It is important to try to corroborate the patient's account of the amount he takes by asking detailed questions about the duration of drug-taking, and the cost and source of drugs; by checking the story for internal consistency, and by external verification whenever possible.

Certain **clinical signs** lead to the suspicion that drugs are being injected. These include needle tracks and thrombosis of veins, especially in the antecubital fossa, wearing garments with long sleeves in hot weather, and scars. Intravenous use should be considered in any patient who presents with subcutaneous abscesses or hepatitis.

Behavioural changes may also suggest drug dependence. These include absence from school or work and occupational decline. Dependent people may also neglect their appearance, isolate themselves from former friends, and adopt new friends in a drug culture. Minor criminal offences, such as petty theft and prostitution, may also be indicators.

Dependent people may come to medical attention in several ways. Some declare that they are dependent on drugs. Others conceal their dependency, and ask for controlled drugs for the relief of pain such as renal colic or dysmenorrhoea. It is important to be particularly wary of such requests from temporary patients. Others present with drug-related complications, such as cellulitis, pneumonia, serum hepatitis, or accidents, or for the treatment of acute drug effects, overdose, withdrawal symptoms, or adverse reactions to hallucinogenic drugs. A few are detected during an admission to hospital for an unrelated illness.

Laboratory diagnosis

Whenever possible, the diagnosis of drug abuse should be confirmed by *laboratory tests*. Most drugs of abuse can be detected in the urine; the notable exception is LSD. Urine specimens should be sent to the laboratory as quickly as possible. An indication should be given of the interval between the last admitted drug dose and the collection of the urine sample. The laboratory should be provided with as complete a list as possible of drugs likely to have been taken, including those prescribed as well as those obtained in other ways.

Prevention, treatment and rehabilitation: general principles

Because treatment is difficult, considerable effort should be given to **prevention**. For many drugs, important preventive measures such as restricting availability and lessening social deprivation depend on government, not medical, policy. The *reduction of overprescribing* by doctors is important, especially with benzodiazepines and other anxiolytic drugs. While *education programmes* by themselves do not seem effective in prevention, it is important that information about the dangers of drug abuse should be available to young people in the school curriculum and through the media. Another aspect of prevention is the identification and *treatment of family problems* that may contribute to drug-taking. In all these preventive measures, the general practitioner has an important role.

When drug abuse has begun, treatment is more effective before dependence is established. At this stage, as at later stages, the essential step is to *motivate* the person to control his drug-taking. This requires a combination of advice about the likely effects of continuing abuse and help with any concurrent psychological or social problems. The techniques of *motivational interviewing*

(Miller and Rollnick 1991) (see p. 456) may be useful here. If the person is associating with others who abuse drugs, he should be urged to leave them and establish new friendships and interests.

The main aim of the treatment of the drug-dependent person is the withdrawal of the drug of dependence. However, drug withdrawal (or *detoxification*) by itself has no effect on long-term outcome (Vaillant 1988), and so this process should be part of a wider treatment programme. If withdrawal cannot be achieved, continued prescribing may be considered as part of a *harm reduction* programme (see below). In addition, psychological treatment and social support are required. At this point in the chapter, the general principles of treatment are outlined. In later sections, treatment specific to individual drugs will be considered.

In the United Kingdom most drug-dependent patients are treated in clinics based on psychiatric units. In some large cities there are **special treatment centres**, mainly for people dependent on narcotics. In-patient care is usually provided within the psychiatric unit of general hospitals, in psychiatric hospitals, or in therapeutic communities run by charitable organizations. It is also possible for general practitioners to manage drug-dependent patients in treatment programmes (Wilson *et al.* 1994).

Treatment of physical complications

The complications of self-injection may need treatment in a general hospital. They include skin infections, abscesses, septicaemia, hepatitis B, and HIV infection. In some cities more than half the intravenous drug users are reported to be HIV positive (Moss 1987).

Principles of withdrawal

The withdrawal of misused drugs is sometimes called *detoxification*. For many drugs, particularly opiates and barbiturates, withdrawal may be most effectively carried out in hospital (see below). Withdrawal from stimulant drugs and benzodiazepines can often be an out-patient procedure provided that the doses are not very large and that

barbiturates are not taken as well. Nevertheless, the risk of depression and suicide should be remembered.

Drug maintenance

Some clinicians undertake to prescribe drugs to dependent people who are not willing to give them up. The usual procedure is to prescribe a drug which has a slower action (and therefore is less addictive) than the 'street' drug. Thus methadone is prescribed in place of heroin. When this procedure is combined with help with social problems and a continuing effort to bring the person to accept withdrawal, it is called **maintenance therapy**. The rationale of this approach is twofold. First, prolonged prescribing will remove the need for the patient to obtain 'street' drugs, and will thereby reduce the need to steal money and associate with other drug-dependent people. Second, social and psychological help will make the person's life more normal so that he will be better able to give up drugs eventually. This form of treatment is used particularly for heroin dependence.

If *maintenance* drug therapy is used, it should be remembered that some drug-dependent people convert tablets or capsules into material for injection, which is a particularly dangerous practice. Also, some attend a succession of general practitioners in search of supplementary supplies of drugs. They may withhold information about attendance at clinics or pose as temporary residents.

Some patients who receive maintenance drugs achieve a degree of social stability, but others continue heavy drug abuse and deteriorate both medically and socially. Subjects on maintenance methadone are more likely to be retained in treatment than those in drug-free programmes. This may be important because the length of time spent in treatment, regardless of type, is the best predictor of favourable outcome (Hser *et al.* 1988). However, it is possible that compliance in treatment may be a reflection of particular personality traits that are the important factor in determining outcome, irrespective of the kind of treatment used. (The use of methadone maintenance treatment in the management of opiate dependence is considered in more detail below.)

Harm reduction

The increase in *HIV* infection has emphasized the importance of *harm-reduction* programmes (Advisory Council on Misuse of Drugs 1988; Strang 1993), of which prescribing maintenance may be one component. Such programmes have the aim of increasing the number of substance abusers who enter and comply with treatment. The aim is to identify intermediate treatment goals, which though short of total abstinence, nevertheless reduce the risk of drug abuse to the individual and society. For example, even if subjects continue to abuse drugs, the risk of HIV infection can be lessened by appropriate *education* and *practical help*. Such interventions may lead to drug abusers using safer routes of drug administration or sterile injection equipment. Counselling and screening for *hepatitis* and *HIV* may also be worthwhile. The impact of these policies on rates of HIV infection are not yet fully clear, but preliminary results suggest that drug abusers who participate in harm-reduction programmes have a lower rate of HIV infection than those who do not (Cremers and Matot 1994).

Psychological treatments

Some drug-dependent patients are helped by simple measures such as counselling. In many units, *group psychotherapy* is provided to help patients develop insight into personal and interpersonal problems. Some patients benefit from treatment in a therapeutic community in which there can be a frank discussion of the effects of drug-taking on the person's character and relationships within the supportive relationship of the group (see p. 619 for an outline of community therapy).

More recently, *cognitive–behavioural* methods of treatment as described for the management of alcohol abuse (see p. 458) have been employed and may be helpful (Robson 1992). The aim of such treatment is to increase recreational and personal skills so that the individual becomes less reliant on drugs and the drug culture as a source of satisfaction. *Involvement of family and partners* is often helpful.

As with alcohol dependence the technique of *relapse prevention* can be used to identify, in advance, situations that contain triggers for drug use; in this way alternative methods of coping can be planned. It has already been mentioned that when a drug abuser is confronted with a situation that contains personal cues for drug use, he can experience acute discomfort associated with a strong desire to use the drug. The technique of *cue exposure* aims, through repeated exposure, to desensitize drug abusers to these effects and thus improve their ability to remain abstinent. This kind of intervention is of theoretical interest, but its efficacy is not yet established (Mattick and Heather 1993).

Rehabilitation

Many drug takers have great difficulty in establishing themselves in normal society. The aim of rehabilitation is to enable the drug-dependent person to leave the drug subculture and to develop new social contacts. Unless he can do this, any treatment is likely to fail.

Rehabilitation is often undertaken after treatment in a therapeutic community (see p. 653). Patients first engage in work and social activities in sheltered surroundings, and then take greater responsibility for themselves in conditions increasingly like those of everyday life. Hostel accommodation is a useful stage in this gradual process. Continuing social support is usually required when the person makes the transition to normal work and living.

Use and abuse of specific types of drug

Opioids

This group of drugs includes morphine, heroin, codeine, and synthetic analgesics such as pethidine, methadone, and dipipanone. The pharmacological effects of opiates are mediated primarily through interaction with specific opiate receptors, with morphine and heroin being quite selective for the μ opiate receptor type (Di Chiara and North

1992). The medical use of opioids is mainly for their powerful analgesic actions; they are abused for their euphoriant effects. In the past morphine was abused widely in Western countries, but has been largely replaced as a drug of abuse by **heroin** which has a particularly powerful euphoriant effect, especially when taken intravenously.

Epidemiology

The NHDSA survey in the United States estimated that at least 1.3 per cent of Americans had used heroin at least once in their lives, with the highest prevalence (1.8 per cent) in the 18–25-year age group. There may be about half a million people with opiate dependence in the United States (Kaplan *et al.* 1994). In the United Kingdom there are about 20 000 subjects registered on the Home Office Addicts Index, the majority of whom are addicted to heroin. Of course, many more opiate abusers are not officially registered (Robson 1992). It is estimated that there are about 15 000–20 000 heroin-dependent subjects in the Netherlands; this number is showing signs of stabilization and decline (Chappell *et al.* 1992).

Methods of use

The data from the United States indicate that many people use heroin without becoming dependent on it. However, there is no doubt that repeated heroin use can lead to the rapid development of *dependence* and marked *physiological tolerance*. As well as the intravenous route, opiate users may employ other methods of administration, for example subcutaneous administration ('skin-popping') or sniffing ('snorting'). Heroin may also be heated on a metal foil and inhaled ('chasing the dragon'). Heroin users may change their customary method of drug administration from time to time. From the perspective of harm reduction, methods that avoid intravenous administration are preferable (Strang 1993).

Clinical effects

As well as *euphoria* and *analgesia*, opiates produce *respiratory depression, constipation, reduced ap-petite, and low libido. Tolerance* develops rapidly, leading to increasing dosage. Tolerance does not develop equally to all the effects, and constipation often continues when the other effects have diminished. When the drug is stopped, tolerance diminishes rapidly so that a dose taken after an interval of abstinence has greater effects than it would have had before the interval. This loss of tolerance can result in dangerous—sometimes fatal—*respiratory depression* when a previously tolerated dose is resumed after a drug-free interval, for example after a stay in hospital or prison.

Withdrawal from opiates

Withdrawal symptoms include intense craving for the drug, restlessness and insomnia, pain in muscles and joints, running nose and eyes, sweating, abdominal cramps, vomiting, and diarrhoea, piloerection, dilated pupils, raised pulse rate, and disturbance of temperature control. These features usually begin about six hours after the last dose, reach a peak after 36–48 hours, and then wane. Withdrawal symptoms rarely threaten the life of someone in reasonable health, though they cause great distress and so drive the person to seek further supplies.

Methadone

Methadone is approximately as potent, weight for weight, as morphine. It causes cough suppression, constipation, and depression of the central nervous system and of respiration. Pupillary constriction is less marked. The withdrawal syndrome is similar to that of heroin and morphine, and is at least as severe. Because methadone has a long half-life (one to two days), symptoms of withdrawal may begin only after 36 hours and reach a peak after three to five days. For this reason, methadone is often used to replace heroin in patients dependent on the latter drug.

The natural course of opiate dependence

Longer-term follow-up studies of opiate abusers have revealed that the disorder appears to run a chronic relapsing and remitting course with a significant mortality (10–15 per cent) over 10 years. Nevertheless, up to 50 per cent of opiate users have been found to be abstinent at 10-year follow-up, which suggests a trend towards natural remission in survivors (Robson 1992). Deaths are usually due to accidental overdosage, often related to loss of tolerance after a period of enforced abstinence. In future, death rates of intravenous drug users are likely to rise as a result of infection with HIV. *Pointers to a good outcome* include substantial periods of employment and marriage (Vaillant 1988).

Abstinence is often related to changed circumstances of life. This point is reflected in the report of 95 per cent abstinence among soldiers who returned to the United States after becoming dependent on opiates during service in the Vietnam War (Robins 1993). This finding is consistent with the observation that a change in residence following discharge from in-patient or residential care is associated with increased rates of abstinence (Robson 1992).

Prevention

Because dependence develops rapidly and treatment of dependent opiate abusers is unsatisfactory, preventive measures (see p. 465) are particularly important with this group of drugs.

Treatment of crisis

Heroin-dependent people present in *crisis* to a doctor in three circumstances. First, when their supplies have run out, they may seek drugs either by requesting them directly or by feigning a painful disorder. Although withdrawal symptoms are very unpleasant, so that the abuser will go to great lengths to obtain more drugs, they are not usually dangerous to an otherwise healthy person. Therefore it is best to offer drugs only as the first step of a planned withdrawal programme. This programme is described in the following sections. In the United Kingdom only specially licensed doctors may legally prescribe heroin or certain related drugs to a drug-dependent person as maintenance treatment. The relevant regulations are available within the National Health Service, and are subject to revision from time to time. The second form of crisis is *drug overdose*. This requires medical treatment, directed particularly to any respiratory depression produced by the drug. The third form of crisis is an *acute complication of intravenous drug usage* such as local infection, necrosis at the injection site, or infection of a distant organ, often the heart or liver.

Doctors in the United Kingdom are required (under the Misuse of Drugs Notification of Supply to Addicts Regulations 1973) to notify the names of people addicted to heroin and certain related drugs to the Chief Medical Officer of the Home Office. Doctors working in other countries are advised to find out the local legal requirements.

Planned withdrawal (detoxification)

The severity of withdrawal symptoms depends on psychological as well as pharmacological factors. Therefore the *psychological management* of the patient during withdrawal is as important as the drug regime. The *speed of withdrawal* should be discussed with the patient to establish a timetable which is neither so rapid that the patient will not collaborate, nor so protracted that the state of dependence is perpetuated. During withdrawal, much personal contact is needed to reassure the patient; the relationship formed in this way can be important in later treatment.

When the dose is low, opioids can be withdrawn rapidly while giving *symptomatic treatment for the withdrawal effects*. These effects are usually abdominal cramps, diarrhoea, anxiety, and insomnia. They can be relieved with a combination of *diphenoxylate* or Lomotil, which is diphenoxylate with atropine, and an anxiolytic–hypnotic such as a *benzodiazepine*. Since the latter drugs have a high potential for dependence, they should be taken only under close supervision. The drug combination is usually given for three days, for example Lomotil two tablets six-hourly for 24 hours, followed by one tablet six-hourly for 48 hours, together with diazepam 5–10 mg twice daily (in the morning and at night) for three days. Other drugs that have been used to reduce withdrawal symptoms include propanalol, thioridazine, and clonidine (see below).

When the daily dose of heroin is high, it may be necessary to prescribe an opioid, reducing the dose gradually. This can be done most effectively if heroin is replaced by *methadone*, which has a more gradual action. The difficulty is to judge the correct dose of methadone because patients often lie about their dosage of heroin, either overstating it in the hope of ensuring that the withdrawal regime will be gradual or understating it in an attempt to avoid censure. Methadone should be given in a liquid form to be taken by mouth.

The initial methadone dose is normally between 20 and 70 mg per day depending on the patient's usual consumption. Doses above 50 mg per day should be given with great caution because this amount of methadone could prove fatal for a non-tolerant individual. Although 10 mg of pharmaceutical heroin is about equivalent to 10 mg of methadone, street heroin varies in potency in different places and at different times. Therefore, if possible, advice about the equivalent dose of methadone should be obtained from a doctor experienced in treating drug dependence. The dose should be reduced by about a quarter every two or three days, according to the patient's response. If withdrawal is to be managed as an out-patient, the dose reduction should be smaller and the intervals longer to reduce the patient's temptation to seek additional supplies. An out-patient regime might last for several weeks; nevertheless a clear time limit should be set from the start. Withdrawal from high doses may be better managed on an in-patient basis.

As described on p. 464, *clonidine* may also be useful in the management of opiate detoxification in either in-patient or out-patient settings, though caution has to be employed in the latter situation because of the risk of *hypotension*. For the purposes of detoxification, clonidine is often given in combination with *naltrexone*, a long-acting opiate antagonist. Combined administration of clonidine and naltrexone is at least as effective as methadone withdrawal in detoxifi-cation and is faster (Charney *et al.* 1986; Kleber *et al.* 1987). However, addition of clonidine to a methadone withdrawal programme probably does not convey additional benefit (Ghodse *et al.* 1994). For a description of the use of clonidine and naltrexone in acute opiate withdrawal, see Charney *et al.* (1986).

Pregnancy and opioid dependence

The babies of opioid-dependent women are more likely than other babies to be *premature* and of *low birth-weight*. They may also show *withdrawal symptoms* after birth, including irritability, restlessness, tremor, and a high-pitched cry. These signs appear within a few days of birth if the mother was taking heroin, but are delayed if she was taking methadone which has a longer half-life in the body. Low birth-weight and prematurity are not necessarily related directly to the drug, since poor nutrition and heavy smoking are common among these women.

Later effects have been reported, with these children being more likely, as toddlers, to be over-active and to show poor persistence. However, these late effects may result from the unsuitable family environment provided by these mothers rather than from a lasting effect of the intrauterine exposure to the drug (for a review see Caviston (1987)).

Maintenance treatment for opiate dependence

As described above, withdrawal from opiates is the preferred treatment option, but if this is not possible, maintenance treatment, usually with methadone, may lessen the physical and social harm associated with the intravenous use of illicit drug supplies. The principle of this treatment has been explained on p. 466. Instead of heroin, methadone is prescribed as a liquid preparation formulated to discourage attempts to inject it, in the United Kingdom as Methadone Mixture Drug Tariff Formula containing 5 mg methadone hydrochloride per 5 ml. This preparation is not to be confused with the cough mixture properly referred to as methadone linctus.

Methadone maintenance treatment has been extensively evaluated (for a review see Farrell *et al.* (1994). There is good evidence from randomized studies that methadone maintenance *decreases the use of illicit opiates and reduces criminal activity.* There are also indications that subjects in methadone maintenance programmes show *less risky injecting behaviours and lower rates of HIV infection.*

Methadone doses of 20–40 mg daily have been widely advocated as appropriate for maintenance treatment, but there is some evidence that higher doses (50–120 mg daily) are associated with lower rates of illicit opiate use and improved retention in the therapeutic programme. The latter is associated with an improved therapeutic outcome. The best approach is probably to have a flexible dosing policy, bearing in mind the potential toxicity of methadone in subjects whose tolerance is unknown or hard to assess.

Evaluation of methadone clinics has shown that the more effective programmes have methadone maintenance treatment rather than abstinence as a primary goal of therapy. The more effective clinics were also characterized by high quality counselling and a wide range of medical services (see Farrell *et al.* 1994).

Therapeutic community methods

These forms of treatment aim to produce abstinence by effecting a substantial change in the patient's attitudes and behaviour. Drug-taking is represented as a way of avoiding pre-existing personal problems and as a source of new ones. Group therapy and communal living are combined in an attempt to produce greater personal awareness, more concern for others, and better social skills. In most therapeutic communities, some of the staff have previously been dependent on drugs and are often better able than other staff to gain the confidence of the patients in the early stages of treatment.

Anxiolytic and hypnotic drugs

The most frequently abused drugs of this groups are now the *benzodiazepines.* The most serious problems are presented by *barbiturates*; although more careful prescribing has limited their therapeutic use, they are available as 'street' drugs. Other drugs of this group that are currently abused include *chlormethiazole* and *glutethimide.* The clinical effects of these drugs are thought to result from their ability to facilitate brain *GABA* function. Benzodiazepines produce these effects by binding to a specific benzodiazepine receptor (Cowen and Nutt 1982).

Barbiturates

Many people dependent on barbiturates take them by mouth. These people are usually middle-aged or elderly and they began taking the drug because it had been prescribed as a hypnotic. In recent years younger people have been using barbiturates intravenously by dissolving capsules. These people generally take other intravenous drugs as well. Intravenous abusers favour short-acting barbiturates, for example pentobarbitone or Tuinal (a mixture of quinalbarbitone and amylobarbitone). In aqueous solution these drugs are highly irritant to tissues, causing periphlebitis, indolent abscesses, sloughing ulcers, and even gangrene. Patients with such complications sometimes need surgical treatment varying from skin grafting to amputation. Overdosage amongst users is common and is difficult to treat.

Tolerance develops less rapidly to barbiturates than to opioids, but when it develops it presents a particular danger. The danger is that tolerance to the sedating effect occurs to a greater extent than it does to the depressant effects on vital centres, so increasing the risk of inadvertent fatal overdosage.

Recognition of abusers of barbiturates depends on recognizing characteristic features. They may appear to be drunk, with slurred speech and incoherence. Dullness and drowsiness are common, and so is *depression. Nystagmus* is a valuable sign, which should always be sought. Pupillary size is not helpful. Younger intravenous users tend to be unkempt and dirty, and often appear malnourished. Blood levels are generally only useful in acute poisoning. Urine should be examined to investigate the possible simultaneous abuse of other drugs.

Prevention of dependence There is no indication now to prescribe barbiturates for insomnia or anxiety. This decline in prescribing should help to prevent barbiturate use and dependence.

Withdrawal Abrupt withdrawal of barbiturates from a dependent person is highly dangerous. It may result in a *delirium*, like that after alcohol withdrawal, and may lead to *seizures* and sometimes to death through *cardiovascular collapse*. The withdrawal syndrome may not appear at its most severe for several days.

The syndrome begins with anxiety, restlessness, disturbed sleep, anorexia, and nausea. These may progress to vomiting, hypotension, pyrexia, tremulousness, major seizures, disorientation, and hallucinations—a picture similar to *delirium tremens*.

Withdrawal should nearly always be an in-patient procedure. If the patient has been taking a small dose of barbiturate, out-patient withdrawal may be considered, but only if there is no history of epilepsy before drug abuse or of recent anti-epileptic medication. In the management of withdrawal, antipsychotic drugs should be avoided as they may lower the fit threshold. Instead, withdrawal can be initiated by giving enough *pentobarbital* in divided doses to maintain the patient between intoxication and withdrawal. Following this, dosage is reduced daily at a rate of about 10 per cent of the total daily dose, as long as withdrawal symptoms do not occur. An alternative is to use a *benzodiazepine*.

If *slow withdrawal* is to be attempted *outside hospital*, short-acting substitutes should be avoided because of the risk of abuse. Instead, *phenobarbitone* should be given. An attempt should be made to calculate the patient's usual daily dosage, allowing for all drugs with similar actions such as alcohol and benzodiazepines. For every 100 mg of a shorter-acting barbiturate or its equivalent, a daily dose of 30 mg phenobarbitone should be given in divided dose up to a maximum of 300 mg daily (exceptionally 400 mg). This is reduced progressively over 10—20 days, with the patient being assessed every second or third day.

Maintenance treatment This may be considered for some elderly patients who have taken barbiturates for a long time. The barbiturate is replaced by a benzodiazepine and continued efforts are made to reduce the dose gradually. In most cases withdrawal can be achieved eventually.

Benzodiazepines

These drugs were in use for many years before it became apparent that their prolonged use could lead to tolerance and dependence with a characteristic **withdrawal syndrome**. This syndrome includes irritability, anxiety, sleep disturbance, increased perceptual sensitivity, and somatic symptoms such as tremor, sweating, palpitations, headache, and muscle pain. Rarely, persistent tinnitus, seizures, confusion, and paranoid psychoses can occur (Petursson and Lader 1984).

Epidemiology Benzodiazepine use is extremely widespread; for example it has been estimated that about 10 per cent of the population of Europe and the United States use benzodiazepines as anxiolytics or hypnotics. One-third of these subjects are likely to be long-term users. While most long-term users are older women, the intravenous abuse of benzodiazepines, particularly temazepam gel capsules, has become more prevalent in young people.

Dependence Dependence on benzodiazepines often results from prolonged medical use but may also result from the availability of benzodiazepines as street drugs because of their euphoriant effects. The *withdrawal syndrome* closely resembles the anxiety symptoms for which the drugs are usually prescribed; hence if symptoms appear after the dose of benzodiazepine has been reduced, the doctor may revert to a higher dosage in the mistaken belief that these symptoms indicate a persistent anxiety disorder. It has been estimated that about one-third of subjects who take a benzodiazepine for more than six months may become dependent (Hallstrom 1985).

Treatment Treatment of dependence consists of *gradual withdrawal* over at least six weeks combined with supportive counselling (Lader 1994). Withdrawal appears to be more severe from benzodiazepines that have short half-lives and high potency at the benzodiazepine receptor. For example Rickels *et al.* (1986) found that about two-thirds of patients who abruptly discontinued diazepam and clorazepate were able to remain abstinent compared with only one-third of patients who stopped lorazepam and alprazolam. For this reason it is often suggested that patients on such compounds should be switched to longer-acting drugs such as diazepam before withdrawal is attempted. For patients who have difficulty withdrawing with these measures, *anxiety management* may be useful (see p. 637).

Many patients experience their most troublesome withdrawal symptoms once the benzodiazepine dose has been fully tapered off. Symptoms usually subside fully over the next few weeks, although the time course can be irregular and some symptoms such as muscle spasm may not appear until other features of withdrawal have largely subsided. A few patients continue to experience withdrawal-like symptoms for months or even years after cessation of benzodiazepines ('prolonged withdrawal syndrome'). The status of this syndrome and the role of personality factors in its pathogenesis is controversial (Ashton 1984; Higgitt *et al.* 1990).

Prevention The prevention of benzodiazepine dependence lies in the *restriction of prescribing*. Psychological treatments are effective for most anxiety disorders (see p. 637) and non-pharmacological approaches to insomnia are also beneficial (Morin *et al.* 1994). If benzodiazepines are prescribed, it should be for the short-term relief of symptoms that are severely disabling or distressing. In some patients who are already long-term users, the balance of benefit and risk will favour continued prescribing, but patients should be regularly reviewed. Advice from a general practitioner can be sufficient to persuade between 20 and 40 per cent of long-term benzodiazepine users to reduce their daily dose or discontinue treatment. See Lader and Morton (1991) and Lader (1994) for further information about benzodiazepine dependence.

Cannabis

Cannabis is derived from the plant *Cannabis sativa*. It is consumed either as the dried vegetative parts in the form known as *marijuana* or *grass*, or as the *resin* secreted by the flowering tops of the female plant. Cannabis contains several pharmacologically active substances of which the most powerful psychoactive member is *delta-l-tetrahydrocannabinol*. It seems likely that the pharmacological effects of cannabinols are mediated through interaction with a specific cannabinoid receptor in the central nervous system (Abood and Martin 1992).

Epidemiology

In some parts of North Africa and Asia, cannabis products are consumed in a similar way to alcohol in Western society. In North America and Britain the intermittent use of cannabis is quite widespread. For example it has been estimated that about one-third of the population of the United States have used cannabis at least once in their lifetimes and 13 per cent are current users. About 60 per cent of the adult population aged 26–34 reported some lifetime use of cannabis (National Institute on Drug Abuse 1991). It appears that most users do not take any other illegal drug, but some are given to high consumption of alcohol.

Clinical effects

The effects of cannabis vary with the *dose*, the person's *expectations and mood*, and the *social setting*. Users sometimes describe themselves as 'high' but, like alcohol, cannabis seems to exaggerate the pre-existing mood, whether exhilaration or depression. Users report an increased enjoyment of aesthetic experiences and distortion of the perception of time and space. There may be reddening of the eyes, dry mouth, tachycardia, irritation of the respiratory tract, and coughing. Cannabis intoxication presents hazards for car drivers.

Adverse effects

No serious adverse effects have been proved among those who use cannabis intermittently at small doses. Although there is no positive evidence of teratogenicity, cannabis has not been proved safe in the first three months of pregnancy. Inhaled cannabis smoke irritates the respiratory tract and is potentially carcinogenic.

The most common adverse psychological effect of acute cannabis consumption is *anxiety*. Mild *paranoid ideation* is also not uncommon. At higher doses *toxic confusional states* and occasionally *psychosis* in clear consciousness may rarely occur. It is not certain how far psychotic reactions are a consequence of a predisposition in the individual, rather than a specific reaction to the pharmacological action of the drug (Thomas 1993).

Cannabis and schizophrenia

The question as to whether chronic cannabis use can predispose to schizophrenia is controversial. Andreasson *et al.* (1987) followed up 45 570 Swedish conscripts for 15 years. They found that the relative risk of developing schizophrenia was 2.5 times greater in subjects who used cannabis, and the relative risk for heavy users was six times greater. While these data suggest that cannabis could be a risk factor for the development of schizophrenia, it is also possible that those predisposed to develop the schizophrenia are also predisposed to abuse cannabis. It is better established that cannabis can modify the course of an established schizophrenic illness, with evidence from a number of studies that users are more likely to experience *psychotic episodes* and *relapse* (Thomas 1993; Martinez-Arevalo *et al.* 1994).

It is also said that chronic use of cannabis can lead to a state of apathy and indolence (an *amotivational state*). However, an objective study of chronic cannabis users failed to demonstrate this (Beaubrun and Knight 1973). It is possible that the symptoms and signs of the amotivational state could reflect chronic intoxication (Thomas 1993).

Tolerance and dependence

There is evidence that *tolerance* to cannabis can occur in subjects exposed to high doses for a prolonged period of time, but it is much less evident in those who use small or intermittent dosing. *Withdrawal* from high doses gives rise to a syndrome of irritability, nausea, insomnia, and anorexia. These symptoms are generally mild in nature. Long-term consumption of high doses can give rise to *dependence,* though usually of the non-physiological type (Abood and Martin 1992). The epidemiological data show that the vast majority of cannabis users do not abuse the drug or become dependent on it. For a review of the pharmacology of cannabis see Abood and Martin (1992).

Stimulant drugs

These drugs include *amphetamines*, related substances such as *phenmetrazine*, and *methylphenidate*. *Cocaine* is also a stimulant drug, but is considered separately in the next section. Amphetamines have been largely abandoned in medical practice, apart from their use for the hyperkinetic syndrome of childhood (p. 694) and for narcolepsy (p. 410). The psychomotor stimulant effects of amphetamines are believed to result from their ability to release and block the re-uptake of *dopamine and noradrenaline.*

Epidemiology

Amphetamines are probably the most commonly used stimulant in the United Kingdom, although cocaine is more commonly used in the United States. A survey in the United States estimated that 7 per cent of the population had used an amphetamine at least once in their lives. The highest level of use was among those aged 18–25, of whom 9 per cent reported using amphetamines at least once, while 1 per cent described themselves as current users. A high rate of use was also apparent in those aged 12–17, where 3 per cent reported use of amphetamines on at least one occasion while 1 per cent were current users. It should be noted that in the United States *3,4-methylenedioxymethamphetamine (MDMA)*, a mixed amphetamine and hallucinogen (see below), is classified as an amphetamine derivative, and a considerable proportion of reported amphetamine use is probably accounted for by this drug (Kaplan *et al.* 1994).

In the past, most addiction to stimulant drugs arose from *injudicious prescribing*. However, these drugs are now sought as *'street drugs'* and are often used intravenously to produce euphoria. They can also be 'snorted' (taken like snuff). A pure form of amphetamine ('ice'), can be smoked or injected. It is said to produce particularly powerful effects.

Clinical effects

Apart from their immediate effect on mood, the drugs produce over-talkativeness, over-activity, insomnia, dryness of lips, mouth, and nose, and anorexia. The pupils dilate, the pulse rate increases, and blood pressure rises.

With large doses there may be *cardiac arrhythmia*, *severe hypertension*, *cerebrovascular accident*, and occasionally *circulatory collapse*. At increasingly high doses, neurological symptoms such as *seizures* and *coma* may occur. Acute adverse psychological effects of amphetamines include dysphoria, irritability, insomnia, and confusion. Anxiety and panic can also be present.

Amphetamine-induced psychosis

Prolonged use of high doses of amphetamines may result in repetitive stereotyped behaviour, for example repeated tidying. A **paranoid psychosis** indistinguishable from paranoid schizophrenia may also be induced by prolonged high doses. The features include persecutory delusions, auditory and visual hallucinations, and sometimes hostile and dangerously aggressive behaviour (Connell 1958). Usually the condition subsides in about a week, but occasionally it persists for months. It is not certain whether these prolonged cases are true drug-induced psychoses, schizophrenia provoked by the amphetamine, or merely coincidental. In experimental animal studies, amphetamines produce *neurotoxicity*.

Tolerance and dependence

Tolerance to amphetamines leads to users taking higher doses of the drug. A *withdrawal syndrome* ('crash') of varying severity follows cessation of amphetamine use. In mild cases it consists mainly of low mood and decreased energy. In occasional *severe cases* depression can be severe, and accompanied by anxiety, tremulousness, lethargy, fatigue, and nightmares. *Craving* for the drug may be intense and suicidal ideation prominent. Dependence on amphetamines can develop quickly. Dependence on stimulant drugs may be recognized from the history of over-activity and high spirits alternating with inactivity and depression. Whenever amphetamine use is at all likely, a urine sample should be taken for analysis as soon as possible because these drugs are quickly eliminated.

Prevention and treatment

Prevention of amphetamine abuse depends on restriction of the drugs and careful prescribing. Doctors should be wary of newly arrived patients who purport to suffer from narcolepsy.

Treatment of acute overdoses requires sedation and management of hyperpyrexia and cardiac arrhythmias. Most toxic symptoms, including paranoid psychoses, resolve quickly when the drug is stopped. An antipsychotic drug may be needed to control florid symptoms, but if this medication can be avoided the differential diagnosis from schizophrenia will be easier.

The **treatment of amphetamine dependence** is difficult because craving for the drug can be intense. To achieve lasting abstinence a full range of social and psychological interventions may be needed (see above). Benzodiazepines may be helpful for managing acute distress occasioned by a severe withdrawal syndrome, and antidepressants may be needed for a persistent depressive disorder.

Cocaine

Cocaine is a central nervous stimulant with effects similar to those of amphetamines (described above). It a particularly powerful positive reinforcer in animals and causes strong *dependence* in humans. These latter effects probably stem from the ability of cocaine to block the re-uptake of *dopamine* into presynaptic dopamine terminals. This leads to substantial increases in extracellular levels of dopamine in the nucleus accumbens and consequent activation of the physiological 'reward system' (Koob 1992; Woolverton and Johnson 1992).

Cocaine is *administered* by injection, by smoking, and by sniffing into the nostrils. The latter practice sometimes causes perforation of the nasal septum. In 'freebasing', chemically pure cocaine is extracted from the 'street' drug to produce *crack* which has a very rapid onset of action, particularly when inhaled (Strang and Edwards 1989).

Epidemiology

Cocaine use became a serious public health problem in the United States in the 1980s, triggering a cycle of crime, social deprivation, homelessness, and unemployment (Michels and Marzuk 1993). However, its use now seems to be declining, and rates of cocaine use for all age groups in 1991 were similar to those found in the early 1970s. In 1991, 12 per cent of Americans reported lifetime use of cocaine, with 3 per cent having used the drug in the previous year but fewer than 1 per cent in the previous month. The highest rate of use in the previous month (2 per cent) was in those aged 18–25 (see National Institute on Drug Abuse 1991; Kaplan *et al.* 1994).

The incidence of cocaine use in the United Kingdom has increased over the last decade, but to a much lesser extent than that seen in the United States. Much of the use has been by those who were already dependent on *opiates*. Customs seizures of cocaine have increased substantially over the last five years, but amphetamines continue to be the most widely used psychostimulant (Strang 1993).

Clinical effects

The *psychological effects* of cocaine include excitement, increased energy, and euphoria. This can be associated with grandiose thinking, impaired judgement, and sexual disinhibition. Higher doses can result in visual and auditory hallucinations. *Paranoid ideation* may lead to *aggressive behaviour*. More prolonged use of high doses of cocaine can result in a *paranoid psychosis* with *violent behaviour*. This state is usually short-lived

but may be more enduring in those with a pre-existing vulnerability to psychotic disorder (Strang 1993). *Formication* ('cocaine bugs'), a feeling as if insects are crawling under the skin, is some-times experienced by cocaine abusers. (This symptom can occur also with amphetamine abuse.)

The physical effects of cocaine include increases in pulse rate and blood pressure. Dilatation of the pupils is often prominent. Severe adverse effects of cocaine use include cardiac arrhythmias, myocardial infarction, myocarditis, and cardiomyopathy. Cocaine use has also been associated with cerebrovascular disease, including cerebral infarction, subarachnoid haemorrhage, and transient ischaemic attacks. Seizures and respiratory arrest have also been reported.

Tolerance and dependence

Tolerance to the effects of cocaine develops and a **withdrawal syndrome** similar to that seen following withdrawal of amphetamines can occur. *After acute cocaine use*, the 'crash' consists of dysphoria, anhedonia, anxiety, irritability, fatigue, and hypersomnolence. If the preceding cocaine use has been relatively mild, such symptoms resolve within about 24 hours. *After more prolonged use*, the symptoms are more severe and extended, and are associated with intense craving, depression, and occasionally severe suicidal ideation. *Craving* for cocaine can re-emerge after months of abstinence, particularly if the subject is exposed to psychological or social cues previously associated with its use.

Treatment

Acute intoxication may require sedation with benzodiazepines or, in severe cases, an anti-psychotic agent such as haloperidol. Concurrent medical crises such as seizures or hypertension should be managed in the usual way.

As with amphetamines, the treatment of cocaine dependence is difficult because of the intense craving associated with abstinence from the drug. For moderate cocaine users it may be sufficient to provide *psychological and social support* on an out-patient basis. Heavy and chaotic users with strong dependence will need more intensive management, perhaps as in-patients (Strang *et al.* 1993). There is some evidence where dependence and craving are severe that treatment with the tricyclic antidepressant *desipramine* may be helpful in promoting abstinence. For example in a controlled trial 59 per cent of desipramine-treated patients were abstinent for at least three weeks, compared with 17 per cent of subjects treated with placebo (Gawin *et al.* 1989). Clearly, longer follow-up studies of this treatment are needed.

Various *psychotherapeutic programmes* may also be effective in subjects with cocaine dependence. The evidence suggests that *cognitive– behavioural* approaches, including *relapse prevention*, produce rather better results than *simple counselling* (Higgins *et al.* 1993). *Cue exposure* (see above) may also have a role (Strang 1993). As far as individual treatment programmes are concerned, it is worth noting that subjects who abuse cocaine often abuse other drugs such as opiates and alcohol.

In the United Kingdom, under the Misuse of Drugs (Notification and Supply of Addicts) Regulations 1973, a doctor is required to notify the Chief Medical Officer of the Home Office if he attends a patient whom he considers to be addicted to cocaine.

MDMA (ecstasy)

The recreational use of *MDMA* or *'ecstasy'* is increasing rapidly. Ecstasy is a synthetic drug which is classified in the DSMIV substance list as a *hallucinogen*. However, it has stimulant as well as mild hallucinogenic properties. It is usually taken in tablet or capsule form in a dose of about 120 mg. Given in this way, its effects last for about four to six hours. Like amphetamines, ecstasy increases the *release of dopamine* but it also *releases 5-hydroxtryptamine 5-HT* which may account for its hallucinogenic properties.

Clinical effects

Ecstasy produces a positive mood state with feelings of *euphoria, sociability*, and *intimacy*. It also produces sensations of *newly discovered insights* and *heightened perceptions* (Solowij *et al.* 1992). The physical effects of ecstasy include *loss of appetite, tachycardia, bruxism,* and *sweating*. Tolerance to successive doses of ecstasy develops quickly, but no clear withdrawal syndrome has yet been described.

Adverse reactions

Rarely, ecstasy can cause severe *adverse reactions* and deaths due to *hyperthermia*, and its complications have been reported in healthy young adults (Henry *et al.* 1992). The hyperthermia probably results from the effect of ecstasy in increasing brain 5-HT release, together with the social setting in which the drug is customarily taken (crowded parties with prolonged and strenuous dancing). Deaths have also been reported through *cardiac arrhythmias*, though pre-existing cardiac disease may have played a role. *Intracerebral haemorrhage* has occurred in ecstasy users, probably as a con-sequence of hypertensive crises. Cases of *toxic hepatitis* could reflect impurities in manufacture (Solowij 1993).

The use of ecstasy has been associated with *acute and chronic paranoid psychoses* but, as with other drug-induced psychotic states, it is not clear how far such disorders represent idiosyncratic reactions of vulnerable individuals. There are also reports of *'flashbacks'* which are the recurrence of abnormal experiences weeks or months following drug ingestion. Such effects have been reported with other hallucinogens (see below).

In experimental animals, including primates, repeated treatment with ecstasy produces *degeneration of 5-HT nerve terminals in cortex and forebrain* (Insel *et al.* 1989). Therefore it is possible that such effects could occur in humans, and one study has shown that subjects who reported repeated ecstasy use had lower cerebrospinal fluid concentrations of the 5-HT metabolite 5-hydroxyindolacetic acid than controls (McCann *et al.* 1994). Whether such a change could be associated with detectable neuropsychological or psychiatric sequelae is not known.

Prevention and harm reduction

Although the risk of serious harm following ecstasy use appears to be low, it is important to inform potential users about the acute risks and the potential long-term hazard of *neurotoxicity*. Consumption of large doses and pre-existing psychiatric disorder are likely to be associated with increased risk of adverse reactions. Education may also help users to avoid heatstroke by encouraging breaks from dancing and the consumption of sufficient replacement fluid during vigorous exercise. For a review of the current use of ecstasy and its adverse effects see Solowij (1993).

Hallucinogens

Hallucinogens are sometimes known as *psychedelics*, but we do not recommend this term because it does not have a single clear meaning. The term *psychotomimetic* is also used because the drugs produce changes that bear some resemblance to those of the functional psychoses. However, the resemblance is not close, and so we do not recommend this term.

The synthetic hallucinogens include *lysergic acid diethylamide* (*LSD*), *dimethyl tryptamine, and methyldimethoxyamphetamine*. Of these drugs, LSD is encountered most often in the United Kingdom. Hallucinogens also occur naturally in some species of *mushroom*, and varieties containing *psilocybin* are consumed for their hallucinogenic effects. The mode of action of hallucinogenic drugs is unclear but an ability to activate brain 5-HT_2 receptors may be important (Pierce and Peroutka 1989).

Epidemiology

The NHSDA study in the United States (National Institute on Drug Abuse 1991) indicated that 8.1 per cent of the population had used a hallucinogen at least once in their lives and 1.2 per cent had used a hallucinogen in the previous year. The highest rate of use was in those aged 26–34, of whom 15.5 per cent had used a hallucinogen on at least one occasion.

Clinical effects

The effects of LSD have been most studied and will be described here. The *physical actions* of LSD are variable. There are initial sympathomimetic effects: *heart rate* and *blood pressure* may *increase* and *pupils dilate*. However, overdosage does not seem to result in severe physiological reactions. In predisposed subjects, the hypertensive effects of hallucinogens can cause adverse *myocardial* and *cerebrovascular* effects. There is no certain evidence linking conventional dosage with chromosomal or teratogenic abnormalities, but nevertheless it is prudent that the drug be avoided during *pregnancy*.

The *psychological effects* develop during a period of two hours after LSD consumption and generally last from 8 to 14 hours. The most remarkable experiences are *distortions or intensifications of sensory perception*. There may be confusion between sensory modalities (*synaesthesia*), with sounds being perceived as visual or movements experienced as if heard. Objects may be seen to merge with one another or move rhythmically. The passage of time appears to be slowed and experiences seem to have a profound meaning.

A distressing experience may be *distortion of the body image*, with the person sometimes feeling that he is outside his own body. These experiences may lead to *panic* with *fears of insanity*. The mood may be *exhilaration, distress*, or acute *anxiety*. According to early reports, *behaviour* could be *unpredictable and extremely dangerous*, with the user sometimes injuring or killing himself through behaving as if he were invulnerable. Since then there may have been some reduction in such adverse reactions, possibly because users are more aware of the dangers and take precautions to ensure support from other people during a 'trip'.

Whenever possible, *adverse reactions* should be managed by 'talking down' the user, explaining that the alarming experiences are due to the drug. If there is not time for this, an anxiolytic such as *diazepam* should be given. Haloperidol can be used for severe intoxication. *Tolerance* to the psychological effects of LSD can occur, but a *withdrawal syndrome* has not been described. *Dependence* may occur in long-term and heavy users, but is rare (Kaplan *et al.* 1994).

It has been argued that abuse of LSD can cause long-term abnormalities in thinking and behaviour or even schizophrenia. The evidence for such an association is extremely dubious (Kaplan *et al.* 1994). However, *flashback* i.e. the recurrence of psychedelic experience weeks or months after the drug was last taken, is a recognized event. This experience may be distressing and occasionally requires treatment with an anxiolytic drug.

Phencyclidine

This drug is sufficiently different from the hallucinogens in its actions to require a separate description. It can be synthesized easily, and is taken by mouth, smoked, or injected. Phencyclidine was developed as a disassociative anaesthetic, but its use was abandoned because of adverse effects such as *delirium* and *hallucinations*. It is related to the currently used anaesthetic agent *ketamine*. Both phencyclidine and ketamine antagonize neurotransmission at N-methyl-D-aspartate (NMDA) receptors, which may account for their hallucinogenic effects. The psychological effects of ketamine in healthy volunteers have been used to model some of the cognitive changes seen in patients with schizophrenia.

Phencyclidine is widely available in the United States but is little used in the United Kingdom. Most users of phencyclidine also use other drugs, particularly alcohol and cannabis. Ingestion of phencylcidine may be inadvertent because it is often added to other 'street' drugs to boost their effects.

Clinical effects

Small doses of this drug produce *drunkenness*, with *analgesia* of fingers and toes, and even *anaesthesia*. Intoxication with the drug is prolonged, with the common features being *agitation, depressed consciousness, aggressiveness* and *psychotic-like* symptoms, *nystagmus*, and *raised blood pressure*. With **high doses** there may be *ataxia, muscle rigidity, convulsions*, and *absence of response* to the environment even though the eyes are wide open. Phencyclidine can be detected in the urine for 72 hours after it was last taken.

With serious overdoses, an adrenergic crisis may occur *with hypertensive heart failure, cerebrovascular accident*, or *malignant hyperthermia. Status*

epilepticus may appear. Fatalities have been reported, due mainly to hypertensive crisis but also to *respiratory failure* or *suicide*. Other people may be attacked and injured. Chronic use of phencyclidine may lead to *aggressive behaviour* accompanied by *memory loss*.

Tolerance to the effects of phencyclidine occurs, though *withdrawal* symptoms are rare in humans. *Dependence* occurs in chronic users (Kaplan *et al.* 1994).

Treatment of phencyclidine intoxication

Treatment of acute intoxication is symptomatic, according to the features listed above. Haloperidol, or diazepam, or both may be given. Chlorpromazine should be avoided because it is much less safe; it may also increase the anticholinergic effects of phencyclidine or any other drugs that may have been taken. Hypertensive crisis should be treated with antihypertensive agents such as phentolamine. Respiratory function needs to be carefully monitored because excessive secretions may compromise the airway in an unconscious patient (Milhorn 1991).

Solvent (inhalant) abuse

Solvent abuse started amongst adolescents in the United States in the 1950s. It came to attention in the United Kingdom in the early 1970s and is now causing serious concern (Ashton 1990). The pharmacological actions of solvents in the central nervous system are not clear but, like alcohol, they may increase the fluidity of neuronal cell membranes and could also increase brain *GABA* function (Kaplan *et al.* 1994).

Epidemiology

In the NHSDA survey about 5 per cent of the American population had used solvents at least once in their lives, but fewer than 1 per cent had used them in the past month (National Institute on Drug Abuse 1991). It has been estimated that about 7 per cent of high school children in the United States have used solvents in the past year (Dinwiddie 1994). Similar findings have been reported in the United Kingdom, where Ramsey *et al.* (1989) estimated that between 3 and 10 per cent of adolescents experimented with solvents and up to 1 per cent abused them.

Solvent use occurs mainly in boys aged 8–19, with a peak in those aged 13–15 (Sourindrin 1985). Most of the young people known to use solvents do so as a *group activity*, and only about 5 per cent are solitary users. There is some evidence that a subgroup of solvent users have *antisocial personalities* and are likely to use and *abuse multiple substances*. However, the epidemiological data suggest that most who use solvents do so only a few times and then abandon the practice (Dinwiddie 1994).

Substances used and methods of use

The *substances* used are mainly solvents and adhesives (hence the name 'glue sniffing'), but also include many other substances such as petrol, cleaning fluid, aerosols of all kinds, agents used in fire extinguishers, and butane. Toluene and acetone are frequently used (Dinwiddie 1994). The *methods* of ingestion depend on the substance: they include inhalation from tops of bottles, beer cans, cloths held over the mouth, plastic bags, and sprays. As explained above, solvent use may be associated with taking other illicit drugs or with tobacco or alcohol consumption, which can be heavy.

Clinical effects

The clinical effects of solvents are similar to those of alcohol consumption. The central nervous system is first stimulated and then depressed. The stages of intoxication are similar to those of alcohol: *euphoria, blurring of vision, slurring of speech, incoordination, staggering gait, nausea, vomiting,* and *coma*. Compared with alcohol intoxication, solvent intoxication develops and wanes rapidly (within a few minutes, or up to two hours). There is early *disorientation* and two-fifths of cases may develop *hallucinations*, which are mainly visual and often frightening. This combination of symptoms may lead to serious accidents.

Adverse effects

Solvent abuse has many severe adverse effects, of which the most serious is *sudden death*. These fatalities occur during acute intoxication, and over 100 such deaths are reported annually in the United Kingdom (Ramsey *et al.* 1989). About half the deaths are due to the direct toxic effects of the solvent, particularly *cardiac arrhythmias* and *respiratory depression*. The rest are due to *trauma, asphyxia* (plastic bag over head), or *inhalation of stomach contents*.

Chronic users may show evidence of *neurotoxic effects*. Severe and disabling peripheral neuropathy has been described in teenagers abusing glues containing these substances. Other neurological adverse effects, particularly associated with toluene, include impaired cerebellar function, encephalitis, and dementia (Lolin 1989). Solvent abuse can also damage *other organs* including liver, kidney, heart, and lungs (Ramsey *et al.* 1989). Gastrointestinal symptoms include nausea, vomiting, and haematemesis.

Tolerance and dependence

Dependence can develop if use is regular, but physical withdrawal symptoms are unusual. When such symptoms occur they usually consist of sleep disturbance, irritability, nausea, tachycardia, and, rarely, hallucinations and delusions. With sustained use over 6–12 months, tolerance can develop.

Diagnosis

The diagnosis of acute solvent intoxication is suggested by several features: glue on the hands, face, or clothes; chemical smell on the breath; rapid onset and waning of intoxication; disorientation in time and space. Chronic abuse is diagnosed mainly on an admitted history of habitual consumption, increasing tolerance, and dependence. A suggestive feature is a facial rash ('glue-sniffers rash') caused by repeated inhalation from a bag.

Treatment

As noted above, for many users experimentation with solvents is a temporary phase which does not appear to lead on to abuse or dependence. Advice and support may well be sufficient for such subjects. However, a significant subgroup of those who abuse solvents also abuse other substances such as alcohol and opiates. Such subjects are more likely to have an antisocial personality disorder and to have experienced a chaotic and abusive family life. Treatment of this group is difficult, and a full range of psychological and social treatments is likely to be needed (see above). There is no specific pharmacotherapy for solvent abuse, but associated psychiatric disorders such as depression may require treatment in their own right.

Prevention

Prevention of solvent use and abuse may best be directed at the large numbers of young people who experiment with solvents through curiosity or peer pressure. Policies include the restriction of sales of volatile substances to children and adolescents. Education, particularly concerning the risk of severe injury and death (which can, of course, occur in occasional or first-time users), seems worthwhile. Wider social measures such as the provision of improved recreational facilities have also been advocated (Ashton 1990). For a review of solvent abuse see Dinwiddie (1994).

Further reading

Gerstein, G.R. and Green, L.W. (1993). *Preventing drug abuse: what do we know?* National Academy Press, Washington, DC.

Heather, N., Wodak, A., Nadelmann, E., and O'Hare, P. (1993). *Psychoactive drugs and harm reduction.* Whurr, London.

Helzer, J.E. and Canino, G.J. (1992). *Alcoholism in North America, Europe and Asia.* Oxford University Press.

Mendelson, J.H. and Mello, N.K. (1992). *Medical diagnosis and treatment of alcoholism.* McGraw-Hill, New York.

Miller, W.R. and Rollnick, S. (1991). *Motivational interviewing: preparing people to change addictive behaviour.* Guilford Press, London.

Robson, P. (1995). *Forbidden drugs.* Oxford University Press.

15

Problems of sexuality and gender identity

This chapter is concerned with four topics related to sexuality and gender: homosexual behaviour, sexual dysfunction, abnormalities of sexual preference, and disorders of gender identity. No account is given here of normal sexual physiology; the reader seeking further information on this subject is referred to a suitable account (e.g. Bancroft 1983). The clinician needs to be aware of the wide variations in sexual behaviour and in social attitudes towards this behaviour. Many behaviours that are regarded as abnormal when they are a preferred form of sexual activity are practised widely as a minor component of sexual activity. For example painful stimulation or the wearing of particular clothes to enhance sexual arousal are practised by many heterosexual couples, but these behaviours are abnormal when, as sexual masochism or fetishism, they become the preferred sexual activity (see pp. 503 and 497). Social attitudes to homosexual behaviour have varied widely between different countries and at different times, and masturbation was regarded as a form of sexual disorder in the late nineteenth century but is now incorporated in some forms of treatment (see p. 505).

For the clinician, awareness of these variations in behaviour and attitudes is important for two reasons. First, such awareness can help him to avoid imposing his own attitudes on his patients. Second, it can help him to recognize that some forms of sexual behaviour are abnormal if habitually preferred to normal sexual intercourse, but within normal limits if performed occasionally as a minor part of normal intercourse.

Homosexuality is discussed in this chapter not because it is considered a disorder, but because homosexual people seek psychological help for problems associated directly or indirectly with their sexual orientation. Sexual dysfunction denotes impaired or dissatisfying sexual enjoyment or performance. Such conditions are common, and find an obvious place in this chapter. Abnormalities of sexual preference are uncommon, but they take many forms and therefore require more space here than would otherwise be justified by their clinical importance. The final part of this chapter deals with gender identity, which is a person's sense of being male or female. When this sense of identity is at variance with the biological sex, the person is said to have a gender identity disorder.

Homosexuality

The term homosexuality denotes erotic thoughts and feelings towards a person of the same sex, as well as any associated sexual behaviour. People cannot be divided sharply into those who are homosexual and those who are heterosexual. There is a continuum, with exclusively heterosexual people at one end and exclusively homosexual people at the other; between these extremes are people who engage in varying degrees of both homosexual and heterosexual behaviour and relationships. The expression of homosexual behaviour varies with age and circumstances. The bisexual potential is greater in adolescence than in adult life, and homosexual behaviour is more likely to be expressed when heterosexual behaviour is unavailable, for example in prisons.

In a much quoted study, Kinsey *et al.* (1948) used a six-point scale to rate degree of homosexuality, and estimated that 10 per cent of men were 'more or less exclusively homosexual' (rating 5 or 6) for at least three years, and that 4 per cent of men were exclusively homosexual throughout their lives. In a subsequent study Kinsey *et al.* (1953) reported that 4 per cent of single women were persistently homosexual from the ages of 20 to 35, whilst Kenyon (1980) concluded that about 2 per cent of the adult female population were predominantly homosexual. Gagnon and Simon (1973) reanalysed Kinsey's data, but excluded respondents who had been in prisons or other institutions that might have increased homosexual activity; they found that 3 per cent of men and 1 per cent of women were exclusively homosexual.

Homosexual young people sometimes ask whether their sexual orientation is likely to change as they grow older. There are no adequate studies of this question, but clinical experience suggests that a person who has reached adult life without experiencing heterosexual feelings is unlikely to develop them later, especially when the person is a male with effeminate social behaviour. If young people have some heterosexual interests and are strongly motivated to develop them, they may change their orientation, especially if they have associated anxiety about heterosexual interest arising understandably from previous experience.

Behaviour and personality of homosexual men

Homosexual behaviour parallels the range of tender feelings, physical intercourse, and social behaviour that makes up heterosexual behaviour. Homosexual physical intercourse includes oral–genital contact, mutual masturbation, and less often anal intercourse. The partners usually change roles in these acts, but with some couples one partner is always passive and the other is always active. Relationships between homosexual men do not usually last as long as relationships between men and women, or those between homosexual women.

Some exclusively homosexual men experience strong feelings of identity with other homosexuals and seek their company, often in clubs or bars. As with heterosexual people, there is a promiscuous minority who seek brief experiences with other promiscuous homosexuals. A small minority adopt feminine mannerisms and dress in women's clothes to attract others.

Most homosexual men have a way of life like that of single heterosexual men, but a few prefer work and leisure activities that would usually be undertaken by a woman.

Most homosexual men are as contented as heterosexual men and have a stable relationship with a partner. For others, homosexuality can lead to difficulties at several times of life. In adolescence there may be distress as sexual orientation is recognized for the first time, and a decision has to be made as to whether to follow or suppress homosexual feelings. With the approach of middle age, sexual partnerships may become difficult to arrange; there may then be loneliness and depression if the man does not have other relationships built on friendship. A few middle-aged homosexuals find it increasingly difficult to obtain sexual partners of their own age, and so turn towards younger homosexual prostitutes. (It is exceptional for these men to turn to prepubertal children; homosexual and heterosexual paedophiles are described on p. 500.)

Homosexual men vary in personality as much as heterosexual men. In homosexual men (as in heterosexual men) disorder of personality is more likely to lead to difficulty with other people or with the law, and more likely to lead to referral to a psychiatrist.

Behaviour and personality of homosexual women

Among homosexual women, tender feelings and social activities shared with the partner are important sources of satisfaction. Physical intercourse between homosexual women includes caressing, breast stimulation, mutual masturbation, and oral–genital contact (cunnilingus). A minority of women practise full body contact with genital friction or pressure (tribadism), or use a vibrator or artificial penis. Active and passive roles are usually exchanged, although one partner may habitually take the active role. Other sexual practices such as sexual sadomasochism occur occasionally between female homosexuals, as between men and women. The **social behaviour** of homosexual women is usually like that of heterosexual women, although some homosexual women seek work and leisure activities more often associated with men. A few female homosexuals dress and behave in a masculine way.

As with men, among women there is a continuum between the exclusively heterosexual and the exclusively homosexual. Most female homosexuals engage in heterosexual relationships at some time, even though they may obtain little satisfaction from them, and some marry. As a group, they are less promiscuous than homosexual men and are less likely to seek transient sexual relationships in bars and other places. Homosexual women are more likely to form lasting relationships, and are less likely to suffer loneliness and depression in middle life (Saghir and Robins 1973).

All kinds of **personality** are represented among homosexual women.

Legal aspects of homosexual behaviour

Laws about homosexual behaviour differ in different countries and states. In England and Wales, there are no laws specifically concerning homosexual behaviour between women. The age of legal consent for male homosexual intercourse is 18 (for the female partner in heterosexual intercourse it is 16). An attempt to obtain a partner in a public place breaks the law. Readers working in other countries are advised to enquire about the local rules of law.

Determinants of homosexual behaviour

The reasons why some people prefer homosexual to heterosexual behaviour are not known. Several causes have been investigated: genetic factors, hormonal influences acting before birth, and psychological and social factors. There have been studies comparing the brains of homosexuals with those of heterosexuals. The results of these various studies will be reviewed briefly.

Genetic causes

It has long been supposed that homosexual behaviour might be determined by heredity. This view was apparently confirmed when Kallmann (1952) reported a small study in which he found 100 per cent concordance of homosexuality in 40 monozygotic (MZ) male twin pairs compared with 12 per cent concordance in 26 dizygotic (DZ) twin pairs. Although this report of complete concordance between MZ pairs has not been confirmed, other investigators have found that MZ twins are more often alike in respect of homosexuality than are DZ twins (Eckhert *et al.* 1986; Bailey and Pillard 1991; King and McDonald 1992). These findings are compatible with a genetic aetiology, but they could arise if the early environment of MZ twins is more similar than that of DZ twins. Adoption studies are needed to exclude this possibility. In the only adoption study reported to date, Eckhart *et al.* (1986) studied a small group of twin pairs in which the proband was homosexual and the twins had been reared apart from infancy. Of the two male twin pairs, one pair was concordant for homosexuality. Of the four female twin pairs, none was concordant. It has been suggested that there might be an abnormality of the **sex chromosomes** in homosexuals, but no convincing evidence has been reported in either males or females. Although a linkage with chromosome Xq28 has been reported (Hamer *et al.* 1993), the methodology of the study has been criticized (Baron 1993) and further studies are required.

Hormonal theories

It has been suggested that homosexuality arises from abnormal development of the brain caused by hormonal abnormalities before birth. Research with animals suggests that the intrinsic pattern of the mammalian brain may be female, and that the development of male brain characteristics depends on androgen production by the fetus. In animals male or female brain characteristics are inferred from differences in reproductive behaviour such as mounting by the male and the lordosis of the sexually receptive female; it is doubtful how far information about the control of these behaviours can be extrapolated to human sexual behaviour. Also, although male animals may engage at times in sexual behaviour with other males, exclusively homosexual behaviour occurs only in human beings. If prenatal hormonal levels are a cause of homosexuality, there should be an excess of homosexuals among males with syndromes of prenatal androgen deficiency or insensitivity, and among females with syndromes involving androgen excess. No such increase has been observed among men with androgen deficiency or insensitivity (Byrne and Parsons 1993). However, an increased rate of homosexuality has been reported among women with congenital virilizing adrenal hyperplasia (Money *et al.* 1984; Byrne and Parsons 1993). The interpretation of this finding is uncertain because these females are born with masculinized external genitalia, and it is possible that sexual orientation in adult life is affected by this feature rather than by a direct effect of hormones on the brain.

Neuroanatomical differences

There have been reports of differences between homosexual and heterosexual men in the structure of the hypothalamus (LeVay 1991) and suprachiasmatic nucleus (Swaab and Hoffman 1990), but these studies have been criticized on technical grounds (Byrne and Parsons 1993) and further investigation is required.

Psychological causes

Studies of the *upbringing* of homosexual men have been mainly based on the recollections of adult homosexuals about their childhood, often as part of psychotherapeutic treatment. This method is unreliable because the homosexuals studied are unlikely to be representative, and because the recall of events in childhood is generally open to error. It was on the basis of this kind of evidence that Bieber (1962) and others concluded that, compared with heterosexuals, homosexual men have more often experienced a poor relationship with or prolonged absence of the father. Other

psychoanalysts have reported that the mothers of homosexual men are overprotective or unduly intimate. This view was not upheld when Siegelman (1974) compared homosexual and heterosexual men (all of whom had normal neuroticism scores) and found no evidence of abnormal parental behaviour. If there is any association with upbringing, it could act by inhibiting the development of heterosexual behaviour rather than by directly causing homosexuality.

Some psychoanalysts have suggested that female homosexuality results from failure to resolve unduly close relationships with the parents in early childhood, with the result that intimate involvement with men is frightening and women become the preferred object of love. In a study of homosexual women, Wolff (1971) concluded that their mothers were rejecting or indifferent. Kenyon (1968) found that, compared with heterosexual women, more homosexual women reported a poor relationship with both mother and father; also, a quarter of the homosexual women had parents who had divorced (compared with 5 per cent in the controls).

In the absence of conclusive scientific data, a scheme that can usefully inform clinical practice is that children develop with the capacity for both heterosexual and homosexual behaviour, and that various experiences determine which behaviour develops more strongly (Bancroft 1975). *Heterosexual* development may be *impeded by* repressive family attitudes towards sex, by frightening early heterosexual experiences, or by a lack of self-confidence with the opposite sex. Male *homosexual* development may be *encouraged by* unusually close and satisfying relationships with men.

Social determinants of homosexuality

The expression of homosexual behaviour may vary widely in relation to its varying acceptance in different societies. Ford and Beach (1952) reported that, among 76 societies described in the literature, homosexuality was socially acceptable, at least for certain people, in 64 per cent.

Psychological problems of homosexual people

Homosexual men may consult doctors about five kinds of problem. The first concerns shy and sexually inexperienced young men who fear relationships with women and conclude that they themselves may be homosexual, but in fact are not. The second problem is presented by young men who have realized, correctly, that they are predominantly homosexual and are bewildered about the implications for their lives. The third problem concerns men who have bisexual inclinations and want to discuss how best to adjust to this situation. The fourth problem concerns the established homosexual who has become depressed or anxious because of personal or social difficulties arising from sexual relationships. In these four groups, the doctor's principal role is to help the patient to clarify his thoughts. The fifth kind of problem concerns the homosexual who is concerned about the possibility of being infected with AIDS or has been found to be HIV positive. Such a person requires suitable medical treatment and counselling.

Sometimes a homosexual person, usually a man, asks if he can be helped to suppress homosexual behaviour and to strengthen any heterosexual feelings that he has. Such a person can help this process by avoiding situations that stimulate his homosexual feelings, while seeking opportunities for social encounters with women by joining a club or study group. He should try to distract himself from homosexual thoughts and to avoid masturbating with homosexual fantasies since this activity reinforces homosexual behaviour. It is not known with certainty whether any form of psychotherapy can achieve more than this simple advice. Nowadays behavioural treatment is concerned mainly with encouraging heterosexual behaviour by employing methods used to treat sexual dysfunction (see p. 494). In the past aversion methods were used to suppress homosexual imagery (Bancroft 1974), but they are no longer employed.

Homosexual women are less likely than homosexual men to ask for help in changing their sexual ideas and behaviour, and more likely to ask for help with problems arising in social relationships. These problems include depression or jealousy when a homosexual relationship is insecure. Many homosexual women also have a male partner and some are married. These women may seek advice about problems in the relationship with the man, or about dysfunction in heterosexual intercourse.

Some male and female homosexual couples with stable social relationships ask for help with problems of achieving satisfying homosexual intercourse. Masters and Johnson (1979) have described techniques for helping them.

The classification of problems of sex and gender identity

In both DSMIV and ICD10 problems of sex and gender identity are classified in three groups: (i) sexual dysfunction; (ii) a group of disorders named paraphilias in DSMIV and disorders of sexual preference in ICD10; (iii) gender identity disorders (Table 15.1). In DSMIV these conditions are grouped together under the heading of sexual and gender identity disorders. In ICD10 they appear in two parts of the classification: sexual dysfunction is part of F5 'physiological dysfunction associated with mental or behavioural factors', while disorders of sexual preference and of gender identity are part of F6 'abnormalities of adult personality and behaviour'.

In both systems each of the three main categories is classified further. *Sexual dysfunctions* are divided according to the stage of the sexual response that is mainly affected: (i) disorders of sexual desire; (ii) disorders of sexual arousal; (iii) disorders of orgasm. There are also categories for the painful conditions vaginismus and dyspareunia. The slight differences in terminology between DSMIV and ICD10 are shown in Table 15.1.

Paraphilias (abnormalities of sexual preference) are subdivided in the two classifications. The only substantial difference is that DSMIV has a category for the uncommon disorder known as frotteurism, a condition which in ICD10 would be classified under 'other abnormalities of sexual preference'.

Table 15.1 *Classification of sexual and gender identity disorders*

DSMIV	ICD10
Sexual dysfunction	**Sexual dysfunction not caused by organic disorders***
Sexual desire disorders	Lack or loss of sexual desire
Hypoactive sexual desire disorder	
Sexual aversion disorder	Sexual aversion and lack of sexual enjoyment
Sexual arousal disorder	
Female sexual arousal disorder	Failure of genital response
Male erectile disorder	
Orgasm disorders	Orgasmic dysfunction
Female orgasmic disorder	
Male orgasmic disorder	
Premature ejaculation	Premature ejaculation
Sexual pain disorders	
Dyspareunia	Non-organic dyspareunia
Vaginismus	Non-organic vaginismus
Sexual dysfunction due to a general medical condition	Excessive sexual drive
Paraphilias	**Disorders of sexual preference†**
Exhibitionism	Exhibitionism
Fetishism	Fetishism
Frotteurism	
Paedophilia	Paedophilia
Sexual masochism	Sadomasochism
Sexual sadism	
Voyeurism	Voyeurism
Transvestic fetishism	Fetishistic transvestism
Gender identity disorders	**Gender identity disorders‡**
In children	
In adolescents and adults	

*In ICD10 sexual dysfunction is a subgroup of F5, behavioural syndromes associated with physiological disturbances and physical factors.
†In ICD10 disorders of sexual preference is a subgroup of F6, disorders of adult personality and behaviour (to aid comparison with DSMIV, the order in which the disorders appear has been changed from that in the text of ICD10).
‡In ICD10 gender identity disorders are a subgroup of F6.

In DSMIV *gender identity disorders* are divided into those in children and those in adolescents and adults.

Sexual dysfunctions

In men sexual dysfunction refers to repeated impairment of normal sexual interest and/or performance. In women it refers more often to a repeated unsatisfactory quality to the experience; sexual intercourse can be completed, but without enjoyment (Bancroft *et al.* 1982). What is regarded as normal sexual intercourse, and therefore what is thought to be impaired or unsatisfactory, depends in part on the expectations of the two people concerned. For example when the woman is regularly unable to achieve orgasm, one couple may regard it as normal whilst another may seek treatment. As explained above, problems of sexual dysfunction are classified into those affecting (1) sexual desire and sexual enjoyment, (2) the genital response (erectile impotence in men, lack of arousal in women), and (3) orgasm (premature or retarded ejaculation in men, orgasmic dysfunction in women). A further group includes problems resulting in pain: vaginismus and dyspareunia in women, and painful ejaculation in men. It should be remembered that sexual function is not always disclosed directly but may be revealed during enquiries about another complaint, such as depression or poor sleep, or gynaecological symptoms.

Prevalence of sexual dysfunctions

The prevalence of sexual dysfunction among men is not known with certainty because surveys have failed to obtain a random sample or to characterize the dysfunctions with sufficient precision. Erectile dysfunction that is partial or temporary, rather than complete and permanent, is not uncommon, especially with increasing age. In one population survey, total and persisting erectile dysfunction was reported by 1.3 per cent of American men aged under 35, 6.7 per cent aged under 50, and 18.4 per cent aged under 60 (cumulative figures from Kinsey *et al.* (1948)).

Erectile dysfunction is the most frequent complaint among men presenting for treatment of sexual dysfunction. The prevalence of *premature ejaculation* is not known exactly because it depends in part on the partner's expectations as well as her speed of sexual response. Inhibited male orgasm appears to be uncommon; it was reported to Kinsey *et al.* (1948) by only six of 4108 men interviewed, and it was reported by 5 per cent of men attending a sex therapy clinic (Hawton *et al.* 1986).

Among 436 women randomly selected for interview in a community survey, at least one kind of sexual dysfunction was reported by more than one in three. *Impaired sexual interest* (17 per cent) and *infrequent orgasm* (16 per cent) were about equally common. Dyspareunia was reported by 8 per cent. Among women seeking help for sexual disorders, impaired sexual interest was described by about half and orgasmic dysfunction by about 20 per cent (Hawton 1985). In about a third of couples seen for treatment, both partners have a problem, usually low libido in the woman and premature ejaculation in the man. Sexual dysfunction is found in about 10 per cent of psychiatric out-patients (Swan and Wilson 1979).

Lack or loss of sexual desire (hyposexual desire disorder)

Complaints of diminished sexual desire are much more common among women than among men. They often reflect general problems in the relationship between the partners. Sometimes there is a specific sexual problem which may be due to long-standing inhibitions about sex or an apparent biological variation of sexual drive that cannot be modified.

Sexual desire is reduced during a depressive disorder. In most cases it returns to the previous level as the depressive disorder resolves, but in a few the impairment persists. This may explain why an increased frequency of previous depressive disorders has been reported among people attending for treatment with a current complaint of impaired sexual desire (Schreiner-Engel and Schiavi 1986).

Sexual aversion disorder

Sexual enjoyment may be replaced by a positive aversion to genital contact. When this aversion is persistent or severe and accompanied by avoidance of almost all genital sexual contact with a sexual partner, the condition is classified as sexual aversion disorder. The causes of the condition are not well understood; they seem to be similar to those of hyposexual desire disorder.

Specific sexual fears

A few women are made extremely anxious by specific aspects of the sexual act, such as being touched on the genitalia, the sight or smell of seminal fluid, or even kissing. Despite these specific fears, they may still enjoy other parts of sexual intercourse.

Sexual arousal disorders

Female sexual arousal disorder

Lack of sexual arousal in the female appears as reduced vaginal lubrication. This reduction may be due to (i) inadequate sexual foreplay by the partner, (ii) lack of sexual interest, or (iii) anxiety about intercourse. After the menopause, hormonal changes may lead to reduced vaginal secretions.

Male erectile disorder

This condition is the inability to reach an erection or to sustain it long enough for satisfactory coitus. It may be present from the first attempt at intercourse (primary) or develop after a period of normal function (secondary). It is more common among older than younger men (in contrast with premature ejaculation (see below)). If a man has more than one sexual partner, he should be asked whether the failure occurs with each partner or only with one. It is also important to ask whether erection occurs on waking or in response to masturbation. If erection occurs in these other circumstances, the erectile failure is likely to be psychological rather than physiological in origin. Erectile failure may be a transient disorder arising at times of stress, or may reflect loss of interest in the sexual partner.

Orgasmic dysfunction

Female orgasmic disorder

Orgasmic dysfunction in women may be related to the man's inexperience as well as the woman's capacity to reach orgasm. Whether it is regarded as a disorder depends on social attitudes and the expectations of the individual. Many women do not regularly have an orgasm during vaginal intercourse but do have orgasm in response to genital caressing. About 25 per cent of women have no orgasm during intercourse for the first year of marriage (Gebhard *et al.* 1970). In the past, absence of orgasm was not generally thought to be abnormal. Attitudes then changed, so that some women regarded themselves as abnormal although previously they would have been content with the intimacy of sexual relations without regular orgasm during intercourse.

Male orgasmic disorder

This term refers to serious delay in or absence of ejaculation. Usually the delay occurs only during coitus, but it may also occur in masturbation. It is usually associated with a general psychological inhibition about sexual relations, but it may be caused by drugs including antipsychotics, monoamine oxidase inhibitors, and specific serotonin uptake inhibitors.

Premature ejaculation

This term refers to habitual ejaculation before penetration or so soon afterwards that the women has not gained pleasure. It is more common among younger than older men, especially during their first sexual relationships.

Sexual pain disorders

Dyspareunia

This term refers to pain on intercourse. Such pain has many causes. Pain experienced after partial penetration may result from impaired lubrication of the vagina, from scars or other painful lesions, or from the muscle spasm of vaginismus. Pain on deep penetration strongly suggests pelvic pathology such as endometriosis, ovarian cysts and tumours, or pelvic infection, though it can be caused by impaired lubrication associated with low sexual arousal.

Vaginismus

This is spasm of the vaginal muscles which causes pain when intercourse is attempted and there is no physical lesion causing pain. The spasm is usually part of a phobic response associated with fears about penetration, and may be made worse by an inexperienced partner. Spasms often begin as soon as the man attempts to enter the vagina; in severe cases it occurs even when the woman attempts to introduce her own finger. Severe vaginismus may prevent consummation of marriage. So-called 'virgin wives' may have a generalized fear and guilt about sexual relationships rather than a specific fear of penetration. Some women with vaginismus are married to passive men who have low libido and are able to accept their wives' refusal to permit full sexual relations (Dawkins 1961; Friedman 1962). Low libido in the man becomes evident when the treatment has reduced the woman's problem.

Pain on ejaculation

This problem is uncommon. The usual causes are urethritis or prostatitis, but sometimes no cause is found.

Aetiology of sexual dysfunction

Factors common to more than one form of sexual dysfunction

Sexual dysfunction arises from varying combinations of a *poor general relationship* with the partner, *low sexual drive, ignorance* about sexual technique, and *anxiety* about sexual performance. Other important factors are physical *illness*, depressive and anxiety disorder,

medication, and *alcohol* or *drug abuse*. Some of these factors will now be considered.

Sexual drive varies between people but the reason for this is not known. Endocrine factors have been suggested because in the male the increasing sexual drive at puberty is related to an increased output of androgens. Castration, treatment with oestrogens, or the administration of anti-androgenic drugs also reduce sexual drive in the male. However, no convincing association has been shown between androgens and low sexual drive in men seeking help for this problem. Also, treatment with androgens does not usually increase sexual drive in men with normal endocrine function. Small doses of androgens increase sexual drive in women (Hawton 1985). However, androgens are not used in treatment because of their side-effects.

Anxiety is an important cause of sexual dysfunction. Sometimes anxiety is an understandable consequence of an earlier frightening experience such as a man's failure in his first attempt at intercourse, or a woman's experience of sexual abuse or assault. Sometimes the anxiety relates to frightening accounts of sexual relationships received from parents or other people. Psychoanalysts suggest that anxiety about sexual relationships originates from even earlier experiences, namely failure to resolve the oedipal complex in boys or the corresponding attachment to the father in girls (see p. 82). Such ideas are difficult to test.

Physical or psychiatric illness and associated **treatment** can cause sexual dysfunction. Sometimes such dysfunction dates from a period of abstinence associated with minor physical illness, pregnancy, or childbirth, or from the debilitating effects of physical illness (Table 15.2). Of the diseases that have a direct effect on sexual performance, diabetes mellitus is particularly important. Between a third and a half of **diabetic** men experience erectile dysfunction as a result of either neuropathology affecting the autonomic nerves mediating erection or vascular disorders. Impaired ejaculation also occurs. Some diabetic women may be affected in a corresponding way, although this is less certain (Fairburn 1981). Sexual dysfunction after **myocardial infarction** may result from anxiety rather than from physical causes. Most other physical causes, including effects of treatment, are self-evident. Nevertheless, doctors often fail to think of the sexual consequences of disease and the (often unexpressed) problems that result. For a comprehensive account of the effects of physical illness on sexual function, see Kolodny *et al.* (1979).

Several **drugs** have side-effects that involve sexual function (Table 15.3). The most important drugs are antihypertensives (especially adrenoceptor antagonists), major tranquillizers (especially thioridazine), monoamine oxidase inhibitors, and specific serotonin re-uptake inhibitors. The role of oral contraceptives is still uncertain; if they cause dysfunction, it is probably only in a minority of women (Hawton and Oppenheimer 1983). Anxiolytics, sedatives, and hormones have more effect on the sexual activity of men than of women. Apart from these prescribed drugs, the excessive use of **alcohol** and street drugs impairs sexual performance.

When assessing patients it is important to recognize that both psychological and physical factors are often present.

The aetiology of particular conditions

Male erectile disorder Primary cases may occur through a combination of low sexual drive and anxiety about sexual performance. Secondary cases may arise from diminishing sexual drive in the middle-aged or elderly, loss of interest in the sexual partner, anxiety, depressive disorder, and organic disease and its treatment. A few cases are due to abnormalities of the vascular supply to the penile erectile tissue, including reduced arterial perfusion, increased venous leakage, and Peyronie's disease (Kirby *et al.* 1991).

Female orgasmic disorder This disorder arises from normal variations in sexual drive, poor sexual technique by the partner, lack of affection

Table 15.2 *Medical and surgical conditions commonly associated with sexual dysfunctions*

Medical	
Endocrine	Diabetes, hyperthyroidism, myxoedema, Addison's disease, hyperprolactinaemia
Gynaecological	Vaginitis, endometriosis, pelvic infections
Cardiovascular	Angina pectoris, previous myocardial infarction
Respiratory	Asthma, obstructive airways disease
Arthritic	Arthritis from any cause
Renal	Renal failure with or without dialysis
Neurological	Pelvic autonomic neuropathy, spinal cord lesions, stroke
Surgical	Mastectomy
	Colostomy, ileostomy
	Oophorectomy
	Episiotomy, operations for prolapse
	Amputation

Modified from Hawton and Oppenheimer 1983.

Table 15.3 *Some drugs that may impair sexual function*

Alcohol	
Antihypertensives	Guanethidine, beta-adrenoceptor antagonists, methyl-dopa
Antidepressants	Tricyclics, monoamine oxidase inhibitors, specific serotonin re-uptake inhibitors
Anxiolytics and hypnotics	Benzodiazepines, barbiturates
Antipsychotics	Especially thioridazine
Anti-inflammatory drugs	Indomethacin
Anticholinergics	Probanthine
Diuretics	Bendrofluazide
Hormones	Steroids, possibly oral contraceptives

Modified from Hawton and Oppenheimer 1983.

for the partner, tiredness, depressive disorder, physical illness, and the effects of medication.

Premature ejaculation This disorder is so common in sexually inexperienced young men that it can be regarded as a normal variation. When it persists, it is often because of fear of failure.

Vaginismus The causes of this disorder are described on p. 490.

Dyspareunia This disorder generally has physical causes, but it may result from vaginismus or from failure of arousal with consequent lack of vaginal lubrication.

The assessment of a patient who presents with sexual dysfunction

Whenever possible the sexual partner should be interviewed as well as the patient. The two should be seen separately, and then together. The first step is to define clearly the **nature of the problem** as it appears to each partner. Details should not be omitted because the interviewer is too embarrassed to make full enquiries. The **origin and course** of the dysfunction is recorded next. It is especially important to discover whether the problem has always been present or whether it started after a period of normal function. Each of the partners should be asked, separately, whether the problem has occurred with other partners. The general strength of sexual **drive** is assessed by

Table 15.4 *Important points in the physical examination of men presenting with sexual dysfunctions*

General examination (directed especially to evidence of diabetes mellitus, thyroid disorder, and adrenal disorder)
Hair distribution
Gynaecomastia
Blood pressure
Peripheral pulses
Ocular fundi
Reflexes
Peripheral sensation

Genital examination
Penis: congenital abnormalities, foreskin, pulses, tenderness, plaques, infection, urethral discharge
Testicles: size, symmetry, texture, sensation

Adapted from Hawton 1985.

asking about frequency of intercourse and masturbation, and about sexual thoughts and feelings of sexual arousal.

Next, an assessment is made of **knowledge** of sexual techniques and then of **anxieties** about sex. Possible sources of misinformation and anxiety are considered by asking about the family's attitude to sex, the kind of sex education received by each partner, and the extent of sexual experience with other partners. Each partner should be asked about the sexual **technique** of the other.

Social relationships with the opposite sex are considered next. The interviewer should find out whether either partner is shy and socially inhibited. If the couple are husband and wife, or otherwise cohabiting, disharmony in their relationship should be enquired into carefully. If the couple lack a loving relationship in their everyday life, it is unlikely that they will achieve a fully satisfying sexual relationship. It is important to remember that some couples ask for help with sexual problems which are the result and not (as they suggest) the cause of marital conflict. The interviewer should also find out why the patient has chosen to seek treatment at this time. The

reason may be that the sexual problem has increased, but there may be another reason such as the partner's increased dissatisfaction and threats to leave.

Careful enquiry should be made for evidence of **psychiatric disorder** in either partner, especially depressive disorder, which might account for the sexual problem. Finally, questions are asked about **physical illness** and its **treatment**, about psychotropic medication, and about **abuse of alcohol or drugs**. If the general practitioner or another specialist has not already done so, a **physical examination** should be carried out (Table 15.4). **Laboratory tests** should be arranged in appropriate cases; for example fasting blood sugar, testosterone, sex-hormone-binding globulin, luteinizing hormone, and prolactin in men with erectile dysfunction.

Physiological methods have been used to assess penile blood flow and pelvic reflexes in cases of erectile dysfunction. Several methods have been used, including Doppler imaging after the induction of intracavernosis smooth relaxation with papaverine (Lue *et al.* 1985). These methods can identify uncommon organic causes (see p. 491–2).

Treatment of sexual dysfunction

Before directing treatment to the sexual problem, it is important to consider whether the couple need marital therapy instead. If it is appropriate to focus treatment on the sexual problem, **advice and education** may be all that is needed. If **specific sex therapy** is appropriate, it should be directed to both partners whenever possible. The usual approach, which owes much to the work of Masters and Johnson (1970), has *four characteristic features*. First, the partners are *treated together*. Second, they are helped to *communicate better*, through words and actions, about their sexual relationship. Third, they receive *education* about the anatomy and physiology of sexual intercourse. Fourth, they are given a series of *'graded tasks'*. Masters and Johnson held that two other factors were important. The first was that treatment should be intensive, for example seeing both partners every day for up to three weeks. The second was that treatment should be carried out by a man and a woman working as co-therapists. It has been shown that neither of these measures is essential; good results can be obtained when treatment is given once a week and when only one therapist sees the couples.

Treatment of couples. Although better results are obtained with treatment of couples, some help can be given to a patient who has no regular partner. Such a patient can at least discuss his difficulties and possible ways of overcoming them. Discussion of this kind can sometimes help to overcome social inhibitions and so aid in developing a relationship with someone of the opposite sex.

Communication is not only the ability to talk freely about specific sexual problems; it is also concerned with increasing understanding of the other person's wishes and feelings. Each partner may believe that the other should know instinctively how to give pleasure during intercourse, so that failure to please is attributed to lack of concern or affection rather than to ignorance. Such failure can be overcome by helping the partners to express their own desires more frankly.

Education stresses the physiology of the sexual response. For example if the problem is anor- gasmia in the woman, the doctor may explain the longer time needed for a woman to reach sexual arousal, and may emphasize the importance of foreplay, including clitoral stimulation, in bringing about vaginal lubrication. Suitably chosen sex education books can reinforce the therapist's advice. Such counselling is often the most important part of the treatment of sexual dysfunctions.

Graded tasks begin with simple tender physical contact. The couples are encouraged to caress any part of the other person's body except the genitalia in order to give enjoyment (Masters and Johnson call this the 'sensate focus'). Next, the couple may engage in mutual masturbation, but not in penetration at this stage. At both stages, the partners are encouraged to discover the experience most enjoyed by the other person and then to provide this experience. They are strongly discouraged from checking their own state of sexual arousal because this checking generally has an inhibiting effect. Such checking is a common habit in people with sexual disorder, and has been called the 'spectator role'. Graded tasks are not only directly beneficial; they also help to uncover hidden fears or areas of ignorance that need to be discussed.

Methods for specific problems

Male erectile dysfunction Urologists have developed several physical methods for patients with impotence due to vascular or neurogenic abnormalities including those secondary to diabetes. These methods include drugs, vacuum devices, the surgical correction of vascular abnormalities, and penile prostheses.

Oral medication: benefits have been reported from opiate antagonists and naltrexone (Fabbri *et al.* 1989) and from the dopamine antagonist bromocyptine, though benefits from the latter are less well supported than the others (Gregoire 1992). Treatment with androgens or with the alpha-2 adrenoreceptor antagonist yohimbine has not been shown to be effective.

Intracavernosal injections of the smooth muscle relaxant papaverine, or the alpha receptor blocker phenoxybenzamine, produce erection and have been used to treat impotence (Virag 1982, Padma-Nathan *et al.* 1987), as have injections of prostaglandins (Lee *et al.* 1988). Small doses are given and increased gradually. When an appropriate dose has been determined patients are taught to inject themselves. (Doses are not given here because the treatment should be learnt from a doctor experienced in its use.) Overdose can lead to prolonged erection which may require the aspiration of blood and the injection of an alpha-1 agonist such as phenylephrine (Kirby 1994).

Vacuum devices can be tried for patients who do not respond to intracavernosal drug injections (Witherington 1989). The penis is placed in a surrounding cylinder in which the pressure is reduced; an erection follows, and is maintained by applying a restricting band to the base of the penis before the cylinder is removed. Although this procedure is generally effective in producing an erection, it is disliked by many patients.

Surgical methods have been used to treat erectile disorder resulting from proven vascular abnormalities. Microsurgery can be used to reduce venous leakage or to revascularize the corpora cavernosa when there are stenoses or occlusions in the arteries. Short-term improvement rates of about 50 per cent have been reported in selected cases (Goldstein 1986). An alternative approach is to insert a *penile prosthesis* which may be semirigid or capable of being inflated before intercourse (Wilson, S. K. *et al.* 1988).

For a review of treatments for erectile dysfunction see NIH Consensus Development Panel on Impotence (1993).

Premature ejaculation This disorder can be treated with the 'squeeze technique'. When the man indicates that he will soon have an orgasm, the woman grips the penis for a few seconds and then releases it suddenly. Intercourse is then continued. An alternative 'start–stop' method has been described in which the woman attempts to regulate the amount of sexual stimulation during intercourse.

Results of treatment The general methods described above are followed by a successful outcome in about a third of cases, and by a worthwhile improvement in a further third (Bancroft and Coles 1976). The outcome is least good for problems associated with low sexual desire. Outcome is better among patients who engage wholeheartedly in treatment. The results are as good in problems of long duration as in others (Hawton *et al.* 1986). See Bancroft *et al.* (1986) for a review of research into sex therapy.

Sexual dysfunction among the physically handicapped

Physically handicapped people have sexual problems arising from several sources: direct effects on sexual function, for example disease of the nervous system affecting the autonomic nerve supply, the general effects of tiredness and pain, fears about the effects of intercourse on the handicapping condition, and lack of information about the sexual activities of other people with the same disability. Much can be done to help disabled people by discussing the forms of sexual activity that are possible despite their disability, and, if appropriate, adapting the methods already described for treating sexual dysfunction. See Stewart (1978) and Crown (1978) for accounts of sexual problems among the disabled.

Paraphilias: abnormalities of sexual preference

The study of these disorders in the past

For centuries, abnormalities of sexual preference were regarded as offences against the laws of religion rather than conditions that doctors should study and treat. The systematic investigation of these disorders began in the 1870s with the work of Krafft-Ebing, Hirschfield, Schrenk-Notzing, and Havelock Ellis. Krafft-Ebing (1840–1902), a professor of psychiatry in Vienna, compiled a systematic account of paraphilias in his book *Psychopathia sexualis*, which was first published in 1886 and later achieved twelve editions and translation into seven languages. In 1899 Magnus Hirschfield founded a journal (*Jahrbuch für sexuelle Zwischenstufen*) devoted to the study of abnormalities of sexual preference. Krafft-Ebing considered that these conditions were due mainly to hereditary causes, though the latter could be modified by social and psychological factors.

About the same time, Schrenck-Notzing developed psychological treatments for both sexual inadequacy and abnormalities of sexual preference, and reported striking successes from the use of therapeutic suggestion (Schrenk-Notzing 1895). In England, the study of sexual disorders was particularly associated with the name of Havelock Ellis (1859–1939).

Freud attempted to explain these conditions as failures of the developmental psychosexual processes that he believed he had identified in normal children. As a consequence, most of the literature on paraphilias has been in the psychoanalytic tradition and until recently treatment has been largely centred on psychoanalytic principles. As described later, these approaches have not been successful either in explaining the conditions or in modifying them. More recently behavioural explanations have been proposed and behavioural treatments have been tried, though with mixed success (see below).

The concept of abnormal sexual preference

This concept has three aspects. The first aspect is social: the behaviour does not conform to some generally accepted view of what is normal. The accepted view is not the same in every society or at every period of history; for example regular masturbation was regarded as abnormal by many medical writers in Victorian England. The second aspect concerns the harm that might be done to the other person involved in the sexual behaviour. Intercourse with young children or extreme forms of sexual sadism are examples. The third aspect is the suffering experienced by the person himself. This suffering is related to the attitudes of the society in which the person lives (for example attitudes to cross-dressing), to conflict between the person's sexual urges and his moral standards, and to the person's awareness of harm or distress caused to others.

General considerations

Abnormalities of sexual preference may come to medical attention in various ways, and the doctor should be aware of the different modes of presentation.

A doctor may be consulted directly by the person with the abnormalities. He may be asked to help by the spouse or other sexual partner, sometimes because the behaviour has just been discovered, and sometimes because known behaviour has become more frequent and can be tolerated no longer. Sometimes the problem is presented as sexual dysfunction, and the abnormality of sexual preference is discovered only in the course of history taking.

A doctor may also be asked for an opinion about a patient charged with an offence arising from an abnormality of sexual preference. Offences of this kind include indecent exposure, the behaviour of a 'peeping Tom', the stealing of clothes by fetishists, appearing in public in clothes of the opposite sex, incest and other assaults upon children, and rape. With two exceptions these offences are discussed below in relation to the corresponding abnormality of sexual preference. The exceptions are incest and rape, with are considered in Chapter 22 (pp. 765 and 766).

There are different opinions about the extent to which doctors should attempt to alter abnormal sexual preferences. There seems to be no reason why doctors should not try to help people who wish to alter unusual patterns of sexual behaviour, but they should not try to impose treatment on people who do not want it. How to decide who really wants help is a difficult point that is taken up later in this chapter.

Pornography

People with abnormal sexual preferences often read pornographic material. Doctors may be asked about the effects of such publications. It is not known whether these publications merely provide a harmless outlet for sexual impulses, including those that might otherwise be inflicted on another person, or whether they increase such impulses and so encourage paraphilias and increase sexual offences. Epidemiological studies have attempted to relate the numbers of sexual offences to changes in the law on pornography (as in Denmark), but they have given inconclusive results. Clinical studies of paraphilias suggest that the reading of pornographic literature increases related fantasies experienced during sexual arousal. However, it is not known whether the strengthening of sexual fantasies in this way increases the likelihood of enacting them. It is possible that pictorial material promotes solitary sexual release and so reduces the involvement of others in the person's paraphilia.

Without more definite evidence it seems appropriate that pornographic material relating sexual activity to violence or to children should not be available to young people whose sexual development is incomplete. The arguments for more general restrictions are that pornographic publications debase women, and sometimes put children at risk of exploitation during the making of the material and in other ways. These important matters of public policy, like many others, have to be decided on limited scientific evidence.

In his work with patients and their families, the doctor is more likely to be asked what effect pornographic material, discovered by a wife or parent, may have on a husband or adolescent son. The doctor should explain the different points of view and the uncertainty of the scientific evidence. He should indicate that the effects are likely to differ in different people, and suggest that the person concerned should speak to him. If the doctor interviews this person and reviews his sexual life thoroughly, some useful advice can generally be given. Such advice should extend to broader aspects of personal relationships and not merely to the effects of the pornographic material.

Abnormalities of sexual preference are sometimes usefully divided into two groups: abnormalities of the object of the person's sexual drives, and abnormalities in the preference of the sexual act. Abnormalities of the sexual 'object' include fetishism, paedophilia, transvestic fetishism, zoophilia, and necrophilia. Abnormalities of preference of the sexual object include exhibitionism, voyeurism, sadism, masochism, and auto-

erotic asphyxia. Some people have more than one type of paraphilia.

Abnormalities in the preference of sexual object

These abnormalities involve preferences for an object other than another adult in the achievement of sexual excitement. The alternative object may be inanimate, as in fetishism and transvestic fetishism, or may be a child (paedophilia) or an animal (zoophilia).

Fetishism

In sexual fetishism, the preferred or only means of achieving sexual excitement are inanimate objects or parts of the human body that do not have direct sexual associations. The disorder shades into normal sexual behaviour; it is not uncommon for men to be aroused by particular items of clothing, such as stockings, or by parts of the female body that do not have direct sexual associations, such as the hair. The condition is abnormal when the behaviour takes precedence over the usual patterns of sexual intercourse.

Prevalence Sexual fetishism as the sole or preferred means of sexual arousal is uncommon, but no exact figures are available.

Description Fetishism usually begins in adolescence. It occurs almost exclusively among men, although a few cases have been described among women (Odlum 1955). Most fetishists are heterosexual but some are homosexual—20 per cent according to Chalkley and Powell (1983). The objects that can evoke sexual arousal are many and varied, but for each person there is usually a small number of objects or classes of objects. Among the more frequent are rubber garments, women's underclothes, and high-heeled shoes. Sometimes the object is an attribute of a person, for example lameness or deformity in a woman, or a part of the human body, such as the hair or foot. The texture and smell of objects is often as important as their appearance, for example furs, velvet, rubber garments, and polished leather. Contact with the object causes sexual excitement, which may be followed by solitary masturbation or by sexual intercourse incorporating the fetish if a willing partner or paid prostitute is available.

Fetishists may spend much time seeking desired objects. Some buy what they require and others steal, for example underclothes from a washing line. When the object is a particular attribute, many hours may be spent in searching for and following a suitable woman, for example a woman with a limp. Inanimate fetish objects are often hoarded; Hirschfeld (1944) recorded a striking example—the hoarding by one man of 31 pigtails of hair, each cut with scissors from women he had followed and each labelled with the date and hour when it had been cut. More often, fetishists collect women's shoes or underclothes, or pictures of such objects.

Aetiology Fetishism was the first sexual disorder for which a theory of association learning was put forward. Binet (1877) suggested that the condition arose by a chance coming together of sexual excitation and the object that becomes the fetish object. Support for this idea was claimed by Rachman (1966) from an experiment in which male volunteers were repeatedly shown pictures of women's boots followed immediately by sexually arousing pictures of women. Rachman reported that this procedure led to sexual arousal when the boots were shown alone; this finding has not been confirmed. The suggestion that fetishism results from faulty imprinting (Wilson 1981) rests entirely on analogy. Imprinting apparently affects sexual behaviour in birds, but there is no evidence that it does so in man.

Psychoanalysts suggest that sexual fetishism arises when castration anxiety is not resolved in childhood, and the man attempts to ward off this anxiety by maintaining in his unconscious mind the idea that women have a penis (Freud 1927). In this view, each fetish is a symbolic representation of a phallus. Although some fetishes can be interpreted in this way, others require tortuous interpretations if the general hypothesis is to be sustained (Stekel 1953). In any case the general idea does not convincingly explain most cases.

Rarely, fetishism has been reported in association with EEG evidence of temporal lobe dysfunction (Epstein 1961) or with frank epilepsy (Mitchell *et al.* 1954). However, there is no evidence of such associations in most cases.

It must be concluded that the cause of sexual fetishism is unknown. A useful model for clinical practice was suggested by Binet—that fetishism arises through chance association, and increases and persists when the usual expression of sexual impulses has been inhibited, for example by shyness or fear of sexual intercourse. On this view, it is as important in treatment to improve social functioning and to deal with sexual fears as to seek the origin of the fetishism.

Prognosis In the absence of reliable follow-up data, the prognosis has to be based on clinical experience. Fetishism in adolescents and young adults is often transient, disappearing when satisfying heterosexual relationships have been established. The prognosis at all ages depends crucially on the extent of social relationships and sexual activities. The prognosis is worse for solitary single men who are shy with women and without a sexual partner than for those who are younger and better adjusted socially. The prognosis is generally worse when the behaviour is frequent and has persistently broken social conventions and legal barriers. Legal proceedings instituted for the first time can increase motivation to control the behaviour.

Treatment There are case reports of treatment by psychoanalysis (Nagler 1957) and by behaviour therapy (see Kilmann (1982) for a review) but no controlled trials. Clinical experience indicates that the general measures outlined later in the chapter are generally as effective as psychoanalytic treatment, and may be as effective as behavioural treatment.

Transvestic fetishism

In ICD10 this condition is known as fetishistic transvestism. The term transvestic fetishism is used here because it indicates a similarity to sexual fetishism. Transvestic fetishism varies from the occasional wearing of a few articles of clothing of the opposite sex to complete cross-dressing. The term transvestic fetishism is also preferred because it makes clear the distinction between this condition and other reasons for cross-dressing (that is for transvestism). These other reasons are transsexualism (see p. 506) and the wishes of some homosexuals (see p. 483). Transvestic fetishism is rare among women (almost all women who cross-dress are transsexual or lesbian). For this reason the description below applies to the condition in men.

Prevalence The prevalence of transvestic fetishism is not known.

Description Cross-dressing usually begins about the time of puberty. The person usually starts by putting on only a few garments, but as time goes by he adds more until eventually he may dress entirely in clothes of the other sex. Transvestic fetishists experience sexual arousal when cross-dressing and the behaviour often terminates with masturbation. Sometimes the clothes are worn in public, either underneath male outer garments or in some cases without such precautions against discovery.

Unlike the transsexuals described later, transvestic fetishists have no doubt that they are men. Most are heterosexuals; exceptionally, after many years of cross-dressing a few begin to believe that they are women. Despite these transitional cases, transvestic fetishists differ in important ways from transsexuals. The former are sexually aroused by cross-dressing and have a correct conviction about their gender; transsexuals are not aroused by cross-dressing and are convinced that they are of the opposite gender to that indicated by their genitalia. Many transvestic fetishists are married, and most hide the behaviour from their wives. If wives discover the behaviour, most express distress and disgust but a few collude and may assist in obtaining the clothing.

Aetiology There is no firm evidence that the chromosomal sex or hormonal make-up of transvestic fetishists is abnormal (Lukianowicz 1959).

Despite a report of three cases in one family (Liakos 1967), transvestic fetishism is not familial and there is no evidence that it is inherited. Although occasional associations with temporal lobe dysfunction have been reported (Davies and Morgenstern 1960; Epstein 1960), there is no evidence for such an association in the majority. Transvestic fetishism resembles fetishism in starting after puberty and being associated with sexual arousal; these findings suggest that the causes of transvestic fetishism may resemble those proposed above for fetishism, namely an impediment to normal sexual development possibly coupled with association learning.

Psychoanalysts use similar explanations for transvestic fetishism as for fetishism, namely that the transvestic fetishist is creating a 'phallic woman' (himself in woman's clothes) to allay castration anxiety (Fenichel 1945). The theory is not convincing.

Prognosis In the absence of reliable information from follow-up of a representative group of transvestic fetishists, statements about prognosis must be based on clinical experience. Most cases appear to continue for years, becoming less severe as sexual drives decline in middle age or later. However, there are wide variations in outcome, and the comments made earlier about the prognosis of fetishism apply here as well. As already noted, a minority of transvestic fetishists gradually develop the idea that they are women and continue to cross-dress without sexual arousal; hence the clinical picture comes to resemble that of transsexualism (see p. 506).

Treatment There is no specific treatment for transvestic fetishism. Management should include the general and behavioural procedures described on p. 504.

Paedophilia

Paedophilia is repeated sexual activity (or fantasy of such activity) with prepubertal children as a preferred or exclusive method of obtaining sexual excitement. It is almost exclusively a disorder of men. It is important to distinguish between paedophilia and intercourse with girls who have passed puberty but not yet reached the legal age of consent. The age of consent differs between legislations. In England and Wales it is 16 for girls and 18 for young men consenting to homosexual intercourse. Paedophilia is not concerned with these borderlines of legal consent but with intercourse between an adult and a prepubertal child.

Prevalence There is no reliable information about the prevalence of paedophilia. From the existence of child prostitution in some countries and the ready sale of pornographic material depicting sex with children, it appears that interest in sexual relationships with children is not rare. Nevertheless, paedophilia as an exclusive form of sexual behaviour is probably uncommon.

Description Paedophiles usually choose a child aged between six years and puberty, but some prefer a very young child. The child may be of the opposite sex (heterosexual paedophilia) or the same sex (homosexual paedophilia). Although most paedophiles seen by doctors are men of middle age, the condition can start in adolescence. It seems to be established early; there is no evidence that an established interest in adult sexual partners changes to an interest in child partners. With younger children fondling or masturbation is more likely than full coitus, but sometimes children are injured by forcible attempts at penetration. These are rare and tragic cases of paedophilia associated with sexual sadism. (Sexual abuse of children is described on p. 720.)

Paedophilia has to be distinguished from exhibitionism towards young girls (in which no attempt is made to engage in direct sexual contact). Sexual contact with children may be sought by people with subnormal intelligence, dementia, and alcoholism.

The child Most females involved in paedophilia are aged between 6 and 12 years, and most males involved in homosexual paedophilia are aged between 12 and 15 years. Of the children involved in cases of paedophilia coming to the attention of the law, two-thirds have cooperated in sexual activity more than once with the same or another adult. Most of these children have been involved through fear rather than interest, but a small minority are promiscuous and delinquent. It has been reported that the mothers have often shown an ambivalent attitude to both discipline and the child's sexual development (Gibbens and Prince 1965).

The *long-term* effects of paedophilia on the child are not certain. In a follow-up of children who had experienced sexual activity with an adult, Bender and Grugett (1952) found that there was no lasting maladjustment provided that the child had developed normally up to the time of the sexual experience. Many adults come for help with problems related to sexual abuse in childhood (see p. 154). The effect on the child is probably much influenced by the reaction of the parents, and by how far the child is involved in legal proceedings (Mohr *et al.* 1964).

Aetiology This is unknown. Paedophiles often have a marked incapacity for relationships with adults and fears of relationships with women. The various aetiological theories have been reviewed by Mohr *et al.* (1964).

Prognosis In the absence of reliable information from follow-up studies, prognosis has to be judged in individual patients by the length of the history, the frequency of the behaviour, the absence of other social and sexual relationships, and the strengths and weaknesses of the personality. Behaviour that has been frequently repeated is likely to persist despite efforts at treatment.

Treatment Both group treatment (Hartmann 1965) and behaviour therapy (Beech *et al.* 1971) have been tried, but there is no convincing evidence that either leads to good results in the majority of paedophiles. The general measures described at the end of this chapter should be tried, although good results should not be expected for most patients.

Other abnormalities in the preference of sexual object

Zoophilia Zoophilia, otherwise called bestiality or bestiosexuality, is the use of an animal as a repeated and preferred or exclusive method of achieving sexual excitement. It is uncommon and rarely encountered by doctors.

Necrophilia In this extremely rare condition sexual arousal is obtained through intercourse with a dead body. Occasionally there are legal trials of men who murder and then attempt intercourse with the victim. No reliable information is available about the causes or prognosis of this extreme form of abnormal sexual preference.

Abnormalities in the preference of sexual act

The second group of abnormalities of sexual preference involves variations in the behaviour that is carried out to obtain sexual arousal. Generally the acts are directed towards other adults, but sometimes children are involved (for example by some exhibitionists).

Exhibitionism

Exhibitionism is the repeated exposing of the genitals to unprepared strangers for the purpose of achieving sexual excitement but without any attempts at further sexual activity with the other person. The name exhibitionism was suggested by Lasègue (1877), and further clinical observations were reported by Krafft-Ebing in 1886 (see Krafft-Ebing 1924). The use of the term in this technical sense is to be distinguished from its everyday sense of extravagant behaviour to draw attention to oneself.

Prevalence This is not known. Exhibitionists make up about one-third of sexual offenders referred for psychiatric treatment, and about a quarter of sexual offenders dealt with in the courts (Rosen 1979). Almost all are men, except for a very few women exhibitionists who repeatedly expose the breasts and an even smaller number who expose the genitalia.

Description Amongst exhibitionists seen by doctors, most are aged 20–40 years and two-thirds are married (Gayford 1981). In some the urge to exhibitionism is persistent; in others it is episodic. The act of exposure is usually preceded by a feeling of mounting tension. When in this state of tension, the exhibitionist characteristically seeks to evoke a strong emotional reaction from the other person, generally surprise and shock. Some exhibitionists are satisfied by any evidence of being noticed, even laughter. Most choose places from which escape is easy, although a few choose places where they risk detection. Whatever the exact pattern of behaviour, the experience at the time is one of intense excitement. As a broad generalization, two groups of exhibitionists can be described. The first group includes men of inhibited temperament who struggle against their urges and feel much guilt after the act; they sometimes expose a flaccid penis. The second group includes men who have aggressive traits, sometimes accompanied by features of antisocial personality disorder. They usually expose an erect penis, often while masturbating. They gain pleasure from any distress they cause and often feel little guilt.

It has been suggested that exhibitionism and voyeurism are related to one another aetiologically, but exhibitionists seldom practice voyeurism (Rooth 1971). There is uncertainty about the relationship between exhibitionism and the making of *obscene phone calls* by men who talk to women about sexual activities while masturbating. It has been suggested (Tollison and Adams 1979) that these obscene callers are also exhibitionists, but it is not easy to identify them in order to study their psychopathology.

In the United Kingdom, if a man is brought to court because of exhibitionism, he is charged with the offence of *indecent exposure* (see Chapter 22, p. 766). About four-fifths of men charged with indecent exposure are exhibitionists (as defined at the beginning of this section).

Aetiology There are several theories, all unsubstantiated. The same explanations have been put forward for this paraphilia as for those described earlier in this chapter, namely a specific failure to resolve Oedipal conflict or a general inhibition of relationships with women. Some exhibitionists describe unduly close relationships with their mothers and a poor relationship with ineffectual fathers (Rickles 1950). However, these accounts are retrospective, and it is not certain how far they reflect the actual circumstances of the patient's upbringing. Also, many people describe similar experiences in childhood, but do not become exhibitionists. To the clinician, the most striking feature of many exhibitionists is a personality characterized by lack of assertion and a striking degree of passivity in everyday relationships. Whatever the original cause of exhibitionism, it has been suggested that the behaviour is perpetuated by the reinforcing effects of sexual release during masturbation that often follows the act (Evans 1970).

In middle-aged or elderly people, the onset of exhibitionism may be caused by organic brain disease. Such disease presumably releases behaviour that is preformed but previously inhibited. Alcoholism may also lead to disinhibition of urges to exhibitionism that have previously been controlled.

Prognosis There is no reliable information about prognosis. Men who exhibit only once do not fall within the definition of the disorder. Clinical experience suggests a variable outcome for the rest. Among men who exhibit repeatedly but only at times of stress, the prognosis depends on the likelihood of the stressors returning. Exhibitionists who repeat often and not solely at times of stress are likely to persist with the behaviour for years despite treatment by psychiatrists or punishment by the courts. In keeping with these clinical impressions, the evidence from the courts is that the reconviction rate for indecent exposure is low after a first conviction but high after a second conviction. Although a history of exhibitionism is sometimes given by men who commit rape, most exhibitionists do not go on to commit violent sexual acts nor do they interfere with children

(Rooth 1973). For a fuller account of exhibitionism see Rooth (1971).

Treatment Any associated psychiatric disorder such as depressive disorder, alcoholism, or dementia should be sought and treated appropriately if found. Many treatments have been tried specifically for exhibitionism, including psychoanalysis, individual and group psychotherapy (Witzig 1968), aversion therapy (Rooth and Marks 1974), and covert sensitization (Maletzky 1974, 1977). There is no satisfactory evidence that any of these treatments is generally effective. A practical approach combines counselling and behavioural techniques. Counselling deals with the effects of exhibitionistic behaviour and with problems in personal relationships, whilst behavioural techniques are concerned with self-monitoring to identify circumstances that trigger the behaviour and to help the person avoid them. Cyproterone acetate and related drugs have been used to reduce sex drive, but are not recommended because of uncertain results and problems of long-term use (see p. 506). The general measures described on p. 504 can be tried, though without high expectation of success.

Sexual sadism

Sadism is named after the Marquis de Sade (1774–1814) who inflicted extreme cruelty on women for sexual purposes. Sexual sadism is achieving sexual arousal, habitually and in preference to heterosexual intercourse, by inflicting pain on another person by bondage or by humiliation.

Prevalence Inflicting pain in fantasy or practice is a not uncommon accompaniment of other forms of sexual behaviour. Sex shops sell chains, whips, and shackles, while some pornographic magazines provide pictures and descriptions of sadistic sexual practices. Sexual sadism as a predominant sexual practice is probably uncommon, but its frequency is not known.

Description Beating, whipping, and tying are common forms of sadistic activity. Repeated acts may be with a partner who is a masochist or a prostitute who is paid to take part. Sadism may be a component of homosexual as well as heterosexual acts. Rare cases of sexual sadism towards animals have been reported (Allen 1969). The acts may be symbolic, with little actual damage, and some involve humiliation rather than injury. Sometimes serious and permanent injuries are caused. Extreme examples are the rare 'lust murders', in which the killer inflicts serious repeated injuries—usually stabbings and mutilations—on the genitalia of his victim. In these rare cases ejaculation may occur during the sadistic act or later by intercourse with the dead body (see necrophilia, p. 501). Further information is given by Hirschfeld (1944).

Aetiology This is not known. Psychoanalytical explanations draw attention to the association of loving and aggressive feelings that is supposed to exist in the young child's early relationship with his parents. Behavioural formulations rely on association learning. Both explanations are unsatisfactory.

Prognosis There is no reliable information, but clinical experience suggests that once established the behaviour is likely to persist for many years.

Treatment There are case reports of the use of behaviour therapy (Davison 1968) but no evidence from adequate clinical trials. In the absence of any proven treatment, men who have committed serious injury must be dealt with by legal means if there is risk of another offence. In deciding this issue it is wise to assume that treatment of any kind is unlikely to alter an established pattern of sadistic behaviour. The risks must not be underestimated when potentially dangerous behaviour has been planned or has already occurred.

Sexual masochism

Sexual masochism is achieving sexual excitement, as a preferred or exclusive practice, through the experience of suffering. As a predominant activity

it differs from the common use of minor painful practices as an accompaniment to sexual intercourse. The condition is named after Leopold von Sacher-Masoch (1836–1905), an Austrian novelist, who described sexual gratification from the experience of pain.

Prevalence Fantasies of being beaten or humiliated are sufficiently common among males to create a demand for pornographic literature, and also for prostitutes who will help the man act out his fantasies. Established sexual masochism is probably uncommon, though no exact information is available.

Description The suffering may take the form of being beaten, trodden upon, bound, or chained, or the enactment of various symbolic forms of humiliation, for example dressing as a child and being punished. Masochism, unlike most other sexual deviations, occurs in women as well as in men, perhaps as a reflection of the more submissive role of the woman in normal sexual relationships. It may occur in homosexual as well as heterosexual relationships.

At times the masochist may allow dangerous forms of assault upon himself, including strangulation, a practice that can increase sexual excitation through the resulting partial anoxia (see also auto-erotic asphyxia, p. 504).

Aetiology This is not known. One theory is that, as a result of beatings delivered to pubertal children, sexual arousal becomes associated by chance with the experience of pain and humiliation. Psychoanalytic theory suggests that masochism is sadism turned inwards, and therefore is explicable in the same way as sadism (see above).

Prognosis There is no reliable information about prognosis. Clinical experience suggests that, once established as a preferred form of sexual behaviour, masochism is likely to persist for many years.

Treatment There are case reports of treatment by psychoanalysis (Stekel 1953) and behavioural treatments (Marks *et al.* 1965), but no satisfactory evidence that either is effective.

Voyeurism

Many men are sexually excited by observing others engaged in intercourse. Voyeurism (sometimes called scopophilia) is observing the sexual activity of others repeatedly as a preferred means of sexual arousal. The voyeur also spies on women who are undressing or without clothes, but does not attempt sexual activity with them. Voyeurism is usually accompanied or followed by masturbation.

Voyeurism is a disorder of heterosexual men whose heterosexual activities are usually inadequate. Although the voyeur usually takes great care to hide from the women he is watching, he often takes considerable risks of discovery by other people. Hence most voyeurs are reported by passers-by, and not by the victim.

Aetiology Among adolescents voyeuristic activities are not uncommon as an expression of sexual curiosity, but they are usually replaced by direct sexual experience. The voyeur continues to watch because he is shy, socially awkward with girls, or prevented from normal sexual expression by some other obstacle. Psychoanalytic explanations follow the general lines described above for other sexual disorders. Behavioural theories seek an explanation in terms of chance associations between a first experience of peeping and sexual arousal.

Prognosis No reliable information is available.

Treatment Psychoanalysis, group therapy (Witzig 1968), and counter-conditioning (Jackson 1969) have been used but, as no systematic trials have been reported, no conclusions can be drawn. It is doubtful whether any treatment is effective, but it is reasonable to try the general measures described later in this chapter.

Other abnormalities of preference of the sexual act

Auto-erotic asphyxia is the practice of inducing cerebral anoxia to heighten sexual arousal while masturbating. Asphyxia is usually induced by partial strangulation with cords or by plastic bags placed over the head. The practice occurs almost exclusively in men. It is hazardous and may lead to death. The act may be accompanied by the use of objects to produce anal stimulation, by using fetish objects, or by cross-dressing. The person may look at himself in a mirror or photograph himself, or he may apply bondage, for example by tying the ankles. Few people who practise auto-erotic asphyxia seek help from doctors. Most information about the condition comes from forensic studies of persons who have died during the act. For further information see Blanchard and Hucker (1991) or Hucker (1990).

In **frotteurism** the preferred form of sexual excitement is applying or rubbing the male genitalia against another person, usually a stranger and an unwilling participant, in a crowded place such as an underground train. In **coprophilia**, sexual arousal is induced by thinking about or watching the act of defecation and this is the preferred sexual activity; in **coprophagia** arousal follows the eating of faeces. In **sexual urethism**, which occurs mainly in women, erotic arousal is obtained by stimulation of the urethra.

Urophilia refers to sexual arousal obtained by watching the act of urination, being urinated upon, or drinking urine.

The prevalence of these disorders is not known, but some are sufficiently common to demand provision from prostitutes. Further information will be found in Allen (1969) and Tollison and Adams (1979).

Assessment and management of abnormalities of sexual preference

In assessment, the first step is to **exclude mental illness.** Abnormal sexual preference is sometimes secondary to dementia, alcoholism, depressive disorder, or mania. These illnesses probably release the behaviour in a person who has previously experienced the corresponding sexual fantasies but not acted on them. It is particularly important to look for mental illness when the abnormal sexual preference comes to notice for the first time in middle age or later.

Detailed enquiry is then made about the patient's sexual practices. It should be borne in mind that not uncommonly patients have more than one form of abnormal sexual preference. The extent and vigour of normal heterosexual interests, both in the present and in the past, are determined. Whenever possible, an interview should be arranged with the patient's regular sexual partner.

It is always important to find out what part the abnormal sexual preference is playing in the patient's life. Apart from being a source of sexual arousal, it may be a comforting activity that helps to ward off feelings of loneliness, anxiety, or depression. Unless other means are found to deal with such feelings, treatment may reduce the patient's abnormal sexual preference but worsen his emotional state.

Motives for seeking treatment

People who request treatment for paraphilias often have mixed motives. Many consult a doctor because their sexual behaviour has become known to the sexual partner, a relative, or the police. People with paraphilias may have little wish to change, and many of them prefer to be told that no treatment will help so as to justify the continuation of their sexual practices. Sometimes people with abnormal sexual preferences seek help when they become depressed and feel guilty about the behaviour and its effects on other people. At these times of low mood, strong wishes for change may be expressed only to fade quickly as normal mood returns. Strong motivation is known to be important whether treatment is by psychoanalysis (Bieber 1962), psychotherapy (Ellis 1956), or behaviour therapy (Feldman and McCulloch 1979). Therefore it is important to assess whether the expressed wishes for change will be maintained.

Planning treatment

The aim of treatment must be discussed with the patient: whether it is to control, or if possible give up, the behaviour, or to adapt better to the behaviour so that less guilt and distress are felt. In considering these aims, the doctor will have to take into account whether any psychological or physical harm is being caused to other people. At this early stage it is important to make clear that, whatever the aim, treatment will require considerable effort on the part of the patient.

If the agreed aim is better adjustment, treatment will be by counselling designed to explore the patient's feelings and to help him to identify the problems caused by his sexual practices and to find ways of reducing them. If the agreed aim is change, the first step is to find ways of encouraging ordinary heterosexual relationships. For this purpose, treatment is directed first to any anxieties that are impeding social relationships with the opposite sex. Attention is then directed to any detected sexual inadequacy using the methods outlined earlier in this chapter. In most cases these two steps are the most important part of treatment.

The problems likely to arise from giving up the paraphilia are considered next. Some patients occupy much of their time in preparing for the sexual act (for example fetishists may spend many hours searching for a particular kind of women's underclothes). As already noted, the behaviour often becomes a way or warding off feelings of loneliness or despair. To safeguard against distress, the patient must be helped to develop leisure activities, to seek new friends, and to find other ways of coping with unpleasant emotions.

Only when these steps have been taken should attention be directed to ways of suppressing the unwanted sexual behaviour. Sometimes the preceding steps are enough to strengthen the patient's capacity to control himself, but additional help is often needed.

Masturbation fantasies appear to play an important part in perpetuating abnormal sexual behaviour. Therefore it is important to encourage the patient to imagine normal heterosexual intercourse while masturbating and to keep any abnormal fantasies out of mind. If he cannot rid his mind of the abnormal fantasies, he should be encouraged to modify them progressively so that the themes become less and less sexually abnormal and increasingly concerned with ordinary heterosexual intercourse.

For men, oestrogens and antiandrogens have been used in an attempt to reduce sexual drive. This approach has limited value; it is most useful as a short-term measure to reduce the intensity of the paraphilia while psychological treatment is starting. Oestrogens may be tried if libido is strong, and if the continuation of the behaviour is likely to have serious consequences outweighing the risks of treatment. *Oestrogens* may be given as depot injections such as oestradiol undecylenate, or as an oestradiol implant. The therapy can cause breast enlargement and nodules, testicular atrophy, osteoporosis, and, rarely, breast tumours. For this reason *antiandrogens* such as cyproterone acetate or medroxyprogesterone acetate (Depot-provera) have been used, although it is not clear how far their effects exceed the placebo response. Cyproterone can produce a reversible atrophy of the seminiferous tubules, but its side-effects include gynaecomastia, sedation, and depression. See Maletzky (1991) for information on the use of drugs for paraphilias.

Behavioural treatment can be used to reduce the paraphilia either directly or indirectly by treating the person as if he had sexual inadequacy. Several direct methods have been tried. Techniques to change masturbation fantasies (described above) are sometimes regarded as a specific behavioural technique, to be used alone or with other methods. *Covert sensitization* is used to interfere with paraphiliac thoughts and urges that lead up to sexual arousal. In this approach, patients list (a) situations in which they are more likely to experience paraphiliac thoughts and (b) all immediate adverse consequences of the behaviour such as discovery by family or police, with resultant humiliation or arrest. The patients then imagine a situation and an adverse outcome. In an alternative method called smell or taste aversion, an unpleasant odour or taste is associated with the arousing fantasies.

Relapse prevention focuses on the situations in which paraphiliac urges are more likely to be experienced. Patients are encouraged to identify these situations and to avoid them. Particular attention is paid to the earliest stages of the chain of behaviour leading up to the paraphilias, for example an action which predictably causes the patient's wife to react angrily, and thus provides an excuse to go to a place where the paraphilia may be stimulated.

Many people with abnormal sexual preferences appear before the courts. Sanctions such as a suspended sentence or probation order can sometimes help the patient to gain control of his own behaviour. However, doctors should not agree to treat patients who are referred against their wishes. For a review of behavioural treatments for sexual offenders see Maletzky (1991). Some violent sexual offenders have been encouraged to consider the victim's distress and to identify with it. None of these treatments has been tested in satisfactory clinical trials.

Abnormalities of gender identity

Transsexualism

Transsexual people are convinced that they are of the gender opposite to that indicated by their external genitalia. In addition, they feel estranged from their bodies, have an overpowering wish to live as a member of the opposite sex, and seek to alter their bodily appearance and genitalia to conform to those of the opposite sex. In the past the condition was called **eonism** because it was exemplified by the Chevalier d'Eon de Beaumont. In psychiatric literature, the condition was mentioned by Esquirol in 1838 and described in more detail by Krafft-Ebing in 1886 (see Krafft-Ebing 1924). In the 1960s the attention of doctors and the public was directed to the condition by a number of striking reports and by a publication by Benjamin (1966).

Prevalence

Epidemiological data are difficult to obtain. Wlinder (1968) estimated the prevalence among Swedish males to be 1 in 37 000 and among Swedish females to be 1 in 103 000. Hoenig and

Kenna (1974) reported similar figures (1 in 34 000 men and 1 in 108 000 women) in Great Britain. Despite these male-to-female ratios, most transsexuals seeking medical help are men.

Description

Among men Patients report a strong conviction of belonging to the other sex, usually starting before puberty. It is sometimes reported by the parents that in childhood the patients preferred the company and pursuits of girls, although such a history is not invariable. Follow-up studies have shown that effeminate boys more often grow up as homosexuals than as transsexuals (see Green 1974).

By the time that medical help is requested, most transsexuals have started dressing as women. In contrast with transvestic fetishists, they cross-dress to feel more like women, not to produce sexual arousal. (They also differ from those homosexuals who dress as women to attract other homosexuals.) For this purpose, make-up is worn and the hair is arranged in a feminine style; facial and body hair are usually removed by electrolysis. Transsexuals try to adopt feminine gestures and to alter the pitch of voice, but few succeed wholly convincingly. Transsexuals also seek changes in social role. They apply for the kind of work that is usually done by women and they enjoy cooking and sewing. They do not show maternal interests. Sex drive is usually low and, unlike transvestic fetishists, these patients do not masturbate when cross-dressed. As Benjamin (1966, p. 21) has written, 'the transvestist looks on his sex organ as an organ of pleasure, and the transsexual turns from it in disgust'.

There is no characteristic type of **personality**, but some transsexuals are self-centred, demanding, and attention-seeking; they are often particularly difficult to treat.

Many transsexual patients are greatly distressed by their predicament. Depression is common, and in one series 16 per cent had made suicide attempts (Wlinder 1967). About a third marry but, not surprisingly, about half of these become divorced (Roth and Ball 1964).

Transsexuals often ask the doctor for help in altering the appearance of the breasts and external genitalia. Usually the first requests are for oestrogens to enlarge the breasts. These requests are often followed by increasingly insistent demands for surgery to the breasts, for surgical castration and removal of the penis, and for operations to create an artificial vagina. The patients are greatly distressed by their condition. Their demands are usually made in a determined and persistent way, and are sometimes accompanied by threats of self-mutilation or suicide. Some patients do attempt to castrate themselves, and may maintain afterwards that the injury was accidental.

Among women Many women who appear to be transsexuals are really homosexuals. Transsexual women resemble transsexual men in having held since childhood a strong conviction that they 'occupy' a body of the wrong sex. Some seek to alter their body by mastectomy or hysterectomy, and a few hope for plastic surgery to create an artificial penis. Transsexual women strive to be like men in dress, voice, gestures, and social behaviour, and in their choice of work and hobbies. They wish to have intercourse not with a female homosexual, but in the role of male with a heterosexual woman.

Aetiology

Many ideas have been put forward to explain this puzzling condition but its cause remains unknown. Transsexuals have normal sex chromosomes, and there is no convincing evidence of a genetic cause. It has been suggested that abnormal acquisition of gender role might be relevant. However, there is no convincing evidence that transsexuals have been brought up in the wrong gender role. Most effeminate boys grow up as heterosexual males; the rest are homosexual more often than transsexual.

Endocrine disorders have also been sought in adult transsexuals but no definite abnormality has been found. It has been suggested that transsexualism might result from hormonal abnormalities during intrauterine development. There is some evidence that when pregnant rhesus monkeys are given large doses of androgens, their female infants behave more like males during play (Young *et al.* 1964). There is no direct parallel in human beings, but it may be relevant that female children with adrenogenital syndrome (involving exposure to large amounts of androgen before and after birth) have been reported to show boyish behaviour in childhood (Ehrhardt *et al.* 1968) but do not grow up as transsexuals.

Although Wlinder (1967) found abnormal EEGs in 28 per cent of his transsexual patients, there is no other evidence of organic brain disorder in such patients.

In an interesting minority of patients, transsexualism begins after many years of transvestic fetishism. These patients start by cross-dressing to obtain sexual excitement, but the resultant arousal gradually diminishes. At the same time the patients gradually become convinced that they are women.

Prognosis

There is no reliable information about the prognosis of untreated transsexuals. Clinical experience suggests that, once established, the condition persists for many years, though it is not known whether it lasts beyond middle age. The rate of suicide among transsexuals is probably increased.

Treatment

Although transsexualism is undoubtedly a psychological disorder in a person whose body is normal, patients usually seek treatment directed to the body not the mind. The most rational treatment would be to alter the patient's conviction that he is of the wrong sex, but attempts to do so by psychotherapy rarely succeed. If treatment involving physical changes is offered, it should be done by carefully planned stages. The evident distress of these patients and their insistent demands should not be allowed to interfere with this plan. Many psychiatrists, including the authors, consider that physical changes are seldom appropriate and that in most cases it is better to use supportive psychotherapy. When other treat-

ment is used it should be supervised by a doctor with special experience of the problem. Treatment generally passes through the stages described in the following paragraphs.

A male transsexual usually has three aims: to take on a woman's appearance, to live as a woman, and to change his genitalia. For the first aim the beard is often removed by electrolysis. The man can learn to speak like a woman, and practise appropriate gestures and ways of walking and sitting. These changes in speech and movement are often particularly difficult, but training with video feedback may aid them. Male patients usually seek breast enlargement, at first by oestrogens and later by mammoplasty. Neither treatment should be provided at an early stage, since suitable clothing can be used instead and the drugs are not without danger (see below).

If the patient persists in his intentions, he may try to live as a woman. He may then become so impressed by the problems of living as a woman as to modify his aims. If after a year the patient persists in demanding surgery and his personality is stable, he may be given a full explanation of what the operation entails. He must understand that no surgery can make a man into a woman; at best it can provide only a poor copy of the female body.

Oestrogen is sometimes given at this stage or earlier to produce some enlargement of the breasts and deposition of fat around hips and thighs. Minor side-effects include nausea and dizziness, while more serious risks are thrombosis and malignant breast tumours (Symmers 1968). Methyl testosterone has been prescribed to female transsexuals who wish to become male, but it carries the risk of liver damage.

If, after these preliminaries, the psychiatrist considers that the patient is one of the few who might benefit from surgery, the opinion of an experienced surgeon can be sought. He will make the final decision about the indications for surgery and the type of operation, though with the psychiatrist's advice. The usual criteria include a persistent wish (for many years) to change gender and a demonstrated ability to live successfully in the desired gender role for at least a year (International Gender Dysphoria Association 1985).

More than a dozen follow-up studies have been reported, but few have included a comparison group. Meyer and Reter (1979) compared patients who received sex reassignment surgery with those who were refused surgery and therefore were not directly comparable. Mate-Kohl *et al.* (1990) studied 40 patients selected for surgery, with alternate cases being allocated to immediate treatment or to treatment after two years on a waiting list. Two years after the operation the first group had lower scores for neurotic symptoms and somewhat better social adjustment than the patients who had not yet received the operation. No study has compared patients allocated randomly to surgery and non-surgery. The uncontrolled studies have shown improvement rates of about 60 per cent after sex reassignment surgery, but this figure must be viewed cautiously because many patients are lost from follow-up. In the comparative study by Meyer and Reter (1979), a similar rate of improvement was found in unoperated controls. This finding should be interpreted with caution since only half the patients were assessed at follow-up, and the follow-up period was longer for operated patients. Nevertheless the improvement rate in unoperated patients is an important warning against uncritical acceptance of the value of surgery. The publicity given to a few successful cases should certainly not be given undue weight. For a review of outcome studies of sex reassignment surgery see Abramowitz (1986). More information about the treatment is given by Green and Money (1969) and by Schapira *et al.* (1979).

Dual-role transvestism

This term is used in ICD10 to describe people who wear clothes of the opposite sex but are neither transvestic fetishists (seeking sexual excitement) nor transsexuals (wishing a change of gender and sexual role). Instead they enjoy cross-dressing in order to gain temporary membership of the opposite sex.

Gender identity disorder of adolescence and adulthood—non-transsexual type

This term is used in DSMIV to denote people who have passed puberty and feel a persistent or recurrent discomfort or sense of inappropriateness about their assigned gender identity. They cross-dress persistently or repeatedly, or imagine themselves doing this, but are not sexually excited by these actions or fantasies nor preoccupied with change in their primary or secondary sex characteristics.

Gender identity disorders in children

Parents more often seek advice about effeminate behaviour in boys than about masculine behaviour in small girls (it is not clear whether such behaviour in girls is less frequent or more socially acceptable). Effeminate boys prefer girlish games and enjoy wearing female clothing. The outcome amongst these boys is variable (Green and Money 1961); some develop normal male interests and activities, but others continue their effeminate ways into adolescence. Further information is given by Green (1974).

Other aspects of sexual behaviour

Rape, incest, and pornography are discussed in Chapter 22 on forensic psychiatry.

Further reading

Bancroft, J. (1989). *Human sexuality and its problems* (2nd edn). Churchill Livingstone, Edinburgh.

Elstein, M. (1980). Sexual medicine. *Clinics in Obstetrics and Gynaecology*, 7(2).

Ford, C.S. and Beach, F.A. (1952). *Patterns of sexual behaviour*. Methuen, London.

Hawton, K. (1985). *Sex therapy: a practical guide*. Oxford University Press.

Kaplan, H.I. and Sadock, B.J. (ed.) (1985). Normal human sexuality and psychosexual disorders. In *Comprehensive textbook of psychiatry* (4th edn), Vol. 1, Chapter 24. Williams and Wilkins, Baltimore, MD.

16 | *Psychiatry of the elderly*

Introduction

It is only in the past 40 years that the psychiatric care of the elderly has attracted special interest. This change of interest largely reflects the increasing numbers of old people in the population. At the beginning of this century only 5 per cent of the population of Western Europe were aged over 65; now the figure is about 15 per cent, a third of whom are aged over 75. It is expected that the proportion of elderly people will continue to rise well into the next century, mainly because of an increase in the numbers of those aged over 85 years. In the United Kingdom there were 700 000 people aged over 85 in 1988 and there are projected to be 1.1 million in 2001 and 1.7 million in 2041 (OPCS 1987).

Since the prevalence of mental disorder, particularly dementia, increases with age, there has been a disproportionate increase in the demand for psychiatric care of the elderly. In developed countries this demand will continue to rise. In developing countries with limited resources, in which there are fewer elderly at present, there are likely to be even greater problems in providing care for increasing numbers. Nevertheless, while larger numbers of the elderly are likely to mean more cases of psychiatric disorder, it should not be forgotten that most old people, even the very old, have good mental health.

Although the psychiatric disorders of the elderly have some special features, they do not differ substantially from the psychiatric disorders of younger adults. It is the needs of elderly psychiatric patients that set them apart from others. Special psychiatric skills are required to meet these needs and it is for this reason that a separate chapter is devoted to the subject. The chapter begins with a brief account of normal ageing. General principles of the provision of psychiatric care for the elderly are discussed next, and then an account is given of psychiatric syndromes in the aged.

Normal ageing

The ageing brain

The weight of the human brain decreases by approximately 5 per cent between the ages of 30 and 70 years, by 10 per cent by the age of 80, and by 20 per cent by the age of 90. As well as these changes, the ventricles enlarge and the meninges thicken. There is some loss of nerve cells, though this is relatively minor and selective. There may also be a reduction in dendritic processes. Magnetic resonance studies have shown a decrease with age in cortical grey matter with white matter relatively unchanged. Cerebral blood flow in the thalamus and in the frontal and temporal lobes also appears to decrease with age (Buchsbaum and Siegel 1994).

In old age, the cytoplasm of nerve cells accumulates a pigment called lipofuscin, which is probably made up of degraded cellular components. There are also changes in the components of the neuronal cytoskeleton. A protein called **tau**, which plays a role in linking neurofilaments and microtubules, can accumulate to produce paired helical filaments that form **neurofibrillary tangles** in some nerve cells. In normal ageing neurofibrillary tangles are usually confined to a small number of cells in the hippocampus and entorhinal cortex.

In addition to neurofibrillary tangles, the normal ageing brain can also contain **senile plaques** which are collections of neuritic processes grouped together in a spherical form, sometimes with a core of extracelluar amyloid at the centre. Senile plaques also contain the paired helical filaments of tau protein. The distribution of senile plaques in the normal ageing brain is rather wider than that of neurofibrillary tangles, and can occur in both the neocortex and amygdala as well as in the hippocampus and entorhinal cortex. A small proportion of brains from normal old subjects contain **Lewy bodies**. These are intracellular inclusion bodies with a laminated appearance, usually confined to the substantia nigra and locus ceruleus.

For a review of the neuropathology of normal ageing see Esiri (1991).

The psychology of ageing

Assessment of cognitive function in the elderly is complicated by the frequent presence of physical ill health, particularly by sensory deficits. Longitudinal studies suggest that intellectual function, as measured by standard intelligence tests, shows a significant decline only in old age. A characteristic pattern of change occurs with **psychomotor slowing** and impairment in the manipulation of new information. In contrast, tests of well-rehearsed skills such as verbal comprehension show little or no age-related decline.

Short-term memory, as measured by the digit span test, for example, does not change in the normal elderly. Tests of **working memory** show a gradual decrease in capacity, so that the elderly perform significantly less well than the young if attention has to be divided between two tasks or if the material has to be processed additionally in some way. The elderly can usually recall remote events of personal significance with great clarity. Despite this, their long-term memory for other remote events shows a decline; for a review see Morris (1991). Overall, there appears to be a balance between losses in flexible problem-solving with age and the benefits of accumulated wisdom derived from experience.

As well as these cognitive and motor changes there are important alterations in personality and attitudes, such as increasing cautiousness and rigidity. It used to be said that elderly people 'disengaged' from social life, but there is little evidence to support this generalization.

Physical health

In addition to a general decline in functional capacity and adaptability with ageing, chronic degenerative conditions are common. As a result, the elderly consult their family doctors frequently and occupy half of all general hospital beds. These demands are particularly large in those aged over 75. Medical management is often made more difficult by the presence of more than one disorder, by increased sensitivity to the side-effects of treatment, and by the frequency of psychiatric and social problems (Kane 1985).

Sensory and motor disabilities are frequent among elderly people. In a study in Jerusalem (Davies and Fleischchman 1981) difficulties in seeing were reported by 54 per cent of 70–74-year-olds, and by 69 per cent of those aged over 80. The corresponding figures for other disabilities were as follows: difficulty in hearing, 34 per cent and 50 per cent; difficulty in walking, 29 per cent and 62 per cent; difficulty in talking, 9 per cent and 22 per cent. Studies by the World Health Organization show that although the reported prevalence of these disorders differs somewhat in other countries, the figures from Jerusalem are fairly representative (Davies 1986).

Social circumstances

Almost all the elderly live at home; in the United Kingdom about a third are alone, about half are with their spouse, and about 10 per cent are with their children. Many see their families, friends, and neighbours regularly, and may provide as much help and support to their younger relatives as they themselves are receiving. Many are satisfied with their way of life and claim to be as happy as when they were younger, but a quarter are over 65 with no children to help them and a considerable number rarely have visitors. In an ageing population in which more middle-aged women work, fewer people visit and help the elderly.

Although many of the elderly are financially secure and there is an increasing trend towards home ownership amongst older people, the aged generally have poorer accommodation and lower incomes than younger people. About a half of the aged in the United Kingdom are believed to be near the official definition of poverty, and many lack basic amenities such as adequate bathrooms and good heating. These problems increase with increasing age.

Although the picture of living alone and of unsatisfactory social circumstances is common in most Western countries, it is not so everywhere. In some cultures (for example the Chinese), the elderly are esteemed, and in most developing countries the elderly live with their children and other relatives. See Grundy and Bowling (1991) for a review of social circumstances of the elderly; and Henderson (1990) for a review of the social aspects of the psychiatry of old age.

General considerations

Until the mid-1950s understanding of the mental disorders of later life was largely based upon accounts written at the beginning of the century when Kraepelin, Bleuler, and others described presenile, senile, and arteriosclerotic psychoses. At that time disorders with a predominantly affective picture were usually ascribed to a supposed underlying organic cause. Doubt about the evidence for such an organic cause, together with the success of new physical treatments for affective disorder, led to a re-examination of the problem. Using information from case notes, Roth (1955) divided patients into five diagnostic groups, and assessed their outcome six months and two years after admission. He found that affective psychosis, late paraphrenia, and acute confusion had better prognoses than arteriosclerotic or senile psychoses. Roth found that organic states appeared to be diagnosed more often in the United States than in the United Kingdom, and this finding was subsequently confirmed in a systematic comparison between the two countries (Copeland and Gurland 1985). Roth's investigation provided the basis for many subsequent studies of the epidemiology, clinical features, prognosis, and treatment of mental disorder in old age.

Epidemiology

Kay *et al.* (1964) carried out the first systematic prevalence study of psychiatric disorder amongst elderly people in the *general population*, including those living at home as well as those living in institutions, in an area of Newcastle-upon-Tyne. The findings, which are shown in Table 16.1 have been broadly replicated in subsequent surveys. In this and subsequent studies there have been problems of case definition and case finding especially for organic mental disorders (Cooper 1991*b*). It is particularly difficult to distinguish mild dementia from the effects of normal ageing or from life-long poor cognitive performance due to low intelligence or lack of education. Nevertheless, it is generally agreed that about 5 per cent of people over the age of 65 suffer from moderate or severe dementia and that the prevalence rises to over 30 per cent of those over 85 (Jorm *et al.* 1987; Skoog *et al.* 1993; Cooper 1991*b*).

Other surveys have shown a high prevalence of psychiatric disorder among elderly people in *sheltered accommodation* and in hospital. A third of the residents in old people's homes have significant cognitive impairment. In *general hospital* wards a third to a half of the patients aged 65 or over suffer from some form of psychiatric illness (Mayou and Hawton 1986)

It has frequently been reported that general practitioners are unaware of many of the psychiatric problems amongst elderly people living in the community. For example Williamson *et al.* (1964) found that general practitioners were unaware of 60 per cent of the elderly patients with neurosis, 76 per cent with depression, and 87 per cent with slight to moderate dementia. Since that time, greater awareness of the problems and the use of screening have increased the detection of disorder. Moreover, the presentation of such disorders to general practitioners and psychiatrists is determined as much by social factors as by a change in the patient's mental state. For example there may be a sudden alteration in the patient's environment, such as illness of a relative, or a bereavement. Sometimes an increasingly exhausted or frustrated family decide that they can no longer continue to care for an old person. At other times there is an element of manipulation by relatives who are trying to rid themselves of an unwanted responsibility.

Epidemiology is discussed further when individual syndromes are considered later in this chapter. (A review of epidemiological methods and findings has been provided by Mann (1991).

Table 16.1 *Estimated prevalence of psychiatric disorder*

Disorder	Prevalence in people aged 65 and over (%)	Ratio of patients living at home to those in institutions
Dementia (severe)	5.6	6 : 1
Dementia (mild)	5.7	10 : 1
Manic depressive	1.4	18 : 1
Schizophrenia (excluding long-stay hospital patients)	1.1	9 : 1
Neurosis and personality disorder	12.5	51 : 1
All disorder	26.3	14 : 1

Adapted from Kay *et al.* 1964.

Services for the elderly

National policies for the provision of services for the elderly differ widely. In the United States emphasis has been placed on care in hospitals and nursing homes. In Europe, Canada, and Australasia there has been varying emphasis on social policies to provide sheltered accommodation and care in the community. In this section, the development of services in the United Kingdom will be described as an example.

After the Second World War, legislation was enacted in the United Kingdom making local authorities responsible for domiciliary, day, and residential services for the elderly. In the National Health Service the role of primary care was emphasized, and a new medical specialty of geriatric medicine was established to develop the special interest and skill necessary to provide adequate community and hospital care for old people. Specialist psychiatric services soon followed.

In the United Kingdom a government policy document *Services for mental illness related to old age* (Department of Health and Social Security 1972) divided elderly patients with psychiatric problems into three broad groups: (1) patients with chronic mental illness in earlier life, some of whom were treated in hospitals for the mentally

ill before modern methods of treatment were available (graduates) (Campbell 1991); (2) elderly patients with functional mental illness; (3) elderly patients with dementia. ('Elderly' is usually taken to mean over 65 years of age.) The first group of old long-stay patients are mostly schizophrenics; they are usually cared for by general psychiatrists. Their numbers are being increased by the 'new long-stay patients' who, in their turn, will require some form of institutional care as they grow old (see Chapter 19). With the exception of these patients who have grown old in hospital, the specialist psychogeriatric services are usually responsible for all elderly psychiatric patients, whether their illness is functional or organic. The case for defining the specialty in this way, rather than confining it to the elderly demented, is twofold. First, there are special problems in the diagnosis and management of functional illness in the elderly; second, staff morale can be maintained better if some recoverable cases are seen.

Government policy in the United Kingdom, more than in some other countries, emphasizes the importance of treatment in the community rather than in hospital (Department of Health and Social Security 1972, 1978*b*, 1985; Health Advisory Service 1983). In trying to achieve this goal, obstacles arise because of the administrative separation in the United Kingdom of health services, provided by central government, from social services and community facilities which are the responsibility of local authorities. In most areas there are shortages of hospital beds, special accommodation, and staff, and an increasing use of private care which has resulted in undue emphasis on the determination of placements being determined by criteria other than need and has hindered development of integrated local services. Also many elderly people require practical services, such as home help or day care, but are not receiving them. For example, among a random sample of 477 people aged 65 and over living at home, only about 10 per cent were receiving domiciliary services, although almost three times as many appeared to need them (Foster *et al.* 1976).

The organization of psychiatric services varies in different places, since it reflects the personal style of doctors, local needs, and the extent of provision for this age group by general psychiatric services, as well as national policies. Nevertheless, there are some general principles of planning (Cooper 1991*a*; Jacoby 1991). The aims should be to maintain the elderly person at home for as long as possible, to respond quickly to medical and social problems as they rise, to ensure coordination of the work of those providing continuing care, and to support relatives and others who care for the elderly at home. There should be close liaison with the departments of geriatric medicine and with social services. A multidisciplinary approach should be adopted with a clinical team that includes psychiatrists, psychologists, community nurses, and social workers. Some members of the team should spend more of their working day in patients' homes and in general practices than in the hospital. The contributions of the various parts of a service will now be considered.

Primary care

In the United Kingdom, general practitioners together with their health visitors and nurses deal with most of the problems of mentally ill old people without referring them to specialists. However as already mentioned, general practitioners do not detect all the psychiatric problems of the elderly at an early stage (Williamson *et al.* 1964), nor do they always provide all the necessary long-term medical supervision. These problems are partly due to lack of awareness of the significance of psychiatric illness among the elderly, and partly to the provision of a service in which doctors respond to requests from patients rather than seeking out their problems. It is probable that the recent obligation for general practitioners in the United Kingdom to screen all those over 75 on their list has improved the recognition of problems.

Hospital care

In the United Kingdom, government policy is for assessment of elderly patients in a district general hospital, and for long-term and relief care in local hospitals. In many countries most elderly psychiatric patients are still treated in the wards of psychiatric hospitals. However, the care provided is more important than the type of institution in which it is given. The basic requirements are opportunities for privacy and the use of personal possessions, together with occupational and social therapy. Provided that these criteria are met, long-term hospital care can be the best provision for very disabled patients.

Geriatric medicine

There is inevitably some overlap in the characteristics of patients treated by units for geriatric medicine and those treated in psychiatry units. In the past there was considerable concern that many patients were 'misplaced' and therefore received poor treatment, stayed too long in hospital, and had an unsatisfactory outcome. Research has generally not confirmed these concerns. For example Copeland *et al.* (1975) found that although 64 per cent of patients admitted to geriatric hospitals were psychiatrically ill, only 12 per cent appeared to be wrongly placed. Moreover the outcome of these misplaced patients did not appear to be affected adversely: it seems that many patients can be cared for equally well in either type of hospital. The optimal placement of the rest depends on the relative predominance of behavioural or physical disorder. Medical and psychiatric teams need to cooperate closely if all patients are to receive appropriate treatment.

Day and out-patient care

In the 1950s *day care* began in geriatric hospitals. A few years later the first psychiatric day hospitals for the elderly were opened. Psychiatric day hospitals should provide a full range of diagnostic services and offer both short-term and continuing care for patients with functional or organic disorders, together with support for relatives. In the United Kingdom, day-care provisions by local authorities include day centres and social clubs. They can assist severely demented patients who do not require medical or nursing care. All arrangements depend crucially on adequate transport facilities.

Out-patient clinics

Out-patient clinics have a smaller part to play in providing care for the elderly than for younger patients because assessment at home is particularly important for old people. However, such clinics are convenient for the assessment and follow-up of mobile patients. There are advantages when these clinics are staffed jointly by medical geriatricians and psychogeriatricians.

Informal carers

Informal carers are those unpaid relatives, neighbours, or friends who look after disabled (usually demented) elderly persons at home. The term differentiates these people from formal carers such as paid home helps and district or community psychiatric nurses. Demographic changes and the move to community care have increased the burden on informal carers. Research has shown that most informal carers are (in descending order of frequency) spouses, adult daughters, or sons. About twice as many women as men are informal carers.

Several studies have shown that patients suffering from dementia place the greatest stress on carers (Baumgarten *et al.* 1994). In general, the more severe the dementia, the greater the strain.

Incontinence, wandering at night, and aggression are the most distressing problems for carers. As assessed by the General Health Questionnaire many carers have symptoms as severe as those of a psychiatric case. Symptoms diminish when the patient has moved to permanent residential care (Morris *et al.* 1988; Henderson 1990).

Levin (1991) recommended the following *ten key requirements for comprehensive support for carers*:

(1) early identification of dementia (the role of general practitioners is vital);

(2) comprehensive medical and social assessment of identified cases;

(3) timely referrals between agencies, for example from general practitioner to old age psychiatrist;

(4) continuing reviews of each patient's needs, and back up for carers;

(5) active medical treatment for any intercurrent illness;

(6) the provision of information, advice, counselling for carers;

(7) regular help with household and personal care tasks;

(8) regular breaks for carers, for example by providing day care and respite care for the patient;

(9) appropriate financial support;

(10) permanent residential care when this becomes necessary.

Residential care

In the United Kingdom, under Part III of the National Assistance Act 1984, local authorities are responsible for providing old people's homes and other sheltered accommodation. When the 1984 Act was passed, there was a legacy of large institutions (Townsend 1962). Since then, accommodation has been provided in small units, but it has proved difficult to achieve an acceptable standard of privacy while encouraging independence and involvement in outside activities.

More recent developments in the United Kingdom have strongly emphasized the role of private care. This change has had some advantages in encouraging a greater variety of facilities, but in some places undue emphasis on the cost rather than the standard of care together with a lack of coordinated planning, has made it difficult to develop integrated local services to meet the needs of the elderly.

There is a need for special housing for the elderly, conveniently sited and easy to run. Ideally, the elderly should be able to transfer to more sheltered accommodation if they become more disabled, without losing all independence or moving away from familiar places. In many communities in the United Kingdom there is still not enough variety of accommodation offering a range of independence. Provision of this kind is better in many parts of Europe, the United States, and Australasia (Grundy 1987).

Other countries have different systems of care. Gurland *et al.* (1979) compared two random samples of patients living in institutions in New York and London. Both cities provided residential care for about 4 per cent of their elderly population. However, in New York at that time 60 per cent of the places for the elderly were in large nursing homes staffed by nurses, whereas in London nearly two-thirds were in small residential unites staffed by a warden and domestic helpers. Thus many elderly people with similar problems were receiving different forms of care in the two cities. There are considerably more hospital beds per unit of population for the elderly in Scandinavia than in the United Kingdom; there are also more residential places which are more varied and of better quality, and generally offer greater privacy and freedom of choice.

Domiciliary services

In addition to medical services, domiciliary services include home helps, meals at home, laundry, telephone, and emergency call systems. In the United Kingdom local authorities provide these services; they also support voluntary organizations and encourage local initiatives such as good neighbour schemes and self-help groups. In a random sample of nearly 500 people aged 65 and over living at home, Foster *et al.* (1976) found that 12 per cent were receiving domiciliary services but a further 20 per cent still needed them. Although these provisions are increasing, so are the numbers of those requiring them. Bergmann *et al.* (1978) have argued that if resources are limited, more should be directed to patients living with their families than to those living alone. This is because the former can often remain at home if they receive such help, while many of the latter require admission before long even when extra help is given.

Community psychiatric nursing

Community psychiatric nurses have had an increasing role in acting as a bridge between primary care and specialist services. The nurses may assess referrals from general practitioners, monitor treatment in collaboration with general practitioners and the psychiatric services, and take part in the organization of home support for the demented elderly.

Services in the United States

In the United States the care of those aged 65 accounts for approximately one-third of all health care expenditure, with two-thirds being financed by the federal government, principally by Medicare. Unfortunately the low rates of reimbursement compared with the fees usually expected mean that it is difficult for the elderly to obtain appropriate out-patient psychiatric care and there are few home visits.

In the past the main emphasis in long-term care was on nursing homes, but this is changing to a wider range of provisions. However, many of these provisions are privately funded and the less well- off elderly are unable to afford them. Further difficulties have arisen from the closure of state psychiatric hospitals, limitation of the number of nursing home beds, and restrictions of Medicare and Medicaid Funding. Despite an increasing interest in the psychiatric care of the elderly, it is unlikely that resources will increase to keep pace with the rate of increase of the number of the elderly.

Some general principles of assessment

In the United Kingdom most psychogeriatricians believe that the first assessment should normally take place in the elderly patient's home, where his functioning can be assessed in its normal setting, and other informants can be interviewed and social conditions can be observed. Since less than half the patients assessed in this way are admitted to hospital, the home visit is an important opportunity to plan treatment with all those concerned. The answers to three general questions should be sought.

(1) Can the patient be managed at home?

(2) If so what additional help does the family need?

(3) Can the patient manage his financial affairs?

During the assessment, emphasis should be placed on the medical history and physical examination as well as on a thorough formulation of social problems. Whenever possible the clinician should interview close relatives or friends who can give information about the patient and may be involved in his continuing care. Such interviews are particularly helpful when the patient has cognitive impairment. In every case the reasons for the referral should be considered carefully, since many emergencies reflect changes in the attitudes of the family and neighbours to the patient's long-standing problems rather than a change in his psychiatric state. In order to answer the three questions listed above, the clinician will need to elicit the following information:

(1) The time and mode of onset of symptoms and their subsequent course.

(2) Any previous medical and psychiatric history.

(3) The patient's living conditions and financial position.

(4) The patient's ability to look after himself.

(5) Any odd or undesirable behaviour that may cause difficulties with neighbours or other people.

(6) Other services already involved in the patient's care.

Usually a diagnosis can be made and provisional plans formulated during the first home visit. Hospital admission may be required either for investigation and treatment or for social reasons. More often it is possible to arrange extra social or medical care in the patient's normal surroundings.

With these general points in mind, consideration can now be given to specific aspects of history taking, physical and mental examination, and psychological assessment.

History taking

Normally the problem should be discussed with the general practitioner before the patient is seen. If there is any likelihood of intellectual impairment in the patient, it is usually best to speak to relatives or other informants first. Details of the onset and time course of the symptoms are of particular value in differential diagnosis. Since a social formulation is important, the history should cover information about the patient's financial state and social circumstances, and about people who may be willing to help him. A description of the patient's behaviour over a typical 24-hour period is often helpful in eliciting symptoms and disabilities, and in obtaining a detailed picture of the patient's way of life and the reactions of other people. Enquiries should also be made about hazards such as poor heating; the capacity to handle fire, gas, and electricity; wandering; and allowing strangers into the home.

Examination

A thorough physical examination should be carried out, including an appropriately detailed neurological assessment with particular attention to vision and hearing. The mental state examination should include a systematic assessment of cognitive functions. It may be necessary to test linguistic, visuo-spatial, and other higher cortical functions.

If the patient is admitted to hospital, systematic observations should be made of his behaviour on the ward. Psychological tests are a useful addition to the clinical assessment of intellectual function (see below).

Physical investigations

On admission to hospital, the minimum routine investigations are a full blood count, a biochemical screen including thyroid functioning, urine analysis and culture, and chest radiography. If physical illness is suspected, additional investigations may be required. In the case of very old patients with dementia, the clinician should use his judgement in deciding how far to pursue physical investigations when there are no clinical signs.

If an organic disorder is suspected, additional investigations are indicated: electrolytes, urea and liver function, plasma calcium, vitamin B_{12}, folate chest radiography, syphilis serology, and ECG. Where it is widely available, a CT scan should be routine; elsewhere it may have to be restricted to where there are specific indications. With older patients, especially when a reasonably confident clinical diagnosis can be made, elaborate investigations are seldom necessary.

Psychological assessment

In skilled hands, psychometric testing has an important though limited role (Hodges 1994; Little 1991). Common obstacles to testing include confusion, lack of motivation, sensory handicaps, and the length of time required for patients to become accustomed to the test procedure. The main uses of psychological assessment are (i) clarification of diagnosis (ii) prediction of outcome, (iii) identifying the need of support and intervention and (iv) monitoring change.

Psychometric testing is now used less frequently than in the past. It is seldom used to differentiate cognitive impairment from functional illness; for this purpose clinical assessment is generally preferred. Psychometric testing may be useful in comparing current functioning with premorbid state, as assessed by educational, employment and social achievements. Schedules such as the Mini Mental State Examination (MMSE, see p. 54) are widely used. These are useful measures of current state that help in the planning of treatment and monitoring of change. Serial measurements can provide better evidence of decline, but the assessor must be aware of the test–retest reliability of his methods. Standardized questionnaires should be an addition to clinical assessment; they cannot wholly replace clinical expertise.

Treatment

Essentially, the psychiatric treatment of the elderly resembles that of other adults, but there are differences in emphasis. If practicable, treatment at home is generally preferable to that in hospital, not only because most elderly people want to be at home, but also because they function best there. Home treatment requires a willingness on the part of the doctor to be flexible and responsive to changing needs, to arrange a plan with the family, to organize appropriate day care or help, and to admit to hospital should it become necessary.

Whenever possible, the underlying cause of a dementia should be treated. The treatment of physical disorders, however minor, can also benefit the mental state; for example a urinary tract infection may cause mild delirium which adds to the features of dementia and improves with treatment. Mobility should be encouraged and physiotherapy is often helpful. A good diet should be arranged.

Use of drugs

Substantially more psychotropic and non-psychotropic drugs are prescribed for the elderly than for younger people. Drug-induced morbidity is a major medical problem, partly because the pharmacokinetics of drugs are different in old people. Most problems arise with drugs used to treat cardiovascular disorders (hypotensives, diuretics, and digoxin) and those acting on the central nervous system (antidepressants, hypnotics, anxiolytics, antipsychotics, and antiparkinsonian drugs). It is essential to restrict the number of drugs to avoid harmful drug interactions and prudent to start with small doses. Medication should be reviewed regularly and kept to a minimum.

Compliance with treatment is a problem in elderly patients, especially in those who live alone, have poor vision or are confused. The drug regime should be as simple as possible, medicine bottles should be labelled clearly, and memory aids, such as packs containing the drugs to be taken on a single day with daily dose requirements, should be provided. If possible drug taking should be supervised and the patient's response watched carefully. In these aspects of treatment, as in many others, domiciliary nursing plays a vital part.

Despite the need for caution in prescribing, elderly patients should not be denied effective drug treatment, especially for depressive disorders. Antidepressant medication should be started cautiously and increased gradually. If patients do not respond, it is sometimes helpful to measure the plasma concentration of antidepressant as a guide to increasing the dose.

Many elderly people sleep poorly, and about 20 per cent of those aged over 70 take hypnotics regularly. Hypnotic drugs often cause adverse side-effects in the elderly, notably daytime drowsiness leading to confusion, falls, incontinence, and hypothermia. Some psychiatrists consider that benzodiazepines are specially liable to cause delirium; chlormethiazole, chloral, dichloralphenazone, and trazodone in low doses are useful as alternatives. Haloperidol may be useful in restless demented patients. If a hypnotic is essential, the minimum effective dose should be used and the effects monitored carefully.

Electroconvulsive therapy

ECT is one of the most effective treatments for serious depressive disorder in the elderly. In patients with cognitive impairment, ECT may be followed by temporary memory impairment and confusion so that longer intervals between treatments are advisable. Particular attention should be paid to the physical health of elderly patients undergoing this treatment, and physically frail patients should be assessed by an experienced anaesthetist before receiving ECT.

Psychological treatment

Supportive therapy with clearly defined aims is often helpful, and joint interviews with the spouse are sometimes required. *Family therapy* is increasingly popular. Interpretative psychotherapy is seldom appropriate for the elderly.

Cognitive and behavioural treatments are increasingly used in the treatment of elderly psychiatric patients (Thompson *et al.* 1987; Fisher and Cartensen 1990). Behavioural methods have been used with demented patients to reduce problems in continence, eating behaviours, or social skills. Memory aids such as notebooks and alarm clocks have been used to assist patients with memory disorder.

Reality orientation therapy (Folsom 1967) is intended to reduce confusion and improve the behaviour of demented patients. In this approach, information is given about orientation in time and place, and is repeated on every contact with the patient. The technique is practised widely, even though the results of evaluative studies are conflicting (Holden and Woods 1988).

Social treatment

Some patients can achieve independence through measures to encourage self-care, and domestic skills, and to increase social contacts. More severely impaired patients can benefit from an environment in which individual needs and dignity are respected and each person retains some personal possessions. Disorientation can be reduced by the general design of the ward and the use of aids such as colour codes on doors. For those living at home, a domiciliary occupational therapist may be helpful in arranging additional stair rails or bathroom aids for example.

Support for relatives

Time should be spent with families in giving advice about the care of patients and discussing their problems. Such support can help families avoid some of the frustration and anxiety of caring for elderly relatives. Published guides such as *The 36-hour day* (Mace and Rabins 1993) are often useful. Other practical help may include day-care or holiday admissions, and laundry and meal services to the home. With such assistance many patients can remain in their own homes without imposing an unreasonable burden on their families. Community psychiatric nurses play essential roles in coordinating these services, supporting relatives, and providing direct nursing care. The first two roles can also be undertaken by social workers.

Delirium

Because delirium has physical causes, the patients are usually under the care of physicians or general practitioners. Surveys of elderly patients admitted to medical and geriatric wards show that up to a fifth suffer delirium. The main predisposing factors are pre-existing dementia, defective hearing and vision, Parkinson's disease, and advanced age. The most frequent precipitating causes include pneumonia, infection, stroke, hypoxia, and the toxic effects of medication. See Lipowski (1990) for a review.

The **clinical features** of delirium are discussed on p. 311 (see also Fairweather 1991). It should be remembered that impairment of consciousness, although invariable, is not always obvious amongst elderly patients, especially when the onset is gradual. When this happens, the condition is sometimes misdiagnosed as an irreversible dementia. In patients who have an established dementia, cognitive function can sometimes be affected further by minor physical disorders such as constipation, dehydration, or mild bronchitis.

Since many of the causes of delirium threaten life, the mortality of delirium, is high. In the Newcastle survey of patients admitted to hospital with delirium half were found to have died within two years of admission (Roth 1955). More recent studies (Van Hemert *et al.* 1993) have confirmed this finding.

The basic requirement in **management** is to search for and treat the underlying cause. While this treatment is taking effect, psychotropic drugs can provide symptomatic relief. Small doses of haloperidol or a phenothiazine are usually effective without increasing confusion. The amount and timing of the dosage should be determined carefully for each patient by reference to the *British national formulary* or a similar source. Failure to respond to a psychotropic is likely to be due to a worsening of the underlying cause of the delirium, and is not normally an indication for increasing the dosage of the psychotropics.

If a hypnotic is needed, chlormethiazole, dichloralphenazone, and medium- or short-acting benzodiazepines are safe and effective. Long-acting benzodiazepines should not be used since they may increase confusion. Other drugs which may exacerbate delirium, especially anti-cholinergics, should be avoided.

For a review of delirium in the elderly see Fairweather 1991.

Dementia in the elderly

In this book, the main account of dementia is given in Chapter 11. The reader is referred to that chapter for a definition of dementia (p. 311), and a description of its clinical features (p. 312) and treatment (p. 322). This section is concerned only with dementia in the elderly.

Dementia in the elderly has been recognized since the French psychiatrist Esquirol described *démence senile* in his textbook *Des maladies mentales* (Esquirol 1838). Esquirol's description of the disorder was in general terms, but it can be recognized as similar to the present-day concept (Alexander 1972). Kraepelin distinguished dementia from psychoses due to other organic causes such as neurosyphilis, and he divided it into presenile, senile, and arteriosclerotic forms. In an important study, Roth (1955) showed that dementia in the elderly differed from affective disorders and paranoid disorders in its poorer prognosis. ICD10 introduced a new category of mild cognitive disorder (MCD) but its validity is not proven (Christensen *et al.* 1995).

Dementia in old age can be divided into three groups according to aetiology and pathology.

(1) *Dementia of the Alzheimer type*: This has the same pathological changes in the brain as presenile dementia of the Alzheimer type. It is the most common kind of dementia in old age.

(2) *Vascular dementia*: this includes multiple infarcts in the brain, resulting from vascular occlusions, Binswanger's disease, and lacunar state.

(3) *Dementia due to other causes*: this disease group includes dementia resulting from a wide range of causes, such as Lewy body disease, parkinsonism, neoplasms, infections, toxins, alcohol abuse, and metabolic disorders, some of which are reversible.

Alzheimer's disease is the most common cause of dementia in the elderly. Meta-analyses (such as those by Jorm *et al.* (1987) and Hofman *et al.* (1991)) have shown a relative excess of Alzheimer's disease in females and of vascular disease in men. There are geographical differences, for example the higher prevalence of vascular dementia in Japan. In the United States alcohol is the third most common cause of dementia. This chapter is concerned only with groups 1 and 2 and with Lewy body disease. The other conditions in group 3 are discussed elsewhere.

The clinical picture of dementia is much the same in all three groups. Some minor differences in clinical features are found fairly consistently, but they cannot be relied on to differentiate between the three groups.

For many years it was believed that vascular disease was the most common cause of dementia. Doubt was cast on this belief when Corsellis (1962) reported that pathological changes of Alzheimer's disease were more common. This finding was subsequently confirmed by several workers. For example in a careful clinicopathological study based on 50 successive post-mortems of demented patients, Tomlinson *et al.* (1970) found the following distribution of changes: definite Alzheimer type,

50 per cent; probable Alzheimer type, 16 per cent; definite arteriosclerotic, 12 per cent; probable arteriosclerotic, 6 per cent; both Alzheimer and arteriosclerotic, 8 per cent; no evident pathology, 8 per cent.

Prevalence of dementia in the elderly

There have been many estimates of the *prevalence* of dementia in the elderly (Henderson 1994), but they are difficult to compare because of differences in the samples studied and the methods used. After reviewing the results of 20 surveys of dementia carried out in Europe, Japan, and North America, Jorm *et al.* (1987) concluded that the prevalence of moderate and severe dementia is about 5 per cent of persons aged 65 years and over, and 20 per cent of those aged over 80 years. A more recent analysis of 12 European studies using comparable methods has confirmed these estimates (Hofman *et al.* 1991). Skoog *et al.* (1993) reported a prevalence of dementia of 30 per cent among 85-year-olds living in Gothenberg, Sweden, of which 43.5 per cent was Alzheimer's disease and 46.9 per cent was vascular dementia.

Bickell and Cooper (1994) re-examined a sample of residents in Mannheim, Germany after an interval of seven to eight years and estimated the annual *incidence* for all forms of dementia as 15.4 per 1000 persons aged 65 and over, of which 8.9 were Alzheimer type and 4.4 were vascular dementia.

About 80 per cent of demented people are in the community rather than institutions. In developing countries the total number of demented people is increasing rapidly as life expectancy improves See Cooper (1991*a*) and Henderson (1994) for reviews.

Alzheimer's disease

Clinical features

The general clinical features of dementia have been described on p. 312, the clinical features of Alzheimer's disease in the elderly are considered here. The condition is common and important in the elderly, and its features are characteristic (though not discriminating enough to allow accurate diagnosis from other forms of dementia).

Alzheimer's disease usually begins after the age of 70, and age-specific incidence rates increase sharply with increasing age. It occurs slightly more commonly in women. Death usually occurs within five to eight years of appearances of the first signs of the disease. In the past doctors were seldom consulted until a later stage of chronic deterioration, or following a sudden worsening in relation to some other physical illness. Greater awareness of the diagnosis and the introduction of routine screening in primary care have resulted in earlier diagnosis.

The first evidence of the condition is often with minor forgetfulness which is difficult to distinguish from normal ageing. The progress is usually gradual for the first two to four years, with increasing memory disturbance and lack of spontaneity. Disorientation is normally an early sign and may first be evident when the person is in unfamiliar surroundings, for example on holiday. The mood may be predominantly depressed (Weiner *et al.* 1994), euphoric, flattened, or labile. Self-care and social behaviour decline, although some patients maintain a good social facade despite severe cognitive impairment.

Changes in behaviour are usual and are of particular concern to carers. Patients may be restless and may wake at night, disorientated and perplexed. There may also be aggressiveness (Patel and Hope 1993), changes in eating habits (Morris *et al.* 1989), and wandering (Hope 1994).

In the early stages of Alzheimer's disease the clinical features are modified by the premorbid personality, and any personality defects tend to be exaggerated. In the middle and later stages of the illness intellectual impairment becomes more obvious, and language and visuospatial disorders are frequent. Focal signs of parietal lobe dysfunction (such as dysphasia or dyspraxia) are usual, and frontal impairment, as indicated by decreased verbal fluency, personality change, and loss of emotional control, is common. Incidental physical illness may cause a superimposed delirium resulting in a sudden deterioration in cognitive function, which usually improves when the physical illness has been treated but may sometimes be permanent.

Diagnosis

The differential diagnosis of dementia is discussed on p. 319. Alzheimer's disease can be distinguished clinically from multi-infarct dementia on the basis of additional features of the latter syndrome, such as stepwise progression, fluctuating course, and focal motor signs. Patients with Lewy body dementia may show hallucinations and confusion at an early state and are more likely to have a concomitant extrapyramidal disorder (see p. 524).

Brain imaging can be helpful in supporting the diagnosis of Alzheimer's disease. CT scanning may show enlarged lateral ventricles and atrophy of medial temporal structures, and serial examination may reveal evidence of disease progression. Structural magnetic resonance imaging may demonstrate atrophy in specific structures such as the hippocampus. Studies of cerebral metabolism and blood flow with PET and SPECT characteristically reveal bilateral deficits in posterior parietal and temporal cortex (Buchsbaum *et al.* 1991; Jobst *et al.* 1992).

Histopathology

The pathological changes in senile Alzheimer's disease are the same as those in the presenile form of the disease. On gross examination the brain is shrunken with widened sulci and enlarged ventricles. There is cell loss, particularly of cortical pyramidal neurones, together with proliferation of astrocytes and increased gliosis. Silver staining shows **senile plaques** throughout the cortical and subcortical grey matter, and also **neurofibrillary tangles** and granulovacuolar degeneration. In comparison with the normal elderly brain, the senile plaques and neurofibrillary tangles are more numerous, particularly in limbic regions, and have a more widespread distribution. For example both plaques and tangles may be found in subcortical structures such as Meynert's nucleus and the locus ceruleus which have diffuse projections to the cortex (Esiri 1991). The presence of neurofibrillary tangles in a particular cell may predict neuronal degeneration and death (Rossor 1993).

The degree of cognitive impairment in Alzheimer's disease correlates with the number of neurofibrillary tangles (Wilcock and Esiri 1982; Arriageda *et al.* 1992). More recently, Förstl *et al.* (1993) found correlations between behavioural disturbance, lower brain weight, density of cortical neurofibrillary tangles, and neuronal loss in the hippocampus and Meynert's nucleus.

Molecular genetics and pathology

First-degree relatives of patients with Alzheimer's disease have a threefold risk of developing the disorder, i.e. three times that of the general population. In certain rare pedigrees, usually those with an early onset of illness, an autosomal dominant mode of inheritance can be discerned. However, most cases appear to be sporadic, with genetic vulnerability interacting with environmental factors (St Clair 1994).

Most progress in elucidating the molecular genetics of Alzheimer's disease has been made in the *presenile form* where the studies have identified two definite sites of genetic linkage, one to chromosome 21 and one to chromosome 14. Particular interest has centred on possible abnormalities in chromosome 21 because patients with trisomy of chromosome 21 (Down's syndrome, see p. 737) have a high risk of early-onset Alzheimer's disease. Furthermore, the gene for the β-amyloid precursor protein, is also located on chromosome 21. β-amyloid precursor protein is a transmembrane glycoprotein and a derivative of this molecule, β-amyloid peptide, is found in the amyloid deposits in senile plaques (Mullan 1993).

Investigations into the role of chromosome 21 led to the finding that in families with autosomal linkage of Alzheimer's disease to chromosome 21, affected or at risk individuals have mutations in the β-amyloid precursor protein gene. The first mutation discovered consisted of a single-nucleotide base-pair substitution such that isoleucine rather than valine was incorporated in the β-amyloid cascade. This change could play a key role in the pathophysiology of Alzheimer's disease, but it is not known how mutations in the gene lead to amyloid deposition and the characteristic histopathological changes (Hardy 1992). Some of the mutations may cause excessive amyloid to be synthesized, while in other cases an abnormal amyloid resistant to degradation may be produced.

Mutations in the β-amyloid gene account for only about 10 per cent of early-onset cases of Alzheimer's disease. A high proportion of the remaining cases are linked to a locus on chromosome 14, but the nature of the gene involved is unknown. It will be of interest to see whether the function of this gene further implicates abnormalities related to amyloid deposition or instead suggests a separate mechanism that leads to Alzheimer's disease. Some families with early-onset disease do not show linkage to either chromosome 14 or chromosome 21, suggesting that other loci are also involved in some cases.

Late-onset Alzheimer's disease does not show the Mendelian pattern of inheritance that can be found in the early-onset cases. However, recent studies have shown a strong association between late onset Alzheimer's disease and the **E4 allele of apolipoprotein** E which is found in both the senile plaques and neurofibrillary tangles. The gene for apoliprotein E is located on chromosome 19 which has shown linkage with late-onset Alzheimer's disease in some family studies (Mullan 1993). The frequency of the E4 allele is about 0.50 in patients with Alzheimer's disease from multiply affected families, about 0.40 in patients unselected for family history, and about 0.12 in controls. People with two E4 alleles have an increased risk of developing Alzheimer's disease and an earlier age of onset. However, about 40 per cent of patients with Alzheimer's disease do not possess a copy of the E4 allele. Also, the E4 allele is a risk factor for hypercholesterolaemia and coronary artery disease as well as for Alzheimer's disease (Owen *et al.* 1994).

Thus, although possession of the apolipoprotein E4 allele is a predisposing factor for Alzheimer's disease, it is neither a necessary nor sufficient cause. How it contributes to the pathological changes is not clear. The presence of apolipoprotein E in senile plaques and neurofibrillary tangles suggests that it may play a direct role in pathogenesis, perhaps by increasing the rate at which amyloid is deposited. The fact that the E4 allele also predisposes to vascular disease is intriguing, since post-mortem studies of demented patients frequently reveal neuropathological features of both vascular and Alzheimer's disease.

The suggestion that **aluminium** may be a contributory environmental cause of Alzheimer's disease is controversial. Studies have failed to agree whether aluminium is present in senile plaques and neurofibrillary tangles, and whether aluminium concentrations are raised in various brain areas of patients with the disease. Patients undergoing renal dialysis are at increased risk of developing dementia. Such patients have elevated brain aluminium levels and increased numbers of senile plaques, but do not have obvious neurofibrillary tangles. However, they do have abnormal processing of tau protein with deposition of insoluble phosporylated tau in grey matter as occurs in Alzheimer's disease (Harrington *et al.* 1994). These latter findings, together with *in vitro* studies showing that aluminum can promote the phosphorylation of tau protein and lead to the formation of amyloid peptide, suggest that this metal could play a role in the development of Alzheimer's disease (Garruto and Brown 1994).

Neurotransmitter changes

Neurochemical studies in Alzheimer's disease have shown a widespread loss of several neurotransmitters, particularly in association areas of cerebral cortex and hippocampus. The cells of origin of several of these neurotransmitters, for example noradrenaline and 5-hydroxytryptamine (5-HT), are located in subcortical nuclei which can themselves be subject to the pathological changes of Alzheimer's disease. However, it is believed that the deficits in these neurotransmitters may be secondary to a primary loss of cortical cells with retrograde degeneration of subcortical neurones.

Acetylcholine is synthesized from choline acetyltransferase, and reductions in this enzyme correlate well with loss of cholinergic nerve cells. Patients with Alzheimer's disease have a widespread **loss of choline acetyltransferase** which correlates with the extent of the cognitive changes before death and the severity of the pathological changes in the brain (Neary *et al.* 1986). There are also reductions in concentrations of **noradrenaline** and 5-HT in the cortex and hippocampus, with a marked *loss of 5-HT_2 receptors* as a common finding. Many 5-HT_2 receptors are located on pyramidal neurones, and presumably the reduction in 5-HT_2 receptor binding reflects the death of these cells (Dewar 1991). It has been proposed that the loss of noradrenaline and 5-HT may underlie some of the non-cognitive symptoms of Alzheimer's disease, for example depression and aggression (Perry and Perry 1994).

The cortical cell loss in Alzheimer's disease is accompanied by reductions in terminal markers of **glutamate,** which is the neurotransmitter utilized by pyramidal cells in both the cortex and the hippocampus. Furthermore the pyramidal cells receive inputs from short cortical interneurones which contain GABA and neuropeptides such as somatostatin. Reductions in **somatostatin** concentrations in patients with Alzheimer's disease may be linked to structural pathology because they are most apparent in temporal regions which have the highest density of plaques and tangles. Clearly, these changes in cortical neurotransmitters would be expected to produce severe disruption of the processing of information, particularly in association areas.

The widespread nature of neurotransmitter deficits in Alzheimer's disease suggests that it will be difficult to devise a replacement treatment. Nevertheless, there is some evidence that increasing acetylcholine function may benefit cognitive function in some patients (see p. 526).

See Terry *et al.* (1994) for reviews of Alzheimer's disease.

Lewy body dementia

Lewy body dementia is a progressive dementing illness which can be distinguished clinically from Alzheimer's disease by its fluctuating course and the occurrence of psychotic symptoms such as hallucinations. Lewy body dementia is associated with signs of parkinsonism and extreme sensitivity to the extrapyramidal effects of antipsychotic drugs. These two phenomena are likely to reflect deficits in nigrostriatal dopamine neurones (Perry and Perry 1994). The characteristic histopathological feature of Lewy body dementia is the presence of Lewy bodies in cerebral cortex and substantia nigra. Senile plaques may be present but neurofibrillary tangles are absent.

Like Alzheimer's disease, Lewy body dementia is associated with widespread reductions in choline acetyltransferase in the neocortex and there is also some loss of dopamine from the caudate nucleus. Similar changes are found in Parkinson's disease. In non-demented parkinsonian patients, Lewy bodies are found predominantly in subcortical regions and the loss of choline acetyltransferase in the cortex is modest. In demented parkinsonian patients, Lewy bodies are present in cortex and there is pronounced loss of cortical choline acetyltransferase (Cross 1991).

Vascular dementia

In the past the dementia caused by cerebrovascular disease was referred to as 'atherosclerotic' psychosis. Following the separation of distinct syndromes of psychiatric disorder in late life (Roth 1955), it became apparent that dementia was often associated with multiple infarcts of varying size, mostly caused by thromboembolism from extra-cranial arteries, and Hachinski *et al.* (1974) suggested the term multi-infarct dementia (MID). Recent research has shown that patients with MID are a subgroup of a larger group of patients with dementia due to vascular disease, and the term vascular dementia is now preferred. The pathogenetic mechanisms are very varied.

Vascular dementia is slightly more common in men than in women. The *prevalence* increases with age, approximately doubling every five years. There appear to be geographical differences, with especially high rates of vascular dementia reported in China, Japan, and the Russian Federation (Henderson 1994).

Clinical picture

The *onset*, which is usually in the late sixties or the seventies, may follow a cerebrovascular accident and is often more acute than that of Alzheimer's disease. Emotional and personality changes may appear first, followed by impairments of memory and intellect that are characteristically fluctuating. Depression is frequent, and episodes of emotional lability and confusion are common, especially at night. Fits or minor episodes of cerebral ischaemia are usual at some stage. Insight is often maintained to a late stage. Behavioural retardation and anxiety are more common than in Alzheimer's disease (Sultzer *et al.* 1993).

The *diagnosis* is difficult to make with certainty unless there is a clear history of strokes or definite localizing signs. Suggestive features are patchy psychological deficits, erratic progression, and relative preservation of the personality. On physical examination there are usually signs of hypertension and of arteriosclerosis in peripheral and retinal vessels, and there may be neurological signs such as pseudobulbar palsy, rigidity akinesia, and brisk reflexes.

The *course* of vascular dementia is usually a stepwise progression, with periods of deterioration that are sometimes followed by partial recovery for a few months. About half the patients die from ischaemic heart disease, and others from cerebral infarction or renal complications. From the time of diagnosis the life-span varies widely but averages about four to five years. Most studies show somewhat shorter survival in vascular dementia than in Alzheimer's disease (Burns 1993).

The *gross pathology* is distinctive. There is localized or generalized brain atrophy and ventricular dilatation, with areas of cerebral infarction and evidence of arteriosclerosis in major vessels. On microscopy, multiple areas of infarction and ischaemia are evident. Tomlinson *et al.* (1970) found that the volume of damaged cerebral cortex at post-mortem was related to the degree of intellectual impairment shortly before death; usually no cognitive impairment was detectable until at least 50 ml of brain tissue had been affected.

Biochemical studies have shown that there is no association between cognitive impairment and levels of choline acetyl-transferase in MID (Perry *et al.* 1978)—a distinguishing feature from dementia of the Alzheimer type. For a review of the pathology of vascular dementia see Esiri (1991).

The assessment of dementia in the elderly

The assessment of dementia in the elderly follows the principles that apply to assessment of dementia at other ages. These principles are described on p. 317.

Dementia has to be differentiated from delirium, depressive disorders, and paranoid disorders. *Delirium* is suggested by impaired and fluctuating consciousness, and by perceptual misinterpretations and hallucinations (see p. 311). The differentiation of dementia from *affective disorders* and *paranoid states* is discussed later in this chapter (pp. 528 and 531). It is important to be aware that subjective complaints of cognitive decline are more likely to be associated with anxiety and depression than with dementia (Jorm *et al.* 1994). It should be remembered that *hypothyroidism* may be mistaken for dementia.

In the assessment of dementia it is important to *look for the treatable causes*, although they are rare. The term treatable in relation to dementia usually means *arrestable* rather than reversible; the effect of treatment may stop the progression of the dementia but it seldom results in a return to premorbid function. Treatable causes include deficiency of vitamin B_{12} and folate, normal pressure hydrocephalus, operable tumours, and neurosyphilis (these conditions are considered in Chapter 11). Whilst B_{12} and folate can cause dementia, this is rare, and in most elderly demented patients low B_{12} and folate levels are caused by poor nutrition. The *investigations* listed on p. 518 are usually sufficient for the investigation of dementia in the elderly. As already mentioned, the intensity of investigation must be judged in relation to the patient's age and general debility.

The assessment of a demented patient should also include a thorough search for treatable medical conditions that are associated with dementia rather than its cause. Such conditions are easily overlooked in demented patients.

The *social assessment* of dementia follows the general principle described earlier in the chapter (p. 517).

The treatment of dementia in the elderly

The first concern is to treat any treatable physical disorder. If the latter is the primary cause, the dementia may be arrested. Even if the physical disorder is not a direct cause of dementia, the treatment of the former may lead to improvement in the patient's condition.

For *dementia of any cause* restlessness by day or night may be reduced by promazine, thioridazine, haloperidol, or related drugs, without causing serious side-effects. In the United States legislation prohibits the use of neuroleptics for the treatment of agitation and restlessness in patients in nursing homes. However reducing restlessness can be an important first step when the patient is cared for at home by the family, since they are likely to be worn down by this behaviour. Antipsychotic drugs may also be required for paranoid delusions, and antidepressants for severe depressive symptoms.

It has been reported that drugs of four kinds may be of specific benefit for *Alzheimer's disease*: cholinergic drugs, vasodilators, neuropeptides, and enhancers of brain metabolism. *Cholinergic* drugs include precursors of acetylcholine such as choline and lecithin, stimulators of acetylcholine release such as piracetam, and inhibitors of acetylcholine hydrolysis such as physostigmine and tetrahydroaminoacridine (THA). *Vasodilators* include isoxsuprine, dihydroergotoxine, hydergine (a mixture of ergot alkaloids), and cyclandelate. *Neuropeptides* include vasopressin and its papaverine analogues which are given intranasally. Putative *enhancers of brain metabolism* include pentifylline and pyritinol.

Most clinical trials of substances in these four groups have shown little or no benefit in the treatment of Alzheimer dementia. Some trials of THA have demonstrated marginal effects, such as postponement of the progression of dementia. In general, however, the use of the drugs mentioned above is not recommended.

There is no specific treatment for vascular dementia apart from the control of blood pressure, low-dose aspirin, and, if indicated, surgical treatment of carotid artery stenosis.

Psychological and social treatment

Psychological and social treatments for elderly demented patients are similar to those for dementia in general (see p. 322). Whenever feasible, patients should continue to live in their own homes, particularly if they have someone to live with. Social care should be planned with all concerned— family, friends, general practitioner, community psychiatric nurse, and social worker. Day care may be needed to provide supervision, occupation, and training for the patient and to relieve the family. Brief hospital admissions may be needed to allow the carers to take a holiday or to tide over a crisis.

If the patient cannot be managed at home, residential care may be provided in an old people's home or other sheltered accommodation. Failing this, long-term care in hospital will be needed. For a review of the management of dementia in the elderly see Council on Scientific Affairs (1986).

Mood disorders

Depressive disorder

Depressive disorders are common in later life, and depressive symptoms are even more frequent. The point prevalence for depression of clinical severity is about 10 per cent for those aged over 65, with 2–3 per cent being severe (Copeland *et al.* 1987). First depressive disorders become less common after the age of 60, and they become rare after the age of 80. Rates in major depression depend on the setting. Koenig and Blazer (1992) reported the following rates: 0.4–1.4 per cent in the community, 5–10 per cent among medical out-patients, 10–15 per cent among medical in-patients, and 15–20 per cent among nursing home patients. It seems that many depressive disorders in elderly patients are not detected by their general practitioners. See Jorm (1995) for a review of epidemiology.

The incidence of **suicide** increases steadily with age, and suicide in the elderly is usually associated with depressive disorder.

Clinical features

There is no clear distinction between the clinical features of depressive disorders in the elderly and those in younger people, but some symptoms are often more striking in the elderly. Post (1972) reported that a third of depressed elderly patients had severe *retardation* and *agitation*. Abas *et al.* (1990) found evidence of *cognitive impairment* in 70 per cent of elderly patients with a depressive disorder. The pattern suggested impairment on tasks demanding effort, for example learning and memory, indicating the importance of *poor concentration* (Austin *et al.* 1992). Depressive *delusions* concerning poverty and physical illness are common, and occasionally there are nihilistic delusions such as beliefs that the body is empty, non-existent, or not functioning (see Cotard's syndrome, p. 199). *Hallucinations* of an accusing or obscene kind may occur. Depression itself is sometimes not conspicuous and may be masked by other symptoms, particularly *hypochondriacal complaints*. Depressive disorder in the elderly, should always be considered when the patient presents with anxiety, hypochondriasis, or confusion.

A small proportion of retarded depressed patients present with '*pseudodementia*'; i.e. they have conspicuous difficulty in concentration and remembering, but careful clinical testing shows that there is no major defect of memory function (Bulbena and Burrows 1986).

Course and prognosis

Before the introduction of ECT, many depressive disorders in the elderly lasted for years. Nowadays considerable improvement within a few months can be expected in about 85 per cent of admitted patients; the remaining 15 per cent do not recover completely. However, long-term follow-up shows a less encouraging picture. Post (1972) reported that patients who recovered in the first few months fell into three groups: one-third remained completely well for three years, another third suffered further depressive disorders with complete remissions; and the remaining third developed a state of chronic invalidism punctuated by depressive disorders. More recent studies have generally found a similar outcome (Baldwin and Jolley 1986; Cole 1990; Ames and Allen 1991).

Factors predicting a good *prognosis* for depression in old age are onset before the age of 70, short duration of illness, good previous adjustment, absence of disabling physical illness, and good recovery from previous episodes. Poor outcome is associated with the severity of the initial illness, poor compliance with antidepressant medication, and severe life events in the follow-up period. However, some other recent studies have reported somewhat more favourable outcomes, at least in the short term (Hinrichsen 1992; Reynolds *et al.* 1992; Hughes *et al.* 1993). Physical illness is an important factor determining the outcome of treatment (Murphy *et al.* 1988; Koenig *et al.* 1992; Reynolds 1992). Murphy *et al.* (1988) found that depressed patients had a significantly higher *mortality* than matched controls and that the difference was not entirely due to differences in physical health between the groups when first seen. As explained in Chapter 13 on suicide and deliberate self-harm, *suicide* is frequent in elderly depressives (Blazer 1986; Lindesay 1986).

Despite the poor outlook for depressive disorder in the elderly, only a small number of the patients develop dementia (Roth 1955; Post 1972).

Aetiology

In general the aetiology of depressive disorders in late life almost certainly resembles the aetiology of similar disorders at younger ages (Katona (1993) for a review). However, *genetic factors* may be of less significance. For first-degree relatives the risk is 4–5 per cent with elderly probands, as against 10–12 per cent with young and middle-aged probands (Mendelwicz 1976). However, the recent evidence is complex in that age of onset and whether the depressive illness is of recurrent or non-recurrent type appear to be independent variables. This means that up to a third of those with late-onset depression may give a family history of depression. It might be expected that the *loneliness and hardship* of old age would be important predisposing factors for depressive disorder. Surprisingly, there is no convincing evidence for such an association (Murphy 1982). Indeed, Parkes *et al.* (1969) even found that the association between bereavement and mental illness no longer held in the aged.

Although neurological and other *physical illnesses* are slightly more frequent among depressed compared with non-depressed elderly patients, there is no evidence that they have a specific aetiological role. Instead, such illnesses appear to act as non-specific precipitants.

Differential diagnosis

It is sometimes difficult to distinguish between depressive pseudodementia and *dementia*. It is essential to obtain a detailed history from other informants and to make careful observations of mental state and behaviour. In depressive pseudodementia a history of mood disturbance usually precedes the other symptoms. The depressed patient's unwillingness to answer questions during mental state examination can usually be distinguished from the demented patient's failure of memory. In dementia impairment is global; depressed patients are likely to have partial deficits. Psychological testing is often said to be useful, but it requires experienced interpretation and it usually adds little to skilful clinical assessment (Miller 1980). At times, dementia and depressive illness coexist. If there is real doubt, there is no harm in a trial of antidepressant treatment.

Less frequently, depressive disorder has to be differentiated from a *paranoid disorder*. When persecutory ideas occur in a depressive disorder, the patient usually believes that the supposed persecution is justified by his wickedness (see p. 199). Particular diagnostic difficulty may occur with the small group of patients who suffer from *schizoaffective* illness in old age (see p. 529).

Treatment

The principles of the treatment of depressive disorders are the same for adults of all ages. They are described in Chapter 8 (pp. 237–42). With elderly patients it is especially important to be aware of the risk of suicide. Any intercurrent physical disorder should be treated thoroughly. Antidepressants are effective, but should be used cautiously, perhaps starting with half the normal dosage and adjusting this in relation to side-effects and response. Some psychiatrists prefer to start treatment with a tricyclic antidepressant; others prefer specific serotonin re-uptake inhibitors antidepressants because they have fewer side-effects and are less cardiotoxic. Although it is appropriate to start drug treatment cautiously, it is equally important to avoid undermedication. For patients who do not respond to the full dose of an antidepressant, it may be necessary to use a combination of drugs as in younger patients (see p. 241). However before changing treatment it should be remembered that the compliance with drugs is often poor in elderly patients.

ECT is usually appropriate for severe and distressing agitation, suicidal ideas and behaviour, life-threatening stupor, or failure to respond to drugs. If the patient is unduly confused after ECT, treatments should be given at longer intervals (Sackeim 1994). If a patient has previously responded to antidepressants or ECT, but does not respond in the present episode, undetected physical illness is a likely cause.

After recovery, full dosage antidepressant medication should be continued for several months as in younger patients (see p. 242). Recent evidence (Old Age Depression Interest Group 1993) suggests that treatment should continue for considerably longer and possibly permanently. See NIH Consensus Development Panel on Depression in the Elderly (1992) for a review of diagnosis and treatment.

Mania

Mania accounts for between 5 and 10 per cent of affective illnesses in old age. Unlike depressive disorder, mania does not increase in incidence with age. Broadhead and Jacoby (1990), in the first prospective study of mania in old age, found that the clinical picture was the same as in younger patients but that a depressive episode occurs immediately after the manic episode more frequently in older patients.

Management is similar to that described for younger patients (p. 243). Lithium prophylaxis is valuable, but the blood levels should be monitored with special care and should be kept at the lower end of the therapeutic range used for younger patients. See Jacoby (1991) for a review of mania in old age.

Schizoaffective disorder

In a study of patients aged over 60 admitted to hospital, Post (1971) found that 4 per cent had schizoaffective disorders (i.e. disorders with a more or less equal mixture of the symptoms of schizophrenia and mood disorder) or a schizophrenic illness followed by a mood disorder or vice versa. Intermediate and long-term outcomes were less favourable for these conditions than for depressive disorder. The treatment of this disorder in the elderly is the same as that for younger patients (see p. 288).

Emotional disorder and personality disorder

In later life, emotional disorders are seldom causes for referral to a psychiatrist. In general practice, Shepherd *et al.* (1966) found that, after the age of 55, the incidence of new cases of neurosis (i.e. emotional disorder) declined; however, the frequency of consultations with the general practitioner for neurosis did not fall—presumably as a result of chronic or recurrent cases. *In the community*, surveys indicate that after the age of 65 some new cases of emotional disorder appear, with a prevalence of cases of at least moderate severity of about 12 per cent (Kay and Bergmann 1980). Probably many of these new cases do not present to general practitioners or psychiatrists. In *differential diagnosis* it is important to differentiate emotional disorders present continuously or intermittently from new symptoms. It is wise to consider that the latter may be due to a depressive disorder until proved otherwise.

Personality disorder is an important predisposing factor in most elderly patients with emotional symptoms. Physical illness is a frequent precipitant, and retirement, bereavement, and change of accommodation are other causes (Eastwood and Corbin 1985).

Emotional disorders among the elderly are usually non-specific, with symptoms of both anxiety and depression. Hypochondriacal symptoms are often prominent. Dissociative and conversion disorder, obsessional disorder, and phobic disorders are less common. Lindesay (1991) and Flint (1994) have suggested that anxiety disorders are clinically more important than had been previously recognized. For reviews see Lindesay (1995).

Personality disorder causes many problems for elderly patients and their families (Cohen 1994) Paranoid traits may become accentuated with the social isolation of old age, sometimes to the extent of being mistaken for a paranoid state (see p.298). Abnormal personality is one of the causes of the *senile squalor syndrome*, in which elderly people become isolated and neglect themselves in filthy conditions. Such gross self-neglect is often associated with social isolation and physical illness, and is associated with a high mortality after hospital admission. It is often difficult to decide when to intervene and when to use compulsory powers (Cybulska and Rucinski 1986). *Criminal behaviour* is unusual in the elderly (Jacoby 1991). In England and Wales in 1989 about 1 per cent of males found guilty of indictable offences were aged 60 and older, whereas 36 per cent were aged under 21.

The *treatment* of emotional disorders and personality disorders in old age is generally similar to that in younger adult life. It is essential to treat any physical disorder. Social measures are usually more important than psychological treatments, but psychotherapy and behavioural treatment should not be ruled out because of age alone.

Alcoholism and drug abuse

Reported prevalence rates for *problem drinking* among the elderly vary widely. Excessive drinking declines with increasing age but, whilst not a major problem, it is still significant among the elderly and is probably substantially under-recognized. Elderly excessive drinkers are of two kinds: those who began earlier in life; and those who began in old age, often as a response to social or other stress (Atkinson 1991).

The prevalence of *drug abuse* in the elderly has been little investigated. Excessive use of hypnotics and misuse of analgesics and laxatives are relatively common (Atkinson 1991) but it is difficult to distinguish between inadvertent and deliberate misuse.

Abuse and neglect of the elderly

Abuse and neglect of the elderly by other family members is often overlooked. The types of abuse are physical, psychological, financial, violation of rights, and acute and passive neglect. Women are more often affected than men, and those who have physical illness or psychiatric disorder are most at-risk. Abuse is usually by a relative and is often repeated. It may take the form of neglect or forced confinement, which may be shown in failure to thrive. Also, property may be misused (Fisk 1991). Dementia and the demands it makes of carers may predispose to abuse (Coyne *et al.* 1993). See Lachs and Pillemer (1995) for a review.

Schizophrenia-like and paranoid states in the elderly

Kraepelin introduced the term *paraphrenia* in 1909 to describe patients with chronic delusions and hallucinations without the characteristic personality deterioration of dementia praecox. Mayer (later known as Mayer-Gross) followed up these patients and, over 10 years later, found that just under half showed a typical schizophrenic decline. In 1955 Roth introduced the term *late paraphrenia* for paranoid conditions starting after the age of 60 'in a setting of well preserved personality and affective response'. Although other terms such as 'persistent persecutory states' and 'late-onset schizophrenia' have been suggested, late paraphrenia is the term that has been adopted by most old-age psychiatrists in the United Kingdom. The use of this term does not prejudge the question of whether late paraphrenia is essentially paranoid schizophrenia of delayed onset or a distinct entity.

Patients with late paraphrenia form approximately 10 per cent of admissions to psychiatric wards for the elderly. There are no good data on the prevalence in the community because it is difficult to identify all cases of a condition in which many sufferers keep their experience to themselves for as long as they can and are unlikely to cooperate with 'doorstep' interviews.

Neither ICD10 nor DSMIV have a clear and unequivocal place for late paraphrenia. In both systems cases have to be fitted into one of several categories according to the pattern and duration of symptoms. Howard *et al.* (1993) found that about 61 of 101 cases of late paraphrenia were classified as paranoid schizophrenia when ICD10 criteria were used. The next most frequent category (31 cases) was delusional disorder. The remaining eight fell into the schizoaffective group.

Aetiology

A number of aetiological factors have emerged from research into late paraphrenia compared with aged matched controls. Premorbid *personality* of people with late paraphrenia is characterized by poor adjustment, as suggested by lower marriage and childbearing rates. Many patients have good work and social records. However, they tend to form fewer close personal relationships than others of their generation and to be prickly, querulous, and unapproachable. Cooper *et al.* (1974) found that conductive *deafness* of onset in earlier middle life, a factor which increases *social isolation*, was significantly more common in late paraphrenia. A consistent finding in late paraphrenia is an excess of females over males in a ratio of about 7 : 1. This compares with younger schizophrenic patients who tend to show a more or less equal sex distribution over the period of risk.

Genetically, patients with late paraphrenia occupy an intermediate position between the unaffected population and those with schizophrenia of earlier onset. In one study (Kay 1972) the risk of schizophrenia in first-degree relatives was found to be 3.4 per cent in late paraphrenics compared with 5.8 per cent in young schizophrenics and less than 1 per cent in the general population. HLA genetic markers have been sought, and Naguib *et al.* (1987) found a statistical association with HLA-B37.

Clinical features

The predominant clinical feature of late paraphrenia is delusional thought. Delusions are mostly of persecution, usually more commonplace and narrowly centred than those found in paranoid schizophrenia of earlier life. Thus, where a younger paranoid schizophrenic patient may assert a plot by the KGB or CIA, an older late paraphrenic is more likely to complain that the neighbours are plotting to kill her or impugning her sexual virtue. Hallucinations are characteristic in late paraphrenia. The most commonly are auditory; tactile and olfactory hallucinations are not infrequent, but visual hallucinations are rare. Naguib and Levy (1987) observed mild cognitive impairment in late paraphrenic patients compared with controls, but this finding may have been due to factors such as medication, poor concentration, or coincidental organic cerebral disease. Holden (1987) followed-up patients with late paraphrenia and found the outcome to be heterogeneous with some developing dementia. Although there are these differences between some cases of late paraphrenia and paranoid schizophrenia, other cases are closely similar. See Howard (1994) for a review.

Treatment

Patients with late paraphrenia usually respond well to antipsychotic drugs, but the premorbid personality factors described above make it necessary to try hard to win patients' trust so that they are more likely to comply with treatment. Many patients have to be admitted to hospital to start treatment. It is uncommon for symptoms to remit completely, but after the acute illness delusional beliefs usually become encapsulated so that patients can often return to the premorbid level of functioning. Follow-up usually requires the help of a community psychiatric nurse who should strike a balance between seeing the patient regularly and persuading him to accept medication, and appearing too intrusive so that the patient refuses treatment.

Further reading

Copeland, J. R. M., Abou-Saleh, M. T., and Blazer, D. G. (1994). *Principles and practice of geriatric psychiatry*. John Wiley, Chichester.

Jacoby, R. and Oppenheimer, C. (ed.) (1991). *Psychiatry in the elderly*. Oxford University Press.

17 | *Drugs and other physical treatments*

This chapter is concerned with the use of drugs, electroconvulsive therapy, and psychosurgical procedures. Psychological treatments are the subject of the next chapter. This separation, although convenient when treatments are described, does not imply that the two kinds of therapy are to be thought of as exclusive alternatives when an individual patient is considered; on the contrary many patients require both. In this book, the ways of combining treatments are considered in other chapters where the treatment of individual syndromes is discussed. It is important to keep this point in mind when reading this chapter and the next.

Our concern is with clinical therapeutics rather than basic psychopharmacology, which the reader is assumed to have studied already. An adequate knowledge of the mechanisms of drug action is essential if drugs are to be used in a rational way, but a word of caution is appropriate. The clinician should not assume that the therapeutic effects of psychotropic drugs are necessarily explained by the pharmacological actions that have been discovered so far. For example, substantial delay in the effects of antidepressant and antipsychotic drugs suggests that their actions on transmitters, which occur rapidly, are only the first steps in a chain of biochemical changes.

This caution does not imply that a knowledge of pharmacological mechanisms has no bearing on psychiatric therapeutics. On the contrary, there have been substantial advances in pharmacological knowledge since the first psychotropic drugs were introduced in the 1950s, and it is increasingly important for the clinician to relate this knowledge to his use of drugs.

History of physical treatments

Physical treatments have been applied to patients with psychiatric disorders since antiquity, though, in retrospect, the most that could be claimed for the best of these interventions is that they were relatively harmless. Of course, the same holds for the management of patients with general medical disorders, for which similar treatments, such as bleeding and purging, were often used regardless of diagnosis. It is wise not to be too censorious about the treatment of disorders of which the aetiology is still largely unknown, but to bear in mind that 'It may well be that in a hundred years current therapies, psychotherapies as well as physical therapies, will be looked upon as similarly uncouth and improbable.' (Kiloh *et al.* 1988*b*).

Historically, physical treatments can be divided into two main classes: (a) those that were aimed at producing a direct change in a pathophysiological process, usually by some alteration in brain function; (b) those that were aimed at producing symptomatic improvement through a dramatic psychological impact. The latter interventions were often based on philosophical theories about the moral basis of madness. For example many physicians appear to have followed the proposal of **Heinroth** (1773–1843) that insanity was the product of evil and personal wrongdoing. Accordingly, **restraint** with chains and corporal punishment were seen as appropriate remedies. Other physical treatments, such as the spinning chair introduced by **Erasmus Darwin** (1731–1802), seemed designed to produce a general 'shock to the system', and perhaps thereby interrupt the morbid preoccupations of the patient. A less arduous regime was the use of **continuous warm baths**, often given in combination with cold packs. This treatment was recommended by clinicians as distinguished as **Connolly** (1794–1866) and **Kraepelin** (1856–1926), and was still in use at the Bethlem Hospital in the 1950s.

Drugs that produce changes in the function of the central nervous system, such as **opiates** and anticholinergic agents, have been used in the treatment of mental disorders for hundreds of years. While some of these drugs may sometimes have had calming effects, they were of no specific value in the treatment of psychiatric disorders. Often a physical treatment was used, not because of proven efficacy, but because it was recommended by an eminent and vigorous physician. Also, the assessment of efficacy depended almost entirely on uncontrolled clinical observation.

In 1933, about ten years after the isolation of insulin by Banting and Best, **Sakel** introduced **insulin coma treatment** for psychosis (Sakel 1938). A suitable dose of insulin was used to produce a coma which was terminated by either tube feeding or intravenous glucose. A course of treatment could include up to 60 comas. Not surprisingly, serious side-effects were common, and a mortality of at least 1 per cent could be expected depending on the standard of the clinic and on the basic physical state of the patient. Insulin coma treatment was rapidly taken up throughout Europe and many specialized treatment units were built. There was a great improvement in the morale of patients and staff because of the belief that this dramatic treatment could cure symptoms of some of the most serious psychiatric disorders.

There were always some doctors who doubted the efficacy of insulin coma treatment. Their doubts were reinforced by a controlled trial by Ackner and Oldham (1962), who found that, in schizophrenic patients, insulin coma was no more effective than a similar period of unconsciousness induced by barbiturates. This study was published about the time when chlorpromazine was introduced, and both factors led to a rapid decline in the use of insulin coma treatment. It should be noted that some controlled studies did not exclude the efficacy of insulin treatment in some circumstances, and a number of workers continued to maintain that it was effective. Therefore it is interesting that recent experimental studies have shown that insulin administration causes striking changes in the release of monoamine neurotransmitters in the brain. Perhaps the main lesson to be learned from insulin coma treatment is that the proposed introduction of a new medical treatment should be preceded by adequate **controlled** trials to determine whether it is therapeutically more effective or safer than current therapies. This lesson is particularly important in psychiatry because the aetiology of some disorders may be obscure and outcome may vary widely, even amongst patients with the same syndrome.

Electroconvulsive therapy (ECT) was introduced about the same time as insulin coma treatment. Unlike the latter, ECT has retained a place in current clinical practice. The rationale for convulsive therapy was a postulated antagonism between schizophrenia and convulsions such that the one would exclude the other. This view is erroneous in so far as schizophrenia-like illnesses are more common in patients with temporal lobe epilepsy than would be expected by chance (see p. 340). Astute clinical observation, in combination with controlled trials, has shown that ECT is effective in the acute treatment of **severe mood disorders**. Thus, even though the rationale for the introduction of ECT was incorrect and its mode of action remains unclear, controlled trials have confirmed that, in carefully defined clinical situations, ECT is a safe and effective treatment (see p. 230).

The action of **lithium** in reducing mania was a chance finding by Cade (1949) who had been investigating the effects of urates in animals and had decided to use the lithium salt because of its solubility. Lithium is a toxic agent, and so Cade's important observations did not make a significant impact on clinical practice until the following decade, when controlled trials showed that lithium was effective in both the acute treatment of mania and the prophylaxis of recurrent mood disorders.

Other agents that revolutionized psychopharmacology were introduced about this time (Table 17.1). Their efficacy and their indications were first recognized through clinical observation, and were subsequently confirmed by controlled clinical trials. None of these agents was introduced on the basis of an aetiological hypothesis. Indeed, such aetiological hypotheses as there are in biological psychiatry have been largely derived from knowledge of the mode of action of effective drugs. Thus the dopamine receptor antagonist properties of **antipsychotic drugs** have given rise to the **dopamine hypothesis** of schizophrenia, whilst the action of tricyclic anti-

Table 17.1. *Introduction of some physical treatments in psychiatry*

1934	Insulin coma treatment (Sakel)
1936	Frontal leucotomy (Moniz)
1936	Metrazole convulsive therapy (Meduna)
1938	Electroconvulsive therapy (Cerletti and Bini)
1949	Lithium (Cade)
1952	Chlorpromazine (Delay and Deniker)
1954	Benzodiazepines (Sternbach)
1957	Iproniazid (Crane and Kline)
1957	Imipramine (Kuhn)
1966	Valpromide (valproate) in bipolar disorder (Lambert *et al.*)
1967	Clomipramine in obsessive–compulsive disorder (Fernandez and Lopez-Ibor)
1971	Carbamazepine in bipolar disorder (Takezaki and Hanaoka)
1988	Clozapine in treatment-resistant schizophrenia (Kane *et al.*)

depressants and monoamine oxidase inhibitors (MAOIs) in facilitating the effects of **noradrenaline** and **5-hydroxytryptamine** (**5-HT**) has led to the various **monoamine hypotheses** of mood disorders.

The last 30 years have brought a period of consolidation in psychopharmacology. Thus clinical trials have been widely used to refine the indications of particular drug treatments and to maximize their risk–benefit ratios. New compounds have continuously become available; most have been derived from previously described agents, and so their range of activity is not strikingly different from that of their predecessors. In general, however, the newer agents are better tolerated and sometimes safer—developments which are important for clinical practice.

There may now be grounds for more optimism about the prospects for advances in psychopharmacology. For example there is rapidly increasing knowledge about chemical signalling in the brain. Numerous neurotransmitters and neuromodulators interact with specific families of **receptors**, many of which exist in a number of different **subtypes**. Several of these receptors have been cloned, and selective ligands for them are becoming available. There is increasing knowledge as to how these chemical messengers may modify behaviour through their interactions with specific brain regions and distributed neuronal circuits.

New compounds are likely to differ from current drugs in their range of behavioural effects. These new preparations are likely to lead to important new developments in psychopharmacology. Given the complex causes of psychiatric disorders, it seems likely that detailed knowledge of aetiology and pathophysiology may lag behind advances in therapeutics. Of course, this disparity, is not uncommon in general medicine. It serves to reinforce the importance of controlled clinical trials in the assessment of new psychopharmacological treatments.

General considerations

The pharmacokinetics of psychotropic drugs

Before psychotropic drugs can produce their therapeutic effects, they must reach the brain in adequate amounts. How far they do so depends on their absorption, metabolism, excretion, and passage across the blood–brain barrier. A short review of these processes is given here. The reader who has not studied them before is referred to the chapter on pharmacokinetics in Lader and Herrington (1990).

In general, psychotropic drugs are easily **absorbed** from the gut because most are lipophilic and are not highly ionized at physiological pH values. Like other drugs, they are absorbed faster from an empty stomach, and in reduced amounts by patients suffering from intestinal hurry, malabsorption syndrome, or the effects of a previous partial gastrectomy.

Most psychotropic drugs are **metabolized** in the liver. This process begins as the drugs pass through the liver in the portal circulation on their way from the gut. This 'first-pass' metabolism reduces the amount of available drug, and is one of the reasons why larger doses are needed when a drug such as chlorpromazine is given by mouth than when it is given intramuscularly. The extent of this liver metabolism differs from one person to another. It is altered by certain other drugs which, if taken at the same time, induce liver enzymes (for example barbiturates) or inhibit them (for example SSRIs). Some drugs, such as chlorpromazine, induce their own metabolism, especially after being taken for a long time. Not all drug metabolites are inactive; for example chlorpromazine is metabolized to a 7-hydroxy derivative which has therapeutic properties, as well as to a sulphoxide which is inactive. Because chlorpromazine, diazepam, and many other psychotropic drugs give rise to many metabolites, measurements of plasma concentrations of the parent drug are a poor guide to therapeutic activity.

Psychotropic drugs are **distributed** in the plasma where most are largely bound to proteins; thus diazepam, chlorpromazine, and amitriptyline are about 95 per cent bound. They pass easily from the plasma to the brain because they are highly lipophilic. For the same reason they enter fat stores, from which they are released slowly long after the patient has ceased to take the drug.

Most psychotropic drugs and their metabolites are **excreted** mainly through the kidney. When kidney function is impaired, excretion is reduced and a lower dose of drug should be given. For basic or acidic drugs, renal excretion depends on the pH of the urine; for example aspirin, a weak acid, is excreted more rapidly when the urine is alkaline rather than acid, whilst amphetamine, a weak base, is excreted more rapidly when the urine is acid rather than alkaline. Lithium is filtered passively and then partly reabsorbed by the same mechanism that absorbs sodium. The two ions compete for this mechanism; hence reabsorption of lithium increases when that of sodium is reduced. Certain fractions of lipophilic drugs such as chlorpromazine are partly excreted in the bile, enter the intestine for the second time, and are then partly reabsorbed, i.e. a proportion of the drug is recycled between intestine and liver.

Measurement of circulating drug concentrations

As a result of individual variations in the mechanisms described above, plasma concentrations after standard doses of psychotropic drugs vary substantially from one patient to another. Tenfold differences have been observed with the antidepressant drug nortriptyline. Therefore it might be expected that measurements of the plasma concentration of circulating drugs would help the clinician.

However, there are several reasons why **plasma concentrations** of psychotropic drugs may not correlate well with their clinical effects. For example many psychotropic drugs are **bound** extensively to plasma proteins; however, only their plasma-free fractions are pharmacologically active. Most assays measure the total concentration of drug in plasma (i.e. free and bound), and therefore do not necessarily provide a reliable estimation of the amount of drug available to produce a therapeutic effect. Another important factor is that, as mentioned above, many drugs have **active metabolites**, some of which have therapeutic effects while others do not. Some assays are too specific, measuring only the parent drug but not its active derivatives; others are too general, measuring active and inactive metabolites alike.

Finally, it seems likely that the therapeutic effects of many psychotropic drugs may be associated with **adaptive responses** of neurones to the presence of the drug. These cellular adaptive responses depend on many factors, particularly perhaps on the properties of the neurones concerned. Therefore it would not be surprising if drug-induced adaptive changes in the neurone did not correlate in a straightforward way with levels of the drug in plasma. Nevertheless, the monitoring of plasma levels of psychotropic drugs can be helpful. For example measurement of **serum lithium** levels is certainly useful. Of course, lithium is a simple ion that has no active metabolites and is not protein bound.

As an alternative to these assays, it may be possible to measure the pharmacological property which is thought to be responsible for the therapeutic effect of a particular drug. For example **positron emission tomography** can be used to measure directly the degree of brain **dopamine receptor blockade** produced by antipsychotic drugs during treatment. Such information has proved valuable in improving dosage regimes of drugs such as haloperidol. However, these pharmacodynamic measures have not yet been able to identify why some patients do not respond to medication. For example the degree of dopamine receptor blockade is the same in patients who respond to antipsychotic drugs as in those who do not (Geaney *et al.* 1992).

Plasma concentrations vary throughout the day, rising immediately after the dose and falling at a rate that differs between individual drugs and individual people. The rate at which a drug level declines after a single dose varies from hours with lithium carbonate to weeks with slow–release preparations of injectable neuroleptics. Knowledge of these differences allows more rational decisions to be made about appropriate intervals between doses.

The concept of **plasma half-life** is useful here. The half-life of a drug in plasma is the time taken for its concentration to fall by a half, once dosing has ceased. With most psychotropic drugs, the amount eliminated over time is proportional to plasma concentration and in this case it will take approximately *five times the half-life for the drug to be eliminated from plasma*. Equally, when dosing with a drug begins, it will take five times the half-life for the concentration in plasma to reach steady state. This can be important when planning treatment. For example MAOIs should not be given with selective serotonin (5-HT) re-uptake inhibitors (SSRIs) because of the danger of drug interaction (see below). For example if a patient is taking sertraline, which has an elimination half-life of about 26 hours, it will be important to leave at least five times the half life (a week is

recommended) before starting MAOI treatment. When sertraline treatment begins, the plasma concentrations will continue to rise for about a week before reaching a steady state.

Drug interactions

When two psychotropic drugs are given together, one may interfere with or enhance the actions of the other. Interference may arise through alterations in absorption, binding, metabolism, or excretion, or by interaction between pharmacodynamic effects.

Interactions affecting drug absorption are seldom important for psychotropic drugs, although it is worth noting that absorption of chlorpromazine is reduced by antacids. Interactions due to *protein binding* are uncommon, although the chloral metabolite trichloroacetic acid may displace warfarin from albumin. Interactions affecting drug *metabolism* are of considerable importance. Examples include the inhibition of the metabolism of sympathomimetic amines by MAOIs and the increase in the metabolism of chlorpromazine and tricyclic antidepressants by barbiturates which induce the relevant enzymes. Interaction affecting renal *excretion* are mainly important for lithium, the elimination of which is decreased by thiazide diuretics.

Pharmacodynamic interactions are exemplified by the antagonism of guanethidine and tricyclic antidepressants.

As a rule, a single drug can be used to produce all the effects required of a combination; for example many tricyclic antidepressants have anti-anxiety effects. It is desirable to avoid combinations of psychotropic drugs whenever possible; if a combination is to be used, it is essential to know about possible interactions.

Drug withdrawal

Many psychotropic drugs do not achieve their full effects for several days, and antidepressants may take up to six weeks. After drugs have been stopped, there is often a comparable delay before their effects are lost. With some drugs, tissues have to readjust when treatment is stopped; this readjustment may appear clinically as a withdrawal syndrome. Among the psychotropic drugs, the hypnotics, and the anxiolytic drugs, readjustments are shown clinically as a sleep disturbance and physiologically in an increase in rapid eye movement (REM) sleep. Unless the symptoms are recognized as being due to drug withdrawal, it may be wrongly concluded from the sleep disturbance that the patient still needs a hypnotic. Withdrawal symptoms also occur when daytime benzodiazepines are stopped abruptly (see p. 472).

General advice about prescribing psychotropic drugs

It is good practice to use well-tried drugs with therapeutic actions and side-effects that are thoroughly understood. The clinician should become familiar with a small number of drugs from each of the main groups—two or three antidepressants, two or three antipsychotics, and so on. In this way he can become used to adjusting the dosage and recognizing side-effects. (Recommendations about drugs of choice will be found in the later part of this chapter.) Well-tried drugs are usually less expensive than new preparations.

Having chosen a suitable drug, the doctor should prescribe it in adequate doses. He should not change the drug or add others without a good reason. In general, if there is no therapeutic response to one established drug, there is no likelihood of a better response to another from the same therapeutic group (provided that the first drug has been taken in adequate amounts). However, since the main obstacle to adequate dosage is usually side-effects, it is sometimes appropriate to change to a drug with a different pattern of side-effects—for example from a tricyclic antidepressant to another with fewer anticholinergic effects such as an SSRI.

Some drug companies market tablets that contain a mixture of drugs, for example tricyclic antidepressants with a small dose of a phenothiazine. These mixtures have little value. In the few cases when two drugs are really required, it is better to give them separately so that the dose of each can be adjusted independently.

Occasionally, drug combinations are given deliberately in the hope of producing interactions that will be more potent than the effects of either drug taken alone in full dosage (for example a tricyclic antidepressant with a MAOI). This practice, if it is to be used at all, should be carried out only by specialists because the adverse effects of combinations are much less easy to predict than those of single drugs.

When a drug is prescribed, it is necessary to determine the dose, the interval between doses, and the likely duration of treatment. The dose ranges for commonly used drugs are indicated later in this chapter. Ranges for others will be found in the manufacturers' literature, the *British national formulary*, or a comparable work of reference. Within the therapeutic range, the correct dose for an individual patient should be decided after considering the severity of symptoms, the patient's age, and weight, and any factors that may affect drug metabolism (for example other drugs being taken or renal disease).

Next, the interval between doses must be decided. Psychotropic drugs have often been given three times a day, even though their duration of action is such that most can be taken once or twice a day without any undesirable fall in plasma concentrations between doses. Less frequent administration has the advantage that out-patients are more likely to be reliable in taking drugs. In hospital, less frequent drug rounds mean that nurses have more time for psychological aspects of treatment. Some drugs, such as anxiolytics, are required for immediate effect rather than continuous action; they should not be given at regular intervals but shortly before occasions on which symptoms are expected to be at their worse. The duration of treatment depends on the disorder under treatment; it is considered in the chapter dealing with the clinical syndrome.

Before giving a patient a first prescription for a drug, the doctor should explain several points. He should make clear what effects are likely to be experienced on first taking the drug, for example drowsiness or dry mouth. He should also explain how long it will be before therapeutic effects appear and what the first signs are likely to be, for example improved sleep after starting a tricyclic antidepressant. He should name any serious effects that must be reported by the patient, such as coarse tremor after taking lithium. Finally, he should indicate how long the patient will need to take the drug. For some drugs such as anxiolytics, the latter information is given to discourage the patient from taking them for too long; for others, such as antidepressants, it is given to deter the patient from stopping too soon.

Compliance with treatment

Many patients do not take the drugs prescribed for them. This problem is greater when treating out-patients, but also occurs in hospital where some patients find ways of avoiding drugs administered by nurses.

If a patient is to comply with medication of any kind, he must be convinced of the need to take it, free from unfounded fears about its dangers, and be aware of how to take it. Each of these requirements presents particular problems when the patient has a psychiatric disorder. Thus schizophrenic or seriously depressed patients may not be convinced that they are ill or they may not wish to recover. Deluded patients may distrust their doctors, and hypochondriacal patients may fear dangerous side-effects. Anxious patients often forget the prescribed dosage and frequency of their drugs. Therefore it is not surprising that many psychiatric patients do not take their drugs in the prescribed way. It is important for the clinician to pay attention to this problem. Time spent in discussing the patient's concerns is time well spent, for it often increases compliance. Written instructions can be a valuable adjunct. For a comprehensive review of patients' compliance with treatment, see Haynes *et al.* (1979).

The overprescribing of psychotropic drugs

In the last 30 years, many safe and effective drugs have been produced for the treatment of psychiatric disorders. Unfortunately, their proven value in severe conditions has led to unnecessary prescribing for mild cases that would recover without medication. Similarly, the safety of these drugs has some-times encouraged prolonged prescribing when brief treatment would be more appropriate. These problems have arisen most often with drugs prescribed for insomnia, anxiety, and depressed mood. All three symptoms are important components of psychiatric illness, but in their mildest form they are also part of everyday life. The extent of usage of anxiolytic and antidepressant drugs was shown by a survey of all prescriptions issued in general practices serving 40 000 people (Skegg *et al.* 1977). It was found that psychotropic drugs were prescribed more often than any others. Amongst patients registered in the practices, nearly 10 per cent of men and over 20 per cent of women received at least one prescription for a psychotropic drug during the course of a year. One-third of women aged 45–49 received such a prescription.

While there are important implications about the amount of drugs consumed through these high rates of prescribing, it is equally important to remember that many prescribed drugs are not taken. Nicholson (1967) collected unused drugs from the houses of about 500 patients in the course of six days. He recovered 36 000 tablets, of which nearly 5000 were sedatives and tranquillizers, over 2000 were hypnotics, and 7500 were antidepressants. Unused drugs are a danger to children and a potential source of self-poisoning either by the patient or by other people. For these reasons patients should not be given more drugs than they need, and care should be taken to enquire whether existing supplies have been used before prescribing more.

Prescribing for special groups

Children seldom require medication for psychiatric problems. When they do require it, doses should be adjusted appropriately by consulting an up-to-date work of reference (such as the *British national formulary*). For **elderly** patients, who are often sensitive to side-effects and may have impaired renal or hepatic function, it is important to start with low doses.

There are special problems about prescribing psychotropic drugs in **pregnancy** because of the risk of teratogenesis. This risk, which varies between drugs, will be considered later in the chapter when the actions of the drugs are described. At this point, some general advice will be given. *Anxiolytic and sedative drugs* are seldom essential in early pregnancy, and psychological treatments can usually be used. If medication is needed, benzodiazepines have not been shown to be teratogenic. If an *antidepressant drug* is required, it is probably better to use a long-established preparation such as imipramine and amitriptyline for which there is no evidence of a teratogenic effect after many years of use. Newer drugs, even if they have not been shown to be unsafe, should be avoided because there has been less time to accumulate evidence. It is seldom necessary to start *antipsychotic* drugs in early pregnancy. When a patient is already receiving them, the risk of relapse if the drugs are stopped must be weighed against the uncertainty about the teratogenic effects of the particular drug.

So far there is no evidence that these drugs damage the fetus but, as noted above, the degree of uncertainty must be greater with newer drugs than with those that have been used for many years. Therefore it is wise to avoid these drugs in the first trimester whenever possible (Edlund and Craig 1984).

It has been reported that, among babies born to mothers who have been receiving tricyclic antidepressants or phenothiazines, there may be withdrawal reactions including tremulousness and vomiting. These symptoms usually settle quickly.

Lithium treatment in pregnancy has been associated with **cardiac abnormalities** in the fetus; therefore women considering pregnancy have been strongly recommended to discontinue lithium before conceiving. Similarly, women who become pregnant whilst taking lithium have usually been advised to stop the treatment. However, a recent prospective study has indicated that the risk of teratogenesis with lithium may be less than was previously thought (Jacobsen *et al.* 1992). Clearly the decision whether or not to discontinue lithium in pregnancy is best made after considering the individual circumstances and a full discussion with the patient.

Because it is difficult to be certain of the teratogenic potential of new psychotropic drugs, it is prudent to avoid them when possible in early pregnancy. It is also often appropriate to advise women of childbearing age who require psycho-

tropics to adopt a reliable method of contraception to avoid pregnancy until the need for the drug is over.

Psychotropic drugs should be prescribed cautiously to women who are **breast-feeding**. Diazepam and other *benzodiazepines* pass readily into breast milk and may cause sedation and hypotonicity in the infant. *Neuroleptics* and *antidepressants* also enter breast milk, although rather less readily than diazepam. **Sulpiride** is excreted in significant amounts and should be avoided. Similarly, **doxepin** and **dothiepin** may accumulate in the infant and cause oversedation. **Fluoxetine** and its metabolites could also accumulate, but **fluvoxamine** is present in only very small amounts in breast milk. *Lithium carbonate* enters the milk freely, and serum concentrations in the infant can approach those of the mother so that breast-feeding requires great caution. A general issue is that, even when the concentration of a particular drug in breast milk is low and no detectable clinical effect upon the infant can be discerned, it is nevertheless possible that *subtle longer-term effects on brain development and behaviour could occur*. For this reason some authorities recommend that women receiving psychotropic medication should not breast-feed at all (Lader and Herrington 1990). For further reviews see Loudon (1987) and the *British national formulary*.

What to do if there is no therapeutic response

The first step is to find out whether the patient has been taking the drug in the correct dose. He may not have understood the original instructions, or may be worried that a full dose will produce unpleasant effects. Some patients fear that they will become dependent if they take the drug regularly. Other patients may have little wish to take drugs for different reasons—schizophrenics because they do not regard themselves as ill, and depressed patients because they do not believe that they can be helped. If the doctor is satisfied that the drug has been taken correctly, he should find out whether the patient is taking any other drug (such as carbamazepine) which could affect the metabolism of the psychotropic agent. Finally, he should review the diagnosis to make sure that the treatment is appropriate before deciding whether to increase the dose.

The evaluation of psychotropic drugs

After being tested in animals, new drugs have to be evaluated for clinical use. This requires two stages. First, the drug is used cautiously at doses sufficient to give therapeutic effects without unwanted effects. Then, controlled clinical trials are carried out in which the drug is compared, under double-blind conditions, with a placebo or standard drug. Readers seeking information about the methodology of clinical trials are referred to Harris and Fitzgerald (1970). This section is concerned only with a few essential points that need to be kept in mind when reading a report of a trial of a new psychotropic drug. These points concern patients, treatments, and measurements.

Patients

In evaluating any clinical trial the clinician has to decide how far the selected patients are typical of all those who have the disorder (for example schizophrenics are a diverse group) and how far they resemble the patients that he wishes to treat (for example chronic schizophrenics undergoing rehabilitation). Part of the selection procedure will have been reported by the research workers, but often other parts are not stated explicitly. Thus, if a trial is restricted to hospital out-patients, it is necessary to consider what kinds of patients are likely to be referred to the particular hospital by the local general practitioners. For example it might be important to know whether most patients with depressive disorders are referred, or only those who have not responded to adequate antidepressant treatment. In the latter case, a hospital-based trial will be dealing with drug-resistant patients.

Important questions to consider are as follows. How were diagnoses made, and were standard and generally accepted diagnostic methods used? Of the patients originally referred to the trial, how many were rejected? Of those accepted, how many dropped out and were they replaced? Another important issue is how the patients would have fared without treatment. This question applies particularly to any trial not including a placebo condition. In such a trial, if patients treated with the new drug improve as much as those treated with a standard preparation, this may merely indicate that the patients selected for the investigation have a high rate of spontaneous recovery. A final point to check is whether allocation of patients to the various treatments has been random.

Treatments

The first questions concern dosage, intervals between doses, whether the same quantity of drugs was given to every patient, and whether additional drugs were allowed. It is important to decide whether the treatments were given for long enough. Nearly all clinical trials include precautions to ensure that neither patients nor staff can tell which treatment any one patient is receiving (the double-blind trial). Identical tablets do not always achieve this aim because side-effects, such as dry mouth, tremor, or postural hypotension, can provide clues to the identity of one or more treatments. Therefore it is important to study the frequency and pattern of side-effects reported by each group of patients.

It is also important to know what precautions have been taken to ensure that patients took the drugs prescribed for them. These precautions include counting any tablets remaining at the end of each treatment period, measuring blood plasma levels, and incorporating in the tablets a marker substance, such as riboflavine, that can be detected in urine more easily than the drug itself.

Measurements

When assessment methods are chosen for a clinical trial, a balance has to be struck between precision and reliability on the one hand, and clinical relevance on the other. Psychological test scores may be reliable and precise, but they seldom relate in a simple way to judgements made in everyday clinical work. Psychiatric rating scales are less reliable and precise but more relevant. It is also important to consider whether the assessments are sensitive within the range of changes to be expected in the trial: thus measures developed for use with severely depressed patients treated in hospital may not be appropriate for patients with minor depressive disorders in general practice. The timing of assessments should also be considered, for example whether they were given sufficiently early and were continued long enough. Finally, the appropriateness of the statistical methods should be reviewed. If statistically significant changes are reported, it is important to decide whether they are large enough to justify a change to the new treatment.

Meta-analysis

When the clinical effect of a drug is large, as with penicillin in pneumoccal pneumonia, the superiority of the drug to placebo or current treatment can be demonstrated using only a small group of patients. However, with many drug treatments the effects are of lesser magnitude. Thus if, in comparison with a conventional tricyclic, a new antidepressant led to the recovery of 10 per cent more patients, such an effect would be worthwhile because depression is common and disabling. To demonstrate such an effect with confidence, however, the patient sample would have to be much larger than the samples usually obtainable in individual controlled trials. Hence it is usually necessary to assess the results of a series of smaller clinical trials and then come to a subjective judgement about the value or otherwise of the new treatment. The problem with this approach is that results from small studies are often inconsistent, and so it is difficult to assess how effective the treatment might be.

The technique of **meta-analysis** has been developed to overcome this problem. Meta-analysis involves a systematic overview of information from all the controlled trials that have been carried out with a treatment. It is important that **data from all trials** should be included, because published studies may be biased towards positive findings. The results from individual trials are usually calculated in the form of an odds ratio, which is based on the number of patients in the experimental group reaching a defined end-point compared with the corresponding number in the control group. The odds ratio from each individual trial is then

summarized using an appropriate test of statistical significance, and an estimation of the magnitude of the treatment effect is made (Thompson and Pocock 1991). For example, in a meta-analysis of 53 trials which compared the efficacy of SSRIs with various other classes of antidepressants in the treatment of major depression, Song *et al.* (1993) found that the mean difference in total Hamilton Depression rating score between SSRIs and the other antidepressants after eight weeks of treatment was 0.13 with a confidence interval of -1.01 to 1.28. Thus, even though individual studies have sometimes found other antidepressants superior to SSRIs (and vice versa), the meta-analysis shows clearly that in the broad range of patients entered for controlled clinical trials there is no consistent difference in efficacy between SSRIs and the other antidepressants studied. Of course, how far the patients entered for the controlled trials are representative of those encountered in clinical practice is another matter, and one which puts some limitation on the applicability of the meta-analysis technique.

The classification of drugs used in psychiatry

Drugs that have effects mainly on mental symptoms are called **psychotropic**. Psychiatrists often use two other groups of drugs: **antiparkinsonian** agents, which are employed to control the side-effects of some psychotropic drugs, and **antiepileptic drugs**.

Psychotropic drugs are conventionally divided into different classes, as shown in Table 17.2, but the therapeutic actions of particular compounds are not confined to one diagnostic category. For example, SSRIs are classified as antidepressants and are effective in the treatment of major depression, but they also produce useful therapeutic effects in anxiety states, obsessive–compulsive disorders, and eating disorders. Of course, this breadth of effect does not mean that the latter syndromes are forms of depression. It merely emphasizes that the neuropsychological consequences of facilitating brain 5-HT function may provide beneficial effects in a variety of psychiatric disorders. While there is

Table 17.2.　*Classification of clinical psychotropic drugs*

Class of drug	Examples of classes	Indications
Antipsychotic	Phenothiazines Butyrophenones Substituted benzamides	Acute treatment of schizophrenia and mania, prophylaxis of schizophrenia
Antidepressant	Tricyclic antidepressants MAOIs SSRIs	Major depression (acute treatment and prophylaxis), anxiety disorders, obsessive–compulsive disorder (SSRIs)
Mood stabilizer	Lithium Carbamazepine	Acute treatment of mania Prophylaxis of recurrent mood disorder
Anxiolytic	Benzodiazepines Azapirones (buspirone)	Generalized anxiety disorder
Hypnotic	Benzodiazepines Cyclopyrrolones (zopiclone)	Insomnia
Psychostimulant	Amphetamine	Hyperkinetic syndrome of childhood Narcolepsy

considerable understanding of the pharmacological actions of psychotropic drugs, little is known about the neuropsychological consequences of these pharmacological actions and about the ways in which neuropsychological changes are translated into clinical benefit in different diagnostic syndromes. At present, therefore, the best plan is to classify drugs according to their major therapeutic use but to bear in mind that the therapeutic effects of different classes of drugs may overlap considerably.

The five main groups of drugs will now be reviewed in turn. For each group, an account will be given of therapeutic effects, pharmacology, principal compounds available, pharmacokinetics, unwanted effects (both those appearing with ordinary doses and the toxic effects of unduly high doses), and contraindications. General advice will also be given about the use of each group in everyday clinical practice, but specific applications to the treatment of individual disorders will be found in the chapters dealing with those conditions.

Drugs that have a limited use in the treatment of a single disorder, for example disulfiram for alcohol problems, are discussed in the chapters dealing with the relevant clinical syndromes.

Anxiolytic drugs

Anxiolytic drugs are prescribed widely and often inappropriately. Before prescribing these drugs it is always important to seek the causes of anxiety and to try to modify them. It is also essential to recognize that a degree of anxiety can motivate patients to take steps to reduce the problems that are causing it. Hence removing all anxiety in the short term is not always beneficial to the patient in the long run. Anxiolytics are most useful when given for a short time either to tide the patient over a crisis or to help him tackle a specific problem. Tolerance is a particular problem with anxiolytic drugs, and physical dependence can develop. Because the benzodiazepines are now the most widely used anxiolytics, they will be considered first. Other compounds will then be described in less detail. When reading this section, it is important to keep in mind that psychological treatment can be used for anxiety (see p. 637).

Benzodiazepines

Pharmacology

Benzodiazepines are anxiolytic, sedative, and, in large doses, hypnotic. They also have muscle relaxant and anticonvulsant properties. Their pharmacological actions are mediated through specific receptor sites located in a supramolecular complex with gamma-aminobutyric acid (GABA) receptors. Benzodiazepines enhance GABA neurotransmission, thereby altering indirectly the activity of other neurotransmitter systems such as those involving noradrenaline and 5-HT.

Compounds available

Many different benzodiazepines are available. They differ both in the potency with which they interact with benzodiazepine receptors and in their plasma half-life. In general, **high-potency** benzodiazepines and those with **short half-lives** are more likely to be associated with **dependence and withdrawal**. Benzodiazepines with short half-lives (less than 12 hours) include oxazepam, lorazepam, temazepam, and lormetazepam. Because of problems with dependence, long-acting compounds are preferable for the management of anxiety, even if such treatment is to be given intermittently on an 'as required basis'. However, **temazepam** and **lormetazepam** are useful **hypnotics** (see below). The long-acting benzodiazepines include drugs such as diazepam, nitrazepam, flurazepam, clobazam, clorazepate, and alprazolam. **Diazepam** is rapidly absorbed and can be used both for the continuous treatment of anxiety and for treatment 'as required'. **Alprazolam**, a high-potency benzodiazepine, is effective in the treatment of panic disorder. This therapeutic efficacy is not confined to alprazolam because equivalent doses of other

high-potency agents such as clonazepam are also effective. **Flumazenil**, *a benzodiazepine receptor antagonist,* has recently become available. This drug produces little pharmacological effect by itself but blocks the actions of other benzodiazepines. Therefore it may be useful in reversing acute toxicity produced by benzodiazepines but carries a risk of provoking acute benzodiazepine withdrawal. Flumazenil is available only for intravenous use.

Pharmacokinetics

Benzodiazepines are rapidly absorbed. They are strongly bound to plasma proteins but, because they are lipophilic, pass readily into the brain. They are metabolized to a large number of compounds, many of which have therapeutic effects of their own; temazepam and oxazepam are among the metabolic products of diazepam. Excretion is mainly as conjugates in the urine. See Schwartz (1973) for a review.

Benzodiazepines with short half-lives, such as temazepam and lorazepam, have a **3-hydroxyl** grouping which allows a one-step metabolism to inactive glucuronides. Other benzodiazepines, such as diazepam and clorazepate, are metabolized to long-acting derivatives, such as desmethyldiazepam, which are themselves therapeutically active. Benzodiazepines are occasionally given parenterally to produce a rapid calming effect. It is worth noting that the absorption of diazepam following intramuscular injection is poor, and lorazepam should be preferred if this route of administration is used.

Unwanted effects

Benzodiazepines are well tolerated. When they are given as anxiolytics, their main side-effects are due to the sedative properties of large doses, which can lead to ataxia and drowsiness (especially in the elderly) and occasionally to confused thinking. Minor degrees of drowsiness and of impaired coordination and judgement can affect driving skills and the operation of potentially dangerous machinery; moreover people affected in this way are not always aware of it (Betts *et al.* 1972). For this reason, when benzodiazepines are prescribed, especially those with a longer action, patients should be advised about these dangers and about the potentiating effects of alcohol. The prescriber should remember that these effects are more common among elderly patients and those with impaired renal or liver function. Although in some circumstances benzodiazepines reduce tension and aggression, in certain doses they lead to a release of aggression by reducing inhibitions in people with a tendency to aggressive behaviour (DiMascio 1973). In this they resemble alcohol. This possible effect should be remembered when prescribing to those judged to be at risk of *child abuse* or to any person with a previous history of *impulsive aggressive behaviour*.

Toxic effects

Benzodiazepines have few toxic effects. Patients usually recover from large overdoses because these drugs do not depress respiration and blood pressure as barbiturates do. Even so, fatal overdoses of benzodiazepines have been reported (Serfaty and Masterton 1993). No convincing evidence of teratogenic effects has been reported, but it is wise to avoid prescribing in the first trimester of pregnancy unless there is a strong indication. Cerebral atrophy as judged by CT scan has been reported in some long-term benzodiazepine users, but it has not been shown to be an effect of the drug rather than an incidental findings (Lader *et al.* 1984).

Drug interactions

Benzodiazepines, like other sedative anxiolytics, potentiate the effects of alcohol and of drugs that depress the central nervous system. Significant respiratory depression has been reported in some patients receiving combined treatment with benzodiazepines and clozapine.

Effects of withdrawal

It is now generally agreed that physical dependence develops after prolonged use of benzodiazepines. The frequency depends on the drug and

the dosage, and has been estimated as between 5 and 50 per cent among patients taking the drugs for more than six months (Hallstrom 1985). Dependence is associated with a withdrawal syndrome characterized by apprehension, insomnia, nausea, and tremor, together with heightened sensitivity to perceptual stimuli. In severe cases epileptic seizures have been reported (Petursson and Lader 1984). Since many of these symptoms resemble those of anxiety disorder, it can be difficult to decide whether the patient is experiencing a benzodiazepine withdrawal syndrome or a recrudescence of the anxiety disorder for which the drug was prescribed originally (Rodrigo and Williams 1986).

Withdrawal symptoms generally begin within two to three days of stopping a short-acting benzodiazepine and within seven days of stopping a long-acting one. The symptoms generally last for three to ten days. Withdrawal symptoms seem to be more frequent after drugs with a short half-life than after those with a long half-life (Tyrer *et al.* 1981). If benzodiazepines have been taken for a long time, it is best to withdraw them gradually over several weeks (Committee on the Review of Medicines 1980). If this is done, withdrawal symptoms can be minimized or avoided.

Azapirones

The only drug in the azapirone class currently marketed for the treatment of anxiety is **buspirone**. This drug has no affinity for benzodiazepine receptors but stimulates a subtype of 5-HT receptor called the 5-HT_{1A} receptor. This receptor is found in high concentration in the raphe nuclei in the brain stem where it regulates the firing of 5-HT cell bodies. Administration of buspirone lowers the firing rate of 5-HT neurones and thereby decreases 5-HT neurotransmission in certain brain regions. This action may be the basis of its anxiolytic effect.

Buspirone has poor systemic availability because it has an extensive first-pass metabolism. Its anxiolytic effects take several days to develop. The side-effect profile differs from that of benzodiazepines. For example buspirone treatment does not cause sedation but instead is often associated with **light-headedness, nervousness,** and **headache** early in treatment. There is little evidence that tolerance and dependence occur during buspirone use, although such a judgement must always be made with circumspection. There is some evidence that patients who have previously responded to treatment with benzodiazepines do not respond well to buspirone (Lader 1991). *It is also important to note that buspirone cannot be used to treat benzodiazepine withdrawal.* Clinical trials have indicated that buspirone has some efficacy in the treatment of major depression, but the compound does not have a product licence for this indication. Unlike high-potency benzodiazepines and certain antidepressant drugs, buspirone does not appear to be useful in the treatment of panic disorder.

Drug interactions

Buspirone is relatively free from significant drug interactions, but in combination with **MAOIs** has been reported to cause raised blood pressure.

Other drugs used to treat anxiety

Antidepressant drugs

Antidepressant drugs usually ameliorate the anxiety that accompanies depressive disorders. Tricyclic antidepressants have also been shown to be as effective as benzodiazepines in the management of both generalized anxiety and panic disorder (Lydiard and Ballenger 1987; Rickels *et al.* 1993). Similarly, both SSRIs and MAOIs are effective in the treatment of panic disorder, but the selective noradrenaline uptake inhibitor maprotiline is not (Lydiard and Ballenger 1987; Den Boer and Westenberg 1988; Nutt and Glue 1989). It is not clear whether antidepressant drugs cause withdrawal syndromes as readily as benzodiazepines in patients with anxiety disorders. However, there is a strong clinical impression that relapse of anxiety is relatively common in patients withdrawn from antidepressant drugs, particularly MAOIs.

Antipsychotic drugs

These drugs are sometimes prescribed for their anxiolytic effects. In doses that do not lead to side-effects, they are generally no more effective than benzodiazepines. Nevertheless, antipsychotic drugs have a small place as anxiolytics in the treatment of two groups of patients—those with persistent anxiety who have become dependent on other drugs, and those with aggressive personalities who respond badly to the disinhibiting effects of other anxiolytics.

Beta-adrenoceptor antagonists

These drugs relieve some of the autonomic symptoms of anxiety, such as tachycardia, almost certainly by a peripheral effect (see Bonn *et al.* 1972). They are best reserved for anxious patients whose main symptom is palpitation or tremor, particularly in social situations. An appropriate drugs is propranolol in a dose of 40 mg three times a day. Contraindications are heart block, systolic blood pressure below 90 mmHg or a pulse rate of less than 60 per minute, a history of bronchospasm, metabolic acidosis, for example in diabetes, and after prolonged fasting, as in anorexia nervosa. Great caution is needed if there are signs of poor cardiac reserve. Beta-adrenoceptor antagonists precipitate heart failure in a few patients and should not be given to those with atrioventricular node block as they decrease conduction in the atrioventricular node and bundle of His. They can cause severe bronchospasm and exacerbate both Raynaud's phenomenon and intermittent claudication. In diabetics they may cause hypoglycaemia. Some drugs interact with beta-blockers and so increase these adverse effects. Therefore, it is important, to find out what other drugs are being taken, and to consult a work of reference to find out whether interactions have been reported.

Barbiturates and other sedative anxiolytics

In the past **barbiturates** were widely used as anxiolytics. Although effective, they readily cause dependency and they should not be used as anxiolytics. **Propanediols** such as meprobamate also have no advantage over benzodiazepines and are more sedative in doses needed to relieve anxiety.

Unwanted effects resemble those of the benzodiazepines and generally appear at doses nearer to the anxiolytic dose. Barbiturates may produce irritability, drowsiness, and ataxia. In large doses the **toxic effects** of sedative anxiolytics are to depress respiration and reduce blood pressure. This is a particular problem with the barbiturates. The **interactions** of these drugs with others resemble those of the benzodiazepines. In addition, barbiturates interact with coumarin drugs and reduce their anticoagulant action. They also increase the metabolism of tricyclic antidepressants and tetracyclin. The **effects of withdrawal** resemble the effects of withdrawing benzodiazepines described above, but are more severe. After stopping barbiturates the effects are particularly marked in the form of psychological tension, sweating, tremor, irritability, and, after large doses, seizures. Hence barbiturates should not be stopped suddenly if the dose has been substantial.

Advice on management

Before an anxiolytic drug is prescribed the cause of the anxiety should always be sought. For most patients, attention to life problems, an opportunity to talk about their feelings, and reassurance from the doctor are enough to reduce anxiety to tolerable levels. If an anxiolytic is needed, it should be given for a short time—seldom more than three weeks—and withdrawn gradually. It is important to remember that dependency is particularly likely to develop among people with alcohol problems. If the drug has been taken for several weeks, the patient should be warned that he may feel tense for a few days when it is stopped.

The drug of choice for the short-term treatment of anxiety is a benzodiazepine. A compound such as diazepam is suitable for both the intermittent treatment of anxiety and continuous treatment throughout the day. The other drugs should be kept for the specific purposes outlined above: beta-adrenoceptor antagonists for control of palpitations and tremor caused by anxiety, phenothiazines for patients who respond badly to the disinhibiting effects of sedative anxiolytics (for example abnormally aggressive patients) or patients who have

become dependent upon them; and antidepressant drugs in difficult and persistent cases that do not respond to psychological treatment alone. MAOIs can be useful in cases of refractory anxiety, but should be used as a last resort because of the risk of food and drug interaction. Whether the new reversible MAOI-A inhibitors, such as moclobemide, will be useful in the management of anxiety disorders remains to be determined.

Hypnotics

Hypnotics are drugs used to improve sleep. Many anxiolytic sedatives also act as hypnotics, and they have been reviewed in the previous section. Hypnotic drugs are prescribed widely and often continued for too long. This reflects the frequency of insomnia as a complaint. Mendelson (1980) found that about a third of American adults reported disturbed sleep, and a third of these described it as a major problem. Insomnia is reported more often by women and the elderly. Effective psychological treatments are available for the management of insomnia (Morin *et al.* 1994).

Pharmacology

The ideal hypnotic would increase the length and quality of sleep without residual effects the next morning. It would do so without altering the pattern of sleep and without any withdrawal effects when the patient ceased to take it. Unfortunately, no drug meets these exacting criteria. It is not easy to produce drugs that affect the whole night's sleep and yet are sufficiently eliminated by morning to leave behind no sedative effects. Moreover, the electrophysiological characteristics of sleep are altered while most drugs are being taken and for some nights after they have been stopped. Thus hypnotic drugs usually affect the pattern of the EEG, they suppress REM sleep while they are taken, and they lead to an increase of REM sleep up to several weeks after they have been stopped. These latter changes are often reflected in reports of disturbed sleep.

Compounds available

Nowadays the most commonly used hypnotics are **benzodiazepines**. Barbiturates should no longer be prescribed for insomnia because of their high toxicity in overdose and their significant potential for inducing dependence. Among other available hypnotic agents, **chloral hydrate** (or its derivatives), **chlormethiazole**, and the cyclopyrlone compound **zopiclone** are commonly used.

Of the benzodiazepines, the shorter-acting compounds such as **temazepam** and **lormetazepam** are appropriate as hypnotics because of their relative lack of hangover effects the next day. Other drugs that are marketed as hypnotics, such as flurazepam and nitrazepam, have a long duration of action and produce significant impairments in tests of *cognitive function* on the day following treatment (Hindmarch 1988).

Zopiclone is a newly introduced short-acting hypnotic compound which binds to a site close to the benzodiazepine receptor and thereby facilitates brain GABA function. Zopiclone produces fewer changes in sleep architecture than benzodiazepine hypnotics. It is also reported to be less liable to produce tolerance and dependence than benzodiazepines, but this report has not yet been fully evaluated. The most common side-effect is a bitter after-taste following ingestion, but behavioural disturbances including confusion, amnesia, and depressed mood have been reported.

Other hypnotic drugs include **chloral hydrate**, which is sometimes prescribed for children and old people. It is a gastric irritant and should be diluted adequately. Chloral is also available in capsular form. **Chlormethiazole edisylate** is a hypnotic drug with anticonvulsant properties. It is often used to prevent withdrawal symptoms in patients dependent on alcohol. For this reason it is sometimes thought, mistakenly, to be a suitable hypnotic for alcoholic patients. This belief is wrong because the drug is as likely as any other hypnotic drug to cause dependency.

Unwanted effects

The most important unwanted effects of hypnotics are their residual effects. These are experienced by the patient on the next day as feelings of being slow and drowsy. Psychological tests of reaction time have shown deficits in the afternoon after a single bedtime dose of a barbiturate or long-acting benzodiazepine (Bond and Lader 1973). A person with these deficits is not always aware of them, which may be serious for work involving potentially dangerous machinery or for driving motor vehicles, trains, or aeroplanes. People who sleep badly often make similar complaints after a poor night in which they did not take hypnotics, but these subjective feelings are not accompanied by comparable impairments of performance on psychological tests. The complaints may reflect the cause of insomnia (for example depression or overindulgence in alcohol on the previous day) rather than the loss of sleep itself.

Contraindications

Barbiturates and dichloralphenazone should not be given to patients suffering from acute intermittent porphyria.

Drug interactions

The most important interaction of hypnotic drugs is with alcohol. At first the two potentiate one another, sometimes to a dangerous extent. After prolonged usage, a degree of cross-tolerance develops; however, persistent abuse of alcohol may damage the liver and so increase sensitivity to hypnotic drugs by reducing their metabolism. With the longer-acting benzodiazepines the alcohol-potentiating effect may last well into the day after the drug was taken (Saario *et al.* 1975). The interaction between *chlormethiazole and alcohol* is particularly dangerous and can result in death from respiratory failure. For this reason there must be adequate supervision when the drug is used during withdrawal of alcohol. It should not be prescribed for alcoholics who continue to drink.

Advice on management

Before prescribing hypnotic drugs it is important to find out whether the patient is really sleeping badly and, if so, why. Many people have unrealistic ideas about the number of hours they should sleep. For example, they may not know that length of sleep often becomes shorter in middle and late life. Others take 'cat naps' in the daytime, perhaps through boredom, and still expect to sleep as long at night. Some people ask for sleeping tablets in anticipation of poor sleep for one or two nights, for example when travelling. Such temporary loss of sleep is soon compensated by increased sleep on subsequent nights, and any supposed advantage in alertness after a full night's sleep is likely to be offset by the residual effects of the drugs. If a drug is justifiable in these circumstances, it should be a short-acting benzodiazepine or zuplicone.

Among the common causes of disturbed sleep are excessive caffeine or alcohol, pain, cough, pruritus, and dyspnoea, and anxiety and depression. When any primary cause is present, this should be treated, not the insomnia. If, after careful enquiry, a hypnotic appears to be essential, it should be prescribed for a few days only. The clinician should explain this to the patient, and should warn him that a few nights of restless sleep may occur when the drugs are stopped, but this restlessness will not be a reason for prolonging the prescription.

The prescription of hypnotics for children is not justified, except for the occasional treatment of night terrors and somnambulism. Hypnotics should also be prescribed with particular care for the elderly, who may become confused and get out of bed in the night, perhaps injuring themselves. Many patients are started on long periods of dependency on hypnotics by the prescribing of 'routine night sedation' in hospital. Prescription of these drugs should *not* be routine; it should be a response only to a real need, and should be stopped before the patient goes home.

Antipsychotic drugs

This term is applied to drugs that reduce psycho-motor excitement and control some symptoms of psychosis. Alternative terms for these drugs are **neuroleptic** and **major tranquillizer**. None of these names is wholly satisfactory. Neuroleptic refers to the side-effects rather than to the therapeutic effects of the drugs and major tranquillizers does not refer to the most important clinical action. The term antipsychotic is used here because it appears in the *British national formulary*.

The main therapeutic uses of antipsychotic drugs are to reduce hallucinations, delusions, agitation, and psychomotor excitement in schizophrenia, psychosis secondary to a medical condition, or mania. The drugs are also used prophylactically to prevent relapses of schizophrenia. The introduction of chlorpromazine in 1952 led to substantial improvements in the treatment of schizophrenia and paved the way to the discovery of the many psychotropic drugs now available.

Pharmacology

Antipsychotic drugs share the property of blocking dopamine receptors. This may account for their therapeutic action, a suggestion supported by the close relationship between their potency in blocking dopaminergic receptors *in vitro*, and their therapeutic strength. It is also supported by the finding that, of the two stereo-isomers of flupenthixol, the alpha isomer blocks dopamine receptors and is not therapeutic (Johnstone *et al.* 1978). Both alpha and beta isomers block noradrenergic and cholinergic receptors. These anti-adrenergic and anticholinergic actions account for many of the side-effects of the drugs, while the antidopaminergic actions on basal ganglia are responsible for the extrapyramidal side-effects.

Dopamine receptors are of several biochemical and morphological subtypes (Schwartz *et al.* 1993; Seeman and Van Tol 1993). *Most antipsychotic drugs bind strongly to dopamine* D_2 *receptors, and this action appears to account for both their antipsychotic activity and their propensity to cause movement disorders.* **Clozapine** is an exception in that it has a relatively weak affinity for D_2 receptors, which may account for its lack of extrapyramidal side-effects. How clozapine produces its antipsychotic effect is not well-established because it has numerous pharmacological actions. Current hypotheses have implicated blockade of dopamine D_4 receptors and 5-HT receptors (Seeman *et al.* 1993; Coward 1992).

Actions at other neurotransmitter receptors may offset the liability of D_2 receptor antagonists to produce movement disorders. For example, thioridazine is a potent antagonist at muscarinic cholinergic receptors, and anticholinergic drugs are known to possess antiparkinsonian effects. Similarly, the lack of extrapyramidal effects associated with **risperidone** treatment has been attributed to the ability of this drug to block 5-HT_2 receptors as well as D_2 receptors.

Compounds available

A large number of antipsychotic compounds have been developed (Table 17.3). Some, like chlorpromazine, are phenothiazines. They differ from one another in the nature of the side-chain, for

Table 17.3 *A list of antipsychotic drugs*

Phenothiazines with aliphatic side-chain	Chlorpromazine Promazine
Phenothiazines with piperidine side-chain	Thioridazine Pipothiazine
Phenothiazines with piperazine side-chain	Trifluoperazine Fluphenazine
Thioxanthines	Flupenthixol Clopenthixol
Butyrophenones	Haloperidol Droperidol
Diphenylbutyl-piperidines	Pimozide Fluspirilene
Dibenzazepines	Clozapine Loxapine
Substituted benzamides	Sulpiride Remoxipride
Benzisoxazole	Risperidone

example, and the radical in the 2 position (see Fig. 17.1). Others are thioxanthenes (for example thiothixine and flupenthixol), butyrophenones (for example haloperidol) (Fig. 17.2) or diphenyl-butylpiperidines (for example pimozide). The various compounds differ more in their side-effects than in their therapeutic properties. An account of the relations between structure and function is given by Shepherd *et al.* (1968).

Phenothiazines fall into three groups, according to the side-chain attached to the 10 position R_2 in (Fig. 17.1) Aminoalkyl compounds such as chlorpromazine are the most sedative and have moderate extrapyramidal side-effects. **Piperidine** compounds such as thioridazine have fewer extrapyramidal effects than this first group. **Piperazine** compounds such as trifluoperazine or fluphenazine are the least sedating and the most likely to produce extrapyramidal effects. They are also the most potent therapeutically.

Thioxanthines are similar in structure to the phenothiazines (Figs. 17.1 and 17.2) differing only in the presence of a carbon rather than a nitrogen atom in the 10 position. Their properties are also similar to those of phenothiazines. **Butyrophenones** have a different base structure (Fig. 17.2). They have powerful antipsychotic effects and are very likely to cause extrapyramidal side-effects, but they have relatively little sedative effect. The **butylpiperidines,** of which pimozide is most often used in clinical work, are related in structure to the butyrophenones. Their most important difference is a longer half-life which allows once daily dosage.

Substituted benzamides such as **sulpiride** and **remoxipride** are highly selective D_2 receptor antagonists which, for reasons that are not well understood, seem less likely to produce extrapyramidal movement disorders. They also lack sedative and anticholinergic properties. *Recently, remoxipride has been associated with an increased incidence of aplastic anaemia,* which precludes its use unless no other antipsychotic drug is suitable for an individual patient. *Risperidone is a benzisoxazole derivative which is a potent antagonist at both $5\text{-}HT_2$ receptors and dopamine D_2 receptors.* The full clinical evaluation of risperidone is still being undertaken, but results from initial controlled studies suggest that it is less likely than haloperidol

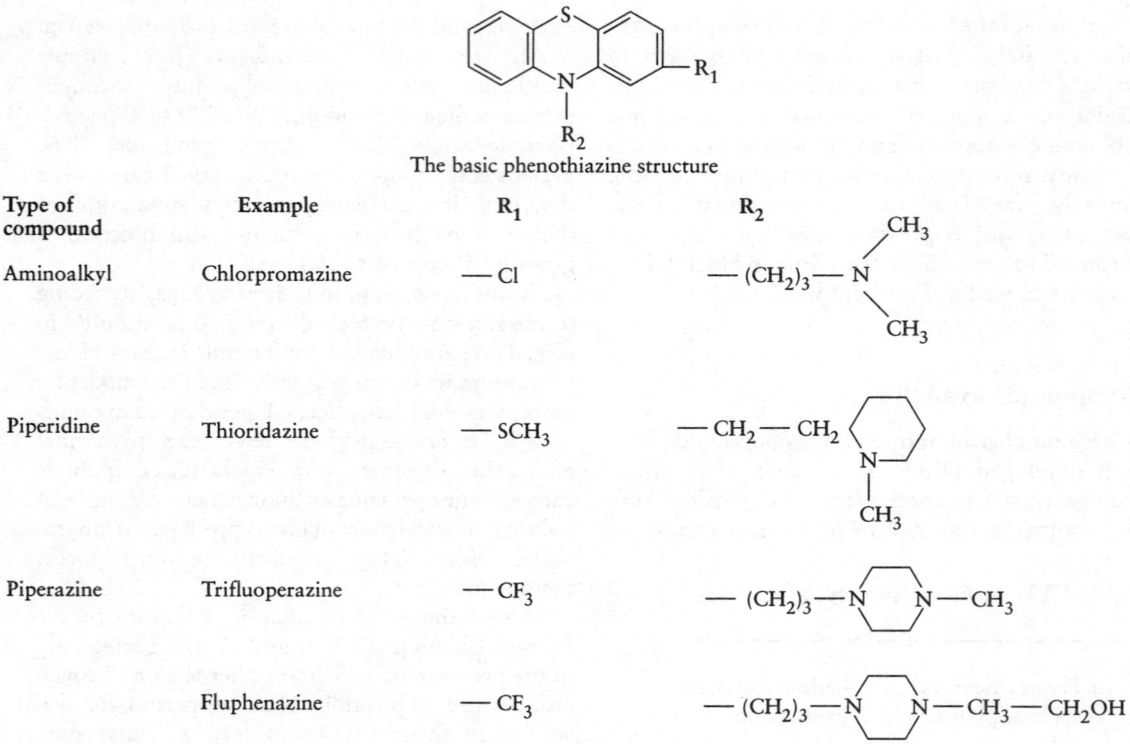

The basic phenothiazine structure

Type of compound	Example	R_1	R_2
Aminoalkyl	Chlorpromazine	— Cl	— $(CH_2)_3$ — N(CH$_3$)(CH$_3$)
Piperidine	Thioridazine	— SCH_3	— CH_2 — CH_2 — (piperidine, N—CH$_3$)
Piperazine	Trifluoperazine	— CF_3	— $(CH_2)_3$ — N(piperazine)N — CH$_3$
	Fluphenazine	— CF_3	— $(CH_2)_3$ — N(piperazine)N — CH$_3$ — CH$_2$OH

Figure 17.1

to induce extrapyramidal movement disorders and that it may also be superior to haloperidol in ameliorating negative symptoms and depressed mood (Chouinard *et al.* 1993).

Clozapine has an unusual pharmacological profile in that it is a weak dopamine D_2 receptor antagonist but binds with a relatively higher affinity to D_4 receptors. Clozapine also binds to a variety of other neurotransmitter receptors including histamine H_1, 5-HT_2, α_1-adrenergic and muscarinic cholinergic receptors (Coward 1992). Controlled studies have shown that a significant proportion of patients (up to about 50 per cent) who are not helped by other antipsychotic drugs will show a useful clinical response to clozapine (Kane *et al.* 1988). Negative symptoms of schizophrenia such as apathy and motivation may also respond. *However, the use of clozapine is associated with a high risk of leucopenia which restricts its use to patients who do not respond to or who are intolerant of other antipsychotic drugs.* The haematological

monitoring of clozapine treatment is discussed below.

The wide range of drugs now available can be seen from Table 17.3, which is not an exhaustive list. Fortunately, the clinician need acquaint himself with only a few of them, as explained below. **Slow-release depot preparations** are used for patients who need to take drugs to prevent relapse but cannot be relied on to take them regularly. These preparations include the esters fluphenazine decanoate, flupenthixol decanoate, clopenthixol decanoate, haloperidol decanoate, and pipothiazine palmitate. All are given in an oily medium. Fluspiriline is an aqueous suspension and has a shorter action than the others. Flupenthixol has been reported to have a mood-elevating effect, but this has not been proved.

Choice of drug

Of the many compounds available, the following are appropriate: chlorpromazine when a more

Flupenthixol
a thioxanthine

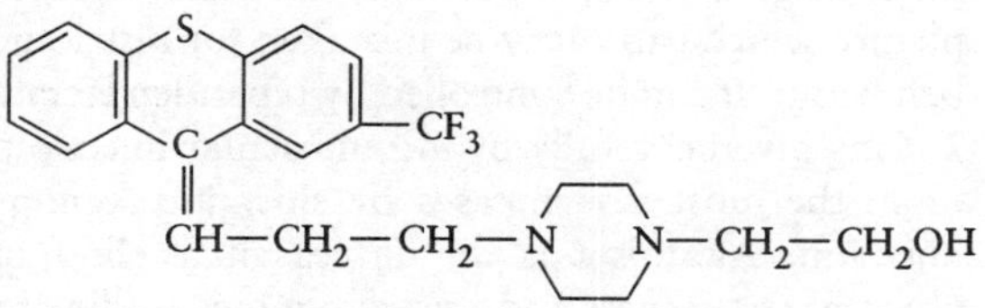

Haloperidol
a butyrophenone

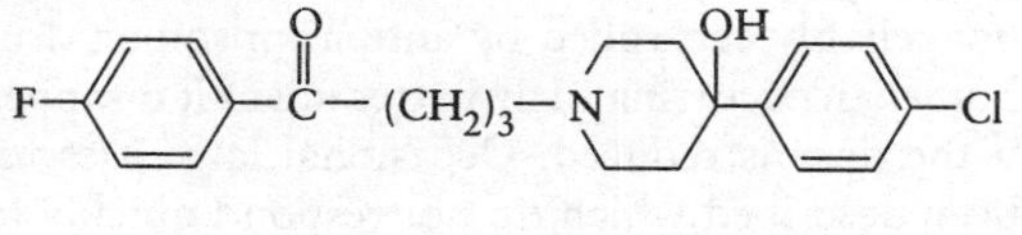

Clozapine

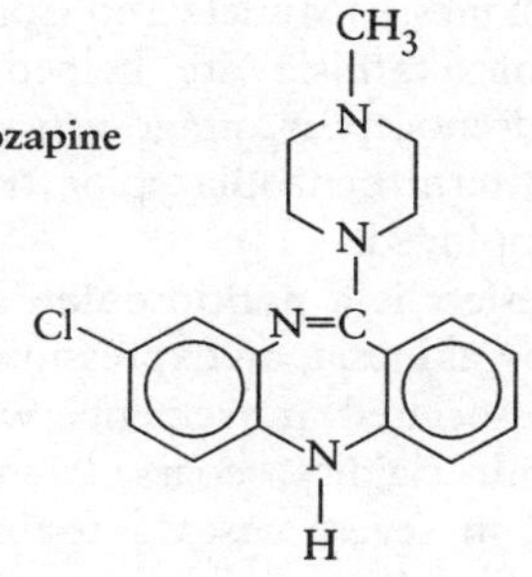

Risperidone

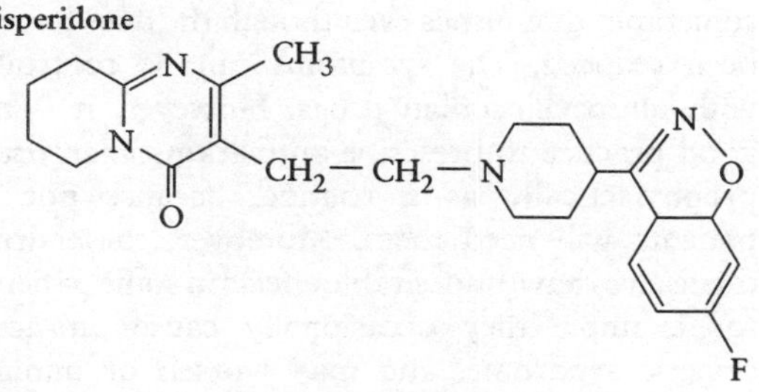

Figure 17.2

sedating drug is required, trifluoperazine or haloperidol when sedation is undesirable, and fluphenazine decanoate when a depot preparation is required. Haloperidol decanoate needs to be given only monthly. Risperidone and sulpiride are also less likely to cause extrapyramidal symptoms. Haloperidol is often preferred for the treatment of mania because it is less sedative than most phenothiazines.

Chlorpromazine, droperidol, and haloperidol can be given by intramuscular injection to produce a rapid calming effect in severely disturbed patients. Clozapine is reserved for patients who have not responded to other antipsychotic drugs in full dosage.

Pharmacokinetics

Antipsychotic drugs are well absorbed, mainly from the jejunum. They are largely metabolized in the liver. When they are taken by mouth, part of this metabolism is completed as they pass through the portal system on their way to the systemic circulation (first-pass metabolism). With chlorpromazine 75 per cent of the drug is metabolized in this way, with fluphenazine the proportion is greater, and with haloperidol and fluphenazine it is even greater, with haloperidol and pimozide it is less. The breakdown of chlorpromazine is complicated, about 75 metabolites have been detected in the blood or urine. The two principal metabolites are 7-hydroxychlorpromazine, which is still therapeutically active, and chlorpromazine sulphoxide which is not. Combinations of active and inactive metabolites also occur with other antipsychotic drugs. They make it difficult to interpret the clinical significance of plasma concentrations; hence the latter are seldom used in everyday clinical work. Chlorpromazine induces liver enzymes that increase its own metabolism; the latter is also increased by barbiturates and some antiparkinsonian drugs (notably orphenadrine). Other drugs (particularly imipramine and amitriptyline) reduce the metabolism of chlorpromazine by competing for relevant enzymes.

Unwanted effects

The many different antipsychotic drugs share a broad pattern of unwanted effects that are mainly related to their antidopaminergic, antiadrenergic, and anticholinergic properties (see Table 17.4). Details of the effects of individual drugs will be found in the *British national formulary* or a similar work of reference. Here we give an account of the general pattern, with examples of the side-effects associated with a few commonly used drugs.

Table 17.4. *Some unwanted effects of antipsychotic drugs*

Antidopaminergic effects
Acute dystonia
Akathisia
Parkinsonism
Tardive dyskinesia

Antiadrenergic effects
Sedation
Postural hypotension
Inhibition of ejaculation

Anticholinergic effects
Dry mouth
Reduced sweating
Urinary hesitancy and retention
Constipation
Blurred vision
Precipitation of glaucoma

Other effects
Cardiac arrhythmias
Weight gain
Amenorrhoea
Galactorrhoea
Hypothermia

Sensitivity reactions
See text

Extrapyramidal effects

These are related to the antidopaminergic action of the drugs on the basal ganglia. As already noted, the therapeutic effects may also derive from the antidopaminergic action, though presumably at a site other than the basal ganglia. Therefore it is not surprising that it has so far proved difficult to produce antipsychotic drugs with no extrapyramidal side-effects.

The effects on the extrapyramidal system fall into four groups. **Acute dystonia** occurs soon after treatment begins, especially in young men. It is observed most often with butyrophenones and with the piperazine group of phenothiazines. The main features are torticollis, tongue protrusion, grimacing, and opisthotonos, an odd clinical picture which can easily be mistaken for histrionic behaviour. It can be controlled by biperiden lactate 2–5 mg given carefully by intramuscular injection, or in the most severe cases by slow intravenous injection. **Akathisia** is an unpleasant feeling of physical restlessness and a need to move, leading to an inability to keep still. It usually occurs in the first two weeks of treatment with neuroleptic drugs, but may begin only after several months. Akathisia is not reliably controlled by antiparkinsonian drugs, but when occurring early in treatment it disappears if the dose is reduced. Occasional late cases have been described which do not respond quickly to a reduction in dose. It is difficult to differentiate these cases from tardive dyskinesia (Munetz and Cornes 1982). Some cases of akathisia are helped by treatment with beta-adrenoceptor antagonists such as propranolol. Short-term benzodiazepine treatment has also been employed.

The common side-effect is a **parkinsonian syndrome** characterized by akinesia, an expressionless face, and lack of associated movements when walking, together with rigidity, coarse tremor, stooped posture, and in severe cases a festinant gait. This syndrome often does not appear until a few months after the drug has been taken, and then sometimes diminishes even though the dose has not been reduced. The symptoms can be controlled with antiparkinsonian drugs. However, it is not good practice to prescribe antiparkinsonian drugs prophylactically as a routine, because not all patients will need them. Moreover, these drugs themselves have undesirable effects in some patients; for example they occasionally cause an acute organic syndrome, and may worsen or unmask concomitant tardive dyskinesia.

This last syndrome, **tardive dyskinesia**, is particularly serious because, unlike the other extrapyramidal effects, it does not always recover when the drugs are stopped. It is characterized by chewing and sucking movements, grimacing, choreo-athetoid movements, and possibly akathisia. The latter usually affect the face, but the limbs and the muscles of respiration may also be involved. The syndrome is seen occasionally among patients who have not taken antipsychotic drugs. Clinical observations suggest that it is much more common among those who have taken antipsychotic drugs for many years. However, a review by an American Psychiatric Association Task Force (1987) concluded that neither size of daily dose nor length of treatment is the main determinant. Tardive dyskinesia is more common among women, the elderly, and patients who have diffuse brain pathology. A diagnosis of affective disorder is also a risk factor (see Jeste and Caligiuri (1993) for a review of prevalence and risk factors). In about half the cases, tardive dyskinesia disappears when the drugs are stopped. Estimates of the frequency of the syndrome vary in different series, but it seems to develop in 20–40 per cent of schizophrenic patients treated with long-term antipsychotic drugs. Whatever the exact incidence, the risk of this syndrome should be a deterrent to the long-term prescribing of antipsychotic drugs unless clearly indicated.

The cause of the syndrome is uncertain, but it could be supersensitivity to dopamine resulting from prolonged dopaminergic blockade. This explanation is consistent with the observations that tardive dyskinesia may be aggravated in three ways: frequently by stopping the antipsychotic drugs, by the action of anticholinergic antiparkinsonian drugs (presumably by upsetting further the balance between cholinergic and dopaminergic systems in the basal ganglia), and by L-dopa and apomorphine in some patients. However, there are other observations that do not readily fit this explanation.

Many *treatments* to tardive dyskinesia have been tried but none is universally effective. Therefore it is important to reduce its incidence as far as possible by limiting long-term treatment to patients who really need it. At the same time a careful watch should be kept for abnormal movements in all patients who have taken antipsychotic drugs for a long time. If dyskinesia is observed, the antipsychotic drug should be stopped if the state of the mental illness allows this. Although the dyskinesia may first worsen after stopping the drug, in many cases it will improve over several months. If the dyskinesia persists after this time or if the continuation of antipsychotic medication is essential, a cautious trial can be made of a drug from one of the groups that have been reported, on the basis of clinical

trials, to reduce the abnormal movements. A drug can be tried from each of the groups in turn. These groups include dopamine receptor antagonists such as sulpiride and pimozide, and dopamine-depleting agents such as tetrabenazine. The reader is referred to Jeste and Caligiuri (1993) for a review of the treatment of tardive dyskinesia, and to Marsden and Jenner (1980) for further information about the pathophysiology of this and other extrapyramidal side-effects of antipsychotic drugs and to Stahl (1986) for a review of the natural history of tardive dyskinesia.

Anti-adrenergic effects

These include sedation postural hypotension with reflex tachycardia, nasal congestion, and inhibition of ejaculation. The effects on blood pressure are particularly likely to appear after intramuscular administration, and may appear in the elderly whatever the route of administration.

Anticholinergic effects

These include dry mouth, urinary hesitancy and retention, constipation, reduced sweating, blurred vision, and, rarely, the precipitation of glaucoma.

Other effects

Cardiac arrhythmias are sometimes reported. ECG changes are more common in the form of prolongation of the QT- and T-wave blurring. The use of pimozide has been associated with serious cardiac arrhythmias. Cautious dose adjustment with ECG monitoring is recommended. Depression of mood had been said to occur, but this is difficult to evaluate because untreated schizophrenic patients may have periods of depression. Some patients gain weight when taking antipsychotic drugs, especially chlorpromazine. Galactorrhoea and amenorrhoea are induced in some women. In the elderly, hypothermia is an important unwanted effect. Some phenothiazines, especially chlorpromazine, increase the frequency of seizures in epileptic patients. Prolonged chlorpromazine treatment can lead to photosensitivity and to accumulation of pigment in the skin, cornea, and lens. Thioridazine in exceptionally high dose (more than 800 mg/day) may cause retinal degeneration. Phenothiazines, particularly chlorpromazine, have been associated with cholestatic jaundice, but the incidence is low (about 0.1 per cent (MacKay 1982)). Blood cell dyscrasias also occur rarely with phenothiazines but are more common with remoxipride and clozapine (see below) (Committee on the Safety of Medicines, 1993).

These drugs have not been shown to be teratogenic, but nevertheless they should be used cautiously in early pregnancy.

Adverse effects of clozapine

The use of clozapine is associated with a significant risk of **leucopenia** (about 2–3 per cent) which can progress to **agranulocytosis** (Krupp and Barnes 1992). Weekly blood counts for the first 18 weeks of treatment and at two-weekly intervals thereafter are mandatory. With this intensive monitoring the early detection of leucopenia can be followed by immediate withdrawal of clozapine and by reversal of the low white cell count. This procedure greatly reduces, but does not eliminate, the risk of progression to agranulocytosis. It is usually recommended that clozapine be used as the sole antipsychotic agent in a treatment regime. Clearly, it is wise to avoid concomitant use of drugs such as carbamazepine which may also lower the white cell count.

Because of its relatively weak blockade of dopamine D_2 receptors, clozapine is less likely than other antipsychotic drugs to cause extrapramidal movement disorders. It does not increase plasma prolactin; hence galactorrhoea does not occur. However, its use is associated with **hypersalivation, drowsiness, postural hypotension, weight gain, and hyperthermia**. Seizures may occur at higher doses. Clozapine is a sedating compound, and cases of respiratory and circulatory embarrassment have been reported during combined treatment with clozapine and benzodiazepines (Sassim and Grohmann 1988). Rarely, **myocarditis** has been reported (Committee on the Safety of Medicines 1993).

The neuroleptic malignant syndrome

This rare but serious disorder occurs in a small minority of patients taking neuroleptics, especially high-potency compounds. Most reported cases have followed the use of neuroleptics for schizophrenia, but in some cases the drugs were used for mania, depressive disorder, and psychosis secondary to a medical condition. Combined lithium and antipsychotic drug treatment may be a predisposing factor. The onset is often, but not invariably, in the first 10 days of treatment. The **clinical picture** includes the rapid onset (usually over 24–72 hours) of severe motor, mental, and autonomic disorders. The prominent *motor* symptom is generalized muscular hypertonicity. Stiffness of the muscles in the throat and chest may cause dysphagia and dyspnoea. The *mental* symptoms include akinetic mutism, stupor, or impaired consciousness. Hyperpyrexia develops with evidence of *autonomic* disturbances in the form of unstable blood pressure, tachycardia, excessive sweating, salivation, and urinary incontinence. In the blood, creatinine phosphokinase (CPK) levels may be raised to very high levels, and the white cells increased. Secondary features may include pneumonia, thromboembolism, cardiovascular collapse, and renal failure. The mortality rate appears to be declining over recent years but may still be of the order of 10 per cent (Shalev *et al.* 1989). The syndrome lasts for one to two weeks after stopping an oral neuroleptic but may last two to three times longer after stopping long-acting preparations. Patients who survive are usually without residual disability.

The **differential diagnosis** includes encephalitis, and in some countries heat stroke. Before the introduction of antipsychotic drugs, a similar disorder was reported as a form of catatonia sometimes called acute lethal catatonia.

The condition can probably occur with any neuroleptic, but in many reported cases the drugs used have been haloperidol or fluphenazine. The cause could be related to excessive dopaminergic blockade, though why this should affect only a minority of patients cannot be explained. **Treatment** is symptomatic; the main needs are to stop the drug, cool the patient, maintain fluid balance, and treat intercurrent infection. No drug treatment is certainly effective. Diazepam can be used for muscle stiffness. Dantrolene, a drug used to treat malignant hyperthermia, has also been tried. Bromocriptine, amantadine, and L-dopa have been used, but with insufficient cases for a definite statement about their value. Some patients who developed the syndrome on one occasion have been given the drug again safely after the acute episode has resolved (Caroff 1980). Nevertheless, if an antipsychotic has to be used again, it is prudent to restart treatment cautiously with a low-potency drug such as thioridazine, used at first in low doses. At least two weeks should elapse before antipsychotic drug treatment is reinstated (Addonizio and Susman 1991).

Contraindications

There are few contraindications and they vary with individual drugs. Before any of these drugs is used, it is important to consult the *British national formulary* or a comparable work of reference. Contraindications include myasthenia gravis, Addison's disease, glaucoma, and evidence of present or past bone marrow depression; all of these conditions can be exacerbated by these drugs. For patients with liver disease chlorpromazine

should be avoided and other drugs used with caution. Caution is also required when there is renal disease, cardiovascular disorder, parkinsonism, epilepsy, or serious infection.

Dosage

Doses of antipsychotic drugs need to be adjusted for the individual patient and changes should be made gradually. Doses should be lower for children, the elderly, patients with brain damage or epilepsy, and the physically ill. The dosage of individual drugs can be found in the *British national formulary* or a comparable work of reference or in the manufacturer's literature.

There is a growing trend for lower doses of antipsychotic drugs to be recommended. This is based in part on recent studies with positron emission tomography which have demonstrated that adequate dopamine D_2 receptor blockade (in the basal ganglia at least) can be obtained with low doses of conventional antipsychotic drugs (about 5 mg of haloperidol for example) (Table 17.5) (Farde *et al.* 1988, 1989). Such doses produce an adequate antipsychotic effect in the majority of patients (McEvoy *et al.* 1991). Higher doses may cause further calming but are also likely to be associated with significant adverse effects, some of

Table 17.5 *Dosage and D_2 receptor blockade of some antipsychotic drugs*

Drug	Relative dose (oral)	Maximum BNF dose (mg)	D_2 receptor occupancy in vivo (%) (daily dose (mg))*
Chlorpromazine	100	1000	80 (200)
Thioridazine	100	800	75 (300)
Trifluoperazine	5	NA	80 (10)
Fluphenazine	2	20	NA
Haloperidol	2	100	80 (4)
Fluphenthixol	1	18	74 (10)
Sulpiride	200	2400	74 (800)
Clozapine	60	900	65 (600)

NA, not available.
*Farde *et al.* 1989.

which may be serious (for example cardiac arrhythmias). A view of growing influence is that the *combination of modest doses of antipsychotic drugs with a benzodiazepine* is a safer and more effective means of producing rapid sedation than high doses of antipsychotic drugs (Pilowsky *et al.* 1992) (see below).

The association of **sudden unexplained death** with antipsychotic drug treatment is a matter of continuing debate. For example it is not established whether the rate of such deaths is greater in patients receiving antipsychotic drugs than in those receiving other treatments, or whether the rate in psychiatric patients is higher than in the general population. However, antipsychotic drugs are known to alter cardiac conduction, and drugs such as chlorpromazine also produce hypotensive effects. An association between high doses of pimozide (often in the setting of a recent dose increase) and sudden unexplained death led to a reduction in the recommended maximum daily dose to 20 mg (Committee on the Safety of Medicines 1990). While the relationship between high dose of antipsychotic drug treatment and sudden death is not established, it is clearly prudent to use as low a dose of an antipsychotic drug as the clinical circumstances permit. For a review of this area and the question of high-dose antipsychotic drug treatment see Royal College Of Psychiatrists (1993).

An indication of the relative dosage of some commonly used drugs taken by mouth is given in Table 17.5. Some practical guidance on the most frequently used drugs is given in the next section.

Advice on management

Use in emergencies

Antipsychotic drugs are used to control psychomotor excitement, hostility, and other abnormal behaviour resulting from schizophrenia, mania, or organic psychosis. If the patient is very excited and is displaying abnormally aggressive behaviour, the aim should be to bring the behaviour under control as quickly and safely as possible. Chlorpromazine (orally or intramuscularly) has previously been recommended for this purpose, but it has the disadvantage of producing autonomic side-effects such as hypotension. *Current practice favours the use of low doses of drugs such as haloperidol (2–10 mg) or droperidol (2–5 mg) with diazepam (5–20 mg) or lorazepam (1–4 mg).* These drugs can be given orally or parenterally (diazepam is poorly absorbed intramuscularly). The intravenous route, if it can be used, has the advantage of allowing small bolus doses to be given so that the amount of drug administered can be titrated against the required effect on behaviour. It is important to check for possible respiratory depression, particularly in the elderly or those with concomitant physical illness. Another possibility is to use a medium-acting depot preparation such as **zuclopenthixol accuphase** which produces sedation shortly after its adminstration and has a duration of action of two to three days. While this regime may be useful in some circumstances, it does not permit the careful titration of dose against clinical response as described above. *For this reason it should not usually be given to patients whose tolerance of antipsychotic drugs is not established* (Royal College of Psychiatrists 1993).

There are several other practical points in the **management of the acutely disturbed patient** that can be dealt with conveniently here. Although it may not be easy in the early stages to differentiate between mania and schizophrenia as causes of the disturbed behaviour, it is necessary to try to distinguish them from psychosis secondary to medical conditions and from outburst of aggression in abnormal personalities. Among medical conditions it is important to consider post-epileptic states, the effects of head injury, transient global amnesia, and hypoglycaemia. People with abnormal personalities may act extremely abnormally when subjected to stressful events, especially if they have taken alcohol or other drugs. When overactive behaviour is secondary to an organic use, it may be necessary to treat it symptomatically, but any drugs must be given cautiously and the primary disorder should be treated whenever possible. If the patient has been drinking alcohol, the danger of potentiating the sedative effects of antipsychotic drugs and benzodiazepines should be remembered. Similarly, antipsychotic drugs that may provoke seizures should be used with caution in postepileptic states.

In order to make a diagnosis, a careful history should be taken from an informant as well as from the patient. It is unwise to be alone with a patient who has already been violent, at least until a diagnosis has been made. The interviewer should do his best to calm the patient. Provided that it seems safe and help remains at hand, he should disengage anyone who is restraining the patient physically. If medication is essential and the patient refuses to accept it, compulsory powers must be acquired by involving the relevant part of the Mental Health Act (Appendix, p 783) before applying treatment. If a calming injection is required, having obtained the necessary legal authority the doctor should assemble enough helpers to restrain the patient effectively. They should act in a swift and determined way to secure the patient; half measures are likely to make him more aggressive. After the patient has become calmer, blood pressure and respiration should be monitored, particularly when the antipsychotic drug has been given by intramuscular injection.

The treatment of the acute episode

When any necessary emergency measures have been taken, or from the beginning in less urgent cases, treatment with moderate doses of one of the less sedating antipsychotic drugs should be started. An appropriate prescription would be trifluoperazine 10–20 mg daily in divided doses, or haloperidol 5–15 mg daily in divided doses. The latter drug is often used for manic patients because it has less sedative side-effects. In the early stages of treatment the amount and timing of doses should be adjusted if necessary from one day to the next, until the most acute symptoms have been brought under control. Thereafter, regular twice-daily dosage is usually appropriate. A careful watch should be kept for acute dystonic reactions in the early days of treatment, especially when large doses are being used. Watch should also be kept for parkinsonian side-effects as treatment progresses; if they appear, an antiparkinsonian drug should be given (see next section). For the elderly or physically ill, appropriate observations of temperature and blood pressure should be made to detect hypothermia or postural hypotension.

While patients often become more settled a few days after starting antipsychotic drugs, improvement in psychotic symptoms is usually slow, with resolution often taking a number of weeks. *Again, current trends are to maintain the dose of antipsychotic drugs at a modest steady level and not to escalate the dose in the hope of speeding up the rate of improvement.* For patients in whom agitation and distress continue to cause concern it may be appropriate to add short-term intermittent treatment with a benzodiazepine rather than increase the dose of antipsychotic agent. If the patient is intolerant of the antipsychotic drug, it may be helpful to change to another drug of a different class, for example by substituting sulpiride for haloperidol if extrapyramidal movement disorders are problematic. A number of patients fail to respond to the initial antipsychotic drug, and it is customary to change the treatment to an antipsychotic drug of a different class. However, with the exception of clozapine and perhaps risperidone, there is little evidence that this practice produces significant benefit. The same is probably true of the practice of combining two antipsychotic drugs. For an account for the clinical management of antipsychotic drug treatment in schizophrenia see Schatzberg and Cole (1991*a*).

Treatment after the acute episode

Episodes of mania and acute psychosis secondary to medical conditions usually subside within weeks. However, schizophrenic patients often require treatment for many months or years. Such maintenance treatment can be a continuation, in a smaller dose, of the oral medication used to bring the condition under control. However, schizophrenic patients frequently fail to take their drugs regularly, and so delayed-release depot preparations are often used. These are given by intramuscular injection. At the start of treatment a test dose is given to find out whether serious side-effects are likely with the full dose; 12.5 mg for fluphenazine decanoate is appropriate. The maintenance is then established by trial and error. It is likely to be between 25 and 50 mg every two to four weeks, and it is appropriate to begin with fluphenazine decanoate 25 mg every three weeks. It is important to find the smallest dose that will control the symptoms; since this may diminish with time, regular reassessment of the remaining symptoms of illness and the extent of side-effects is needed. It is not necessary to give antiparkinsonian drugs routinely; if they are needed, it may be only for a few days after the injection of the depot preparation (when the drug plasma concentrations are highest).

Alternative sustained-action injectable preparations are haloperidol decanoate, flupenthixol decanoate, and clopenthixol decanoate. It has been reported that the former leads to less depression of mood than fluphenazine preparations, but this has not been substantiated.

Depot preparations have long half-lives, and therefore it may take several weeks for maximum plasma concentrations to be reached. This has implications for the rate at which dose increases and decreases should be made, and also for the tapering of doses of oral antipsychotic medication once depot treatment has started.

Antiparkinsonian drugs

Although these drugs have no direct therapeutic use in psychiatry, they are often required to control the extrapyramidal side-effects of antipsychotic drugs.

Pharmacology

Of the drugs used to treat idiopathic parkinsonism, the anticholinergic compounds are used for drug-induced extrapyramidal syndromes.

Preparations available

Many anticholinergic drugs are available and there is no rational reason for choosing any particular compound. Those most often used in psychiatric practice are the synthetic anticholinergics benzhexol, benztropine mesylate, and procylidine, and the antihistaminic orphenadrine. Orphenadrine is said to have a mood-elevating effect. An injectable preparation of biperiden is useful for the treatment of acute dystonias.

Unwanted effects

In large doses these drugs may cause an acute organic syndrome, especially in the elderly. Their anticholinergic activity can summate with those of antipsychotic drugs so that glaucoma or retention of urine in men with enlarged prostates may be precipitated. Drowsiness, dry mouth, and constipation also occur. These effects tend to diminish as the drug is continued. Some studies have found that concomitant treatment with anticholinergic drugs can attenuate the therapeutic effect of antipsychotic drug treatment (Johnstone *et al.* 1983). Anticholinergic drugs can also exacerbate tardive dyskinesia but are probably not a predisposing factor in its development (Jeste and Caligiuri 1993).

Drug interactions

Antiparkinsonian drugs can induce drug metabolizing enzymes in the liver, so that plasma concentrations of antipsychotic drugs are sometimes reduced.

Advice on management

As noted already, anticholinergic drugs should not be given routinely because they may increase the risk of tardive dyskinesia. It has also been pointed out that patients receiving injectable long-acting antipsychotic preparations usually require anticholinergic drugs for only a few days after injection, if at all. There have been reports of dependence on benzhexol, possibly resulting from a mood-elevating effect (Harrison 1980). Benzhexol 5–15 mg daily in divided doses or orphenadrine 50–100 mg three times daily is appropriate for routine use.

Antidepressants

Currently used antidepressant drugs can be divided into three main classes, depending on their acute pharmacological properties. The first class consists of compounds that *inhibit the reuptake of noradrenaline and/or 5-HT*. This class includes the **tricyclic antidepressants** and the **SSRIs**. The second class consists of drugs that inhibit *monoamine oxidase* (**MAOIs**). The third class consists of drugs with complex effects on monoamine mechanisms, for example **mianserin** and **trazodone,** which cannot easily be categorized under the first two headings. In the broad range of major depression, these drugs are of equivalent efficacy. The main distinctions between them are in their **adverse effects, toxicity,** and **cost** (Table 17.6). These three classes of drugs will be considered in turn after some comments on the possible mechanism of action of antidepressants.

Mechanism of action

The acute effect of re-uptake inhibitors and of MAOIs is to *enhance the functional activity of noradrenaline and/or 5-HT.* These actions can be detected within hours of the start of treatment starting and yet the antidepressant effects of drug treatment can be delayed for several weeks. For example it has been suggested that at least six weeks should elapse before an assessment of the effects of an antidepressant drug can be made in an individual patient.

To some extent this delay in the onset of therapeutic activity may be due to pharmacokinetic factors. For example the half-life of most tricyclic antidepressants is around 24 hours, which means that steady state in plasma drug levels will be reached only after five to seven days. However, it seems unlikely that this can account completely for the lag in antidepressant activity.

In animal experimental studies, the acute effects of antidepressants in facilitating noradrenaline and

Table 17.6 *Groups of antidepressant drugs*

Drug	Advantages	Disadvantages
Tricyclic antidepressants*	Well studied No serious long-term toxicity Useful sedative effect in selected patients Inexpensive	Cardiotoxic, dangerous in overdose Anticholinergic side-effects Cognitive impairment Weight gain during longer-term treatment
SSRIs	Lack cardiotoxicity, relatively safe in overdose Not anticholinergic No cognitive impairment Relatively easy to give effective dose	Long-term toxicity not fully evaluated Gastrointestinal disturbance May worsen sleep and anxiety symptoms initially Greater risk of drug interaction Expensive
Trazodone Mianserin	Lack cardiotoxicity[†], relatively safe in overdose Not anticholinergic Useful sedative effect in selected patients	Wide dose range Cognitive impairment Less well-established efficacy in severe depression Expensive

*Lofepramine has important differences from conventional tricyclic antidepressants (see p. 562).
[†]Cardiac arrhythmias have rarely been reported with trazodone.

5-HT neurotransmission are followed by a cascade of secondary biochemical changes in noradrenaline and 5-HT pathways (Heninger *et al.* 1983*a*). These **adaptive changes** are of interest because their appearance is delayed for several days and thus parallels the lag in onset of the clinical antidepressant effects of drug treatment. In addition, many different kinds of antidepressant treatment may produce common changes in certain neurotransmitter receptors, although it is not known which of these changes may be relevant to clinical antidepressant activity.

An important feature of both noradrenaline and 5-HT pathways is that the cell bodies in the midbrain possess inhibitory autoreceptors, stimulation of which decreases cell firing. Drugs that acutely increase synaptic neurotransmitter levels of noradrenaline and 5-HT (such as tricyclics and MAOIs) indirectly activate these autoreceptors through dendritic release of noradrenaline and 5-HT. This action diminishes cell body firing and attenuates the increase in neurotransmission caused by the antidepressant drug.

Biochemical and behavioural studies have indicated that, as antidepressant treatment is continued for several days, the autoreceptors on noradrenaline and 5-HT cell bodies become subsensitive (Green *et al.* 1986). The effect of this is to free noradrenergic and 5-HT neurones from inhibitory feedback control, and to restore the firing rate of the cell bodies to normal levels despite the presence of increased synaptic concentrations of noradrenaline and 5-HT. This would be expected to increase further the ability of antidepressant drugs to augment noradrenaline and 5-HT function. *Thus the clinical effects of antidepressant treatment may result from an increasing potentiation of noradrenaline and 5-HT neurotransmission over time.*

Some support for this view has come from a recent series of clinical investigations in the United States. It is possible to produce a rapid temporary decrease in brain 5-HT function by a dietary manipulation that diminishes the availability of L-tryptophan, the amino acid precursor of 5-HT, to the brain. Amongst depressed patients who have recently recovered from a depressive illness and are maintained on antidepressant drug treatment, a substantial proportion show acute clinical relapse when exposed to this dietary manipulation. Such a relapse is particularly apparent in patients receiving treatment with drugs whose primary pharmacological action is exerted upon 5-HT mechanisms, for example SSRIs (Delgado *et al.* 1992). In contrast, when patients receive treatment with antidepressant drugs whose main effect is to inhibit the re-uptake of noradrenaline, for example the tricyclic antidepressant desipramine, they tend to show clinical relapse when noradrenaline neurotransmission is interrupted by the noradrenaline synthesis inhibitor alpha-methyl-tyrosine (Salomon *et al.* 1993). These findings support the proposal that the therapeutic effect of antidepressant drug treatment depends on sustained increases in noradrenaline and/or 5-HT neurotransmission.

Tricyclic antidepressants

Structure and pharmacological properties

Tricyclic antidepressants have a three-ringed structure with an attached side-chain. Modification of either the central ring or the side-chain produces derivatives with differing pharmacological properties. In general, these differing properties are associated with changes in adverse effects rather than in efficacy as treatment for major depression. A useful distinction is between compounds that have a terminal methyl group on the side-chain (tertiary amines) and those that do not (secondary amines). In general, compared with the secondary amines, tertiary amines have a higher affinity for the 5-HT uptake site and are more potent antagonists of α_1-adrenoceptors and muscarinic cholinergic receptors. *Therefore, in clinical use tertiary amines are more sedating and cause more anticholinergic effects than secondary amines.*

Tricyclic antidepressants inhibit the re-uptake of both 5-HT and noradrenaline. They also have antagonist activities at a variety of neurotransmitter receptors. In general, these receptor-blocking actions have been thought to cause adverse effects (Table 17.7), though some investigators have argued that the ability of some tricyclics to antagonize brain 5-HT$_2$ receptors may also mediate some of their therapeutic effects. *Tricyclics*

Table 17.7 *Some adverse effects of tricyclic antidepressants*

Pharmacological action	Adverse effect
Muscarinic receptor blockade (anticholinergic)	Dry mouth, tachycardia, blurred vision, glaucoma, constipation, urinary retention, sexual dysfunction, cognitive impairment
α_1-Adrenoceptor blockage	Drowsiness, postural hypotension, sexual dysfunction, cognitive impairment
Histamine H_1-receptor blockade	Drowsiness, weight gain
Membrane-stabilizing properties	Cardiac conduction defects, cardiac arrhythmias, epileptic seizures
Other	Rash, oedema, leucopenia, elevated liver enzymes

have quinidine-like membrane-stabilizing effects, and this may explain why they impair cardiac conduction and cause high toxicity in overdose.

Pharmacokinetics

Tricyclic antidepressants are well absorbed from the gastrointestinal tract, with the result that peak plasma levels occur two to four hours after ingestion. Tricyclics are subject to significant first-pass metabolism in the liver and are highly protein bound. The free fraction is widely distributed in body tissues. In general, the elimination half-life of tricyclics is such that it is unnecessary to give them more than once daily. Tricyclics are metabolized in the liver by hydroxylation and demethylation; it is noteworthy that demethylation of tricyclics with a tertiary amine structure gives rise to significant plasma concentrations of the corresponding secondary amine. There can be substantial (10–40-fold) differences in plasma tricyclic antidepressant levels between individual subjects when fixed-dose regimes are employed.

Plasma monitoring of tricyclic antidepressants

Despite a considerable research effort the role of plasma-level monitoring in the use of tricyclics is not well established. In general it has been difficult to show a consistent relationship between plasma level and therapeutic response. The possible reasons for this have been discussed above. However, there is some agreement (though the studies are not all in accord) that plasma levels of **nortriptyline** demonstrate a **curvilinear** relationship with clinical outcome. *The highest response rates occur with plasma concentrations in the range 50–150 ng/ml, and above this level the response rate may actually decline.* However, the relationship between clinical response during amitriptyline treatment and total plasma levels of amitriptyline and nortriptyline is not clear, with different studies reporting variously a linear relationship, a curvilinear relationship, and no relationship at all. For reviews of the plasma monitoring of tricyclic antidepressants see the American Psychiatric Association (1985) and Preskorn (1993).

There is some evidence that **high levels of tricyclic antidepressants** are more likely to be associated with toxic side-effects such as delirium, seizures, and cardiac arrhythmias. *The risk of such side-effects is minimized if total plasma levels of tricyclic antidepressants are lower than 300 ng/ml* (Preskorn 1993). In this context it is worth noting that a small proportion of patients who metabolize drugs slowly may develop significantly increased plasma levels of tricyclics while taking routine clinical doses (Preskorn *et al.* 1989).

Overall, plasma-level monitoring has a useful but rather limited role in the management of tricyclic antidepressant treatment (Table 17.8). Plasma monitoring may be useful to assess **compliance** and is often helpful in patients who have **not responded to what are usually adequate tricyclic doses**, particularly if increases in dose above 200 mg daily are contemplated. Finally, plasma-level monitoring is useful in patients with **co-existing medical disorders**, especially if there is a possibility of drug interaction. For example, in patients with seizure disorders for example, it is prudent to maintain plasma tricyclic levels within the usual range for the particular compound being used because tricyclics **lower the seizure threshold**. An additional level of complexity is added by the effects of coadministered antiepileptic drugs which can increase or lower plasma tricyclic levels through metabolic interactions.

Compounds available

These include amitriptyline, amoxapine, clomipramine, desipramine, dothiepin, doxepin, imipramine, iprindole, lofepramine, nortriptyline, protriptyline,

Table 17.8 *Indications for plasma monitoring of tricyclic antidepressants*

To check compliance
Toxic side-effects at low dose
Lack of therapeutic response
Coexisting medical disorder (e.g. epilepsy)
Possibility of drug interaction

---

and trimipramine. Some of these are sufficiently distinct from amitriptyline and imipramine to be worth separate mention. **Amoxapine** is a fairly selective inhibitor of noradrenaline uptake but, unusually for a tricyclic antidepressant, produces significant **blockade of dopamine D_2 receptors.** The combined effect of amoxapine to increase noradrenaline neurotransmission and antagonize D_2 receptors has led to suggestions that this compound may be particularly useful in the treatment of depressive psychosis when combined treatment with antidepressant and antipsychotic drugs is often required. However, the use of a single preparation to produce a combined pharmacological effect limits prescribing flexibility. Furthermore, as might be expected, the D_2-receptor-blocking properties of amoxapine may result in **extrapyramidal disorders** (Rudorfer and Potter 1989).

Clomipramine is the most potent of the tricyclic antidepressants in inhibiting the re-uptake of 5-HT; however, its secondary amine metabolite, desmethylclomipramine, is an effective noradrenaline re-uptake inhibitor. In studies of depressed inpatients the antidepressant effect of clomipramine was found to be superior to that of the SSRIs, citalopram and paroxetine (Danish University Antidepressant Group 1990). *Unlike other tricyclic antidepressants, clomipramine is also useful in ameliorating the symptoms of obsessive–compulsive disorder (whether or not there is a coexisting major depressive disorder)* (p. 185). Clomipramine is also available as an intravenous infusion. In general, this form of administration does not appear to produce better therapeutic effects than the oral route (Pollock *et al.* 1989).

Lofepramine is a tertiary amine which is metabolized to desipramine; however, during lofepramine treatment desipramine levels are probably too low to contribute significantly to the therapeutic effect. Lofepramine is a fairly selective inhibitor of noradrenaline re-uptake, and has fewer anticholinergic and antihistaminic properties than amitriptyline. Lofepramine has not been widely tested against placebo in depressed patients, but in two published studies it was more effective (Lancaster and Gonzalez 1990). It has been widely compared with other tricyclic antidepressants and in general its antidepressant efficacy appears equivalent. Lofepramine is not sedating; early in treatment it can be experienced as activating, an effect which some depressed patients find unpleasant. Similarly, impaired sleep does not usually improve until the underlying depression remits. *The most important feature of lofepramine is that in overdose it is not cardiotoxic and it is much safer than conventional tricyclic antidepressants.* Therefore lofepramine is likely to be safer than other tricyclics for patients with cardiovascular disease, though caution is still recommended. There have been reports of **hepatitis** in association with lofepramine, but it is not clear whether the incidence is greater than with other tricyclic antidepressants.

Maprotiline is often referred to as a quadricyclic antidepressant because the tricyclic nucleus is supplemented by an ethylene bridge across the middle ring. It is the most selective noradrenaline uptake inhibitor of the tricyclic antidepressants currently available, and has moderate antihistaminic properties but rather less anticholinergic effects than imipramine. It is not well established whether maprotiline is more effective than placebo for depression, but in comparative studies with reference tricyclics its therapeutic activity appears equivalent. *The use of maprotiline at doses above 200 mg has been associated with a higher incidence of seizures than is usual during tricyclic treatment* (Skowron and Stimmel 1992). Therefore a dose range of 75–150 mg daily has been recommended, and the co-prescription of other drugs that may lower the seizure threshold, such as phenothiazines, should be approached with caution. Maprotiline has effects on the heart that are similar to those of conventional tricyclics, and in overdose it is at least as toxic (Rudorfer and Potter 1989).

Unwanted effects of tricyclic antidepressants

These are numerous and important (Table 17.7). They can be divided conveniently into five groups.

1. **Autonomic:** dry mouth, disturbance of accommodation, difficulty in micturition leading to retention, constipation leading rarely to ileus, postural hypotension, tachycardia, and increased sweating. Retention of urine, especially in elderly men with enlarged prostates, and worsening of glaucoma are the most serious of these effects; dry mouth and accommodation difficulties are the most common. Iprindole and lofepramine are least likely to produce these anticholinergic side-effects.

2. **Psychiatric:** tiredness and drowsiness with amitriptyline and other sedative compounds; insomnia with desipramine and lofepramine; acute organic syndromes; mania may be provoked in manic depressive patients.

3. **Cardiovascular effects:** tachycardia and hypotension occur commonly. The electrocardiogram frequently shows prolongation of PR and QT intervals, depressed ST segments, and flattened T waves. Ventricular arrhythmias and heart block develop occasionally, more often in patients with pre-existing heart disease.

4. **Neurological:** fine tremor (commonly), incoordination, headache, muscle twitching, epileptic seizures in predisposed patients, and, rarely, peripheral neuropathy.

5. **Other:** allergic skin rashes, mild cholestatic jaundice, and, rarely, agranulocytosis; weight gain and sexual dysfunction are also common. Teratogenic effects have not been recorded in women, but nevertheless antidepressant drugs should be used cautiously in the first trimester of pregnancy.

Antidepressants should be withdrawn slowly. Sudden cessation may be followed by nausea, anxiety, sweating, and insomnia.

Toxic effects

In overdosage, tricyclic antidepressants produce a large number of effects, of which some are extremely serious. Therefore urgent expert treatment in a general hospital is required, but the psychiatrist should know the main signs of overdosage. These can be listed as follows. The **cardiovascular** effects include ventricular fibrillation, conduction, disturbances, and low blood pressure. Heart rate may be increased or decreased depending partly on the degree of condition disturbance. The **respiratory** effects lead to respiratory depression. The resulting hypoxia increases the likelihood of cardiac complications. Aspiration pneumonia may develop. The **central nervous system** complications include agitation, twitching, convulsions, hallucinations, delirium, and coma. **Parasympathetic** effects include dry mouth, dilated pupils, blurred vision, retention of urine, and pyrexia. Most patients need only supportive care, but cardiac monitoring is important and arrhythmias require urgent treatment by a physician in an intensive care unit. Tricyclic antidepressants delay gastric emptying, and so gastric lavage is valuable for several hours after the overdose. Lavage must be carried out with particular care to prevent aspiration of gastric contents; if necessary a cuffed endotracheal tube should be inserted before lavage is attempted.

Antidepressants and heart disease

The cardiovascular side-effects of tricyclic drugs noted above, coupled with their toxic effects on the heart when these drugs are taken in overdose, have led to the suggestion that tricyclic antidepressant drugs may be dangerous in patients with heart disease. The evidence is conflicting: a British drug-monitoring system linked cardiac deaths with amitriptyline (Coull *et al.* 1970), but a similar system in the United States did not confirm such a link (Boston Collaborative Drug Surveillance Program 1972). Tricyclic antidepressants have anticholinergic and quinidine-like effects, and they decrease myocardial contractility. Therefore the drugs could impair cardiac function. However, Veith *et al.* (1982) found no effect of tricyclic antidepressants on left ventricular function at rest or after exercise in depressed patients with chronic heart disease. However, patients with abnormal cardiac function do seem to be more at risk of orthostatic hypotension and heart block during treatment.

As noted above some of the newer compounds, such as mianserin and the SSRIs, appear safer than tricyclic antidepressants. Orme (1984) concluded that any antidepressant drug is probably safe for patients with only mild heart disease, but tricyclic antidepressants should be used very cautiously for patients with severe heart disease, such as recent myocardial infarction, heart failure, or electrocar-diographic evidence of bundle branch block or heart block.

Interactions with other drugs

The metabolism of tricyclic drugs is reduced competitively by phenothiazines and increased by barbiturates (though not by benzodiazepines). Tricyclic compounds potentiate the pressor effects of noradrenaline, adrenaline, and phenylephrine by preventing re-uptake (Boakes *et al.* 1973), and this is a potential hazard when local anaesthetics are used for dental surgery or other purposes. Tricyclic antidepressants also interfere with the effects of the antihypertensive agents bethanidine, clonidine, debrisoquine, and guanethidine. However, they do not interact with the β-adrenoceptor antagonists used to treat hypertension. Tricyclics may increase the action of warfarin. Interactions of tricyclic drugs with MAOIs are considered later.

Contraindications

Contraindications include agranulocytosis, severe liver damage, glaucoma, and prostatic hypertrophy. The drugs must be used cautiously in epileptic patients, in the elderly, and after coronary thrombosis.

Clinical use of tricyclic antidepressants

In the use of tricyclics the old adage is recommended that it is best to get to know one or two drugs well and stick to them. It is probably sufficient to be familiar with one sedating compound (for example amitriptyline) and one less sedating drug (for example nortriptyline). Other tricyclics can then be reserved for special purposes; for example *lofepramine can be reserved for patients who present the risk of overdose or of cardiotoxicity, whilst clomipramine can be reserved for patients in whom a depressive disorder is related to an obsessional illness.*

The prescribing of amitriptyline can be taken as an example. At the outset it is important to explain to patients that, while side-effects may be noticed early in treatment, any improvement in mood may be delayed for a week or more, and therefore it is important to persist. *Early signs of improvement may include better sleep and a lessening of tension.* Common side-effects should be mentioned because a forewarned patient is more likely to continue with medication.

The usual practice of starting with a low dose of amitriptyline and building up is probably wise, because side-effects are generally milder and patients are more likely to develop tolerance to them. The starting dose will depend to some extent on the patient's age, weight, physical condition, and history of previous exposure to tricyclics; daily doses of 25–50 mg for an out-patient and 50–75 mg for an in-patient would be reasonable. *The whole dose can be given at night about one to two hours before bedtime because the sedative effects of the drug will help sleep.*

Patients should be reviewed frequently in the first few weeks of treatment when support and advice are helpful both to maintain morale and to ensure compliance with medication. Often the clinician can detect improvements in rapport and initiative early in treatment. It can then be useful to discuss these changes with the patient. The dose of amitriptyline to be aimed at is about 150 mg daily. With careful monitoring and encouragement this dose can usually be reached over two weeks. In some patients, side-effects limit the rate of dosage increase, but if there is clinical improvement it is reasonable to settle for lower doses. In general, side-effects should not be greater than the patient can comfortably tolerate.

For patients who show little or no improvement, it is usually advisable to continue amitriptyline for four weeks at the maximum tolerated dose before deciding that the drug is ineffective. Some patients respond only to higher doses (up to 300 mg daily) (Quitkin 1985), and cautious increases towards this level are warranted provided that side-effects are tolerable. *In doses above 200 mg daily it is* wise to monitor plasma tricyclic levels and the electrocardiogram before each further dosage increase.

Plasma tricyclic concentrations have to be interpreted in the context of a patient's clinical condition, but generally *levels above 450 ng/ml are more likely to be associated with severe toxic reactions. In patients with concomitant medical disorders, a limit* of *300 ng/ml may be more appropriate* (Preskorn 1993). In the ECG it is important to note any evidence of impaired cardiac conduction, for example lengthening of the QT interval and the appearance of bundle branch block or arrhythmias. Because of the half-life of amitriptyline, each dose increase will take about a week to reach steady state. If the patient has not improved, and if he cannot tolerate an increase in dose or fails to respond to higher doses, then other treatments should be considered. Some possible strategies are outlined in Chapter 8 on mood disorders (see p. 239).

Maintenance and prophylaxis

If patients respond to amitriptyline, they should be maintained on treatment for at least three months because continuation therapy greatly reduces the risk of early relapse. The same dose of amitriptyline should be maintained if possible, but if side-effects become a problem the dose can be lowered until tolerance is again satisfactory. It is often not clear when antidepressant drug treatment should be withdrawn, because in some patients depression is a recurrent disorder. Long-term prophylactic treatment may then be justified. Obviously the risk of recurrence increases with the number of episodes that the patient suffers, but other clinical and biochemical predictors of relapse are not well established. It has been reported that, if patients have been entirely free of depressive symptoms for at least 16 weeks, they are most likely to do well when drug treatment is withdrawn (Prien and Kupfer 1986).

Selective serotonin reuptake inhibitors (SSRIs)

Structure and pharmacological properties

Four SSRIs—**fluoxetine, fluvoxamine, paroxetine,** and **sertraline**—are available at present for clinical use in the United Kingdom. SSRIs are a structurally diverse group, but they all inhibit the reuptake of 5-HT with **high potency** and **selectivity.** None of them has an appreciable affinity for the noradrenaline uptake site, and present data suggest that they have a very low affinity for monoamine neurotransmitter receptors.

Pharmacokinetics

The SSRIs are absorbed slowly and reach peak plasma levels after about four to eight hours. The half-lives of fluvoxamine, paroxetine, and sertraline are about 24 hours, while the half-life of fluoxetine is 48–72 hours. *The SSRIs are primarily eliminated by hepatic metabolism.* Fluoxetine is metabolized to norfluoxetine, which is also a potent 5-HT uptake blocker and has a half-life of seven to nine days. Sertraline is converted to desmethylsertraline, which has a half-life of two to three days and is 5–10 times less potent than the parent compound in inhibiting the reuptake of 5-HT. The contribution of desmethylsertraline to the antidepressant effect of sertraline during treatment is unclear.

Efficacy of SSRIs in depression

The SSRIs have been extensively compared with placebo and with reference tricyclic antidepressants. The SSRIs are all clearly superior to placebo and are generally as effective as tricyclics in the treatment of major depression. Most comparative studies have been of moderately depressed outpatients. There has been concern that SSRIs may be less effective than conventional tricyclic anti-

depressants for more severely depressed patients. For example the Danish University Antidepressant Group (1990) found that **clomipramine** was significantly more effective than either paroxetine or citalopram. However, meta-analyses usually show that SSRIs are at least as effective as tricyclic antidepressants for patients with 'melancholic' depression and for those with high scores on the Hamilton Depression Rating Scale (Montgomery 1992). To some extent this issue is clouded by the lack of consensus among clinicians as to the definition of 'severe depression'. It is worth noting that conventional tricyclic antidepressants as sole therapy may be of only limited benefit in patients with severe depression, particularly when there are psychotic symptoms (Spiker and Kupfer 1988).

Unwanted effects of SSRIs

The adverse effects of SSRIs differ significantly from those of tricyclic antidepressants (Table 17.9). A major difference is that *SSRIs are less cardiotoxic than TCAs and are much safer in overdose.* SSRIs also lack anticholinergic effects and are not sedating. Side-effects can be grouped as follows (Table 17.9). **Gastrointestinal:** nausea (in about 20 per cent of patients, though it may resolve with continued administration), dyspepsia, bloating, flatulence, and diarrhoea. Unlike tricyclic antidepressants, SSRIs are not usually associated with weight gain but may

Table 17.9   *Side-effects of SSRIs*

Gastrointestinal	**Common:** nausea, appetite loss, dry mouth, diarrhoea, constipation, dyspepsia **Uncommon:** vomiting, weight loss
Central nervous system	**Common:** headache, insomnia, dizziness, anxiety, fatigue, tremor, somnolence **Uncommon:** extrapyramidal reaction, seizures, mania
Other	**Common:** sweating, delayed orgasm, anorgasmia **Uncommon:** rash, pharyngitis, dyspnoea, serum sickness, hyponatraemia, alopecia

be associated with weight loss despite improvement in depression. **Neuropsychiatric:** insomnia, daytime somnolence, agitation, tremor, restlessness, and irritability. SSRIs have also been associated with confusion and mania. **Extrapyramidal side-effects** such as parkinsonism and akathisia may be more common during treatment with SSRIs than with tricyclics. Paroxetine has been associated with **acute dystonias** in the first few days of treatment. SSRIs have also been associated with **seizures.** It is possible that some SSRIs such as sertraline are less likely than tricyclics to be associated with seizures, but the data do not allow a firm conclusion at present.

There have been anecodotal reports that fluoxetine treatment may be associated with **hostile and suicidal behaviour.** A meta-analysis of controlled trials of fluoxetine found no increase in suicidal ideation or suicidal acts, as compared with placebo or tricyclic antidepressants. It is likely that some of the reported cases of restlessness and suicidal behaviour associated with fluoxetine are attributable to **akathisia** (Power and Cowen 1992). **Sexual dysfunction** including ejaculatory delay and anorgasmia are fairly common during SSRI treatment. **Sweating, headache,** and **dry mouth** are also reported. **Cardiovascular** side-effects are rare with SSRIs, but some reduction in pulse rate may occur and postural hypotension has been reported. Fluoxetine has been associated with skin rashes and, rarely, a more generalized allergic reaction with arthritis. SSRIs have been associated with a low sodium state secondary to inappropriate ADH secretion. As with tricyclic antidepressants, elevation of liver enzymes can occur but is generally reversible on treatment withdrawal.

Interactions with other drugs

The most serious interaction yet reported is that simultaneous administration of SSRIs and MAOIs has provoked a **5-HT toxicity syndrome** with hyperpyrexia, rigidity, myoclonus, coma, and death (Beasley *et al.* 1993) (for further details see the section on MAOIs). Other drugs that increase brain 5-HT function must be used with caution in combination with SSRIs; they include **lithium** and **tryptophan** which have been reported to be associated with mental state changes, myoclonus, and seizures. SSRIs may potentiate the induction of extrapyramidal movement disorders by antipsychotic drugs, although this effect could be partly due to a pharmacokinetic interaction whereby SSRIs increase plasma levels of certain antipsychotic drugs (see below). *SSRIs have been reported to inhibit the hepatic metabolism of some other drugs,* including **tricyclic antidepressants, antipsychotic drugs,** and **anticonvulsants.** Sertraline may cause fewer such interactions (Preskorn 1993). The anticoagulant effect of **warfarin** may be increased by SSRIs.

The clinical use of SSRIs in depression

Some authorities recommend that SSRIs should be used as the initial treatment in major depression; whilst others believe that (unless specifically contraindicated) tricyclic antidepressants should be used for this purpose because more is known about their efficacy and adverse effects. There is controversy as to whether drop out due to adverse effects is significantly less frequent in patients taking SSRIs than in those taking conventional tricyclic antidepressants (Song *et al.* 1993). In general, it seems reasonable to use tricyclic antidepressants as initial treatment for severely depressed patients, particularly in-patients and those with depressive psychosis. SSRIs are suitable alternatives to tricyclics for patients who may be vulnerable to anticholinergic effects or cardiotoxicity. *Particularly when the risk of suicide is high and when supervision of medication is not assured, the lower acute toxicity of SSRIs makes them preferable to conventional tricyclic antidepressants.* The lack of sedation makes SSRIs useful for out-patients who are striving to maintain their usual social and work activities. SSRIs may also be preferable for patients who gain excessive weight with tricyclic antidepressants. As mentioned above, *depression in association with an obsessional disorder is an indication for treatment with an SSRI or clomipramine* (Table 17.10).

At present it is difficult to draw clear clinical distinctions between the different SSRIs. Fluoxetine may be the most activating SSRI as judged from adverse event data, but this impression has

Table 17.10 *Indications for SSRI treatment in depression*

Concomitant cardiac disease*
Intolerance of anticholinergic effects
Significant risk of deliberate overdose
Likelihood of excessive weight gain
Sedation undesirable
Depression and obsessive–compulsive
  disorder

---

*SSRIs are safer than tricyclic antidepressants but should be used with caution.

not yet been confirmed by controlled comparative trials. Fluoxetine also has a distinctive pharmacokinetic profile in relation to its long-acting metabolite (see above) which may be involved in drug interactions several weeks after fluoxetine has been stopped. *For example at least five weeks should elapse between stopping fluoxetine and starting an MAOI.*

In treating depressive disorder, dosing is easier with SSRIs than with tricyclic antidepressants because most SSRIs can be started at a standard dose that can often be maintained throughout treatment. For example, although fluoxetine can be given in doses of up to 80 mg daily, there is little evidence of increasing therapeutic efficacy above the 20 mg dose. For example maintaining the 20 mg dose for 6 weeks is as effective in most patients as increasing it to 40 or 60 mg over a similar period (Power and Cowen 1992).

As with tricyclic antidepressant treatment, patients starting SSRIs should be warned about likely side-effects including **nausea** and some **restlessness during sleep**. A number of patients become more anxious and agitated during SSRI treatment; therefore it is important to explain that such effects are sometimes experienced during treatment but do not mean that the underlying depression is worsening. If patients persist with treatment, anxiety and agitation usually diminish, but short-term treatment with a benzodiazepine may be helpful, particularly if sleep disturbance is a problem. Small

doses of trazodone (50–150 mg) may also help sleep. *As with tricyclic antidepressants, when patients respond to SSRIs there is good evidence that continuing treatment for several months lowers the rate of relapse.* In addition, placebo-controlled studies have shown that SSRIs are effective in the prophylaxis of recurrent depressive episodes (Montgomery and Montgomery 1992). SSRIs should not be stopped suddenly as there have been reports of **withdrawal reactions** (insomnia, nausea, agitation, dizziness) after the cessation of paroxetine treatment.

Monoamine oxidase inhibitors (MAOIs)

MAOIs were introduced just before the tricyclic antidepressants but their use has been less widespread because of both troublesome interactions with foods and drugs and uncertainty about their therapeutic efficacy. Recent controlled studies have shown that in adequate doses MAOIs are useful antidepressants, *often producing clinical benefit in depressed patients who have not responded to other medication or ECT.* In addition MAOIs can be useful in refractory anxiety states (Nutt and Glue 1989). These beneficial effects have to be weighed against the need to adhere to strict dietary and drug restrictions in order to avoid reactions with tyramine and other sympathomimetic agents. *In practice this means that MAOIs are very rarely used as first-line treatment.* It remains to be seen whether this approach will be altered by the recent availability of MAOIs, such as moclobemide, that do not potentiate tyramine.

Pharmacological actions

MAOIs inactivate enzymes that oxidize noradrenaline, 5-HT, tyramine, and other amines that are widely distributed in the body as transmitters, or are taken in food and drink or as drugs. Monoamine oxidase (MAO) exists in a number of forms that differ in their substrate and inhibitor specificities. From the point of view of psychotropic drug treatment it is important to recognize that there are two forms of MAO (type A and type B) which are encoded by separate genes. In general, MAO-A metabolizes intraneuronal noradrenaline and 5-HT, while both MAO-A and MAO-B metabolize dopamine and tyramine.

Compounds available

Phenelzine is the most widely used and widely studied compound. **Isocarboxazid** is reported to have fewer side-effects than phenelzine, and can be useful for patients who respond to the latter drug but suffer from its side-effects of hypotension or sleep disorder. **Tranylcypromine** differs from the other compounds in combining the ability to inhibit MAO with an amphetamine-like stimulating effect which many patients welcome. Indeed, the drug is partly metabolized to amphetamine. It is sometimes used in combination with trifluoperazine as the proprietary preparation Parstelin (tranylcypromine 10 mg plus trifluoperazine 1 mg in each tablet), but there is no good reason to use this mixture. Some patients become dependent on the stimulant effect of tranylcypromine. Moreover, compared with phenelzine, tranylcypromine is more likely to give rise to hypertensive crises, though less likely to damage the liver. For these reasons, tranylcypromine should be prescribed with particular caution. **Moclobemide** is the most recently developed MAOI to be marketed. It differs from the other compounds in selectively binding to MAO-A, which it inhibits in a reversible way. This results in a lack of significant interactions with foodstuffs and a quick offset of action (see below).

Pharmacokinetics

Phenelzine, isocarboxazid, and tranylcypromine are rapidly absorbed and widely distributed. They have short half-lives (about two to four hours), as they are quickly metabolized in the liver by acetylation, oxidation, and deamination. People differ in their capacity to acetylate drugs; for example in the United Kingdom approximately 60 per cent of the population are 'fast acetylators' who would be expected to metabolize hydrazine MAOIs more quickly than 'slow acetylators'. Some studies have shown a better clinical response to phenelzine in 'slow acetylators', but this finding has not been consistently replicated (Paykel 1990).

However, *it may underlie the observation that the best response rate with MAOIs occurs in studies that have used higher dose ranges*, presumably because even patients who metabolize MAOIs quickly will receive an adequate dose. Phenelzine, isocarboxazid, and tranylcypromine bind irreversibly to MAO-A and MAO-B by means of a covalent linkage. Hence, the enzyme is permanently deactivated and MAO activity can be restored only when new enzyme is synthesized. *Thus, despite their short half-lives, irreversible MAOIs cause a long-lasting inhibition of MAO.*

In contrast with these compounds, moclobemide binds reversibly to MAO-A. This compound has a short half-life (about two hours), and therefore its inhibition of MAO-A is brief, declining to some extent even during the latter periods of the thrice-daily dosing regime. *Full MAO activity is restored within 24 hours of stopping moclobemide*; with the irreversible MAOIs, two weeks or more may be needed for synthesis of new MAO.

Efficacy of MAOIs in depression

For many years MAOIs were in relative disuse because several studies, in particular a large controlled trial by the Medical Research Council (Clinical Psychiatry Committee 1965), found them no better than placebo in the treatment of depressive disorders. It seems likely that the doses of MAOIs were too low in these early investigations; in the Medical Research Council study the maximum dose of phenelzine was 45 mg daily as against the current practice of doses up to 90 mg daily if side-effects permit. *Subsequent studies have shown that in this wider dose range MAOIs are superior to placebo and are generally equivalent to tricyclic antidepressants in their therapeutic activity* (Paykel 1990).

Recent investigations in the United States have confirmed early clinical impressions that MAOIs may be of particular value in the treatment of **atypical depression** (Quitkin *et al.* 1989) (see p.200). It also seems that MAOIs are more effective than tricyclic antidepressants for patients with **bipolar depression** if the clinical features include hypersomnia and anergia (Himmelhoch *et al.* 1991). Partly because of the concept that MAOIs are effective in atypical depression, it is often said that they are less effective than tricyclics in the treatment of more typical depressive disorders with characteristic vegetative changes. However, if adequate doses are given, MAOIs seem to be effective in the treatment of endogenous depression (Paykel 1990). There is also good evidence that they may be beneficial for depressed patients who do not respond to tricyclics, whether or not the depression has endogenous features (Cowen 1988).

Unwanted effects

These include dry mouth, difficulty in micturition, postural hypotension, headache, dizziness, tremor, paraesthesia of the hands and feet, constipation, and oedema of the ankle. Hydrazine compounds can give rise to hepatocellular jaundice (Table 17.11).

Interactions with foodstuffs and drugs

Some foods contain tyramine, a substance that is normally inactivated by MAO in the liver and gut wall. When MAO is inhibited, tyramine is not broken down and is free to exert its antihypertensive effects. These effects are due to release of noradrenaline with a consequent elevation in blood pressure. This may reach dangerous levels and may occasionally result in subarachnoid haemorrhage. Important early symptoms of such a crisis include a severe and usually throbbing headache.

The incidence of hypertensive reactions is about 8 per cent in patients taking MAOIs, even in those who have received dietary counselling (Davidson 1992). Therefore regular reminders about dietary restrictions may be helpful, particularly in patients on longer-term treatment. There have been reports of many foods being implicated in hypertensive reactions with MAOIs, but many of these have cited single cases and hence are of uncertain validity. Another complication is that the tyramine content of a particular food item may vary, as may the susceptibility of an individual patient to a hypertensive reaction. If a forbidden food has been consumed on one occasion without adverse effects, this does not preclude a future reaction. Davidson (1992) concluded that the following foods and drinks should be avoided:

- *all cheeses* except cream, cottage, and ricotta cheeses
- *red wine, sherry, beer, and liquors*
- *pickled or smoked fish*
- *brewer's yeast products* (for example Marmite, Bovril, and some packet soups)
- *broad bean pods* (such as Italian green beans)
- *beef or chicken liver*
- *fermented sausage* (for example, bologna, pepperoni, salami)
- *unfresh, overripe, or aged food* (for example pheasant, venison, unfresh dairy products).

Table 17.11 *Adverse effects of MAOIs*

Central nervous system	Insomnia, drowsiness, agitation, headache, fatigue, weakness, tremor, mania, confusion, convulsions (rare)
Autonomic	Blurred vision, difficulty in micturition, sweating, dry mouth, postural hypotension, constipation
Other	Sexual dysfunction, weight gain, peripheral neuropathy (pyridoxine deficiency), oedema, rashes, hepatocellular toxicity (rare), leucopenia (rare)

It is notable that about four-fifths of all reported reactions between foodstuffs and MAOIs, and nearly all the deaths, have followed the consumption of cheese (McCabe 1986). Hypertensive reactions should be treated with parenteral adminstration of an α_1-adrenoceptor antagonist, such as phentolamine. If this drug is not available, chlorpromazine can be used. Recently, the use of oral nifedipine has been advocated. Whatever treatment is given, blood pressure must be monitored carefully.

Moclobemide and tyramine reactions

Tyramine is metabolized by both MAO-A and MAO-B. Experimental studies have shown that *the hypertensive effect of oral tyramine is potentiated much less by moclobemide than by non-selective MAOIs*. In patients taking moclobemide the dose of tyramine required to produce a significant pressor response is above 100 mg. Even a five-course meal with wine would be unlikely to result in a tyramine intake of more than 40 mg (Simpson and De Leon 1989). It seems likely that the tyramine has relatively little effect in patients receiving moclobemide because MAO-B (present in the gut wall and liver) is still available to metabolize much of the tyramine ingested. Another factor may be that the interaction between moclobemide and MAO-A is reversible, thus allowing displacement of moclobemide from MAO when tyramine is present in excess.

Drugs

Patients taking MAOIs must not be given drugs whose metabolism depends on enzymes that are affected by the MAOI. These drugs include sympathomimetic amines such as adrenaline, noradrenaline, and amphetamine, as well as phenylpropanolamine and ephedrine (which may be present in proprietary cold cures). Dopa and dopamine may also cause hypertensive reactions. Antihypertensive drugs, such as methyldopa and guanethidine, and antihistamines are also affected. Local anaesthetics often contain a sympathomimetic amine and should also be avoided. Morphine, pethidine, procaine, cocaine, alcohol, barbiturates, and insulin can also be involved in dangerous interactions. Sensitivity to oral antidiabetic drugs is increased, with consequent risk of hypoglycaemia. The metabolism of barbiturates, phenytoin, and other drugs broken down in the liver may be slowed.

The 5-HT syndrome

A number of drugs that potentiate brain 5-HT function can produce a severe neurotoxicity syndrome when combined with MAOIs. The main features of this syndrome are as follows:

- **neurological symptoms:** myoclonus, nystagmus, headache, tremor, rigidity, seizures
- **mental state changes:** irritability, confusion, agitation, coma
- **other symptoms:** hyperpyrexia, cardiac arrhythmias, death.

Some of these symptoms resemble the **neuroleptic malignant syndrome** (see p. 554) with which 5-HT neurotoxicity is occasionally confused (Sternbach 1991; Beasley *et al.* 1993). In view of the interactions between dopamine and 5-HT pathways it is possible that similar mechanisms may be involved.

Current clinical data suggest that combination of MAOIs with **SSRIs, clomipramine,** and **fenfluramine** is contraindicated. The combination of MAOIs with **L-tryptophan** has also been reported to cause 5-HT toxicity. Adverse reactions have been reported between the $5-HT_{1A}$ receptor agonist, **buspirone,** and MAOIs. In general, the use of **lithium** with MAOIs seems safe and is often effective in patients with resistant depression (Cowen 1988). If a 5-HT syndrome develops, all medication should be stopped and supportive measures instituted. In theory, drugs with 5-HT receptor antagonist properties such as cypropeptadine or propranolol may be helpful, but formal studies have not been carried out.

Combination of MAOIs with tricyclic antidepressants

The combined use of MAOIs and tricyclic antidepressants fell into disuse because of the severe reactions associated with the 5-HT syndrome. Current views are that combination therapy is safe provided that the following rules are followed.

- *Clomipramine and imipramine are not used.* The most favoured tricyclics in combination with MAOIs are amitriptyline and trimipramine.

- *the MAOI and tricyclic are started together at low dosage, or the MAOI is added to the tricyclic.*

Adding tricyclics to MAOIs is more likely to provoke dizziness and postural hypotension.

The advantages and disadvantages of combined tricyclic and MAOI therapy have not been fully established. On the one hand, patients taking tricyclics with MAOIs are less likely to suffer from MAOI-induced insomnia; on the other hand, they are more likely to experience **postural hypotension** and troublesome **weight gain**. The combination is said to be useful in patients with resistant depression. Although formal studies have not been carried out in this patient group, there are case reports of patients for whom combined MAOI–tricyclic treatment was successful when either treatment alone had not been helpful. In the United States, **trazodone** is often used to ameliorate MAOI-induced insomnia; present experience suggests that this drug has the advantage that, in modest dosage, it can be safely added to MAOI treatment (Schatzberg and Cole 1991*b*). However, this combination is not recommended in the current United Kingdom data sheet for trazodone.

Contraindications

These include liver disease, phaeochromocytoma, congestive cardiac failure, and conditions that require the patient to take any of the drugs that react with MAOI.

Clinical use of MAOIs in depression

Because of the potential danger of drug interactions and the need for a tyramine-free diet, irreversible MAOIs are rarely used as first-line antidepressant agents. The exception may be when patients have previously shown a favourable response to these drugs as against other classes of antidepressants. Even in atypical depression, for which MAOIs may well be superior to tricyclic antidepressants, it is probably better to try a tricyclic or SSRI first because many patients will respond to this approach.

The clinical use of phenelzine can be taken as an example. Treatment should start with 15 mg daily increasing to 30 mg daily in divided doses (with the final dose not later than 5 p.m.) in the first week. *Patients should be given clear written instructions about foods to be avoided (see below) and should be warned to take no other medication unless it has been specifically checked with a pharmacist or doctor who knows that the patient is taking MAOIs.* As always, patients should be warned about the delay in therapeutic response (up to six weeks) and about common side-effects (sleep disturbance, dizziness).

TREATMENT CARD

Carry this card with you at all times. Show it to any doctor who may treat you other than the doctor who prescribed this medicine, and to your dentist if you require dental treatment.

INSTRUCTIONS TO PATIENTS

Please read carefully

While taking this medicine and for 10 days after your treatment finishes you must observe the following simple instructions: –

1 Do not eat CHEESE, PICKLED HERRING OR BROAD BEANS PODS.
2 Do not eat or drink BOVRIL, OXO, MARMITE or ANY SIMILAR MEAT OR YEAST EXTRACT.
3 Do not take any other MEDICINES (including tablets, capsules, nose drops, inhalations or suppositories) whether purchased by you or previously prescribed by your doctor, without first consulting your doctor or your pharmacist.
NB *Treatment for coughs and colds, pain relievers and tonics are medicines.*
4 Drink ALCOHOL only in moderation and avoid CHIANTI WINE completely.

Report any severe symptoms to your doctor and follow any other advice given by him.

M.A.O.I.

Preparared by the Pharmaceutical Society and the British Medical Association on behalf of the Health Departments of the United Kingdom.

Figure 17.3

In the second week the dose of phenelzine can be increased to 45 mg daily. At this stage a greater increase to 60 mg may produce a quicker response, but it is also associated with more adverse effects. Accordingly, if feasible, it is better to find out whether an individual patient will respond to lower doses (about 45 mg) before increments are made (up to 90 mg daily). If patients do not respond to 45 mg, the dose can be increased by 15 mg weekly if side-effects permit.

The response to MAOIs can often be sudden; over the course of a day or two the patient suddenly feels better. If there are signs of over-activity or excessive buoyancy in mood, the dose can be reduced and the patient monitored for signs of developing hypomania. Side-effects likely to be particularly troublesome are insomnia and postural hypotension. Insomnia is best managed by lowering the dose of MAOI if feasible. Otherwise, the addition of a benzodiazepine or trazodone (50–150 mg at night) can be helpful, although the latter drug can sometimes increase problems of dizziness and postural hypotension.

Postural hypotension can be a disabling problem with MAOIs. Again, dose reduction is worth considering. Various measures have been suggested, for example the use of support stockings, an increase in salt intake, or even the use of a mineralocorticoid. Of course, the latter two measures have their own adverse effects.

Withdrawal from MAOIs

Patients who respond to MAOIs have often suffered from disabling depression for many months or even years. For such patients the usual practice is to continue therapy for at least six months to a year. With MAOIs (but not tricyclic antidepressants), it is wise to lower the dose if the patient can tolerate the reduction without relapsing. Sudden cessation of MAOIs can lead to anxiety and dysphoria. Even gradual withdrawal can be associated with increasing anxiety and depression. *Clinical experience indicates that it is more difficult to stop MAOI than tricyclic antidepressant treatment.* An explanation for this difference may be that MAOIs may produce physical dependence in some patients; another possible explanation is that MAOIs are given to patients with chronic disabling disorders who frequently relapse. *It is emphasized that, because of the time taken to synthesize new MAO, two weeks should elapse between the cessation of irreversible MAOI treatment and the easing of dietary and drug restrictions.*

Treatment with reversible type A MAOI inhibitors

In their freedom from tyramine reactions and their quick offset of activity, the reversible type A MAOIs (such as moclobemide) have clear advantages over conventional MAOIs. As with all newer antidepressants, however, the therapeutic efficacy of moclobemide, particularly in severely ill patients, is not as well established as that of phenelzine or tranylcypromine. Also, it is not yet known whether moclobemide will prove effective for patients with the various forms of atypical depression and tricyclic-resistant depression for which conventional MAOIs can be useful. Issues such as long-term efficacy, toxicity, and withdrawal will also need to be fully examined. If future studies of these issues are favourable, it is likely that reversible type A MAOIs will replace the older irreversible non-selective inhibitors.

The starting dose of moclobemide is 150–300 mg daily, which can be increased to 600 mg over a number of weeks. Moclobemide is better tolerated than tricyclic antidepressants or irreversible MAOIs, but *side-effects such as nausea and insomnia occur in about 20–30 per cent of patients.*

Drug interactions of moclobemide At present it is recommended that moclobemide should not be combined with SSRIs or clomipramine. As 5-HT is metabolized by MAO-A, the combination of moclobemide with 5-HT-potentiating drugs could lead to a 5-HT syndrome (see above). Like the irreversible MAOIs, moclobemide may react adversely with opiates. Similarly, moclobemide may potentiate the pressor effects of sympathomimetic amines; therefore combined use should be avoided. It is not yet clear whether moclobemide can be safely combined with L-DOPA in patients with Parkinson's disease.

Other antidepressant drugs

Other antidepressant drugs are available for use in the United Kingdom. Their mechanism of action is such that they cannot easily be grouped with tricyclic antidepressants, SSRIs, or with MAOIs. These drugs also have differing adverse-event profiles. Therefore they are discussed individually below.

Mianserin

Mianserin is a quadricyclic compound with complex pharmacological actions. It has weak noradrenaline reuptake inhibiting effects, and is a fairly potent antagonist at several 5-HT receptor subtypes, particularly 5-HT$_2$ receptors. Mianserin is also a competitive antagonist at histamine H$_1$ receptors and α_1 and α_2-adrenoceptors. It is not a muscarinic cholinergic antagonist and is not cardiotoxic. *Because of these various actions mianserin has a sedating profile, but it is not anticholinergic and is relatively safe in overdose.*

Pharmacokinetics Mianserin is rapidly absorbed, and the peak plasma concentration occurs after two to three hours. Its half-life is 10–20 hours, and the entire daily dose can usually be given in a single administration at night.

Efficacy Controlled trials have shown that mianserin is superior to placebo in the management of depression, and comparative studies against imipramine and clomipramine have shown no difference in effect. These studies are difficult to assess because of the wide range of doses that have been used. Many early studies of mianserin used doses of 30–60 mg daily, whereas much higher doses of up to 200 mg daily have sometimes been advocated for in-patients.

Unwanted effects The main adverse effects of mianserin are **drowsiness** and **dizziness**, though these effects can be lessened by starting at a modest dosage and then increasing gradually. Significant cognitive impairment is more likely with mianserin than with SSRIs. **Weight gain** is a common problem. Dyspepsia and nausea have also been reported. Like tricyclics, mianserin appears to **lower seizure threshold** to some extent. Postural hypotension occurs occasionally. *The most serious adverse effect of mianserin is lowering of the white cell count, and fatal agranulocytosis has been reported.* These adverse reactions occur more commonly in elderly patients, though some authorities have suggested that the incidence of leucopenia with mianserin is no different to that seen with tricyclics (Cowen 1992). It is recommended that a blood count be obtained before starting mianserin treatment, and that the white cell count be monitored monthly for three months after treatment has started. Rare side-effects of mianserin include **arthritis** and **hepatitis**.

Drug interactions Mianserin can potentiate the effect of other central sedatives. Other significant drug interactions are rare, although in theory the α_2-adrenoceptor antagonist effects of mianserin could offset the hypotensive effects of α_2-adrenoceptor agonists such as clonidine.

Trazodone

Trazodone is a triazolopyridine derivative with complex actions on 5-HT pathways. Studies *in vitro* suggest that trazodone has some weak 5-HT reuptake inhibiting properties which are probably not manifest during clinical use; for example repeated administration of trazodone does not lower platelet 5-HT content. Trazodone has antagonist actions at 5-HT$_2$ receptors but its active metabolite *m*-chlorophenylpiperazine (*m*-CPP), is a 5-HT receptor agonist. Therefore the precise balance of effects on 5-HT receptors during trazodone treatment is difficult to determine and may depend on relative blood levels of the parent compound and metabolite. Trazodone also blocks postsynaptic α_1-adrenoceptors. Overall it has a distinct sedating profile.

Pharmacokinetics Trazodone has a short half-life (about 4–14 hours). It is metabolized by hydroxylation and oxidation, with the formation of a number of metabolites including *m*-CPP. During treatment plasma levels of *m*-CPP may exceed those of trazodone itself.

Efficacy Several placebo-controlled studies have shown that trazodone in doses of 150–600 mg is superior to placebo in the treatment of depressed patients. Trazodone also appears to have equivalent antidepressant activity to reference compounds such as imipramine. Many of these studies were carried out in moderately depressed out-patients, and the efficacy of trazodone in depressed in-patients is not as well established (Rudorfer and Potter 1989). Some workers have maintained that the efficacy of trazodone is improved if treatment is started at low doses (50 mg) and increased slowly to 300 mg over two to three weeks. Despite the short half-life of trazodone, once-daily adminstration of the drug is often sufficient. The drug is usually given in the evening to take advantage of its sedative properties. Doses above 300 mg daily are usually better given in divided amounts.

Unwanted effects The major unwanted effect of trazodone is **excessive sedation,** which can result in significant cognitive impairment. **Nausea** and **dizziness** are also reported, particularly if the drug is taken on an empty stomach. The α_1-adrenoceptor antagonist properties of trazodone may lower blood pressure to some extent, and postural hypotension has been reported. Trazodone is less cardiotoxic than conventional tricyclics, but there are reports that *cardiac arrhythmias may be worsened in patients with cardiac disease.* Nevertheless, trazodone is much less toxic in overdose than tricyclic antidepressants. The most serious side-effect of trazodone is **priapism.** This reaction is seen rarely (about 1 in 6000 male patients). It can cause considerable problems, requiring the local injection of noradrenaline agonists such as adrenaline or even surgical decompression. Long-term sexual dysfunction has sometimes resulted. It is recommended that male patients be warned of this potential side-effect and advised to seek medical help urgently if persistent erection occurs.

Drug interactions As with all sedative antidepressants trazodone may potentiate the sedating effects of alcohol and other central tranquillizing drugs. Studies in animals have raised the possibility that trazodone could attenuate the hypotensive effect of clonidine, but it is not known whether such an interaction occurs in humans.

Nefazodone

Nefazodone is related to trazodone but lacks α_1-adrenoceptor antagonist properties and is therefore not sedating. Like trazodone it has rather modest 5-HT reuptake blocking properties and is metabolized to the 5-HT receptor agonist, *m*-CPP.

Pharmacokinetics Nefazodone is rapidly absorbed and undergoes extensive first pass metabolism. It is highly protein bound. The principal metabolites are *m*-CPP and hydroxynefazodone which has similar pharmacological properties to nefazodone. Both nefazodone and hydroxynefazodone have relatively short half-lives (2–4 hours) which makes twice daily dosing necessary.

Efficacy Controlled trials in patients with major depression have shown that in doses of 400 mg and greater, nefazodone is more effective than placebo and generally equal in therapeutic activity to comparator drugs (Rickels *et al.* 1994). As with trazodone, these studies have focused on out-patients with moderate depressive disorders. Nefazodone is usually given in two divided doses starting at 200 mg daily with titration to 400 mg daily after about a week. The maximum dose is 600 mg daily.

Unwanted effects Nefazodone is generally well tolerated with the most common side-effects being headache, loss of energy, dizziness, dry mouth, nausea, and somnolence. It appears less cardiotoxic than tricyclic antidepressants and is probably safer in overdose. Nefazodone is less likely than the SSRIs to cause anxiety, insomnia, and sexual dysfunction.

Drug interactions Limited data are available on the potential of nefazodone to cause drug interactions. Co-administration of nefazodone and propranolol led to an increase in plasma *m*CPP levels. Nefazodone is reported to increase haloperidol levels and may also elevate plasma benzodiazepine concentrations.

Venlafaxine

Venlafaxine is a phenylethylamine derivative which produces a potent blockade of both 5-HT and noradrenaline re-uptake. In this respect the pharmacological properties of venlafaxine resemble those of clomipramine (see p. 562). However, unlike clomipramine and other tricyclic antidepressants, venlafaxine has a negligible affinity for other neurotransmitter receptor sites and so lacks sedative and anticholinergic effects (Feighner 1994). Venlafaxine has therefore been classified as a selective serotonin and noradrenaline re-uptake inhibitor (SNRI).

Pharmacokinetics Venlafaxine is well absorbed achieving peak plasma levels about 1.5–2 hours after oral administration. The half-life of venlafaxine is 3–7 hours but it is metabolized to O-desmethylvenlafaxine which has essentially the same pharmacodynamic properties as the parent compound and a half-life of 8–13 hours.

Efficacy Venlafaxine has been studied in both in-patients and out-patients with major depression and compared with placebo and active comparators. Current studies suggest that it is more effective than placebo and at least of equal efficacy to other available antidepressant drugs (Feighner 1994). Venlafaxine also appears to be effective in depressed in-patients, perhaps more so than fluoxetine (Clerc *et al.* 1994). This finding is of interest in view of the fact that clomipramine, another potent 5-HT and noradrenaline reuptake inhibitor, seems more effective than SSRIs in this patient group (Danish University Antidepressant Group 1990).

Venlafaxine has a wider dosage range than SSRIs, from 75–375 mg daily in two divided doses. The antidepressant efficacy of venlafaxine appears more robust at higher doses which may also confer a faster onset of action. However, higher doses are associated with a greater incidence of adverse effects. The usual starting dose of venlafaxine is 75 mg daily which may be sufficient for many patients. Upward titration can be considered where there is insufficient response, or if a faster onset of therapeutic activity is needed. It is also possible that venlafaxine may be of value in treatment-resistant patients (Nierenberg *et al.* 1994). In these circumstances it may provide a simpler approach than the combination of SSRI and tricyclic (see p. 241).

Unwanted effects The adverse effect profile of venlafaxine resembles that of SSRIs, with the most common adverse effects being nausea, headache, somnolence, dry mouth, dizziness, and insomnia. Anxiety and sexual dysfunction may also occur. Venlafaxine occasionally causes postural hypotension but in addition, dose-related increases in blood pressure can also occur. Blood pressure monitoring may be advisable in patients receiving more than 200 mg venlafaxine daily. Sudden discontinuation of venlafaxine has been associated with symptoms of fatigue, nausea, and dizziness. It is recommended that patients who received venlafaxine for 6 weeks or more should have the dose reduced gradually over at least one-week period. Preliminary evidence suggest that venlafaxine is less toxic in overdose than tricyclic antidepressants.

Drug interactions Unlike the SSRIs, venlafaxine appears to produce little effect on hepatic drug metabolizing enzymes and therefore should be less likely to inhibit the metabolism of co-administered drugs. Like other drugs that potently inhibit the uptake of 5-HT, venlafaxine should not be given concomitantly with monoamine oxidase inhibitors (MAOIs) because of the danger of a toxic serotonin syndrome. It is also recommended that 14 days should elapse after the end of MAOI treatment before venlafaxine is started and that 7 days should elapse after venlafaxine cessation before MAOIs are given.

Viloxazine

Viloxazine is a bicyclic tetrahydroxazine. It has modest inhibitory effects on noradrenaline re-uptake and may also potentiate some of the effects of 5-HT; however, inhibition of 5-HT reuptake does not seem to be involved. Viloxazine has no peripheral anticholinergic or antihistaminic effects.

Pharmacokinetics Peak plasma concentrations of viloxazine are reached about two hours after oral administration. The drug has a short half-life of two to five hours, and is extensively metabolized by elimination and conjugation. Viloxazine is usually given in two or three daily doses.

Efficacy Viloxazine has been studied in doses of 150–400 mg daily. A number of studies comparing viloxazine with imipramine have shown equivalent antidepressant activity; however, by current standards these studies were flawed by lack of clear diagnostic criteria and by inadequate ratings of clinical change. The efficacy of viloxazine against placebo is not clearly established.

Unwanted effects The major adverse effects of viloxazine are **nausea** and **vomiting**. Dry mouth can occur, although viloxazine has fewer anticholinergic effects than the tricyclics. **Drowsiness** and **headache** are also reported fairly frequently. Viloxazine appears less cardiotoxic than tricyclics and is probably safer in overdose. In animal studies viloxazine has some anticonvulsive properties, but in humans there are occasional reports of seizures being associated with its use.

Drug interactions Viloxazine may potentiate the sedative actions of other centrally active drugs. It also increases plasma levels of phenytoin and therefore may delay the hepatic metabolism of other drugs.

L-Tryptophan

L-Tryptophan is a naturally occurring amino acid, present in the normal diet; about 500 mg of tryptophan is consumed daily in the typical Western diet. Most ingested tryptophan is used for protein synthesis and the formation of nicotinamide nucleotides; only a small proportion (about 1 per cent) is synthesized to 5-HT via 5-hydroxtryptophan. Tryptophan hydroxylase, the enzyme that catalyses the formation of 5-HTP from L-tryptophan, is normally unsaturated with tryptophan. Accordingly, increasing tryptophan availability to the brain increases 5-HT synthesis.

Pharmacokinetics L-Tryptophan is rapidly absorbed, with plasma levels peaking about one to two hours after ingestion. It is extensively bound to plasma albumin. The amount of L-tryptophan available for brain 5-HT synthesis depends on several factors, including the proportion of L-tryptophan free in plasma, the activity of tryptophan pyrrolase, and the concentration of other plasma amino acids that compete with L-tryptophan for brain entry (for a review of these mechanisms see Bender (1982)).

Efficacy There is only weak evidence that L-tryptophan has antidepressant activity when given alone, though it may be superior to placebo in moderately depressed out-patients (Boman 1988). L-Tryptophan may be therapeutically effective only in a subgroup of depressed patients with low plasma tryptophan concentrations before treatment. There is rather better evidence that *L-tryptophan combined with MAOI treatment can enhance the antidepressant effects of MAOIs*. Similar synergistic effects have been reported in some studies of L-tryptophan combined with tricyclics, though overall the therapeutic benefit of this combination seems weak and inconsistent (Chalmers and Cowen 1990).

Unwanted effects L-Tryptophan is generally well tolerated, although **nausea** and **drowsiness** soon after dosing are not unusual. Recently, however, the prescription of L-tryptophan has been associated with the development of a severe scleroderma-like illness, the **eosinophilia–myalgia syndrome (EMS)**, in which there is a **very high circulating eosinophil count** (about 20 per cent of peripheral leucocytes) with severe **muscle pain, oedema, skin sclerosis and peripheral neuropathy**. Fatalities have been reported. It is now reasonably well established that EMS is not caused by L-tryptophan itself but rather by a contaminant formed in the manufacturing process used by a particular manufacturer (Waller *et al.* 1991). L-Tryptophan remains available for the treatment of severe refractory depression, when it can be used as an adjunct to other antidepressant medication (Barker *et al.* 1987). Patients receiving L-tryptophan require close supervision, including monitoring for possible symptoms of EMS and regular blood eosinophil counts. L-tryptophan should be withdrawn if there is any evidence that EMS may be developing.

Drug interactions The only significant drug interactions of L-tryptophan are *with drugs that also increase brain 5-HT function.* Thus, while administration of TRP with MAOIs may produce clinical benefit, there are also reports that this combination may lead to 5-HT neurotoxicity as described above. Similarly the combination of L-tryptophan with SSRIs has been reported to cause myoclonus, shivering, and mental state changes (Sternbach 1991).

Mood-stabilizing drugs

Three agents are grouped under this heading: **lithium, carbamazepine, and sodium valproate.** *These three drugs are effective in the prevention of recurrent affective illness and also in the acute treatment of mania.* Lithium also has useful antidepressant effects in some circumstances, but the antidepressant activity of carbamazepine and sodium valproate is less well established.

Lithium

The antidepressant effect of lithium alone is not striking, and is most readily demonstrated in bipolar depressed patients. However, lithium can be of *significant therapeutic benefit in combination with other antidepressant drugs such as tricyclic antidepressants and MAOIs when the latter have proved ineffective alone* (see Price (1989) for a review of the antidepressant effects of lithium given alone and in combination with other drugs). Lithium has also been shown to lower the frequency of aggressive behaviour in violent offenders and in patients with learning difficulties (for reviews see Cowen (1990) and Nilsson (1993).

Animal studies have shown that lithium has important effects on the intracellular signalling molecules or 'second messengers' that are activated when a neurotransmitter or agonist binds to a specific receptor. *At clinically relevant doses lithium inhibits the formation of cyclic adenosine monophosphate (cAMP) and also attenuates the formation of various inositol lipid derived mediators.* Through these actions lithium could exert profound effects on a wide range of neurotransmitter pathways, many of which use the above messenger systems. It has been proposed that the effects of lithium may be particularly marked when the turnover and recycling of second messengers is increased, and accordingly lithium may act preferentially to inhibit over-active neurotransmitters systems. Viewed from a higher level of organization, lithium produces striking enhancements in some aspects of brain 5-HT function, an action which has been related to its antidepressant and anti-aggressive activity (Lithium Mechanisms Study Group 1993).

Pharmacokinetics

Lithium is rapidly absorbed from the gut and diffuses quickly throughout the body fluids and cells. Lithium moves out of cells more slowly than sodium. It is removed from plasma by renal excretion and by entering cells and other body compartments. Therefore there is a rapid excretion of lithium from the plasma, and a slower phase reflecting its removal from the whole-body pool. Like sodium, lithium is filtered and partly reabsorbed in the kidney. When the proximal tubule absorbs more water, lithium absorption increases. Therefore dehydration causes plasma lithium concentrations to rise. Because lithium is transported in competition with sodium, more is reabsorbed when sodium concentrations fall. Thiazide diuretics increase sodium excretion without increasing that of lithium; hence they can lead to toxic concentrations of lithium in the blood.

Dosage and plasma concentrations

Because the therapeutic and toxic doses are close together, it is essential to measure plasma concentrations of lithium during treatment. Measurements should first be made after four to seven days, then weekly for three weeks, and then, provided that a satisfactory steady state has been achieved, once every six weeks. Subsequently lithium levels are often very stable, and plasma monitoring can be carried out at intervals of two to three months unless there are clinical indications for more frequent monitoring. After an oral dose, plasma lithium levels rise by a factor of two or three within about four hours. For this reason, concentrations are normally measured 12 hours after the last dose, usually just before the morning dose which can be delayed for an hour or two if necessary. It is important to follow this routine because published information about lithium concentrations refers to the level twelve hours after the last dose, and not to the 'peak' reached in the four hours after that dose. If an unexpectedly high concentration is found, it is important to establish whether the patient has inadvertently taken the morning dose before the blood sample was taken.

Previously, the accepted range for prophylaxis was 0.7–1.2 mmol/1 measured 12 hours after the last dose. However, Srinivasan and Hullin (1980) proposed that these levels were unnecessarily high. They suggested that concentrations in the range 0.5–0.8 mmol/1 were sufficient for prophylaxis, and that higher levels were required only for treatment of acute illness. This view is now accepted by many.

A study in the United States found that patients with lower lithium levels (0.4–0.6 mmol/1) had more affective illness during maintenance treatment than did patients with higher levels (0.8–1.0 mmol/1). However, they also experienced more side-effects (Gelenberg *et al.* 1989). In practice, it seems that many patients can be managed satisfactorily if their lithium levels are kept in the 0.5–0.8 mmol/1 range. *However, if a patient's course is unstable it may be worthwhile maintaining slightly higher lithium levels if side-effects permit.*

In the treatment of acute mania, plasma concentrations below 0.8 mmol/1 appear to be ineffective and a range of 0.8–1.2 mmol/1 is probably required (Prien *et al.* 1972). Serious toxic effects appear with concentrations above 2.0 mmol/1, though early symptoms may appear between 1.5 and 2.0 mmol/1.

A number of delayed-release preparations of lithium are now available, but their pharmacokinetics *in vivo* do not differ significantly from those of standard lithium carbonate preparations. A liquid formulation of lithium citrate is available for patients who have difficulty in taking tablets. Lithium may be administered once or twice daily. *Frequency of administration does not appear to affect urine volume* (O'Donovan *et al.* 1993). In general it is more convenient to take lithium as a single dose at night, but patients who experience gastric irritation on this regime may be helped by divided daily dosage.

Unwanted effects (Table 17.12)

A mild diuresis due to sodium excretion occurs soon after the drug is started. Other common effects include tremor of the hands, dry mouth, a metallic taste, feelings of muscular weakness, and fatigue. After the initial sodium diuresis, many patients develop poor renal concentrating ability, resulting in polyuria and polydipsia. A few patients develop a diabetes insipidus syndrome (see below). Polyuria can lead to dehydration with the risk of lithium intoxication so that patients should be advised to drink enough water to compensate for the fluid loss. Some patients, especially women, gain some weight when taking the drug. Persistent fine tremor, mainly affecting the hands, is common, but coarse tremor suggests that the plasma concentration of lithium has reached toxic levels. Most patients adapt to the fine tremor; for those who do not, propanolol 10 mg three times daily often reduces the symptom. Occasional cases of partial hair loss have been reported in the absence of hypothyroidism (Mortimer and Dawber 1984), and some cases of coarsening of the hair (McCreadie and Farmer 1985).

Thyroid gland enlargement occurs in about 5 per cent of patients taking lithium. The thyroid shrinks again if thyroxine is given while lithium is continued and it returns to normal a month or two after lithium has been stopped (Schou *et al.* 1968). Lithium interferes with thyroid production, and **hypothyroidism** occurs in up to 20 per cent of women patients (Lindstedt *et al.* 1977) with a compensatory rise in thyroid-stimulating hormone. Tests of thyroid function should be performed every six months to help to detect these changes, but these intermittent tests are no substitute for a continuous watch for suggestive clinical signs, particularly lethargy and substantial weight gain.

Table 17.12 *Some adverse effects of lithium and carbamazepine*

	Lithium	*Carbamazepine*
Neurological	Tremor, weakness, dysarthria, ataxia, impaired memory, seizures (rare)	Dizziness, weakness, drowsiness, ataxia, headache, visual disturbance
Renal/fluid balance	Increased urine output with decreased urine-concentrating ability, thirst, diabetes insipidus (rare), oedema	Acts to increase urine concentrating ability, low sodium states, oedema
Gastrointestinal/hepatic	Altered taste, anorexia, nausea, vomiting, diarrhoea, weight gain	Anorexia, nausea, constipation, hepatitis
Endocrine	Decreased thyroxine with increased TSH, goitre, hyperparathyroidism (rare)	Decreased thyroxine with normal TSH
Haematological	Leucocytosis	Leucopenia, agranulocytosis (rare)
Dermatological	Acne, exacerbation of psoriasis	Erythematous rash
Cardiovascular	ECG changes (usually clinically benign)	Cardiac conduction disturbance

TSH, thyroid-stimulating hormone.

If hypothyroidism develops and the reasons for lithium treatment are still strong, thyroxine treatment should be added. Lithium has been associated with elevated serum calcium levels in the context of hyperparathyroidism. Whether this is a chance finding is unclear.

Reversible **ECG changes** also occur. These may be due to displacement of potassium in the myocardium by lithium for they resemble those of hypokalaemia, with T-wave flattening and inversion or widening of the QRS. Other changes include a reversible **leucocytosis** and occasional papular or maculopapular rashes. There is some uncertain evidence that prolonged treatment may lead to osteoporosis in women.

Long-term effects on the kidney In 10 per cent of cases there is a persistent impairment of concentrating ability, and in a small number the syndrome of nephrogenic diabetes insipidus develops due to interference with the effect of antidiuretic hormone. This syndrome does not respond to antidiuretic treatments but usually recovers when the drug is stopped, though there are reports of persisting cases (Simon *et al.*1977). Structural changes have been reported in the kidneys of animals receiving toxic doses of lithium (Radomski *et al.* 1950), but these are much higher than the equivalent doses in humans. There have also been reports of tubular damage in patients on prolonged treatment (Herstbech *et al.* 1977). The incidence of pathological findings varies according to criteria for selection of the patients. Among those with impaired renal function, most show pathological changes in the tubules; in other patients about one in six show such changes (Bendz 1983).

Several follow-up studies have examined the effect of longer-term lithium maintenance treatment on glomerular function. Schou (1988) concluded that long-term lithium treatment, in the absence of toxic blood levels, does not result in a lowering of glomerular filtration rate. However, while lithium may not significantly lower mean glomerular filtration rate in groups of patients with bipolar illness, there are case reports of increases in plasma creatinine in lithium-treated subjects when other causes of nephrotoxicity appear to be absent. Frank renal failure, however, appears to be very rare, with only two probable cases being reported in the world literature (Gitlin 1993). Even so, with the current trends towards long-term prophylaxis of mood disorders, it is wise to monitor plasma creatinine levels regularly. It seems likely that the risk of nephrotoxicity will be minimized by maintaining plasma lithium levels at the lower end of the therapeutic range provided that they are therapeutically effective for the individual patient.

Effects on memory are sometimes reported by patients, who complain particularly of everyday lapses of memory such as forgetting well-known names. It is possible that this impairment of memory is caused by the mood disorder rather than by the drug itself, but there is also evidence that lithium can be associated with impaired performance on certain cognitive tests. For example with normal volunteers lithium has been shown to cause minor but reliably detectable memory deficits and to impair certain kinds of information processing (Glue *et al.* 1987).

Toxic effects These are related to dose. They include ataxia, poor coordination of limb movements, muscle twitching, slurred speech, and confusion. They constitute a serious medical emergency for they can progress through coma and fits to death. If these symptoms appear, lithium must be stopped at once and a high intake of fluid provided, with extra sodium chloride to stimulate an osmotic diuresis. In severe cases renal dialysis may be needed. Lithium is rapidly cleared if renal function is normal so that most cases either recover completely or die. However, a few cases of permanent neurological damage despite haemodialysis have been reported (von Hartitzsch *et al.* 1972).

Lithium crosses the placenta. Retrospective studies have found increased rates of abnormalities in the babies of mothers receiving lithium in pregnancy. For example a rate of 7 per cent has been reported, with most abnormalities affecting the baby's heart (Kallen and Tandberg 1983). However, a recent prospective study of 148 women found no increase in congenital malformations in patients exposed to lithium in the first trimester of pregnancy compared with matched controls (2.8 versus 2.4 per cent). The authors concluded that lithium did not appear to be an important human teratogen (Jacobsen *et al.* 1992). However, this conclusion was based on relatively few patients. *Clearly, it is desirable for patients to be medication free during the first trimester of pregnancy, and the decision whether or not to continue with lithium treatment must be carefully weighed.* Important factors include the likelihood of affective relapse if lithium is withheld and the difficulty that could be experienced in managing an episode of affective illness in the individual woman. If pregnant patients continue with lithium, plasma levels should be monitored closely. Ultrasound examination and fetal echocardiography are valuable screening tests as the pregnancy progresses. *Patients with a history of bipolar disorder have a substantially increased risk of psychotic relapse in the post-partum period.* In such patients it may be worth considering the introduction of lithium shortly after delivery to provide a prophylactic effect. However, it should be noted that lithium is secreted into breast milk and that significant concentrations of lithium can be measured in the plasma of breast-fed infants. Therefore bottle feeding is usually advisable.

Drug interactions

Plasma lithium levels may be increased by several drugs, which include the following:

- **thiazide diuretics** (potassium-sparing and loop diuretics seem less likely to increase lithium levels but should still be used with caution)

- **non-steroidal anti-inflammatory agents**

- **antibiotics** (metronidazole and spectinomycin)

- **antihypertensive agents** (angiotensin-converting enzyme inhibitors and methyldopa).

Lithium appears to potentiate **antipsychotic drugs in producing extrapyramidal movement disorders.** Occasionally confusion and delirium have been reported. Therefore, when combining lithium and antipsychotic drugs, it is prudent to use lower doses of the latter compounds. Lithium increases brain 5-HT function and in combination with SSRIs has led to 5-HT neurotoxicity, as shown by myoclonus, seizures, and hyperthermia. There are reports that the continuation of lithium during ECT may lead to **neurotoxicity.** If feasible, lithium treatment should be suspended or plasma levels reduced during ECT because the customary overnight fast beforehand may leave patients relatively dehydrated the following morning. If feasible, lithium treatment should be discontinued before major surgery because the effects of muscle relaxants may be potentiated.

Neurotoxicity has been reported following the combined use of **lithium and carbamazepine** and of **lithium- and calcium-channel blockers,** although plasma lithium levels are within the normal therapeutic range.

The effects of **digoxin** on cardiac conduction may be potentiated by lithium.

Contraindications

These include renal failure or recent renal disease, current cardiac failure or recent myocardial infarction, and chronic diarrhoea sufficient to alter electrolytes. It is advisable not to use lithium for children or, as explained above, in early pregnancy. Lithium should not be prescribed if the patient is judged unlikely to observe the precautions required for its safe use.

The management of patients on lithium

A careful routine of management is essential because of the effects of therapeutic doses of lithium on the thyroid and kidney, and the toxic effects of excessive dosage. The following routine is one of several that have been proposed and can be adopted safely. Successful treatment requires attention to detail, and so the steps are set out below at some length.

Before starting lithium a physical examination should be carried out, including the measurement of blood pressure. It is also useful to weigh the patient. Blood should be taken for estimation of electrolytes, urea, serum creatinine, haemoglobin, and a full blood count. When a particularly thorough evaluation is indicated, creatinine clearance is carried out, with an 18-hour collection usually being adequate. Thyroid function tests are also necessary. If indicated, an ECG and pregnancy tests should be performed as well.

If these tests show no contraindication to lithium treatment, the doctor should check that the patient is not taking any drugs that might interact with lithium. A careful explanation should then be given to the patient. He should understand the possible early toxic effects of an unduly high blood level, and also the circumstances in which this can arise, for example during intercurrent gastroenteritis, renal infection, or the dehydration secondary to fever. He should be advised that if any of these arise, he should stop the drug and seek medical advice. It is usually appropriate to include another member of the family in these discussions. Providing printed guidelines on these points is often helpful (either written by the doctor, or in one of the forms provided by pharmaceutical firms). In these discussions a sensible balance must be struck between alarming the patient by overemphasizing the risks and failing to give him the information that he needs to take a responsible part in the treatment.

Starting treatment: lithium should normally be prescribed as the carbonate, and treatment should begin and continue with a single daily dose unless there is gastric intolerance, in which case divided doses can be given. If the drug is being used for prophylaxis, it is appropriate to begin with 600–800 mg daily in divided doses, taking blood for lithium estimations every week and adjusting the dose until an appropriate concentration is achieved. A lithium level of 0.5 mmol/l (in a sample taken 12 hours after the last dose) may be adequate for prophylaxis, as explained above; if this is not effective, the previously accepted higher range of 0.8–1.0 mmol/l should be used. In judging response, it should be remembered that several months may elapse before lithium achieves its full effect.

As treatment continues, lithium estimations should be carried out every 6–12 weeks. It is important to have some means of reminding patients and doctors about the times at which repeat investigations are required. If a doctor is treating many patients with lithium, it is useful to keep a card index arranged in order of date to ensure that tests are not overlooked. Every six months, blood samples should be taken for electrolytes, urea, and creatinine, a full blood count, and the thyroid function tests listed above. The results should be recorded in tabular form in the patient's notes so that the results of successive estimations can be compared easily. If two consecutive thyroid function tests a month apart show hypothyroidism, lithium should be stopped or L-thyroxine prescribed. Troublesome polyuria is a reason for attempting a reduction in dose, while severe persistent polyuria is an indication for specialist renal investigation including tests of concentrating ability. A persistent leucocytosis is not uncommon and is apparently harmless. It reverses soon after the drug is stopped.

While lithium is given, the doctor must keep in mind the interactions that have been reported with psychotropic and other drugs (see above). It is also prudent to watch for toxic effects with extra care if ECT is being given. If the patient requires an anaesthetic for any reason, the anaesthetist should be told that the patient is taking lithium; this is because, as noted above, there is some evidence that the effects of muscle relaxant may be potentiated.

Lithium is usually continued for at least a year, and often for much longer. The need for the drug should be reviewed once a year, taking into account any persistence of mild mood fluctuations which suggest the possibility of relapse if treatment is stopped. Continuing medication is more likely to be needed if the patient has previously had several episodes of mood disorder within a short time, or if previous episodes were so severe that even a small risk of recurrence should be avoided. Some patients have taken lithium continuously for 15 years or more, but there should always be compelling reasons for continuing treatment for more than five years.

In some studies abrupt lithium withdrawal has been associated with the rapid onset of mania (Mander and Loudon 1988). Undoubtedly there is an increased risk of recurrent mood disorder after lithium discontinuation, probably because lithium is an effective prophylactic agent and because it is used for disorders with a high risk of recurrence. It is unclear whether there is also a distinct lithium withdrawal syndrome with 'rebound' mania (Schou 1993). *The risk of rapid relapse is lessened if lithium is discontinued slowly over a period of several weeks* (Faedda *et al.* 1993). Even patients who have remained entirely well for many years may experience a further episode of affective disorder after lithium discontinuation. There are worrying case reports of such patients not responding to the reintroduction of lithium treatment (Post *et al.* 1992).

Carbamazepine

Carbamazepine was originally introduced as an anticonvulsant and was found to have useful effects on mood in certain patients. Subsequently it was found to be beneficial in many bipolar patients, *including those who had proved refractory to lithium*. There is good evidence that carbamazepine is effective in the management of **acute mania** and also in the **prophylaxis of bipolar disorder.** Carbamazepine may also have some benefit in the treatment of drug-resistant depression, but it does not have an established role in the prophylaxis of recurrent unipolar depression. For a review of the role of carbamazepine in mood disorders see Post (1991).

Like certain other anticonvulsants, carbamazepine blocks neuronal sodium channels. It is unclear whether this action plays a role in the mood-stabilizing effects. Like lithium, carbamazepine facilitates some aspects of brain 5-HT function in human beings (Elphick *et al.* 1990).

Pharmacokinetics

Carbamazepine is slowly but completely absorbed and widely distributed. It is extensively metabolized, with at least one metabolite being therapeutically active. The half-life during long-term treatment is about 20 hours. Carbamazepine is a strong inducer of hepatic microsomal enzymes and can lower the plasma concentrations of other drugs.

Dosage and plasma concentrations

The dosage of carbamazepine in the treatment of mood disorders is similar to that used in the treatment of epilepsy, within the range of 400–1600 mg daily. Treatment is usually given in divided doses two or three times daily, because this practice may improve tolerance. No clear relationship has been established between plasma carbamazepine concentrations and therapeutic response, but it seems prudent to monitor levels (about 12 hours after the last dose) and to maintain them in the usual anticonvulsant range as a guard against toxicity.

Unwanted effects

Side-effects are common at the beginning of treatment. They include **drowsiness, dizziness, ataxia, diplopia, and nausea.** Tolerance to these effects usually develops quickly. A potentially serious side-effect of carbamazepine is **agranulocytosis**, though this complication is very rare (variously estimated from 1 in 10 000 to 1 in 125 000 patients). A relative leucopenia is more common, with the white cell count often falling in the first few weeks of treatment, though usually remaining within normal levels. **Rashes** occur in about 5 per cent of patients. **Elevations in liver enzymes** may also occur and, rarely, hepatitis has been reported. Carbamazepine can cause disturbances of **cardiac conduction** and therefore is contraindicated in patients with pre-existing abnormalities of cardiac conduction. Carbamazepine **lowers plasma thyroxine concentrations**, but thyroid-stimulating hormone levels are not elevated and clinical hypothyroidism is unusual.

Carbamazepine has also been associated with **low sodium states** (Post 1991). The unwanted effects of carbamazepine are compared with those of lithium in Table 17.12.

Drug interactions

Carbamazepine increases the metabolism of other drugs including tricyclic antidepressants, benzodiazepines, haloperidol, oral contraceptive agents, thyroxine, warfarin, and other anticonvulsants. A similar mechanism may underlie the decline in plasma carbamazepine levels that occur after the first few weeks of treatment. Carbamazepine levels may be increased by **SSRIs, erthyromycin, and isoniazid.** The pharmacodynamic effects and plasma levels of carbamazepine may be increased by some calcium-channel blockers such as **diltiazem** and **verapamil.** Conversely, carbamazepine may decrease the effect of certain other calcium-channel antagonists such as **felodipine** and **nicardipine.** **Neurotoxicity** has been reported when **carbamazepine** and **lithium** have been combined even in the presence of normal lithium levels. On theoretical grounds the manufacturers of carbamazepine recommend that combination of **carbamazepine** with **MAOIs** be avoided. However, there are case reports of these drugs being used safely together. It is possible that some MAOIs may increase plasma carbamazepine levels.

Clinical use of carbamazepine

The usual indication for carbamazepines is the prophylactic management of bipolar illness in patients for whom lithium treatment is ineffective or poorly tolerated. In particular, carbamazepine may be more effective than lithium in the treatment of patients with frequent mood swings, the so-called 'rapid cyclers'. Carbamazepine can be added to lithium treatment in patients who have shown a partial response to the latter drug; in these circumstances it is important to remember that this combination can cause neurotoxicity. Carbamazepine may also be useful in the acute treatment of mania, again usually as an alternative to lithium (Gerner and Stanton 1992).

If clinical circumstances permit, it is preferable to start treatment with carbamazepine slowly at a dose of 100–200 mg daily, increasing in steps of 100–200 mg twice weekly. Patients show wide variability in the blood levels at which they experience adverse effects; accordingly it is best to titrate the dose against the side-effects and the clinical response. Because of the risk of a lowered white cell count, it is prudent to monitor the count in the first three months of treatment. Patients should be instructed to seek help urgently if they develop a fever or other sign of infection. When patients have responded to the addition of carbamazepine to lithium, it is possible subsequently to attempt a cautious lithium withdrawal. However, the current clinical impression is that, for many patients, the maintenance of mood stability requires continuing treatment with both drugs.

Sodium valproate

Like carbamazepine, **sodium valproate** was first introduced as an anticonvulsant. In recent years there has been increasing interest in using the drug in the management of mood disorders. There have been controlled studies suggesting that valproate is effective in the **acute management of mania**, but there are no controlled trials of its efficacy in the prophylaxis of bipolar disorder (Gerner and Stanton 1992). There have been numerous case studies and open studies that have reported useful prophylactic effects of valproate in patients unresponsive to lithium and carbamazepine, including those with rapid cycling mood disorders (McElroy *et al.* 1987; Post 1991).

Valproate is a simple branch-chain fatty acid with a mode of action that is unclear. However, there is some evidence that it can slow the breakdown of the inhibitory neurotransmitter GABA. This action could account for the anticonvulsant properties of valproate, but whether it also underlies the psychotropic effects is unclear.

Pharmacokinetics

Valproate is rapidly absorbed, with the peak plasma concentrations occurring about two hours after ingestion. It is widely and rapidly distributed and has a half-life of 8–18 hours. Valproate is metabolized in the liver to produce a wide variety of metabolites, some of which have anticonvulsant activity. Unlike carbamazepine, valproate does not induce hepatic microsomal enzymes and, if anything, tends to delay the metabolism of other drugs.

Dosing and plasma concentrations

Valproate can be started at a dose of 400–600 mg daily, which may be increased once or twice weekly to a range of 1–2 g daily. Plasma levels of valproate do not correlate well with either the anticonvulsant or the mood-stabilizing effects, but it has been suggested that efficacy in the treatment of mood disorders is usually apparent when plasma levels are greater than 50 μg/ml (McElroy *et al.* 1992).

Unwanted effects

Common side-effects with valproate include **gastrointestinal disturbances, tremor, sedation, and tiredness.** Other troublesome side-effects include weight gain and transient hair loss with changes in texture on regrowth. Patients taking valproate may have some **elevation in hepatic transaminase enzymes;** provided that this increase is not associated with hepatic dysfunction the drug can be continued while enzyme levels and liver function are carefully monitored. However, there have been several reports of **fatal hepatic toxicity** associated with the use of valproate; thus far, these reports have been confined to children taking multiple anticonvulsant drugs. Valproate must be withdrawn immediately if vomiting, anorexia, jaundice, or sudden drowsiness occur. Valproate may also cause **thrombocytopenia** and may **inhibit platelet aggregation. Acute pancreatitis** is another rare but serious side-effect, and increases in **plasma ammonia** have also been reported. Other possible side-effects include oedema, amenorrhoea, and rashes.

Drug interactions

Valproate **potentiates** the effects of **central sedatives**. It has been reported to increase the side-effects of other anticonvulsants (without necessarily improving anticonvulsant control). It may increase plasma levels of **phenytoin**.

Clinical use

At present valproate can be considered for patients with refractory bipolar disorder who do not respond to either lithium or carbamazepine or who are intolerant of these drugs. Valproate does not have a product licence for the treatment of mood disorders. Before prescribing valproate there should be both a careful review of other treatment options and also appropriate discussion with the patient and family. In general, *valproate seems more effective in the prophylaxis of manic than depressive episodes.* It has not been established whether valproate is valuable in the management of resistant depression. It has often been used in combination with lithium for patients who have shown a partial response to lithium, and this combination appears to be safe. Valproate has also been used in combination with carbamazepine. For patients who continue to show episodes of mood disturbance, valproate can be combined with antidepressant or antipsychotic drugs. For reviews of the use of valproate in the treatment of mood disorders see Post (1991) and McElroy *et al.* (1992).

Antiepileptic drugs

In the United Kingdom, the treatment of epilepsy is usually undertaken by neurologists. Psychiatrists need to be aware of the general lines of drug management for two reasons: first, patients with epilepsy can have coexisting psychiatric disorders; second, antiepileptic drugs can produce changes in mood and behaviour. The following section summarizes some properties of major drugs currently used to treat epilepsy.

Compounds available

Carbamazepine and **valproate** have been reviewed in the section on mood stabilizers. **Phenytoin** is a hydantoin, which was introduced many years ago but is still important in the management of epilepsy. The use of barbiturates is declining because of their numerous adverse effects. However, **phenobarbitone** and **primodone**, are still occasionally used. **Ethosuximide** is effective in the treatment of absence seizures. **Benzodiazepines**, such as **clonazepam** and **clobazam**, are generally used as second-line drugs when partial seizures or secondary generalized seizures are difficult to control. They may also be used in atypical absence, atonic, and tonic seizures. Three recently introduced compounds—**vigabatrin, lamotrogine,** and **gabapentin**—are available for the treatment of generalized and partial seizures that do not respond well to first-line drugs.

The mode of action of antiepileptic drugs is not well understood. Drugs such as benzodiazepines, valproate, and vigabatrin may act by *increasing brain GABA function.* Lamotrogine may decrease the activity of the *excitatory amino acid neurotransmitter glutamate.* Other compounds such as carbamazepine and phenytoin appear to act at the level of neuronal membranes, perhaps via *blockade of sodium channels.* The mode of action of gabapentin is unclear. Despite its structural resemblance to GABA, this drug does not clearly enhance GABA function. It appears to bind to a specific protein on neurones, but the nature of this receptor is not known.

Choice of drug and type of seizure (Table 17.13)

In treating epilepsy, the choice of drug is based more on freedom from adverse effects than on any differences in effectiveness in controlling seizures. Carbamazepine is the drug of choice for partial (otherwise called focal) seizure, whether complex or simple in type, and phenytoin or sodium valproate are the main alternatives. Phenytoin has a narrow optimal dosage range for seizure control and is more likely than carbamazepine to give rise to adverse effects. For tonic–clonic

Table 17.13 *Classification of seizures and drugs of choice*

Type of seizure	First choice	Others
Partial or focal (whether simple or complex)	Carbamazepine	Phentoin or sodium valproate
Generalized tonic–clonic	Carbamazepine or sodium valproate	Phenytoin
Absence seizures	Sodium valproate	Ethosuximide
Myoclonic and atonic	Sodium valproate	Clonazepam

generalized seizures, the first choice lies between carbamazepine and sodium valproate. Sodium valproate is the first choice for absence seizures, with ethosuximide as the alternative. Until recently myoclonic and atonic seizures did not respond well to antiepileptic drugs, but some can now be controlled with sodium valproate or clonazepam.

At present lamotrogine, vigabatrin, and gabapentin are used as add-on therapies in patients with partial seizures and secondarily generalized tonic–clonic seizures that are not well controlled by first-line agents.

Drugs used in status epileptics

Diazepam given intravenously is the drug of first choice. Care must be taken to avoid respiratory depression and venous thrombophlebitis. As a rule, diazepam is not effective in status epilepticus when injected intramuscularly, but it can be given effectively by rectal infusion when entry to a vein is difficult (Munthe-Kaas 1980). If diazepam fails, an intravenous infusion of chlormethiazole should be used. Paraldehyde was the mainstay of treatment in the past, but until recently was out of fashion. It is now being used increasingly when diazepam fails. It can be given intramuscularly, rectally, or by intravenous infusion, though it is recommended that this latter route be used only in intensive care settings. If a plastic syringe is used, the drug must be given as soon as it has been drawn up. If status epilepticus persists despite these measures, intravenous phenytoin, with ECG

monitoring (because of the danger of cardiac arrhythmia), or phenobarbitone may be tried. These latter measures should not be taken without advice from a neurologist unless the circumstances are exceptional. Details of dosage will be found in the *British national formulary* or comparable handbooks; a useful discussion of the treatment of status epilepticus is given by Rimmer and Richens (1982).

Unwanted effects

All antiepileptic drugs are potentially harmful and must be used with care. Because adverse effects differ between the many compounds in use, only general guidance can be given here. Before prescribing, it is important to study carefully a work of reference such as the *British national formulary*.

The unwanted effects of carbamazepine and valproate have been described above. **Phenytoin** has many adverse effects. If commonly causes gum hypertrophy. Acne, hirsutism, and coarsening of the facial features are sufficiently frequent to demand caution in its use. In the nervous system cerebellar signs occur (ataxia, dysarthria, nystagmus) and indicate overdosage; among children intoxication may occur without these signs and therefore may be missed. High plasma concentrations (above 40 $\mu g/$ml) may result in an acute organic mental disorder. According to Glaser (1972) phenytoin can cause an encephalopathy, of which one feature is an increase in seizure frequency. Uncommon haematological effects include a megaloblastic anaemia related to folate deficiency, leucopenia, thrombocytopenia, and agranulocytosis. Serum calcium may be lowered. Reynolds (1968) has suggested that the mental side-effects of phenytoin are due to folate deficiency. However, the evidence for this is not convincing (Grant and Stores 1970).

In the treatment of epilepsy the unwanted effects of **phenobarbitone** include drowsiness and irritability and, in larger doses, slurred speech and ataxia. In children hyperactivity and emotional upsets are frequent, and impaired learning and skin rashes can occur. *For these reasons the drug should be avoided if practicable.* **Primodone** is metabolized to **phenobabarbitone** in the liver, and so their adverse effect profiles are similar.

Ethosuximide has been reported to cause a wide range of side-effects including gastrointestinal disturbances, drowsiness, ataxia, and mood changes. Changes in hepatic function and blood dyscrasia may occur rarely. **Vigabatrin** is associated with a high incidence of behavioural and mood disturbances, including drowsiness, fatigue, irritability, and depression. Less commonly, confusion, aggression, and psychotic reactions have been described. **Lamotrogine** may cause diplopia, blurred vision, and drowsiness, as well as irritability and aggression. The use of **gabapentin** has been associated with somnolence, dizziness, and ataxia. Fatigue, nystagmus, and headache have also been reported.

Teratogenic effects

The use of antiepileptic drugs is associated with an increased risk of teratogenesis. However, if the patient is clearly helped by such agents, the benefits of continued treatment generally outweigh the risk to the fetus. The risk is least if a single drug is used. Screening during pregnancy is advisable, and mothers taking phenytoin and carbamazepine should receive folate supplements. The use of phenytoin and valproate is associated with an increased risk of bleeding in the newly born child. In general antiepileptic drugs are not considered to be a contraindication to breast-feeding, except perhaps for barbiturates. *The baby should be monitored carefully for signs of sedation.*

Drug interactions

The drug interactions of carbamazepine and sodium valproate have been discussed in detail above. *Antiepileptic agents often interact significantly with other drugs.* Carbamazepine, phenytoin, and barbiturates have the effect of **inducing hepatic microsomal enzymes** which accelerate the metabolism of other drugs and can lower their plasma levels. Examples include tricyclic antidepressants and clozapine, as well as exogenously administered hormones such as corticosteroids, thyroxine, and oestrogen (which may render the contraceptive pill ineffective). Conversely, plasma levels of phenytoin can be increased by drugs such as fluoxetine, cimetidine, and tolbutamide. Antiepileptic drugs also potentiate the central effects of other psychotropic agents such as lithium, with resultant neurotoxicity. *Given together, antiepileptic drugs can produce an increased risk of toxicity and sedation without improving the anticonvulsant effect.* Monitoring plasma levels of antiepileptic drugs is important, especially when the introduction of new drugs may alter plasma levels. If a clinician is not fully familiar with the possible drug interactions of a particular antiepileptic agent, it is advisable to consult a reference work such as the *British national formulary* before prescribing this agent in combination with another drug.

Contraindications

These are few in number and depend on the particular drug. They should be checked carefully before prescribing a compound with which the doctor is not already familiar. It should be noted especially that phenobarbitone has a limited place in treating epilepsy, particularly among children and psychiatric patients, because it frequently leads to disturbed behaviour. In patients who have renal or hepatic disease, antiepileptic drugs must be given cautiously.

Management

The psychiatrist is more likely to be involved in maintaining established treatment for a patient with epilepsy than in starting treatment for a newly diagnosed patient. In some patients the epilepsy and the psychiatric disorder will be unrelated. Other patients will have psychiatric symptoms that are secondary to the epilepsy or its treatment. Adverse behavioural effects of treatment occur particularly with barbiturates but also with overdosage of any epileptic drug. The psychiatrist will usually be taking over treatment of an established condition from a general practitioner or neurologist, and he should normally discuss the case with them before making any changes. If the psychiatrist initiates treatment of a new case he should remember that the treatment is likely to continue for years; hence discussion with a specialist as well as the family doctor will usually be appropriate. Drug treatment is not indicated for a single seizure (though the cause must be investigated).

It is good practice to prescribe only one anti-epileptic drug at a time and to adjust its dose carefully. Sudden changes in dosage are potentially hazardous, for they may cause status epilepticus. The drug chosen should be known to be effective for the particular type of epilepsy presented by the patient (see above). If this first choice fails, a second can be tried, again given on its own. With the range of preparations now available it should be uncommon to combine two drugs, and most exceptional to use more than two. Whenever combinations are used, careful consideration must be given to possible interactions.

Throughout treatment a careful watch should be kept for the particular side-effects of the drug in use. At the same time the doctor should make sure that the patient is continuing to comply with the dosage schedule. He should warn the patient about the dangers of suddenly stopping taking the tablets (an important cause of status epilepticus). If it becomes necessary to change from one drug to another, the new drug should be introduced gradually until its full dosage is reached. Only then should the old one be phased out.

Once an effective regime is established, it should be continued until there has been freedom from seizures for at least two years. Plasma concentration should be monitored regularly, except in the case of valproate when it is not considered routinely helpful. Additional plasma estimations may be useful if there are changes in seizure control or mental state, or if another drug is added to the treatment regime. When drugs are eventually withdrawn this should be done gradually.

In the United Kingdom epileptic patients may drive a private motor vehicle, but not a public service or heavy goods vehicle, provided that they have experienced no epileptic attacks for at least two years or epileptic attacks only while asleep during a period of at least three years. However, patients whose epilepsy can be controlled only at the expense of drowsiness should not drive. If there is doubt, advice should be obtained from a consultant with special experience in the treatment of epilepsy. For a review of epilepsy and driving see O'Brien (1986).

Psychostimulants

This class of drugs includes mild stimulants, of which the best known is **caffeine,** and more powerful stimulants such as **amphetamine** and **methylphenidate. Pemoline** has intermediate effects. **Cocaine** is a powerful psychostimulant with a particularly high potential for inducing dependence (see p. 475). It is useful as a local anaesthetic but has no other clinical indications. Psychostimulants increase the release and block the reuptake of dopamine and noradrenaline.

Amphetamines were used for numerous conditions in the past, but they are now prescribed much less frequently because of the high risk of dependence. They are not appropriate for the treatment of obesity. In adults the agreed indication for amphetamines is **narcolepsy.** Amphetamines are also used to treat the **hyperkinetic syndrome** of childhood. In the past amphetamines were widely prescribed for the treatment of depression, but they have been superseded by the antidepressant drugs. Some specialists, mainly in the United States, believe that psychostimulants may have a role either as sole agent or in combination with other antidepressant drugs for patients with **refractory depressive disorder.** Also, there is some interest in using psychostimulants for **elderly depressed patients** with **concomitant medical illness** (Gurevitch *et al.* 1991). A blanket proscription of psychostimulant treatment in depression is unjustified. Psychostimulants should be used only by practitioners with special experience in the psychopharmacological management of resistant depression.

The main **preparations** are dexamphetamine sulphate, given for narcolepsy in divided doses of 10 mg daily increasing to a maximum of 50 mg daily in steps of 10 mg each week, and methylamphetamine hydrochloride which has similar effects.

Unwanted effects

These include restlessness, insomnia, poor appetite, dizziness, tremor, palpitations, and cardiac arrhythmias. **Toxic effects** from large doses include disorientation and aggressive behaviour, hallucinations, convulsions, and coma. Persistent abuse can lead to a paranoid state similar to paranoid schizophrenia (Connell 1958). Amphetamines **interact** dangerously with MAOIs. They are **contraindicated** in cardiovascular disease and thyrotoxicosis.

Electroconvulsive therapy (ECT)

Convulsive therapy was introduced in the late 1930s on the basis of the mistaken idea that epilepsy and schizophrenia do not occur together. It seemed to follow that induced fits should lead to improvement in schizophrenia. However, when the treatment was tried it became apparent that the most striking changes occurred not in schizophrenia but in severe depressive disorders, in which it brought about a substantial reduction in chronicity and mortality (Slater 1951). At first, fits were produced either by using cardiazol (Meduna 1938) or by passing an electric current through the brain (Cerletti and Bini 1938). As time went by, electrical stimulation became the rule. The subsequent addition of brief anaesthesia and muscle relaxants made the treatment safe and acceptable.

Indications

This section summarizes the indications for ECT. Further information about the efficiency of the procedure will be found in the chapters dealing with the individual psychiatric syndromes.

ECT is a rapid and effective treatment for severe **depressive disorders**. In the Medical Research Council trial (Clinical Psychiatry Committee 1965) it acted faster than imipramine or phenelzine, and was more effective than imipramine in women and more effective than phenelzine in both sexes. These findings accord with the impression of many clinicians, and with the recommendations of this book, that ECT should be used mainly when it is essential to bring about improvement quickly. Therefore, the strongest indications are an immediate high risk of suicide, depressive stupor, or danger to physical health because the patient is not drinking enough to maintain adequate renal function.

In the United Kingdom recent placebo-controlled studies have confirmed that ECT is particularly effective for patients with **depressive psychosis** or definite **psychomotor disturbance** (Buchan *et al.* 1992). In addition, there is evidence that treatment with ECT can be beneficial in depressed patients for whom **antidepressant medication** has been unsuccessful (Chalmers and Cowen (1990).

ECT is also effective in the treatment of the **affective psychoses** that follow **childbirth**. These puerperal psychoses often present with mixed affective features that can be difficult to resolve quickly with psychotropic drug treatment. ECT may often prove rapidly effective for such patients, a matter of some importance in the early development of the relationship between mother and baby. ECT is also effective in the treatment of **mania** (Small *et al.* 1988), but is generally reserved for patients who do not respond to drug treatment or for those whose manic illness is severe, requiring high doses of antipsychotic drugs.

On the basis of clinical case studies it has long been held that ECT is useful in the treatment of **acute catatonic states** and **schizoaffective disorders**. Recent controlled studies have also shown that ECT is effective in patients with **acute schizophrenia** with predominantly positive symptoms. In these studies ECT is effective not only for affective symptoms but also for positive symptoms such as delusions and thought disorder (Brandon *et al.* 1985). In general, however, ECT adds little to the effects of adequate doses of antipsychotic drugs, though it probably produces a greater rate of symptomatic improvement in the short-term. The role of ECT is unclear for schizophrenic patients with positive symptoms who do not respond to antipsychotic medication. Several older studies suggest that ECT does not improve the negative symptoms of schizophrenia. The current evidence for the efficacy of ECT for different psychiatric disorders is well discussed in the *ECT handbook* of the Royal College of Psychiatrists (1995).

Mode of action

Presumably, the specific therapeutic effects of ECT must be brought about through physiological and biochemical changes in the brain. The first step in identifying the mode of action must be to find out whether the therapeutic effect depends on the seizure, or whether other features of the treatment are sufficient, such as the passage of the current through the brain and the use of anaesthesia and muscle relaxants. Clinicians have generally been convinced that the patient does not improve unless a convulsion is produced during ECT procedure. This impression has been confirmed by several double-blind trials which, taken together, show that ECT is strikingly more effective than a full placebo procedure that includes anaesthetic and muscle relaxant (Freeman *et al.* 1978; Johnstone *et al.* 1980; West 1981; Brandon *et al.* 1984; Gregory *et al.* 1985). *This evidence does not support the notion that a full seizure is the sufficient and necessary therapeutic component of ECT*, and recent studies have shown that this notion is incorrect. Modern ECT machines deliver brief pulses of electrical current that enable a seizure to be induced by administration of relatively low doses of electrical energy. With this mode of administration, both electrode placement and electrical dosage can have profound effects on the therapeutic efficacy of ECT. A recent study by Sackheim *et al.* (1993) examined the effect of four ECT treatment regimes in severely depressed patients. These treatment regimes were as follows:

- low electrical dose with right unilateral electrode placement

- high electrical dose with right unilateral electrode placement
- low electrical dose with bilateral electrode placement
- high electrical dose with bilateral electrode placement.

The results showed that *low-dose right unilateral treatment was significantly less effective than high-dose right unilateral treatment,* even though the two procedures produced identical seizures as determined by EEG monitoring. *Both the bilateral electrode placements were more therapeutically effective than the unilateral placements, and bilateral high-dose treatment was marginally more effective than low-dose bilateral treatment in speed of effect.* These findings have important implications for the practical management of ECT when the clinician's aim is to find the best balance between therapeutic efficacy and cognitive side-effects (see below). From the theoretical viewpoint, however, it can be concluded that an important determinant of ECT efficacy is how far the applied electrical energy exceeds the seizure threshold of the individual patient. Also it is clear that bilateral electrical stimulation is more effective than unilateral stimulation even though both produce bilateral convulsions.

Electrical seizures in animals produce many biochemical and electrophysiological changes, and therefore it is difficult to identify the processes that are important in the antidepressant effect of ECT. It is of interest that some of the changes in noradrenaline pathways found in rodents after ECT (for example the down-regulation of α_2- and β_1-adrenoceptors) are also found after antidepressant drug treatment (Green *et al.* 1986). *However, ECT also produces striking changes in some aspects of brain dopamine function which are not seen after treatment with antidepressants* (Grahame-Smith *et al.* 1978). It may be that the effect of ECT on dopamine neurotransmission is responsible for its beneficial effects in retarded depression. It is notable that ECT may also have a useful, though somewhat transient, effect in patients with Parkinson's disease (Abrams 1989).

During a course of ECT the seizure threshold of patients tends to increase (Sackheim *et al.* 1993). This change occurs more with bilateral than with unilateral electrode placement, suggesting that the processes underlying the seizure threshold change could be important in the therapeutic action of ECT in depression. From this viewpoint ECT can be regarded as an anticonvulsant, and it is therefore intriguing that various antiepileptic compounds, such as carbamazepine and sodium valproate, are now known to be useful in the treatment of severe mood disorders.

Physiological changes during ECT

If ECT is given without atropine premedication, the pulse first slows and then rises quickly to 130–190 beats/minute, falling to the original resting rate or beyond towards the end of the seizure before a final less marked tachycardia lasting several minutes. It is generally believed that atropine abolishes both these periods of slowing, although a controlled trial by Wyant and MacDonald (1980) did not confirm this. If no muscle relaxant is given, there are corresponding changes in blood pressure; if a relaxant is given, blood pressure changes are less although systolic pressure can still rise to 200 mmHg. Cerebral blood flow also increases by up to 200 per cent. More details of these physiological changes are given by Perrin (1961). There is an increased output of prolactin and neurophysin during and soon after the seizure (Whalley *et al.* 1982).

Unilateral or bilateral ECT

Results from several controlled studies (Squire 1977; Sackheim *et al.* 1993) indicate that the temporary post-ictal confusion and memory impairment after ECT are significantly less if the stimulating electrodes are applied **unilaterally** to the right cerebral hemisphere. However, it has been disputed whether this reduction in cognitive side-effects is at the expense of therapeutic efficacy. The older ECT machines used modified sine wave current and delivered a dose of electricity that was substantially beyond seizure threshold. In these circumstances the therapeutic advantages of bilateral compared with unilateral ECT are seldom of much clinical significance (Gregory *et al.* 1985). However, the cognitive impairment is substantially greater after bilateral sine wave ECT than after right unilateral treatment. Hence right unilateral electrode placement was the preferred treatment with these older machines.

As discussed above, with the newer brief pulse machines there are clinically important differences in antidepressant effects between low-dose right unilateral ECT and low-dose bilateral ECT. The current advice of the Royal College of Psychiatrists (1995) is that *low-dose treatment with bilateral electrode placement* provides the best combination of therapeutic efficacy and minimal post-treatment cognitive impairment.

As mentioned above, an important factor in the efficacy of ECT is not the absolute dose of electrical current but the amount by which the dose exceeds seizure threshold for the particular patient. Since seizure threshold is not routinely measured before ECT (and there are up to 40-fold differences in threshold between subjects) it is not easy to predict how far a particular ECT dose will be above threshold in a given individual. Therefore, some authorities have proposed that the best treatment regime for routine clinical practice is *right unilateral treatment given at doses substantially above seizure threshold*, for example 350–400 millicoulombs (Abrams *et al.* 1991). This dose is significantly higher than the dose used in the study by Sackheim *et al.* (1993) in which patients treated with 'high-dose' right unilateral ECT received an average dose of 175 millicoulombs. Clearly, further studies of this issue are needed. At present, it is recommended that high-dose right unilateral ECT be reserved for patients in whom low-dose bilateral ECT produces significant cognitive side-effects. (The technique for determining handedness and electrode placement is described on p. 544).

Unwanted effects after ECT

Subconvulsive shock may be followed by *anxiety and headache*. ECT can cause a brief retrograde *amnesia* as well as loss of memory for up to 30 minutes after the fit. Brief disorientation can occur, particularly with bilateral electrode placement. Headache can also occur. Some patients complain of *confusion, nausea,* and *vertigo* for a few hours after the treatment, but with modern methods these unwanted effects are mild and brief. A few patients complain of *muscle pain,* especially in the jaws, which is probably attributable to the relaxant. There have been a few reports of sporadic major *seizures* in the months after ECT (Blumenthal 1955), but these may have had other causes. If they occur at all, it is only during the first year after treatment.

Occasional *damage to the teeth, tongue, or lips* can occur if there have been problems in positioning the gag or airway. Poor application of the electrodes can lead to small electrical *burns.* Fractures, including *crush fractures* of the vertebrae, have occurred occasionally when ECT was given without muscle relaxants. All these physical consequences are rare provided that a good technique of anaesthesia is used and the fit is modified adequately. *Other complications* of ECT are rare and mainly occur in people suffering from physical illness. They include arrhythmia, pulmonary embolism, aspiration pneumonia, and cerebrovascular accident. *Prolonged apnoea* is a rare complication of the use of muscle relaxants. Rarely, status epilepticus may occur in predisposed subjects or in those taking medication that prolongs seizure duration.

Since the introduction of ECT there has always been concern as to whether it may cause brain damage. When ECT is given to animals in the usual clinical regime there is no evidence that brain damage occurs. Also, structural imaging studies in patients have been reassuring on this point (Coffey *et al.* 1991).

Memory disorder after ECT

As already mentioned, the immediate effects of ECT include loss of memory for events shortly before the treatment, and impaired retention of information acquired soon after the treatment. Loss of more remote personal memories may also be experienced. These effects depend on both electrode placement (unilateral versus bilateral) and electrical design; electrode placement appears to be the more

important factor. However, two months after treatment there is no difference between treatment regimes (Sackheim *et al.* 1993). These memory changes are experienced by nearly all patients receiving ECT, and they disappear within a few weeks of the end of the treatment.

Many patients fear that there will be lasting memory change, and some complain of it after ECT. However, studies have revealed either no differences in memory or some improvements in memory some weeks after ECT (Cronholm and Molander 1964; Sackheim *et al.* 1993). Depressive disorders substantially impair cognitive function, and many patients report their memory as subjectively much improved after ECT (Sackheim *et al.* 1993). Also, several studies have found no significant differences in memory tests between ECT-treated patients and controls who had not received ECT (Johnstone *et al.* 1980; Weeks *et al.* 1980). However, in a study of former patients who were complaining that they had suffered permanent harm to memory from ECT given in the past, Freeman *et al.*(1980) found that these patients did worse than controls on some tests in a battery designed to test memory. They also had residual depressive symptoms, and so it is possible that continuing depressive disorder accounted for the memory problems. It seems reasonable to conclude that, when used in the usual way, ECT is not followed by persisting memory disorder, except perhaps in a small minority, and that even in this group it is still uncertain whether the impairment is due to the effects of ECT or to a continuation of the original depressive disorder.

The mortality of ECT

The death rate attributable to ECT was estimated to be 3–4 per 100 000 treatments by Barker and Barker (1959). A survey of all ECT treatments given with anaesthesia in Denmark found a similar death rate of 1 in 22 210 treatments, i.e. 4–5 per 100 000 treatments (Heshe and Roeder 1976). The risks are related to the anaesthetic procedure and are greatest in patients with cardiovascular disease. When death occurs it is usually due to ventricular fibrillation or myocardial infarction.

Contraindications

The contraindications to ECT are any medical illnesses that increase the risk of anaesthetic procedure by an unacceptable amount, for example respiratory infections, serious heart disease, and serious pyrexial illness. Other contraindications are diseases likely to be made worse by the changes in blood pressure and cardiac rhythm that occur even in a well-modified fit; these include serious heart diseases, recent coronary thrombosis, cerebral or aortic aneurysm, and raised intracranial pressure. Mediterranean and Afro-Caribbean patients who might have sickle cell trait need additional care that oxygen tension does not fall. Extra care is also required with diabetic patients who take insulin. Although risks rise somewhat in old age, so do the risks of untreated depression and drug treatment.

ECT should not be given to patients taking reserpine (Crammer *et al.* 1982, p. 233). Adverse effects, including increased cognitive impairment, have been reported when ECT has been given with **lithium**. **SSRIs** have been associated with prolonged seizures during ECT. Some anaesthetists prefer not to anaesthetize patients taking MAOIs, but ECT can be given safely to patients receiving MAOI therapy.

Technique of administration

In this section we outline the technical procedures used at the time of treatment. Although the information in this account should be known, it is important to remember that ECT is a practical procedure that must be learnt by apprenticeship as well as by reading. Much useful information is given in the *ECT handbook* of the Royal College of Psychiatrists (1995) and by the American Psychiatric Association (1990).

ECT should be given in pleasant safe surroundings. Patients should not have to wait where they can see or hear treatment given to others. There should be waiting and recovery areas separate from the room in which treatment is given, and adequate emergency equipment should be available including a sucker, endotracheal tubes, adequate supplies of oxygen, and facilities to carry out full resuscitation. The nursing and medical staff who give ECT should receive special training.

The first step in giving ECT is to put the patient at ease and to check his identity. The case notes should then be seen to make sure that there is a valid consent form. The drug sheet should be checked to ensure that the patients is not receiving any drugs, such as MAOIs, that might interfere with anaesthetic procedures. It is also important to check for evidence of drug allergy or adverse effects of previous general anaesthetics. The drug sheet should be available for the anaesthetist to see. A full physical evaluation should have been carried out by the patient's treating doctor. Specialist advice should be sought when there may be medical contraindications to ECT. The next steps are to make sure that the patient has taken nothing by mouth for at least five hours, and then, with the anaesthetist, to remove dentures and check for loose or broken teeth. Finally, the record of any previous ECTs should be examined for evidence of delayed recovery from the relaxant (due to deficiency in pseudocholinesterase) or other complications.

An anaesthetist should be present when ECT is given (though this cannot always be achieved in developing countries). Suction apparatus, a positive-pressure oxygen supply, and emergency drugs should always be available. A tilting trolley is also valuable. As well as the psychiatrist and anaesthetist, at least one nurse should be present.

There is some debate as to whether atropine pretreatment helps to prevent vagal arrhythmias and dry excess bronchial secretion. There is little evidence that atropine in usual dosage is useful for these purposes, and the current advice of the Royal College of Psychiatrists (1995) is that atropine should not be given routinely. If an anticholinergic agent is required to dry secretions, **glycopyrrolate** should be used because it does not cross the blood–brain barrier. Anaesthesia for ECT can be induced with a short-acting agent such as **methohexitone**. **Propofol** should not be used because it substantially shortens seizure duration. Methohexitone is followed immediately by a muscle relaxant (often suxamethonium chloride) from a separate syringe, although the same needle can be used. The anaesthetist is responsible for the choice of drugs.

He ensures that the lungs are well oxygenated before a mouth gag is inserted.

While the anaesthetic is being given, the psychiatrist checks both the dose of electricity and the electrode placement that has been prescribed for the patient. For unilateral ECT it is important to establish handedness as a guide to cerebral dominance; handedness should be determined at least by asking which hand is used to catch and throw, and which foot to kick. In right-handed people the left hemisphere is nearly always dominant; in left-handed people either hemisphere may be dominant. Hence, if there is evidence that the patient is not right-handed, it is usually better to use bilateral electrode placements. In all cases the patient should be watched carefully after the first allocation of ECT. Marked confusion, especially with dysphasia, for more than five minutes after the return of consciousness suggests that the dominant side has been chosen inadvertently. In such an event either the opposite side should be stimulated subsequently, or bilateral placement should be used instead. If there is any doubt about handedness, bilateral positioning should be chosen.

With the current recommendations, bilateral treatment will be prescribed for most patients. While the standard electrode placement for bilateral treatment is bitemporal, some workers believe that a suitable bifrontal position produces therapeutic effects at least as good as those of bitemporal placement, with less cognitive impairment (Fig. 17.4) (Lawson *et al.* 1990; Letemendia *et al.* 1993).

The skin is cleaned in the appropriate areas and moistened electrodes are applied. (If good electrical contact is to be obtained, it is also important that grease and hair lacquer are removed by ward staff before the patient is sent for ECT.) While dry electrodes can cause skin burns, it is also important to remember that excessive moisture causes shorting and may prevent a seizure response. (This can happen more readily with unilateral placement because the electrodes are closer to one another.) Although enough muscle relaxant should have been given to ensure that convulsive movements are minimal, a nurse or other assistant should be ready to restrain the patient gently if necessary. The electrodes are now secured firmly. For unilateral ECT the first electrode is placed on the non-dominant side, 4 cm above the midpoint between the external angle of the orbit and the external auditory meatus. The second is 10 cm away from the first, vertically above the meatus of the same side (see Fig 17.4). For bilateral ECT, electrodes are on opposite sides of the head, each 4 cm above the midpoint of the line joining the eternal angle of the orbit to the external auditory meatus—usually just above the hairline. The shock is then given.

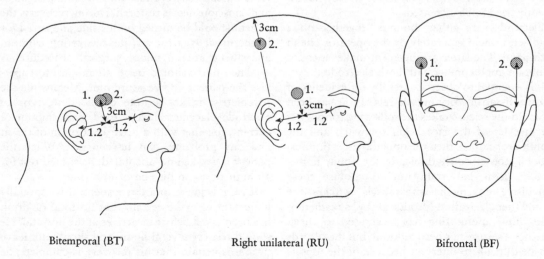

Bitemporal (BT)   Right unilateral (RU)   Bifrontal (BF)

Figure 17.4

ECT machines that deliver sine wave current should be abandoned because of the increased risk of cognitive side-effects. Modern machines deliver a **constant current** which takes into account factors such as skull impedance. Hence they can be preset to deliver a fixed amount of electrical charge, which is best measured in millicoulombs. As discussed above, when using bilateral electrode placement the present practice is to administer a dose of electricity that is only modestly (50–100 per cent) above the seizure threshold for the individual. Seizure thresholds differ considerably between individuals, but in two-thirds of the population are between 100 and 200 millicoulombs. Seizure threshold is higher in men than in women, and increases with age. For example a reasonable starting dose for a patient under 40 who is to receive bilateral ECT, would be 150 millicoulombs. Subsequently this dose could be adjusted depending on the length of seizure, the cognitive side-effects, and clinical response. The dose might be increased if the seizure were short or absent, or if there were no improvement after several treatments. Conversely, troublesome post-ECT cognitive disturbance would indicate that the dose of electricity should be reduced. It is important to note that seizure duration may decrease during a course of ECT because repeated treatment tends to increase seizure threshold. Thus, if a dose of electrically initially produced a seizure of satisfactory duration, it may subsequently need to be increased.

Electrodes are either mounted together on a head-set, or held separately by the operator one in each hand. The latter arrangement makes it easier to obtain good contact with both electrodes.

It is essential to observe carefully for **evidence of seizure**. If satisfactory muscle relaxation has been achieved the seizure takes the following form. First the muscles of the face begin to twitch and the mouth drops open; then the upper eyelids, thumbs, and big toes jerk rhythmically for about half a minute. It is important not to confuse these convulsive movements with muscle twitches due to the depolarization produced by suxamethonium. EEG monitoring has been used to check whether a seizure has been induced, but the records can be difficult to interpret because of the muscle artefact produced by direct stimulation of the frontalis muscle. An alternative is to isolate one forearm from the effects of the muscle relaxant. This can be done by blowing up a blood pressure cuff to above systolic pressure before the relaxant is injected; this pressure is maintained during the period in which the seizure should occur and then released. Seizure activity can then be observed in the muscles of the isolated part of the arm. When judging the appropriate cuff pressure it is important to remember that systolic pressure rises during the seizure; if the cuff is not at sufficient pressure the relaxant will pass into the forearm at this stage.

There is no direct correlation between treatment outcome and duration of seizure activity, but it is recomended that the dose of electricity be adjusted to achieve a *seizure duration of between 20 and 50 seconds*. This duration is recommended because short seizures are likely to be therapeutically ineffective while long seizures are more likely to be associated with cognitive disturbance. *Psychotropic drugs may alter seizure threshold and seizure duration*. For example most antidepressant and antipsychotic drugs lower seizure threshold, whilst benzodiazepines and carbamazepine have the reverse effect.

After the seizure, the lungs are oxygenated thoroughly with an airway in place. The patient remains in the care of the anaesthetist and under close nursing observation until breathing resumes and consciousness is restored. During recovery, the patient should be turned on his side and cared for in the usual way for anyone recovering from an anaesthetic after a minor surgical procedure. A qualified nurse should be in attendance to supervise the patient and reassure him. Meanwhile the psychiatrist makes a note of the date, type of electrode placement, drugs used, and amount of current, together with a brief description of the fit and any problems that have arisen. When the patient is awake and orientated, he should rest for an hour or so on his bed or in a chair.

If ECT is given to a day patient, it is especially important to make certain that no food or drink has been taken before he arrives at the hospital. He should rest for several hours and should not leave until it is certain that his recovery is complete; he should leave in the company of a responsible adult, preferably by ambulance, and certainly not riding a bicycle or driving a car. The authors believe that it is usually preferable to admit patients to hospital for ECT.

The most important problem, apart from those relating to the anaesthetic procedure, is failure to produce a clonic convulsion (a tonic jerk produced by the current must not be mistaken for a seizure). If it is certain that no seizure has appeared, checks should be made of the machine, electrodes, and contact with the skin. The possibility of shorting due to excess moisture on the scalp should also be considered. If all these are excluded, the patient may have either an unusually high resistance to the passage of current through the extracranial tissues and skull, or a high convulsive threshold. The charge can then be increased by 50 per cent and a further stimulus given.

Frequency and number of treatments

In the United Kingdom bilateral ECT is usually given twice a week, although in urgent cases three applications may be used in the first week. In general, thrice-weekly ECT has little therapeutic advantage over a twice-weekly regime, and may produce more cognitive impairment.

Decisions about the length of a course of ECT have to depend on clinical experience since relevant information is not available from clinical trials. A course of ECT is usually from six to a maximum of 12 treatments. Progress should be reviewed at least once a week; there is usually little response until two or three treatments have been given, after which increasing improvement takes place. If the response is more rapid than this, fewer treatments may be given. If there has been no response after six to eight treatments, the course should usually be abandoned since it is unlikely that more ECT will produce useful change.

It is important that, while ECT may produce striking benefit in depressed patients, there is a *high relapse rate unless continuation therapy with antidepressant medication is undertaken.* Sometimes the choice of antidepressant drug can be difficult because recent studies have shown that, if a patient does not respond to an adequate dose of

an antidepressant drug prior to ECT, little prophylactic effect may be obtained if the same drug is continued after the course of ECT is completed (Sackheim *et al.* 1990). Thus, if a patient has required ECT because of non-response to antidepressant medication, it is good practice to consider a different class of antidepressant drug or else lithium carbonate in the continuation and prophylactic phases of drug treatment. A few patients respond well to ECT but continually relapse even when maintained on multiple drug therapy. In these circumstances some practitioners give **maintenance ECT** at a reduced frequency, such as fortnightly or monthly. Patients receiving continuing ECT require careful monitoring of both mood and cognitive state (Scott *et al.* 1991).

Medicolegal issues including consent to ECT

Before a patient is asked to agree to ECT, it is essential to explain the procedure and indicate its expected benefits and possible risks (especially the possible effects on memory). The importance of this step is underlined by the finding (Freeman and Kendell 1980) that only one-fifth of patients receiving ECT thought that they had received adequate explanation. Many patients expect severe and permanent memory impairment after treatment and some even expect to receive unmodified fits. Once the doctor is sure that the patient understands what he has been told, the latter is asked to sign a standard form of consent. The patient should understand that consent is being sought for the whole course of ECT and not just for one treatment (although he can of course withdraw consent at any time). All this is the doctor's job—he should not delegate it to other staff.

If a patient refuses consent or is unable to give it because he is in a stupor or for other reasons, and if the procedure is essential, further steps must be considered in the United Kingdom. The first is to decide whether there are grounds for involving the appropriate section of the Mental Health Act (see Appendix). The section does not allow anyone to give consent on behalf of the patient, but it does establish formally that he is mentally ill and in need of treatment. In England and Wales the opinion of a second independent consultant is required by the Mental Health Act 1983. The requirements of the Act are outlined in the Appendix, p. 786. Readers working elsewhere should find out the relevant legal requirements. If the decision is made in this careful way, it is rare for patients to question the need for treatment once they have recovered. Instead, most acknowledge that treatment has helped, and they understand why it was necessary to give it without their expressed consent.

Bright light treatment

There is increasing interest in the use of **bright light treatment for seasonal affective disorder** (p. 206). Several studies have compared the effect of bright light (above 2500 lux) and of dim light in patients with winter depression. Most studies agree that bright-light treatment produces a rapid antidepressant effect, though this effect tends to wane when treatment is withdrawn (Blehar and Rosenthal 1989). There is disagreement as to the schedule of light administration that produces the best antidepressant effect. On the hypothesis that the circadian rhythms of patients with seasonal affective disorder are abnormally phase delayed, some workers have maintained that early morning light treatment (given between about 6 a.m. and 8 a.m.) should be most beneficial. *Some studies support this hypothesis, while others suggest that light given in the evening, or even at midday, can also be effective.*

Mechanism of action

Light is an important organiser (*Zeitgeber*) of circadian rhythms in human beings. Therefore it is possible that, if abnormally delayed circadian rhythms are involved in the aetiology of seasonal affective disorder, appropriately timed light treatment could be beneficial. An argument against this hypothesis is that bright light given in the evening, or even at midday, may also be effective in seasonal affective disorder (Wehr *et al.* 1986). It is important to note that, while bright light treatment is more effective than dim light, there is the difficulty of constructing a really convincing placebo condition with which to compare bright light.

Adverse effects

In general, bright light treatment is well tolerated, and if subjects follow the advice to sit at the side of the light box and glance at the screen from time to time, adverse effects are uncommon. However, headaches are sometimes reported. Light therapy would not be expected to have adverse effects on the eye because the outdoor illumination on a bright spring day (above 5000 lux) is significantly greater than that produced by conventional light boxes. Of course, it is important to make sure that a light box is soundly constructed and electrically safe. Several reliable machines are now commercially available.

Psychosurgery

Psychosurgery refers to the use of neurosurgical procedures to modify the symptoms of psychiatric illness by operating on either the nuclei of the brain or the white matter. Psychosurgery began in 1936 with the work of Moniz whose operation consisted of an extensive cut in the white matter of the frontal lobes (**frontal leucotomy**). This extensive operation was modified by Freeman and Watts (1942) who made smaller coronal incisions in the frontal lobes through lateral burr holes. Although their so-called **standard leucotomy** was far from standardized anatomically, and although it produced unacceptable side-effects (see below), the procedure was widely used in the United Kingdom and other countries. There was enthusiasm for the initial improvements observed in patients, but this was followed by growing evidence of adverse effects including intellectual impairment, emotional lability, disinhibition, apathy, incontinence, obesity, and epilepsy. This led to a search for more restricted lesions capable of producing the same therapeutic benefits without these adverse consequences. Some progress was made, but at the same time advances in pharmacology made it possible to use drugs to treat the disorders for which surgery was intended.

There have been no controlled trials to test the value of these operations. If such surgery is used at all, it should be only after the most thorough and persistent attempts to produce improvement with other forms of treatment. If this is done, the requirement for psychosurgery will be extremely small. It is not easy to judge how widely the operations are used. *Psychosurgery appears to be declining* in frequency, and in many countries it is not carried out at all. In the United Kingdom the major centre for psychosurgery is the Geoffrey Knight Unit, London, which now carries out about 15 procedures annually (Poynton *et al.* 1988; Bridges 1992). In Australia and New Zealand the number of operations declined through the 1980s, and by the end of the decade only two or three operations a year were conducted (Hay and Sachdev 1992).

More recent follow-up studies of patients with obsessive–compulsive disorder suggest low rates of improvement (about 40 per cent). Nevertheless, some patients show significant benefit which is not usually apparent in subjects matched for severity of illness and followed up naturalistically (Hay *et al.* 1993).

The indications for psychosurgery are considered to be intractable mood disorder, obsessive–compulsive disorder, and anxiety disorder.

Types of operation

As operations on the frontal lobe became anatomically less extensive, the most commonly used procedures in the United Kingdom became the restricted undercutting of the medial third of the orbital cortex and a bimedial operation aimed at the frontothalamic bundle. At the same time it became clear that the frontal cortex has complex connections with the hypothalamus, temporal cortex, hippocampus, amygdala, and mamilliary bodies. Therefore the surgical approach was therefore directed to some of these connections as well as to the frontal lobe.

Nowadays the older 'blind' operations have been replaced by stereotactic procedures that allow the lesions to be placed more accurately. These stereotactic operations are tractotomy, limbic leucotomy, and amygdalotomy. In **stereotactic tractotomy**, the target is the posterior part of the area incised in orbital undercutting. The lesion is produced by implanting radioactive yttrium 'seeds' (Knight 1972). In **stereotactic limbic leucotomy** small bilateral lesions are placed in the lower medial quadrant of the frontal lobe to interrupt two of the frontolimbic pathways, and in the cingulum (Richardson 1973). In **amygdalotomy** bilateral lesions are placed in the amygdala, usually in an attempt to control aggressive behaviour (Small *et al.* 1977).

The operation should never be carried out until the effects of several years of vigorous treatment have been observed. If this rule is followed, the operation will hardly be used. If the operation is to be considered at all, it should only be for chronic intractable obsessional disorder and severe chronic depressive disorders in older patients. There is no clear justification for psychosurgery for anxiety neuroses or schizophrenia. For a less conservative view the reader should consult Bartlett *et al.* (1981) and Cobb and Kelly (1990).

Further reading

Abrams, R. (1992). *Electroconvulsive therapy*. Oxford University Press.

Cooper, J. R., Bloom, F. E. and Roth, R. H. (1991). *The biochemical basis of neuropharmacology* (6th edn). Oxford University Press.

Grahame-Smith, D. G. and Aronson, J. K. (1992). *Oxford textbook of clinical pharmacology and drug therapy*, (2nd edn). Oxford University Press.

Kiloh, L. G., Smith, J. S., and Johnson, G. F. (1988). *Physical treatments in psychiatry*. Blackwell Scientific Publications, Melbourne.

Lader, M. and Herrington, R. (1990). *Biological treatments in psychiatry*. Oxford University Press.

Leonard, B.E. (1992). *Fundamentals of psychopharmacology*. Wiley, Chichester.

Schatzberg, A. F. and Cole, J. O. (1991). *Manual of clinical psychopharmacology* (2nd edn). American Psychiatric Press, Washington, DC.

18 | *Psychological treatment*

This chapter is concerned with various kinds of counselling, psychotherapy, behavioural and cognitive therapies, and some related techniques. The subject is large, and will be easier to follow if the reader's attention is drawn to four points about the organization of the chapter. First, it is recommended that this account be read in conjunction with the general advice on planning treatment (Chapter 17) and on the provision of services (Chapter 19). Second, this chapter includes general comments on the value of different treatments, but specific advice about these treatments is given in the chapters concerned with particular syndromes. Third, the chapter covers many different techniques of treatment, and therefore cannot consider each in detail; for this reason suggestions for further reading are given. Finally, and most important, it is emphasized that psychological treatments cannot be learnt by reading alone; they must also be learnt by appropriate and adequate supervised experience. Psychological treatments are important, and are part of the management of most patients, sometimes in combination with drug treatment. The ways of combining psychological and drug treatments for particular disorders are described in other chapters.

The word psychotherapy is used in two ways. In the first usage, psychotherapy denotes any psychological treatment, including for example counselling and cognitive and behavioural treatments. In the second usage psychotherapy does not include counselling and cognitive and behavioural treatments. Here, the second usage is adopted because it is more convenient for the purposes of the present chapter. The treatments considered in this chapter are listed in Table 18.1.

General considerations

General indications for the use of counselling and psychotherapy

It is helpful to have a broad framework in which to place the many forms of psychotherapy. One practical scheme divides psychotherapy along two dimensions, the first concerned with the complexity of the procedures and the second with the number of patients taking part. On the first dimension treatments can be divided into counselling that is appropriate in primary care; short- or long-term psychotherapy that can be provided by all psychiatrists, and methods that are best provided by specialists in psychotherapy. On the second dimension, treatments can be divided into those used with a single patient; those used with couples and families, and those used with small or large groups of patients.

In primary care and hospital practice, counselling may be considered for patients with emotional problems of short duration and for patients who need help in coming to terms with the effects of incurable illness, whether physical or mental. Counselling may also be provided for special groups such as mothers with postnatal depression, the bereaved, and people with alcohol problems.

Psychiatrists are likely to use brief psychotherapy as the main treatment of younger patients with moderately severe but recurrent depressive or anxiety symptoms arising mainly from intrapsychic problems such as persistent low self-esteem or from difficulties in interpersonal relationships. (Phobic, panic, and obsessional disorders are more likely to be treated by cognitive and behavioural methods.) For depressive disorders, brief psychotherapy is often combined with antidepressant drug treatment and with social measures. Counselling and psychotherapy are an important part of the treatment of many other disorders; the indications for these treatments are discussed in other chapters.

General psychiatrists are likely to refer two kinds of difficult problem to specialist psychotherapists: certain kinds of personality disorder, and unusually complicated cases of the conditions mentioned in the previous paragraph. Specialist psychotherapists are more likely to use psychotherapy as the sole means of treatment

Table 18.1. *Psychological treatments considered in this chapter*

than to combine it with medication. (In practice, many patients suitable for treatment by general psychiatrists receive psychotherapy from trainee psychiatrists under the supervision of a specialist therapist.)

With regard to the choice between individual, group, marital, or family therapy, it can be said at this stage that individual therapy is generally suitable for specific problems that can be the focus of short-term treatment. Individual therapy may also be chosen for people who could be treated in a group but would feel unduly awkward there, either because of shyness or because of the nature of their problems (for example a sexual deviation). Individual therapy may also be chosen for patients with more complex psychopathology. For patients whose problems mainly concern their relationships with other people, individual and group therapy are usually equally effective. (The indications for group therapy are considered further on p. 619.)

Marital therapy is appropriate when emotional problems are mainly determined by the relationship between the two partners of a marriage. Family therapy may be used when the difficulties of an older child or adolescent involve members of the family.

These preliminary generalizations are provided as a background to the chapter; exceptions will be considered later in the chapter.

Common factors in psychotherapy

Jerome Frank (1967) pointed out that all forms of psychotherapy share certain basic processes. He went on to suggest that these common factors may be more important in bringing about change than are the factors specific to each treatment. At first this view was controversial, but now it is widely accepted. These shared factors are listening and talking, release of emotion, giving information and providing a rationale, restoring morale, prestige suggestion, and therapeutic relationships.

1. *Listening and talking* In psychotherapy the patient talks and the therapist listens. By listening intently, the therapist signals his

concern for the patient's problems and begins to develop a helping relationship. He detects common themes and revealing omissions in the patient's remarks. When the therapist speaks it is usually to comment on these matters and to clarify ideas that have not been put into words before. Thus patient and therapist seek links between aspects of feeling and behaviour that have previously been unrecognized.

2. *Release of emotion* In the early stages of treatment patients may release much emotion about their problems. This emotional release may be helpful at this stage, but is not generally useful if evoked repeatedly. (The term abreaction refers to a procedure in which a particularly intense and rapid release of feeling is encouraged.)

3. *Giving information* When giving information the therapist should bear in mind that distressed patients may remember little of what they have been told. There are two reasons for such lack of recall. First, doctors often use technical language that is unnecessarily complicated; second, they often underestimate the patient's capacity to understand the nature and significance of illness when explained in simple language. Important points may need to be explained more than once, and may be usefully put in writing so that the patient can study them.

4. *Providing a rationale* All forms of psychological treatment include a rationale that makes the patient's disorder more intelligible. This rationale may be described by the therapist (as in short-term psychotherapy), or it may be discovered by the patient with the help of the therapist's interpretations (as in psychoanalytically oriented treatments). Whatever the method of imparting the rationale, the effect is to make problems more understandable to the patient and thereby to give him more confidence that he can solve them.

5. *Restoration of morale* is an important part of psychotherapy because patients have mostly become demoralized through repeated failure, and have lost the conviction that they can help themselves. Unless this conviction is restored little progress can be made in psychotherapy.

6. *Suggestion* All psychotherapy contains an element of suggestion. This process can bring about change, but generally the effects are not lasting. For this reason suggestion is not used in most forms of psychotherapy (the exception is hypnotherapy). Nevertheless in the early stages of treatment suggestion may bring about some temporary improvement until more lasting changes are effected.

7. *Guidance and advice* There is an element of guidance and advice in all psychotherapies, but in the more intensive methods there is usually a requirement that the patient discover the answers for himself.

8. *The therapeutic relationship* This is a most important element in all psychotherapy. It will be discussed next under the heading of transference and counter-transference.

Transference and counter-transference

In all psychological treatment the relationship between the patient and therapist may help or hinder the patient's progress. This relationship is present from the start and grows more intense as treatment lasts longer or is carried out more frequently. Even in the shortest forms of psychotherapy this relationship forms the cornerstone of treatment, helping to sustain the patient through his difficulties and to motivate him to overcome his problems.

The relationship has a 'realistic' and an 'unrealistic' component. The realistic component is known as the therapeutic alliance; the unrealistic element results from transference. As treatment progresses the unrealistic elements become increasingly strong. These transference elements arise largely because the therapist listens more than he talks. As a result, the patient reveals personal problems that in other circumstances would be revealed only (if at all) to a close relation or intimate friend. The psychotherapist, however, reveals little of his own background or personal beliefs. The intimacy of the situation is such that the patient reacts to the therapist as if they were closely related. Knowing little about the therapist, the patient imagines what he is like; in doing so the patient draws on past experience of other significant persons with whom he has experienced a comparable intimacy, usually one or both parents. The patient transfers to the therapist some of the feelings and attitudes that originated in relation to parents or other intimate relatives; hence the process is called transference. When the therapist is conceived as a good figure, transference is said to be positive; when the therapist is conceived as a bad figure, it is said to be negative. Usually transference has both positive and negative components, and these may involve a complex mixture of love, hate, envy, and reverence for the therapist.

In psychotherapy, therapists have a role unlike that of their everyday relationships. They have to remain impartial and professional and yet be genuinely concerned about their patients' most intimate problems. Despite their training, therapists cannot always achieve this ideal combination of detachment and concern. They may then respond in ways that are not simply a reflection of their patients' personal qualities but also a displacement onto their patients of ideas and feelings related to other figures in the therapists' own lives. This process is called counter-transference.

Both transference and counter-transference can be impediments to treatment, but both can be turned to advantage. One disadvantage of transference is that it may induce behaviour that distracts from the main plan of treatment; for example the patient may attempt to prolong interviews, request extra appointments, and show dramatic behaviour demanding urgent action, such as threats of suicide. Transference may also make it difficult to bring treatment to an end; thus there may be a recrudescence of symptoms that have improved, and also demands for further treatment. If these behaviours are noticed early and discussed, further difficulties can be prevented and the patients can learn more about themselves by understanding how their emotions affect their behaviour.

Counter-transference causes difficulties when therapists become inappropriately involved in their patients' problems or inappropriately angry with the patients. However, counter-transference be turned to advantage. If therapists recognize these feelings and examine their origins, they will learn more not only about the patient but also about themselves.

Transference and counter-transference are most developed in the long-term forms of treatment. In psychoanalysis, transference is encouraged for therapeutic use. However, it is important to realize that transference and counter-transference are relevant in every form of psychological treatment.

How psychotherapy developed

The use of psychological healing is as old as the practice of medicine. Parallels have been drawn between aspects of modern psychotherapy and the ceremonials carried out in some of the ancient Greek temples of healing. However, in the history of psychiatric treatment, psychotherapy starts in the nineteenth century. At that time, the most important developments were in hypnosis, which had come to the attention of doctors through the activities of Anton Mesmer (1734—1815), a Viennese physician. Mesmer believed that cures could be obtained by altering the function of the body by magnetic forces. These forces could arise from actual magnets and, according to Mesmer, also from 'animal magnetism' present in the body of the therapist. Mesmer believed that the curative effect of animal magnetism could be enhanced by the use of physical magnets (Block 1980). Mesmer's theories were not generally accepted by the medical profession, but a Manchester doctor, James Braid, gave a more physiologically plausible explanation of the phenomena. Braid drew parallels between the phenomenon and sleep, and suggested the name hypnosis. He showed that therapeutic effects could be obtained without the use of magnets (Braid 1843).

Treatment with hypnotism became popular in France, where its use was championed particularly by A. A. Liebeault (1823–1904) and Hippolyte Bernheim (1837–1919) in Nancy, and by Jean

Martin Charcot (1825–1893) who practised at the Salpêtrière hospital in Paris. Bernheim's book *Suggestive therapeutics* (Bernheim 1890) described the use of suggestion to treat a variety of conditions including hysteria and other neurotic conditions as well as a variety of painful afflictions and gastro-intestinal problems. Bernheim considered that hypnosis was a normal state, allied to sleep, that could be induced in most patients and used therapeutically. However, Charcot maintained that hypnosis was a pathological state occurring in patients with hysteria and lacked a wider value in therapeutics.

At that time the main alternative psychotherapy was a method that relied on rational discussion rather than suggestion. In contrast with hypnosis, there was no attempt to increase suggestibility; instead symptoms and other problems were discussed in the expectation that greater understanding would enable patients to control them and to become more self-reliant. This alternative treatment was known as 'persuasion'. It was based not on any special theory of psychological development, but rather on the physician's good sense and understanding of people and disease. The important proponent of this suggestive therapy was a Swiss professor of neuropathology, Paul Dubois, whose book *The psychic treatment of nervous disorders*, published in 1904 (see Dubois 1909) was widely influential. Dubois taught that hypnosis had some limited value for hysteria, but an educative treatment was needed for other kinds of neurosis. He taught that educative treatment required more time than hypnosis, but was more effective because it 'changes the point of view of the patient' (see Dubois 1909, p.xiii).

Freud's interest in psychological treatment began with hypnosis (Freud 1892). He visited Bernheim in Nancy and also went to Paris to see Charcot's demonstrations of hypnosis in hysterical patients. When he returned to Vienna, Freud tried hypnosis with some of his neurotic patients and at first was pleased by its success. However, his early successes were not always repeated. Therefore he began to try alternative ways of using hypnosis. At first, Freud used hypnosis to suppress symptoms, but by 1889, in the case of Emmy von N, he was using it to release the emotion associated with the repressed ideas that he now believed to be the root cause of the symptoms (Greenson 1967). Remembering Bernheim's demonstration that patients could recall forgotten events under the influence of waking suggestion, Freud tried a new method. Instead of hypnotizing patients he asked them simply to shut their eyes while he placed his hands on their forehead (Breuer and Freud 1893, pp. 109, 279). Then Freud discovered that it was equally effective for the patient to lie on a couch and talk freely while the therapist kept out of sight. In this way the method of *free association* developed. Freud used various methods to encourage the flow of associations and to comment on them, and to regulate the intensity of the relationship between patient and analyst. These methods made up the basic technique of psychoanalysis and subsequently of much psychotherapy. They are described briefly later in this chapter. The interested reader is recommended to read one of the accounts written by Freud himself (for example Freud 1895*a,b*, 1923). The reader is referred also to the account of Freud's theories of mental functioning in Chapter 4, p. 82.

Interest in psychoanalytic methods increased and the use of educational methods declined. Freud continued to publish striking accounts of his new treatment and to elaborate his theories. He built up a group of colleagues, but not all agreed with the developments of his ideas. Some parted from him and formed their own 'schools' of psychotherapy. A brief outline of some of these developments will now be given. The reader will find more information in the books by Brown and Pedder (1979) and Munroe (1955).

Adler and Jung were the first important figures to leave Freud and develop their own theories and methods of treatment. **Adler**, who left Freud in 1910, rejected the libido theory (see p. 83) and emphasized instead the influence of social factors in development. In keeping with this approach, Adler's therapeutic technique, known as *individual analysis*, was an attempt to bring about greater understanding of how the patient's lifestyle had developed, and also placed considerable emphasis on current problems (Ansbacher and Ansbacher 1964). Adler's theories lacked the ingenuity and interest of Freud's ideas. His methods have never been used widely, although they laid the foundation for the development of the influential dynamic-cultural school of American analysts (see below).

While Adler emphasized the real problems in the patient's life, **Jung** was more concerned with the inner world of fantasy. As a result his technique of psychotherapy relies more on the interpretation of unconscious material as represented in dreams and artistic production, although contemporary problems were by no means neglected. In interpreting these problems, reference was made not only to the past experience of the individual but also to aspects of the 'collective' unconscious, which Jung believed to be common to all mankind and to be revealed through universal images called archetypes (Fordham 1990). In contrast with Freudian analysis, in Jungian analysis the relationship between therapist and patient is less one-sided because the therapist is more willing to be active and to reveal information about himself.

Several other developments followed and were important in the evolution of psychotherapy. All shared the basic analytic technique; they differed mainly in their theories of mental development and consequently in the kind of response made by the therapist to the patient's account of himself.

The so-called **dynamic-cultural** or neo-Freudian school of analysis shared some of Adler's concerns with social causes of neurosis. The most important figures in this movement were Karen Horney and Erich Fromm (two refugees from Nazi Germany who settled in the United States in 1930s) and the American, Harry Stack Sullivan. All three emphasized social factors in the development of personality and in the aetiology of neurosis, and all three considered that the stages of development, which Freud attributed to the unfolding of biological influences, were determined much more by family influences.

Horney accepted Freud's idea that abnormal anxiety was the basis of all neuroses, that other symptoms represented mechanisms of defence acting to reduce this anxiety, and that the origins of neurosis were in childhood. However, she rejected Freud's ideas about the origins of anxiety. Horney did not accept Freud's theory of instinctual forces and the stages of libido development, and

she particularly took issue with his view of the psychological development of women, including the notion of 'penis envy'. For Horney, anxiety was inevitable in childhood, arising from the experience of being insignificant, helpless, and threatened. She held that anxiety was normally overcome by the experience of being brought up by loving parents. In some children who have lacked this experience, anxiety persists and defences develop against it. Horney's defences (which she called 'neurotic trends') were not identical with Freud's defence mechanisms; they included striving for affection, striving for power, and submissiveness. Horney also emphasized the social context outside the family, and pointed out that behaviour could be neurotic in one society but adaptive in another. For Horney, neurotic symptoms were not an essential feature of neurosis; the important features were the neurotic trends and the characteristic structure that grew around them. She summarized the difference between her treatment and Freud's method as follows: 'I differ from Freud in that after the recognition of the neurotic trends, while he primarily investigates their genesis I primarily investigate their actual functions and their consequences. My contention is that by working through the consequences the patient's anxiety is so much lessened and his relation to self and others so much improved, that he can dispense with the neurotic trends.' (Horney 1939, p.282)

Fromm also rejected Freud's theory of instinctual development in causing neurosis, and instead emphasized relationships between the individual and society. Fromm's account of psychopathology includes psychic mechanisms which are broadly similar to those of Freud but differ in important detail. They include 'moral masochism' (a need to be helpless and dependent on others), sadism (a need to exploit others or make them suffer), and 'automaton conformity' (excessive conformity with and submission to other people). Fromm accepted the importance of family influences in shaping character, but also drew attention to wider cultural influences that produce features of personality shared by all members of a society. Fromm was generally more concerned with social than with clinical issues, and partly for this reason his ideas have not had a great influence on therapy.

Fromm's views on character structure and the interaction between psychological and social factors are described in *The fear of freedom* (Fromm 1942).

Sullivan was also concerned with the patient's relationships with other people in adult life. For him, sexual problems were only one aspect of the patient's problems; they lacked the central importance ascribed to them in psychoanalysis. His treatment centred on the relationship between analyst and patient, and the discussion of everyday social encounters. In this process, patient and therapist were more equal than in Freudian analysis, and Sullivan preferred pointed questions and provocative statements to interpretations based on theory (Sullivan 1953).

Melanie Klein, working in the United Kingdom in the 1930s and 1940s, enlarged on some of the biological and psychoanalytical aspects of Freud's theories. Her work has been influential among analysts in the United Kingdom, where it has grown into the '**object-relations**' school expanded by Fairburn, Guntrip, and Winnicot. Her developments of technique were particularly related to play therapy with children, in which she made extensive use of interpretations. For a time her ideas were applied to the treatment by psychotherapy of schizophrenia and severe ('psychotic') depression, but such a usage is not to be recommended—indeed it is generally contraindicated. An essential concept in Klein's theory is the 'object', a term that refers to a person who is emotionally important to the patient (for example a parent) and to an internal psychological representation of such a person. Klein's theory of personality is much concerned with the earliest development of the infant, with the way in which 'objects' are dealt with at this time, and with the instinctual feelings of love and hatred that accompany them. The theory has been described as 'fanciful projections of a theoretically based therapist' (Wolberg 1977, p. 186). However, a substantial group of psychotherapists use interpretations about object relations whilst employing a technique broadly similar to psychoanalysis. For an outline of Klein's theories see Segal (1963), and

for an account of Klein's technique of analysis of children see Klein (1963).

Bowlby, a British analyst, developed attachment theory, based on the idea that infants need a secure relationship with their parents. When this attachment is insecure, emotional problems may develop in later life.

A further line of development can be traced to **Ferenczi**. By the early 1920s psychoanalysis, which began as a brief treatment, had grown increasingly long. Ferenczi tried to shorten it while keeping broadly within Freud's methods. He set time limits to treatment, adopted a less passive role as therapist, and planned the way in which the main themes were to be dealt with in treatment. Many of these innovations have found their way into the brief psychotherapy used today.

Counselling

In everyday usage, the word counselling denotes the giving of advice. As a technical term, it denotes a wider procedure concerned with emotion as well as with giving information. One form of counselling, developed by Carl Rogers, is wholly *non-directive*; the counsellor takes a passive role, giving little information and largely restricting his interventions to comments on the emotional content of the client's utterances ('reflection of feelings'). Although in vogue for some years, this technique has generally been replaced by several more structured procedures, adapted to particular purposes. Counselling is based on the general factors present in all kinds of psychotherapy (see p. 601). The central feature is the relationship between the counsellor and the person counselled; the relative importance of giving information, allowing the release of emotion, and thinking afresh about the situation vary according to the purpose of counselling. The following account outlines some of these procedures; it is illustrative and not comprehensive.

Counselling about risks has several elements: giving information about a problem, providing an opportunity for reflection on the impact of the problem on daily life, and working out ways of lessening this impact or of coming to terms with it. Examples of this kind of procedure include counselling about genetic risks, about the excessive use of alcohol or drugs, and about AIDS.

Counselling to relieve distress some counselling is directed to relieving distress among people who are reacting to difficult circumstances. For this purpose the giving of information is less important than listening, providing an opportunity for emotional release, and discussing ways of coping with problems. Examples of this approach include counselling for university students, as well as counselling for patients with adjustment disorders (see p. 145), acute reactions to stress (p. 139), minor affective disorders (see p. 156), and post-natal depression (Holden *et al.* 1989). This form of counselling is important in general hospital practice, for example for women with breast cancer or for parents who have a handicapped child.

An alternative approach is *interpersonal counselling*, which was developed by Klerman *et al.* (1987) from interpersonal therapy (see p. 608). Up to six sessions of half an hour are focused on changes in life events, sources of persistent distress in the family or place of work, and current difficulties in relationships. Patients are encouraged to consider whether there may be better ways of coping with these difficulties.

Marriage guidance counselling is directed to helping couples to talk constructively about problems in their relationships; this approach focuses on the need for each partner to understand the point of view and feelings of the other, and to identify positive aspects of the relationship as well as those causing conflict.

Bereavement counselling focuses on working through the stages of grief. It combines an opportunity for emotional release (including the expression of despair and anger), information about the normal course of grieving, and advice about the practical problems of living without the deceased person. (This form of counselling is considered further on p. 153.)

Problem-solving counselling is the most structured of the various forms of counselling, and is particularly suitable for patients with reactions to stress and with minor affective disorder. The basic counselling techniques are combined with a systematic approach to the resolution of problems. The patient is helped to do the following:

(i) identify and list problems that are causing distress;

(ii) consider what practicable courses of action might solve or reduce each problem;

(iii) select one problem, and try out the course of action that appears most feasible and likely to succeed;

(iv) review the results of the attempt to solve the problem and then either (a) choose another problem for solution if the first action has succeeded; or (b) choose another course of action if the first has not succeeded.

These steps are carried out in the context of a caring relationship. In clinical trials in general practice this approach has been shown to be effective for reactions to stress, for minor affective disorders, and for less severe forms of major depressive disorder (Catalan *et al.* 1991).

Crisis intervention

Crisis intervention is a form of counselling designed to help patients to adapt to the immediate effects of severe life events, whilst acquiring better ways of dealing with any future stressful circumstances. This approach is used when there is an acute disruption of personal affairs such as the break-up of a relationship, or the aftermath of disastrous circumstances such as floods and earthquakes. The ideas of crisis intervention originated in the work of Lindemann (1944) and Caplan (1961). Although these and subsequent authors have used technical terms to describe their ideas and methods, the essential notions are straightforward. Crisis intervention seeks to restore **coping**, i.e. behaviour used to deal with a difficulty or threatened difficulty. Coping can take four forms: **problem-solving behaviour**, which is the satisfactory adaptive form, **regression**, which is use of behaviours that were appropriate at an earlier time of life but are no longer adaptive, and **denial** and **inertia** which are self-explanatory. Denial is adaptive in the early stages of a crisis but may be maladaptive later. The response to a crisis can be thought of as passing through *four stages*: emotional arousal with efforts to solve the problem; if these fail, greater arousal and distress accompanied by disorganization of behaviour; then trials of alternative ways of dealing with the problem; finally, if there is still no resolution, exhaustion and 'decompensation' (Caplan 1961).

Although no two problems are exactly alike, it is useful to recognize four groups according to their themes (Bancroft 1986): **loss problems** which include not only the loss of a person through death or separation but also the loss of a body part or the function of an organ; **role changes** such as entering marriage, parenthood, or a new job with added responsibilities; **problems in relationships** such as those between sexual partners, or between parent and child; **conflict problems**, which are usually difficulties in choosing between two undesirable alternatives.

Treatment starts by attempting to restore emotional arousal to a near normal level. This is an important first step for a patient in crisis because high emotional arousal interferes with problem solving. To achieve a near normal level, reassurance and an opportunity to express emotions are usually enough, although occasionally anxiolytic drugs may be required for a few days. Patients are encouraged to help themselves, but early in treatment arrangements sometimes have to be made for them, for example over the care of children.

The next stage of crisis intervention resembles the problem-solving counselling described above. The patient's problems and assets are assessed carefully. The patient is encouraged to suggest alternative solutions and to choose the most promising. The therapist's role is to encourage, prompt, and question. He does not formulate problems or suggest solutions directly but helps the patient to do so himself. In crisis intervention, an important aim is that the patient should recognize that he has learnt a general method that can be used for solving future problems. Treatment is usually short but intensive.

Indications

Clinical experience suggests that crisis intervention may be most valuable for well-motivated people with stable personalities who are facing major but transitory difficulties; in other words, those who are most likely eventually to cope on their own. The generally accepted indications have been mentioned already. They include emotional reactions or disturbed behaviour (for example, deliberate self-harm) arising in response to social crises such as the breakdown of marriage or other relationships, traumatic events such as rape or other forms of assault, severe disruptions in life such as unexpected bereavement, and natural disasters such as floods.

Interpersonal therapy

Interpersonal therapy was developed as a structured psychological treatment for problems in relation to personal roles and interpersonal relationships of depressed patients (Klerman *et al.* 1984). This therapy has a wider application to disorders in which such problems are maintaining behaviour, for example eating disorders. The number and content of treatment sessions are planned carefully. Interpersonal problems are categorized as involving: (i) bereavement and other loss, (ii) interpersonal disputes, (iii) role transitions, and (iv) 'interpersonal deficits' such as loneliness. Problems are considered in relation to specific situations and to alternative ways of coping. Clear goals are set and progress towards them is monitored. New coping strategies are tried out in homework assignments.

Treatment makes use of the general therapeutic techniques that have been described above. Specific methods are added for each of the four kinds of problem listed above. For grief and other problems of loss of relationships, the methods resemble those described above under bereavement counselling. For interpersonal disputes patients are helped to identify clearly the issues in the dispute, as well as any differences between their own values and those of the other person. They are helped to negotiate with the other person and to recognize their own contributions to problems that they ascribe to that person. Problems of role transition are dealt with in a similar way. Interpersonal deficits are discussed by analysing present relationship problems and the patient's previous attempts to overcome them, and by discussing alternatives.

Interpersonal therapy has been used particularly in the treatment of depressive disorders (see p. 232). For an account of interpersonal therapy see Klerman *et al.* (1984).

Supportive psychotherapy

Supportive psychotherapy is used to relieve distress or to help a person to persist despite difficulties, when all opportunities for problem solving have been tried. It is helpful for patients with chronic mental or physical illness or handicap when other treatment is unhelpful. The treatment is specially suitable for the dying (see p. 149).

Supportive therapy uses the basic procedures shared by all forms of psychotherapy (see p. 601). Its basic elements are a therapeutic relationship, listening, explaining, and encouraging hope. Supportive therapy is often prolonged and carried out in distressing circumstances, and therefore attention to the therapeutic relationship is important.

1. **The therapeutic relationship** This relationship should be an alliance in which patients feel more able to struggle with their problems without becoming unduly dependent on the therapist. Supportive therapy may need to be lengthy; if the patient has severe handicap or has a dependent personality, it may be difficult to avoid a degree of dependence. If dependency develops, it should be directed as far as possible to the group of staff working with the patient in hospital or general practice and not to an individual carer. From the beginning, limits should be agreed with the patient about the amount of time that can be given to his needs.

2. **Listening** is important in supportive therapy. As in all forms of psychological treatment patients should feel that they have their doctors' full attention and concern, and that their worries are being taken seriously.

3. **Emotional release** In the early stages of supportive treatment patients may be helped by emotional release. With the progress of treatment further release may be needed as new problems have to be surmounted, for example in a person with a progressive physical illness.

4. **Explanation and advice** are important in supportive therapy, and their timing should be considered carefully. Patients with serious and incurable illness may need to receive information gradually, coping with part before they face the whole. Information should always be accurate, but it is not necessary to explain everything at the first session. The patient's requests for information are generally the best guide.

5. **Reassurance** is valuable, but premature or unrealistic reassurance can destroy a patient's confidence in the doctor. Reassurance should be offered only when the patient's concerns have been fully understood. Reassurance must be truthful; if a patient asks about prognosis, it may be appropriate to give the most optimistic outcome of those that can be foreseen because hope should not be destroyed. However, if patients find that they have been deceived, they will lose the basic trust on which supportive therapy depends. Even with the most difficult problems, a positive approach can often be maintained by encouraging patients to build on their few remaining assets and opportunities.

6. **Suggestion:** In supportive treatment, patients should generally be encouraged to reach a realistic adjustment to their situation. Nevertheless there are times when it is appropriate for doctors to use their authority as experts to persuade the patients to take some necessary step, for example to continue to cope despite a temporary exacerbation of their condition.

Supportive treatment need not always be given by a health professional. Self-help groups can support some patients and their relatives. This form of support is sometimes more appropriate than individual supportive psychotherapy given by a doctor. A useful account of supportive treatment is given by Bloch (1986).

Psychodynamic supportive psychotherapy

This approach has aims similar to those of other supportive therapy, but places more emphasis on the relationship between patient and therapist, and also on an understanding of the influence of the patient's past experience on his current reactions in treatment. These ideas are useful in supportive therapy for patients with severe emotional disorders (including borderline personality disorder) in which intensive transference may arise and interfere with treatment unless the therapist is aware of the potential problem and is equipped to avoid it.

Dynamic supportive therapy uses methods of other supportive treatment, including reassurance, prestige suggestion, information, advice, and emotional release. Adaptive mechanisms of defence (see p. 136) are encouraged, but there is avoidance of techniques that enhance transference and dependency (such as free association and dream analysis). Patients are encouraged to accept causes for problems that are outside their control ('benign projection'), whilst interpretations of unconscious motivations are generally avoided. Goals and limits are set. A brief account of these methods has been given by Rockland (1987), a longer account by Rockland (1989), and an account of applications to borderline personality by Rockland (1992).

Brief insight-oriented psychotherapy

Brief eclectic psychotherapy

This kind of psychotherapy aims to produce limited but worthwhile changes within a short time; the duration of treatment is seldom more than six months and is often considerably less. The treatment is focused upon specific problems; hence the term *focal psychotherapy* is sometimes used.

The term eclectic (meaning selecting freely from numerous sources) is used to indicate that the treatment is not linked to any special theory such as psychoanalysis.

The basic *procedures* of brief psychotherapy are mainly the techniques common to all kinds of psychotherapy (see p. 601) rather than the techniques specific to any of the 'schools' of therapy which are more important in the long-term therapies described on p. 613). The following account summarizes the main elements of brief psychotherapy.

After taking a full psychiatric history, the therapist begins the first phase of treatment by discussing with the patient what aspects of the problems are to be dealt with in treatment, what aims are realistic, and how long treatment will last. In this sort of treatment, from five to twenty sessions may be allocated depending on the complexity of the problems. The therapist emphasizes that patients will be helped to find their own solutions to their problems; the therapist's role is not to provide the solutions but to help patients towards them.

Patients are then asked to talk about one of the problems selected for consideration. They are encouraged to give specific examples of events that can be examined in detail with the intention of finding out how they thought, felt, and acted at the time. Various prompts and other interviewing techniques are required for this, but they are essentially similar to those described in Chapter 2. To encourage patients to think aloud about their difficulties, the therapist says little. He encourages patients to talk about emotionally painful subjects rather than to avoid them, to review their own part in any difficulties that they ascribe to other people, and to look for common themes in what they are describing. At times the therapist helps patients to look back on their lives to see how present patterns of behaviour began. Patients are asked to consider whether they are maintaining behaviour that served a purpose in the past, although it is no longer appropriate. Finally, they are encouraged to consider alternative ways of thinking and behaving in situations that cause difficulties.

Throughout treatment, the therapist pays as much attention to patients' non-verbal behaviour as to their words, because discrepancies between the two often point to problems that have not yet been expressed directly. The therapist also watches for behaviour that suggests undue emotional attachment. If such attachment is suspected, it is discussed with the patient. At the same time therapists must be sensitive to their own emotional reactions to patients, and must make sure that they are neither over-involved nor rejecting. If they are over involved or rejecting, they should try to find out why—if necessary through discussion with a colleague. (These emotional reactions, which are known respectively as transference and counter-transference, are discussed further on p. 612.)

In the middle phase of treatment, patients continue to talk about the problems that they identified at the start and they examine current examples. The therapist points out any patterns of behaviour that are being repeated, and then relates them to patients' accounts of childhood experiences. The therapist also comments on the patient's emotional reactions during the interview.

As the end of treatment approaches, patients should feel that they have a better understanding of the problems selected at the beginning and should be more confident of dealing with them. At this stage patients should no longer feel dependent on the therapist because the potential problems of termination of treatment have been worked through already. However, it is sometimes useful to ease the separation by arranging a few follow-up appointments spaced over two or three months.

Indications

Brief psychotherapy is mainly helpful for patients who have difficulties in personal relationships but are free from serious disorder of personality. Suitable patients are intelligent, interested in gaining psychological understanding of their own behaviour, and well motivated to change by their own efforts. Particularly suitable patients are those who have problems in relationships leading to unhappiness and anguish in the absence of a specific neurotic syndrome. Patients with obsessional or hypochondriacal neurosis are much less likely to respond to such brief treatment.

An account of the eclectic type of brief psychotherapy is given by Garfield (1980).

Brief psychodynamic psychotherapy

Psychodynamic psychotherapy is intended to examine the effects of past experience on present behaviour, including the current mechanisms of defence. The present attitudes and expectations that give meaning to events are formed by past experience; psychodynamic therapy aims to alter these organizers of behaviour. To bring this change about, the basic shared techniques of psychotherapy (see p. 601) are used, together with the methods described above as eclectic psychotherapy. These methods are supplemented by free association and by interpretations of defences and transference. Treatment is generally once a week for six to nine months. See Ursano *et al.* (1991) for a brief account of the theories and methods of brief analytically oriented therapy.

In *free association* the patient is encouraged to talk freely about whatever comes to mind about his present and past life without considering whether the content is true or logical. This procedure has two effects. First, patients express thoughts and feelings which seem illogical or shameful, and which they might otherwise have suppressed. Second, difficulties in free association are pointers to mechanisms of defence whereby patients exclude from their own consciousness material within their unconscious store of memories.

Interpretations are hypotheses about causal links between events and behaviours. Interpretations about defence mechanisms are hypotheses about the reasons for blocks in the process of free association or about the ways in which the patient appears to be unconsciously protecting himself from unpleasant feelings in his daily life. Interpretations about transference are hypotheses about the ways in which the patient responds to the therapist or to other people; they suggest ways in which the patient responded to close relations in earlier life.

At the start of brief treatment, it is appropriate to explain to the patient the general aims of linking past and present behaviour patterns through his own recollections and through interpretations. It should also be explained that the patient's role is to work out his own answers with help from the therapist; the latter will be relatively inactive and will not give information or advice once the initial explanation of the treatment process has been completed. An atmosphere should be created in which the patients feels safe to speak about ideas and fantasies not previously revealed to anyone.

The therapist responds as much to the emotional content as to the intellectual components of the patient's utterances (for example 'it sounds as though you felt angry when this happened'). In this way the patient is helped to examine feelings that he may have denied, and to think back to past situations in which similar feelings were experienced.

Throughout treatment therapists should examine their own emotional responses to their patients. *Counter-transference* (the therapist's response to patients based on the therapist's own past experience) can involve inappropriate responses to the patient's emotional state or inappropriate identification with another person in the patient's life, for example one of the parents. Awareness of such reactions can help the therapist to understand the patient's situation better; lack of awareness can reduce his objectivity. Because such insight is difficult to achieve, therapists using dynamic psychotherapy often work with a supervisor who can help to identify the counter-transference. Some therapists undergo a period of psychotherapy to become more aware of events in the past which may determine their present counter-transference reactions to patients.

When links with past and present have been identified, patients are encouraged to use their new knowledge to change their current behaviour and feelings, and to deal more adaptively in relationships with others. Treatment should allow sufficient time for the patient to try out these new behaviours; otherwise patients may have greater insight but be left with their original problems.

In brief insight-oriented treatment it is important to discuss termination from an early stage; otherwise the patient may become dependent and the treatment may be prolonged. To achieve the planned time for termination, realistic goals should be chosen and retained, and the focus should remain on these original goals and should not be widened. In most cases, some problems will remain after focal treatment; it is important to make this point clear from the beginning so that any disappointment can be dealt with in the early stage of therapy rather than in the final stage.

The *selection criteria* for brief psychodynamic psychotherapy are a willingness to be introspective and honest, a problem that can be conceptualized readily in psychodynamic terms, and the absence of a severe mood disorder, schizophrenia, or a personality disorder leading to acting out of problems.

Further information about brief psychotherapy can be obtained from Mann (1973), Malan (1976), Sifneos (1979), and Davanloo (1980).

Cognitive analytic therapy

The intention of this treatment is to apply cognitive therapy techniques within a framework of psychoanalytic understanding. The patient is helped to recognize his own contributions to his problems by keeping diaries of moods, behaviour, and intrusive thoughts. In this way he can identify common patterns of behaviour such as avoidance or inappropriate efforts to please. Through a discussion of these behaviours, he is helped to identify maladaptive assumptions about relationships, for example if one person cares for another, the first must always give in to the demands of the other. Present patterns of behaviour and thinking are formulated as patterns that were adaptive when the person was younger but have become maladaptive as circumstances have changed. These behaviours can safely be changed since the original problems no longer exist.

At the start of treatment, goals are agreed with the patient. The relation of these goals to the patient's attitudes and values is examined so that possible resistances to change are identified. Goals are recorded on a 'target problems list', procedures are agreed that could bring about desired changes, and a time limit is set for therapy. The procedures are written down in an appropriate formulation that links them with the target problems; for example if the problem is that the person is trying to please everyone, the general procedure would be 'to assert needs appropriately'. Specific examples of general procedures are identified from the patient's diaries, and homework is arranged whereby these specific procedures can be tried out. At the end of treatment, the therapist may write a letter to the patient summarizing what has been learnt and encouraging him to continue to practice the new behaviours.

The general similarities of this approach to cognitive therapy will be apparent (see p. 635 for a description of cognitive therapy). The treatment differs from cognitive therapy in focusing less on specific patterns of thinking and more on interpersonal behaviours and the formulation of the assumptions that lie behind them. The sessions differ from psychoanalytic psychotherapy mainly in focusing less on transference interpretations and on identifying defence mechanisms, and more on the use of behavioural homework to test out formulations, a sequence sometimes described as description followed by demonstration. For an account of cognitive analytic therapy see Ryle (1990).

Long-term psychotherapy

Long-term psychotherapy differs from the brief methods in several ways. First, long-term treatments may begin in an unstructured way; instead of choosing a focus for discussion, the therapist asks patients to talk about anything that occurs to them as important, with the expectation that themes and goals will emerge later. Second, treatments differ in the number and kind of explanations offered by the therapist. In the brief treatment described above, explanations are based mainly on a common-sense understanding of human life. In more intensive treatment, they are based on some theory of psychological development, for example that of psychoanalysis, which holds that complex unconscious processes may determine how a person thinks, feels, and acts. As a rule, formulations based on a theoretical framework are not presented as a whole but are revealed piecemeal in the form of comments about the origins of behaviours and feelings as the latter gradually emerge during interviews. (These comments are one form of interpretation (see p. 611). Third, treatments differ in how far attention is given to matters other than conscious memories and attitudes. In some treatments unconscious material in the form of fantasies and dreams, or as revealed through paintings, is used to encourage patients to examine aspects of their mental life of which they were previously unaware. Fourth, treatments differ in how far the relationship between patient and therapist is encouraged to develop into a transference, which can then be utilized to put patients more in touch with their own feelings and reactions. Transference is discussed further on p. 602.)

Psychoanalysis

Psychoanalysis is the most intensive and prolonged form of long-term psychotherapy. Its practitioners receive lengthy training which includes personal analysis as well as supervised experience in treating patients. For these reasons and because results have not been shown to be better than those of shorter forms of treatment, psychoanalysis is not widely available as part of the health services of most countries.

The basic psychoanalytic techniques are those described above under brief insight-oriented psychotherapy, namely free association and interpretation. In psychoanalysis use is also made of dream analysis, which is thought to allow access to unconscious processes. The analyst asks questions to make the material clearer, confronts the patient with any contradictions, and makes interpretations. Otherwise he remains non-directive. As this procedure continues, the patient usually begins to avoid certain topics and may show other forms of **resistance** to treatment such as rejecting the therapist's interpretations. As in brief therapy, these evasions or rejections are examined for evidence about the patient's use of mechanisms of defence.

Gradually the patient's behaviour and talk begin to give direct or indirect evidence that he is developing intense but distorted ideas and feelings about the analyst. These distortions result from the **transference** to the analyst of ideas and feelings related to earlier experiences in the patient's life. As in brief therapy the analyst maintains a **treatment alliance**, i.e. a realistic relationship between patient and analyst reflecting the former's wish to change. In analysis, however, transference is deliberately encouraged by seeing the patient frequently (up to five times a week) and often by placing the patient on a couch instead of a chair.

As explained on p. 602, **negative transference** denotes the patient's hostile feelings to the therapist, whilst **positive transference** denotes the opposite feelings to the therapist such as dependency, idealization, or erotic feelings. If transference develops to such an extent that many of the patient's neurotic problems are re-experienced in relation to the analyst, this development is called **transference neurosis**. The analysis of the latter is an essential part of treatment. If transference becomes intense, the patient may express ideas and feelings not in words (as requested) but in actions (acting out). Such behaviour, like transference neurosis, is interpreted to reveal what light it throws on the underlying problems.

Interpretations may be rejected, sometimes because they are inaccurate but also because ingrained habits of thought can be changed only slowly and therefore require repeated **working through**. As the patient begins to accept interpretations he is said to gain **insight**.

As treatment progresses the analyst's feelings towards the patient change in ways that are partly realistic and partly distorted by his own previous experiences (**counter-transference** (see p. 603)). As already described, **counter-transference** refers to the feelings of the therapist towards the patient that are highly subjective and so a potential interference in treatment. In recent years the term counter-transference has been extended to all the analyst's feelings towards the patient, whether 'unrealistic' or not. According to Heimann (1950), some of these feelings provide the analyst with valuable insights into the patient's problems. For an account of the basic concepts of psychoanalysis the reader is referred to Greenson (1967), Symington (1986), and Sandler *et al.* (1992).

Other forms of insight-oriented psychotherapy

These vary in the nature of interpretations, in the relative emphasis put on present problems and early experience, and in the attention given to dreams. The variants of treatment proposed by *Jung, Klein,* and the *neo-Freudians* were referred to in the historical section earlier in this chapter. Readers who wish to obtain further information are referred to Munroe (1955).

Existential psychotherapy stems originally from the existential movement in philosophy. It is concerned with patients' ways of dealing with the fundamental issues of human existence—the meaning and purpose of life, isolation, freedom, and the inevitability of death. In this method of treatment, increased awareness of the self is more important than exploration of the unconscious, but many of the techniques are borrowed from brief psychoanalytic therapy. An account of existential psychotherapy has been written by Yalom (1980).

Results and indications of psychodynamic psychotherapy

It is uncertain how far the results of long-term psychotherapy are superior to those of shorter methods. In general the research literature supports the impression of experienced clinicians that most patients can be helped as effectively with short-term dynamic therapy as with longer methods. Clinical experience suggests that long-term therapy is more appropriate than shorter treatment for patients who have long-lasting and complicated emotional difficulties, or significant maturational problems in their personal development. In general, specific psychiatric symptoms respond less well than do problems of personal relationships. The use of these methods in schizophrenia and manic depressive disorder is not recommended. Long-term methods are contraindicated in patients with marked paranoid personality traits. Histrionic and schizoid personality disorders, while not contraindications, are particularly difficult to treat.

Small-group psychotherapy

This section is concerned with psychotherapy carried out with a group of patients, usually about eight in number. Treatment in larger groups is considered later. Small-group psychotherapy can be used with the intention of bringing about substantial change in symptoms, personal problems, or difficulties in interpersonal relationships, as a form of supportive treatment, or to encourage limited adjustments to specific problems including those of disabling physical or mental illness.

The development of group psychotherapy

Group therapy is often said to originate from the work of Joseph Pratt, an American physician who used 'class methods' to treat patients with pulmonary tuberculosis (Pratt 1908). However, Pratt's classes had little resemblance to modern group therapy, for they combined supportive conversations with instruction about the effects of the disease. A more obvious precursor of modern group psychotherapy was the work of J. L. Moreno, a Romanian who worked in Vienna before emigrating to the United States (he also laid the foundations of psychodrama). Trigant Burrow, an American, experimented with analysis in small groups (Burrow 1927), and his combination of analytic enquiry and comment on each patient's 'social image' has an obvious similarity to modern group therapy. Despite these developments and others up to the 1940s, it was the experience of treating neuroses in wartime in the United Kingdom that led to the further development of group therapy.

Pioneering steps were taken in the Northfield Military Hospital in England, where S. H. Foulkes developed the methods of group analysis which he had first tried in his civilian practice in 1941 (Foulkes and Lewis 1944). His approach was based on psychoanalysis; the therapist or group leader was relatively passive, and much use was made of analytic interpretations addressed both to individuals and to the group as a whole (Foulkes 1948, p.136). A different approach was developed by W.R. Bion, a Kleinian analyst whose interest in groups also grew from wartime experience at the Northfield Hospital (Bion 1961). Bion focused specifically on the unconscious defences of the group as a whole.

Other experiential group methods include sensitivity groups and encounter groups. These methods can be traced back to the **sensitivity groups**, which were started in the National Training Laboratories in the United States and were intended to teach community workers about group process by direct experience, and to **training groups** (T groups) in which people seeking 'personal awareness' (rather than treatment) talked frankly about their experiences and were willing to receive frank comments from other members of the group.

Encounter groups were developed as an intense form of experiential group. The 'encounter' might be entirely verbal, though the words were direct and emotive, or it might include physical contact, usually touching the other participants. In some groups, members' experience was intensified by prolonging it for a day or longer without interruption except for meals and sleep (**marathon** groups). These groups were popular in the United States in the 1960s and early 1970s. Although many participants reported that encounter groups were helpful, a systematic study by Lieberman *et al.* (1973) indicated that emotional problems increased in a minority. This increase was greater in members with substantial emotional disorders at the start, and in groups with the most direct and attacking methods.

Psychodrama was developed by Moreno (see above) as an intense form of encounter. The group enacts events from the life of one member, in scenes reflecting either current relationships or those of the family in which the person grew up. The enactment usually provokes strong feelings in the person represented, and may also reflect the problems of the other members. In these enactments members sometimes exchange roles so as to understand better the other person's point of view. The drama is followed by a discussion in which everyone takes part. Instead of building a drama round the personal experiences of one member, the action may be concerned with problems that the participants share, for example how to deal with authority. This method is called sociodrama. For an account of psychodrama see Goldman and Morrison (1984) and Holmes and Karp (1991).

Action techniques are now used mainly in combination with other group methods. A session of psychodrama can provide topics for discussion when a group using other methods is failing to make progress. Role reversal can help some patients to view their problems more objectively and perhaps for the first time from the standpoint of other people. An account of action techniques is given by Blatner (1973).

Therapeutic factors in small-group treatments

Group treatments share the therapeutic factors common to all kinds of psychological treatment (p. 601), namely listening and talking, release of emotion, giving information, providing a rationale, restoring morale, and prestige suggestion. In group treatment there are additional factors not present in individual therapies but common to all kinds of group psychotherapy. These factors are sharing experience, support to and from group members, socialization, imitation, and interpersonal learning.

Sharing experience, sometimes called universality, helps patients to realize that they are not isolated and that others have similar experiences and problems. Hearing from other patients that they have shared experiences is often more convincing and helpful than reassurance from a therapist to this effect.

Support to and from group members Offering understanding to other members of the group ('altruism') can increase the self-esteem of the person making the offer. Receiving help from other group members can be supportive to the person helped. The shared action of being mutually supporting is an aspect of *group cohesiveness* that can provide an important sense of belonging for patients who feel isolated in their everyday lives. Mutual support can also bind the group together when problems of individual members might otherwise cause divisions.

Socialization is the acquisition of social skills within a group through comments that members provide about one another's deficiencies in social skills, for example avoiding eye contact. This process can be helped by trying out new ways of interacting within the safety of the group.

Imitation or vicarious learning is learning from observing and adopting the behaviours of other group members. If the group is run well, patients imitate the adaptive behaviours of other group members; if it is not run well, they may imitate maladaptive behaviours such as extravagant displays of emotion, acting out problems, undue defensiveness, or talking in a way that deflects attention from their own emotional problems.

Interpersonal learning refers to learning about difficulties in relationships by examining the interaction of individuals with the other members of the group. As the group proceeds each person begins to interact with the other members as he would interact with people in daily life. If these interactions are characterized by transference, distortions, and maladaptive behaviour, the group provides a setting in which these characteristics can be pointed out to the patient, their origins considered, and alternatives tried out. Interpersonal learning is an important component of group therapy.

An outline of group treatment

In small groups the most widely used methods are dynamic interactional techniques which concentrate on present problems in relationships and how these problems are reflected within the group. The past is discussed only in so far as it helps to make sense of the present problems. The therapist seeks to utilize the therapeutic factors mentioned in the preceding paragraph, and to help each member to correct his false assumptions about other people's views of him.

Stages of group treatment

Group therapy of this kind goes through predictable stages. In the *first stage* the group tends to depend on the therapist, seeking expert advice about everyday problems and about appropriate ways of behaving within the group. Before long, some patients miss meetings or come late, either because they are anxious about talking in the group, or because they are angry and resentful about lack of immediate progress. In the *second stage* members begin to know each other better, they become used to discussing each other's problems, and they begin to seek answers. This is the stage in which most change can be expected. The therapist encourages the examination of current problems and relationships, and he comments on the dynamics of the group. In the *third stage* of treatment the group may be dominated by the residual problems of the members who have made least progress and still show the most dependency. This development should be anticipated by starting to discuss problems of termination several months before the group is due to end.

Preparation for the group

It is useful to prepare patients for their experience in a group by emphasizing the following points.

1. The confidentiality of the proceedings of the group.

2. The requirement for self-disclosure and for concern with problems disclosed by other members.

3. The need to attend regularly and remain throughout the sessions.

4. A warning about common reactions in the early stages of treatment, such as disappointment at the lack of rapid change, or frustration at the requirement that each member shared with others the time for speaking.

5. The duration of the group.

6. Whether the group is open or closed. Groups for dynamic psychotherapy are usually closed, thus helping to build support and sharing of responsibility. Open groups usually have supportive or educational purposes.

7. The requirement that members should not meet outside the group; if this rule is broken, it should be reported to the next meeting of the group.

Setting up the group

A therapeutic group usually has six to eight members. They should have problems in common, and should be able to empathize with each other's difficulties and to assist each other. No member should have problems that set him aside from everyone else in the group. Meetings are best held in a room of adequate size, with chairs in a circle so that all members can see one another. Meetings usually last for an hour or 90 minutes to allow adequate time for every member to take part; they are usually held once a week and generally continue for 12–18 months.

Groups may be open or closed. A *closed group* does not accept any new members, however many drop out. An *open group* accepts new members. Therapeutic groups are usually closed, though sometimes one or two new members are accepted if there are drop-outs in the first few sessions. Thereafter it is difficult to integrate new members into the group.

One or two therapists? Some groups are run by one therapist, others by co-therapists. The advantage of employing two therapists is that one can help the other to recognize and deal with countertransference problems (which are as important in group as in individual psychotherapy). The potential disadvantages are that the two therapists may develop different views about the running of the group, or may behave defensively with one another. If the therapists trust each other's judgements and if these differences are discussed as they arise, they can gain further insight into the group process. At the start, differences between co-therapists may be discussed outside the group. When the group is well integrated, such differences can be discussed within the group as a way of encouraging members to gain more understanding of interpersonal processes.

Managing the group

The therapist should be aware of five basic issues that are likely to appear in the members of a group and to require attention:

(1) the conflict between the wish to be helped and the requirement to help others;

(2) the conflict between a wish to gain the therapist's approval and the desire to be approved by the other members of the group;

(3) the process by which members of a group establish a hierarchy of dominance, and the rivalries that this may produce;

(4) the risk that one person may become a scapegoat when the group is dissatisfied;

(5) the risk that members may be excessively passive and dependent on the therapist for advice instead of working out solutions themselves.

If a therapeutic group is working well, giving and taking are realized to be complementary, emotional support is balanced by a readiness to challenge and be challenged, and problems are accepted rather than projected on to other people.

To supplement this outline of group therapy the reader is referred to Vinogradov and Yalom (1989) for a brief guide, and to Yalom (1985) for a longer account.

Problems in group therapy

Some members may form a coalition based on age, shared values, social class, or some other characteristic. This subgrouping is undesirable and may disrupt the therapeutic process. Early signs of such alliances should be looked for and discouraged, and the reasons for their formation should be discussed by the whole group.

Members who talk too much

In its early stages the group may welcome a talkative member who relieves the others of the need to speak about themselves. As meetings continue the group is likely to become dissatisfied with this member for monopolizing time that could be better shared. The therapist should draw attention to this problem at an early stage before the group becomes rejecting. Attention can some-

times be drawn most effectively by asking the group why they let one person absolve them from the need to speak about themselves when they could be addressing the talkative member.

Members who talk too little

The therapist should assist silent members to speak and should therefore understand the reasons for silence. Some patients are generally awkward in company; some are afraid to reveal a specific problem to the group and fear that this problem will be uncovered if they speak at all, and some are silent because they are angry and dissatisfied with the progress of the group.

Conflict between individual members

Conflicts can impede the therapeutic process. The therapist should not take sides in such conflicts but should encourage the whole group to discuss the issue in a way that leads them to understand why the conflict has arisen, for example because a hostile transference has developed.

Focus on current problems

In most therapeutic groups the focus is mainly on the current problems of the members in their daily lives as reflected in their interactions within the group. As the group progresses, attention to group interactions increases and is used to understand the individual members' ways of feeling, thinking, and behaving. Members are encouraged to try out these new insights in daily life and to report the result to the group. When patients persist in talking about the past, questions and interpretations can be used to bring discussion back to relationships in the group. For example if a woman mentions difficulties with her overpowering father many years ago, she may be asked whether she raises this issue because one or more of the men in the group seem to be behaving in a similar way. Similarly questions and interpretations can be used to link accounts of recent problems outside the group to interactions with other group members. For example if a patient says he has been unfairly criticized by colleagues at work, he may be asked whether he feels unfairly criticized in the group, and by whom.

The therapist should help the group to examine their interactions and feelings in a constructive way, so that the members feel helped by the opportunity to learn more about themselves rather than criticized or embarrassed. Members are more likely to benefit from this when specific rather than general issues are discussed; for example it is more useful to discuss why a woman feels angry when a particular man in the group offers her advice than to discuss why she has difficulties with men in general. Consideration of specific instances can lead to understanding of wider problems. Other members are encouraged to work on the problem, and the therapist can help by commenting on their reactions and on his own feelings about the interaction.

Group analytic therapy

This treatment aims to bring psychoanalytic understanding to group work. It differs from the eclectic method described above mainly in the greater use of transference interpretations and interpretations of the unconscious context of the members' discussions. The therapist generally speaks less than his counterpart in an eclectic group. He encourages the group to speak and interact, and he uses his own feelings and experiences to identify and point out the unspoken messages and emotional reactions within the group. The therapist is alert to detect shared preoccupations and conflicts within the group, and he interprets them to the group. In analytic group therapy particular attention is given to transferences to the therapist and between members. The group processes are thought to stimulate members to experience feelings and responses that stem from their families of origin. These experiences are interpreted in terms of relationships in earlier life with parents and siblings. Examination of transference in the group can help patients to understand how it may be disturbing relationships outside the group in terms of earlier experiences. This process can lead to questioning of other preconceptions, and a greater understanding of aspects of the self that have been disowned. The method has been described by Foulkes and Anthony (1957).

Indications for group therapy

There is no evidence that the results of group therapy differ in general from those of individual psychotherapy of the same duration, or that the results of any one form of group therapy differ from those of the rest. In particular there is no evidence that encounter groups or action techniques are superior to other methods, and there is some evidence that they can increase symptoms in some patients (Yalom *et al.* 1973).

Long-term group therapy (as opposed to long-term individual psychotherapy) has no specific proven indications, but it is generally thought to be well suited to patients whose problems are mainly in relationships with other people. As in individual psychotherapy, results are better in patients who are young, well-motivated, able to express themselves fluently in words, and free from severe personality disorder. Groups are often suitable for patients with moderate degrees of social anxiety, presumably because such patients benefit from the opportunity to rehearse social behaviour. Severe social anxiety is a contraindication. Other contraindications are similar to those for long-term individual psychotherapy, with the additional point (noted above) that a group should never include a solitary patient whose problem is so unlike those of other members that he may become an outsider or scapegoat.

Other forms of group therapy

Therapy in large groups

This form of therapy is characteristic of therapeutic communities and is also part of the daily programme of many psychiatric wards. Large groups usually include all the patients in a treatment unit together with some or all of the staff, with the total membership varying from 20 to 50. At the simplest level, large groups allow patients to express problems of living together. At a more ambitious level, such groups can attempt to change their members through social learning. Change is attempted by presenting each member with examples of his disordered behaviour or irrational responses. At the same time support is provided by other members who share similar problems and opportunities for social learning. The group is sometimes made into a kind of governing body that formulates rules and seeks to enforce them. Because large groups can evoke much anxiety in patients and staff, care should be taken to prepare new members for the experience. It is also important to protect vulnerable people from attacks by other group members and to decide when patients are too unwell to participate.

Supportive groups

Groups led by a professional therapist may be used to support patients with chronic psychiatric or physical illness, or people with shared problems such as the parents of handicapped children.

Self-help groups

All the groups described so far have been led by a professional therapist. Self-help groups are organized and led by patients or ex-patients who have learnt ways of overcoming or adjusting to their difficulties. The other group members benefit from this experience, from the opportunity to talk about their own problems and express feelings, and from mutual support. There are self-help groups for people who suffer from many kinds of disorders; among the best known are Alcoholics Anonymous (see p. 459) and groups to help people lose weight (Weight Watchers). Other self-help groups are helpful to patients with chronic physical conditions such as colostomy, to people facing special problems such as single parents or those with a handicapped child, and to the bereaved (Cruse Clubs). If they are well run, such self-help groups can be of value. To ensure good practice, some groups have professional advisers.

A study of self-help group treatment for recently bereaved women found that it was as effective as brief dynamic psychotherapy (Marmar *et al.* 1988), although the value of self-help for weight reduction seems to be more limited (Stunkard *et al.* 1970). For a review of self-help groups see Lieberman (1990).

Marital therapy

The term marital therapy implies treatment for a husband and wife; hence the term *couple therapy* is sometimes used as being applicable to couples who are living together, whether married or not. In this section the term marital therapy will be used because most of the reported work is with spouses. Treatment of this kind is usually given either because marital conflict appears to be the cause of emotional disorder in one of the partners, or because the marriage appears likely to break up and both partners wish to save it. Treatment for the latter reason is sometimes called marriage guidance rather than marital therapy. (Family therapy, which is discussed later, differs in including one or more other family members, usually children.)

In the apparently simple step from treating an individual person to treating a couple, there is an important conceptual issue—that the problem is not confined to one person but shared between the two. The problem is conceived as resulting from the way that the partners interact, and treatment is directed to this interaction. In assessing the interaction it is useful to examine issues that are important in all marriages and cohabitations, for example the sharing of values, concern for the welfare and personal development of the partner, tolerance of differences, and an agreed balance of dominance and decision-making. It is useful also to bear in mind the stages through which a marriage passes: first living together, then bringing up children, and finally readjusting when the children leave home. To avoid imposing values, the clinician adopts a 'target problem' approach, whereby couples are required to identify the difficulties that they would like to put right.

The development of marital therapy

This therapy is a relatively recent development which, in the United Kingdom, owes much to the work of Henry Dicks. In his book *Marital tensions* Dicks (1967) proposed that psychoanalytic ideas were useful in understanding and treating marital problems. In the United States, an important influence was that of Bateson's group who studied modes of communication in families in Palo Alto (Haley 1963; Watzlawick *et al.* 1968). Another development was the introduction of behavioural principles. Each of these three approaches to marital therapy will be described briefly.

Types of marital therapy

Analytic methods

These methods employ concepts from psychoanalysis. A central idea is that the behaviour of a married couple is largely determined, from the moment that they choose each other, by unconscious forces. Each person selects a spouse who is perceived as completing unfulfilled parts of himself. When the selection is successful the couple complement one another, but sometimes one partner fails to live up to the (unconscious) expectations of the other. For example a wife may criticize her husband for failing to show the independence and self-reliance that she lacks herself. Also, each partner may attribute to the other unwanted aspects of the self which are split off, denied, and projected onto the other. For example a husband may project on to his wife a vulnerability that he feels but cannot accept.

Therefore the aim of this kind of treatment is to help each person to understand his own emotional needs and how they relate to those of the partner. Various combinations of patients and therapists are used. One therapist may see two partners together, two therapists may see them together (each therapist having a primary concern with one of them), or separate therapists may see the patients separately but meet regularly to coordinate their treatments. Opinions differ about the value of and indications for each of these methods. In marital therapy of this kind, therapists take a more active part than they would in the analytic treatment of a single patient. Therapists are also less likely to make interpretations about transferences towards themselves and more likely to comment on the relationship between the partners and how it reflects the childhood experience of each, and how each may be transferring needs or fears on to the other.

Transactional methods

In these methods, one or two therapists may take part but the partners are always seen together. The focus of treatment is on the hidden rules that govern the behaviour of the couple towards one another, on disagreements about who makes these rules, and on inconsistencies between these two 'levels' of interaction. These issues are discussed around conflicts arising in the everyday life of the couple, for example who decides where to go on holiday, and how they decide who is to decide this. In this way it is hoped to arrive at a more balanced and more cooperative relationship. A lively account of the method has been given by Haley (1963).

Behavioural therapy

This form of marital therapy is brief and highly structured. It uses so-called operant-interpersonal techniques. The therapist tries to identify ways in which undesired behaviour between the couple is being reinforced unwittingly by one of its consequences. Each partner is then asked to say what alternative behaviours would be desirable in the other person. These behaviours must be described in specific terms, for example 'talk to me for half an hour when you come in from work' rather than 'take more notice of me'. Each partner then has to agree a way of rewarding the other when the desired behaviour is carried out. The rewarding behaviour could be the expression of approval and affection, or the carrying out of some behaviour that the partner desires. This exchange is often called 'give to get'. The couple are helped to communicate more directly, to listen to one another, and to express their own wishes more clearly. Described as briefly as this, the treatment may seem a crude form of bargaining that is remote from a loving relationship. In practice it can enable a couple to co-operate and give up habits of criticism and nagging, with consequent improvement in their feelings for one another. The method has been described by Stuart (1980).

Eclectic methods

As well as these formal methods, there is an important place for less formal treatment directed to specific problems as part of a wider plan of treatment. For example many depressed patients have some marital difficulties that are contributing to their problems. A few sessions can then be undertaken with the couple, with attention directed to specific and limited goals.

Results and indications

Gurman (1979) reviewed the literature and found fairly strong evidence that marital therapy is better than no treatment, and that the behavioural form of marital therapy is followed by improvement in about 60 per cent of cases. Crowe (1973) compared a behavioural form of marital treatment with two other methods, one combining elements of systems theory and interpretation, and the other a non-directive approach. The non-directive approach was followed by least improvement, but the other two methods did not differ in their effects. Clinical experience supports the value of the eclectic methods just described, particularly if they are incorporated in a wider plan of treatment.

Family therapy

Several or all of the family members take part in this treatment. Usually both parents are involved, often together with a child whose problems have made the family seek help. At times they are joined by other children, and by grandparents or others who make up the family unit. The aim of treatment is to alleviate the problems that led to the disorder in the identified patient and to improve family functioning, rather than to achieve some ideal state of a healthy family. Since success depends on the collaboration of several people drop-out rates are understandably high.

Family therapy is a recent development dating from the 1950s. It can be traced to two sources: an influential book by Ackerman (1958) on *The psychodynamics of family life*, and the work on communication by Bateson and his colleagues mentioned above. Ackerman's work led to psychodynamic methods of treatment, whilst Bateson's treatments were directed to improving communications between family members. At the same time Minuchin developed methods suitable for use with disadvantaged poor families, with the emphasis being on a practical approach to resolving problems—an approach he called structural family therapy.

In the following account four kinds of family therapy will be described: psychodynamic methods, structural family therapy, systemic family therapy, and eclectic approaches. The interested reader will find a comprehensive account of family therapy in Gurman and Kriskern (1991).

Whatever their methods, family therapists have the following goals for the family: (a) improved communication, (b) improved autonomy for each member, (c) improved agreement about roles, (d) reduced conflict, and (e) reduced distress in the member who is the patient.

Types of family therapy
Psychodynamic methods

These methods use concepts taken from the psychoanalytical treatment of individual patients. It is assumed that current problems in the family originate in the separate past experiences of its individual members, particularly those of the parents. Present problems arise in part from unconscious conflicts within individual members. Family members need to gain insight into these conflicts in order to change their behaviour. The therapist's task is to help members gain these insights into their own unconscious, and into the ways in which one person's problems may interlock with those of another. The therapist does this by examining the relationship between himself and the different family members to throw light on the unconscious aspects of the problems. The therapist may give interpretations but is not directive. A group analytic approach to family therapy has been developed by Skynner (1991) who uses a Kleinian object relationship approach (see p. 606).

Skynner drew attention to the ways in which the childhood experience of each of the patients affects the ways that they relate to their own children; for example a mother who received no adequate mothering herself may have developed no internal model to guide her and may develop a 'projective system', i.e. current expectations shaped by childhood experiences. Projective systems affect the parents' ways of relating to one another as well as to the children. Sometimes a conflict between the projective systems of the parents is resolved by diverting the projections onto the child. The therapist must experience these projective systems in order to understand them and help the family resolve the conflicts between them.

Structural family therapy

The term family structure refers to a set of unspoken rules that organize the ways in which family members relate to one another. Some rules determine the hierarchy in the family, for example that parents have more authority and more responsibility than children. Some rules determine cooperation in the family, for example that father and mother share certain tasks and responsibilities, and take on others individually. In some families both parents set rules for behaviour and admonish children when the rules are broken; in other families the father is the strict parent. Rules also determine boundaries; sometimes these are broken, for example when an unhappy wife involves her son in her problems with her husband. In structural family therapy, hypotheses about these rules are often presented to the family in a paradoxical way; for example, 'you seem to be very dependent on your wife; what does she do to make you feel less competent?' Such interventions may increase family tension in the short term until a new set of rules is established.

Systemic family therapies

These methods originated in the work of Haley (1963) and Satir (1967), both members of the Mental Research Institute in Palo Alto, California, and in the work of Minuchin (1974), an Argentinian psychoanalyst who worked in New York and Philadelphia. Strategic therapy attempts to change the present functioning of the family, rather than to explore the past. It is assumed that current symptoms result from a maladaptive attempt to change a problem. The attempt is maladaptive because of problems in the family system, i.e. unspoken rules of behaviour, disagreements about who makes these rules, and distorted communication. The therapist's role is to identify the rules, to help the family to modify them, and to improve communication.

Systemic family therapy was developed by a group of Italian psychiatrists in Milan (Palazzoli *et al.* 1978). In this approach the family is asked to provide information from which 'hypotheses' are constructed about the functioning of the family system, and the way in which family rituals operate. These 'hypotheses' concern the breakdown of the boundary between parent and child subsystems, for example that a parent has formed a strong alliance with one child. Another 'hypothesis' is that symptoms have a positive side, aiding some aspect of family cohesion. Such family cohesion hypotheses are presented to the family, who consider them in and between sessions. The therapist may use challenges or paradoxical injunctions; for example he may tell the family of an anorexic patient that she must remain underweight. This injunction is intended to unite the parents against the therapist and thereby reduce the patient's concern that they may separate; the therapist's assumption is that the family will react against the injunction and act in a contrary way. The therapist may also use 'circular questioning' in which each person is asked to comment on a single event; the purpose is to discover and clarify confused or conflicting views. The therapy is brief, generally consisting of five to ten sessions, but the interval between sessions is often a month or more ('long brief therapy'). In a review of ten outcome studies of Milan systemic family therapy (five of the studies included comparison groups), this approach was found to be as effective as other kinds of family therapy. There was symptomatic improvement in about two-thirds of patients, and improved functioning in about half the families. Generally these results were obtained in less than ten sessions (Carr 1991).

Eclectic approaches

In everyday clinical work, especially with adolescents, it is practicable to use a simple short-term method designed to bring about limited changes in the family. For this purpose, it is appropriate to concentrate on the present situation of the family and to examine how the members communicate with one another. Specific goals are agreed with the members of the family, who are asked to consider how any changes will affect themselves and others, and what has prevented the family from making the changes. The number of family members taking part should be decided on practical rather than theoretical grounds; for example some children may be too young, whilst others may be away as students for much of the time. Sessions can be at varying intervals, for example weekly at first and then every three weeks to allow time for the family to work on the problems raised in treatment.

Indications and contraindications

Family therapy is used mainly in the treatment of problems presented by young people living with their parents. Skynner (1969) has suggested that conjoint family therapy is most useful in two circumstances: when the parents cannot cope with the behaviour of a child or adolescent, or when the family is making one member a scapegoat for shared problems. These problems are often related to difficulties in communication between members of the family, or to role problems such as the adoption of a parental role by one of the children (who may be looking after other children or one of the parents).

If one member of a family has a serious illness requiring treatment in its own right (for example a severe depressive disorder) family therapy may be combined with other treatment. For example antidepressant medication can be given for the depressive disorder, whilst family therapy can be used for any family conflicts that appear to be prolonging the disorder. Similarly, family therapy may help in the treatment of anorexia nervosa after appropriate steps have been taken to restore weight. The use of special kinds of family treatment to reduce relapses in schizophrenia is described on pp. 285 and 289.

Results of family therapy

In a meta-analysis of the results of 19 studies of family therapy, the effect was found to be comparable to that of other forms of psychotherapy. About 75 per cent of patients receiving family therapy had a better outcome than similar patients receiving minimal or no treatment. The size of the effect increased during the first year after treatment, remained unchanged for a further six months, and then decreased (Markus *et al.* 1990).

For an account of the techniques of family therapy see Barker (1992).

Psychotherapy for children

The kinds of psychotherapy discussed so far do not lend themselves to the treatment of young children who lack the necessary verbal skills. In practice there are fewer difficulties than might be expected, because many emotional problems of younger children are secondary to those of their parents and it is often appropriate to direct psychotherapy mainly to the parents.

Some psychotherapists believe that it is possible to use the child's play as equivalent to the words of the adult in psychotherapy. **Klein** developed this approach extensively by making frequent analytic interpretations of the symbolic meaning of the child's actions during play, and by attempting to relate these actions to the child's feelings towards his parents. Although ingenious, this approach is highly speculative since there is almost no evidence against which the interpretations can be checked. **Anna Freud** developed child psychotherapy by a less extreme adaptation of her father's techniques to the needs of the child. She recognized the particular difficulty for child analysis of the child's inability or unwillingness to produce free associations in words. However, she considered that neither play with toys, nor drawing and painting, nor fantasy games could be an adequate substitute. Moreover, she cautioned against the use of uncontrolled play which may lead to the acting out of aggressive urges in a destructive way. Anna Freud accepted that non-analytic techniques could be helpful for many disorders of development. These techniques include reassurance, suggestion, the giving of advice, and acting a role model (an 'auxiliary ego'). However, for neurotic disorders in childhood, and for the many mixed disorders, she advocated analytic techniques to identify the unconscious content of the disorder and to interpret it in a way that strengthens ego functions. A concise account of Anna Freud's views on child analysis is contained in A. Freud (1966, Chapters 2 and 6).

In the United Kingdom most psychotherapy for children is *eclectic*; the therapist tries to establish a good relationship with the child and to learn about his feelings and thoughts, partly through talking and listening, and partly through play. Older children can communicate verbally with adults but younger children can communicate better through actions, including play. The therapist can help children to find words that express thoughts and feelings, and thus can make it easier for them to control and change these thoughts and feelings. However, it is important, to ensure that the therapist's interpretations are reflecting the child's own thoughts and not implanting new ideas. Child psychotherapy is discussed further in Chapter 20 on child psychiatry (p. 678).

Research in psychotherapy

The aim of medicine must be to use only treatments that have proved valuable in clinical trials, but many treatments have not yet been evaluated in this way. Thus many common

surgical procedures can be judged only on clinical grounds because they have not been subjected to comparative trials.

There are several reasons why a treatment continues to be used without complete evaluation. Sometimes it has been used for so long with such obvious benefit that testing it seems unethical, for example appendicectomy. For other procedures, the desired outcome is so complicated that it is difficult to devise valid and reliable ways of assessing it objectively, for example the use of physiotherapy for rheumatoid arthritis. Psychotherapy is used largely on the basis of clinical opinion rather than scientific evaluation. One reason for this limitation is the real problem of measuring the changes that psychotherapists aim to achieve (this and other research problems are discussed below). Another reason is the belief, held by some clinicians, that some kinds of psychotherapy are obviously beneficial. In reading this chapter, it is important to remember that many recommendations about the indications for various kinds of psychotherapy are based only on clinical experience and therefore are subject to revision when evaluative studies are performed.

Although many investigations of psychotherapy have been carried out, definite results are few. Hence only a brief account will be given here. Further information on the extensive research literature can be found in the review edited by Bergin (1994).

When first attempted, psychotherapy research suffered from overambitious attempts to investigate complex problems although the available methods of assessment were suited only to simple issues. There was an understandable wish to establish the value of psychoanalytically based treatments. This problem cannot be answered because there are no sufficiently valid and reliable methods for the assessment of basic psychodynamic variables. Moreover, there is no unambiguous specification of the psychoanalytical method that would allow one study to be replicated by another research team. No amount of complicated experimental design or elaborate statistics can overcome these fundamental problems.

Psychotherapy research has been most informative when concerned with the less complex forms of treatment. Carl Roger's pioneering studies of client-centred therapy showed that research was feasible (Rogers and Dymond 1954), but these methods are seldom used in clinical practice. In subsequent enquiries into short-term dynamic therapies, a group at Johns Hopkins Hospital led by Jerome Frank showed that clinically relevant results could be obtained (Frank *et al.* 1978). These investigations and others in psychotherapy can be divided into studies of outcome and studies of the processes occurring in the therapy sessions.

Many of the published studies of **outcome** are difficult to interpret because they lack the untreated control groups that are required if treatment effects are to be separated from spontaneous recovery. In long-term treatments this problem is difficult to overcome because untreated controls are seldom available, as most patients are unwilling to wait long without treatment. Thus most information concerns brief treatment.

Several attempts have been made to summarize the information from the various studies of psychotherapy. There have been two approaches. The first approach, which was adopted many years ago by Luborsky *et al.* (1975), is to accept for review only investigations that reach certain minimal scientific standards. The second approach, which was first adopted by Smith and Glass (1977), is to include all studies having a comparison group and to subject them to a statistical analysis designed to yield a composite estimate of change. The two approaches give similar results: psychotherapy leads to greater change than is found in comparable patients who are untreated, but no differences can be detected between the results of different forms of psychotherapy. More recent reviews have reached similar conclusions (Svartberg and Stiles 1991; Crits-Cristoph 1992).

Even this finding has to be qualified. Frank's research group found that the difference between treated and control groups grew less with the passage of time after treatment had ended, so that five years later no differences existed. The explanation appeared to be that the control groups continued to improve slowly for several years and eventually caught up with the treated patients. This finding suggested that psychotherapy had

merely accelerated natural processes of change in these patients (Stone *et al.* 1961). This finding must be treated cautiously because Frank's group studied patients with disorders that were not severe or long-standing; hence the results may not apply to patients with disorders of greater severity or duration.

In assessing any outcome study it is important to recognize that hidden selection processes operate before a patient is accepted for psychotherapy. This applies particularly to comparisons between American and British studies. As Goldberg and Huxley (1980) point out, in the United Kingdom a patient with a psychiatric disorder has to pass through three 'filters' before he can be treated with psychotherapy. First, he has to decide that his problems are appropriate to take to a general practitioner. Second, the general practitioner has to recognize that the problem requires treatment from a psychiatrist. Third, the psychiatrist has to decide to use psychotherapy rather than some other treatment. In the United States, one of these filters is often removed because patients refer themselves directly to psychiatrists. Shepherd *et al.* (1966) have shown that, in England, general practitioners refer only about 5 per cent of their patients with identified psychiatric disorder to psychiatrists. This difference in referral methods is likely to alter substantially the types of patient receiving psychotherapy in the two countries, possibly resulting in less severe disorders reaching psychotherapists in the United States.

The investigations reviewed so far were designed to measure possible beneficial effects of psychotherapy. The idea that psychotherapy could harm some patients was first discussed fully in the book *Psychotherapy for better or worse* by Strupp *et al.* (1977) and soon afterwards in the review by Bergin and Lambert (1978). The latter authors found nine well-conducted investigations in which some worsening of symptoms seemed to have occurred in some patients. Clinical experience indicates that, when harm results from psychotherapy, it is usually in the form of excessive preoccupation with emotional problems, increase of symptoms, and 'acting out'.

Research into the **process** of psychotherapy has added little to the results of outcome studies. There have been several attempts to find which techniques and what aspects of the therapist's personality are associated with good results. Among investigators concerned with the personality of the therapist, Whitehorn and Betz (1954) reported that they could identify two types of therapist whose results differed when treating schizophrenic patients with psychotherapy. However, other workers, have not confirmed these claims consistently (Parloff *et al.* 1978). It has also been argued that the results of treatment vary with measurable qualities of empathy, warmth, and genuineness of the therapist (Truax and Carkhuff 1967). This notion has also not been confirmed by others (Shapiro 1976).

The study of patients who **drop out** of treatment has been more fruitful. Dropping out depends on factors in the patient, the therapist, and the treatment (Frank *et al.* 1957). Patients who drop out are more likely to be of lower social class, less educated, less integrated into society, less ready to talk about their feelings, to have persevered less in any previous treatment, and to be receiving a treatment that does not match their expectations.

The factors that determine whether patients stay in treatment are not necessarily the same as those that lead to improvement in those who remain. Every psychiatrist knows of patients who have persisted for years with a treatment that has not helped them at all. According to Frank's group, the likelihood of patients dropping out of psychotherapy can be reduced by making their expectations about treatment more realistic (Hoehn-Sarik *et al.* 1964). Although such methods are attractive, their value is not firmly established. When Yalom *et al.* (1967) prepared patients in a similar way for group therapy, attendance did not improve although the patients learnt more quickly what was required of them during group sessions.

Behaviour and cognitive therapies

The term **behaviour therapy** is applied to psychological treatments based on experimental psychology and intended to change symptoms and behaviour. Two other terms are used to describe these methods: **behaviour modification** is employed both as a synonym for behaviour therapy and to refer to a particular group of procedures based on operant conditioning; **behavioural psychotherapy** generally refers to behaviour therapies other than operant methods. The term **cognitive therapy** is applied to psychological treatments intended to change maladaptive ways of thinking and thereby bring about improvement in psychiatric disorders.

How behaviour therapy developed

Behaviour therapy can be traced to Janet's (1925) methods of **re-education** which were used for disorders with a behavioural element. These early methods arose from practical experience rather than from any formal theory. It was the well-known experiments on conditioning by Pavlov and on reward learning by Thorndike and others (Thorndike 1913) that provided a theoretical basis for a treatment based on experimental psychology. The practical application of these findings can be traced to the experiments of Watson and Rayner (1920). These workers showed, for example, that fear responses in a healthy child could become associated with a previously neutral stimulus by Pavlovian conditioning. This experiment suggested that naturally occurring fears might be removed by comparable methods. However, although behaviourism continued as a dominant force in psychology throughout the 1920s and 1930s, especially in the United States, there were few applications to treatment except in the (now largely abandoned) use of aversion therapy for alcoholism.

The development of modern behaviour therapy dates from the 1950s when it grew from three separate beginnings. In the United Kingdom, psychologists working at the Maudsley Hospital applied learning principles to the treatment of individual patients, especially those with phobic disorders. In South Africa, Wolpe developed a treatment based on his experimental work with animals. He subsequently described it in an influential book *Psychotherapy by reciprocal inhibition* (Wolpe 1958). This book was a landmark because, for the first time, the clinician was offered a practical and widely applicable behavioural treatment procedure backed by a reasonably convincing theory and supported by results.

The third strand of development began with Skinner's *Science and human behaviour* (1953) in which he argued that normal and abnormal behaviours are governed by the laws of operant conditioning and that similar principles could be used to change them. These beginnings explain the subsequent course of development of behaviour therapy in the United Kingdom and the United States. Wolpe's ideas were introduced to the United Kingdom by Eysenck (1960), and were soon adopted because they were easily assimilated to the methods which the Maudsley Hospital psychologists had started. In the United States however, there was a vigorous development of methods based on operant conditioning. Over the subsequent years there has been a convergence of approach, and similar methods are now used in the two countries.

How cognitive therapy developed

Cognitive therapy followed two distinct but related lines of development. The first began with the work of A. T. Beck, a psychiatrist, who became dissatisfied with the results of psychoanalytic psychotherapy for depressed patients and sought an alternative approach. He was struck by the recurring themes in the thinking of depressed patients (for example about personal failure), and suggested that these themes should be regarded as part of the primary disorder rather than secondary to either underlying unconscious conflicts or biochemical abnormalities. Beck went on to develop a treatment designed to alter these recurring thoughts (see p. 639).

The second line in the development of cognitive therapy arose from the work of psychologists who were dissatisfied with the behavioural approach to treatment originating from work on operant conditioning. The contribution of one of the psychologists, D. H. Meichenbaum, will serve as

an example. He examined the thinking of people with minor emotional disorders and noticed that certain kinds of recurrent thoughts were described frequently by these people (for example thoughts about social embarrassment). He then set out to develop ways of controlling these thoughts. Subsequently, this approach to treatment has been taken up by many other psychologists and psychiatrists, and has been applied to conditions such as bulimia nervosa, panic attacks, and hypochondriasis. These applications will be discussed later in this chapter.

Cognitive and behavioural approaches to treatment

All psychiatric disorders have cognitive and behavioural components and, for the patient to recover, both elements must change. Behaviour therapy is designed to bring about a direct change in the behavioural component of psychiatric disorders, but if the patient is to recover, the other cognitive components of the disorder must change indirectly. Equally, if a patient is to recover with cognitive therapy, a primary change in cognitions must be followed by a secondary change in behaviour. This does not mean that cognitive and behavioural treatments are interchangeable; in some conditions it is more effective to start by changing behaviour, in others by changing cognitions. For example obsessional disorders are characterized by ritual behaviours and obsessional thoughts; in these conditions improvement is greater when rituals are treated with behaviour therapy than when obsessional thoughts are treated with cognitive therapy. However, it is becoming clear that in many disorders the best results are obtained by combining the two approaches. Thus in most phobic disorders behaviour therapy given alone produces incomplete improvement, and a combined cognitive–behaviour treatment is more effective (see p. 637). Because these two approaches are often combined, the term cognitive–behaviour therapy (abbreviated to CBT) is often used. Some general principles of behavioural and cognitive treatment are outlined below. This account will be followed by brief descriptions of the more common procedures.

Principles of behavioural treatment

In behavioural treatment the therapist attempts to alter a prominent behavioural element in a psychiatric disorder by encouraging the patient to carry out an incompatible behaviour. For example in phobic patients avoidance behaviour is treated by encouraging them to enter situations that they fear ('exposure'), and in depressed patients inactivity and social withdrawal are treated by participation in enjoyable activities ('activity scheduling').

Although these simple procedures are central to behaviour therapy, it is not enough just to instruct the patient to carry them out. Two difficulties must be overcome. First, patients are often unaware of all the occasions on which the abnormal behaviour is present and consequently do not attempt the therapeutic behaviours at the right times. Second, the therapeutic behaviours have to be practised frequently and for a long time, and most patients lack the motivation to persevere. Each of these difficulties can be overcome.

Patients can be helped to identify the problem behaviours by keeping detailed records of the occurrence of symptoms, the events that provoked them, and their attempts to cope. This kind of enquiry is called a *behavioural analysis*. The analysis of social phobia can serve as an example; although patients with this disorder are aware of anxiety in social situations, they can seldom report which features of the situation are most stressful and which are avoided. A behavioural analysis will usually establish that the patient is anxious only in specific aspects of social encounters, such as initiating conversation when speaking to an older man. Graduated exposure to the specific elements, one at a time, usually reduces anxiety more effectively than a less targeted approach.

It is essential to *overcome poor motivation* if good results are to be obtained with behaviour therapy. Although most patients have a general wish to recover, they usually have inadequate motivation to practise specific behavioural proce-

dures repeatedly for weeks or months. Lack of motivation is often explained by demoralization consequent upon previous unsuccessful attempts to control the abnormal behaviour. Usually motivation can be restored by explaining how the proposed programme differs from the patient's previously unsuccessful attempts at overcoming the problems. Motivation can be maintained more easily if each procedure is presented as an experiment in which the completion of a task is valuable, whilst inability to complete is not a failure but an opportunity to learn more about the disorder.

Principles of cognitive treatment

In cognitive treatment, the therapist attempts to change one or more of the disordered ways of thinking that characterize the disorder, for example the irrational fears of a phobic patient or the unreasonably pessimistic ideas of a depressed patient. The aim is to change these ways of thinking directly with the expectation that other changes will follow. Several techniques are used to produce changes in thinking.

The first step is to help the patient to become aware of the irrational ideas. Although patients can describe some of these ideas, they are usually unaware of others. These unacknowledged ideas are usually important in maintaining the disorder. Irrational ideas can be elicited by careful interviewing, in which the patient is asked about the reasons for his own actions and what could be the worse outcome of his behaviour; for example what would happen if a social phobic were to join a social group instead of avoiding it. Irrational ideas can also be identified by asking patients to keep a daily record of thoughts experienced at times when the symptoms increase (for example what they were thinking just before they felt depressed).

Next, an attempt is made to change the irrational ideas. Two kinds of technique are used, verbal and behavioural. It may at first seem paradoxical that behavioural procedures are part of cognitive therapy; however, cognitive therapy is characterized by its aims rather than its methods, and these techniques are intended to bring about a direct change in thinking.

Techniques to suppress thoughts are used in two ways: with guidance from the therapist in therapy sessions, and during everyday activities. Techniques to be practised outside the sessions must be easy to remember and to carry out at times of distress. They are of two kinds. The first kind are techniques to distract the patient from the cognitions (for example an anxious patient's thoughts that he will die of a heart attack). Distraction is achieved either by focusing attention on something in the patient's immediate environment (for example by counting the objects in the window of a shop) or by focusing on a normal mental activity such as counting or some other exercise in mental arithmetic. Alternatively, a sudden sensory stimulus may be used to distract the patient, for example the snapping of a rubber band on the wrist; this method is sometimes called *thought stopping*. The second kind of technique is intended to counterbalance the effect of irrational thoughts by helping the patient to formulate a rational response; for example, 'my heart is beating fast because I feel anxious, not because I have heart disease'. It is difficult for the patient to focus on reassuring thoughts at times when the irrational thoughts are most intrusive; hence it is useful for him to carry a 'prompt card' listing these thoughts for quick consultation.

Techniques to change thoughts include giving information and questioning the logical basis of the thoughts. The therapist helps patients to identify illogical ways of thinking which allow the intrusive thoughts to persist despite evidence that they are ill founded. These *logical errors* include unjustified wide generalization from single instances, and focusing on evidence that supports the irrational ideas while failing to take account of evidence that negates them. The therapist also tries to identify irrational beliefs that make the patient experience excessive anxiety or depression in response to minor problems. Beck (1976) has called these irrational beliefs *underlying assumptions*. An example is the belief that to be free from depression, people must be successful in everything they do. Such a belief will lead to excessive response to minor failures in work or in personal

relationships. Specific examples of cognitive techniques will be given later in the chapter. Behavioural methods are described next.

Behavioural techniques

Relaxation training

The simplest behavioural treatment for anxiety disorders is **relaxation training**. The original method, *progressive relaxation*, was an elaborate procedure intended to reduce the tonus in individual muscle groups one by one and to regulate breathing (Jacobson 1938). Subsequently the procedure has been simplified and shortened (Bernstein and Borkovec 1973). A particularly intensive method is called applied relaxation (Öst 1987). Part of the training can be acquired from tape-recorded instructions or by patients learning together in a group; both procedures can reduce the therapist's time. Outside the training sessions, relaxation should be practised at regular times of the day in stressful situations. Applied relaxation is a rapid method particularly suitable for the latter purpose.

There is a surprising lack of controlled studies of the efficacy of relaxation training. The only technique to have been evaluated in controlled trials is applied relaxation, which was found to be an effective treatment for panic anxiety. The most common uses of relaxation are for less severe anxiety disorders and insomnia. There have been no satisfactory trials comparing relaxation directly with drugs, but clinical experience suggests that relaxation, when practised persistently, produces improvement equivalent to that produced by moderate doses of benzodiazepines. However, it often difficult to persuade patients to practise diligently enough to produce these useful effects.

Relaxation training has been used in the treatment of mild hypertension (see p. 383) and in other conditions in which stressful events may worsen a physical disorder by producing excessive arousal. Clinical trials have shown relaxation training to have specific effects in lowering blood pressure, greater than those of non-specific supportive contact with the therapist (Brauer *et al.*

1979). For a review of relaxation training see Glaister (1982).

Exposure

Exposure techniques are used in the treatment of phobic disorders to reduce avoidance behaviour directly and phobic anxiety indirectly. For simple phobias, exposure techniques give good results when used alone. For social phobia and agoraphobia, exposure techniques alone are less effective and therefore are usually combined with cognitive procedures (see p. 637). Exposure treatments are described in this section; combined cognitive and exposure treatments are considered on p. 635.

Exposure can be carried out in two ways: in the actual situations that provoke anxiety (*exposure in practice*) or in the clinic by helping the patient to imagine the phobic situations vividly so that anxiety is experienced as it would be in the actual situation (*exposure in imagination*). In either procedure exposure can be gradual, starting with situations that provoke little anxiety and building up slowly through increasingly more difficult ones (a procedure called *desensitization*); alternatively, exposure can be sudden, starting with situations that provoke severe anxiety (a procedure called *flooding*).

Desensitization in imagination is a widely used method first developed by Wolpe (1958). First a list is compiled of situations that provoke increasing amounts of anxiety—a *hierarchy*. There are usually about 10 items on the list, chosen so that each step introduces an equal increment of anxiety. This grading may be difficult to achieve; hence it may be necessary to adjust the severity of anxiety by introducing modifying factors. For example the anxiety response to a situation may be smaller if the person is with a friend and greater if he is with a critical stranger. Sometimes the anxiety-provoking situations seem so diverse that they cannot form a single hierarchy; for example an agoraphobic patient may fear visiting the hairdresser, going to the cinema, and attending a teacher–parent evening at a school. In such cases a unifying fear should be sought; for example the fear of being unable to leave a situation without embarrassment. In other cases it may be necessary to construct two hierarchies and to deal with each separately in treatment.

Relaxation is taught and used for two purposes. The first is to improve mental imagery, since relaxed patients can more easily imagine situations vividly. The second is to reduce the anxiety evoked by the imagery. Patients begin by imagining repeatedly the first situation on the hierarchy; at first anxiety is experienced, but as the procedure is repeated the anxiety diminishes until eventually the situation can be imagined without any anxiety. When this has been achieved the procedure is repeated with the second item on the hierarchy, then with the third, and so on. Wolpe believed that it was important to neutralize completely the anxiety response to the imagined phobic stimuli; it is now agreed that it is sufficient to reduce the anxiety substantially. Until recently, desensitization was the most frequently used technique of behaviour therapy, but desensitization in imagination is now used mainly in the treatment of simple phobias and as an adjunct to exposure in actual situations.

In **flooding treatment**, patients encounter and remain in real or imagined situations that produce severe anxiety. They remain in the situation until the anxiety diminishes; the process is then repeated with another near-maximum stimulus. When successful this procedure leads to a rapid reduction of anxiety, but many patients find the experience very unpleasant. This technique is not recommended because there is no evidence that inducing severe anxiety is more effective than desensitization. (The term *implosion* is sometimes applied to a particularly intense form of flooding in which severe anxiety is induced by imagining exceptionally frightening scenes that may be beyond anything the patient has experienced or is ever likely to experience. For example a spider phobic patient may imagine an encounter with a mass of giant spiders.)

Exposure treatment is usually carried out in a way that lies between the extremely cautious approach of desensitization and the rapid and intensive approach of flooding. The chosen situations cause moderate anxiety when first encountered, and can be encountered repeatedly.

Exposure sessions last about 45 minutes and, if feasible, are repeated daily by the patient working alone or with a relative or friend. The exposure sessions are planned carefully in advance to ensure that the effects can be predicted, and particularly that there will be time for anxiety to subside before the patient leaves the situation. Some of the phobic situations are brief; for example a socially phobic patient may order and drink a cup of coffee in a café. In such a case several brief assignments may be planned for the day, rather than a single treatment period. Some patients disengage from the situations by thinking of other things; this habit may have enabled them to cope with anxiety in the past, but during treatment sessions it is important that they are fully engaged in the situation and experience anxiety at the start of the session.

The outcome of these attempts at exposure is reviewed during weekly treatment sessions, and new goals are set for the next week. If anxiety has diminished during the previous week's sessions, the next item on the hierarchy is chosen and specific situations are selected. If anxiety has not diminished, the reasons are considered and, if necessary, a new situation is chosen to be intermediate in severity between the level on the hierarchy that was dealt with successfully in previous sessions and the level of the present unsuccessful sessions.

Exposure and response prevention

This combination of techniques is used to treat obsessional disorders in which there are rituals. (Obsessional thoughts without rituals need different techniques as described on p. 185.) The basic procedure is to help patients to refrain from rituals whilst experiencing strong urges to carry them out. As treatment progresses, these urges are deliberately increased by encouraging the patient to enter the situations that provoke rituals and have usually been avoided. When rituals are prevented, anxiety increases at first but then declines after 30 minutes to an hour; after experiencing this decline, patients feel a less strong urge to carry out rituals when they next encounter a situation that provokes them. At first, patients need to be accompanied and reassured while they strive to prevent

the rituals, but with practice they can manage on their own. Sometimes the therapist demonstrates the exposure procedure, for example by picking up an object that the patient cannot touch without experiencing anxiety to hand-washing urges ('modelling').

Treatment has to be agreed with the patient. The rationale is explained and targets for exposure are decided; for example, a short-term behavioural target for a patient with obsessions concerned with contamination, might be to touch an object that he thinks of as contaminated (such as a door handle) and not to wash his hands during the next hour. A long-term target might be to do all the household dusting and not to wash himself until the end of the work. It is important that all stages in treatment are agreed in advance so that patients are confident that they will not be faced with unexpected situations. The tasks usually cause substantial anxiety at first, but patients can generally tolerate this if they are convinced that it will eventually decline. Sessions should be planned so that the therapist can be present until anxiety has diminished; for this purpose the therapist has to allocate up to two hours at a time in the early sessions of treatment. Later sessions can be shorter, and eventually patients can carry out the procedures alone.

Obsessional thoughts accompanying rituals generally improve as the rituals are brought under control. Obsessional thoughts without rituals are more difficult to treat. Two approaches can be tried, neither of which is certainly effective. In *habituation training* patients are required to dwell on the obsessional thoughts for long periods, or to listen repeatedly to a tape-recording of the thoughts spoken aloud for an hour or more each time. *Thought stopping* is a form of distraction powerful enough to control intrusive obsessional thoughts. An effective distracting stimulus has to be sudden and intense, and also unobtrusive. A common technique is to snap a strong elastic band worn on the wrist.

Social skills training

This technique derives from work suggesting that social behaviour can be regarded as a set of skills that can be changed by new learning (Trower *et al.* 1978). Video recordings are used to define and rate elements of the patient's behaviour in standard social encounters. Patients are then taught more appropriate behaviours by a combination of direct instruction, modelling, video-feedback, and role reversal. The procedure has been applied mainly to socially inadequate people, though it is sometimes helpful for patients with schizophrenia as part of a wider programme of rehabilitation.

Assertiveness training

Assertiveness training is a particular kind of social skills training designed for people who have difficulty in being appropriately assertive. It was described originally by Salter (1949) and developed by Wolpe (1958). The essence of the treatment is that patients enact social encounters in which a degree of self-assertion would be appropriate, for example being ignored by a gossiping shop assistant. By a combination of coaching, modelling, and role reversal, patients are encouraged to practise appropriate verbal and non-verbal behaviour. The latter might take into account eye contact, facial expression, and posture. An account of these methods has been given by Rimm and Masters (1974).

Self-control techniques

All behavioural treatments encourage patients to learn to control their own behaviour and feelings. Such learning is the principal aim in self-control techniques. The methods can be traced to Goldiamond (1965), who suggested that operant conditioning principles could be applied for this purpose, and to Bandura (1969), who pointed out the importance of self-reward in controlling social behaviour. Self-control techniques attempt to increase patients' ability to make common-sense efforts at altering their behaviour. For this reason they have usually been employed when the goals of treatment are obvious to the patient but the effort required to achieve them is great. Over-

eating and excessive smoking are examples of such conditions.

There are two stages of treatment: self-monitoring and self-reward. **Self-monitoring** refers to the keeping of daily records of the problem behaviour and the circumstances in which it appears. For example patients who overeat can be asked to record what they eat, when they eat, and any associations between eating, stressful events, and mood states. Keeping such records can itself act as a powerful stimulus to self-control, because patients have often avoided facing the true extent of their problem and the factors that make it worse. **Self-evaluation** refers to making daily records of the behaviour to assess progress. This procedure also helps to bring about change, and it provides information for **self-reward** when behaviour has been controlled enough to reach a planned target. Thus a woman who is trying to diet might buy herself new shoes on reaching a pre-agreed weight.

Self-control methods are generally used as part of a wider programme, for example as part of a cognitive behaviour programme in the treatment of eating disorders (see p. 640).

Contingency management

The term contingency management refers to a group of procedures based on the principle that, if any behaviour persists, it is being reinforced by certain of its consequences. If these consequences can be altered, the behaviour will change. If the consequences are lessened, the behaviour will be weakened; if the consequences increase, the behaviour will be strengthened. In practice, the positive reinforcers of human behaviour are mainly social. They include expressions of approval and disapproval from other people, and actions that are enjoyable and rewarding for the patient.

Contingency management has four stages. First, the behaviour to be changed is defined and another person (usually a nurse, spouse, or parent) is trained to record it; for example a mother might count the number of times a child with learning difficulties shouts loudly. Second, the events that immediately follow (and therefore are presumed to reinforce) the behaviour are identified; for example the parents may pay attention to the child when he shouts but ignore him at other times. Third, reinforcements are devised for alternative behaviours, for example being approved or earning points by refraining from shouting for an agreed time. Staff or relatives are trained to provide the chosen reinforcements immediately after the desired behaviour and to withhold them at other times. As treatment progresses, records are kept of the frequency of the problem behaviours and of the desired behaviours. Although treatment is mainly concerned with the consequences of behaviour, attention is also given to changing any events that might be provoking the behaviour. For example in a psychiatric ward the abnormal behaviour of one patient may be provoked on each occasion by the actions of another patient. Contingency management has some value as part of a programme for some of the behaviour problems of people with severe learning difficulties, and of autistic children. [Practical details of contingency management have been described by Rimm and Masters (1979).]

A **token economy** is a form of contingency management that can be used for a group of patients living together in a ward or hostel. The aim is to reinforce behaviour by providing tokens that can be exchanged for privileges or goods. The method is seldom used because of the problems. The first problem is ethical; it may be difficult to devise effective privileges or rewards to be earned with tokens without first depriving the patient of some amenity that he should have by right. The second problem is practical: any changes of behaviour during the token economy may disappear when the patient leaves it; the results do not generalize. The third problem is verification: it has not been demonstrated that the changes brought about in a token economy result directly from the use of tokens rather than from increased interest in the patients and from the planning of goals.

Other behavioural techniques

The pad-and-bell method The pad-and-bell method is a special procedure, first developed in the 1930s for the treatment of enuresis. In the original method two metal plates separated by a pad are placed under the sheets of the bed. If the child passes urine in his sleep, the pad becomes moist and its resistance falls, allowing electric contact between the metal plates which are wired to a battery and a bell. The noise of the bell wakes the child, who must then rise to empty his bladder. After this has been repeated on several nights, the child does not pass urine in his sleep but wakens to empty the bladder. Eventually he sleeps through the night without being enuretic. The waking from sleep before passing urine can be understood as the result of classical conditioning. It is less easy to understand how the treatment leads to the child passing an uninterrupted dry night. Readers who wish to know more about this treatment and its rationale are referred to Dische *et al.* (1983). The method is mentioned again on p. 708.

Biofeedback In biofeedback techniques patients attempt to gain control of bodily functions, such as blood pressure, over which they otherwise have little or no control. Such lack of control occurs normally with functions regulated by the autonomic nervous system, and as an abnormal state after injury or disease has interfered with neural pathways.

The essentials of treatment are simple. A physiological monitor is used to provide information about the function to be controlled, and this information is presented to the person in an easily understood form such as a tone of varying pitch or a visual display. The person then tries to alter the display by regulating the function in some way, for example by relaxing. Biofeedback has not been shown to add usefully to the ability that people already have to control autonomic functions indirectly by relaxing without feedback. However, the technique can be of some value when there has been interference with normal sensory information, for example after spinal injury (Brudny *et al.* 1974). Biofeedback of galvanic skin response or pulse rate has been used as an aid to relaxation training, but most patients relax as well without this additional help. For a longer account of biofeedback and a more positive assessment of its value, see Basmajian (1983) or Stroebel (1985).

Aversion therapy In this technique, negative reinforcement is used to help patients suppress inappropriate behaviour that they wish to control. There are two problems with the use of negative reinforcement, one technical and the other ethical. The technical problem is that the effects of negative reinforcement on behaviour are usually temporary. The ethical problem is that negative reinforcement requires the use of stimuli that are unpleasant or mildly painful, and agents with these effects are used in other circumstances as a form of punishment. Although there is a distinct difference between negative reinforcement agreed to by a person to help suppress behaviour that he wishes to control and punishment imposed by others for wrongdoing, it is nevertheless important to make sure that the boundary between the two is maintained. This is particularly important when aversion therapy is being considered for people with conditions, such as disorders of sexual preference, which may be the subject of legal proceedings.

Nowadays, aversion therapy is seldom used because it has not been shown to produce better results than less unpleasant treatments. In the 1940s, however, the procedures were used to treat alcohol dependence by classical conditioning in which the sight, smell, and taste of alcohol were linked with nausea and vomiting induced by apomorphine. After early enthusiasm (Lemere *et al.* 1942) the method was given up in all but a few centres because the uncertain results did not justify the unpleasantness of the procedure. Instead of externally applied stimuli, unpleasant mental images have been used ('covert sensitization'), but this method has not been shown to have specific therapeutic effects.

Negative practice Negative practice derives from the work of Dunlap (1932) who suggested that tics, stammering, thumb-sucking, and nail-biting could be reduced when the patient deliberately repeated the behaviour. Some theoretical support for this idea was provided by experiments showing that inhibition accumulates during massed practice (Hull 1943); on repetition, reactive inhibition becomes associated with the behaviour which is then reduced. Negative practice has been used mainly to treat tics. Although short-term benefit has been reported (Walton 1961), there is no convincing evidence of sustained improvement.

Techniques for treating sexual dysfunction

These are described in Chapter 15.

Cognitive and cognitive–behaviour therapies

Most cognitive treatments include some behavioural elements and so the term cognitive–behaviour therapy is generally applicable. These treatments follow the same general principles (see p. 629), but are designed for particular disorders, each of which requires a different programme.

What happens in cognitive therapy

Cognitive therapy uses specific techniques designed to identify and change patterns of thinking that maintain psychiatric disorder. These methods are used with the common factors of psychotherapy such as restoring morale, listening to distress, giving information, and providing a rationale for treatment (see p. 601). Therefore cognitive therapy is similar to other kinds of psychotherapy in the general approach, and also in the preliminary agreement about the duration and number of sessions. This account outlines the features that are common to the several forms of cognitive therapy; features specific to a single kind of cognitive therapy (for example the kind used for depressive disorder) are described later.

The three characteristic features of cognitive therapy are as follows.

1. A detailed *analysis of the problem*, focusing on ways of thinking and how they precede or otherwise relate to the onset of symptoms or abnormal behaviours.

2. A *formulation of the problem* that can be tested by the patient in everyday life.

3. The aim of the analysis of the problem is to *identify the ABC of the problem*: Antecedents,

Behaviour and beliefs, and Consequences (O'Leary and Wilson 1975).

The focus of enquiry is on specific examples of general problems; for example patients with social phobia are asked to describe a specific occasion when they felt anxious in company, and to consider what they were thinking immediately before and after the anxiety began. By considering sequences of events on several occasions, common themes are identified in which there is a regular pattern of thinking and responding. This information is obtained in three ways: daily monitoring of thoughts and behaviours, observation within treatment sessions, and homework assignments.

1. *Daily monitoring of thoughts and behaviours* Monitoring is most useful when an immediate record is made of the thoughts; important details may be forgotten if the recording is made at the end of the day. The record of thoughts should include the date and time of day, the symptoms experienced, the situation and events at the time, and thoughts before, during, and after the onset symptoms.

2. *Observations within treatment sessions* Patients may be asked to describe thoughts arising while they imagine the situations in which problems arise. Symptoms may be provoked by exercise or overbreathing, and by the associated thoughts reported to the therapist. (The latter technique is used especially for panic symptoms (see p. 638)).

3. *Homework assignments* Between treatment sessions specific tasks may be arranged in order to obtain more information. For example patients may be asked to enter places that provoke symptoms and avoidance, and to record thoughts occurring before, during, and after these experiences. Homework should follow logically from the work of the session; it should be defined clearly so that patients understand the purpose of the exercise and how the results relate to their thoughts. For example a patient with panic disorder may be

asked to note how far physical exercise provokes fears of dying from a heart attack.

Information is also collected about *general beliefs* that may be relevant to the disorder, for example the belief that popularity depends on slimness. Some patients are aware of these beliefs and able to express them directly; others are only partly aware of them and may need to be helped by a technique known as *laddering*, in which a series of questions is asked about the consequences of each previous statement. For example agoraphobic patients may be asked what would happen if they became very anxious; they may reply that they would faint; they may then be asked the consequence of fainting; they may then reply that they would recover in a group of unsympathetic and mocking people. In this way general ideas may be revealed from a single instance.

Inquiries should lead to information about *events that provoke* symptoms (for example opening a conversation) and about *contexts* in which events are most likely to have this effect (for example speaking to a woman of the same age). Inquiry should also be made about *maintaining factors* (for example problems in relationships.)

As in other forms of therapy, beliefs about the proposed treatment and any previous treatment should be explored in order to anticipate problems in compliance.

There are several reasons why abnormal cognitions persist despite evidence that they are unfounded. First, patients *attend selectively* to evidence that confirms their beliefs, and they ignore evidence that contradicts them. For example patients with social phobias attend to other people's behaviour that might indicate criticism, while ignoring behaviour that signals approval. Second, these patients commit *errors in logic* when assessing evidence; for example they generalize too widely from single examples ('he does not love me so no-one else will ever love me'). Third, they engage in *safety behaviours* that give immediate relief from distress but prevent the logical examination of the belief. For example a socially phobic person may remain silent in company because he thinks that others will consider him foolish if he talks; this behaviour reduces his immediate anxiety about speaking, but prevents him from finding out that other people may help him to enter the conversation.

Continuous assessment of progress is an important feature of cognitive therapy. As treatment progresses, assessment consists increasingly of homework assignments. Homework should be designed to provide challenges to the patient's inappropriate beliefs and should have a clearly specified purpose. As far as possible goals should be positive; the patient should be trying to behave adaptively rather than to stop behaving maladaptively. Homework should be undertaken as an experiment that is informative whatever its outcome. Thus if the homework changes cognitions, it is successful; if the homework does not change cognitions, it provides a valuable starting point for further enquiry into the factors that maintain the cognitions, and then for another better-planned experiment.

Structure of the sessions Every session begins with the setting of an agreed agenda with the patient. Next, the outcome of the previous session is reviewed; for example the therapist may say, 'now that you have had more time for reflection, what is your reaction to our last discussion?' The results of homework are discussed, followed by one or more new topics chosen for the session, the next week's homework is planned, and the main points of the session are summarized.

Cognitive–behaviour therapy for phobic disorders Although, as explained on p. 631, most simple phobias can be treated effectively with exposure (a form of behaviour therapy), social phobia and agoraphobia generally respond better to a combined cognitive and behavioural approach. In this combined treatment behavioural methods are used to reduce avoidance, whilst cognitive techniques are used to interrupt intrusive thoughts and neutralize their effects. Three kinds of intrusive thought are modified: general concerns about the effects of being anxious ('fear of fear'); concerns about specific symptoms, and concerns that other people will react unfavourably to the patient ('fear of negative evaluation'). In social phobia, fears of 'negative evaluation' are particularly important, whilst concerns about symptoms

focus mainly on blushing and trembling. In agoraphobia 'fear of fear' is particularly important (usually as thoughts that the person will faint, die, or lose control), and fears of symptoms are also common (especially when there are associated panic attacks). Treatment combines exposure, a thorough explanation of the physiology of anxiety, questioning the logical basis of the fears, and instruction in distraction techniques. Information about the physiology of anxiety helps patients to attribute symptoms such as dizziness and palpitations to the correct cause instead of to physical illness such as a brain tumour or heart disease (which are common concerns of phobic patients). Questioning about the logical basis of the fear follows the general lines described on p. 636, namely reviewing the patient's evidence for his beliefs and pointing out alternative ways of thinking. Distraction techniques should be used selectively. Phobic patients often use distraction in situations that they fear. This use of distraction prevents exposure to the situation and provides short-term relief. Instead, distraction should be used to keep anxiety-provoking thoughts out of mind when the patient is planning new exposure tasks, or as a short-term measure if anxiety is greater than anticipated when exposure is carried out. In the latter circumstance patients can pay close attention to things around them instead of focusing on cues for anxiety (for example a patient might count the objects in a shop window).

Anxiety management for generalized anxiety disorders

This treatment combines relaxation training with cognitive techniques designed to reduce the effects of anxiety-provoking thoughts. Patients with generalized anxiety disorder have a combination of anxiety-provoking thoughts of the kind that occur in a more focused way in phobic and panic disorders, namely fears of the effects of being anxious and fears about specific physical symptoms. Anxiety-provoking beliefs are identified by the general methods described on p. 630. Patients are given an account of the physiology of fear responses and of the origins of anxiety symptoms.

The aim here is to help patients to ascribe their symptoms to anxiety rather than to physical illness (fears of physical illness are common in anxious patients).

Patients are helped to reduce the effects of intrusive thoughts in three ways. The first method is distraction (described in the previous section). The second method is repeating a reassuring statement that negates intrusive thoughts; for example if the intrusive thought is that dizziness indicates incipient loss of control, patients can repeat to themselves that anxiety does not lead to loss of control. The third technique is to challenge anxiety-producing thoughts. Among patients with generalized anxiety disorder, many have beliefs that make minor life-events highly stressful. These beliefs usually concern social situations (beliefs similar to those of patients with social phobia (see p. 637)), the safety of others (for example that a person who is late has had an accident), and health (beliefs similar to those of patients with panic disorder (see below)). The therapist challenges these ideas by asking questions such as: (i) How might other people think in the situation? (ii) Are patients adopting standards that they would not apply to others? (iii) Are they thinking in all or none terms? (iv) Are they over-estimating the extent of their own responsibility for the welfare of others? (v) What is the worst that could happen, and what is so bad about that outcome? Patients can be helped to reassess their role by constructing pie charts that show, for example, the importance of their role in ensuring that their family is content on holiday; thus they can be asked to write down factors that might affect the children's enjoyment (facilities, availability of friends, the weather) and then assign segments of the pie chart to each. Patients enter all these factors before recording their own contribution; they generally find that there is less space for the latter factor than they had assumed.

For further information about cognitive treatment of generalized anxiety disorder see Butler *et al.* (1987) and Clark (1994).

Cognitive techniques for panic disorder

Patients with frequent panic attacks have particularly strong convictions that physical symptoms of anxiety are evidence of serious physical disease (see p. 178). These convictions, which often relate to heart disease, create a vicious circle in which anxiety symptoms such as tachycardia generate more anxiety. In the usual cognitive behavioural technique, physical symptoms like those feared by the patient are induced in a benign way (usually by voluntary hyperventilation or by taking exercise). These symptoms trigger the anxious thoughts typical of the disorder, for example thoughts about heart disease. This sequence is pointed out to the patient with the suggestion that similar sequences, in which autonomic symptoms induce anxiety-provoking thoughts, occur in panic disorder. Patients are encouraged to monitor the thoughts that precede naturally occurring panic attacks, and to compare them with the thoughts that precede panic attacks induced during treatment sessions.

The general approach to treatment is similar to that described above for generalized anxiety disorder, but attention is focused more on the fears of physical symptoms that are the specific feature of panic disorder. Behavioural experiments are particularly important for panic patients whose beliefs may be resistant to the verbal techniques of cognitive therapy. Hyperventilation can demonstrate that palpitations induced in a harmless way may trigger fearful mental imagery. Focusing on the imagery of an attack can demonstrate convincingly that thinking may provoke physical symptoms of anxiety such as palpitations. In this way the two parts of the vicious circle of panic anxiety (see p. 178) can be demonstrated and the link between them can be easily explained.

Patients with panic attacks often engage in 'safety behaviours' that prevent the disconfirmation of their beliefs. For example, patients who fear that they will faint during a panic attack may be asked what has prevented that outcome so far. They will usually describe an action that they have taken, such as tensing the leg muscles or holding a support. They should be persuaded to try the effect of giving up these safety behaviours, and thus learn that panic attacks do not end in fainting whether or not they take these actions.

Some panic disorder patients experience anxiety because of assumptions that make ordinary situations stressful. They may assume, for example, that other people find them unattractive and boring. People with this assumption may act a part in social situations, striving to appear lively and witty and yet constantly fearing that others will see behind the facade. In treatment patients with these assumptions are encouraged to list their strengths and assets, to consider whether there is any evidence that people find them unattractive or boring, and to test the effect of meeting people without adopting this stressful role.

Cognitive–behaviour therapy for depressive disorders

Cognitive therapy for depressive disorders is a complex procedure combining behavioural and cognitive techniques, with emphasis on procedures for changing ways of thinking. It was the first effective form of cognitive therapy, and was developed by A. T. Beck (1976).

Cognitive therapy for depression is intended to alter three aspects of the thinking of depressed patients: negative intrusive thoughts, assumptions that render ordinary situations stressful, and errors of logic that allow these assumptions to persist despite evidence to the contrary. A combination of behavioural and cognitive procedures is used to bring about these changes. Behavioural techniques are also used to encourage a general increase in activity among the most depressed patients who have become so inactive as to be deprived of the everyday rewarding experiences that help to maintain normal mood, such as walking in pleasant surroundings or spending time on a hobby such as gardening.

Depressive cognitions have to be identified before they can be changed. Patients are encouraged to record their thoughts as soon as possible after a change of mood. Negative intrusive thoughts ('automatic thoughts') can be identified most readily, for example the thought 'I am a failure'. Assumptions, which are less obvious to the patient, can usually be discovered by asking four questions: (i) What is your reason for thinking in this way, what evidence do you have? (ii) What are the alternative ways of thinking about the issue? (iii) What are the consequences of thinking in this way? (iv) Are you thinking logically? Each of these questions will be considered more fully. These questions usually reveal one or more dysfunctional assumptions such as 'unless I always do what pleases other people, they will not like me'.

The next step is to try to challenge these automatic thoughts and assumptions by considering the answers to these questions. Meanwhile patients may gain some control over intrusive thoughts by using distraction techniques (see p. 630) or by repeating to themselves an alternative reassuring thought such as 'because I think something it will not necessarily take place'. To give these reassuring thoughts greater priority, it is useful to write them on a card that can be read when intrusive thoughts are severe.

Reviewing evidence Depressed patients pay more attention to evidence that supports their depressive ideas than to evidence that contradicts them. The therapist helps patients to assemble a complete set of evidence, positive and negative, and to review it in a balanced way.

Considering alternatives The therapist does not state alternatives directly since statements are likely to be rejected by the patient. Instead, the therapist helps that patient to consider alternatives by asking questions: What do you think that another person would think about this situation; What would you think if another person had done what you have done?

Considering consequences Patients are helped to see the consequences of thinking negative thoughts; for example the thought that everything is hopeless may prevent patients from attempting small changes that could be beneficial.

Considering errors of logic The patient is asked to consider a series of questions about himself: (i) Am I thinking in black and white terms? (ii) Am I drawing too wide conclusions from a single event? (iii) Am I blaming myself for something for which I am not responsible? (iv) Am I exaggerating the importance of events? These questions should be asked about specific ideas and situations so that they relate particularly to the patient's thinking.

Behavioural experiments A prediction is based on the patient's current thoughts. An action is planned that will test the prediction, and the outcome is reviewed. Illogical modes of thinking are identified by questioning patients about the reasoning behind their intrusive thoughts and by considering their replies. Several kinds of distorted logic allow maladaptive thoughts to persist despite evidence to the contrary. Common examples are generalizing too widely from single occurrences (which Beck calls 'over-generalization'), focusing on a single unfavourable aspect of a situation while ignoring favourable aspects ('selective abstraction'), and inappropriate self-blame for the consequences of the actions of other people ('personalization'). When these errors of logic have been identified, they are pointed out repeatedly to the patient. He is encouraged to recognize the errors, to think more logically, and to reach more adaptive conclusions.

For a more detailed account of cognitive therapy for the depressed patient, the reader is referred to Fennell (1994).

Cognitive-behaviour therapy for bulimia nervosa

The treatment of bulimia nervosa by cognitive–behaviour therapy is based on the idea that the central problem is overconcern about shape and weight, often associated with low self-esteem. This concern leads to extreme dieting, which makes the control of eating more difficult and leads to binge eating. Self-induced vomiting, sometimes combined with abuse of laxatives and diuretics, prevents weight gain and also reduces restraint on overeating, leading to more binges and a greater tendency to diet. In this way a vicious circle is set up which can be interrupted by restoring a regular pattern of eating three meals a day, increasing restraint on binge eating, and discussing ideas about shape, weight, and self-esteem.

This cognitive model is explained to patients. They are then asked to keep records of what they eat, when they eat, and when they induce vomiting or take laxatives. The situations that provoke binge eating are noted. Patients are encouraged to take three meals a day, so that they do not have long periods without food, or to take snacks between meals. They are assured that this pattern of eating will make it easier to control the urge to overeat and therefore will not lead to weight gain. These patients have often imposed a strict regime of dieting upon themselves, and they may require considerable time to become convinced of the need to change this. They are more likely to control their eating successfully if (a) meals are always eaten in a dining room or other place used exclusively for eating, (b) a limited amount of food is available at each meal, for example two slices of bread are put on the table but not a whole loaf, (c) a small amount of food is left on the plate and then thrown away, in order to mark the end of the meal, (d) food is stored out of sight of the eating area, and (e) a shopping list is made in advance and purchases are limited strictly to the list.

Binges often occur when the patient is unhappy, lonely, or bored. Patients should be encouraged to think of other behaviours that could relieve these feelings, for example seeking out friends or playing music. Efforts should be focused on preventing binges since vomiting is likely to stop when binges are under control. The dangers of abuse of laxatives and diuretics should be explained, and patients should be encouraged strongly to throw away all such drugs.

When eating is under better control, attention is given to cognitions. These are recorded on the diary sheets that are used to chart eating behaviour. The relevant cognitions are concerned not only with body shape and weight but also with self-esteem, for example to be fat is to be a failure and to be out of control, dieting is a sign of strong will and self-control, and it is necessary to be thin to be happy and successful. These beliefs persist because of the same errors of logic that maintain depressive cognitions, namely selective abstraction of evidence, overgeneralization from limited examples, overestimation of the significance of events, and all or none reasoning.

Questions are asked to identify cognitions and to question their logical status, using the techniques that have been described under cognitive therapy for depression (see p. 639). Patients with bulimia nervosa may have a distorted body image, which cannot usually be changed directly by cognitive procedures. This distortion of self-perception often diminishes as the other symptoms are brought under control.

Other uses of cognitive therapy

Currently the cognitive approach is being applied to several other disorders. As yet there is insufficient evidence to decide the value of these new applications, which include hypochondriasis and personality disorder.

Cognitive–behaviour therapy for hypochondriasis

The approach is twofold: to identify behaviours that maintain the disorder, and to change hypochondriacal ideas directly. The relevant behaviours are repeatedly seeking reassurance, which relieves anxiety briefly but reinforces the concerns in the longer term, and checking bodily functions (for example counting the pulse rate) or structure (for example palpitating for lumps). Hypochondriacal ideas are approached by asking the patient to explain the basis of the belief, by presenting evidence to counter this belief, and by considering alternative explanations, for example that the pulse rate is increased because the person is aroused and not because the heart is diseased. Preliminary results (Warwick *et al.* 1995) indicate that this combination of techniques has a beneficial effect. Controlled trials are being carried out.

Cognitive therapy for personality disorder

This application derives from the work of A. T. Beck, who developed cognitive therapy for depression. For each type of personality disorder, the approach is to identify characteristic beliefs and ways of thinking. Each disorder can be considered in terms of the person's self-view, views of others, general beliefs, major perceived threat, main strategy for coping, and primary affective responses.

In treatment, 'schemas' are inferred from the prominent 'automatic' thoughts (see p. 639). These schemas are then examined using the general techniques of cognitive therapy (see p. 629). Schemas are given for each personality disorder. For example the schema for histrionic personality disorder includes the following statements, which can be the basis of treatment: 'unless I captivate people, I am nothing'; 'to be happy, I need other people to admire me'; 'I must show people that they have hurt me'.

The general cognitive approach is often useful because of its focus on specific problems and limited goals, together with the keeping of records of thoughts. The methods have not been worked out in sufficient detail to form the basis of a clinical trial. For more information about the general use of cognitive therapy for personality disorders see Beck and Freeman (1990); for information about the specific use for borderline personality disorder see Layden *et al.* (1993).

Other forms of psychological treatment

Hypnosis

Hypnosis is a state in which the person is relaxed and drowsy, and unusually suggestible. Enhanced suggestibility can be shown by inducing various experiences such as diminished sensitivity to painful stimuli, vivid mental imagery, hallucinations, failures of memory, and 'age regression' (behaving as the person might have done when younger). Although enhanced suggestibility is characteristic of hypnosis, it is not confined to this method. Some suggestible subjects can produce the above phenomena in response to direct suggestion when in a state of full alertness (Barber 1962). There seem to be no phenomena peculiar to the hypnotic trance.

Hypnosis can be induced in many ways. The main requirements are that the subject should be willing to be hypnotized and should be convinced that hypnosis will occur. Most hypnotic procedures contain some combination of a task to focus attention (such as watching a moving object), rhythmic monotonous instructions, and the use of a graduated series of suggestions, for example that the person's arm will rise. The therapist uses the suggestible state either to implant direct suggestions of improvement or to encourage recall of previously repressed memories.

Indications for hypnosis

In psychiatry, hypnosis can be used in several ways. The first and simplest use, which requires only a light trance, is as a *form of relaxation*. For this purpose, hypnosis has not been shown to be better than methods that leave patients with more control over their actions. The second use, which requires a deeper trance, is to *enhance suggestion* in order to relieve symptoms, especially those of hysteria. Although this procedure is often effective, at least in the short term, it has not been proved better than more gradual forms of suggestion with no trance. Moreover, the sudden removal of symptoms by hypnosis is sometimes followed by an intense emotional reaction of anxiety or depression. The third use of hypnosis is as an *aid to psychotherapy* by bringing about the recall of repressed memories, but there is no evidence that this improves the effects of treatment. For all these reasons the authors do not recommend the use of hypnosis in clinical psychiatry. Readers seeking more information about hypnosis are referred to Wolberg (1977) for a brief account, or Wolberg (1948) for an extended account.

Autogenic training

This technique derives from the work of Oskar Vogt who, at the turn of the century, studied the psychophysiological changes brought about by hypnosis and autosuggestion. Shortly afterwards, in 1905, Schultz developed from this work a clinical procedure known as autogenic training, which he used to treat physical symptoms caused by emotional disorder (Schultz 1932).

In autogenic training 'standard exercises' are used to induce feelings of heaviness, warmth, or cooling in parts of the body, and to slow respiration. 'Meditative exercises' follow, in which colours or objects are imagined as vividly as

possible. Practising these two kinds of exercise is supposed to induce changes in autonomic nervous activity and thus to benefit patients with diverse disorders of the cardiovascular, endocrine, and urogenital systems, as well as patients with neuroses and habit disorders. It has not been established that autonomic changes after autogenic training differ substantially from those after simple relaxation, nor is there any good evidence about the therapeutic effects of the procedure. These methods have not been used widely in the United Kingdom or the United States, although they are employed more in Europe and Canada. The interested reader is referred to the short account by Schultz and Luthe (1959).

Techniques of meditation

In recent years several techniques of meditation have attracted popular attention, and some have been used to treat neurotic patients. Although the individual methods are based on different systems of belief, they have certain common features. First, they include some kind of instruction about relaxation and about the regulation of the speed and depth of breathing. Second, they include some mental process to direct the person's attention away from the external world and from the stream of thoughts that would otherwise occupy his mind. Often this requires concentration on a repeated word or phrase (a mantra). Third, emphasis is placed on setting aside from the day's activities periods when calm can be restored. Fourth, the person joins a group of people who believe strongly in the method and encourage each other to practise it. Such group pressure is often lacking from hospital-based programmes of relaxation or meditation, and this may explain why hospital patients often fail to persist with the exercises.

One review of the literature concluded that the level of physiological arousal in meditating subjects was not significantly different from that in resting subjects (Holmes 1987). There is no satisfactory evidence about the value of these methods. Clinical experience suggests that the less extreme forms have some value for patients with neurotic symptoms that result from a style of life that is too stressful and hurried. For a review of the various forms of meditation and their effects see West (1990).

Abreaction

It has long been known that abreaction (the unrestrained expression of emotion) often leads to temporary relief of mental disorder. Abreaction is part of many forms of religious healing and has also been used in medicine. The method is of most value in acute neuroses caused by extreme stress, and therefore has been used mainly for war neuroses. After Sargant and Slater (1940) used abreaction in the treatment of acute neurosis in soldiers evacuated from Dunkirk, the method was widely employed in the front line of battle to bring rapid relief to soldiers and so enable them to return to combat quickly. In civilian practice, abreaction has much less value because comparable neuroses are seldom seen a few hours after emotional trauma. Abreaction can be brought about by strong encouragement to relive the traumatic events; the procedure can be facilitated by giving a sedative drug intravenously. The use of abreaction in peacetime is not recommended. Details of the procedure have been given by Sargant and Slater (1963).

Further reading

Bloch, S. (1986). *An introduction to the psychotherapies* (2nd edn). Oxford University Press.

Brown, D. and Pedder, J. (1991). *Introduction to psychotherapy: an outline of psycho-dynamic principles and practice.* Tavistock/Routledge, London.

Frank, J. D. and Frank J. B. (1991). *Persuasion and healing* (3rd edn). Johns Hopkins Press, Baltimore, MD.

Gurman, A. S. and Kriskern, D. P. (1991). *Handbook of family therapy* (2nd edn). Brunner-Mazel, New York.

Hawton, K., Salkovskis, P. M., Kirk, J. W., and Clark, D. M. (1989). *Cognitive behavioural approaches for adult psychiatric disorders: a practical guide.* Oxford University Press.

Storr, A. (1979). *The art of psychotherapy.* Secker and Warburg with Heinemann Medical Books, London.

Wolberg, L. R. (1988). *The technique of psychotherapy* (4th edn). Harcourt Brace Jovanovich, Nre York.

Yalom, I. (1991). *The theory and practice of group psychotherapy* (3rd edn). Harper Collins, New York.

The last two chapters dealt with the treatment of individual patients. This chapter is concerned with the provision of psychiatric care for populations. It deals mainly with the needs of and provisions for people aged 18–65 (services for children are described on pp. 677–81; services for the elderly are described on pp. 513–17; and services for patients with severe learning difficulties are described on pp. 741–7). The organization of psychiatric services in any country inevitably depends on the organization of general medical services in that country. This chapter will refer specifically to services in the United Kingdom, but the principles of these services apply widely. The chapter begins with an account of the historical development of psychiatric services. This is followed by descriptions of the commonly available psychiatric services and of the problems encountered with these provisions. The chapter ends with a consideration of some innovations designed to overcome these problems.

The development of psychiatric services

Until the middle of the eighteenth century, there were hardly any special provisions for the mentally ill. In England, for example, the only hospital for these patients was the Bethlem Hospital, founded in 1247. In most of continental Europe there was a similar lack of hospital provision; in the Middle Ages hospitals in Spain were a notable exception (Chamberlain 1966). Nearly all mentally ill people lived in the community, often with help from Poor Law provisions, or else they were in prison. In England the Vagrancy Act of 1744 made the first legal distinction between paupers and lunatics, and made provision for the treatment of the latter. In response, private provisions for the mentally ill ('madhouses'—later to be called private asylums) were developed mainly for those who could pay for care, but also for some paupers supported by their parishes (Parry-Jones 1972). About the same time,

a few hospitals or wards were established through private benefaction and public subscription. The Bethel Hospital in Norwich was founded in 1713. In London, the lunatic ward at Guy's Hospital was founded in 1728, and in 1751 St. Luke's Hospital was founded as an alternative to the overcrowded Bethlem Hospital. Then, as now, the value of psychiatric wards in general hospitals was debated (Allderidge 1979).

Moral management

At the end of the eighteenth century, public concern in many countries about the poor standards of private and public institutions led to renewed efforts to improve the care of the mentally ill. In Paris in 1793 Pinel gave an important lead by releasing patients from the chains that were used for restraint. Subsequently he introduced other changes to make the care of patients more humane. In England similar reforming ideas were proposed by William Tuke, a Quaker philanthropist who founded the Retreat in York in 1792. The Retreat provided pleasant surroundings and adequate facilities for occupation and recreation. Treatment was based on 'moral' (i.e. psychological) management and respect for patients' wishes, in contrast with the physical treatments (usually bleeding and purging) and authoritarian approach favoured by most doctors at that time. This enlightened form of care was described several years later by William Tuke's grandson, Samuel, in *A description of the Retreat*, published in 1813. These humane methods were adopted in other hospitals as it became clear that many mentally ill patients could exert self-control and did not require physical restraint and drastic medical treatment.

The asylum movement

Despite such pioneering efforts, in the early years of the nineteenth century many mentally ill people received no care and lived as vagrants or as inmates of workhouses and gaols. Also, there was public concern in England about the welfare of those in care, following reports of scandalously low standards in some private madhouses. This concern led to the County Asylum Act of 1808 which provided for the building of mental hospitals in each of the English counties. Unfortunately little was done by the county authorities, and in 1845 it was necessary to enact the Lunatics Act which required the building of an asylum in every county. At first the new asylums provided good treatment in spacious surroundings. Moral management was championed especially by members of the Non-restraint Movement, which had started with the work of Gardiner Hill at the Lincoln Asylum in 1837 and was developed further by John Conolly at the Middlesex County Asylum, Hanwell. In 1856 Conolly published a significant book, *The treatment of the insane without mechanical restraints*.

Unfortunately these liberal steps were soon followed by a new restrictive approach. Increasing public intolerance led to the transfer of more and more patients from the community and prisons to the new asylums. Initial optimism about the curability of psychiatric disorder dissipated as the limitations of moral treatment became apparent. More gloomy organic and hereditary views prevailed. By the 1850s, the problems of overcrowded asylums were evident. Attempts were made to house patients with chronic illness in less restrictive and more domestic surroundings in detached annexes or houses in the grounds of the asylum. Other hospitals returned patients to the community either by boarding them out with a family (a form of care which was practised most successfully at Gheel in Belgium) or by returning them to workhouses. The Lunacy Commissioners, whose role was to oversee the care of the mentally ill, were concerned that these arrangements could lead to abuse, and were opposed to them. Nevertheless nineteenth-century asylums, even when overcrowded, provided a standard of care for the mentally ill that was lacking elsewhere. Thus the mentally ill were protected from exploitation, and were provided with shelter, food, and general health care. These benefits were counterbalanced by the disadvantages of loss of personal choice and autonomy, and of a mono-

tonous and overprotective regime that could lead to institutionalism (see p. 284).

Under the increasing pressures of overcrowding and staff shortages, there was less and less time for moral management. Again, a custodial approach was adopted. This change to custodial care was endorsed by the Lunacy Act 1890, which imposed restrictions on discharge from hospital. These custodial arrangements continued into the twentieth century, and their legacy is still seen in the size and structure of the large Victorian hospitals in which most psychiatry was practised until recently. See Jones (1972) and Rothman (1971) for accounts of psychiatric hospitals in the United Kingdom and the United States in the nineteenth century.

Early treatment

In the United Kingdom the start of a return to more liberal policies was signalled, shortly before the First World War, by a substantial gift of money by Henry Maudsley, a wealthy psychiatrist, to provide for a hospital devoted to early treatment. Unfortunately the war interfered with the project, and the opening of the Maudsley Hospital was delayed until 1923. The hospital provided an out-patient service and voluntary in-patient treatment in surroundings in which teaching and research were carried out.

In the post-war years, the impetus for change increased. The Mental Treatment Act 1930 repealed many of the restrictions on discharge of patients imposed by the Lunacy Act 1890, and allowed county asylums to accept patients for voluntary treatment. The 1930 Act also encouraged local authorities to set up out-patient clinics and to establish facilities for after-care. Therapeutic optimism increased further as two new treatments were discovered: insulin coma (later abandoned) and electroconvulsive therapy. At the same time, efforts were made to improve conditions in hospitals, to unlock previously locked wards, and to encourage occupational activities. Similar changes took place in other countries.

In most countries these reforms were halted by the Second World War. Psychiatric hospitals became understaffed as doctors and nurses were recruited to the war effort. They also became overcrowded as some were allocated to the care of the war injured, with the result that their patients had to be relocated among the rest. The effects of the war on an English county asylum have been described by Crammer (1990).

Social psychiatry and the beginnings of community care

After the Second World War, several influences led to further changes in psychiatric hospitals. Social attitudes had become more sympathetic towards disadvantaged people. Among psychiatrists, wartime experience of treating 'battle neuroses' had encouraged interest in the early treatment of mental disorder and in the use of group treatment and social rehabilitation. In the United Kingdom, the advent of the National Health Service led to a general reorganization of medical services including psychiatry. The introduction of chlorpromazine in 1952 made it easier to manage disturbed behaviour, and therefore easier to open wards which had been locked, to engage patients in social activities, and to discharge some of them into the community.

Despite these changes, services continued to be concentrated on a single site, often remote from centres of population. In the United States, Goffman (1961) argued that State Hospitals were 'total institutions', i.e. segregated communities isolated from everyday life. He described such institutions as impersonal, inflexible, and authoritarian. In the United Kingdom, Wing and Brown (1970) found that some of the large mental hospitals were characterized by 'clinical poverty' and 'social poverty'. Vigorous methods of social rehabilitation were used to improve conditions in hospital and to reduce the effects of long years of institutional living. Occupational and industrial therapies were used to prepare chronically disabled patients for the move from hospital to sheltered accommodation or to ordinary housing (Bennett 1983). Many long-stay patients were responsive to these vigorous new methods. There was optimism that newly admitted patients could also be helped in these ways.

For patients in the community, day hospitals were set up to provide continuing treatment and rehabilitation, and hostels were opened to provide sheltered accommodation. As a result of all these changes, the numbers of patients in psychiatric hospitals fell substantially in the United Kingdom and in other countries. The changes were particularly rapid in the United States. Despite these changes, services were still based in large mental hospitals that were often far from patients' homes. Unfortunately, in many places the provision of community facilities was insufficient for the needs of all newly discharged patients.

Hospital closure

After the initial success of discharging many institutionalized patients, it was optimistically proposed that large asylums could be closed and replaced by small psychiatric units in general hospitals, with support from community facilities. In most countries the programme of hospital closure took place gradually. A notable exception was Italy, which at first lagged behind most other countries but later made rapid changes. In 1978 the Italian Parliament passed Law 180, which aimed to abolish the mental hospitals and replace them by a comprehensive system of community care. Admission to psychiatric hospitals was prohibited, and there were requirements that psychiatric units be set up in general hospitals and that community services be developed in defined areas. The scheme was based on the work of Franco Basaglia in hospitals in northeast Italy, and on the proposals of the professional and political movement he founded. This movement—*Psichiatria Democratica*—combined an extreme left-wing political view that patients in psychiatric hospitals were the victims of oppression by the capitalist system with the conviction that severe mental illness was induced by social conditions and not by biological causes. Basaglia's forceful personality and qualities of leadership helped him to succeed in finding new ways of caring for patients in the community. Other workers found it difficult to repeat his successes. The consequences of this sudden change have been varied. In those parts of Italy where the reforms were financed adequately and were implemented by enthusiastic staff, the new provisions were successful. In areas where the provision of new facilities was inadequate, there were many problems for patients and their families (Bollini and Mollica 1989).

In the United Kingdom and elsewhere the pace of change was slower, but similar problems arose. Some patients could not manage in the community without intensive support, and even then required repeated readmission to hospital so that the arrangements became known as the 'revolving door policy'. The rehabilitation services had been expected to discharge patients in an improved state, but they found it necessary to provide continuing care for so many that it was difficult to take on new patients. Some discharged patients attended day hospitals for years without further improvement (Gath *et al.* 1973). It became clear that earlier views of the benefits of 'de-institutionalization' had been over optimistic, and that services outside hospital were inadequate to provide the help needed by discharged patients and their families. Attempts were made to develop more adequate community facilities, a policy known as community care.

The rise of community care

The term 'community care' applies to two distinct approaches to treatment. The first is concerned with the treatment of chronic psychiatric disorder outside hospital, and the second is concerned with the prevention or early treatment of acute psychiatric disorder. The two approaches have certain principles in common: responsibility for a defined population, provision of treatment close to the patient's home, comprehensive services, continuity of care, and a multidisciplinary team approach. However, there are important differences between the two approaches. These are illustrated by the ways in which community care developed in the United Kingdom and in the United States.

In the United Kingdom community care developed originally as an alternative to long-term hospital care for patients with serious psychiatric disorders. In the United States more emphasis was given to the prevention and early treatment of mental disorder to avoid the need for admission to hospital. The US Federal Government established a Joint Commission on Mental Illness and Health which issued a report in 1961 strongly criticizing the State hospital system and recommending community treatment, delivered from community mental health centres (CMHCs). These centres had staff from several disciplines who offered psychological and social care. In most centres emphasis was placed more on early intervention for psychosocial problems (crisis intervention) than on the care of patients with chronic psychiatric disorders. There was no requirement that the centres should provide sheltered accommodation for patients with chronic disorders. This limited role of the centres led to dissatisfaction as patients discharged from long-term hospital care found their way into private hospitals or prisons, or joined the homeless population of large cities (Goldman and Morrisey 1985).

Emerging principles

Some commonly agreed principles about services needed to enable the closure of the old asylums emerged from these experiences.

Hospital care: hospital admissions were to be brief, and as far as possible to psychiatric units in general hospitals rather than to psychiatric hospitals. Whenever practicable patients were to be treated as out-patients or day patients.

Rehabilitation: was to be provided, initially in the hope that most patients would progress to independent living, but subsequently with the more modest aim of preventing further deterioration.

Community care: was to be offered to patients by staff working outside hospital. Two factors gradually became apparent: first that this kind of care required considerable resources, and second, that some of the most vulnerable patients were unwilling to make use of the available care. (These problems are discussed further on p. 654–5.)

Care was to be provided by *multidisciplinary teams* in which members worked closely together. These teams usually included psychiatrists, community nurses, clinical psychologists, and social workers, all of whom worked in collaboration with members of voluntary groups.

In many countries *legal reforms* were introduced to limit the uses of compulsory treatment and to encourage alternatives to in-patient care. These reforms also reflected a greater public concern for the rights of the individual.

These principles have led to a widely adopted pattern of service provision in the community which will be described next.

The provision of psychiatric services for a community

To determine what psychiatric services are required for a community, it is necessary to know how frequent mental disorders are in the population, and how these disorders become known to the medical services. These issues will be described first in relation to primary care services, and then in relation to specialist provisions.

Services in primary care

Psychiatric disorder in the community and in primary care

Psychiatric disorder in the community It is difficult to determine the exact frequency of mental disorders in the community (for methods of epidemiological research see p. 86), but approximate estimates are usually sufficient for service planning. Table 19.1 shows one set of figures, which were derived from a large population survey in the United States. Psychological disorders are common; amongst persons at risk, about one in five experience one of these disorders in the course of one year. Amongst conditions recorded in such surveys, many are brief anxiety and depressive reactions to stressful circumstances rather than definite psychiatric disorders. Amongst the psychiatric disorders, anxiety disorders are most frequent followed by abuse of drugs or alcohol. Affective disorders come next in frequency. Obsessional disorders and schizophrenia are much less frequent.

Psychiatric disorder in primary care Amongst people identified in a community survey as having a psychiatric disorder, many do not seek medical advice. Thus many people with acute reactions to stress or with adjustment disorders seek no help from doctors, but are supported by family, friends, the clergy, or non-medical counsellors. Many people with problems of substance abuse do not consider themselves as needing help of any kind. Nevertheless, when primary care services are well developed (as in the United Kingdom) about nine in ten people with a definite psychiatric disorder attend a general practitioner. Whether a person

with a psychiatric disorder consults a general practitioner depends on several factors:

- The *severity and duration* of the disorder.
- The *person's attitude* to psychiatric disorder; some people feel ashamed and embarrassed to ask for help.
- The *attitudes* and knowledge of *family and friends*; if these people are unsympathetic, the affected person may be less likely to seek help.
- *Knowledge* about possible help; if people do not know what help can be provided, they are less likely to seek help.
- The person's *perception of the doctor's attitude* to psychiatric disorder; if the doctor is viewed as unsympathetic, the person is less likely to ask for help.

Detection of psychiatric disorder by general practitioners

People with psychiatric disorder who consult their general practitioner may not mention their psychiatric problem but may complain instead of physical symptoms, which may be symptoms of

Table 19.1. *Approximate prevalence of psychiatric disorder in the community (one year prevalence per 100 population)* *

All disorders	20.0
Anxiety disorders	13.5
Substance abuse	8.8
Affective disorders	4.3
Obsessive–compulsive disorder	1.7
Schizophrenia	1.0
Somatization	0.1

*Data are based on an American survey reported by Robins and Regier (1991). Note that the rates for schizophrenia are higher than those reported in the United Kingdom.

an associated minor physical illness, or somatic symptoms of anxiety (such as palpitations) or depression (such as tiredness). Some of these patients are aware that the physical symptoms are part of an emotional disorder, but are uncertain whether the doctor will respond sympathetically to a request for treatment for psychological symptoms. Other patients are unaware that the physical symptoms have a psychological cause.

General practitioners vary in their ability to detect undeclared psychiatric symptoms. Their success depends on two factors: ability to gain the patients' confidence and so to enable them to disclose psychiatric symptoms; and skill in assessing the relationship between physical and psychiatric symptoms.

Treatment of psychiatric disorder in primary care

Most psychiatric disorders in general practice attenders can be treated successfully in primary care. Examples are most adjustment disorders, the less severe anxiety and depressive disorders, and many problems of alcohol abuse. Only about 5–10 per cent of patients whom general practitioners have identified as having psychiatric disorders need referral to psychiatrists. This small group includes patients with severe depressive disorders, schizophrenia, and dementia. General practitioners are more likely to refer patients with other disorders when the *diagnosis is uncertain,* the *condition is severe,* there is a significant *suicide risk* the condition is *chronic,* necessary *treatment cannot be provided* by a member of the practice team, psychiatric services are accessible and responsive, and the patient is willing to attend.

Treatment resources provided by the general practitioner As mentioned above, most patients presenting with psychiatric disorders in primary care have adjustment disorders, affective disorders, or problems of substance abuse. Consequently the provisions required for psychiatric treatment in primary care are mainly counselling, simple behavioural treatment, and the prescription of drugs. Since few general practitioners have time to provide counselling for all patients who need it, a well-staffed primary care team should include a coun-

sellor. With some additional training, practice nurses can assist in the counselling of patients with less severe affective disorders (Wilkinson *et al.* 1993).

General practitioners have a role in the aftercare of patients with chronic psychiatric disorder. For such patients it is good practice to define clearly the role of the general practitioner in treatment, and the role of the psychiatric team. For example, when a general practitioner prescribes lithium for a patient with recurrent affective disorder, there should be an agreed plan that includes a specified frequency of estimations of lithium concentration and of renal and thyroid function. General practitioners may play a part in the community care of patients with chronic schizophrenia by assessing progress, encouraging compliance with treatment, and supporting families (Regier *et al.* 1985; Sibbald *et al.* 1991).

The psychiatric team working in primary care It is now common for *psychiatrists* to visit primary care centres to support the general practitioner in his work. There are four ways in which a psychiatrist can work in primary care:

1. The psychiatrist can act as the doctor of *first contact*, instead of the general practitioner. This way of working is appropriate in countries with poorly developed primary care services, but less appropriate in countries (such as the United Kingdom) where general practitioners are widely available to make the first assessment of the patient's needs.

2. The psychiatrist can *advise* the general practitioner about the management of patients in primary care. The psychiatrist does not see the patient but bases his advice on the general practitioner's assessment.

3. The psychiatrist can *assess patients* when the general practitioner is uncertain about diagnosis or treatment. Patients identified as needing specialist treatment are then referred to a psychiatric out-patient clinic in the usual way.

4. The psychiatrist can work mostly *in primary care,* seeing patients at the health centre or at home and thus curtailing or closing the hospital out-patient clinic.

The most appropriate arrangement depends on the needs of the general practitioners, the accessibility of hospital out-patient clinics to patients (generally greater in an urban than in a rural area), and the number of psychiatrists available (if less time is spent in travelling, more time is available for patients).

Clinical psychologists and *psychiatric nurses* may work with the primary care team, providing counselling or behaviour treatment. It is not certain whether community psychiatric nurses are more effective than primary care in treating minor disorders. One study showed no benefit when a group of patients treated by a community psychiatric nurse was compared with a control group treated by the general practitioner (Gournay and Brooking 1994). However, it is generally agreed that community psychiatric nurses can play an important part in caring for patients with severe or chronic psychiatric disorders such as schizophrenia, manic depressive disorder, and substance abuse.

Specialist psychiatric services

Patients treated by the psychiatric services are a highly selected subgroup of people with mental disorder. In countries where the general practitioner acts as the 'gatekeeper' to specialist services, the numbers and types of patient reaching the psychiatric services depend largely on the willingness of general practitioners to treat psychiatric disorder, on the available resources, and on the patients' willingness to seek specialist psychiatric advice. In the United Kingdom, most of the patients in contact with the psychiatric services have severe and chronic neurosis, severe affective disorder, schizophrenia, or dementia. Among patients particularly likely to be cared for by specialists are those who are suicidal and those who are dangerous to others. Several kinds of provision are required to manage these conditions: in-patient care, day care, out-patient care, and community care. Originally it was hoped that rates of admission to hospital would be reduced by increased work in the community, but it is not certain that such a reduction can be achieved. Reductions have been reported (Williams and Ballestieri 1989) but, in a careful study of the effects of introducing community health teams to primary care, no decrease was found in the use of in-patient resources whilst out-patient referrals declined in the expected way (Jackson *et al.* 1993).

In the United Kingdom and some other countries, the organization of specialist services is made difficult by a division of responsibilities between different funding agencies. In the United Kingdom, for example, central government provides funds for hospital-based community services, whilst local authorities employ social workers and provide day activities and sheltered accommodation in hostels and group homes. Voluntary organizations also play a part in providing care. Unless there is close liaison between these various providers, services become uncoordinated and deficiencies develop in the provisions needed by patients. Different provisions are required by patients with acute disorder as against those with chronic disorders. These provisions will be described separately, starting with specialist services for patients with acute psychiatric disorders.

Acute specialist care

Specialist care of acute psychiatric disorder requires out-patient, day-patient, and in-patient provisions, as well as working with primary care teams (described above).

Out-patient clinics

Although it is often helpful for a psychiatrist to work in general practice in order to assess and treat patients with acute psychiatric disorders, working in a central clinic has two important *advantages*: first, the time of the professional staff is used more effectively because they spend less time travelling; second, a senior person is available immediately to give advice to less experienced staff. The *disadvantages* of out-patient clinics are that patients may have to travel long distances and family members may find if difficult to attend the clinic if necessary.

Day hospitals

Day hospitals are valuable in two ways in the treatment of acute psychiatric disorder: first, day hospital care may avoid admission to hospital for patients with less severe disorders; second, in-patients may transfer to day care when they no longer need the full facilities of an in-patient unit.

In the treatment of patients with acute disorders, day care is most appropriate for those who can be with their families in the evening and at weekends. Suitable psychiatric conditions include depressive disorders of moderate intensity and without substantial risk of suicide, anxiety disorders, and obsessional disorders, and some eating disorders. Day care should be planned as carefully as in-patient care, and there should be an active treatment programme specific to each patient's needs. If there is no active programme, patients can become dependent on the general supportive function of a day hospital without gaining benefit from the specific treatment procedures that could be offered.

In-patient units

Although it was originally hoped that well-resourced provision of community care would greatly reduce the need for in-patient care, it is now recognized that every psychiatric service requires an in-patient unit capable of treating the most severely ill patients promptly in an emergency. The number of beds required for the support of a community psychiatric service depends on several factors,

1. The facilities for treatment of acute psychiatric disorder outside hospital; the willingness of families to care for acutely ill relatives; the availability of psychiatric nurses and other community staff to provide 24-hour intensive care in the home; and the provision of staffed hostels for patients without a family home.

2. The facilities for early discharge of patients from hospital after the acute phase of the disorder; these facilities resemble those listed above, although it is generally easier to discharge early than to avoid admission.

A reasonable balance between in-patient and community care can usually be achieved with the number of hospital places shown in Table 19.2. The figure of 100 places for acute hospital care per 250 000 total population includes the requirements

Table 19.2. *Estimated need for specialist residential provision per 250 000 total population (excluding dementia)*

Type of provision	No. of places
Hospital provision	
Acute care	100
Intensive care unit	10
Secure unit	4
Intensive continuing care ('hostel ward')	50
Hostels	
Hostels—staff sleeping in	75
Hostels—staff by day only	50
Staff visiting	
Group homes	45
Other supported placements	60
Total	394

Source: Wing J. K. 1994*a*.

for both acute disorders and acute exacerbations of chronic disorders in people up to 65 years of age; for patients over this age the figure excludes dementia. (Provisions for the elderly are discussed on p. 513.) Special provision is required for patients who require intensive nursing during episodes of disturbed behaviour, and for those who may be dangerous to others (some provision for the latter may be provided in a secure unit serving a wider area). See Hirsch (1987) for a review of the needs for beds and other resources in general psychiatry.

In-patient units for acutely ill patients should strike a balance between the patients' needs for privacy and the staff's requirement to observe them. There is a need for secure areas for the most disturbed patients, areas where patients can be alone, and areas where they can interact with others. There should be provisions for occupational therapy, the practice of domestic skills, and recreation. Outdoor space is desirable.

General hospital units

In many places in-patient care for acute psychiatric disorders is provided in general hospitals. These units have the *advantages* of lack of stigma and easy access to medical services when required. Also, such units are usually nearer than specialist psychiatric units to patients' homes. The *disadvantages* of such general hospital units include the difficulties of providing adequate space for occupational activities and of creating an informal environment suitable for psychiatric care in a hospital designed primarily for the different needs of physically ill patients. Some of these problems can be overcome if the psychiatric unit occupies a separate building within the general hospital complex.

Long-term care

Characteristics of patients needing long-term care

Diagnosis With the exception of the elderly (considered in Chapter 16), most psychiatric patients requiring long-term care have schizophrenia, chronic affective disorders, presenile dementia, or personality disorders associated with aggressive behaviour or substance abuse. Patients who need care in hospital for more than a year are often referred to as the 'new long-stay' (in contrast with the 'old long-stay' who had been resident in hospital for many years before hospital closure programmes were initiated).

Problems There are several ways of classifying the problems of patients who need long-term psychiatric care. The first was suggested by Wing and Morris (1981):

(1) **impairments** due directly to psychiatric disorder, for example persistent hallucinations, social withdrawal, underactivity, and slowness;

(2) **secondary social disadvantages** such as unemployment, poverty, and homelessness, as well as the stigma still attached to chronic psychiatric disorder;

(3) **adverse personal reactions** to illness and social disadvantage, such as low self-esteem and hopelessness.

A second method of classification is based on patient characteristics that lead to difficulties in management (Wing and Furlong 1986):

(1) *risk of harm* to self and others;

(2) *unpredictable behaviour* and liability to relapse;

(3) *poor motivation* and poor capacity for self-management or for performance of social roles;

(4) *lack of insight* into the need for treatment;

(5) *low public acceptability*.

If patients present severe and persistent problems of this kind, they cannot be given adequate care in the community; instead they require continuous residential care in hospital or in a well-staffed hostel. Groups of patients of this type have been identified in community services in the United Kingdom (Wing 1986; Wing and Furlong 1986), the United States (Bachrach 1986), and other countries (Häfner 1987*a*).

A third way of classifying the problems of patients needing long-term care is related to their *needs for rehabilitation*. Wing and Morris (1981) suggested the following classification:

(1) persistent symptoms, both positive (such as hallucinations) and negative (such as lack of drive);

(2) unusual behaviour, especially if likely to be disapproved of socially, for example shouting obscenities;

(3) activities of daily living, such as the capacity to wash and dress;

(4) occupational or domestic skills, such as shopping and cooking;

(5) personal attitudes and expectations;

(6) social circumstances to which the patient is likely to return.

Rehabilitation for patients needing long-term care

In psychiatry, the term rehabilitation denotes measures for helping patients to reach and maintain their best level of functioning. The *procedures* used in rehabilitation are medical, psychological, occupational, social, and residential. These procedures may be provided in an in-patient unit, day hospital, or rehabilitation centre.

Medical Most patients in rehabilitation programmes require drug treatment to control symptoms of schizophrenia or of chronic affective disorder.

Psychological Psychological methods include supportive therapy, behavioural programmes, and social skills training (Liberman *et al*. 1986).

Occupational Occupational therapy helps to structure the day and to provide an opportunity for interaction with other people. Good results can be a source of self-esteem, and payment is a further incentive. In the past occupational therapy was often intended to prepare patients for simple industrial work (Carstairs *et al*. 1956; Wing *et al*. 1964), but in recent years unemployment has increased among healthy people and the chances of employment for the handicapped have fallen. For this reason alternative activities such as gardening, crafts, home repairs, and cooking are provided to instil a sense of achievement and to help patients to use leisure time constructively.

Social Although handicapped people should be encouraged whenever practicable to join social groups attended by healthy people whenever practicable, some need special clubs and social centres where they can be with other people who have similar difficulties in coping with the demands of normal life.

Provisions needed for long-term care

In-patient care

For patients with chronic psychiatric disorders, in-patient care may be needed for acute treatment at times of relapse, for intensive rehabilitation, and occasionally for long stay. The basic requirements for an in-patient unit for patients with chronic disorders are broadly similar to those for patients with acute illness (see p. 651). However, since the pace and intensity of treatment of patients is usually slower for those with chronic disorders than for those with acute disorders, it is generally better to separate the care of the two groups. There are needs for secure areas for patients with acute and severely disturbed behaviour, single rooms, areas for observation and other areas for privacy, and provisions for occupational and social activities including outdoor space. Sheltered work and recreational facilities need not be in the same building, but should be within walking distance. The provisions can be in buildings of a more domestic type than those generally available for patients with acute illness.

Day hospitals

Day hospitals can play an important part in the care of patients with chronic psychiatric disorder. Patients may attend for assessment and for the supervision of drug treatment, and for occupational and social activities. Since it is difficult to provide simultaneously for the needs of patients with acute disorder and of patients with long-term disorders, it is better to separate activities in different areas of the building or to arrange them at different times.

Out-patient clinics

Out-patient clinics play a smaller part in the care of patients with chronic psychiatric disorders than in the care of those with acute disorders. This is because patients with chronic disorder are more likely to miss booked appointments, and because it is often important to visit their family or other carers. For these reasons, follow-up by a community nurse is generally more effective than out-patient care.

Provisions required in the community

If patients with chronic psychiatric disorder are to be treated in the community it is necessary to provide all the elements of care that they would have received in hospital. In the community seven provisions are required to replace long-term care in hospital:

- suitable well-supported carers
- suitable accommodation
- suitable occupation
- arrangements to ensure the patient's collaboration with treatment
- regular reassessment, including assessment of physical health
- effective collaboration amongst carers
- continuity of care and rapid response to crises.

Complicated and expensive arrangements are required to make these seven elements available as readily in the community as in hospital. Lack of these arrangements may leave patients homeless, without constructive occupation, inadequately treated, and without a carer. Failure of community care may also leave carers unsupported and family life disrupted. The first five of the above list of seven provisions will be discussed next.

The carers When patients live at home, **family and friends** are the main carers, and they provide much of the help that would be provided by nurses if patients were to remain in hospital. For example they may encourage patients to get up in the morning, maintain personal hygiene, eat regular meals, and occupy themselves constructively. Carers also encourage compliance with treatment. If patients have many problem behaviours, prolonged involvement in care is stressful. Carers may then need support and advice, and sometimes periods of respite.

Voluntary carers play an important part in many systems of community care. Trained volunteers can help to support patients and families, and some charitable organizations employ professional carers such as hostel staff.

Community psychiatric nurses play an essential part in community care by supporting patients and their families, evaluating the patients' clinical state, supervising drug therapy, and encouraging social interaction.

Accommodation Patients discharged from hospital have obvious needs for food and shelter. Many patients live with their **families**. Some can care for themselves in rented accommodation. Others need more help, which can be provided in three ways.

In lodgings: some people are willing to receive patients with mental disorders as lodgers and to provide them with extra care. This practice works well in some countries but has not been adopted widely in the United Kingdom.

In group homes: some patients are able to live in group homes, i.e. houses in which four or five patients live together. The houses may be owned by social or health services or by a charity. The patients are often chronic schizophrenics with social handicaps but few positive symptoms. They are chosen as being able to perform the essential tasks of running the house together, with each using his or her remaining abilities, even though separately unable to complete all the tasks. Patients living in group homes receive regular support and supervision, usually from a community nurse who ensures that arrangements are working well and who encourages patients to take on as much responsibility as possible.

In hostels: much long-term residential care is in hostels. In the 1950s the first hostels for psychiatric patients were intended to be half-way houses from which patients would soon move to more independent living. Experience showed that many residents could not leave the half-way houses, which thus became long-term hostels (Wing and Hailey 1972). This change should not be seen as a failure, since long-term hostels serve a useful purpose. For example, in a study of hostel residents, Hewett and Ryan (1975) found that half had remained in the hostel for over two years and had reached a plateau in recovery, but most had little behavioural impairment and were working. Although most hostel residents live fairly independent lives, a few of the most disabled require additional care. Levels of supervision can be varied according to the needs of the residents; for example staff may sleep at night, or they may remain awake as in a hospital ward. The latter arrangement is sometimes called a *hospital hostel*.

Occupation Some patients with chronic psychiatric disorders can take on normal employment. Other patients require specially arranged *sheltered work*, in which they can work productively but more slowly than would be acceptable elsewhere. Such work includes horticulture or the making of craft items. Some patients who cannot meet the requirements of sheltered work may undertake voluntary activities. For those who are more severely handicapped, there is a need for *occupational therapy* to avoid boredom, understimulation, and lack of social contacts. These occupational activities may be provided in day hospitals or *day centres*.

Encouraging collaboration with treatment In hospital the continuous presence of nurses can ensure compliance with treatment. In community care it is much more difficult to ensure the compliance of poorly motivated patients. It is often the *relatives* who undertake this role, and it is important that they understand the plan of treatment (for example the dose and timing of medication) and that they know who to inform if the patient has departed from the plan. It is also important that the patient understands the treatment plan. *Community nurses* have an important role in encouraging compliance with treatment. Patients who object to their plan of treatment may be helped to express their views by an *advocate* who can speak for them

Some patients with chronic psychiatric disorder do not recognize their need for continuing treatment and relapse because they stop taking antipsychotic drugs. In the United Kingdom there are no compulsory legal powers to require compliance with treatment from patients who are not in-patients. This is a concern of some psychiatrists because failure to comply with treatment is an important cause of relapse.

Reassessment Patients living outside hospital require the same regular reassessments that they would have received in a long-stay hospital. Regular reappraisal of the mental state and of compliance with treatment is usually performed by community nurses, with planned but less frequent reassessment by a psychiatrist. Some patients with chronic disorders forget appointments; therefore it is important to have a *recall system* whereby prompt steps can be taken to re-establish contact as soon as possible. To ensure this arrangement the psychiatric team needs to work closely with the carers and the general practitioner. It is important to review *physical health* as well as the psychiatric disorder, because patients with chronic psychiatric disorder may not seek help for physical illness or may not comply with the care that is offered (Brugha *et al.* 1989). Despite thorough arrangements for community assessment, the frequency of assessment can never be as great as in hospital. Hence it can be very difficult to anticipate dangerous behaviour in patients receiving community care.

Working with the family and volunteers

Community care is costly, and in most countries public funds are limited with the result that arrangements often depend on contributions by families and voluntary groups. It is important that these families and voluntary groups are involved in the planning of services, and that there is agreement about their responsibilities and those of professional staff. Without this agreement family members may believe that they are required to take on overdemanding tasks, and professionals may be concerned that volunteers are taking on tasks beyond their capabilities. It is good practice to involve families and voluntary groups in the evaluation of services.

Effective collaboration

Effective community care requires complicated arrangements involving several professionals as well as relatives and voluntary organizations. To ensure that the necessary arrangements are in place before patients are discharged from hospital, it is good practice to agree a care plan with all those involved. It is usual to appoint one person (a *key worker*) who is responsible for ensuring that the planned services are delivered and for checking regularly whether the patient's needs have changed. Sometimes a case manager is appointed to ensure that the whole care plan is carried out (see p. 660).

Continuity of care and response to crisis

Community care staff need to gain the confidence of their patients to ensure treatment and ask for help if their problems increase. Staff need to know their patients well enough to be able to predict their response to stress and to detect small changes in behaviour that may indicate relapse. These aims cannot be achieved if staff change frequently. Continuity of care is important, and staff should be extra vigilant when care has passed to a new worker.

Community care staff need to respond quickly to crises. Staffing levels must be adequate for a quick response, preferably by staff who know the patient or else by an emergency team.

The evaluation of psychiatric services

Policies of community care were originally based on optimistic views about the possibility of managing outside hospital both acute episodes of illness and the long-term problems associated with chronic mental illness. When the policy was initiated, it applied mainly to patients who, after many years in hospital, had become institutionalized and compliant. These patients could be managed in the community without much difficulty. It then became evident that less compliance could be expected from patients who had spent less time in hospital and were at an earlier stage of illness. It also became clear that the disorders of these patients were more unstable and thus difficult to manage in the community. The consequent problems have led to questioning about the effectiveness of community care, and have increased the need for evaluation.

There have been two main approaches to the evaluation of community care: (a) studies of whole services, and (b) studies of particular elements of service, for example day hospital care. These two approaches will be considered in turn.

Studies of whole services

Wing and Hailey (1972) suggested six questions that should be asked about an area psychiatric service for a population.

1. How many patients are in contact with the service?

2. What are their needs and those of the relatives?

3. Are the services meeting these needs at present?

4. How many others, not in contact with the service, also have needs?

5. What new services, or modifications to existing services, are required to cater for unmet needs?

6 Having introduced the new or modified services, are the needs met?

These questions can be answered by using case registers, by specific studies (for example evaluation of the effects of closing a hospital), by using routine records, and by seeking the opinions of those who use the service and those who contribute to it.

Case registers enable continuous monitoring of the use of services within a defined area by enumerating contacts with the various parts of the service. Such an approach is exemplified in reports of the services in Camberwell, South London, by Wing and Hailey (1972) and Wing (1982).

Systematic studies of the consequences of closing a mental hospital, resettling its patients, and caring for new cases in the community have been reported in the United Kingdom (Leff 1993*b*) and Italy (Tansella 1991). The results of such surveys are generally reassuring in that they suggest that well-resourced community provision can replace hospital care, but the results should be applied cautiously to less well provided services.

Clinical records can be used for routine monitoring of services. Such monitoring requires simple indicators that are sensitive to the wider working of the system of care. Ideally, these indicators should be of **outcome** (the effects of the service) rather than of input (the resources used). Outcome indicators are difficult to identify. Suggested indicators are rates at which patients are lost to follow-up, rates of suicide among recently discharged patients, and the number of homeless mentally ill. The value of the last two indicators is limited since both are affected by factors other than the quality of the psychiatric services. In the absence of good outcome indicators, most community services are monitored by measuring inputs, for example the number of community nurses employed or the number of hostel places provided. **Process** variables are difficult to monitor, but simple indices can be measured such as the amount of time spent by community nurses with patients who have chronic schizophrenia. However, quality of care depends not only on time spent with each patient but also on what is provided during the sessions.

Surveys of users and providers of the service can give useful information. Patients and their relatives, general practitioners, social services, and voluntary agencies may have an important perspective on the effectiveness of the services and the need for change.

Studies of the elements of a service

Trials of day care and of intensive community support have shown that, for acute disorders, brief in-patient care followed by day care or support at home can be as effective as more prolonged in-patient care (Braun *et al.* 1981). The most comprehensive studies (Stein and Test 1980; Hoult *et al.* 1983) compared hospital admission with community care in the management of psychiatric problems from defined populations. The results of these and related studies are described on p. 660, where future developments in services are considered.

Patients with special needs

Clinical work and research have identified several groups of patients whose needs cannot be met easily by the usual psychiatric services. These groups will be discussed next.

Members of ethnic minorities

Use of the services Members of ethnic minorities are less likely than other people to use services provided for the majority of the population. They are less likely to consult general practitioners when they have a psychiatric disorder, less likely to accept referral to psychiatric services, and less likely to comply with psychiatric treatment. When members of ethnic minorities ask for help, professional staff are less likely to identify psychiatric

disorder, and less able to explain illness and treatment in terms that take account of the beliefs and cultural background of the patients.

In the United Kingdom, these problems have been studied particularly among people of Asian and Afro-Caribbean origin. People of Asian origin consult their general practitioners more frequently about most conditions than do members of the general population, but consult less about psychiatric symptoms (Murray and Williams 1986; Gillam *et al.* 1989). Some Asian people prefer to seek treatment from a traditional healer when they have a psychiatric disorder (Bhopal 1986). Others consult a general practitioner, but present physical rather than psychiatric symptoms. In some ethnic minorities, referral to a psychiatrist may be avoided because it could affect marriage prospects.

Under-recognition There are two reasons why general practitioners and psychiatric staff may fail to recognize mental disorder in members of ethnic minorities. First, there may be problems of communication, which could partly be overcome by the provision of doctors who are members of the ethnic minority or by providing interpreters to aid other doctors. Second, the presentation of psychiatric disorders may differ between members of ethnic minorities and the general population. As mentioned, people from the Indian subcontinent are more likely than the general population of the United Kingdom to complain of physical symptoms when anxious or depressed (Bal 1987).

Problems of diagnosis People of Afro-Caribbean origin are more likely than members of the general population to be diagnosed as schizophrenic when their behaviour is acutely disturbed (Flaskerud and Hu 1992). It is not known whether this difference reflects a true difference in rates of schizophrenia, or whether it is due to misdiagnosis of disturbed behaviour that has other causes, for example a reaction to stress.

Problems of management In England an excess of black people has been observed among patients who are admitted to hospital on a compulsory order (Dunn and Fahy 1990). This finding raises the possibility that different criteria for admission are applied to this group.

Homeless mentally ill people

When hospital closures are planned carefully, few discharged patients become homeless (Leff 1993*a*; Harrison *et al.* 1994). Nevertheless, concern has been expressed in the United States and other countries about the numbers of psychiatric patients living on the streets. Surveys have found high rates of chronic psychiatric disorder among the residents of hostels for the homeless in the United Kingdom (Priest 1976; Marshall 1989; Marshall and Reed 1992), the United States (Bassuk 1984; Susser *et al.* 1989), and Australia (Herrman *et al.* 1989). Similar high rates of psychiatric disorder were found among people sleeping on the streets (Reed *et al.* 1992). Alcohol abuse is also common in people who live in hostels and among the homeless.

When people with psychiatric disorder have adopted a vagrant way of life it is difficult to persuade them to accept treatment or help with accommodation, and many remain on the streets or move to unsatisfactory housing (Marshall and Gath 1992; Caton *et al.* 1993). Patience and persistence are required if this outcome is to be prevented.

For a review of the problems of the homeless mentally ill, especially as seen in the United States, see Lamb (1989).

Young patients with chronic disorder

Most psychiatric patients can be treated by means of a short stay in a psychiatric unit followed by intensive community care. However, there is an important group, composed mainly of young schizophrenic men, who have unremitting illness and need prolonged care. Such patients do not fit well into an acute admission ward and are helped more by treatment in a less stressful environment. Many can be cared for in well-staffed hostels (hostel wards or hospital hostels), provided that they can be admitted to an acute psychiatric unit if their disorder relapses (Garety and Morris 1984; Creighton *et al.* 1991). In planning community services it has to be recognized that most members of this group are likely to require intensive care throughout their lives.

Patients with aggressive or other difficult behaviour

In the past this small group of patients remained for many years in hospital, where their abnormal behaviour could be managed quickly. Most of these patients are now cared for in the community, where it is difficult to meet their needs. Most of these patients have schizophrenia, often accompanied by personality disorder or the misuse of alcohol or drugs. A few have brain damage. In the community these patients require intensive supervision, which is expensive to provide and may not be accepted readily by the patients. In the long-stay hospital periods of aggressive behaviour can be identified and treated quickly, but such behaviour may be difficult to predict for a patient in the community, however close the supervision. Since aggressive behaviour may endanger other people, the public is likely to judge the effectiveness of the service on its ability to care for this small group of patients. These patients are not managed well in a general psychiatric unit, where their aggressive or unpredictable behaviour may alarm other patients. It may be better to provide a separate unit designed to meet their special needs.

Not all violent or otherwise difficult behaviour results from chronic psychiatric disorder. In some patients this behaviour is related to associated personality disorder or substance abuse. It is particularly difficult to prevent or predict episodes of violence in such patients.

Doctors with psychiatric problems

Although doctors have striven to remove stigma from psychiatric disorders, many do not seek help if they develop such a disorder themselves. It is helpful to provide special arrangements to enable psychiatrically ill doctors to obtain treatment away from their place of work. If doctors are to seek treatment appropriately, a greater acceptance of psychiatric disorder is required within the medical profession. This issue should be discussed openly during medical education and training. Occasionally, chronic or recurrent psychiatric disorder raises problems of fitness to practise.

Other problems of community care

The burden on relatives

Community care may make great demands on relatives. If members of the family are to take responsibility for housing the patient, encouraging adaptive behaviour, supervising medication, and reporting signs of relapse, they need to be well informed, adequately supported, and able to obtain help in an emergency. Such support is time-consuming and expensive, and is not available unless adequate community care is well resourced and the needs of the family are given high priority. There have been many reports of poor communication between psychiatric services and patients' relatives (for example about discharge plans). Other reports have highlighted the difficulties that carers experience in coping with the negative symptoms and socially embarrassing or aggressive behaviour of patients whom they support. It is especially important to consider the effects on children of psychiatric disorder in a care-giver.

Problems in the distribution of resources

There are problems in the distribution of care between patients with acute and patients with chronic psychiatric disorder.

Staff working in the community may have the responsibility of supporting primary care teams in the early treatment of psychiatric disorder, and also of supporting patients with chronic psychiatric disorder. Such staff may devote more time than was originally planned to the former group who are generally more demanding of care and more responsive to it. This trend has been shown in studies of community nursing (Wooff and Goldberg 1988; Burns *et al.* 1991), but it is not likely to be confined to this professional group.

A second problem of resource allocation is that, during the move from long-stay hospital care to community care, there is a stage when the costs of both kinds of provision must be incurred simultaneously until the long-stay hospital provision can be closed. There may be a temptation to shorten this period of transition by accelerating the discharge of patients to a degree that exceeds the tolerance of families and voluntary agencies to take on the additional burdens involved.

Problems in the coordination of services

In most countries long-stay hospital care for patients with chronic psychiatric disorder is provided by a single agency (a hospital authority), while community care requires coordinated action by several agencies, each of which usually has other responsibilities (for example social services departments have responsibilities for mentally healthy children and elderly people as well as for psychiatric patients). These arrangements present problems of two kinds—the allocation of resources to the service as a whole, and the availability of specific items of care required by individual patients (for example sheltered housing). The former problem can be reduced by a coordinating committee with members from all the agencies involved in the funding of community care. The latter problem can be reduced by designating one member of staff as a case manager responsible for ensuring that the necessary components of care are provided for individual patients (case management is discussed further on p. 660).

Developments in the provision of services

The problems encountered in the provision of services, especially for patients with chronic psychiatric disorder, have led to new developments. Some of these developments have been incorporated into practice, though not always on the basis of sound evidence of effectiveness. These innovations will be considered next in relation to primary care, acute care, and long-term care.

Developments in primary care services

It is now common practice for one or more members of the psychiatric community care team (usually a community nurse, a clinical psychologist, or a psychiatrist) to work part-time with primary care teams. A new development is to extend the skills of members of the primary care team by providing additional training in the management of psychiatric problems. General practitioners have been trained to increase their ability to identify psychiatric disorder among their patients (Gask 1992), and to counsel patients with acute reactions to stress and adjustment disorders (Mynors-Wallis *et al.* 1995).

Developments in acute specialist care

Alternatives have been sought to the usual in-patient treatment of patients with acute psychiatric disorder, either by greatly shortening length of stay or by avoiding in-patient admission. These alternatives have included the greater use of day hospitals and the provision of intensive care at home.

Greater use of day hospitals

Studies have compared admission to a day hospital with admission to an in-patient unit. One study showed that about 40 per cent of acutely ill psychiatric patients can be cared for in day hospitals (Creed *et al.* 1991). However, successful day care has not been shown to be less expensive than in-patient care when the full costs are included, for example costs of transport (Creed *et al.* 1990). Attempts to direct care from hospitals to less well provided day hospitals may place an undue burden on families.

Intensive community treatment

Several controlled studies have compared hospital treatment with intensive community treatment for acute psychiatric disorder. In the United States, Stein and Test (1980) developed a form of management which they called assertive community treatment, involving assessment and treatment at home by a multidisciplinary team. They compared this approach with standard treatment in hospital. Over 14 months, symptoms, social functioning, and satisfaction were better in the community group, and bed use was reduced. In Australia, Hoult *et al.* (1983) obtained comparable results with a similar kind of intensive community treatment, and reported that this treatment was less costly than

hospital treatment. In the United Kingdom, Marks *et al.* (1994) compared routine hospital care with a form of intensive community treatment which they called the daily living programme. The study lasted for three years, and was concerned with acutely ill patients who had not previously been admitted to hospital. About three-quarters of the community treatment group required an initial brief admission to hospital (average six days) before they could be managed in the community. Their stay in hospital was significantly less than that of the control group (average 53 days), and the outcomes for symptoms and social adjustment were slightly better. Deaths from self-harm were not reduced.

Similar findings have been reported for somewhat different kinds of intensive community treatment by Merson *et al.* (1992) and Dean and Gadd (1990) in the United Kingdom, and by Wright *et al.* (1989) in the United States. However, there are reasons to doubt whether the results of such demonstration projects, staffed by highly motivated staff working for a limited period, can be obtained continuously in routine practice (Audini *et al.* 1994). All the studies confirmed the need for some beds for the treatment of the acute stage of illness; intensive home care can reduce the number of beds needed, but not the basic requirement.

Assertive community treatment has also been compared with less intensive community treatment. As in the comparisons of assertive community treatment with in-patient care, this second group of studies showed that assertive community treatment reduced the use of hospital beds. Patients were more satisfied with the more intensive care, but mental state and social functioning were not different in the two groups (Jerrell and Hu 1989).

Developments in care for patients with chronic mental disorder

Case management

Case management (also known as care management) is intended to improve the coordination of the care provided to patients by the various agencies involved in community care. The approach was developed first in the United States in the early 1980s.

The term case management has been used to describe several rather different procedures. All have four *elements in common*: patients' needs are assessed, a plan is made to provide services to meet the needs, services are provided, and the patient and the service delivery are monitored. The *different elements* of the various forms of case management can be characterized by dividing them into the following three groups: (i) *general management* methods, in which the managers organize services provided by others but do not provide direct clinical services; (ii) *clinical management*, in which the managers not only arrange the services of other agencies but also provide some treatment for patients, and so have a more direct relationship with them; (iii) *rehabilitation management*, in which the emphasis is on preserving or developing normal areas of functioning rather than identifying deficits. (Case management forms a part of the assertive community treatment methods discussed above.)

The method of case management adopted in the United Kingdom is an amalgam of the three types described above; it varies from one place to another but in most is closest to 'general management'. Case management is a plausible way of improving service delivery and has been adopted widely. In England and Wales it forms part of the requirements of the Community Care Act (1990). Nevertheless, it is appropriate to ask (a) whether the patients' needs for services are met more effectively under case management than under other arrangements, and (b) if better provision of services is achieved, whether this leads to improvement in the symptoms or social functioning of patients and a lessening of the burden on their families.

To date only one trial of the *general management* approach has been reported (Franklin *et al.* 1987). In comparison with controls, patients receiving this kind of case management received more services and were admitted to hospital more often; even so, their quality of life was not better. *Clinical case management* has not been tested in a controlled trial at the time of writing. *Rehabilitation management* has been tested in one randomized controlled trial (Modcrin *et al.* 1988). Compared with the control group, patients receiving this form of case management improved more in social functioning but not in

quality of life. (Assertive community treatment has been tested more extensively (see p. 660) however, it consists of more than the case management).

Further research is needed to establish the costs and benefits of case management. Since the services to be coordinated in different countries differ in their nature and accessibility, the results of a study in one country should be applied with caution to the planning of services in another.

Supervision registers

Supervision registers are an extension of the case management approach. The intention is to identify, among the patients cared for in the community, those who are specially likely to be aggressive or at high risk for suicide. These and other patients are included in the register when more than usual efforts should be made to respond quickly to crises, to maintain active follow-up, and to seek patients without delay if they fail to keep appointments. Supervision registers alone do little to improve care; they need to be backed by funding sufficient to supply all the needs that have been identified for those on the register. Usually the main need is for intensive individual care and supervision by a member of staff who has a small case load.

Transcultural aspects of service provision

In developing countries, the prevalence and nature of psychiatric disorders is broadly similar to that in developed countries. However, in the developing countries there is more psychiatric morbidity associated with untreated or inadequately treated physical illness, and there are some differences in the presentation of illnesses, for example more presentations with physical symptoms. In each country the organization of care should take account of the stage of development of primary care services, and the number of trained psychiatrists and psychiatric nurses. Further account should be taken of the attitudes and beliefs of the population concerning mental disorder. For example in Zimbabwe a 'defeat depression' campaign integrated local beliefs and culturally appropriate counselling with scientific knowledge of psychiatric disorder. The campaign emphasized local causes such as the low status of women and the breakdown of traditional family support systems (Abas *et al.* 1994).

In many developing countries there is limited provision of specialist psychiatric resources, and so it is not possible to establish the broad range of psychiatric services available in more developed countries. Therefore, it is essential to identify priorities and to make the maximum use of other resources, including traditional healers when appropriate.

The World Health Organization (WHO) (1984) has identified four priorities for the provision of psychiatric services in developing countries: rapid response to psychiatric emergencies, provisions for chronic severe psychiatric disorder, care for psychiatric disorders associated with general medical illness, and care for any high risk groups found in the country (for example drug abuse). The WHO has stressed the need for the training of auxiliary workers who can supplement the efforts of fully trained staff. In countries with few trained psychiatrists, resources may be used most effectively by improving the skills of general practitioners and nurses in the first-line management of psychiatric emergencies, by making use of a limited list of psychiatric drugs, and by undertaking the care of medically ill patients who have associated psychiatric problems. In keeping with these aims, the WHO has recommended that the care of psychiatric problems should be an important part of the work of primary care teams (World Health Organization 1978*b*).

Further reading

Bhugra, D. and Leff, J. (1993). *Principles of social psychiatry*. Blackwell, Oxford.

Liberman, R. P. (ed.) (1992). *Handbook of psychiatric rehabilitation*. Macmillan, New York.

Tyrer, P. and Creed, F. (ed.) (1995). *Community psychiatry in action*. Cambridge University Press.

20 | Child psychiatry

The practice of child psychiatry differs from that of adult psychiatry in several important ways. It is seldom that children initiate a consultation with the clinician. Instead, they are brought by adults—usually the parents—who think that some aspect of behaviour or development is abnormal. Much depends on the attitudes and tolerance of these adults, and how they perceive the child's behaviour. Healthy children may be brought to the doctor by overanxious and solicitous parents or teachers, whilst in other circumstances severely disturbed children may be left to themselves. A related factor is that psychiatric problems in a child may be a manifestation of disturbance in other members of his family. When a child's problems have previously been contained in a family or school, the child may be referred when another problem arises in the family or school and reduces their capacity to cope with the child.

Another difference from adult psychiatry is that, in deciding what is normal and what is abnormal, greater attention must be paid to the stage of development of the patient and the duration of the disorder. For example repeated bed-wetting can be regarded as normal in a three-year-old child but abnormal in a child aged seven, whilst separation from the parents is more likely to affect a toddler than an older child.

The practice of child psychiatry differs from adult psychiatry in two other ways. First, children are generally less able to express themselves in words. Therefore evidence of disturbance is based more on observations of behaviour made by parents, teachers, and others. Epidemiological studies have shown that there is only moderate agreement between different informants. This is partly because the child's behaviour may differ in different circumstances, and partly because informants may have different criteria for abnormality. The assessment of informants' accounts requires skills in taking a developmental history, assessing behaviour, evaluating the emotional involvement of informants, and understanding the home and

school background. It is important to ask for specific examples of general problems.

The second difference between child and adult psychiatry is that the treatment of children makes less use of medication or other methods of individual treatment. Instead, the main emphasis is on changing the attitudes of parents, reassuring and retraining children, working with the family, and coordinating the efforts of others who can help children, especially at school. Achieving these changes usually requires teamwork from doctors, psychologists, and social workers so that multidisciplinary working is even more important in child than in adult psychiatry.

The first part of this chapter is concerned with a number of general issues concerning psychiatric disorder in childhood, including its frequency, causes, assessment, and management. The second part of the chapter contains information about the principal syndromes encountered in the practice of child psychiatry. The chapter does not provide a comprehensive account of child psychiatry. It is an introduction to the main themes for the psychiatrist who is starting his specialist general training. It is expected that he will follow it by reading a specialist text such as the textbook by Graham (1991) or one of those listed in the further reading at the end of the chapter. **In this book, childhood mental retardation is dealt with in Chapter 21.** This is a convenient arrangement, but the reader should remember that many aspects of the study and care of mentally retarded children are closely related to child psychiatry.

Normal development

The practice of child psychiatry calls for knowledge of the normal process of development from a helpless infant into an independent adult. In order to judge whether any observed emotional, social, or intellectual functioning is abnormal, it has to be compared with the corresponding normal range for the age group. This section provides a summary of the main aspects of development that concern the psychiatrist. A textbook of paediatrics should be consulted for details of these developmental phases. A useful review of psychological and social development has been provided by Rutter (1980).

The first year of life

This is a period of rapid development of motor and social functioning. Three weeks after birth, the baby smiles at faces; selective smiling appears by six months, fear of strangers by eight months, and anxiety on separation from the mother shortly after.

Bowlby (1980) emphasized the importance in the early years of life of a general process of **attachment** of the infant to the parents and of more selective emotional **bonding**. Although bonding to the mother is most significant, important attachments are also made to the father and other people who are close to the infant. Research has stressed the reciprocal nature of this process and the probable importance of early contacts between the mother (or other carers) and the newborn infant in initiating bonding (Rutter 1980).

By the end of the first year, the child should have formed a close and secure relationship with the mother or other close carer. There should be an ordered pattern of sleeping and feeding, and weaning has usually been accomplished. The child has begun to learn about objects outside himself, simple causal relationships, and spatial relationships. By the end of the first year, the child enjoys making sounds and may say 'mama', 'dada', and perhaps one or two other words.

Year two

This is also a period of rapid development. The child begins to wish to please the parents and appears anxious when they disapprove. He begins to learn to control his behaviour. By now, attachment behaviour should be well established. Temper tantrums occur, particularly if exploratory wishes are frustrated. These tantrums do not last long, and should lessen as the child learns to accept constraints. By the end of the second year he should be able to put two or three words together as a simple sentence.

Pre-school years (two to five years)

This phase brings a rapid increase in intellectual abilities, especially in the complexity of language. Social development occurs as the child learns to live within the family. He begins to identify with the parents and adopt their standards in matters of conscience. Social life develops rapidly as he learns to interact with siblings, other children, and adults. Temper tantrums continue, but diminish and should disappear before the child starts school. At this age, the child has much curiosity about the environment and may ask a great number of questions.

In children aged two to five, fantasy life is rich and vivid. It can form a temporary substitute for the real world, enabling desires to be fulfilled regardless of reality. Special objects such as teddy bears or pieces of blanket become important to the child. They appear to comfort and reassure the child, and help sleep. They have been called 'transitional objects'.

The child begins to learn about his own sexual identity. He realizes the differences between males and females in their appearance, clothes, behaviour, and anatomy. Sexual play and exploration are common at this stage.

According to psychodynamic theory, at this stage defence mechanisms develop to enable the child to cope with anxiety arising from unacceptable emotions. These defence mechanisms have been described on p. 136. They include repression, rationalization, compensation, and displacement.

Common problems in early childhood

In children from birth to the beginning of the fifth year, common problems include difficulties in feeding and sleeping, as well as clinging to the parents (separation anxiety), temper tantrums, oppositional behaviour, and minor degrees of aggression.

Middle childhood

By the age of five, the child should understand his or her identity as boy or girl, and his position in the family. He has to learn to cope with school, and to read, write, and acquire numerical concepts. The teacher becomes an important person in the child's life. At this stage the child gradually learns what he can achieve and what his limitations are. Defence mechanisms, conscience, and standards of social behaviour develop further. According to psychoanalytic theory, this is a period in which psychosexual development is quiescent (the latent period). This notion has been questioned (Rutter 1971), and it now seems that in the five- to ten-year period sexual interest and activities are present, although they may be concealed from adults.

Common problems in middle childhood

Common problems in children aged 5 to 11 years include fears, nightmares, minor difficulties in relationships with peers, disobedience, and fighting.

Adolescence

Adolescence is the growing-up period between childhood and maturity. Among the most obvious features are the physical changes of puberty. The age at which these changes occur is quite variable, usually between 11 and 13 in girls, and 13 and 17 in boys. The production of sex hormones precedes these changes, starting in both sexes between the ages of eight and ten. Adolescence is a time of increased awareness of personal identity and individual characteristics. At this age, young people become self-aware, are concerned to know who they are, and begin to consider where they want to go in life. They can look ahead, consider alternatives for the future, and feel hope and despair. It is popularly but wrongly believed that emotional turmoil and alienation from the family are characteristic of adolescence (see p. 713).

Peer group relationships are important and close friendships often develop, especially among girls. Membership of a group is common, and this can help the adolescent in moving towards autonomy. Adolescence brings a marked increased in heterosexual interest and activity. At first, tentative approaches are made to the opposite sex. Gradually these become more direct and confident. In late adolescence, there is a capacity for affection towards the opposite sex as well as sexual feelings. How far and in what way sexual feelings are expressed depends greatly on the standards of society and on rules in the family.

Common problems in later childhood and early adolescence

Common problems among children aged from 12 to 16 years include moodiness, anxiety, minor problems of school refusal, difficulties in relationships with peers, disobedience and rebellion including truancy, experimenting with illicit drugs, fighting, and stealing.

Developmental psychopathology

In child psychiatry it is important to adopt a developmental approach for three reasons. First, as explained on p. 663, the child's stage of development determines whether behaviour is normal or pathological; for example bed-wetting is normal at the age of three years but abnormal at the age of seven years. Second, the effects of life events differ as the child develops; for example infants aged under six months can move to a new caretaker with little disturbance, but children aged six months to three years of age show great distress when separated from an original caretaker, because an attachment relationship has been formed. After the age of three years, attachment bonds are still strong but the child's ability to use language and to understand reduce the effect of a change of caretaker provided that it is arranged sensitively.

The third reason for a developmental approach is that psychopathology may change as the child grows older: anxiety disorders in childhood tend to improve as the child develops; depressive disorders often recur and continue into adult life; conduct disorders frequently continue into adolescence as aggressive and delinquent behaviour, and also commonly as substance abuse—a problem that is uncommon in younger children.

These continuities and discontinuities from childhood to adulthood depend not only on maturation of the children but also on changes in their environment, especially on changes in the family. An adverse family environment, particularly persistent marital discord, appears to be important, but personality disorder in the parents may be important in causing the marital conflict and the persistent disorder in the child. Conversely, there may be protective factors in the child's environment that reduce some of the continuities between childhood and adult disorder, for example a close relationship with a caring grandparent. Long-term studies of childhood pathology have been reviewed by Robins and Rutter (1990).

The classification of psychiatric disorders in children and adolescents

Both DSMIV and ICD10 contain a scheme for classifying the psychiatric disorders of childhood. Disorders of adolescence are classified partly with this scheme, and partly with the categories used in adult psychiatry.

Seven main groups of childhood psychiatric disorders are generally recognized by clinicians and are supported by studies using multivariate analysis (Quay and Werry 1986). The terms used in this book for the seven groups are listed below, with some alternatives in parentheses:

(1) adjustment reactions;
(2) pervasive developmental disorders (psychoses of childhood);
(3) specific developmental disorders;
(4) conduct (antisocial or externalizing) disorders;
(5) hyperkinetic (attention-deficit) disorders;
(6) emotional (neurotic or internalizing) disorders;
(7) symptomatic disorders.

Many child psychiatric disorders cannot be classified in a satisfactory way by allocating them to a single category. Therefore *multiaxial systems* have been proposed. A widely adopted system has axes for (i) clinical psychiatric syndromes, (ii) specific delays in development, (iii) intellectual level, (iv) medical conditions, and (v) abnormal social situations. Further information is given by Rutter and Gould (1985). This scheme is easy to use, and allows clinicians to record systematically the different kinds of information required in categorizing children's problems.

The DSMIV and ICD10 classifications for child psychiatric disorders are shown in Table 20.1. Both schemes are complicated, and so only the main categories are shown in the table. (In this book mental retardation in childhood is considered in Chapter 21.) In most ways the two systems are similar; both have categories for specific and pervasive developmental disorders, with the former divided into disorders affecting motor skills, speech and language (communication), and scholastic skills (learning). Both systems have categories for disorders of behaviour, which are divided into conduct (or disruptive behaviour) disorder and attention-deficit (hyperkinetic) disorder. Both systems have a category for anxiety (emotional) disorder (DSMIV does not provide a separate category for childhood anxiety disorder but uses the adult category instead). Both systems have a category for tic disorders. DSMIV also has a category for eating and elimination disorders, but ICD10 classifies these conditions with sleep disorders under 'other behavioural and emotional disorders'. For further information about classification in child psychiatry see Cantwell and Rutter (1994).

Epidemiology

Behavioural and emotional disorders occur frequently in the general population of children. Estimates vary according to the diagnostic criteria and other methods used, but it appears that rates in the different developed countries are similar. The limited evidence suggests that rates of emotional and behavioural disorders in developing

Table 20.1

DSMIV	ICD10
	F8 Disorders of psychological development
Learning disorders	Specific developmental disorders of scholastic skills
Motor skills disorders	Specific developmental disorder of motor function
Communication disorders	Specific developmental disorders of speech and language
Pervasive developmental disorders	Pervasive developmental disorders
	F9 Behavioural and emotional disorders with onset usually occurring in childhood and adolescence
Attention-deficit and disruptive behaviour disorders	Hyperkinetic disorders Conduct disorders
	Mixed disorders of conduct and emotions
	Emotional disorders with onset specific to childhood
Tic disorders	Tic disorders
Feeding and eating disorders of infancy and childhood	
	Disorders of social functioning with onset specific to childhood and adolescence
Elimination disorder	
Other disorders of infancy, childhood, and adolescence	Other behavioural and emotional disorders with onset usually in childhood and adolescence (include elimination disorders and feeding disorders)

countries are similar to those in developed ones. In the United Kingdom the prevalence of child psychiatric disorder in ethnic minority groups has usually been found to be similar to that in the rest of the population. The exception is a high prevalence of conduct disorder found among West Indian girls (Rutter *et al.* 1974).

The frequency of psychiatric problems varies with age. Richman *et al.* (1982) reported that 7 per cent of three-year-olds had symptoms amounting to a moderate or severe problem, and a further 15 per cent had mild problems such as disobedience. In the middle years of childhood, rates of psychiatric problems vary in different kinds of areas, being twice as high in urban areas (about 25 per cent) than in rural areas (about 12 per cent) (Rutter *et al.* 1975*b*).

Evidence about adolescence was provided by a four-year follow-up of the Isle of Wight study described below (Rutter *et al.* 1976*a*). At the age of 14, the one-year prevalence rate of significant psychiatric disorder was about 20 per cent. Similar findings have been reported from other countries. Less is known about prevalence among older adolescents, but the rates are probably similar to those in mid-adolescence.

The most detailed findings come from a study of physical health, intelligence, education, and psychological difficulties in all the 10- and 11-year-olds attending state schools in the Isle of Wight—a total of 2199 children (Rutter *et al.* 1970*a*). Screening questionnaires were completed by parents and teachers. Children identified in this way were given psychological and educational tests and their parents were interviewed. The one-year prevalence rate of psychiatric disorder was about 7 per cent, with the rate in boys being twice that in girls. There was no correlation with social class, but prevalence increased as intelligence decreased. Psychiatric disorder was associated with physical handicap and especially with evidence of organic brain damage. There was also a strong association between reading retardation and conduct disorder. Several years later the same methods were used to survey an inner London borough (Rutter *et al.* 1975*b*,*c*). It was found that the rates of all types of disorder were twice those in the Isle of Wight.

Follow-up studies

Mild symptoms and behavioural or developmental problems are usually brief. However, this is not so for the conditions severe enough to be diagnosed as childhood psychiatric disorders. These conditions often persist for several years. Thus in the Isle of Wight Study three-quarters of children with conduct disorder and half of those with emotional disorders at age 10 were still handicapped by these problems four years later (Rutter *et al.* 1976*a*).

The prognosis for adult life can be established only by long follow-up, which is difficult to arrange. The outstanding study is that of Robins (1966) who followed up people who had attended a child guidance clinic 30 years previously and compared them with a control group who had attended the same schools but had not been referred to the clinic. She found that emotional disorders had a good prognosis. When these disorders did continue they took the form, in the adult, of neurosis or depression. In contrast, children with conduct disorder had a poor outcome. As adults they were likely to develop antisocial personality disorder or alcoholism, have problems with employment or marriage, or commit offences. More recent research has confirmed that the outcomes of neurotic and conduct disorders are very different. It has also been shown that definite over-activity syndromes have a poor prognosis and psychoses a worse one (Robins 1979; Zeitlin 1986). Zeitlin examined the records of patients who attended the same hospital for psychiatric treatment both as children and as adults. He found considerable continuity in the types of symptoms reported, especially when the original problems were depressive or obsessive symptoms or conduct disorders. For a review of outcome studies see Rutter (1995).

Aetiology

In discussing the causes of child psychiatric disorders, much the same principles apply as those described in the earlier chapter on the aetiology of adult disorders. In child psychiatry, there are fewer disease entities and more reactions to environmental factors, notably those in the family, school, and neighbourhood. Even more than in adult life, the determinants of childhood disturbance are usually multiple. There is also a developmental aspect; children mature psychologically and socially as they grow up, and their disorders reflect this maturation. In the following paragraphs four interacting groups of factors will be considered briefly. These are inheritance, temperament, physical impairment, with special reference to brain damage, and environmental, family, social, and cultural causes. Aetiology is reviewed in the textbooks edited by Graham (1991) and Rutter *et al.* (1994).

Genetic factors

Children with psychiatric problems often have parents who suffer from a psychiatric disorder. There is much evidence that environmental factors account for a large part of this association (see below) and it has been difficult to determine what contribution, if any, is made by genetic factors. Early studies were flawed by lack of standardized diagnoses, problems in sampling, and lack of adequate controls, but some progress has been made recently. This progress has been greater in studies of defined psychiatric disorders than in studies of minor emotional disturbances.

The results of the studies are reviewed later in this chapter; some general comments will be made here. First, recent studies show the value of using standard criteria for diagnosis, but they also suggest that standard diagnostic categories do not all coincide closely with the genetic factors in aetiology. For example the phenotype of autism seems to include some kinds of developmental language disorder, whilst the phenotype of Gilles de la Tourette's syndrome seems to include some cases of obsessional disorder. Second, many children have symptoms that qualify for more than one psychiatric diagnosis, for example depressive disorders, conduct disorder, or eating disorder. It is not yet clear whether this overlap arises because one disorder predisposes to another (as, for example, infection with HIV predisposes to other infections) or whether they are two

manifestations of the same genetic predisposition. Third, it has become clear that single-gene effects may be operating when there is no obvious Mendelian pattern of inheritance. This can arise when many of the affected persons do not have children (for example in autism), and when the phenotype is very variable so that it is not at first recognized as such.

In child psychiatry, genetic research has so far been most productive in the study of autism. (For a review of genetic factors in child psychiatry see Rutter *et al.* (1990).) The hereditary factors of importance in child psychiatry are polygenic. As in adult life they interact with psychosocial factors, and genetic investigations may include estimates of environmental factors. There are also indirect genetic influences acting through polygenic control of intelligence and temperament which in part determine whether situations are experienced as stressful. Methods of genetic investigation are discussed on pp. 93–7, and examples of genetic findings in child psychiatry are given at relevant points in the chapter.

Temperament and individual differences

In a longitudinal study in New York, Thomas *et al.* (1968) found that certain temperamental factors detected before the age of two might predispose to later psychiatric disorder. In the first two years, one group of children ('difficult children') tended to respond to new environmental stimuli by withdrawal, slow adaptation, and an intense behavioural response. Another group ('easy children') responded to new stimuli with positive approach, rapid adaptation, and a mild behavioural response. This group was less likely than the first to develop behavioural disorders later in childhood. The investigators thought that these early temperamental differences were determined both genetically and by environmental factors. The validity of the methods in this study, and the significance of the findings, have been questioned (Graham and Stevenson 1987).

Brain disorder

Although serious physical disease of any kind can predispose to psychological problems in childhood, brain disorders are the most important. In the Isle of Wight Study about 7 per cent of physically healthy children aged 10–11 years were classified as having psychiatric problems, compared with about 12 per cent of physically ill children of the same age and 34 per cent of children with brain disorders (Rutter *et al.* 1976*a*). The high prevalence in the latter group was not explained by the adverse social factors known to be associated with the risk of brain disorder. Nor is it likely to have been due to physical disability as such, because rates of psychiatric disorder are less in children equally disabled by muscular disorders. The rate of psychiatric disorder among children with brain damage is related to the severity of the damage, though not closely to the site. It is as common among brain-injured girls as boys, a finding which contrasts with the higher rate of psychiatric disorder among boys in the general population.

Children with brain injury are more likely to develop psychiatric disorder if they encounter adverse psychosocial influences of the kind that provoke psychiatric disorder in children without brain damage (see for example the study of children with head injury by Rutter *et al.* (1983)).

The theory of minimal brain dysfunction

The observation that defined brain damage was associated with psychiatric disorder led to the suggestion that lesser degrees of damage, insufficient to cause definite neurological signs, could account for otherwise unexplained disorders. The term minimal brain damage was suggested, but was later changed to minimal brain dysfunction after repeated failures to find evidence of any structural changes. It was suggested that this brain dysfunction originated in damage at birth, and for a time the phrase 'a continuum of reproductive casualty' (Pasamanick and Knobloch 1966) was used to express this notion. There is an association between histories of abnormal pregnancy, prematurity, and birth asphyxia on the one hand, and psychiatric disorder on the other, but the former factors are also associated with social disadvantage which could be the real cause of the

psychiatric disorder. The theory of minimal brain damage (or dysfunction) is no longer influential, although the role of established brain damage is well recognized. See Taylor (1991*a*) for a review.

Maturational changes and delayed effects

The effects of brain lesions are more complex in childhood because the brain is developing. The immature brain seems more able than the adult brain to compensate for localized damage. For example even complete destruction of the left hemisphere in early childhood can be followed by normal development of language (Goodman 1987).

Early damage may not become manifest as a disorder until a later stage of development when the damaged area takes up a key function. It is known that, when epilepsy follows brain injury at birth, seizures may not begin until many years later; similarly it has been suggested that there may be delays in the behavioural consequences of brain injury.

Lateralized processes and psychopathology

The behavioural effects of lateralized brain damage are less specific in childhood than in adult life. Therefore attempts to explain, for example, educational problems in terms of left- and right-sided damage or dysfunction are unwise, and attempts to use left-brain or right-brain training are unlikely to be helpful.

The consequences of head injury in childhood

Head injury is a common cause of neurological damage in childhood. The form of the disorder is not very specific, partly because the effects of head injury are seldom localized to one area of the brain. Common consequences of severe injury are intellectual impairment and social inhibition. Intellectual impairment is proportional to the severity of the injury, but the relationship of behaviour disorder to the injury is less direct.

Epilepsy as a cause of childhood psychiatric disorder

The relationship between epilepsy and psychiatric disorder in adults is considered on pp. 334–41. In childhood there is a strong association between recurrent seizures and psychiatric disorder. As in adult life, the causes of the relationship are complex. The brain lesion causing the epilepsy may cause psychological disorder, the psychological and social consequences of recurrent seizures may lead to emotional and behavioural disorder, and may affect school performance, and the side-effects of anti-epileptic drugs may have the same effects. The site and the type of epilepsy seem generally less important, with the exception that temporal lobe epilepsy seems more likely to be associated with psychological disorder. The age of onset of seizures also determines the child's response to epilepsy.

Lead intoxication as a cause of psychological problems

There has been concern that chronic exposure to lead may impair the developing brain and may produce intellectual impairment and behaviour disorder. It is now generally accepted that there is an association between IQ and body lead as measured in blood or tooth dentine. However, the cause of the association is still uncertain. Lead may damage the brain and may cause low IQ, but psychosocial factors could explain the association. Children brought up in poor social conditions could ingest more lead because they play in streets polluted by car exhausts; also, low IQ is known to be related to social adversity. When social factors are allowed for, it seems that there is still a small but consistent negative association between high lead levels and IQ (Taylor 1991*b*). Behavioural disorder cannot be measured as precisely as IQ; hence an association of such a disorder with body levels of lead is even more difficult to establish. To date, cohort studies in several cities in the United States have not given a clear answer to this question (see Taylor (1991*b*) for a review).

Environmental factors: the effect of life events

The concept of *life events* is useful in child psychiatry as well as in the psychiatry of adult life (see p. 89). Life events may be *undesirable* or *protective* in nature, and may vary in severity. Events can be classified by their social characteristics, for example family problems, divorce, or the death of a parent. They can also be classified according to their general significance; for example they may be exit events (i.e. separations) or entrance events (i.e. additions such as the birth of a sibling).

The study of life events is not as advanced in child psychiatry as in adult psychiatry, but in child psychiatry it has been shown that life events have an additive effect, that exit events are associated with anxious and depressive disorders (Goodyer *et al.* 1985), and that events cluster about four months before the onset of symptoms (Goodyer *et al.* 1987).

Family

As a child progresses from complete dependence on others to independence, he needs a stable and secure family background with a consistent pattern of emotional warmth, acceptance, help, and constructive discipline. Prolonged separation from or loss of parents can have a profound effect on psychological development in infancy and childhood. Poor relationships in the family may have similar adverse effects; overt conflict between the parents seems especially important (Jenkins and Smith 1990).

The well-known work of Bowlby (1951) led to widespread concern with the effects of 'maternal deprivation'. Bowlby originally suggested that prolonged separation from the mother was a major cause of juvenile delinquency. Subsequently he argued that the early experience or threat of separation from the mother is associated with anxiety or depression in later years (Bowlby 1973, 1980). Since the original formulation of the consequences of maternal deprivation, it has become apparent that the effect of separation depends on many factors. These include the age of the child at the time of separation, his previous relationship with his mother and father, the reasons for the separation, the way in which separation was managed, and the quality of the new care. It is also apparent that the various consequences of parental deprivation have different long-term effects. An unstimulating environment and lack of encouragement to learn in infancy is associated with educational underachievement in later years. Poor emotional attachments in early life may result in difficulties in social relationships.

Other studies have confirmed that *family risk factors* for psychiatric disorder in childhood are multiple and additive. Thus the risk increases in children of families with (i) severe marital discord, (ii) low social status, (iii) large size or overcrowding, (iv) paternal criminality, and (v) maternal psychiatric disorder. The risk is also increased in children placed in care away from the family. In children with one risk factor, rates of psychiatric disorder are not significantly greater than in children with no risk factors. Children with two risk factors have a fourfold increase in rate of disorder. One in five of children with four risk factors have psychiatric disorder (Garmezy and Mastern 1994).

In addition to risk factors, *protective factors* can be identified which reduce the rate of psychiatric disorder associated with a given level of risk factors. Protective factors include good mothering, strong affectionate ties within the family, sociability and the capacity for problem solving in the child, and support outside the family from individuals, or from the school or church (Werner and Smith 1982; Rutter 1985b). Patterns of child rearing are not clearly related to psychiatric disturbance in the child except when they involve abuse of the child (Rutter and Madge 1976).

Physical and sexual abuse are important risk factors which are considered on pp. 716 and 720.

Effects of parental mental disorder

Rates of psychological problems are higher in the children of parents with mental illness than in the children of healthy parents (Rutter 1966). These problems usually involve poor adjustment at home or at school, and often include disruptive behaviour (Orvaschel 1983; Radke-Yarrow *et al.* 1993). The

causes are complex, but it is known that depressed mothers involve themselves less with their children, thus providing a less stimulating environment (Gordon *et al.* 1989; Stein *et al.* 1991). Older children of depressed mothers may be depressed, an association that could be determined genetically or through environmental factors. Schizophrenia impairs mothering skills when the mother has distracting or psychotic symptoms, or negative symptoms, or impairment of affect.

Recently there has been increasing concern about the psychological development of the children of mothers who are addicted to alcohol or drugs. Such children suffer from a series of disadvantages, which are related to the effects of drug taking on the mother's care of the child, as well as to features of the mother's personality that led her to take drugs and to associated social problems (for reviews see Singer *et al.* (1991) and Von Knorring (1991)).

Divorce

The children of divorced parents have more psychological problems than the children of parents who are not divorced. It is not certain how far these problems precede the divorce and are related to disharmony between the parents, or to the behaviour of one or both parents that contributed to the decision to divorce.

Distress and dysfunction in the children are greatest in the year after the divorce; after two years these problems are still present but are generally less severe than those of children remaining in conflictual marriages. There is some evidence that among children brought up by a divorced mother, boys have more problems than girls (Hetherington *et al.* 1985). For a review of the long term effects of divorce on children, see Wallerstein (1991).

Death of a parent

In children the response to the death of a parent varies with age. *Children aged below four to five years* do not have a complete concept of death as causing permanent separation. Such children react with despair and anxiety to separation, however caused, and their reaction to bereavement is not different.

Children aged 5–11 years have increasing understanding of death. They usually become depressed and over-active, and may show disorders of conduct. Some have suicidal thoughts and ideas that death would unite them with the lost parent. Suicidal actions are infrequent. In *children over the age of 11 years* the response is increasingly like that of adults (see p. 151).

Bereavement may have long-term effects on development, especially if the child was young at the time of the parent's death and if the death was sudden or violent. Outcome probably depends largely on the effects of the bereavement on the surviving parent. Most studies of bereavement in children have been concerned with the death of a parent; few have concerned the death of a sibling. For a review see Kane (1979).

Social and cultural factors

Although the family is undoubtedly the part of the child's environment with most effect on his development, wider social influences are important as well, particularly in the aetiology of conduct disorder. In the early years of childhood these social factors act indirectly through their influence on the patterns of family life. As the child grows older and spends more time outside the family, they have a direct effect as well. These factors have been studied by examining the associations between psychiatric disorder and type of neighbourhood and school.

Rates of childhood psychiatric disorder are higher in areas of social disadvantage. For example, as already noted (p. 668), the rates of both emotional and conduct disorder were found to be higher in a poor inner London borough than in the Isle of Wight. The important features of inner city life may be lack of play space, inadequate social amenities for teenagers, overcrowded living conditions, and lack of community involvement. Rates of child psychiatric referral and delinquency also vary between schools (Power *et al.* 1972; Gath *et al.* 1977). These differences do not seem to be due to the size of the school or the age of its buildings, but rather to social environment.

Psychiatric assessment of children and their families

The aims of assessment are to obtain a clear account of the presenting problem, to find out how this problem is related to the child's past development and his present life in its psychological and social context, and to plan treatment of the child and family.

The psychiatric assessment of children differs in several ways from that of adults. With children it is often difficult to follow a set routine: a flexible approach to interviewing is required, though it is still important that information and observations are recorded systematically. Both parents should be asked to attend the assessment interview, and it is often helpful to have other siblings present. Time can be saved by asking permission to obtain information from teachers before the child attends the clinic. This information should be concerned with the child's behaviour in school and his educational attainments.

Child psychiatrists vary in their methods of assessment. All agree that it is important to see the family together at some stage. Some prefer to do this from the start. Others believe that they can make a better assessment by interviewing the parents and the child separately, and then proceeding to a joint interview at which family interactions can be observed. If separate interviews are used, it may be advisable to start by seeing the adolescent patient on his own before seeing the parents. In the case of younger children the main informants are usually the parents, but children over the age of six should usually be seen on their own at some stage. Of course, in the special case of suspected child abuse, the interview with the child is particularly important. Whatever the problem, the parents should be made to feel that the interview is supportive and does not undermine their confidence.

Interviewing the parents

Parents should be encouraged to talk spontaneously before systematic questions are asked. The methods of interviewing are similar to those used in adult psychiatry (see Chapter 2). The items to be included in the history are listed in the scheme described by Graham (1991) which appears in the appendix to this chapter. It is important to obtain specific examples of general problems, to elicit factual information, and to assess feelings and attitudes. There are several standardized schedules for interviewing parents; they are described in reference books such as Rutter *et al.* (1994).

Interviewing and observing the child

Because younger children may not be able or willing to express ideas and feelings in words, observations of their behaviour and interaction with the interviewer are especially important; with very young children, drawing and the use of toys may be helpful. With older children, it may be possible to follow a procedure similar to that used with adults, provided that care is taken to use words and concepts appropriate to the child's age and background. Several standardized systems of observation have been described for specific purposes. They are described in books of reference such as Rutter *et al.* (1994).

It is essential to begin by establishing a friendly atmosphere and winning the child's confidence. It is appropriate to ask the child what he likes to be called. It is usually better to begin with a discussion of neutral topics such as pets, games, or birthdays before turning to the presenting problem. When a friendly relationship has been established, the child can be asked about the problem, his likes and dislikes, and his hopes for the future. It is often informative to ask what he would request if given three wishes. Younger children may be given the opportunity to express their concerns and feelings in paintings or play. Children can generally recall events accurately, but they are more suggestible than adults; therefore it is particularly important not to use leading questions in interviewing, and not to suggest actions or interpretations to a child who is being observed at painting or play.

Observations of the child's behaviour and mental state should be recorded. The items to be included are listed in the appendix to this chapter (p. 722). When assessing the mental state it should be remembered that children who are seeing a psychiatrist for the first time may be silent and withdrawn; this behaviour should not be misinterpreted as evidence of depression. At some stage, preferably late in the consultation, a physical examination may be performed, with particular attention to the central nervous system (see appendix). By the end of the interview an assessment should have been made of the child's development relative to other children of his age.

Interviewing the family

A family interview helps the clinician to evaluate interactions between family members, but it is not a good way to obtain factual information; the latter is generally obtained more effectively in interviews with individual family members. Of the various aspects of family life, discord and disorganization have been shown to be most closely associated with the development of psychiatric disorder; hence it is important to assess these characteristic processes when interviewing the family. Patterns of communication between family members may also be important.

If a family interview is to be part of the assessment, it is generally best held in the early stages; for example it can usefully be in the second part of the first assessment before the interviewer has formed close relationships with the patient or an informant which may impede the interviewing of the others.

The main aim of the interview is to observe how members of the family interact. The interviewer may begin by asking: 'Who would be the best person to tell me about the problems?' If one member presents a monologue, it should be brought to an end, for example by asking another member to comment on what has been said. Another useful question is: 'How do you think that your wife (husband, son) would see the problem?' The interviewer can then ask the wife (husband, son) how they see the problem. As an alternative, family members who are present can be asked what they think that an absent member would think about the issues.

While observing the family's ways of responding to these and other questions, the interviewer should consider the following.

(1) Who is the spokesman for the family?
(2) Who seems most worried about the problem?
(3) What are the 'subsystems' in the family, i.e. who is allied with whom?
(4) What is the hierarchy in the family, i.e. who is most dominant?
(5) How well do the family members communicate with one another?
(6) How do they seem to deal with conflict?

Psychological assessment

Measures of intelligence and educational achievement are often valuable. Thus if mental development and achievement are inconsistent with chronological or mental age, or with the expectations of parents or teachers, this may indicate a generalized or specific disorder of development or may indicate a source of stress in disorders of other kinds. Some of the more commonly used procedures are listed in Table 20.2. Some psychologists also use one or more of the projective techniques, which are difficult to score, and not of established validity. Sometimes, however, such techniques provide a useful way of discovering the child's feelings about the members of his family and other issues. Used in this way, projective techniques resemble clinical methods (for example asking the child to make up a story) rather than psychological tests.

Other information

The most important additional informants are the child's teachers. They can describe his classroom behaviour, educational achievements, and relationships with other children. They may also make useful comments about the family and home circumstances. It is often helpful for a social worker to visit the home. This can provide useful information about material circumstances in the home, the relationship of family members, and the pattern of their life together.

Ending the assessment

At the end of the assessment the psychiatrists should explain to the parents (and the child, depending on age) how he plans to proceed. He should explain to them that he will contact the general practitioner and that he should seek their permission to contact other people involved, such as teachers, or social workers. He should give clear information about any proposed treatment, and should encourage questions and discussion.

Formulation

A formulation should be made in every case. This starts with a brief statement of the current problem. The diagnosis and differential diagnosis are discussed next. Aetiological factors are then considered. The developmental stages of the child should be noted, as well as any particular strengths and achievements. An assessment of the problems and the strengths of the family is also recorded. Any further assessments should be specified, a treatment plan drawn up, and the expected outcome recorded.

Court reports

Psychiatrists may be asked to prepare court reports in relation to children. These reports are usually undertaken by specialists in child psychiatry; hence only an outline will be given here. If required to make such a report, psychiatrists should ensure that they are thoroughly aware of the relevant legislation (in England and Wales, the Children Act 1989) and should study a more detailed account of the child psychiatrist's role in court, for example the account by Wolkind (1994). Whenever practicable, the psychiatrist should seek the advice of a colleague experienced in this kind of work.

Courts concerned with children obtain evidence from several sources, including social workers,

Table 20.2 *Notes on some psychological measures in use with children and the mentally retarded*

Intelligence tests

Stanford–Binet intelligence scale	A revision of the original intelligence test; now seldom used. Provides mental age. Weighted to verbal abilities and this may result in cultural bias. More useful for middle-class patients and for low ability
Weschler intelligence scale for children (WISC III)	Provides a profile of specific verbal and performance ability as well as IQ for children aged 6–14 years; widely used and well standardized; cannot be used for IQ below 40
Weschler pre-school and primary scale of intelligence (WPPSI)	A version of WISC for used with younger children (4–6.5 years) and with the mentally retarded
British ability scales	Twenty-four subscales suitable for 2.5–17 years, and covering six areas; speed of information processing, reasoning, spatial imagery, perceptual matching, short-term memory, retrieval, and application of knowledge; analysis can be general or specific
Goodenough–Harris drawing test	A brief test of non-verbal intelligence for children aged 3–10 years

Social development assessments

Vineland social maturity scale	The original development scale recently revised, which has psychometric limitations; covers general self-help, self-help in dressing, self-help in eating, locomotion, communication, self-direction, social isolation, and occupation; provides 'social age'
Adaptive behaviour scales (Nihira)	Rating scales to evaluate abilities and habits in 10 behavioural areas
Gunzburg progress assessment charts	Provides a clear visual display of self-help, communication, social and occupational abilities

Other developmental assessment

Denver development scale	Assessment of gross and fine motor skills, language, and social development; used for children up to 2 years of age
Bayley scales of infant development	Range of items which can be scored on mental and psychomotor development indices; comprehensive and reliable to ages 2 months to 2.5 years

Educational attainment

Neale analysis of reading	Graded test of reading ability, accuracy, comprehension, and rate for age 6 upwards
Schonell graded word reading test	The child reads words of increasing difficulty
Schonell graded word spelling test	The child spells words of increasing difficulty
Tests of mathematical ability	No satisfactory test; arithmetic subtests of WISC-R, WPPSI, and British ability scales

probation officers, community nurses, psychologists, and psychiatrists. In their reports psychiatrists should focus on matters within their expertise, such as (a) the child's stage of development and its relevance to the case, (b) whether the child has a psychiatric disorder, (c) the child's own wishes, and (d) the parenting skills of the carers, and other relevant aspects of the family.

Generally a court seeks information about the following issues (which are referred to in the Children's Act 1989 of England and Wales (Harrison 1991)):

(1) the child's physical, emotional, and educational needs;

(2) the ascertainable wishes of the child about his future, considered in the light of his age and understanding;

(3) the likely effect on the child of any change of circumstances (for example removal from home, or living with one or other parent after a divorce);

(4) matters concerned with the child's age, personal characteristics and background that are relevant to the case;

(5) any harm that the child has suffered or is likely to suffer;

(6) how far each parent, or any relevant other person, is capable of meeting the child's needs.

The *wishes of the child* should be considered in relation to his age and ability to understand the present situation and possible future arrangements, and to relevant factors in the present situation, for example some abused children maintain strong attachments with the abuser.

Parenting skills are judged partly on the history and reports of other people; they are also judged partly on direct observations of the interactions between the parents and child, including the parent's attachment to the child, sensitivity to cues from the child, and ability to meet the child's needs.

The report is similar in structure to a court report for an adult (see p. 778), and is presented under the following headings:

(1) the qualifications of the writer;

(2) who commissioned the report, and what questions were asked;

(3) what written information was available, and who was interviewed;

(4) a summary of the findings from the interview (it is not necessary to repeat information contained in social enquiry reports, but it may be necessary to add observations that supplement or contradict this information);

(5) the writer's interpretation of the information from the interviews and written material;

(6) (in the light of the findings) comments on the options before the court, remembering always that it is for the court and not the psychiatrist to determine which option is selected.

(See Wolkind (1994) for a more comprehensive account of the preparation of a Court Report.)

Psychiatric treatment for children and families

The role of the general practitioner

General practitioners spend much of their time in advising and reassuring parents about children, but they refer only a small proportion of these children to a child psychiatrist or a paediatrician (Bailey *et al.* 1978). General practitioners are more likely to refer certain problems to a paediatrician, for example developmental difficulties, physical symptoms with a probable psychological cause, and psychological complications of physical illness. Emotional and conduct disorders are more likely to be referred to a child psychiatry clinic; however, many of the disorders referred are no more severe than those which the general practitioner manages himself (Gath *et al.* 1977).

The psychiatric team

Although the members of the team (doctors, social workers, and psychologists) have special skills, they do not confine themselves to their traditional professional roles when working with children and families. Instead, they take whatever role seems most likely to be helpful in the particular case.

It is usual to adopt a family approach, and to maintain close liaison with other people or agencies working with the child or his family, including paediatricians, child health and social services, teachers, and educational psychologists. Since many childhood problems are obvious at school, or lead to educational difficulties, the child's teachers usually need to be involved in treatment. Teachers may require advice about the best way to manage behavioural disturbance, remedial teaching may be required, or some change may be needed in the child's school timetable. Occasionally a change of school is indicated.

In the following sections, brief general descriptions are given of the main kinds of treatment. In the second part of the chapter further information is given about the management of individual disorders. Further information about treatment in child psychiatry can be obtained from Graham (1991) or from one of the textbooks listed at the end of this chapter.

Drug treatment

Drugs have a limited but important place in child psychiatry. The main indications are in the treatment of epilepsy, depressive disorders, obsessional disorders, over-activity syndromes, Gilles de la Tourette's syndrome, and, occasionally, nocturnal enuresis (these uses are considered later in this chapter). In all cases dosages should be checked carefully in a standard reference book, with particular attention to the allowances required for the child's age and body weight.

Individual psychotherapy with the child

This treatment originated in the separate methods developed by Anna Freud and Melanie Klein. These methods differed, especially in the emphasis placed by Klein on the psychodynamic interpretation of child's play. Nowadays lengthy intensive treatment of this kind is uncommon. Instead, most psychotherapy with children is brief and aims to help current problems. The main indications are

emotional disorders and physical complaints related to important psychological factors.

The psychotherapist tries to make a warm and accepting relationship with the child. He uses the relationship to encourage the child to express feelings and to find alternative ways of behaving. Toys are usually provided to help the child to relate to the therapist and to express in play the problems and feelings that cannot be put into words. Acceptance of the child is important and criticism should be avoided, but there should be no implication that every aspect of the child's behaviour is approved. At first, the child often perceives the psychotherapist as an agent of his parents and expects him to share their attitudes. For this reason, it is advisable to delay discussion of the presenting problems until the child's confidence has been gained by talking about neutral things that interest him.

There have been few well-conducted evaluations of individual psychotherapy for children. Reviews have generally concluded that most studies are marred by methodological flaws, that the results of the better investigations indicate that psychotherapy is more effective than no treatment, and that behavioural methods generally have effects greater than other techniques (Casey and Berman 1985; Barnett *et al.* 1991).

Family therapy

This is a specific form of treatment which should be distinguished from the general family approach to treatment described above. In family therapy, the child's symptoms are considered as an expression of the functioning of the family, which is the primary focus of treatment. Several approaches have been used based on behavioural or psychoanalytical systems, or on communication or structural theories. These kinds of therapy are described on pp. 622–4. In practice, most therapists adopt an eclectic approach.

The indications for family therapy are still debated. Such treatment may be appropriate under the following conditions:

(1) the child's symptoms appear to be part of a disturbance of the whole family;
(2) individual therapy is not proving effective;
(3) family difficulties arise during another kind of treatment.

Family therapy may be contraindicated when the parents' marriage is breaking up, or the child's problems do not seem to be closely related to family function. It is important that a therapist's interest in family therapy should not prevent a thorough evaluation of the case and the use of other treatments when indicated. The practice of family therapy in child and adolescent psychiatry has been reviewed by Barker (1988).

Uncontrolled studies of family therapy have led to claims that it has substantial effects. Controlled evaluations suggest more modest benefits for children with a wide range of emotional and behavioural disorders. In a study of a mixed group of children with emotional and behavioural disorders, the benefits of Milan Family Therapy for the child were no greater than those of eclectic treatment (Simpson 1990). However, when dynamic family therapy was used for a specific disorder—eating disorders in adolescents—the benefits were greater than those of a control treatment (Russell *et al.* 1987). Research on family therapy has been reviewed by Gurman and Kriskern (1991).

Group therapy

Group therapy can be used for the child or the parents. Older children and adolescents may be helped by the sharing of problems, discussions, and modelling that form part of group therapy. Parents may be helped by the opportunity to discuss shared problems of child management or other difficulties. The principles of group therapy are described in Chapter 18.

Cognitive and behaviour therapy

Behavioural methods have several applications in child psychiatry. They can be used to encourage new behaviour by positive reinforcement and modelling. This is often done by first rewarding behaviour that approximates to the desired beha-

viour (shaping), and then giving reinforcement in a more discriminating way. Thus with autistic and retarded children, shaping has been used for behavioural problems such as temper tantrums and refusal to go to bed, and for problems in toilet training. Punishment is not used in shaping because its effects are temporary and it is ethically unacceptable. Instead, efforts are made to identify and remove any factors in the child's environment that are reinforcing unwanted behaviour. It is often found that undesired behaviour is reinforced unwittingly by extra attention given to the child when the behaviour occurs. If this child is ignored at such times and attended to when his behaviour is more normal, beneficial changes often take place. More specific forms of behaviour therapy can be used for enuresis (see p. 708) or phobias. Social skills training in a group or in individual sessions may be used for children who have difficulty in relationships with other children and adults. The methods generally resemble those used with adults (see Chapter 18).

Cognitive therapy is useful for older children who can describe and learn to control ways of thinking that give rise to symptoms and problem behaviours. Older children and adolescents with anxiety disorders and eating disorders can be treated with methods devised for adults. Special techniques have been devised for children with aggressive behaviour (see p. 697).

Parent training

Parent training is used to improve the skills of parents with deficient parenting skills, including those who abuse or neglect their children and those with low intelligence. It is also used to assist parents of children with behaviour problems that require special parenting skills, for example the parents of children with conduct disorder or hyperactivity. Most research has been with parents of children who are oppositional or defiant or have conduct disorders (see for example the earlier studies of Patterson (1982), and the more recent investigations of Webster-Stratton (1991)). Parent training can be carried out with an *educational approach*, in which skills of general importance are taught, or with a *behavioural approach*, in which the problems of the particular parent and child are analysed and corrected by selected techniques. In the behavioural approach, use may be made of video-tape vignettes showing desirable and undesirable parental responses to children's behaviour. These responses are discussed with the parents of one child, or with a group of parents. Whatever approach is adopted, it is important to take account of the stage of development of the child and of the changing needs of children of different ages.

Studies of the behavioural training of parents have generally found it more effective in the short term than less structured approaches, but the long-term benefits have not been established with certainty.

Social work

Social workers play an important role in the care of children with psychiatric disorders and of their families. They have statutory duties for the protection of children who are at risk within the family, and who require special care or special supervision. They help parents to improve their skills in caring for their children and to solve problems with finances or accommodation. Social workers carry out family assessments and family therapy, and also provide individual counselling for the child and members of the family.

Occupational therapy

Occupational therapists can play a valuable part in assessment of the child's development, in psychological treatment, and in devising measures to improve parent–child interaction. Their work is mainly in hospital day and in-patient units, though increasingly they carry out some work in the community, for example visiting children at home. They work closely with teachers both in assessing and providing therapeutic activities for children.

Special education

Children attending as out-patients, as well as the smaller number who are day patients or in-patients, often benefit from additional educational arrange-ments. Special teaching may be needed to remedy backwardness in writing, reading, and arithmetic, which is common among children with conduct disorders as well as those with specific develop-mental disorders.

Substitute care

Residential care

Residential care can be valuable for children with symptoms that result from a severely unstable home environment or from extreme parental rejection. When children are considered for residential placement, they often have conduct disorders and severe educational problems. Removal of a child from home should be considered only after every practical effort has been made to improve the circumstances of the family. Residential care can be arranged in a foster home, a children's home (in which a group of about 10 children live in circumstances as close as possible to those of a large family), or a boarding school.

Residential care is seldom arranged for children under five years of age because they have a special need for attachment to parental figures. In general, children who have been in residential care have high rates of psychosocial problems in later child-hood and in adult life. As men they most often have problems with the law, whilst as women they most often have unmarried pregnancy and parent-ing problems. It is not clear how far these problems relate to the experience of residential care or to previous adverse experiences that led to the residential placement. Reports of the abuse of children placed within care are reminders of the need to ensure good training and supervision of the staff of children's homes and residential schools. (For a review of residential care see Wolkind and Rushton (1994).)

Fostering

Foster care may be of three kinds:

(1) **short-term** emergency care, for example when a care giver is ill or when the parents of an autistic child need respite;

(2) **medium-term** care, which may be followed by return home, for example if the care-giver is receiving treatment for problems that led to neglect or abuse of the child;

(3) **long-term** care, in which the child remains until able to leave independently.

Children in long-term foster care have more problems than children who have been adopted from origin, but it is difficult to determine how far these are related to experiences before fostering and how far they are due to the lesser security of fostering as against adoption. Problems seem to be greater when the fostered child is older and when children in the fostering family are of the same age as the fostered child. Children in foster care usually retain some contact with their biological parents. It is not helpful to the child to have sporadic and distressing contacts or to have contacts when there is no prospect of returning home. For a review of fostering see Wolkind and Rushton (1994).

In-patient and day-patient care

Child psychiatric in-patients units require easy access to paediatric advice, adequate space for play, easy access to schooling, and an informal design that still allows close observation. There should be some provisions for mothers to stay with their children.

Admission for **in-patient** treatment is usually arranged for any of three reasons. First, a **behaviour disorder** may be too **severe** to treat in any other way: examples include extreme hyperactivity, severe pervasive developmental disorder, life-threatening anorexia nervosa, and school refusal resistant to out-patient treatment. Second, the child may be admitted for observation when the **diagnosis is uncertain**. Third, in-patient treatment is one way of providing a period away from a severely disturbing home environment, for example when there is child abuse or gross overprotection.

Sometimes the mother is admitted as well as the child, thus helping to maintain the bonds between the two. This arrangement also allows close observation of the ways in which the mother responds to the child, for example in cases of child abuse. Once the nature of the problem is clear, the mother can be helped to learn new parenting skills by taking an increasing part in the child's care while both remain in hospital.

Day hospital treatment for a child provides many of the advantages of in-patient care without removing him from home. Unless there is any danger that the child may be abused, remaining at home has the advantage that relationships with other family members are maintained. Day care can relieve the family from some of the stressful effects of managing an over-active or autistic child.

REVIEW OF SYNDROMES

The review of syndromes begins with the problems encountered in preschool children. Specific and pervasive developmental disorders are then described. An account is then given of the main psychiatric disorders of childhood, in the order in which they appear in the major systems of classification. Other psychiatric disorders of childhood are described next. An account is given of disorders of adolescence (which are generally similar to those of childhood or adult life). The review ends with a consideration of child abuse, which is an important cause of physical and psychiatric disorder in children. Table 20.3 shows the arrangement of the remaining sections of this chapter.

Problems of preschool children and their families

It has already been noted that in the preschool years children are learning several kinds of social behaviour. They are acquiring sphincter control. They are learning how to behave at mealtimes, to go to bed at an appropriate time, and to control angry feelings. They are also becoming less dependent. All these behaviours are learnt within the family. The psychiatric problems of preschool children centre around these behaviours and they often reflect factors in the family as well as factors in the child. Many psychological problems at this age are brief, and can be thought of as delays in

Table 20.3 *Arrangement of sections on disorder of children in this and other chapters*

Problems of preschool children

Specific developmental disorder

Pervasive developmental disorders

Principal psychiatric disorders of
 childhood
Hyperkinetic disorder
Conduct disorder
(Juvenile delinqency)
Anxiety and obsessional disorders
Somatization disorders
Depressive disorders
(School refusal)

Other childhood psychiatric disorders
Functional enuresis
Functional encopresis
Elective mutism
Stammering
Dementia
Schizophrenia
Gender identity disorders
Suicide and self-harm

Psychiatric aspects of physical illness

Psychiatric problems of adolescence

Child abuse
Physical abuse
Emotional abuse
Child neglect
Non-organic failure to thrive
Sexual abuse

Conditions considered in other chapters
Tic disorders (p. 400)
Suicide and deliberate self-harm (p. 414)
Munchausen's syndrome by proxy
 (p. 357)

---

normal development. Most of these problems are treated by general practitioners and paediatricians. The more serious problems may be referred to child psychiatrists.

Prevalence

Richman *et al.* (1982) studied a sample of 705 families with a three-year-old child in a London borough. The most frequent abnormalities of behaviour in these children were bed-wetting at least three times a week (present in 37 per cent), wetting by day at least once a week (17 per cent), over-activity (14 per cent), soiling at least once a week (13 per cent), difficulty in settling at night (13 per cent), fears (13 per cent), disobedience (11 per cent), attention-seeking (10 per cent), and temper tantrums (5 per cent).

Whether these behaviours are reported as problems depends on the attitudes of the parents as well as on the nature, severity, and frequency of the behaviour. Richman *et al.* overcame this difficulty by making their own ratings of the extent of problems. They based this assessment on the effects on the child's well-being and the consequences for the other members of the family. They used common-sense criteria to decide whether the problems were mild, moderate, or severe. Seven per cent of three-year-olds in their survey had behaviour problems of marked severity and 15 per cent had mild problems. The behaviours most often rated as problems were temper tantrums, attention-seeking, and disobedience.

Prognosis

As explained above, many psychological problems of preschool children are brief. However, Richman *et al.* (1982) found that certain problems detected in three-year-old children were still present at the age of eight; these included over-activity, conduct disorder, speech difficulty, effeminacy, and autism.

Aetiology

Aetiological factors are related to the stage of development, the child's temperament, and influences in the family. There are wide individual variations in the rate at which normal development proceeds, particularly in sphincter control and language acquisition. As noted above (p. 669), a child's temperamental characteristics are evident from the earliest weeks. These characteristics are capable of affecting the mother's response—how much time she spends with him, how often she picks him up, and so on. These maternal responses may in turn affect the child's development. Behaviour problems at this age are also associated with poor marital relationships, maternal depression, rivalry with siblings, and inadequate parental behaviour. Richman *et al.* (1982) for a review of the evidence.

Some common problems of preschool children

Temper tantrums

Occasional temper tantrums are normal in toddlers, and only persistent or very severe tantrums are abnormal. The immediate cause is often unwitting reinforcement by excessive attention and inconsistent discipline on the part of the parents. When this arises it is often because the parents have emotional problems of their own or because their relationship is unsatisfactory. Temper tantrums usually respond to kind but firm and consistent setting of limits. In treatment it is first necessary to discover why the parents have been unable to set limits in this way. They should be helped with any problems of their own and also advised on how to respond to the tantrums.

Sleep problems

The commonest sleep difficulty is *wakefulness* at night, which is most frequent between the ages of one and four years. About a fifth of children of this age take at least an hour to get to sleep or are wakeful for long periods during the night. When wakefulness is an isolated problem and not overdistressing to the family, it is enough to reassure parents about the prognosis. When sleep disturbances are severe or persistent, two possible causes should be considered. First, the problems may

have been made worse by physical or emotional disorders. Second, they may have been exacerbated by the parents' overconcern or inability to provide reassurance. If either of these causes is found, it should be treated by attending to the reasons for the overconcern and by helping the parents to feel more secure. Some parents overstimulate their child in the evening, or condone crying in the night by taking the child into their own bed. A behavioural approach to these problems is generally useful (Richman *et al.* 1985). Otherwise, it is usually sufficient to reassure the parents and the child.

Hypnotic medication may be useful for special occasions but is unlikely to be effective in the long term. The handbook by Douglas and Richman (1984) is useful for parents.

Other difficulties, such as *nightmares* and *night terrors*, are common among healthy toddlers but they seldom persist for long. They are discussed on pp. 412 and 413.

Feeding problems

Minor food fads or food refusal are common in preschool children, but do not usually last long. In a minority the behaviour is severe or persistent, although not accompanied by signs of poor nourishment. When this happens it is often because the parents are overattentive and obsessional, and unwittingly reinforce the child's behaviour. Treatment is directed to the parents' management of the problem. They should be encouraged to ignore the feeding problem and refrain from offering the child special foods or otherwise attempting to do anything unusual to persuade him to eat. Instead, he should be offered a normal meal and left to decide whether to eat it or not.

Pica

Pica is the eating of items generally regarded as inedible, for example soil, paint, and paper. It is often associated with other behaviour problems. Cases should be investigated carefully because some are due to brain damage, or autism, or mental retardation. Some are associated with emotional

Table 20.4   *Specific disorders of psychological development*

DSMIV	ICD10
Learning disorder	**Specific developmental disorders of scholastic skills**
Reading disorder	Specific reading disorder
Disorder of written expression	Specific spelling disorder
Mathematics disorder	Specific disorder of arithmetic skills
	Mixed disorder of scholastic skills
Communication disorders	**Specific developmental disorders of speech and language**
Phonological disorder	Specific speech articulation disorder
Expressive language disorder	Expressive language disorder
Mixed receptive–expressive language disorder	Receptive language disorder
	Acquired aphasia with epilepsy*
Stuttering	(Stuttering†)
Motor skills disorder	**Specific developmental disorder of motor function**
Developmental coordination disorder	

*In DSMIV acquired aphasia with epilepsy is classified as an (acquired) receptive–expressive language disorder.
†Classified in ICD10 under F98, other behavioural and emotional disorders of childhood, and *not* as a developmental language disorder.

distress, which should be reduced if possible. Otherwise, treatment consists of common-sense precautions to keep the child away from the abnormal items of diet. Pica usually diminishes as the child grows older. For a review of the history of ideas about pica see Parry-Jones and Parry-Jones (1992).

Reactive attachment disorder of infancy and early childhood

This term denotes a syndrome starting before the age of five years and associated with grossly abnormal care-giving. Affected children may appear inhibited and hypervigilant, and may show 'frozen watchfulness' (the inhibited subtype) which is a combination of vigilance and fearfulness. These children are difficult to console, miserable, and sometimes aggressive. Some fail to thrive. Such behaviour has been described most

clearly in children who have been abused (see p. 717). Other children are indiscriminately sociable; they relate to people irrespective of their closeness and are excessively familiar with strangers (the disinhibited subtype). Such behaviour has been described most clearly in children raised in institutions. In DSMIV the diagnosis is made when the disturbance of relationships appears to be a direct result of abnormal care-giving. ICD10 does not use this criterion but instead requires that the behaviour is present in various situations.

It seems that these syndromes are characteristic of the type of care-giving (abusive or institutional) rather than of the child. Insecure attachment in infancy is often followed by conflicts with care-givers and impulsive behaviour later in childhood. Nevertheless, considerable improvement can occur if the child experiences a secure attachment to a care-giver, for example as a result of fostering or adoption. These observations have not been made specifically in relation to attachment disorder as defined in ICD10 and DSMIV.

Assessment and treatment

In assessing problems in preschool children, the psychiatrist usually has to rely largely on information from the parents. As mentioned above, it is important to distinguish between a primary disorder in the child and a disorder that reflects the difficulties of the mother or the entire family. It is necessary to make a careful assessment of the particular behaviour, the child's general level of development, and the functioning of the family as a whole.

Apart from particular points already mentioned under the specific disorders, treatment includes counselling for the mother (and sometimes for other family members as well) and advice about child rearing. Little is known about the value of specific treatments. Behavioural methods are probably useful, language delays may benefit from educational measures, and occasionally medication is required to control extreme over-activity (see p. 695). It is often helpful to arrange for the child to spend part of the day away from the family in a playgroup or nursery school.

Specific developmental disorders

Both DSMIV and ICD10 contain categories for specific developmental disorders, which are circumscribed developmental delays that are not attributable to another disorder or to lack of opportunity to learn (Table 20.4). It is debatable whether these conditions should be classified as mental disorders at all, since many children meeting the criteria have no other signs of psychopathology. In ICD10 these developmental disorders are divided into specific developmental disorders of scholastic skills, speech and language, and motor function. In DSMIV the same disorders are called learning disorder, communication disorders, and motor skill disorder respectively.

Specific developmental disorders of scholastic skills are divided further into specific reading disorder, specific spelling disorder, and specific

arithmetic disorder. In DSMIV these conditions are called reading disorder, disorder of written expression, and mathematics disorder respectively.

Specific reading disorder

In DSMIV this condition is named *reading disorder*. It is defined by a reading age well below (usually 1.5–2 standard deviations) the level expected from the child's age and IQ (Yule 1967). Defined in this way, the disorder was found in about 4 per cent of 10–11 year olds in the Isle of Wight, and about twice that percentage in London (Yule and Rutter 1985).

Clinical features

Specific reading disorders should be clearly distinguished from general backwardness in scholastic achievement resulting from low intelligence or inadequate education. They should also be distinguished from poor reading due to lack of opportunity to learn at home or at school, or due to poor visual acuity. The child presents with a history of serious delay in learning to read, which has been evident from the early years of schooling and has sometimes been preceded by delayed acquisition of speech and language. Errors in reading include omissions, substitutions or distortions of words, slow reading, long hesitations, and reversals of words or letters. There may also be poor comprehension. Writing and spelling are impaired, and in older children these problems may be more obvious than the reading problems. There may be associated emotional problems, but development in other areas is not affected. Compared with children with general backwardness at school, those with specific reading retardation are much more often boys; they are also more likely to have minor neurological abnormalities and are likely to come from socially disadvantaged homes.

Specific reading retardation is associated with *conduct disorder* more often than would be expected through chance (Rutter *et al.* 1970*a, b*). The association may arise in part because the two conditions have common neurodevelopmental or temperamental origins; in part because reading retardation leads to conduct problems at school when the child is frustrated by failures; and in part because conduct disorder gives rise to problems in learning to read.

Aetiology

Reading is a complex skill which depends on more than one psychological process and is learnt in several stages. Therefore it is not surprising that no single cause has been identified for specific reading disorder. A widely held theory of the learning of reading is that children first use visual methods; they learn the appearance of whole words, and cannot decipher new words. The next stage of learning is alphabetic; children become able to decode new words from the sounds associated with the letters. In the final stage, reading becomes automatic and flexible in combining visual and alphabetic methods. (This model of reading, although not accepted by all, is useful in clinical practice.)

The frequent occurrence of reading disorder in family members suggests a *genetic cause*, but more direct evidence is lacking. The family patterning shows that, if there is a genetic cause, it cannot be on the basis of any single mode of inheritance (Rutter *et al.* 1990). Children with cerebral palsy and epilepsy have increased rates of specific reading disorder; therefore it has been suggested that children with specific reading disorder, but no obvious neurological disease, may have minor neurological abnormalities. The evidence does not support this idea. The most likely cause appears to be disorder of *brain maturation* affecting one or more of the perceptual and language skills required in reading. This explanation is consistent with the following findings: difficulties in verbal coding and sequencing in many cases, confusion between right and left, and general improvement with age.

Social factors may add to these psychological problems. It seems that minor innate difficulties may not be sufficient in themselves to cause reading retardation, but may do so when children are brought up in a large family or attend a school where they receive little personal attention. Frequent changes of school and an illiterate home background are also associated with specific read-

ing disorder and could act in a similar way. As already noted, children with reading disorder have a high rate of conduct disorder. For a review of the acquisition of reading skills and of reading disorder see Snowling (1991).

Assessment and treatment

It is important to identify the disorder early. *Assessment* is carried out by an educational or clinical psychologist using an individually administered standardized test of reading accuracy and comprehension. *Treatment* is educational unless there are additional medical or behavioural problems requiring separate intervention. Treatment should be started as early as possible before the child has a sense of failure. Several educational approaches are used but it is most important to reawaken the interest of a child with a long experience of failure. Continued extra teaching and parental interest seem to be helpful, but there is no evidence that any one method of teaching is better than others (Gittelman 1985). If there are behavioural problems secondary to frustration caused by the reading difficulty, they may lessen as reading improves; others may need separate attention.

Prognosis

Prognosis varies with the severity of the condition. Among children with a mild problem in mid-childhood only about a quarter achieve normal reading skills by adolescence. Very few with severe problems in mid-childhood overcome them by adolescence. Whilst there is insufficient evidence to be certain what happens to these people as adults, those with substantial difficulties in adolescence seem likely to retain them (Maughan *et al.* 1985). For a review of developmental reading disorder see Snowling (1991).

Mathematics disorder (specific arithmetic disorder)

The first of these names is used in DSMIV; the term in parentheses is used in ICD10. Difficulty with arithmetic is probably the second most common specific disorder. Problems include failure to understand simple mathematical concepts, failure to recognize numerical symbols or mathematical signs, difficulty in carrying out arithmetic manipulations, and inability to learn mathematical tables. These problems are not due simply to lack of opportunities to learn and are evident from the time of the child's first attempts to learn mathematics.

There has been no study of its *epidemiology*, although it is thought to be quite common. Although it causes less severe handicap in everyday life than reading difficulties, it can lead to secondary emotional difficulties when the child is at school.

The *causes* are uncertain. Dyscalculia occurs in some adults with parietal lobe lesions, but no brain damage has been found in children with specific arithmetic disorder. It seems unlikely that there is a single cause. **Assessment** is usually based on the arithmetic subtests of the Wechsler Intelligence Scale for Children (WISC) and the Wechsler Adult Intelligence Scale (WAIS) and on specific tests. **Treatment** is by remedial teaching but it is not known whether it is effective. The **prognosis** is not known. For a review see O'Hare *et al.* (1991).

Communication disorders (developmental disorders of speech and language)

Children vary widely in their achievement of speech and language. Half of all children use words with meanings by 12.5 months and 97 per cent do so by 21 months. Half form words into simple sentences by 23 months (Neligan and Prudham 1969). Vocabulary and complexity of language develop rapidly during the preschool years. However, when children start school, 1 per cent are seriously retarded in speech and 5 per cent have difficulty in making themselves understood by strangers. The process by which language is acquired is complex and is still not fully understood.

No cause can be found in most children who present with speech and language disorders. However, it is most important to detect primary conditions when these are present. The most common cause of delay in the development of normal speech is *mental retardation*. Other important causes are *deafness, cerebral palsy,* and *pervasive developmental disorder. Social deprivation* can cause mild delays in speaking or add to the effects of the other causes. The remaining cases are attributed to the specific developmental speech and language disorder. Children with *developmental language disorder* have a marked delay in acquiring normal speech, in the absence of any primary cause.

Three categories of speech and language disorder are recognized in ICD10 and DSMIV. The first is known as phonological disorder in DSMIV and speech articulation disorder in ICD10. The second is expressive language disorder, and the third is mixed receptive–expressive disorder in DSMIV (receptive language disorder in ICD10). ICD10 (but not DSMIV) has a fourth category of acquired aphasia with epilepsy, and DSMIV includes stuttering with these other conditions (and widens the title of the group to communication disorders). ICD10, which uses the title specific developmental disorders of speech and language, has to exclude stuttering (which is not a disorder of this kind); it is coded instead under behavioural disorders of childhood (see Table 20.4).

Phonological disorder (speech articulation disorder)

In this condition use of speech sound is below the level appropriate for the child's mental age but language skills are normal. Of course, errors in making speech sounds are normal in children up to about the age of four years, but by age seven most speech sounds should be made normally, though a few may still be produced imperfectly. By age 12 years nearly all speech sounds should be made normally. Children with specific speech articulation disorder make errors of articulation so severe that it is difficult for others to understand their speech. Speech sounds may be omitted or distorted, or other sounds substituted. In assessing speech production, appropriate allowance should be made for regional accents and dialects. The sounds affected most often are those developing later in the normal sequence of development (l, r, s, z, th, and ch for English speakers).

Prevalence depends on the criteria used to determine when speech production is abnormal; a rate of 2–3 per cent has been cited among six- to seven-year-olds (American Psychiatric Association 1994).

Expressive language disorder

In this disorder the ability to use expressive spoken language is markedly below the level appropriate for the child's mental age; language comprehension is within normal limits but there may be abnormalities in articulation. Language development varies considerably among normal children, but the absence of single words by two years of age and of two-word phrases by three years of age signifies abnormality. Signs at later ages include restricted vocabulary, difficulties in selecting appropriate words, and immature grammatical usage. Non-verbal communication, if impaired, is not affected as severely as spoken language, and the child tries to communicate. Disorders of behaviour are often present.

Some children speak rapidly and with an erratic rhythm such that the grouping of words does not reflect the grammatical structure of their speech. This abnormality, which is known as **cluttering**, is classified as an associated feature of expressive language disorder in DSMIV while in ICD10 it is classified (with stammering) among other behavioural disorders of childhood.

The *prevalence* of expressive language disorder depends on the method of assessment; a rate of 3–5 per cent of children has been proposed (American Psychiatric Association 1994).

Prognosis: it is reported that about half of the children meeting DSMIV criteria develop normal speech by adult life, while the rest have long-lasting difficulties (American Psychiatric Association 1994).

Receptive–expressive (or receptive) language disorder

In this disorder the understanding of language is below the level appropriate to the child's mental age. In almost all cases expressive language is also disturbed (a fact recognized in DSMIV by the term receptive–expressive language disorder).

The development of receptive language ability varies considerably among normal children. However, failure to respond to familiar names, in the absence of non-verbal cues, by the beginning of the second year of age, or failure to respond to simple instructions by the end of the second year are significant signs suggesting receptive language disorder (provided that deafness, mental retardation and pervasive developmental disorder have been excluded). Associated social and behavioural problems are particularly frequent in this form of language disorder.

The *prevalence* depends on the criteria for diagnosis, but a frequency of up to 3 per cent of school age children has been suggested (American Psychiatric Association 1994).

The *prognosis* is variable.

Acquired aphasia with epilepsy (Landau–Kleffner syndrome)

In this disorder, a child whose language has so far developed normally loses both receptive and expressive language but retains general intelligence. There are associated EEG abnormalities, nearly always bilateral and temporal and often with more widespread disturbances. Most of the affected children develop seizures either before or after the change in expressive language. The disorder starts usually between three and nine years of age. In most cases the loss of language occurs over several months but it may be more rapid. In the early stages the severity of the impairment may fluctuate.

The *aetiology* is unknown but an inflammatory encephalitis has been proposed. The prognosis is variable: about two-thirds are left with a receptive language deficit, but the other third recover completely.

Assessment: early investigation is essential both to determine the nature and severity of the speech and language disorder and to exclude mental retardation, deafness, cerebral palsy, and pervasive developmental disorder. The speech-

Table 20.5 *Pervasive developmental disorders*

DSMIV	ICD10
Autistic disorder	Childhood autism
Rett's syndrome	Rett's syndrome
Childhood disintegrative disorder	Other childhood disintegrative disorder Overactive disorder with mental retardation and stereotyped movements
Asperger's disorder	Asperger's syndrome
Pervasive developmental disorder not otherwise specified (including atypical autism)	Atypical autism Pervasive developmental disorder not otherwise specified

producing organs should be examined. It is particularly important to detect deafness at an early stage.

Parents can give some indication of the child's speech and language skills, especially if they complete a standardized inventory. With younger children it may be necessary to rely on this information, but children about the age of three years can be tested by a standard test of language appropriate to the child's age. If possible, such a test should be carried out by a speech therapist or a psychologist specializing in the subject.

Treatment depends partly on the cause but usually includes a programme of speech training carried out through play and social interaction. In milder cases this help is best provided at home by the parents who are given information on what to do. More severe difficulties are likely to require specialized help in a remedial class or a special school. Treatment should start early. For a review the development of speech and language and of their disorders see Yule and Rutter (1987).

Motor skills disorder

Some children have delayed motor development, which results in clumsiness in school-work or play.

In ICD10 this condition is called **specific developmental disorder of motor function**. It is also known as clumsy child syndrome or specific motor dyspraxia. The children can carry out all normal movements, but their coordination is poor. They are late in developing skills such as dressing, walking, and feeding. They tend to break things and are poor at handicrafts and organized games. They may also have difficulty in writing, drawing, and copying. IQ testing usually shows good verbal but poor performance scores.

These children are sometimes referred to a psychiatrist because of secondary emotional disorder. An explanation of the nature of the problem should be given to the child, the family, and the teachers. Special teaching may improve confidence. It may be necessary to exempt the child from organized games or other school activities involving motor coordination. There is usually some improvement with time. Further information is given by Cantwell and Baker (1985) and Henderson (1987).

Pervasive developmental disorders

The term pervasive developmental disorder refers to a group of disorders characterized by abnormalities in communication and social interaction and by restricted repetitive activities and interests. These abnormalities occur in a wide range of situations. Usually, development is abnormal from infancy and most cases are manifest before the age of five years.

Six conditions are included under this rubric in ICD10 (see Table 20.5); two of these do not appear in DSMIV, namely atypical autism and over-active disorder with mental retardation and stereotyped movements.

Childhood autism (autistic disorder)

This condition was described by Kanner (1943) who suggested the name infantile autism. The term childhood autism is used in ICD10 and autistic disorder in DSMIV.

The prevalence of autism is probably about 30–40 per 100 000 children, it is four times as common in boys as in girls (Rutter 1985*a*).

Clinical features

In his original description, Kanner (1943) identified the main features, which are still used to make the diagnosis. In both DSMIV and ICD10 three kinds of abnormality are required to make the diagnosis of autism. These are abnormalities of communication, abnormalities of social development, and a restriction of behaviour and interests. Of these, the abnormalities of social development are the most specific to autism. These and other features will be described more fully.

Autistic aloneness is an abnormality of social development in which the child is unable to make warm emotional relationships with people. Autistic children do not respond to their parents' affectionate behaviour by smiling or cuddling. Instead, they appear to dislike being picked up or kissed. They are no more responsive to their parents than to strangers and do not show interest in other children. There is little difference in their behaviour towards people and inanimate objects. A characteristic sign is gaze avoidance, that is the absence of eye-to-eye contact.

Speech and language disorder is another important feature. Speech may develop late or never appear. Occasionally, it develops normally until about the age of two years and then disappears in part or completely. This lack of speech is a manifestation of a severe cognitive defect. As autistic children grow up, about half acquire some useful speech, although serious impairments usually remain, such as the misuse of pronouns and the inappropriate repeating of words spoken by other people (echolalia). Some autistic children are talkative, but their speech is a repetitive monologue rather than a conversation with another person.

This **cognitive defect** also affects non-verbal communication and play; autistic children do not take part in the imitative games of the first year of life, and later they do not use toys in an appropriate way. They show little imagination or creative play.

Obsessive desire for sameness is a term applied to stereotyped behaviour together with evidence of distress if there is any change in the environment. For example autistic children may prefer the same food repeatedly, insist on wearing the same clothes, or engage in repetitive games. They are often fascinated by spinning toys.

Bizarre behaviour and mannerisms are common. Some autistic children engage in odd motor behaviour such as whirling round and round, twiddling their fingers repeatedly, flapping their hands, or rocking. Others do not differ obviously in motor behaviour from normal children.

Other features: autistic children may suddenly show anger or fear without apparent reason. They may be over-active and distractible, sleep badly, or soil or wet themselves. Some injure themselves deliberately. About 25 per cent of autistic children develop seizures, usually about the time of adolescence.

Kanner originally believed that the intelligence of autistic children was normal. Later research has shown that three-quarters have IQ scores in the retarded range, and this finding appears to represent true intellectual impairment (Rutter and Lockyer 1967). Some autistic children show areas of ability despite impairment of other intellectual functions, and in some cases they have exceptional but restricted powers of memory or mathematical skill (Hermelin and O'Connor 1983).

Aetiology

The cause of childhood autism is unknown. It is likely that the central abnormality is cognitive, affecting particularly symbolic thinking and language (Rutter 1983), and that the behavioural abnormalities are secondary to this cognitive defect.

Genetic influences have a major importance, which is so large that environmental factors are likely to be unimportant. The condition is 50 times more frequent in the siblings of affected persons than in the general population (Rutter *et al.* 1990). Several twin studies have shown a much higher concordance between monozygotic than between dizygotic twins (Folstein and Rutter 1971; Steffenburg *et al.* 1989; Rutter *et al.* 1993). Cognitive abnormalities are more frequent among the siblings of autistic patients than in the general population, and more frequent among the siblings of people with Down's syndrome (August *et al.* 1981). These findings suggest that the phenotype may be wider than the syndrome of autism as currently defined. Another finding supports this idea: although autistic people rarely have children, the condition does not die out. This persistence could be explained if the genetic disorder leads not only to autism but to other less extreme manifestations which do not prevent reproduction.

Organic brain disorder has been suspected as a cause of autism because there is an increased frequency of complications in pregnancy and childbirth among these patients, and an association of autism with epilepsy (in 2 per cent of cases). When an autistic patient has an unaffected identical twin, the autistic twin is more likely to have had obstetric complications at birth. However, the birth complications are strongly associated with minor congenital abnormalities. This finding suggests that the obstetric complications may have resulted from an abnormality in the fetus. If so, they may be a result not a cause of autism (Rutter *et al.* 1993). Also some patients have non-localizing neurological abnormalities ('soft signs').

Reports of reduced size of neurones in the limbic system and reduced cell number in the cerebellum (Bauman 1991) need to be confirmed. **Neuroimaging** studies have so far been inconclusive.

There is no evidence for a relationship between childhood autism and schizophrenia, and autistic children do not grow up to be schizophrenic.

Theory of mind in autism By the age of 4 years, normal children are able to form an idea of what others are thinking. For example if a normal child watches while another normal child is shown the location of a hidden object and is then sent out of the room while the object is moved to a new hiding place, the observer who has remained in the room will conclude that the other child will expect the object to be in the original position when he returns to the room. An autistic child tends to lack this appreciation of what information others possess and what they are likely to be thinking; in this example an autistic child is likely to say that the other child will think that the object has been moved and is in its new place. It is not certain how specific to autism is this difficulty in appreciating what others know and expect, or how central it is to the psychopathology. In any case, its cause is not known. For a review of the theory of mind in autism see Frith (1989).

Abnormal parenting Kanner (1943) suggested that autism was a response to **abnormal parents** who were characterized as cold, detached, and obsessive. Kanner's idea has not been substantiated (Koegel *et al.* 1983). It is now thought that any psychological abnormalities in the parents are likely to be either a response to the problems of bringing up the autistic child or, alternatively a manifestation in the parents of the genes that have produced autism in the child. The possible biological causes of autism are reviewed in Gillberg and Coleman (1992).

Prognosis

Between 10 and 20 per cent of children with childhood autism begin to improve between the ages of about four and six years, and are eventually able to attend an ordinary school and obtain work. A further 10–20 per cent can live at home but cannot work and need to attend a special school or training centre. The remainder, at least 60 per cent, improve little and are unable to lead an independent life; many need long-term residential care (Russell 1970). Those who improve may continue to show language problems, emotional coldness, and odd behaviour. As noted already, a substantial minority develop epilepsy in adolescence.

Differential diagnosis

It is more usual to encounter partial syndromes than the full syndrome of childhood autism. These partial syndromes must be distinguished from the **childhood disintegrative disorders** arising after the age of 30 months (see p. 693) and **Asperger's syndrome** (autistic psychopathy) (see p. 693).

Deafness should be excluded by appropriate tests of hearing. **Developmental language disorder** (see p. 687) differs from autism in that the child usually responds normally to people and has good non-verbal communication. **Mental retardation** can be differentiated because, although the child has general intellectual retardation, responses to other people are more normal than those of an autistic child. Also, an autistic child has more impairment of language relative to other skills than is found in a mentally retarded child of the same age.

Assessment

Assessment should be concerned with more than the diagnosis of autism. The following additional factors need to be considered (Lord and Rutter 1994):

(1) cognitive level;

(2) language ability;

(3) communication skills, social skills and play, and repetitive or other abnormal behaviour;

(4) stage of social development in relation to age, mental age, and stage of language development;

(5) associated medical conditions;

(6) psychosocial factors.

Treatment

Treatment has three main aspects: management of the abnormal behaviour, arrangements for social and educational services, and help for the family. Individual psychotherapy has been used in the hope of effecting more fundamental changes but there is no evidence that it succeeds. Nor is there evidence that any form of medication is effective in childhood autism, except in the short-term management of behaviour problems when an antipsychotic drug may be used.

Behavioural methods using contingency management (see p. 633) may control some of the abnormal behaviour of autistic children. Behavioural treatment is often carried out at home by the parents, instructed and supervised by a clinical psychologist. It is not known whether these methods have any lasting benefit, but in autism even temporary changes are often worthwhile for the patient and the family.

Most autistic children require *special schooling*. It is generally thought better for them to live at home and to attend special day schools. If the condition is so severe that the child cannot stay in the family, residential schooling is necessary even though the characteristic social withdrawal may be increased by an institutional atmosphere. In some places, the educational and residential needs of autistic children are best provided through the services for the mentally retarded. Older adolescents may need vocational training.

The *family* of an autistic child needs considerable help to cope with the child's behaviour, which is often bewildering and distressing. Although little can be done to treat many of these patients, the doctor must not withdraw from the family, who need continuing support and encouragement in their efforts to help the child to realize any potential for normal development. Some parents request genetic counselling and it seems that the risk of a further autistic child is about 3 per cent (Lord and Rutter 1994). Many parents find it helpful to join a voluntary organization in which they can meet other parents of autistic children and discuss common problems.

A general review of treatment is given by Rutter (1985c).

Rett's disorder

Rett's disorder (or Rett's syndrome) is a rare condition which has to date been reported only in girls. The reported prevalence is 0.8 per 10 000 girls (Kerr and Stevenson 1985). After a period of normal development in the first months of life, head growth slows and over the next two years there is arrest of cognitive development and loss of purposive skilled hand movements. Stereotyped movements develop with hand clapping and hand-wringing movements. Ataxia of the legs and trunk may develop. Interest in the social environment diminishes in the first few years of the disorder, but may increase again later. Expressive and receptive language development is impaired severely and there is psychomotor retardation. Some patients develop severe mental retardation. Pedigree studies show that there is rarely a family history of the disorder. See Olsson and Rett (1990) for a review.

Over-active disorder with mental retardation and stereotyped movements

This condition is included in ICD10 but not in DSMIV. It is an ill-defined syndrome which occurs in children with severe mental retardation (IQ less than 50) who have hyperactivity, inattention, and stereotyped movements. In adolescence, over-activity may be replaced by reduced activity (an unusual outcome in hyperkinetic syndrome). It is not certain whether this combination of features defines a distinct entity.

Childhood disintegrative disorder

The term *childhood disintegrative disorder* is used in ICD10; an alternative is Heller's disease. The condition begins after a period of normal development usually lasting for more than two years. It is unclear how far the childhood disintegrative disorder is distinct from childhood autism. It resembles childhood autism in the marked loss of cognitive functions, abnormalities of social behaviour and communication, and unfavourable outcome. It differs from childhood autism in the loss of motor skills and of bowel or bladder control. The condition may arrest after a time, or progress to a severe neurological condition.

Asperger's syndrome

This rare condition, first described by Asperger (1944), is sometimes called *autistic psychopathy*. (Asperger's original paper in German has been translated into English (see Frith 1991).) The condition is characterized by severe and sustained abnormalities of social behaviour similar to those of childhood autism, with stereotyped and repetitive activities and motor mannerisms such as hand and finger twisting or whole body movements. It differs from autism in that there is no general delay or retardation of cognitive development or language. The disorder is more common in boys than girls. The children develop normally until about the third year when they begin to lack warmth in their relationships and speak in monotonous ways. They are solitary, and spend much time in narrow interests and routines. They are often clumsy and eccentric. They do not share interest or pleasures with others and are without friends.

The *cause* of autistic psychopathy is unknown. It is uncertain whether the condition is a variant of childhood autism or a separate disorder. Usually the abnormalities persist into adult life. Most people with the disorder can work, but few form successful relationships and marry. See Frith (1991) for a review of Asperger's syndrome.

Hyperkinetic and conduct disorders

About a third of children are described by their parents as over-active, and 5–20 per cent of schoolchildren are so described by teachers. These reports encompass behaviour varying from normal high spirits to a severe and persistent disorder. This over-activity often varies in different situations. Hyperkinetic disorders are more severe forms of over-activity, associated with marked inattention (hence the alternative name attention-deficit hyperactivity disorder).

Hyperkinetic disorder (attention-deficit hyperactivity disorder)

Clinical features

The cardinal features of this disorder are extreme and persistent restlessness, sustained and prolonged motor activity, and difficulty in maintaining attention. Children with the disorder are often impulsive, reckless, and prone to accidents. There are learning difficulties which result in part from poor attention and lack of persistence with tasks. Minor forms of antisocial behaviour are common, particularly disobedience, temper tantrums, and aggression. However, these antisocial behaviours do not occur early. These children are often socially disinhibited and unpopular with other children. Mood fluctuates, but low self-esteem and depressive mood are common.

Restlessness, over-activity, and related symptoms often start before school age. Sometimes the child was over-active as a baby, but more often significant problems begin when the child begins to walk; he is constantly on the move, interfering with objects and exhausting his parents.

Diagnostic criteria

In both ICD10 and DSMIV the cardinal features for the diagnosis of the disorder are impaired attention and hyperactivity, but the two systems differ in the details of the criteria for diagnosis. DSMIV requires that these symptoms have been present for six months; the clinical diagnostic guidelines for ICD10 do not specify a time limit. Both systems require that symptoms should have started in childhood: ICD10 specifies before six years of age; DSMIV specifies before seven years of age. An important difference is that ICD10 requires both hyperactivity *and* impaired atten-

tion, while for DSMIV the diagnosis can be made if there is *either* inattention *or* hyperactivity–impulsivity.

In ICD10 the disorder can be further classified as (a) disturbance of activity and attention and (b) hyperkinetic conduct disorder. The latter term is used when criteria for both hyperkinetic disorder and conduct disorder are met. (The division is made because the presence of associated aggression, delinquency, or antisocial behaviour is associated with a less good outcome—see below.)

Autistic children are often hyperactive and inattentive, but these features are regarded as part of the syndrome of childhood autism; hyperactivity disorder is not diagnosed in addition to autism.

Epidemiology

Estimates of the prevalence of hyperkinetic disorder vary according to the criteria for diagnosis. Using DSMIV criteria, a prevalence of 3–5 per cent is suggested (American Psychiatric Association 1994). Using ICD10 criteria, a prevalence of 1.7 per cent was found among primary school boys (Taylor *et al.* 1991). Rates are about four times higher in boys than in girls (Ross and Ross 1982).

Aetiology

Signs suggesting **neurodevelopmental impairment** or delay are found in children with hyperkinetic disorder, for example clumsiness, language delay, and abnormalities of speech (Schachar 1991). Although these signs are associated with birth complications, it is not certain whether they result from factors acting at birth or at an earlier stage of development of the brain.

Genetic factors are suggested by a study of adopted children (see Cantwell 1975) and by comparisons of monozygotic and dizygotic twins (Goodman and Stevenson 1989). However, the results of both studies are inconclusive because the first used wide criteria and the second used questionnaire measures. Biederman *et al.* (1992), using DSMIIIR criteria, found that, compared with controls, probands with attention-deficit hyperactivity disorder had more first-degree relatives with the same disorder but also had more with antisocial and major depressive disorders. Therefore the data do not allow a firm conclusion to be drawn at present about the role of genetic factors in hyperkinetic disorder.

It is possible that **social factors** increase an innate tendency to hyperactivity since over-active behaviour is more frequent among young children living in poor social conditions (Richman *et al.* 1982). However, social factors are unlikely to be the sole cause of hyperkinetic disorder. **Lead intoxication** (Needleman *et al.* 1979) and **food additives** (Feingold 1975) have been suggested as causes of hyperkinetic syndrome, but there is no convincing evidence for either (Taylor 1984).

Prognosis

Over-activity usually lessens gradually as the child grows older, especially when it is mild and not present in every situation. It usually ceases by puberty. The prognosis for any associated learning difficulties is less good, while antisocial behaviour has the worst outcome. When the over-activity is severe, and is accompanied by learning failure or associated with low intelligence, the prognosis is poor and the conditions may persist into adult life, usually as antisocial disorder and drug abuse rather than continued hyperactivity. Mood and anxiety disorders are not increased in adult life (Mannuzza *et al.* 1993).

Treatment

A hyperactive child exhausts his parents who need support from the start of treatment, particularly as it may be difficult to reduce the child's behaviour. The child's teachers need advice about management, which may include remedial teaching. Methods of behaviour modification may help to reduce the inadvertent reinforcement of over-activity by parents and teachers.

Stimulant drugs may be tried, especially when attention deficits are severe. The usual drugs are methylphenidate or dexamphetamine. Dosage should be related to body weight. When methyl-phenidate is used, it is appropriate to start with 2.5 mg in the morning, adding after four days a further 2.5 mg at midday, and, depending on the response and side-effects, increasing cautiously to a maximum of 10 mg in the morning and 10 mg at midday for a five-year-old of average weight (or a lower dose for a younger child). The side-effects include irritability, depression, insomnia, and poor appetite. With high doses there may be some slowing of growth but adult stature and weight do not seem to be affected (Klein and Mannuzza 1988; Taylor, E. 1994). The drug may be needed for several months or even a year or more, and careful monitoring is essential. The drug may be stopped from time to time in an attempt to minimize side-effects and to confirm that medication is still needed. In clinical trials, short-term benefits of the drug have been shown in about two-thirds of children with hyperkinetic syndrome (Ottenbacher and Cooper 1983), but the long-term benefits are uncertain. It seems best to reserve drug treatment for severe cases which have not responded to other treatment. Surprisingly, there is no report of children treated in this way becoming addicted to the drug.

For a review of the hyperkinetic syndrome see Schachar (1991).

Conduct disorders

Conduct disorders are characterized by severe and persistent antisocial behaviour. They form the largest single group of psychiatric disorders in older children and adolescents. Because conduct disorders vary widely in their clinical features, attempts have been made to classify them. One of the earliest classifications was into socialized, unsocialized, and overinhibited groups (Hewett and Jenkins 1946). In DSMIV, conduct disorders are divided into childhood onset type (onset before 10 years of age) and adolescent onset type (with onset after this age). DSMIV has an additional category, 'oppositional defiant disorder', for persistently hostile defiant provocative and disruptive behaviour outside the normal range but without aggressive or dyssocial behaviour. This disorder occurs mainly in children below ten years of age. ICD10 has four subdivisions of conduct disorder:

socialized conduct disorder, unsocialized conduct disorder, conduct disorders confined to the family context, and oppositional defiant disorder.

Clinical features

The essential feature of conduct disorder is persistent abnormal conduct which is more serious that ordinary childhood mischief. In the preschool period the disorder usually manifests as aggressive behaviour in the home, often with over-activity. In later childhood it usually first begins in the home as stealing, lying, and disobedience, together with verbal or physical aggression. Later, the disturbance often becomes evident outside as well as inside the home, especially at school, or as truanting, delinquency, vandalism, and reckless behaviour, or as alcohol or drug abuse.

In children above the age of seven years persistent stealing is abnormal. Below that age, children seldom have a real appreciation of other people's property. Many children steal occasionally, so that minor or isolated instances need not be taken seriously. A small proportion of children with conduct disorder present with sexual behaviour that incurs the disapproval of adults. In younger children, masturbation and sexual curiosity may be frequent and obtrusive. Promiscuity is a particular problem in adolescent girls. Fire-setting is rare, but obviously dangerous (see p. 768).

Prevalence

The prevalence of conduct disorders is difficult to estimate because the dividing line between them and normal rebelliousness is arbitrary. Rutter *et al.* (1970*a*) found the prevalence of 'antisocial disorder' to be about 4 per cent among 10–11 year olds on the Isle of Wight; in a subsequent study in London about twice this rate was found (Rutter *et al.* 1976*b*). In a study in a more disadvantaged area of London the rate was about twice that in the Isle of Wight (Rutter *et al.* 1975*b*). A subsequent study in a province of Canada found a rate of 5.5 per cent (Offord *et al.* 1987). Studies in the community, in psychiatric practice, and in the juvenile courts all indicate that conduct

disorders are about four times more common in boys than girls (Rutter *et al.* 1970*a*; Gath *et al.* 1977).

Aetiology

Environmental factors are important; conduct disorders are commonly found in children from unstable, insecure, and rejecting families living in deprived areas. Antisocial behaviour is frequent among children from broken homes, those from homes in which family relationships are poor, and those who have been in residential care in their early childhood. Conduct disorders are also related to adverse factors in the wider social environment of the neighbourhood and school (Power *et al.* 1972; Rutter *et al.* 1975*c*; Gath *et al.* 1977).

As well as these environmental causes, certain factors in the child may predispose to conduct disorder. Adoption studies suggests that *genetic factors* play only a small part in the aetiology of conduct disorder (Rutter *et al.* 1990). This contrasts with the stronger genetic influences identified in adult antisocial behaviour but accords with the observation that most childhood conduct disorders improve with time. It is possible that the persistent cases have a stronger genetic aetiology. Alcoholism and personality disorder in the father is reported to be strongly associated with conduct disorder (Earls *et al.* 1988); the association could be through genetic or environmental mechanisms.

Children with brain damage and epilepsy are prone to conduct disorder, as they are to other psychiatric disorders. An important finding in the Isle of Wight survey was a strong association between antisocial behaviour and specific reading retardation (see p. 668). It is not known whether antisocial behaviour and reading retardation result from common predisposing factors, or whether one causes the other.

See Rutter and Giller (1983) for a review of the aetiology of conduct disorder.

Prognosis

Conduct disorders usually run a prolonged course in childhood (Rutter *et al.* 1976*b*). The long-term outcome varies considerably with the nature and extent of the disorder. Robins (1966) found that almost half of people who had attended a child guidance clinic for conduct disorder in adolescence showed some form of antisocial behaviour in adult life. No cases of sociopathic disorder were found in adult life among those with diagnoses other than conduct disorder in adolescence. Follow-up of conduct-disordered children cared for in children's homes and of controls led to similar conclusions: about 40 per cent of the conduct-disordered children had DSMIII antisocial personality disorder in their twenties, and many of the rest had persistent and widespread social difficulties below the threshold for diagnosis of a personality disorder. Pervasive social difficulty in adult life was also related to upbringing in a children's home, but conduct disorder had an additional effect when this variable was controlled for (Zoccolillo *et al.* 1992).

There are no good indicators of the long-term outcome of individual cases. The best available predictors seem to be the extent of the childhood antisocial behaviour and the quality of relationships with other people (Robins 1978). Other research indicates that the presence of hyperactivity or inattention and poor peer relations predict a poor outcome of childhood conduct disorder (Farrington *et al.* 1990), although it is not certain whether these are specific factors or merely indices of the severity of the conduct disorder. In the study by Zoccolillo *et al.* (1992) there were few changes until adult life, when some people improved substantially following marriage to a supportive and non-deviant partner.

Treatment

Mild conduct disorders often subside without treatment other than common-sense advice to the parents. For more severe disorders, treatment is mainly directed to the family. It usually takes the form of social casework or family therapy. There is no convincing evidence that treatment affects the long-term outlook. However, awareness of a generally poor long-term prognosis should not lead to inadequate attention to the immediate

problems that can be modified in the individual case. In many cases some of the immediate distress of patients and their families can be reduced. In some cases it is possible to modify adverse social and family factors and thereby perhaps improve the long-term outlook.

Parent training programmes

These programmes use behavioural principles (see p. 679). Parents are taught how the child's antisocial behaviour may be reinforced unintentionally by their attention to it, and how it may be provoked by interactions with members of the family. Parents are also taught how to reinforce normal behaviour by praise or rewards and how to set limits on abnormal behaviour, for example by removing the child's privileges such as an hour less time to play a game. As aids to learning, parents are provided with written information and videotapes showing other parents applying behavioural procedures. For a review of parent training programmes see Webster-Stratton (1991).

Anger management

Young people who are habitually aggressive have been shown to misperceive hostile intentions in other people who are not in fact hostile. They also tend to underestimate the level of their own aggressive behaviour, and choose inappropriate behaviours rather than more appropriate verbal responses (Dodge *et al.* 1990). Anger management programmes seek to correct these ideas by teaching how to inhibit sudden inappropriate responses to angry feelings, (for example Stop! What should I do!), and how to reappraise the intentions of other people and use socially acceptable forms of self-assertion. Kazdin *et al.* (1987) showed that these methods reduce the problems of aggressive children who do not have conduct disorder. Similar benefits have been reported with conduct-disordered children (Kendall *et al.* 1991).

Other methods

Group therapy for the child by using peer pressures to reduce antisocial behaviour is sometimes helpful. If there are associated reading difficulties, remedial teaching should be arranged. Medication is of little value. Some families are difficult to help by any means, especially where there is material deprivation, chaotic relationships, and poorly educated parents.

Residential care

Occasionally, residential placement may be necessary in a foster home, group home, or special school. This should be done only for compelling reasons. There is no evidence that institutional care improves the prognosis for conduct disorder.

Truancy

The treatment of truancy requires separate consideration. A direct and energetic approach is called for. Pressure should be brought to bear upon the child to return to school and, if possible, the support of the family should be enlisted. At the same time, an attempt should be made to resolve any educational or other problems at school. In all this it is essential to maintain good communications between clinician, parents, and teachers. If other steps fail, court proceedings may need to be initiated.

Juvenile delinquency

A juvenile delinquent is a young person who has been found guilty of an offence that would be categorized as a crime if committed by an adult. In most countries the term applies only to a young person who has attained the age of criminal responsibility—at present 10 years in the United Kingdom, but ranging widely in other places. Thus delinquency is not a psychiatric diagnosis but a legal category. However, juvenile delinquency may be associated with psychiatric disorder, especially conduct disorder. For this reason it is appropriate to interrupt this review of the syndromes of child psychiatry to consider juvenile delinquency. The majority of adolescent boys, when asked to report their own behaviour, admit to offences against the

law and a fifth are convicted at some time (West and Farrington 1973); most of the offences are trivial. Amongst boys who are convicted, only a half are reconvicted. Few juvenile delinquents continue to offend in adult life. Of these, a few who offend repeatedly are very difficult to manage. Many more boys than girls are delinquent, and the peak age of contact with the police is 15–16 years. In considering these figures it has to be remembered that crime statistics may be misleading. Nevertheless, there seems to be a substantial similarity in the characteristics of self-reported offenders and of convicted offenders (West and Farrington 1973).

Delinquency is often equated with conduct disorder. This is wrong, for although the two categories overlap, they are not the same. Many delinquents do not have conduct disorder or any other psychological disorder. Equally, many of those with conduct disorder do not offend. Nevertheless, in an important group persistent lawbreaking is preceded and accompanied by abnormalities of conduct, such as truancy, aggressiveness, and attention-seeking, and by poor concentration.

Causes

The causes of juvenile delinquency overlap with those of conduct disorder. However, greater emphasis must necessarily be given to social explanations, since delinquency is defined by the provisions of the law and by the way it is operated. For a review see Rutter and Giller (1983).

Social factors

Delinquency is related to low social class, poverty, poor housing, and poor education. There are marked differences in delinquency rates between adjacent neighbourhoods which differ in these respects. Rates also differ between schools. Many social theories have been put forward to explain the origins of crime, but none offers a completely adequate explanation.

Family factors

Many studies have found that crime runs in families. For example, about half of boys with criminal fathers are convicted, compared with a fifth of those with fathers who are not criminals (West and Farrington 1977). The reasons for this are poorly understood. They may include poor parenting and shared attitudes to the law.

In a retrospective study, Bowlby (1944) examined the characteristics of 'juvenile thieves' and argued that prolonged separation from the mother during childhood was a major cause of their problems. More recent work has not confirmed such a precise link (see p. 671). Although delinquency is particularly common among those who come from broken homes, this seems to be largely because separation often reflects family discord in early and middle childhood (Rutter and Madge 1976). Other family factors correlated with delinquency are large family size and child-rearing practices, including erratic discipline and harsh or neglecting care.

Factors in the child

Genetic factors appear to be less significant among the causes of delinquency than in the more serious criminal behaviour of adult life (see p. 754). (The possible role of genetic factors in conduct disorder has been considered already (p. 696).) There are important relationships between delinquency and slightly below average IQ as well as educational and reading difficulties (Rutter *et al.* 1976b). There are at least two possible explanations for the latter finding. Temperament or social factors may predispose to both delinquency and reading failure. Alternatively, reading difficulties may result in frustration and loss of self-esteem at school, and these may in turn predispose to antisocial behaviour.

Physical abnormalities probably play only a minor role among the causes of delinquency, even though brain damage and epilepsy predispose to conduct disorder.

Assessment

When the child is seen as part of an ordinary psychiatric referral and the delinquency is accompanied by a psychiatric syndrome, the latter should be assessed in the usual way. Sometimes the child psychiatrist is asked to see a delinquent specifically to prepare a court report. In these circumstances, as well as making enquiries among the parents and teachers it is essential to consult any social worker or probation officer who has been involved with the child.

Psychological testing of intelligence and educational achievements can also be useful. The form of the report is similar to that described in Chapter 22 (p. 778). It should include a summary of the history and present mental state together with recommendations about treatment.

Treatment

When considering the treatment of delinquent children and adolescents, psychiatrists need to understand the legal system in the country in which they work. In England and Wales the provisions are largely covered by the Children Act 1989. The legal responses include a fine, the requirement that the parent or guardian take proper control, supervision by a social worker, a period at a special centre, or an order committing the child to the care of the local authority. The exact provisions vary from one country to another, and readers should enquire about the arrangements in the places in which they work. Since delinquent behaviour is common, mainly not serious, and usually a passing phase, it is generally appropriate to treat first offences with minimal intervention coupled with firm disapproval. The same applies to minor offences that are repeated. A more vigorous response is required for more serious recurrent delinquency. For this purpose a community-based programme is usually preferred, with the main emphasis on improving the family environment, reducing harmful peer group influences, helping the offender to develop better skills for solving problems, and improving educational and vocational accomplishments. When this approach fails, custodial care will be considered.

In recent years the main aim of the law as it applies to children and young persons has been treatment rather than punishment or even deter-

Table 20.6 *Anxiety disorders in childhood*

DSMIV	ICD10
	F93 Emotional disorders with specific onset in childhood
Separation anxiety disorder	Separation anxiety disorder of childhood
Phobic anxiety disorder*	Phobic anxiety disorder of childhood
Social phobia*	Social anxiety disorder of childhood
(sibling relationship problems)†	Sibling rivalry disorder
	Other anxiety disorders*
Post-traumatic stress disorder*	Post-traumatic stress disorder
Obsessive–compulsive disorder*	Obsessive–compulsive disorder
	Other emotional disorders*
	Dissociative disorder
	Somatization disorder

*There is no separate category for these disorders in childhood; the adult categories are used (see text).
†Listed under 'other conditions that may be the focus of clinical attention'.

rence. There has been extensive criminological research to determine the effectiveness of the measures used. The general conclusions are not encouraging, though not altogether surprising since delinquency is strongly related to factors external to the child, including family disorganization, antisocial behaviour among the parents, and poor living conditions. The risk of reconviction seems to be greater among children who have had any court appearance or period of detention than among children who have committed similar offences without any official action having been taken (West and Farrington 1977).

There have been many attempts to establish and evaluate treatments that might be effective. One of the earliest, the Highfields Project, compared group treatment in a small well-staffed unit with the usual custodial sentence. Modest benefits were found for the former. A larger study, the Pilot Intensive Counselling Organization (PICO) project, found some evidence that nine months' counselling was of more benefit to 'amenable' boys in a medium-security unit that to more difficult and uncooperative ('non-amenable') boys. An elaborate investigation known as the Community Treatment Project of the Californian Youth Authority (Warren 1973) found that community treatment was generally at least as effective as institutional care.

These and other studies suggest the need to match the type of treatment to the type of offender. Some delinquents seem to respond better to authoritative supervision, and others to more permissive counselling. Unfortunately, it is not yet possible to provide any satisfactory practical guidelines about the choice of treatment for the individual delinquent. See Mulvey *et al.* (1993) for a review of research on the prevention and treatment of juvenile delinquency.

Anxiety disorders

In ICD10 anxiety disorder in childhood are classified as emotional disorders with onset specific to childhood (Table 20.6). DSMIV does not contain this category and with one exception classifies childhood anxiety disorders in the same way as anxiety disorders in adult life. The exception is separation anxiety disorder which is listed under the heading 'other disorders of infancy, childhood or adolescence'. ICD10 has a diagnosis of sibling rivalry disorder. DSMIV does not have this diagnosis in the main classification, but sibling relationship problems can be coded under 'other conditions that may be the focus of clinical attention'.

Prevalence

The prevalence of anxiety disorders in childhood is uncertain because epidemiological studies have usually employed the wider category of emotional disorder, or asked about symptoms rather than syndromes of anxiety (Orvaschel and Weissman 1986). In their survey of the Isle of Wight, Rutter *et al.* (1970*a*) found a prevalence of emotional disorders of 2.5 per cent in both boys and girls. In a London suburb the corresponding figure was doubled. (Rutter *et al.* 1975*b*). (In both places, the rate of conduct disorder was about twice that of emotional disorder.) More recent surveys of the general population suggest rates of anxiety disorders of 6–9 per cent among seven- to eleven-year-olds, of which about half was separation anxiety disorder (see below) (Anderson *et al.* 1987; Benjamin *et al.* 1990). A survey of children referred to a child guidance clinic in London showed that one-third were diagnosed as having emotional disorders (Gath *et al.* 1977).

Normal anxiety in childhood

Anxiety is common in childhood, but its nature changes as the child grows older: infants pass through a stage of fear of strangers; during preschool years separation anxiety and fears of animals, imaginary creatures, and the dark are common; in early adolescence these fears are replaced by anxiety about social situations and personal adequacy. Anxiety disorders in childhood resemble these normal anxieties and follow the same developmental sequence though they are more severe and more prolonged. Phobias and separation anxiety disorder usually start in early childhood, and social anxiety disorder starts in adolescence.

There is no clear dividing line between normal anxiety and anxiety disorders in childhood and there is often overlap, i.e. an individual child's disorder may fulfill the criteria for more than one disorder, for example phobic disorder and separation anxiety disorder.

Separation anxiety disorder

Separation anxiety disorder is a fear of separation from people to whom the child is attached which is clearly greater than normal separation anxiety of toddlers or preschool children, or persists beyond the usual preschool period, and is associated with significant problems of social functioning. The onset is before the age of six years. The diagnosis is not made when there is a generalized disturbance of personality development.

Clinical picture

Children with this disorder are excessively anxious when separated from parents or other attachment figures, and unrealistically concerned that harm may befall these persons or that they will leave the child. They may refuse to sleep away from these persons or, if they agree to separate, may have disturbed sleep with nightmares. They cling to their attachment figures by day, demanding attention. Anxiety is often manifested as physical symptoms of stomach ache, headache, nausea, and vomiting, and may be accompanied by crying, tantrums, or social withdrawal. Separation anxiety disorder is one cause of school refusal (see p. 706).

Epidemiology

Community surveys suggest that rates of separation anxiety disorder are about 3–4 per cent among seven- to eleven-year-olds (Anderson *et al.* 1987; Benjamin *et al.* 1990).

Aetiology

Separation anxiety disorder is sometimes precipitated by a frightening experience. This may be brief, for example admission to hospital, or prolonged, for example conflict between the parents. In some cases separation anxiety disorder develops in children who react with excessive anxiety to a large number of everyday stressors and who are therefore said to have an anxiety-prone temperament. Sometimes the condition appears to be a response to anxious or overprotective parents.

Treatment

Account should be taken of the whole range of possible aetiological factors including stressful events, previous actual separation, an anxiety-prone temperament, and the behaviour of the parents. Stressors should be reduced if possible, and the children should be helped to talk about their worries. It is more important to involve the family, helping them to understand how their own concerns or overprotection effect the child. Anxiolytic drugs may be needed occasionally when anxiety is extremely severe, but they should be used for short periods only. When separation anxiety is worse in particular circumstances, the child may benefit from the simple behavioural techniques used for phobias as described in the next section.

Phobic anxiety disorders

This diagnosis for children corresponds to specific phobia for adults. Minor phobic symptoms are common in childhood. They usually concern animals, insects, the dark, school, and death. The prevalence of more severe phobias varies with age. Severe and persistent fears of animals usually begin before the age of five, and nearly all have declined by the early teenage years. At age 11 a rate of only 2–4 per cent has been found (Anderson *et al.* 1987).

Most childhood phobias improve without specific treatment provided that the parents adopt a firm and reassuring approach. For phobias that do not improve, simple behavioural treatment can be combined with reassurance and support. The child is encouraged to encounter feared situations in a graded way, as in the treatment of phobias in adult life. Dynamic psychotherapy has also been used, but it is not obviously more effective than simple behavioural treatment. A full account of childhood phobias is given by Johnson (1985).

Social anxiety disorder of childhood

This term is used in ICD10 to describe disorders starting before the age of six years in which there is anxiety with strangers greater or more prolonged than the fear of strangers which normally occurs in the second half of the first year of life.

These children are markedly anxious in the presence of strangers and avoid them. The fear, which may be mainly of adults or of other children, interferes with social functioning. It is not accompanied by severe anxiety on separation from the parents.

Aetiology and treatment resemble those of other anxiety disorders of childhood.

Sibling rivalry disorder

This category is listed in ICD10 for children who show extreme jealousy or other signs of rivalry of a sibling, starting during the months following the birth of that sibling. The signs are clearly more severe than the emotional upset and rivalry which is common in such circumstances, and they are persistent and cause social problems. When the disorder is severe there may be hostility and even physical harm to the sibling. The child may regress in behaviour, for example losing previously learned control of bladder or bowels, or act in a way appropriate for a younger child. There is usually opposition to the parents and behaviour intended to obtain their attention, often with temper tantrums. There may be sleep disturbance and problems at bedtime.

In treatment parents should be helped to divide their attention appropriately between the two children, to set limits for the older child, and to help him or her feel valued. For a review see Dunn and Kendrick (1982).

Post-traumatic stress disorder

Although not included in ICD10 among the anxiety disorders with onset usually in childhood, post-traumatic stress disorder can occur in childhood life. The *clinical picture* resembles that of the same disorder in adult life with disturbed sleep, nightmares, and flashbacks (p. 140). Children with post-traumatic stress disorder often have irrational separation anxiety, and young children may show regressive behaviour (Yule and Williams 1990). The *prognosis* of the disorder has not been studied systematically in childhood, but severe reactions have been reported to last for six months to a year (Yule 1994). As in adults, the *cause* is an encounter with exceptionally severe stressors, for example those encountered by children caught up in war, civil unrest, or natural disasters. Physical and sexual abuse may also provoke post-traumatic stress disorder in children (Goodwin 1988). *Treatment* resembles that for adults (see p. 143) with the same condition, adapting counselling to the child's stage of development.

Obsessive–compulsive disorders

Obsessive–compulsive disorders are rare in childhood. However, several related forms of repetitive behaviour are common, particularly between the ages of four and ten years. These repetitive behaviours include preoccupation with numbers and counting, the repeated handling of certain objects, and hoarding. Normal children commonly adopt rituals such as avoiding cracks in the pavement or touching lamp-posts. These behaviours cannot be called compulsive because the child does not struggle against it (see p. 15 for the definition of obsessive and compulsive symptoms). The preoccupations and rituals of obsessive–compulsive disorder are more extreme than these behaviours of healthy children and take up an increasing amount of the child's time, for example rechecking schoolwork many times or frequently repeated hand-washing.

Clinical picture

Obsessional disorder rarely appears in full form before late childhood, though the first symptoms may appear earlier. The onset may be rapid or gradual.

Obsessional disorders in childhood generally resemble those in adult life (see p. 181). The presenting symptoms are more often rituals than obsessional thoughts. Washing rituals are the most frequent, followed by repetitive actions and checking. Obsessional thoughts are most often concerned with contamination, accidents or illness affecting the patient or another person, and concerns about orderliness and symmetry. The content of symptoms often change as the child grows older. The obsessional symptoms may be provoked by external cues such as unclean objects. Children with obsessional symptoms usually try to conceal them, especially outside the family. Obsessional children often involve their parents by asking them to take part in the rituals or give repeated reassurance about the obsessional thoughts. See Swedo *et al.* (1989) for an account of the clinical picture.

Aetiology

Genetic factors are suggested by the observation that obsessive–compulsive disorder is more frequent among the first-degree relatives of children and adolescents with obsessive–compulsive disorder than among the general population (Lenane *et al.* 1990.) However, the probands in this study had been referred for treatment, and it is possible that this biased the sample since parents with obsessive–compulsive disorder may be more likely to seek treatment for their affected children. Also, familial aggregation of cases might indicate *social learning* rather than genetic inheritance.

There is an association between obsessive–compulsive disorder in childhood and conditions thought to arise from *dysfunction of the basal ganglia*. Some children with obsessive–compulsive disorder have tics or choreiform movements. Conversely, children with Gilles de la Tourette's syndrome have obsessional and compulsive symptoms, and these symptoms have also been described in children with Sydenham's chorea. These observations have led to the suggestion that childhood obsessive–compulsive disorder arises from a disorder of some part of the basal ganglia (see Rappoport 1991 for a review).

Associated disorders

Severe and persistent obsessional thoughts and compulsive rituals in childhood are often accompanied by anxiety and depressive symptoms. In some cases there is an associated anxiety or depressive disorder. As noted above, abnormal choreiform movements occur in some children with obsessive–compulsive disorder. Children with Gilles de la Tourette's syndrome have obsessional symptoms, and it is important to make the distinction between this condition (see p. 400) and obsessive–compulsive disorder.

Prognosis

Clinical observations suggest that less severe forms of the disorder have a generally good outcome, but severe forms have a poor prognosis. In a two- to seven-year follow-up of patients initially treated in drug trials, 43 per cent still met diagnostic criteria for obsessive–compulsive disorder and only 6 per cent were in full remission (Leonard *et al.* 1993).

Treatment

When obsessional symptoms occur as part of an anxiety or depressive disorder, treatment is directed to the primary disorder. True obsessional disorders of later childhood are treated along similar lines to an anxiety disorder with the addition of behavioural methods similar to those used with adults. It is important to involve the family in treatment.

Clomipramine has been found to be more effective than placebo in obsessive–compulsive disorder of children aged 10 and over (de-Veaugh-Geiss *et al.* 1992). As in adults, the symptoms are reduced but not removed by this treatment (Leonard *et al.* 1991).

Somatization disorders

Children with a psychiatric disorder often complain of somatic rather than psychological symptoms. These complaints include abdominal pain, headache, cough, and limb pains. Most of these children are treated by family doctors. The minority who are referred to specialists are more likely to be sent to paediatricians than to child psychiatrists.

Abdominal pain

Abdominal pain is the symptom that has been studied most thoroughly. It has been estimated to occur in between 4 and 17 per cent of all children, and is a common reason for referral to a paediatrician. In most cases abdominal pain is associated with headache, limb pains, and sickness (Apley and Hale 1973). Physical causes for the abdominal pain are seldom found and psychological causes are often suspected. Some of these unexplained abdominal pains are related to anxiety and, as discussed on p. 701, others have been ascribed to 'masked' depressive disorder. Some appear to be a direct symptomatic response to stressful events. Treatment is similar to that for other emotional disorders. Follow-up suggests that a quarter of cases severe enough to require investigation by a paediatrician develop chronic psychiatric problems.

Conversion and dissociative disorder

Conversion and dissociative disorder are more common in adolescence than in childhood, both in individual patients and in its epidemic form (see p. 191). In childhood, symptoms are usually mild and seldom last long. The most frequent symptoms include paralyses, abnormalities of gait, and inability to see or hear normally. As in adults, dissociative and conversion symptoms can occur in the course of organic illness as well as in a dissociative or conversion disorder. As with adults, organically determined physical symptoms are sometimes misdiagnosed as conversion disorder when the causation physical pathology is difficult to detect and stressful events coincide with the onset of the symptom (Rivinus *et al.* 1975). For this reason, the diagnosis of dissociative (conversion) disorder should be made only after the most careful search for organic disease.

Epidemiology

Dissociative disorders were encountered rarely in the Isle of Wight study of children in the community (Rutter *et al.* 1970*a*). Among children referred to paediatricians, these disorders have been reported in 3–13 per cent. In a survey of prepubertal children referred to a psychiatric hospital, Caplan (1970) found that conversion disorder was diagnosed in about 2 per cent. In almost half of this 2 per cent, organic illness was eventually detected either near the time or during the four- to eleven-year follow-up. Amblyopia was the symptom of organic disorder most likely to be diagnosed as psychogenic.

Treatment

Dissociative (conversion) and other somatization disorders should be treated as early as possible before secondary gains accumulate. Treatment is directed mainly at reducing any stressful circumstances and encouraging the child to talk about the problem. Symptoms may subside with these measures, or may need management comparable to that used for conversion disorder in adults (see p. 354). Physiotherapy and behavioural methods may be valuable for motor symptoms (Dubowitz and Hersov 1976). For a review of child psychiatric syndromes with somatic presentation see Goodyer and Taylor (1985) and Garralda (1992).

Depressive disorders

Healthy children are understandably unhappy in distressing circumstances, for example when a parent is seriously ill or a grandparent has died. Some of these children are tearful, lose interest and concentration, and may eat and sleep badly. This section is not concerned with these normal forms of unhappiness, but with depressive disorders.

Psychiatrists have disagreed about the boundaries of the syndromes of depressive disorder in childhood. Some believe that the condition resembles that in adult life with only minor modifications accounted for by age. Thus young children do not experience guilt in an adult form and may have difficulty in describing feelings as sadness or despair. Other psychiatrists maintained that in childhood depressive disorders present not only with symptoms seen in adults but also in a form with little or no depressed mood but with a variety of other symptoms including unexplained abdominal pains, headache, anorexia, and enuresis, as well as anxiety (Brady and Kendall 1992). It is not unreasonable to suggest that in childhood, as in adult life, depressive symptoms can come to light because of associated physical or behavioural symptoms. However, in children, as in adults, further examination of such cases will reveal evidence of depressed mood and the diagnosis of depressive disorder should be made only when there is clear evidence of the principal features of the syndrome seen in adults (see p. 197).

Depressive disorder should be distinguished clearly from depressive symptoms occurring as a component of an emotional or conduct disorder. It seems that bipolar disorder does not occur before puberty.

Epidemiology

Although depressive symptoms are common in middle and late childhood, depressive disorders are infrequent. Thus Rutter *et al.* (1970*a*) found depressive disorder in only three of the girls and none of the boys among 2000 10- to 11-year-olds, though depressive symptoms were common as part of other disorders. Among 14-year-olds, Rutter *et al.* (1976*a*) found a depressive disorder in 1.5 per cent. More recent estimates give higher figures: 1 per cent of children in middle childhood, and 2–5 per cent in mid-adolescence (Harrington 1994). Rates are about equal among males and females (Kashani and Simonds 1979), and this finding contrasts with the excess of females among adults with this disorder.

Aetiology

Two findings suggest that the causes of depressive disorder in childhood may be similar to those of depressive disorder in adult life. First, the rates of depressive disorder among first-degree relatives of children with depressive disorder (Harrington *et al.* 1993) are greater than the rates in the general population, suggesting *genetic* factors. Second, there are strong continuities between depressive disorders in childhood and adult life. Harrington *et al.* (1990) followed-up people who, when children, had been treated for a depressive disorder diagnosed by criteria similar to those in use today. Of the index group, 58 per cent had a depressive disorder in adult life compared with 31 per cent of a control group. However, members of the index group were no more likely than controls to have other kinds of psychiatric disorder in adult life.

Some findings point to differences in the aetiology of depressive disorders in children and adults. As noted above, bipolar disorders do not seem to occur in childhood (they appear first in adolescence), and the sex ratio in childhood is about 1 : 1 (see above) whilst in adults the ratio of women to men is about 2 : 1 (see p. 211).

Treatment

Any distressing circumstances should be reduced, if this is possible, while the child is helped to talk about feelings. Antidepressant drugs are usually reserved for older children with definite symptoms of a severe depressive disorder. Clinical trials of antidepressant drugs for children with depressive disorder suggest that they may be less effective than for adults. However, there are methodological problems in the trials which make conclusion uncertain (Ambrosini *et al.* 1993). It is important in management to deal with the whole range of problems of a depressed child, including difficulties at school or in the family, problems in relationships with peers, and low self-esteem.

Suicide *among children is considered on p. 419.*

For a review of depressive disorders in childhood see Kopelwicz and Klass (1993).

School refusal

School refusal is not a psychiatric disorder but a pattern of behaviour that can have many causes. It is convenient to consider it at this point in the chapter because of its association with anxiety and depressive disorder. School refusal is one of many causes of repeated absence from school. Physical illness is the most common. A small number of children miss school repeatedly because they are deliberately kept at home by parents to help with domestic work or for company. Some are truants who could go to school but choose not to, often as a form of rebellion. An important group stay away from school because they are anxious or miserable when there. These are the school-refusers. The important distinction between truancy and school refusal was first made by Broadwin (1932). Later, Hersov (1960) studied 50 school-refusers and 50 truants, all referred to a child psychiatric clinic. Compared with the truants, the school-refusers came from more neurotic families, were more depressed, passive, and overprotected, and had better records of schoolwork and behaviour.

Prevalence

Temporary absences from school are extremely common, but the prevalence of school refusal is uncertain. In the Isle of Wight school refusal was reported in rather less than 3 per cent of 10- and 11-year-olds with psychiatric disorder (Rutter *et al.* 1970a). It is most common at three periods of school life, between five and seven years, at 11 years with the change of school, and especially at 14 years and older.

Clinical picture At times, the first sign to the parents that something is wrong is the child's sudden and complete refusal to attend school. More often there is an increasing reluctance to set out, with signs of unhappiness and anxiety when it is time to go. These children complain of somatic symptoms of anxiety such as headache, abdominal pain, diarrhoea, sickness, or vague complaints of feeling ill. These complaints occur on school days but not at other times. Some children appear to want to go to school but become increasingly distressed as they approach it. The final refusal can arise in several ways. It may follow a period of gradually increasing difficulty of the kind just described. It may appear after an enforced absence for another reason, such as a respiratory tract infection. It may follow an event at school such as a change of class. It may occur when there is a problem in the family such as the illness of a grandparent to whom the child is attached. Whatever the sequence of events, the children are extremely resistant to efforts to return them to school and their evident distress makes it hard for the parents to insist that they go.

Aetiology

Several causes have been suggested. Johnson *et al.* (1941) emphasized the general role of separation anxiety, a mechanism also stressed by Eisenberg (1958). More recent observations suggest that separation anxiety is particularly important in younger children. In older children there may be a true school phobia, i.e. a specific fear of certain aspects of school life including travel to school, bullying by other children, or failure to do well in class. Other children have no specific fears but feel inadequate and depressed. Some older children have a depressive disorder.

Prognosis

Clinical experience suggests that most younger children eventually return to school. However, a proportion of the most severely affected adolescents do not return before the time when their compulsory school attendance ceases. There have been few studies of the longer prognosis of school refusal. Berg and Jackson (1985) followed-up 168 teenage school-refusers who had been treated as in-patients. After 10 years, about half still suffered from emotional or social difficulties or had received further psychiatric care. This study was concerned with severe cases and the general prognosis may be rather better.

Treatment

Except in the most severe cases, arrangements should be made for an early return to school. There should be discussion with the school-teachers, who should be given advice about any difficulties that are likely to be encountered. It is sometimes more satisfactory for someone other than the mother to accompany the child to school at first. In a few cases a more elaborate graded behavioural plan is necessary. In the most severe cases admission to hospital may be required to reduce anxiety before a return to school can be arranged. Occasionally a change of school is appropriate.

Any depressive disorder should be treated. It has been reported that antidepressants are effective for school refusal even when there is no depressive disorder, but this view is not generally accepted. In all cases the child should be encouraged to talk about his feelings and the parents given support. See Hersov and Berg (1980) and Berg (1984) for reviews of school refusal.

Other childhood psychiatric disorders

Functional enuresis

Functional enuresis is the repeated involuntary voiding of urine occurring after an age at which continence is usual (see below) in the absence of any identified physical disorder. Enuresis may be **nocturnal** (bed-wetting) or **diurnal** (daytime wetting) or both. Most children achieve daytime and night-time continence by three or four years of age. Nocturnal enuresis is often referred to as **primary** if there has been no preceding period of urinary continence. It is called **secondary** if there has been a preceding period of urinary continence.

Nocturnal enuresis can cause great unhappiness and distress, particularly if the parents blame or punish the child. This unhappiness may be made worse by limitations imposed by enuresis on activities such as staying with friends or going on holiday.

Epidemiology

In the United Kingdom, the prevalence of nocturnal enuresis is about 10 per cent at five years of age, 4 per cent at eight years, and 1 per cent at 14 years. Similar figures have been reported from the United States. Nocturnal enuresis occurs more frequently in boys. Daytime enuresis has a lower prevalence and is more common in girls than boys. About half of daytime wetters also wet their beds at night.

Aetiology

Nocturnal enuresis occasionally results from physical conditions but more often appears to be caused by delay in maturation of the nervous system, either alone or in combination with environmental stressors. There is some evidence for a *genetic* cause; about 70 per cent of children with enuresis have a first-degree relative who has been enuretic (Bakwin 1961). Also, concordance rates for enuresis are twice as high in monozygotic as in dizygotic twins (Hallgren 1960).

Although most enuretic children are free from psychiatric disorder, the proportion with psychiatric disorder is greater than that of other children. *Psychological factors* can contribute to aetiology, for example unduly rigid toilet training, negative or indifferent attitudes of parents, and stressful events leading to anxiety in the child.

Assessment

A careful history and appropriate physical examination is required to exclude undetected physical disorder, particularly urinary infection, diabetes, or epilepsy, and to assess possible precipitating factors and the child's motivation.

Psychiatric disorder should be sought. If none is found, an assessment should be made of any distress caused to the child. An evaluation is made of the attitudes of the parents and siblings to the bed-wetting. Finally, the parents should be asked how they have tried to help the child.

Treatment

Any physical disorder should be treated. If the enuresis is functional, an explanation should be given to the child and the parents that the condition is common and the child is not to blame. It should be explained to the parents that punishment and disapproval are inappropriate and unlikely to be effective. The parents should be encouraged not to focus attention on the problem but to reward success without drawing attention to failure. Many younger enuretic children improve spontaneously soon after an explanation of this kind, but those over six years of age are likely to need more active measures.

Treatment begins with advice about *restricting fluid* before bedtime, *lifting* the child during the night, and the use of *star charts* to reward success.

Children who do not improve with these simple measures may be treated with *enuresis alarm* methods. In the original pad and bell method two perforated metal plates, separated by a cotton sheet, were incorporated in a low-voltage circuit including a battery, a switch, and a bell or buzzer. The resistance of the cotton sheet prevented current from flowing in the circuit. When the bed was made, the plates were placed under the position in which the child's pelvis will rest. When the child began to pass urine, the circuit was complete and the bell or buzzer sounded. The child turned off the switch, and rose to complete the emptying of the bladder. The bed was remade and a dry sheet was put between the metal plates before the child returned. Although effective, this method was cumbersome and has now been largely replaced by a modern version in which a sensor to detect the voiding of urine is attached to the child's pyjama trousers and an alarm is carried on the wrist or in a pocket (Schmitt 1986). Either method requires about six to eight weeks of treatment and some families break off before this has been completed.

The enuresis alarm seldom succeeds with children under the age of six, or those who are uncooperative. For the rest the original pad and bell method was found to be effective within a month in about 70–80 per cent of cases (Shaffer *et al.* 1968), although about a third relapsed within a year (Forsythe and Butler 1989). Similar results have been reported with modern sensor methods (Butler *et al.* 1990) provided that families can be persuaded to persist long enough with the treatment. It has been suggested that children with associated psychiatric disorder do less well than the rest even if they complete treatment.

Enuresis can be treated with a *tricyclic antidepressant*, usually imipramine or amitriptyline, although this treatment is not generally the first choice. These drugs are given in a dose of 25 mg at night increasing to 50 mg if necessary. Their beneficial effect has been demonstrated in clinical trials. Most bed-wetters improve initially, and about a third recover completely. However, among those who improve or recover, most relapse when the drug is stopped. Because of this high relapse rate, the side-effects of tricyclics, and the danger of accidental overdose, the drugs have limited value in treating enuresis. They are most useful when it is important to control the enuresis for a short time—for example when the child goes on holiday.

The synthetic *antidiuretic hormone* desamino-D-arginine vasopressin (desmopressin) has a more prolonged action than natural vasopressin. It has been used in the treatment of nocturnal enuresis, when it is often given intranasally, though an oral preparation is also available. The results appear comparable with those of tricyclic antidepressants. In one clinical trial about half the enuretic children treated with intranasal hormone became dry (Miller and Klauber 1990). When treatment is stopped, most patients relapse. Side-effects include rhinitis, nasal pain, nausea, and abdominal pain. See Shaffer (1985) for a review of enuresis and its treatment.

Functional encopresis

Encopresis is the repeated voluntary or involuntary passing of faeces into inappropriate places after the age at which bowel control is usual in the absence of known organic cause. The diagnosis should not be made unless the chronological and mental ages are greater than four years. Encopresis may have been present continuously from birth (primary) or have started after a period of continence (secondary).

At the age of three years, 6 per cent of children are still incontinent of faeces at least once a week; at seven years the figure is 1.5 per cent. Among children over the age of three years, loss of bowel control is more often secondary to constipation, and true encopresis is less common. The condition is three to four times more frequent among boys than girls. See Hersov (1985) for a review.

Faeces may be passed into clothing or deposited in inappropriate places such as the floor of the living room. Children who soil their clothes may deny what has happened and try to hide the dirty clothing. Some children smear faeces on walls or elsewhere. Most of these children have associated psychological problems of various kinds.

Aetiology

Repeated faecal soiling may be secondary to chronic constipation which may be associated with several causes including mental retardation, conditions that cause pain on defecation (for example anal fissure), and Hirschsprung's disease. The causes of true encopresis are less understood. In some cases, parental attitudes to toilet training seem important. Some parents have unrealistic expectations about the age at which control can be achieved and are unduly punitive; others fail to adopt a consistent approach. Emotional disorder is common among children with encopresis, and may sometimes be a contributory cause. For example soiling sometimes begins after an upsetting event such as the illness of a parent or the birth of a sibling. In some cases soiling develops when the child has a poor relationship with one or both parents and appears to be rebelling.

Assessment and treatment

The first step is to exclude physical causes of chronic constipation (see above). For this purpose joint assessment by a paediatrician and psychiatrist is often helpful. The next step is to assess parental attitudes and emotional factors in the child.

Treatment begins with finding out what the child thinks and feels about the problem, and providing him with an explanation of the disorder and reassurance. The parents need similar explanation and reassurance. The most successful approach is usually a *behavioural programme* in which the child is encouraged to sit on the toilet for about 10 minutes after each meal, and is rewarded for doing this and for succeeding in passing a motion into the toilet. When there are associated emotional problems or conflicts with the parents, individual or family psychotherapy may be helpful. When out-patient treatment fails, the child may respond to similar behavioural management in hospital. If the child is admitted, the parents need to be closely involved in the treatment to reduce the risk of relapse when the child returns home.

Prognosis

Whatever the cause, it is unusual for encopresis to persist beyond the middle teenage years, although associated problems (especially aggressive behaviour) may continue. When treated, most cases improve within a year.

Elective mutism

In this condition, a child refuses to speak in certain circumstances, although he does so normally in others. Usually speech is normal in the home but lacking in school. There is no defect of speech or language, only a refusal to speak in certain situations. Often there is other negative behaviour such as refusing to sit down or to play when invited to do so. The condition usually begins between three and five years of age after normal speech has been acquired.

Although reluctance to speak is not uncommon among children starting school, clinically significant elective mutism is rare, probably occurring in about 1 per 1000 children. Assessment is difficult because the child often refuses to speak to the psychiatrist so that diagnosis depends to a large extent on the parents' account. In questioning them it is important to ask whether speech and comprehension are normal at home. Although psychotherapy, behaviour modification, and speech therapy have been tried, there is no evidence that any treatment is generally effective.

In some cases, elective mutism lasts for months or years. A five- to ten-year follow-up of a small group showed that only about half had improved (Kolvin and Fundudis 1981).

Stammering

Stammering (or stuttering) is a disturbance of the rhythm and fluency of speech. It may take the form of repetitions of syllables or words, or of blocks in the production of speech. Stammering is four times more frequent in boys than in girls. It is usually a brief problem in the early stages of language development. However, 1 per cent of children suffer from stammering after they have entered school.

The cause of stammering is not known, although many theories exist. It seems unlikely that all cases have the same causes; genetic factors, brain damage, and anxiety may all play a part in certain cases but do not seem to be general causes. Stammering is not usually associated with a psychiatric disorder even though it can cause embarrassment and distress. Most children improve whether treated or not. Many kinds of psychiatric treatment have been tried, including psychotherapy and behaviour therapy, but none has been shown to be effective. The usual treatment is speech therapy.

Tic disorders

Tic disorders including Gilles de la Tourette's syndrome are considered on p. 400.

Dementia

Dementing disorders are rare in childhood. They result from organic brain diseases such as lipidosis, leucodystrophy, or subcaudate sclerosing panencephalitis. Some of the causes are genetically determined and may affect other children in the family. The prognosis is variable. Many cases are fatal, others progress to profound mental retardation.

Schizophrenia

Schizophrenia is almost unknown before seven years of age, and seldom begins before late adolescence. When it occurs in childhood, the onset may be acute or insidious. The whole range of symptoms that characterize schizophrenia in adult life may occur (see Chapter 9), and in both DSMIV and ICD10 the criteria for diagnosis in children are the same as those used with adults; there is no separate category of childhood schizophrenia. Before symptoms of schizophrenia appear, many of these children are odd, timid, or sensitive, and their speech development is delayed. Early diagnosis is difficult particularly when these non-specific abnormalities precede the characteristic symptoms. Treatment is with antipsychotic drugs as in the management of schizophrenia in adults, though with appropriate reductions in dosage. The child's educational needs should be met and support given to the family. See Werry (1992) for a review of schizophrenia in childhood.

Gender identity disorders

Effeminacy in boys

Some boys prefer to dress in girls' clothes and to play with girls rather than boys. Some have an obvious effeminate manner and say that they want to be girls. The cause of this condition is unknown. There is no evidence of any endocrine basis for these behaviours. Various family influences have been suggested, including the encouragement of feminine behaviour by the parents, a lack of boys as companions in play, a girlish appearance, and a lack of an older male with whom the child can identify. However, many children experience these influences without being effeminate.

In treatment it is difficult to know how far intervention is appropriate. Associated emotional disturbance in the child may require help, and it may be useful to investigate and discuss any family behaviours which seem to be contributing to or maintaining the child's behaviour. The prognosis is uncertain. Adult males with transvestism and transsexualism frequently recall enjoying feminine play as children, but follow-up studies of effeminate behaviour in early childhood show that the condition is more likely to proceed to homosexuality or bisexuality in adult life than to transsexualism or transvestism (Zuger 1984; Green 1985).

Tomboyishness in girls

In girls the significance of marked tomboyishness for future sexual orientation is not known. It is usually possible to reassure the parents, and sometimes necessary to discuss their attitudes to the child and their responses to her behaviour.

Suicide and deliberate self-harm

Both deliberate self-harm and suicide are rare amongst children less than 12 years of age (though more common in adolescence). *These problems are discussed in the chapter on suicide and deliberate self-harm (pp. 419 and 433).*

Psychiatric aspects of physical illness

The associations between physical and psychiatric disorders in children resemble those in adults (see Chapters 11 and 12). There are three main groups of association which are met at least as frequently in paediatric as in child psychiatric practice. The first group comprises the psychological and social consequences of physical illness. The second consists of psychiatric disorders presenting with physical symptoms, for example abdominal pain. The third consists of physical complications of psychiatric disorders; for example eating disorders and encopresis.

Most medical disorders of children are discussed in the chapter on psychiatry and medicine (Chapter 12). In this section we consider only some special problems in childhood.

The consequences of physical illness

Psychiatric disorder provoked by physical illness

When physically ill, children are more likely than adults to develop *delirium*. A familiar example is delirium caused by febrile illness.

Some chronic physical illnesses have psychological consequences for the child. In the Isle of Wight study of children (Rutter *et al.* 1970*a*) the prevalence of psychiatric disorder was only slightly increased with physical illnesses that do not affect the brain (for example asthma or diabetes); however, the prevalence was considerably higher with organic brain disorder or epilepsy. Chronic illness may impair reading ability and general intellectual development (Rutter *et al.* 1970*a*; Eiser 1986), and sometimes also self-esteem and ability to form relationships.

Effect on parents

The effects on parents are particularly important when the child's physical illness is chronic. Parents are naturally distressed by learning that their child has a chronic disabling physical illness. The effects on the parents depend on many factors including the nature of the physical disorder, the temperament of the child, the parents' emotional resources, and the circumstances of the family. The parents may experience a sequence of emotional reactions like those of bereavement, and their marital and social lives may be affected.

Most parents eventually develop a warm loving relationship with a handicapped child and cope successfully with the difficulties. A few manage less well and may have unrealistic expectations, or they may be rejecting or overprotective. See reviews by Breslau *et al.* (1981), Romans-Clarkson *et al.* 1986).

Effect on siblings

The brothers and sisters of children with physical problems may develop emotional or behavioural disturbances. They may feel neglected, irritated by restrictions on their social activities, or resentful of having to spend so much time helping in the case of the handicapped child. Although some studies have shown more emotional and behavioural disturbances in siblings than would be expected by chance (Ferrari 1984; Breslau and Prabucki 1987), most siblings manage well and may even benefit through increased abilities to cope with stress and to show compassion for others.

Management

Everyone involved in the care of physically disabled children should be aware of the psychological difficulties commonly experienced by these children and their families. When giving distressing information to families, it is particularly important to take time. It may be necessary to see the family many times, providing continuing advice and support. There should be regular liaison between the paediatrician and the child psychiatrist. There is a need for good communication with teachers and any social workers or others involved with the welfare of the child. Short periods of relief care can enable a family to continue with the care of a handicapped child.

Advice on imparting to parents the diagnosis of life-threatening illness affecting their children is given by Wooley *et al.* (1989).

Children in hospital

The admission of a child to hospital has important psychological consequences for the child and family. In the past, most hospitals discouraged families from visiting children. Bowlby (1951) suggested that this separation could have adverse immediate and long-term psychological effects; he identified successive stages of protest, despair, and detachment in the child during admission to hospital. These ideas were influential. It is now general policy to encourage parents to visit and take part in the care of their child and, if the child is young, to sleep in the hospital if family circumstances allow this. It is also recognized as important to prepare children for admission by explaining in simple terms what will happen, and by introducing to them the members of staff who will care for them in hospital.

It has been shown that repeated admission to hospital in early or middle childhood is associated with behavioural and emotional disturbances in adolescence (Rutter 1981). It is possible that these long-term consequences are being reduced with the improvements in hospital care mentioned above (Shannon *et al.* 1984).

Adolescence

There are no specific disorders of adolescence. However, special experience and skill are required to apply the general principles of psychiatric diagnosis and treatment to patients at this time of transition between childhood and adult life. It is often particularly difficult to distinguish psychiatric disorder from the normal emotional reactions of the teenage years. For this reason, this section begins by discussing how far emotional disorder is an inevitable part of adolescence. For a general review of problems in adolescence and their treatment, the reader is referred to Steinberg (1982).

Psychological changes in adolescence

Considerable changes—physical, psychosexual, emotional, and social—take place in adolescence. In the 1950s and 1960s it was widely assumed that these changes were commonly accompanied by emotional upset of such a degree that it could be considered to be a psychiatric disorder. Indeed, Anna Freud (1958) regarded 'disharmony within the psychic structure' as a 'basic fact' of adolescence. Others described alienation, inner turmoil, adjustment reactions, and identity crises as common features of this time of life. Recently, a more cautious view has prevailed. Rutter *et al.* (1976*b*) have reviewed the evidence, including their own findings in 14-year-olds on the Isle of Wight. They concluded that rebellion and parental alienation are uncommon in mid-adolescence, although inner turmoil, as indicated by reports of misery, self-depreciation, and ideas of reference, is present in about half of all adolescents. However, this turmoil seldom lasts for long and usually goes unnoticed by adults. Rebellious behaviour is more

common among older adolescents and many become estranged from school during their last year of compulsory attendance. Other problems include excessive drinking of alcohol and the use of drugs and solvents (discussed in Chapter 14), problems in relationships and sexual difficulties, and irresponsible behaviour in driving cars and motorcycles (Spicer 1985; Bewley 1986).

Epidemiology of psychiatric disorder in adolescence

Although psychiatric disorders are only a little more common in adolescence than in the middle years of childhood, the pattern of disorder is markedly different, being closer to that of adults. In adolescence the sexes are affected equally, anxiety is less common than in earlier years, and depression and school refusal are more frequent. An epidemiological study of adolescents in the United States (Whitaker *et al.* 1990) has found that the most common disorders are dysthymic disorder, major depression, and generalized anxiety disorder, followed by bulimia and anorexia nervosa, obsessive–compulsive disorder, and panic disorder. The precise estimates are difficult to interpret in relation to DSMIV criteria because DSMIII was used, the population was confined to school attenders, and the information was solely from self-report.

Psychiatric disorders of adolescence

Clinical features

Anxiety disorders

School refusal is common between 14 years of age and the end of compulsory schooling, and at this age is often associated with other psychiatric disorders. Generalized anxiety states are less common in adolescence than in childhood. Social phobias begin to appear in early adolescence; agoraphobia appears in the later teenage years.

Conduct disorders

About half the cases of conduct disorder seen in adolescents have started in childhood. Those which begin in adolescence differ in being less strongly associated with reading retardation and family pathology. Among younger children, aggressive behaviour is generally more evident in the home or at school. Among adolescents, it is more likely to appear outside these settings as offences against property. Truancy also forms part of the conduct disorders occurring at this age.

Mood disorders

Depressive symptoms are more common in adolescence than in childhood. In the Isle of Wight study they were 10 times more frequent among 14-year-olds than among 10-year-olds. In depressive disorder of adolescence, depressive mood is often less immediately obvious than anger, alienation from parents, withdrawal from social contact with peers, and underachievement at school.

The classification and aetiology of mood disorder are the same for adolescents as for adults, as described in Chapter 8. In adolescents the clinical features of depressive disorders are similar to those in adults, but less frequently include sleep disturbances, delusions, and hallucinations. Contrary to early beliefs, bipolar affective disorder is now thought to occur in adolescence. From retrospective studies of the adolescent disorder of adults diagnosed as manic depressive, it appears that the first episodes of these illnesses may manifest in adolescence as abnormal behaviour, which may be misdiagnosed because the associated mood disorder is not detected.

The treatment of affective disorder is as described in Chapter 8, but with appropriate reductions in drug doses. Lithium is usually an effective prophylactic when the illness is recurrent. See Ryan and Puig-Antich (1986) for a review of affective disorder in adolescence.

Schizophrenia

Schizophrenia in adolescence is more common in boys than in girls. Usually the diagnosis presents little difficulty. When there is difficulty it is usually in detecting characteristic symptoms, especially in patients whose main features are gradual deterioration of personality, social withdrawal, and decline in social performance. The prognosis may be good for a single acute episode with florid symptoms, but is poor when the onset is insidious.

Eating disorders

Problems with eating and weight are common in adolescence. They are discussed in Chapter 12 (p. 372) since they resemble closely the same conditions in adult life. It is particularly important to involve the parents and perhaps other family members in the treatment of an adolescent patient with eating disorders. Formal family therapy has been shown to be of value for anorexia nervosa in adolescents. For a review see Hodes *et al.* (1991).

Suicide and deliberate self-harm

In recent years there has been a marked increase in suicide and deliberate self-harm among adolescents. These subjects are discussed in Chapter 13 (pp. 419 and 433).

Alcohol and substance abuse

Problems of substance abuse in adolescence are similar to those in adults, as described in Chapter 14. There is increasing evidence that the frequency of excessive drinking among adolescents is increasing. Most adolescent heavy drinkers seem to reduce their drinking as they grow older, but a few progress to more serious drinking problems in adult life.

Occasional drug taking is common in adolescence and is often a group activity. Cigarette smoking and the use of cannabis are especially frequent. Solvent abuse is largely confined to adolescence and is usually of short duration. Abuse of drugs such as amphetamines, barbiturates, opiates, and cocaine is less common but more serious, since most drug-dependent adults have experimented with these drugs during adolescence. There is a strong association between conduct disorder in childhood and drug-taking in adolescence (Robins 1966).

Most adolescents experiment with drugs for short periods and do not become regular users. Those who persist in taking drugs are more likely to come from discordant families or broken homes, to have failed at school, and to be members of a group of persistent drug users. Feelings of alienation and low self-esteem may also be important.

Since regular drug-taking starts less often in adult life than in adolescence, limiting drug-taking among adolescents is an important preventive measure, but there is no evidence that it can be achieved. Drug dependency clinics specifically for adolescents have been provided in the United States for example, but their effectiveness has not been demonstrated convincingly.

See Swadi (1992*a*) for a review of alcoholism and substance abuse in adolescents.

Sexual problems

Concern about sexuality is normal in adolescence. Excessive worry about masturbation and sexual identity and orientation may lead to medical consultation. Sexual abuse is increasingly a cause of referral to psychiatrists (see p. 720).

It is probable that at least two-thirds of teenage pregnancies are terminated and some of the remainder are unwanted. There is a raised incidence of prenatal complications as compared with older mothers. Very young mothers frequently have substantial difficulties as parents and there is a poor outlook for many teenage marriages. The psychological and social problems of teenage pregnancy show that there is a need for access to continuing medical and social services during and after pregnancy (Black 1986).

Assessment

There are special skills in interviewing adolescents. In general, young adolescents require an approach similar to that used for children, while with older adolescents it is more appropriate to employ that used with adults. It must always be remembered that a large proportion of adolescents attending a psychiatrist do so somewhat unwillingly and also that most have difficulty in expressing their feelings in adult terms. Therefore the psychiatrist must be willing to spend considerable time establishing a relationship with an adolescent patient. To do this, he must show interest in the adolescent, respecting his point of view and talking in terms that he can understand. As in adult psychiatry, it is important to collect systematic information and describe symptoms in detail, but with adolescents the psychiatrist must be prepared to adopt a more flexible approach to the interview.

It is usually better to see the adolescent before interviewing the parents. In this way, the psychiatrist makes it clear that he regards the adolescent as an independent person. Later, other members of the family may be interviewed and the family seen as a whole. As well as the usual psychiatric history, particular attention should be paid to information about the adolescent's functioning at home, in school, or at work, and about his relationship with peers. A physical examination should be carried out unless the general practitioner has performed one recently and reported the results.

Such an assessment should allow allocation of the problem to one of three classes. In the first, no psychiatric diagnosis can be made and reassurance is all that is required. In the second, there is no psychiatric diagnosis but anxious parents or a disturbed family need additional help. In the third, there is a psychiatric disorder requiring treatment.

Treatment

Treatment methods are intermediate between those employed in child and adult psychiatry. As in the former, it is important to work with relatives and teachers. It is necessary to help, reassure, and support the parents and sometimes extend this to other members of the family. This is especially important when the referral reflects the anxiety of the family about minor behavioural problems rather than the presence of a definite psychiatric disorder. However, it is also important to treat the adolescent as an individual who is gradually becoming independent of the family. In these circumstances family therapy as practised in child psychiatry is usually inappropriate and may at times be harmful.

Services for adolescents

The proportion of adolescents in the population who are seen in psychiatric clinics is less than the proportion of other age groups. Of those referred, some of the less mature adolescents can be helped more in a child psychiatry clinic. Some of the older and more mature adolescents are better treated in a clinic for adults. Nevertheless for the majority the care can be provided most appropriately by a specialized adolescent service provided that close links are maintained with child and adult psychiatry services and with paediatricians. There are variations in the organization of these units and the treatment that they provide, but most combine individual and family psychological treatment with the possibility of drug treatment for severe disorders. Most units accept out-patient referrals not only from doctors but also from senior teachers, social workers, and the courts. When the referral is non-medical the general practitioner should be informed and the case discussed with him. All adolescent units work with schools and social services. In-patient facilities are usually limited in extent, so that it is important to agree with social services what kinds of problems need admission to a health service unit and which should be cared for in residential facilities provided (in the United Kingdom) by social services. Reasons for admission to a health service in-patient unit include the following:

(i) severe or very unusual mental symptoms requiring that the person's mental state be observed carefully, investigations carried out, or treatment monitored closely;

(ii) behaviour that is dangerous to the self or others *and* that is due to psychiatric disorder.

When dangerous behaviour relates to personality and circumstances and not to illness, a hospital unit is not more effective than secure residential accommodation, and the behaviour of such ado-

lescents may be stressful for others with mental disorders.

Child abuse

In recent years, the concept of child abuse has been widened to include the overlapping categories of physical abuse (non-accidental injury), emotional abuse, sexual abuse, and neglect. Most of the literature on child abuse refers to developed countries, rather than to developing countries in which children commonly face poor nutrition, other hardships such as severe physical punishment, abandonment, and employment as beggars and prostitutes.

The term *fetal abuse* is sometimes applied to various behaviours detrimental to the fetus including physical assault and the taking by the mother of substances likely to cause fetal damage. *Munchausen's syndrome by proxy* is the name given to apparent illness in children which has been fabricated by the parents and to conditions induced by parents, for example by partly smothering the child. It is discussed on p. 357.

For reviews of child abuse the reader is referred to Mrazek and Mrazek (1985) and Cicchetti and Carlson (1989).

Physical abuse (non-accidental injury)

Estimates of the prevalence of physical abuse vary with the criteria used. A survey of children under four years of age in an English county suggested an annual rate of 1 per 1000 children for injuries of such severity that there was evidence of bone fracture or bleeding around the brain (Baldwin and Oliver 1975). It has been reported that in 1989 in England, 3.5 per 1000 children below the age of 18 were on Child Protection Registers and about 1 in 4 of these had suffered physical abuse (Browne and Saqui 1987). Less severe injury is probably much more frequent, but often does not come to professional attention.

Clinical features

Parents may bring an abused child to the doctor with an injury said to have been caused accidentally. Alternatively, relatives, neighbours, or other people may become concerned and report the problem to police, social workers, or voluntary agencies. The most common forms of injury are multiple bruising, burns, abrasions, bites, torn upper lip, bone fractures, subdural haemorrhage, and retinal haemorrhage. Some infants are smothered, usually with a pillow, and the parents report an apnoeic attack. Suspicion of physical abuse should be aroused by the pattern of the injuries, a previous history of suspicious injury, unconvincing explanations, delay in seeking help, and incongruous parental reactions. The psychological characteristics of abused children vary but include fearful responses to the parents, other evidence of anxiety or unhappiness, and social withdrawal. Such children often have low self-esteem, may avoid adults and children who make friendly approaches, and may be aggressive.

Aetiology

Child abuse is more frequent in neighbourhoods in which family violence is common, schools, housing, and employment are unsatisfactory, and there is little feeling of community.

In *parents* the factors associated with child abuse include youth, abnormal personality, psychiatric disorder, lower social class, social isolation, disharmony and breakdown in marriage, and a criminal record. When a parent has a psychiatric disorder, it is most often a personality disorder; only a few parents have disorders such as schizophrenia or affective disorder. Many parents give a history of having themselves suffered abuse or deprivation in childhood. In abusing families relationships between the parents are harsher and colder than in matched controls (Jones and Alexander 1978). Although child abuse is much more common in families with other forms of social pathology, it is certainly not limited to such families.

In the children, risk factors include premature birth, early separation, need for special care in the neonatal period, congenital malformations, chronic illness, and a difficult temperament.

Management

Doctors and others involved in the care of children should always be alert to the possibility of child abuse. They need to be particularly aware of the risks to children who have some of the characteristics described above, or are cared for by parents with the predisposing factors listed.

Doctors who suspect abuse should refer the child to hospital and inform a paediatrician or casualty officer of their suspicions. In the hospital emergency department, in-patient admission should be arranged for all children in whom non-accidental injury is suspected. If possible, the doctor's concerns should be discussed with the parents, and in any case they should be told that admission is necessary to allow further investigations. If the parents refuse admission, it may be necessary in England and Wales to apply to a magistrate for a Place of Safety Order; similar action may be appropriate in other countries. During admission assessment must be thorough and include photographs of injuries and skeletal radiography. Radiological examination may show evidence of previous injury or, occasionally, of bone abnormalities such as osteogenesis imperfecta. A CT scan may be needed if subdural haemorrhage is suspected. All findings must be fully documented.

Once it has been decided that non-accidental injury is probable, senior doctors should talk to the parents. Other children in the family should be seen and examined. In assessing the parents the following points should be considered (Skuse and Bentovim (1994)).

(1) Do they acknowledge their part in the abuse?
(2) Do they accept the need to change their behaviour?
(3) Do they show a willingness to try new approaches to the child?
(4) Will they accept help with their personal or relationship problems?

The subsequent procedure will vary according to the administrative arrangements in different countries. In the United Kingdom, the social services staff should be notified so that they can organize a case conference for the exchange of information and opinions between various representatives of hospital and community. It may be decided to put the child's name on a child abuse register, thereby making the Social Services Department responsible for visiting the home and checking the problem regularly.

In some cases, the risk of returning the child to the parents is too great and separation is required. If the parents do not agree to separation, a care order can be sought by the Social Services Department. When abuse is severe, prolonged, or permanent, separation may be necessary and parents may face criminal charges. Because there are known cases of injury or death in children returned to their parents, it is vitally important that most careful assessment be made before physically abused children are returned. Countries vary in the requirements and procedures for reporting and monitoring possible physical abuse in children, and readers should inform themselves of the arrangements in the area in which they work.

Prognosis

Children who have been subjected to physical abuse are at high risk of further problems. For example the risk of further severe injury is probably between 10 and 30 per cent, and sometimes the injuries are fatal. Abused children are likely to have subsequent high rates of physical disorder, delayed development, and learning difficulties. There are also increased rates of behavioural and emotional problems in later childhood and adult life even when there has been earlier therapeutic intervention (Lynch and Roberts 1982). As adults, many former victims of abuse have difficulties in rearing their own children. The outcome is better for abused children who can establish a good relationship with an adult and can improve their self-esteem; and for those without brain damage (Lynch and Roberts 1982; Rutter 1985*b)*.

Emotional abuse

The term emotional abuse usually refers to persistent neglect or rejection sufficient to impair a child's development. However, the term is sometimes applied to gross degrees of overprotection, verbal abuse, or scapegoating which impair development. Emotional abuse often accompanies other forms of child abuse.

Emotional abuse has various effects on the child, including failure to thrive physically, impaired psychological development, and emotional and conduct disorders (Rutter 1985*b*; Garbarino *et al.* 1986). Diagnosis depends on observations of the parents' behaviour towards the child, which may include frequent belittling or sarcastic remarks about him during the interview. One or both parents may have a disorder of personality, or occasionally a psychiatric disorder. The parents should be interviewed separately and together to discover any reasons for the abuse of this particular child; for example he may fail to live up to their expectations, or may remind them of another person who has been abusive to one of them. The parents' mental state should be assessed.

Treatment

In treatment, the parents should be offered help with their own emotional problems and with the day-to-day interactions with the child. It is often difficult to persuade parents to accept such help. If they reject help and if the effects of emotional abuse are serious, it may be necessary to involve the social services and to consider the steps described above for the care of children suffering physical abuse. The child may need individual help.

Child neglect

Child neglect may take several forms including emotional deprivation, neglect of education, physical neglect, lack of appropriate concern for physical safety, and denial of necessary medical or surgical treatment. These forms of neglect may lead to physical or psychological harm.

Child neglect is more common than physical abuse, and it may be detected by various people including relatives, neighbours, teachers, doctors, or social workers. Child neglect is associated with adverse social circumstances, and is a common reason for a child to need foster care (Fanshel 1981).

Non-organic failure to thrive and deprivation dwarfism

Paediatricians recognize that some children fail to thrive for no apparent organic cause. In children under three, this condition is called non-organic failure to thrive (NOFT); in older children it is called psychosocial short stature syndrome (PSSS) or deprivation dwarfism.

Clinical picture

Non-organic failure to thrive is caused by the deprivation of food and close affection. There is usually evidence of problems in the parent–child relationship since the child's early infancy; these include rejection and, in extreme cases, expressed hostility towards the child. There may be physical or sexual abuse as well. The infant may present either with recent weight loss or a weight persistently below the third percentile for chronological age. Height (or length) may be reduced. Head circumference may eventually be affected, and there may be cognitive and developmental delay. The infant may be irritable and unhappy, or in more severe cases lethargic and resigned. There is a clinical spectrum ranging from infants with mild feeding problems to those with all the severe features described above (Skuse 1985). If treated with food and care, the infants usually grow and develop quickly (Kempe and Goldbloom 1987).

Psychosocial short stature or 'deprivation dwarfism' was first reported by Powell *et al.* (1967). They described 13 children with abnormally short stature, unusual eating patterns, retarded speech development, and temper tantrums. Since this original account, the syndrome has been widely recognized. Although short in stature, the children may be of normal weight when seen by the doctor or even slightly overweight for their height. Growth hormone secretion is abnormal, with diminished 24-hour circulating levels due to diminished pulse amplitude (Stan-

hope *et al.* 1988). In severe cases the head circumference is reduced. Emotional and behavioural disorders occur, and may include food searching, scavenging, hoarding, disturbed sleep, and sometimes urination or defecation in inappropriate places (McCarthy 1981). There is often cognitive and developmental delay with impairment of language skills. They have low self-esteem and are commonly depressed. There is usually a history of deprivation or of psychological maltreatment. Away from the deprived environment these children eat ravenously, sometimes until they vomit through overindulgence.

Treatment

In treating either syndrome the first essential is to ensure the child's safety, which often means admission to hospital. Subsequently some children can be managed at home, but some need foster care. Some parents can be helped to understand their child's needs and to plan for them; other parents are too hostile to be helped. If help is feasible, it should be intensive and should probably focus on changing patterns of parenting (Kempe and Goldbloom 1987). It is unusual for the parents to be psychiatrically ill, but some have severe post-partum depression or other psychiatric disorder.

Prognosis

With both syndromes, the prognosis for severe cases is poor for psychological development and physical growth. The mortality rate is significant (Oates *et al.* 1985). Some of the less severe cases improve when removed from the abusing environment. Some children have to be placed permanently in foster care because family patterns are resistant to change. The abnormal behaviour is usually lost quickly and mental development follows physical growth (Skuse 1989).

Munchausen syndrome by proxy

This condition, in which a parent brings a child for treatment of fabricated symptoms, is discussed on p. 357.

Sexual abuse

The term sexual abuse refers to the involvement of children in sexual activities which they do not fully comprehend and to which they cannot give informed consent, and which violate generally accepted cultural rules. The term covers various forms of sexual contact with or without varying degrees of violence. The term also covers some activities not involving physical contact, such as exhibitionism and posing for pornographic photographs or films. The abuser is commonly known to the child and is often a member of the family (incest). A minority of children are abused by groups of paedophiles (sex rings).

Prevalence

The prevalence of sexual abuse has been estimated from criminal statistics or from surveys, but differences in definition and thoroughness of reporting making it difficult to interpret published figures. It is agreed that children are more often female—probably about 2–3 : 1 (Finkelhor 1986). The offender is usually male. Much sexual abuse takes place within the family, and stepfathers are overrepresented among abusers (Russell 1984). The extent of sexual abuse by women is not known. It has been suggested that women may carry out about 10 per cent of the sexual abuse of children (Glaser 1991). Faller (1987) reported that four-fifths of the women involved were the mothers of at least one of their victims, and McCarty (1986) found that many of the women abusers reported having been sexually abused themselves as children.

Retrospective studies suggest that between 20 and 50 per cent of women in populations surveyed recall some experience of abuse in childhood (Peters *et al.* 1986). These figures include a wide range of experiences, ranging from minor touching to repeated intercourse. Effects on adults of abuse in childhood are described on p. 154.

Clinical features

The presentation of child sexual abuse depends on the type of sexual act and the relationship of the offender to the child. Children are more likely to report abuse when the offender is a stranger. Sexual abuse may be reported directly by the child or a relative, or it may present indirectly with unexplained problems in the child, such as physical symptoms in the urogenital or anal area, pregnancy, behavioural or emotional disturbance, or precocious or otherwise inappropriate sexual behaviour. In adolescent girls, running away from home or unexplained suicide attempts should raise the suspicion of sexual abuse. When abuse occurs within the family, marital and other family problems are common (Furniss *et al.* 1984).

Effects of sexual abuse

Early emotional consequences of sexual abuse include anxiety, fear, depression, anger, and inappropriate sexual behaviour, as well as reactions to any unwanted pregnancy. As noted above, there may be inappropriate sexual behaviour and aggressive acts. A sense of guilt and responsibility is common. Some children show signs of post-traumatic stress disorder (see p. 703).

It is not certain how common these reactions are, or how they relate to the nature and circumstances of the abuse. Long-term effects are said to include depressed mood, low self-esteem, self-harm, difficulties in relationships, and sexual maladjustment in the form of either hypersensitivity or sexual inhibition. Effects of abuse are generally greater when the abuse has involved physical violence and penetrative intercourse. Some of the long-term effects are probably related to the events surrounding the disclosure of the abuse, including any legal proceedings.

In assessing these long-term effects, it should be remembered that sexual abuse often occurs in families with severe and chronic problems which are likely to have their own adverse long-term consequences. See Conte (1985), Alter-Reid *et al.* (1986), and Browne and Finkelhor (1986) for reviews of sexual abuse.

Aetiology

There is little reliable information about sexual abuse of children. It occurs in all socio-economic groups, but is more frequent among socially deprived families. Finkelhor (1984) suggests that there are several preconditions which make sexual abuse more likely: in the abuser, deviant sexual motivation, impulsivity, a lack of conscience, and a lack of external restraints (for example cultural tolerance); in the child, a lack of resistance (through insecurity, ignorance, or other causes of vulnerability).

Assessment

It is important to be ready to detect sexual abuse and to give serious attention to any complaint by a child of being abused in this way. When abuse has been established, it is important to assess whether it is likely to continue if the child remains at home and, if so, how dangerous it is likely to be. It is also important not to make the diagnosis without adequate evidence, which requires social investigation of the family as well as psychological and physical examination of the child. The child should be interviewed sympathetically and encouraged to describe what has happened; drawings or toys may help younger children to give a description, but great care must be taken to ensure that they are not used in a way that suggests to the child events that have not taken place. Young children can recall events accurately, but they are more suggestible than adults (see Jones (1992) for advice on interviewing). At an appropriate time it is often necessary to arrange a physical examination, including inspection of the genitalia and anal region and, if intercourse may have taken place within 72 hours, the collection of specimens from the genital and other regions (Kingman and Jones 1987). Usually this physical examination should be carried out by a paediatrician or police surgeon with special experience in the problem (see Royal College of Physicians (1991) for advice about physical examination).

The final decision as to whether abuse has taken place should be made after collecting information from the child, the physical examination, and social enquiries about the family.

Treatment

The initial management and the measures to protect the child are similar to those for physical abuse (see p. 717), including a decision about separating the child from the family. There are particular difficulties involved in intervening with families in which sexual abuse has occurred. These include a marked tendency to deny the seriousness of the abuse and of other family problems, and in some cases deviant sexual attitudes and behaviour of other family members, possibly including other children (Furniss *et al.* 1984; Mrazek and Mrazek 1985). Individual and group treatment has been used for the offenders with the general aim of enabling the person to reduce denial and consider the effects of the abuse on the child. If the mother has a history of abuse, this needs to be discussed to help her understand how this may have affected her response to her child's abuse.

Sexually abused children may often have highly abnormal sexual development for which they require help. They also need counselling to help them to deal with the emotional impact of the abuse, to come to terms with it, and to improve low self-esteem. Help should be in the form of a staged programme of rehabilitation for the whole family rather than a brief intervention. The treatment of child sexual abuse has been reviewed by Glaser (1991).

Appendix: History taking and examination in child psychiatry

The format and extent of an assessment will depend on the nature of the presenting problem. The following scheme is taken from the book by Graham (1991), which should be consulted for further information. Graham suggests that clinicians with little time available should concentrate on the items in bold type.

1. **Nature and severity of presenting problem(s). Frequency. Situations in which it occurs. Provoking and ameliorating factors. Stresses thought by parents to be important.**

2. Presence of other current problems or complaints.
   (a) Physical. Headaches, stomach ache. Hearing, vision. Seizures, faints, or other types of attacks.
   (b) Eating, sleeping, or elimination problems.
   (c) **Relationship with parents and siblings. Affection, compliance.**
   (d) Relationships with other children. Special friends.
   (e) Level of activity, attention span, concentration.
   (f) Mood, energy level, sadness, misery, depression, suicidal feelings. General anxiety level, specific fears.
   (g) Response to frustration. Temper tantrums.
   (h) Antisocial behaviour. Aggression, stealing, truancy.
   (i) **Educational attainments, attitude to school attendance.**
   (j) Sexual interest and behaviour.
   (k) Any other symptoms, tics, etc.

3. Current level of development.
   (a) Language: comprehension, complexity of speech.
   (b) Spatial ability.
   (c) Motor coordination, clumsiness.

4. Family structure.
   (a) Parents. Ages, occupations. **Current physical and emotional state.** History of physical or psychiatric disorder. Whereabouts of grandparents.
   (b) Siblings. Ages, presence of problems.
   (c) Home circumstances: sleeping arrangements.

5. Family function.
   (a) **Quality of parental relationship. Mutual affection. Capacity to communicate about and resolve problems. Sharing of attitudes over child's problems.**
   (b) **Quality of parent–child relationship. Positive interaction: mutual enjoyment. Parental level of criticism, hostility, rejection.**
   (c) Sibling relationships.
   (d) Overall pattern of family relationships. Alliance, communication. Exclusion, scapegoating. Intergenerational confusion.

6. Personal history.
   (a) Pregnancy—complications. Medication. Infectious fevers.
   (b) Delivery and state at birth. Birth-weight and gestation. Need for special care after birth.
   (c) Early mother–child relationship. Post-partum maternal depression. Early feeding patterns.
   (d) Early temperamental characteristics. Easy or difficult, irregular, restless baby and toddler.
   (e) Milestones. Obtain exact details only if outside range of normal.
   (f) **Past illnesses and injuries. Hospitalizations**
   (g) Separations lasting a week or more. Nature of substitute care.
   (h) Schooling history. Ease of attendance. Educational progress.

7. Observation of a child's behaviour and emotional state.
   (a) **Appearance. Signs of dysmorphism. Nutritional state. Evidence of neglect, bruising, etc.**
   (b) **Activity level. Involuntary movements. Capacity to concentrate.**
   (c) **Mood. Expression of signs of sadness, misery, anxiety, tension.**
   (d) **Rapport, capacity to relate to clinician. Eye contact. Spontaneous talk. Inhibition and disinhibition.**
   (e) **Relationship with parents. Affection shown. Resentment. Ease of separation**
   (f) Habits and mannerisms.
   (g) Presence of delusions, hallucinations, thought disorder.
   (h) Level of awareness. Evidence of minor epilepsy.

8. Observation of family relationships.

(a) Patterns of interaction—alliances, scape-goating.

(b) Clarity of boundaries between generations: enmeshment.

(c) Ease of communication between family members.

(d) Emotional atmosphere of family. Mutual warmth. Tension, criticism.

9. Physical examination of child.

10. Screening neurological examination.

(a) Note any facial asymmetry

(b) Eye movements. Ask the child to follow a moving finger and observe eye movement for jerkiness, incoordination.

(c) Finger–thumb apposition. Ask the child to press the tip of each finger against the thumb in rapid succession. Observe clumsiness, weakness.

(d) Copying patterns. Drawing a man.

(e) Observe grip and dexterity in drawing.

(g) Jumping up and down on the spot.

(h) Hopping.

(i) Hearing. Capacity of child to repeat numbers whispered two metres behind him.

Further reading

Graham, P. (1991). *Child psychiatry: a developmental approach* (2nd edn). Oxford University Press.

Quay, A. C. and Werry, J. S. (1986). *Psychopathological disorders of childhood* (3rd edn). Wiley, New York.

Rutter, M., Taylor, E., and Hersov, L. (1994). *Child psychiatry: modern approaches* (3rd edn). Blackwell, Oxford.

Shaffer, D., Ehrhardt, A. A., and Greenhill, L. L. (1985). *The clinical guide to child psychiatry*. Free Press, New York.

Steinberg, D. (1982). *The clinical psychiatry of adolescence*. Wiley, Chichester.

Until recent years, many mentally retarded adults lived in large isolated hospitals under the care of psychiatrists and mental handicap nurses. Nowadays hospital admission is for specialized assessment and treatment, and almost everyone lives in the community supported by services organized by social workers. Medical care is provided by family doctors and by specialists according to need. Nevertheless, the psychiatrist still has an important role in the prevention, assessment, and treatment of psychiatric disorders in mentally retarded children and adults and in research. This chapter is concerned with a general outline of the features, epidemiology, and aetiology of retardation, the organization of services and, more specifically, with psychiatric disorder in both children and adults. Many of the psychiatric problems of mentally retarded children are similar to those of children of normal intelligence; an account of these problems is given in Chapter 20 on child psychiatry.

Terminology

Over the years several terms have been applied to people with intellectual impairment from early life. In the nineteenth and early twentieth centuries, the word idiot was used for people with severe intellectual impairment, and imbecile for those with moderate impairment. The special study and care of such people was known as the field of *mental deficiency*. When these words came to carry stigma, they were replaced by the terms *mental subnormality* and *mental retardation*. The term *mental handicap* has also been widely used, but recently the terms *developmental disability* or *learning disability* have often been preferred. In this chapter, the term mental retardation is used because it is used in ICD10 and DSMIV.

The concept of mental retardation

A fundamental distinction has to be made between intellectual impairment starting in early childhood (mental retardation) and intellectual impairment developing later in life (dementia). In 1845 Esquirol made this distinction when he wrote:

Idiocy is not a disease, but a condition in which the intellectual faculties are never manifested; or have never been developed sufficiently to enable the idiot to acquire such an amount of knowledge as persons of his own age and placed in similar circumstances with himself are capable of receiving. (Esquirol 1845, pp. 446–7).

At the end of the nineteenth century a significant advance was made when methods of measuring intellectual capacity became available. Early in the twentieth century Binet's introduction of tests of intelligence provided quantitative criteria for ascertaining mental retardation. These tests also made it possible to identify mild intellectual retardation that might not be obvious otherwise (Binet and Simon 1905). Unfortunately, those responsible for the mentally retarded began to assume that people with such mild degrees of intellectual impairment were socially incompetent and required institutional care (Corbett 1978).

Similar views were reflected in the legislation of the time. For example in England and Wales in 1886 the Idiots Act had made a simple distinction between idiocy (more severe) and imbecility (less severe). In 1913 the Mental Deficiency Act added a third category for people who 'from an early age display some permanent mental defect coupled with strong vicious or criminal propensities in which punishment has had little or not effect'. As a result of this legislation, people of normal or near-normal intelligence were admitted to hospital for long periods simply because their behaviour offended against the values of society. Although some of these people had committed crimes, others were girls whose repeated illegitimate pregnancies were interpreted as a sign of the 'criminal' propensities mentioned in the Act.

In the past, the use of social criteria clearly led to abuse. Nevertheless, it is unsatisfactory to define mental retardation in terms of intelligence alone. Social criteria must be included, since a clinical distinction must be made between people who can lead a normal or near-normal life and those who cannot. In practice, the most useful modern definition is probably the one used by the American Association for Mental Deficiency (AAMD), which defines mental retardation as 'sub-average general intellectual functioning which originated during the development period and is associated with impairment in adaptive behaviour' (Heber 1981).

DSMIV defines mental retardation as a 'significantly sub-average general intellectual functioning, that is accompanied by significant limitations in adaptive functioning in at least two of the following skill areas: communication, self-care, home living, social/interpersonal skills used for community resources, self direction, functional academic skills, work, leisure, health and safety' and having an onset before the age of 18.

Both ICD10 and DSMIV have the following four subtypes: mild (IQ 50–70); moderate (IQ 35–49); severe (IQ 20–34); profound (IQ below 20).

Educationalists use other terms and these differ between countries. In the United Kingdom, the term is special needs, whereas in the United States, three groups are recognized: educable mentally retarded (EMR), trainable mentally retarded (TMR), and severely mentally retarded (SMR) (for a review see Howlin (1994)).

Epidemiology

Mental retardation

In 1929, in an important survey of schoolchildren in six areas of the United Kingdom, E. O. Lewis found that the total prevalence of mental retardation was 27 per 1000, and the prevalence of moderate and severe retardation (IQ less than 50)

was 3.7 per 1000. Subsequent studies in many countries have generally shown that, in the population aged 15–19, the prevalence of moderate and severe retardation is about 3.0–4.0 per 1000. The prevalence of moderate and severe retardation has changed little since the 1930s. The incidence of severe retardation has fallen substantially, but the prevalence has not changed because patients are living longer, particularly those with Down's syndrome. The incidence has fallen because antenatal and neonatal care have improved, and the numbers of children born to mothers aged over 35 are smaller. The age distribution of severely retarded people in the population has been changing, so that the numbers of adults (particularly the middle-aged) have increased (see Scott (1994) for a review of epidemiology).

Tizard (1964) drew attention to the distinction between 'administrative' prevalence and 'true' prevalence. He defined administrative prevalence as 'the numbers for whom services would be required in a community which made provision for all who needed them'. It has become standard practice for district services to carry out local censuses in terms of administrative prevalence. The use of standardized scales of adaptive behaviours (see p. 740) has made it easier to make these assessments.

If the true prevalence of all levels of retardation (IQ less than 70) is taken to be 20–30 per 1000 of the population of all ages (Broman *et al.* 1987), then the administrative prevalence is about 10 per 1000 of all ages. In other words, less than half of all retarded people require special provision. Administrative prevalence is higher in lower socio-economic groups and in childhood. It falls after the age of 16 (Richardson 1992) because there is continuing slow intellectual development and gradual social adjustment, as well as lower expectations of some of the more severely retarded.

Psychiatric disorder

Because the definition of mental retardation is imprecise and because it is difficult to identify psychiatric disorder among mentally retarded people, estimates of the prevalence of *dual diagnoses* (i.e. mental retardation plus another psychiatric disorder) are likely to be inaccurate. However, it is undoubtedly greater than in the general population (Borthwick-Duffy 1994). Rates of psychiatric disorder are high among mentally retarded people in hospital, presumably because psychiatric disorder is a common reason for admission and because long-term institutional care may lead to behavioural disturbance. The few general population studies have also shown high rates of psychiatric disorder and disturbed behaviour among the mentally retarded.

In a survey of all children aged 9–11 years with an IQ under 70, Rutter *et al.* (1970*a*) found that almost a third were rated as 'disturbed' by their parents, whilst about 40 per cent were so rated by their teachers. These rates were three to four times higher than the rates among intellectually normal children. In those with severe retardation, most surveys using standardized instruments have found that about half have psychiatric disorder, with the highest rates among those most profoundly retarded. See Scott (1994) for a review.

Among people with mild mental retardation, the range of psychiatric syndromes is similar to that in people of normal intelligence. Among people with moderate and severe retardation, certain kinds of disorder are especially frequent, including hyperactivity, pervasive developmental disorders, stereotypes, and self-injury.

Clinical features of mental retardation

General description

The most frequent manifestation of mental retardation is uniformly low performance on all kinds of intellectual task including learning, short-term memory, the use of concepts, and problem-solving. Specific abnormalities may lead to particular difficulties. For example lack of visuospatial skills may cause practical difficulties, such as inability to dress, or there may be disproportionate difficulties with language or social interaction, both of which are strongly associated with

behaviour disorder. Among retarded children, the common behaviour problems of childhood tend to occur when they are older and more physically developed than the normal child and they last longer. Such behaviour problems usually improve slowly as the child grows older.

Mild mental retardation (IQ 50–70)

People with mild retardation account for about 85 per cent of the mentally retarded. Usually their appearance is unremarkable and any sensory or motor deficits are slight. Most people in this group develop more or less normal language abilities and social behaviour during the preschool years, and their mental retardation may never be formally identified. In adult life most people with mild retardation can live independently in ordinary surroundings, though they may need help in coping with family responsibilities, housing, and employment, or when under unusual stress.

Moderate retardation (IQ 35–49)

People in this group account for about 10 per cent of the mentally retarded. Most can talk, or at least learn to communicate, and most can learn to care for themselves albeit with some supervision. As adults, they can usually undertake simple routine work and find their way about, but very few can lead an independent life.

Severe retardation (IQ 20–34)

People with severe retardation account for about 3–4 per cent of the mentally retarded. In the preschool years their development is usually greatly slowed. Eventually many of them can be helped to look after themselves under close supervision and to communicate in a simple way. As adults they can undertake simple tasks and engage in limited social activities, but they need supervision and a clear structure to their lives. Among the severely retarded, a small number of *idiots savants* have highly specific cognitive abilities (for example a musical or artistic talent) that are normally associated with superior intelligence (Hermelin 1994).

Profound retardation (IQ below 20)

People in this group account for 1–2 per cent of the mentally retarded. Few of them learn to care for themselves completely. Some eventually achieve some simple speech and social behaviour.

Physical disorders among the mentally retarded

The most important physical disorders in the mentally retarded are sensory and motor disabilities, epilepsy, and incontinence. Severely retarded people (especially children) usually have such problems, often as multiple disorders. Only a third are continent, ambulant, and without severe behaviour problems; a quarter are highly dependent on other people. Among the mildly retarded, similar problems occur, but less frequently, and determine whether special educational programmes are needed. Any **sensory disorders** add an important additional obstacle to normal cognitive development. **Motor disabilities** are frequent, and include spasticity, ataxia, and athetosis. Increasing age results in a greater likelihood of physical problems in both mild and severe retardation (Moss *et al.* 1993; Day and Jancar 1994).

Epilepsy is common among the mentally retarded, especially the severely retarded. Corbett *et al.* (1975) surveyed all the severely retarded children (whether in hospital or outside) originating from a London suburb. One-third of these children had experienced seizures at some time, and one-fifth had had at least one seizure in the year preceding the enquiry. Epilepsy is most common when mental retardation is due to cerebral damage, and is uncommon when the retardation is due to chromosomal abnormalities. It becomes less prevalent with increasing age, partly because those with severe cortical damage tend to die early, and partly because epilepsy tends to improve with age irrespective of intelligence level.

The types of epilepsy found in the mentally retarded are generally the same as those found in people of normal intelligence. However, certain rare syndromes are particularly associated with mental retardation. An example is *infantile spasms* in which seizures start in the first year and take the form of so-called salaam attacks with tonic flexion of the neck and body and movement of the arms outward and forward. The episodes last for a few seconds (Corbett and Pond 1979). The condition has also been reported in association with autism.

Psychiatric disorders among the mentally retarded

Psychiatric disorder amongst the mentally retarded has been seen as very different to that seen in people of normal intelligence. One view was that individuals who had retardation were unable to develop emotional disorders, whilst another view was that such disorders were common but that they differed from those of people with normal intelligence and would usually have a biological origin. Recently, it has been generally agreed that people with mental retardation have very similar psychiatric disturbances to the general population.

All varieties of *dual diagnoses* (of psychiatric disorder and mental retardation) occur, but the symptoms are often greatly modified by low intelligence (Borthwick-Duffy 1994; Fraser and Nolan 1994). Delusions, hallucinations, and obsessions may not be easily recognized in people with severe retardation and limited language development. Other symptoms may be difficult to detect because patients need a minimum verbal fluency (probably at an IQ level of about 50) to describe their experiences. Hence in diagnosing psychiatric disorder among the mentally retarded, more emphasis has to be given to behaviour and less to reports of mental phenomena than would be the case in people of normal intelligence.

Reported prevalences for psychiatric disorder are much higher than those in the general population, but the range of estimates is very wide, in part because of methodological problems (Borthwick-Duffy 1994; Scott 1994). These include problems in defining and recognizing both the mental retardation (and assessing the severity) and the psychiatric disorder, as well as problems of sampling and drawing control groups from the general population. Rates of psychiatric disorders are related to intellectual level, and they are lowest in mentally retarded people living at home.

Schizophrenia

The point prevalence of schizophrenia in people with mental retardation is 3 per cent compared with 1 per cent in the general population. The clinical picture of schizophrenia in the mentally retarded is characterized by poverty of thinking. Delusions are less elaborate than in schizophrenics of normal intelligence. Hallucinations have a simple and repetitive content. It may be difficult to distinguish between the motor disorders of schizophrenia and the motor disturbances common among the retarded, especially when the IQ is below 45. It is difficult to make a definite diagnosis of schizophrenia in mentally retarded people, but the diagnosis should be considered when there is a distinct worsening of intellectual or social functioning without evidence of an organic cause, especially if any new behaviour is odd and out of keeping with the patient's previous behaviour. When there is continuing doubt, a trial of antipsychotic drugs is often appropriate. Although a response to such medication is helpful to the patients, it does not prove that schizophrenia is the correct diagnosis.

In the past, psychiatrists (including Kraepelin) described a syndrome called *Pfropfschizophrenie*. This disorder was said to begin in mentally retarded children and adolescents and to be characterized by mannerisms and stereotypes. It now appears that these features were related more to severe mental retardation than to schizophrenia (see Turner (1989) for a review).

The principles of treatment of schizophrenia in mentally retarded patients are essentially the same as in patients of normal intelligence (see Chapter 9).

Affective disorder

When suffering from a **depressive disorder**, mentally retarded people are less likely than those of normal intelligence to complain of mood changes or to express depressive ideas. Diagnosis has to be made mainly on an appearance of sadness, changes in appetite and sleep, and behavioural changes of retardation or agitation. Severely depressed patients with adequate verbal abilities may describe hallucinations or delusions. A few make attempts at suicide (which are usually poorly planned). Differential diagnosis involves thyroid dysfunction (which is especially prevalent in those with Down's syndrome). **Mania** has to be diagnosed mainly on over-activity and behavioural signs such as excitement, irritability, and nervousness. The principles of treatment of affective disorders among the mentally retarded are essentially the same as among people of normal intelligence.

Emotional disorders

Emotional symptoms and disorders commonly occur among the less severely retarded, especially adjustment disorders when they are facing changes in the routine of their lives, but establishing prevalence is very difficult. The clinical picture is often mixed. *Anxiety disorders* are frequent, especially at times of stress; phobic disorders are especially likely to be overlooked. *Obsessive–compulsive disorders* are considerably more frequent than in the general population. *Conversion and dissociative* symptoms may sometimes be florid. *Somatoform disorders* and other causes of functional somatic symptoms can result in persistent requests for medical attention. Treatment is usually directed more to bringing about adjustments in the patient's environment than to discussion of problems.

Eating disorders

Although overeating and unusual dieting preferences are frequent among people with mental retardation, specific eating disorders appear to be less common than in the general population. Overeating and obesity are features of the Prader–Willi syndrome, a genetic cause of retardation.

Personality disorder

Personality disorder is common among the mentally retarded, but is difficult to diagnose. Sometimes it leads to greater problems in management than those caused by the retardation itself. The general approach is as described on p. 128, though with more emphasis on finding an environment to match the patient's temperament and less on attempts to bring about self-understanding (Reid and Ballinger 1987).

Cognitive psychiatric disorders

These are common among mentally retarded people. *Delirium* is relatively frequent, probably because underlying cerebral disorders predispose to this kind of response to infection, medication, and other precipitating factors. It is more common in childhood and old age than at other ages. Disturbed behaviour due to delirium is sometimes the first indication of physical illness. Delirium is frequently due to side-effects of drugs (anticonvulsants, antidepressants, and psychotropic medication). Similarly, a progressive decline in intellectual and social functioning may be the first indication of *dementia*. Care is necessary to distinguish dementia from conditions such as depression and delirium which give the impression of loss of intellectual capacity. The syndrome known as *childhood disintegrative disorder* (p. 693) is a form of dementia occurring in early life, often associated with lipidoses or other progressive brain pathology. As the life expectation of mentally retarded patients is increasing, dementia in later life is becoming more common. There is a particular association between Alzheimer's disease and Down's syndrome (see p. 738).

Disorders which are usually first diagnosed in childhood and adolescence

Many of the disorders in this category are more frequent in mentally retarded children than in the general population and they are more likely to continue into adult life. It is important to be aware that specific developmental disorder of language, learning, and motor skills may occur alongside general intellectual retardation.

Autism and over-activity syndromes

Both autism and over-activity syndromes are more common among the mentally retarded than among the general population. These disorders are discussed in Chapter 20 on child psychiatry (pp. 690 and 694) and will not considered further here. For a review of autism see Wing (1994b).

Behaviour disorders

Stereotyped or repetitive and apparently purposeless activities, such as mannerisms, head banging, and rocking, are common in the severely retarded, occurring in about 40 per cent of children and about 20 per cent of adults. *Repeated self-injurious* behaviour is less frequent but may be even more persistent, and there is a specific association with the Lesch–Nyhan syndrome (see p. 735) Many severely retarded children are *over-active, distractible, and impulsive*, but not to an extent that would indicate a diagnosis of a hyperkinetic syndrome. Another common disturbance is *emotional lability* (including temper tantrums).

The term *challenging behaviour* is often used to describe behaviours which are of an intensity or frequency sufficient to impair the physical safety of the retarded person or to pose a danger to others (Emerson 1993; Emerson *et al.* 1994). It is probable that around 20 per cent of mentally retarded children and adolescents and 15 per cent of mentally retarded adults have some form of behaviour which challenges the services providing care and support. The causes of such behaviour include pain and discomfort, difficulties in communication, epilepsy, the side-effects of medication, and psychiatric disorders. Behavioural treatment (see p. 748) carried out in the places in which the behaviour most often appears or, in severe cases, in a residential unit may be helpful (Emerson 1993).

Forensic problems

Mildly mentally retarded people have higher rates of criminal behaviour than the general population (see also Chapter 22, p. 757). The causes of this excess are complex, with predisposing influences in the family and social environment being of most importance. Impulsivity, suggestibility, exploitability, and desire to please are other important reasons for involvement in crime. Compared with the general population mentally retarded people who commit offences are more likely to be detected and, once apprehended, may be more likely to confess. Arson and sexual offences (usually exhibitionism) are said to be particularly common.

Increased suggestibility, with the risk of false confessions, requires special care in questioning a mentally retarded person about an alleged offence. Once convicted, psychiatric supervision and specialized education may be needed (see Turk (1989) for a review).

Sleep disorders

Serious sleep problems appear to be common among mentally retarded people and can be a source of considerable distress to the people themselves and to their carers. These problems have not received much attention, but it seems that treatments in use for the general population may also be effective for the mentally retarded (Stores 1992).

Sexual problems

In the past, much concern was expressed about the risk that mentally retarded people would have sexual intercourse and produce mentally retarded children. It is now known that many kinds of severe retardation are not inherited, and those which are inherited are often associated with infertility. A more important concern is that, even if their children are of average intelligence, the severely mentally retarded are unlikely to make good parents. With modern contraceptive methods, the risk of unplanned pregnancy is much reduced.

Public masturbation is the most frequent forensic problem. Some of the mentally retarded show a child-like curiosity about other people's bodies which can be misunderstood as sexual.

Effects of mental retardation on the family

The effects on the family change over the years following the initial shock of being told that the child is mentally retarded. When a newborn child is found to be mentally retarded, the parents are inevitably distressed. Feelings of rejection are common, but seldom last long. Frequently, the diagnosis of mental retardation is not made until after the first year of life, and the parents then have to make great changes in their hopes and expectations for the child. They often experience prolonged depression, guilt, shame, or anger, and inability to cope with substantial practical and financial problems. A few reject their children, while others become overinvolved in their care, sacrificing other important aspects of family life (Floyd and Phillippe 1993). The majority achieve a satisfactory adjustment, although the temptation to overindulge the child remains. However well they adjust psychologically, the parents are still faced with a long prospect of hard work, frustration, and social problems. If the child also has a physical handicap, these problems are increased (see Benson and Gross (1989) for a review).

There have been several studies on the effect of a child's mental retardation on the family. In an early study, Ann Gath (1978) compared two groups of families, those with a Down's syndrome child at home and controls with a normal child of the same age. It was concluded that 'despite the understandable emotional reactions to the fact of the baby's abnormality, most of the families in the study have adjusted well, and two years later are providing a home environment that is stable and enriching for both the normal and handicapped children' (Gath 1978, p. 116). However, it seemed likely that siblings were often at some disadvantage because of the time and effort that had to be devoted to the retarded child.

Other investigators have surveyed the needs of the mothers of older mentally retarded children who need continuing care at an age at which normal children are achieving independence. In a widely influential study, Tizard and Grad (1961) compared two groups of families, those with a mentally retarded child living at home and those with a comparable child in an institution. The former were found to be preoccupied with the 'burden of care' for their child, while the latter were living nearly normal lives. Psychiatric and mental problems usually cause even more distress than physical care (Quine and Pahl 1987).

More recent studies have consistently found that mothers with a mentally retarded child at home received help from their husbands but little help from other people, and many professionals were seen as lacking interest and expertise. Many parents report difficulties in obtaining help in looking after the children in school holidays and during weekends or evenings. For a review of the problems of families with a mentally retarded member see Beresford (1994).

Maltreatment and abuse

Many mentally retarded children and their families are characterized by factors known to be associated in the general population with maltreatment of children (see p. 717). Nevertheless, although it is often said that abuse of mentally retarded children is frequent, there is no convincing epidemiological evidence to support this assertion.

Aetiology of mental retardation

Introduction

Lewis (1929) distinguished two kinds of mental retardation: **subcultural**, i.e. the lower end of the normal distribution curve of intelligence in the population, and **pathological**, i.e. retardation due to specific disease processes. In a study of the 1280 mentally retarded people living in the Colchester Asylum, Penrose (1938) found that most cases

were due not to a single cause but to an interaction of inherited and environmental factors. Subsequent research has confirmed that mental retardation has multiple causes. This is particularly true for mild mental retardation which is usually due to a combination of genetic and adverse environmental factors, and which is more common in the lower social classes.

Among the severely retarded, physical causes are found in 55–75 per cent of cases (Broman *et al.* 1987). Prenatal causes predominate, of which the most frequent are idiopathic cerebral palsy, Down's syndrome, and fragile X syndrome; the proportion with fragile X syndrome is increasing steadily, accounting for 10–15 per cent. Another three causes—other chromosomal anomalies, single-gene disorders, and idiopathic epilepsy—each account for 5–10 per cent of cases.

It should be noted that increasing success in identifying specific causes of severe mental retardation does not remove the need to consider all the additional social and other factors in each case. Until recently, severe retardation was thought to be evenly distributed in the population; now it is known to be more common in the lower socioeconomic groups, possibly because preventive measures have been less effective in this group.

Whilst the aetiology of mental retardation in developed countries is predominantly due to genetic and perinatal causes, perinatal and postnatal factors (hypothyroidism, infection, trauma, toxicity) are much more significant in developing countries. See Scott (1994) for a review of aetiology.

Inheritance

There is good evidence from family, twin, and adoption studies that polygenic inheritance is important in determining intelligence within the normal range, and that much mild mental retardation represents the lower end of the distribution curve of intelligence (Plomin 1994, 1995; Thapar *et al.* 1994). Fragile X syndrome is the most common known inherited cause of retardation, and is particularly associated with moderate and mild syndromes. In severe retardation, many genetic abnormalities (including about 180 single-gene defects) are responsible for metabolic disorders and other anomalies that cause mental disorder (see Tables 21.1 and 21.2).

Table 21.1 summarizes the most frequent causes, but it is not exhaustive. Many of these causes are rare, and they will not be described in this chapter. Information about some of the less rare conditions is summarized in Table 21.2. These syndromes are more likely to be dealt with by paediatricians than by psychiatrists. If psychiatrists take over the care of a patient suffering from any of these rare syndromes, they should work closely with the paediatrician and family doctor, and should acquaint themselves with up-to-date knowledge of the particular syndrome by studying a textbook of paediatrics and discussing the case with the appropriate specialist.

Specific genetic syndromes require some separate comment. Five groups may be recognized.

1. **Dominant conditions** These are rare. Examples are the phakomatoses, including neurofibromatosis.
2. **Recessive conditions** This is the largest group of specific gene disorders. It includes most of the inherited metabolic conditions such as phenylketonuria (the commonest inborn error of metabolism), homocystinuria, and galactosaemia.
3. **Chromosal abnormalities**
   (a) **Sex-linked conditions** The prevalence of mental retardation is 25 per cent greater in males than in females. Lehrke (1972) was the first to suggest that the excess among males might be due to X-chromosome-linked causes. Recent research suggests that up to a fifth of retardation in males is due to X-linked causes. Many rare specific X-linked syndromes have been identified, for example glucose dehydrogenase deficiency and the Lesch–Nyhan syndrome. However, in most cases there is no metabolic abnormality, for example the 'fragile X syndrome' (see p. 738). Sex chromosome abnormalities, such as Klinefelter's syndrome (XXY) and Turner's syndrome (XO), may also cause retardation.

(b) **Autosomal chromosome abnormalities** The most common is Down's syndrome (see p. 737).
4. **Conditions with partial and complex inheritance** such as anencephaly. This group is poorly understood.

Social factors

Studies of the general population suggest that factors in the social environment may account for variation in IQ of as much as 20 points. The evidence comes from two kinds of enquiry (Rutter 1980; Clarke *et al.* 1985). The first is epidemiological. Low IQ is related to lower social class, poverty, poor housing, and an unstable family environment. Such social factors may be the effects of low intelligence and do not necessarily exclude a genetic cause. Thus mentally retarded people might drift into an adverse social environment and bring up their children there. Follow-up studies from birth to adolescence have found marked differences in measured IQ according to psychosocial predictors (Sameroff *et al.* 1987).

The second source of evidence is from attempts to enrich the environment of deprived children in special residential care (O'Connor 1968) and to provide special education. In one early experiment, children from large and unsatisfactory institutions were transferred to small well-staffed children's homes or given more stimulating education. Twenty years later they were found to have higher IQs than those who, as children, had remained in their original institutions (Skeels 1966). More recent studies have confirmed that well-planned and prolonged intervention can be beneficial for socially deprived children (Garber 1988).

Other environmental factors

These include intrauterine infection (such as rubella), environmental pollutants (such as lead), maternal alcoholism in pregnancy (see pp. 448 and 462), severe malnutrition, iodine deficiency, and excessive irradiation to the womb. There may be vulnerable periods of brain development during which damage is particularly likely to follow

Table 21.1. *The aetiology of mental retardation*

Genetic
Chromosome abnormalities

Fragile site	Fragile X syndrome
Trisomy 21	Down's syndrome
Trisomy 13	Patau's syndrome
Trisomy 18	Edwards' syndrome
Terminal deletion 5	Cri du chat
Microdeletions	Williams' syndrome
	Prader–Willi syndrome
	Angelman's syndrome

Metabolic disorders affecting
Amino acids (e.g. phenylketonuria, homocystinuria, Hartnup disease)
The urea cycle (e.g. citrullinuria, aminosuccinic aciduria)
Lipids (Tay–Sachs, Gaucher's, and Niemann–Pick diseases)
Carbohydrate (galactosaemia)
Purines (Lesch–Nyhan syndrome)
Mucopolysaccahridoses (Hurler's, Hunter's, Sanfillipo's, and Morquio's syndromes)

With gross disease of the brain
Tuberose sclerosis
Neurofibromatosis

With brain malformations
Neural tube defects
Hydrocephalus
Microcephalus

Antenatal damage
Infections (rubella, cytomegalovirus, syphilis, toxoplasmosis)
Intoxications (lead, certain drugs, alcohol)
Physical damage (injury, radiation, hypoxia)
Placental dysfunction (toxaemia, nutritional growth retardation)
Endocrine disorders (hypothyroidism, hypoparathyroidism)

Perinatal
Birth asphyxia
Complications of prematurity
Kernicterus
Intraventricular haemorrhage

Postnatal damage:
Injury (accidental, child abuse)
Intoxication (lead mercury)
Infections (encephalitis, meningitis)
Autism
Impoverished environment

Table 21.2 *Notes on some causes of mental retardation*

Syndrome	Aetiology	Clinical features	Comments
Chromosome abnormalities (for Down's syndrome and X-linked retardation, see text)			
Triple X	Trisomy X	No characteristic feature	Mild retardation
Trisomy 18 (Edward's syndrome)	Trisomy 18	Growth deficiency, abnormal skull shape and facial features, clenched hands, rocker bottom feet, cardiac and renal abnormalities	Most die within a few weeks of birth
Trisomy 13 (Patau's syndrome)	Trisomy 13	Structural abnormalities of the brain, lip, and palate, polydactyly	
Cri du chat	Deletion in chromosome 5	Microcephaly, hyperteleorism, typical cat-like cry, failure to thrive	
Inborn errors of metabolism			
Phenylketonuria	Autosomal recessive causing lack of liver phenylalanine hydroxylase. Commonest inborn error of metabolism	Lack of pigment (fair hair, blue eyes); retarded growth; associated epilepsy, microcephaly, eczema, and hyperactivity	Detectable by postnatal screening of blood or urine; treated by controlling the intake of phenylalanine from the diet during early years of life
Homocystinuria	Autosomal recessive causing lack of cystathione synthetases	Ectopia lentis, fine and fair hair, joint enlargement, skeletal abnormalities similar to Marfan's syndrome; associated with thromboembolic episodes	Retardation variable; sometimes treated by methionine restriction
Galactosaemia	Autosomal recessive causing lack of galactose 1-phosphate uridyl transferase	Presents following introduction of milk into diet; failure to thrive, hepatosplenomegaly, cataracts	Detectable by postnatal screening for the enzymic defect; treatable by galactose-free diet; toluidine blue test on urine
Tay–Sachs disease	Autosomal recessive resulting in increased lipid storage (the earliest form of the cerebromacular degenerations)	Progressive loss of vision and hearing; spastic paralysis; cherry-red spot at macula of retina; epilepsy	Death at 2–4 years

Hurler's syndrome	Autosomal recessive affecting mucopolysaccharide storage	Grotesque features; protuberant abdomen; hepatosplenomegaly; associated cardiac abnormalities	Death before adolescence
Lesch–Nyhan syndrome	X-linked recessive leading to enzyme defect affecting purine metabolism. Excessive uric acid production and excretion	Normal at birth. Development of choreo-athetoid movements, scissors position of legs, and self-mutilation	Diagnosed pre-natally, by sampling amniotic fluid and enzyme estimation. Post-natal diagnosis by enzyme estimation in hair roots. Death in second or third decade from renal failure or infection. Self-mutilation may be reduced by treatment with hydroxy-tryptophan
Other inherited disorders			
Neurofibromatosis (Von Recklinghausen's syndrome)	Autosomal-dominant inheritance	Neurofibromata, café au lait spots, vitiligo; associated with symptoms determined by site of neurofibromata; astrocytomas, menigioma	Retardation in a minority
Tuberose sclerosis (epiloia)	Autosomal dominant (very variable penetrance)	Epilepsy, adenoma sebaceum on face, white skin patches, shagreen skin, retinal phakoma, subungal fibromata; associated multiple tumours in kidney, spleen, and lungs	Retardation in about 70 per cent
Lawrence–Moon–Biedl syndrome	Autosomal recessive	Retinitis pigmentosa, polydactyly, sometimes with obesity and impaired genital function	Retardation usually not severe
Infection			
Rubella embryopathy	Viral infection of mother in first trimester	Cataract, microphthalmia, deafness, microcephaly, congenital heart disease	If mother infected in first trimester, 10–15 per cent infants are affected (infection may be subclinical)
Toxoplasmosis	Protozoal infection of mother	Hydrocephaly, microcephaly, intracerebral calcification, retinal damage, hepatosplenomegaly, jaundice, epilepsy	Wide variation in severity

continued

Table 21.2 *Continued*

Syndrome	Aetiology	Clinical features	Comments
Cytomegalovirus	Virus infection of mother	Brain damage; only severe cases are apparent at birth	
Congenital syphilis	Syphilitic infection of mother	Many die at birth; variable neurological signs, 'stigmata' (Hutchinson teeth and rhagades often absent)	Uncommon since routine testing of pregnant women; infant's WR positive at first but may become negative
Cranial malformations			
Hydrocephalus	Sex-limited recessive; inherited developmental abnormality, e.g. atresia of aqueduct, Arnold–Chiari malformation, meningitis, spina bifida	Rapid enlargement of head. In early infancy, symptoms of raised CSF pressure; other features depend on aetiology	Mild cases may arrest spontaneously; may be symptomatically treated by CSF shunt; intelligence can be normal
Microcephaly	Recessive inheritance, irradiation in pregnancy, maternal infections	Features depend upon aetiology	Evident in up to a fifth of institutionalized mentally retarded patients
Miscellaneous			
Spina bifida	Aetiology multiple and complex	Failure of vertebral fusion; *spina bifida cystica* is associated with meningocele or, in 15–20 per cent, myelomeningocele; latter causes spinal cord damage with lower limb paralysis, incontinence, etc.	Hydrocephalus in four-fifths of those with myelomeningocele; retardation in this group
Cerebral palsy	Perinatal brain damage; strong association with prematurity	Spastic (common), athetoid and ataxic types; variable in severity	Majority are below average intelligence; athetoid are more likely to be of normal IQ
Hypothyroidism (cretinism)	Iodine deficiency or (rarely) atrophic thyroid	Appearance normal at birth; abnormalities appear at 6 months; growth failure, puffy skin, lethargy	Now rare in UK; responds to early replacement treatment
Hyperbilirubinaemia	Haemolysis, rhesus incompatibility, and prematurity	Kernicterus (choreoathetosis), opsthotonus, spasticity, convulsions	Prevention by anti-rhesus globulin; neonatal treatment by exchange transfusion

exposure to such environmental hazards (Davison 1984; Holland 1994*b*). *Malnutrition* in the first two years of life is probably the most common cause of retardation in the world as a whole, but is much less frequent in developed countries. *Iodine deficiency* is an important cause in many developing countries.

There is no doubt that severe **lead intoxication** can cause an encephalopathy with consequent intellectual impairment. It is much less certain whether the moderate levels of lead found in some British children (resulting in part from air pollution with lead additives in petrol) can cause intellectual retardation. It is known that children absorb lead more readily than adults, and therefore are at greater risk from environmental pollution. However, most of the studies have been of children from lower socio-economic groups, and it is impossible to be certain how far findings of low intelligence (compared with children in other areas) are due to slightly raised lead levels in their blood and how far to social influences (see Pocock *et al.* (1994) for a review).

Birth injury

This is an important cause of mental retardation Early studies estimated that clinically recognizable birth injuries accounted for about 10 per cent of mental retardation. Pasamanick and Knobloch (1966) suggested a 'continuum of reproductive casualty' in which additional cases of mild retardation resulted from less obvious brain lesions sustained *in utero* or perinatally. Although prematurity and low birth-weight are associated with mental retardation (Lukeman and Melvin 1993; Hack *et al.* 1994), the theory of a continuum of reproductive causality is not otherwise supported.

Some specific causes of mental handicap

Down's syndrome

In 1866, Langdon Down tried to relate the appearance of certain groups of patients to the physical features of ethnic groups. One of his groups had the condition originally called mongolism, and now generally known as Down's syndrome. This condition is a frequent cause of mental retardation, occurring in 1 in every 650 live births. It is more frequent among older women, occurring in about 1 in 2000 live births for mothers aged 20–25 and 1 in 30 for those aged 45. The incidence of Down's syndrome has decreased because of reduced birth rates among older women, and increases in detection of the condition by amniocentesis and subsequent termination of pregnancy. Usually the retardation is mild or moderate, but occasionally it is severe.

The **clinical picture** is made up of a number of features, any one of which can occur in a normal person. Four of these features together are generally accepted as strong evidence for the syndrome. The most characteristic signs are (a) a small mouth and teeth, furrowed tongue, and high-arched palate, (b) oblique palpebral fissures and epicanthic folds, (c) flat occiput, (d) short and broad hands, a curved fifth finger and a single transverse palmar crease, and (e) hyperextensibility or hyperflexibility of joints and hypotonia.

There are often other associated abnormalities, and 10 per cent are multiply handicapped. Congenital heart disease (especially septal defects) occurs in about 5 per cent. Intestinal abnormalities are common, especially duodenal obstruction. Hearing may be impaired. An immunological defect predisposes to infections. There are also increased risks of acute leukaemia, hypothyroidism, and atlantoaxial instability.

There is considerable variation in the degree of mental retardation; the IQ is generally between 20 and 50, and in 15 per cent it is above 50. Mental abilities usually develop fairly quickly in the first six months to a year of life but then increase more slowly. The temperament of children with Down's syndrome has often been described as lovable and easygoing, but there is a wide individual variation. Emotional and behaviour problems are less frequent than in forms of retardation associated with clinically detectable brain damage.

In the past, the infant mortality of Down's syndrome was high, but with improved medical care survival into adult life is usual. About a quarter of people with Down's syndrome now live beyond 50 years of age. In middle life, people with Down's syndrome develop Alzheimer-like changes in the brain (Oliver and Holland 1986; Berg *et al.*

1995). However, clinical decline occurs in a smaller proportion of people and much later than would be expected from the neuropathological data (Wisniewski *et al.* 1994).

Many of the older people with Down's syndrome live with their families or in large residential groups and have led restricted lives. In the future, the social outlook may be better for those who have been involved in intensive early teaching and it is likely that most will require long-term support of a kind more vigorous than is currently available if the quality of their lives is to be maintained. See Carr (1994) for a review.

Aetiology In 1959, Down's syndrome was found to be associated with the chromosomal disorder of trisomy (three chromosomes instead of the usual two). About 95 per cent of cases are due to trisomy 21. These cases result from failure of disjunction during meiosis and are associated with increasing maternal age. The risk of recurrence in a subsequent child is about 1 in 100. The remaining 5 per cent of cases of Down's syndrome are attributable either to translocation involving chromosome 21 or to mosaicism. The disorder leading to translocation is often inherited, and the risk of recurrence is about 1 in 10. Mosaicism occurs when non-disjunction takes place during any cell division after fertilization. Normal and trisomic cells occur in the same person, and the effects on cognitive development are particularly variable (Thapar *et al.* 1994). Down's pathology is presumed to be due to the increased 'dosage' of genes on chromosome 21. This could account for the excess of early-onset Alzheimer's (see p. 523).

Fragile X syndrome

Fragile X syndrome is the second most common specific cause of mental retardation after Down's syndrome and is the most common inherited cause. It occurs in up to 1 in 1000 males and in a milder form in 1 in 200 females. It accounts overall for about 10 per cent of those with learning difficulties, i.e. for about 7 per cent of moderate and 4 per cent of mild retardation amongst males, and about 2.5 per cent of moderate and 3 per cent of mild retardation in females.

There are a number of characteristic but somewhat variable clinical features, none of which is diagnostic. These features include enlarged testes, large ears, a long face, and flat feet. Psychological disturbances are said to include abnormalities of speech and language, autism and other social impairments, and disorders of attention and concentration. There are increased rates of several psychiatric disorders, including attention-deficit hyperactivity disorder. The inheritance of the condition is unusual and complex, and has been difficult to unravel. During the last few years there has been a major advance in understanding the molecular genetic basis of the condition, and a gene referred to as *FMR-1* has been identified. This gene contains an amplified CGG repeat sequence which constitutes the fragile X anomaly.

The identification of the genetic abnormality has transformed the advice that can be given to individuals, since it is now possible to identify affected heterozygous females who are clinically and even cytogenetically normal. These women can be advised of the high risk of affected offspring and the need for prenatal testing. Conversely, many women who are at risk of being carriers can be reassured they do not have the condition. Similarly, it is possible to determine whether men are transmitting carriers.

The psychological and psychiatric complications in the syndrome mean that there is a need for regular review of affected people and for the use, when appropriate, of the medical psychological interventions described later in this chapter. See Hagerman (1992), Dykens *et al.* (1994), and Turk *et al.* (1994) for reviews of the fragile X syndrome.

Causes of psychiatric disorder and behaviour problems in the mentally retarded

The diversity of psychiatric disorders among the mentally retarded makes it unlikely that they have a single aetiology. As with mental disorder among people of normal intelligence, several causes have to be considered: biological, psychosocial, and developmental (Dosen 1993; Matson and Sevin 1994). Overall there is no reason to expect that particular mental disorders co-morbid with mental retardation have aetiologies which differ from those in people with normal intelligence.

Biological factors include **specific** associations such as those between the fragile X syndrome and attention-deficit hyperactivity disorder and anxiety disorder, between the Lesch–Nyhan syndrome and self-injury, and between Down's syndrome and Alzheimer's disease.

As noted already, most severely retarded people have some **organic brain pathology** and so do a smaller number of those with moderate and mild retardation. As mentioned in the chapter on child psychiatry, psychiatric disorder is associated with brain damage in children of normal intelligence (Rutter *et al.* 1970*a*). There is also a known association between cerebral pathology on the one hand and schizophrenia and affective disorders on the other (Davison and Bagley 1969; Davison 1983). Therefore it is likely that some psychiatric disorders (including major psychiatric illnesses) in mentally retarded people, are related to brain pathology. There is an especially close association between **epilepsy** and behaviour disorder in the mentally retarded (see Matson and Sevin 1994). Such behaviour disorder may be due not only to the direct effects of epilepsy but also to the side-effects of anticonvulsant drugs (see p. 587).

Psychosocial factors such as bereavement or a disrupted family may cause mental disorder and behaviour problems in the mentally retarded just as in people of normal intelligence. The mentally retarded may develop adjustment reactions or mental disorder when the arrangements for their care are disrupted or if they are treated badly or exploited. Such factors have, in part, been seen as mediated by behavioural processes, classical conditioning, operant conditioning, and social learning. Operant models point to the importance of inadequate environmental reinforcement of adaptive behaviours and the reinforcement of maladaptive behaviours.

Developmental factors associated with mental retardation include abnormalities of temperament, difficulty in acquiring language and social skills, low self-esteem, and educational failure.

It should not be forgotten that **iatrogenic** factors can contribute to the causes of psychiatric disorder among the mentally retarded. As mentioned above, these include the side-effects of drugs, especially those used to treat epilepsy, and over- or under-stimulating environments within an institution or in the community.

The assessment of the mentally retarded

Assessment of the mentally retarded is directed towards four main areas:

(1) aetiology of mental retardation;

(2) associated biomedical conditions;

(3) intellectual and social skills development;

(4) psychological and social functioning.

Severe retardation can usually be diagnosed in infancy, especially as it is often associated with detectable physical abnormalities or with retardation of motor development. Some mentally retarded people have specific developmental disorders, i.e. impairment of specific functions greater than would be expected from the general level of retardation. The clinician should be cautious in diagnosing less severe mental retardation on the basis of delays in development. Although routine examination of a child may reveal signs of developmental delay, suggesting possible mental retardation, confident diagnosis often requires a second opinion from a specialist in mental retardation. Full assessment has several stages: history taking, physical examination, developmental testing, functional behavioural assessment, analysis of the interaction between the retarded person and the social care and support systems, and examination of the mental state and emotional adjustment. These will be considered in turn. Although this section is concerned mainly with the assessment of children, similar principles apply in adolescence and adult life.

History taking

In the course of obtaining a full history, particular attention should be given to any family history suggesting an inherited disorder and to abnormalities in the pregnancy or the delivery of the child. Dates of passing developmental milestones should be ascertained (see p. 664). A full account of any behaviour disorders should be obtained. Details of any associated medical conditions, such as congenital heart disease, epilepsy, and cerebral palsy, should be documented.

Physical examination

A systematic physical examination should include the recording of head circumference. It is important to be alert for the physical signs of the many specific syndromes (see Table 21.2). The neurological examination should include particular attention to vision and hearing.

Developmental assessment

This assessment is based on a combination of clinical experience and standardized methods of measuring intelligence, language, motor performance, and social skills. Although the IQ is the best general index of intellectual development, it is not reliable in the very young. Some commonly used developmental tests are listed below. More information is provided about some of these scales in Table 20.2 (p. 676).

Tests used in developmental assessment

The following standardized assessment instruments are widely used in screening, diagnosing, or assessing the severity of disorders of psychological development. Several do not require special training for their correct administration.

The first four instruments provide a general assessment over a range of developmental domains.

The other four focus, either entirely or principally, on specific aspects of development.

1. **Vineland Social Maturity Scale** The Vineland Scale is useful in the assessment of children who do not cooperate in testing since it can be completed by interview with a reliable informant. An overall social age can be derived from the scale, which may be usefully compared with mental and chronological age.
2. **The British Ability Scales and Differential Ability Scales** These scales measure a range of functions and educational attainments, and can be used to calculate an overall IQ.
3. **The Adaptive Behaviour Scales** These scales are probably the most widely used method for the assessment of social functioning in mentally retarded adults.
4. **The Portage Guide to Early Education** This assessment has been adopted for use internationally. It provides a broad-based developmental assessment including socialization, language, self-help, and cognitive and motor domains. Only a brief period of training is required for the assessor, together with the active involvement of the parent or other carer.
5. **Autism Behaviour Check List** This questionnaire is completed by the parent or other carer. It is a reliable indicator of problems in the area of autism and delay in social development.
6. **The Disability Assessment Scale** This scale focuses on autistic and related social developmental, language disorders, and behavioural problems.
7. **British (Peabody) Picture Vocabulary Test** This is a test of language comprehension, suitable for non-speaking children. The test booklet is largely pictorial and the age range covered is 3–19 years. Although professional training is not required, the test is used by psychologists and speech therapists.
8. **Reynell Scales of Language Development** These scales assess comprehension and expressive language in the age range one month to six years. The test is of particular use with non-verbal children.

Functional behavioural assessment

This is based on accounts by family and carers and the observations by the clinical team of the person's ability to care for himself, his social abilities including his ability to communicate, his sensory motor skills, and his relationships with others.

Psychological assessment

This is directed at the interaction between the individual and people who are closely involved in care, determining the correct needs and wishes for the future. It should examine opportunities for learning new skills, making relationships, and achieving maximum choice about way of life. If the mentally retarded person has reasonable language ability, it is possible to carry out a standard psychiatric interview whilst making appropriate allowance for any difficulty in concentration or abstract concepts. When language is less well developed, an account has to be obtained mainly from informants. It is particularly important to obtain a complete description of any change from the usual pattern of behaviour. It is often necessary to ask parents, teachers, or care staff to keep records of behaviours such as eating, sleeping, and general activity. The interviewer should keep in mind the possible causes of psychiatric disorder outlined above, including unrecognized epilepsy (Dosen 1993).

The care of the mentally retarded

A historical perspective

In the last few decades the aims and methods of care for the mentally retarded have been transformed. These changes and the remaining unsolved problems can best be understood in relation to the history of the development of services.

Although actual distinctions between inborn 'idiocy' and dementia have been made for many hundreds of years, the distinctions were not clearly described and related to psychiatric classification as a whole until the beginning of the nineteenth century. At this time there were numerous reports of treatments, most remarkably the efforts of Itard, physician-in-chief at the Asylum for the Deaf and Dumb in Paris, to train the 'wild boy' found in Aveyron in 1801. This child was thought to have grown up in the wild, isolated from human beings. Itard made great efforts to educate the boy, but after persisting for six years he concluded the training had been a failure. Nevertheless, his work had important and lasting consequences, one of which was the development of special educational methods for the mentally retarded. These methods were most conspicuously developed by Seguin, director of the School for Idiots at the Bicêtre in Paris, who in 1842 published his *Theory and nature of the education of idiots*. Seguin believed that the mentally retarded had latent abilities which could be encouraged by special training. Therefore he advised an educational programme of physical exercises, moral instruction, and graded tasks (Seguin 1864, 1866).

These ideas were particularly taken up in Switzerland and Germany. Another pioneer was the Swiss physician Guggenbühl, who founded the first special residential institution for the mentally retarded at Abendberg in 1841. Similar institutions were soon opened in other parts of Europe to provide a training that would enable their pupils to live as independently as possible. However, it was recognized that some mentally retarded people needed long-term care.

At the end of the nineteenth century, several influences led to a more custodial approach to the care of the mentally retarded. These influences included the development of the science of genetics, the measurement of intelligence, the beliefs embodied in the eugenics movement, and a general decrease in public tolerance of abnormal behaviour. In England and Wales, such ideas were reflected in the Mental Deficiency Act 1913 which empowered local authorities to provide for the confinement of the intellectually and morally defective and imposed upon them a responsibility to provide training and occupation. As a result, the total number of in-patients of this kind rose from 6000 in 1916 to 50 000 in 1939.

In the 1960s the need for reform was recognized in a number of developed countries, partly because of changes that had already been effected in psychiatric hospitals (see p. 647), partly because of improved psychological research, partly because of campaigning by groups of parents, and partly because of public concern about the generally poor conditions in which the mentally retarded were housed. Surveys of hospitals for the retarded showed that the mean IQ of their patients was over 70. Many residents had only mild retardation, and many did not need hospital care. About the same time, it was shown that simple training could help many patients, both the mildly and severely retarded (O'Connor 1968). Further investigations showed the advantages of residential care in small homely units (Tizard 1964). However, public concern in the United Kingdom and other countries was aroused less by research findings than by scandals about the conditions in hospitals for the mentally retarded.

The last 30 years have seen the acceptance in all developed countries (especially Scandinavia) of the need for methods of care with a less medical approach and for greater integration of the disabled into society. Unfortunately, there have been divergent views about the methods to achieve this, and resources have been inadequate and progress slow. In the United States deinstitutionalization has been carried out at great speed with both successes and failures.

Among the newer concepts of care, the main principle has been 'normalization', an idea developed in Scandinavia in the 1960s. This term refers to the general approach of providing a pattern of life as near normal as possible (Nirje 1970). Normalization implies that almost all mentally retarded people will live in the community, participating in normal activities and relationships, experiencing choice, and having full access to social opportunities. Children are brought up in their own homes, and adults are encouraged to lead independent lives wherever possible. For the few who need special social and health care, accommodation and activities are designed to be as close as possible to those of family life. The concept of normalization has been greatly developed in the United States (Wolfensberger 1980) and elsewhere, and now denotes a system of care that is more elaborate and sophisticated than the word itself might suggest. Increasingly, disabled people are organizing themselves into self-advocacy groups. Those who are unable to speak for themselves about the services they need require an *advocate* to speak on their behalf.

General provisions

The precise model for the care of the mentally retarded in a community matters less than the detail in which it is planned and the enthusiasm with which it is carried out. Good planning requires an estimate of the needs of the population to be served, but must also be based on the summation of individual assessments since each person has an individual profile of needs. To achieve this, local case registers and linked developmental records are important.

The general approach to care is educational and psychosocial. The family doctor and paediatrician are mainly responsible for the early detection and assessment of mental retardation. The team providing continuing health care also includes psychologists, speech therapists, nurses, occupational therapists, and physiotherapists. Volunteers can often play a valuable part, and it is useful to encourage self-help groups for parents.

The main role of the psychiatrist specializing in the problems associated with mental retardation is the assessment and treatment of emotional and behavioural problems and psychiatric disorder. In the United Kingdom most specialist psychiatrists work as members of a community-based multidisciplinary team seeing referrals from colleagues, family doctors, and other services, and being available for consultation with other team members. (See King *et al.* (1994) for an account of psychiatric consultation.) They have access to day or overnight assessment and treatment facilities and to a hospital-based multidisciplinary team particulary skilled in the care of psychiatric disorder and of associated epilepsy in liaison with community services.

The mildly retarded

The number of mildly retarded people in the population is not known accurately. Few need specialist services. Some have additional problems such as physical disability, minor emotional disorders, and psychiatric illness. A few mildly retarded children require fostering, boarding-school placements, or residential care, either because of such additional problems or because of difficulties in the family. Mildly retarded adults may need support when they are facing extra problems; for example they may need help with housing and employment or with the special problems of old age.

Severely retarded children

The population of severely retarded children is about 90 in 100 000. Some require special services throughout their lives, and appropriate planning should begin as soon as the diagnosis is certain. The practical help may include a sitting service, day respite during school holidays, or overnight stays in a foster family or residential care when the parents need a holiday or when another family member is ill. About a quarter of severely retarded children living at home have behavioural problems, difficulty in walking, or incontinence. In the United Kingdom education can be provided from the age of two and must be available until the age of 19.

Severely retarded adults

In the care of the mentally retarded, it is often difficult to arrange a smooth transition during adolescence from children's services to adult services and to make the best use of normal facilities and special provisions. At this stage the coordination of services passes from the school to the social services. Care at home becomes less easy to manage and less appropriate, and residential provision is often preferable. Provisions are required for work, occupation, housing, adult education, etc. The main principle now guiding the provision of resources is that the retarded person should be given sufficient help to be able to use the usual community services rather than to provide specialist segregated services. Some severely retarded people will continue to need treatment for behaviour disorders, psychiatric illness, epilepsy, or physical disability. A few will need to remain in health-service-supported accommodation. Because of the marked increase in the life expectancy of severely retarded people, large unforeseen demands are now being placed on services and it is difficult to predict the scale of future needs.

Specific services

An account will now be given of the main elements in a comprehensive service for people with mental retardation and their families or carers:

(1) the prevention and early detection of mental handicap;

(2) regular assessment of the mentally retarded person's attainments and disabilities;

(3) advice, support, and practical measures for families;

(4) provision for education, training, occupation, or work appropriate for each handicapped person;

(5) housing and social support to enable self-care;

(6) medical, nursing, and other services for those who require them as out-patients, day patients, or in-patients;

(7) psychiatric and psychological services.

Preventive services

Primary prevention depends largely on genetic counselling, early detection of fetal abnormalities during pregnancy, and safe childbirth. *Secondary prevention* aims to prevent the progression of disability by either medical or psychological means. The latter include 'enriching' education and early attempts to reduce behavioural problems. In developed countries there remains considerable scope for reduction of the genetic causes of severe retardation, but it is unlikely that it will be possible to affect the incidence of mild retardation significantly. In developing countries, incidence could be substantially reduced by general measures to improve the health of mothers during pregnancy, and by better perinatal care. See Murphy (1994) for a review.

Genetic screening and counselling This begins with assessment of the risk of an abnormal child being born. Such an assessment is based on study of the family history, on knowledge of the genetics of conditions that give rise to mental impairment, and on awareness of the possibilities for genetic screening. The parents are given an explanation of the risks of screening and encouraged to discuss them. Most parents seek advice only after a first abnormal child has been born. Some do so before starting to have children because there is a mentally retarded person on one or other side of the family. The distress following a positive diagnosis of abnormality leading to termination and following a false-positive result of screening may be considerable (Iles and Gath 1993; Marteau 1994). Therefore it is important that those involved in screening are alert to psychological issues and have the appropriate counselling skills. For a general account of genetic counselling, see p. 365.

Prenatal care Prenatal care begins even before conception with advice on diet, alcohol, and smoking, and by the provision of immunization against rubella for girls who lack immunity.

Prenatal diagnosis overlaps with genetic screening. It is becoming available for an increasing number of conditions with the aim of providing information to those at risk of having abnormal children, reassurance to others, and appropriate treatment of affected infants through early diagnosis. Amniocentesis, fetoscopy, and ultrasound scanning of the fetus in the second trimester can reveal chromosomal abnormalities, most open neural tube defects, and about 60 per cent of inborn errors of metabolism. Amniocentesis carries a small but definite risk, and so is usually offered only to women who have carried a previous abnormal fetus, women with a family history of congenital disorder, and women over 35 years of age.

Rhesus incompatibility is now largely preventable. Sensitization of a rhesus-negative mother can usually be avoided by giving anti-D antibody. An affected fetus can be detected by amniocentesis and treated if necessary by exchange transfusion. For pregnant women with *diabetes mellitus*, special care can improve the outlook for the fetus. Further information about these aspects of care will be found in current textbooks of obstetrics and paediatrics.

Postnatal prevention In the United Kingdom all infants are routinely tested for phenylketonuria, and routine testing for hypothyroidism and galactosaemia is becoming increasingly common. Universal screening for elevated levels of lead has been advocated, but recent evidence suggests that it is more appropriate to target screening in areas where lead levels are known to be high (Diermayer *et al.* 1994). Intensive care units and improved methods of treatment for premature and low-birth-weight infants can prevent mental retardation in some who would previously have suffered brain damage. However, the methods also enable the survival of some retarded children who would otherwise have died.

'Compensatory' education 'Compensatory' education is intended to provide optimal conditions for the mental development of the retarded child. This was the aim of the Head Start programme in the United States, which provided extra education for deprived children. Its methods varied from nursery schooling to attempts to teach specific skills (Rutter and Madge 1976). Many of the results were disappointing. A more intensive programme with similar aims has been carried out in Milwaukee (Garber 1988). Skilled teachers educated children living in slum areas with mothers who had a low IQ (under 75). This additional education started when the child was three months old and continued until school age. At the same time the mothers were trained in a variety of domestic skills. These children were compared with control children of the same age who came from similar families but who had not received additional education and whose mothers had not been trained. At the age of four and a half, the trained children had a mean IQ 27 points higher than that of the controls. Although this study can be faulted because the selection of children was not strictly random and because some of the changes in test scores could have been due to practice, the main findings, that substantial effort by trained staff can produce worthwhile improvement in children of low intelligence born to socially disadvantaged mothers, probably stand. The findings also indicate a need to train the parents as well as the children.

Overall, there is increasing evidence that early interventions can be effective, especially if they are family centred. However, many uncertainties about the components and delivery of such help remain (Murphy 1994).

Assessment

Severe retardation is usually obvious from an early stage. Lesser degrees may become apparent only when the child starts school. Family doctors and teachers should be able to detect possible retardation, but a full assessment may require attendance at a special centre where the child can be observed in many different activities. The methods have been described earlier (p. 739).

Once mental retardation has been diagnosed, regular reviews are required. For children these reviews will usually be carried out by a multi-specialist child health team in cooperation with teachers and social workers. The child psychiatrist has a liaison role with the team and takes children referred to him with emotional, behavioural, and psychiatric problems.

It is important to arrange a thorough review before the child leaves school. This review should assess his need for further education, prospects for employment, occupation, independent living, and specialist physical and psychological health care. Mentally retarded adults also need to be assessed regularly to make sure that they are continuing to achieve their potential and still receiving appropriate care. This is usually carried out by a multi-

professional community team which includes a specialist psychiatrist.

Help for families

Help for families is needed from the time that the diagnosis is first made. It is not enough to give worried parents an explanation on just one occasion. They may need to hear the explanation several times before they can absorb all its implications. Adequate time must be allowed to explain the prognosis, indicate what help can be provided, and discuss the part that the parents can play in helping their child to achieve full potential.

Thereafter, the parents need continuing support. When the child starts school, they should not only be kept informed about progress, but should feel involved in the planning and provision of care. They should be given help with practical matters, such as day care for the child during school holidays, baby-sitting, or arrangements for family holidays. In addition to practical matters, the parents need continuing psychological support, which may be provided as a community programme for parents, siblings, and whole families (Murphy 1994; Petronko *et al.* 1994).

Families are likely to need extra help when their mentally retarded child is approaching puberty or leaving school. Making the transition from child to adult services is extremely stressful, especially as adult services are usually neither as well resourced or as clearly focused at helping the disabled person and his carers. Both day and overnight care is often required to relieve carers and to encourage the retarded person to become more independent.

Education, training, and occupation

In 1929, the British Mental Deficiency Committee made the following comment on schools for the mentally retarded:

If the majority of children for whom these schools are intended are to lead the lives of ordinary citizens . . . these schools must be brought into close relation with the public elementary school system and presented to parents not as something distinct and humiliating but as a helpful variation of an ordinary school.

Progress in achieving this aim has been slow and there are considerable geographical variations and improvisions. The aim is that as many mentally retarded children as possible are educated in ordinary schools either in normal classes or in special classes, but with social integration outside the classroom. Health and local authorities are expected to identify all mentally retarded children and to make written statements of their needs.

Research has consistently shown the value of an early start. Such a start can be made in a special preschool nursery class or playgroup, or occasionally in a paediatric day centre. When normal school age is reached, the least handicapped children can attend remedial classes in ordinary schools. Others need to attend special educational programmes for children with learning difficulties. It is still not certain which retarded children benefit from ordinary schooling, and particularly whether the severely retarded do so. Education in an ordinary school offers the advantages of more normal social surroundings and the expectation of progress, but may carry the disadvantage of lack of special teaching skills and equipment. There are advantages in having disabled children in ordinary schools so that other pupils learn to accept their integration into society as the norm.

Traditionally, education for the more severely retarded was based on the sensory training methods started by Itard and Seguin (see p. 741). It is only recently that the content of the curriculum has been reconsidered. The first change was towards an approach similar to that of an ordinary primary school with an emphasis on self-expression. However, methods of this kind may be inappropriate. There is now an increasing use of more specialist teaching and a variety of innovative procedures for teaching language and other methods of communication (Howlin 1994).

Before retarded children leave school, they need reassessment and vocational guidance. Most mildly retarded young people are able to take normal jobs or enter sheltered employment. The severely retarded are likely to transfer to adult day centres, which should provide a wide range of activities if the abilities of each attender are to be developed as much as possible. A minority of moderately and severely retarded people require intensive care programmes with input from a team of psychologists, occupational therapists, specialist nurses, speech therapists, physiotherapists, dieticians, and social workers.

Residential care

It is now widely accepted that parents should be supported in caring for their retarded children at home or, if they are too heavy a burden for their parents, the child should be in an alternative family and the adult supported in ordinary housing, a family placement, lodgings, or a small residential group. Early support for this view came from an important study by Tizard (1964), who compared two groups of children who had moderate or severe mental retardation but no serious additional physical handicaps. One group was reared in a large hospital, and the other in a small residential unit where care was provided in small family-like groups. The children brought up in this small unit developed better verbal abilities, emotional relationships, and personal independence. Studies such as those of Landesman-Dyer (1981) confirmed that merely moving mentally retarded children to smaller living units is not beneficial unless the staff encourage residents to develop the skills and to live as normally as possible.

Specialist medical services

In most developed countries, official policy is to replace all large isolated hospitals by smaller domestic units in the community that they serve. Various alternatives have been tried. Evaluation of small residential units suggests that they are generally as good as or better than large hospitals. Nevertheless, advocates of 'normalization' criticize these units as still being segregated. Mentally retarded people should have the same access to health services as all other citizens, but require extra individual support if they are to obtain full benefit. People with multiple and complex needs are usually best served by a specialist team who can ensure that they receive coordinated care by medical specialists which also takes account of their social and educational needs.

Retarded children and adults often have physical handicaps or epilepsy for which continuing medical care is needed. This medical care is usually obtained from the ordinary medical services and can work well, but there may be difficulties for both the mentally retarded person and the family if doctors and nurses are insufficiently aware of how to deal with an uncomprehending patient. Extra knowledge, skills, and resources are needed for a high-quality service, and families and carers are always keen for care to be coordinated by a single person so that they can avoid visiting various specialists who offer conflicting advice and care.

Psychiatric services

Expert psychiatric care is an essential part of a comprehensive community service for the mentally retarded. In the past there have been arguments about whether this care should be provided within generic mental health services or by specialists in mental retardation. Specialist psychiatric services are increasingly seen as the more appropriate answer in most countries.

Treatment of psychiatric disorder in the mentally retarded

Treatment of psychiatric disorder in the mentally retarded follows the principles described elsewhere in this book, taking account of the special problems described on p. 742.

Behavioural problems

Psychiatric disorder in the mentally retarded usually comes to notice through changes in behaviour. It should be remembered that behavioural change can also result from physical illness or from stressful events, both of which should be carefully excluded. In the most retarded, and especially those with sensory deficits, behavioural disturbance may be due to understimulation and

frustration at the inability to communicate wishes and needs. Once the cause is clear, the treatment follows. Physical illness should be treated promptly, stressful events reduced if possible, or a more stimulating environment provided when appropriate. If the disturbed behaviour results from a psychiatric disorder, the treatment is similar in many ways to that for a patient of normal intelligence with the same disorder (see below).

It is important to advise and support the parents or others who are caring for the patient during the period of treatment. Community-based care may involve use of behavioural assessment and treatment methods involving the carer (whether parent or paid worker) in observation and in implementing the programme.

The most serious and persistent disorders may require hospital admission for more intensive behavioural management which may be combined with pharmacotherapy. See Petronko *et al.* (1994) for a review of community-based methods and Spreat and Behar (1994) for a review of in-patient treatment.

Drugs

Although antipsychotic drugs are widely used to control abnormal behaviour in the mentally retarded, there have been few controlled trials of their effects. The indications for these drugs are similar to those for patients of normal intelligence. Chlorpromazine or haloperidol are suitable preparations. A particularly careful watch should be kept for side-effects because the patient may not be able to draw attention to them himself. Although antipsychotic drugs may be used for the short-term control of behaviour problems, whenever possible social measures or behavioural treatment should be used for long-term management.

Many mentally retarded patients suffer from epilepsy and require anticonvulsant treatment. Special care is needed in arriving at a drug and a dosage that controls seizures without producing unwanted effects (see p. 587 for the side-effects of anticonvulsant drugs).

Counselling

The patients' limited understanding of language sets obvious limitations to the use of psychotherapy. However, simple discussion can help. As noted already, counselling for parents is an important part of treatment.

Behaviour modification

Behavioural methods have become widely used since they were first introduced in the United States in the 1960s, and can be potentially helpful to the severely retarded since some methods do not require language. They can be used to encourage basic skills such as washing, toilet training, and dressing. Often parents and teachers are taught to carry out the training so that it can be maintained in the patient's everyday environment (Petronko *et al.* 1994). First, the behaviour to be modified is specified. If the problem is an undesired behaviour, a search is made for any environmental factors that seem regularly to provoke it or reinforce it. If possible, these environmental factors are changed. In this way, problem behaviours are reduced or eliminated by ensuring that they are not rewarded inadvertently and by reinforcing alternative responses. Aggressive behaviour is sometimes dealt with by withdrawing all reinforcement; in so-called 'time-out' the patient is ignored or secluded until the behaviour subsides. Punishment and aversive techniques raise important ethical issues, and if used at all require very careful supervision and regulations (Matson and Taras 1989). In general, methods of reward have more lasting effects than aversive techniques and have fewer ethical problems. If the problem is the lack of some socially desirable behaviour, attempts are made to reinforce any such behaviour with material or social rewards, if necessary by 'shaping' the final behaviour from simpler components. Reward should be given immediately after the desired behaviour has taken place (for example using the toilet). For training in skills such as dressing, it is often necessary to provide modelling and prompting in the early stages, and to reduce them gradually later (Yule and Carr 1987; Petronko *et al.* 1994; Spreat and Behar 1994).

Consent to treatment

The important general issues relating to informed consent for physical and psychiatric treatment are discussed on pp. 750–1. Many of the more severely mentally retarded are unable to give informed consent, and it is essential to be aware of local legislation and practice. There is an increasingly important role for advocacy by others on the patient's behalf.

The relevant provisions of the legislation in England and Wales are discussed in the Appendix to this volume, pp. 781–7.

Further reading

Bouras, N. (1994). *Mental health in mental retardation.* Cambridge University Press.

Bregman, J. D. and Harris, J. C. (1995). Mental retardation. In *Comprehensive textbook of psychiatry* (ed. H. I. Kaplan and B. J. Sadock) (6th edn), pp. 2207–42. Williams and Wilkins, Baltimore, MD.

Matson, J. L. and Mulick, J. A. (eds) (1991). *Handbook of mental retardation.* Pergamon Press, New York.

22 | *Forensic psychiatry*

The term *forensic psychiatry* is used in two different senses, one narrow and one broad. In its narrow sense the term is applied only to the branch of psychiatry that deals with the assessment and treatment of mentally abnormal offenders. In its broad sense the term is applied to all legal aspects of psychiatry including the civil law and laws regulating psychiatric practice as well as the subspecialty concerned with mentally abnormal offenders. Forensic psychiatrists also assess and treat people with violent behaviour who have not at the time committed an offence in law. In the title of this chapter the term is used in the broad sense.

The clinical psychiatrist needs a working knowledge of two sets of laws: those relating to patients seen in ordinary clinical practice, and those relating to mentally abnormal offenders. The first set of laws (concerned with ordinary patients) consists of two main groups. First are the laws regulating clinical practice, particularly the compulsory detention of patients in hospital and consent to treatment. Second are civil laws dealing with issues such as the patient's capacity to make a will or to care for his own property.

The second set of laws are those relevant to the management of mentally abnormal offenders, i.e. criminal offenders who suffer from mental disorder, mental retardation, or severe personality disorder. This group form a small minority of all offenders, but they present many difficult problems in psychiatry and the law. These problems include legal issues, such as criminal responsibility and fitness to plead, and practical questions, such as whether an offender needs psychiatric treatment and whether such treatment should be provided in the community, in a psychiatric hospital or special hospital, or in prison. For the management of such problems the psychiatrist needs knowledge not only of the law but also of the relationship between particular kinds of crime and particular kinds of psychiatric disorder. It is important to remember that concepts and evidence are markedly influenced by national patterns of criminal law and their application (R. Smith 1989; Gunn and Taylor 1993, pp. 118–66).

The chapter begins with a brief discussion of the law in relation to ordinary psychiatric practice, with particular reference to confidentiality, informed consent, and compulsory admission to hospital. Next comes a short section dealing with the civil law in relation to issues such as fitness to drive and the care and disposal of patients' property.

The main part of the chapter is concerned with the mentally abnormal offender. A brief review of the general causes of crime is followed by discussion of the relationship between crime and the various psychiatric diagnostic categories. An account is then given of the types of offence (violence, sexual offences, and property offences) most likely to be associated with psychological factors. Next the role of the psychiatrist is described, with particular reference to the offender's fitness to plead, mental state at the time of the offence, and diminished responsibility, and the psychiatric treatment of mentally abnormal offenders. This is followed by some guidelines for the psychiatrist on the work of courts, and on interviewing defendants and preparing psychiatric court reports. Dangerousness and violence are then discussed. The main provisions of the Mental Health Act in England and Wales are given in the Appendix. In reading this chapter, two important points need to be borne in mind.

1. There are substantial differences between the laws of different countries. For this reason, the chapter deals largely with general principles rather than the details of the law.

2. There are differences between the *legal* and the *psychiatric* concepts of mental abnormality. These differences are made more complicated

because the concept of mental abnormality varies between different parts of the law. In this chapter it is not possible to review all these diverse concepts, but a few examples will be given in relation to such issues as fitness to plead, testamentary capacity (fitness to make a will), and the legal defence of insanity. If a psychiatrist is called upon to give a psychiatric opinion on a legal issue, he should acquaint himself with the relevant legal concepts and clarify the points to be addressed.

The law in relation to ordinary psychiatric practice

The principles concerning *confidentiality* and *informed consent to treatment* are the same in psychiatry as in general medicine, but certain points need to be stressed. Issues of compulsory treatment which are special to psychiatry are considered next, before a discussion of selected aspects of civil law.

Confidentiality

Confidentiality is particularly important in psychiatry because information is collected about private and sometimes highly sensitive matters. In general, the psychiatrist should not collect information from other informants without the patient's consent. If the patient is too mentally disturbed to give an account of himself, the psychiatrist should use his own discretion about seeking information from someone else. Sometimes such information is of vital importance to assessment and management. The guiding principle should be to try to act in the patient's best interests, and to obtain information as far as possible from close relatives rather than employers. The same principles apply when the psychiatrist needs to give information or an opinion to relatives or other people. In most countries there are legal requirements that medical information be disclosed in certain circumstances, usually to the courts. In the United Kingdom the General Medical Council has guidelines on situations in which confidentiality may be breached by doctors; they

include concerns for public safety and threats to an identified person. Similar considerations apply in most other countries, and it is important that psychiatrists are aware of the ethical and legal requirements in the country in which they are working.

Consent to medical treatment

The patient should have a clear and full understanding of the nature of a treatment procedure and its probable side-effects, and should freely agree to receive the treatment. For most treatments, such as established forms of medication, it is sufficient for the psychiatrist to explain the nature of the treatment and probable side-effects. However, there are, national differences in the degree of explanation required. For example, the concept of *informed consent* involves a more detailed account of side-effects of treatment in the United States than in the United Kingdom. (See Katz *et al.* (1995) for an account of psychiatric consultation with those refusing medical treatment.)

In general, **competent adults have a right to refuse medical treatment,** even if this refusal results in death or permanent disablement. However, there are several situations in which consent is not needed (although the precise nature of these exceptions depends on local law and precedent).

1. **Implied consent** such as where a patient is unconscious and a reasonable person would consent.
2. **Necessity** in which grave harm or death are likely to occur without intervention and there is doubt about the patient's competence.
3. **Emergency:** in order to prevent *immediate* serious harm to the patient or to others, or to prevent a crime.

If the patient is refusing medical treatment, the doctor needs to make two judgements before accepting that the patient has the right to refuse.

1. Is the patient **competent** and does he have the mental capacity to refuse treatment?

2. Has the patient been **influenced by others** to the extent that a refusal has been coerced and cannot be relied upon?

Issues about competence to consent to treatment arise particularly in three groups of patients: mentally retarded people, children and adolescents, and patients with mental illness.

The judgement of competence is usually not defined precisely; it depends upon the patient's being able to comprehend and retain information about the treatment, to believe this information, and to be able to use it to make an informed choice. Judgements of competence are specific to the particular decision that is to be made about treatment. A patient with a mental disorder may not be competent in other respects, but none the less may be competent to decide upon a particular treatment. In the United Kingdom the Mental Health Acts distinguish between consent to medical treatment and consent to psychiatric treatment. They provide for the compulsory treatment of mental disorder, but do not allow compulsory treatment for physical conditions.

Advance directives (living wills) are accepted in many countries as a means of ensuring that those who previously had the capacity to take decisions but have lost it, for example because of dementia, are treated in the way that they would have wished. Where patients do not have the required capacity to consent and have not made an advance directive, others have to decide whether treatment should be given or withdrawn. This decision must be made on the basis of the patient's **best interests** as determined by the responsible clinicians on the basis of their clinical judgement in accordance with general medical opinion. It is wise to consult relatives (although they cannot give or withhold consent) and to discuss the case with other professional staff. Detailed notes should be kept of the reasons for the decision and the consultations that took place. Current requirements for consent to treatment in England and Wales are given in the Appendix.

Compulsory admission and treatment

In developed countries there are laws to protect mentally disordered persons and to protect society from the consequences of their mental disorder, although the form of these laws differs between countries. Special legal provision is needed for people who are a danger to themselves or others because of mental disorder, and who refuse to accept the treatment that they require. Such people usually have little or no insight into their own psychiatric condition. They present a difficult ethical dilemma: they have a right to be at liberty but they also have a need for care and treatment, and society has a right to be protected. Countries vary widely in their approach to this dilemma. In some Scandinavian countries, for example, procedures for compulsory care are simple, whilst in some states of the United States, a court hearing may be required. In England and Wales, provisions for compulsory admission and treatment are embodied in the Mental Health Act 1983. The relevant sections of this Act are explained in the Appendix.

An experienced psychiatrist can often avoid the use of compulsory legal powers by patiently and tactfully discussing with the patient and relatives the reasons why admission is necessary. If this fails and compulsory treatment is justified, a further discussion takes place with family members to seek their support for the patient's admission to hospital. When talking to the family the doctor should understand their feelings of anxiety and guilt. When the patient is in hospital, restrictions should be kept to the minimum required for safety and adequate treatment. Usually the patient and the family soon realize that compulsory hospital care is little different from that of a voluntary patient, and they may feel relieved that treatment has started. If the hospital staff are patient, understanding, and adaptable, it is usually possible to establish an effective therapeutic relationship between staff, patient, and relatives. However, a patient admitted under a compulsory order sometimes refuses to accept restrictions or treatment. Such refusal can usually be overcome by skilful nursing with firmness tempered by sympathy, patience, and flexibility.

Another problem arises *when a voluntary patient who is judged to need electroconvulsive therapy (ECT)* for a severe and life-threatening psychiatric disorder (for example extreme depression with dangerous refusal of food and drink) refuses to consent to this treatment. As explained in the Appendix, in England and Wales the procedure is to discuss the problem fully with relatives, to seek an independent psychiatric opinion, and to complete a compulsory treatment order with the relatives' agreement.

In most mental health legislation, safeguards against unnecessary detention are provided and patients are entitled to easy access to appeal procedures. The type of safeguard varies from country to country. The system in England and Wales is reviewed in the Appendix (pp. 781–7). See Hoggett (1990) for a review of principles and practice, and Unsworth (1987) for a review of the development of mental health legislation in the United Kingdom.

Civil law

As explained in the introduction to this chapter, civil law is concerned with property, inheritance, and contracts. In other words, it deals with the *rights and obligations of individuals to one another* and includes family law (which is a concern of child psychiatrists. In this respect it differs from criminal law, which is concerned with offences against the state (not necessarily against an individual). Proceedings in civil law are undertaken by individuals or groups who believe that they have suffered a breach of the civil law, rather than by an agency of the state as with criminal law.

In matters of civil law, psychiatrists may be involved in issues such as *fitness to drive, testamentary capacity, receivership, marriage contracts, divorce, guardianship, torts and contracts, and compensation.* These matters are outlined below. The psychiatrist may be asked to submit a written report on a patient's state of mind in relation to these issues. The report should be prepared only after full discussion with the patient and only with the patient's full consent. In preparing such a report, the psychiatrist should follow the principles of writing a court report (described on p. 778). As with all psychiatric reports for legal purposes, the report should be concise and factual, and should give the reasons for any opinions.

Since the law on these issues is complicated, it is often advisable for the psychiatrist to seek legal guidance, particularly in relation to local legal concepts of abnormal mental state relevant to the issue in question. Sometimes international law is relevant, for example laws determined by the European Court of Human Rights.

Testamentary capacity

This term refers to the capacity to make a valid will. If someone is suffering from mental disorder at the time of making a will, its validity may be in doubt and other people may challenge it. However, the will may still be legally valid if the testator is of 'sound disposing mind' at the time of making it. Psychiatrists may be asked to report in relation to two issues: (1) testamentary capacity; (2) the possibility that the testator was subjected to undue influence.

In order to decide whether or not a testator is of sound disposing mind, the doctor should use four legal criteria:

(1) whether the testator understands what a will is and what its consequences are;

(2) whether he knows the nature and extent of his property (though not in detail);

(3) whether he knows the names of close relatives and can assess their claims to his property;

(4) whether he is free from an abnormal state of mind that might distort feelings or judgements relevant to making the will (a deluded person may legitimately make a will, provided that the delusions are unlikely to influence it).

In conducting an examination, the doctor should see the testator alone, but should also see relatives and friends to check the accuracy of factual statements.

Assessment of undue influence is more complex and requires assessment of the relationship between testator and beneficiary, the mental state of the testator, and what is know of earlier intentions. See Spar and Garb (1992) for a review of the issues and of assessment procedures.

Power of attorney and receivership

If a patient is incapable of managing his possessions by reason of mental disorder, alternative arrangements must be made, particularly if the incapacity is likely to last a long time. Such arrangements may be required for patients living in the community as well as those in hospital. In English law two methods are available—power of attorney and receivership.

Power of attorney is the simpler method, only requiring the patient to give written authorization for someone else to act for him during his illness. In signing such authorization, the patient must be able to understand what he is doing. He may revoke it at any time.

Receivership is the more formal procedure and is likely to be more in the patient's interests. In England and Wales an application is made to the Court of Protection, which may decide to appoint a receiver. The procedure is most commonly required for the elderly. The question of receivership is one that places special responsibility on the psychiatrist. If a patient is capable of managing his affairs on admission to hospital, but later becomes incapable by reason of intellectual deterioration, then it is the doctor's duty to advise the patient's relatives about the risks to property. If the relatives are unwilling to take action, then it is the doctor's duty to make an application to the Court of Protection. The doctor may feel reluctant to act in this way, but any actions taken subsequently are the Court's responsibility and not the doctor's.

Family law

A marriage contract is not valid if at the time of marriage either party was so mentally disordered as not to understand its nature. If mental disorder of this degree can be proved, a marriage may be decreed null and void by a divorce court. If a marriage partner becomes of 'incurably unsound mind' later in a marriage, this may be grounds for divorce.

A doctor may be asked for an opinion about the capacity of parents or a guardian to care adequately for a child. (See p. 718.)

Torts and contracts

Torts are wrongs for which a person is liable in civil law as opposed to criminal law. They include negligence, libel, slander, trespass, and nuisance. If such a wrong is committed by a person of unsound mind, then any damages awarded in a court of law are usually only nominal. In this context the legal definition of unsound mind is restrictive, and it is advisable for a psychiatrist to take the advice of a lawyer on it.

If a person makes a contract and later develops a mental disorder, then the contract is binding. If a person is of unsound mind when the contract is made, a distinction is made between the 'necessaries' and 'non-necessaries' of life. Necessaries are legally defined as goods (or services) 'suitable to the condition of life of such person and to his actual requirements at the time' (Sale of Goods Act 1893). In a particular case it is for the court to decide whether any goods or services are necessaries within this definition. A contract made for necessaries is always binding. In the case of a contract for non-necessaries made by a person of unsound mind, the contract is binding unless it can be shown both (a) that he did not understand what he was doing and (b) that the other person was aware of the incapacity.

Fitness to drive

Although not strictly part of civil law, it is convenient to discuss fitness to drive at this point. Questions of fitness to drive may arise in relation to many psychiatric disorders, but particularly the major mental disorders. Reckless driving may result from suicidal inclinations or manic disinhibition, panicky or aggressive driving may result from persecutory delusions, and indecisive or inaccurate driving may be due to dementia. Concentration on driving may be impaired in severe anxiety or depressive disorders. The question of fitness to drive also arises in relation to psychiatric drugs, particularly those with sedative effects such as anxiolytic or antipsychotic drugs in high dosage.

A doctor giving an opinion on fitness to drive should consider whether any medical condition or its treatment is liable to cause loss of control, impair perception or comprehension, impair judgement, reduce concentration, or affect motor functions involved in handling the vehicle. Doctors should determine the criteria for fitness to drive in the places in which they are working; these criteria may differ somewhat for drivers of cars and drivers of heavy goods vehicles.

Compensation

Psychiatrists may be asked to write reports in relation to claims for compensation by patients with post-traumatic stress disorder or other psychological sequelae of accidents. These conditions are described on p. 140 and p. 402 respectively. Reports should clearly state the sources of information, the history of the trauma, the psychiatric and social history, and the post-accident course. It should include a detailed functional assessment and examine the relationship between the trauma and subsequent symptoms and disability (Hoffman 1986; Hoffman and Spiegel 1989).

Patterns and causes of crime

Patterns of crime

It is possible to give only a brief outline of patterns of crime here. Crime is predominantly an activity of young men. In England and Wales half of all indictable offences are committed by males aged under 21, and a quarter by males aged under 17. In recent decades crime has increased in amount. Since the Second World War there has been a steady rise in the rates of crimes against property

and of violent crimes. This rise has included a sharp increase in the numbers of offences committed by women. Four-fifths of all crimes are against property. For further information on criminology see Wilson and Herrnstein (1985), Gunn and Taylor (1993, pp. 252–84), and McGuire (1994).

General causes of crime

The theory of degeneracy

The causes of crime have been seen in terms of the causes within the individuals and social causes. In the nineteenth century criminologists were interested in causes within the individual, especially the idea that criminals were degenerate. In 1876 Lombroso published an important book, *L'uomo delinquente*, in which he described characteristic physical stigmata in criminals. These ideas were the forerunners of interest, in the present century, in possible genetic causes for criminal behaviour.

Genetic studies

Early studies of twins suggested that concordance rates for criminality were substantially greater in monozygotic twins than in dizygotic twins (Lange 1931). Later adoption studies in Sweden and in Denmark have confirmed the genetic influence but shown that it is more modest than Lange supposed. It is mainly significant for severe and persistent criminality. The relationship is stronger for crimes against property than for violent crimes (see Brennan and Mednick 1993).

It is uncertain how genetic factors could lead to criminality. Prospective studies have shown a relationship between measures of physiological arousal at age 15 years and criminal acts at age 24 years (Loeb and Mednick 1977; Raine *et al.* 1990). It is possible that genetic factors determine arousability, which is related to the speed of learning to inhibit antisocial behaviour (Brennan and Mednick 1993).

Chromosomal abnormalities have also been suggested as causes of criminal behaviour. It was originally reported that the XYY chromosomal abnormality was more frequent in patients in maximum security hospitals than in the general population, but surveys suggest that the XYY constitution and other chromosomal abnormalities are not associated with criminal behaviour or with aggression in particular (Witkin *et al.* 1976).

Social studies

Social studies have drawn attention to *social and economic causes of crime* in the family, peer group, and subculture, and in conditions of poverty, poor schooling, and unemployment. Since many of the proposed predisposing social and individual factors are interrelated, simple conclusions are not possible. However, it is widely held that social causes of crime are generally more important than psychological causes (Downes and Rock 1988).

Psychological causes

There is a small but important group of offenders whose criminal behaviour seems to be partly explicable by psychological abnormalities. It is this group that particularly concerns the psychiatrist and is discussed next (Wilson and Herrnstein 1985; West 1988; Gunn and Taylor 1993, pp. 252–84; Farrington 1994).

Associations between psychiatric disorders and crime

Frequency of the association

It is difficult to obtain a reliable estimate of the numbers of mentally disordered offenders. An unknown but probably considerable number of mentally abnormal offenders bypass the courts; some of these are offenders whose offences are known to their doctors but not to the police, and others are known to the police but dealt with in ways other than prosecution.

Much of the evidence on prevalence of mental disorder relates to highly selected groups of offenders, particularly those in prison. Prisons contain many psychiatrically disturbed people.

Gunn *et al.* (1991) described the prevalence of psychiatric disorder among male sentenced prisoners in England and Wales, and their treatment needs, on the basis of a 5 per cent sample. Using standardized measures they concluded that overall 37 per cent had diagnosable disorder (with some having more than one year of disorder), 0.6 per cent had mental retardation, 2 per cent psychosis, 6 per cent neurotic disorder, 10 per cent personality disorder, 12 per cent alcohol dependence, and 12 per cent drug dependence. They considered that 3 per cent required treatment in a psychiatric hospital and 5 per cent could benefit from treatment in a prison therapeutic community. A further 10 per cent were thought to require further psychiatric assessment treatment within prison.

Whilst there is little evidence of large or growing numbers of people with serious mental illnesses in prison either in the United Kingdom or in other countries, there appears to be an important small group of such people who are not appropriately placed or treated.

The nature of the association

Much social or criminological evidence has suggested that criminality and mental illness are only weakly associated, and that apparent relationships may be coincidental (Monahan and Steadman 1994). An alternative psychiatric view has been that there is a real relationship between mental illness and crime (Wessely and Taylor 1991; Gunn and Taylor 1993, pp. 330–5; Taylor 1993), for example in the Epidemiologic Catchment Area Study. The psychiatric disorders most likely to be associated with crime were personality disorders, alcohol and drug dependence, and mental retardation. In addition to these categories, there is a sizeable group of recidivist offenders who are socially isolated, and often homeless and unemployed. They are frequently of low intelligence, and some have chronic schizophrenia. In this group, criminality is just one manifestation of all-round incompetence. Recent evidence suggests an association between violence and psychosis, especially those cases with paranoid ideation (Beck 1994; Mulvey 1994; Torrey 1994). However, only a small minority of all people who commit violent acts are psychotic, and the vast majority of patients with serious mental illness are no more dangerous than members of the general population.

Associations of mental disorder and crime among women

In the United Kingdom, five times as many men as women are convicted of indictable offences, and about 30 times as many men are imprisoned. Although this difference may be due in part to lower identification and reporting of female crime, it reflects a major difference in behaviour.

The most common offence committed by women is stealing. Shoplifting accounts for half of all convictions of women for indictable offences. In contrast, violent and sexual offences are uncommon (Heidensohn 1994*a*). It has been argued that women indulge in forms of antisocial behaviour that are regarded less severely by the law than those for which men are prosecuted, such as soliciting and some forms of social security fraud. However, in general, it would appear that women are more law-abiding than men. There are differences in the way men and women are treated by the criminal justice system, generally they are sentenced more leniently and they are more likely to be seen as 'sick' (Heidensohn 1994*b*; Maden *et al.* 1994).

It is widely accepted that a substantial proportion of crime by women is associated with mental disorder. Psychiatric disorder is frequent amongst women admitted to prison (Maden 1994), with personality disorder and drug abuse being especially common. Rates for self-harm before and during imprisonment are also high. The premenstrual syndrome is often suggested as an aetiological factor by defence lawyers and has occasionally been accepted as such in a number of recent court decisions. It is possible that premenstrual symptoms may complicate or exacerbate pre-existing social and psychological difficulties, but it is unlikely that they are ever a primary cause of offences (Gunn and Taylor 1993, pp. 598–623).

Specific psychiatric disorders as causes of crime

A brief review will now be given of the associations between crime and the various psychiatric diagnostic categories. It should be borne in mind that such associations are not necessarily causal. Moreover, if an association of this kind is to have any potential significance, the psychiatric diagnoses must be based on independent evidence and must not be deduced solely from the criminal behaviour, however bizarre.

Personality disorder

There are close associations between crime and personality disorder, particularly antisocial personality disorder. (The features, aetiology, and treatment of antisocial personality disorders are discussed in Chapter 5.) Offenders with such personality problems are often more susceptible to social than to psychiatric care, but there are sometimes indications for psychological treatment, such as therapeutic community techniques, or treatment for sexual problems or anxiety disorders.

In this book the term antisocial personality disorder is used in preference to 'psychopathic disorder'. However, the latter term is in current legal use in some countries, and when working with the courts in such places the psychiatrist may be required to use it. In the Mental Health Act 1983 of England and Wales, the term psychopathic disorder is employed and defined as a 'persistent disorder or disability of mind (whether or not including significant impairment of intelligence), which results in abnormally aggressive or seriously irresponsible conduct'. If a compulsory order is to be made on the grounds of psychopathy, then the Act requires that there be evidence that treatment 'is likely to alleviate or prevent a deterioration of (the patient's) condition' as well as the requirement 'that it is necessary for the health and safety of the patient, or the protection of other persons'. It is widely held that this legal concept of psychopathy is unsatisfactory. There is a danger that the diagnosis will be based on the nature of the crime rather than on any independent evidence of psychiatric disorder, and that it will be used as a label for people whose behaviour is not acceptable to conventional society. A further criticism is that the stipulation concerning psychiatric treatment is unrealistic, since most patients with psychopathic disorder are not susceptible to psychiatric treatment.

Alcohol dependence

There are close relationships between substance abuse and crime which have substantially affected legislation, enforcement, and national policies. Individual offences can only be understood in the wider context of attitudes to drugs and alcohol and legal policy (South 1994).

Alcohol and crime are related in three important ways:

(1) alcohol intoxication may lead to charges related to public drunkenness or to driving offences;

(2) intoxication reduces inhibitions and is strongly associated with crimes of violence, including murder;

(3) the neuropsychiatric complications of alcoholism (see Chapter 14) may also be linked with crime.

For example offences may be committed during alcoholic amnesias or 'blackouts' (periods of several hours or days which the heavy drinker cannot subsequently recall, although at the time he appeared normally conscious to other people and was able to carry out complicated actions). However, the association is complex, and social factors related to drinking may be as important as alcohol itself.

Drug dependence

Intoxication with drugs may lead to criminal behaviour including violent offences. Drug abusers, especially those dependent on heroin and cocaine, commit repeated offences against both property and people to pay for their drugs. Some of the offences involve violence. Rates of drug abuse are increased among prisoners, and some succeed in continuing to obtain drugs in prison. Involvement in criminal activity and with other criminals may lead to drug usage. For a review of the relationship between drug dependence and crime see South (1994) and Gunn and Taylor (1993, p. 435–81).

Mental retardation

Contrary to former early beliefs, there is no evidence that most criminals are of markedly low intelligence. Recent surveys have shown that most delinquent youths are within the lower part of the normal range of intelligence, but only about 3 per cent are mentally retarded (Lund 1990). There is no reason to suppose that the distribution of intelligence is any different amongst adult criminals. However, compared with other offenders, the mentally retarded are more likely to be caught.

Mentally retarded people may commit offences because they do not understand the implications of their behaviour, or because they are susceptible to exploitation by other people. The closest association between mental retardation and crime is a high incidence of sexual offences, particularly indecent exposure by males (Craft 1984). The exposer is often known to the victim and therefore the rate of detection is high. There is also said to be an association between mental retardation and arson. Apart from sexual offences and arson, no other crimes are closely associated with mental retardation.

Mood disorder

Depressive disorder Depressive disorder is sometimes associated with shoplifting (see p. 767). Much more seriously, severe depressive disorder may lead to homicide. When this happens, the depressed person usually has delusions, for example that the world is too dreadful a place for him and his family to live in; he then kills his spouse or children to spare them from the horrors of the world. The killer often commits suicide afterwards. A mother suffering from post-partum disorder may sometimes kill her newborn child or her older children. Rarely, a person with severe depressive disorder may commit homicide because of a persecutory belief, for example that the victim is conspiring against the patient. Occasionally, ideas of guilt and unworthiness lead depressed patients to confess to crimes that they did not commit.

Mania Manic patients may spend excessively on expensive objects such as jewellery or cars that they cannot pay for. They may hire cars and fail to return them, or steal cars for their own use. They may be charged with fraud or false pretences. Manic patients are also prone to irritability and aggression, which may lead to offences of violence, though the violence is seldom severe.

Schizophrenia

There is conflicting evidence about the relationship between schizophrenia and crime, probably because of differences in the populations studied. Although most research has been concerned with homicide and other violent crimes, recent evidence suggests that schizophrenic patients are more likely to commit non-violent as well as violent offences. A twin study by Coid *et al.* (1993) found increased rates of criminality in schizophrenics and showed that most of the criminal behaviour followed the onset of schizophrenia. Frequently, criminality is a result of personality difficulties and an all-round social incompetence rather than psychiatric symptoms. These people are usually apathetic and lacking in judgement. Until recently most lived as long-stay patients in mental hospitals; nowadays, most live in the community, and some become destitute. Their crimes are usually petty, but because they are repeated may lead to prison sentences (Valdiserri *et al.* 1986).

In a large study of mentally abnormal offenders in Germany, the risk of homicide was found to be moderately increased in schizophrenia compared with the general population (Böker and Häfner 1977). In Sweden, a study of 790 schizophrenic patients aged over 15 years found that the overall crime rate of male schizophrenics was similar to that of the general population but that the rate of

violent offences was four times higher. However, the violence was almost always of minor severity (Lindqvist and Allebeck 1990). In the United States, the Epidemiological Catchment Area Study found that the one-year prevalence of reported violence by schizophrenics was considerably higher than in the general population.

In some schizophrenics, violence results from delusions and hallucinations. According to a hospital case record study by Planansky and Johnston (1977), violence in schizophrenics may be associated with any of the following features: great fear and loss of self-control associated with non-systematized delusions, systematized paranoid delusions including the conviction that enemies must be defended against, irresistible urges, instructions from hallucinatory voices, and unaccountable frenzy. More recently, Buchanan *et al.* (1993) concluded that the risk of violence is greatest where delusions are accompanied by strong affect and when the person has made efforts to try and confirm the truth of the delusions. Taylor (1985) interviewed remanded psychotic men and concluded that psychiatric symptoms accounted for most of the very violent behaviour, and that in about 40 per cent there was a direct psychiatric drive to commit the offence. As mentioned in the section on dangerousness (p. 779), violent threats in schizophrenics should be taken very seriously (especially in those with a history of previous violence); most serious violence occurs in those already known to psychiatrists. The change to community care has made minor criminality and rare violent offences more conspicuous and has resulted in increased public disquiet. Psychiatric services need the resources to minimize difficulties and to identify and manage serious threats of violence. See Gunn and Taylor (1993, pp. 329–72) and Taylor (1993) for reviews of the risk of violence in psychotic illness.

Organic mental disorders

Acute organic mental disorders are occasionally associated with criminal behaviour. Diagnostic problems may arise if the mental disturbance improves before the offender is examined by a doctor. Dementia is sometimes associated with offences, though crime is otherwise uncommon among the elderly and violent offences are rare. A few elderly men commit sexual offences, usually in the form of indecency with children. Such men have usually had lifelong sexual difficulties but no previous offences of any kind. Whenever an elderly man is charged with a sexual offence, it is important to consider the possibility of dementia. Violent and disinhibited behaviour may also occur after traumatic damage to the brain following head injury.

Epilepsy

It is uncertain whether criminality is more common among epileptics than among non-epileptics (see p. 340). The uncertainty results from the use of selected populations and different definitions of epilepsy in different surveys. It is known that there are more epileptics in prison than would be expected in relation to the general population (Gunn 1977*a*). This finding holds particularly for young persons and those convicted of violent offences. It appears that fits themselves are not a significant cause of crime but that there are other explanations. Some epileptics suffer from *brain disorder* that induces both seizures and criminal behaviour, some carry out crimes because of the *low self-esteem and social problems* associated with the epilepsy, and some carry out crimes because of an associated mental disorder. It is possible also that epileptic automatism is a rare cause of crime. It has been suggested that violent behaviour is sometimes associated with EEG abnormalities in the absence of clinical epilepsy.

Episodic dyscontrol syndrome

The episodic dyscontrol syndrome was described by Bach-y-Rita *et al.* (1971) in patients who had repeated unprovoked episodes of violence. These authors considered that the syndrome had more than one cause and their series included patients with epilepsy. Maletzky (1973) excluded patients with epilepsy, schizophrenia, pathological intoxication with alcohol, and acute intoxication with drugs, and described a residual group of patients whose unexplained episodes of violence were preceded by a sequence of aura, headache, and drowsiness. About half reported amnesia for the episode and half had EEG abnormalities, usually in the temporal lobes. Maletzky reported improvement with the antiepileptic drug phenytoin, but there was no placebo control group against which to assess this finding. These findings have not been confirmed by subsequent studies. The evidence for a distinct syndrome of episodic dyscontrol is not convincing (Fenton 1986).

Impulse control disorders

DSMIV contains a rubric for 'impulse control disorders not otherwise classified' which brings together in a convenient way four conditions relevant to forensic psychiatry: intermittent explosive disorder, pathological gambling, pyromania, and kleptomania. (The rubric also contains another condition, trichotillomania, which is considered on p. 389.) In ICD10 these conditions are classified under abnormalities of adult personality and behaviour as 'habit and impulse disorders'.

Intermittent explosive disorder This term is used to describe repeated episodes of seriously aggressive behaviour directed to people or property that is out of proportion to any provoking events and is not accounted for by another psychiatric disorder (for example antisocial personality disorder, substance abuse, or schizophrenia). The aggression may be preceded by tension and followed by relief of this tension. Later the person feels remorse. If care is taken to exclude other causes, the condition is rare. Some psychiatrists doubt whether this condition exists as a distinct entity.

Pathological gambling Pathological ('compulsive') gambling is not itself illegal but may lead to behaviours that bring the gambler to the attention of the courts, for example fraud or stealing to obtain money to pay for the habit. Gambling is pathological when it is repeated frequently and dominates the person's life; the gambling persists when the person can no longer afford to pay his debts. The person lies, steals, or defrauds in order to obtain money or avoid repayment and to continue the habit. Family life may be damaged, other social relationships impaired, and employment put at risk. The pathological gambler has an intense urge to gamble which is difficult to control. He is preoccupied with thoughts of gambling, much as a person dependent on alcohol is preoccupied with drink. Often, increasing sums of money are gambled, either to increase the excitement or in an attempt to recover previous losses. Gambling continues despite inability to repay debts and despite awareness of the resulting social and legal problems. If gambling is prevented, the person becomes irritable and even more preoccupied with the behaviour. Similarities between patterns of behaviour and those of people dependent on drugs have led to the suggestion that pathological gambling is itself a form of addictive behaviour.

The *prevalence* of pathological gambling is not known. It is probably more frequent among males. Most gamblers seen by psychiatrists are adults, but there is concern that young people are increasingly being involved, usually with gambling machines in amusement arcades and other places. The *causes* of pathological gambling are not known.

In the absence of randomized controlled comparative trials, it is impossible to assess the value of psychiatric *treatment*. The usual management is similar to that for dependence on alcohol. First, the gambler takes part in a thorough review of the effects of the habit on himself and his family; then he is given strong encouragement to abstain from gambling. There are self-help groups (Gamblers Anonymous) which resemble Alcoholics Anonymous in providing a combination of individual confession and catharsis, and group support. Help is given to the family and advice is provided about any legal consequences of excessive gambling. Unfortunately, the prognosis is often poor. For a review of pathological gambling see Gunn and Taylor (1993, pp. 481–5).

Pyromania Pyromania is one cause of fire-setting (see p. 768). The term pyromania refers to repeated episodes of deliberate fire-setting which are not carried out for monetary gain, to conceal a crime, as an act of vengeance, for social or political motives, or as a consequence of hallucinations, delusions, or impaired judgement (resulting from intoxication, dementia, or mental retardation for example). The diagnosis is not made when there is an associated antisocial personality disorder, a manic episode, or (among children or adolescents) a conduct disorder. In this rare condition, the act of fire-setting is preceded by tension or arousal, and is followed by relief of tension. Patients with pyromania have a preoccupying interest in fires and in matters related to them such as firefighting equipment. They enjoy watching fires. They may plan the fire-setting in advance, taking no account of the danger to other people from their actions. When causes other than those listed above are excluded carefully, pyromania is rare; indeed, some writers doubt its existence. For a review of pyromania see Geller (1992) and Barker (1994).

Kleptomania

The term kleptomania refers to repeated failure to resist impulses to steal objects that are not needed, either for use or for their monetary value. The impulses are not associated with delusions or hallucinations, or with motives of anger or vengeance. Before the act of stealing there is increased tension; after the act there is relief of tension. The diagnosis is not made when there is an associated antisocial personality disorder, a manic episode, or (among children or adolescents) a conduct disorder, nor when the stealing results from sexual fetishism. The objects stolen may be of little value and could have been afforded; they may be hoarded, thrown away, or returned later to the owner. The patient knows that the stealing is unlawful, and may feel guilty and depressed after the immediate pleasurable sensations that follow the act. The disorder occurs more often among women. Associations with anxiety and eating disorders have been described (Goldman 1991; McElroy *et al.* 1991). The behaviour may be sporadic with long intervals of remission, or may persist for years despite repeated prosecutions.

When other causes of repeated stealing are excluded, the condition is rare; indeed, some writers doubt its existence, pointing out that diagnosis depends on the accused persons' descriptions of their own motives.

Post-traumatic stress disorder

Post-traumatic stress disorder has been claimed to be a cause of offending in some cases. However, there may be confounding factors of drug or alcohol abuse, and the strength of the association with post-traumatic stress disorder is uncertain (Sparr and Atkinson 1986).

Psychiatric aspects of specific crimes

The following sections are concerned with offences of the types that are most likely to be associated with psychological factors. These offences can be divided into crimes of violence, sexual offences, and offences against property.

Crimes of violence

There have been many attempts to understand the cause of violence. The main hypotheses have been as follows: (1) aggression as a fundamental instinct which may be expressed or in undesirable aggressive acts as well as in socially acceptable ways; (2) aggression as the result of a frustration in goal-directed behaviour; (3) aggression as learned behaviour resulting either from experiences in which aggression was rewarded, or from observation and modelling of the aggressive behaviour of other people. Several factors determine whether a person is aggressive in a particular situation: personality, the immediate social group, the behaviour of the victim, disinhibiting factors such as alcohol or drugs, general environmental factors such as noise and social pressure, physiological factors such as fatigue, hunger, and lack of sleep, and the presence of mental abnormality.

Amongst mentally abnormal offenders, violence is associated more often with personality disorder than with psychiatric disorder (Taylor 1993). Violence is particularly common in people with antisocial personality traits who abuse alcohol or drugs, or who have marked paranoid or sadistic traits. Violence is often part of a persistent pattern of impulsive and aggressive behaviour, but it may be a sporadic response to stressful events in 'overcontrolled' personalities (Megargee 1966). There are cultural norms concerning the expression of violence. It is still uncertain whether depictions of violence in the media increase violent behaviour among the viewers. Several kinds of violent offence will be discussed next. The assessment of dangerousness and the management of violence are discussed on p. 779.

Homicide

There are several legal categories of homicide: murder, manslaughter, and infanticide. In England and Wales murder and manslaughter are defined by historical precedent and not by statute (Smith and Hogan 1988). According to a widely quoted definition put forward by Lord Coke in 1797, **murder** occurs

when a man of sound memory and of the age of discretion unlawfully killeth within any country of the realm any reasonable creature in rerum natura under the King's peace with malice aforethought, either expressed by the party or implied by law, so as the party wounded or hit, etc, die of the wound or hit within a year a day after the same.

The phrase 'malice aforethought' is important, although it has no statutory definition and can be interpreted only from case law.

According to Smith and Hogan (1988), **manslaughter**

is a diverse crime covering all unlawful homicides which are not murder. A wide variety of types of homicide fall within this category, but it is customary and useful to divide manslaughter into two main groups which are designated 'voluntary' and 'involuntary' manslaughter respectively. The distinction is that in voluntary manslaughter the defendant may have malice aforethought of murder but the presence of some defined mitigating circumstances reduces his crime to a less serious grade of criminal homicide.

In involuntary manslaughter, there is no malice aforethought; it includes, for example, causing death by gross negligence.

As mentioned elsewhere in this chapter, the category of manslaughter resulting from diminished responsibility is defined in England not by common law but by statute, namely the Homicide Act 1957. This Act also provides that the survivor of a genuine suicide pact should be guilty only of manslaughter.

Normal and abnormal homicide It is common practice to divide homicide into 'normal' and 'abnormal' according to the legal outcome. Homicide is 'normal' if there is a conviction of murder or common law manslaughter; it is 'abnormal' if there is a finding of insane murder, suicide murder, diminished responsibility, or infanticide.

'Normal' homicide accounts for half to two-thirds of all homicides occurring in the United Kingdom. In countries such as the United States, where the overall homicide rate is much higher than in the United Kingdom, the excess is largely made up of 'normal' homicide. 'Normal' homicide is most likely to be committed by young men of low social class. In the United Kingdom, the victims are mainly family members or close acquaintances, and they are less often killed in the course of a robbery or sexual offence. In countries with high homicide rates a greater proportion of killings are associated with robbery or sexual offences. Sexual homicide may result from panic during a sexual offence or may be a sadistic killing, sometimes committed by a shy man with bizarre sadistic and other violent fantasies.

'Abnormal' homicide accounts for a third to half of all homicides in the United Kingdom. It is usually committed by older people. Homicide by women is much less frequent than by men; when it occurs, it is nearly always 'abnormal' and the most common category is infanticide. The victims of abnormal homicide are usually family members. In those who commit 'abnormal' homicide, the most common psychiatric diagnosis is depressive disorder, especially in those who kill themselves afterwards. Other associated diagnoses are schizo-phrenia, personality disorder, and alcoholism. The syndrome of pathological jealousy may be associated with any of the above diagnoses; it has been identified in 12 per cent of insane male murderers and 3 per cent of insane women murderers. It is particularly dangerous because of the risk of the offence being repeated (see p. 301).

A large proportion of all murderers are *under the influence of alcohol* at the time of the crime (Vieweg *et al.* 1974). In a survey of 400 people charged with murder in Scotland, 58 per cent of the men and 30 per cent of the women were found to have been intoxicated at the time of the offence (Gillies 1976). Drug abuse is also an important factor (Tardiff *et al.* 1994).

Multiple homicide Multiple murders are rare, although they attract great public interest. They include (a) those who kill several people at once, sometimes a family killing which is often followed by suicide, (b) killings attributable to a psychotic illness in which the killer aims to save himself or his family from a perceived threat, and (c) serial killings taking place over a period of time. Serial murders may be 'normal' (for example, killings by terrorists), psychotic, or motivated by sexual sadism or necrophilia.

Homicide followed by suicide Homicide is followed by suicide in about 10 per cent of homicides in England and Wales. Marzuk *et al.* (1992) reviewed epidemiological findings and concluded that in the United States murder followed by suicide presents 1.5 per cent of all suicides and 5 per cent of all homicides. In the United Kingdom, West (1965) studied 78 cases occurring in the London area over the years 1954–1961. The offenders were much more likely to be women, were of higher social class, and had fewer previous convictions than other convicted homicide offenders. The victims were usually children. Half the homicides were 'abnormal' in the sense defined above; in most cases the offender was severely depressed at the time of the offence. In most of the 'normal' offences, the killer appeared to have felt driven to suicide by illness or distressing circumstances.

In Marzuk's study in the United States, the four most common clinical types were jealousy (50–75 per cent), killing of elderly spouses in poor health, killing of a child often attributable to depression in the parent, and multiple murders by a depressed, paranoid, or intoxicated person. Approximately a quarter to a fifth of all jealous men who killed their spouses also committed suicide.

Parents who kill their children A quarter of all victims of murder or manslaughter in the United Kingdom are under the age of 16. Most of them are killed by a parent who is mentally ill, usually the mother (d'Orban 1979). The classification of child murder is difficult, but useful categories suggested by Scott (1973) are mercy killing, psychotic murder, and killing as the end result of battering or neglect. This last category is discussed further on pp. 716 and 718. Infanticide, which is a special legal category, is discussed next.

Infanticide A woman who kills her child may be charged with murder or manslaughter, but under special circumstances the charge may be infanticide. In England and Wales the Infanticide Act 1922, afterwards amended by the Infanticide Act 1938, defined a category of offence which can now be seen as a special case of the later and wider concept of diminished responsibility. Section 1 of the Act provides that

where a woman causes the death of her child under the age of 12 months, but at the time the balance of her mind was disturbed by reason of her not having fully recovered from the effects of childbirth or lactation consequent upon the birth of the child, she shall be guilty not of murder but infanticide.

The judge has the same freedom of sentencing for a conviction of infanticide as for a conviction of manslaughter. The legal concept of infanticide is unusual in that the accused is required to show only that her mind was disturbed as a result of birth or lactation, but not that the killing was a consequence of her mental disturbance.

Resnick (1969) found that two types of infanticide could be discerned. When the killing occurred within the first 24 hours after birth, in most cases the child was unwanted, and the mother was young and unequipped to care for the child but not psychiatrically ill. When the killing occurred more than 24 hours after childbirth, in most cases the mother had a depressive disorder and killed the child to save it from the suffering she anticipated for it; about a third of the mothers also tried to take their own lives. Marks and Kumar (1993) examined records of all infants aged under a year when the victims of homicide during 1982–1988. Infants were most at risk on the first day of life, and the relative risk decreased steadily thereafter until by the final quarter of the first year of life the risk of homicide was the same as that of the general population. Parents were the most frequent perpetrators, mothers on the first day and thereafter fathers were slightly more likely to be recorded as the prime suspect. Mothers received less severe convictions and less severe sentences than fathers. It would seem that, after the first day, fathers are generally as likely to be responsible as mothers. Contrary to wide belief, puerperal psychotic illness was a relatively infrequent cause of homicide by the mother. International comparisons are difficult, but rates of infant homicide in the United States are higher than in England and Wales, and are especially so amongst black Americans (Christoffel *et al.* 1989).

Victims of homicide The statistics about the victims of homicide are of interest. A quarter of all homicide *victims are aged under 16*; their deaths usually result from 'abnormal' homicide or repeated child abuse by the parents. Amongst *adult victims*, women outnumber men by three to two. Nearly half of women victims are killed by their husbands, and the rest mainly by relatives or intimate friends. In contrast, nearly half of male victims are killed by strangers or chance associates. In one study that about a third of homicide victims were probably intoxicated with alcohol at the time of the crime (Gillies 1976).

Violence within the family

This subject has received increasing attention in recent years (Smith 1989). Family violence is strongly associated with excessive drinking. Some people are violent only within their family, whilst others are also violent outside the family. Violence in the family can have long-term deleterious effects on the psychological and social development of the children as well as on the mental health of the spouse (see Chapter 20). Violence in the family, may be directed to the spouse, to children, (*child abuse* is reviewed on p. 716), or to elderly relatives (*violence to the elderly* is referred to on p. 530). Any of these forms of violence may result in homicide.

Violence to spouses Violence by men towards their wives is much more conspicuous than violence to husbands. The latter is less frequent, physically less serious, and much less often reported. It appears that most of the perpetrators of wife-battering are men with aggressive personalities, whilst a few are violent only when suffering from psychiatric illness, usually a depressive disorder. Other common features in the men are morbid jealousy and heavy drinking. Such men may have suffered violence in childhood, and often come from backgrounds in which violence is frequent and tolerated. In general, the role of the victim in contributing to and provoking violence is important but is often unclear in individual cases.

When psychiatric advice is requested in the management of domestic violence, it is important to gain an understanding of the interaction between the participants and of the help that is acceptable to the couple. It is rarely helpful to apportion blame. Marital therapy and family therapy are sometimes helpful, but individual psychotherapy is usually of little benefit (Rosenfeld 1992). In some cases, a frightened wife may need practical help to leave the home. It is important to keep in mind that family violence may result in homicide.

Sexual offences

In the United Kingdom sexual offences account for less than 1 per cent of all indictable offences recorded by the police. Sexual offenders make up a relatively large proportion of offenders referred to psychiatrists, though only a small proportion of people charged with sexual offences are assessed by psychiatrists. Most sexual offences are committed by men; apart from soliciting for purposes of prostitution, women seldom commit such offences. This section applies almost entirely to men. As a group, sexual offenders are older than other offenders. Reconviction rates of sexual offenders are generally lower than those of other offenders, but a minority of recidivist sexual offenders are extremely difficult to manage (Gunn 1985b).

The most common sexual offences are indecent assault against women, indecent exposure, and unlawful intercourse with girls aged under 16. Some sexual offences do not involve physical violence (for example indecent exposure, voyeurism, and most sexual offences involving children); others may involve considerable violence (for example rape). The nature and treatment of non-violent sexual offences are discussed in Chapter 15, but their forensic aspects are considered here (for reviews see Bancroft (1991) and Gunn and Taylor (1993, pp. 522–66)). The psychological consequences for victims of sexual offences are discussed on p. 145, and sexual abuse of children is discussed on p. 720.

Sexual offences against children

The age of consent varies in different countries. In England and Wales it is illegal to have any heterosexual activity with a person aged under 16, or homosexual activity with people aged under 18. Sexual offences involving children are reported commonly, amounting to over half of all reported sexual offences in the United Kingdom. It is probable that many more offences are not reported, particularly those occurring within families. The offences vary in severity from mild indecency to seriously aggressive behaviour, but the large majority do not involve violence.

Adults who commit sexual offences against children are known as *paedophiles*. They are almost always male. They may be homosexual or heterosexual. They are rarely mentally ill. Victims are most commonly female. It is difficult to classify paedophiles, but the following groups have been recognized: timid and sexually inexperienced, the mentally retarded and untrained, those who have experienced normal sexual relationships but prefer sexual activity with children, and a predatory group who repeatedly seek out vulnerable children. In rare cases paedophile sexual activities end in murder (Quinsey 1986).

The *victims* are known to the offender in four-fifths of cases, and belong to the offender's family in a third (Mohr *et al.* 1964). Girl victims outnumber boys by about two to one. Many child victims suffer emotional difficulties later (Constantine 1981). (Sexual abuse of children by family members is considered on p. 716.) The *prognosis* is difficult to determine with certainty. Among those who receive a prison sentence, the recidivism rate is about one in three. Most offenders do not progress from less serious to more serious activities, although an important minority progress to violent sexual offences. For this reason psychiatrists may be asked to give an opinion on an offender's dangerousness.

Assessment Generally, a different person is asked to assess the perpetrator and the victim. The interviewing of children after sexual abuse is considered on p. 721; the interviewing of the adult is considered here. In trying to decide whether an offence is likely to be repeated and whether there is likely to be a progression to more serious offences, the psychiatrist should consider the duration and frequency of the particular sexual activity in the past. Paedophiles often deny their offending and the psychiatrist should study the depositions and the victim's statement carefully (Kennedy and Grubin 1992). The offender's predominant sexual orientation should be considered; exclusively paedophile inclinations and behaviour indicate greater risk of repetition. Older paedophiles are less likely to be aggressive. It is important to determine whether alcohol or drugs played any part in the offence, and if so whether the person is likely to continue using them, and whether the offender feels regret or guilt. Relevant environmental factors include stressful circumstances associated with the offence (and the likelihood that these will continue) and the degree of access to children. Finally, evidence should be sought of any

psychiatric disorder or personality features such as lack of self-control. In drawing conclusions it is important to be aware of the limitations of psychiatric knowledge of this form of behaviour. For an assessment of adults who have committed sexual offences against children see Becker and Quinsey (1993).

Treatment Treatment is directed towards any associated psychiatric disorder. Direct treatment of the sexual behaviour is difficult. Group therapy run jointly by a psychiatrist or psychologist and a probation officer may be tried. Behavioural treatment has been directed towards encouraging desirable sexual behaviour, but the evidence for effectiveness is unconvincing. The use of sex hormones or drugs to reduce sex drive has been advocated, but its value is uncertain and its use raises ethical issues.

Sexual abuse of children is discussed further on p. 716.

Incest

It is difficult to determine the frequency of incest because this kind of behaviour is particularly unlikely to be revealed to an interviewer. Community studies report that many adults recall sexual interference by a family member during childhood. Most reported cases involve a father and daughter, but brother–sister relationships may be more common, and relationships between mothers and sons may not be as uncommon as reported cases suggest. Incest between father and daughter often starts as the girl reaches puberty. Several social factors may contribute. There may be a history of marital breakdown such that the daughter replaces the mother. The family is often socially isolated and share bedrooms in crowded accommodation. About a third of the fathers have antisocial personalities and many drink excessively.

It is likely that some cases are known to the medical or social services but not to the police. Among cases known to the police, only half are prosecuted. When the girl is a young child, most prosecutions result in imprisonment of the father. The family needs considerable psychological and social support, particularly if there is publicity about the prosecution or the father is imprisoned.

The long-term consequences of incest on the family are uncertain. The long-term consequences of child sexual abuse are discussed on p. 154.

Indecent exposure

The term indecent exposure is the legal name for the offence of indecently exposing the genitals to other people. It is applied to all forms of exposure; exhibitionism is by far the most frequent form, but exposure may also occur as an invitation to intercourse, as a prelude to sexual assault, or as an insulting gesture. Exhibitionism, as explained on p. 501, is the medical name for the behaviour of men who gain sexual satisfaction from repeatedly exposing to women. In England and Wales, indecent exposure is one of the most frequent sexual offences. It is most common in men aged between 25 and 35.

Indecent exposers rarely have a history of psychiatric disorder or other criminal behaviour. The reconviction rate is low, and few of the offenders proceed to more serious offences.

Indecent assault

The term indecent assault refers to a wide range of behaviour from attempting to touch a stranger's buttocks to sexual assault without attempted penetration. The psychiatrist is most commonly asked to give a psychiatric opinion on adolescent boys and on men who have assaulted children. Many adolescent boys behave in ways that could be construed as 'indecent'. More serious indecent behaviour is associated with aggressive personality, ignorance and lack of social skills, personal unattractiveness, and occasionally subnormal intelligence. Treatment depends on the associated problems.

Rape

In England and Wales the Sexual Offences Act 1956 states that 'a man commits rape if (a) he has unlawful sexual intercourse with a woman who at the time of the intercourse does not consent to it and (b) at the time he knows that she does not consent to the intercourse or he is reckless as to whether she consents to it'. The Act refers to unlawful sexual intercourse *per vaginam*; male rape is excluded and so is forcible anal intercourse with a woman. Although forced sexual intercourse within marriage is excluded from the 1956 Act, there is case law recognizing rape in marriage. Rape varies from the use of deception without violence to extreme brutality. In over half the recorded cases, the offender was known to the victim and the most common place for rape to occur is in the home (Lewis 1994). Rape and other forms of sexual aggression towards women are probably much more frequent in the population than the number reported to the police would suggest. Amir (1971) found that *group rape* accounted for a quarter of all rapes.

Male rape may be motivated by a wish to degrade or dominate the other person as well as sexual motives. Male rape is more common in prisons where heterosexual activity is impossible. Homosexual or, less commonly, heterosexual men may be assaulted. There are rare cases of sexual assault on vulnerable men by women (Sarrell and Masters 1982; Mezey and King 1987).

Explanations of rape are principally sociocultural in terms of male attitudes to women and women's likely responses. *Rapists* have been classified by psychiatrists in various ways. Most rapists are young and sexually frustrated, with little experience of sexual intercourse, and most have a record of previous criminal offences and of aggression on slight provocation. Many try to persuade themselves that the victim was a willing partner. Major psychiatric disorder is rare among rapists. The following types of rapist have been described clinically (but have no proven validity): (i) aggressive antisocial men who have a history of general criminal behaviour but do not have a psychiatric disorder; (ii) aggressive sadistic men who wish to humiliate and hurt women; (iii) so-called explosive rapists who are often timid and inhibited, and who carry out the act as a deliberate plan to relieve their frustration; (iv) mentally ill rapists, who most often suffer from mania (this is the least common group).

The *reconviction* rate is fairly low. Gibbens *et al.* (1977) found that, among men charged with rape, 12 per cent of those convicted and 14 per cent of those acquitted were convicted of a further sexual offence during a 12-year follow-up. In interpreting these findings, it should be remembered that most of the convicted men were in prison and therefore not at risk during part of the follow-up period.

Victims of rape In about a third of cases the victim is an acquaintance of the rapist, and in a fifth she appears to have participated initially in the events leading up to the offence. It is sometimes said that women who are raped have frequently encouraged the man initially, or else submitted without much resistance. There is little justification for this view. Amir (1971) found that half of rape victims were threatened with injury either verbally or with weapons, and about a third were handled roughly or violently. In such a dangerous situation submission without much physical resistance is understandable.

There is much evidence that rape victims may suffer long-term psychological effects (see also p. 145). Nadelson *et al.* (1982) interviewed 41 women who attended a clinic in a general hospital shortly after being raped. At follow-up 12 to 18 months later it was found that half the women were afraid of being alone, and three-quarters were still suspicious of other people. Many women reported depression and sexual difficulties which they attributed to the rape. More recent research has shown very high levels of intrusive thoughts and other post-traumatic symptoms in the week following rape. At one month 65 per cent met the criteria for post-traumatic stress syndrome and at 6 months 15 per cent met these criteria. In one study, 17 per cent of rape victims were still suffering severe distress over 10 years after the rape (Kilpatrick 1987).

Serious distress may also be experienced by the partners and families of rape victims. In the United States many crisis intervention centres staffed by multidisciplinary teams have been set up for rape victims. For a review of the psychological reactions and treatment of rape victims see Rothbaum and Foa (1993).

Sexual violence

Some men who commit rape, homicide, or other violent offences have considerable sexual problems or suffer sexual jealousy, and these may have contributed to their dangerousness (Gunn 1985*a*). A small group of men obtain sexual pleasure from sadistic assaults on unwilling partners (MacCulloch *et al.* 1985). Frequently, the only evidence of psychiatric abnormality is the deviant sexual desire itself.

Child abduction

Child abduction is rare. A child may be abducted by one of the parents, by a man with a sexual motive, or by an older child. Babies are usually abducted by women who may have one of three kinds of motives (d'Orban 1976): to achieve comfort, to manipulate another person, and on impulse by psychiatrically disturbed women. Fortunately, most stolen babies are well cared for and are found quickly (see Stephenson 1995).

Offences against property

Shoplifting

Many adolescents admit occasional shoplifting, but few admit persistent shoplifting. Both observational studies (Buckle and Farrington 1984) and the reports of huge losses from shops suggest that shoplifting is common among adults. It is generally accepted that most shoplifters are covetous rather than psychiatrically disturbed. However, an important minority have a psychiatric disorder. This minority was identified in an important study by Gibbens *et al.* (1971) who followed-up over 500 women who had been convicted of shoplifting 10 years previously. Most reconvictions were for further shoplifting. During the follow-up period the admission rate to mental hospital was three times the expected figure for women of similar age in the population. The authors distinguished two subgroups of women. One consisted of women who were persistently and widely deviant and had committed numerous previous and subsequent offences including theft, violence, and drunkenness. The other subgroup had suffered, at the time of the original offence, from a depressive disorder or from the effects of chronic background difficulties or recent life events. These women had generally been of good previous character and most had no further convictions. A more recent study in Canada identified a similar group of shoplifting women suffering from depressive disorder (Bradford and Balmaceda 1983).

Apart from depressive disorders, various other psychiatric diagnoses may be associated at times with shoplifting. Patients with drug or alcohol dependence may steal because of economic necessity. Patients with mania or acute or chronic schizophrenia occasionally steal from shops, and patients with anorexia nervosa occasionally steal food. In other conditions, shoplifting may result from distractibility; examples are organic mental disorders, when the person is confused or forgetful, and panic attacks when the patient may run out of the shop without paying. *Kleptomania* (see p. 768) is a rare cause of repeated stealing without obvious motive in the absence of another psychiatric cause. Some writers doubt the validity of this psychiatric syndrome.

The *assessment* of a person charged with shoplifting is similar to that for any other forensic problem. If the accused has a depressive disorder at the time of the examination, the psychiatrist should try to establish whether the disorder was present at the time of the offence or whether it developed after the charge was brought. The timing is important because, in England and Wales, a successful prosecution under the Theft Act 1968 requires that there should have been intention to steal (*mens rea*). The psychiatrist's report should include opinions on the prognosis and the need for treatment. See Gudjonsson (1990) for a review of psychiatric aspects of shoplifting.

Arson

This offence is generally regarded extremely seriously, not only because it threatens life but also because it can result in great damage to property. Most arsonists are males. Although the courts refer many arsonists for psychiatric assessment, the psychiatric literature on arson is small

(see Geller (1992) for a historical review). As often in forensic psychiatry, it is difficult to make a behavioural classification (Lewis and Yarnell 1951), but certain groups can be recognized. First, there are arsonists who are free from psychiatric disorder and who start fires for financial or political reasons or for revenge; they are sometimes referred to as *motivated* arsonists. Second, there are so-called *pathological* arsonists, who suffer from mental retardation, mental illness, or alcoholism. In a consecutive series of men remanded in custody, Taylor and Gunn (1984*a*) found an association between psychotic disorder and arson. However, psychotic fire-setters are reported to account for only 10–15 per cent of arsons. A third group resembles the DSM criteria for *pyromania* (see p. 760), although the validity of this diagnostic criteria is unsubstantiated. These individuals (who sometimes join conspicuously in firefighting) obtain intense satisfaction and tension relief from fire-setting.

The risks of further offences were assessed in a 20-year follow-up by Soothill and Pope (1973) who found that only 4 per cent of arsonists were reconvicted for arson, but about half of them were charged with offences of other kinds. An important guideline is that a person convicted of arson a second time is at a much greater risk of further offences. Apart from this guideline, certain other factors point to an increased risk of a further offence: antisocial personality disorder, mental retardation, persistent social isolation, and evidence that fire-raising was done for sexual gratification or relief of tension. The scope for psychiatric intervention is limited. Management of arsonists within hospital requires a secure setting and close observation.

Children also present with problems of fire-raising (Showers and Pickrell 1987). Sometimes the behaviour represents extreme mischievousness in psychologically normal children, at times as a group activity, and sometimes it springs from psychiatric disturbance. Most of 104 child fire-setters referred to a child psychiatric clinic in London had shown marked antisocial and aggressive behaviour before the fire-setting. The most frequent diagnosis was conduct disorder (Jacobson 1985). Among children charged with fire-setting, the recurrence rate in the following two years is reported to be under 10 per cent (Strachan 1981).

For reviews of arson see Gunn and Taylor (1993, pp. 587–98) and Barnett and Spitzer (1994).

Victims of crime

It is only relatively recently that criminology and society have paid attention to the role and needs of victims (see Zedner (1994) for a review). Recent surveys of general populations indicate that the experience of being a victim of crime is frequent and is related to geographical area, sex, age, and social habits. Much violence, especially sexual and domestic assaults, is unreported. There are differences between men and women in the experience as victims of crime. Young men are particularly at risk of personal violence by reason of their ways of life, whilst women are more likely to suffer domestic violence. Women express greater fear of crime and are more likely to avoid areas that may be dangerous (Heidensohn 1994*a*). In some cases, the victim's behaviour is perceived by the offender as a provocation. After an offence, the response of the victim is important in determining reporting to the police and enforcement of the law (Walker 1987).

Crime victims suffer a variety of early and late psychological problems. (Weaver and Clum 1995.) These include the immediate distress following the crime and the subsequent distress associated with investigation and court hearings. Post-traumatic stress disorder (see p. 140) is frequently reported (Davidson and Foa 1992). These consequences are more common and severe immediately after the crime, but they may persist for many years (Kilpatrick 1985). In general rape victims report more problems than victims of other kinds of assault, and victims of assault describe more than those of robbery. Norris and Kaniasty (1994) compared the psychological distress reported by victims of violent crime and property crime three months after the crime and for a year afterwards. Symptoms improved substantially between three and nine months, but changed little thereafter. Throughout the follow-up victims of violent crime were more distressed than victims of property crime. The severity of the reaction could not be accounted for by precrime differences in social status or psychological functioning. Some of those with lasting distress had suffered further crimes. In addition to psychological effects, financial, social, family, and other consequences are frequently very considerable.

The severity and persistence of psychological problems indicate the need for help for victims. In some places victim support groups have been set up, as well as counselling for rape victims (see Hampton 1995). Victims may also need information and practical and financial assistance. Victims frequently seek compensation from those responsible for road traffic accidents and other trauma, and from national criminal injury compensation schemes. The psychiatric aspects of compensation are discussed on p. 403.

Psychiatric issues in the assessment of offenders

Although mentally abnormal offenders are only a small minority of all offenders, both forensic and general psychiatrists can play an important role by helping to identify, assess, and manage them. The psychiatrist may be asked to give advice in relation to the following issues: fitness to plead, mental state at the time of the offence, diminished criminal responsibility, the validity of confessions, and the psychiatric management of offenders. Each of these issues will be discussed in turn. The discussion will be based on the law in England and Wales, but the principles apply more widely. For further information on the criminal law of England and Wales, the reader is referred to the standard legal texts by Smith and Hogan (1988) and Hoggett (1990), and to the textbooks of forensic psychiatry by Gunn and Taylor (1993) and Bluglass and Bowden (1990).

In England and Wales, a prison doctor carries out a psychiatric assessment of every person charged with murder. This doctor may ask for a second psychiatric opinion, often from a specialist

Table 22.1 *The involvement of psychiatrists in the stages of the legal process*

Stage 1 Arrest	Stage II Pre-trial	Stage III At the trial	Stage IV After the trial
Removal to a place of safety (S136)*	Court report	Special problems: Fitness to plead diminished responsibility	Treatment under hospital orders (S37, 38, 41) or guardianship (S37)
Assessment after arrest	Remand for inpatient assessment (S35) or treatment (S36)	Advice about disposal (S37/38/41)	Transfer from prison (S47, 49)
Court diversion schemes	Transfer from prison for assessment (S48, 49)		Decisions about release
			Treatment in the community

*Numbers in the table refer to sections of the Mental Health Act of England and Wales. (See Appendix, pp. 781–7.)

in forensic psychiatry but sometimes from a general psychiatrist. The defence lawyers often seek independent psychiatric advice. It is good practice for the doctors involved, whether engaged by the prosecution or defence lawyers, to discuss the case. When this is done, disagreement is unusual. Copies of the reports are distributed to the judge and to the prosecution and defence lawyers. Similar arrangements apply to other offences in which a psychiatric opinion is required.

The psychiatric report should be based on full psychiatric and social examination. It is essential that the psychiatrist read all the depositions by witnesses, statements by the accused, and any previous medical notes and social reports. Family members should be interviewed. When evidence about previous offences is not admissible (as is the case in English Law) the psychiatrist's report should not include these facts. This may cause problems for the psychiatrist, whose opinion is often based in large part on the previous behaviour of the offender. The writing of the court report follows the usual format (see p. 778) and should include discussions of mental state at the time of the alleged offence and of fitness to plead. The involvement of the psychiatrist at

various stages of the legal process is shown in Table 22.1.

Fitness to plead

English law requires that the defendant must be in a fit condition to defend himself. The issue may be raised by the defence, the prosecution, or the judge. It cannot be decided in a magistrates' court, but only by a jury. If the accused is found unfit to plead and the charge is murder, an order is made committing him to any hospital specified by the Home Secretary where he may be detained without limit of time and can be discharged only at the discretion of the Home Secretary. If discharged from hospital, he is returned to the penal system for trial. In England and Wales when the charge is other than murder a range of disposals are available to the judge including orders for hospital or community treatment (Criminal Procedures (Insanity and Fitness to Plead) Act 1991).

In determining fitness to plead, it is necessary to determine how far the defendant can

(i) understand the nature of the charge,

(ii) understand the difference between pleading guilty and not guilty,

(iii) instruct counsel,

(iv) challenge jurors,

(v) follow the evidence presented in court.

A person may be suffering from severe mental disorder but still be fit to stand trial. An American study of 85 people judged incompetent to stand trial showed that most had been charged with serious offences. Plans for psychiatric care after release were generally inadequate (Lamb 1987).

Mental state at the time of the offence

The issue of criminal responsibility is relevant at every trial. In a small proportion of cases a psychiatric opinion is sought on the influence of psychiatric disorder on criminal responsibility. Underlying this issue is the principle that a person should not be regarded as culpable unless he was able to control his own behaviour and to choose whether to commit an unlawful act or not. It follows from this principle that, in determining whether or not a person is guilty, it is necessary to consider his mental state at the time of the act. Before anyone can be convicted of a crime, the prosecution must prove the following:

(1) he carried out an unlawful act (*actus reus*);

(2) he had at the time the state of mind necessary to commit a crime (*mens rea*).

The latter is a technical term which is often loosely translated as meaning a 'guilty mind'. However, this translation can be misleading since a person may commit a legal offence whilst completely confident that he is morally right. The various categories of *mens rea* are not precisely defined. They vary from crime to crime, and are interpreted in the light of the precedents of case law. The categories are as follows.

1. *Intent* Intent has various meanings but the main principle is that the person perceives and intends that his act of omission will produce unlawful consequences.

2. *Recklessness* 'Recklessness is the deliberate taking of an unjustifiable risk. A man is reckless with respect to the consequence of his act, when he foresees it may occur but does not desire it. Recklessness with respect to circumstances means the realisation that the circumstances may exist, without either knowing or hoping that they do. D points a gun at P and pulls the trigger; if he does not know that it is loaded but realises that it may be, he is reckless with respect to that circumstance, whether he hopes it is unloaded or just does not care.' (Smith and Hogan 1988)

3. *Negligence* 'A man acts negligently when he brings about a consequence which a reasonable and prudent man would have foreseen and avoided.' (Smith and Hogan 1988)

4. *Blameless inadvertence* 'A man may reasonably fail to foresee a consequence of his act, as when a slight slap causes the death of an apparently healthy person: or reasonably fail to consider the possibility of the existence of a circumstance, as when goods, which are in fact stolen, are bought in the normal course of business from a trader of high repute.' (Smith and Hogan 1988)

These definitions are themselves the subject of debate, and the state of the law is complicated. A psychiatrist who has been asked to give an opinion on these matters should discuss with the lawyers what the relevant psychiatric contribution might be.

Children under 10 are excluded because they are deemed incapable of criminal intent (*doli incapax*). Children aged over 10 and under 14 are excluded unless it can be proved that they knew the nature of their act and knew it to be morally and legally wrong (mischievous discretion); in other words the law assumes that the children in this age group do not have *mens rea* unless it can be proved otherwise.

The degree of *mens rea* required for a conviction varies from crime to crime. For murder it is necessary to establish 'specific intent', for manslaughter it is sufficient to establish gross negligence, and for some types of offence, such as traffic offences, it is not necessary to establish any degree of *mens rea* at all. For most offences it is necessary to establish some degree of intent.

When a person is charged with an offence, the defence can be made that he is not culpable because he did not have a sufficient degree of *mens rea*. This defence can be raised in several ways:

(1) not guilty by reason of insanity (under the McNaghten rules);
(2) diminished responsibility (not guilty of murder, but guilty of manslaughter, which requires a lesser degree of criminal intent);
(3) incapacity to form an intent because of an automatism.

A further example is that if a mother kills her child in the first year of its life, she is not usually held legally responsible for murder but only for the less serious crime of infanticide (see p. 763).

The types of defence listed above will now be considered in turn. For further information the reader is referred to the textbooks by Bluglass and Bowden (1990) and Gunn and Taylor (1993).

Not guilty by reason of insanity

This concept is embodied in the McNaghten Rules. In 1843 Daniel McNaghten, a wood turner from Glasgow, shot and killed Edward Drummond, private secretary to the Prime Minister, Sir Robert Peel. In the trial at the Old Bailey, a defence of insanity was presented on the grounds that McNaghten had suffered from delusions for many years. He believed that he was persecuted by spies, and had gone to the police and other public figures seeking help. His delusional system gradually focused on the Tory Party, and he decided to kill their leader, Sir Robert Peel. He killed Peel's secretary but was prevented from firing a second shot at the Prime Minister (West 1974). In accordance with suggestions made by the judge in summing up, McNaghten was found not guilty on the grounds of insanity and was admitted to Bethlem Hospital. This verdict outraged public opinion and was debated urgently in the House of Lords. At the request of the Lords, the judges drew up rules which were not enacted in the law but provided guidance as follows:

To establish a defence on the ground of insanity, it must be clearly proved that, at the time of committing the act, the party accused was labouring under such a defect of reason, from disease of the mind, as not to know the nature and quality of the act he was doing, or, if he did know it, that he did not know what he was doing was wrong.

The McNaghten Rules have no statutory basis, but they are accepted by the courts as having the same status as statutory law. If an offender is found 'not guilty by reason of insanity', the court must order his admission to a hospital specified by the Home Secretary (Criminal Procedure (Insanity) Act 1964).

The rules are more restrictive than the summing up in the McNaghten trial. They have been strongly criticized as providing a concept of insanity that is much too narrow. Critics have argued that insanity affects not only cognitive faculties, but also emotions and willpower. Both for this reason and because of the increasing concern about capital punishment, the defence of diminished responsibility for murder was introduced in 1957. Since then a defence of insanity in terms of the McNaghten Rules is seldom raised.

The McNaghten Rules are used in several other jurisdictions, and have led to other formulations of the insanity defence. In the United States the American Psychiatric Association reviewed procedures after public outrage about the finding that, by reason of insanity, John Hinckley was not guilty of the attempted murder of President Reagan. It concluded that further legislation was needed to cover the confinement, treatment, review, and release of such persons (Insanity Defense Work Group 1983). See Restak (1993) for a review of neurological disorder as a defence against criminal charges.

Diminished responsibility

Diminished responsibility may be pleaded as a defence to the charge of murder. If the defence is upheld, the accused is found guilty only of manslaughter. The concept of diminished responsibility is based on a definition of mental abnormality that is much wider than that embodied in the McNaghten Rules. This point is illustrated by the following extract from the Homicide Act 1957 (Section 2):

where a person kills or is party to a killing of another, he shall not be convicted of murder if he was suffering from such abnormality of mind (whether arising from a condition of arrested or retarded development of mind or any inherent causes or induced by disease or injury) as substantially impaired his mental responsibility for his acts and omissions in doing or being party to the killing.

In practice, if a person is charged with murder, he may plead that he is not guilty of murder but guilty of manslaughter on the grounds of diminished responsibility. If this plea is acceptable to the prosecution and to the judge, there is no trial but a hearing of evidence (without a jury) and a sentence for manslaughter is passed. However, if the plea is not acceptable to the prosecution or the judge, a trial is held. The jury must then consider the evidence, both medical and non-medical, to decide whether at the material time the accused was suffering from abnormality of mind and, if so, whether the abnormality was such as substantially to impair his responsibility. If the accused is convicted of manslaughter, the judge may pass whatever sentence he deems appropriate (which may include life imprisonment) on the grounds of dangerousness. In contrast with this discretion in the sentence for manslaughter, there is a statutory sentence of life imprisonment for a conviction of murder.

Diminished responsibility has been widely interpreted and has made the insanity defence virtually obsolete. Successful pleas have been based on conditions such as 'emotional immaturity', 'mental instability', 'psychopathic personality', 'reactive depressed state', 'mixed emotions of depression, disappointment, and exasperation', and 'premenstrual tension'.

Automatism

If a person has no control over an act, he cannot be held responsible for it. For this reason, verdicts of not guilty have been returned when acts of violence were judged to be committed as 'sane automatisms'. Such circumstances are rare, but have occurred.

Automatism thought to arise from a 'disease of the mind', is referred to as 'insane automatism'; the appropriate defence is then insanity, and the McNaghten Rules apply. In legal practice there have been varying interpretations of 'disease of the mind' in this context. In English law epileptic automatism, hypoglycaemia, hyperglycaemia, and sleep-walking are not recognized as sane automatisms but caused by 'disease of the mind' because the conditions may recur. Four types of association have been described between *sleep disorders* and violence: confusion on sudden awakening (sleep drunkenness), sleep-walking, night terrors, and rapid eye movement sleep behaviour disorder (Howard and d'Orban 1987).

Intoxication

The English law relating to alcohol and drug addiction is complicated. It can be summarized as follows.

1. Involuntary intoxication (as when someone unwittingly takes a drink to which a drug has been added) or automatism occurring as a side-effect to medical treatment, constitutes a valid defence.

2. Self-induced intoxication is not a defence unless (a) it is itself evidence of 'disease of the mind' under the McNaghten Rules or (b) it is evidence of lack of intent in relation to those crimes for which 'specific intent' must be proved (for example murder, theft, and burglary). Self-induced intoxication is not a defence to those crimes for which evidence of 'specific intent' is not required (for example manslaughter, rape, indecent assault, and common assault).

Amnesia

Over a third of those charged with serious offences, especially homicide, report some degree of amnesia for the offence and inadequate recall of what happened. It has sometimes been argued that loss of memory should be regarded as evidence of unfitness to plead, but such arguments have been unsuccessful. The factors most commonly associated with claims of amnesia are extreme emotional arousal, alcohol abuse and intoxication, and severe depression. Amnesia has to be distinguished from malingering in an attempt to avoid the consequences of the offence. However, there appear to be instances of true amnesias for offences, just as there is impaired recall by victims and witnesses of offences. Moreover, the factors associated with amnesia are similar in offenders and victims. In the absence of organic disease, the presence of amnesia is unlikely to carry any legal implications (see Gunn and Taylor (1993, pp. 291–9).

False confessions

Accounts of trials and other descriptive evidence indicate that false confessions to criminal deeds are sometimes made. However, the frequency of such confessions is unknown. Gudjonsson (1992) suggested that there are three main types of false confession: voluntary, coerced–compliant, and coerced–internalized. *Voluntary* confessions may arise from a morbid desire for notoriety, from difficulty in distinguishing fact from fantasy, from a wish to expiate guilt feelings, or from a desire to protect another person. *Coerced–compliant* confessions result from forceful interrogation and are usually retracted subsequently. *Coerced–internalized* confessions are made when the technique of interrogation undermines suspects' own memories and recollections so that they come to believe that they may have been responsible for the crime. Factors making a person more likely to make a false confession include a history of substance abuse, head injury, a bereavement, current anxiety, or guilt.

Particular difficulties arise in the questioning of children whether as victims, witnesses, or possible perpetrators of crime. Children are more suggestible than adults during questioning, but even young children can provide a reliable account if questioned in a way appropriate to their ages (Ceci and Bruck 1993).

The assessment of possible false confessions is difficult. It requires a thorough review of the circumstances of arrest, custody, and interrogation, as well as an assessment of the personality and the current mental and physical state of the suspect. Usually the assistance of a clinical psychologist will be required to carry out a neuropsychological assessment and in some cases an assessment of suggestibility (for a review see Lancet 1994).

Treatment of mentally abnormal offenders

When sentence is passed in court, the need for psychiatric treatment may be taken into account. After conviction, an offender may be treated on a compulsory or a voluntary basis (Hoggett 1990). Facilities for mentally abnormal offenders are reviewed in the books by Gunn and Taylor (1993) and Bluglass and Bowden (1990). The role of forensic psychiatric services is described by Faulk (1994).

In the United Kingdom, special treatment for mentally abnormal offenders is, in principle, provided by the Home Office (the prison medical service and the probation service) and by the Department of Health (special hospitals, specialist forensic services, and general psychiatry services). However, many mentally abnormal offenders do not receive the psychiatric treatment that they require (Gunn *et al.* 1991).

Much work with offenders is carried out by general psychiatrists who assess patients and prepare court reports. General psychiatrists as well as forensic psychiatrists treat offenders given non-custodial sentences. Forensic psychiatrists work in separate units and undertake specialized assessment and court work. In many places there are community forensic services to provide assessment and treatment. Forensic psychiatrists may work to provide care for patients needing security in ordinary psychiatric hospitals.

The mentally abnormal in prison

Surveys have shown that about a third of sentenced prisoners have a psychiatric disorder and 2 per cent have a psychosis (see Gunn *et al.* 1991). Most of these disorders can be treated in prison, but a few need transfer to a hospital.

Reasons for transfer include unpredictable violence, life-threatening self-harm, and failure to improve with treatment in prison.

The medical services have to provide psychiatric care under extremely difficult conditions, and it has been argued that there should be a substantial increase in the contribution of psychiatrists to the provision of medical care within prisons. A few prisons offer psychiatric treatment, usually of personality disorders and sexual offences, as a main part of their work; one such prison is at Grendon Underwood in England (Gunn and Robertson 1982). Although there is an undoubted need for psychiatric care within prisons, there would be disadvantages in a system which encouraged the courts to send the mentally abnormal to prison rather than to hospital services.

Offenders in hospital

In England and Wales, a convicted offender may be committed to hospital for compulsory psychiatric treatment under a Mental Health Act hospital order (see Appendix). There is also provision in law for a prisoner to be transferred from prison to a psychiatric hospital. An important point is that hospital orders may have no time limit, while most prison sentences are of fixed length. The length of stay in a psychiatric hospital may be shorter or longer than a prison sentence.

Committal is usually to a local psychiatric hospital, but may be to a medium security unit or to a high security hospital or special prison. In England the first special provision for the criminally insane was made in 1800. Following a trial in which Hadfield was found not guilty by reason of insanity for shooting at King George III, a special criminal wing was established at the Bethlem Hospital. In 1863 Broadmoor, the oldest of the special hospitals, opened under the management of the Home Office. There are now four high-security special hospitals in England and Wales which are the responsibility of the Department of Health.

The detention of patients in special hospitals is for an indeterminate length of stay. Dell *et al.* (1987) showed that, for those with mental illness (mostly schizophrenia), length of detention at Broadmoor Special Hospital was associated with the severity or chronicity of the psychiatric disorder rather than the nature of the offence. In contrast, for patients suffering from psychopathic disorder, the main determinant of length of stay was the nature of the offence.

The closure of the larger mental hospitals has had unforeseen consequences for the care of mentally abnormal offenders. There is less physical security in the new psychiatric units than was available in the old long-stay hospitals and less willingness by hospital staff to tolerate severely disturbed behaviour. As a result it has become increasingly difficult to arrange admission to hospital for offenders, particularly those who are severely disturbed. Also, length of stay in hospital has shortened, making it more difficult to arrange treatment for patients with chronic disorders and severe behaviour disorder. Two alternative provisions have been developed. One is the provision of well-staffed secure areas in ordinary psychiatric hospitals in which the less dangerous of these patients can be treated. The second provision (recommended in England and Wales by the Committee on Mentally Abnormal Offenders (1975)) is the setting up of special secure units associated with psychiatric hospitals to provide a level of security between that of an ordinary hospital and a special hospital. Problems have arisen about the criteria for selecting patients for these special secure units, and about their role in relation to both ordinary psychiatric hospitals and the special hospitals.

Violent incidents in hospitals

Violent incidents are not confined to patients with forensic problems, but this is a convenient place to consider their management. Although not frequent, violent incidents in hospitals are increasing (Haller and Deluty 1988), especially in the United States (Davis 1991). The reasons for this increase appear to include changes in mental health policies that have made dangerousness a relatively more common reason for an admission (since non-violent patients are more likely to be treated in the community), overcrowding, lack of experienced staff, and, in some hospitals, an increased tolerance of violence.

It is important that the staff have a clear policy for managing incidents of violence and are trained to carry it out. Such a policy calls for attention to the design of wards, arrangements for summoning assistance, and suitable training of the staff. When violence is threatened or takes place, staff should be available in adequate numbers, and emergency medication such as intramuscular chlorpromazine should be unobtrusively available. The emphasis should be on the prevention of violence. Dangerous people can often be calmed by sympathetic discussion or reassurance, preferably given by someone whom the patient knows and trusts. It is important not to challenge the patient. It is inappropriate to reward violent or threatening behaviour by making concessions in treatment or ward rules, but every effort should be made to allow the patient to withdraw from confrontation without loss of face.

After an incident has occurred, the clinical team should meet to consider the future care of the patient and also any possible changes in the general policy of the ward. For mentally disordered patients, there should be a review of the drugs prescribed and their dosage. When violence occurs in a person with a personality disorder, medication may be required in an emergency, but it is usually best to avoid maintenance medication. Other measures include trying to reduce factors that provoke violence or to provide the patient with more constructive ways of managing tension, such as taking physical exercise or asking a member of staff for help.

Treatment in the community

When a non-custodial sentence is passed, the court may require social and psychological care by the probation service. In addition, psychiatric treatment as an in-patient or out-patient may be made a condition of probation under the Powers of the Criminal Courts Act 1973. When treatment is made a condition of probation, the offender must state that he is willing to comply. The psychiatric treatment provided for a mentally abnormal offender is similar to that for a patient with the same psychiatric disorder who has not broken the law.

It is often difficult to provide psychiatric care for offenders with chronic psychiatric disorders who commit repeated petty offences. In the past they would have been long-stay patients in a psychiatric hospital, but now they are treated in the community where they may be unwilling to collaborate with treatment and may be difficult to follow-up because they change address or become homeless.

The psychiatrist and the court

Psychiatrists need some knowledge about the workings of the courts and the legal system. They should also be familiar with the role of psychiatrists in relation to the courts, the psychiatric examination of defendants, and the preparation of a court report. In the courts psychiatrists may be working outside the familiar conventions of the doctor–patient relationship. It is essential that they have a clear understanding of their role, and of the ethical issues concerning their relationship with the alleged offender.

The workings of the courts

In England and Wales, **Magistrates' Courts** deal with 98 per cent of all criminal prosecutions. They also deal with civil cases. Most magistrates are non-stipendiary laymen who receive some basic legal training and are advised on legal points by the Clerk to the Justices.

Indictable offences are those that can be tried by a judge and jury in a Crown Court, though they may also be tried in a Magistrates' Court. The Magistrates' Court tries non-indictable offences and some indictable offences. Magistrates can impose a sentence on a person found guilty of an indictable offence, but if the offence seems to merit a more severe penalty than they are empowered to imposed they refer the case to the Crown Court for sentencing. If a defendant is charged with a serious indictable offence and he chooses to be tried in a Crown Court, the magistrates must first decide whether the prosecution has established a prima-facie case.

In the **Crown Courts**, indictable offences are tried by jury and the sentence is passed by a judge.

Appeals against conviction in a Magistrates' Court are heard either in the Crown Court or the Appeal Court, and appeals against convictions in Crown Court are heard in the Appeal Court.

The role of the psychiatrist in relation to the court

The psychiatrist's role is to draw on his special knowledge to help the court. He should not attempt to tell the court what to do. In the United Kingdom an expert medical witness is expected to remain neutral and not to favour either the accused or the defendant. The psychiatrist should be aware that the court will see the report and that it may be read out in open court. Reports commissioned and paid for by lawyers are the property of the lawyers.

The psychiatrist's interview with the defendant

Psychiatrists should prepare themselves as thoroughly as possible before the interview. They should have a clear idea as to the purpose of the examination, and particularly as to any question of fitness to plead. They should have details of the present charge and past convictions, together with copies of any statements made by the defendant and witnesses. Psychiatrists should also study any available reports of the defendant's social history; during their subsequent interview they should go through this report with the defendant and check its accuracy.

Psychiatrists should begin by explaining to the client the source of the referral and why the referral was made. They should explain that the psychiatrist's opinion may be given in court and that the defendant is under no obligation to answer any questions if he chooses not to. The interview should be carried out in strict confi-

dence. Detailed notes should be made, recording any significant comments in the defendant's own words.

At some stage in the interview (not necessarily at the start) the alleged crime should be discussed. The defendant may or may not admit guilt. A detailed history of physical illnesses should be taken; particular attention should be paid to neurological disorders including head injury and epilepsy. A careful history of previous psychiatric disorder and treatment should be obtained. If there has been a previous psychiatric opinion or treatment, further information should be sought. Full examination of the present mental state is made in the usual way. Special investigations should be requested if suitable. If the defendant's intelligence level is under question, an assessment should be made by a clinical psychologist who normally submits a separate report.

It is important to obtain further information from relatives and other informants. If the defendant is remanded in custody, the staff may have long periods of contact with the prisoner and may be able to give particularly useful information.

Preparing a psychiatric court report

In preparing a court report, the psychiatrist should remember that it will be read by non-medical people. Therefore the report should be written in simple English and should avoid jargon. If technical terms are used, they should be defined as accurately as possible. The report should be concise and set out as follows.

1. A *statement* of the psychiatrist's full name, qualifications, present appointment, and (in England and Wales) whether approved under Section 12 of the Mental Health Act.

2. *Where and when the interview was conducted* and whether any third person was present.

3. *Sources of information* including documents that have been examined.

4. *Family and personal history of the defendant* Usually this need not be given in great detail,

particularly if a social report is available to the court. The focus should be on information relevant to the diagnosis and disposal.

5. *The account of the crime given by the accused* This will depend on whether the defendant is pleading guilty or not guilty. If the accused admits to the crime, comment may be made on his attitude to it such as degree of remorse. If he is pleading not guilty, any reference to the alleged crime is inadmissible.

6. *Other behaviour* It may be relevant to mention other items of behaviour, even if not directly involved in the crime, such as alcohol or drug abuse, quality of relationships with other people, tolerance of frustration, and general social competence.

7. *Present mental state* Only the salient positive findings should be stated and negative findings should be omitted. A general diagnosis should be given in the terms of the Mental Health Act (mental illness, mental impairment, or psycho-pathic disorder). A more specific diagnosis can then be given, but the court will be interested in a categorical statement rather than the finer nuances of diagnosis.

8. *Mental state at the time of the crime* This is often a highly important issue and yet it can be based only on retrospective speculation. The assessment can be helped by accounts given by eye-witnesses who saw the offender at the time of the crime or soon after. A current psychiatric diagnosis may suggest the likely mental state at the time of the crime. For example if the accused suffers from chronic schizophrenia or a chronic organic mental syndrome, the mental state may well have been the same at the time of the crime as at the examination. However, if the accused suffers from a depressive disorder (now or recently) or from an episodic disorder such as epilepsy, it is more difficult to infer what the mental state is likely to have been at the material time. To add to the difficulty, even if it is judged that the defendant was suffering from a mental disorder, a further judgement is needed as to his *mens rea* at the time of the crime.

9. *Fitness to plead* It is often helpful for the psychiatrist to include a statement of fitness to plead (the criteria for deciding this are given on p. 771).

Advice on medical treatment

One of the psychiatrist's main functions is to give an opinion as to whether or not psychiatric treatment is indicated. The psychiatrist should make sure that any recommendations on treatment are feasible, if necessary by consulting colleagues, social workers, or others. If hospital treatment is recommended, the court should be informed whether or not a suitable placement is available. The assessment of dangerousness is important here (see next section).

The psychiatrist should not recommend any form of disposal other than treatment. However, the court often welcomes tactfully worded comments on the suitability of possible sentences, particularly in the case of young offenders.

The psychiatrist appearing in court

The psychiatrist appearing in court should be fully prepared and should have well-organized copies of all reports and necessary documents. It is helpful to speak to the lawyer involved beforehand, in order to clarify any points that may be raised in court. When replying to any questions in court, it is important to be brief and clear, to restrict the answers to the psychiatric evidence, and to avoid speculation.

Dangerousness

The psychiatrist may need to assess dangerousness in everyday psychiatric practice and also in forensic work. In everyday practice, both out-patients and in-patients may appear to be dangerous, and careful assessment may be required so that the most appropriate steps can be taken in the interests of the patient and of other people. Dangerousness is an important reason for recommending compulsory detention in hospital. In forensic work the court may ask for the psychiatrist's advice on the defendant's dangerousness so that a suitable sentence can be passed. The psychiatrist may also be asked to comment on offenders who are detained in institutions and who are being considered for release. In both kinds of circumstance there is an ethical dilemma between the need to protect the community from someone who might show violent behaviour and the obligation to respect the human rights of the offender.

There are no fixed rules for assessing dangerousness. Criminologists have identified predictive factors in offender populations, but the low correlations between predicted and observed dangerousness mean that they are unhelpful in individual predictions. Psychiatrists have tried to identify factors associated with dangerousness in an individual patient, but no reliable predictors of violence have been established. Actuarial prediction has been attempted but has proved to be inaccurate (Monahan 1984; Gunn and Taylor 1993, pp. 624–45). There are a few guidelines, as shown in Table 22.2; these guidelines apply to offenders, but the same principles hold for non-offenders. A thorough review should be made of the history of previous violence, the characteristics of the current offence and the circumstances in which it occurred, and the mental state. In making the review it is helpful to consider certain key questions: whether any consistent pattern of behaviour can be discerned, whether any circumstances have provoked violence in the past and are likely to occur again in the future, whether there is any good evidence that the defendant is willing to change his behaviour, and whether there is likely to be any response to treatment. Of these predictors, the most useful is a history of past violence.

Particular difficulties may arise in the assessment of dangerousness in people of antisocial personality or in the mentally retarded, both of whom may be poorly motivated to comply with care. Another difficult problem is presented by the person who threatens to commit a violent act such as homicide. Here the assessment is much the same as for suicide threats (Gunn and Taylor 1993 p. 632). The psychiatrist should ask the threatener about his intent, motivation, and the potential victim, and should make a full assessment of

Table 22.2 *Factors associated with dangerousness*

History
One or more previous episodes of violence
Repeated impulsive behaviour
Evidence of difficulty in coping with stress
Previous unwillingness to delay gratification
Sadistic or paranoid traits

The offence
Bizarre violence
Lack of provocation
Lack of regret
Continuing major denial

Mental state
Morbid jealousy
Paranoid beliefs plus a wish to harm others
Deceptiveness
Lack of self-control
Threats to repeat violence
Attitude to treatment

Circumstances
Provocation or precipitant likely to recur
Alcohol or drug abuse
Social difficulties and lack of support

mental state. Some patients who make threats can often be helped by out-patient support and treatment, but sometimes hospital admission is required if the risk is high. It may be necessary to warn potential victims.

It is a valuable principle for the psychiatrist not to rely entirely on his own evaluation of dangerousness, but to discuss the problem with other colleagues, including psychiatrists, general practitioners, social workers, and relatives. For a review see Gunn and Taylor (1993, pp. 624–45) or Tardiff (1992).

Further reading

Bluglass, R. S. and Bowden, P. (ed.) (1990). *Principles and practice of forensic psychiatry*. Routledge, London.

Faulk, M. (1994). *Forensic psychiatry* (2nd edn). Blackwell, Oxford.

Gunn, J. and Taylor, P. J. (1993). *Forensic psychiatry: clinical, legal and ethical issues*. Butterworth-Heinemann, London.

Halleck, S. L. (1986). *The mentally disordered offender*. National Institute of Mental Health, Rockville, MD.

McGuire, M., Morgan, R., and Reiner, R. (eds) (1994). *The Oxford handbook of criminology*. Oxford University Press.

Appendix: The law in England and Wales

This appendix is an introduction to the principal sections of the law of England and Wales relating to psychiatric practice. It begins with a review of the Mental Health Act. More detailed information can be obtained from the works listed under further reading (p. 787) and, of course, from the Act itself. Psychiatrists practising elsewhere than in England and Wales will need to consult guides to their local legislation and its application in clinical practice.

The Mental Health Act

The Mental Health Act 1983 regulates the care of mentally abnormal persons. It consolidates the Mental Health Act 1959 and the Mental Health (Amendment) Act 1982, which made provisions for compulsory treatment and about consent to treatment. It also set up the Mental Health Act Commission, which is an independent multidisciplinary body appointed by the Secretary of State. The Commission has powers to safeguard the interests of detained patients. Its duties are to visit such patients, investigate complaints, receive reports on patients' treatment, and appoint doctors and others to give opinions to consent to treatment.

Parts II–V of the Mental Health Act provide the legal basis for compulsory admission and detention of psychiatric patients. Provision for compulsory detention is also made by the Criminal Procedure (Insanity) Act 1964, which relates to people found 'not guilty by reason of insanity' or 'unfit to plead'.

Under the Mental Health Act there are three main groups of compulsory order for assessment and treatment:

(a) admission for assessment (Sections 2, 4, 5, 135, 136);
(b) treatment orders (Sections 3 and 7);
(c) admission and transfer of patients concerned with criminal proceedings (Sections 35–37, 41, 47–49).

The short-term orders listed in (a) above apply to any mental disorder, which need not be specified. For the long-term orders listed under (b) and (c), it must be stated that the patient suffers from one of four types of mental disorder: mental illness, psychopathic disorder, mental impairment, and severe mental impairment.

The Act does not define 'mental illness', but it states that no one should be 'treated as suffering from mental disorder by reason only of promiscuity, or other immoral conduct, sexual deviancy or dependence on alcohol or drugs'.

The Act gives the following definitions of the three other types of mental disorder:

(a) **severe mental impairment** means a state of arrested or incomplete development of mind which includes severe impairment of intelligence and social functioning and is associated with abnormally aggressive or seriously irresponsible conduct on the part of the person concerned.

(b) **mental impairment** means a state of arrested or incomplete development of mind (not amounting to severe mental impairment) which includes significant impairment of intelligence and social functioning and is associated with abnormally aggressive or seriously irresponsible conduct on the part of the person concerned.

(c) **psychopathic disorder** means a persistent disorder or disability of mind (whether or not including significant impairment of intelligence) which results in abnormally aggressive or seriously irresponsible conduct on the part of the person concerned.

The Act also specifies the various people who may be involved in procedures for admission and treatment.

Responsible medical officer: The doctor in charge of treatment.

Nearest relative. Nearest adult relative (in the order of spouse, son or daughter, father or mother, sibling, grandparent, grandchild, uncle or aunt, nephew or niece). The elder or eldest of relatives of the same kind (for example siblings) is preferred, and full siblings have precedence over half-siblings. Also preference is given to a relative with whom the patient lives or who cares for him. The definition includes cohabitees (who may be of the same sex as the patient) who have lived with the patient for not less than six months. However, a cohabitee cannot claim precedence over a husband or wife other than through agreement under an Order of Court or by permanent separation by desertion.

Approved social worker Social workers approved by the local authority as having appropriate competence in dealing with mentally disordered persons.

Approved doctor. A doctor approved under Section 12 of the Act by the Secretary of State as having special experience in the diagnosis or treatment of mental disorder.

The three main groups of compulsory order will now be reviewed in turn.

Admission for assessment

Whilst a full understanding of the proper use of compulsory admission can be gained only from clinical experience, the clinician needs to be aware of the following general conditions governing compulsory detention.

Section 2: Application for admission for assessment (28 days)

This is the usual procedure for compulsory admission when informal admission is not appropriate in the circumstances. Detention is for assessment, or for assessment followed by medical treatment. The following grounds must be satisfied.

1. The patient suffers from a mental disorder which warrants the patient's detention in hospital for assessment (assessment followed by treatment).

2. Admission is necessary in the interests of the patient's own health or safety or for the protection of others.

The procedure requires the following.

(a) *Application* by the patient's nearest relative, or an approved social worker who must have seen the patient within the last 14 days. The approved social worker should, so far as it is practicable, consult the nearest relative.

(b) *Medical recommendations* by two doctors, one of whom must be approved under Section 12 of the Act. The two doctors should not be on the staff of the same hospital unless it would cause undesirable delay to find a doctor from elsewhere. There should not be more than five days between the examinations.

Section 4: Emergency order for assessment (72 hours)

This section allows a simpler procedure than Section 2 and provides power to detain patients in emergencies. It is usually completed in the patient's home by the family doctor but is also occasionally used in general hospital Accident and Emergency departments. Section 4 should be used only when there is insufficient time to obtain the opinion of an approved doctor who could complete Section 2. The grounds are as for Section 2. It is expected that a Section 4 order will be converted into a Section 2 order as soon as possible after the patient has arrived in hospital. The procedure for a Section 4 order requires the following:

(a) Application by an approved social worker who must have seen the patient within the previous 24 hours, or the nearest relative.

(b) Medical recommendation by one doctor, who need not be approved under Section 12 of the Act. The patient must be admitted within 24 hours of the examination (or of the application if made earlier).

Section 5: Change to compulsory detention (72 hours)

This is an order for the emergency detention of a patient who is already in hospital as a voluntary patient but wishes to leave, and for whom the doctor'believes an application should be made for compulsory admission under the Act. It requires a single medical recommendation by the doctor in charge of the patient's care or by another doctor who is on the staff of the hospital and nominated by the doctor in charge. (This power applies to a patient in any hospital.) It is usual to consider a change to a Section 2 or 3 order as soon as possible.

If a Section 5 order cannot be obtained immediately, a registered mental nurse or registered nurse for the mentally subnormal may invoke a six-hour **holding order**. The nurse must record that the patient is suffering from mental disorder such that, in the interests of the patient's health or safety or for the protection of others, the patient should be restrained from leaving the hospital. The holding order applies only when the patient is already under treatment for a mental disorder. It lapses as soon as the doctor signs Section 5. (This power applies only to a patient in a psychiatric ward.)

Section 115: Powers of entry and inspection

An approved social worker can enter and inspect any premises (within the area of the local authority in which the patient lives) if he has reason to believe that a mentally disordered patient is not under proper care. The social worker must be able to provide authenticated documentation of his status.

Section 135: Warrant to search for and remove patients

Any approved social worker who believes that someone is suffering from a mental disorder and is unable to care for himself or is being ill-treated or neglected may apply to a magistrate for a warrant for that person's removal to a place of safety.

Section 136: Mentally disordered person found in a public place

Any police constable who finds in a public place someone who appears to be suffering from a mental disorder may take that person to a place of safety (which usually means a police station or a hospital), if the person appears to be in immediate need of care or control, or if the police constable thinks that it is necessary to do so in the person's interest, or for the further protection of other persons. The person is detained so that he can be examined by a doctor and any necessary arrangements can be made for his treatment or care. The authority under Section 136 expires when these arrangements have been completed or within 72 hours, whichever is the shorter.

Treatment orders

Section 3: Admission for treatment (six months)

The grounds for this longer-term order are that the patient:

(a) is suffering from mental illness, severe mental impairment, psychopathic disorder or mental impairment, being a mental disorder of a nature or degree which makes it appropriate for him to receive medical treatment in a hospital; and

(b) in the case of psychopathic disorder or mental impairment, that such treatment is likely to alleviate or prevent a deterioration of his condition; and

(c) that it is necessary for the health or safety of the patient or for the protection of other persons that he should receive such treatment and that it cannot be provided unless he is detained under this section.

The procedure requires the following.

1. *Application.* This is made by the patient's nearest relative or an approved social worker. The latter must, if practicable, consult the nearest relative before making an application and cannot proceed if the nearest relative objects.

2. *Medical recommendation.* As for Section 2. In addition the recommendations must state the particular grounds for the doctor's opinion, specifying whether any other methods of dealing with the patient are available and, if so, why they are not appropriate. The doctor must specify one of the four forms of mental disorder.

3. *Renewal.* The order may be renewed on the first occasion for a further six months and subsequently for a year at a time.

Section 7: Reception into guardianship

Guardianship is more appropriate than the provisions of Section 3 for the long-term treatment of patients living in the community. The application, medical recommendation, duration, and renewal procedure are similar to those for Section 3. The guardian, who is usually but not always the local Social Services Department, is given authority (Section 8) for supervision in the community, including power to:

(a) require the patient to live at a place specified by the guardian;

(b) require the patient to attend places specified by the guardian for medical treatment, occupation, training, or education;

(c) ensure that a doctor, social worker, or other person specified by the guardian can see the patient at his home.

Admission to hospital of those appearing before the courts

These sections of the Mental Health Act allow the Courts to order psychiatric care for those charged with or convicted of an offence punishable by imprisonment. Medical recommendations are required together with an assurance that a hospital place is available.

Remands to hospital and interim hospital orders

Persons on remand (but not in custody) may be treated as voluntary patients. Sometimes psychiatric care may be made a condition of the granting of bail. In addition, the Mental Health Act 1983 gave the courts powers to:

(a) remand an accused person to a hospital for medical reports (Section 35);

(b) remand an accused person to hospital for treatment (except for murder cases) (Section 36);

(c) make an interim hospital order on a convicted person to assess suitability for a hospital order (Section 38).

Procedure (a) requires a medical recommendation by an approved doctor that there is reason to suspect mental disorder. Procedures (b) and (c) require medical recommendations by two doctors (one of whom must be approved) that the person is suffering from mental disorder.

Section 37: Hospital order

A court may impose a hospital order, which commits an offender to hospital on a similar basis to that of a patient admitted for treatment under the civil provisions of Section 3 of the Act (see above). The duration of the order is six months.

Medical recommendation: two doctors, one of whom must be approved.

Section 41: Restriction order

When a Section 37 hospital order is made by a Crown Court, the Court may also make an order under Section 41 of the Act restricting the person's discharge from hospital. The restriction order may be either without limit of time or for a specified period. If it is for a fixed term, once that term expires or otherwise ceases to have effect, the patient will still be detained under a hospital order but without restriction, i.e. Section 37.

Section 47: Transfer to hospital from prison

This section authorizes the Home Secretary to transfer a person serving a sentence of imprisonment to a local National Health Service hospital or special hospital (Section 48 covers other prisoners not serving sentences for criminal offences). A direction for transfer has the same effect as a hospital order. The patient's status changes to that of a notional Section 37 at the time of the 'earliest date of release'.

The Home Secretary can make the direction with or without special restriction or discharge (Section 49).

Medical recommendation: two doctors, one of whom must be approved.

Discharge of patients

Patients on emergency orders (Sections 4, 5, 135, 136) can be discharged by the responsible medical officer. Patients on a Section 2 or 3 order can be discharged by the responsible medical officer, the hospital managers, the nearest relative, or a Mental Health Review Tribunal. The same applies to patients on Section 3, except that the responsible medical officer may register an objection to discharge by relatives if he considers that the patient is a danger to himself or others. Patients under guardianship are in the same position as patients on Section 3 except that the local Social Services replace the hospital managers.

The nearest relative has no rights to discharge patients on Section 37 or 47 orders. Patients on Section 41 and 49 restriction orders can be discharged only by the responsible medical officer with the consent of the Secretary of State for Home Affairs, or by a Mental Health Review Tribunal.

Mental Health Review Tribunals

These are regional tribunals that provide an appeal procedure for patients subject to longer-term orders. They hear appeals against compulsory orders and automatically review certain patients under Sections 3 and 37 (see list below). Review tribunals may order immediate or delayed discharge. The members of a panel are appointed by the Lord Chancellor and include a lay member, a doctor, and a lawyer who is the chair. When the patient is subject to a restriction order, the chair of the panel is a judge. Patients are entitled to be provided with legal representation.

For the various sections of the Act, the timing of application for appeal is specified:

Section 4 and 5:	No appeal
Section 2:	Application must be made within 14 days.
Section 3:	Application can be made in the first six months, in the second six months, and then annually. Review is automatic if there has been no appeal either in the first six months or in any three-year period.
Section 37 (with or without Section 41)	Application can be made in the second six months, then annually. Review is automatic if there has been no appeal in any three-year period.
Guardianship	An application can be made in each period of detention. No automatic review.

Consent to treatment

Under common law, no treatment can be given to a voluntary patient without his valid consent. This requires that the patient voluntarily (that is without being subjected to coercion or unreasonable influence) agrees to the treatment, and is capable of making that decision. Doctors should not give treatment without such consent unless the treatment is essential to safeguard the health or preserve the life of the patient.

The Mental Health Act 1983 introduced provisions to serve two purposes: to give authority for certain treatments to be given without consent, and to safeguard psychiatric patients' interests in relation to treatment procedures.

The Act specifies certain emergency conditions under which treatment can be given without consent to a detained patient. Any treatment (provided that it is not irreversible or hazardous) can be given to such a patient without his consent if it is immediately necessary to save the patient's life, to prevent a serious deterioration in his condition, to alleviate serious suffering, or to prevent violence or danger to the patient himself or to others.

The Act also defines various groups of treatments according to the type of consent required for them. The allocations of particular treatments to these groups are specified in the Act, in Regulations (which are compulsory), and in a Code of Practice (which is advisory). Certain patients are excluded from the stipulations concerning consent to treatment, namely those detained under Sections 4, 5, 135, and 136, those remanded to hospital for reports, and those subject to Section 41 orders but conditionally discharged by the Home Secretary. For these patients the doctor has only 'common-law' rights and duties when giving treatment.

For other patients, three groups of treatments are stipulated, of which all three apply to detained patients, and only the first to voluntary patients.

(1) Treatments which give rise to special concern

This group applies to both voluntary and detained patients. It includes psychosurgery and other treatments which are yet to be specified, but does not include ordinary medication or electro-convulsive therapy (ECT). For treatments included in this group the patient must consent, *and* there must be a second opinion. The second opinion must be provided by an independent doctor who

will be required to consult two people (one a nurse, one neither a doctor nor a nurse) who have been professionally concerned with the patient's treatment. He considers both the treatment proposed and the patient's ability to give consent. In addition two independent people must certify as to the patient's ability to give consent.

Approval of treatment by the second-opinion procedure may cover a plan of care including more than one form of treatment.

(2) Other treatments listed in Regulations

This group applies to detained patients. It consists of other treatments specified in the Act or in Regulations. It includes some forms of medication and ECT. For these treatments, a second opinion must be obtained as in (1) above if the patient does not consent, cannot give consent, or withdraws consent. Again, the second opinion may cover a plan of treatment. However, for most forms of medication these procedures do not apply during the first three months of treatment.

(3) Other forms of treatment

Treatments not referred to in (1) or (2) above can be given to detained patients without their consent. It should be noted that, according to the act, medical treatment 'includes nursing and also includes care, habilitation and rehabilitation under medical supervision'.

The Court of Protection

This Court has a very long history and is responsible for the mentally ill or impaired. Most of its functions are carried out by the Master and other officers appointed by the Lord Chancellor. They are assisted by medical, legal, and general panels of Lord Chancellor's Visitors, who visit patients to review their capacities and the implementation of procedures approved by the Court. Applications to the Court may be made by the nearest relative or any interested party. They should include a medical certificate from a doctor concerned in the patient's care and an affidavit of the patient's family and property. After considering the evidence the judge may appoint a receiver to administer the patient's affairs and also to 'do or secure the doing of all such things as appear necessary or expedient'. During the management of a patient's affairs by the Court, medical opinion may be sought about the patient's ability to make a will and about any application to end the Court management. (See Gostin (1983), for a general review, and MacFarlane (1985) for guidance on providing medical evidence.)

Hospitals and the police

The police are entitled to question any person, whether suspected or not, whom they think may be able to provide useful information. The person need not say anything and cannot be compelled to go to a police station except by arrest. If the police wish to interview a psychiatric patient, it may sometimes be necessary for the doctor to give a medical opinion that the patient's mental condition is such that it would be inappropriate for him to be interviewed.

The Judges' Rules state that, as far as practicable, children and young persons under the age of 17 (whether suspected of a crime or not) should be interviewed only in the presence of a parent or guardian or, in their absence, some person who is not a police officer and who is of the same sex as the child. This recommendation applies to juveniles who are hospital patients. If the parents or guardians cannot be present, the hospital should act *in loco parentis* and an appropriate member of hospital staff of the same sex as the child should be present.

The law and child psychiatry

In England and Wales, the legislation relating to the welfare of children is contained in the Children's Act 1989. The provisions of the Act as they apply to the practice of child psychiatry have been summarized by Graham (1991, pp. 441–3). The law of England and Wales relating to consent to treatment during childhood has been reviewed by Pearce (1994). Doctors working in other parts of the United Kingdom and in other countries should seek advice from a colleague who is well informed about the relevant aspects of law relating to the practice of child psychiatry, and should study the relevant legislation.

Further reading

Bluglass, R. S. (1983). *A guide to the Mental Health Act 1983*. Churchill Livingstone, Edinburgh

Graham, P. (1991). *Child psychiatry: a developmental approach* (2nd edn), pp. 441–3. Oxford University Press

Hoggett, B. M. (1984). *Mental health law* (2nd edn). Sweet and Maxwell, London.

Pearce, J. (1994). Consent to treatment during childhood. *British Journal of Psychiatry*, 165, 713–16.

References

Abas, M. A., Sahakian, B. J., and Levy, R. (1990). Neuropsychological deficits and CT scan changes in elderly depressives. *Psychological Medicine* 20(3), 507–20.

Abas, M., Broadhead, J.C., Mbape, P., and Khumalo-Satakukwa, G. (1994). Defeating depression in the developing world. *British Journal of Psychiatry* 164, 293–6.

Abood, M. E. and Martin, B. R. (1992). Neurobiology of marijuana abuse. *Trends in Pharmacological Sciences* 13, 201–6.

Abou-Saleh, M. T. (1992). Lithium. In *Handbook of affective disorders* (ed. E. S. Paykel), pp. 369–85. Churchill Livingstone, Edinburgh.

Abraham, K. (1911). Notes on the psychoanalytic investigation and treatment of manic-depressive insanity and allied conditions. In *Selected papers on psychoanalysis*, pp. 137–56. Hogarth Press and Institute of Psychoanalysis, London (1927).

Abrahamson, L. Y., Seligman, M. E. P., and Teasdale, J. (1978). Learned helplessness in humans: critique and reformulation. *Journal of Abnormal Psychology* 87, 49–74.

Abramowitz, S. I. (1986). Psychosocial outcomes of sex reassignment surgery. *Journal of Consulting and Clinical Psychology* 54, 183–9.

Abrams, R. (1989). ECT for Parkinson's disease. *American Journal of Psychiatry* 146, 1391–3.

Abrams, R., Swartz, C. M., and Vedak, C. (1991). Antidepressant effects of high dose right unilateral electroconvulsive therapy. *Archives of General Psychiatry* 48, 746–8.

Achté, K. A., Hillbom, E., and Aalberg, V. (1969). Psychoses following war brain injuries. *Acta Psychiatrica Scandinavica* 45, 1–18.

Ackerman, N. W. (1958). *The psychodynamics of family life*. Basic Books, New York.

Ackner, B. (1954a). Depersonalization: I. Aetiology and phenomenology. *Journal of Mental Science* 100, 939–53.

Ackner, B. (1954b). Depersonalization: II. The clinical syndromes. *Journal of Mental Science* 100, 954–72.

Ackner, B. and Oldham, A. J. (1962). Insulin treatment of schizophrenia. A three year follow up of a controlled study. *Lancet* i, 504–6.

Ackner, B., Cooper, J. E., Gray, C. H., and Kelly, M. (1962). Acute porphyria, a neuro-psychiatric and biochemical study. *Journal of Psychosomatic Research* 6, 1–24.

Addonizio, G. and Susman, V.L. (1991). *Neuroleptic malignant syndrome: a clinical approach*. In Mosby Year Book, St Louis, MO.

Ader, R. (1976). Psychosomatic research in animals. In *Modern trends in psychosomatic medicine* (ed. O. W. Hill). Butterworths, London.

Adler, A. (1943). Neuropsychiatric complications in victims of Boston's Coconut Grove disaster. *Journal of the American Medical Association* 123, 1098–111.

Advisory Council on Misuse of Drugs (1988). *AIDS and drug misuse*. HMSO, London.

Akhtar, S. and Thomson, A. J. (1982). Overview: narcissistic personality disorder. *American Journal of Psychiatry* 139, 12–20.

Alanen, Y. O. (1958). The mothers of schizophrenic patients. *Acta Psychiatrica Neurologica Scandinavica* 33(Suppl.), 124.

Alanen, Y. O. (1972). The families of schizophrenic patients. *Proceedings of the Royal Society of Medicine* 63, 227–30.

Albert, M. L., Feldman, R. G., and Willis, A. L. (1974). The 'sub-cortical dementia' of progressive supranuclear palsy. *Journal of Neurology, Neurosurgery and Psychiatry* 37, 121–30.

Alexander, F. (1950). *Psychosomatic medicine*. Norton, New York.

Alexander, D. A. (1972). 'Senile dementia'. A changing perspective. *British Journal of Psychiatry* 121, 207–14.

Alexander, P. C. and Lupfer, S. L. (1987). Family characteristics and long term consequences associated with sexual abuse. *Archives of Sexual Behaviour* 16, 235–45.

Alexopoulos, G. S., Young, R. C., and Shindledecker, R. D. (1992). Brain computer tomographic findings in geriatric depression and primary degenerative dementia. *Biological Psychiatry* 31, 591–9.

Allderidge, P. (1979). Hospitals, madhouses and asylums: cycles in the care of the insane. *British Journal of Psychiatry* 134, 321–4.

Allebeck, P. and Wisledt, B. (1986). Mortality in schizophrenia. *Archives of General Psychiatry* 43, 650–3.

Allebeck, P., Bolund, C., and Ringback, G. (1989). Increased suicide rate in cancer patients. *Journal of Clinical Epidemiology* 42, 611–16.

Allen, C. (1969). *A textbook of psychosexual disorders* (2nd edn). Oxford University Press, London.

Allodi, F. A. (1991). Assessment and treatment of torture

victims: a critical review. *Journal of Nervous and Mental Disease* 179, 4–11.

Alter-Reid, K. A., Gibbs, M. S., Lachenmeyer, J. R., *et al.* (1986). Sexual abuse of children: a review of the empirical findings. *Clinical Psychology Review* 6, 249–66.

Alzheimer, A. (1897). Beiträge zur pathologischen Anatomie der Hirnrinde und zur anatomischen Grundlage einiger Psychosen. *Monatsschrift für Psychiatrie und Neurologie* 2, 82–120.

Ambrosini, P. J., Bianchi, M. D., Rabinovich, H., and Elia, J. (1993). Antidepressant treatments in children and adolescents 1. Affective disorders. *Journal of American Academy of Child and Adolescent Psychiatry* 32, 1–6.

American Psychiatric Association (1980). *Diagnostic and statistical manual of mental disorders* (3rd edn, revised). American Psychiatric Association, Washington, DC.

American Psychiatric Association (1985). *Task force on ECT: the practice of ECT: recommendations for training and privileging.* American Psychiatric Association, Washington, DC.

American Psychiatric Association (1987). *Diagnostic and statistical manual of mental disorders* (3rd edn). American Psychiatric Association, Washington, DC.

American Psychiatric Association (1990). *Task force on ECT. The practice of ECT: recommendations for training and privileging.* American Psychiatric Association, Washington, DC.

American Psychiatric Association (1994). *Diagnostic and statistical manual of mental disorders* (4th edn). American Psychiatric Association, Washington, DC.

American Psychiatric Association (1995). *Diagnostic and statistical manual of mental disorders* (4th edn), International version. American Psychiatric Press, Washington, DC.

Ames, D. and Allen, N. (1991). The prognosis of depression in old age: good, bad or indifferent? *International Journal of Geriatric Psychiatry* 6, 477–81.

Amies, P. L., Gelder, M. G., and Shaw, P. M. (1983). Social phobia: a comparative clinical study. *British Journal of Psychiatry* 142, 174–9.

Amir, M. (1971). *Patterns in forcible rape.* Chicago University Press, Chicago, IL.

Ammerman, R. T., Van Hasselt, V. B., and Herson, M. (1986). Psychological adjustment of visually handicapped children and youth. *Clinical Psychological Review* 6, 67–85.

Andersen, B. L. (1992). Psychological interventions for cancer patients to enhance the quality of life. *Journal of Consulting and Clinical Psychology* 60(4), 552–68.

Anderson, G. M., Pollak, E. S., Chatterjee, D., *et al.* (1992). Postmortem analysis of subcortical monoamines and amino acids in Tourette syndrome. In *Tourette syndrome: genetics, neurobiology and treatment advances*

in neurology (ed. T. N. Chase, A. J. Friedhoff, and D. J. Cohen), pp. 123–33. Raven Press, New York.

Anderson, J. C., Williams, S., McGee, R., and Silva, P. A. (1987). DSMIII disorder in pre-adolescent children; prevalence from a large sample in the general population. *Archives of General Psychiatry* 44, 69–76.

Anderson, I. M., Parry-Billings, M., Newsholme, E. A., *et al.* (1990). Decreased plasma tryptophan concentration in major depression: relationship to melancholia and weight loss. *Journal of Affective Disorders* 20, 185–91.

Andreasen, N. J. C. (1982). Negative versus positive schizophrenia: definition and validation. *Archives of General Psychiatry* 36, 1325–30.

Andreasen, N. J. C. (1985). Post-traumatic stress disorder. In *Comprehensive textbook of psychiatry* (4th edn), Vol. 3 (ed. H. I. Kaplan and B. J. Sadock). Williams and Wilkins, Baltimore, MD.

Andreasen, N. C. and Carpenter, W. T. (1993). Diagnosis and the classification of schizophrenia. *Schizophrenia Bulletin* 2, 199–212.

Andreasen, N. J. C. and Norris, A. S. (1972). Management of emotional reactions in severely burned adults. *Journal of Nervous and Mental Disease* 154, 352–62.

Andreasen, N. C., Flaum, M., Swayze, V. W., *et al.* (1990a). Positive and negative symptoms in schizophrenia. *Archives of General Psychiatry* 47, 615–21.

Andreasen, N. C., Swayze, V. W., Flaum, M., *et al.* (1990b). Ventricular enlargement in schizophrenia evaluated with computed tomographic scanning. *Archives of General Psychiatry* 47, 1008–15.

Andreasen, N. C., Rezai, K., Alliger, R., *et al.* (1992). Hypofrontality in neuroleptic-naive patients and in patients with chronic schizophrenia. *Archives of General Psychiatry* 49, 943–58.

Andreasson, S., Allebeck, P., Engstrom, A., *et al.* (1987). Cannabis and schizophrenia: a longitudinal study of Swedish conscripts. *Lancet* ii, 1483–5.

Andrykowski, M. A. and Jacobsen, P. B. (1993). Anticipatory nausea and vomiting with cancer chemotherapy. In *Psychiatric aspects of symptom management in cancer patients* (ed. W. Breitbart), pp. 107–28. American Psychiatric Press, Washington, DC.

Angst, J. (1992). How recurrent and predictable is depressive illness. In *Long-term treatment of depression* (ed. S. Montgomery and F. Rouillon), pp. 1–13. Wiley, Chichester.

Angst, J. and Dobler-Mikola, A. (1985). The Zurich Study. VI A continuum from depression to anxiety disorders? *European Archives of Psychiatry and Neurological Sciences* 235, 179–86.

Angst, J., Baastrup, P., Grof, P., *et al.* (1973). The course of monopolar depression and bipolar psychosis. *Psychiatrica, Neurologica, Neurochirurgia* 76, 486–500.

Angst, J., Scheidegger, P., and Stabl, M. (1993). Efficacy of moclobemide in different patient groups. *Clinical Neuropharmacology* **16** (Suppl. 2), S55–S62.

Ansbacher, H. and Ansbacher, R. (1964). *The individual psychotherapy of Alfred Adler*. Basic Books, New York.

Apley, J. and Hale, B. (1973). Children with recurrent abdominal pain: how do they grow up? *British Medical Journal* ii, 7–9.

Appleby, L., Fox, H., Shaw, M., and Kumar, R. (1989). The psychiatrist in the obstetric unit establishing a liaison service. *British Journal of Psychiatry* **154**, 510–15.

Appleby, L. (1993). Parasuicide: features of repetition and the implications for intervention. *Psychological Medicine* **23**, 13–16.

Arieti, S. (1974). Individual psychotherapy for schizophrenia. In *American handbook of psychiatry* (ed. S. Arieti), Vol. III, Chapter 27. Basic Books, New York.

Arieti, S. (1977). Psychotherapy of severe depression. *American Journal of Psychiatry* **134**, 864–8.

Armistead, L., Klein, K., and Forehand, R. (1995). Parental physical illness and child functioning. *Clinical and Psychological Reviews* **15**(5), 409–22.

Armor, D. J., Polich, J. M., and Stambul, H. B. (1976). *Alcoholism and treatment*. Rand Corporation and Interscience, Santa Monica, CA.

Arnetz, B. B., Horte, L. G., Hedberg, A., *et al.* (1987). Suicide patterns among physicians related to other academics as well as to the general population. *Acta Psychiatrica Scandinavica* **75**, 139–43.

Arora, R. C. and Meltzer, H. Y. (1989). Serotonergic measures in the brains of suicide victims: 5-HT binding sites in the frontal cortex of suicide victims and control subjects. *American Journal of Psychiatry* **146**, 730–6.

Arriageda, P. V., Growdon, J. H., Hedley Whyte, E. T., and Hyman, R. T. (1992). Neurofibrillary tangles but not senile plaque parallel the duration and severity of Alzheimer's disease. *Neurology* **42**, 63–5.

Arthur, A. Z. (1964). Theories and explanations of delusions: a review. *American Journal of Psychiatry* **121**, 105–15.

Asher, R. (1949). Myxoedematous madness. *British Medical Journal* ii, 555–62.

Asher, R. (1951). Munchausen's syndrome. *Lancet* i, 339–41.

Asherson, P., Parfitt, E., Sargeant, M., *et al.* (1992). No evidence for a pseudoautosomal locus for schizophrenia: linkage analysis of multiply affected families. *British Journal of Psychiatry* **161**, 63–8.

Ashton, C. H. (1990). Solvent abuse. *British Medical Journal* **300**, 135–6.

Ashton, H. (1984). Benzodiazepine withdrawal: an unfinished story. *British Medical Journal* **288**, 1135–40.

Asperger, H. (1944). Die 'Autistischen Psychopathien' Kindesalter. *Archives für Psychiatrie und Nervenkrankheiten* **117**, 76–136.

Astrup, C. and Ødegaard, Ø. (1960). Internal migration and mental disease in Norway. *Psychiatric Quarterly* **34** (Suppl. 116).

Audini, B., Marks, I. M., Lawrence, R. E., *et al.* (1994). Home-based versus out-patient/in-patient care for people with serious mental illness. Phase II of a controlled trial. *British Journal of Psychiatry* **165**, 204–10.

August, G. J., Stewart, M. A., and Tsai, L. (1981). The incidence of cognitive disabilities in the siblings of autistic children. *British Journal of Psychiatry* **144**, 461–9.

Austin, M.-P., Ross, M., Murray, C., *et al.* (1992). Cognitive function in major depression. *Journal of Affective Disorders* **25**, 21–30.

Austoker, J. (1994). Reducing alcohol intake. *British Medical Journal* **308**, 1549–52.

Axelson, D. A., Doraiswamy, P. M., McDonald, W. N., *et al.* (1993). Hypercortisolaemia and hippocampal changes in depression. *Psychiatry Research* **47**, 163–73.

Bachrach, L. L. (1986). De-institutionalization: What do the numbers mean? *Hospital and Community Psychiatry* **37**, 118–21.

Bach-y-Rita, G., Lion, J. R., Climent, C. E., and Ervin, F. R. (1971). Episodic dyscontrol: a study of 130 violent patients. *American Journal of Psychiatry* **127**, 1473–8.

Baddeley, A. D. (1990). *Human memory: theory and practice*. Erlbaum, London.

Bailey, J. M. and Pillard, R. C. (1991). A genetic study of male sexual orientation. *Archives of General Psychiatry* **48**, 1089–96.

Bailey, V., Graham, P., and Boniface, D. (1978). How much child psychiatry does a general practitioner do? *Journal of the Royal College of General Practitioners* **28**, 621–6.

Bakwin, H. (1961). Enuresis in children. *Journal of Paediatrics* **58**, 806–19.

Bal, S. S. (1987). Psychological symptomatology and health beliefs of Asian patients. In *Clinical psychology: research and development* (ed. H. Dent), pp. 101–10. Croom Helm, London.

Baldwin, J. A. and Oliver, J. E. (1975). Epidemiology and family characteristics of severely abused children. *British Journal of Preventive and Social Medicine* **29**, 205–21.

Baldwin, R. C. and Jolley, D. J. (1986). The prognosis of depression in old age. *British Journal of Psychiatry* **149**, 574–83.

Bale, R. N. (1973). Brain damage in diabetes mellitus. *British Journal of Psychiatry* **122**, 337–41.

Ball, J. R. B. and Kiloh, L. G. (1959). A controlled trial of imipramine in the treatment of depressive states. *British Medical Journal* ii, 1052–5.

Ballard, C. G., Davis, C., Cullen, P. C., Mohan, R. N., and Dean, C. (1994). Prevalence of postnatal psychiatric morbidity in mothers and fathers. *British Journal of Psychiatry* **164**, 782–8.

Ballinger, C. B. (1990). Psychiatric aspects of the menopause. *British Journal of Psychiatry* **156**, 773–87.

Ban, T. A. (1982). Chronic schizophrenias: a guide to Leonhard's classification. *Comprehensive Psychiatry* **23**, 155–69.

Bancroft, J. H. J. (1974). *Deviant sexual behaviour: modification and assessment*. Oxford University Press.

Bancroft, J. H. J. (1975). Homosexuality in the male. In *Contemporary psychiatry* (ed. T. Silverstone and B. Barraclough). *British Journal of Psychiatry*, Special Publication No. 9.

Bancroft, J. H. J. (1983). *Human sexuality and its problems*. Churchill Livingstone, Edinburgh.

Bancroft, J. H. J. (1986). Crisis intervention. In *An introduction to the psychotherapies* (ed. S. Bloch) (2nd edn). Oxford University Press.

Bancroft, J. H. J. (1991). The sexuality of sexual offending: the social dimension. *Criminal Behaviour and Mental Health* **1**, 181–92.

Bancroft, J. H. J. and Coles, L. (1976). Three years experience in sexual problems clinic. *British Medical Journal* i, 1575–7.

Bancroft, J. H. J., Reynolds, F., Simkin, S., and Smith, J. (1975). Self-poisoning and self-injury in the Oxford area. *British Journal of Preventive and Social Medicine* **29**, 170–7.

Bancroft, J. H. J., Skrimshire, A. M., Casson, J., et al. (1977). People who deliberately poison or injure themselves: their problems and their contacts with helping agencies. *Psychological Medicine* **7**, 289–303.

Bancroft, J. H. J., Simkins, S., Kingston, B., et al. (1979). The reasons people give for taking overdoses: a further inquiry. *British Journal of Medical Psychology* **52**, 353–65.

Bancroft, J. H. J., Tyrer, G., and Warner, P. (1982). The classification of sexual problems in women. *British Journal of Sexual Medicine* **9**, 30–7.

Bancroft, J. H. H., Dickerson, M., Fairburn, C. G., et al. (1986). Sex therapy outcome research: a reappraisal of methodology. *Psychological Medicine* **16**, 851–63.

Bandura, A. (1969). *Principles of behaviour modification*. Holt, Rinehart and Winston, New York.

Banki, C. M., Bissette, G., Arato, M., and Nemeroff, C. B. (1988). Elevation of immunoreaction CSF TRH in depressed patients. *American Journal of Psychiatry* **145**, 1526–31.

Banks, M. H. and Jackson, P. R. (1982). Unemployment and risk of minor psychiatric disorder in young people: cross-sectional and longitudinal evidence. *Psychological Medicine* **12**, 789–98.

Bannister, D. (1962). The nature and measurement of schizophrenic thought disorder. *Journal of Mental Science* **108**, 825–42.

Bannister, D. and Fransella, F. (1966). A grid test of schizophrenic thought disorder. *British Journal of Social and Clinical Psychology* **5**, 95–102.

Barber, T. X. (1962). Towards a theory of hypnosis: posthypnotic behaviour. *Archives of General Psychiatry* **1**, 321–42.

Barker, A. (1994). *Arson*. Oxford University Press.

Barker, J. C. and Barker, A. A. (1959). Deaths associated with electroplexy. *Journal of Mental Science* **105**, 339–48.

Barker, M. G. (1968). Psychiatric illness after hysterectomy. *British Medical Journal* ii, 91–5.

Barker, P. (1988). *Basic child psychiatry* (5th edn). Blackwell Scientific Publications, Oxford.

Barker, P. (1992). *Basic family therapy* (3rd edn). Blackwell Scientific Publications, Oxford.

Barker, W. A., Scott, J., and Eccleston, D. (1987). The Newcastle chronic depression study: results of a treatment regime. *International Clinical Psychopharmacology* **2**, 261–72.

Barnett, R. J., Docherty, J. P., and Frommelt, S. M. (1991). A review of child psychotherapy research since 1963. *Journal of the American Academy of Child and Adolescent Psychiatry* **30**, 1–14.

Barnett, W. and Spitzer, H. (1994). Pathological firesetting 1951–1991. A review. *Medical Scientific Law* **34**, 4–20.

Baron, M. (1993). Genetic linkage and male homosexual orientation. *British Medical Journal* **307**, 337–8.

Baron, M., Gruen, R., and Ranier, J. D. (1985). A family study of schizophrenic and normal control probands: implications for the spectrum concept of schizophrenia. *American Journal of Psychiatry* **142**, 447–55.

Barr, L. C., Goodman, W. K., Price, L. H., McDougle, C. J., and Charney, D. S. (1992). The serotonin hypothesis of obsessive compulsive disorder: implications of pharmacologic challenge studies. *Journal of Clinical Psychiatry* **53** (suppl.), 17–28.

Barraclough, B. M. (1973). Differences between national suicide rates. *British Journal of Psychiatry* **122**, 95–6.

Barraclough, B. M. (1987). The suicide rate of epilepsy. *Acta Psychiatrica Scandinavica* **76**, 339–45.

Barraclough, B. M. and Shea, M. (1970). Suicide and Samaritan clients. *Lancet* ii, 868–70.

Barraclough, B. M. and Shepherd, D. M. (1976). Public interest: private grief. *British Journal of Psychiatry* **129**, 109–13.

Barraclough, B. M. Bunch, J., Nelson, B., and Sainsbury, P. (1974). A hundred cases of suicide: clinical aspects. *British Journal of Psychiatry* **125**, 355–73.

Barrett, C. J. (1978). Effectiveness of widow's groups in facilitating change. *Journal of Consulting and Clinical Psychology* **46**, 20–31.

Barrowclough, C., Johnston, M., and Tarrier, N. (1994). Attributions, expressed emotion, and patient relapse: an attributional model of relatives' response to schizophrenic illness. *Behaviour Therapy* **25**, 67–88.

Barsky, A. J., Wyshak, G., Latham, K. S., and Klerman, G. L. (1991). The relationship between hypochondriasis and medical illness. *Archives of Internal Medicine* 151, 84–8.

Bartlett, J., Bridges, P., and Kelly, D. (1981). Contemporary indications for psychosurgery. *British Journal of Psychiatry* 138, 507–11.

Baruk, H. (1959). Delusions of passion. Reprinted in *Themes and variations in European psychiatry* (ed. S. R. Hirsch and M. Shepherd), pp. 375–84. Wright, Bristol (1974).

Basmajian, J. V. (ed.) (1983). *Biofeedback: principles and practice for clinicians*. Williams and Wilkins, Baltimore, MD.

Bass, C. and Mayou, R. A. (1995). Chest pain and palpitations. In *Treatment of functional somatic symptoms* (ed. R. A. Mayou, C. Bass, and M. Sharpe), pp. 328–52. Oxford University Press, Oxford.

Bassuk, E. L. (1984). Is homelessness a mental health problem? *American Journal of Psychiatry* 141, 1546–50.

Bateson, G., Jackson, D., Haley, J., and Weakland, J. (1956). Towards a theory of schizophrenia. *Behavioral Science* 1, 251–64.

Bauer, M. S. and Whybrow, P. C. (1991). Rapid cycling bipolar disorder: clinical features, treatment and etiology. In *Refractory depression* (ed. J. D. Amsterdam), pp. 191–208. Raven Press, New York.

Bauman, M. (1991). Microscopic neuroanatomic abnormalities. *Paediatrics* 31, 791–6.

Baumgarten, M., Hanley, J. A., Infante-Rivard, C., *et al.* (1994). Health of family members caring for elderly persons with dementia. *Annals of Internal Medicine* 120, 126–32.

Baxter, L. R., Schwartz, J. M., Bergman, K. S., *et al.* (1992). Caudate glucose metabolic rate changes with both drug and behavior therapy for obsessive–compulsive disorder. *Archives of General Psychiatry* 49, 681.

Bearison, D. J. and Mulhern, R. K. (1994). *Pediatric psychooncology: psychological perspectives on children with cancer*. Oxford University Press.

Beasley, C. M., Sayler, M. E., Cunningham, G. E., *et al.* (1990). Fluoxetine in tricyclic refractory major depressive disorder. *Journal of Affective Disorders* 20, 193–200.

Beasley, C. M., Masica, D. N., Heiligenstein, J. H., *et al.* (1993). A possible monoamine oxidase inhibitor–serotonin uptake inhibitor interaction: fluoxetine clinical data and preclinical findings. *Journal of Clinical Psychopharmacology* 13, 312–20.

Beaubrun, M. H. and Knight, F. (1973). Psychiatric assessment of 30 chronic users of cannabis and 30 matched controls. *American Journal of Psychiatry* 130, 309–11.

Beaurepaire, J. E., *et al.* (1992). The acute appendicitis syndrome: psychological aspects of the inflamed and non-inflamed appendix. *Journal of Psychosomatic Research* 36(5), 425–37.

Bebbington, P. E., Sturt, E., Tennant, C., and Hurry, J. (1984). Misfortune and resilience: a community study of women. *Psychological Medicine* 14, 347–63.

Bebbington, P., Wilkins, S., Jones, P., *et al.* (1993). Life events and psychosis. *British Journal of Psychiatry* 162, 72–9.

Beck, A. T. (1967). *Depression: clinical experimental and theoretical aspects*. Harper and Row, New York.

Beck, A. T. (1976). *Cognitive therapy and the emotional disorders*. International Universities Press, New York.

Beck, A. T. and Freeman, A. (1990). *Cognitive therapy for personality disorders*. Guilford Press, New York.

Beck, A. T., Ward, C. H., Mendelson, M., *et al.* (1961). An inventory for measuring depression. *Archives of General Psychiatry* 4, 561–85.

Beck, A. T., Laude, R., and Bohnert, M. (1974a). Ideational components of anxiety neuroses. *Archives of General Psychiatry* 31, 319–25.

Beck, A. T., Schuyler, D., and Herman, I. (1974b). Development of suicide intent scales. In *The prediction of suicide* (ed. A. T. Beck, H. L. P. Resaik, and D. J. Lettie). Charles Press, Maryland.

Beck, A. T., Steer, R. A., Kovacs, M., and Garrison, B. (1985). Hopelessness and eventual suicide: a 10-year prospective study of patients hospitalised with suicidal ideation. *American Journal of Psychiatry* 145, 559–63.

Beck, J. C. M. (1994). Epidemiology of mental disorder and violence: beliefs and research findings. *Harvard Review of Psychiatry* 2(1), 1–6.

Becker, J. and Quinsey, V. (1993). Assessing suspected child molesters. *Child Abuse and Neglect* 17, 169–74.

Beech, H. R., Watts, F., and Poole, A. D. (1971). Classical conditioning of a sexual deviation: a preliminary note. *Behaviour Therapy* 2, 400–2.

Beitchman, J. H., Zucker, K. J., and Hood, J. E. (1992). A review of the long term effects of child sexual abuse. *Child Abuse and Neglect* 16, 101–18.

Bekkers, M. J. T. M. *et al.* (1995). Psychosocial adaption to stoma surgery: a review. *Journal of Behavioral Medicine* 18(1), 1–31.

Benaim, S., Horder, J., and Anderson, J. (1973). Hysterical episode in a classroom. *Psychological Medicine* 3, 366–73.

Bench, C. J., Dolan, R. J., Friston, K. J., and Frackowiak, R. S. J. (1990). Positron emission tomography in the study of brain metabolism in psychiatric and neuropsychiatric disorders. *British Journal of Psychiatry* 157 (Suppl. 9), 82–95.

Bender, D. A. (1982). Biochemistry of tryptophan in health and disease. *Molecular Aspects of Medicine* 6, 101–97.

Bender, L. and Grugett, A. A. (1952). A follow up report on

children who had atypical sexual experiences. *American Journal of Orthopsychiatry* 22, 825–37.

Bendz, H. (1983). Kidney function in lithium treated patients: a literature survey. *Acta Psychiatrica Scandinavica* 68, 303–24.

Benedetti, G. (1952). *Die Alkoholhalluzinosen.* Thieme, Stuttgart.

Benedetti, G. (1987). *Psychotherapy of schizophrenia.* University Press, London.

Benedict, R. H. B. (1989). The effectiveness of cognitive remediation strategies for victims of traumatic head-injury: a review of the literature. *Clinical Psychology Review* 9, 605–26.

Benjamin, H. (1966). *The transsexual phenomenon.* Julian Press, New York.

Benjamin, R. S., Costello, E. J., and Warren, M. (1990). Anxiety disorders in a pediatric sample. *Journal of Anxiety Disorders* 4, 293–316.

Benjamin, S., Mawer, J., and Lennon, J. (1992). The knowledge and beliefs of family care givers about chronic pain patients. *Journal of Psychosomatic Research* 36(3), 211–17.

Benjamin, S. and Main, C. J. (1995). Psychiatric and psychological approaches to the treatment of chronic pain: concepts and individual treatments. In *Treatment of functional somatic symptoms* (ed. R. Mayou, C. Bass, and M. Sharpe), p. 188. Oxford University Press, Oxford.

Bennett, D. H. (1983). The historical development of rehabilitation services. In *The theory and practice of rehabilitation* (ed. F. N. Watts and D. H. Bennett). Wiley, Chichester.

Benson, B. A. and Gross, A. M. (1989). The effect of a congenitally handicapped child upon the marital dyad: a review of the literature. *Clinical Psychology Review* 9, 747–58.

Berelowicz, M. and Tarnopolsky, A. (1993). Borderline personality disorder. In *Personality disorder reviewed* (ed. P. Tyrer and G. Stein), pp. 90–112. Gaskell, London.

Beresford, B. A. (1994). Resources and strategies: how parents cope with the care of a disabled child. *Journal of Child Psychology and Psychiatry* 35(1), 171–209.

Berg, I. (1984). School refusal. *British Journal of Hospital Medicine* 31, 59–62.

Berg, I. and Jackson, A. (1985). Teenage school refusers grow up: a follow-up study of 168 subjects, ten years on average after in-patient treatment. *British Journal of Psychiatry* 147, 366–70.

Berg, J. M., Karlinsky, H., and Holland, A. J. (ed.) (1995). *Alzheimer disease, Down syndrome and their relationship.* Oxford University Press, Oxford.

Bergen, A. L. M., Dahl, A. A., Guldberg, C., and Hansen, H. (1990). Langfelt's schizophreniform psychoses fifty years later. *British Journal of Psychiatry* 157, 351–4.

Berger, M. (1985). Temperament and individual differences. In *Child and adolescent psychiatry: modern approaches* (ed. M. Rutter and L. Hersov) (2nd edn). Blackwell Scientific, Oxford.

Bergin, A. E. (1994). *Handbook of psychotherapy and behaviour change* (4th edn). Wiley, New York.

Bergin, A. E. and Lambert, M. J. (1978). The evaluation of therapeutic outcomes. In *Handbook of psychotherapy and behaviour change* (ed. S. L. Garfield and A. E. Bergin) (2nd edn). Wiley, New York.

Berglund, M. and Nordstrom, G. (1989). Mood disorders in alcoholism. *Current Opinion in Psychiatry* 2, 428–33.

Bergmann, K., Foster, E. M., Justice, A. W., and Matthews, V. (1978). Management of the demented patient in the community. *British Journal of Psychiatry* 132, 441–9.

Berlin, R. M. (1984). Sleep disorders in a psychiatric consultation service. Often complicated by physical illness and common but frequently overlooked in liaison psychiatry. *American Journal of Psychiatry* 141, 582–4.

Berman, K. F., Torrey, E. F., Daniel, D. G., and Weinberger, D. R. (1992). Regional cerebral blood flow in monozygotic twins discordant and concordant for schizophrenia. *Archives of General Psychiatry* 49, 927–34.

Berman, K. F., Doran, A. R., Pickar, D., and Weinberger, D. R. (1993). Is the mechanism of prefrontal hypofunction in depression the same as schizophrenia. *British Journal of Psychiatry* 162, 183–92.

Berman, S. M. and Noble, E. P. (1993). Childhood antecedents of substance misuse. *Current Opinion in Psychiatry* 6, 382–7.

Bernheim (1890). *Suggestive therapeutics* (2nd edn). Young J. Pentland, Edinburgh and London.

Bernstein, D. A. and Borkovec, T. D. (1973). *Progressive, relaxation training: a manual for the helpful professions.* Research Press, Champaign, IL.

Berrios, G. E. (1981a). Stupor: a conceptual history. *Psychological Medicine* 11, 677–88.

Berrios, G. E. (1981b). Delirium and confusion in the 19th century. *British Journal of Psychiatry* 139, 439–49.

Berrios, G. E. (1985). Delusional parasitosis and physical disease. *Comprehensive Psychiatry* 26, 395–403.

Berrios, G. E. (1994). Dementia: historical review. In *Dementia* (ed. A. Burns and R. Levy), pp. 5–20. Chapman & Hall Medical, London.

Berson, R. J. (1983). Capgras' syndrome. *American Journal of Psychiatry* 140, 969–78.

Bertelsen, A., Harvald, B., and Hauge, M. (1977). A Danish twin study of manic depressive disorders. *British Journal of Psychiatry* 130, 330–51.

Besson, J. A. O. (1993). Structural and functional brain imaging in alcoholism and drug misuse. *Current Opinion in Psychiatry* 6, 403–10.

Betts, T. (1990). Pseudoseizures: seizures that are not epilepsy. *Lancet* ii, 9–10.

Betts, T. A., Clayton, A. B., and Mackay, G. M. (1972). Effects of four commonly-used tranquillizers on low-speed driving performance tests. *British Medical Journal* iv, 580–4.

Bewley, B. (1986). The epidemiology of adolescent behaviour problems. *British Medical Bulletin* 42, 200–3.

Bhopal, R. S. (1986). The inter-relationship of folk, traditional and Western medicine within an Asian community in Britain. *Social Scientific Medicine* 22, 99–105.

Bibring, E. (1953). The mechanism of depression. In *Affective disorders* (ed. P. Greenacre), pp. 14–47. International Universities Press, New York.

Bickell, H. and Cooper, B. (1994). Incidence and relative risk of dementia in an urban elderly population: findings of a prospective field study. *Psychological Medicine* 24, 179–92.

Bieber, I. (1962). *Homosexuality: a psychoanalytic study of male homosexuals*. Basic Books, New York.

Biederman, J., Faraone, S. V., Keenan, K., *et al.* (1992). Further evidence for family-genetic risk factors in attention deficit hyperactivity disorder. *Archives of General Psychiatry* 49, 728–38.

Biehl, H. (1989). The WHO Psychological Impairments Rating Schedule. *British Journal of Psychiatry* 155(7), 78–80.

Biggins, C. A., Boyd, J. L., Harrop, F. M., *et al.* (1992). A controlled, longitudinal study of dementia in Parkinson's disease. *Journal of Neurology, Neurosurgery and Psychiatry* 55, 566–71.

Binet, A. (1877). Le fétishisme dans l'amour. *Revue Philosophique* 24, 143.

Binet, A. and Simon, T. (1905). Méthodes nouvelles pour le diagnostic du niveau intellectuel des normaux. *L'Année Psychologique* 11, 193–244.

Binswanger, O. (1894). *Münchener Medizinische Wochenschrift* 52, 252.

Bion, W. R. (1961). *Experiences in groups*. Tavistock Publications, London.

Birnbaum, K. (1908). *Psychosen mit Wahnbildung und wahnhafte Einbildungen bei Degenerativen*. Marhold, Halle.

Bishay, N. R., Petersen, M., and Tarrier, N. (1989). An uncontrolled study of cognitive therapy for morbid jealousy. *British Journal of Psychiatry* 154, 386–9.

Bitondo Dyer, C., Ashton, C. M., and Teasdale, T. A. (1995). Postoperative delirium. *Archives of Internal Medicine* 155, 461–5.

Black, D. (1986). Schoolgirl mothers. *British Medical Journal* 293, 1047.

Black, D., Winokur, G., and Hasrallah, A. (1987). Suicide in subtypes of major affective disorders. *Archives of General Psychiatry* 44, 878–80.

Black, D. W., Wesner, R., Bowers, W., and Gabel, J. (1993). A comparison of fluvoxamine, cognitive therapy, and placebo in the treatment of panic disorder. *Archives of General Psychiatry* 50, 44.

Blackburn, I. M., Bishop, S., Glen, A. I. M., *et al.* (1981). The efficacy of cognitive therapy on depression: a treatment trial using cognitive therapy and pharmacotherapy, each alone and in combination. *British Journal of Psychiatry* 139, 181–9.

Blackburn, I. M., Eunson, K. M., and Bishop, S. (1986). A two-year naturalistic follow-up of depressed patients treated with cognitive therapy, pharmacotherapy and a combination of both. *Journal of Affective Disorders* 10, 67–75.

Blackwood, D. H. R. and Muir, W. J. (1990). Cognitive brain potentials and their application. *British Journal of Psychiatry* 157 (Suppl. 9), 96–101.

Blake, F., Gath, D., and Salkovskis, P. M. (1995). Psychological aspects of the premenstrual syndrome: developing a cognitive approach. In *Treatment of functional somatic symptoms* (ed. R. Mayou). Oxford University Press.

Blanchard, E. B. (1992). Psychological treatment of benign headache disorders. *Journal of Consulting and Clinical Psychology* 60(4), 537–51.

Blanchard, R. and Hucker, S. J. (1991). Age, transvestism, bondage, and concurrent paraphilic activities in 117 fatal cases of autoerotic asphyxia. *British Journal of Psychiatry* 159, 371–7.

Blanchard, E. B., Hickling, E. J., Taylor, A. E., Loos, W. R., and Gerardi, R. J. (1994). Psychological morbidity associated with motor vehicle accidents. *Behavioural Research and Therapy* 32(3), 283–90.

Blanchard, E. B., Hickling, E. J., Taylor, A. E., Loos, W. R., Forneris, C. A. and Jaccard, J. (1995). Who develops PTSD from motor vehicle accidents. *Behaviour Research and Therapy* (in press).

Blank, A. S. (1992). The longitudinal course of posttraumatic stress disorder. In *Posttraumatic stress disorder: DSMIV and beyond* (ed. J. R. T. Davidson and E. B. Foa), pp. 3–22. American Psychiatric Press, Washington, DC.

Blatner, H. A. (1973). *Acting-in: practical aspects of psychodramatic methods*. Springer, New York.

Blazer, D. G. (1986). Suicide in late life: review and commentary. *Journal of the American Geriatric Society* 34, 519–25.

Blehar, M. C. and Rosenthal, N. E. (1989). Seasonal affective disorders and phototherapy: report of the National Institute of Mental Health-Sponsored Workshop. *Archives of General Psychiatry* 46, 469–74.

Bleuler, E. (1906). *Affektivität, Suggestibilität, und Paranoia*. Marhold, Halle.

Bleuler, E. (1911). (English edn 1950). *Dementia praecox or the group of schizophrenias*. International University Press, New York.

Bleuler, M. (1972). (English edn 1978). *The schizophrenic*

disorders: long term patient and family studies. Yale Universities Press, New Haven, CT.

Bleuler, M. (1974). The long term course of the schizophrenic psychoses. *Psychological Medicine* 4, 244–54.

Bliss, E. L. (1984). A symptom profile of patients with multiple personalities, including MMPU results. *Journal of Nervous and Mental Disease* 172, 197–202.

Bliss, E. L., Clark, L. D., and West, C. D. (1959). Studies of sleep deprivation: relationship to schizophrenia. *Archives of Neurology and Psychiatry* 81, 348–59.

Bloch, S. (1986). Supportive psychotherapy. In *An introduction to the psychotherapies* (ed. S. Bloch) (2nd edn). Oxford University Press.

Bloch, S. and Chodoff, P. (1981). *Psychiatric ethics.* Oxford University Press.

Block, G. J. (1980). *Mesmerism.* William Kaufmann, Los Altos, California.

Bloom, F. E. (1993). Advancing a neurodevelopmental origin for schizophrenia. *Archives of General Psychiatry* 50, 224–7.

Bluglass, R. and Bowden, P. (1990). *Principles and practice of forensic psychiatry.* Churchill Livingstone, Edinburgh.

Blumenthal, E. J. (1955). Spontaneous seizures and related electroencephalographic findings following shock therapy. *Journal of Nervous Mental Disease* 122, 581–8.

Blumenthal, J. A. and Mark, D. B. (1994). Quality of life and recovery after cardiac surgery. *Psychosomatic Medicine* 56, 213–15.

Blumer, D. and Heilbronn, M. (1982). Chronic pain as a variant of depressive disease. The pain prone disorder. *Journal of Nervous and Mental Disease* 170, 381–406.

Blurton-Jones, N. G. (1972). Non-verbal communication in children. In *Non-verbal communication* (ed. R. A. Hinde). Cambridge University Press.

Boakes, A. J., Laurence, D. R., Teoh, P. C., *et al.* (1973). Interactions between sympathomimetic amines and antidepressant agents in man. *British Medical Journal* i, 311–15.

Bogerts, B., Meertz, E., and Schonfield-Bausch, R. (1985). Basal ganglia and limbic system pathology in schizophrenia: a morphometric study. *Archives of General Psychiatry* 42, 784–91.

Bogren, L. Y. (1983). Couvade. *Acta Psychiatrica Scandinavica* 68, 55–63.

Bohman, M. (1978). Some genetic aspects of alcoholism and criminality. *Archives of General Psychiatry* 35, 269–76.

Bohman, M., Sigvardsson, S., and Cloninger, C. R. (1981). Maternal inheritance of alcohol abuse: cross fostering analysis of adopted women. *Archives of General Psychiatry* 38, 965–9.

Bohnen, N. and Jolles, J. (1992). Neurobehavioural aspects of postconcussive symptoms after mild head injury. *Journal of Nervous and Mental Disorder* 180, 683–92.

Böker, W. and Häfner, H. (1977). Crimes of violence by mentally disordered offenders in Germany. *Psychological Medicine* 7, 733–6.

Bollini, P. and Mollica, R. F. (1989). Surviving without the asylum: an overview of the studies on the Italian Reform Movement. *Journal of Nervous and Mental Disease* 177, 607–15.

Bond, A. J. and Lader, M. H. (1973). Residual effects of flurazepam. *Psychopharmacologia* 32, 223–35.

Bonhoeffer, K. (1909). Exogenous psychoses. *Zentralblatt für Nervenheilkunde* 32, 499–505. Translated by H. Marshall in *Themes and variations in European psychiatry* (ed. S. R. Hirsch and M. Shepherd). Wright, Bristol (1974).

Bonn, J., Turner, P., and Hicks, D. C. (1972). Beta-adrenergic receptor blockade with practolol in the treatment of anxiety. *Lancet* i, 814–15.

Böök, J. A. (1953). A genetic and neuropsychiatric investigation of a North-Swedish population with special regard to schizophrenia and mental deficiencies. *Acta Genetica et Statistica Medica* 4, 1–100.

Bools, C. N., Neale, B. A., and Meadow, S. R. (1993). Follow up of victims of fabricated illness (Munchausen syndrome by proxy). *Archives of Disease in Childhood* 69, 625–30.

Boman, B. (1988). L-Tryptophan: a rational antidepressant or a natural hypnotic? *Australia and New Zealand Journal of Psychiatry* 22, 83–97.

Borthwick-Duffy, S. A. (1994). Epidemiology and prevalence of psychopathology in people with mental retardation. *Journal of Consulting and Clinical Psychology* 62(1), 17–27.

Boston Collaborative Drug Surveillance Program (1972). Adverse reactions to tricyclic antidepressant drugs. *Lancet* i, 529–31.

Bowlby, J. (1944). Forty-four juvenile thieves. Their characters and home life. *International Journal of Psychoanalysis* 25, 19–53.

Bowlby, J. (1946). *Forty-four juvenile thieves: their characters and home-life.* Baillière, Tindall and Cox, London.

Bowlby, J. (1951). *Maternal care and maternal health.* World Health Organization, Geneva.

Bowlby, J. (1969). Psychopathology of anxiety: the role of affectional bonds. In *Studies in anxiety* (ed. M. H. Lader). *British Journal of Psychiatry*, Special Publication No. 3.

Bowlby, J. (1973). *Attachment and loss*, Vol. 2, *Separation, anxiety and anger.* Hogarth Press, London.

Bowlby, J. (1980). *Attachment and loss*, Vol. 3. *Loss, sadness and depression.* Basic Books, New York.

Boyd, J. H. and Weissman, M. M. (1982). Epidemiology. In *Handbook of affective disorders* (ed. E. S. Paykel). Churchill Livingstone, Edinburgh.

Bradbury, E. (1994). The psychology of aesthetic plastic surgery. *Aesthetic Plastic Surgery* 18, 301–5.

Bradbury, T. N. and Miller, G. A. (1985). Season of birth in schizophrenia: a review of evidence, methodology and etiology. *Psychological Bulletin* 98, 569–594.

Braddock, L. (1986). The dexamethasone suppression test: fact and artefact. *British Journal of Psychiatry* 148, 363–74.

Bradford, J. and Balmaceda, R. (1983). Shoplifting: is there a specific psychiatric syndrome? *Canadian Journal of Psychiatry* 28, 248–54.

Bradley, C., Parsons, P. E., and Clarke, A. J. (1994). Contributions of psychology to diabetes management. *British Journal of Clinical Pharmacology* 33, 11–21.

Bradley, D., *et al.* (1993). Experience with screening newborns for Duchenne muscular dystrophy in Wales. *British Medical Journal* 306, 357–60.

Bradwejn, J., Koszcki, D., and Shriqui, C. (1991). Enhanced sensitivity to cholecysokinin tetrapeptide in panic disorder. *Archives of General Psychiatry* 48, 603.

Brady, E. U. and Kendall, P. C. (1992). Comorbidity of anxiety and depression in children and adolescents. *Psychological Bulletin* 111, 244–55.

Braid, J. (1843). *Neurohypnology: or the rationale of nervous sleep, considered in relation with animal magnetism.* Churchill, London.

Brandon, S. (1991). The psychological aftermath of war. *British Medical Journal* 302, 305–6.

Brandon, S., Cowley, P., McDonald, C., *et al.* (1984). Electroconvulsive therapy: results in depressive illness from the Leicestershire Trial. *British Medical Journal* 288, 22–5.

Brandon, S., Cowley, P., McDonald, C., *et al.* (1985). Leicester ECT trial: results in schizophrenia. *British Journal of Psychiatry* 146, 177–83.

Brauer, A., Horlick, L. F., Nelson, E., *et al.* (1979). Relaxation therapy for essential hypertension: a Veterans Administration out-patient study. *Journal of Behavioural Medicine* 2, 21–9.

Braun, P., Kuchansky, G., Shapiro, R., *et al.* (1981). Overview; reinstitutionalization of psychiatric patients, a critical review of outcome studies. *American Journal of Psychiatry* 138, 736–49.

Breitbart, W. and Holland, J. C. (ed.) (1993). *Psychiatric aspects of symptom management in cancer patients.* American Psychiatric Press, Washington, DC.

Brennan, P. A. and Mednick, S. A. (1993). Genetic perspectives on crime. *Acta Psychiatrica Scandinavica* 370 (Suppl.), 19–36.

Breslau, N. and Prabucki, M. A. (1987). Siblings of disabled children: effects of chronic stress in the family. *Archives of General Psychiatry* 44, 1040–46.

Breslau, N., Weitzman, M., and Messenger, K. (1981). Psychologic functioning of the siblings of disabled children. *Pediatrics* 67, 344–57.

Breuer, J. and Freud, S. (1893–5). *Studies on hysteria.* In *The standard edition of the complete psychological works,* Vol. 2. Hogarth Press, London (1955).

Bridges, K. W. and Goldberg, D. P. (1985). Somatic presentation of DSMIII psychiatric disorders in primary care. *Journal of Psychosomatic Research* 29, 563–9.

Bridges, P. (1992). Resistant depression and psychosurgery. In *Handbook of affective disorders* (ed. E. S. Paykel), pp. 437–51. Churchill Livingstone, Edinburgh.

Briere, J. (1988). *Annals of the New York Academy of Sciences* 528, 327–33.

Brietbart, W. and Holland, J. C. (ed.) (1993). *Psychiatric aspects of symptom management in cancer patients.* American Psychiatric Press, Washington, DC.

British Medical Association (1993). *The boxing debate.* British Medical Association, London.

British Medical Formulary (1981). British Medical Association and the Pharmaceutical Society of Great Britain, London.

Broadbent, D. E. (1981). Chronic effects from the physical nature of work. In *Working life: a social science contribution to work reform* (ed. B. Gardell and G. Johansson). Wiley, London.

Broadbent, D. E. and Gath, D. H. (1979). Chronic effects of repetitive and non-repetitive work. In *Response to stress: occupational aspects* (ed. C. G. McKay and T. R. Cox). Independent Publishing Company, London.

Broadhead, J. and Jacoby, R. J. (1990). Mania in old age: a first prospective study. *International Journal of Geriatric Psychiatry* 5, 215–22.

Broadwin, I. T. (1932). A contribution to the study of truancy. *American Journal of Orthopsychiatry* 2, 253–9.

Brockington, I. (1986). Diagnosis of schizophrenia and schizoaffective psychoses. In *The psychopharmacology and treatment of schizophrenia* (ed. P. B. Bradley and S. R. Hirsch). Oxford University Press.

Brockington, I. F., Kendell, R. E., Kellett, J. M., *et al.* (1978). Trials of lithium, chlorpromazine and amitriptyline on schizoaffective patients. *British Journal of Psychiatry* 133, 162–8.

Broman, S., Nichols, P. L., Shaughnessy, P., and Kennedy, W. (1987). *Retardation in young children: a developmental study of cognitive deficit.* Lawrence Erlbaum, Hillsdale, NJ.

Brooks, N. (1984). *Closed head injury.* Oxford University Press.

Brooks, N. (1991). The head-injured family. *Journal of Clinical and Experimental Neuropsychology* 13, 155–88.

Brown, A. S. and Gershon, S. (1993). Dopamine and depression. *Journal of Neural Transmission* 91, 75–109.

Brown, D. and Pedder, J. (1979). *Introduction to psychotherapy: an outline of psychodynamic principles and practice.* Tavistock Publications, London.

Brown, F. W. (1942). Heredity in the psychoneuroses.

Proceedings of the Royal Society of Medicine 35, 785–90.

Brown, G. L. and Linnoila, M. I. (1990). CSF serotonin metabolite (5-HIAA) studies in depression, impulsivity, and violence. *Journal of Clinical Psychiatry* 51 (Suppl. 4), 31–41.

Brown, G. W. and Birley, J. L. T. (1968). Crisis and life change at the onset of schizophrenia. *Journal of Health and Social Behaviour* 9, 203–24.

Brown, G. W. and Harris, T. O. (1978). *Social origins of depression*. Tavistock, London.

Brown, G. W. and Harris, T. O. (1986). Stressor, vulnerability, and depression: a question of replication. *Psychological Medicine* 16, 739–44.

Brown, G. W. and Harris, T. O. (1993). Aetiology of anxiety and depressive disorders in an inner-city population. 1 Early adversity. *Psychological Medicine* 23, 143–54.

Brown, G. W. and Prudo, R. (1981). Psychiatric disorder in a rural and urban population. 1. Aetiology of depression. *Psychological Medicine* 11, 581–99.

Brown, G. W., Carstairs, G. M., and Topping, G. G. (1958). *Lancet* ii, 685–9.

Brown, G. W., Monck, E. M., Carstairs, G. M., and Wing, J. K. (1962). Influence of family life on the cause of schizophrenic illness. *British Journal of Preventive and Social Medicine* 16, 55–68.

Brown, G. W., Bone, M., Dalison, B., and Wing, J. K. (1966). *Schizophrenia and social care*. Maudsley Monograph 17, Oxford University Press, London.

Brown, G. W., Harris, T. O., and Peto, J. (1973a). Life events and psychiatric disorders: the nature of the causal link. *Psychological Medicine* 3, 159–76.

Brown, G. W., Sklair, F., Harris, T. O., and Birley, J. L. T. (1973b). Life events and psychiatric disorder: some methodological issues. *Psychological Medicine* 3, 74–87.

Brown, G. W., Harris, T. O., and Eales, M. J. (1993). Aetiology of anxiety and depressive disorders in an inner-city population. 2. Comorbidity and adversity. *Psychological Medicine* 23, 155–65.

Brown, J. (1992). Pick's disease. In *Baillière's clinical neurology*, pp. 535–53. Baillière Tindall, London.

Brown, R., Colter, N., Corsellis, J. A. N., *et al.* (1986). Post-mortem evidence of structural brain changes in schizophrenia. *Archives of General Psychiatry* 43, 36–42.

Brown, T. M. and Brown, R. L. S. (1995). Neuropsychiatric consequences of renal failure. *Psychosomatics* 36 244–53.

Brown, W. (1934). *Psychology and psychotherapy* (3rd edn). Edward Arnold, London.

Browne, A. and Finkelhor, D. (1986). Impact of child sexual abuse: a review of research. *Psychological Bulletin* 99, 66–77.

Browne, K. and Saqui, S. (1987). Parent child interaction in abusing families and its possible causes and consequences. In *Child abuse: the educational perspective* (ed. P. Maher). Blackwell, Oxford.

Brownell, K. D. and Fairburn, C. G. (1995). *Eating disorders and obesity: a comprehensive handbook*. Guilford Press, New York.

Bruce, M., Scott, N., Shine, P., and Lader, M. (1992). Anxiogenic effects of caffeine in patients with anxiety disorders. *Archives of General Psychiatry* 49, 867–9.

Bruch, H. (1974). *Eating disorders: anorexia nervosa and the person within*. Routledge & Kegan Paul, London.

Brudny, J., Korein, J., Levidow, A., and Friedman, L. W. (1974). Sensory feedback therapy as a modality of treatment in central nervous system disorders of voluntary movement. *Neurology* 24, 925–32.

Brugha, T., Wing, J. K., and Smith, B. L. (1989). Physical health of the long-term mentally ill in the community: Is there unmet need? *British Journal of Psychiatry* 155, 777–81.

Bruton, C. J., Crow, T. J., Frith, C. D., *et al.* (1990). Schizophrenia and the brain: a prospective clinico-neuropathological study. *Psychological Medicine* 20, 285–304.

Bryant, B., Trower, P., Yardley, K., Urbieta, H., and Letemendia, F. J. J. (1976). A survey of social inadequacy among psychiatric outpatients. *Psychological Medicine* 6, 101–12.

Bryant, R. A. and Harvey, A. G. (1995). Avoidant coping style and post-traumatic stress following motor vehicle accidents. *Behaviour Research and Therapy* 33(6), 631–5.

Bryer, J. B., Nelson, B. A., and Miller, J. B. (1992). Childhood sexual and physical abuse as factors in adult psychiatric illness. *American Journal of Psychiatry* 131, 981–7.

Bryer, J. B., Nelson, B. A., Miller, J. B., and Krol, P. A. (1987). Childhood sexual and physical abuse as factors in adult psychiatric illness. *Americal Journal of Psychiatry* 144, 1426–30.

Buchan, H., Johnstone, E., McPherson, K., *et al.* (1992). Who benefits from electroconvulsive therapy? Combined results of the Leicester and Northwick Park Trials. *British Journal of Psychiatry* 160, 355–9.

Buchanan, A., Reed, A., Wessely, S., *et al.* (1993). The phenomenological correlates of acting on delusions. *British Journal of Psychiatry* 163, 77–81.

Buchsbaum, M. S. and Siegel, B. V. (1994). Neuroimaging and the aging process in psychiatry. *International Review of Psychiatry* 6, 109–18.

Buchsbaum, M. S., Kesslak, J. P., Lynch, G., *et al.* (1991). Temporal and hippocampal metabolic rate during an olefactory memory task assessed by positron emission tomography in patients with dementia of the Alzheimer type and controls. *Archives of General Psychiatry* 48, 840–7.

Buchsbaum, M. S., Haier, R. J., Potkin, S. G., *et al.* (1992). Frontostriatal disorder of cerebral metabolism in never-medicated schizophrenics. *Archives of General Psychiatry* 49, 935–42.

Buckle, A. and Farrington, D. P. (1984). An observational study of shop-lifting. *British Journal of Criminology* 24, 63–73.

Bucknill, J. C. and Tuke, D. H. (1858). *A manual of psychological medicine.* John Churchill, London.

Buglass, D. and Duffy, J. C. (1978). The ecological pattern of suicide and parasuicide in Edinburgh. *Social Science and Medicine* 12, 241–53.

Buglass, D. and Horton, J. (1974). The repetition of parasuicide: a comparison of three cohorts. *British Journal of Psychiatry* 125, 168–74.

Buglass, D., Clarke, J., Henderson, A. S., *et al.* (1977). A study of agoraphobic housewives. *Psychological Medicine* 7, 73–86.

Bulbena, A. and Burrows, G. E. (1986). Pseudodementia: facts and figures. *British Journal of Psychiatry* 148, 87–94.

Bunch, J. (1972). Recent bereavement in relation to suicide. *Journal of Psychosomatic Research* 16, 361–6.

Burgess, A. and Holmstrom, L. (1979*a*). Rape: sexual disruption and recovery. *American Journal of Orthopsychiatry* 49, 648–57.

Burgess, A. and Holmstrom, L. (1979*b*). Adaptive strategies and recovery from rape. *American Journal of Psychiatry* 136, 1278–82.

Burns, A. (1993). *Ageing and dementia: a methodological approach.* Arnold, London.

Burns, A. and Levy, R. (1994). *Dementia.* Chapman & Hall Medical, London.

Burns, B. H. and Howell, J. B. (1969). Disproportionately severe breathlessness in chronic bronchitis. *Quarterly Journal of Medicine* 38, 277–94.

Burns, T., Paykell, E. S., and Lemon, S. (1991). Care of chronic neurotic out-patients by community psychiatric nurses. *British Journal of Psychiatry* 158, 685–90.

Burrow, T. (1927). The group method of analysis. *Psychoanalytic Review* 14, 268–80.

Burvill, P. W., Johnson, G. A., Jamrozik, K. D., Anderson, C. S., Stewart-Wynne, E. G., and Chakera, T. M. H. (1995). Anxiety disorders after stroke: results from the Perth Community Stroke Study. *British Journal of Psychiatry* 166, 328–32.

Bushnell, J. A., Wells, J. E., and Oakley Browne, M. (1993). Long term effects of intrafamilial sexual abuse in childhood. *Acta Psychiatrica Scandinavica* 85, 136–42.

Bussing, R. and Johnson, S. B. (1992). Psychosocial issues in hemophilia before and after the HIV crisis: a review of current research. *General Hospital Psychiatry* 14, 387–403.

Butler, G., Cullington, A., Munby, M., *et al.* (1984). Exposure and anxiety management in the treatment of social phobia. *Journal of Consulting and Clinical Psychology* 52, 642–50.

Butler, G., Cullington, A., Hibbert, G. A., *et al.* (1987) Anxiety management for persistent generalized anxiety. *British Journal of Psychiatry* 151, 535–42.

Butler, R. J., Forsythe, W. I., and Robertson, J. (1990). The body worn alarm in treatment of childhood enuresis. *British Journal of Child Psychiatry* 44, 237–41.

Butler, G., Gelder, M., Hibbert, G., Cullington, A., and Klimes, I. (1987*b*). Anxiety management: developing effective strategies. *Behaviour Research and Therapy* 25, 517–22.

Buysse, D. J., *et al.* (1994). Diagnostic concordance for DSM-IV sleep disorders: a report from the APA/NIMH DSM-IV Filed Trial. *American Journal of Psychiatry* 151(9), 1351.

Bynum, W. F. (1983). Psychiatry in its historical context. In *Handbook of psychiatry*, Vol. 1 (ed. M. Shepherd and O. L. Zangwill). Cambridge University Press.

Bynum, W. F. (1985). The nervous patient in eighteenth- and nineteenth-century Britain: the psychiatric origins of British neurology. In *The anatomy of madness*. Vol. I, *People and ideas* (ed. W. F. Bynum, R. Porter, and M. Shepherd). Tavistock Publications, London.

Byrne, W. and Parsons, B. (1993). Human sexual orientation. *Archives of General Psychiatry* 50, 228–39.

Cade, J. F. (1949). Lithium salts in the treatment of psychotic excitement. *Medical Journal of Australia* 2, 349–52.

Cadoret, R. J. (1978*a*). Evidence of genetic inheritance of primary affective disorder in adoptees. *American Journal of Psychiatry* 135, 463–6.

Cadoret, R. J. (1978*b*). Psychopathology in adopted-away offspring of biologic parents with antisocial behaviour. *Archives of General Psychiatry* 35, 176–84.

Cadoret, R. J. and Gath, A. (1978). Inheritance of alcoholism in adoptees. *British Journal of Psychiatry* 132, 252–8.

Cadoret, R. J., Cunningham, L., Loftus, R., and Edwards, J. (1975). Studies of adoptees from psychiatrically disturbed biological parents — II Temperament, hyperactive, antisocial and developmental variables. *Journal of Pediatrics* 87, 301–6.

Caine, E. D. (1994). Amnestic disorders. In *DSM-IV sourcebook*, Vol. 1 (ed. T. A. Widiger, A. J. Frances, H. A. Pincus, *et al.*), pp. 237–42. American Psychiatric Association, Washington, DC.

Calabrese, L. V. and Stern, T. A. (1995). Neuropsychiatric manifestations of systemic lupus erythematosus. *Psychosomatics* 36, 344–59.

Calabrese, J. R., Markovitz, P., Kimmel, S. E., and Wagner, S. C. (1992). Spectrum of efficacy of valproate in 78 rapid cycling bipolar patients. *Journal of Clinical Psychopharmacology* 12, 53S–6S.

Calhoun, K. S. and Atkeson, B. M. (1991). *Treatment of*

rape victims: facilitating psychosocial adjustment. Pergamon Press, New York.

Cameron, N. (1938). Reasoning, regression and communication in schizophrenia. *Psychological Monographs* 50, 1–34.

Cameron, N. (1963). *Personality development and psychopathology: a dynamic approach.* Houghton Mifflin, Boston, MA.

Campbell, E., Cope, S., and Teasdale, J. (1983). Social factors and affective disorder: an investigation of Brown and Harris's model. *British Journal of Psychiatry* 143, 548–53.

Campbell, P. G. (1991). Graduates. In *Psychiatry in the elderly* (ed. R. Jacoby and C. Oppenheimer), pp. 779–818. Oxford University Press.

Campo, J. V. and Fritsch, S. L. (1994). Somatization in children and adolescents. *Journal of the Academy for Child and Adolescent Psychiatry* 33(9), 1223–5.

Cannon, T. D., Mednick, S. A., Parnas, J., *et al.* (1993). Developmental brain abnormalities in the offspring of schizophrenic mothers. (1) Contributions of genetic and perinatal factors. *Archives of General Psychiatry* 50, 551–64.

Cantwell, D. (1975). Genetics of hyperactivity. *Journal of Child Psychology and Psychiatry* 16, 261–4.

Cantwell, D. P. and Baker, L. (1985). Coordination disorder. In *Comprehensive textbook of psychiatry* (ed. H. I. Kaplan and B. J. Sadock) (4th edn). Williams and Wilkins, Baltimore, MD.

Cantwell, D. P. and Rutter, M. (1994). Classification: conceptual issues and substantive findings. In *Child and adolescent psychiatry: modern approaches* (3rd edn) (ed. M. Rutter, E. Taylor, and L. Hersov), pp. 3–21. Blackwell Scientific Publications, Oxford.

Capgras, J. and Reboul-Lachaux, J. (1923). L'Illusion des sosies dans un délire systématisé chronique. *Bulletin de la Société Clinique de Médicine Mentale* 11, 6–16.

Caplan, A. L., Engelhardt, T., and McCartney, J. J. (1981). *Concepts of health and disease: interdisciplinary perspectives.* Addison-Wesley, Reading, MA.

Caplan, G. (1961). *An approach to community mental health.* Tavistock Publications, London.

Caplan, H. L. (1970). Hysterical conversion symptoms in childhood. M. Phil. Dissertation, University of London. (See the account in *Child psychiatry: modern approaches* (ed. M. L. Rutter and L. Hersov) (2nd edn). Blackwell, Oxford (1985).

Carey, G., Gottesman, I. I., and Robins, E. (1980). Prevalence rates among neuroses, pitfalls in the evaluation of familiarity. *Psychological Medicine* 10, 437–43.

Carlson, G. A. and Goodwin, F. K. (1973). The stages of mania: a longitudinal analysis of the manic episode. *Archives of General Psychiatry* 28, 221–8.

Carlsson, A. and Lindquist, M. (1963). Effect of chlorpromazine and haloperidol on formation of methoxytyr-amine and normetanephrine in mouse brain. *Acta Pharmacologia et Toxicologia* 20, 140–4.

Carlsson, M. and Carlsson, A. (1990). Interactions between glutamatergic and monoaminergic systems within the basal ganglia — implications for schizophrenia and Parkinson's disease. *Trends in Neurosciences* 13, 272–5.

Carney, M. W. P., Roth, M., and Garside, R. F. (1965). The diagnosis of depressive syndromes and the prediction of ECT response. *British Journal of Psychiatry* 111, 659–74.

Carnwath, T. C. M. and Johnson, D. A. W. (1987). Psychiatric morbidity among spouses of patients with stroke. *British Medical Journal* 294, 409–11.

Caroff, S. N. (1980). The neuroleptic malignant syndrome. *Journal of Clinical Psychiatry* 41, 79–83.

Carone, B. J., Harrow, N., and Westermeyer, J. F. (1991). Posthospital course and outcome in schizophrenia. *Archives of General Psychiatry* 48, 247–53.

Carpenter, W. T., Strauss, J. S., and Muleh, S. (1973). Are there pathognomonic symptoms of schizophrenia? An empiric investigation of Schneider's first rank symptoms. *Archives of General Psychiatry* 28, 847–52.

Carr, A. (1991). Milan systemic family therapy: a review of ten empirical investigations. *Journal of Family Therapy* 13, 237–63.

Carr, J. (1994). Annotation: long term outcome for people with Down's syndrome. *Journal of Child Psychology and Psychiatry* 35, 425–39.

Carstairs, G. M., O'Connor, N., and Rawnsley, K. (1956). Organisation of a hospital workshop for chronic psychiatric patients. *British Journal of Preventive and Social Medicine* 10, 136–40.

Casey, P. R. and Tyrer, P. J. (1986). Personality functioning and symptomatology. *Journal of Psychiatric Research* 20, 363–74.

Casey, P. R., Tyrer, P. J., and Dillon, S. (1984). The diagnostic status of patients with conspicuous psychiatric morbidity in primary care. *Psychological Medicine* 14, 673–81.

Casey, R. and Berman, J. (1985). The outcome of psychotherapy with children. *Psychological Bulletin* 98, 388–400.

Cassidy (1986). Emotional stress in terminal cancer: discussion paper. *Journal of the Royal Society of Medicine* 79, 717–20.

Castillo, C. S., Starkstein, S. E., Fedoroff, J. P., *et al.* (1993). Generalized anxiety disorder after stroke. *Journal of Nervous and Mental Disease* 181, 100–6.

Castle, D. J. and Murray, R. M. (1991). The neurodevelopmental basis of sex differences in schizophrenia. *Psychological Medicine* 21, 565–75.

Castle, D. J., Scott, K., Wessely, S., and Murray, R. M. (1993). Does social deprivation during gestation and early life predispose to later schizophrenia? *Social Psychiatry in Psychiatric Epidemiology* 28, 1–4.

Catalan, J., Gath, D., Edmonds, G., and Ennis, J. (1984). The effects of non-prescribing of anxiolytics in general practice: I. Controlled evaluation of psychiatric and social outcome. *British Journal of Psychiatry* 144, 593–602.

Catalan, J., Gath, D. H., Anastasiades, P., *et al.* (1991). Evaluation of a brief psychological treatment for emotional disorders in primary care. *Psychological Medicine* 21, 1013–19.

Catalan, J., Klimes, I., and Burgess, A. (1995). *Psychological medicine of HIV infection*. Oxford University Press, Oxford.

Caton, C. L. M., Wyatt, R. J., Felix, A., *et al.* (1993). Follow-up of chronically homeless mentally ill men. *American Journal of Psychiatry* 150, 1639–42.

Cattell, R. B. (1963). *The Sixteen Personality Factor Questionnaire*. Institute for Personality and Ability Testing, Chicago, IL.

Caviston, P. (1987). Pregnancy and opiate addiction. *British Medical Journal* 295, 286–6.

Cay, E. I. (1984). Psychological problems in relation to coronary care. In *Recent advances in cardiology* (ed. D. J. Rowlands). Churchill Livingston, Edinburgh.

Ceci, S. J. and Bruck, M. (1993). Suggestibility of the child witness: a historical review and synthesis. *Psychological Bulletin* 113(3), 403–39.

Central Health Services Council (1968). *Hospital treatment of acute poisoning*. HMSO, London.

Cerletti, U. and Bini, I. (1938). Un nuovo metodo di shokterapia; 'Tettroshock'. *Bulletin Accademia Medica di Roma* 64, 136–8.

Chadwick, D. (1990). Diagnosis of epilepsy. *Lancet* ii, 15–18.

Chadwick, P. and Birchwood, M. (1994). The omnipotence of voices: a cognitive approach to auditory hallucinations. *British Journal of Psychiatry* 164, 190–201.

Chalkley, A. J. and Powell, G. (1983). The clinical description of forty-eight cases of sexual fetishism. *British Journal of Psychiatry* 142, 292–5.

Chalmers, J. (1990). Mania: whaif neuroleptics don't work? In *Dilemmas and difficulties in the management of psychiatric patients* (ed. K. Hawton and P. Cowen), pp. 31–40. Oxford University Press.

Chalmers, J. S. and Cowen, P. J. (1990). Drug treatment of tricyclic resistant depression. *International Review of Psychiatry* 2, 239–48.

Chamberlain, A. S. (1966). Early mental hospitals in Spain. *American Journal of Psychiatry* 123, 143–9.

Chaplin, J. E., Yepez Lasso, R., Shorvon, S. D., and Floyd, M. (1992). National general practice study of epilepsy: the social and psychological effects of a recent diagnosis of epilepsy. *British Medical Journal* 304, 1416–18.

Chappell, D., Reitsma, R., O'Connell, D., and Strang, H. (1992). Law enforcement as a harm-reducing strategy in Rotterdam and Merseyside. In *Psychoactive drugs and harm reduction* (ed. N. Heather, A. Wodak, E. Nadelmann, and P. O'Hare), pp. 118–36. Whurr, London.

Charlton, J., Kelly, S., Dunnell, K., *et al.* (1992). Trends in suicide deaths in England and Wales. *Population Trends* 69, 10–16.

Charlton, J., Dunnell, K., Evans, B., and Jenkins, R. (1993). Suicide deaths in England and Wales: Trends in factors associated with suicide deaths. *Population Trends* 71, 34–42.

Charney, D. S., Heninger, G. R., and Breier, A. (1984). Noradrenergic function in panic patients. *Archives of General Psychiatry* 41, 751–62.

Charney, D. S., Heninger, G. R., and Kleber, H. D. (1986). Combined use of clonidine and naltrexone as a rapid, safe and effective treatment of abrupt withdrawal from methadone. *American Journal of Psychiatry* 143, 831–7.

Charney, D. S., Deutch, A. Y., Krystal, J. H., *et al.* (1993). Psychobiologic mechanisms of posttraumatic stress disorder. *Archives of General Psychiatry* 50, 294.

Checkley, S. A. (1992). Neuroendocrinology. In *Handbook of affective disorders* (ed. E. S. Paykel), pp. 255–66. Churchill Livingstone, Edinburgh.

Checkley, S. A., Murphy, D. G. M., Abbas, M., *et al.* (1993). Melatonin rhythms in seasonal affective disorder. *British Journal of Psychiatry* 163, 332–7.

Chick, J. (1987). Early interventions in the general hospital. In *Helping the problem drinker* (ed. T. Stockwell and S. Clements), pp. 105–17. Croom Helm, London.

Chick, J. (1992). Doctors with emotional problems: How can they be helped? In *Practical problems in clinical psychiatry* (ed. K. Hawton and P. Cowen), pp. 242–52. Oxford University Press.

Chick, J. (1993). Brief interventions for alcohol misuse. *British Journal of Psychiatry* 307, 1374.

Chick, J. (1994). Alcohol problems in the general hospital. *British Medical Bulletin* 50, 200–10.

Chick, J., Lloyd, G., and Crombie, E. (1985). Counselling problem drinkers in medical wards: a controlled study. *British Medical Journal* 290, 965–7.

Chick, J., Ritson, B., Connaughton, J., *et al.* (1988). Advice versus extended treatment for alcoholism: a controlled study. *British Journal of Addiction* 83, 159–70.

Chick, J., Gough, K., Falkowski, W., *et al.* (1992). Disulfiram treatment of alcoholism. *British Journal of Psychiatry* 161, 84–9.

Chodoff, P. and Lyons, H. (1958). Hysteria: the hysterical personality and hysterical conversion. *American Journal of Psychiatry* 114, 734–40.

Chou, J. C. (1991). Recent advances in the treatment of mania. *Journal of Clinical Psychopharmacology* 11, 3–21.

Chouinard, G., Ross-Chouinard, A., Annable, L., and Jones, B. D. (1980). Extrapyramidal symptom rating scale. *Canadian Journal of Neurological Science* 7(3), 233.

Chouinard, G., Jones, B., Remington, G., *et al.* (1993). A Canadian multicentre placebo-controlled study of fixed doses of risperidone and haloperidol in the treatment of chronic schizophrenic patients. *Journal of Clinical Psychopharmacology* 13, 25–40.

Chowdhury, N., Hicks, R. C., and Kreitman, N. (1973). Evaluation of an after-care service for parasuicide ('attempted suicide') patients. *Social Psychiatry* 8, 67–81.

Christensen, H., Henderson, A. S., Jorm, A. F., Mackinnon, A. J., Scott, R., and Korten, A. E. (1995). *Psychological Medicine* 25, 105–20.

Christie, J. E., Whalley, L. J., Dick, H., *et al.* (1986). Raised plasma cortisol concentrations a feature of drug-free psychotics and not specific for depression. *British Journal of Psychiatry* 148, 58–65.

Christodoulou, G. N. (1976). Delusional hyper-identifications of the Frégoli type. *Acta Psychiatrica Scandinavica* 54, 305–14.

Christodoulou, G. N. (1977). The syndrome of Capgras. *British Journal of Psychiatry* 130, 556–64.

Christoffel, K. K., Anzinger, N. K., and Merrill, D. D. (1989). Age-related patterns of violent death, Cook County, Illinois, 1977 through 1982. *American Journal of Disease in Childhood* 143, 1403–9.

Chua, S. E. and McKenna, P. J. (1995). Schizophrenia a brain disease? *British Journal of Psychiatry* 166, 563–82.

Cicchetti, D. and Carlson, V. (ed.) (1989). *Child maltreatment. Theory and research on the causes and consequences of child abuse and neglect.* Cambridge University Press, New York.

Ciompi, L. (1980). The natural history of schizophrenia in the long term. *British Journal of Psychiatry* 136, 413–20.

Clancy, J., Noyes, R., Noenk, P. R., and Slymen, D. J. (1978). Secondary depression in anxiety neurosis. *Journal of Nervous and Mental Disease* 166, 846–50.

Clare, A. (1979). The disease concept in psychiatry. In *Essentials of postgraduate psychiatry* (ed. P. Hill, R. Murray, and A. Thorley). Academic Press/Grune & Stratton, New York.

Clare, A. W. and Tyrrell, J. (1994). Psychiatric aspects of abortion. *Irish Journal of Psychological Medicine* 11(2), 92–8.

Claridge, G. and Hewitt, J. K. (1987). A biometrical study of schizotypy in a normal population. *Personality and Individual Differences* 8, 303–12.

Clark, D. B. and Agras, W. S. (1991). The assessment and treatment of performance anxiety in musicians. *American Journal of Psychiatry* 148, 598–605.

Clark, D. M. (1986). A cognitive approach to panic. *Behaviour Research and Therapy* 24, 461–70.

Clark, D. M. (1990). Cognitive therapy for depression and anxiety: Is it better than drug treatment in the long-term? In *Dilemmas and difficulties in the management of psychiatric patients* (ed. K. Hawton and P. Cowen), pp. 55–64. Oxford University Press.

Clark, D. M. (1994). Anxiety states. In *Cognitive behaviour therapy for psychiatric patients: a practical guide* (ed. K. Hawton, P. M. Salkovskis, J. Kirk, and D. M. Clark). Oxford University Press.

Clark, D. M. and Teasdale, J. D. (1982). Diurnal variation in clinical depression and accessibility of memories of positive and negative experiences. *Journal of Abnormal Psychology* 91, 87–95.

Clark, D. M., Salkovskis, P. M., and Chalkley, A. J. (1985). Respiratory control as a treatment for panic attacks. *Journal of Behaviour Therapy and Experimental Psychiatry* 16, 23–30.

Clark, D. M., Salkovskis, P. M., Hackmann, A., *et al.* (1994). A comparison of cognitive therapy, applied relaxation and imipramine in the treatment of panic disorder. *British Journal of Psychiatry* 164, 759–69.

Clarke, A. M., Clarke, A. D. B., and Berg, J. M. (ed.) (1985). *Mental deficiency: the changing outlook* (4th edn). Methuen, London.

Clarkin, J. F., Marziali, E., and Munroe-Blum, H. (1991). Group and family treatments for borderline personality disorder. *Hospital and Community Psychiatry* 42(10), 1038–42.

Clausen, J. A. and Kohn, M. L. (1959). Relation of schizophrenia to the social structure of a small town. In *Epidemiology of mental disorder* (ed. B. Pasamanick). American Association for the Advancement of Science, Washington, DC.

Clayton, P. J. (1979). The sequelae and non-sequelae of conjugal bereavement. *American Journal of Psychiatry* 136, 1530–4.

Clayton P. J. (1990). The comorbidity factor: establishing the primary diagnosis in patients with mixed symptoms of anxiety and depression. *Journal of Clinical Psychiatry* 51 (suppl.), 35–9.

Clayton, P. J. and Darvish, H. S. (1979). Course of depressive symptoms following the stress of bereavement. In *Stress and mental disorder* (ed. J. Bartlett, R. M. Rose, and G. L. Klerman), pp. 121–39. Raven Press, New York.

Clayton, P. J., Herjanic, M., and Murphy, G. E. (1974). Mourning and depression: their similarities and differences. *Canadian Psychiatric Association Journal* 19, 309–13.

Clayton, P. J., Herjanic, M., Murphy, G. E., and Woodruff, R. (1974). Mourning and depression: their similarities and differences. *Canadian Psychiatric Association Journal* 19, 309–12.

Cleckley, H. M. (1964). *The mask of sanity: an attempt to*

clarify issues about the so-called psychopathic personality (4th edn). Mosby, St Louis, MO.

Clerc, G. E., Ruimy, P., and Verdeac-Pailles, J. (1994). A double-blind comparison of venlafaxine and fluoxetine in patients hospitalised for major depression and melancholia. *International Clinical Psychopharmacology* 9, 139–43.

Clifford, C. A., Hopper, J. L., Fulker, D. W., and Murray, R. M. (1984). A genetic and environmental analysis of a twin family study of alcohol use, anxiety and depression. *Genetics and Epidemiology* 1, 63–79.

Clinical Psychiatry Committee (1965). Clinical trials of the treatment of depressive illness: report to the Medical Research Council. *British Medical Journal* i, 881–6.

Clinical Research Centre (Division of Psychiatry) (1984). The Northwick Park ECT trial: predictors of response to real and simulated ECT. *British Journal of Psychiatry* 144, 227–37.

Clomipramine Collaborative Study Group (1991). Clomipramine and the treatment of patients with obsessive-compulsive disorder. *Archives of General Psychiatry* 48, 730–8.

Cloninger, C. R. (1986a). Classification of the somatoform disorders: a critique of DSM-III. In *Diagnosis and classification in psychiatry* (ed. G. Tischler), pp. 243–59. Cambridge University Press, New York.

Cloninger, C. R. (1986b). A unified biosocial theory of personality and its role in the development of anxiety states. *Psychiatric Developments* 3, 167–226.

Cloninger, C. R., Bohman, M., and Sigvardsson, S. (1981). Inheritance of alcohol abuse: cross fostering analysis of adopted men. *Archives of General Psychiatry* 38, 861–8.

Clouston, P. D., De Angelis, L. M., and Posner, J. B. (1992). The spectrum of neurological disease in patients with systemic cancer. *Annals of Neurology* 31, 268–73.

Clozapine Study Group (1993). The safety and efficacy of clozapine in severe treatment-resistant schizophrenic patients in the UK. *British Journal of Psychiatry* 163, 155–65.

Cobb, J. and Kelly, D. (1990). Psychosurgery: is it ever justified? In *Dilemmas and difficulties in the management of psychiatric patients* (ed. K. Hawton and P. Cowen), pp. 219–30. Oxford University Press.

Cobb, J. P. and Marks, I. M. (1979). Morbid jealousy featuring as obsessive compulsive neurosis. Treatment by behavioural psychotherapy. *British Journal of Psychiatry* 134, 301–5.

Cobb, S. and Rose, R. M. (1973). Hypertension, peptic ulcer and diabetes in air traffic controllers. *Journal of the American Medical Association* 224, 489–92.

Coccaro, E. F., Siever, L. J., Clar, H. M., et al. (1989). Serotonergic studies in patients with affective and personality disorders. *Archives of General Psychiatry* 46, 587–99.

Cochran, E., Robins, E., and Grote, S. (1976). Regional serotonin levels in the brain: comparison of depressive suicides and alcoholic suicides with controls. *Biological Psychiatry* 11, 283–94.

Cochrane, R. and Bal, S. (1990). The drinking habits of Sikh, Hindu, Muslim and White men in the West Midlands: a community survey. *British Journal of Addiction* 85, 759–69.

Cody, M. (1990). Depression and the use of antidepressants in patients with cancer. *Palliative Medicine* 4, 271–8.

Coffey, B. J., Euripides, C. M., Savage, C. R., and Rauch, S. L. (1994). Tourette's disorder and related problems: a review and update. *Harvard Review of Psychiatry* 2, 121–32.

Coffey, C. E., Rweiner, R. D., Djang, W. T., et al. (1991). Brain anatomic effects of electroconvulsive therapy: a prospective magnetic resonance imaging study. *Archives of General Psychiatry* 48, 1013–21.

Coffey, C. E., Wilkinson, W. E., Weiner, R. D., et al. (1993). Quantitative cerebral anatomy in depression: a controlled magnetic resonance imaging study. *Archives of General Psychiatry* 50, 7–16.

Cohen, B. J., et al. (1994). Personality disorder in later life: a community study. *British Journal of Psychiatry* 165, 493–9.

Cohen, F. and Lazarus, R. (1973). Active coping processes, coping dispositions and recovery from surgery. *Psychosomatic Medicine* 35, 375–89.

Cohen, J. (1961). A study of suicide pacts. *Medicolegal Journal* 29, 144–51.

Cohen, S. (1980). Cushing's syndrome: a psychiatric study of 29 patients. *British Journal of Psychiatry* 136, 120–4.

Cohen, S. D., Monteiro, W., and Marks, I. M. (1984). Two-year follow-up of agoraphobics after exposure and imipramine. *British Journal of Psychiatry* 144, 276–81.

Cohen-Cole, S. A. (1991). *The medical interview: the three function approach*. Mosby Year Book, St Louis, MO.

Coid, B., Lewis, S. W., and Reveley, A. M. (1993). A twin study of psychosis and criminality. *British Journal of Psychiatry* 162, 87–92.

Cole, J. D., Goldberg, S. C., and Klerman, G. L. (1964). Phenothiazine treatment in acute schizophrenia. *Archives of General Psychiatry* 10, 246–61.

Cole, M. G. (1990). The prognosis of depression in the elderly. *Canadian Medical Association Journal* 143, 633–9.

Collinge, J., Delisi, L. E., Boccio, A., et al. (1991). Evidence for a pseudo-autosomal locus for schizophrenia using the method of affected sibling pairs. *British Journal of Psychiatry* 158, 624–9.

Collis, I. and Lloyd, G. (1992). Psychiatric aspects of liver disease. *British Journal of Psychiatry* 161, 12–22.

Collis, I., Burroughs, A., Rolles, K., and Lloyd, D. (1995). Psychiatric and social outcome of liver transplantation. *British Journal of Psychiatry* 166, 521–4.

Committee on the Review of Medicines (1980). Systematic

review of the benzodiazepines: guidelines for data sheets on diazepam, chlordiazepoxide, medazepam, temazepam, triazolam, nitrazepam and flurazepam. *British Medical Journal* i, 910–12.

Committee on the Safety of Medicines (1993). *Current Problems in Pharmacovigilance* 19, 9–11.

Committee on the Safety of Medicines (1995). Cardiac arrhythmias with pimozide. *Current Problems in Pharmacovigilance* 21, 2.

Connell, P. H. (1958). *Amphetamine psychosis*. Maudsley Monograph No. 5, Oxford University Press, London.

Connolly, J. F., Gruzelier, J. H., Marchanda, R., and Hirsch, S. R. (1983). Visual evoked potentials in schizophrenia. *British Journal of Psychiatry* 142, 152–5.

Conolly, J. (1856). *The treatment of the insane without mechanical restraints* (reprinted 1973). Dawson, London.

Conrad, K. (1958). *Die beginnende Schizophrenie: versuch einer gestaltanalyse des Wahns*. Thieme, Stuttgart.

Constantine, L. L. (1981). The effects of early sexual experience. In *Children and sex* (ed. L. L. Constantine and F. M. Martinson). Little Brown, Boston, MA.

Conte, J. R. (1985). The effects of sexual abuse on children: a critique and suggestions for future research. *Victimology* 10, 110–30.

Cooper, A. F., Kay, D. W. K., Curry, A. R., Garside, R. F., and Roth, M. (1974). Hearing loss in paranoid and affective psychoses of the elderly. *Lancet* 2, 851–61.

Cooper, A. F. (1984). Psychiatric aspects of sensory deficits. In *Handbook of studies on psychiatry and old age* (ed. D. W. K. Kay and G. D. Burrows). Elsevier, Amsterdam.

Cooper, B. (1978). Epidemiology. In *Schizophrenia. Towards a new synthesis* (ed. J. K. Wing). Academic Press, London.

Cooper, B. (1986). Mental disorder as reaction: the history of a psychiatric concept. In *Life events and psychiatric disorder: controversial issues* (ed. H. Katchnig). Cambridge University Press.

Cooper, B. (1991*a*). Psychiatric services for the elderly: principles of service provision in old age psychiatry. In *Psychiatry in the elderly* (ed. R. Jacoby and C. Oppenheimer), pp. 274–300. Oxford University Press.

Cooper, B. (1991*b*). The dementias of old age: epidemiology of dementia. In *Psychiatry in the elderly* (ed. R. Jacoby and C. Oppenheimer), pp. 574–85. Oxford University Press.

Cooper, B. (1993). Single spies and battalions: the clinical epidemiology of mental disorders. *Psychological Medicine* 23, 891–907.

Cooper, J. E. (1990). The classification of mental disorders for use in general medical settings. In *Psychological disorders in general medical settings* (ed. N. Sartorius, D. Goldberg, D. de Girolamo, *et al.*), pp. 49–59. Hogrefer and Huber, Göttingen.

Cooper, J. E., Kendell, R. E., Gurland, B. J., *et al.* (1972).

Psychiatric diagnosis in New York and London. Maudsley Monograph No. 20. Oxford University Press, London.

Cooper, P. J., Gath, D., Rose, N., and Fieldsend, R. (1982). Psychological sequelae to elective sterilisation: a prospective study. *British Medical Journal* 284, 461–3.

Cooper, P. J., Campbell, G. A., Day, A., *et al.* (1988). Nonpsychotic psychiatric disorder after childbirth: a prospective study of prevalence, incidence, course and nature. *British Journal of Psychiatry* 152, 799–806.

Cooper, P. J. and Murray, L. (1995). Course and recurrence of postnatal depression. Evidence for the specificity of the diagnostic concept. *British Journal of Psychiatry* 166, 191–5.

Cooper, Z. (1995). The development and maintenance of eating disorders. In *Eating disorders and obesity: a comprehensive handbook* (ed. K. D. Brownell and C. G. Fairburn). Guilford Press, New York.

Cope, H., David, A., Pelosi, A., and Mann, A. (1994). Predictors of chronic "postviral" fatigue. *Lancet* 344, 864–8.

Copeland, J. R. M. and Gurland, B. J. (1985). International comparative studies. In *Recent advances in psychogeriatrics* (ed. T. Arie). Churchill Livingstone, Edinburgh.

Copeland, J. R. M., Kelleher, M. J., Kellett, J. M., *et al.* (1975). Evaluation of a psychogeriatric service: the distinction between psychogeriatric and geriatric patients. *British Journal of Psychiatry* 126, 21–9.

Coppen, A. J. and Shaw, D. M. (1963). Mineral metabolism in melancholia. *British Medical Journal* ii, 1439–44.

Coppen, A. J., Montgomery, S. A., Gupta, R. K., and Bailey, J. (1976). A double blind comparison of lithium carbonate or maprotiline in the prophylaxis of affective disorder. *British Journal of Psychiatry* 128, 479–85.

Coppen, A., Standish-Barry, H., Bailey, J., *et al.* (1991). Does lithium reduce the mortality of recurrent mood disorders. *Journal of Affective Disorders* 23, 1–7.

Corbett, J. A. (1978). The development of services for the mentally handicapped: a historical and national review. In *The care of the handicapped child* (ed. J. Apley). Heinemann, London.

Corbett, J. A. and Pond, D. A. (1979). Epilepsy and behaviour disorder in the mentally handicapped. In *Psychiatric illness and mental handicap* (ed. F. F. James and R. P. Snaith). Gaskell, Ashford, Kent.

Corbett, J. A. and Turpin, G. (1985). Tics and Tourette's syndrome. In *Child and adolescent psychiatry: modern approaches* (ed. M. Rutter and L. Hersov) (2nd edn). Blackwell, Oxford.

Corbett, J. A., Harris, R., and Robinson, R. G. (1975). Epilepsy. In *Mental retardation and developmental disabilities: an annual review*, Vol. VII (ed. J. Wortis). Brunner-Mazel, New York.

Corbin, S. L. and Eastwood, M. R. (1986). Sensory deficits

and mental disorders of old age: causal or coincidental associations? *Psychological Medicine* 16, 251–6.

Corcoran, T. and Frith, C. D. (1993). Neuropsychology and neurophysiology in schizophrenia. *Current Opinion in Psychiatry* 6, 74–9.

Cornelius, J. R., Day, N. L., Fabrega, H., *et al.* (1991). Characterizing organic delusional syndrome. *Archives of General Psychiatry* 48, 749–53.

Corr, A. and Corr, D. M. (1983). *Hospice care, principles and practice.* Springer, New York.

Corsellis, J. A. N. (1962). *Mental illness and the ageing brain.* Maudsley Monographs No. 9. Oxford University Press, London.

Corsellis, J. A. N., Bruton, C. J., and Freeman-Browne, D. (1973). The aftermath of boxing. *Psychological Medicine* 3, 270–303.

Coryell, W. and Winokur, G. (1992). Course and outcome. In *Handbook of affective disorders* (ed. E. S. Paykel), pp. 89–108. Churchill Livingstone, Edinburgh.

Coryell, W. and Zimmerman, M. (1989). Personality disorder in the families of depressed schizophrenic and never-ill probands. *American Journal of Psychiatry* 146, 469–502.

Coryell, W., Noyes, R., and Clancy, J. (1982). Excess mortality in panic disorder: comparison with primary unipolar depression. *Archives of General Psychiatry* 39, 701–3.

Costello, G. G. (1982). Social factors associated with depression: a retrospective community study. *Psychological Medicine* 12, 329–39.

Cotard, M. (1882). Du délire de négations. *Archives de Neurologie, Paris* 4, 152–70 and 282–96. (Trans. by M. Rohde in S. R. Hirsch and M. Shepherd (ed.). *Themes and variations in European psychiatry*, pp. 353–73. Wright, Bristol.)

Coull, D. C., Crooks, J., Dingwell-Fordyce, I., *et al.* (1970). Amitriptyline and cardiac disease: risk of sudden death identified by monitoring system. *Lancet* ii, 590–1.

Council on Scientific Affairs (1986). Dementia. *Journal of the American Medical Association* 256, 2234–8.

Courbon, P. and Fail, G. (1927). Syndrome 'd'illusion de Frégoli' et schizophrénie. *Bulletin de la Société Clinique de Médecine Mentale* 15, 121–4.

Covinsky, K. E., Goldman, L., Francis Cook, E., *et al.* (1994). The impact of serious illness on patients' families. *Journal of the American Medical Association* 272(23), 1839–44.

Coward, D. M. (1992). General pharmacology of clozapine. *British Journal of Psychiatry* 160, (Suppl. 17), 5–11.

Cowdrey, R. and Gardner, D. L. (1988). Pharmacotherapy of borderline personality disorder. *Archives of General Psychiatry* 45, 111–19.

Cowen, P. J. (1988). Depression resistant to tricyclic antidepressants. *British Medical Journal* 297, 435–6.

Cowen, P. J. (1992). New antidepressants: have they superseded tricyclics? In *Practical problems in clinical psychiatry* (ed. K. Hawton and P. J. Cowen), pp. 22–32. Oxford University Press.

Cowen, P. J. (1990). Personality disorders: are drugs useful? In *Dilemmas and difficulties in the management of psychiatric patients* (ed. K. Hawton and P. J. Cowen), pp. 105–16. Oxford University Press.

Cowen, P. J. (1991). Serotonin receptor subtypes: implications for psychopharmacology. *British Journal of Psychiatry* 159 (Suppl. 12), 7–14.

Cowen, P. (1992). Drugs in focus: 5. Buspirone. *Prescribers' Journal* 32(5), 201–5.

Cowen, P. J. and Anderson, I. M. (1991). Abnormal 5-HT neuroendocrine function in depression: association or artefact. In *5-Hydroxytryptamine in psychiatry: a spectrum of ideas* (ed. M. Sandler, A. Coppen, and S. Harnett), pp. 124–56. Oxford University Press.

Cowen, P. J. and Nutt, D. J. (1982). Abstinence symptoms after withdrawal of tranquillising drugs: is there a common neurochemical mechanism. *Lancet* ii, 360–2.

Cowie, V. (1961). The incidence of neurosis in the children of psychotics. *Acta Psychiatrica Scandinavica* 37, 37–71.

Cox, D. J. and Gonder-Frederick, L. (1992). Major developments in behavioral diabetes research. *Journal of Consulting and Clinical Psychology* 60, 628–38.

Cox, D. J., Gonder-Frederick, L., Julian, D. M., and Clarke, W. (1994). Long-term follow-up evaluation of blood glucose awareness training. *Diabetes Care* 17, 1–5.

Cox, S. M. and Ludwig, A. (1979). Neurological soft signs and psychopathology: 1. Findings in schizophrenia. *Journal of Nervous and Mental Disease* 167, 161–5.

Coyne, A. C., Reichmann, W. E., and Berbig, L. T. (1993). The relationship between dementia and elder abuse. *American Journal of Psychiatry* 150(4), 643–6.

Craft, M. (1965). *Ten studies in psychopathic personality.* Wright, Bristol.

Craft, M. (1984). Low intelligence, mental handicap and crime. In *Mentally abnormal offenders* (ed. M. Craft and A. Craft). Baillière Tindall, London.

Crammer, J. (1990). *Asylum history: Buckinghamshire County Pauper Lunatic Asylum—St Johns.* Gaskell, London.

Crammer, J., Barraclough, B., and Heine, B. (1982). *The use of drugs in psychiatry.* Gaskell, London.

Craven, J. and Rodin, G. M. (1992). *Psychiatric aspects of organ transplantation.* Oxford University Press, New York.

Creed, F. (1981). Life events and appendectomy. *Lancet* i, 1381–5.

Creed, F. (1995). Psychological treatment of the irritable bowel syndrome and abdominal pain. In *Treatment of functional somatic symptoms* (ed. R. A. Mayou, C. Bass, and M. Sharpe), pp. 255–70. Oxford University Press.

Creed, F. and Ash, G. (1992). Depression in rheumatoid

arthritis: aetiology and treatment. *International Review of Psychiatry* 4, 23–34.

Creed, F., Black, D., Anthony, P., *et al.* (1991). Randomised controlled trial of day and in-patient psychiatric treatment. 2: Comparison of two hospitals. *British Journal of Psychiatry* 158, 183–9.

Creed, F., Black, D., Anthony, P., *et al.* (1990). Randomised controlled trial of day patient versus inpatient psychiatric treatment. *British Medical Journal* 300, 1033–7.

Creer, C. (1978). Social work with patients and their families. In *Schizophrenia: towards a new synthesis* (ed. J. K. Wing). Academic Press, London.

Creer, C. and Wing, J. K. (1975). Living with a schizophrenic patient. *British Journal of Hospital Medicine* 14, 73–82.

Creighton, F. J., Hyde, C. E., and Farragher, B. (1991). Douglas House: Seven years' experience of a community hostel ward. *British Journal of Psychiatry* 159, 500–4.

Cremers, L. and Matot, J-P. (1994). Dimensions of drug and alcohol use and misuse in HIV risk behaviour. *Current Opinion in Psychiatry* 7, 285–90.

Creutzfeldt, H. E. (1920). Über eine eigenartige herdformige. Erkrankung des zentral Nevensystems. *Zeitschrift für die gesamte Neurologie und Psychiatrie* 57, 1–18.

Crisp, A. H. (1977). Diagnosis and outcome of anorexia nervosa: the St George's view. *Proceedings of the Royal Society of Medicine* 70, 464–70.

Critchley, M. (1953). *The parietal lobes*. Edward Arnold, London.

Critchley, M. (1984). The history of Huntington's chorea. *Psychological Medicine* 14, 725–7.

Crits-Christoph, P. (1992). The efficacy of brief dynamic psychotherapy: a meta-analysis. *American Journal of Psychiatry* 149, 151–8.

Crombie, I. K. (1989). Trends in suicide and unemployment in Scotland, 1976–86. *British Medical Journal* 298, 782–4.

Crombie, I. K. (1990). Suicide in England and Wales and in Scotland. An examination of divergent trends. *British Journal of Psychiatry* 157, 529–32.

Cronholm, B. and Molander, L. (1964). Memory disturbance after electroconvulsive therapy. *Acta Psychiatrica Scandinavica* 40, 211–16.

Cross, A. J. (1991). Neurotransmitter systems in Alzheimer's disease. In *Neurobiology and psychiatry*, Vol. 1 (ed. R. Kerwin, D. Dawbarn, J. McCulloch, and C. Tamminga). Cambridge University Press, Cambridge.

Crow, T. J. (1980). Molecular pathology of schizophrenia; more than one disease process? *British Medical Journal* 280, 66–8.

Crow, T. J. (1985). The two-syndrome concept: origins and current status. *Schizophrenia Bulletin* 11, 471–85.

Crow, T. J. (1994a). Aetiology of schizophrenia. *Current Opinion in Psychiatry* 7, 39–42.

Crow, T. J. (1994b). The demise of the Kraepelin binary system as a prelude to genetic advance. In *Genetic approaches to mental disorders* (ed. E. S. Gershon and R. Cloninger), pp. 163–92. American Psychiatric Press, Washington, DC.

Crow, T. J. (1994c). Prenatal exposure to influenza as a cause of schizophrenia. *British Journal of Psychiatry* 164, 588–92.

Crow, T. J., Ball, J., Bloom, S. R., *et al.* (1989). Schizophrenia as an anomaly of development of cerebral asymmetry. *Archives of General Psychiatry* 46, 1145–50.

Crowe, M. J. (1973). Conjoint marital therapy: advice or interpretation. *Journal of Psychosomatic Research* 17, 309–15.

Crowe, R. R. (1974). An adoption study of antisocial personality. *Archives of General Psychiatry* 31, 785–91.

Crowe, R. R., Noyes, R., Pauls, D. L., and Slymen, D. (1983). A family study of panic disorder. *Archives of General Psychiatry* 40, 1065–9.

Crown, S. (1978). *Psychosexual counselling*. Academic Press, London.

Cryns, A. G., Gorey, K. M., and Goldstein, M. Z. (1990). Effects of surgery on the mental status of older persons: a meta-analytic review. *Journal of Geriatric Psychiatry and Neurology* 3, 184–91.

Cullen, Gulielmo (1814). *Synopsis: nosologiae methodicae*. John Anderson, Edinburgh/Longman, Hurse, Reese, Orme, and Brown, London.

Cummings, J. L. (1992). Depression and Parkinson's disease: a review. *American Journal of Psychiatry* 149, 443–54.

Cummings, J. L. and Frankel, M. (1985). Gilles de la Tourette syndrome and the neurological basis of obsessions and compulsions. *Biological Psychiatry* 20, 1117–26.

Curran, D. (1937). The differentiation of neuroses and manic-depressive psychosis. *Journal of Mental Science* 83, 156–74.

Cutting, J. (1978). The relationship between Korsakov's syndrome and 'alcoholic dementia'. *British Journal of Psychiatry* 132, 240–51.

Cutting, J. (1987). The phenomenology of acute psychosis. *British Journal of Psychiatry* 151, 324–32.

Cutting, J., Cowen, P. J., and Mann, A. H. (1986). Personality and psychosis: the use of the Standardized Assessment of Personality. *Acta Psychiatrica Scandinavica* 73, 86–92.

Cutting, J., Cowen, P. J., Mann, A. H., and Jenkins, R. (1986). Personality and psychosis: use of the standardised assessment of personality. *Acta Psychiatrica Scandinavica* 73, 87–92.

Cybulska, E. and Rucinski, J. (1986). Gross self-neglect in old age. *British Journal of Hospital Medicine* 31, 21–6.

Da Costa, J. M. (1871). An irritable heart: a clinical study of functional cardiac disorder and its consequences. *American Journal of Medical Science* 61, 17–52. (See extracts in S. Jarcho (1959). On irritable heart. *American Journal of Cardiology* 4, 809–17.)

Dahl, A. A. (1985). A critical examination of empirical studies of the diagnosis of borderline disorders in adults. *Psychiatric Developments* 3, 1–29.

Dahl, A. A. (1993). The personality disorders: a critical review of family, twin and adoption studies. *Journal of Personality Disorders* Supplement, 86–99.

Dalbiez, R. (1941). *Psychoanalytic method and the doctrine of Freud*, Vols 1 and 2. Longmans Green, London.

Dalton, K. (1964). *The premenstrual syndrome*. Heinemann, London.

d'Amato, T., Campion, D., Gorwood, M., *et al.* (1992). Evidence for pseudoautosomal locus for schizophrenia (2): replication of a non-random segregation of alleles at the DXYS14 locus. *British Journal of Psychiatry* 161, 59–62.

Danish University Antidepressant Group (1990). Paroxetine: a selective serotonin reuptake inhibitor showing better tolerance but weaker antidepressant effect than clomipramine in a controlled multicentre study. *Journal of Affective Disorders* 18, 289–99.

Davanloo, H. (1980). *Short-term dynamic psychotherapy*. Aronson, New York.

David, A. S. and Wessely, S. C. (1995). The legend of Camelford: medical consequences of a water pollution accident. *Journal of Psychosomatic Research* 39, 1–10.

Davidson, J., Kudler, H., and Smith, S. L. (1990). Treatment of post-traumatic stress disorder with amitriptyline and placebo. *Archives of General Psychiatry* 47, 259–66.

Davidson, J., Turnbull, C. D., and Miller, R. D. (1980). A comparison of inpatients with primary unipolar depression and depression secondary to anxiety. *Acta Psychiatrica Scandinavica* 621, 377–86.

Davidson, J. R. T. (1992). Monoamine oxidase inhibitors. In *Handbook of affective disorders* (ed. E. S. Paykel), pp. 345–58. Churchill Livingstone, Edinburgh.

Davidson, J. R. T. and Foa, E. A. (1992). *Posttraumatic stress disorder: DSMIV and beyond*. American Psychiatric Press, Washington, DC.

Davidson, J. R. T., Giller, E. L., Zisook, S., and Overall, J. E. (1988). An efficacy study of isocarboxazid and placebo in depression, and its relationship to depressive nosology. *Archives of General Psychiatry* 45, 120–7.

Davidson, J. R. T., Hughes, D., Blazer, D. G., and George, L. K. (1991). Post-traumatic stress disorder in the community: an epidemiological study. *British Journal of Psychiatry* 21, 713–21.

Davidson, J. R. T., Hughes, D. L., George, L. K., and Blazer, D. (1993). The epidemiology of social phobia: findings from the Duke Epidemiological Catchment Area Study. *Psychological Medicine* 23, 709–18.

Davidson, K. and Ritson, E. B. (1993). The relationship between alcohol dependence and depression. *Alcohol and Alcoholism* 28, 147–55.

Davies, A. M. (1986). *Epidemiological data on the health of the elderly: a review of the present state of research* (ed. H. Häfner, G. Moschel, and N. Sartorius). Springer-Verlag, Berlin.

Davies, A. M. and Fleischman, R. (1981). Health status and the use of health services as reported by older residents of the Baka neighbourhood, Jerusalem. *Israeli Medical Sciences* 17, 138–44.

Davies, B. M. and Morgenstern, F. S. (1960). A case of cysticercosis, temporal lobe epilepsy and transvestism. *Journal of Neurology, Neurosurgery and Psychiatry* 23, 247–9.

Davies, D. L. (1962). Normal drinking in recovered alcohol addicts. *Quarterly Journal of Studies on Alcohol* 23, 94–104.

Davies, K. and Wardle, J. (1994). Body image and dieting in pregnancy. *Journal of Psychosomatic Research* 38(8), 787–99.

Davis, J. M., Schaffer, C. B., Killian, G. A., Kinard, C., and Chan, C. (1980). Important issues in the drug treatment of schizophrenia. *Schizophrenia Bulletin* 6, 70–87.

Davis, S. (1991). Violence by psychiatric inpatients: a review. *Hospital and Community Psychiatry* 42(6), 585–90.

Davison, A. N. (1984). Neurobiology and neurochemistry of the developing brain. In *Scientific studies in mental retardation* (ed. J. Dobbing, A. D. B. Clarke, J. A. Corbett, and R. O. Robinson). Royal Society of Medicine and Macmillan, London.

Davison, G. (1968). Elimination of a sadistic fantasy by a client-controlled counter-conditioning technique: a case study. *Journal of Abnormal Psychology* 73, 84–90.

Davison, K. (1983). Schizophrenia-like psychoses associated with cerebral disorders: a review. *Psychiatric Developments* 1, 1–34.

Davison, K. and Bagley, C. R. (1969). Schizophrenia-like psychoses associated with organic disorders of the central nervous system: a review of the literature. In *British Journal of Psychiatry* Special Publication No. 4, *Current problems in neuropsychiatry* (ed. R. N. Herrington). Headley, Ashford, Kent.

Dawkins, S. (1961). Non-consummation of marriage. *Lancet* ii, 1029–33.

Daws, D. and Boston, M. (1988). *The child psychotherapist*. Karnac Books, London.

Day, K. and Jancar, J. (1994). Mental and physical health and ageing in mental handicap: a review. *Journal of Intellectual Disability Research* 38, 241–56.

Deakin, J. F. W. (1991a). Serotonin subtypes and affective disorders. In *Serotonin, sleep and mental disorder* (ed. C. Idzikowski and P. J. Cowen), pp. 161–78. Wrighton Biomedical, Petersfield, Hants.

Deakin, J. F. W. (1991*b*). Depression and 5-HT. *International Clinical Psychopharmacology* **6** (Suppl. 3), 23–31.

de Alarcón, R. D. and Franchesini, J. A. (1984). Hyperparathyroidism and paranoid psychosis; case report and review of the literature. *British Journal of Psychiatry* **145**, 477–86.

Dean, C. and Gadd, E. M. (1990). Home treatment for acute psychiatric illness. *British Medical Journal* **301**, 1021–3.

Deary, I. J. and Wilson, J. A. (1994). Problems in treating globus pharyngis. *Clinical Otolaryngology* **19**, 55–60.

De Clérambault, G. (1921). Les délires passionels. Erotomanie, revendication, jalousie. *Bulletin de la société Clinique de Médicine Mentale* 61–71.

De Clérambault, G. G. (1987). Psychoses of passion (English translation). In *The clinical roots of the schizophrenia concept* (ed. J. Cutting and M. Shepherd). Cambridge University Press.

de Leen, J., Verghese, C., Tracy, J. I., Jusiassen, R. L., and Simpson, G. M. (1993). Polydipsia and water intoxication in psychiatric patients: a review of the epidemiological literature. *Biological Psychiatry* **35**, 408–19.

Dedman, P. (1993). Home treatment for acute psychiatric disorder. *British Medical Journal* **306**, 1359–60.

Dee Higley, J., Mehlman, P. T., Taub, D. M., *et al.* (1992). Cerebrospinal fluid monoamine and adrenal correlates of aggression in free-ranging rhesus monkeys. *Archives of General Psychiatry* **49**, 436–41.

Déjerine, J. and Gauckler, E. (1913). *Psychoneurosis and psychotherapy* (trans. S. E. Jelliffe and J. B. Lippincott). Reissued by Arno Press, New York.

De Jesus Mari, J. and Streiner, D. L. (1994). An overview of family interventions and relapse on schizophrenia: meta-analysis of research findings. *Psychological Medicine* **24**, 565–78.

Delgado, P. L., Price, L. H., Heninger, G. R., and Charney, D. S. (1992). Neurochemistry. In *Handbook of affective disorders* (ed. E. S. Paykel), pp. 219–53. Churchill Livingstone, Edinburgh.

Dell, S., Robertson, G., and Parker, E. (1987). Detention in Broadmoor: factors in length of stay. *British Journal of Psychiatry* **150**, 824–7.

Dement, W. C. and Mitler, M. M. (1993). It's time to wake up to the importance of sleep disorders. *Journal of the American Medical Association* **269**, 1548–50.

de Montigny, C., Cournoyer, G., and Morissette, R. (1983). Lithium carbonate addition in tricyclic antidepressant-resistant unipolar depression: correlations with the neurobiological actions of tricyclic drugs and lithium ion on the serotonin system. *Archives of General Psychiatry* **40**, 1327–34.

de Girolamo, G. and Reich, J. H. (1993). *Personality disorders*. World Health Organization, Geneva.

Den Boer, J. A. and Westenberg, H. G. M. (1988). Effect of a serotonin and nor-adrenalin uptake inhibitor in panic disorder: a double blind comparative study with fluvoxamine and maprotiline. *International Clinical Psychopharmacology* **3**, 59–74.

Denicoff, K. D., Joffe, R. T., Lakshman, M. C., *et al.* (1990). Neuropsychiatric manifestations of altered thyroid state. *American Journal of Psychiatry* **147**, 94–9.

Dening, T. R. (1989). Wilson's disease. Psychiatric symptoms in 195 cases. *Archives of General Psychiatry* **46**, 1126–34.

Denko, J. D. and Kaelbling, R. (1962). The psychiatric aspects of hypoparathyroidism. *Acta Psychiatrica Scandinavica* Suppl. **164**, 1–70.

Denmark, J. C. (1985). A study of 250 patients referred to a department of psychiatry for the deaf. *British Journal of Psychiatry* **146**, 282–6.

Department of Health and Social Security (1984). *The management of deliberate self harm*. HM (84) 25, DHSS, London.

Depression Guideline Panel (1993). *Depression in primary care*, Vol. 2, *Treatment of major depression. Clinical Practice Guideline No. 5*. US Department of Health and Human Services, Rockville, MD.

Derby, I. M. (1933). Manic-depressive 'exhaustion' deaths. *Psychiatric Quarterly* **7**, 435–9.

Derogatis, L. R. *et al.* (1985). Prevalence of psychiatric disorders among cancer patients. *Journal of the American Medical Association* **249**, 751–7.

Detera-Wadleigh, S. D., Berrettini, W. H., Goldini, R., *et al.* (1987). Close linkage of the C-Harvey-ras-1 and the insulin gene to affective disorders is ruled out in three North American pedigrees. *Nature, London* **325**, 306–7.

Deutsch, H. (1937). Absence of grief. *Psychoanalytic Quarterly* **6**, 12–22.

De Veaugh-Geiss, J., Moroz, G., Biederman, J., *et al.* (1992). Clomipramine hydrochloride in childhood and adolescent obsessive–compulsive disorder: a multicenter trial. *Journal of American Academy of Child and Adolescent Psychiatry* **31**, 45–9.

Dew, M. A., Ragni, M. V., and Nimorwicz, P. (1990). Infection with human immunodeficiency virus and vulnerability to psychiatric distress. *Archives of General Psychiatry* **47**, 737–44.

Dewar D. (1991). Neurotransmitter systems in Alzheimer's disease. In *Neurobiology and psychiatry*, Vol. 1 (ed. R. Kerwin, D. Dawbarn, J. McCulloch, and C. Tamminga). Cambridge University Press.

Dewhurst, K. (1980). *Thomas Willis's Oxford lectures*. Sandford Publications, Oxford.

Di Chiara, G. and North, R. A. (1992). Neurobiology of opiate abuse. *Trends in Pharmacological Sciences* **13**, 185–93.

Dicks, H. (1967). *Marital tensions: clinical studies towards a psychological theory of interaction*. Routledge & Kegan Paul, London.

Diekstra, R. (1994). The prevention of suicidal behaviour: a review of evidence for the efficacy of community-based programs.

Diermayer, M., Hedberg, K., and Fleming, D. (1994). Backing off universal childhood lead screening in the USA: opportunity or pitfall? *Lancet* 344, 1587–8.

DiMascio, A. (1973). The effects of benzodiazepines on aggression: reduced or increased? In *The benzodiazepines* (ed. S. Garattini, E. Mussini, and L. O. Randall). Raven Press, London.

DiMascio, A., Weissman, M. M., Prusoff, B. A., *et al.* (1979). Differential symptom reduction by drugs and psycho-therapy in acute depression. *Archives of General Psychiatry* 36, 1450–6.

Dinan, T. and Barry, S. (1989). A comparison of electroconvulsive therapy with a lithium and tricyclic combination among depressed tricyclic non-responders. *Acta Psychiatrica Scandinavica* 80, 97–100.

Dingemanse, J. (1993). An update of recent moclombemide interaction data. *International Clinical Psychopharmacology* 7, 167–80.

Dinwiddie, S. H. (1994). Abuse of inhalants: a review. *Addiction* 89, 925–39.

Dische, S., Yule, W., Corbett, J., and Hand, D. (1983). Childhood nocturnal enuresis: factors associated with the outcome of treatment with an enuretic alarm. *Developmental Medicine and Child Neurology* 25, 67–80.

Ditunno, J. F. Jr and Formal, C. (1994). Chronic spinal cord injury. *New England Journal of Medicine* 330, 550–6.

Dodge, K. A., Price, J. M., Bachorowski, J., and Newman, J. P. (1990). Hostile attributional biases in severely aggressive adolescents. *Journal of Abnormal Psychology* 99, 385–92.

Dolan, B. and Coid, J. (1993). *Psychopathic and antisocial personality disorders*. Gaskell, London.

Dollard, J. and Miller, N. E. (1950). *Personality and psychotherapy*. McGraw-Hill, New York.

Done, D. J., Crow, T. J., Johnstone, E. C., and Sacker, A. (1994). Childhood antecedence of schizophrenia and affective illness: social adjustment at ages 7 and 11. *British Medical Journal* 309, 699–703.

Done, D. J., Johnstone, E. C., Frith, C. D., *et al.* (1991). Complications of pregnancy and delivery in relation to psychosis in adult life: data from the British perinatal mortality survey sample. *British Medical Journal* 302, 1576–80.

d'Orban, P. T. (1976). Child stealing: a typology of female offenders. *British Journal of Criminology* 16, 275–9.

d'Orban, P. T. (1979). Women who kill their children. *British Journal of Psychiatry* 134, 560–71.

Dosen, A. (1993). Diagnosis and treatment of psychiatric and behavioural disorders in mentally retarded individuals: the state of the art. *Journal of Intellectual Disability Research* 37, 1–7.

Douglas, J. and Richman, N. (1984). *My child won't sleep: a handbook for management for parents*. Penguin, Harmondsworth.

Douglas, N. J. (1993). The sleep apnoea/hypopnoea syndrome and snoring. *British Medical Journal* 306, 1057–60.

Downes, D. and Rock, P. (1988). *Understanding deviance* (2nd edn). Clarendon Press, Oxford.

Drake, R. E. and Cotton, P. G. (1986). Depression, hopelessness and suicide in chronic schizophrenia. *British Journal of Psychiatry* 148, 554–9.

Drake, R. E., Gates, C., Cotton, P. G., and Whitaker, A. (1984). Suicide among schizophrenics: who is at risk? *Journal of Nervous and Mental Disease* 172, 613–17.

Dreifuss, F. E., Bancaud, J., Henricksen, O., *et al.* (1981). Proposal for a revised clinical and electroencephalographic classification of epileptic seizures. *Epilepsia* 22, 489–503.

Drife, J. O. (1987). Pseudocyesis. *Integrative Psychiatry* 5, 194–200.

Driver, H. S. (1993). Parasomnias. *British Medical Journal* 306, 921–4.

Drossman, D. A. (1994). Irritable bowel syndrome: the role of psychosocial factors. *Stress Medicine* 10, 49–55.

Drossman, D. A. (1995). *The functional gastrointestinal disorders: diagnosis, pathophysiology and treatment: a multinational consensus*. Little Brown, London.

Drugs and Therapeutics Bulletin (1983). Drugs which can be given to nursing mothers. *Drugs and Therapeutics Bulletin* 21, 5–8.

Dubois P. (1909). *The psychic treatment of nervous disorders* (6th edn). Funk and Wagnalls Company, New York and London.

Dubowitz, V. and Hersov, L. (1976). Management of children with non-organic (hysterical) disorders of motor function. *Developmental Medicine and Child Neurology* 18, 358–68.

Duffy, J. (1977). Estimating the proportion of heavy drinkers. In *The Ledermann curve* (ed. D. L. Davies). Alcohol Education Centre, London.

Dunbar, H. F. (1954). *Emotions and bodily changes*. Columbia University Press, New York.

Dunbar, J. A., Ogston, S. A., Ritchie, A., *et al.* (1985). Are problem drinkers dangerous drivers? An investigation of arrest for drinking and driving, serum gamma glutyltranspeptidase activities, blood alcohol concentrations and road accidents: the Tayside safe-driving project. *British Medical Journal* 290, 827–30.

Dunlap, K. (1932). *Habits: their making and unmaking*. Liverheight, New York.

Dunn, J. and Fahy, T. A. (1990). Police admissions to a psychiatric hospital: demographic and clinical differences between ethnic groups. *British Journal of Psychiatry* 156, 373–8.

Dunn, J. and Kendrick, C. (1982). *Siblings: love, envy and understanding*. Cambridge University Press.

Dunne, F. J. (1993). Subcortical dementia. *British Medical Journal* 307, 1–2.

Dunner, D. L., Ishiki, D., Avery, D. H., *et al.* (1986). Effect of alprazolam and diazepam on anxiety and panic attacks in panic disorder: a controlled trial. *Journal of Clinical Psychiatry* 47, 458–60.

Dupont, R. M., Jernigan, T. L., Butters, N., *et al.* (1990). Subcortical abnormalities detected in bipolar affective disorder using magnetic resonance imaging: clinical and neuropsychological significance. *Archives of General Psychiatry* 47, 55–9.

Durkheim, E. (1951). *Suicide: a study in sociology* (trans. J. A. Spaulding and G. Simpson). Free Press, Glencoe, IL.

Dworkin, S. F. (1990*a*). Epidemiology of signs and symptoms in temporomandibular disorders: clinical signs in cases and controls. *Journal of the American Dental Association* 120, 273–81.

Dworkin, S. F. (1990*b*). Multiple pains and psychiatric disturbance. *Archives of General Psychiatry* 47, 239–44.

Dworkin, S. F., Von Korff, M., and LeResche, L. (1990). Multiple pains and psychiatric disturbance. *Archives of General Psychiatry* 47, 239–44.

Dykens, E. M., Hodapp, R., and Leckman, J. F. (1994). Behaviour and development in fragile X syndrome. *Developmental Clinical Psychology and Psychiatry* 28.

Eagles, J. M. and Whalley, L. J. (1985). Decline in the diagnosis of schizophrenia among first admissions to Scottish mental hospitals from 1969 to 1978. *British Journal of Psychiatry* 146, 151–4.

Eapen, V., Pauls, D. L., and Robertson, M. M. (1993). Evidence for autosomal dominant transmission in Gilles de la Tourette syndrome — United Kingdom cohort. *British Journal of Psychiatry* 162, 593–6.

Earls, F., Reich, W., Jung, K., and Cloninger, C. R. (1988). Psychopathology in children of alcoholic and antisocial parents. *Alcoholism: Clinical and Experimental Research* 12, 481–7.

Eastwood, R. and Corbin, S. (1985). Epidemiology of mental disorders in old age. In *Recent advances of psychogeriatrics* (ed. T. Arie). Churchill Livingstone, Edinburgh.

Eaton, J. W. and Weil, R. J. (1955). *Culture and mental disorders: a comparative study of the Hutterites and other populations*. Free Press, Glencoe, IL.

Eaton, W. W. and Keyl, P. M. (1990). Risk factors for the onset of diagnostic interview schedule/DSM-III agoraphobia in a prospecctive population-based study. *Archives of General Psychiatry* 47, 819.

Eaton, W. W., Mortenson, P. B., Herrman, H., *et al.* (1992). Long-term course of hospitalisation for schizophrenia: (1) risk for hospitalisation. *Schizophrenia Bulletin* 18, 217–28.

Ebmeier, K. P., Calder, S. A., Crawford, J. R., *et al.* (1991). Dementia in idiopathic Parkinson's disease: prevalence and relationship with symptoms and signs of parkinsonism. *Psychological Medicine* 21, 69–76.

Eckhert, E. D., Bouchard, T. J., Bohlen, J., and Heston, L. L. (1986). Homosexuality in monozygotic twins reared apart. *British Journal of Psychiatry* 148, 421–5.

Edlund, M. J. and Craig, T. J. (1984). Antipsychotic drug use and birth defects: an epidemiologic assessment. *Comprehensive Psychiatry* 25, 32–7.

Edwards, G. (1985). A later follow-up of a classic case series: D. L. Davies' 1962 report and its significance for the present. *Journal of Studies on Alcohol* 46, 181–90.

Edwards, G., *et al.* (1977). Alcoholism: a controlled trial of 'treatment' and 'advice'. *Journal of Studies on Alcohol* 38, 1004–31.

Egeland, J. A., *et al.* (1987). Bipolar affective disorders linked to DNA markers on chromosome 11. *Nature, London* 325, 783–7.

Ehlers, A., Stangier U., and Gieler, U. (1994). Treatment of atopic dermatitis: a comparison of psychological and dermatological approaches to relapse prevention. *Journal of Consulting and Clinical Psychology* 63, 624–35.

Ehrhardt, A. A., Epstein, R., and Money, J. (1968). Fetal androgens and female gender identity in the early-treated adrenogenital syndrome. *Johns Hopkins Medical Journal* 122, 160–7.

Eisenberg, D. M., Kessler, R. C., Foster, C., *et al.* (1993). Unconventional medicine in the United States — prevalence, costs, and patterns of use. *New England Journal of Medicine* 328, 246–52.

Eisenberg, L. (1958). School phobia — a study in the communication of anxiety. *American Journal of Psychiatry* 114, 712–18.

Eisenberg, L. (1986). Does bad news about suicide beget bad news? *New England Journal of Medicine* 315, 705–7.

Eiser, C. (1986). Effects of chronic illness on the child's intellectual development. *Journal of the Royal Society of Medicine* 79, 2–3.

Eitinger, L. (1960). The symptomatology of mental disease among refugees in Norway. *Journal of Mental Science* 106, 947–66.

Elkin, I., Shea, T., Watkins, J. T., *et al.* (1989). National Institute of Mental Health Treatment of Depression Collaborative Research Programme: general effectiveness of treatments. *Archives of General Psychiatry* 46, 971–82.

Ellenberg, J. H. , Hirtz, D. G., and Nelson, K. B. (1986). Do seizures in children cause intellectual deterioration? *New England Journal of Medicine* 314, 1085–8.

Ellenberger, H. F. (1970). *The discovery of the unconscious.* Basic Books, New York.

Ellis, A. (1956). The effectiveness of psychotherapy in individuals who have severe homosexual problems. *Journal of Consulting Psychology* 20, 191–5.

Elphick, M., Yang, J.-D., and Cowen, P. J. (1990). Effects of carbamazepine on dopamine and serotonin mediated neuroendocrine responses. *Archives of General Psychiatry* 47, 135–40.

Emanuel, E. J. (1994). Euthanasia: historical, ethical, and empiric perspectives. *Archives of Internal Medicine* 154, 1890–1901.

Emerson, E. (1993). Challenging behaviours and severe learning disabilities: recent developments in behavioural analysis and intervention. *Behavioural and Cognitive Psychotherapy* 21, 171–98.

Emerson, E., McGill, P., and Mansell, J. (1994). *Severe learning disabilities and challenging behaviours: designing high quality services.* Chapman & Hall, London.

Endicott, J. and Spitzer, R. L. (1978). A diagnostic interview: the schedule for affective disorders and schizophrenia. *Archives of General Psychiatry* 35, 837–44.

Engel, G. (1962). *Psychological development in health and disease.* Saunders, Philadelphia, PA.

Engel, G. and Romano, J. (1959). Delirium, a syndrome of cerebral insufficiency. *Journal of Chronic Diseases* 9, 260–77.

Enoch, M. D. and Trethowan, W. H. (1979). *Uncommon psychiatric syndromes.* Wright, Bristol.

Epstein, A. W. (1960). Fetishism: a study of its psychopathology with particular reference to a proposed disorder in brain mechanisms as an aetiological factor. *Journal of Nervous and Mental Disease* 130, 107–19.

Epstein, A. W. (1961). Relationship of fetishism and transvestism to brain and particularly to temporal lobe dysfunction. *Journal of Nervous and Mental Disease* 133, 247–53.

Epstein, L., Valoski, A., Wing, R. R., and McCurley, J. (1994). Ten-year outcomes of behavioral family-based treatment for childhood obesity. *Health Psychology* 13(5), 373–83.

Epstein, L. H. (1992). Role of behavior theory in behavioral medicine. *Journal of Consulting and Clinical Psychology* 60(4), 493–8.

Eriksson-Mangold, M. and Carlsson, S. G. (1991). Psychological and somatic distress in relation to perceived hearing disability, hearing handicap and hearing measurements. *Journal of Psychosomatic Research* 35, 729–40.

Errera, P. (1962). Some historical aspects of the concept, phobia. *Psychiatric Quarterly* 36, 325–36.

Escobar, J. I., Rubiio-Stipec, M., Canino, G., and Karno, M. (1989). Somatic Symptom Index (SSI): a new and abridged somatization construct. Prevalence and epide-miological correlates in two large community samples. *Journal of Nervous and Mental Disease* 177, 140.

Esiri, M. (1991). Neuropathology. In *Psychiatry in the elderly* (ed. R. Jacoby and C. Oppenheimer), pp. 113–47. Oxford University Press.

Esquirol, E. (1838). *Des maladies mentales.* Baillière, Paris. (Reprinted in 1976 by Arno Press, New York.)

Esquirol, E. (1845). *Mental maladies, a treatise on insanity* (transl. E. K. Hunt). Lea and Blanchard, Philadelphia, PA.

Essau, C. A. and Wittchen, H. U. (1993). An overview of the composite. International Diagnostic Interview (CID). *International Journal of Methods in Psychiatric Research* 3, 79–85.

Essen-Möller, E. (1971). Suggestions for further improvement of the international classification of mental disorders. *Psychological Medicine* 1, 308–11.

Evans, D. R. (1970). Exhibitionism. In *Symptoms of psychopathology* (ed. C. G. Costello), pp. 7–59. Wiley, New York.

Evans, M. D., Hollon, S. D., Derubeis, R. J., *et al.* (1992). Differential relapse following cognitive therapy and pharmacotherapy for depression. *Archives of General Psychiatry* 49, 802–8.

Eysenck, H. J. (1960). *Behaviour therapy and the neuroses.* Pergamon Press, Oxford.

Eysenck, H. J. (1970a). *Crime and personality.* Paladin Press, London.

Eysenck, H. J. (1970b). A dimensional system of psycho-diagnosis. In *New approaches to personality classification* (ed. A. R. Mahrer), pp. 169–207. Columbia University Press, New York.

Eysenck, H. J. (1970c). *The structure of human personality.* Methuen, London.

Eysenck, H. J. (1976). The learning theory model of neurosis: a new approach. *Behaviour Research and Therapy* 14, 251–67.

Eysenck, H. J. and Eysenck, S. B. G. (1976). *Psychoticism as a dimension of personality.* Hodder and Stoughton, London.

Fabbri, A., Jannini, E. A., Gnessi, L., *et al.* (1989). Endorphine in male impotence: evidence for naltrexone stimulation of erectile activity in patient therapy. *Psychoneuroendocrinology* 14, 103–11.

Fabrega, H. (1987). Psychiatric diagnosis, a cultural perspective. *Journal of Nervous and Mental Disease* 175, 383–94.

Facione, N. C. (1993). Delay versus help seeking for breast cancer symptoms: A critical review of the literature on patient and provider delay. *Social Science and Medicine* 36(12), 1521–34.

Faedda, G. L., Tondo, L., Baldessarini, R. J., *et al.* (1993). Outcome after rapid versus gradual discontinuation of lithium treatment in bipolar disorders. *Archives of General Psychiatry* 50, 448–58.

Faergeman, P. M. (1963). *Psychogenic psychoses.* Butterworths, London.

Fahy, T. A. (1991). Eating disorders in pregnancy. *Psychological Medicine* 21, 577–80.

Fairburn, C. (1981). A cognitive behavioural approach to the treatment of bulimia. *Psychological Medicine* 11, 707–11.

Fairburn, C. G. (1995). *Overcoming binge eating.* Guilford Press, New York.

Fairburn, C. G. and Beglin, S. J . (1990). Studies of the epidemiology of bulimia nervosa. *American Journal of Psychiatry* 147, 401–8.

Fairburn and Brownell (1995).

Fairburn, C., Peveler, R. C., Davies, B., *et al.* (1991). Eating disorders in young adults with insulin-dependent diabetes mellitus: a controlled study. *British Medical Journal* 303, 17–20.

Fairburn, C. G., Marcus, M. D., and Wilson, G. T. (1993). Cognitive-behavioral therapy for binge eating and bulimia nervosa: a comprehensive treatment manual. In *Binge eating: nature, assessment and treatment* (ed. C. G. Fairburn and G. T. Wilson). Guilford Press, New York.

Fairburn, C. G., Jones, R., Peveler, R. C., *et al.* (1993). Psychotherapy and bulimia nervosa. Longer-term effects of interpersonal psychotherapy, behavior therapy, and cognitive behaviour therapy. *Archives of General Psychiatry* 50, 419–28.

Fairweather, D. S. (1991). Delirium. In *Psychiatry in the elderly* (ed. R. Jacoby and C. Oppenheimer), pp. 647–75. Oxford University Press.

Falkai, P. and Bogerts, S. (1993). Brain development and schizophrenia. In *Neurobiology and psychiatry*, Vol. 2 (ed. R. Kerwin), pp. 43–70. Cambridge University Press.

Faller, K. C. (1987). Women who sexually abuse children. *Violence and Victims* 2, 263–76.

Fallowfield, L. J., *et al.* (1994). Psychological effects of being offered choice of surgery for breast cancer. *British Medical Journal* 309, 448.

Falret, J. P. (1854). Mémoire sur la folie circulaire. *Bulletin de l'Academie de Médicine* 19, 382–415. (Trans. into English in M. J. Sedler and E. C. Dessain (1983). Falret's discovery: the origin of the concept of bipolar affective illness. *American Journal of Psychiatry* 140, 1227–33.)

Fanshel, D. (1981). Decision-making under uncertainty: foster care for abused or neglected children? *American Journal of Public Health* 71, 685–6.

Farde, L., Wiesel, F.-A., Hall, H., *et al.* (1987). No D_2 receptor increase in PET study of schizophrenia. *Archives of General Psychiatry* 44, 671–2.

Farde, L., Wiesel, F. A., Halldin, C., and Sedfall, G. (1988). Central C_2-dopamine receptor occupancy in schizophrenic patients treated with antipsychotic drugs. *Archives of General Psychiatry* 45, 71–6.

Farde, L., Wiesel, F. A., Nordstrom, A. L., and Sedvall, G.

(1989). D_1 and D_2 dopamine receptor occupancy during treatment with conventional and atypical neuroleptics. *Psychopharmacology* 99, S28–31.

Farde, L., Wiesel, F. A., Stone-Elander, S., *et al.* (1990). D_2 dopamine receptors in neuroleptic naive patients. *Archives of General Psychiatry* 47, 213–19.

Faris, R. E. L. and Dunham, H. W. (1939). *Mental disorders in urban areas.* Chicago University Press.

Farmer, A., Jackson, R., McGuffin, P., and Storey, P. (1987). Cerebral ventricular enlargement in chronic schizophrenia: consistencies and contradictions. *British Journal of Psychiatry* 150, 324–30.

Farmer, A. E., Wessely, S., Castle, D., and McGuffin, P. (1992*a*). Methodological issues in using a polydiagnostic approach to define psychotic illness. *British Journal of Psychiatry* 161, 824–30.

Farmer, R., Tranah, T., O'Donnell, I., and Catalan, J. (1992*b*). Railway suicide: the psychological effect on drivers. *Psychological Medicine* 22, 407–14.

Farrell, M., Ward, J., Mattick, R., *et al.* (1994). Methadone maintenance treatment in opiates dependents: a review. *British Medical Journal* 309, 997–1001.

Farrington, D. P. (1994). Human development and criminal careers. In *The Oxford handbook of criminology* (ed. M. Maguire), pp. 511–84. Clarendon Press, Oxford.

Farrell, B. A. (1979). Mental illness: a conceptual analysis. *Psychological Medicine* 9, 21–35.

Farrell, B. A. (1981). *The standing of psychoanalysis.* Oxford University Press.

Farrington, D. P., Loeber, R., and VanLammen, W. B. (1990). Long-term criminal outcomes of hyperactivity–impulsivity–attention deficit and conduct problems in childhood. In *Straight and devious pathways from childhood to adulthood* (ed. L. N. Robins and M. Rutter), pp. 62–81. Cambridge University Press, New York.

Faulk, M. (1994). *Basic forensic psychiatry* (2nd edn). Blackwell Scientific Publications, Oxford.

Fava, G. A., Morphy, M. A., and Sonino, N. (1994). Affective prodromes of medical illness. *Psychotherapy and Psychosomatics* 62, 141–5.

Favazza, A. R. and Rosenthal, F. J. (1993). Diagnosic issues in self-mutilation. *Hospital and Community Psychiatry* 44(2), 134–40.

Fawcett, J., Scheftner, W., Clark, D., *et al.* (1987). Clinical predictors of suicide in patients with major affective disorders: a controlled prospective study. *American Journal of Psychiatry* 144, 35–40.

Fawzy, F. I. (1993). Malignant melanoma: effects of an early structured psychiatric intervention, coping and affective state on recurrence and survival 6 years later. *Archives of General Psychiatry* 50(9), 681–9.

Fawzy, F. I., Fawzy, N. W., Arndt, L. A., and Pasnau, R. O. (1995). Critical review of psychosocial interventions in cancer care. *Archives of General Psychiatry* 52, 100–13.

Fedoroff, J. P., Starkstein, S. E., Forrester, A. W., *et al.* (1992). Depression in patients with acute traumatic brain injury. *American Journal of Psychiatry* **149**, 918–23.

Feighner, J. P. (1994). The role of venlafaxine in rational antidepressant therapy. *Journal of Clinical Psychiatry* **59**(9 Suppl. A), 62–8.

Feighner, J. P., Robins, E., Guze, S. B., *et al.* (1972). Diagnostic criteria for use in psychiatric research. *Archives of General Psychiatry* **26**, 57–63.

Feingold, B. F. (1975). Hyperkinesis and learning difficulties linked to artificial food and colors. *American Journal of Nursing* **75**, 797–803.

Feinmann, C. and Harris, M. (1984). Psychogenic facial pain. Part 1: The clinical presentation. *British Dental Journal* **156**, 165.

Feldman, M. P. and McCulloch, M. J. (1979). *Homosexual behaviour, therapy and assessment*. Pergamon Press, Oxford.

Feldman, E., Mayou, R. A., Hawton, K., Ardern, M., and Smith, E. D. O. (1988). Psychiatric disorder in medical inpatients. *Quarterly Journal of Medicine* **241**, 405–12.

Fenichel, O. (1945). *The psychoanalytic theory of neurosis*. Kegan Paul, Trench and Trubner, London.

Fennell, M. J. V. (1994). Depression. In *Cognitive behaviour therapy for psychiatric problems* (ed. K. Hawton, P. M. Salkovskis, J. Kirk, and D. M. Clark), pp. 169–234. Oxford University Press.

Fenton, G. W. (1983). Epilepsy, personality and behaviour. In *Research progress in epilepsy* (ed. F. C. Rose). Pitman, Bath.

Fenton, G. W. (1986). Epilepsy and hysteria. *British Journal of Psychiatry* **149**, 28–37.

Fenton, G., McClelland, R., Montgomery, A., *et al.* (1993). The postconcussional syndrome: social antecedents and psychological sequelae. *British Journal of Psychiatry* **162**, 493–7.

Fenton, W. S. and McGlashan, T. H. (1991). Natural history of schizophrenia subtypes. (1) Longitudinal study of paranoid, hebephrenic and undifferentiated schizophrenia. *Archives of General Psychiatry* **48**, 969–77.

Ferrari, M. (1984). Chronic illness: psychosocial effects on siblings. *Journal of Child Psychology and Psychiatry* **25**, 459–76.

Ferreira, A. J. and Winter, W. D. (1965). Family interaction and decision making. *Archives of General Psychiatry* **13**, 214–23.

Ferrier, I. N. (1985). Water intoxication in patients with psychiatric illness. *British Medical Journal* **291**, 1594–5.

Ferrier, I. N., Roberts, G. W., Crow, T. J., *et al.* (1983). Reduced cholecystokinin-like and somatostatin-like immunoreactivity in limbic lobe is associated with negative symptoms of schizophrenia. *Life Sciences* **3**, 475–82.

Fielding, D., Moore, B., Dewey, M., *et al.* (1985). Children with end-stage renal failure: psychological effects on patients, siblings and parents. *Journal of Psychosomatic Research* **29**, 457–65.

Fifer, S. K., *et al.* (1994). Untreated anxiety among adult primary care patients in a health maintenance organization. *Archives of General Psychiatry* **51**, 740–50.

Fink, P. (1992). Physical complaints and symptoms of somatizing patients. *Journal of Psychosomatic* **36**, 125–36.

Fink, P. (1995). Psychiatric illness in patients with persistent somatisation. *British Journal of Psychiatry* **166**, 93–9.

Finkelhor, D. (1984). *Child sexual abuse: new theory and research*, pp. 53–68. Free Press, London.

Finkelhor, D. (1986). *A sourcebook of child sexual abuse*. Sage, Beverley Hills, CA.

Finlay-Jones, R. and Brown, G. W. (1981). Types of stressful life event and the onset of anxiety and depressive disorders. *Psychological Medicine* **11**, 803–16.

Fiordelli, E., Beghi, E., Boglium, G., and Crespi, V. (1993). Epilepsy and psychiatic disturbance. A cross-sectional study. *British Journal of Psychiatry* **163**, 446–50.

Firth, M. A. (1983). Diagnosis of Duchenne muscular dystrophy: experience of parents of sufferers. *British Medical Journal* **286**, 700–1.

Fischer, M. (1973). Genetic and environmental factors in schizophrenia: a study of twins and their families. *Acta Psychiatrica Scandinavica* **238** (Suppl.).

Fish, B., Marcus, J., Hans, S. L., *et al.* (1992). Infants at risk for schizophrenia: sequelae of a genetic neurointegrative defect. *Archives of General Psychiatry* **49**, 221–35.

Fishbain, D. A. and Aldrich, T. E. (1985). Suicide pacts: international comparisons. *Journal of Clinical Psychiatry* **46**, 11–15.

Fisher, J. E. and Carstensen, L. L. (1990). Behavior management of the dementias. *Clinical Psychology Review* **10**, 611–29.

Fisher, S. and Greenberg, R. P. (1977). *The scientific credibility of Freud's theories and therapy*. Basic Books, New York.

Fisk, J. (1991). Medico-legal issues in old age psychiatry: Abuse of the elderly. In *Psychiatry in the elderly* (ed. R. Jacoby and C. Oppenheimer), pp. 901–14. Oxford University Press, Oxford.

Fitzpatrick, R. and Hopkins, A. (1981). Referrals to neurologists for headaches not due to structural disease. *Journal of Neurology, Neurosurgery and Psychiatry* **44**, 1061–7.

Fitzpatrick, R., Fletcher, A., Gore, S., *et al.* (1992). Quality of life measures in health care. I: Applications and issues in assessment. *British Medical Journal* **305**, 1074–7.

Flaskerud, J. H. and Hu, L. T. (1992). Relationship of ethnicity to psychiatric diagnosis. *Journal of Nervous and Mental Disease* **180**, 296–303.

Fletcher, A., Gore, S., Jones, D., *et al.* (1992). Quality of life measures in health care. II Design, analysis and interpretation. *British Medical Journal* 305, 1145–8.

Flint, A. J. (1994). Epidemiology and comorbidity of anxiety disorders in the elderly. *American Journal of Psychiatry* 151(5), 640–9.

Flor-Henry, P. (1969). Psychosis and temporal lobe epilepsy: a controlled investigation. *Epilepsia* 10, 363–95.

Floyd, F. and Phillippe, K. (1993). Parental interactions with children with and without mental retardation: behavior, management, coerciveness, and positive exchange. *American Journal of Mental Retardation* 97, 673–84.

Foa, E. B., Steketee, G., Kozak, M. J., and Dugger, D. (1987). Imipramine and placebo in the treatment of obsessive compulsives: their effect on depression and on obsessional symptoms. *Psychopharmacology Bulletin* 23, 8–11.

Foa, E. B., Rothbaum, B. O., Riggs, D. S., and Murdock, T. B. (1991). Treatment of posttraumatic stress disorder in rape victims: a comparison between cognitive–behavioural procedures and counselling. *Journal of Consulting and Clinical Psychology* 59, 715–23.

Foa, E. B., Riggs, D. S., and Gershvny, B. S. (1995). Arousal, numbing and intrusion: Symptom structure of PTSD following assault. *American Journal of Psychiatry* 152, 116–20.

Folks, D. G. and Kinney, F. C. (1992*a*). The role of psychological factors in dermatologic conditions. *Psychosomatics* 33(1), 45–54.

Folks, D. G. and Kinney, F. C. (1992*b*). The role of psychological factors in gastrointestinal conditions. *Psychosomatics* 33(3), 257–70.

Folsom, J. C. (1967). Intensive hospital therapy for psychogeriatric patients. *Current Psychiatric Therapy* 7, 209–15.

Folstein, S. and Rutter, M. (1971). Infantile autism: a genetic study of 21 twin pairs. *Journal of Child Psychology and Psychiatry* 18, 297–321.

Ford, D. E. and Kamerow, D. B. (1989). Epidemiologic study of sleep disturbances and psychiatric disorders. *Journal of the American Medical Association* 262(11), 1479–84.

Ford, C. S. and Beach, F. A. (1952). *Patterns of sexual behaviour.* Eyre and Spottiswoode, London.

Fordham, F. (1990). *An introduction to Jung's psychology.* Penguin, Harmondsworth.

Forrest, G. C. and Standish, E. (1984). Supporting bereaved parents after perinatal death. In *Recent research in developmental psychopathology* (ed. J. E. Stevenson). Pergamon Press, Oxford.

Förstl, H., Almeida, O. P., Owen, A. M., *et al.* (1991). Psychiatric, neurological and medical aspects of mis-identification syndromes: a review of 260 cases. *Psychological Medicine* 21, 905–10.

Förstl, H., Burns, A., Levy, R., *et al.* (1993). Neuropathological correlates of behavioural disturbance in confirmed Alzheimer's disease. *British Journal of Psychiatry* 163, 364–8.

Forsythe, W. I. and Butler, R. J. (1989). Fifty years of enuretic alarms. *Archives of Disease in Childhood* 64, 879–85.

Foster, E. M., Kay, D. W. K., and Bergmann, K. (1976). The characteristics of old people receiving and needing domiciliary services. *Age and Ageing* 5, 345–55.

Foulkes, S. H. (1948). *Introduction to group-analytic psychotherapy.* Heinemann, London.

Foulkes, S. H. and Anthony, E. J. (1957). *Group psychotherapy: the psychoanalytic approach.* Penguin, Harmondsworth.

Foulkes, S. H. and Lewis, E. (1944). Group analysis: a study in the treatment of groups on psychoanalytic lines. *British Journal of Medical Psychology* 20, 175–82.

Fowler, R. C., Rich, C. L., and Young, D. (1986). San Diego suicide study. II. Substance abuse in young cases. *Archives of General Psychiatry* 43, 962–5.

Frank, E., Kupfer, D. J., Perel, J. M., *et al.* (1990). Three year outcomes of maintenance therapies in recurrent depression. *Archives of General Psychiatry* 47, 1093–9.

Frank, E., Kupfer, D. J., Perel, J. M., *et al.* (1993). Comparison of full dose versus half-dose pharmacotherapy in the maintenance treatment of recurrent depression. *Journal of Affective Disorders* 27, 139–41.

Frank, J. D. (1967). *Persuasion and healing.* Johns Hopkins Press, Baltimore.

Frank, J. D., Bliedman, L. H., Imber, S. D., *et al.* (1957). Why patients leave psychotherapy. *Archives of Neurology and Psychiatry* 77, 283–99.

Frank, J. D., Hoehn-Sarik, R., Imber, S. D., *et al.* (1978). *Effective ingredients of successful psychotherapy.* Brunner-Mazel, New York.

Frank, R. G., Kashani, J. H., Kashani, S. R., *et al.* (1984). Psychological response to amputation as a function of age and time since amputation. *British Journal of Psychiatry* 144, 493–7.

Frankel, F. H. (1993). Adult reconstruction of childhood events in the multiple personality literature. *American Journal of Psychiatry* 150, 954–8.

Franklin, J., Solovitz, B., Mason, M., *et al.* (1987). An evaluation of case management. *American Journal of Public Health* 77, 674–8.

Fras, I., Litin, E. M., and Pearson, J. S. (1967). Comparison of psychiatric symptoms in carcinoma of the pancreas with those in some other intra-abdominal neoplasms. *American Journal of Psychiatry* 123, 1553–62.

Fraser, R. (1947). The incidence of neurosis among factory workers. *Industrial Health Research Board Report*, No. 90. HMSO, London.

Fraser, W. and Nolan, M. (1994). Psychiatric disorders in mental retardation. In *Mental health in mental retarda-*

tion: *recent advances and practices* (ed. N. Bouras), pp. 79–92. Cambridge University Press.

Frasure-Smith, N., Lesperance, F., and Talajic, M. (1993). Depression following myocardial infarction. *Journal of the American Medical Association* 270, 1819–25.

Frasure-Smith, N., Lesperance, F., and Talajic, M. (1995). Depression and 18-month prognosis after myocardial infarction. *Circulation* 91, 999–1005.

Frederiks, J. A. M. (1969). Disorders of the body schema. In *Handbook of clinical neurology* (ed. P. J. Vinken and G. W. Bruyn), Vol. 4, pp. 207–40. North-Holland, Amsterdam.

Freedland, K. E., Carney, R. M., Krone, R. J., *et al.* (1991). Psychological factors in silent myocardial ischemia. *Psychosomatic Medicine* 53, 13–24.

Freeman, C. P. L. and Kendell, R. E. (1980). ECT: patients' experiences and attitudes. *British Journal of Psychiatry* 137, 8–16.

Freeman, C. P. L., Basson, J. V., and Crighton, A. (1978). Double blind controlled trial of electroconvulsive therapy (ECT) and simulated ECT in depressive illness. *Lancet* i, 738–40.

Freeman, C. P. L., Weeks, D., and Kendell, R. E. (1980). ECT: II Patients who complain. *British Journal of Psychiatry* 137, 17–25.

Freeman, H. (ed.) (1984). *Mental health and the environment.* Churchill Livingstone, Edinburgh.

Freeman, H. (1994). Schizophrenia and city residents. *British Journal of Psychiatry* 164, 39–50.

Freeman, W. and Watts, J. W. (1942). *Psychosurgery.* Thomas, Springfield, IL.

Fremming, K. H. (1951). The expectation of mental infirmity in a sample of the Danish population. *Occasional Papers on Eugenics*, No. 7. Cassell, London.

French, S. A. and Jeffery, R. W. (1994). Consequences of dieting to lose weight: effects on physical and mental health. *Health Psychology* 13(3), 195–212.

Freud, A. (1936). *The ego and the mechanisms of defence.* Hogarth Press, London.

Freud, A. (1958). Adolescence. I. Adolescence in the psychoanalytic theory. In *The psychoanalytic study of the child* (ed. A. Freud), Vol. XIII. International University Press, New York.

Freud, A. (1966). *Normality and pathology in childhood: assessments of development.* Hogarth Press and Institute of Psychoanalysis, London.

Freud, S. (1892). *The standard edition of the complete psychological works* (ed. J. Strachey), Vol. 1. Hogarth Press, London.

Freud, S. (1893). *The standard edition of the complete psychological works* (ed. J. Strachey), Vol. 2. Hogarth Press, London.

Freud, S. (1893). On the psychical mechanisms of hysterical phenomena. *The standard edition of the complete psychological works* (ed. J. Strachey), Vol. 3, pp. 25–42. Hogarth Press, London.

Freud, S. (1895a). Obsessions and phobias, their psychical mechanisms and their aetiology. In *The standard edition of the complete psychological works* (ed. J. Strachey), Vol. 3. Hogarth Press, London.

Freud, S. (1895b). The justification for detaching from neurasthenia a particular syndrome: the anxiety neurosis. *Neurologisches Zentralblatt* 14, 50–66. (Reprinted (transl. J. Riviere) in *Collected papers* 1, 76–106 (1940).)

Freud, S. (1911). Psychoanalytic notes upon an autobiographic account of cases sof paranoia. (Schreber). In *The standard edition of the complete psychological works*, Vol. 12, pp. 1–82. Hogarth Press, London.

Freud, S. (1914). On narcissism: an introduction. In *Collected papers*, vol. 4 (1925), pp. 30–59 (transl. J. Riviere). Hogarth Press and Institute of Psychoanalysis, London.

Freud, S. (1917). Mourning and melancholia. *The standard edition of the complete psychological works*, Vol. 14, pp. 243–58. Hogarth Press, London.

Freud, S. (1923). Psychoanalysis. In *The standard edition of the complete psychological works*, Vol. 18, pp. 235–54. Hogarth Press, London.

Freud, S. (1924a). *Neurosis and psychosis.* Reprinted in Penguin Freud Library, Vol. 10, pp. 209–18. Penguin, Harmondsworth.

Freud, S. (1924b). *The loss of reality in neurosis and psychosis.* Reprinted in Penguin Freud Library, Vol. 10, pp. 219–29. Penguin, Harmondsworth.

Freud, S. (1927). Fetishism. *International Journal of Psychoanalysis* 9, 161–6. Also in *The standard edition of the complete psychological works*, Vol. 21, pp. 147–57. Hogarth Press, London.

Freud, S. (1935). *An autobiographic study.* Hogarth Press, London.

Friedman, A. S. (1975). Interaction of drug therapy with marital therapy for depressed patients. *Archives of General Psychiatry* 32, 619–37.

Friedman, L. J. (1962). *Virgin wives: a study of unconsummated marriage.* Tavistock Publications, London.

Friedman, M. and Rosenman, R. H. (1959). Association of specific behaviour pattern with blood and cardiovascular findings. *Journal of the American Medical Association* 169, 1286–96.

Friedman, M., Thorensen, C. E., and Gill, J. J. (1986). Alteration of Type A behaviour and its effect on cardiac recurrences in post myocardial infarction patients: summary results of the recurrent coronary prevention project. *American Heart Journal* 112, 653–65.

Friedman, N., Abel, L. A., Jesberger, J. A., *et al.* (1992). Saccadic intrusions into smooth pursuit in patients with schizophrenia or affective disorder and normal controls. *Biological Psychiatry* 31, 1110–18.

Friedman, T. and Gath, D. (1989). The psychiatric consequences of spontaneous abortion. *British Journal of Psychiatry* 155, 810–30.

Frierson, R. L. (1991). Suicide attempts by the old and the very old. *Archives of Internal Medicine* 151, 141–4.

Frischer, M. (1992). Estimated prevalence of injecting drug use in Glasgow. *British Journal of Addiction* 87, 235–44.

Frith, C. D. (1979). Consciousness information processing in schizophrenia. *British Journal of Psychiatry* 134, 225–35.

Frith, C. D. (1992). *The cognitive neuropsychology of schizophrenia*. Lawrence Erlbaum, Hillsdale, NJ.

Frith, U. (1989). *Autism: explaining the enigma*. Blackwell, Oxford.

Frith, U. (1991). Autistic psychopathy in childhood. In *Autism and Asperger syndrome* (ed. U. Frith), pp. 37–92. Cambridge University Press.

Fromm, E. (1942). *The fear of freedom*. Kegan Paul, London.

Fromm-Reichmann, F. (1948). Notes on the development of treatment of schizophrenia by psychoanalytic psychotherapy. *Psychiatry* 11, 263–73.

Fuhrer, M. J., Rintala, D. H., Hart, K. A., *et al.* (1993). Depressive symptomatology in persons with spinal cord injury who reside in the community. *Archives of Physical Medicine and Rehabilitation* 74, 255–60.

Fukuda, K., Straus, S. E., Hickie, I., *et al.* (1994). The chronic fatigue syndrome: a comprehensive approach to its definition and study. *Annals of Internal Medicine* 121, 953–9.

Fulford, K. W. M. (1989). *Moral theory and medical practice*. Cambridge University Press.

Fulton, M. and Winokur, G. (1993). A comparative study of paranoid and schizoid personality disorders. *British Journal of Psychiatry* 150, 1363–7.

Furniss, T., Bingley-Miller, L., and Bentovim, A. (1984). Therapeutic approach to sexual abuse. *Archives of Disease in Childhood* 59, 865–70.

Fusco, L., Iani, C., Faedda, M. T., *et al.* (1990). Mesial frontal lobe epilepsy: a clinical entity not sufficiently described. *Journal of Epilepsy* 3, 123–35.

Fyer, A. J., Mannuzza, S., Chapman, T. F., *et al.* (1993). A direct interview family study of social phobia. *Archives of General Psychiatry* 50, 286.

Fyer, M. R., Frances, A. J., Sullivan, T., *et al.* (1988). Co-morbidity of borderline personality disorder. *Archives of General Psychiatry* 45, 348–52.

Gagnon, J. and Simon, W. (1973). *Sexual conduct: the social sources of human sexuality*. Aldine, Chicago, IL.

Gale, E. and Ayer, W. A. (1969). Treatment of dental phobias. *Journal of the American Dental Association* 78, 1304–7.

Ganser, S. J. (1898). Über einen eigenartigen hysterischen Dämmerzustand. *Archiv für Psychiatrie und Nervenkrankheiten* 30, 633–40. (Trans. by C. E. Schorer in *British Journal of Criminology* 5, 120–6 (1965).)

Ganz, L. I. and Friedman, P. L. (1995). Medical progress: supraventricular tachycardia. *New England Journal of Medicine* 332, 162–73.

Garbarino, J., Guttman, E., and Seeley, J. W. (1986). *The psychologically battered child*. Jossey Bass, London.

Garber, H. J., Ananth, J. V., Chin, L. C., *et al.* (1989). Nuclear magnetic resonance study of obsessive compulsive disorder. *American Journal of Psychiatry* 146, 1001–5.

Garber, H. L. (1988). *The Milwaukee Project: preventing mental retardation in children at risk*. American Association on Mental Retardation, Washington, DC.

Gardner, D. L. and Cowdrey, R. W. (1986). Positive effects of carbamazepine on behavioural dyscontrol in borderline personality disorders. *American Journal of Psychiatry* 143, 519–22.

Gardner, R., Hanka, R., O'Brien, V. C., *et al.* (1977). Psychological and social evaluation in cases of deliberate self-poisoning admitted to a general hospital. *British Medical Journal* ii, 1567–70.

Garety, P. and Morris, I. (1984). A new unit for psychiatric patients: organization, attitudes and quality of care. *Psychological Medicine* 14, 183–92.

Garfield, S. L. (1980). *Psychotherapy: an eclectic approach*. Wiley, New York.

Garfinkel, P. E. and Garner, D. N. (1982). *Anorexia nervosa: a multidimensional perspective*. Brunner-Mazel, New York.

Garmezy, N. and Mastern, A. S. (1994). Chronic adversities. In *Child and adolescent psychiatry: modern approaches* (3rd edn) (ed. M. Rutter, E. Taylor, and L. Hersov), pp. 191–208. Blackwell Scientific Publications, Oxford.

Garner, D. M. (1993). Binge eating in anorexia nervosa. In *Binge eating: nature, assessment and treatment* (ed. C. G. Fairburn and G. T. Wilson). Guilford Press, New York.

Garralda, M. E. (1992). A selective review of child psychiatric syndromes with a somatic presentation. *British Journal of Psychiatry* 161, 759–73.

Garruto, R. M. and Brown, P. (1994). Tau protein, aluminium and Alzheimer's disease. *Lancet* 343, 989.

Gask, L. (1992). Training general practitioners to detect and manage emotional disorders. *International Review of Psychiatry* 4, 293–300.

Gastaut, M. (1969). Clinical and electroencephalographic classification of epileptic seizures. *Epilepsia* 10 (Suppl.), 2–21.

Gath, A. (1978). *Down's syndrome and the family*. Academic Press, London.

Gath, D. and Iles, S. (1990). Depression and the menopause. *British Medical Journal* 300, 1287–8.

Gath, D., Cooper, P., Gattoni, F., and Rockett, D. (1977). *Child guidance and delinquency in a London Borough.* Maudsley Monograph No. 24. Oxford University Press, London.

Gath, D., Cooper, P., and Day, A. (1982*a*). Hysterectomy and psychiatric disorder: 1. Levels of psychiatric morbidity before and after hysterectomy. *British Journal of Psychiatry* 140, 335–42.

Gath, D., Cooper, P., Bond, A., and Edmonds, G. (1982*b*). Hysterectomy and psychiatric disorder: II. Demographic psychiatric and physical factors in relation to psychiatric outcome. *British Journal of Psychiatry* 140, 343–50.

Gath, D., Hassal, C., and Cross, K. W. (1973). Whither psychotic day patients? A study of day patients in Birmingham. *British Medical Journal* 1, 94–8.

Gath, D., Osborn, M., Bungay, G., *et al.* (1987). Psychiatric disorder and gynaecological symptoms in middle-aged women: a community survey. *British Medical Journal* 24, 213–18.

Gaupp, R. (1914). The scientific significance of the case of Ernst Wagner. In *Themes and variations in European psychiatry* (ed. S. R. Hirsch and M. Shepherd), pp. 121–33 (1974). Wright, Bristol.

Gawin, F. H., Kleber, H. D., Byck, R., *et al.* (1989). Desipramine facilitation of initial cocaine abstinence. *Archives of General Psychiatry* 46, 117–21.

Gayford, J. J. (1981). Indecent exposure: a review of the literature. *Medicine, Science and the Law* 21, 233–42.

Gazzard, R. G., Davis, M., Spooner, J., and Williams, R. (1976). Why do people use paracetamol for suicide? *British Medical Journal* i, 212–13.

Geaney, D. P. (1994). Single photon emission tomography. In *Dementia* (ed. A. Burns and R. Levy). Chapman & Hall Medical, London.

Geaney, D. P. and Abou-Saleh, M. T. (1990). The use and applications of single photon emission computerised tomography in dementia. *British Journal of Psychiatry* 157 (Suppl. 9), 66–75.

Geaney, D. P., Ellis, P. M., Soper, N., *et al.* (1992). Single photon emission tomography assessment of cerebral dopamine D_2 receptor blockade in schizophrenia. *Biological Psychiatry* 32, 293–5.

Gebhard, P. H., Raboch, J., and Giese, H. (1970). *The sexuality of women* (transl. C. Bearne). André Deutsch, London.

Gelder, M. G. (1991). Adolf Meyer and his influence on British psychiatry. In *150 Years of British psychiatry 1841–1991* (ed. G. E. Berrios and H. Freeman), pp. 419–35. Gaskell, London.

Gelenberg, A. J., Kane, J. M., Keller, M. B., *et al.* (1989). On maintenance treatment of bipolar disorder. *New England Journal of Medicine* 3, 1489–93.

Gelernter, C. S., Uhde, T. W., Cimbolic, P., *et al.* (1991). Cognitive-behavioural and pharmacological treatments of social phobia—a controlled study. *Archives of General Psychiatry* 49, 938.

Geller, J. L. (1992). Pathological firesetting in adults. *International Journal of Law and Psychiatry* 15, 283–302.

General Register Office (1968). A glossary of mental disorders. *Studies on Medical and Population Subjects* 22. HMSO, London.

Gentil, V., Lotufo-Neto, F., Andrade, L., *et al.* (1993). Clomipramine, a better reference drug for panic/agoraphobia. I. Effectiveness comparison with imipramine. *Journal of Pharmacology* 7(4), 316–24.

Gerner, R. H. and Stanton, A. (1992). Algorithm for patient management of acute manic states: lithium, valproate or carbamazepine. *Journal of Clinical Psychopharmacology* 12, 57S–63S.

Gershon, E. S., Mark, A., Cohen, N., *et al.* (1975). Transmitted factors in the morbidity of affective disorders: a controlled study. *Journal of Psychiatric Research* 12, 283–99.

Gersons, B. P. R. and Carlier, I. V. E. (1992). Post-traumatic stress disorder: the history of a recent concept. *British Journal of Psychiatry* 161, 742–8.

Ghodse, H., Myles, J., and Smith, S. E. (1994). Clonidine is not a useful adjunct to methadone gradual detoxification in opioid addiction. *British Journal of Psychiatry* 165, 370–4.

Gibb, W. R. (1989). Dementia and Parkinson's disease. *British Journal of Psychiatry* 154, 596–614.

Gibbens, T. C. N. and Prince, J. (1965). *Child victims of sex offences.* Institute for the Study and Treatment of Delinquency. London.

Gibbens, T. C. N., Pond, D. A., and Stafford Clark, D. A. (1959). A follow-up study of criminal psychopaths. *Journal of Mental Science* 105, 108–15.

Gibbens, T. C. N., Palmer, C., and Prince, J. (1971). Mental health aspects of shoplifting. *British Medical Journal* iii, 612–15.

Gibbens, T. C. N., Way, C., and Soothill, K. L. (1977). Behavioural types of rape. *British Journal of Psychiatry* 130, 32–42.

Gibbs, C. J., Gajdusek, D. C., Asher, D. M., *et al.* (1968). Creutzfeldt–Jacob disease (spongiform encephalopathy): transmission to the chimpanzee. *Science* 161, 388–9.

Giles, D. E., Biggs, M. M., Rush, A. J., and Roffwarg, H. P. (1988). Risk factors in families of unipolar depression 1. Psychiatric illness and reduced REM latency. *Journal of Affective Disorders* 14, 51–60.

Gilchrist *et al.* (1995). Termination of pregnancy and psychiatric morbidity. *British Journal of Psychiatry* 167, 243–8.

Giles, D. E., Jarrett, R. B., Rush, A. J., *et al.* (1993). Prospective assessment of electroencephalographic sleep in remitted major depression. *Psychiatry Research* **46**, 269–84.

Gillam, S. J., Jarman, B., White, P., and Law, R. (1989). Ethnic differences in consultation rates in urban general practice. *British Medical Journal* **299**, 958–60.

Gillberg, C. and Coleman, N. (1992). *The biology of the autistic syndromes* (2nd edn). MacKeith Press, London.

Gilles de la Tourette (1885). Etude sur une affection nerveuse characterisée par l'incoordination motrice accompagnee d'echolalie et de coprolalie. *Archives de Neurologie* **9**, 19–42.

Gillies, N. (1976). Homicide in the west of Scotland. *British Journal of Psychiatry* **128**, 105–27.

Girela, E., Villanueva, E., Hernandez-Cueto, C., and Luna, J. D. (1994). Comparison of the CAGE questionnaire versus some biochemical markers in the diagnosis of alcoholism. *Alcohol and Alcoholism* **29**, 337–43.

Gitlin, M. J. (1993). Lithium-induced renal insufficiency. *Journal of Clinical Psychopharmacology* **13**, 276–9.

Gittelman, R. (1985). Controlled trials of remedial approaches to reading disability. *Journal of Child Psychology and Psychiatry* **6**, 843–6.

Gjessing, R. (1947). Biological investigations in endogenous psychoses. *Acta Psychiatrica* (Kbh). Suppl. 47.

Glaister, B. (1982). Muscle relaxation training for fear reduction of patients with psychological problems: a review on controlled studies. *Behaviour Research and Therapy* **20**, 493–504.

Glaser, D. (1991). Treatment issues in child sexual abuse. *British Journal of Psychiatry* **159**, 769–82.

Glaser, G. H. (1972). Diphenylhydantoin toxicity. In *Antiepileptic drugs* (ed. M. Dixon, J. Woodbury, J. Kiffin Penry, and R. P. Schmidt), Chapter 20. Raven Press, London.

Glassman, A. H., Roose, S. P., and Bigger, J. T. (1993). The safety of tricyclic antidepressants in cardiac patients. *Journal of the American Medical Association* **269**(20), 2673–5.

Glen, A. I. M., Johnson, A. L., and Shepherd, M. (1984). Continuation therapy with lithium and amitriptyline in unipolar depressive illness: a randomized, double-blind controlled trial. *Psychological Medicine* **14**, 37–50.

Glover, G. R. (1992). *CAPSE-10. Computer assisted PSE10.* World Health Organization, Geneva.

Glover, L. and Pearce, S. (1995). Chronic pelvic pain. In *Treatment of functional somatic symptoms* (ed. R. A. Mayou, C. Bass, and M. Sharpe), pp. 313–27. Oxford University Press, Oxford.

Glue, P. W., Nutt, D. J., Cowen, P. J., and Broadbent, D. (1987). Selective effects of lithium on cognitive performance in man. *Psychopharmacology* **91**, 109–11.

Goffman, E. (1961). *Asylums: essays on the social situation of mental patients and other inmates.* Doubleday, New York.

Göktepe, E. O., Young, L. B., and Bridges, P. K. (1975). A further review of the results of stereotactic tractotomy. *British Journal of Psychiatry* **126**, 270–81.

Gold, J. N. and Weinberger, D. R. (1991). Frontal lobe structure, function, and connectivity in schizophrenia. In *Neurobiology in psychiatry*, Vol. 1 (ed. R. Kerwin), pp. 39–59. Cambridge University Press.

Goldacre, M. Seagroatt, V., and Hawton, K. (1993). Suicide after discharge from psychiatric inpatient care. *Lancet* **342**, 283–6.

Goldberg, D. (1972). *The detection of psychiatric illness by questionnaire.* Maudsley Monograph No. 21. Oxford University Press, London.

Goldberg, D. (1990). Reasons for misdiagnosis. In *Psychological disorders in general medical settings* (ed. N. Sartorius, D. Goldberg, D. de Girolamo, *et al.*), pp. 139–45. Hogrefe and Hubert, Göttingen.

Goldberg, D. and Hillier, V. P. (1979). A scaled version of the General Health Questionnaire. *Psychological Medicine* **9**, 139–45.

Goldberg, D. and Huxley, P. (1980). *Mental illness in the community.* Tavistock Publications, London.

Goldberg, E. M. and Morrison, S. L. (1963). Schizophrenia and social class. *British Journal of Psychiatry* **109**, 785–802.

Goldberg, D., Richels, J., Downing, R., and Hesbacher, P. (1976). A comparison of two psychiatric screening tests. *British Journal of Psychiatry* **129**, 61–7.

Goldberg, D., Steele, J., and Smith, J. (1980). Teaching psychiatric interview techniques to family doctors. *Acta Psychiatrica Scandinavica* **62** (Suppl. 285), 41–7.

Goldberg, D., Jenkins, L., Millar, T., and Faragher, B. (1993). The ability of *trainee* general practitioners to identify psychological distress among their patients. *Psychological Medical* **23**(1), 185–93.

Goldberg, S. C., Schulz, S. C., Schulz, P. M., *et al.* (1986). Borderline and schizotypal personality disorders treated with low-dose thiothixine or placebo. *Archives of General Psychiatry* **43**, 680–6.

Goldiamond, I. (1965). Self-control procedures in personal behaviour problems. *Psychological Reports* **17**, 851–68.

Goldman, D. (1993). The DRD$_2$ dopamine receptor and the candidate gene approach in alcoholism. In *Advances in biomedical alcohol research* (ed. P. V. Taberner and A. A. Badawy), pp. 27–30. Pergamon Press, Oxford.

Goldman, E. and Morrison, D. (1984). *Psychodrama: experience and process.* Kendall/Hunt, Iowa.

Goldman, H. and Morrisseym, J. P. (1985). The alchemy of mental health policy: homelessness and the fourth cycle of reform. *American Journal of Public Health* **75**, 727–31.

Goldman, M. J. M. (1991). Kleptomania: making sense of the nonsensical. *American Journal of Psychiatry* **148**(8), 986–96.

Goldstein, I. (1986). Arterial revascularisation procedures. *Seminars in Urology* **4**, 252–8.

Goldstein, K. (1944). Methodological approach to the study of schizophrenic thought disorder. In *Language and thought in schizophrenia* (ed. J. S. Kasanin). University of California Press, Berkeley, CA.

Goldstein, K. (1975). Functional disturbance in brain damage. In *American handbook of psychiatry* (ed. S. Arieti and M. F. Reisser) (2nd edn), Vol. 4. Basic Books, New York.

Goldstein, L. H. (1990). Behavioural and cognitive–behavioural treatments for epilepsy: a progress review. *British Journal of Clinical Psychology* 29, 257–69.

Goldstein, R. B., Black, D. W., Nasrallah, A., and Winokur, G. (1991). The prediction of suicide-sensitivity, specificity, and predictive value of a multivariate model applied to suicide among 1906 patients with affective disorders. *Archives of General Psychiatry* 48, 418–22.

Good, B. J. (1977). The heart of what's the matter: the semantics of illness in Iran. *Culture, medicine and psychiatry* 1, 25–58.

Goodacre, T. E. and Mayou, R. (1995). Dysmorphophobia in plastic surgery and its treatment. In *Treatment of functional somatic symptoms* (ed. R. Mayou, M. Sharpe, and C. Bass), pp. 231–54. Oxford University Press.

Goodman, J. E. and McGrath, P. J. (1991). The epidemiology of pain in children and adolescents: a review. *Pain* 46, 247–64.

Goodman, R. (1987). The developmental neurobiology of language. In *Language development and disorders* (ed. W. Yule and M. Rutter), pp. 129–45. MacKeith Press, London.

Goodman, R. and Stevenson J. (1989). A twin study of hyperactivity: I. An examination of hyperactivity scores and categories derived from Rutter Teacher and Parent Questionnaires. II. The aetiological role of genes, family relationships, and perinatal adversity. *Journal of Child Psychology and Psychiatry* 30, 671–710.

Goodman, W. K., Price, L. H., and Rasmussen, S. A. (1989a). The Yale–Brown Obsessive Compulsive Scale. *Archives of General Psychiatry* 46, 1006–11.

Goodman, W. K., Price, L. H., Rasmussen, S. A., et al. (1989b). Efficacy of fluvoxamine in obsessive–compulsive disorder. *Archives of General Psychiatry* 46, 36–44.

Goodwin, F. K. and Jamison, K. R. (1990). Medical treatment of acute bipolar depression. In *Manic depressive illness* (ed. F. K. Goodwin and K. R. Jamison). Oxford University Press.

Goodwin, F. K., Schulsinger, F., Hermansen, L., et al. (1973). Alcohol problems in adoptees raised apart from alcoholic biological parents. *Archives of General Psychiatry* 28, 238–43.

Goodwin, G. M. (1994). Recurrence of mania after lithium withdrawal. *British Journal of Psychiatry* 164, 149–52.

Goodwin, J. (1988). Post-traumatic symptoms in abused children. *Journal of Traumatic Stress* 4, 475–88.

Goodyer, I. and Taylor, D. C. (1985). Hysteria. *Archives of Diseases in Childhood* 60, 680–1.

Goodyer, I. M., Kolvin, I., and Gatzanis, S. (1985). Recent undesirable life events and psychiatric disorder in childhood and adolescence. *British Journal of Psychiatry* 147, 517–23.

Goodyer, I. M., Kolvin, I., and Gatzanis, S. (1987). The impact of recent life events in psychiatric disorders of childhood and adolescence. *British Journal of Psychiatry* 151, 179–85.

Gordon, D., Burge, D., Hammen, C., et al. (1989). Observations of interactions of depressed women with their children. *American Journal of Psychiatry* 146, 50–5.

Gortmaker, S. L., Must, A., Perrin, J. M., et al. (1993). Social and economic consequences of overweight in adolescence and young adulthood. *New England Journal of Medicine* 329, 1008–12.

Gossop, M. (1981). *Theories of neurosis*. Springer, Berlin.

Gostin, L. (1983). *The court of protection*. Mind, London.

Gottesman, I. and Shields, J. A. (1967). A polygenic theory of schizophrenia. *Proceedings of the National Academy of Sciences of the USA* 58, 199–205.

Gottesman, I. and Shields, J. A. (1972). *Schizophrenia and genetics: a twin study vantage point*. Academic Press, New York.

Gould, M. S., Wallenstein, S., and Kleinman, M. (1990). Time–space clustering of teenage suicide. *American Journal of Epidemiology* 131, 71–8.

Gournay, K. and Brooking, J. (1994). Community psychiatric nurses in primary health care. *British Journal of Psychiatry* 165, 231–8.

Graham, P. (1991). *Child psychiatry: a developmental approach* (2nd edn). Oxford University Press.

Graham, P. and Rutter, M. (1968). Organic brain dysfunction and child psychiatric disorder. *British Medical Journal* iii, 695–700.

Graham, P. and Rutter, M. (1970). Psychiatric aspects of physical disorder. In *Education, health and behaviour* (ed. M. Rutter, J. Tizard, and K. Whitmore). Longman, London.

Graham, P. and Stevenson, J. (1987). Temperament and psychiatric disorder: the genetic contribution to behaviour in childhood. *Australian and New Zealand Journal of Psychiatry* 21, 267–74.

Grahame-Smith, D. G., Green, A. R., and Costain, D. W. (1978). Mechanisms of the antidepressant action of electroconvulsive therapy. *Lancet* i, 254–6.

Gralnick, A. (1942). Folie à deux. The psychosis of association. *Psychiatric Quarterly* 16, 230–63.

Gram, L. (1990). Epileptic seizures and syndromes. *Lancet* ii, 7–8.

Grant, I., Heaton, R. K., McSweeny, J., et al. (1982). Neuropsychological findings in hypoxemic chronic obstructive pulmonary disease. *Archives of Internal Medicine* 142, 1470–6.

Gray J. A. (1982). *The neuropsychology of anxiety: an enquiry into the functions of the septo-hippocampal system.* Clarendon Press, Oxford.

Gray, J., Feldon, V., Rawlins, J., *et al.* (1990). The neuropsychology of schizophrenia. *Behavioural and Brain Science* 14, 1–84.

Green, A. I. and Austin, C. P. (1993). Psychopathology of pancreatic cancer: a psychobiologic probe. *Psychosomatics* 34(3), 208–21.

Green, A. R., Heal, D. J., and Goodwin, G. M. (1986). The effect of electroconvulsive therapy and antidepressant drugs on monoamine receptors in rodent brain—similarities and differences. In *Antidepressants and receptor function*, Ciba Foundation Symposium 123, pp. 246–67. Wiley, Chichester.

Green, R. (1974). *Sexual identity conflict in children and adults.* Duckworth, London.

Green, R. (1985). Atypical psychosexual development. In *Child and adolescent psychiatry* (2nd edn) (ed. M. Rutter and L. Hersov). Blackwell Scientific Publications, Oxford.

Green, R. and Money, J. (1961). Effeminacy in prepubertal boys: summary of eleven cases and recommendations for case management. *Pediatrics* 27, 286–91.

Green, R. and Money, J. (1969). *Transsexualism and sex reassignment.* Johns Hopkins Press, Baltimore, MD.

Greene, B. and Blanchard, E. B. (1994). Cognitive therapy for irritable bowel syndrome. *Journal of Consulting and Clinical Psychology* 62, 576–82.

Greenson, R. R. (1967). *The techniques and practice of psychoanalysis.* Hogarth Press, London.

Greer, S. (1969). The prognosis of anxiety states. In *Studies in anxiety* (ed. M. H. Lader), pp. 151–7. Royal Medicopsychological Association, London.

Greer, S., Marcus, T., and Pettingale, K. W. (1979). Psychological response to breast cancer: effect on outcome. *Lancet* ii, 785–7.

Gregoire, A. (1992). New treatments for erectile impotence. *British Journal of Psychiatry* 160, 315–26.

Gregory, S., Shawcross, C. R., and Gill, D. (1985). The Nottingham ECT study: a double blind comparison of bilateral, unilateral and simulated ECT in depressive illness. *British Journal of Psychiatry* 146, 520–4.

Griesinger, W. (1867). *Mental pathology and therapeutics* (2nd edn) (transl. C. Lockhart Robertson and J. Rutherford). New Sydenham Society, London.

Griffin, J. and Wyles, M. (1991). *Epilepsy, towards tomorrow.* Office of Health Economics, London.

Gross, M. M., Rosenblatt, S. M., Lewis, E., *et al.* (1971). Hallucinations and clouding of sensorium during alcohol withdrawal. *Quarterly Journal of Studies on Alcohol* 32, 1061–9.

Grove, W. M. and Andreasen, N. C. (1992). Concept

diagnosis and the classification. In *Handbook of affective disorders* (ed. E. S. Paykel), pp. 25–42. Churchill Livingstone, Edinburgh.

Grundy, E. (1987). Community care for the elderly 1976–84. *British Medical Journal* 294, 626–9.

Grundy, E. and Bowling, A. (1991). The sociology of ageing. In *Psychiatry in the elderly* (ed. R. Jacoby and C. Oppenheimer), pp. 35–37. Oxford University Press.

Gudjonsson, G. H. (1990). Psychological and psychiatric aspects of shoplifting. *Medicine, Science and Law* 30, 45–51.

Gudjonsson, G. H. (1992). *The psychology of interrogations, confessions and testimony.* Wiley, Chichester.

Guerrant, J., Anderson, W. W., Fischer, A., *et al.* (1962). *Personality in epilepsy.* Thomas, Springfield, IL.

Gunderson, J. G. and Kolb, J. E. (1978). Discriminating features of borderline patients. *American Journal of Psychiatry* 135, 792–6.

Gunn, J. (1977a). *Epileptics in prison.* Academic Press, London.

Gunn, J., Maden, A., and Swinton, M. (1985a). Sexual offenders. In *Current themes in psychiatry* (ed. R. N. Gaind, A. Fawzy, B. I. Hudson, and R. O. Pasnau), Vol. 4. Spectrum, New York.

Gunn, J. (1985b). The role of psychiatry in prisons and the right to punishment. In *Psychiatry, human rights and the law* (ed. M. Roth and R. Bluglass). Cambridge University Press, Cambridge.

Gunn, J., Maden, A., and Swinton, M. (1991). Treatment needs of prisoners with psychiatric disorders. *British Medical Journal* 303, 338–41.

Gunn, J. and Fenton, G. W. (1971). Epilepsy, automatism and crime. *Lancet* i, 1173–6.

Gunn, J. and Robertson, G. (1982). An evaluation of Grendon Prison. In *Abnormal offenders, delinquency and the criminal justice system* (ed. J. Gunn and D. P. Farrington). Wiley, Chichester.

Gunn, J. and Taylor, P. J. (1993). *Forensic psychiatry: clinical, legal and ethical issues.* Butterworth Heinemann, Oxford.

Gunn, J., Maden, A., and Swinton, M. (1991). Treatment needs of prisoners with psychiatric disorders. *British Medical Journal* 303, 338–40.

Gunnell, D. and Frankel, S. (1994). Prevention of suicide: aspirations and evidence. *British Medical Journal* 308, 1227–33.

Gupta, M. A., Gupta, A. K., and Haberman, H. F. (1987). The self-inflicted dermatoses: a critical review. *General Hospital Psychiatry* 9, 45–52.

Gurevitch, D., Bagne, C. A., Perl, E., and Dumlao, M. S. (1991). A review of psychostimulants in elderly patients with refractory depression. In *Advances in neuropsychiatry and psychopharmacology*, Vol. 2, *Refractory depression* (ed. J. D. Amsterdam), pp. 167–75. Raven Press, New York.

Gurland, B., *et al.* (1979). A cross-national comparison of the institutionalised elderly in the cities of New York and London. *Psychological Medicine* 9, 781–8.

Gurman, A. S. (1979). Research on marital and family therapy: progress, perspective and prospect. In *Handbook of psychotherapy and behaviour change* (ed. S. L. Garfield and A. E. Bergin) (2nd edn). Wiley, New York.

Gurman, A. S. and Kriskern, D. P. (1991). *Handbook of family therapy*. Brunner-Mazel, New York.

Gusella, J. F. and MacDonald, M. E. (1994). Huntington's disease and repeating trinucleotides. *New England Journal of Medicine* 330(20), 1450–1.

Gusella, J. F., Wexler, N. S., Conneally, P. M., *et al.* (1983). A polymorphic DNA marker genetically linked to Huntington's disease. *Nature, London* 306, 234–8.

Gustafson, L., Brun, A., and Passant, U. (1992). Frontal lobe degeneration of non-Alzheimer type. In *Baillière's clinical neurology*. Baillière Tindall, London.

Guthrie, E., Creed, F., Dawson, D., and Tomenson, B. (1991). A controlled trial of psychological treatment for the irritable bowel syndrome. *Gastroenterology* 100, 450–7.

Guttman, E. and Maclay, W. S. (1936). Mescalin and depersonalization; therapeutic experiments. *Journal of Neurology and Psychopathology* 16, 193–212.

Guze, S. B., Woodruff, R. A., and Clayton, P. J. (1971). 'Secondary' affective disorder: a study of 95 cases. *Psychological Medicine* 1, 426–8.

Guze, S. B., Cloninger, C. R., Martin, R. L., and Clayton, P. J. (1986). A follow-up and family study of Briquets' syndrome. *British Journal of Psychiatry* 149, 17–23.

Haaga, D. A. F. and Beck, A. T. (1992). Cognitive therapy. In *Handbook of affective disorders* (ed. E. S. Paykel), pp. 511–23. Churchill Livingstone, Edinburgh.

Hachinski, V., Lassen, N. A., and Marshall, J. (1974). Multi-infarct dementia. *Lancet* ii, 207–9.

Hack, M., Taylor, H. G., Klein, N., *et al.* (1994). School-age outcomes in children with birth weights under 750 g. *New England Journal of Medicine* 331, 753–9.

Hackett, T. P. and Weissman, A. (1962). The treatment of the dying. *Current Psychiatric Therapy* 2, 121–6.

Häfner, H. (1987a). Do we still need beds for psychiatric patients? *Acta Psychiatrica Scandinavica* 75, 113–26.

Häfner, H. (1987b). The concept of disease in psychiatry. *Psychological Medicine* 17, 11–14.

Häfner, H. and Reimann, H. (1970). Spatial distribution of mental disorders in Mannheim. In *Psychiatric epidemiology* (ed. E. H. Hare and J. K. Wing). Oxford University Press, London.

Häfner, H., Riecher, A., Maurer, K., *et al.* (1989). How does gender influence age at first hospitalisation for schizophrenia. *Psychological Medicine* 19, 903–18.

Häfner, H., Maurer, K., Loffler, W., and Riecher-Rossler, A. (1993). The influence of age and sex on the onset and early course of schizophrenia. *British Journal of Psychiatry* 162, 80–6.

Hagerman, R. J. (1992). Annotation: fragile X syndrome: advances and controversy. *Journal of Child Psychology and Psychiatry* 33(7), 1127–39.

Hagnell, O. (1966). *A prospective study of the incidence of mental disorder*. Scandinavian University Books, Copenhagen, Denmark.

Hagnell, O. (1970). Incidence and duration of episodes of mental illness in a total population. In *Psychiatric epidemiology* (ed. E. H. Hare and J. K. Wing). Oxford University Press, London.

Haig, R. A. (1992). Management of depression in patients with advanced cancer. *Medical Journal of Australia* 156, 499–503.

Hakim, S. and Adams, R. D. (1965). The special problem of symptomatic hydrocephalus with normal cerebrospinal fluid pressures: observations on cerebrospinal fluid hydrodynamics. *Journal of Neurological Sciences* 2, 307–27.

Haley, J. (1963). *Strategies of psychotherapy*. Grune and Stratton, New York.

Hall, G. S. and Lindzey, G. (1980). *Theories of personality* (3rd edn). Wiley, Chichester.

Hall, J. (1983). Ward based rehabilitation programmes. In *Theory and practice of psychiatric rehabilitation* (ed. F. N. Watts and D. H. Bennett). Wiley, Chichester.

Haller, R. M. and Deluty, R. H. (1988). Assaults on staff by psychiatric in-patients: a critical review. *British Journal of Psychiatry* 152, 174–9.

Hallgren, B. (1960). Nocturnal enuresis in twins. *Acta Psychiatrica Scandinavica* 35, 73–90.

Hallstrom, C. (1985). Benzodiazepines: clinical practice and central mechanisms. In *Recent advances in psychiatry* (ed. K. Granville-Grossman), Vol. 5. Churchill Livingstone, Edinburgh.

Hamburg, D. A., Artz, P., Reiss, E., *et al.* (1953). Clinical importance of emotional problems in the care of patients with burns. *New England Journal of Medicine* 248, 355–9.

Hamer, D. H., Hu, S., Magnuson, V. L., *et al.* (1993). A linkage between DNA markers on the X chromosome and male sexual orientation. *Science* 261, 321–7.

Hamilton, M. (1959). The assessment of anxiety states by rating. *British Journal of Medical Psychology* 32, 50–5.

Hamilton, M. (1967). Development of a rating scale for primary depressive illness. *British Journal of Social and Clinical Psychology* 6, 278–96.

Hamilton, M. (ed.) (1984) *Fish's Schizophrenia* (3rd edn). Wright, Bristol.

Hamilton, M. (ed.) (1984). *Fish's Clinical psychopathology* (2nd edn). Wright, Bristol.

Hammer, T. (1992). Unemployment and use of drugs and

alcohol among young people: a longitudinal study in the general population. *British Journal of Addiction* 87, 1571–81.

Hampton, H. L. (1995). Current concepts: care of the woman who has been raped. *New England Journal of Medicine* 332, 234–7.

Hanks, G. W. (1992). Cancer pain: management. *Lancet* 339, 1031–6.

Hanson, J. W., Jones, K. L., and Smith, D. W. (1976). Fetal alcohol syndrome: an experiment with 41 patients. *Journal of the American Medical Association* 235, 1458–60.

Harding, A. E. (1993). The gene for Huntington's disease. *British Medical Journal* 307, 397.

Harding, C. M., Zubin, J., and Strauss, J. S. (1987). Chronicity in schizophrenia: fact, partial fact, or artifact? *Hospital and Community Psychiatry* 38, 477–86.

Harding, T. W., DeAiango, M. V., Baltasar, J., *et al.* (1980). Mental disorders in primary health care: a study of their frequency in diagnosis in four developing countries. *Psychological Medicine* 10, 231–41.

Hardy, J. A. (1992). An anatomical cascade hypothesis for Alzheimer's disease. *Trends in Neuroscience* 15, 200–1.

Hare, E. H. (1956a). Mental illness and social conditions in Bristol. *Journal of Mental Science* 102, 349–57.

Hare, E. H. (1956b). Family setting and the urban distribution of schizophrenia. *Journal of Mental Science* 102, 753–60.

Hare, E. H. (1959). The origin and spread of dementia paralytica. *Journal of Mental Science* 105, 594–626.

Hare, E. H. (1973). A short note on pseudo-hallucinations. *British Journal of Psychiatry* 122, 469–76.

Hare, E. H. (1975). Season of birth in schizophrenia and neurosis. *American Journal of Psychiatry* 132, 1168–71.

Hare, E. H. and Shaw, G. K. (1965). *Mental health on a new housing estate: a comparative study of two districts of Croydon.* Maudsley Monograph No. 12. Oxford University Press, London.

Harper, C. and Kril, J. (1985). Brain atrophy in chronic alcoholic patients: a quantitative pathological study. *Journal of Neurology, Neurosurgery and Psychiatry* 48, 211–17.

Harper, C., Gold, C., Rodgriguez, M., and Perdices, M. (1989). The prevalence of the Wernicke–Korsakoff syndrome in Sydney, Australia: a prospective necropsy study. *Journal of Neurology, Neurosurgery and Psychiatry* 52, 282–5.

Harper, P. S. (1993). Clinical consequences of isolating the gene for Huntington's disease. *British Medical Journal* 307, 397–8.

Harrington, C. R., Wischik, C. M., McArthur, F. K., *et al.* (1994). Alzheimer's-disease-like changes in tau protein processing: association with aluminium accumulations in brain and renal dialysis patients. *Lancet* 343, 993–7.

Harrington, R. (1994). Affective disorders. In *Child and adolescent psychiatry: modern approaches* (ed. M. Rutter, E. Taylor and L. Hersov), pp. 330–50. Blackwell Scientific Publications, Oxford.

Harrington, R., Fudge, H., Rutter, M., *et al.* (1990). Adult outcomes of childhood and adolescent depression. *Archives of General Psychiatry* 47, 465–73.

Harrington, R. C., Fudge, H., Rutter, M. L., *et al.* (1993). Child and adult depression: a test of continuities with data from a family study. *British Journal of Psychiatry* 162, 627–33.

Harris, E. L. and Fitzgerald, J. D. (1970). *The principles and practice of clinical trials.* Livingstone, Edinburgh.

Harris, P. (1982). The symptomatology of abnormal appearance: an anecdotal survey. *British Journal of Plastic Surgery* 35, 312–13.

Harris, E. C. and Barraclough, B. M. (1995). Suicide as an outcome for medical disorders. *Medicine* 73, 281–96.

Harrison, F. (1991). The Children's Act 1989. *Journal of the Medical Defence Union* 4, 82–3.

Harrison, G. (1980). The abuse of anti-cholinergic drugs in adolescents. *British Journal of Psychiatry* 137, 494–6.

Harrison, G. Owens, D., Holton, A., *et al.* (1988). A prospective study of severe mental disorder in Afro-Caribbean patients. *Psychological Medicine* 18, 643–57.

Harrison, G., Mason, P., Glazebrook, C., *et al.* (1994). Residence of incident cohort of psychotic patients after 13 years of follow-up. *British Medical Journal* 308, 813–19.

Harrison Jr, G. P. and Katz, D. L. (1994). Psychiatric and medical effects of anabolic–androgenic steroid use. *Archives of General Psychiatry* 51, 375–82.

Harrison, P. J. and Pearson, R. C. A. (1989). Gene expression and mental disease. *Psychological Medicine* 19, 813–19.

Harrison, P., McLaughlin, D., and Kerwin, R. W. (1991). Decreased hippocampal expression of a glutamate receptor gene in schizophrenia. *Lancet* 337, 450–2.

Harrison, J. and Maguire, P. (1994). Predictors of psychiatric morbidity in cancer patients. *British Journal of Psychiatry* 165, 593–8.

Hartmann, H. (1964). *Essays on ego psychology.* Hogarth Press, London.

Hartmann, V. (1965). Notes on group therapy with pedophiles. *Canadian Psychiatric Association Journal* 10, 283–8.

Harvey Smith, E. A. and Cooper, B. (1970). Patterns of neurotic illness in the community. *Journal of the Royal College of General Practitioners* 19, 132–9.

Harwood, D. D., Hanumanthus, S., and Stoudemire, A. (1992). Pathophysiology and management of phantom limb pain. *General Hospital Psychiatry* 14, 107–18.

Haug, J. O. (1962). Pneumoencephalographic studies in mental disease. *Acta Psychiatrica Scandinavica* Suppl. 165, 1–114.

Hauser, W. A. and Annegers, J. F. (1993). Epidemiology of epilepsy. In *A textbook of epilepsy* (ed. J. Laidlaw, A. Richens, and D. Chadwick), pp. 23–46. Churchill Livingstone, Edinburgh.

Hawker, A. (1978). *Adolescents and alcohol*. Edsall, London.

Hawton, K. E. (1985). *Sex therapy: a practical guide*. Oxford University Press.

Hawton, K. E. (1986). *Suicide and attempted suicide among children and adolescents*. Sage, Beverley Hills, CA.

Hawton, K. (1988). Controlled studies of psychosocial interventional following attempted suicide. In *Current research on suicide and parasuicide* (ed. S. Platt and N. Kreitman), pp. 180–95. Edinburgh University Press.

Hawton, K. (1990). Self-cutting: can it be prevented? In *Dilemmas and difficulties in the management of psychiatric patients* (ed. K. Hawton and P. Cowen). Oxford University Press.

Hawton, K. E. and Catalan, J. (1987). *Attempted suicide: a practical guide to its nature and management* (2nd edn). Oxford University Press.

Hawton, K. and Fagg, J. (1988). Suicide and other causes of death following attempted suicide. *British Journal of Psychiatry* 152, 359–66.

Hawton, K. and Fagg, J. (1990). Deliberate self-poisoning and self-injury in older people. *International Journal of Geriatric Psychiatry* 5, 367–73.

Hawton, K. and Fagg, J. (1992a). Trends in deliberate self poisoning and self injury in Oxford, 1976–90. *British Medical Journal* 304, 1409–11.

Hawton, K. and Fagg, J. (1992b). Deliberate self-poisoning and self-injury in adolescents. A study of characteristics and trends in Oxford, 1976–89. *British Journal of Psychiatry* 161, 816–23.

Hawton, K. and Kirk, J. (1989). Problem solving. In *Cognitive behaviour therapy for psychiatric problems. A practical guide* (ed. K. Hawton, P. Salkovskis, J. Kirk, and D. J. Clark), pp. 406–26. Oxford University Press.

Hawton, K. E. and Oppenheimer, C. (1983). Women's sexual problems. In *Women's problems in general practice* (ed. A. Anderson and A. McPherson). Oxford University Press.

Hawton, K. and Rose, N. (1986). Unemployment and attempted suicide among men in Oxford. *Health Trends* 18, 29–32.

Hawton, K. E., Gath, D., and Smith, E. (1979). Management of attempted suicide in Oxford. *British Medical Journal* ii, 1040–2.

Hawton, K. E., Fagg, J., and Marsack, P. (1980). Association between epilepsy and attempted suicide. *Journal of Neurology, Neurosurgery and Psychiatry* 43, 168–70.

Hawton, K. E., Bancroft, J., Catalan, J., et al. (1981). Domiciliary and out-patient treatment of self-poisoning patients by medical and non-medical staff. *Psychological Medicine* 116, 169–77.

Hawton, K. E., Catalan, J., Martin, P., and Fagg, J. (1986). Long-term outcome of sex therapy. *Behaviour Research and Therapy* 24, 665–75.

Hawton, K. E., McKeown, S., Day, A., et al. (1987). Evaluation of out-patient counselling compared with general practitioner care following overdoses. *Psychological Medicine* 17, 751–62.

Hawton, K., Fagg, J., and Simkin, S. (1988). Female unemployment and attempted suicide. *British Journal of Psychiatry* 152, 632–7.

Hawton, K., Fagg, J., and McKeown, P. (1989). Alcoholism, alcohol and attempted suicide. *Alcohol and Alcoholism* 24, 3–9.

Hawton, K., Fagg, J., Platt, S., and Hawkins, M. (1993). Factors associated with suicide after parasuicide in young people. *British Medical Journal* 306, 1641–4.

Hawton, K., Haigh, R., Simkin, S., and Fagg, J. (1995a). Attempted suicide in Oxford University students, 1976–1990. *Psychological Medicine* 25, 179–88.

Hawton, K., Simkin, S., Fagg, J., and Hawkins, M. (1995b). Suicide in Oxford University students, 1976–1990. *British Journal of Psychiatry* 166, 44–50.

Hay, E. M., Huddy, A., Black, D., et al. (1994). A prospective study of psychiatric disorder and cognitive function in systemic lupus erythematosus. *Annals of Rheumatic Disease* 53, 298–303.

Hay, G. G. (1970a). Psychiatric aspects of cosmetic nasal operations. *British Journal of Psychiatry* 116, 85–97.

Hay, G. G. (1970b). Dysmorphophobia. *British Journal of Psychiatry* 116, 399–406.

Hay, G. G. and Heather, B. B. (1973). Changes in psychometric test results following cosmetic nasal operations. *British Journal of Psychiatry* 122, 81–90.

Hay, P. J. and Sachdev, P. S. (1992). The present status of psychosurgery in Australia and New Zealand. *Medical Journal of Australia* 157, 17–19.

Hay, P. J., Sachdev, P. S., Cummings, S., et al. (1993). Treatment of obsessive–compulsive disorder by psychosurgery. *Acta Psychiatrica Scandinavica* 87, 197–207.

Haynes, R. B., Taylor, D. W., and Sackett, D. L. (ed.) (1979). *Compliance in health care*. Johns Hopkins University Press, Baltimore, MD.

Hays, J. C. (1994). The course of psychological distress following threatened and actual conjugal bereavement. *Psychological Medicine* 24, 917–27.

Head, H. (1920). *Studies in neurology*, Vol. 2. Oxford University Press.

Heather, N. (1993). Application of harm-reduction principles to the treatment of alcohol problems. In *Psychoactive drugs and harm reduction* (ed. N. Heather, A. Wodak, E. Nadelmann, and P. O'Hare), pp. 168–83. Whurr, London.

Heaton, K. (1992). What makes people with abdominal pain consult their doctor? In *Medical symptoms not explained by organic disease* (ed. F. Creed), pp. 1–8. Royal College of Psychiatrists, London.

Heber, R. (1981). A manual on terminology and classification in mental retardation. *American Journal of Mental Deficiency*, Suppl. 64.

Hecker, E. (1871). Die Hebephrenie. *Virchows Archiv für Pathologie and Anatomie* 52, 394–429. (See *American Journal of Psychiatry* 142, 1265–71.)

Heidensohn, F. (1994a). Women as perpetrators and victims of crime: a sociological perspective. *British Journal of Psychiatry* 158 (Suppl. 10), 50–4.

Heidensohn, F. (1994b). Gender and crime. In *The Oxford handbook of criminology* (ed. M. Maguire), pp. 997–1040. Clarendon Press, Oxford.

Heilig, M., Koob, G. F., Eckman, R., and Britton, K. T. (1994). Corticotropin-releasing factor and neuropeptide Y: role in emotional integration. *Trends in Neurological Sciences* 17, 80–5.

Heimann, P. (1950). On countertransference. *International Journal of Psychoanalysis* 31, 81–4.

Heimberg, R. G. (1989). Cognitive and behavioural treatments for social phobia: a critical analysis. *Clinical Psychology Review* 9, 107–28.

Helgason, T. (1964). Epidemiology of mental disorders in Iceland. A psychiatric and demographic investigation of 5395 Icelanders. *Acta Psychiatrica Scandinavica*, Suppl. 173.

Helmsley, D. R. and Garety, P. A. (1986). The formation and maintenance of delusions. *British Journal of Psychiatry* 149, 51–6.

Helweg-Larsen, P., Hoffmeyer, H., Kieler, J., *et al.* (1952). Famine disease in German concentration camps: complications and sequels. *Acta Psychiatrica et Neurologica Scandinavica* Suppl. 83, 1–460.

Helzer, J. E. and Canino, G. J. (1992). Comparative analysis of alcoholism in ten cultural regions. In *Alcoholism in North American, Europe and Asia* (ed. J. E. Helzer and G. J. Canino), pp. 289–308. Oxford University Press.

Helz, J. W. and Templeton, B. (1990). Evidence of the role of psychosocial factors in diabetes mellitus: a review. *American Journal Psychiatry* 147, 1275–82.

Helzer, J. E., Robins, L. N., and McEvoy, L. (1987). Posttraumatic stress disorder in the general population: findings of the Epidemiological Catchment Area Survey. *New England Journal of Medicine* 317, 1630–4.

Helzer, J. E., Kendell, R. E., and Brockington, I. F. (1983). Contribution of the six-month criteria to the predictive validity of the DSMIII definition of schizophrenia. *Archives of General Psychiatry* 40, 1277–80.

Helzer, J. E., Chammas, S., Norland, C. C., *et al.* (1984). A study of the association between Crohn's disease and psychiatric illness. *Gastroenterology* 86, 324–30.

Hemsley, D. R. (1993). A simple (or simplistic?) cognitive model for schizophrenia. *Behaviour Research Therapeutics* 31, 633–45.

Henderson, A. S. (1990). The social psychiatry of later life. *British Journal of Psychiatry* 156, 645–53.

Henderson, A. S. (1994). *Dementia*. World Health Organization, Geneva.

Henderson, D. K. (1939). *Psychopathic states*. Chapman & Hall, London.

Henderson, D. K. and Gillespie, R. D. (1930). *Textbook of psychiatry for students and practitioners* (2nd edn). Oxford University Press, London.

Henderson, S., Duncan-Jones, P., McAuley, H., and Ritchie, K. (1978). The patient's primary group. *British Journal of Psychiatry* 132, 74–86.

Henderson, S., Byrne, D. G., and Duncan-Jones, P. (1982). *Neurosis and the social environment*. Academic Press, London.

Henderson, S. E. (1987). The assessment of 'clumsy' children: old and new approaches. *Journal of Child Psychology and Psychiatry* 28, 511–27.

Hendin, H. and Haas, A. P. (1991). Suicide and guilt as manifestations of PTSD in Vietnam combat veterans. *American Journal of Psychiatry* 148, 586–91.

Hendriksen, C. and Binder, V. (1980). Social prognosis in patients with ulcerative colitis. *British Medical Journal* ii, 581–3.

Heninger, G. R., Charney, D. S., and Menkes, D. B. (1983a). Receptor sensitivity and the mechanism of action of antidepressant treatment. In *Treatment of depression: old approaches and new controversies* (ed. P. J. Clayton and J. E. Barrett). Raven Press, New York.

Heninger, G. R., Charney, D. S., and Sternberg, D. E. (1983b). Lithium carbonate augmentation of antidepressant treatment. *Archives of General Psychiatry* 40, 1335–42.

Henry, J. P., Meehan, J. P., and Stephens, P. M. (1967). Use of psychosocial stimuli to induce prolonged systolic hypertension in mice. *Psychosomatic Medicine* 29, 408–32.

Henry, J. A., Jeffreys, K. J., and Dawling, S. (1992). Toxicity and deaths from 3,4-methylenedioxymethamphetamine ('ecstasy'). *Lancet* 340, 384–7.

Herbert, T. B. and Cohen, S. (1993). Stress and immunity in humans: a meta-analytic review. *Psychosomatic Medicine* 55, 364–79.

Hermelin, B. (1994). Visual and motor functions in graphically gifted savants. *Psychological Medicine* 24, 673–80.

Hermelin, B. and O'Connor, N. (1983). The idiot savant: flawed genius or clever Hans? *Psychological Medicine* 13, 479–81.

Herrman, H., McGorry, P., Bennett, P., *et al.* (1989). Prevalence of severe mental disorders in disaffiliated and homeless people in inner Melbourne. *American Journal of Psychiatry* 146, 1179–84.

Hershon, H. I. (1977). Alcohol withdrawal symptoms and

drinking behaviour. *Journal of Studies on Alcoholism* 38, 953–71.

Hersov, L. (1960). Refusal to go to school. *Journal of Child Psychology and Psychiatry* 1, 137–45.

Hersov, L. (1985). Encopresis. In *Child psychiatry: modern approaches* (ed. L. Hersov and M. Rutter) (2nd edn). Blackwell, Oxford.

Hersov, L. and Berg, I. (eds) (1980). *Out of school.* Wiley, Chichester.

Herstbech, S., Hansea, H. E., Amdisen, A., and Olsen, S. (1977). Chronic renal lesions following long term treatment with lithium. *Kidney International* 12, 205–13.

Heshe, J. and Roeder, E. (1976). Electroconvulsive therapy in Denmark. *British Journal of Psychiatry* 128, 241–5.

Heston, L. J. (1966). Psychiatric disorders in foster home reared children of schizophrenic mothers. *British Journal of Psychiatry* 112, 819–25.

Hetherington, E. M., Cox, M., and Cox, R. (1985). Long-term effects of divorce on the adjustments of children. *Journal of the American Academy of Child Psychiatry* 24, 518–30.

Hewett, L. E. and Jenkins, R. L. (1946). *Fundamental patterns of maladjustment: the dynamics of their origin.* Thomas, Springfield, IL.

Hewett, S. H. and Ryan, P. J. (1975). Alternatives to living in psychiatric hospitals—a pilot study. *British Journal of Hospital Medicine* 14, 65–70.

Hibbert, G. A. (1984a). Ideational components of anxiety, their origin and content. *British Journal of Psychiatry* 144, 618–24.

Hibbert, G. A. (1984b). Hyperventilation as a cause of panic attacks. *British Medical Journal* 288, 263–4.

Hibbert, G. and Pilsbury, D. (1988). Hyperventilation in panic attacks. *British Journal of Psychiatry* 153, 76–80.

Hietala, J., Syvalahti, E., Vuorio, K., *et al.* (1994). Striatal D^2 dopamine receptor characteristics in neuroleptic-naive schizophrenic patients studied with positron emission tomography. *Archives of General Psychiatry* 51, 116–23.

Higgins, S. R., Budney, A. J., Bickel, W. K., *et al.* (1993). Achieving cocaine abstinence with a behavioural approach. *American Journal of Psychiatry* 150, 763–9.

Higgitt, A. and Fonagy, P. (1993). Psychotherapy in borderline and narcissistic personality disorder. In *Personality disorder review* (ed. P. Tyrer and G. Stein), pp. 225–61. Gaskell, London.

Higgit, A., Fonangy, P., Toone, B., and Shine, P. (1990). The prolonged benzodiazepine withdrawal syndrome: anxiety or hysteria. *Acta Psychiatrica Scandinavica* 82, 165–8.

Hill, D. (1952). EEG in episodic psychotic and psychopathic behaviour: a classification of data. *Electroencephalography and Clinical Neurophysiology* 4, 419–42.

Hill, D. (1953). Psychiatric disorders of epilepsy. *Medical Press* 229, 473–5.

Hill-Beuff, A. and Porter, J. D. R. (1984). Children coping with impaired appearance: social and psychologic influences. *General Hospital Psychiatry* 6, 294–301.

Hillbrand, M. (1994). Clinical predictors of self-mutilation in hospitalized forensic patients. *Journal of Nervous and Mental Disease* 182(1), 9–13.

Hiller, W., Zaudig, M., and Bose, M. V. (1989). The overlap between depression and anxiety on different levels of psychopathology. *Journal of Affective Disorders* 16, 223–31.

Hilton, M. E. (1988). Trends in US drinking patterns: further evidence from the past twenty years. *British Journal of Addiction* 83, 269–78.

Himmelhoch, J. M., Thase, M. E., Mallinger, A. G., and Houck, P. (1991). Tranylcypromine versus imipramine in anergic bipolar depression. *American Journal of Psychiatry* 148, 910–16.

Hinde, R. A. (1985). Ethiology in relation to psychiatry. In *Handbook of psychiatry*, Vol. 5 (ed. M. Shepherd). Cambridge University Press.

Hindmarch, I. (1988). The psychopharmacological approach: effects of psychotropic drugs on car handling. *International Clinical Psychopharmacology* 3 (Suppl. 1), 73–9.

Hingson, R. and Hawland, J. (1993). Alcohol and non-traffic unintended injuries. *Addiction* 88, 877–83.

Hinrichsen, G. A. (1992). Recovery and relapse from major depression in the elderly. *American Journal of Psychiatry* 149, 1575–9.

Hirsch, S. R. (1986a). Clinical treatment of schizophrenia. In *The psychopharmacology and treatment of schizophrenia* (ed. P. Bradley and S. R. Hirsch). Oxford University Press.

Hirsch, S. R. (1987). Planning for bed needs and resource requirements in acute psychiatry. *Bulletin of the Royal College of Psychiatrists* 11, 398–407.

Hirsch, S. R. and Bristow, M. F. (1993). Psychological, family, ethnic and community factors affecting the course and treatment of schizophrenia. *Current Opinion in Psychiatry* 6, 53–7.

Hirsch, S. and Leff, J. (1975). *Abnormalities in parents of schizophrenics.* Maudsley Monograph No. 22. Oxford University Press, London.

Hirsch, S. R., Gaind, R., Rohde, P. D., *et al.* (1973). Outpatient maintenance of chronic schizophrenic patients with long acting fluphenazine: double blind placebo trial. *British Medical Journal* i, 633–7.

Hirschfeld, M. (1944). *Sexual anomalies and perversions: physical and psychological development and treatment.* Aldor, London.

Hoare, P. and Kerley, S. (1991). Psychosocial adjustment of children with chronic epilepsy and their families.

Developmental Medicine and Child Neurology 33, 201–15.

Hobbs, N. (1994). Childhood sexual abuse: how women can be helped to overcome its long-term effects. In *Dilemmas and difficulties in the management of psychiatric patients* (ed. K. Hawton and P. Cowen), pp. 183–96. Oxford University Press.

Hobson, R. F. (1953). Prognostic factors in electric convulsive therapy. *Journal of Neurology, Neurosurgery and Psychiatry* 16, 275–81.

Hoch, P. H. and Polantin, P. (1949). Pseudoneurotic forms of schizophrenia. *Psychiatric Quarterly* 23, 249–96.

Hodes, M., Eisler, I., and Dare, C. (1991). Family therapy for anorexia in adolescence: a review. *Journal of the Royal Society of Medicine* 84, 359–62.

Hodges, J. R. (1994). *Cognitive assessment for clinicians.* Oxford University Press.

Hodgkinson, S., *et al.* (1987). Molecular genetic evidence for heterogeneity in manic depression. *Nature, London* 325, 805–6.

Hoehn-Sarik, R., Frank, J. D., Imber, S. D., *et al.* (1964). Systematic preparation of patients for psychotherapy I: Effects on therapy behaviour and outcome. *Journal of Psychiatric Research* 2, 267–81.

Hoehn-Saric, R., Pearlson, G. D., Harris, G. J., *et al.* (1991). Effects of fluoxetine on regional cerebral blood flow in obsessive–compulsive patients. *American Journal of Psychiatry* 148, 1243–5.

Hoenig, J. and Kenna, J. C. (1974). The prevalence of transsexualism in England and Wales. *British Journal of Psychiatry* 124, 181–90.

Hoffman, B. F. (1986). How to write a psychiatric report for litigation following a personal injury. *American Journal of Psychiatry* 143, 164–9.

Hoffman, B. F. and Spiegel, H. (1989). Legal principles in the psychiatric assessment of personal injury litigants. *American Journal of Psychiatry* 146, 304–10.

Hofman, A., Rocca, W. A., Brayne, C., *et al.* (1991). The prevalence of dementia in Europe: a collaborative study of 1980–1990 findings. *International Journal of Epidemiology* 20, 736–48.

Hogarty, G. E. and Ulrich, R. (1977). Temporal effects of drug and placebo in delaying relapse in schizophrenic out-patients. *Archives of General Psychiatry* 34, 297–301.

Hogarty, G. E., Goldberg, S. C., and Schooler, N. (1974). Drugs and sociotherapy in the aftercare of schizophrenic patients II. Two year relapse rates. *Archives of General Psychiatry* 31, 603–8.

Hoggett, B. (1984). *Mental health law* (2nd edn). Sweet and Maxwell, London.

Hoggett, B. M. (1990). *Mental health law* (3rd edn). Sweet & Maxwell, London.

Holden, N. (1987). Late paraphrenia or the paraphrenias? *British Journal of Psychiatry* 150, 635–9.

Holden, U. P. and Woods, R. T. (1988). *Reality orientation: psychological approaches to the "confused elderly"* (2nd edn). Churchill Livingstone, Edinburgh.

Holden, J. M., Sagovsky, R., and Cox, J. L. (1989). Counselling in a general practice setting: controlled study of health visitor intervention in treatment of postnatal depression. *British Medical Journal* 298, 223–6.

Holding, T. A., Buglass, D., Duffy, J. C., and Kreitman, N. (1977). Parasuicide in Edinburgh—a seven year review, 1968–1974. *British Journal of Psychiatry* 130, 534–43.

Holland, A. J. (1994a). *Neurobiology and psychiatry*, Vol. 2 (2nd edn). Cambridge University Press.

Holland, A. J. (1994b). Down's syndrome and Alzheimer's disease. In *Mental health in mental retardation: recent advances and practices* (ed. N. Bouras), pp. 154–67. Cambridge University Press.

Holland, J. C. and Rowland, J. H. (1989). *Handbook of psychooncology. Psychological care of the patient with cancer.* Oxford University Press, New York.

Holland, A. J., Hall, A., Murray, R., *et al.* (1984). Anorexia nervosa: a study of 34 twin pairs and a set of triplets. *British Journal of Psychiatry* 145, 414–19.

Holland, J. C., *et al.* (1986). Comparative psychological disturbance in patients with pancreatic and gastric cancer. *American Journal of Psychiatry* 143, 982–6.

Hollander, E., Neville, D., Frenkel, M., *et al.* (1992). Body dysmorphic disorder. Diagnostic issues and related disorders. *Psychosomatics* 33, 156–65.

Hollingshead, A. B. and Redlich, F. C. (1958). *Social class and mental illness: a community study.* Wiley, New York.

Hollister, J. N., Mednick, S. A., Brenan, P., and Cannon, T. D. (1994). Impaired autonomic nervous system habituation in those at genetic risk for schizophrenia. *Archives of General Psychiatry* 51, 552–8.

Hollon, S. D., Derubeis, R. J., Evans, M. D., *et al.* (1992). Cognitive therapy and pharmacotherapy for depression. *Archives of General Psychiatry* 49, 774–81.

Holmes, D. B. (1987). The influence of meditation versus rest on physiological arousal: a second examination. In *The psychology of meditation* (ed. M. A. West). Clarendon Press, Oxford.

Holmes, N., Shah, A., and Wing, L. (1982). The disability assessment schedule: a brief assessment device for use with the mentally retarded. *Psychological Medicine* 12, 879–90.

Holmes, P. and Karp, M. (1991). *Psychodrama, inspiration and technique.* Tavistock Publications and Routledge, London.

Holmes, T. and Rahe, R. H. (1967). The social adjustment rating scale. *Journal of Psychosomatic Research* 11, 213–18.

Holsboer, F. (1992). The hypothalamic–pituitary–adrenocortical system. In *Handbook of affective disorders* (ed.

E. S. Paykel), pp. 267–87. Churchill Livingstone, Edinburgh.

Holzman, P. S., Kringlen, E., Matthysse, S., *et al.* (1988). A single dominant gene can amount for eye tracking dysfunctions in schizophrenia in offspring of discordant twins. *Archives of General Psychiatry* 45, 641–7.

Home Office, Department of Health (1975). *Report of the Committee on Mentally Abnormal Offenders* (Chairman: Lord Butler), Cmnd 6244. HMSO, London.

Hooley, J. M., Orley, J., and Teasdale, J. D. (1986). Levels of expressed emotion and relapse in depressed patients. *British Journal of Psychiatry* 148, 642–7.

Hope, T. (1994). The structure of wandering in dementia. *International Journal of Geriatric Psychiatry* 9, 149–55.

Hopkins, A. (1992). The management of patients with chronic headache not due to obvious structural disease. In *Medical symptoms not explained by organic disease* (ed. F. Creed), pp. 34–46. Royal College of Psychiatrists, London.

Hopwood, P. and Maguire, P. (1992). Priorities in the psychological care of cancer patients. *International Review of Psychiatry* 4, 35–44.

Horney, K. (1939). *New ways in psychoanalysis.* Kegan Paul, London.

Horowitz, M. J. (1976). *Stress response syndromes.* Aronson, New York.

Horowitz, M. J. (1986). *Stress response systems* (2nd edn). Jason Aronson, New Jersey.

Horton, R. W. (1992). The neurochemistry of depression: evidence derived from studies of post-mortem brain tissue. *Molecular Aspects of Medicine* 13, 191–203.

Horwath, E., Lish, J., Johnson, J., *et al.* (1993). Agoraphobia without panic: clinical reappraisal of an epidemiologic finding. *American Journal of Psychiatry* 150, 1496–1501.

Hotopf, M. F. and Wessely, S. (1994). Viruses, neurosis and fatigue. *Journal of Psychosomatic Research* 38(6), 499–514.

Hoult, J. (1986). Community care of the acutely mentally ill. *British Journal of Psychiatry* 149, 137–44.

Hoult, J., Reynolds, I., Charbonneau-Powis, M., *et al.* (1983). Psychiatric hospital versus community treatment: the results of a randomised trial. *Australia and New Zealand Journal of Psychiatry* 17, 160–7.

House, A. D. and Andrews, H. B. (1988). Life events and difficulties preceding the onset of functional dysphonia. *Journal of Psychosomatic Research* 32, 311–19.

House, A. and Hodges, J. (1988). Persistent denial of handicap after infarction of the right basal ganglia: a case study. *Journal of Neurology, Neurosurgery and Psychiatry* 51, 112–15.

House, A., Dennis, M., Molyneux, A., *et al.* (1989). Emotionalism after stroke. *British Medical Journal* 298, 991–4.

House, A., Dennis, M., Mogridge, L., *et al.* (1991). Mood disorders in the year after first stroke. *British Journal of Psychiatry* 158, 83–92.

Housekamp, B. M. and Foy, D. W. (1991). The assessment of posttraumatic stress disorder in battered women. *Journal of Interpersonal Violence* 6(3), 367–75.

Houston, F. and Royse, A. B. (1954). Relationship between deafness and psychotic illness. *Journal of Mental Science* 100, 900–3.

Howard, C. and d'Orban, P. T. (1987). Violence in sleep: medico-legal issues and two case reports. *Psychological Medicine* 17, 915–25.

Howard, L. M. and Wessely, S. (1993). The psychology of multiple allergy. *British Medical Journal* 307, 747–8.

Howard, L. M., Williams, R., and Fahy, T. A. (1994). The psychiatric assessment of liver transplant patients with alcoholic liver disease: a review. *Journal of Psychosomatic Research* 38(7), 643–53.

Howard, R. (1994). Phenomenology, demography and diagnosis in late paraphrenia. *Psychological Medicine* 24, 397–410.

Howard, R., Castle, D., Wessely, S., and Murray, R. (1993). A comparative study of 470 cases of early onset and late onset schizophrenia. *British Journal of Psychiatry* 163, 352–7.

Howlin, P. (1994). Special educational treatment. In *Child and adolescent psychiatry: modern approaches* (3rd edn) (ed. M. Rutter, E. Taylor, and L. Hersov), pp. 1071–88. Blackwell Scientific Publications, Oxford.

Hser, Y.-I., Anglin, M. D., and Chou, C.-P. (1988). Evaluation of drug abuse treatment—a repeated measures design assessing methadone maintenance. *Evaluation Review* 12, 547–70.

Hsu, L. K. G. (1990). *Eating disorders.* Guilford Press, New York.

Huber, G., Gross, G., and Schuttler, R. (1975). A long-term follow up study of schizophrenia: psychiatric course of illness and prognosis. *Acta Psychiatrica Scandinavica* 52, 49–57.

Hucker, S. J. (1990). Sexual asphyxia. In *Principles and practice of forensic psychiatry* (ed. P. Boeden and R. Bluglass). Churchill Livingston, Edinburgh.

Hughes, A. L. (1992). The prevalence of illicit drug use in six metropolitan areas in the United States: results from the 1991 National Household Survey on Drug Abuse. *British Journal of Addiction* 87, 1481–5.

Hughes, D. C., Demallie, D., and Blazer, D. G. (1993). Does age make a difference in the effects of physical health and social support on the outcome of a major depressive episode? *American Journal of Psychiatry* 150, 728–33.

Hull, C. L. (1943). *Principles of behaviour.* Appleton, New York.

Humphreys, M. S., Johnstone, E. C., MacMillan, J. F., and Taylor, P. J. (1992). Dangerous behaviour preceding first admissions for schizophrenia. *British Journal of Psychiatry* 161, 501–5.

Hunter, R. and MacAlpine, I. (1963). In *Three hundred years of psychiatry 1535–1860*, pp. 441–4. Oxford University Press.

Huntington's Disease Collaborative Research Group (1993). A novel gene containing a trinucleotide repeat that is expanded and unstable on Huntington's disease chromosomes. *Cell* 72, 971–83.

Hussebye, D. G., Westlie, L., Thomas, J. S., and Kjellstrand, C. M. (1987). Psychological, social and somatic prognostic indicators in old patients undergoing long-term dialysis. *Archives of Internal Medicine* 147, 1921–4.

Hyde, T. H., Nawroz, S., Goldberg, T. E., *et al.* (1994). Is there a cognitive decline in schizophrenia? A cross sectional study. *British Journal of Psychiatry* 164, 494–500.

Hyler, S., Reider, R., and Spitzer, R. (1987). *Personality Diagnostic Questionnaire PDQ*. New York State Psychiatric Institute, New York.

Iles, S. and Gath, D. (1993). Psychiatric outcome of termination of pregnancy for foetal abnormality. *Psychological Medicine* 23, 407–13.

Imboden, J. B., Canter, A., and Cluff, L. E. (1961). Convalescence from influenza: a study of the psychological and clinical determinants. *Archives of Internal Medicine* 108, 393–9.

Insanity Defense Work Group (1983). American Psychiatric Association statement of the insanity defense. *American Journal of Psychiatry* 140, 681–8.

Insel, T. R. (1991). Serotonin in obsessive compulsive disorder: a causal connection or more monomania about a major monoamine. In *5-Hydroxytryptamine in psychiatry: a spectrum of ideas*. (ed. M. Sandler, A. Coppen, and S. Harnett). Oxford University Press.

Insel, T. R. (1992). Towards a neuroanatomy of obsessive compulsive disorder. *Archives of General Psychiatry* 49, 739–44.

Insel, T. R., Battaglia, G., Johannsseen, J. N., and De Souza, E. B. (1989). 3,4-methylenedioxymethamphetamine; ('ecstasy') selectively destroys brain serotonin terminals in rhesus monkeys. *Journal of Pharmacology and Experimental Therapeutics* 249, 713–20.

International Gender Dysphoria Association (1985). Standards of care: the hormonal and surgical sex reassignment of gender dysphoric persons. *Archives of Sexual Behaviour* 14, 79–90.

Irvine, D., Brown, B., Crooks, D., *et al.* (1991). Psychosocial adjustment in women with breast cancer. *Cancer* 67, 1097–1117.

Isacsson, G., Holmgren, P., Wasserman, D., and Bergman, U. (1994). Use of antidepressants among people committing suicide in Sweden. *British Medical Journal* 308, 506–9.

Isometsa, E. T., Henriksson, M. M., Aillevi, M. A., *et al.* (1994). Suicide in major depression. *American Journal Psychiatry* 151(4), 530–6.

Jablensky, A. (1986). Epidemiology of schizophrenia: a European perspective. *Schizophrenia Bulletin* 12, 52–73.

Jablensky, A. (1993). The epidemiology of schizophrenia. *Current Opinion in Psychiatry* 6, 43–52.

Jablenksy, A., Korten, A., Ernberg, G., *et al.* (1986). Manifestations and first-contact incidence of schizophrenia in different cultures. *Psychological Medicine* 16, 909–28.

Jablensky, A., Sartorius, N., Ernberg, G., *et al.* (1992). Schizophrenia: manifestations, incidence and course in different cultures. A World Health Organisation 10-Country Study. *Psychological Medicine Monograph* Suppl. 20.

Jackson, B. M. (1969). A case of voyeurism treated by counterconditioning. *Behaviour Research and Therapy* 7, 133–4.

Jackson, G., Gater, R., Goldberg, D., *et al.* (1993). A new community mental health team based in primary care. *British Journal of Psychiatry* 162, 375–84.

Jackson, M. and Cawley, R. (1992). Psychodynamics and psychotherapy on an acute psychiatric ward. *British Journal of Psychiatry* 160, 41–50.

Jacobs, D. and Silverstone, T. (1986). Dextroamphetamine-induced arousal in human subjects as a model for mania. *Psychological Medicine* 16, 323–9.

Jacobs, P. A., Brunton, M., Melville, M. M., *et al.* (1965). Aggressive behaviour and subnormality. *Nature, London* 208, 1351–2.

Jacobs, S. (1993). *Pathological grief — maladaption to loss*. American Psychiatric Press, Washington, DC.

Jacobs, S. and Myers, J. (1976). Recent life events and acute schizophrenic psychosis: a controlled study. *Journal of Nervous and Mental Disease* 162, 75–87.

Jacobs, S., Prusoff, B. A. and Paykel, E. S. (1974). Recent life events in schizophrenia and depression. *Psychological Medicine* 4, 444–52.

Jacobs, S. C., Hansen, F., and Berkman, L. (1989). Depressions of bereavement. *Comprehensive Psychiatry* 30, 218–24.

Jacobsen, J. L., Jacobsen, S. W., Sokol, R. J., *et al.* (1993). Teratogenic effects of alcohol on infant development. *Alcohol Clinical and Experimental Research* 17, 174–83.

Jacobson, A. M., Hauser, S. T., Lavori, P., *et al.* (1994). Family environment and glycemic control: a four-year prospective study of children and adolescents with insulin-dependent diabetes mellitus. *Psychosomatic Medicine* 56, 401–9.

Jacobson, E. (1938). *Progressive relaxation*. Chicago University Press.

Jacobson, E. (1953). Contribution to the metapsychology of cyclothymic depression. In *Affective disorders* (ed. P. Greenacre). International Universities Press, New York.

Jacobson, L. K., Rabinowitz, I., Popper, M. S., *et al.* (1994). Interviewing prepubertal children about suicidal ideation and behavior. *Journal of the Academy of Child and Adolescent Psychiatry* 33(4), 439.

Jacobson, S. J., Jones, K., Johnson, K., et al. (1992). Prospective multicentre study of pregnancy outcome after lithium exposure during the first trimester. *Lancet* 339, 530–3.

Jacobson, R. R. (1985). Child firesetters: a clinical investigation. *Journal of Child Psychology and Psychiatry* 26, 759–68.

Jacobson, R. R. (1995). The post-concussional syndrome: physiogenesis, psychogenesis and malingering: an integrative model. *Journal of Psychosomatic Research* (in press).

Jacoby, A. (1992). Epilepsy and the quality of everyday life. *Social Science and Medicine* 34, 657–66.

Jacoby, R. (1991). Affective disorders: manic illness. In *Psychiatry in the elderly* (ed. R. Jacoby and C. Oppenheimer), pp. 720–34. Oxford University Press.

Jaffe, J. H. (1989). Addictions: what does biology have to tell? *International Review of Psychiatry* 1, 51–61.

Jahanshahi, M. (1991). Psychosocial factors and depression in torticollis. *Journal of Psychosomatic Research* 35, 493–507.

Jahoda, G. and Cramond, J. (1972). *Children and alcohol. A developmental study in Glasgow*, Vol. 1. HMSO, London.

Jakob, A. (1921). Über eingenarte Erkrankungen des Zentralnervensystems mit bemerkenswerten anatomischen Befunde. *Zeitschrift für die gesamte Neurologie und Psychiatrie* 64, 147–228.

James, I. P. (1967). Suicide and mortality among heroin addicts in Britain. *British Journal of Addictions* 62, 391–8.

Jamison, K. R. (1992). Manic depressive illness: what role does psychotherapy have in management? In *Practical problems in clinical psychiatry* (ed. K. Hawton and P. Cowen), pp. 33–50. Oxford University Press.

Janca, A., Üstür, T. B., and Sartorius, N. (1994). New versions of World Health Organization instruments for the assessment of mental disorders. *Acta Psychiatrica Scandinavica* 90, 73–83.

Janet, P. (1925). *Psychological healing*. Allen and Unwin, London.

Janicak, P. G., Davis, J. M., Gibbons, R. D., et al. (1985). Efficacy of ECT: a meta-analysis. *American Journal of Psychiatry* 142, 297–302.

Janis, I. L. (1958). *Psychological stress: psychoanalytic and behavioural studies of surgical patients*. Wiley, New York.

Janoff-Bulman, R. (1985). The aftermath of victimization: rebuilding shattered assumptions. In *Trauma and its wake: the study and treatment of post traumatic stress disorder* (ed. C. R. Figley), pp. 15–25. Brunner-Mazel, New York.

Janoff-Bulman, R. and Frieze, I. H. (1983). A theoretical perspective for understanding reactions to victimization. *Journal of Social Issues* 39, 1–17.

Janssen, R., Nwanyanwu, O., Selik, R., and Stehr-Green, J. (1992). Epidemiology of HIV encephalopathy in the US. *Neurology* 42, 1472–6.

Jaspers, K. (1913). *Allgemeine Psychopathologie*. Springer, Berlin.

Jaspers, K. (1963). *General psychopathology (Allgemeine Psychopathologie* (7th edn), 1959, trans. J. Hoenig and M. W. Hamilton). Manchester University Press.

Jeavons, P. M. (1983). Non-epileptic attacks in childhood. In *Research progress in epilepsy* (ed. F. C. Rose). Pitman, Bath.

Jellinek, E. M. (1960). *The disease concept of alcoholism*. University Press, New Haven, CT.

Jenike, M. A. and Rauch, S. L. (1994). Managing the patient with treatment resistant obsessive compulsive disorder: current strategies. *Journal of Clinical Psychiatry* 55 (Suppl.), 11–17.

Jenike, M. A., Hyman, S., Baer, L., et al. (1990). A controlled trial of fluvoxamine in obsessive-compulsive disorder: implications for a serotonergic theory. *American Journal of Psychiatry* 147, 1209–15.

Jenike, M. A., Baer, L., Ballantine, H. T., et al. (1991). Cingulotomy for refractive obsessive-compulsive disorder. *Archives of General Psychiatry* 48, 548.

Jenkins, C. D. (1994). Quantifying and predicting recovery after heart surgery. *Psychosomatic Medicine* 56, 203–12.

Jenkins, J. M. and Smith, M. A. (1990). Factors protecting children living in disharmonious homes: maternal reports. *Journal of the American Academy of Child and Adolescent Psychiatry* 29, 60–9.

Jenkins, L., Tarnopolsky, A., and Hand, D. (1981). Psychiatric admissions and aircraft noise from London Airport: four year, three-hospitals' study. *Psychological Medicine* 11, 765–82.

Jenkins, S. and Wingate, C. (1994). Who cares for young carers? *British Medical Journal* 308, 733–4.

Jennings, C., Barraclough, B. M., and Moss, J. R. (1978). Have the Samaritans lowered the suicide rate? A controlled study. *Psychological Medicine* 8, 413–22.

Jensen, M. P. (1994). Relationship of pain-specific beliefs to chronic pain adjustment. *Pain* 57(3), 301–9.

Jerrell, J. and Hu, T. (1989). Cost-effectiveness of intensive clinical and case management compared with an existing system of care. *Inquiry* 26, 224–34.

Jeste, D. V. and Caligiuri, M. P. (1993). Tardive dyskinesia. *Schizophrenia Bulletin* 19, 303–15.

Jobst, K. A., Smith, A. D., Barker, D. S., et al. (1992). Association of atrophy of the medical temporal lobe with reduced blood blow in the posterior parietotemporal cortex in patients with a clinical and pathological diagnosis of Alzheimer's disease. *Journal of Neurology, Neurosurgery and Psychiatry* 55, 190–4.

Joffe, R. T., Singer, W., Levitt, A. J., and MacDonald, C. (1993). A placebo-controlled comparison of lithium and triiodothyronine augmentation of tricyclic antidepressants in unipolar refractory depression. *Archives of General Psychiatry* 50, 387–93.

Johnson, A. M., Falstein, E. K., Szorek, S. A., and Svendsen, M. (1941). School phobia. *American Journal of Orthopsychiatry* 11, 702–11.

Johnson, D. A. W. (1986). Depressive symptom in schizophrenia: some observations on frequency, morbidity and possible causes. In *Contemporary issues in schizophrenia* (ed. A. Kerr and P. Snaith). Gaskell, London.

Johnson, G. A. (1991). Research into psychiatric disorder after stroke: the need for further studies. *Australia and New Zealand Journal of Psychiatry* 25, 358–70.

Johnson, J. (1984). Stupor: a review of 25 cases. *Acta Psychiatrica Scandinavica* 70, 376–7.

Johnson, J., Weissman, M. M., and Klerman, G. L. (1992). Service utilization and social morbidity associated with depressive symptoms in the community. *Journal of the American Medical Association* 267(11), 1478–83.

Johnson, S. B. (1985). Situational fears and objects phobias. In *The clinical guide to child psychiatry* (ed. D. Shaffer, A. A. Ehrhardt, and L. L. Greenhill). Free Press, New York.

Johnston, L. D., O'Malley, P. M., and Bachman, J. G. (1991). *Drug use among American high school seniors, college students and young adults, 1975–1990.* National Institute on Drug Abuse, Washington, DC.

Johnston, M. (1986). Preoperative emotional status and post operative recovery. *Advances in Psychosomatic Medicine* 15, 1–22.

Johnston D. W. (1993). The current status of the coronary prone behaviour. *Journal of the Royal Society of Medicine* 86, 406–9.

Johnston, D. W., Gold, A., Kentish, J. *et al.* (1993). Effect of stress management on blood pressure in mild primary hypertension. *British Medical Journal* 306, 963–6.

Johnstone, E. C. (1991). Disabilities and circumstances of schizophrenic patients: a follow-up study. *British Journal of Psychiatry* 159 (Suppl. 13), 5–46.

Johnstone, E. C., Crow, T. J., Frith, C. D., *et al.* (1976). Cerebral ventricular size and cognitive impairment in chronic schizophrenia. *Lancet* ii, 924–6.

Johnstone, E. C., Crow, T. J., Frith, C. D., *et al.* (1978). Mechanism of the antipsychotic effect in the treatment of acute schizophrenia. *Lancet* i, 848–51.

Johnstone, E. C., Crow, T. J., Frith, C. D., *et al.* (1978). The dementia of dementia praecox. *Acta Psychiatrica Scandinavica* 57, 305–24.

Johnstone, E. C., Cunningham-Owens, D. G., Rith, C. D., *et al.* (1980). Neurotic illness and its response to anxiolytic and antidepressant treatment. *Psychological Medicine* 10, 321–8.

Johnstone, E. C., Deakin, J. F. W., Lawler, P., *et al.* (1980). The Northwick Park electroconvulsive therapy trial. *Lancet* ii, 1317–20.

Johnstone, E., Crow, T. J., Ferrier, I., *et al.* (1983). Adverse effects of anticholinergic medication on positive schizophrenic symptoms. *Psychological Medicine* 13, 513–77.

Johnstone, E. C., Owens, D. G. C., Gold, A., *et al.* (1984). Schizophrenia patients discharged from hospital—a follow-up study. *British Journal of Psychiatry* 145, 586–90.

Johnstone, E. C., Crow, T. J., Johnstone, A. L., and MacMillan, J. F. (1986). The Northwick Park study of first episodes of schizophrenia. I. Presentation of the illness. Problems relating to admission. *British Journal of Psychiatry* 148, 115–20.

Johnstone, E. C., Crow, T. J., Frith, C. D., and Owens, D. G. C. (1988). The Northwick Park 'functional' psychosis study: diagnosis and treatment response. *Lancet* ii, 120–5.

Jolley, A. G., Hirch, S. R., Morrison, E., *et al.* (1990). Trial of brief intermittent neuroleptic prophylaxis for selected schizophrenic outpatients: clinical and social outcome at two years. *British Medical Journal* 301, 837–41.

Jones, D. (1992). *Interviewing children who have been sexually abused* (4th edn). Gaskell Press, London.

Jones, D. P. H. and Alexander, H. (1978). Treating the abusive family within the family care system. In *The battered child* (ed. R. E. Helfer and R. S. Kempe) (4th edn). University of Chicago Press, London.

Jones, K. (1972). *A history of the mental health services.* Routledge & Kegan Paul, London.

Jones, K. and Smith D. W. (1973). Recognition of the fetal alcohol syndrome in early infancy. *Lancet* ii, 999–1001.

Jones, M. (1952). *Social psychiatry: a study of therapeutic communities.* Tavistock, London.

Jones, P. and Murray, R. M. (1991). The genetics of schizophrenia is the genetics of neurodevelopment. *British Journal of Psychiatry* 158, 615–23.

Jorge, R. E., Robinson, R. G., and Arndt, S. (1993). Are there symptoms that are specific for depressed mood in patients with traumatic brain injury? *Journal of Nervous and Mental Disease* 181, 91.

Jorm, A. F. (1995). The epidemiology of depressive states in the elderly: implications for recognition, intervention and prevention. *Social Psychiatry and Psychiatric Epidemiology* 30, 53–9.

Jorm, A. F., Korten, A. E., and Henderson, A. F. (1987). The prevalence of dementia: a quantitative integration of the literature. *Acta Psychiatrica Scandinavica* 76, 465–79.

Jorm, A. F., Christensen, H., Henderson, A. S., *et al.* (1994). Complaints of cognitive decline in the elderly: a comparison of reports by subjects and informants in a community survey. *Psychological Medicine* 24, 365–74.

Judd, F. K. and Brown, D. J. (1992). Psychiatric consultation in a spinal injuries unit. *Australia and New Zealand Journal of Psychiatry* 26, 218–22.

Kahlbaum, K. (1863). *Die Gruppirung der psychichen Krankheiten.* Kafemann, Danzig.

Kahn, E. (1928). Die psychopäthischen Persönlichkeiten. In *Handbuch der Geisteskrankheiten*, Vol. 5, p. 227. Springer, Berlin.

Kahn, R. J., McNair, D. M., Lipman, R. S., *et al.* (1986). Imipramine and chlordiazepoxide in depressive and anxiety disorders. II Efficacy in anxious outpatients. *Archives of General Psychiatry* 43, 79–85.

Kales, A., Soldatos, C. R., and Kales, J. D. (1987). Sleep disorders: insomnia, sleepwalking, night terrors, nightmares, and enuresis. *Annals of Internal Medicine* 106, 582–92.

Kalichman, S. C. and Sikkema, K. J. (1994). Psychological sequelae of HIV infection and aids: review of empirical findings. *Clinical Psychology Review* 14, 611–32.

Kallen, B. and Tandberg, A. (1983). Lithium and pregnancy—a cohort study of manic depressive women. *Acta Psychiatrica Scandinavica* 62, 134–9.

Kallmann, F. J. (1938). *The genetics of schizophrenia.* Augustin, New York.

Kallmann, F. J. (1946). The genetic theory of schizophrenia: an analysis of 691 schizophrenic twin index families. *American Journal of Psychiatry* 103, 309–22.

Kallmann, F. J. (1952). Study on the genetic affects of male homosexuality. *Journal of Nervous and Mental Disease* 115, 1283–98.

Kalucy, R. S., Crisp, A. H., and Harding, B. (1977). A study of 56 families with anorexia nervosa. *British Journal of Medical Psychology* 50, 381–95.

Kamarck, T. and Jennings, J. R. (1991). Biobehavioral factors in sudden cardiac death. *Psychological Bulletin* 109, 42–75.

Kaminski, M., Rumeau-Rouquette, C., and Schwartz, D. (1976). Consommation d'alcool chez les femmes enceintes et issue de la grossesse. *Revue d'Epidemiologie et de Santé Publique* 24, 27–40.

Kane, B. (1979). Children's concepts of death. *Journal of Genetic Psychology* 134, 141–53.

Kane, J., Honigfeld, G., Singer, J., and Meltzer, H. Y. (1988). Clozapine for the treatment-resistant schizophrenic: a double blind comparison with chlorpromazine. *Archives of General Psychiatry* 45, 789–96.

Kane, J. M. (1992). Clinical efficacy of clozapine in treatment-refractory schizophrenia: an overview. *British Journal of Psychiatry* 160 (Suppl. 17), 41–5.

Kane, J. M. and Marder, S. R. (1993). Psychopharmacologic treatment of schizophrenia. *Schizophrenia Bulletin* 19, 287–302.

Kane, R. L. (1985). Special needs of the elderly. In *Oxford textbook of public health* (ed. W. W. Holland), Vol. 4. Oxford University Press.

Kanner, L. (1943). Autistic disturbance of affective contact. *Nervous Child* 2, 217–50.

Kantor, J. S., Zitrin, C. M., and Zeldis, S. M. (1980). Mitral valve prolapse in agoraphobic patients. *American Journal of Psychiatry* 137, 467–9.

Kaplan, H. I., Sadock, B. J., and Grebb, J. A. (1994). Substance related disorders. In *Synopsis of psychiatry*, pp. 383–456. Williams and Wilkins, Baltimore, MD.

Karasu, T. B. (1979). Psychotherapy of the medically ill. *American Journal of Psychiatry* 136, 1–11.

Karlehagen, S., Malt, U. F., Hoff, H., Tibell, E., Herrstromer, U., Hildingson, K., and Leymann, H. (1993). *Journal of Psychosomatic Research* 37(8), 807–17.

Karush, A., Daniels, G. E., O'Connor, J. F., and Stern, L. O. (1977). *Psychotherapy in chronic ulcerative colitis.* Saunders, Philadelphia, PA.

Kasanin, J. (1933). The acute schizoaffective psychoses. *American Journal of Psychiatry* 13, 97–126.

Kashani, J. H. and Simonds, J. F. (1979). The incidence of depression in children. *American Journal of Psychiatry* 136, 1203–5.

Katerndahl, D. A. (1993). Lifetime prevalence of panic states. *American Journal of Psychiatry* 150, 246–9.

Kathol, R. G. and Delahunt, J. W. (1986). The relationship of anxiety and depression to symptoms of hyperthyroidism using operational criteria. *General Hospital Psychiatry* 8, 23–8.

Katon, W. J., Ries, R. K., Bokan, J. A., and Kleinman, A. (1980). Hyperemesis gravidarum: a biopsychosocial perspective. *International Journal of Psychiatry in Medicine* 10, 151–62.

Katona, C. (1993). The aetiology of depression in old age. *International Review of Psychiatry* 5, 407–16.

Katz, M., Abbey, S., Rydall, A., and Lowy, F. (1995). Psychiatric consultation for competency to refuse medical treatment. A retrospective study of patient characteristics and outcome. *Psychosomatics* 36, 33–41.

Kavanagh, D. J. (1992). Recent developments in expressed emotion in schizophrenia. *British Journal of Psychiatry* 160, 601–20.

Kavka, J. (1949). Pinel's conception of the psychopathic state. *Bulletin of the History of Medicine* 23, 461–8.

Kay, D. W. K. (1972). Schizophrenia and schizophrenia-like states in the elderly. *British Journal of Hospital Medicine* 8, 369–76.

Kay, D. W. K. and Bergmann, K. (1980). Epidemiology of mental disorder among the aged in the community. In *Handbook of mental health and ageing* (J. E. Birren and R. B. Sloane). Prentice-Hall, Englewood Cliffs, NJ.

Kay, D. W. K. and Roth, M. (1961). Environmental and

hereditary factors in the schizophrenias of old age ('late paraphrenia') and their bearing on the general problem of causation in schizophrenia. *Journal of Mental Science* 107, 649–86.

Kay, D. W. K., Beamish, P., and Roth, M. (1964). Old age mental disorders in Newcastle-upon-Tyne: 1: a study in prevalence. *British Journal of Psychiatry* 110, 146–58.

Kay, D. W. K., Cooper, A. F., Garside, R. F., and Roth, M. (1976). The differentiation of paranoid and affective psychoses by patients' premorbid characteristics. *British Journal of Psychiatry* 129, 207–15.

Kay, S. R., Fiszbein, A., and Opler, L. A. (1987). The Positive and Negative Syndrome Scale (PANSS) for schizophrenia. *Schizophrenia Bulletin* 13, 261–76.

Kazdin, A. E., Esveldt-Dawson, K., French, N. H., and Unis, A. S. (1987). Problem solving skills and relationship therapy in the treatment of antisocial child behaviour. *Journal of Consulting and Clinical Psychology* 55, 76–85.

Kedward, H. B. and Cooper, B. (1966). Neurotic disorders in urban practice: a 3 year follow-up. *Journal of the Royal College of General Practitioners* 12, 148–63.

Keefe, R. S., Silverman, J. M., Siever, L. J., and Cornblatt, B. A. (1991). Refining phenotype characterization in genetic linkage studies of schizophrenia. *Social Biology* 38, 197–218.

Keller, M. B., Beardslee, W. R., Dorer, D. J., *et al.* (1986). Impact of severity and chronicity of parental affective illness on adaptive functioning and psychopathology in children. *Archives of General Psychiatry* 43, 930–7.

Kellner, C., Jolley, R., Holgate, R., *et al.* (1991). Brain MRI in obsessive compulsive disorder. *Psychiatry Research* 36, 45–9.

Kellner, R. (1990). Somatization: the most costly comorbidity? In *Comorbidity of mood and anxiety disorders* (ed. J. D. Maser and C. R. Cloninger), pp. 239–52. American Psychiatric Press, Washington, DC.

Kellner, R. (1992). Diagnosis and treatments of hypochondriacal syndromes. *Psychosomatics* 33, 278–89.

Kellner, R. (1994). Psychosomatic syndromes, somatization and somatoform disorders. *Psychotherapy and Psychosomatics* 61, 4–24.

Kelly, W. F., Checkley, S. A., Bender, D. A., and Mashifer, K. (1985). Cushing's syndrome and depression—a prospective study of 26 patients. *British Journal of Psychiatry* 142, 16–19.

Kelsoe, J. R., Ginns, E. I., Egeland, J. A., *et al.* (1989). Re-evaluation of the linkage relationship between chromosome 11p loci and the gene for bipolar affective disorders in the old order Amish. *Nature, London* 342, 238–43.

Kemmer, F. W., Bisping, R., Steingruber, H. J., *et al.* (1986). Psychological stress and metabolic control in patients with type 1 diabetes mellitus. *New England Journal of Medicine* 314, 1078–84.

Kempe, R. S. and Goldbloom, R. B. (1987). Malnutrition and growth retardation (failure to thrive) in the context of child abuse and neglect. In *The battered child* (ed. R. E. Helfer and R. S. Kempe), pp. 315–35. University of Chicago Press, London.

Kendall, P. C., Ronan, K. R., and Epps, J. (1991). Aggression in children and adolescents: cognitive behavioural treatment perspective. In *The development of treatment of childhood aggression* (ed. D. J. Pepler and K. H. Rubin), pp. 341–60. Lawrence Erlbaum, Hillsdale, NJ.

Kendell, R. E. (1968). *The classification of depressive illness*. Maudsley Monograph No. 18. Oxford University Press, London.

Kendell, R. E. (1975). *The role of diagnosis in psychiatry*. Blackwell, Oxford.

Kendell, R. E. (1985) Emotional and physical factors in the genesis of puerperal mental disorders. *Journal of Psychosomatic Research* 29, 3–11.

Kendell, R. E. (1991). Suicide in pregnancy and the puerperium. *British Medical Journal* 302, 126–7.

Kendell, R. E., de Roumanie, M., and Ritson, E. B. (1983). Influence of an increase in excise duty on alcohol consumption and its adverse effects. *British Medical Journal* 287, 809–11.

Kendell, R. E., Chalmers, J. C., and Platz, C. (1987). Epidemiology of puerperal psychoses. *British Journal of Psychiatry* 150, 662–73.

Kendell, R. E., Malcolm, D. E., and Adams, W. (1993). The problem of detecting changes in the incidence of schizophrenia. *British Journal of Psychiatry* 162, 212–18.

Kendler, K. S. (1982). Demography of paranoid psychosis (delusional disorder). *Archives of General Psychiatry* 39, 890–902.

Kendler, K. S. (1986). Genetics of schizophrenia. In *American Psychiatric Association Annual Review* (ed. A J. Frances and R. E. Hales), Vol. 5. American Psychiatric Press, Washington, DC.

Kendler, K. S. (1991). Mood incongruent affective illness. *Archives of General Psychiatry* 48, 362–9.

Kendler, K. S. and Gruenberg, A. M. (1984). An independent analysis of the Danish adoption study of schizophrenia. VI. The relationship between psychiatric disorders as defined by DSMIII in the relatives and adoptees. *Archives of General Psychiatry* 41, 555–64.

Kendler, K. S. and Tsuang, M. T. (1981). Nosology of paranoid schizophrenia and other paranoid psychoses. *Schizophrenia Bulletin* 7, 594–610.

Kendler, K. S., Gruenberg, A. M., and Strauss, J. S. (1981). An independent analysis of the Copenhagen sample for the Danish adoption study of schizophrenia. The relationship between schizotypal personality disorder and schizophrenia. *Archives of General Psychiatry* 38, 982–7.

Kendler, K. S., Masterson, C. C., Ungaro, R., and Davis, K. L. (1984). A family history study of schizophrenia-related personality disorders. *American Journal of Psychiatry* 141, 424–7.

Kendler, K. S., Gruenberg, A. M., and Tsuang, M. T. (1985a). Psychiatric illness in first degree relatives of schizophrenic and surgical control patients: a family study using DSMIII criteria. *Archives of General Psychiatry* 42, 770–9.

Kendler, K. S., Masterson, C. C., and Davis, K. L. (1985b). Psychiatric illness in first-degree relatives of patients with paranoid psychosis, schizophrenia and medical illness. *British Journal of Psychiatry* 47, 524–31.

Kendler, K. S., Neale, M. C., Kessler, R. C. et al. (1992a). The genetic epidemiology of phobias in women. *Archives of General Psychiatry* 49, 273.

Kendler, K. S., Neale, M. C., Kessler, R. C., et al. (1992b). Generalized anxiety disorder in women. *Archives of General Psychiatry* 43, 267–72.

Kendler, K. S., McGuire, M., Gruenberg, A. M., et al. (1993a). The Roscommon family study. (1) Methods, diagnosis of probands and risk of schizophrenia in relatives. *Archives of General Psychiatry* 50, 527–40.

Kendler, K. S., McGuire, M., Gruenberg, A. M., et al. (1993b). The Roscommon family study. (2) The risk of non-schizophrenic non-affective psychosis in relatives. *Archives of General Psychiatry* 50, 645–52.

Kendler, K. S., McGuire, M., Gruenberg, A. M., et al. (1993c). The Roscommon family study. (3) Schizophrenia-related personality disorders in relatives. *Archives of General Psychiatry* 50, 781–8.

Kendler, K. S., McGuire, M., Gruenberg, A. M., et al. (1993d). The Roscommon family study. (4) Affective illness, anxiety disorders and alcoholism in relatives. *Archives of General Psychiatry* 50, 952–60.

Kendler, K. S., Gruenberg, A. N., and Kinney, D. K. (1994a). Independent diagnosis of adoptees and relatives as defined by DSM-III in the provincial and national samples of the Danish Adoption Study of Schizophrenia. *Archives of General Psychiatry* 51, 456–68.

Kendler, K. S., Neale, M. C., Heath, A. C., et al. (1994b). A twin-family study of alcoholism in women. *American Journal of Psychiatry* 151, 707–15.

Kendrick, T., Sibbald, B., Burns, T., and Freeling, P. (1991). Role of general practitioners in the care of long-term mentally ill patients. *British Medical Journal* 302, 508–10.

Kennedy, A. and Neville, J. (1957). Sudden loss of memory. *British Medical Journal* ii, 428–33.

Kennedy, H. and Grubin, D. (1992). Patterns of denial in sex offenders. *Psychological Medicine* 22, 191–6.

Kennedy, P. and Kreitman, N. (1973). An epidemiological survey of parasuicide ('attempted suicide') in general practice. *British Journal of Psychiatry* 123, 23–34.

Kennerley, H. and Gath, D. (1989). Maternity blues:

associations with obstetric, psychological and psychiatric factors. *British Journal of Psychiatry* 155, 367–73.

Kenyon, F. E. (1968). Studies in female homosexuality: social and psychiatric aspects: sexual development, attitudes and experience. *British Journal of Psychiatry* 114, 1337–50.

Kenyon, F. E. (1980). Homosexuality in gynaecological practice. *Clinics in Obstetrics and Gynaecology* 1, 363–86.

Kernberg, O. F. (1975). *Borderline conditions and pathological narcissism.* Aronson, New York.

Kernberg, O. F. (1993). *Severe personality disorders: psychotherapeutic strategies.* Yale University Press, New Haven, CT.

Kerr, A. M. and Stevenson, J. B. P. (1985). Rett's syndrome in the West of Scotland. *British Medical Journal* 291, 579–82.

Kerr, T. A., Roth, M., Shapira, K., and Gurney, C. (1972). The assessment and prediction of outcome in affective disorders. *British Journal of Psychiatry* 121, 167–74.

Kerr, T. A., Roth, M., and Shapira, K. (1974). Prediction of outcome in anxiety states and depressive illness. *British Journal of Psychiatry* 124, 125–31.

Kershner, P. and Wang-Cheng, R. (1989). Psychiatric side effects of steroid therapy. *Psychosomatics* 30, 135–9.

Kerwin, R. W., Patel, S., Meldrum, B. S., et al. (1988). Asymmetrical loss of glutamate receptor subtype in left hippocampus in schizophrenia. *Lancet* 334, 583–4.

Kessel, N. and Grossman, G. (1965). Suicide in alcoholics. *British Medical Journal* ii, 1671–2.

Kessler, R. C., McGonagle, K. A., Zhao, S., et al. (1994). Lifetime and 12-month prevalence of DSM-III-R psychiatric disorders in the United States: results from the National Comorbidity Survey. *Archives of General Psychiatry* 51, 8–20.

Kety, S. (1983). Mental illness in the biological and adoptive relatives of schizophrenic adoptees: findings relevant to genetic and environmental factors in etiology. *American Journal of Psychiatry* 140, 720–7.

Kety, S., Rosenthal, D., Wender, P. H., et al. (1975). Mental illness in the biological and adoptive families of adopted individuals who have become schizophrenic. In *Genetic research in psychiatry* (ed. R. R. Fieve, D. Rosenthal, and H. Bull). Johns Hopkins University Press, Baltimore, MD.

Kety, S. S., Wender, P. A., Jacobsen, B., et al. (1994). Mental illness in the biological and adoptive relatives of schizophrenic adoptees. *Archives of General Psychiatry* 51, 442–55.

Khoo, C. T. K. (1982). Cosmetic surgery—where does it begin? *British Journal of Plastic Surgery* 35, 277–80.

Kidson, M. A. (1973). Personality and hypertension. *Journal of Psychosomatic Research* 17, 35–41.

Kiev, A. (1972). *Transcultural psychiatry.* Penguin, Harmondsworth.

Kilmann, P. R. (1982). The treatment of sexual paraphilias: a review of outcome research. *Journal of Sex Research* 18, 193–252.

Kiloh, L. G. and Garside, R. F. (1963). The independence of neurotic depression and endogenous depression. *British Journal of Psychiatry* 109, 451–63.

Kiloh, L. G., Ball, J. R. B., and Garside, R. F. (1962). Prognostic factors in treatment of depressive states with imipramine. *British Medical Journal* i, 1225–7.

Kiloh, L. G., Andrews, G., Nielson, M., and Bianchi, G. N. (1972). The relationship between the syndromes called endogenous and neurotic depression. *British Journal of Psychiatry* 121, 183–96.

Kiloh, L. G., Andrews, G., and Neilson, M. (1988a). The long-term outcome of depressive illness. *British Journal of Psychiatry* 153, 752–7.

Kiloh, L. G., Smith, J. S., and Johnson, G. F. (1988b). *Physical treatments in psychiatry*. Blackwell Scientific Publications, Oxford.

Kilpatrick, D. (1987). Victim and crime factors associated with PTSD. *Behaviour Therapy* 20, 199–214.

Kilpatrick, D. G. and Resnick, H. S. (1993). Posttraumatic stress disorder associated with exposure to criminal victimization in clinical and community populations. In *Posttraumatic stress disorder. DSM-IV and beyond* (ed. J. R. T. Davidson and E. B. Foa), pp. 113–46. American Psychiatric Press, Washington, DC.

King, B. H., DeAntonio, C., McCracken, J. T., Forness, S. R., and Ackerland, V. (1994). Psychiatric consultation in severe and profound mental retardation. *American Journal of Psychiatry* 151, 1802–8.

King, M. and McDonald, E. (1992). Homosexuals who are twins: a study of 46 probands. *British Journal of Psychiatry* 160, 407–9.

King, M. B. (1993). *Aids, HIV and mental health*. Cambridge University Press.

Kingdon, D., Turkington, D., and John, C. (1994). Cognitive behaviour therapy of schizophrenia. *British Journal of Psychiatry* 164, 581–7.

Kingman, R. and Jones, D. P. H. (1987). Incest and other forms of sexual abuse. In *The battered child* (ed. R. E. Helfer and R. S. Kempe) (4th edn). University of Chicago Press, London.

Kinsey, A. C., Pomeroy, W. B., and Martin, C. E. (1948). *Sexual behavior in the human male*. Saunders, Philadelphia, PA.

Kinsey, A. C., Pomeroy, W. B., Martin, C. E., and Gebhard, P. H. (1953). *Sexual behaviour in the human female*. Saunders, Philadelphia, PA.

Kirby, R. S. (1994). Impotence: diagnosis and management of male erectile dysfunction. *British Medical Journal* 308, 957–61.

Kirby, R. S., Carson, C., and Webster, G. D. (1991). *Impotence: diagnosis and management*. Butterwroth Heinemann, Oxford.

Klaf, F. S. and Hamilton, J. G. (1961). Schizophrenia — a hundred years ago and today. *Journal of Mental Science* 107, 819–28.

Kleber, H. D., Topazian, N., Gaspari, J., *et al.* (1987). Clonidine and naltrexone in the outpatient treatment of heroin withdrawal. *American Journal of Drug and Alcohol Abuse* 13, 1–17.

Klein, D. F. (1964). Delineation of two drug-responsive anxiety syndromes. *Psychopharmacologia* 5, 397–408.

Klein, D. F. (1993). False suffocation alarms, spontaneous panics, and related conditions: an integrative hypothesis. *Archives of General Psychiatry* 50, 306.

Klein, M. (1952). Notes on some schizoid mechanisms. In *Developments in psychoanalysis* (ed. J. Jacobs and J. Riviere). Hogarth Press, London.

Klein, M. (1934). A contribution to the psychogenesis of manic-depressive states. Reprinted in *Contributions to psychoanalysis 1921–1945: developments in child and adolescent psychology*, pp. 282–310. Hogarth Press, London (1948).

Klein, M. (1963). *The psychoanalysis of children* (translated by A. Strachey). Hogarth Press and Institute of Psycho-analysis, London.

Klein, R. G. and Mannuzza, S. (1988). Hyperactive boys almost grown up: III. Methylphenidate effects on ultimate height. *Archives of General Psychiatry* 45, 1131–4.

Klein, R. H. and Nimorwicz, P. (1982). Psychosocial aspects of hemophilia in families: assessment strategies and instruments. *Clinical Psychology Review* 2, 153–69.

Kleinknecht, R. A., Klepac, R. K., and Alexander, L. D. (1973). Origin and characteristics of fear of dentistry. *Journal of the American Dental Association* 86, 842–8.

Kleinman, A. (1982). Neurasthenia and depression: a study of somatization and culture in China. *Culture Medicine and Psychiatry* 6, 117–96.

Kleinman, A. (1986). *Social origins of distress and disease: depression neurosthenia and pain in modern China*. Yale University Press, New Haven, CT.

Kleinman, A., Wang, W. Z., Li, S. C., Chang, X. M., Dai, X. Y., Li, K. T., and Kleinman J. (1995). The social cause of epilepsy: chronic illness as social experience in interior China. *Social Science and Medicine* 40, 1319–30.

Kleist, K. (1928). Cycloid paranoid and epileptoid psychoses and the problem of the degenerative psychosis. Reprinted in *Themes and variations in European psychiatry* (ed. S. R. Hirsch and M. Shepherd). Wright, Bristol (1974).

Kleist, K. (1930). Alogical thought disorder: an organic manifestation of the schizophrenic psychological deficit. In *The clinical roots of the schizophrenic concept* (ed. J. Cutting and M. Shepherd). Cambridge University Press (1987).

Klerman, G. L. and Weissman, M. M. (1989). Increasing

rates of depression. *Journal of the American Medical Association* **261**, 2229–35.

Klerman, G. L., Weissman, M. M., Rounsaville, B. J., and Chevron, E. S. (1984). *Interpersonal psychotherapy of depression*. Basic Books, New York.

Klerman, G. L., Budman, S., Berwick, D., *et al.* (1987). Efficacy of a brief psychosocial intervention for symptoms of stress and distress among patients in primary care. *Medical Care* **25**, 1078–88.

Klimes, I., Mayou, R. A., and Pearce, J. (1989). An evaluation of a psychological treatment of persistent atypical chest pain. *Quarterly Journal of Medicine* **73**, 970–1.

Kline, M. (1993). Using field trials to evaluate proposed changes in DSM diagnostic criteria. *Hospital and Community Psychiatry* **44**, 621–3.

Kluft, R. P. (1985). *Childhood antecedents of multiple personality disorder*. American Psychiatric Press, Washington, DC.

Knight, G. (1972). Neurosurgical aspects of psychosurgery. *Proceedings of the Royal Society of Medicine* **65**, 1099–104.

Knights, A. and Hirsch, S. R. (1981). Revealed depression and drug treatment for schizophrenia. *Archives of General Psychiatry* **38**, 806–11.

Knott, D. G. and Beard, J. D. (1971). In *Treatment of the alcohol withdrawal syndrome* (ed. F. A. Seixas), p. 29. National Council on Alcoholism, New York.

Koch, J. L. A. (1891). *Die Psychopathischen Minderwertigkeiter*. Dorn, Ravensburg.

Kocsis, J. H., Croughan, J. L., Katz, M. N., *et al.* (1990). Response to treatment with antidepressants of patients with severe or moderate non-psychotic depression and of patients with psychotic depression. *American Journal of Psychiatry* **147**, 621–4.

Koegel, R., Schreibman, L., O'Neil, R. E., and Burke, J. C. (1983). The personality and family interaction characteristics of parents with autistic children. *Journal of Consulting and Clinical Psychology* **51**, 683–92.

Koenig, H. G. and Blazer, D. G. (1992). Epidemiology of geriatric affective disorders. *Clinics in Geriatric Medicine* **8**, 235–51.

Koenig, H. G. *et al.* (1988). Depression in the elderly hospitalized patient with medical illness. *Archives of Internal Medicine* **148**, 1929–36.

Kolle, K. (1931). *Die primare Verrucktheit: psychopathologische, klinische und genealogische Untersuchungen*. Thieme, Leipzig.

Kolodny, R. C., Masters, W. H., and Johnson, V. E. (1979). *Textbook of sexual medicine*. Little Brown, Boston, MA.

Kolvin, I. and Fundudis, T. (1981). Elective mute children: psychological development and background factors. *Journal of Child Psychology and Psychiatry* **22**, 219–32.

Koo, J. Y. M. and Pham, C. T. (1992). Psychodermatology. *Archives of Dermatology* **128**, 381–8.

Koob, G. F. (1992). Drugs of abuse: anatomy, pharmacology and function of reward pathways. *Trends in Pharmacological Sciences* **13**, 177–93.

Kopelman, M. D. (1986). Clinical tests of memory. *British Journal of Psychiatry* **148**, 517–625.

Kopelman, M. D. (1987). Amnesia: organic and psychogenic. *British Journal of Psychiatry* **150**, 428–42.

Kopelman, M. D. (1995). The Korsakoff syndrome. *British Journal of Psychiatry* **166**, 154–73.

Kopelwicz, H. S. and Klass, E. (1993). *Depression in children and adolescents*. Harwood Academic, Chur, Switzerland.

Koranyi, R. K. and Potoczny, W. M. (1992). Physical illnesses underlying psychiatric symptoms. *Psychotherapy and Psychosomatics* **58**, 155–60.

Kornhuber, J., Riederer, P., Reynolds, G. P., *et al.* (1989). [3H]-Spiperone binding sites in post-mortem brains from schizophenic patients: relationship to neuroleptic drug treatment, abnormal movements, and positive symptoms. *Journal of Neural Transmission* **75**, 1–10.

Korsakov, S. S. (1889). Translated and reprinted as 'Psychic disorder in conjunction with multiple neuritis'. *Neurology* **5**, 394–406.

Koslow, S. H., Maas, J. W., Bowden, C. L., *et al.* (1983). CSF and urinary biogenic amines and metabolites in depression and mania. *Archives of General Psychiatry* **40**, 999–1010.

Kotin, J. and Goodwin, F. K. (1972). Depression during mania. Clinical observations and theoretical implications. *American Journal of Psychiatry* **129**, 679–86.

Kraepelin, E. (1897). Dementia praecox. In *Clinical roots of the schizophrenia concept* (ed. J. Cutting and M. Shepherd). Cambridge University Press (1981).

Kraepelin, E. (1904). *Clinical psychiatry: a textbook for students and physicians* (edited and translated from 7th edition of Kraepelin's *Textbook* by A. R. Diefendof). Macmillan, New York.

Kraepelin, E. (1912). Über paranoide Erkrankungen. *Zentralblatt für die gesamte Neurologie und Psychiatrie* **11**, 617–38.

Kraepelin, E. (1915). Der Verfolgungswahn der Schwerhörigen. *Psychiatrie*, Vol. 8, Part 4. Barth, Leipzig.

Kraepelin, E. (1919). *Dementia praecox and paraphrenia*. Livingstone, Edinburgh.

Kraepelin, E. (1921). Manic depressive insanity and paranoia (translated by R. M. Barclay from the 8th edition of *Lehrbuch der Psychiatrie*, Vols III and IV). Livingstone, Edinburgh.

Krafft-Ebing, R. (1888). *Lehrbuch der Psychiatrie*. Enke, Stuttgart.

Krafft-Ebing, R. (1924). *Psychopathic sexuality with special reference to contrary sexual instinct*. Authorized translation of the 7th German edition by C. G. Chaddock. F. A. Davis, Philadelphia, PA.

Kramlinger, K. G. and Post, R. M. (1989*a*). Adding lithium carbonate to carbamazepine: anti-manic efficacy in treatment resistant mania. *Acta Psychiatrica Scandinavica* 79, 378–85.

Kramlinger, K. G. and Post, R. M. (1989*b*). The addition of lithium to carbamazepine. *Archives of General Psychiatry* 46, 794–800.

Kreitman, N. (ed.) (1977). *Parasuicide*. Wiley, London.

Kreitman, N. (1989). Can suicide and parasuicide be prevented? *Journal of the Royal Society of Medicine* 82, 648–52.

Kreitman, N. (1993). Suicide and parasuicide. In *Companion to psychiatric studies* (ed. E. Kendell and A. K. Zealley), pp. 743–60. Churchill Livingstone, Edinburgh.

Kreitman, N. and Dyer, J. A. T. (1980). Suicide in relation to parasuicide. *Medicine* 2nd series, 1826–30.

Kreitman, N. and Foster, J. (1991). The construction and selection of predictive scales, with special reference to parasuicide. *British Journal of Psychiatry* 159, 185–92.

Kreitman, N., Collins, J., Nelson, B., and Troop, J. (1970). Neurosis and marital interaction. *British Journal of Psychiatry* 117, 33–46, 47–58.

Kretschmer, E. (1921). *Physique and character* (transl. from German). Harcourt Brace, New York.

Kretschmer, E. (1927). Der sensitive Beziehungswahn. Reprinted and translated as Chapter 8 in *Themes and variations in European psychiatry* (ed. S. R. Hirsch and M. Shepherd). Wright, Bristol (1974).

Kretschmer, E. (1936). *Physique and character* (2nd edn) (transl. W. J. H. Sprott and K. P. Trench). Trubner, New York.

Kretschmer, E. (1961). *Hysteria, reflex and instinct* (transl. V. Baskin and W. Baskin). Peter Owen, London.

Kringlen, E. (1965). Obsessional neurosis: a long term follow up. *British Journal of Psychiatry* 111, 709–22.

Kringlen, E. (1967). *Heredity and environment in the functional psychoses*. Heinemann, London.

Kripke, D. F., Risch, S. C., and Janowsky, D. S. (1983). Bright light alleviates depression. *Psychiatry Research* 10, 105–12.

Kroenke, K. and Mangelsdorff, D. (1989). Common symptoms in ambulatory care: incidence, evaluation, therapy and outcome. *American Journal of Medicine* 86, 262–6.

Kroenke, K. and Price, R. K. (1993). Symptoms in the community. Prevalence, classification, and psychiatric comorbidity. *Archives of Internal Medicine* 153, 2474–80.

Krupp, P. and Barnes, P. (1992). Clozapine-associated agranulocytosis: risk in aetiology. *British Journal of Psychiatry* 160 (Suppl. 17), 38–40.

Krystal, J. H., Karper, L. P., Seibyl, J. P., *et al.* (1994). Subanesthetic effects of the noncompetitive NMDA antagonist, ketamine, in humans. *Archives of General Psychiatry* 51, 199–214.

Kubler-Ross, E. (1969). *On death and dying*. Macmillan, New York.

Kuczmaski, R. J. (1994). Increasing prevalence of overweight among US adults. *Journal of the American Medical Association* 272(3), 205–11.

Kuhn, R. (1957). Uber die Behandlung depressiver Zustände mit einem Iminodibenzylderivat. *Schweizerische medizinische Wochenschrift* 36, 1135–40.

Kuller, J. A. and Laifer, S. A. (1994). Preconceptional counseling and intervention. *Archives of Internal Medicine* 154, 2273–9.

Kumar, R., Marks, M., Platz, C., and Yoshida, K. (1995). Clinical survey of a psychiatric mother and baby unit: characteristics of 100 consecutive admissions. *Journal of Affective Disorders* 33, 11–22.

Kupfer, D. J., Buysse, D. J., Nofzinger, E. A., and Reynolds, C. F. (1994). Sleep disorders. In *DSM-IV sourcebook*, Vol. 1 (ed. T. A. Widiger, A. J. Frances, H. A. Pincus, *et al.*), pp. 597–606. American Psychiatric Association, Washington, DC.

Lachs, M. S. and Pillemer, K. (1995). Abuse and neglect of elderly persons. *New England Journal of Medicine* 332, 437–43.

Lacks, P. and Morin, C. M. (1992). Recent advances in the assessment and treatment of insomnia. *Journal of Consulting and Clinical Psychology* 60(4), 586–94.

Lader, M. H. (1969). Psychophysiological aspects of anxiety. In *Studies of anxiety* (ed. M. H. Lader). *British Journal of Psychiatry* Special Publication, No. 3.

Lader, M. (1991). Can buspirone induce rebound dependent or abuse? *British Journal of Psychiatry* 159 (Suppl. 12), 45–51.

Lader, M. (1994). Anxiolytic drugs: dependence, addiction and abuse. *European Neuropsychopharmacology* 4, 85–91.

Lader, M. and Herrington, R. (1990). *Biological treatments in psychiatry*. Oxford University Press.

Lader, M. and Morton, S. (1991). Benzodiazepine problems. *British Journal of Addiction* 86, 823–8.

Lader, M. H. and Sartorius, N. (1968). Anxiety in patients with hysterical conversion symptoms. *Journal of Neurology, Neurosurgery and Psychiatry* 31, 490–7.

Lader, M. H. and Wing, L. (1966). *Physiological measures, sedative drugs and morbid anxiety*. Maudsley Monograph No. 14. Oxford University Press, London.

Lader, M. H., Ron, M., and Petursson, H. (1984). Computed axial brain tomography in long-term benzodiazepine users. *Psychological Medicine* 14, 203–6.

Laidlaw, J., Richens, A., and Chadwick, D. (1993). *A textbook of epilepsy* (4th edn). Churchill Livingstone, Edinburgh.

Lam, D. (1991). Psychological family intervention in

schizophrenia: a review of empirical studies. *Psychological Medicine* 21, 423–41.

Lamb, H. R. (1987). Incompetency to stand trial: appropriateness and outcome. *Archives of General Psychiatry* 44, 754.

Lamb, H. R. (1989). *The homeless mentally ill: a task force of the American Psychiatric Association.* American Psychiatric Association, Washington, DC.

Lambourn, J. and Gill, D. (1978). A controlled comparison of simulated and real ECT. *British Journal of Psychiatry* 133, 514–19.

Lancaster, S. G. and Gonzalez, J. P. (1990). Lofepramine: a review of its pharmacodynamic and pharmacokinetic properties and therapeutic efficacy in depressive illness. *Drugs* 37, 123–40.

Lancet (1981). Epilepsy and violence (Editorial). *Lancet* ii, 966–7.

Lancet (1982). Trials of coronary heart disease prevention (Editorial). *Lancet* ii, 803–4.

Lancet (1987). Non-convulsive status epilepticus (Editorial). *Lancet* i, 958–9.

Lancet (1992a). Psychogenic vomiting—a disorder of gastrointestinal motility? (Editorial). *Lancet* 339, 279.

Lancet (1992b). How can one assess damage caused by treatment of childhood cancer? (Editorial). 340, 758–9.

Lancet (1994). Guilty innocents: the road to false confessions. *Lancet* 344, 1447–50.

Landesmann-Dyer, S. (1981). Living in the community. *American Journal of Mental Deficiency* 86, 223–34.

Lane, R. D. (1990). Successful fluoxetine treatment of pathologic jealousy. *Journal of Clinical Psychology* 51, 345–6.

Lange, J. (1931). *Crime as destiny* (transl. C. Haldane). George Allen, London.

Langer, K. G. and Padrone, F. J. (1992). Psychotherapeutic treatment of awareness in acute rehabilitation of traumatic brain injury. *Neuropsychological Rehabilitation* 2, 59–70.

Langfeldt, G. (1961). The erotic jealousy syndrome. A clinical study. *Acta Psychiatrica Scandinavica* Suppl. 151.

Lapensée, M. A. (1992). A review of schizoaffective disorder. (1) Current concepts. *Canadian Journal of Psychiatry* 37, 335–46.

Lasègue, C. (1877). Les exhibitionnistes. *Union Medicale* 23, 709–14.

Lawson, J. S., Inglis, J., Delva, N. J., et al. (1990). Electroplacement in ECT: cognitive effects. *Psychological Medicine* 20, 335–44.

Layden, M. A., Newman, C. F., Freeman, A., and Morse, S. B. (1993). *Cognitive therapy of borderline personality disorder.* Allyn & Bacon, Boston, MA.

Lazarus, R. S. (1966). *Psychological stress and the coping processes.* McGraw-Hill, New York.

Lazarus, R. S. (1993). Coping theory and research: past, present and future. *Psychosomatic Medicine* 55, 234–47.

Ledermann, S. (1956). *Alcool, alcoolisme, alcoolisation.* Presses Universitaires de Paris, Paris.

Lee, A. S. and Murray, R. M. (1988). The long-term outcome of Maudsley depressives. *British Journal of Psychiatry* 153, 741–51.

Lee, L. M., Stevenson, R. W., and Szasz, G. (1988). Prostaglandin E_1 versus phentolamine/papaverine for the treatment of erectile impotence: a double-blind comparison. *Journal of Urology* 141, 54–7.

Lee, S. (1993). How abnormal is the desire for slimness? A survey of eating attitudes and behaviour among Chinese undergraduates in Hong Kong. *Psychological Medicine* 23, 437–51.

Lee, S. (1995). Neurasthenia and Chinese psychiatry in the 1990's. *Journal of Psychosomatic Research* (in press).

Leff, J. (1978). Social and psychological causes of the acute attack. In *Schizophrenia: towards a new synthesis* (ed. J. K. Wing). Academic Press, London.

Leff, J. (1981). *Psychiatry around the globe: a transcultural view.* Dekker, New York.

Leff, J. (1993a). All the homeless people—where do they all come from? *British Medical Journal* 306, 669–70.

Leff, J. (1993b). The Taps Project: evaluating community placement of long-stay psychiatric patients. *British Journal of Psychiatry* 162 (Suppl. 19), 1–56.

Leff, J. (1994). Working with the families of schizophrenic patients. *British Journal of Psychiatry* 164 (Suppl. 23), 71–6.

Leff, J. and Isaacs, A. D. (1978). *Psychiatric examination in clinical practice.* Blackwell, Oxford.

Leff, J. and Vaughn, C. (1981). The role of maintenance therapy and relative expressed emotion in relapse of schizophrenia: a two year follow up. *British Journal of Psychiatry* 139, 102–4.

Leff, J. and Wing, J. K. (1971). Trial of maintenance therapy in schizophrenia. *British Medical Journal* iii, 599–604.

Leff, J., Kuipers, L., Berkowitz, R., et al. (1982). A controlled trial of social intervention in the families of schizophrenic patients. *British Journal of Psychiatry* 141, 121–34.

Leff, J. P., Kuipers, L., Berkowitz, R., and Sturgeon, D. (1985). A controlled trial of intervention in the families of schizophrenic patients: two year follow-up. *British Journal of Psychiatry* 146, 594–600.

Lehrer, P. M., Sargunaraj, D., and Hochron, S. (1992). Psychological approaches to the treatment of asthma. *Journal of Consulting and Clinical Psychology* 60, 639–43.

Lehrke, R. (1972). A theory of X-linkage of major intellectual traits. *American Journal of Mental Deficiency* 76, 611–19.

Leibenluft, E. and Wehr, T. A. (1992). Is sleep deprivation

useful in the treatment of depression? *American Journal of Psychiatry* 149, 159–68.

Lelliott, P. T., Noshirvani, H. F., Basoglu, M., *et al.* (1988). Obsessive–compulsive beliefs and treatment outcome. *Psychological Medicine* 18, 697–702.

Lemere, F., Voegtlin, W. L., Broz, W. R., *et al.* (1942). Conditioned reflex treatment of chronic alcoholism. VIII: a review of six years experience with this treatment of 1526 patients. *Journal of the American Medical Association* 120, 269–70.

Lemert, E. (1951). *Social pathology: a systematic approach to the theory of sociopathic behaviour.* McGraw-Hill, New York.

Lemoine, P., Harousseau, H., Borteyru, J.-P., and Menuet, J.-C. (1968). Les enfants de parents alcooliques: anomalies observées à propos de 127 cas. *Ouest Médical* 25, 477–82.

Lenane, M. C., Swedo, S. E., Leonard, H., *et al.* (1990). Psychiatric disorders in first degree relatives of children and adolescents with obsessive compulsive disorder. *Journal of the American Academy of Child and Adolescent Psychiatry* 29, 407–12.

Leonard, H. L., Swedo, S., Lenane, M. C., *et al.* (1993). A 2- to 7-year follow-up study of 54 obsessive–compulsive children and adolescents. *Archives of General Psychiatry* 50, 429.

Leonard, H. L., Swedo, S. E., Lenane, M. C., *et al.* (1991). A double-blind desipramine substitution during long-term clomipramine treatment in children and adolescents with obsessive–compulsive disorder. *Archives of General Psychiatry* 48, 922–7.

Leonhard, K. (1957). *The classification of endogenous psychoses.* English translation of the 8th German edition of *Aufteilung der Endogenen Psychosen* by R. Berman. Irvington, New York (1979).

Leonhard, K., Korff, I., and Schultz, H. (1962). Die Temperamente und den Familien der monopolaren und bipolaren phasishen Psychosen. *Psychiatrie und Neurologie* 143, 416–34.

Lernan, C. and Croyle, R. (1994). Psychological issues in genetic testing for breast cancer susceptibility. *Archives of Internal Medicine* 154, 609–16.

Lesage, A. D., Boyer, R., Grunberg, F., *et al.* (1994). Suicide and mental disorders: a case-controlled study of young men. *American Journal of Psychiatry* 151(7), 1063–8.

Lesko, L. M. (1993). Oncology. In *Psychiatric care of the medical patient* (ed. A. Stoudemire and B. S. Vogel). Oxford University Press, New York.

Lesko, L. M., Ostroff, J. S., Mumma, G. H., *et al.* (1992). Long-term psychological adjustment of acute leukemia survivors: Impact of bone marrow transplantation versus conventional chemotherapy. *Psychosomatic Medicine* 54, 30–47.

Letemendia, F. J., Delva, N. J., Rodenburg, M., *et al.*

(1993). Therapeutic advantage of bifrontal electroplacement in ECT. *Psychological Medicine* 23, 349–60.

LeVay, S. (1991). A difference in hypothalamic structure between heterosexual and homosexual men. *Science* 253, 1034–7.

Levenson, J. and Bernis, C. (1991). The role of psychological factors in cancer onset and progression. *Psychosomatics* 32, 124–32.

Levenson, J. L. and Glocheski, S. (1991). Psychological factors affecting end-stage renal disease. *Psychosomatics* 32, 382–9.

Levenson, J. L., Hamer, R. M., Myers, T., *et al.* (1987). Psychological factors predict symptoms of severe recurrent genital herpes infection. *Journal of Psychosomatic Research* 31, 153–9.

Levin, E. (1991). Carers—problems, strains and services. In *Psychiatry in the elderly* (ed. R. Jacoby and C. Oppenheimer), pp. 301–12. Oxford University Press, Oxford.

Levin, R., Banks, S., and Berg, B. (1988). Psychosocial dimensions of epilepsy: a review of the literature. *Epilepsia* 29, 805–16.

Levy, N. B. (1994). Psychological aspects of renal transplantation. *Psychosomatics* 35(5), 427–33.

Lewin, B., Robertson, I. H., Cay, E. L., *et al.* (1992). Effects of self-help post-myocardial-infarction rehabilitation on psychological adjustment and use of health services. *Lancet* 339, 1036–40.

Lewin, J. and Lewis, S. (1995). Organic and psychosocial risk factors for duodenal cancer. *Journal of Psychosomatic Research* (in press).

Lewis, A. (1974). Psychopathic personality: a most elusive category. *Psychological Medicine* 4133–40.

Lewis, A. J. (1934). Melancholia: a clinical survey of depressive states. *Journal of Mental Science* 80, 277–8.

Lewis, A. J. (1936a). Melancholia: prognostic study and case material. *Journal of Mental Science* 82, 488–558.

Lewis, A. J. (1936b). Problems of obsessional neurosis. *Proceedings of the Royal Society of Medicine* 29, 325–36.

Lewis, A. J. (1938). States of depression: their clinical and aetiological differentiation. *British Medical Journal* ii, 875–8.

Lewis, A. J. (1942). Discussion on differential diagnosis and treatment of post-confusional states. *Proceedings of the Royal Society of Medicine* 35, 607–14.

Lewis, A. J. (1953a). Hysterical dissociation in dementia paralytica. *Monatsschrift für Psychiatrie und Neurologie* 125, 589–604.

Lewis, A. J. (1953b). Health as a social concept. *British Journal of Sociology* 4, 109–24.

Lewis, A. J. (1956). Psychological medicine. In *Price's Textbook of the practice of medicine* (ed. D. Hunter) (9th edn). Oxford University Press, London.

Lewis, A. J. (1957). Obsessional illness. *Acta Neuropsiquià-*

trica Argentina 3, 325–35. Reprinted as Chapter 7 in *Inquiries in psychiatry: clinical and social investigations*. Routledge & Kegan Paul, London.

Lewis, A. J. (1970). Paranoia and paranoid: a historical perspective. *Psychological Medicine* 1, 2–12.

Lewis, A. J. (1976). A note on classification of phobia. *Psychological Medicine* 6, 21–2.

Lewis, C., O'Sullivan, C., and Barraclough, J. (ed.) (1994). *The psychoimmunology of cancer: mind and body in the fight for survival*. Oxford University Press.

Lewis, C. E. and Bucholz, K. K. (1991). Alcoholism, antisocial behaviour and family history. *British Journal of Addiction* 86, 177–94.

Lewis, E. O. (1929). Report on an investigation into the incidence of mental deficiency in six areas. 1925–27. In *Report of the mental deficiency committee*, Part IV. HMSO, London.

Lewis, N. D. S. and Yarnell, P. (1951). *Pathological firesetting*. Monograph No. 82. Nervous and Mental Diseases Publishing Company, New York.

Lewis, R. (1993). The merits of a structured settlement: the plaintiffs' perspective. *Oxford Journal of Legal Studies* 13, 530–47.

Lhermitte, J. (1951). Visual hallucinations of the self. *British Medical Journal* i, 431–4.

Liakos, A. (1967). Familial transvestism. *British Journal of Psychiatry* 113, 49–51.

Liberman, R. P., Mueser, K. T., Wallace, C. J., *et al.* (1986). Training skills in the psychiatrically disabled: learning coping and competence. *Schizophrenia Bulletin* 12, 631–47.

Liddle, P. F. (1987). The symptoms of chronic schizophrenia: a re-examination of the positive and negative dichotomy. *British Journal of Psychiatry* 151, 145–51.

Liddle, P. F. (1994). Neurobiology of schizophrenia. *Current Opinion in Psychiatry* 7, 43–6.

Liddle, P. F. and Morris, D. (1991). Schizophrenic syndromes and frontal lobe performance. *British Journal of Psychiatry* 158, 340–5.

Liddle, P. F., Friston, K. J., Frith, C. D., *et al.* (1992). Patterns of cerebral blood flow in schizophrenia. *British Journal of Psychiatry* 160, 179–86.

Lidz, R. W. and Lidz, T. (1949). The family environment of schizophrenic patients. *American Journal of Psychiatry* 106, 332–45.

Lidz, T., Fleck, S., and Cornelison, A. (1965). *Schizophrenia and the family*. International Universities Press, New York.

Lieberman, J. A. and Alvir, J. M. J. (1992). A report of clozapine-induced agranulocytosis in the United States. Incidence and risk factors. *Drug Safety* 7 (Suppl.), 1–2.

Lieberman, J. A. and Sobel, S. N. (1993). Predictors of treatment response in course of schizophrenia. *Current Opinion in Psychiatry* 6, 63–9.

Lieberman, M. A. (1990). A group therapist perspective on

self-help groups. *International Journal of Group Psychotherapy* 40, 251–77.

Lieberman, M. A. and Yalom, I. (1992). Brief group psychotherapy for the spousally bereaved: a controlled study. *International Journal of Group Psychotherapy* 42, 117–32.

Lieberman, M. A., Yalom, I. D., and Miles, M. B. (1973). *Encounter groups: first facts*. Basic Books, New York.

Liebowitz, M. R., Gorman, J. M., Fyer, A. J., and Klein, D. F. (1985). Social phobia: review of a neglected anxiety disorder. *Archives of General Psychiatry* 42, 729–36.

Liebowitz, M. R., Gorman, J. M., and Fyer, A. J. (1988). Pharmacotherapy of social phobia: an interim report of a placebo controlled comparison of phenelzine and atenolol. *Journal of Clinical Psychiatry* 49, 252–7.

Liebowitz, M. R., Schneier, F. R., and Campeas, R. (1992). Phenelzine versus atenolol in social phobia: a placebo controlled comparison. *Archives of General Psychiatry* 49, 290.

Liem, J. H. (1980). Family studies of schizophrenia: an update and commentary. *Schizophrenia Bulletin* 6, 429–55.

Lindemann, E. (1944). Symptomatology and management of acute grief. *American Journal of Psychiatry* 101, 141–8.

Lindesay, J. (1986). Suicide and attempted suicide in old age. In *Affective disorders in the elderly* (ed. E. Murphy). Churchill Livingstone, Edinburgh.

Lindesay, J. (1991). Anxiety disorders in the elderly. In *Psychiatry in the elderly* (ed. R. Jacoby and C. Oppenheimer), pp. 735–57. Oxford University Press, Oxford.

Lindesay, J. (1995). *Neurotic disorders in the elderly*. Oxford University Press.

Lindqvist, P. and Allebeck, P. (1990). Schizophrenia and crime. *British Medical Journal* 157, 345–50.

Lindsay, M. (1985). Emotional management. In *Care of the child with diabetes* (ed. J. D. Baum and A. L. Kimmonth). Churchill Livingstone, Edinburgh.

Lindstedt, G., Nilsson, L. A., Walinder, J., *et al.* (1977). On the prevalence, diagnosis and management of lithium-induced hypothyroidism in psychiatric patients. *British Journal of Psychiatry* 130, 452–8.

Lineberger, H. P. (1981). Social characteristics of a haemophiliac clinic population. *General Hospital Psychiatry* 3, 157–63.

Linehan, M. M., Tutek, D. A., Heard, H. L., and Armstrong, H. E. (1994). Interpersonal outcome of cognitive behavioral treatment for chronically suicidal borderline patients. *American Journal of Psychiatry* 151(12), 1771–6.

Lingjaerde, O., Edlund, A. H., Gorinsen, C. A., *et al.* (1974). The effects of lithium carbonate in combination with tricyclic antidepressants in endogenous depression. *Acta Psychiatrica Scandinavica* 50, 233–42.

Linn, M. W., Caffey, E. M., Klett, J., *et al.* (1979). Day treatment and psychotropic drugs in the aftercare of schizophrenic patients. *Archives of General Psychiatry* 36, 1055–66.

Linnoila, M. I. and Virkkunen, M. (1992). Aggression, suicidality and serotonin. *Journal of Clinical Psychiatry* 53 (Suppl. 10), 46–51.

Linton, S. J., Hellsing, A.-L., and Anderson, D. (1993). A controlled study of the effects of an early intervention on acute musculoskeletal pain problems. *Pain* 54, 353–9.

Lipowski, Z. J. (1980). Organic mental disorders: introduction and review of syndromes. In *Comprehensive textbook of psychiatry* (ed. H. I. Kaplan, A. M. Freedman, and B. J. Sadock) (3rd edn). Williams and Wilkins, Baltimore, MD.

Lipowski, Z. J. (1988). An in-patient programme for persistent somatizers. *Canadian Journal of Psychiatry* 33, 275–8.

Lipowski, Z. J. (1990). *Delirium: acute confusional states.* Oxford University Press, New York.

Liptzin, B., Levkoff, S. E., Gottlieb, G. L., and Johnson, J. C. (1994). Delirium. In *DSMIV sourcebook*, Vol. 1 (ed. T. A. Widiger, A. J. Frances, H. A. Pincus, *et al.*), pp. 199–212. American Psychiatric Association, Washington, DC.

Lishman, W. A. (1968). Brain damage in relation to psychiatric disability after head injury. *British Journal of Psychiatry* 114, 373–410.

Lishman, W. A. (1978). Research into the dementias. *Psychological Medicine* 8, 353–6.

Lishman, W. A. (1987). *Organic psychiatry* (2nd edn). Blackwell Scientific Publications, Oxford.

Lishman, W. A. (1988). Physiogenesis and psychogenesis in the post-concussional syndrome. *British Journal of Psychiatry* 153, 460–9.

Lithium Mechanisms Study Group (1993). Mechanisms of lithium a ction. *Reviews in Contemporary Pharmacotherapy* 4, 287–317.

Littman, A. B. (1993). Review of psychosomatic aspects of cardiovascular disease. *Psychotherapy and Psychosomatics* 60, 148–67.

Litz, B. T. and Keane, T. M. (1989). Information processing in anxiety disorders: application to the understanding of post-traumatic stress disorder. *Clinical Psychology Review* 9, 243–57.

Ljungberg, L. (1957). Hysteria. *Acta Psychiatrica Scandinavica* Suppl. 12.

Lloyd, C. E., Matthews, K. A., Wing, R. R., and Orchard, T. J. (1992). Psychosocial factors and complications of IDDM. *Diabetes Care* 15, 166.

Lloyd, K. G., Farley, I. J., Deck, J. H. N., and Hornykiewicz, O. (1974). Serotonin and 5-hydroxyindolacetic acid in discrete areas of the brain stem of suicide victims and control patients. *Advances in Biochemical Psychopharmacology* 11, 387–97.

Lock, T., Abou-Saleh, M. T., and Edwards, R. H. T. (1990). Psychiatry and the new magnetic resonance era. *British Journal of Psychiatry* 57 (Suppl. 9), 38–55.

Lockwood, A. H. (1989). Medical problems of musicians. *New England Journal of Medicine* 320, 221–7.

Loeb, J. and Mednick, S. A. (1977). A prospective study of predictors of criminality: electrodermal response patterns. In *Biosocial bases of criminal behaviour* (ed. S. A. Mednick and K. O. Christiansen), pp. 245–54. Gardner, New York.

Loebel, J. A., Lieberman, J. A., Alvir, J. M. J., *et al.* (1992). Duration of psychosis and outcome in first episode schizophrenia. *American Journal of Psychiatry* 149, 1183–8.

Loehlin, J. C., Willerman, L., and Horn, J. M. (1988). Human behaviour genetics. *Annual Review of Psychology* 39, 101–33.

Lolin, Y. (1989). Chronic neurological toxicity associated with exposure to volatile substances. *Human Toxicology* 8, 293–300.

Lord, C. and Rutter, M. (1994). Autism and pervasive developmental disorder. In *Child and adolescent psychiatry: modern approaches* (3rd edn) (ed. M. Rutter, E. Taylor and L. Hersov), pp. 569–93. Blackwell Scientific Publications, Oxford.

Loudon, J. B. (1987). Prescribing in pregnancy: psychotropic drugs. *British Medical Journal* 293, 167–9.

Lowman, R. L. and Richardson, L. M. (1987). Pseudoepileptic seizures of psychogenic origin: a review of the literature. *Clinical Psychology Review* 7, 363–89.

Luborsky, L., Singer, B., and Luborsky, L. (1975). Comparative studies psychotherapies. *Archives of General Psychiatry* 31, 995–1008.

Lucas, A. R., Beard, C. M., O'Fallon, W. M., and Kurland, L. T. (1991). 50 year trends in the incidence of anorexia nervosa in Rochester, Minnesota: a population-based study. *American Journal of Psychiatry* 148, 917–22.

Lue, T. F., Hricak, H., Marich, K. W., and Tanagho, E. A. (1985). Vasogenic impotence evaluated by high resolution ultrasonography and pulsed Doppler spectrum analysis. *Radiology* 155, 777–81.

Lukeman, D. and Melvin, D. (1993). Annotation: the preterm infant: psychological issues in childhood. *Journal of Child Psychology and Psychiatry* 34(6), 837–49.

Lukianowicz, N. (1958). Autoscopic phenomena. *Archives of Neurology and Psychiatry* 80, 199–220.

Lukianowicz, N. (1959). Survey of various aspects of transvestism in the light of our present knowledge. *Journal of Nervous and Mental Disease* 128, 36–64.

Lund, J. (1990). Mentally retarded criminal offenders in Denmark. *British Journal of Psychiatry* 156, 726–31.

Lundberg, S. G. and Guggenheim, F. (1986). Sequelae of

limb amputation. *Advances in Psychosomatic Medicine* 15, 199–210.

Lundquist, G. (1945). Prognosis and course in manic depressive psychosis. A follow-up study of 319 first admissions. *Acta Psychiatrica Scandinavica* Suppl. 35.

Lustman, P. J. *et al.* (1986). Psychiatric illness in diabetes mellitus. Relationship to symptoms and glucose control. *Journal of Nervous and Mental Disorders* 174, 736–42.

Luxenberger, H. (1928). Vorläufiger Bericht über psychiatrische Serienuntersuchungen an Zwillingen. *Zeitschrift für die gesamte Neurologie und Psychiatrie* 116, 297–326.

Luxon, L. M. (1993). Tinnitus: its causes, diagnosis and treatment. *British Medical Journal* 306, 1490–1.

Lydiard, R. B. and Ballenger, J. C. (1987). Antidepressants in panic disorder and agoraphobia. *Journal of Affective Disorders* 13, 153–68.

Lynch, M. and Roberts, J. (1982). *Consequences of child abuse*. Academic Press, London.

MacAlpine, I. and Hunter, R. (1966). The 'insanity' of King George III: a classic case of porphyria. *British Medical Journal* i, 65–71.

Mackay, A. V. P. (1982). Antischizophrenic drugs. In *Drugs in psychiatric practice*. (ed. P. J. Tyrer), pp. 42–81. Butterworths, London.

MacKay, R. I. (1982). The causes of severe mental handicap. *Developmental Medicine and Child Neurology* 24, 386–93.

McCabe, B. J. (1986). Dietary tyramine and other pressoramines in MAOI regimens: a review. *Journal of the American Dietetic Association* 86, 1059–64.

McCann, U. D., Ridenour, A., Shahan, Y., and Ricaurte, G. A. (1994). Serotonin neurotoxicity after 3,4-methylenedioxymethamphetamine (MDMA; ecstasy): a controlled study in humans. *Neuropsychopharmacology* 10, 129–38.

MacCarthy, D. (1981). The effects of emotional disturbance and deprivation and somatic growth. In *Scientific foundations of paediatrics* (ed. J. A. Davis and J. Dobbing), pp. 54–73. Heinemann, London.

McCarthy, G., Blamire, A. M., Rothman, D. L., *et al.* (1993). Echo-planar magnetic resonance imaging studies of frontal cortex activation during word generation in humans. *Proceedings of the National Academy of Sciences of the United States of America* 90, 495–6.

McCarthy, P. D. and Walsh, D. (1975). Suicide in Dublin: 1. The under-reporting of suicide and the consequences for national statistics. *British Journal of Psychiatry* 126, 301–8.

McCarty, L. M. (1986). Mother–child incest: characteristics of the offender. *Child Welfare* 65, 447–59.

McClure, G. M. G. (1984). Suicide in England and Wales 1975–84. *British Journal of Psychiatry* 150, 309–14.

McClure, G. M. G. (1994). Suicide in children and adolescents in England and Wales 1960–1990. *British Journal of Psychiatry* 165, 510–14.

McCrady, B. S., Stout, R., Noel, N., *et al.* (1991). Effects of three types of spouse-involved behavioural alcoholism treatment. *British Journal of Addiction* 86, 1415–24.

McCreadie, R. G. and Farmer, J. G. (1985). Lithium and hair texture. *Acta Psychiatrica Scandinavica* 72, 387–8.

McCreadie, R. G. and Ohaeri, J. U. (1994). Movement disorder in never and minimally treated Nigerian schizophrenic patients. *British Journal of Psychiatry* 164, 184–9.

McCulloch, D. K., Young, R. J., Prescott, R. J., *et al.* (1984). Natural history of impotence in diabetic men. *Diabetologia* 26, 437–40.

MacCulloch, M. J., Snowden, P. R., Wood, P. J. W., and Mills, H. E. (1985). Sadistic fantasy, sadistic behaviour, and offending. *British Journal of Psychiatry* 143, 20–9.

McDaniel, J. S., Musselmann, D. L., Porter, M. R., Reed, D. A., and Nemeroff, C. B. (1995). Depression in patients with cancer. *Archives of General Psychiatry* 52, 89–99

McDougal, W. S. and Nickeleit, V. (1995). A 71-year-old man with masses in the pancreas, presacral region, and left kidney. *New England Journal of Medicine* 332, 174–80.

McDougall, W. (1926). *An outline of abnormal psychology*. Methuen, London.

McDougle, C. J., Goodman, W. C., Price, L. H., *et al.* (1990). Neuroleptic addition in fluvoxamine refractory obsessive compulsive disorder. *American Journal of Psychiatry* 147, 652–4.

McElroy, S. L., Keck, P. E., and Pope, H. G. (1987). Sodium valproate: its use in primary psychiatric disorders. *Journal of Clinical Psychopharmacology* 7, 16–24.

McElroy, S. L., Hudson, J. I., Pope, H. G., and Keck, P. E. (1991). Kleptomania: clinical characteristics and associated psychopathology. *Psychological Medicine* 21, 93–108.

McElroy, S. L., Keck, P. E., Pope, H. G., and Hudson, J. I. (1992). Valproate in the treatment of bipolar disorder: literature review and clinical guideline. *Journal of Clinical Psychopharmacology* 12, 42S–52S.

McEvoy, J. P., Hogarty, G. E., and Steinjard, S. (1991). Optimal dose of neuroleptic in acute schizophrenia. *Archives of General Psychiatry* 48, 739–45.

MacFarlane, A. B. (1985). Medical evidence in the Court of Protection. *Bulletin of the Royal College of Psychiatrists* 9, 26–8.

McFarlane, A. C. (1988). The longitudinal course of posttraumatic morbidity. *Journal of Nervous and Mental Disease* 176, 30–9.

McFarlane, A. C. (1989). The aetiology of post-traumatic

morbidity: predisposing, precipitating and perpetuating factors. *British Journal of Psychiatry* 154, 221–8.

McGauley, G. A. (1989). Quality of life assessment before and after growth hormone treatment in adults with growth hormone deficiency. *Acta Paediatrica Scandinavica* Suppl. 356, 70–2.

McGee, R. (1994). Depression and the development of cancer: a meta-analysis. *Social Science and Medicine* 38(1), 187–92.

McGorry, P. D., Singh, B. S., Connell, S., *et al.* (1992). Diagnostic concordance in functional psychosis revisited: a study of inter-relationships between alternative concepts of psychotic disorder. *Psychological Medicine* 22, 367–78.

McGuffin, P. (1984). Principles and methods in psychiatric genetics. In *The scientific principles of psychopathology* (ed. P. McGuffin, M. F. Shanks, and R. J. Hodgson). Academic Press, London.

McGuffin, P. (1988). Genetics of schizophrenia. In *Schizophrenia. The major issues* (ed. P. Bebbington and P. McGuffin), pp. 107–26. Heinemann, Oxford.

McGuffin, P. and Thapar, A. (1992). The genetics of personality disorder. *British Journal of Psychiatry* 160, 12–23.

McGuffin, P., Farmer, A. E., Gottesman, M., *et al.* (1984). Twin concordance for operationally defined schizophrenia. *Archives of General Psychiatry* 49, 541–5.

McGuffin, P., Farmer, A. E., and Gottesman, I. I. (1987). Is there really a split in schizophrenia? The genetic evidence. *British Journal of Psychiatry* 150, 581–92.

McGuffin, P., Asherson, P., Owen, M., and Farmer, A. (1994). The strength of the genetic effect: is there room for an environmental influence in the aetiology of schizophrenia. *British Journal of Psychiatry* 164, 593–9.

McGuiness, I. (1980). An econometric analysis of total demand for alcoholic beverages in the UK: 1956–1975. *Journal of Industrial Economics* 29, 85–109.

McGuire, P. K., Shah, G. M. S., and Murray, R. M. (1993). Increased blood flow in Broca's area during auditory hallucinations in schizophrenia. *Lancet* 342, 703–6.

McGuire, M., Morgan, R., and Reiner, R. (ed.) (1994). *Oxford handbook of criminology.* Oxford University Press.

McHugh, P. R. and Slavney, P. R. (1986). *The perspectives of psychiatry.* Johns Hopkins University Press, Baltimore, MD.

MacKay, A. V. P. (1982). Antischizophrenic drugs. In *Drugs in psychiatric practice* (ed. P. J. Tyrer), pp. 42–81. Butterworths, London.

McKeigue, P. M. and Karmi, G. (1993). Alcohol consumption and alcohol related problems in Afro-Caribbeans and South Asians in the United Kingdom. *Alcohol and Alcoholism* 28, 1–10.

McKenna, P. J. (1984). Disorders with overvalued ideas. *British Journal of Psychiatry* 145, 579–85.

McKenna, P. J. and Bailey, P. E. (1993). The strange story of clozapine. *British Journal of Psychiatry* 162, 32–7.

McKenna, P., Willison, J. R., Lowe, D., and Neil-Dwyer, G. (1989). Cognitive outcome and quality of life one year following subarachnoid hemorrhage. *Neurosurgery* 24, 361–7.

MacKenzie, T. B. and Popkin, M. K. (1987). Suicide in the medical patient. *International Journal of Psychiatry in Medicine* 17, 3–22.

MacKinnon, D. P., Pentz, M. A., and Stacy, A. W. L. (1993). The alcohol warning label and adolescents: the first year. *American Journal of Public Health* 83, 585–7.

MacMahon, B. and Pugh, T. F. (1965). Suicide in the widowed. *American Journal of Epidemiology* 81, 23–31.

McPherson, H., Herbison, P., and Romans, S. (1993). Life events and relapse in established bipolar affective disorder. *British Journal of Psychiatry* 163, 381–5.

McSweeney, A. J., Grant, I., Medlen, R. K., *et al.* (1982). Life quality of patients with chronic obstructive pulmonary disease. *Archives of Internal Medicine* 142, 473–8.

McWilliam, T. M. (1991). Post-traumatic stress disorder and severe head injury. *British Journal of Psychiatry* 159, 431–3.

Mace, C. J. (1992). Hysterical conversion I: a history. *British Journal of Psychiatry* 161, 369–77.

Mace, N. L. and Rabins, P. V. (1993). *The 36-hour day.* Johns Hopkins University Press, Baltimore, MD.

Machlin, S., Harris, G., Pearlson, G., *et al.* (1991). Elevated medical frontal cerebral blood flow in obsessive compulsive patients: a SPECT study. *American Journal of Psychiatry* 148, 1240–2.

Maden, T., Swinlon, M., and Gonn, J. (1994). Psychiatric disorder in women serving a prison sentence. *British Journal of Psychiatry* 164, 44–54.

Magni, G., Bernasconi, P., and Mauro, P. (1991). Psychiatric diagnoses in ulcerative colitis: a controlled study. *British Journal of Psychiatry* 158, 413–19.

Magni, G. (1993). General considerations on psychiatric illness in relation to musculoskeletal pain. In *Progress in fibromyalgia and myofascial pain* (ed. H. Veroy). Elsevier, Amsterdam.

Mahendra, B. (1981). Where have all the catatonics gone? *Psychological Medicine* 11, 669–71.

Mahl, G. F. (1953). Physiological changes during chronic fear. *Annals of the New York Academy of Science* 56, 240–9.

Maier, W., Lichtermann, D., Minges, J., *et al.* (1993). Continuity and discontinuity of affective disorders and schizophrenia. *Archives of General Psychiatry* 50, 871–83.

Main, C. J. and Benjamin, S. (1995). Psychological treatment and the health care system: The chaotic case of back pain. Is there a need for a paradigm shift? *Treatment of functional somatic symptoms,* (ed. R. Mayou, C. Bass, and M. Sharpe), p.214. Oxford University Press.

Maj, M. (1990). Psychiatric aspects of HIV infection and AIDS. *Psychological Medicine* 20, 547–63.

Maj, M. and Perris, C. (1990). Patterns of course in patients with a cross-sectional diagnosis of schizoaffective disorder. *Journal of Affective Disorders* 20, 71–7.

Makanjuola, R. O. A. (1982). Manic disorder in Nigerians. *British Journal of Psychiatry* 141, 459–63.

Malan, D. (1976). *The frontier of brief psychotherapy.* Plenum Medical, New York.

Maletzky, B. M. (1973). The episode dyscontrol syndrome. *Diseases of the Nervous System* 34, 178–84.

Maletzky, B. M. (1974). 'Assisted' covert sensitization in the treatment of exhibitionism. *Journal of Consulting and Clinical Psychology* 42, 34–40.

Maletzky, B. M. (1976). The diagnosis of pathological intoxication. *Journal of Studies on Alcoholism* 37, 1215–20.

Maletzky, B. M. (1977). 'Booster' sessions in aversion therapy: the permanency of treatment. *Behaviour Therapy* 11, 655–7.

Maletzky, B. M. (1991). *Treating the sexual offender.* Sage, Newbury Park, CA.

Malmberg, A. K. and David, A. S. (1993). Positive and negative symptoms in schizophrenia. *Current Opinions in Psychiatry* 6, 58–62.

Malt, U., Myhrer, T., Bikra, G., and Hoivik, B. (1987). Psychopathology and accidental injuries. *Acta Psychiatrica Scandinavica* 76, 261–71.

Maltzman, I. and Schweiger, A. (1991). Individual and family characteristics of middle-class adolescents hospitalised for alcohol and other drug abuse. *British Journal of Addiction* 86, 1435–7.

Malzberg, B. and Lee, E. S. (1956). *Migration and mental disease: a study of first admission to hospitals for mental disease in New York 1939–41.* Social Science Research Council, New York.

Mander, A. J. and Loudon, J. B. (1988). Rapid recurrence of mania following abrupt discontinuation lithium. *Lancet* ii, 15–17.

Mankanjuola, R. A. O. (1982). Manic disorder in Nigerians. *British Journal of Psychiatry* 141, 459–63.

Mann, A. H. (1977). The psychological effect of a screening programme and clinical trial for hypertension upon the participants. *Psychological Medicine* 7, 431–8.

Mann, A. H. (1986), Psychological aspects of essential hypertension. *Journal of Psychosomatic Research* 30, 527–41.

Mann, A. (1991). Epidemiology. In *Psychiatry in the elderly* (ed. R. Jacoby and C. Oppenheimer), pp.89–112. Oxford University Press.

Mann, A. (1992). Psychiatric symptoms and low blood pressure: more evidence for an association. *British Medical Journal* 304, 64–5.

Mann, A. H., Jenkins, R., Cutting, J. C., and Cowen, P. J. (1981). The development and use of a standardized assessment of abnormal personality. *Psychological Medicine* 11, 839–47.

Mann, D. M. A., South, P. W., Snowden, J. S., and Neary, D. (1993). Dementia of frontal lobe type: neuropathology and immunohistochemistry. *Journal of Neurology, Neurosurgery and Psychiatry* 56, 605–14.

Mann, J. (1973). *Time limited psychotherapy.* Commonwealth Fund/Harvard University Press, Cambridge, MA.

Mann, K., Batra, A., Gunther, A., and Schroth, G. (1992). Do women develop alcoholic brain damage more readily than men? *Alcohol: clinical and experimental research* 16, 1052–6.

Mann, R. E. and Smart, R. G. (1990). Alcohol problems, preventions and epidemiology: looking for the next questions. *British Journal of Addiction* 85, 1385–7.

Mannuzza, S., Klein, R. G., Bessler, A., *et al.* (1993). Adult outcome of hyperactive boys. *Archives of General Psychiatry* 50, 565–76.

Manschreck, T. C., Maher, B. A., Rucklos, M. E., and Vereen, D. R. (1982). Disturbed voluntary motor activity in schizophrenic disorder. *Psychological Medicine* 12, 73–84.

Mapother, E. (1926). Manic depressive psychosis. *British Medical Journal* ii, 872–9.

Marcé, L. V. (1858). *Traité de la folie des femmes enceintes, des nouvelles accouchés et des nourrices.* Baillière, Paris.

Marcovitz, P. J. and Schulz, S. C. (1993). Drug treatment of personality disorder (letter). *British Journal of Psychiatry* 162, 122.

Marcus, M. D. (1995). Binge eating in obesity. In *Binge eating: nature, assessment and treatment* (ed. C. G. Fairburn and G. T. Wilson). Guilford Press, New York.

Marcus, J., Hans, S. L., Auerbach, J. G., and Auerbach, A. G. (1993). Children at risk for schizophrenia: the Jerusalem infant developmental study. *Archives of General Psychiatry* 50, 797–809.

Marder, S. R. and Meibach, R. C. (1994) Risperidone in the treatment of schizophrenia. *American Journal of Psychiatry* 151, 825–35.

Marks, I. M. (1969). *Fears and phobias.* Heinemann, London.

Marks, I. M. and Gelder, M. G. (1966). Different ages of onset of varieties of phobia. *American Journal of Psychiatry* 123, 218–21.

Marks, I. M., Rachman, S., and Gelder, M. G. (1965). Method for the assessment of aversion therapy in fetishism with narcissism. *Behaviour Research and Therapy* 3, 253–8.

Marks, I. M., Swinson, R. P., and Basoglu, M. (1993a). Alprazolam and exposure alone and combined in panic disorder and agoraphobia: a controlled study in London and Toronto. *British Journal of Psychiatry* **162**, 776–87.

Marks, I. M., Swinson, R. P., Basoglu, M., *et al.* (1993*b*). Reply to comment on the London/Toronto study. *British Journal of Psychiatry* **162**, 790–4.

Marks, I. M., Connolly, J., Muijen, M., *et al.* (1994). Home-based versus hospital based care for people with serious mental illness. *British Journal of Psychiatry* **165**, 179–94.

Marks, M. N. and Kumar, R. (1993). Infanticide in England and Wales. *Medical Scientific Law* **33**(3), 329–39.

Marks, V. and Rose, F. C. (1965). *Hypoglycaemia*. Blackwell, Oxford.

Markus, E., Lange, A., and Pettigrew, T. F. (1990). Effectiveness of family therapy: a meta-analysis. *Journal of Family Therapy* **12**, 205–21.

Marmar, C. R. (1991). Brief dynamic psychotherapy of post-traumatic stress disorder. *Psychiatric Annals* **21**, 405–14.

Marmar, C. R., Horowitz, M. J., Weiss, D. S., *et al.* (1988). A controlled trial of brief psychotherapy and mutual help group treatment of conjugal bereavement. *American Journal of Psychiatry* **145**, 203–9.

Marsden, C. D. and Jenner, R. (1980). Pathophysiology of extrapyramidal side-effects of neuroleptic drugs. *Psychological Medicine* **10**, 55–72.

Marshall, E. J., Edwards, G., and Taylor, C. (1994). Mortality in men with drinking problems: a 20-year follow-up. *Addiction* **89**, 1293–8.

Marshall, E. J. and Murray, R. M. (1991). The familial transmission of alcoholism. *British Medical Journal* **303**, 72–3.

Marshall, E. J. and Reed, J. L. (1992). Psychiatric morbidity in homeless women. *British Journal of Psychiatry* **160**, 761–8.

Marshall, M. (1989). Collected and neglected: are Oxford hostels for the homeless filling up with disabled psychiatric patients. *British Medical Journal* **299**, 706–9.

Marshall, M. and Gath, D. (1992). What happens to homeless mentally ill people? Follow-up of residents of Oxford hostels for the homeless. *British Medical Journal* **304**, 79–80.

Marteau, R. M. (1994). Psychology and screening: narrowing the gap between efficacy and effectiveness. *British Journal of Clinical Psychology* **33**, 1–10.

Martin, P. R., Marie, G. V., Nathan, P. R. (1994). Psychophysiological mechanisms of chronic headaches: investigation using pain induction and pain reduction procedures. *Journal of Psychosomatic Research* **36**(2), 137–48.

Martinez-Arevalo, M. J., Calcedo-Ordonez, A., and Varo-Prieto, J. R. (1994). Cannabis consumption as a prognostic factor in schizophrenia. *British Journal of Psychiatry* **164**, 679–81.

Marttunen, M. J., Hillevi, M. S., Henriksson, M. M., and Lonnqvist, J. K. (1991). Mental disorders in adolescent suicide: DSMIII-R axes I and II diagnoses in suicides among 13 to 19 year olds in Finland. *Archives of General Psychiatry* **48**, 834–9.

Marttunen, M. J., Hillevi, M. A., and Lonnqvist, J. K. (1993). Adolescence and suicide: a review of psychological autopsy studies. *European Child and Adolescent Psychiatry* **2**, 10–18.

Marzuk, P. M., Tierney, H., Tardiff, K., *et al.* (1988). Increased risk of suicide in persons with AIDS. *Journal of the American Medical Association* **259**(9), 1333–7.

Marzuk, P. M., Tardiff, K., and Hirsch, C. S. (1992). The epidemiology of murder-suicide. *Journal of the American Medical Association* **267**, 3179–81.

Marzuk, P. M., Tierney, H., Tardiff, K. *et al.* (1988). Increased risk of suicide in persons with AIDS. *Journal of the American Medical Association* **259**(9), 1333–7.

Masters, W. H. and Johnson, V. E. (1970). *Human sexual inadequacy*. Churchill, London.

Masters, W. H. and Johnson, V. E. (1979). *Homosexuality in perspective*. Little Brown, Boston, MA.

Mate-Kole, C., Freschi, M., and Robin, A. (1990). A controlled study of psychological and social change after surgical gender reassignment in selected male transsexuals. *British Journal of Psychiatry* **157**, 261–4.

Matson, J. L. and Sevin, J. A. (1994). Theories of dual diagnosis in mental retardation. *Journal of Consulting and Clinical Psychology* **62**(1), 6–16.

Matson, J. L. and Taras, M. E. (1989). A 20 year review of punishment and alternative methods to treat problem behaviors in developmentally delayed persons. *Research in Developmental Disabilities* **10**, 85–104.

Matthews, A. (1990). Why worry? The cognitive function of anxiety. *Behaviour Research and Therapy* **28**, 455–68.

Matthews, A., Gelder, M. G., and Johnson, D. (1981). *Agoraphobia: nature and treatment*. Tavistock Publications, London.

Mattick, R. P. and Heather, N. (1993). Developments in cognitive and behavioural approaches to substance misuse. *Current Opinion in Psychiatry* **6**, 424–9.

Mattick, R. P. and Peters, L. (1988). Treatment of severe social phobia: effects of guided exposure with and without cognitive restructuring. *Journal of Clinical Psychology* **56**, 251–60.

Maudsley, H. (1879). *The pathology of mind*. Macmillan, London.

Maudsley, H. (1885). *Responsibility in mental disease*. Kegan Paul and Trench, London.

Maughan, B., Gray, G., and Rutter, M. (1985). Reading retardation and antisocial behaviour: a follow-up into employment. *Journal of Child Psychology and Psychiatry* **25**, 741–58.

Mavissakalian, M. and Perel, J. M. (1992). Clinical experiments in maintenance and discontinuation of imipramine therapy in panic disorder with agoraphobia. *Archives of General Psychiatry* **49**, 318.

Mawson, D., Marks, I. M., and Ramm, L. (1981). Guided mourning for morbid grief: a controlled study. *British Journal of Psychiatry* **138**, 185–93.

May, P. R. A. (1968). *Treatment of schizophrenia*. Science House, New York.

Mayer, W. (1921). Über paraphrene psychosen. *Zentralblatt für die gesamte Neurologie und Psychiatrie* **71**, 187–106.

Mayer-Gross, W. (1932). Die Schizophrenie. In *Bumke's Handbuch der Geisteskrankheiten*, Vol. 9. Springer, Berlin.

Mayer-Gross, W. (1935). On depersonalization. *British Journal of Medical Psychology* **15**, 103–26.

Mayou, R. A. (1992). Psychiatric aspects of road traffic accidents. *International Review of Psychiatry* **4**, 45–54.

Mayou, R. A. (1995). Medico-legal aspects of road traffic accidents. *Journal of Psychosomatic Research* **39** (in press).

Mayou, R. (1996). Accident neurosis revisited. *British Journal of Psychiatry* (in press).

Mayou, R. A. and Bryant, B. (1996). The effects of road traffic accidents on travel. *Injury* **25**, 457–60.

Mayou, R. A. and Sharpe, M. (1995). Psychiatric illnesses associated with physical disease. *Baillière's clinical psychiatry* **1**, 2.

Mayou, R. A., Peveler, R., Davies, B., Mann, J., and Fairburn, C. (1991). Psychiatric morbidity in young adults with insulin-dependent diabetes mellitus. *Psychological Medicine* **21**, 639–45.

Mayou, R. A. and Hawton, K. E. (1986). Psychiatric disorder in the general hospital. *British Journal of Psychiatry* **149**, 172–90.

Mayou, R. A., Bryant, B., and Duthie, R. (1993). Psychiatric consequences of road traffic accidents. *British Medical Journal* **307**, 647–51.

Mayou, R. A., Bass, C., and Sharpe, M. (ed.) (1995). Chronic fatigue chronic fatigue syndrome and fibroyalgia. In *Treatment of functional somatic symptoms* (ed. S. Wessely and M. Sharpe). Oxford University Press.

Meadow, R. (1985). Management of Munchausen syndrome by proxy. *Archives of Diseases of Childhood* **60**, 385–93.

Mechanic, D. (1978). *Medical sociology* (2nd edn). Free Press, Glencoe.

Medical Research Council Drug Trials Subcommittee (1981). Continuation therapy with lithium and amitriptyline in unipolar depressive illness: a controlled clinical trial. *Psychological Medicine* **11**, 409–16.

Medina-Mora, M. E. (1992). Epidemiology of substance misuse. *Current Opinion in Psychiatry* **5**, 403–7.

Mednick, S. A. and Schulsinger, F. (1968). In *The transmission of schizophrenia* (ed. D. Rosenthal and S. Kety). Pergamon Press, Oxford.

Meduna, L. (1938). General discussion of cardiazol therapy. *American Journal of Psychiatry* **94**, Suppl. 40.

Meecham, W. C. and Smith, N. (1977). Effects of jet aircraft noise on mental hospital admissions. *British Journal of Audiology* **11**, 81–5.

Megargee, E. I. (1966). Uncontrolled and overcontrolled personality type in extreme antisocial aggression. *Psychological Monographs* **80**, No. 3.

Mehlum, L., Friis, S., Irion, T., *et al.* (1991). Personality disorders 2–5 years after treatment: a prospective follow-up study. *Acta Psychiatrica Scandinavica* **84**, 72–7.

Mellor, C. S. (1982). The present status of first-rank symptoms. *British Journal of Psychiatry* **140**, 423–4.

Meltzer, H. Y. (1992). Dimensions of outcome with clozapine. *British Journal of Psychiatry* **160** (Suppl. 17), 46–53.

Mendelewicz, J. (1976). The age factor in depressive illness: some genetic considerations. *Journal of Gerontology* **31**, 300–3.

Mendels, J. (1965). Electroconvulsive therapy and depression I: the prognostic significance of clinical features. *British Journal of Psychiatry* **111**, 675–81.

Mendelson, G. (1995). "Compensation neurosis" revisited: outcome studies of the effects of litigation. *Journal of Psychosomatic Research* (in press).

Mendelson, M. (1992). Psychodynamics. In *Handbook of affective disorders* (ed. E. S. Paykel), pp. 195–207. Churchill Livingstone, Edinburgh.

Mendelson, W. B. (1991). Neurotransmitters, sleep and affective disorder. In *Serotonin, sleep and mental disorder* (ed. C. Idzikowski and P. J. Cowen), pp. 277–88. Wrightson Biomedical, Petersfield, Hants.

Mendelson, W. B. (1980). *The use and misuse of sleeping pills: a clinical guide*. Plenum Press, New York.

Mendelwicz, J. and Rainer, J. D. (1977). Adoption study supporting genetic transmission of manic depressive illness. *Nature, London* **268**, 327–9.

Mendez, M. F., Grau, R., Doss, R. C., and Taylor, J. L. (1993). Schizophrenia in epilepsy: seizure and psychosis variables. *Neurology* **43**, 1073–7.

Merikangas, K. R., Wicki, W., and Angst, J. (1994). Heterogeneity of depression classification of depressive subtypes by longitudinal course. *British Journal of Psychiatry* **164**, 342–8.

Merskey, H. and Spear, F. G. (1967). *Pain, psychological and psychiatric aspects*. Baillière Tindall and Cassell, London.

Merson, S., Tyrer, P., Onyett, S., *et al.* (1992). Early intervention in psychiatric emergencies: a controlled clinical trial. *Lancet* **339**, 1311–14.

Meyer, J. K. and Reter, D. J. (1979). Sex reassignment: follow up. *Archives of General Psychiatry* **36**, 1010–15.

Meyers, S., Walfish, J. S., Sachar, D. B., *et al.* (1980).

Quality of life after surgery for Crohn's disease: a psychosocial survey. *Gastroenterology* 78, 1–6.

Mezey, G. and King, M. (1987). Male victims of sexual assault. *Medicine, Science and Law* 27, 122–4.

Micale, M. S. (1990). Hysteria and its historiography: the future perspective. *History of Psychiatry* i, 33–124.

Michael, R. P. and Gibbons, J. L. (1963). Interrelationships between the endocrine system and neuropsychiatry. *International Review of Neurobiology* 5, 243–302.

Michels, R. and Marzuk, P. M. (1993). Progress in psychiatry. *New England Journal of Medicine* 329, 628–38.

Miles, P. (1977). Conditions predisposing to suicide: a review. *Journal of Nervous and Mental Diseases* 164, 231–46.

Milhorn, H. T. (1991). Diagnosis and management of phencyclidine intoxication. *American Family Physician* 43, 1293–1302.

Miller, E. (1980). Psychological intervention in the management and rehabilitation of neuropsychological impairments. *Behaviour Research and Therapy* 18, 527–35.

Miller, G. H. and Agnew, N. (1974). The Lederman model of alcohol consumption. *Quarterly Journal of Studies on Alcoholism* 35, 877–98.

Miller, H. (1961). Accident neurosis. *British Medical Journal* i, 919–25, 992–8.

Miller, K. and Klauber, G. T. (1990). Desmopressin acetate in children with severe primary nocturnal enuresis. *Clinical Therapeutics* 12, 357–66.

Miller, R. J., Horn, A. S., and Iversen, L. L. (1974). The action of neuroleptic drugs on dopamine-stimulated adenosine cyclic 3'5' monophosphate production in rat neostriatum and limbic forebrain. *Molecular Pharmacology* 10, 759–66.

Miller, W. R. and Rollnick, S. (1991). *Motivational interviewing: preparing people to change addictive behaviour*. Guilford Press, London.

Milstein, V., Small, J. G., Klapper, M. H., *et al.* (1987). Unilateral versus bilateral ECT in the treatment of mania. *Convulsive Therapy* 3, 1–9.

Mindham, R. H. S., Bagshaw, A., Howland, C., and Shepherd, M. (1973). An evaluation of continuation therapy with tricyclic antidepressants in depressive illness. *Psychological Medicine* 3, 5–17.

Minichiello, W. E. and O'Sullivan, R. C. (1994). Trichotillomania: clinical aspects and treatment strategies. *Harvard Review of Psychiatry* 1(6), 336–44.

Minski, L. and Guttmann, E. (1938). Huntington's chorea: study of thirty four families. *Journal of Mental Science* 84, 21–96.

Minuchin, S. (1974). *Families and family therapy*. Tavistock Publications, London.

Minuchin, S., Rosman, B., and Baker, L. (1978). *Psychosomatic families: anorexia nervosa in context*. Harvard University Press, Cambridge, MA.

Mitchell, W., Falconer, M. A., and Hill, D. (1954). Epilepsy with fetishism relieved by temporal lobectomy. *Lancet* ii, 626–30.

Mitchell-Heggs, N., Kelly, D., and Richardson, A. (1976). Stereotactic limbic leucotomy — a follow-up after 16 months. *British Journal of Psychiatry* 128, 226–41.

Modcrin, M., Rapp, C., and Poertner, J. (1988). The evaluation of case management studies with the chronically mentally ill. *Evaluation and Program Planning* 11, 307–14.

Modigh, K., Wetenberg, P., and Eriksson, E. (1992). Superiority of clomipramine over imipramine in the treatment of panic disorder: a placebo controlled trial. *Journal of Clinical Psychopharmacology* 12, 251–61.

Mohr, J. W., Turner, R. E., and Jerry, M. B. (1964). *Pedophilia and exhibitionism*. Toronto, University Press.

Mohr, J., Turner, E. R., and Jerry, M. (1964). *Pedophilia and exhibitionism*. Toronto University Press, Toronto.

Monahan, J. (1984). The prediction of violent behaviour: toward a second generation of theory and policy. *American Journal of Psychiatry* 141, 263–6.

Monahan, J. S. (1994). Crime and mental disorder: an epidemiological approach. In *Crime and justice: an annual review of research* (ed. N. Morris and M. Tonry), Vol. 3, pp. 145–89.

Monahan, J. and Steadman, H. (1994). Crime and mental disorder: an epidemiological approach. In *Crime and justice: an annual review of research* (ed. N. Morris and M. Tonry) 3, 145–89.

Money, J., Schwartz, M., and Lewis, V. G. (1984). Adult erotosexual status and fetal hormonal masculinization and demasculinization: 46 XX congenital virilizing adrenal hyperplasia and 46 XY androgen-insensitivity syndrome compared. *Psychoneuroendocrinology* 9, 405–14.

Monteiro, W., Marks, I. M., and Ramm, E. (1985). Marital adjustment and treatment outcome in agoraphobia. *British Journal of Psychiatry* 146, 383–90.

Montgomery, S. A. (1992). The advantages of paroxetine in different subgroups of depression. *International Clinical Psychopharmacology* 6 (Suppl. 4), 91–100.

Montgomery, S. A. and Asberg, M. (1979). A new depression scale designed to be sensitive to change. *British Journal of Psychiatry* 134, 382–9.

Mongtomery, S. A. and Montgomery, D. B. (1992). Prophylactic treatment in recurrent unipolar depression. In *Long-term treatment of depression* (ed. S. A. Montgomery and F. Rouillon), pp. 53–79. Wiley, Chichester.

Moore, B. E. and Fine, B. A. (1990). *Psychoanalytic terms and concepts*. American Psychiatric Press and Yale University Press, New Haven and London.

Moorey, S. and Greer, S. (1989). *Psychological therapy for patients with cancer: a new approach*. Heinemann Medical, Oxford.

Moran, M. G. (1993). Connective tissue diseases. In *Psychiatric care of the medical patient* (ed. A. Stoudemire), pp. 739. Oxford University Press.

Moran, M. G. and Stoudemire, A. (1992). Sleep disorders in the medically ill patient. *Journal of Clinical Psychiatry* 56(6), 29–36.

Morel, B. A. (1860), *Traité des malades mentales*. Masson, Paris.

Morgan, H. G. (1979). *Death wishes? The understanding and management of deliberate self-harm*. Wiley, Chichester.

Morgagni, G. B. (1769). *The seats and causes of diseases investigated by anatomy* (transl. B. Alexander). Millar, London.

Morgan, H. G., Burns-Cox, C.J., Pocock, H., and Pottle, S. (1975). Deliberate self-harm: clinical and socio-economic characteristics of 368 patients. *British Journal of Psychiatry* 127, 564–74.

Morin, C. M., Culbert, J. P., and Schwartz, S. N. (1994). Nonpharmacological interventions for insomnia: a meta-analysis of treatment efficacy. *American Journal of Psychiatry* 151(8), 1172–80.

Morris, C. H., Hope, R. A., and Fairburn, C. G. (1989). Eating habits in dementia: a descriptive study. *British Journal of Psychiatry* 154, 801–6.

Morris, C. (1991). Non-ulcer dyspepsia. *Journal of Psychosomatic Research* 35, 129–40.

Morris, J. B. and Beck, A. T. (1974). The efficacy of antidepressant drugs. A review of research (1958–1972). *Archives of General Psychiatry* 30, 667–74.

Morris, R. G. (1991). Cognition and aging. In *Psychiatry in the elderly* (ed. R. Jacoby and C. Oppenheimer), pp. 58–88. Oxford University Press.

Morris, R. G., Morris, L. W., and Britton, P. G. (1988). Factors affecting the emotional wellbeing of the caregivers of dementia sufferers. *British Journal of Psychiatry* 153, 147–56.

Morrison, J. (1989). Childhood sexual histories of women with somatization disorder. *American Journal of Psychiatry* 146, 239–41.

Morselli, E. (1886). Sulla dismorfofobia e sulla tabefobia. *Bolletin Academica Medica* VI, 110–19.

Morstyn, R., Duffy, F. H., and McCarley, R. W. (1983). Altered P_{300} topography in schizophrenia. *Archives of General Psychiatry* 40, 729–34.

Mortimer, P. S. and Dawber, R. P. R. (1984). Hair loss and lithium, commentary. *International Journal of Dermatology* 23, 603–4.

Mosher, I., and Keith, S. (1980). Psychosocial treatment: individual, group, family and community support approaches. *Schizophrenia Bulletin* 6, 10–41.

Moss, A. R. (1987). AIDS and intravenous drug use: the real heterosexual epidemic. *British Medical Journal* 294, 389–90.

Moss, P. D. and McEvedy, C. P. (1966). An epidemic of overbreathing among schoolgirls. *British Medical Journal* ii, 1295–1300.

Moss, S., Goldberg, D., Patel, P., and Wilkin, D. (1993). Physical morbidity in older people with moderate, severe and profound mental handicap, and its relation to psychiatric morbidity. *Social Psychiatry and Psychiatric Epidemiology* 28, 32–9.

Mowat, R. R. (1966). *Morbid jealousy and murder*. Tavistock Publications, London.

Mowrer, O. H. (1950). *Learning theory and personality dynamics*. Ronald Press, New York.

Mrazek, D. and Mrazek, P. (1985). Child maltreatment. In *Child and adolescent psychiatry* (ed. M. Rutter and L. Hersov) (2nd edn). Blackwell, Oxford.

Muir, W. J., St Claire, D. N., Blackwood, D. H. R., *et al.* (1992). Eye-tracking dysfunction in the affective psychoses in schizophrenia. *Psychological Medicine* 22, 573–80.

Mukherjee, S., Sackheim, H. A., and Lee, C. (1988). Unilateral ECT in the treatment of manic episodes. *Convulsive Therapy* 4, 74–80.

Mukherjee, S., Sackheim, H. A., and Schnur, D. B. (1994). Electroconvulsive therapy of acute manic episodes: a review of fifty years experience. *American Journal of Psychiatry* 151, 169–76.

Mullan, M. J. (1993). Molecular pathology of Alzheimer's disease. *International Review of Psychiatry* 5, 351–62.

Mullen, P. E. and Maack, L. H. (1985). Jealousy, pathological jealousy and aggression. In *Aggression and dangerousness* (ed. D. P. Farington and J. Gunn). Wiley, Chichester.

Mullen, P. E. and Martin, J. (1994). Jealousy: a community study. *British Journal of Psychiatry* 164, 35–43.

Mullen, P. E., Linsell, C. R., and Parker, D. (1986). Influence of sleep disruption and calorie restriction on biological markers for depression. *Lancet* ii, 1051–5.

Mullen, P. E., Martin, J. L., Anderson, J. C., *et al.* (1993). Childhood sexual abuse and mental health in adult life. *British Journal of Psychiatry* 163, 721–32.

Muller-Oerlinghausen, B., Ahrens, B., Grof, E., *et al.* (1992). The effect of long-term lithium treatment on the mortality of patients with manic depressive and schizoaffective illness. *Acta Psychiatrica Scandinavica* 86, 218–22.

Mulrow, C. D., *et al.* (1990). Quality of life changes and hearing improvement: a randomized trial. *Annals of Internal Medicine* 113, 188–94.

Mulvey, E. P. (1994). Assessing the evidence of a link between mental illness and violence. *Hospital and Community Psychiatry* 45, 663–8.

Mulvey, E. P., Arthur, M. W., and Reppucci, A. (1993). The prevention and treatment of juvenile delinquency: a review of the research. *Clinical Psychology Review* 13, 133–67.

Munetz, M. R. and Cornes, C. L. (1982). Akathisia, pseudo-akathisia and tardive dyskinesia. *Comprehensive Psychiatry* 23, 345–52.

Munro, A. (1980). Monosymptomatic hypochondriacal psychosis. *British Journal of Hospital Medicine* 24, 34–8.

Munro, A. (1984). Excellent response of pathologic jealousy to pimozide. *Canadian Medical Association Journal* 131, 852–3.

Munroe, R. L. (1955). *Schools of psychoanalytic thought.* Hutchinson Medical, London.

Munthe-Kaas, A. (1980). Rectal administration of diazepam: theoretical basis and clinical experience. In *Antiepileptic therapy: advances in drug monitoring* (ed. S. L. Johannensen *et al.*). Raven Press, New York.

Murphy, E., Smith, R., Lindesay, J., and Slatter, J. (1988). Increased mortality rates in late-life depression. *British Journal of Psychiatry* 152, 347–53.

Murphy, G. E. (1982). Social origins of depression in old age. *British Journal of Psychiatry* 141, 135–42.

Murphy, G. (1994). Services for children and adolescents with severe learning difficulties (mental retardation). In *Child and adolescent psychiatry: modern approaches* (3rd edn) (ed. M. Rutter, E. Taylor, and L. Hersov), pp. 1023–39. Blackwell Scientific Publications, Oxford.

Murphy, G. E. and Guze, S. B. (1960). Setting limits: the management of the manipulative patient. *American Journal of Psychotherapy* 14, 30–47.

Murphy, G. E. and Wetzel, R. D. (1990). The lifetime risk of suicide in alcoholism. *Archives of General Psychiatry* 47, 383–92.

Murphy, G. E., Simons, A. D., Wetzel, R. D., and Lustman, P. J. (1984). Cognitive therapy and pharmacotherapy: singly and together in the treatment of depression. *Archives of General Psychiatry* 41, 33–41.

Murphy, G. E., Wetzel, R. D., Robins, E., and McEvoy, L. (1992). Multiple risk factors predict suicide and alcoholism. *Archives of General Psychiatry* 49, 459–63.

Murphy, H. B. M. (1968). Cultural factors in the genesis of schizophrenia. In *The transmission of schizophrenia* (ed. D. Rosenthal and S. S. Kety). Pergamon Press, Oxford.

Murphy, H. B. M. (1977). Migration, culture and mental health. *Psychological Medicine* 7, 677–84.

Murphy, H. B. M. and Raman, A. C. (1971). The chronicity of schizophrenia in indigenous tropical people. *British Journal of Psychiatry* 118, 489–97.

Murray, J. and Williams, P. (1986). Self-reported illness and general practice consultations in Asian born and British born residents of West London. *Social Psychiatry* 21, 139–45.

Murray, R. M. and Lewis, S. W. (1987). Is schizophrenia a neurodevelopmental disorder? *British Medical Journal* 295, 681–2.

Murray, R. M. and McGuffin, P. (1993). Genetic aspects of psychiatric disorders. In *Companion to psychiatric*

studies (ed. E. Kendell and A. K. Zealley), pp. 227–61. Churchill Livingstone, Edinburgh.

Murray R. M. and Reveleley, A. (1981). The genetic contribution to the neuroses. *British Journal of Hospital Medicine* 25, 185–90.

Murtagh, D. R. R. and Greenwood, K. M. (1995). Identifying effective psychological treatments for insomnia: a meta-analysis. *Journal of Consulting and Clinical Psychology* 63, 79–89.

Mutryn, C. S. (1993). Psychological impact of cesarean section on the family: a literature review. *Social Science and Medicine* 37(10), 1271–81.

Mynors-Wallis, L. M., Gath, D. H., Lloyd-Thomas, A. R., and Tomlinson, D. (1995). Randomised controlled trial comparing problem solving treatment with amitriptyline and placebo for major depression in primary care. *British Medical Journal* 310, 441–5.

Nadelson, C. (1989). Consequences of rape: clinical treatment and aspects. *Psychotherapy and psychosomatics* 51, 187–92.

Nadelson, C. C., Notman, M. T., Zackson, H., and Garnick, J. (1982). A follow-up study of rape victims. *American Journal of Psychiatry* 139, 1266–70.

Nagler, S. H. (1957). Fetishism. *Psychiatric Quarterly* 31, 713–41.

Naguib, M. and Levy, R. (1987). Late paraphrenia—neuropsychological impairment and structural brain abnormalities on computed tomography. *International Journal of Geriatric Psychiatry* 2, 83–90.

Nagy, L. M., Morgan, C. A., Southwick, S. M., and Charney, D. S. (1993). Open prospective trial of fluoxetine for posttraumatic stress disorder. *Journal of Clinical Psychopharmacology* 13, 107–13.

Naranjo, C. A. and Bremner, K. E. (1993). Behavioural correlates of alcohol intoxication. *Addiction* 88, 31–41.

National Institute of Mental Health (1985). Consensus Development Conference statement. Mood disorders: pharmacological prevention of recurrence. *American Journal of Psychiatry* 142, 469–76.

National Institute on Drug Abuse (1991). *National Household Survey on Drug Abuse: Highlights, 1991.* US Government Printing Office, Washington, DC.

Navia, B. and Price, R. (1987). AIDS dementia complex and the presenting or sole manifestation of HIV infection. *Archives of Neurology* 44, 65–9.

Naylor, G. J., Dick, D. A. T., Dick, E. G., *et al.* (1973). Electrolyte membrane cation carrier in depressive illness. *Psychological Medicine* 3, 502–8.

Naylor, G. T., Worrall, E. P., Peet, M., and Dick, P. (1976).

Whole blood adenosine triphosphate in manic depressive illness. *British Journal of Psychiatry* 129, 233–5.

Ndetei, D. M. and Muhangi, J. (1979). The prevalence and clinical presentation of psychiatric illness in a rural setting in Kenya. *British Journal of Psychiatry* 135, 269–72.

Neary, D., Snowden, J. S., and Mann, D. M. A. (1986). Alzheimer's disease: a correlative study. *Journal of Neurology, Neurosurgery and Psychiatry* 49, 229–37.

Neary, D., Snowden, J. S., Northern, B., and Goulding, P. (1988). Dementia of frontal lobe type. *Journal of Neurology, Neurosurgery and Psychiatry* 51, 353–61.

Needleman, H., Gunnoe, C., Leviton, A., et al. (1979). Deficits in psychologic and classroom performances of children with elevated dentine lead levels. *New England Journal of Medicine* 300, 689–95.

Neligan, G. and Prudham, D. (1969). Norms for four standard developmental milestones by sex, social class and place in the family. *Developmental Medicine and Child Neurology* 11, 413–22.

Nemiah, J. C. and Sifneos, P. E. (1970). Psychosomatic illness: a problem of communication. *Psychotherapy and Psychosomatics* 18, 154–60.

Newman, N. J. (1993). Neuro-ophthalmology and psychiatry. *General Hospital Psychiatry* 15, 102–14.

Newson-Smith, J. G. B. and Hirsch, S. R. (1979a). Psychiatric symptoms in self-poisoning patients. *Psychological Medicine* 9, 493–500.

Newson-Smith, J. G. B. and Hirsch, S. R. (1979b). A comparison of social workers and psychiatrists in evaluating suicide. *British Journal of Psychiatry* 134, 335–42.

Nicholson, A. N. and Pasco, E. (1991). Monoaminergic transmission and sleep in man. In *Serotonin, sleep and mental disorder* (ed. C. Idzikowski and P. J. Cowen), pp. 215–26. Wrightston Biomedical, Petersfield, Hants.

Nicholson, W. A. (1967). Collection of unwanted drugs from private homes. *British Medical Journal* iii, 730–1.

Nierenberg, A. A., Feighner, J. P., Rudolph, R., Cole, J. O., and Sullivan, J. S. (1994). Venlafaxine for treatment-resistant depression. *Jorunal of Clinical Psychopharmacology* 41, 419–23.

NIH Consensus Development Panel on Depression in the Elderly (1992). *Journal of the American Medical Association,* 268(8), 1018–24.

NIH Consensus Development Panel on Impotence (1993). Impotence. *Journal of the American Medical Association* 270, 83–9.

Nilsson, A. (1993). The anti-aggressive actions of lithium. *Reviews in Contemporary Pharmacotherapy* 4, 269–85.

NiNuallain, M., O'Hare, A., and Walsh, D. (1987). Incidence of schizophrenia in Ireland. *Psychological Medicine* 17, 943–8. ·

Nirje, B. (1970). Normalisation. *Journal of Mental Subnormality* 31, 62–70.

Nolen, W. A., Putte, J. J., van de Dijken, W. A., et al. (1988). Treatment strategy in depression. 2. MAO inhibitors in depression resistant tricyclic antidepressants: two controlled studies with tranylcypromine versus 1-5-hydroxytryptophan and nomifensine. *Acta Psychiatrica Scandinavica* 78, 676–83.

Nordmeyer, J. P. (1994). An internist's view of patients with factitious disorders and factitious clinical sympatomatology. *Psychotherapy and Psychosomatics* 62, 30–40.

Norman, R. M. G. and Malla, A. K. (1993). Stressful life events and schizophrenia. (1): A review of research. *British Journal of Psychiatry* 162, 161–6.

Norris, F. H. and Kaniasty, K. (1994). Psychological distress following criminal victimization in the general population: cross-sectional, longitudinal, and prospective analyses. *Journal of Consulting and Clinical Psychology* 62(1), 111–23.

North, C. S., Clouse, R. E., Spitznagel, E. L., and Alpers, D. H. (1990). The relation of ulcerative colitis to psychiatric factors: a review of findings and methods. *American Journal of Psychiatry* 147(8), 974–81.

North, C. S., Ryall, J. M., and Wetzel, R. D. (1993). *Multiple personalities, multiple disorders.* Oxford University Press, New York.

Noyes, R. and Clancy, J. (1976). Anxiety neurosis: a 5 year follow up. *Journal of Nervous and Mental Disease* 162, 200–5.

Noyes, R. and Kathol, R. G. (1986). Depression and cancer. *Psychiatric Developments* 2, 77–100.

Noyes, R., Kathol, R. G., Crowe, R., et al. (1978). The familial prevalence of anxiety neurosis. *Archives of General Psychiatry* 35, 1057–9.

Noyes, R. Jr., Reich, J., Christiansen, J., Suelzer, M., Pfohl, B., and Coryell, W. A. (1990). Outcome of panic disorder. Relationship to diagnostic subtypes and comorbidity. *Archives of General Psychiatry* 47, 809–18.

Noyes, R., Kathol, R. G., Fisher, M. M., et al. (1994). One-year follow-up of medical outpatients with hypochondriasis. *Psychosomatics* 35, 533–45.

Nuffield Council on Bioethics (1993). *Genetic screening: ethical issues.* London.

Nurnberger, J. I. and Gershon, E. S. (1992). Genetics. In *Handbook of affective disorders* (ed. E. S. Paykel), pp. 131–48. Churchill Livingstone, Edinburgh.

Nutt, D. J. and Glue, P. W. (1989). Monoamine oxidase inhibitors: rehabilitation from recent research. *British Journal of Psychiatry* 154, 287–91.

Nutt, D. and Lawson, C. (1992). Panic attacks: a neurochemical overview of models and mechanisms. *British Journal of Psychiatry* 160, 165–78.

Oates, R. K., Peacock, A., and Forrest, D. (1985). Long-term effects of non-organic failure to thrive. *Paediatrics* 75, 36–40.

O'Brien, S. J. (1986). The controversy surrounding epilepsy and driving: a review. *Public Health* 100, 21–7.

O'Callaghan, E., Gibson, T., Colohan, H. A., *et al.* (1991). Season of birth in schizophrenia. *British Journal of Psychiatry* 158, 764–9.

O'Callaghan, E., Gibson, T., Colohan, A. J., *et al.* (1992). Risk of schizophrenia in adults born after obstetric complications and their association with early onset of illness: a controlled study. *British Medical Journal* 305, 1256–9.

O'Connor, N. (1968). Psychology and intelligence. In *Studies in psychiatry* (ed. M. Shepherd and D. L. Davis). Oxford University Press, London.

Ødegaard, Ø. (1932). Emigration and insanity. *Acta Psychiatrica Scandinavica* Suppl. 4.

Odlum, D. (1955). Fetishism. *British Medical Journal* i, 302.

O'Donovan, C., Hawkes, J., and Bowen, R. (1993). Effect of lithium dosing schedule on urinary output. *Acta Psychiatrica Scandinavica* 87, 92–5.

Offord, D. R., Boyle, M. H., Szatmari, P., and Rae-Grant, N. I. (1987). Ontario Child Health Study. I Six month prevalence of disorder and service utilization. *Archives of General Psychiatry* 44, 832–6.

O'Hare, A. E., Brown, J. K., and Aitken, K. (1991). Dyscalculia in children. *Developmental Medicine and Child Neurology* 33, 356–61.

Old Age Depression Interest Group (1993). How long should the elderly take antidepressants? A double-blind placebo-controlled study of continuation/prophylaxis therapy with dothiepin. *British Journal of Psychiatry* 162, 175–82.

Oldham, J. M., Skodol, A. E., Kellman, H. D., *et al.* (1992). Diagnosis of DSMIIIR personality disorders by two structured interviews: patterns of co-morbidity. *American Journal of Psychiatry* 149, 213–20.

Oldridge, N., Guyatt, G. H., Jones, N., *et al.* (1991). Effects on quality of life with comprehensive rehabilitation after acute myocardial infarction. *American Journal of Cardiology* 67, 1084–9.

O'Leary, K. D. and Beach, S. R. H. (1990). Marital therapy: a viable treatment for depression and marital discord. *American Journal of Psychiatry* 147, 183–6.

O'Leary, K. D. and Wilson, G. T. (1975). *Behaviour therapy: application and outcome*. Prentis Hall, Englewood Cliffs, NJ.

Oliver, C. and Holland, A. J. (1986). Down's syndrome and Alzheimer's disease: a review. *Psychological Medicine* 16, 307–22.

Oliver, J. E. (1970). Huntington's chorea in Northamptonshire. *British Journal of Psychiatry* 116, 241–53.

Olsson, B. and Rett, A. (1990). A review of the Rett syndrome with a theory of autism. *Brain and Development* 12, 11–15.

Onstad, S., Skre, J., Torgersen, S., and Kringlen, E. (1991). Subtypes of schizophrenia—evidence from a twin family study. *Acta Psychiatrica Scandinavica* 84, 203–6.

Onstad, S., Skre, I., Torgersen, S., and Kringlen, E. (1992). Birth weight and obstetric complications in schizophrenic twins. *Acta Psychiatrica Scandinavica* 85, 70–3.

Orford, J. and Edwards, G. (1977). *Alcoholism*. Maudsley Monographs No. 26. Oxford University Press, London.

Orme, M. L. E. (1984). Antidepressants and heart disease. *British Medical Journal* 289, 1–2.

Ormel, J., Oldehinkel, T., Brilman, E., and van den Brink, W. (1993). Outcome of depression and anxiety in primary care. *Archives of General Psychiatry* 50, 759–66.

Ornish, D., Brown, S., Scherwitz, L. W., *et al.* (1990). Can lifestyle changes reverse coronary heart disease? *Lancet* 336, 129–33.

Orvaschel, H. (1983). Maternal depression and child dysfunction: children at risk. In *Advances in clinical psychology*, Vol. 6 (ed. B. B. Lahey and A. E. Kazdun), pp. 167–97. Plenum Press, New York.

Orvaschel, H. and Weissman, M. M. (1986). The epidemiology of anxiety disorders. In *Anxiety disorders in childhood* (ed. R. Gittleman). Guilford Press, New York.

Osberg, J. W., Meares, G. J., McKee, D. C. M., and Burnett, G. B. (1982). Intellectual functioning in renal failure and chronic dialysis. *Journal of Chronic Diseases* 35, 445–57.

Osler, W. (1910). Angina pectoris. *Lancet* i, 697–702, 839–44.

Öst, L-G. (1987). Applied relaxation: description of a coping technique and review of controlled studies. *Behaviour Research and Therapy* 25, 397–409.

Ottenbacher, K. J. and Cooper, H. M. (1983). Drug treatment of hyperactivity in children. *Developmental Medicine and Child Neurology* 25, 358–66.

Ovenstone, I. M. K. and Kreitman, N. (1974). Two syndromes of suicide. *British Journal of Psychiatry* 124, 336–45.

Overall, J. E. and Gorham, D. R. (1962). The Brief Psychiatric Rating Scale. *Psychological Reports* 10, 799–812.

Owen, F., Cross, A. J., Crow, T. J., *et al.* (1978). Increased dopamine receptor sensitivity in schizophrenia. *Lancet* ii, 223–5.

Owen, M., Liddell, M., and McGuffin, P. (1994). Alzheimer's disease. *British Medical Journal* 308, 672–3.

Owen, M. J. (1992). Will schizophrenia become a graveyard for molecular geneticists? *Psychological Medicine* 22, 289–93.

Owen, R. R., Gutierrez-Esteinou, R., Hsiao, J., *et al.* (1993). Effects of clozapine and fluphenazine treatment on responses to *m*-chlorophenylpiperazine infusions in schizophrenia. *Archives of General Psychiatry* 50, 636–44.

Owens, D. G. C. and Johnstone, E. C. (1982). Spontaneous

involuntary disorders of movement. *Archives of General Psychiatry* 39, 452–61.

Paddison, P. (1993). *Treatment of adult survivors of incest.* American Psychiatric Press, Washington, DC.

Padma-Nathan, H., Goldstein, I., Payton, T., and Krane, R. J. (1987). Intracavernosal pharmacotherapy: the pharmacological erection program. *World Journal of Urology* 5, 160–5.

Page, A. C. (1994), Blood-injury phobia. *Clinical Psychology Review* 14(5), 443–61.

Palazzoli, M., Boscolo, L., Cecchin, G., and Prata, G. (1978). *Paradox and counterparadox.* Aronson, New York.

Palmer, M. S. and Collinge, J. (1992). Human prion diseases. In *Baillière's clinical neurology*, pp. 627–47. Baillière Tindall, London.

Palmer, R. L., Chaloner, D. A., and Oppenheimer, R. (1992). Childhood sexual experiences with adults reported by female psychiatric patients. *British Journal of Psychiatry* 160, 261–5.

Pardes, H., Kaufmann, C. A., Pincus, H. A., and West, A. (1989). Genetics and psychiatry: past discoveries, current dilemmas and future directions. *American Journal of Psychiatry* 135, 435–43.

Parker, G. (1979). Parental characteristics in relation to depressive disorders. *British Journal of Psychiatry* 134, 138–47.

Parker, G. (1992). Early environment. In *Handbook of affective disorders* (ed. E. S. Paykel), pp. 171–83. Churchill Livingstone, Edinburgh.

Parkes, C. M. (1965). Bereavement and mental illness. Part 2: a classification of bereavement reactions. *British Journal of Medical Psychology* 38, 13–26.

Parkes, C. M. (1971). The first year of bereavement: a longitudinal study of the reaction of London widows. *Psychiatry* 33, 444–6.

Parkes, C. M. (1978). Psychological reactions to loss of a limb. In *Modern perspectives in the psychological aspects of surgery* (ed. J. G. Howells). Macmillan, London.

Parkes, C. M. and Brown, R. J. (1972). Health after bereavement: a controlled study of young Boston widows and widowers. *Psychosomatic Medicine* 34, 449–61.

Parkes, C. M. and Weiss, R. S. (1983). *Recovery from bereavement.* Basic Books, New York.

Parkes, C. M., Benjamin, B., and Fitzgerald, R. G. (1969). Broken heart: a statistical study of increased mortality among widowers. *British Medical Journal* i, 740–3.

Parkes, J. D. (1985). *Sleep and its disorders.* Saunders, London.

Parkes, J. D. (1993). Daytime sleepiness. *British Medical Journal* 306, 772–5.

Parkes, K. R. (1982). Occupational stress among student nurses: a natural experiment. *Journal of Applied Psychology* 67, 784–96.

Parloff, M. B., Waskow, I. E., and Wolfe, B. E. (1978). Research on therapist variables in relation to process and outcome. In *Handbook of psychotherapy and behaviour change* (ed. S. L. Garfield and A. E. Bergin) (2nd edn). Wiley, New York.

Parnas, J., Schulsinger, F., Teasdale, T. W., *et al.* (1982). Perinatal complications and clinical outcome within the schizophrenia spectrum. *British Journal of Psychiatry* 140, 416–20.

Parnas, J., Cannon, T. D., Jacobsen, B., *et al.* (1993). Lifetime DSM–III–R diagnostic outcomes in the offspring of schizophrenic mothers. *Archives of General Psychiatry* 50, 707–14.

Parry-Jones, B. and Parry-Jones, W. L. (1992). Pica: symptom or eating disorder? A historical assessment. *British Journal of Psychiatry* 160, 341–54.

Parry-Jones, W. Ll. (1972). *The trade in lunacy.* Routledge & Kegan Paul, London.

Parry-Jones, W. L.l. (1973). Criminal law and complicity in suicide and attempted suicide. *Medicine, Science and the Law* 13, 110–19.

Parry-Jones, W. L., Santer-Westrate, H. C., and Crawley, R. C. (1970). Behaviour therapy in a case of hysterical blindness. *Behaviour Research and Therapy* 8, 79–85.

Parsons, T. (1951). *The social system.* Free Press, Glencoe.

Pasamanick, B. and Knobloch, H. (1966). Retrospective studies on the epidemiology of reproductive casualty: old and new. *Merril-Palmer Quarterly of Behavioral Development* 12, 7–26.

Pasmanick, B., Scarpitti, F. R., and Lefton, M. (1964). Home versus hospital care for schizophrenics. *Journal of the American Medical Association* 187, 177–81.

Patchell, R. A. (1994). Neurological complications of organ transplantation. *Annals of Neurology* 36, 688–703.

Patel, V. and Hope, T. (1993). Aggressive behaviour in elderly people with dementia: a review. *International Journal of Geriatric Psychiatry* 8, 457–72.

Pato, C. N., Macciardi, F., Pato, M. T., *et al.* (1993). Review of the putative association of dopamine D_2 receptor and alcoholism: a meta-analysis. *American Journal of Medical Genetics* 48, 78–82.

Patrick, C., Padgett, D. K., Schlesinger, H. J., *et al.* (1992). Serious physical illness as a stressor: effects on family use of medical services. *General Hospital Psychiatry* 14, 219–27.

Patterson, D. R. (1993). Psychological effects of severe burn injuries. *Psychological Bulletin* 113(2), 362–78.

Patterson, E. M. (1977). *The experience of dying.* Prentice-Hall, London.

Patterson, G. R. (1982). *Coercive family process.* Castalia Eugene, OR.

Patterson, D. R., *et al.* (1993). Psychological effects of severe burn injuries. *Psychological Bulletin* 113(2), 362–78.

Pauls, D. and Leckman, J. (1988). The genetics of Tourette syndrome. In *Tourette's syndrome and tic disorders: clinical understanding and treatment* (ed. D. Cohen, R. Bruun, and J. Leckman), pp. 91–102. Wiley, New York.

Paykel, E. S. (1978). Contribution of life events to causation of psychiatric illness. *Psychological Medicine* 8, 245–53.

Paykel, E. S. (1983). Methodological aspects of life event research. *Journal of Psychosomatic Research* 27, 341–52.

Paykel, E. S. (1989). Treatment of depression: the relevance of research to clinical practice. *British Journal of Psychiatry* 155, 754–63.

Paykel, E. S. (1990). Monoamine oxidase inhibitors: when should they be used. In *Dilemmas and difficulties in the management of psychiatric patients* (ed. K. Hawton and P. J. Cowen), pp. 17–30. Oxford University Press.

Paykel, E. S. and Cooper, Z. (1992). Life events and social stress. In *Handbook of affective disorders* (ed. E. S. Paykel), pp. 149–70. Churchill Livingstone, Edinburgh.

Paykel, E. S., Myers, J. K., Dienelt, M. N., *et al.* (1969). Life events and depression: a controlled study. *Archives of General Psychiatry* 21, 753–60.

Paykel, E. S., Prusoff, B. A., and Myers, J. K. (1975a). Suicide attempts and recent life events: a controlled comparison. *Archives of General Psychiatry* 32, 327–33.

Paykel, E. S., Di Mascio, A., Haskell, D., and Prusoff, B. A. (1975b). Effects of maintenance amitriptyline and psychotherapy on symptoms of depression. *Psychological Medicine* 5, 67–77.

Paykel, E. S., Emms, E. M., Fletcher, J., and Rassaby, E. S. (1980). Life events and support in puerperal depression. *British Journal of Psychiatry* 136, 339–46.

Paykel, E. S., Hollyman, J. A., Freeling, P., and Sedgwick, P. (1988). Predictors of therapeutic benefit from amitriptyline in mild depression. A general practice placebo controlled trial. *Journal of Affective Disorders* 14, 83–95.

Payne, R. W. and Friedlander, D. (1962). A short battery of simple tests for measuring over-inclusive thinking. *Journal of Mental Science* 108, 362–7.

Peachey, J. E. and Loh, E. (1994). Validity of alcohol and drug assessment. *Current Opinion in Psychiatry* 7, 252–7.

Pearce, J. (1994). Consent to treatment during childhood. *British Journal of Psychiatry* 165, 713–16.

Pearce, J., Hawton, K., and Blake, F. (1995). Psychological and sexual symptoms associated with the menopause and the effects of hormone replacement therapy. *British Journal of Psychiatry* 167, 163–73.

Pedersen, N. L., Plomin, R., McLearn, G. E., and Friberg, L. (1988). Neuroticism, extroversion and related traits in adult twins reared apart and reared together. *Journal of Personality and Social Psychology* 55, 950–7.

Penrose, L. (1938). *A clinical and genetic study of 1280 cases of mental deficiency.* HMSO, London.

Perley, M. J. and Guze, S. B. (1962). Hysteria — the stability and usefulness of clinical criteria. *New England Journal of Medicine* 266, 421–6.

Perlmutter, L. C., Hakami, M.K., Hodgson-Harrington, C., *et al.* (1984). Decreased cognitive function in aging non-insulin-dependent diabetic patients. *American Journal of Medicine* 77, 1043–8.

Perrin, G. M. (1961). Cardiovascular aspects of electric shock therapy. *Acta Psychiatrica Scandinavica* 36 (Suppl. 152), 1–45.

Perry, E. K. and Perry, R. H. (1994). Alzheimer's disease, neurobiology and therapy. *International Review of Psychiatry* 5, 363–80.

Perry, E. K., Tomlinson, B. E., and Blessed, G. (1978). Correlation of cholinergic abnormalities with senile plaques and mental test scores in senile dementia. *British Medical Journal* ii, 1457–9.

Perry, E. K., McKeith, I., Thompson, P., *et al.* (1991). Topography, extent, and clinical relevance of neuro-chemical deficits in dementia of Lewy body type, Parkinson's disease, and Alzheimer's disease. *Annals of the New York Academy of Science* 640, 197–202.

Perry, P. S., Morgan, D. E., Smith, R. E., and Tsuang, M. T. (1982). Treatment of unipolar depression accompanied by delusions. *Journal of Affective Disorders* 4, 195–200.

Perry, T. L., Hansen, S., and Kloster, M. (1973). Huntington's chorea: deficiency of gamma-aminobutyric acid in brain. *New England Journal of Medicine* 288, 337–42.

Peselow, E. D., Fieve, R. R., Difiglia, C., and Sanfilipo, M. P. (1994). Lithium prophylaxis of bipolar illness: the value of combination treatment. *British Journal of Psychiatry* 164, 208–14.

Peselow, E. D., Robins, C., Bloch, P., *et al.* (1990). Dysfunctional attitudes in depressed patients before and after clinical treatment and in normal control subjects. *American Journal of Psychiatry* 147, 439–44.

Peters, A. A., van Dorst, E., Jellis, B., *et al.* (1991). A randomized clinical trial to compare two different approaches in women with chronic pelvic pain. *Obstetrics and Gynecology* 77(5), 740–4.

Peters, S. D., Wyatt, G. E., and Finkelhor, D. (1986). Prevalence. In *A source book on child sexual abuse* (ed. D. Finkelhor). Sage, London.

Petersen, P. (1968). Psychiatric disorders in primary hyperparathyroidism. *Journal of Clinical Endocrinology and Metabolism* 28, 1491–5.

Petronko, M. R., Harris, S. L., and Kormann, R. J. (1994). Community based behavioral training approaches for people with mental retardation and mental illness. *Journal of Consulting and Clinical Psychology* 62(1), 49–54.

Petursson, H. and Lader, M. H. (1984). *Dependence on tranquillizers.* Oxford University Press.

Peyser, C. E. and Folstein, S. E. (1993). Depression in Huntington disease. In *Depression in neurologic disease* (ed. S. E. Starkstein and R. G. Robinson), pp. 117–38. Johns Hopkins University Press, Baltimore, MD.

Pfeffer, C. R., Normandin, L., and Kakuma, T. (1994). Suicidal children grow up: demographic and clinical risk factors for adolescent suicide attempts. *Journal of American Academy of Child and Adolescent Psychiatry* 33(8), 1087.

Phillips, H. C., Grant, L., and Berkowitz, J. (1991). The prevention of chronic pain and disability: a preliminary investigation. *Behaviour Research and Therapy* 29, 443–50.

Phillips, K. A. (1991). Body dysmorphic disorder: the distress of imagined ugliness. *American Journal of Psychiatry* 148(9), 1138–49.

Phillips, K. A., McElroy, S. L., Keck, P. E., *et al.* (1993). Body dysmorphic disorder: 30 cases of imagined ugliness. *American Journal of Psychiatry* 150(2) (abstract), 302–8.

Piccinelli, N. and Wilkinson, G. (1994). Outcome of depression in psychiatric settings. *British Journal of Psychiatry* 164, 297–304.

Pichot, P. (1982). The diagnosis and classification of mental disorders in French speaking countries: background, current view and comparison with other nomenclature. *Psychological Medicine* 12, 475–92.

Pichot, P. (1984). The French approach to classification. *British Journal of Psychiatry* 144, 113–18.

Pichot, P. (1994). Nosological models in psychiatry. *British Journal of Psychiatry* 164, 232–40.

Pick, A. (1892). Über die Beziehungen der senilen Hirnatrophie zur Aphasie. *Prager medizinische Wochenschrift* 17, 165–7.

Pickens, R. W., Svikis, D. S., McGue, M., *et al.* (1991). Heterogeneity in the inheritance of alcoholism: a study of male and female twins. *Archives of General Psychiatry* 48, 19–29.

Pierce, P. A. and Peroutka, S. J. (1989). Hallucinogenic drug interaction with neurotransmitter receptor binding sites in human cortex. *Psychopharmacology* 97, 118–22.

Piggot, T. A., Pato, M. T., Bernstein, S. E., Grover, G. N., Hill, J. L., Tolliver, T. J., and Murphy, D. L. (1990). Controlled comparisons of clomipramine and fluoxetine in the treatment of obsessive-compulsive disorder. *Archives of General Psychiatry* 47, 926–32.

Pilowsky, L. S. (1992). Understanding schizophrenia. *British Medical Journal* 305, 327–8.

Pilowsky, L. S., Ring, H., Shine, P. J., *et al.* (1992). Rapid tranquillisation: a survey of emergency prescribing in a general psychiatric hospital. *British Journal of Psychiatry* 160, 831–5.

Pincus, J. H. and Tucker, G. J. (1985). *Behavioural neurology* (3rd edn). Oxford University Press, New York.

Pitres, A. and Régis, E. (1902). *Les obsessions et les impulsions*. Doin, Paris.

Pitts, F. N. and McClure, J. N. (1967). Lacate metabolism in anxiety neurosis. *New England Journal of Medicine* 25, 1329–36.

Pitts, F. N. and Winokur, G. (1964). Affective disorders III: diagnostic correlates and incidence of suicide. *Journal of Nervous and Mental Disease* 139, 176–81.

Pizella, V. and Romani, G. L. (1990). In *Advances in neurology*, Vol. 54, *Magnetoencephalography* (ed. S. Sato). Raven Press, New York.

Planansky, K. and Johnston, R. (1977). Homicidal aggression in schizophrenic men. *Acta Psychiatrica Scandinavica* 55, 65–73.

Plant, M. A. (1975). *Drug takers in an English town*. Tavistock Publications, London.

Plant, M. A., Peck, D. F., and Samuel, E. (1985). *Alcohol, drugs and school leavers*. Tavistock Publications, London.

Plasky, P. (1991). Antidepressant usage in schizophrenia. *Schizophrenia Bulletin* 17, 649–57.

Platt, S. (1987). Suicide trends in 24 European countries, 1972–1984. In *Current issues in suicidology* (ed. H. J. Moller, A. Schmidtke, and R. Welz). Springer-Verlag, Berlin.

Platt, S., Hawton, K., Kreitman, N., Fagg, J., and Foster, J. (1988). Recent clinical and epidemiological trends in parasuicide in Edinburgh and Oxford: a tale of two cities. *Psychological Medicine* 18, 405–18.

Plomin, R. (1994). *Genetics and experience*. Sage, Thousand Oaks, CA.

Plomin, R. (1995). Genetics and children's experiences in the family. *Journal of Child Psychology and Psychiatry* 36(1), 33–68.

Pocock, S. J., Smith, M., and Baghurst, M. (1994). Environmental lead and children's intelligence: a systematic review of the epidemiological evidence. *British Medical Journal* 309, 1189–97.

Poirier, J., Davignon, J., Bouthillier, D., *et al.* (1990). Apoliprotein E polymorphism and Alzheimer's disease. *Lancet* 342, 697–9.

Pokorny, A. (1964). Suicide rates in various psychiatric disorders. *Archives of General Psychiatry* 139, 499–506.

Polak, P. B., Egan, D., and Bandenbergh, R. (1975). Prevention in mental health: a controlled study. *American Journal of Psychiatry* 132, 146–9.

Pollin, W. and Stabenau, J. (1968). Biological, psychological, and historical differences in a series of monozygotic twins discordant for schizophrenia. In *Transmission of schizophrenia* (ed. D. Rosenthal and S. Kety). Pergamon Press, London.

Pollitt, J. (1957). Natural history of obsessional states. *British Medical Journal* i, 194–8.

Pollitt, J. (1960). Natural history studies in mental illness: a discussion based upon a pilot study of obsessional states. *Journal of Mental Science* 106, 93–113.

Pollock, B. G., Perel, J. M., Nathan, S., and Kupfer, D. J. (1989). Acute antidepressant effect following pulse loading with intravenous and oral clomipramine. *Archives of General Psychiatry* 46, 29–35.

Pollock, C., Freemantle, N., Sheldon, T., *et al.* (1993). Methodological difficulties in rehabilitation research. *Clinical Rehabilitation* 7, 63–72.

Pond, D. A. (1957). Psychiatric aspects of epilepsy. *Journal of the Indian Medical Profession* 3, 1441–51.

Pond, D. A. and Bidwell, B. H. (1960). A survey of epilepsy in fourteen general practices. II. Social and psychological aspects. *Epilepsia* 1, 285–99.

Pope, H. G., Jonas, J. M., Hudson, J. I., *et al.* (1983). The validity of DSM III borderline personality disorder: a phenomenological, family history, treatment response, and long term follow-up study. *Archives of General Psychiatry* 40, 23–30.

Popkin, M. K. and Tucker, G. J. (1994). Mental disorders due to a general medical condition and substance-induced disorders. Mood, anxiety, psychotic, catatonic and personality disorders. In *DSM-IV sourcebook*, Vol. I (ed. T. A. Widiger, A. J. Frances, H. A. Pincus, *et al.*), pp. 243–76. American Psychiatric Association, Washington, DC.

Porter, R. (1995). Psychosomatic disorder: historical perspectives. In *Treatment of functional somatic symptoms* (ed. R. A. Mayou, C. Bass, and M. Sharpe), pp. 17–41. Oxford University Press, Oxford.

Post, F. (1971). Schizo-affective symptomatology in late life. *British Journal of Psychiatry* 118, 437–45.

Post, F. (1972). The management and nature of depressive illnesses in late life: a follow-through study. *British Journal of Psychiatry* 121, 393–404.

Post, R. M. (1991). Anticonvulsants as adjuncts or alternatives to lithium in refractory bipolar illness. In *Advances in neuropsychiatry and psychopharmacology*, Vol. 2, *Refractory depression* (ed. J. D. Amsterdam), pp. 155–65. Raven Press, New York.

Post, R. M., Uhde, T. W., Roy-Byrne, P. P., and Joffe, R. T. (1986). Antidepressant effects of carbamazepine. *American Journal of Psychiatry* 143, 29–34.

Post, R. M., Leverich, G. S., Rosoff, A. S., and Altschuler, L. L. (1990). Carbamazepine prophylaxis in refractory affective disorders: a focus on long-term follow-up. *Journal of Clinical Psychopharmacology* 10, 318–22.

Post, R. M., Leverich, G. S., Altshuler, L., and Mikalauskas, K. (1992). Lithium discontinuation-induced refractoriness: preliminary observations. *American Journal of Psychiatry* 149, 1727–9.

Potter, W. Z. and Manji, H. K. (1993). Are monoamine metabolites in cerebrospinal fluid worth measuring? *Archives of General Psychiatry* 50, 653–6.

Potter, W. Z., Rudorfer, M. V., and Manji, H. (1991). The pharmacological treatment of depression. *New England Journal of Medicine* 325, 633–42.

Powell, G. F., Brasel, J. A., and Blizzard, R. M. (1967). Emotional deprivation and growth retardation simulating idiopathic hypopituitarism. *New England Journal of Medicine* 276, 1271–83.

Power, A. C. and Cowen, P. J. (1992). Fluoxetine and suicidal behaviour: some clinical and theoretical aspects of a controversy. *British Journal of Psychiatry* 161, 735–41.

Power, D. J., Benn, R. T., and Homes, J. N. (1972). Neighbourhood, school and juveniles before courts. *British Journal of Criminology* 12, 111–32.

Poynton, A., Bridges, P. K., and Bartlett, J. R. (1988). Psychosurgery in Britain now. *British Journal of Neurosurgery* 2, 297–306.

Pratt, J. H. (1908). Results obtained in treatment of pulmonary tuberculosis by the class method. *British Medical Journal* ii, 1070–1.

Preskorn, S. H. (1993). Pharmacokinetics of antidepressants: why and how they are relevant to treatment. *Journal of Clinical Psychiatry* 54(9 Suppl.), 14–34.

Preskorn, S. H., Gerkovish, G. S., Baber, J. H., and Widener, P. (1989). Therapeutic drug monitoring of tricyclic antidepressants: a standard of care issue. *Psychopharmacology Bulletin* 25, 281–4.

Preskorn, S. H., Alderman, J., Chung, M., *et al.* (1994). The pharmacokinetics of desipramine co-administered with sertraline or fluoxetine. *Journal of Clinical Psychopharmacology* 14, 90–8.

Price, L. H. (1989). Lithium augmentation in tricyclic resistant depression. In *Treatment of tricyclic-resistant depression* (ed. I. L. Extein), pp. 51–79. American Psychiatric Press, Washington, DC.

Price, L. H., Charney, D. S., and Heninger, G. R. (1985). Efficacy of lithium-tranylcypromine treatment in refractory depression. *American Journal of Psychiatry* 142, 619–23.

Prichard, J. C. (1835). *A treatise on insanity*. Sherwood Gilbert and Piper, London.

Prien, R. F. (1992). Maintenance treatment. In *Handbook of affective disorders* (ed. E. S. Paykel), pp. 419–35. Churchill Livingstone, Edinburgh.

Prien, R. F. and Kupfer, D. J. (1986). Continuation drug therapy for major depressive episodes: how long should it be maintained? *American Journal of Psychiatry* 143, 18–23.

Prien, R. F., Caffey, E. M., and Glett, C. J. (1972). Comparison of lithium carbonate and chlorpromazine in the treatment of mania. *Archives of General Psychiatry* 26, 146–53.

Priest, R. G. (1976). The homeless person and the psychiatric services: an Edinburgh survey. *British Journal of Psychiatry* 128, 128–36.

Prigatano, G. P. (1992). Personality disturbances associated with traumatic brain injury. *Journal of Consulting and Clinical Psychology* **60**, 360–8.

Prigatano, G. P., Wright, E. C., and Levin, D. (1984). Quality of life: its predictors in patients with mild hypoxaemia and chronic obstructive pulmonary disease. *Archives of Internal Medicine* **144**, 1613–19.

Prince, M. (1908). *Dissociation of personality, a biographical study in abnormal psychology*. Longmans Green, New York.

Prins, H. (1993). *Fire-raising: its motivation and management*. Routledge, London.

Protheroe, C. (1969). Puerperal psychoses: a long term study, 1927–1961. *British Journal of Psychiatry* **115**, 9–30.

Prusoff, B. A., Weissman, M. M., Klerman, G. L., and Rounsaville, B. J. (1980). Research diagnostic criteria subtypes of depression as predictors of differential response to psychotherapy and drug treatment. *Archives of General Psychiatry* **37**, 796–801.

Prusiner, S. B. (1993). Genetic and infectious prion diseases. *Archives of Neurology* **50**, 1129–53.

Prusiner, S. B. and Hsiao, K. K. (1994). Human prion diseases. *Annals of Neurology* **35**, 385–95.

Pugh, R., Jerath, B. K., Schmidt, W. M., and Reed, R. B. (1963). Rates of mental disease related to child rearing. *New England Journal of Medicine* **22**, 1224–8.

Putnam, F. W., Guroff, J. J., and Silberman, E. K. (1986). *Journal of Clinical Psychiatry* **47**, 285–93.

Quality Assurance Project (1991). Treatment outlines for borderline, narcissistic and histrionic personality disorders. *Australia and New Zealand Journal of Psychiatry* **25**, 392–403.

Quay, H. C. and Werry, J. S. (1986). *Psychopathological disorders of childhood* (3rd edn). Wiley, New York.

Querido, A. (1959). Forecast and follow-up. An investigation into the clinical, social and mental factors determining the results of hospital treatment. *British Journal of Preventive and Social Medicine* **13**, 334–9.

Quine, L. and Pahl, J. (1987). First diagnosis of severe handicap: a study of parental reactions. *Developmental Medicine and Child Neurology* **9**, 232–42.

Quinsey, V. L. (1986). Men who have sex with children. In *Law and mental health* (ed. D. Weisstub). Pergamon Press, Oxford.

Quitkin, F. M. (1985). The importance of dosage in prescribing antidepressants. *British Journal of Psychiatry* **147**, 593–7.

Quitkin, F., Rifkin, A., and Klein, D. (1976). Neurologic soft signs in schizophrenia and character disorders. *Archives of General Psychiatry* **33**, 845–53.

Quitkin, F. M., McGrath, P. J., Stewart, J. W., *et al.* (1989).

Phenelzine and imipramine in mood reactive depressives. *Archives of General Psychiatry* **46**, 787–93.

Quitkin, F. M., McGrath, P. J., Stewart, J. W., *et al.* (1989). Phenelzine and amitriptyline in mood reactive depressives. *Archives of General Psychiatry* **46**, 787–93.

Rachman, S. (1966). Sexual fetishism—an experimental analogue. *Psychological Record* **16**, 293–6.

Rachman, S. (1974). Primary obsessional slowness. *Behaviour Research and Therapy* **11**, 463–71.

Rachman, S. and Hodgson, R. J. (1980). *Obsessions and compulsions*. Prentice-Hall, Englewood Cliffs, NJ.

Radke-Yarrow, M., Nottelmann, E., Martinez, P., and Fox, M. B. (1993). Young children of affectively ill parents: a longitudinal study of social development. *Journal of the American Academy of Child and Adolescent Psychiatry* **31**, 68–77.

Radomski, J. K., Fuyat, H. N., Belson, A. A., and Smith, P. K. (1950). The toxic effects, excretion and distribution of lithium chloride. *Journal of Pharmacology and Experimental Therapeutics* **100**, 429–44.

Rahe, R., Gunderson, E. K. E., and Arthur, R. J. (1970). Demographic and psychosocial factors in acute illness reporting. *Journal of Chronic Diseases* **23**, 245–55.

Rahe, R. H., McKean, J. D., and Ransom, J. A. (1967). A longitudinal study of life-changes and illness patterns. *Journal of Psychosomatic Research* **10**, 355–66.

Raine, A., Venables, P. H., and Williams, M. (1990). Relationships between central and autonomic measures of arousal at age 15 years and criminality at age 24 years. *Archives of General Psychiatry* **47**, 1003.

Ramsay, B. and O'Reagan, M. (1988). A survey of the social and psychological effects of psoriasis. *British Journal of Dermatology* **118**, 195–201.

Ramsay, N. (1992). Referral to a liaison psychiatrist from a palliative care unit. *Palliative Medicine* **6**, 54–60.

Ramsay, R., Gorst-Unsworth, C., and Turner, S. (1993). Psychiatric morbidity in survivors of organised state violence including torture. *British Journal of Psychiatry* **162**, 55–9.

Ramsey, J., Anderson, H. R., Bloor, K., and Flanagan, R. J. (1989). An introduction to the practice, prevalence and chemical toxicology of volatile substance abuse. *Human Toxicology* **8**, 261–9.

Rao, S. M., Huber, S. J., and Bornstein, R. A. (1992). Emotional changes with multiple sclerosis and Parkinson's disease. *Journal of Consulting and Clinical Psychology* **60**, 369–78.

Raphael, B. (1977). Preventive intervention with the recently bereaved. *Archives of General Psychiatry* **34**, 1450–4.

Raphael, B. (1986). The problems of mental health and adjustment. In *When disaster strikes: a handbook for the caring profession* (ed. B. Raphael). Hutchinson, London.

Rapoport, R. N. (1960). *Community as doctor*. Tavistock Publications, London.

Rappoport, J. L. (1991). Recent advances in obsessive compulsive disorder. *Neuropsychopharmacology* 5, 1–9.

Ratnasuriva, R. H., Eisler, I., Szmukler, G. I., and Russell, G. F. (1991). Anorexia nervosa: outcome and prognostic factors after 20 years. *British Journal of Psychiatry* 158, 495–502.

Rauch, S. L., Jenike, M. A., Alpert, N. M., *et al.* (1994). Regional cerebral blood flow measured during symptom provocation in obsessive–compulsive disorder using oxygen 15-labeled carbon dioxide and positron emission tomography. *Archives of General Psychiatry* 51, 62–70.

Razani, J., White, J., Simpson, G., *et al.* (1983). The safety and efficacy of combined amitriptyline and tranylcypromine antidepressant treatment. *Archives of General Psychiatry* 40, 657–61.

Rebok, G. W. and Folstein, M. F. (1994). Dementia. In *DSM-IV sourcebook*, Vol. 1 (ed. T. A. Widiger, A. J. Frances, H. A. Pincus, *et al.*), pp. 213–35. American Psychiatric Association, Washington, DC.

Reed, A., Ramsden, S., Marshall, J., *et al.* (1992). Psychiatric morbidity and substance abuse among residents of a cold weather shelter. *British Medical Journal* 304, 1028–9.

Reed, G. F. and Sedman, G. (1964). Personality and depersonalization under sensory deprivation conditions. *Perceptual and Motor Skills* 18, 659–60.

Regier, D. A., Burke, J. D., Manderscheid, R. W., and Burns, B. J. (1985). The chronically mentally ill in primary care. *Psychological Medicine* 15, 265–73.

Regier, D. A., Boyd, J. H., Burke, J. D., *et al.* (1988). One month prevalence of mental disorders in the United States. *Archives of General Psychiatry* 45, 977–86.

Regier, D. A. *et al.* (1994). The ICD-10 Clinical Field Trial for Mental and Behavioral Disorders: Results in Canada and the United States. *American Journal of Psychiatry* 151(9), 1340–50.

Regier, D. A., Narrow, W. E., Rae, D. S., *et al.* (1994). The *de facto* US Mental and Addictive Disorders Service System. *Archives of General Psychiatry* 50, 85–94.

Reich, P. and Gottfried, L. A. (1983). Factitious disorders in a teaching hospital. *Annals of Internal Medicine* 99, 240–7.

Reich, P., Regestein, Q. R., Murawski, B. J., *et al.* (1983). Unrecognized organic mental disorders in survivors of cardiac arrest. *American Journal of Psychiatry* 140, 1194–7.

Reid, A. H. and Ballinger, B. R. (1987). Personality disorder in mental handicap. *Psychological Medicine* 17, 983–7.

Reilly, P. A. (1995). "Repetitive strain injury": from Australia to the UK. *Journal of Psychosomatic Research* (in press).

Resnick, P. J. (1969). Child murder by parents. *American Journal of Psychiatry* 126, 325–34.

Restak, R. (1993). The neurological defense of violent crime: "insanity defense" retooled. *Archives of Neurology* 50, 869–71.

Reynolds, C. F. (1992). Treatment of depression in special populations. *Journal of Clinical Psychiatry* 53(9S), 45–53.

Reynolds, C. F., Frank, E., and Perel, J. M. (1992). Combined pharmacotherapy and psychotherapy in the acute and continuation treatment of elderly patients with recurrent major depression: a preliminary report. *American Journal of Psychiatry* 149, 1687–92.

Reynolds, E. H. (1968). Mental effects of anticonvulsants and folic acid metabolism. *Brain* 91, 197–214.

Reynolds, G. P. (1983). Increased concentration and lateral asymmetry of amygdala dopamine in schizophrenia. *Nature, London* 305, 527–9.

Reynolds, G. P. and Mason, S. L. (1994). Are striatal D4 receptors increased in schizophrenia? *Journal of Neurochemistry* 63, 1576–7.

Ribeiro, S. C. M., Tandon, R., Grunhaus, L., and Greden, J. F. (1993). The DST as a predictor of outcome in depression: a meta-analysis. *American Journal of Psychiatry* 150, 1618–29.

Richardson, A. (1973). Stereotactic limbic leucotomy: surgical technique. *Postgraduate Medical Journal* 49, 860.

Richardson, S. K. and Koller, H. (1992). Vulnerability and resilience in adults who were classified as mildly mentally handicapped in childhood. In *Vulnerability and resilience in human development* (ed. B. Tizard and V. Varma), pp. 102–23. JKP, London.

Richman, N., Stevenson, J., and Graham, P. (1982). *Preschool to school: a behavioural study*. Academic Press, London.

Richman, N., Douglas, J., Hunt, H., *et al.* (1985). Behavioural methods in the treatment of sleep disorders—a pilot study. *Journal of Child Psychology and Psychiatry* 26, 581–90.

Rickels, K., Csanalosi, I., and Chung, H. R. (1974). Amitriptyline in anxious-depressed outpatients: a controlled study. *American Journal of Psychiatry* 130, 25–30.

Rickels, K., Case, W. G., Schweizer, E. E., *et al.* (1986). Low-dose dependence in chronic benzodiazepine users: a preliminary report on 119 patients. *Psychopharmacology Bulletin* 22, 407–15.

Rickels, K., Downing, R., Schweizer, E., and Hassman, H. (1993). Antidepressants for the treatment of generalised anxiety disorder. *Archives of General Psychiatry* 50, 884–95.

Rickels, K., Schweizer, E., Clary, C., Fox, I., and Weise, C. (1994). Nefazodone and imipramine in major depression: a placebo-controlled trial. *British Journal of Psychiatry* 164, 802–5.

Rickles, N. K. (1950). *Exhibitionism*. Lippincott, Philadelphia, PA.

Ridgeway, V. and Mathews, A. (1982). Psychological preparation for surgery: a comparison of methods. *British Journal of Clinical Psychology* 21, 271–80.

Riether, A. M. and McDaniel, J. S. (1993). Surgery and trauma: general principles. In *Psychiatric care of the medical patient* (ed. A. Stoudemire and B. S. Fogel), pp. 759–81. Oxford University Press, New York.

Rimm, D. C. and Masters, J. C. (1974). *Behaviour therapy: techniques and empirical findings*. Academic Press, New York.

Rimm, D. C. and Masters, J. C. (1979). *Behaviour therapy: techniques and empirical findings*. Academic Press, New York.

Rimmer, E. M. and Richens, A. (1982). Clinical pharmacology and medical treatment. In *A textbook of epilepsy* (ed. J. Laidlaw, A. Richens, and J. Oxley) (3rd edn). Churchill Livingstone, Edinburgh.

Ring, H. (1993). Psychological and social problems of Parkinson's disease. *British Journal of Hospital Medicine* 49, 111–16.

Ring, H. A. and Trimble, M. R. (1993). Depression in epilepsy. In *Depression in neurologic disease* (ed. S. E. Starkstein and R. G. Robinson), pp. 63–83. Johns Hopkins University Press, Baltimore, MD.

Ritson, B. (1977). Alcoholism and suicide. In *Alcoholism: new knowledge and new responses* (ed. G. Edwards and M. Grant). Croom Helm, London.

Ritson, B. and Patience, D. (1994). Alcohol-related physical and psychiatric disorders. *Current Opinion in Psychiatry* 7, 258–61.

Ritson, B., Chick, J. D., and Strang, J. (1993). Dependence on alcohol and other drugs. In *Companion to psychiatric studies* (ed. R. E. Kendall and A. K. Zealley), pp. 359–95. Churchill Livingstone, Edinburgh.

Rivers, W. H. (1920). *Instinct and the unconscious*. Cambridge University Press.

Rivinus, T. M., Jamison, D. L., and Graham, P. J. (1975). Childhood organic neurological disease presenting as psychiatric disorder. *Developmental Medicine and Child Neurology* 23, 747–60.

Rix, B., Pearson, D. J., and Bentley, S. J. (1984). A psychiatric study of patients with supposed food allergy. *British Journal of Psychiatry* 145, 121–6.

Robbins, T. W., Joyce, E. M., and Sahakian, B. J. (1992). Neuropsychology and imaging, In *Handbook of affective disorders* (ed. E. S. Paykel), pp. 289–309. Churchill Livingstone, Edinburgh.

Roberts, A. H. (1969). *Brain damage in boxers*. Pitman, London.

Roberts, G. W., Allsop, D., and Bruton, C. (1990a). The occult aftermath of boxing. *Journal of Neurology, Neurosurgery and Psychiatry* 53, 373–8.

Roberts, G. W., Done, D. J., Bruton, C., and Crow, T. J.

(1990b). A 'mock-up' of schizophrenia: temporal lobe epilepsy and schizophrenia-like psychosis. *Biological Psychiatry* 28, 127–43.

Roberts, J. and Hawton, K. (1980). Child abuse and attempted suicide. *British Journal of Psychiatry* 137, 319–23.

Robertson, M. M. (1994). Annotation: Gilles de la Tourette syndrome—an update. *Journal of Child Psychology and Psychiatry* 35, 597–611.

Robertson, M. M., Trimble, M. R., and Lees, A. J. (1988). The psychopathology of the Gilles de la Tourette syndrome. *British Journal of Psychiatry* 152, 383–90.

Robins, E., Gassner, S., Kayes, J., *et al.* (1959). The communication of suicidal intent: a study of 134 successful (completed) suicides. *American Journal of Psychiatry* 155, 724–33.

Robins, L. N. (1966). *Deviant children grown up*. Williams and Wilkins, Baltimore, MD.

Robins, L. N. (1978). Sturdy childhood predictors of adult antisocial behaviour: replications from longitudinal studies. *Psychological Medicine* 8, 611–22.

Robins, L. N. (1979). Follow-up studies. In *Pathological disorders of childhood* (ed. H. C. Quay and J. S. Werry), pp. 483–513. Wiley, New York.

Robins, L. N. (1993). Vietnam veterans' rapid recovery from heroin addiction: a fluke or normal expectation? *Addiction* 88, 1041–54.

Robins, L. N. and Kulbok, P. A. (1988). Epidemiologic studies in suicide. In *Annual Review of Psychiatry*, Vol. 7 (ed. A. J. Frances and R. E. Hales), pp. 289–306. American Psychiatric Association, Washington, DC.

Robins, L. N. and Regier, D. A. (1991). *Psychiatric disorder in America: the epidemiological catchment area study*. Free Press, New York.

Robins, L. N. and Rutter, M. (1990). *Straight and devious pathways from childhood to adulthood*. Cambridge University Press.

Robins, L. N., Helzer, J. E., Croughan, J., and Ratcliff, K. S. (1981). National Institutes of Mental Health Diagnostic Interview Schedule. *Archives of General Psychiatry* 38, 381–9.

Robins, L. N., Helzer, J. E., Ratcliff, K. S., and Seyfried, W. (1982). Validity of the Diagnostic Interview Schedule, version II: DSMIII diagnoses. *Psychological Medicine* 12, 855–70.

Robins, L. N., Helzer, J. E., Weissman, M. M., *et al.* (1984). Lifetime prevalence of specific psychiatric disorder in three sites. *Archives of General Psychiatry* 41, 949–58.

Robinson, D. (1979). *Talking out of alcoholism: the self-help process of Alcoholics Anonymous*. Croom Helm, London.

Robinson, R. G. (1993). Pathological laughing and crying following stroke: validation of a measurement scale and a double blind trial treatment study. *American Journal of Psychiatry* 150, 286–93.

Robinson, R. G., Kubos, K. L., Storr, L. B., *et al.* (1984). Mood disorders in stroke patients: importance of location of lesion. *Brain* 107, 87–94.

Robinson, R. G. *et al.* (1993). Pathological laughing and crying following stroke: Validation of a measurement scale and a double blind trial treatment study. *American Journal of Psychiatry* 150, 286–93.

Robson, P. (1992). Opiate misusers: are treatments effective? In *Practical problems in clinical psychiatry* (ed. K. Hawton and P. Cowen), pp. 141–58. Oxford University Press.

Rochford, J. M., Detre, T., Tucker, G. J., and Harrow, M. (1970). Neuropsychological impairments in functional psychiatric disease. *Archives of General Psychiatry* 22, 114–19.

Rockland, L. H. (1987). A supportive approach: psychodynamically oriented supportive treatment of borderline patients who self-mutilate. *Journal of Personality Disorders* 1, 350–3.

Rockland L. H. (1989). *Supportive therapy: a psychodynamic approach*. Basic Books, New York.

Rockland L. H. (1992). *Supportive therapy for borderline patients: a psychodynamic approach*. Guilford Press, New York.

Rodrigo, E. K. and Williams, P. (1986). Frequency of self-reported 'anxiolytic withdrawal' symptoms in a group of female studies experiencing anxiety. *Psychological Medicine* 16, 467–72.

Rogers, C. R. and Dymond, R. F. (ed.) (1954). *Psychotherapy and personality change*. University of Chicago Press.

Rogers, S. C. and May, P. M. (1975). A statistical review of controlled trials of imipramine and placebo in the treatment if depressive illness. *British Journal of Psychiatry* 127, 599–603.

Romans-Clarkson, S. E., Clarkson, J. E., and Dittmer, I. D. (1986). Impact of a handicapped child on mental health of patients. *British Medical Journal* 293, 1395–417.

Ron, M. A. (1989). Psychiatric manifestations of frontal lobe tumours. *British Journal of Psychiatry* 155, 735–8.

Ron, M. A. (1993). The brain and the mind: the case of multiple sclerosis. In *Clinical psychiatry* (ed. K. Granville-Gross). Churchill Livingstone, New York.

Ron, M. A. (1994). Somatisation in neurological practice. *Journal of Neurology, Neurosurgery and Psychiatry* 57, 1161–4.

Ron, M. A. and Feinstein, A. (1992). Multiple sclerosis and the mind. *Journal of Neurology, Neurosurgery and Psychiatry* 55, 1–3.

Ron, M. A. and Logsdail, S. J. (1993). Psychiatric morbidity in multiple sclerosis: a clinical and MRI study. *Psychological Medicine* 19, 887–95.

Ron, M. A., Toone, B. K., Garralda, M. E., and Lishman, W. A. (1979). Diagnostic accuracy in presenile dementia. *British Journal of Psychiatry* 134, 161–8.

Ron, M. A., Callanan, M. M. and Warrington, E. K. (1991). Cognitive abnormalities in multiple sclerosis: a psychometric and MRI study. *Psychological Medicine* 21, 59–68.

Roose, S. P. and Glassman, A. H. (1994). Antidepressant choice in the patient with cardiac disease: lessons form the cardiac arrhythmia suppression trial (CAST) studies. *Journal of Clinical Psychiatry* 55(9) (Suppl. A), 83–7.

Rooth, F. G. (1971). Indecent exposure and exhibitionism. *British Journal of Hospital Medicine* 5, 521–33.

Rooth, F. G. (1973). Exhibitionism, sexual violence and paedophilia. *British Journal of Psychiatry* 122, 705–10.

Rooth, F. G. and Marks, I. M. (1974). Persistent exhibitionism: short-term response to aversion self regulation and relaxation treatment. *Archives of Sexual Behaviour* 3, 227–43.

Rosanoff, A. J., Handy, L. M., and Rosanoff, I. A. (1934). Criminality and delinquency in twins. *Journal of Criminal Law and Criminology* 24, 923–34.

Rosch, P. J. (1993). Stressful life events and Graves' disease. *Lancet* 342, 566–7.

Rosen, B. K. (1981). Suicide pacts: a review. *Psychological Medicine* 11, 525–33.

Rosen, I. (1979). Exhibitionism, scopophilia and voyeurism. In *Sexual deviations* (ed. I. Rosen) (2nd edn). Oxford University Press.

Rosenbaum, M. (1983). Crime and punishment—the suicide pact. *Archives of General Psychiatry* 40, 979–82.

Rosenfeld, B. D. (1992). Court-ordered treatment of spouse abuse. *Clinical Psychology Review* 12, 205–26.

Rosengren, A., Orth-Gomer, K., Wedel, H., and Wilhelmsen, L. (1993). Stressful life events, social support, and mortality in men born in 1933. *British Medical Journal* 307, 1102–5.

Rosenman, R. H., Brand, R. J., Jenkins, C. D., *et al.* (1975). Coronary heart disease: a Western Collaborative Group study. Final follow up experience of eight and a half years. *Journal of the American Medical Association* 233, 872–7.

Rosenthal, A. (1993). Adults with tertralogy of Fallot—repaired, yes; cured, no. *New England Journal of Medicine* 329, 655–6.

Rosenthal, D., Wender, P. H., Kety, S. S., and Welner, J. (1971). The adopted-away offspring of schizophrenics. *American Journal of Psychiatry* 128, 307–11.

Rosenthal, N. E., Sack, D. A., Gillin, J. C., *et al.* (1984). Seasonal affective disorder. *Archives of General Psychiatry* 41, 72–80.

Rosenthal, N. E., Sack, D. A., Carpenter, C. J., *et al.* (1985). Antidepressant effect of light in seasonal affective disorder. *American Journal of Psychiatry* 142, 163–70.

Rosenthal, P. A. and Rosenthal, S. (1984). Suicide behaviour by pre-school children. *American Journal of Psychiatry* 141, 520–5.

Ross, C. A., McInnis, M. G., Margolis, R. L., and Li, S.-H.

(1993). Genes with triplet repeats: candidate mediators of neuropsychiatric disorders. *Trends in Neurosciences* 16, 254–60.

Ross, D. M. and Ross, S. A. (1982). *Hyperactivity: current issues, research and theory.* Wiley, New York.

Ross, G. A., Miller, S. D., Reagon, P., *et al.* (1990). Structured interview data on 102 cases of multiple personality disorder for four centers. *American Journal of Psychiatry* 147, 596–601.

Rossor, M. N. (1994). Management of neurological disorders: dementia. *Journal of Neurology, Neurosurgery and Psychiatry* 57, 1451–6.

Roth, M. (1955). The natural history of mental disorder in old age. *Journal of Mental Science* 101, 281–301.

Roth, M. (1959). The phobic anxiety-depersonalization syndrome. *Proceedings of the Royal Society of Medicine* 52, 587–95.

Rothbaum, B. O. and Ninan, P. T. (1994). The assessment of trichotillomania. *Behaviour Research and Therapy* 32, 651–62.

Roth, M. and Ball, J. R. B. (1964). Psychiatric aspects of intersexuality. In *Intersexuality in vertebrates including man* (ed. C. N. Armstrong and A. J. Marshall). Academic Press, London.

Roth, M. and Kroll, J. (1987). *The reality of mental illness.* Cambridge University Press.

Rothbaum, B. O., Foa, E. B., Riggs, D. S., Murdock, T., and Walsh, W. (1992). A prospective examination of post-traumatic stress disorder in rape victims. *Journal of Traumatic Stress* 5(3), 455–76.

Rotheram-Borus, M. J. (1994). Brief cognitive–behavioral treatment for adolescent suicide attempters and their families. *Journal of the American Academy of Child and Adolescent Psychiatry* 33(4), 508.

Rothlind, J. C., Bylsma, F. W., Peyser, C., *et al.* (1993). Cognitive and motor correlates of everyday functioning in early Huntington's disease. *Journal of Nervous and Mental Disease* 181, 194–8.

Rothman, D. (1971). *The discovery of the asylum.* Little Brown, Boston, MA.

Rowan, P. R., Paykel, E. S., and Parker, R. P. (1982). Phenelzine and amitriptyline effects on symptoms of neurotic depression. *British Journal of Psychiatry* 140, 475–83.

Rowland, J. H., *et al.* (1993). Psychological response to breast reconstruction. *Psychosomatics* 34(3), 241–50.

Roy, A. (1982). Suicide in chronic schizophrenia. *British Journal of Psychiatry* 141, 171–7.

Roy, A. (1985). Suicide in doctors. *Psychiatric Clinics of North American* 8, 377–87.

Roy, A., Virkkunen, M., and Linnoila, M. (1990). Serotonin in suicide, violence and alcoholism. In *Serotonin in major psychiatric disorders* (ed. E. F. Coccaro and D. L. Murphy), pp. 187–208. American Psychiatric Press, Washington, DC.

Royal College of Physicians (1987). *A great and growing evil: the medical consequences of alcohol abuse.* Tavistock Publications, London.

Royal College of Physicians (1991). *Physical signs of sexual abuse in children.* Royal College of Physicians, London.

Royal College of Psychiatrists (1986). *Alcohol: our favourite drug: new report on alcohol and alcohol related problems.* Tavistock Publications, London.

Royal College of Psychiatrists (1995). *ECT handbook.* Royal College of Psychiatrists, London.

Royal College of Psychiatrists (1993). *Consensus statement on the use of high dose antipsychotic medication.* Council report CR26. Royal College of Psychiatrists, London.

Roy-Byrne, P., Post, R. M., Uhde, T. B., *et al.* (1985). The longitudinal course of recurrent affective illness: life chart data from research patients at the NIMH. *Acta Psychiatrica Scandinavica* 71 (Suppl. 317), 1–34.

Royston, M. C. and Simpson, M. D. C. (1991). Post-mortem neurochemistry of schizophrenia. In *Neurobiology and psychiatry*, Vol. 1 (ed. R. Kerwin), pp. 1–13. Cambridge University Press.

Rubin, R. T., Villanueva-Meyer, J., Ananth, J., *et al.* (1992). Regional xenon 133 cerebral blood flow and cerebral technetium Tc99m-HMPAO uptake in unmedicated patients with obsessive compulsive disorder and match normal control subjects. *Archives of General Psychiatry* 49, 695–702.

Rubinow, D. R., Post, R. M., Savard, R., and Gold, P. W. (1984). Cortisol hypersecretion and cognitive impairment in depression. *Archives of General Psychiatry* 41, 279–83.

Rubinsztein, D. C., Barton, D. E., and Ferguson-Smith, M. A. (1994). Issues in Huntington's disease testing. *Quarterly Journal of Medicine* 87, 71–3.

Rüdin, E. (1916). Studien über Vererbung und Entstehung geistiger Störungen: I. *Zur Vererbung und Neuentstehung der Dementia Praecox.* Springer, Berlin.

Rüdin, E. (1953). Ein Beitrag zur Frage der Zwangskrankheit, unsbesondere ihrer hereditären Beziehungen. *Archiv für Psychiatrie und Nervenkrankheiten* 191, 14–54.

Rudorfer, M. V. and Potter, W. Z. (1989). Antidepressants: a comparative review of the clinical pharmacology and therapeutic use of the 'newer' versus the older drugs. *Drugs* 37, 713–38.

Rundell, J. R., Ursano, R. J., Holloway, H. C., and Silberman, E. K. (1989). Psychiatric responses to trauma. *Hospital and Community Psychiatry* 40(1), 68–73.

Rush, A. J., Beck, A. T., Kovacs, M., and Hollon, S. (1977). Comparative efficacy of cognitive therapy and imipramine in the treatment of depressed out-patients. *Cognitive Therapy and Research* 1, 17–31.

Rush, A. J., Kain, J. W., Rease, J., *et al.* (1991).

Neurological bases for psychiatric disorders. In *Comprehensive neurology* (ed. R. N. Rosenberg), pp. 555–603. Raven Press, New York.

Russell, D. E. H. (1984). The prevalence and seriousness of incestuous abuse: stepfathers versus biological fathers. *Child Abuse and Neglect* 8, 15–22.

Russell, G. F. M. (1979). Bulimia nervosa: an ominous variant of anorexia nervosa. *Psychological Medicine* 9, 429–48.

Russell, G. F. M., Szmulker, G., Dare, C., and Eisler, I. (1987). An evaluation of family therapy in anorexia nervosa and bulimia nervosa. *Archives of General Psychiatry* 44, 1047–56.

Russell, O. (1970). Autistic children: infancy to adulthood. *Seminars in Psychiatry* 2, 435–40.

Rutter, M. (1966). *Children of sick parents: an environmental and psychiatric study*. Institute of Psychiatry, Maudsley Monographs No. 16, Oxford University Press.

Rutter, M. (1971). Parent–child separation: psychological effects on the children. *Journal of Child Psychology and Psychiatry* 14, 201–8.

Rutter, M. (1972). Relationships between child and adult psychiatric disorders. *Acta Psychiatrica Scandinavica* 48, 3–21.

Rutter, M. (ed.) (1980). *Scientific foundations of developmental psychiatry*. Heinemann, London.

Rutter, M. (1981). *Maternal deprivation reassessed*. Penguin, Harmondsworth.

Rutter, M. (1983). Cognitive deficits in the pathogenesis of autism. *Journal of Child Psychology and Psychiatry* 24, 513–32.

Rutter, M. (1985a). Infantile autism and other pervasive developmental disorders. In *Child and adolescent psychiatry: modern approaches* (ed. M. Rutter and L. Hersov) (2nd edn). Blackwell, Oxford.

Rutter, M. (1985b). Resilience in the face of adversity: protective factors and resistance to psychiatric disorder. *British Journal of Psychiatry* 147, 598–611.

Rutter, M. (1985c). The treatment of autistic children. *Journal of Child Psychology and Psychiatry* 2, 193–214.

Rutter, M. (1995). Relationships between mental disorders in childhood and adulthood. *Acta Psychiatrica Scandinavica* 91, 73–85.

Rutter, M. and Giller, H. (1983) *Juvenile delinquency: trends and perspectives*. Penguin, Harmondsworth.

Rutter, M. and Gould, M. (1985). Classification. In *Child and adolescent psychiatry: modern approaches* (ed. M. Rutter and L. Hersov) (2nd edn). Blackwell, Oxford.

Rutter, M. and Lockyer, L. (1967). A five to fifteen year follow-up study of infantile psychosis: I. Description of sample. *British Journal of Psychiatry* 113, 1169–82.

Rutter, M. and Madge, N. (1976). *Cycles of disadvantage: a review of research*. Heinemann, London.

Rutter, M., Graham, P., and Birch, H. G. (1970a). A neuropsychiatric study of childhood. Clinics in Developmental Medicine No. 35/36. Heinemann, London.

Rutter, M., Tizard, J., and Whitmore, K. (eds) (1970b). *Education, health and behaviour*. Longmans, London.

Rutter, M., Yule, W., Berger, M., *et al.* (1974). Children of West Indian immigrants. I. Rates of behavioural deviance and of psychiatric disorder. *Journal of Child Psychology and Psychiatry* 15, 241–62.

Rutter, M., Shaffer, D., and Shepherd, M. (1975a). *A multiaxial classification of child psychiatric disorders*. World Health Organization, Geneva.

Rutter, M. L., Cox, A., Tupling, C., *et al.* (1975b). Attainment and adjustment in two geographical areas: I. Prevalence of psychiatric disorders. *British Journal of Psychiatry* 126, 493–509.

Rutter, M., Yule, B., Quinton, D., *et al.* (1975c). Attainment and adjustment in two geographical areas III: Some factors accounting for area differences. *British Journal of Psychiatry* 126, 520–33.

Rutter, M., Tizard, J., Yule, W., *et al.* (1976a). Isle of Wight Studies 1964–1974. *Psychological Medicine* 6, 313–32.

Rutter, M., Graham, P., Chadwick, O., and Yule, W. (1976b). Adolescent turmoil: fact or fiction. *Journal of Child Psychology and Psychiatry* 17, 35–56.

Rutter, M., Chadwick, O., and Shaffer, D. (1983). Head injury. In *Developmental neuropsychiatry* (ed. M. Rutter), pp. 83–111. Churchill Livingstone, Edinburgh.

Rutter, M. L., MacDonald, H., LeCouteur, A., *et al.* (1990). Genetic factors in child psychiatric disorder. II Empirical findings. *Journal of Child Psychology and Psychiatry* 31, 39–83.

Rutter, M., Bailey, A., Bolton, P., and Le Couter, A. (1993). Autism: syndrome definition and possible genetic mechanisms. In *Nature, Nurture and Psychology* (ed. R. Plomin, and G. E. McClearn), pp. 269–84. American Psychiatric Association, Washington, DC.

Rutter, M., Taylor, E., and Hersov, L. (ed.) (1994). *Child and adolescent psychiatry: modern approaches* (3rd edn). Blackwell Scientific Publications, Oxford.

Rutz, W., von Knorring, L., and Walinder, J. (1989). Frequency of suicide on Götland after systematic postgraduate education of general practitioners. *Acta Psychiatrica Scandinavica* 80, 151–4.

Rutz, W., von Knorring, L., and Walinder, J. (1992). Long-term effects of an educational program for general practitioners given by the Swedish Committee for the prevention and treatment of depression. *Acta Psychiatrica Scandinavica* 85, 83–8.

Ryan, N. D. and Puig-Antich, J. (1986). Affective illness in adolescence. *American Psychiatric Association Annual Review*, Vol. 5 (ed. A. J. Frances and R. E. Hales). American Psychiatric Association, Washington, DC.

Ryle, A. (1990). *Cognitive analytic therapy: active participation in change*. Wiley, Chichester.

Saario, I., Linnoila, M., and Maki, M. (1975). Interaction of drugs with alcohol on human psychomotor skills related to driving: effects of sleep deprivation or two weeks treatment with hypnotics. *Journal of Clinical Pharmacology* 15, 52–9.

Sackeim, H. A. (1994). Use of electroconvulsive therapy in late-life depression. In *Diagnosis and treatment of depression in late life* (ed. L. S. Schneider), pp. 259–77. American Psychiatric Press, Washington, DC.

Sackheim, H. A., Prudic, J., Devanand, D. P., *et al.* (1990). The impact of medication resistance and continuation of pharmacotherapy on relapse following response to electroconvulsive therapy in major depression. *Journal of Clinical Psychopharmacology* 10, 96–104.

Sackheim, H. A., Prudie, J., Devanand, D. P., *et al.* (1993). Effects of stimulus intensity and electrode placement on the efficacy and cognitive effects of electroconvulsive therapy. *New England Journal of Medicine* 328, 839–46.

Sacks, O. (1973). *Awakenings.* Duckworth, London.

Saghir, M. T. and Robins, E. (1973). *Male and female homosexuality: a comprehensive investigation.* Williams and Wilkins, Baltimore, MD.

Sainsbury, P. (1955). *Suicide in London.* Maudsley Monograph No. 1. Chapman & Hall, London.

Sainsbury, P. (1962). Suicide in later life. *Gerontologia Clinica* 4, 161–70.

Sainsbury, P. (1986). The epidemiology of suicide. In *Suicide* (ed. A. Roy). Williams and Wilkins, Baltimore, MD.

Sainsbury, P. and Barraclough, B. (1968). Differences between suicide rates. *Nature, London* 220, 1252–3.

St Clair, D. (1994). Genetics of Alzheimer's disease. *British Journal of Psychiatry* 164, 153–6.

Sakel, M. (1938). *The pharmacological shock treatment of schizophrenia.* Nervous and Mental Diseases Monograph Series No. 62. Nervous and Mental Diseases Publications, New York.

Salkovskis, P. M. and Warwick, H. M. C. (1986). Morbid preoccupations, health anxiety and reassurance: a cognitive–behavioural approach to hypochondriasis. *Behaviour Research and Therapy* 24, 597–602.

Salkovskis, P. M. and Clark, D. M. (1993). Panic disorder and hypochondriasis. *Advances in Behaviour Research and Therapy* 15, 23–48.

Salkovskis, P. M., Atha, C., and Storer, D. (1990). Cognitive–behavioural problem solving in the treatment of patients who repeatedly attempt suicide. A controlled trial. *British Journal of Psychiatry* 157, 871–6.

Salminen, J. K., Saarijarvi, S., and Aarela, E. (1995). Two decades of alexithymia. *Journal of Psychosomatic Research* (in press).

Salomon, R. M., Miller, H. L., Delgado, P. L., and Charney, D. (1993). The use of tryptophan depletion to evaluate central serotonin function in depression and other neuropsychiatric disorders. *International Clinical Psychopharmacology* 8 (Suppl. 2), 41–6.

Salter, A. (1949). *Conditioned reflex therapy.* Farrar Strauss, New York.

Sameroff, A., Seifer, R., Barocas, R., *et al.* (1987). IQ scores of 4-year-old children: social-environmental risk factors. *Pediatrics* 79, 343–50.

Sanders, R. D., Keschaven, M. S., and Schooler, N. R. (1994). Neurological examination abnormalities in neuroleptic-naive patients with first break schizophrenia: preliminary result. *American Journal of Psychiatry* 151, 1231–3.

Sanderson, W. C. and Barlow, D. H. (1990). A description of patients diagnosed with DSM-III-R generalized anxiety disorder. *Journal of Nervous and Mental Disease* 178, 588–91.

Sandifer, M. G., Hordern, A., Timbury, G. C., and Green, L. M. (1968). Psychiatric diagnosis: a comparative study in North Carolina. *British Journal of Psychiatry* 114, 1–9.

Sandler, J., Dare, C., and Holder, A. (1992). *The patient and the analyst.* Karnac, London.

Sapolsky, R. M. (1992). *Stress, the ageing brain and the mechanisms of neuron death.* MIT Press, Cambridge, MA.

Sapolsky, R. M., Uno, H., Rebert, C., and Finch, C. (1990). Hippocampal damage associated with prolonged glucocorticoid exposure in primates. *Journal of Neuroscience* 10, 2897–904.

Sargant, W. and Dally, P. (1962). Treatment of anxiety state by antidepressant drugs. *British Medical Journal* i, 6–9.

Sargant, W. and Slater, E. (1940). Acute war neuroses. *Lancet* ii, 1–2.

Sargant, W. and Slater, E. (1963). *An introduction to physical methods of treatment in psychiatry.* Livingstone, Edinburgh.

Sarrell, P. M. and Masters, W. H. (1982). Sexual molestation of men by women. *Archives of Sexual Behaviour* 11, 117–31.

Sartorius, N., Kaelber, C. T., Cooper, J. E., *et al.* (1993). Progress toward achieving a common language in psychiatry: results from the field trial of the clinical guidelines accompanying the WHO classification of mental and behavioral disorders in ICD-10. *Archives of General Psychiatry* 50, 115.

Sassim, N. and Grohmann, R. (1988). Adverse reactions with clozapine and simultaneous application of benzodiazepines. *Pharmacopsychiatry* 21, 306–7.

Satir, V. (1967). *Conjoint family therapy.* Science and Behaviour Books, Palo Alto, CA.

Saykin, A. J., Shtasel, D. J., Gur, R. E., *et al.* (1994). Neuropsychological deficits in neuroleptic naive patients with first-episode schizophrenia. *Archives of General Psychiatry* 51, 124–31.

Schachar, R. (1991). Childhood hyperactivity. *Journal of Child Psychology and Psychiatry* 32, 155–91.

Schade, D. S., Drumm, D. A., Eaton, R. P., and Sterling, W. A. (1985). Factitious brittle diabetes mellitus. *American Journal of Medicine* 78, 777–83.

Schapira, K., Davison, K., and Brierley, H. (1979). The assessment and management of transsexual problems. *British Journal of Hospital Medicine* 22, 63–9.

Schapira, K., Roth, M., Kerr, T. A., and Gurney, C. (1972). The prognosis of affective disorders. The differentiation of anxiety states from depressive illness. *British Journal of Psychiatry* 121, 175–81.

Scharfetter, C. (1980). *General psychopathology: an introduction* (transl. H. Marshall). Cambridge University Press.

Schatzberg, A. F. and Cole, J. O. (1991*a*). Combination and adjunctive treatments. In *Manual of clinical psychopharmacology* (ed. A. F. Schatzberg and J. O. Cole), pp. 263–79. American Psychiatric Press, Washington, DC.

Schatzberg, A. F. and Cole, J. O. (1991*b*). Anxiolytic drugs. In *Manual of clinical psychopharmacology* (ed. A. F. Schatzberg and J. O. Cole), pp. 85–143. American Psychiatric Press, Washington, DC.

Scheff, T. J. (1963). The role of the mentally ill and the dynamics of mental disorder: a research framework. *Sociometry* 26, 436–53.

Schiffer, R. B. and Wineman, N. M. (1990). Antidepressant pharmacotherapy with multiple sclerosis. *American Journal of Psychiatry* 147, 1493–7.

Schilder, P. (1935). *The image and appearance of the human body*. International Universities Press, New York.

Schittecatte, M., Charles, G., Machowski, R., and Wilmott, E. (1989). Tricyclic washout and growth hormone response to clonidine. *British Journal of Psychiatry* 154, 858–63.

Schmideberg, M. (1947). The treatment of psychopaths and borderline patients. *American Journal of Psychotherapy* 1, 45–70.

Schmidtke, A. and Häfner, H. (1988). The Werther effect after television films. *Psychological Medicine* 18, 665–76.

Schmitt, B. D. (1986). New enuresis alarms: safe successful and child operable. *Archives of Disease in Childhood* 43, 665–71.

Schneider, K. (1950). *Psychopathic personalities* (9th edn) (transl. M. W. Hamilton). Cassel, London.

Schneider, K. (1959). *Clinical psychopathology*. Grune and Stratton, New York.

Schneier, F. R., Jonson, J., Hornig, C. D., *et al.* (1992). Social phobia: comorbidity and morbidity in an epidemiologic sample. *Archives of General Psychiatry* 49, 282.

Schneier, F. R., Spitzer, R. L., Gibbon, M., *et al.* (1991). The relationship of social phobia subtypes and avoidant personality disorder. *Comprehensive Psychiatry* 32(6), 496–502.

Schooler, N. R., Levine, J., Severe, J. B., *et al.* (1980). Prevention of relapse in schizophrenia. *Archives of General Psychiatry* 37, 16–24.

Schor, J. D., Levkoff, S. E., Lipsitz, L. A., *et al.* (1992). Risk factors for delirium in hospitalized elderly. *Journal of the American Medical Association* 267, 827–31.

Schou, M. (1988). Effects of long-term lithium treatment on kidney function: an overview. *Journal of Psychiatric Research* 22, 287–96.

Schou, M. (1993). Is there a lithium withdrawal syndrome? An examination of the evidence. *British Journal of Psychiatry* 163, 514–18.

Schou, M., Amisden, A., Jensen, S. E., and Olsen, T. (1968). Occurrence at goitre during lithium treatment. *British Medical Journal* iii, 710–13.

Schreiner-Engel, P., and Schiavi, R. C. (1986). Lifetime psychopathology in individuals with low sexual desire. *Journal of Nervous and Mental Disease* 174, 646–51.

Schrenck-Notzing, A. von (1895). *The use of hypnosis in psychopathia sexualis with special reference to contrary sexual instinct* (trans. C. G. Chaddock). Institute of Research in Hypnosis Publication Society and the Julian Press, New York (1956).

Schuckit, M. A., Smith, T. L., Anthenellic, R. A., and Irwin, M. (1993). A clinical course of alcoholism in 636 male inpatients. *American Journal of Psychiatry* 150, 786–92.

Schulsinger, F. (1982). Psychopathy: heredity and environment. *International Journal of Mental Health* 1, 190–206.

Schultz, J. H. (1932). *Das autogene training*. Thieme, Liepzig.

Schultz, J. H. and Luthe, W. (1959). *Autogenic training: a psychophysiological approach*. Grune and Stratton, New York.

Schultz P. M., Schulz, S. C., Golberg, S. C., Ettigi, P., Resnick, R. J., and Friedel, R. O. (1986). Diagnoses of relatives of schizotypal outpatients. *Journal of Nervous and Mental Disease* 174, 457–63.*et al.* (1986).

Schwartz, A. (1993). The epidemiology of suicide among students at colleges and universities in the United States. In *College student suicide* (ed. L. C. Whitaker and R. E. Slimak), pp. 25–44. Haworth Press, New York.

Schwartz, J. C., Levesque, D., Martres, M.-P., and Sokoloff, P. (1993). Dopamine D_3 receptor: basic and clinical aspects. *Clinical Neuropharmacology* 16, 295–314.

Schwartz, M. A. (1973). Pathways of metabolism of diazepines. *The benzodiazepines* (ed. S. Garrattini, E. Mussini, and L. O. Randall). Raven Press, New York.

Schwarz, S. P. and Blanchard, E. B. (1991). Evaluation of a psychological treatment for inflammatory bowel disease. *Behaviour Research and Therapy* 29(2), 167–77.

Scott, A. I. F., Weeks, D. J., and McDonald, C. F. (1991). Continuation of electroconvulsive therapy: preliminary guidelines and an illustrative case report. *British Journal of Psychiatry* 159, 867–70.

Scott, P. D. (1960). The treatment of psychopaths. *British Medical Journal* i, 1641–6.

Scott, P. D. (1973). Parents who kill their children. *Medicine, Science and the Law* 13, 120–6.

Scott, S. (1994). Mental retardation. In *Child and adolescent psychiatry: modern approaches* (3rd edn) (ed. M. Rutter, E. Taylor, and L. Hersov), pp. 616–46. Blackwell Scientific Publications, Oxford.

Seager, C. P. and Flood, R. A. (1965). Suicide in Bristol. *Bristol Journal of Psychiatry* 111, 919–32.

Sedler, M. J. (1985). The legacy of Ewald Hecker: a new translation of 'Die Hebephrenie'. *American Journal of Psychiatry* 142, 1265–71.

Sedman, G. (1966). A phenomenological study of pseudo-hallucinations and related experiences. *British Journal of Psychiatry* 113, 1115–21.

Sedman, G. (1970). Theories of depersonalization: a reappraisal. *British Journal of Psychiatry* 117, 1–14.

Sedvall, G., Farde, L., Persson, A., and Wiesel, F. A. (1986). Imaging of neurotransmitter receptors in the living human brain. *Archives of General Psychiatry* 43, 995–1005.

Seeman, P. and Van Tol, H. M. (1993). Dopamine receptor pharmacology. *Current Opinion in Neurology and Neurosurgery* 6, 602–8.

Seeman, P., Guan, H-C., and Van Tol, H. M. M. (1993). Dopamine D_4 receptors elevated in schizophrenia. *Nature, London* 365, 441–5.

Segal, H. (1963). *Introduction to the work of Melanie Klein.* Heinemann Medical, London.

Seguin, E. (1864). Origin of the treatment and training of idiots. In *History of mental retardation* (ed. M. Rosen, G. R. Clark, and M. S. Kivitz), Vol. 1. University Park Press, Baltimore, MD (1976).

Seguin, E. (1866). *Idiocy and its treatment by the physiological method.* Brandown, Albany, NY.

Seligman, M. E. P. (1975). *Helplessness: on depression, development and death.* Freeman, San Francisco, CA.

Sell, L. A., Cowen, P. J., and Robson, P. J. (1995). Ondansetron and opiate craving; a novel pharmacological approach to addiction. *British Journal of Psychiatry* 166, 511–14.

Selye, H. (1950). *Stress.* Acta, Montreal.

Serfaty, M. and Masterton, G. (1993). Fatal poisonings attributed to benzodiazepines in Britain during the 1980s. *British Journal of Psychiatry* 163, 286–393.

Serieux, P. and Capgras, J. (1987). Misinterpretation delusion states (English translation). In *The clinical roots of the schizophrenia concept* (ed. J. Cutting and M. Shepherd). Cambridge University Press.

Shaffer, D. (1974). Suicide in childhood and early adolescence. *Journal of Child Psychology and Psychiatry* 15, 275–91.

Shaffer, D. (1985). Enuresis. In *Child psychiatry: modern approaches* (ed. M. Rutter and L. Hersov) (2nd edn). Blackwell. Oxford.

Shaffer, D. and Piacentini, J. (1994). Suicide and attempted suicide. In *Child and adolescent psychiatry: modern approaches* (3rd edn) (ed. M. Rutter, E. Taylor, and L. Hersov), pp. 407–24. Blackwell Scientific Publications, Oxford.

Shaffer, D., Costello, A. J., and Hill, I. D. (1968). Control of enuresis with imipramine. *Archives of Diseases of Childhood* 43, 665–71.

Shaffer, D., Garland, A., Gould, M., et al. (1988). Preventing teenage suicide: a critical review. *Journal of the American Academy of Child and Adolescent Psychiatry* 27, 675–87.

Shakin, E. J. and Thompson, A. (1993). Hematologic disorders. In *Psychiatric care of the medical patient* (ed. A. Stoudemire). Oxford University Press.

Shalev, A., Hermesh, H., and Munitz, H. (1989). Mortality from neuroleptic malignant syndrome. *Journal of Clinical Psychiatry* 50, 18–25.

Shallice, T., Burgess, B. W., and Frith, C. D. (1991). Can the neuropsychological case study approach be applied to schizophrenia? *Psychological Medicine* 21, 661–73.

Sham, P. C., O'Callaghan, E., Takei, N., et al. (1992). Schizophrenia following pre-natal exposure to influenza epidemics between 1939 and 1960. *British Journal of Psychiatry* 160, 461–6.

Shannon, F. T., Fergusson, D. M., and Dimond, M. E. (1984). Early hospital admissions and subsequent behaviour problems in six-year-olds. *Archives of Diseases of Childhood* 59, 815–19.

Shaper, A. G. (1990). Alcohol and mortality: a review of prospective studies. *British Journal of Addiction* 85, 837–47.

Shapiro, C. M. and Dement, W. C. (1993). Impact and epidemiology of sleep disorders. *British Medical Journal* 306, 1604–7.

Shapiro, D. (1976). The effects of therapeutic conditions: positive results revisited. *British Journal of Medical Psychology* 49, 315–23.

Sharan, S. N. (1965). Family interaction with schizophrenia and their siblings. *Journal of Abnormal Psychology* 71, 345–53.

Sharma, T. and Murray, R. M. (1993). Aetiological theories in schizophrenia. *Current Opinion in Psychiatry* 6, 80–4.

Sharp, C. W. and Freeman, P. L. (1993). The medical complications of anorexia. *British Journal of Psychiatry* 162, 452–62.

Sharpe, M. (1994). Cognitive behavioural therapy. In *Chronic fatigue syndrome* (ed. S. Straus). Dekker, New York.

Sharpe, M., Peveler, R., and Mayou, R. A. (1992). Invited review. The psychological treatment of patients with functional somatic symptoms: a practical guide. *Journal of Psychosomatic Research* **36**, 515–29.

Sharpe, M., Hawton, K., Seagroatt, V., *et al.* (1994). Depressive disorders in long-term survivors of stroke. Associations with demographic and social factors, functional status, and brain lesion volume. *British Journal of Psychiatry* **164**, 380–6.

Shavitt, R. G., Gentil, V., and Mandetta, R. (1992). The association of panic/agoraphobia and asthma. Contribution factors and clinical implications. *General Hospital Psychiatry* **14**(6), 420–3.

Sheldon, W. H., Stevens, S. S., and Tucker, W. B. (1940). *The varieties of human physique.* Harper, London.

Sheldon, W. H., Stevens, S. S., and Tucker, W. B. (1942). *The varieties of temperament.* Harper, London.

Shepherd, M. (1961). Morbid jealousy: some clinical and social aspects of a psychiatric symptom. *Journal of Mental Science* **107**, 687–753.

Shepherd, M., Cooper, B., Brown, A. C., and Kalton, G. W. (1966). *Psychiatric illness in general practice.* Oxford University Press, London.

Shepherd, M., Lader, M., and Rodnight, R. (1968). *Clinical psychopharmacology.* English Universities Press, London.

Shepherd, M., Harwin, B. G., Depla, C., and Cairns, V. (1979). Social work and primary care of mental disorder. *Psychological Medicine* **9**, 661–70.

Sherman, R. A., Sherman, C. J., and Bruno, G. M. (1987). Psychological factors influencing chronic phantom limb pain: an analysis of the literature. *Pain* **28**, 285–95.

Sherman, D. I. N., Ward, R. J., Warren-Perry, M., *et al.* (1993). Association of restriction fragment length polymorphism in alcohol dehydrogenase 2 gene with alcohol induced liver damage. *British Medical Journal* **307**, 1388–90.

Shestatzky, M., Greenberg, D., and Lehrer, B. (1988). A controlled trial of phenelzine in post-traumatic stress disorder. *Psychiatry Research* **24**, 149–55.

Shibuya, A. and Yoshida, A. (1988). The genotypes of alcohol-metabolising enzymes in Japanese with alcohol liver disease. *American Journal of Human Genetics* **43**, 744–8.

Shields, J. (1962). *Monozygotic twins brought up apart and brought up together.* Oxford University Press, London.

Shields, J. (1980). Genetics and mental development. In *Scientific foundations of developmental psychiatry* (ed. M. Rutter). Heinemann Medical, London.

Shillito, F. H., Drinker, C. K., and Shaughnessy, T. J. (1936). The problem of nervous and mental sequelae of carbon monoxide poisoning. *Journal of the American Medical Association* **106**, 669–74.

Shore, J. H., Vollmer, W. M., and Tatum, E. L. (1989). Community patterns of post-traumatic stress disorder. *Journal of Nervous and Mental Disease* **177**, 681–5.

Shorvon, S. D. (1990). Epidemiology, classification, natural history and genetics of epilepsy. *Lancet* ii, 3–6.

Shorvon, H. J., Hill, J. D. N., Burkitt, E., and Hastead, H. (1946). The depersonalization syndrome. *Proceedings of the Royal Society of Medicine* **39**, 779–92.

Showers, J. and Pickrell, E. (1987). Child firesetters: a study of three populations. *Hospital and Community Psychiatry* **38**, 495–501.

Shulman, R. (1967). Vitamin B_{12} deficiency and psychiatric illness. *British Journal of Psychiatry* **113**, 252–6.

Siegelman, M. (1974). Parental background of male homosexuals and heterosexuals. *Archives of Sexual Behaviour* **3**, 3–18.

Sifneos, P. (1979). *Short term dynamic therapy.* Plenum Medical, New York.

Sigal, M., Gelkopf, M., and Meadow, R. (1989). Munchausen by proxy syndrome: the triad of abuse, self-abuse and deception. *Comprehensive Psychiatry* **30**, 527–33.

Silver, J. M., Yudofsky, S. C., and Hales, R. E. (ed.) (1994). *Neuropsychiatry of traumatic brain injury.* American Psychiatry Press, Washington, DC.

Silverstone, T. and Cookson, J. (1982). The biology of mania. In *Recent advances in clinical psychiatry* (ed. K. Granville-Grassman), pp. 201–41. Churchill Livingstone, Edinburgh.

Simon, G. E., Danniell, W., Stockbridge, H., *et al.* (1993). Immunologic, psychological, and neuropsychological factors in multiple chemical sensitivity: a controlled study. *Annals of Internal Medicine* **119**(2), 97–103.

Simon, N. M., Garber, E., and Arieff, A. J. (1977). Persistent nephrogenic diabetes insipidus after lithium carbonate. *Annals of Internal Medicine* **86**, 446–7.

Simons, R. C. and Hughes, C. C. (1985). *Culture bound syndromes: folk illnesses of psychiatric and anthropological interest.* Reidel, Dordrecht.

Simpson, D. M. and Tagliatti, M. (1994). Neurologic manifestations of HIV infection. *Annals of Internal Medicine* **121**, 769–85.

Simpson, G. M. and De Leon, J. (1989). Tyramine and new monoamine oxidase inhibitor drugs. *British Journal of Psychiatry* **155** (Suppl. 6), 32–7.

Simpson, L. (1990). The comparative efficacy of Milan Family Therapy for disturbed children and their families. *Journal of Family Therapy* **13**, 267–84.

Sims, A. (1995). *Symptoms in the mind; an introduction to descriptive psychopathology*, (2nd edn). Baillière Tindall, London.

Singer, L. T., Garber, R., and Kliegman, R. (1991). Neurobehavioural sequelae of fetal cocaine exposure. *Journal of Pediatrics* **119**, 667–71.

Singer, M. T. and Wynne, L. C. (1965). Thought disorder and family relations of schizophrenics: IV. Results and implications. *Archives of General Psychiatry* **12**, 201–12.

Siris, S. G., Morgan, V., Fagerstrom, R., *et al.* (1987). Adjunctive imipramine in the treatment of post-psychotic depression. *Archives of General Psychiatry* **42**, 533–9.

Siris, S. G., Bermanzohn, P. C., Gonzales, A., Mason, S. E., White, C. V., and Shuwall, M. A. (1991). Use of antidepressants for negative symptoms in a subset of schizophrenic patients. *Psychopharmacology Bulletin* **27**, 331–5.

Sjöbring, H. (1973). Personality structure and development: a model and its applications. *Acta Psychiatrica Scandinavica* Suppl. 244.

Sjögren, T., Sjögren, H., and Lindgren, A. G. H. (1952). Morbus Alzheimer and morbus Pick. A genetic, clinical and patho-anatomical study. *Acta Psychiatrica Neurologica Scandinavica* Suppl. 82.

Skeels, H. (1966). Adult status of children with contrasting life experiences: a follow-up study. *Monograph of the Society for Research into Child Development* **31**(3).

Skegg, D. C. G., Doll, R., and Perry, J. (1977). Use of medicines in general practice. *British Medical Journal* i, 1561–3.

Skegg, K. (1993). Multiple sclerosis presenting as a pure psychiatric disorder. *Psychological Medicine* **23**, 909–14.

Skegg, K., Corwin, P. A., and Skegg, D. C. G. (1988). How often is multiple sclerosis mistaken for a psychiatric disorder? *Psychological Medicine* **18**, 733–6.

Skevington, S. M. (1986). Psychological aspects of pain in rheumatoid arthritis: a review. *Social Science and Medicine* **23**, 567–75.

Skinner, B. F. (1953). *Science and human behaviour.* Macmillan, New York.

Skodol, A. E. and Oldham, J. M. (1991). Assessment and diagnosis of borderline personality disorder. *Hospital and Community Psychiatry* **42**(10), 1021–8.

Skoog, I., Nilsson, L., Palmertz, B., *et al.* (1993). A population-based study of dementia in 85-year olds. *New England Journal of Medicine* **328**, 153–8.

Skowron, D. M. and Stimmel, G. L. (1992). Antidepressants and the risk of seizures. *Pharmacotherapy* **12**, 18–22.

Skuse, D. (1985). Non-organic failure to thrive. *Archives of Disease in Childhood* **60**, 173–8.

Skuse, D. (1989). Emotional abuse and delay in growth. In *ABC of child abuse* (ed. R. Meadow), pp. 23–5. British Medical Association, London.

Skuse, D. and Bentovim, A. (1994). Physical and emotional maltreatment. In *Child and adolescent psychiatry: modern approaches* (3rd edn) (ed. M. Rutter, E. Taylor, and L. Hersov), pp. 209–29. Blackwell Scientific Publications, Oxford.

Skynner, A. C. R. (1969). Indications for and against conjoint family therapy. *International Journal of Social Psychiatry* **15**, 245–9.

Skynner, A. C. R. (1991). Open-systems group-analytic approach to family therapy. In *Handbook of family therapy* (ed. A. S. Gurman and D. P. Kriskern). Brunner Mazel, New York.

Slater, E. (1951). Evaluation of electric convulsion therapy as compared with conservative methods in depressive states. *Journal of Mental Science* **97**, 567–9.

Slater, E. (1953). *Psychotic and neurotic illness in twins.* HMSO, London.

Slater, E. (1958). The monogenic theory of schizophrenia. *Acta Genetica Statistica Medica* **8**, 50–6.

Slater, E. (1961). Hysteria 311. *Journal of Mental Science* **107**, 359–81.

Slater, E. (1965). The diagnosis of hysteria. *British Medical Journal* i, 1395–9.

Slater, E. and Cowie, V. (1971). *The genetics of mental disorders.* Oxford University Press, London.

Slater, E. and Glithero, E. (1965). A follow-up of patients diagnosed as suffering from hysteria. *Journal of Psychosomatic Research* **9**, 9–13.

Slater, E. and Shields, J. (1969). Genetical aspects of anxiety. In *Studies of anxiety* (ed. M. H. Lader). *British Journal of Psychiatry* Special Publication No. 3.

Slater, E. Beard, A. W., and Glithero, E. (1963). The schizophrenia-like psychoses of epilepsy. *British Journal of Psychiatry* **109**, 95–150.

Slavney, P. R. and McHugh, P. R. (1974). The hysterical personality: a controlled study. *Archives of General Psychiatry* **30**, 325–9.

Slocumb, J., Kellner, R., Rosenfeld, R. C., and Palthak, D. (1989). Anxiety and depression in patients with the abdominal pelvic pain syndrome. *General Hospital Psychiatry* **11**, 58–53.

Small, G. W. (1986). Pseudocyesis: an overview. *Canadian Journal of Psychiatry* **31**, 452–7.

Small, I. F., Heimburger, R. F., Small, J. G., *et al.* (1977). Follow up of stereotaxic amygdalotomy for seizure and behaviour disorders. *Biological Psychiatry* **12**, 401–11.

Small, J. G., Klapper, M. H., Kellams, J. J., *et al.* (1988). Electroconvulsive treatment compared with lithium in the management of manic states. *Archives of General Psychiatry* **45**, 727–32.

Small, J. G., Klapper, M. H., Milstein, V., *et al.* (1991). Carbamazepine compared with lithium in the treatment of mania. *Archives of General Psychiatry* **48**, 915–21.

Smart, R. G. and Cutler, R. E. (1976). The alcohol advertising ban in British Columbia: problems and effects on beverage consumption. *British Journal of Addiction* **71**, 13–21.

Smith, A. L. and Weissman, M. M. (1992). Epidemiology. In *Handbook of affective disorders* (ed. E. S. Paykel), pp. 111–29. Churchill Livingstone, Edinburgh.

Smith, C. A. and Walston, K. A. (1992). Adaptation in patients with chronic rheumatoid arthritis: application of a general model. *Health Psychology* **11**(3), 151–62.

Smith, E. M., North, C. S., McCool, R. E., and Shea, J. M.

(1990). Acute postdisaster psychiatric disorders: identification of persons at risk. *American Journal of Psychiatry* 147, 202–6.

Smith, G. R. (1995). Treatment of patients with multiple symptoms. In *Treatment of functional somatic symptoms* (ed. R. A. Mayou, C. Bass, and M. Sharpe), pp. 175–87. Oxford University Press, Oxford.

Smith, G. R., Hanson, R. A., and Ray, D. C. (1986). Patients with multiple unexplained symptoms. *Archives of Internal Medicine* 146, 69–72.

Smith, J. C. and Hogan, B. (1988). *Criminal law* (6th edn). Butterworths, London.

Smith, J. S. and Brandon, S. (1973). Morbidity from acute carbon monoxide poisoning at 3 year follow up. *British Medical Journal* i, 318–21.

Smith, L. J. F. (1989). *Domestic violence: an overview of the literature*. Home Office Research Study No. 107. HMSO, London.

Smith, M. L. and Glass, G. V. (1977). Meta-analysis of psychotherapy outcome studies. *American Psychologist* 32, 752–60.

Smith, R. (1981). Alcohol, women, and the young: the same old problem? *British Medical Journal* 283, 1170–2.

Smith, R. (1985). Occupationless health. *British Medical Journal* 291, 1024–7, 1191–5, 1338–41, 1409–12.

Smith, R. (1989). Expertise, procedure and the possibility of a comparative history of forensic psychiatry in the nineteenth century. *Psychological Medicine* 19, 289–300.

Snaith, R. P., Baugh, S. J., Clayden, A. D., *et al.* (1982). The clinical anxiety scale: an instrument derived from the Hamilton Anxiety Scale. *British Journal of Psychiatry* 141, 518–23.

Sneddon, I. B. (1983). Simulated disease: problems in diagnosis and management. *Journal of the Royal College of Physicians* 17, 199–205.

Snowling, M. J. (1991). Developmental reading disorders. *Journal of Child Psychology and Psychiatry* 32, 49–77.

Soloff, P. H. and Millward, J. W. (1983). Psychiatric disorders in families of borderline patients. *Archives of General Psychiatry* 40, 37–44.

Soloff, P. H., Cornelius, J., Anselm, G., *et al.* (1993). Efficacy of phenelzine and haloperidol in borderline personality disorder. *Archives of General Psychiatry* 30, 377–86.

Soloff, P. H., George, A., Nathan, R. S., *et al.* (1986). Progress in the pharmacotherapy of borderline disorders: a double-blind study of amitriptyline, haloperidol and placebo. *Archives of General Psychiatry* 43, 691–7.

Solomon, S. D., Gerrity, E. T., and Muff, A. M. (1992). Efficacy of treatments for posttraumatic stress disorder. *Journal of the American Medical Association* 268, 633–8.

Solomon, Z. and Bromet, E. (1982). The role of social factors in affective disorder: an assessment of the vulnerabilty model of Brown and his colleagues. *Psychological Medicine* 12, 123–30.

Solowij, N. (1993). Ecstasy (3,4-methylenedioxymethanphetamine). *Current Opinion in Psychiatry* 6, 411–15.

Solowij, N., Hall, W., and Lee, N. (1992). Recreational MDMA use in Sidney: a profile of 'Ecstasy' users and their experience with the drug. *British Journal of Addiction* 87, 1161–72.

Song, F., Freemantle, N., Sheldon, T. A., *et al.* (1993). Selective serotonin reuptake inhibitors: a meta-analysis of efficacy and acceptability. *British Medical Journal* 306, 683–7.

Soothill, K. L. and Pope, P. J. (1973). Arson: a twenty-year cohort study. *Medicine, Science and the Law* 13, 127–38.

Sourindrin, I. (1985). Solvent abuse. *British Medical Journal* 290, 94–5.

South, N. (1994). Drugs: control, crime, and criminological studies. In *The Oxford handbook of criminology* (ed. M. Maguire, R. Morgan, and R. Reiner), pp. 393–440. Clarendon Press, Oxford.

Southard, E. E. (1910). A study of the dementia praecox group in the light of certain cases showing anomalies or sclerosis in particular brain regions. *American Journal of Insanity* 67, 119–76.

Southwick, S. M., Krystal, J. H., Morgan, C. A., *et al.* (1993). Abnormal noradrenergic function in posttraumatic stress disorder. *Archives of General Psychiatry* 50, 266.

Soyka, M., Naber, G., and Volcker, A. (1991). Prevalence of delusional jealousy in different psychiatric disorders. *British Journal of Psychiatry* 158, 549–53.

Spar, J. E. M. and Garb, A. S. (1992). Assessing competency to make a will. *American Journal of Psychiatry* 149(2), 169–74.

Sparr, L. and Atkinson, R. (1986). PTSD as an insanity defence. *American Journal of Psychiatry* 143, 608–13.

Speilberger, C. D., Gorsuch, R. L., and Lushene, R. (1970). *State-Trait Anxiety Inventory manual*. Consulting Psychologists Press, Palo Alto, CA.

Spicer, R. F. (1985). Adolescents. In *Oxford textbook of public health* (ed. W. Holland, R. Detels, and G. Knox), Vol. 4. Oxford University Press.

Spiegel, D. (1990). Can psychotherapy prolong cancer survival? *Psychosomatics* 31, 361–6.

Spiegel, D. (1994). Health caring: psychosocial support for patients with cancer. *Cancer* 74, 1453–7.

Spiegel, D., Bloom, J., and Kraemer, H. (1989). Effect of psychosocial treatment on survival in metastatic breast cancer. *Lancet* ii, 889–91.

Spiker, D. G. and Kupfer, D. J. (1988). Placebo response rate in psychotic and non-psychotic depression. *Journal of Affective Disorders* 14, 21–3.

Spiker, D. G., Weiss, J., Dealy, R., *et al.* (1985). The

pharmacological treatment of delusional depression. *American Journal of Psychiatry* 142, 430–6.

Spitzer, R. L. (1990). *User's guide to the structured clinical interview for DSMIIIR*. American Psychiatric Press, Washington, DC.

Spitzer, R. L. and Endicott, J. (1968). DIAGNO: a computer programme for psychiatric diagnosis utilizing the differential diagnostic procedures. *Archives of General Psychiatry* 18, 746–56.

Spitzer, R. L. and Williams, J. B. W. (1985). Classification in psychiatry. In *Comprehensive textbook of psychiatry* (ed. H. I. Kaplan and B. J. Sadock) (4th edn). Williams and Wilkins, Baltimore, MD.

Spitzer, R. L., Endicott, J., and Robins, E. (1978). Research diagnostic criteria: rationale and reliability. *Archives of General Psychiatry* 35, 773–82.

Spitzer, R. L., Endicott, J., and Gibson, M. (1979). Research diagnostic criteria: rationale and reliability. *Archives of General Psychiatry* 36, 17–24.

Spitzer, R. L., Williams, J. B. W., Gibbon, M., and First, M. B. (1990). *User's guide for the structured clinical interview for DSMIIIR*. American Psychiatric Association, Washington, DC.

Spitzer, R., First, M. B., Williams, J. B. W., *et al.* (1992). Now is the time to retire the term 'organic mental disorders'. *American Journal of Psychiatry* 149(2), 240–4.

Spohr, H. L., Willms, J., and Steinhausen, H.-C. (1993). Prenatal alcohol exposure and long-term developmental consequences. *Lancet* 341, 907–10.

Spreat, S. and Behar, D. (1994). Trends in the residential (inpatient) treatment of individuals with a dual diagnosis. *Journal of Consulting and Clinical Psychology* 62(1), 43–8.

Squire, L. R. (1977). ECT and memory loss. *American Journal of Psychiatry* 134, 997–1001.

Srinivasan, D. P. and Hullin, R. P. (1980). Current concepts of lithium. *British Journal of Hospital Medicine* 24, 466–75.

Stabenau, J. R. (1992). Is risk for substance abuse unitary? *Journal of Nervous and Mental Disorders* 180, 583–8.

Stahl, S. (1986). Tardive dyskinesia: natural history studies assist the pursuit of preventive therapies. *Psychological Medicine* 16, 491–4.

Stanhope, R., Adlard, P., Hamill, G., *et al.* (1988). Psychological growth hormone secretion during the recovery from psychosocial dwarfism: a case report. *Clinical Endocrinology* 28, 335–9.

Starkstein, S. E. and Robinson, R. G. (ed.) (1993). *Depression in neurologic disease*. The Johns Hopkins University Press, Baltimore, MD.

Stedeford, A. and Bloch, S. (1979). The psychiatrist in the terminal care unit. *British Journal of Psychiatry* 135, 7–14.

Stedeford, A. and Regnard, C. (1991). Confusional states in advanced cancer — a flow diagram. *Palliative Medicine* 5, 256–61.

Steffenburg, S., Gillberg, C., and Helgren, L. (1989). A twin study of autism in Denmark, Finland, Iceland, Norway and Sweden. *Journal of Child Psychology and Psychiatry* 30, 405–416.

Stein, A. and Fairburn, C. G. (1989). Children of mothers with bulimia nervosa. *British Medical Journal* 299, 777–8.

Stein, A., Gath, D. H., Bucher, J., *et al.* (1991). The relationship between post-natal depression and mother child interaction. *British Journal of Psychiatry* 158, 46–52.

Stein, A., Stein, J., Walters, E. A., and Fairburn, C. G. (1995). Eating habits and attitudes among mothers of children with feeding disorders. *British Medical Journal* 310, 228.

Stein, G. (1992). Drug treatment of personality disorder. *British Journal of Psychiatry* 161, 167–84.

Stein, L. J. and Test, M. A. (1980). Alternative to mental hospital treatment. 1. Conceptual model, treatment program and clinical evaluation. *Archives of General Psychiatry* 37, 392–7.

Stein, Z. and Susser, M. (1969). Widowhood and mental illness. *British Journal of Preventative and Social Medicine* 23, 106–10.

Steinberg, A. (1991). Issues in providing mental health services to hearing impaired persons. *Hospital and Community Psychiatry* 42(4), 380.

Steinberg, D. (1982). Treatment, training, care or control. *British Journal of Psychiatry* 141, 306–9.

Steinhausen, H. C., Rauss-Mason, C., and Seidel, R. (1991). Follow-up studies of anorexia nervosa: a review of four decades of outcome research. *Psychological Medicine* 21, 447–54.

Stekel, W. (1953). *Sadism and masochism*, Vols 1 and 2. Liveright, London.

Stenager, E. N., Stenager, E., Koch-Henriksen, N., *et al.* (1992). Suicide and multiple sclerosis: an epidemiological investigation. *Journal of Neurology, Neurosurgery and Psychiatry* 55, 542–5.

Stengel, E. (1941). On the aetiology of fugue states. *Journal of Mental Science* 87, 572–99.

Stengel, E. (1952). Enquiries into attempted sucide. *Proceedings of the Royal Society of Medicine* 45, 613–20.

Stengel, E. (1959). Classification of mental disorders. *Bulletin of the World Health Organization* 21, 601–63.

Stengel, E. and Cook, N. G. (1958). *Attempted suicide: its social significance and effects*. Maudsley Monograph No. 4. Chapman & Hall, London.

Stenstedt, A. (1952). A study of manic depressive psychosis: clinical, social and genetic investigations. *Acta Psychiatrica et Neurologica Scandinavica* 79 (Suppl.), 3–85.

Stephens, J. H. (1978). Long term prognosis and follow up in schizophrenia. *Schizophrenia Bulletin* 4, 25–47.

Stephenson, T. (1995). Abduction of infants from hospital. *British Medical Journal* 310, 754–5.

Steptoe, A. (1984). Psychological aspects of bronchial asthma. In *Contributions to medical psychology*, Vol. 3 (ed. S. Rachman). Pergamon Press, Oxford.

Steptoe, A. and Wardle, J. (1994). *Psychosocial process and health: a reader*. Cambridge University Press.

Stern, R. S., Lipsedge, M. A., and Marks, I. M. (1973). Thought-stopping of neutral and obsessional thoughts: a controlled trial. *Behaviour Research and Therapy* 11, 659–62.

Sternbach, H. (1991). The serotonin syndrome. *American Journal of Psychiatry* 148, 705–13.

Stevens, J. (1987). Brief psychoses: do they contribute to the good prognosis and equal prevalence of schizophrenia in developing countries? *British Journal of Psychiatry* 151, 393–6.

Stewart, A. L., Greenfield, S., Wells, K., *et al.* (1989). Functional status and well-being of patients with chronic conditions. *Journal of the American Medical Association* 262, 907–13.

Stewart, D. E. and Streiner, D. (1994). Alcohol drinking in pregnancy. *General Hospital Psychiatry* 16, 406–12.

Stewart, W. F. R. (1978). Sexual fulfilment for the handicapped. *British Journal of Hospital Medicine* 22, 676–80.

Stolerman, I. (1992). Drugs of abuse: behavioural principles, methods and terms. *Trends in Pharmacological Sciences* 13, 170–6.

Stone, A. R., Frank, J. D., Nash, E. H., and Imber, S. D. (1961). An intensive five year follow up study of treated psychiatric outpatients. *Journal of Nervous and Mental Disease* 133, 410–22.

Stone, M. (1985). Shellshock and the psychologists. In *The anatomy of madness*, Vol. 2, *Institutions and society* (ed. W. F. Bynum, R. Porter, and M. Shepherd), pp. 242–71. Tavistock Publications, London.

Stone, M. H. (1993). Long-term outcome in personality disorders. *British Journal of Psychiatry* 162, 299–313.

Stone, M. H., Hurt, S. W., and Stone, D. K. (1987). The PI-500: long term follow-up of borderline inpatients meeting DSMIII criteria I: global outcome. *Journal of Personality Disorders* 1, 291–8.

Stores, G. (1981). Problems of learning and behaviour in children with epilepsy. In *Epilepsy and psychiatry* (ed. E. H. Reynolds and M. R. Trimble). Churchill Livingstone, Edinburgh.

Stores, G. (1986). Psychological aspects of nonconvulsive status epilepticus and children. *Journal of Child Psychology and Psychiatry* 27, 575–82.

Stores, G. (1992). Annotation: sleep studies in children with a mental handicap. *Journal of Child Psychology and Psychiatry* 33(8), 1303–17.

Stores, G., Zaiwalla, Z., and Bergel, N. (1991). Frontal lobe complex partial seizures in children: a form of epilepsy at particular risk of misdiagnosis. *Developmental Medicine and Child Neurology* 33, 998–1009.

Stoudemire, A. and Fogel, B. S. (1993). *Psychiatric care of the medical patient*. Oxford University Press, New York.

Strachan, J. G. (1981). Conspicuous firesetting in children. *British Journal of Psychiatry* 138, 26–9.

Strang, J. (1993). Drug use and harm reduction: responding to the challenge. In *Psychoactive drugs and harm reduction* (ed. N. Heather, A. Wodak, E. Nadelmann, and P. O'Hare), pp. 3–20. Whurr, London.

Strang, J. and Edwards, G. (1989). Cocaine and crack. *British Medical Journal* 229, 337–8.

Strauss, J. S. and Carpenter, W. T. (1974). The prediction of outcome of schizophrenia. *Archives of General Psychiatry* 31, 37–42.

Strauss, J. S. and Carpenter, W. T. (1977). Prediction of outcome in schizophrenia III. Five-year outcome and its predictors. *Archives of General Psychiatry* 34, 159–63.

Strober, M. (1995). Family-genetic perspectives on anorexia nervosa and bulimia nervosa. In *Eating disorders and obesity: a comprehensive handbook* (ed. K. D. Brownell and C. G. Fairburn). Guilford Press, New York.

Stroebe, M. S. and Stroebe, W. (1993). The mortality of bereavement: a review. In *Handbook of bereavement* (ed. M. S. Stroebe and R. O Hansson), pp. 175–95. Cambridge University Press.

Stroebel, C. F. (1985). Biofeedback and behavioural medicine. In *Comprehensive textbook of psychiatry* (ed. H. I. Kaplan and B. J. Sadock) (4th edn), pp. 1467–73. Williams and Wilkins, Baltimore, MD.

Strömgren, E. (1968). Psychogenic psychoses. In *Themes and variations in European psychiatry* (ed. S. R. Hirsch and M. Shepherd), pp. 97–120. Wright, Bristol (1974).

Strömgren, E. (1985). World-wide issues in psychiatric diagnosis and classification and the Scandinavian point of view. In *Mental disorders, alcohol and drug related problems*. Excerpta Medica, Amsterdam.

Strömgren, E. (1986). The development of the concept of reactive psychoses. *Psychopathology* 20, 62–7.

Strunk, R. C., Mrazek, D. A., Fuhimann, G. S. W., and La Breque, J. F. (1985). Physiologic and psychological characteristics associated with death due to asthma in childhood. *Journal of the American Medical Association* 254, 1193–8.

Strupp, H. H., Hadley, S. W., and Gomes-Schwartz, B. (1977). *Psychotherapy for better or worse*. Aronson, New York.

Stuart, R. B. (1980). *Helping couples change: a social learning approach to marital therapy*. Guilford Press, New York.

Stunkard, A. J. and Wadden, T. A. (1993). *Obesity theory and therapy*.

Stunkard, A., Levine, H., and Fox, S. (1970). The management of obesity: patient self help and medical treatment. *Archives of Internal Medicine* 125, 1067–72.

Stürup, G. K. (1968). *Treating the 'untreatable': chronic criminals at Herstedvester*. Johns Hopkins University Press, Baltimore, MD.

Suddath, R. L., Christison, G. W., Torrey, E. F., *et al.* (1990). Anatomical abnormalities in the brains of monozygotic twins discordant for schizophrenia. *New England Journal of Medicine* 322, 789–94.

Sugarman, P. A. and Crawford, D. (1994). Schizophrenia in the Afro-Caribbean community. *British Journal of Psychiatry* 164, 474–80.

Sullivan, C., Grant, M. Q., and Grant, J. D. (1959). The development of interpersonal maturity: applications to delinquency. *Psychiatry* 20, 373–85.

Sullivan, H. (1953). *The interpersonal theory of psychiatry*. Norton, New York.

Sullivan, M., Kalon, W., Rosso, J., *et al.* (1993). A randomized trial of nortriptyline for severe chronic tinnitus: effects on depression, disability, and tinnitus symptoms. *Archives of Internal Medicine* 153(19), 2251–9.

Sullivan, M. J. L., Reesar, K., Mikail, S., and Fisher, R. (1992). The treatment of depression in chronic low back pain: review and recommendations. *Pain* 50, 5–13.

Sulloway, F. J. (1979). *Freud: biologist of the mind*. Fontana, London.

Sultzer, D. L., Levin, H. S., Mahler, M. E., *et al.* (1993). A comparison of psychiatric symptoms in vascular dementia and Alzheimer's disease. *American Journal of Psychiatry* 150, 1806–12.

Summerskill, W. H. J., Davidson, E. A., Sherlock, S., and Steiner, R. E. (1956). The neuropsychiatric syndrome associated with hepatic cirrhosis and an extensive portal collateral circulation. *Quarterly Journal of Medicine* 25, 245–66.

Surawy, C., Hackman, A., Hawton, K., and Sharpe, M. (1995). Chronic fatigue syndrome: a cognitive approach. *Behaviour Research and Therapy* 33(5), 535–44.

Surman, O. S. (1994). Psychiatric aspects of liver transplantation. *Psychosomatics* 35, 297–307.

Surtees, P. G. and Barkley, C. (1994). Future imperfect: the long-term outcome of depression. *British Journal of Psychiatry* 164, 327–41.

Susser, E., Struening, E. L., and Conover, S. (1989). Psychiatric problems in homeless men. *Archives of General Psychiatry* 46, 845–50.

Susser, M. (1990). Disease, illness, sickness; impairment, disability and handicap. *Psychological Medicine* 20, 471–3.

Sutherland, A. J. and Rodin, G. M. (1990). Factitious disorders in a general hospital setting: Clinical features and a review of the literature. *Psychosomatics* 31, 392–414.

Svartberg, M. and Stiles, T. C. (1991). Comparative effects of short-term psychodynamic psychotherapy: a meta-analysis. *Journal of Consulting and Clinical Psychology* 59, 704–14.

Swaab, D. F. and Hoffman, M. A. (1990). An enlarged suprachiasmatic nucleus in homosexual men. *Brain Research* 537, 141–8.

Swadi, H. (1988). Drugs and substance use among 3333 London adolescents. *British Journal of Addiction* 83, 935–42.

Swadi, H. (1992a). A longitudinal perspective on adolescent substance abuse. *European Child and Adolescent Psychiatry* 1, 156–70.

Swadi, H. (1992b). Relative risk factors in detecting adolescent drug abuse. *Drug and Alcohol Dependence* 29, 253–4.

Swan, W. and Wilson, L. J. (1979). Sexual and marital problems in a psychiatric outpatient population. *British Journal of Psychiatry* 135, 310–15.

Swartz, M. S., Blazer, D. G., George, L. K., and Landerman, R. (1986). Somatization disorder in a community population. *American Journal of Psychiatry* 143, 1403–8.

Swedo, S., Rappoport, J. L., Leonard, H. L., *et al.* (1989). Obsessive compulsive disorder in children and adolescents. *Archives of General Psychiatry* 46, 335–41.

Swedo, S. E., Pietrini, P., Leonard, H. L., *et al.* (1992). Cerebral glucose metabolism in childhood-onset obsessive-compulsive disorder: revisualization during pharmacotherapy. *Archives of General Psychiatry* 49, 690–4.

Symington, N. (1986). *The analytic experience*. Free Association Books.

Symmers, W. St C. (1968). Carcinoma of breast in transsexual individuals after surgical and hormonal interference with primary and secondary sex characteristics. *British Medical Journal* ii, 83–5.

Szasz, T. S. (1960). The myth of mental illness. *American Psychology* 15, 113–18.

Talley, N. J., McNeill, D., Heyden, A., *et al.* (1987). Prognosis of chronic unexplained dyspepsia. A prospective study of potential predictor variables in patients with endoscopically diagnosed nonulcer dyspepsia. *Gastroenterology* 92, 1060–6.

Tamlyn, D., McKenna, P. J., Mortimer, A. M., *et al.* (1992). Memory impairment in schizophrenia: its extent, affiliations, neuropsychological character. *Psychological Medicine* 22, 105–15.

Tan, E., Marks, I. M., and Marset, P. (1971). Bimedial leucotomy in obsessive compulsive neurosis: a controlled serial enquiry. *British Journal of Psychiatry* 118, 155–64.

Tansella, M. (1991). Community based psychiatry: long-term patterns of care in South Verona. *Psychological Medicine* **19** (Suppl.).

Tardiff, K. (1992). The current state of psychiatry in the treatment of violent patients. *Archives of General Psychiatry* **49**, 493–9.

Tardiff, K., Marzule, P. M., Leon, A. C., et al. (1994). Homicide in New York City: cocaine use and firearms. *Journal of the American Medical Association* **272**, 43–6.

Tarnopolsky, A., and Berelowitz, M. (1987). Borderline personality: a review of recent research. *British Journal of Psychiatry* **151**, 724–34.

Tarnopolsky, A., Watkins, G. V., and Hand, D. J. (1980). Aircraft noise and mental health: I. Prevalence of individual symptoms. *Psychological Medicine* **10**, 683–98.

Tarnowski, K. J., Rasnake, L. K., Gavaghan-Jones, H. P., and Smith, L. (1991). Psychosocial sequelae of pediatric burn injuries: a review. *Clinical Psychological Review* **11**, 371–98.

Tarrier, N., Beckett, R., Harwood, S., et al. (1993). A trial of two cognitive–behavioural methods of treating drug resistant psychotic symptoms in schizophrenic patients. (1) Outcome. *British Journal of Psychiatry* **162**, 524–32.

Task Force of the American Psychiatric Association (1989). *Delirium: definition and diagnostic criteria*, pp. 804–915. American Psychiatric Press, Washington, DC.

Tattersall, R. B. (1981). Psychiatric aspects of diabetes — a physician's view. *British Journal of Psychiatry* **139**, 485–93.

Tattersall, R., Gregory, R., Selby, C., et al. (1991). Course of brittle diabetes: 12 year follow up. *British Medical Journal* **302**, 1240–3.

Taylor, E. A. (1984). Diet and behaviour. *Archives of Disease of Childhood* **59**, 97–8.

Taylor, E. (1991a). Developmental neuropsychiatry. *Journal of Child Psychology and Psychiatry* **32**, 3–47.

Taylor, E. (1991b). *Biological risk factors for psychosocial disorders*. Cambridge University Press.

Taylor, E. (1994). Physical treatments. In *Child and adolescent psychiatry* (3rd edn), (ed. M. Rutter, E. Taylor, and L. Hersov). Blackwell Scientific Publications, Oxford.

Taylor, E., Sandberg, S., Thorley, G., and Giles, S. (1991). *The epidemiology of childhood hyperactivity*. Maudsley Monograph No. 3. Oxford University Press.

Taylor, F. H. (1966). The Henderson therapeutic community. In *Psychopathic disorders* (ed. M. Craft). Pergamon Press, Oxford.

Taylor, F. K. (1979). *Psychopathology: its causes and symptoms*. Quartermaine House, Sunbury on Thames.

Taylor, F. K. (1981). On pseudo-hallucinations. *Psychological Medicine* **11**, 265–72.

Taylor, M. (1994). Madness and Maastricht: a review of reactive psychoses from a European perspective. *Journal of the Royal Society of Medicine* **87**, 683–6.

Taylor, M. A. and Abrams, R. (1984). Cognitive impairment in schizophrenia. *American Journal of Psychiatry* **141**, 196–201.

Taylor, P. J. (1985). Motives for offending among violent and psychotic men. *British Journal of Psychiatry* **147**, 491–8.

Taylor, P. J. (1993). *Violence in society*. Royal College of Physicians, London.

Taylor, P. J. and Fleminger, J. J. (1980). ECT for schizophrenia. *Lancet* **i**, 1380–2.

Taylor, P. J. and Gunn, J. (1984a). Violence and psychosis. 1 — Risk of violence among psychotic men. *British Medical Journal* **288**, 1945–9.

Taylor, P. J., Mahandra, B., and Gunn, J. (1983). Erotomania in males. *Psychological Medicine* **13**, 645–50.

Taylor, S. J. L. and Chave, S. (1964). *Mental health and environment*. Longman, London.

Teasdale, J. D. (1983). Changes in cognition during depression: psychopathological implications. *Journal of the Royal Society of Medicine* **76**, 1038–44.

Teasdale, J. D. and Bancroft, J. (1977). Manipulation of thought content as a determinant of mood and corrugator electromyographic activity in depressed patients. *Journal of Abnormal Psychology* **86**, 235–41.

Teasdale, J. D. and Fogarty, S. (1979). Differential aspects of induced mood on the retrieval of pleasant events from episodic memory. *Journal of Abnormal Psychology* **88**, 248–57.

Tellenbach, R. (1975). Typologische untersuchungen zur prämorbiden Persönlichkeit von Psychotikern unter besonderer Berucksichtigung manisch-depressiver. *Confinia Psychiatrica* **18**, 1–15.

Tennant, C. (1985). Female vulnerability to depression. *Psychological Medicine* **15**, 733–7.

Tennant, C. (1988). Parental loss in childhood: its effect in adult life. *Archives of General Psychiatry* **45**, 1045–50.

Tennant, C. and Bebbington, P. (1978). The social causation of depression: a critique of the work of Brown and his colleagues. *Psychological Medicine* **8**, 565–75.

Tennant, C., Bebbington, P., and Hurry, J. (1981a). The short-term outcome of neurotic disorders in the community: the relation of remission to clinical factors and to 'neutralizing' life events. *British Journal of Psychiatry* **139**, 213–20.

Terry, R. D., Katzmann, R., and Bick, K. L. (1994). *Alzheimer disease*. Raven Press, New York.

Thapar, A., Irving, I., Gottesman, I., et al. (1994). The genetics of mental retardation. *British Journal of Psychiatry* **164**, 747–58.

Theorell, T. and Lind, E. (1973). Systolic blood pressure, serum cholesterol, and smoking in relation to sociological factors and myocardial infarctions. *Journal of Psychosomatic Research* **17**, 327–32.

Thigpen, C. H., Thigpen, H., and Cleckley, H. M. (1957). *The three faces of Eve*. McGraw-Hill, New York.

Thomas, A., Chess, S., and Birch, H. G. (1968). *Temperament and behaviour disorders in children*. University Press, New York.

Thomas, A. J. (1981). Acquired deafness and mental health. *British Journal of Medical Psychology* 54, 219–29.

Thomas, C., Madden, F., and Jehu, D. (1987). Psychological effects of stomas — I. Psychosocial morbidity one year after surgery. *Journal of Psychosomatic Research* 31, 311–16.

Thomas, H. (1993). Psychiatric symptoms in cannabis users. *British Journal of Psychiatry* 163, 141–9.

Thompson, C. (1989). *The instruments of psychiatric research*. Wiley, Chichester.

Thompson, C., Franey, C., Arendt, J., and Checkley, S. A. (1988). A comparison of melatonin secretion in depressed patients and normal subjects. *British Journal of Psychiatry* 147, 389–93.

Thompson, J. (1814). *Synopsis: Nosologiae methodicae. Auctore Guilelmo Cullen M. D. to shich is added an appendix and a translation of Cullen's Nosology*. John Anderson, Edinburgh.

Thompson, L. W., Gallagher, D., and Breckenridge, J. S. (1987). Comparative effectiveness of psychotherapies for depressed elders. *Journal of Consulting and Clinical Psychology* 55, 385–90.

Thompson, S. G. and Pocock, S. J. (1991). Can meta-analyses be trusted? *Lancet* 338, 1127–30.

Thorndike, E. L. (1913). *Educational psychology*, Vol. 2. *The psychology of learning*. Teachers College, Columbia University, New York. (Also Kegan Paul, Trench, and Trubner, London (1923).)

Tienari, P. (1968). Schizophrenia in monozygotic male twins. In *The transmission of schizophrenia* (ed. D. Rosenthal and S. S. Kety). Pergamon Press, New York.

Tizard, B. (1962). The personality of epileptics: discussion of the evidence. *Psychological Bulletin* 59, 196–210.

Tizard, J. (1964). *Community services for the mentally handicapped*. Oxford University Press, London.

Tizard, J. and Grad, J. C. (1961). *Mentally handicapped children and their families*. Oxford University Press, London.

Tollison, C. D. and Adams, H. E. (1979). *Sexual disorders: treatment, theory and research*. Gardner Press, New York.

Tomlinson, B. E., Blessed, G., and Roth, M. (1970). Observations on the brains of demented old people. *Journal of the Neurological Sciences* 11, 205–42.

Toone, B. (1985). Sexual disorders in epilepsy. In *Recent advances in epilepsy*, Vol. 2 (ed. T. A. Pedley and B. S. Meldrum). Churchill Livingstone, Edinburgh.

Torgersen, S. (1984). Genetic and nosological aspects of schizotypal and borderline personality disorders: a twin study. *Archives of General Psychiatry* 41, 546–54.

Torrey, F. E. M. (1994). Violent behavior by individuals with serious mental illness. *Hospital Community Psychiatry* 45(7), 653.

Townsend, P. (1962). *The last refuge*. Routledge & Kegan Paul, London.

Tracy, J. I., Gorman, D. M., and Leventhal, E. A. (1992). Reports of physical symptoms in alcohol use: findings from a primary health care sample. *Alcohol and Alcoholism* 27, 481–91.

Treiman, D. M. (1993). Epilepsy and the law. In *A textbook of epilepsy* (ed. J. Laidlaw, A. Richens, and D. Chadwick), pp. 645–60. Churchill Livingstone, Edinburgh.

Treiman, D. M. and Delgado-Escueta, A. V. (1983). Violence and epilepsy: a critical review. In *Recent advances in epilepsy*, Vol. 1 (ed. T. A. Pedley and B. S. Meldrum). Churchill Livingstone, Edinburgh.

Trijsburg, R. W., van Knippenberg, F. C. E., and Rigma, S. C. (1992). Effects of psychological treatment on cancer patients: a critical review. *Psychosomatic Medicine* 54, 489–517.

Trimble, M. R. (1992). The schizophrenia-like psychosis of epilepsy. *Neuropsychiatry, Neuropsychology and Behavioural Neurology* 5, 103–7.

Trower, P. E., Bryant, B., and Argyle, M. (1978). *Social skills and mental health*. Methuen, London.

Truax, C. B. and Carkhuff, R. R. (1967). *Towards effective counselling and psychotherapy*. Aldine, Chicago.

True, W. R., Rice, J., Eisen, S. A., *et al.* (1993). A twin study of genetic and environmental contributions to liability of posttraumatic stress symptoms. *Archives of General Psychiatry* 50, 257.

Trzepacz, P. T. (1994). The neuropathogenesis of delirium: a need to focus our research. *Psychosomatics* 35, 374–91.

Trzepacz, P. T., Klein, I., Roberts, M., *et al.* (1989). Graves' disease: an analysis of thyroid hormone levels and hyperthyroid signs and symptoms. *American Journal of Medicine* 87, 558–61.

Tseng, W.-S., Kan-Ming, M., Hsu, J., *et al.* (1988). A sociocultural study of koro epidemics in Guangdong, China. *American Journal of Psychiatry* 145, 1538–43.

Tseng, W.-S., Asai, M., Kitanishi, K., *et al.* (1992). Diagnostic patterns of social phobia: comparisons in Tokyo and Hawaii. *Journal of Nervous and Mental Disease* 180, 380–5.

Tsoi, W. F. and Wong, K. E. (1991). A fifteen-year follow-up study of Chinese schizophrenic patients. *Acta Psychiatrica Scandinavica* 84, 217–20.

Tsuang, J. W. (1994). Anabolic steroids withdrawal, dependence and abuse. In *DSM-IV sourcebook*, Vol. 1 (ed. T. A. Widiger, A. J. Frances, H. A. Pincus, *et al.*), pp. 135–43. American Psychiatric Association, Washington, DC.

Tsuang, M. T., Woolson, R. F., and Fleming, J. A. (1979). Long-term outcome of major psychosis: I. Schizophrenia and affective disorder compared with psychiatrically symptom free surgical controls. *Archives of General Psychiatry* 36, 1295–301.

Tucker, G. J. (1994). Regional syndromes. In *DSM-IV sourcebook*, Vol. 1 (ed. T. A. Widiger, A. J. Frances, H. A. Pincus, *et al.*), pp. 277–86. American Psychiatric Association, Washington, DC.

Tucker, G. J., Caine, E. D., and Popkin, M. K. (1994). Delirium, dementia and amnestic and other cognitive disorders. In *DSM-IV sourcebook*, Vol. 1 (ed. T. A. Widiger, A. J. Frances, H. A. Pincus, *et al.*), pp. 185–98. American Psychiatric Association, Washington, DC.

Tuinier, S. and Verhoeven, W. M. A. (1994). Pharmacological advances in mental retardation: a need for reconceptualization. *Current Opinion in Psychiatry* 7, 380–6.

Tuke, D. H. (1872). *Illustrations of the influence of the mind upon the body in health and disease*. J. and A. Churchill, London.

Tuke, S. (1813). *A description of the Retreat*. Dawson, London (1964).

Tunbridge, R. J., Murray, P. A., Kinsella, A. N. and Galasko, C. B. (1990). *The cost of long-term disability resulting from road traffic accidents: interim report (Contractors report 212)*. Department of Transport (Transport and road research), HMSO, London.

Turk, J. (1989). Forensic aspects of mental handicap. *British Journal of Psychiatry* 155, 591–4.

Turk, J., Hagerman, R. J., Barnicott, A., and McEvoy, J. (1994). The fragile X syndrome. In *Mental health in mental retardation: recent advances and practices* (ed. N. Bouras), pp. 135–53. Cambridge University Press.

Turner, T. H. (1989). Schizophrenia and mental handicap: an historical review, with implications for further research. *Psychological Medicine* 19, 301–14.

Tyc, L. T. (1992). Psychosocial adaptation of children and adolescents with limb deficiencies: a review. *Clinical Psychology Review* 12, 275–91.

Tyrer, P. and Alexander, J. (1979). Reliability of a schedule for rating personality disorders. *British Journal of Psychiatry* 135, 168–74.

Tyrer, P. and Steinberg, D. (1975). Symptomatic treatment of agoraphobia and social phobias: a follow-up study. *British Journal of Psychiatry* 127, 163–8.

Tyrer, P., Rutherford, D., and Huggett, T. (1981). Benzodiazepine withdrawal symptoms and propanolol. *Lancet* i, 520–2.

Tyrer, S. P. (1986). Learned pain behaviour. *British Medical Journal* 292, 1–2.

Tyrer, S. (1994). Repetitive strain injury. *Journal of Psychosomatic Research* 38(6), 493–8.

Unsworth, C. (1987). *The politics of mental health legislation*. Oxford University Press.

Urbach, J. R. and Culbert, J. P. (1991). Head-injured parents and their children. Psychosocial consequences of a traumatic syndrome. *Psychosomatics* 32, 24–33.

Ursano R. J., Sonnenberg, S. M., and Lazar, S. G. (1991). *Concise guide to psychodynamic psychotherapy*. American Psychiatric Press, Washington, DC.

Urwin, P. and Gibbons, J. L. (1979). Psychiatric diagnosis in self-poisoning patients. *Psychological Medicine* 9, 501–8.

Vaillant, G. E. (1988). What can long-term follow-up teach us about relapse and prevention in addiction? *British Journal of Addiction* 83, 1147–57.

Valdiserri, E. V., Carroll, K. R., and Hartl, A. J. (1986). A study of offenses committed by psychotic inmates in a county jail. *Hospital and Community Psychiatry* 37, 163–6.

Vance, M. L. (1994). Medical progress: hypopituitarism. *New England Journal of Medicine* 330, 1651–62.

Van der Hart, O. and Horst, R. (1989). The dissociation theory of Pierre Janet. *Journal of Traumatic Stress* 2, 397–412.

Van der Plate, C. and Aral, S. O. (1987). Psychosocial aspects of genital herpes virus infection. *Health Psychology* 6, 57–72.

van Gijn, J. (1993). Treating uncontrolled crying after stroke. *Lancet* 342, 816–17.

Van Hemert, A. M., Hengeveld, M. W., Bolk, J. J., *et al.* (1993). Psychiatric disorders in relation to medical illness among patients of a general medical out-patient clinic. *Psychological Medicine* 23, 167–73.

Van Loon, F. H. G. (1927). Amok and latah. *Journal of Abnormal and Social Psychology* 21, 434–44.

Vaughn, C. E. and Leff, J. P. (1976). The influence of family and social factors as the course of psychiatric illness. *British Journal of Psychiatry* 129, 125–37.

Vaukhonen, K. (1968). On the pathogenesis of morbid jealousy. *Acta Psychiatrica Scandinavica* Suppl. 202.

Veith, I. (1965). *Hysteria: the case history of a disease*. University of Chicago Press.

Veith, R. C., Raskind, M. A., Caldwell, J. H., *et al.* (1982). Cardiovascular effects of tricyclic antidepressants in depressed patients with chronic heart disease. *New England Journal of Medicine* 306, 954–9.

Versiani, M., Nardi, A. E., and Mundim, F. D. (1992). Pharmacology of social phobia: a controlled study of moclobemide and phenelzine. *British Journal of Psychiatry* 161, 353–60.

Victor, M. (1964). Observations on the amnesic syndrome in man and its anatomical basis. In *Brain function: RNA and brain function, memory and learning*, Vol. 2 (ed. M. A. B. Brazier). University of California Press, Berkeley, CA.

Victor, M. and Adams, R. D. (1953). The effect of alcohol on the nervous system. *Proceedings of the Association for Research in Nervous and Mental Diseases* 32, 526–73.

Victor, M., Adams, R. D., and Collins, G. H. (1971). *The Wernicke-Korsakoff syndrome*. Blackwell, Oxford.

Vieweg, W. W. R., David, J. J., Rowe, W. T., *et al.* (1974). Alcohol as a factor precipitating aggression and conflict behaviour leading to homicide. *British Journal of Addiction* 69, 149–54.

Vinogradov, S. and Yalom, I. D. (1989). *Concise guide to group psychotherapy*. American Psychiatric Press, Washington, DC.

Virag, R. (1982). Intracavernosus injection of papaverine for erectile failure. *Lancet* ii, 938.

Vitousek, K. and Manke, F. (1994). Personality variables and disorders in anorexia nervosa and bulimia nervosa. *Journal of Abnormal Psychology* 101, 137–47.

Volkan, V. D. (1994). Identification with a therapist's functions in ego-building in the treatment of schizophrenia. *British Journal of Psychiatry* 164 (Suppl. 23), 77–82.

Von Ameringen, M., Mancini, C., and Streiner, D. L. (1993). Fluoxetine efficacy in social phobia. *Journal of Clinical Psychiatry* 54, 27–32.

Von Economo, C. (1929). *Encephalitis lethargica: its sequelae and treatment* (trans. K. O. Newman). Oxford University Press (1931).

Von Hartitzsch, B., Hoenich, N. A., Leigh, R. J., *et al.* (1972). Permanent neurological sequelae despite haemodialysis for lithium intoxication. *British Medical Journal* iv, 757–9.

Von Knorring, A. V. (1991). Children of alcoholics. *Journal of Child Psychology and Psychiatry* 132, 411–21.

Von Korff, M. (1992). Case definitions in primary care: the need for clinical epidemiology. *General Hospital Psychiatry* 14, 293–5.

Von Korff, M., Barlow, W., Charkin, D., and Deyo, R. A. (1994). Effects of practice style in managing back pain. *Annals of Internal Medicine* 121, 187–95.

Von Korff, M., Dworkin, S. F., Le Resche, L. L., and Kruger, A. (1988). An epidemiologic comparison of pain complaints. *Pain* 32, 173–83.

Von Korff, R., Eaton, W. W., and Keyl, P. M. (1985). The epidemiology of panic attacks and panic disorder: results in three community surveys. *American Journal of Epidemiology* 122, 970–81.

von Zerssen, D. (1976). Physique and personality. In *Human behaviour genetics* (ed. A. R. Kaplan), pp. 230–78. Thomas, Springfield, IL.

Wade, D. T. (1992). Stroke: rehabilitation and long-term care. *Lancet* 339, 791–5.

Wade, S. L., Monroe, S. M., and Michelson, L. K. (1993). Chronic life stress and treatment outcome in agoraphobia with panic attacks. *American Journal of Psychiatry* 150, 1491–5.

Wade, S. T., Legh-Smith, J., and Hewer, R. L. (1986). Effects of living with and looking after survivors of a stroke. *British Medical Journal* 293, 418–20.

Wålinder, J. (1967). *Transsexualism: a study of 43 cases*. Akademi-Förlaget Göteborg.

Walker, E. A., Katon, W., Jemelka, R. P., and Roy-Byrne, P. P. (1992). Comorbidity of gastrointestinal complaints, depression, and anxiety in the epidemiologic catchment area (ECA) study. *American Journal of Medicine* 92, 26S–9S.

Walker, L. G. *et al.* (1994). How distressing is attendance for routine breast cancer screening? *Psycho-Oncology* 3, 299–304.

Walker, N. (1987). *Crime and criminology*. Oxford University Press.

Walker, V. and Beech, H. R. (1969). Mood state and the ritualistic behaviour of obsessional patients. *British Journal of Psychiatry* 115, 1261–3.

Wall, P. D. and Melzach, R. (1994). *Textbook of pain* (3rd edn). Churchill Livingstone, Edinburgh.

Wall, T. L., Thomasson, H. R., Schuckit, M. A., and Ehlers, C. L. (1992). Subjective feelings of alcohol intoxication in Asians with genetic variations of ALDH2 alleles. *Alcohol Clinical and Experimental Research* 16, 991–5.

Wallace, C., Nelson, C., and Liberman, R. (1980). A review and critique of social skills and training with schizophrenic patients. *Social Bulletin* 6 42–63.

Wallace, P. and Jarman, B. (1994). Alcohol: strengthening the primary care response. *British Medical Bulletin* 50, 211–20.

Wallach, J. (1994). Laboratory diagnosis of factitious disorders. *Archives of Internal Medicine* 154, 1690–6.

Waller, P., Wood, S., Breckenridge, A., and Rawlings, N. (1991). Eosinophilia-myalgia syndrome associated with prescribed L-tryptophan in the United Kingdom. *Health Trends* 53–55.

Wallerstein, J. A. (1991). The long-term effects of divorce on children: a review. *Journal of the American Academy of Child and Adolelscent Psychiatry* 30, 349–60.

Walsh, B. W. and Rosen, P. R. (1985). Self mutilation and contagion: an empirical test. *American Journal of Psychiatry* 142, 119–20.

Walton, D. (1961). Experimental psychology and the treatment of the ticquer. *Journal of Child Psychology* 2, 148–55.

Walton, J. N. (ed.) (1985). *Brain's Diseases of the nervous system* (9th edn). Oxford University Press, Oxford.

Wang, Z. W., Crowe, R. R., and Noyes, R. (1992). Adrenergic receptor genes as candidate genes for panic disorder: a linkage study. *American Journal of Psychiatry* 149, 470–4.

Ward, C. H., Beck, A. T., Mendelson, M., *et al.* (1962). The psychiatric nomenclature. *Archives of General Psychiatry* 7, 198–205.

Warr, P. and Jackson, P. (1985). Factors influencing the psychological impact of prolonged unemployment and of re-employment. *Psychological Medicine* 15, 795–808.

Warren, M. Q. (1973). Correctional treatment in community settings. *Proceedings of the International Congress of Criminology, Madrid*.

Warrington, E. K. and Weiskrantz, L. (1970). Amnesic syndrome—consolidation or retrieval? *Nature, London* 228, 628–30.

Warwick, H. M. C., Clark, D. M., and Cobb, A. (1995). A controlled trial of cognitive behaviour therapy for hypochondriasis. Unpublished data.

Watson, J. B. and Rayner, R. (1920). Conditioned emotional reactions. *Journal of Experimental Psychology* 3, 1–14.

Watson, M. (1991). *Cancer patient care: psychosocial treatment methods*. British Psychological Society, Leicester, and Cambridge University Press.

Watzlawick, P., Bearn, J. H., and Jackson, D. D. (1968). *Pragmatics of human communication*. Faber, London.

Weatherall, D. J. (1991). *The new genetics and clinical practice*. Oxford University Press.

Weatherall, D., Ledingham, J. G. G., and Warrell, D. A. (1995). *Oxford textbook of medicine*. Oxford University Press.

Weaver, T. L. and Clum, G. A. (1995). Psychological distress associated with interpersonal violence: a meta-analysis. *Clinical Psychology Reviews* 15(2), 115–20.

Webster-Stratton, C. (1991). Annotation strategies for helping families with conduct disordered children. *Journal of Child Psychology and Psychiatry* 32, 1047–61.

Wechsler, D. (1945). A standardized memory scale for clinical use. *Journal of Psychology* 19, 87–95.

Weeks, D., Freeman, C. P. L., and Kendell, R. E. (1980). ECT:III Enduring cognitive deficits. *British Journal of Psychiatry* 137, 26–37.

Wehr, T. A., Jacobsen, F. M., Sack, D. A., *et al.* (1986). Phototherapy of seasonal affective disorder: time of day and suppression of melatonin are not critical for antidepressant effects. *Archives of General Psychiatry* 43, 870–5.

Wehr, T. A., Sack, D. A., Rosenthal, N. E., and Cowdry, R. W. (1988). Rapid cycling affecting disorder: contributing factors and treatment responses in 51 patients. *American Journal of Psychiatry* 145, 179–84.

Weilburg, J. B., Rosenbaum, J. F., Biederman, J., *et al.* (1989). Fluoxetine added to non-MAOI antidepressants converts non-responders to responders: a preliminary report. *Journal of Clinical Psychiatry* 50, 447–9.

Weilburg, J. B., Rosenblaum, J. F., Meltzer-Brody, S., and Shushtari, J. (1991). Tricyclic augmentation of fluoxetine. *Annals of Clinical Psychiatry* 3, 209–13.

Weinberger, D. R. (1987). Implications of normal brain development for the pathogenesis of schizophrenia. *Archives of General Psychiatry* 44, 660–9.

Weinberger, D. R. and Kleinman, J. E. (1986). Observations on the brain in schizophrenia. *Psychiatry update: the American Psychiatric Association Annual Review*, Vol. 5 (ed. A. J. Frances and R. E. Hales). American Psychiatric Press, Washington, DC.

Weinberger, D. R., Berman, K. F., Suddath, R. L., and Torrey, E. (1992). Evidence of dysfunction of a prefrontal-limbic network in schizophrenia: a magnetic resonance imaging and regional cerebral blood flow study of discordant monozygotic twins. *American Journal of Psychiatry* 149, 890–7.

Weiner, H. (1977). *Psychobiology and human disease*. Elsevier, New York.

Weiner, M. F., Edland, S. D., and Luszczynska, I. I. (1994). Prevalence and incidence of major depression in Alzheimer's disease. *American Journal of Psychiatry* 151(7), 1006–9.

Weinstein, E. A. and Kahn, R. L. (1955). *Denial of illness: symbolic and physiological aspects*. Thomas, Springfield, IL.

Weissman, A. D. (1974). The epidemiology of suicide attempts, 1960–1971. *Archives of General Psychiatry* 30, 737–46.

Weissman, M. M. (1993). The epidemiology of personality disorders: an update. *Journal of Personality Disorders* Suppl., 44–62.

Weissman, M. M. and Klerman, G. L. (1978). Epidemiology of mental disorder: emerging trends in the US. *Archives of General Psychiatry* 35, 705–12.

Weissman, M. M. and Merikangas, K. R. (1986). The epidemiology of anxiety and panic disorders. *Journal of Clinical Psychiatry* 47 (Suppl.), 11–17.

Weissman, M., Pottenger, M., Kleber, H., *et al.* (1977). Symptom patterns in primary and secondary depression: a comparison of primary depressives, with depressed opiate addicts, alcoholics and schizophrenics. *Archives of General Psychiatry* 34, 854–62.

Weissman, M., Prusoff, B. A., DiMascio, A., *et al.* (1979). The efficacy of drugs and psychotherapy in the treatment of acute depressive episodes. *American Journal of Psychiatry* 136, 555–8.

Weissman, M. M., Merikangas, K. R., John, K., *et al.* (1986). Family-genetic studies of psychiatric disorders. *Archives of General Psychiatry* 43, 1104–16.

Weissman, M. M., Bland, R. C., Canino, G. J., *et al.* (1994). The gross national epidemiology of obsessive compulsive disorder. *Journal of Clinical Psychiatry* 55, 5–11.

Wells, K. B., Golding, J. M., and Burnam, M. A. (1988). Psychiatric disorder in a sample of the general population with and without chronic medical conditions. *American Journal of Psychiatry* 145, 976–81.

Wells, K. B., Stewart, A., Hays, R. D., *et al.* (1989). The functioning and well-being of depressed patients. *Journal of the American Medical Association* 262(7), 914–19.

Welner, J. and Strömgren, E. (1958). Clinical and genetic studies on benign schizophreniform psychoses based on a follow up. *Acta Psychiatrica Neurologica Scandinavica* 33, 377–99.

Wender, P., Rosenthal, D., Kety, S. S., *et al.* (1974). Cross-fostering: a research strategy for clarifying the role of genetic and experimental factors in the aetiology of schizophrenia. *Archives of General Psychiatry* 30, 121–8.

Wender, P. H., Kety, S. S., Rosenthal, D., *et al.* (1986). Psychiatric disorder in the biological and adoptive families of adopted individuals with affective disorders. *Archives of General Psychiatry* 43, 923–9.

Werner, E. E. and Smith, R. S. (1982). *Vulnerable but invincible: a study of resilient children.* McGraw-Hill, New York.

Wernicke, C. (1881). *Lehrbuch der Gehirnkrankheiten*, Part 2, p. 229. Kassel, Berlin.

Wernicke, C. (1900). *Grundriss der Psychiatrie.* Thieme, Leipzig.

Werry, J. S. (1992). Child and adolescent (early onset) schizophrenia: a review in the light of DSMIIIR. *Journal of Autism and Development Disorders* 22, 601–24.

Wertheimer, A. (1992). *A special scar. The experiences of people bereaved by suicide* (2nd edn). Routledge, London.

Wessely, S., Nickson, J., and Cox, B. (1990). Symptoms of low blood pressure: a population study. *British Medical Journal* 301, 362–5.

Wessely, S. (1991). History of postviral fatigue syndrome. *British Medical Bulletin* 47(9), 919–41.

Wessely, S. C. and Lewis, G. H. (1989). The classification of psychiatric morbidity in attenders at a dermatology clinic. *British Journal of Psychiatry* 155, 686–91.

Wessely, S. and Taylor, P. (1991). Madness and crime: criminology versus psychiatry. *Criminal Behaviour and Mental Health* 1, 193–228.

Wessely, S., Castle, D., Der, G., and Murray, R. (1991). Schizophrenia in Afro-Caribbeans. A case controlled study. *British Journal of Psychiatry* 159, 795–801.

West, D. (1965). *Murder followed by suicide.* Heinemann, London

West, D. (1974). Criminology, deviant behaviour and mental disorder. *Psychological Medicine* 4, 1–3.

West, D. J. (1988). Psychological contributions to criminology. *British Journal of Criminology* 28(2), 77–92.

West, D. and Farrington, D. P. (1973). *Who becomes delinquent?* Heinemann Educational, London.

West, D. and Farrington, D. P. (1977). *The delinquent way of life.* Heinemann, London.

West, E. D. (1981). Electric convulsion therapy in depression: a double blind controlled trial. *British Medical Journal* 282, 355–7.

West, D. J. (1988). Psychological contributions to criminology. *British Journal of Criminology* 28(2), 77–92.

West, M. A. (1990). *The psychology of meditation.* Clarendon Press, Oxford.

Westphal, C. (1872). Die agoraphobie, eine neuropatische Erscheinung. *Archiv für Psychiatrie und Nervenkrankheiten* 3, 209–37.

Wexler, L., Weissman, N. M., and Kasl, S. V. (1978). Suicide attempts 1970–75. Updating a United States study and comparisons with international trends. *British Journal of Psychiatry* 132, 180–5.

Weyerer, S. (1990). Relationships between physical and psychological disorders. In *Psychological disorders in general medical settings* (ed. N. Sartorius, D. Goldberg, G. de Girolamo, *et al.*), pp. 34–46. Hogrefe & Huber, Toronto.

Whalley, L. J., Borthwick, N., Copolov, D., *et al.* (1986). Corticosteroid receptors and depression. *British Medical Journal* 292, 859–61.

Whalley, L. J., Rosie, R., Dick, H., *et al.* (1982). Immediate increases in plasma prolactin and neurophysin but not other hormones after electroconvulsive therapy. *Lancet* ii, 1064–8.

Wheeler, E. O., White, P. D., Reed, E. W., and Cohen, M. E. (1950). Neurocirculatory asthenia (anxiety neurosis, effort syndrome, neurasthenia). A twenty year follow up of one hundred and seventy three patients. *Journal of the American Medical Association* 142, 878–89.

Wheeler, K., Leiper, A. D., Jannoun, L., and Chessells, J. M. (1988). Medical cost of curing childhood acute lymphoblastic leukaemia. *British Medical Journal* 296, 162–6.

Whiffen, V. E. (1992). Is postpartum depression a distinct diagnoses? *Clinical Psychology Review* 12, 485–508.

Whitaker, A., Johnson, J., Shaffer, D., Rapoport, J. L., Kalikow, K., Walsh, B. T., *et al.* (1990). Uncommon troubles in young people: prevalence estimates of selected psychiatric disorders in a nonreferred adolescent population. *Archives of General Psychiatry* 47, 487–96.

Whitaker, L. C. and Slimak, R. E. (1990). *College student suicide.* Haworth Press, New York.

White, K. P. and Nielson, W. R. (1995). Cognitive behavioural treatment of fibromyalgia syndrome: a following assessment. *Journal of Rheumatology* 22(4), 717–21.

White, P. D., Thomas, J. M., Amess, J. *et al.* (1995). The existence of a fatigue syndrome after glandular fever. *Psychological Medicine* 25, 907–16.

Whitehead, W. E. (1992). Behavioural medicine approaches

to gastrointestinal disorders. *Journal of Consulting and Clinical Psychology* 60(4), 605–12.

Whitehorn, J. C. and Betz, B. J. (1954). A study of psychotherapeutic relationship between physicians and schizophrenic patients. *American Journal of Psychiatry* 111, 321–31.

Whiteley, S. (1975). The psychopath and his treatment. In *Contemporary psychiatry* (ed. T. Silverstone and R. B. Barraclough). *British Journal of Psychiatry* Special Publication No. 9.

Whitlock, F. (1961). The Ganser syndrome. *British Journal of Psychiatry* 113, 19–29.

Whitwam, J. G. (1991). Flumazenil. *Prescribers' Journal* 31, 57–61.

Wieck, A., Kumar, R., Hirst, A. D., *et al.* (1991). Increased sensitivity of dopamine receptors and recurrence of affective psychosis after childbirth. *British Medical Journal* 303, 613–16.

Wiggins, S., Whyte, P., Huggins, M., *et al.* (1992). The psychological consequences of predictive testing for Huntington's disease. *New England Journal of Medicine* 327, 1401–5.

Wilcock, H. K. and Esiri, M. M. (1982). Plaque, tangles and dementia: a quantitative study. *Journal of Neurological Science* 56, 343–56.

Wilkinson, G., Allen, P., Marshall, E., *et al.* (1993). The role of the practice nurse in the management of depression in general practice: treatment adherence to antidepressant medication. *Psychological Medicine* 23, 229–37.

Williams, D. (1969). Neural factors related to habitual aggression: consideration of the differences between those habitual aggressive and others who have committed crimes of violence. *Brain* 92, 503–20.

Williams, J. B. W. (1985). The multiaxial system of DSMIII: where did it come from and where should it go? *Archives of General Psychiatry* 42, 175–86.

Williams, J. M. G., Watts, F. N., McLeod, C., and Matthews, A. (1988). *Cognitive psychology and emotional disorders*. Wiley, Chichester.

Williams, P. and Balestrieri, M. (1989). Psychiatric clinics in general practice: do they reduce admissions? *British Journal of Psychiatry* 154, 67–71.

Williamson, D. F. (1993). Descriptive epidemiology of body weight and weight change in US adults. *Annals of Internal Medicine* 119, 646–9.

Williamson, J., Stokoe, I. H., Gray, S., *et al.* (1964). Old people at home: their unreported needs. *Lancet* i, 1117–20.

Williamson, P. D. and Spencer, S. S. (1986). Clinical and EEG features of complex partial seizures of extra temporal origin. *Epilepsia* Suppl. 2, 546–63.

Willich, S. N., Maclure, M., Mittleman, M., Arntz, H. R., and Muller, J. E. (1994). Sudden cardiac death. Support for a role of triggering in causation. *Circulation* 87, 1442–9.

Willner, P., Muscat, R., and Papp, M. (1992). Chronic mild stress-induced anhedonia: a realistic animal model of depression. *Neuroscience and Biobehaviour Review* 16, 525–34.

Wilson, B. (1989). Injury to the central nervous system. In *The practice of behavioural medicine* (ed. S. Pearce and J. Wardle), pp. 51–81. Oxford University Press.

Wilson, G. (1994). Behavioural treatment of childhood obesity: theoretical and practical implications. *Health Psychology* 13(5), 371–2.

Wilson, G. D. (1981). *Love and instinct*. Temple Smith, London.

Wilson, G. T. (1993). Behavioral treatment of obesity: thirty years and counting. *Advances in Behaviour Research and Therapy* 16, 31–75.

Wilson, G. T. (1994). Behavioral treatment of childhood obesity: theoretical and practical implications. *Health Psychology* 13, 371–2.

Wilson, J. A., Deary, I. J., and Maran, A. G. D. (1988). Is globus hystericus? *British Journal of Psychiatry* 153, 335–9.

Wilson, J. Q. and Herrnstein, •. (1985). *Crime and human nature*. Simon and Schuster, New York.

Wilson, M. (1993). DSM-III and the transformation of American psychiatry: a history. *American Journal of Psychiatry* 150, 399–410.

Wilson, P., Watson, R., and Ralston, G. E. (1994). Methadone maintenance in general practice: patients, workload and outcome. *British Medical Journal* 309, 641–4.

Wilson, S. K., Wahman, G. E., and Lange, J. L. (1988). Eleven years' experience with the inflatable penile prosthesis. *Journal of Urology* 139, 951–2.

Wilson, S. R., Scamagas, P., and German, D. F. (1993). A controlled trial of two forms of self-management education for adults with asthma. *American Journal of Medicine* 94, 564–76.

Wimmer, A. (1916). Psykogene sindssygdomsformer. (Psychogenic varieties of mental diseases.) In *St Hans Hospital 1816–1916. Jubilee Publication*, pp. 85–216. Gad, Copenhagen.

Wing, J. K. (ed.) (1982) Long term community care: experience in a London borough. *Psychological Medicine Monograph* Suppl. 2.

Wing, J. K. (1986). The cycle of planning and evaluation. In *The provision of mental health services in Britain: the way ahead* (ed. G. Wilkinson and H. Freeman). Gaskell, London.

Wing, J. K. (1994a). Mental illness. In *Health care needs assessment* (ed. A. Stevens and J. Raferty). Radcliffe Medical Press, Abingdon, Oxon.

Wing, L. (1994b). The autistic continuum. In *Mental health in mental retardation: recent advances and practices* (ed. N. Bouras), pp. 108–25. Cambridge University Press.

Wing, J. K. and Brown, G. W. (1970). *Institutionalism and schizophrenia*. Cambridge University Press, London.

Wing, J. K. and Fryers, T. (1976). *Psychiatric services in Camberwell and Salford*. MRC Social Psychiatry Unit, London.

Wing, J. K. and Furlong, R. (1986). A haven for the severely disabled within the context of a comprehensive psychiatric community service. *British Journal of Psychiatry* 149, 449–57.

Wing, J. K. and Hailey, A. M. (ed.) (1972). *Evaluating a community psychiatric service*. Oxford University Press, London.

Wing, J. K. and Morris, B. (1981). *Handbook of rehabilitation practice*. Oxford University Press.

Wing, J. K., Bennett, D. H., and Denham, J. (1964). *The industrial rehabilitation of long stay psychiatric patients*. Medical Research Council Memorandum No. 42. HMSO, London.

Wing, J. K., Cooper, J. E., and Sartorius, N. (1974). *Measurement and classification of psychiatric symptoms*. Cambridge University Press.

Wing, R. R., Blair, E., Marcus, M., *et al.* (1994). Year-long weight loss treatment for obese patients with type II diabetes: does including an intermittent very-low-calorie diet improve outcome? *American Journal of Medicine* 97, 354.

Winokur, G., Black, D. W., and Nasrallah, A. (1988). Depressions secondary to other psychiatric disorders and medical illnesses. *American Journal of Psychiatry* 145, 233–7.

Winokur, G., Coryell, W., Keller, M., *et al.* (1993). A prospective follow-up of patients with bipolar and primary unipolar affective disorder. *Archives of General Psychiatry* 50, 457–65.

Winterborn, M. H. (1987). Growing up with chronic renal failure. *British Medical Journal* 295, 870.

Wisniewski, H. M., Silverman, J., and Wegiel, J. (1994). Ageing, Alzheimer disease and mental retardation. *Journal of Intellectual Disability Research* 38, 233–9.

Witherington (1989). Vacuum constriction device for the management of erectile impotence. *Journal of Urology* 141, 320–2.

Witkin, H. A., Mednick, S. A., and Schulsinger, F. (1976). Criminality and XYY and XXY man. *Science* 193, 547–8.

Witte, R. A. (1985). The psychosocial impact of a progressive physical handicap and terminal illness (Duchenne muscular dystrophy) on adolescents and the families. *British Journal Medical Psychology* 58, 179–87.

Witzig, J. S. (1968). The group treatment of male exhibitionists. *American Journal of Psychiatry* 125, 179–85.

Woerner, M. G., Pollack, M., and Klein, D. F. (1973). Pregnancy and birth complications in psychiatric patients: a comparison of schizophrenic and personality disorder patients with their siblings. *Acta Psychiatrica Scandinavica* 49, 712–21.

Wolberg, L. R. (1948). *Medical hypnosis*. Grune and Stratton, New York.

Wolberg, L. R. (1977). *The techniques of psychotherapy*. Grune and Stratton, New York.

Wolberg, L. R. (1988). *The technique of psychotherapy* (4th edn). Harcourt Brace Jovanovich, New York.

Wolf, S. and Wolff, H. G. (1947). *Human gastric function*. Oxford University Press, New York.

Wolfe, F. (1990). Fibromyalgia. *Rheumatic Disease Clinics of North America* 16, 681–98.

Wolfensberger, W. (1980). The definition of normalisation — update, problems, disagreements, and misunderstandings. In *Normalisation, social integration and community services* (ed. R. J. Flynn and K. E. Nitsch). University Park Press, Baltimore, MD.

Wolff, C. (1971). *Love between women*. Duckworth, London.

Wolff, H. G. (1962). A concept of disease in man. *Psychosomatic Medicine* 24, 25–30.

Wolff, H. G. and Curran, D. (1935). Nature of delirium and allied states. *Archives of Neurology and Psychiatry* 35, 1175–215.

Wolkind, S. (1994). Legal aspects of child care. In *Child and adolescent psychiatry: modern approaches* (3rd edn) (ed. M. Rutter, E. Taylor, and L. Hersov). Blackwell Scientific Publications, Oxford.

Wolkind, S. N. and Rushton, A. (1994). Residential and foster family care. In *Child and adolescent psychiatry: modern approaches* (3rd edn) (ed. M. Rutter, E. Taylor, and L. Hersov), pp. 252–66. Blackwell Scientific Publications, Oxford.

Wolpe, J. (1958). *Psychotherapy by reciprocal inhibition*. Stanford University Press, Stanford, CA.

Wong, D. F., Wagner, H. N., Tune, L. E., *et al.* (1986). Positron emission tomography reveals elevated D_2 dopamine receptors in drug-naive schizophrenics. *Science* 234, 1558–62.

Wood, A. J., Aronson, J. K., Cowen, P. J., and Grahame-Smith, D. G. (1989). The measurement of transmembrane cation transport *in vivo* in acute manic illness. *British Journal of Psychiatry* 155, 501–4.

Wood, A. J., Smith, C. E., Clarke, E. E., *et al.* (1991). Altered *in vitro* adaptive responses of lymphocyte Na^+, K^+-ATPase in patients with manic depressive psychosis. *Journal of Affective Disorders* 21, 199–206.

Wood, P. (1941). Da Costa's syndrome (or effort syndrome). *British Medical Journal* 1, 767–74, 805–11, 846–51.

Wooff, K. and Goldberg, D. (1988). Further observations on the practice of community care in Salford: differences between community psychiatric nurses and mental health social workers. *British Journal of Psychiatry* 163, 30–7.

Wooley, H., Stein, A., Forrest, G. C., and Baum, J. D. (1989). Imparting the diagnosis of life threatening illness in children. *British Medical Journal* 298, 1623–6.

Wooley, S. C. and Garner, D. M. (1994). Dietary treatments for obesity are ineffective. *British Medical Journal* 309, 655–6.

Woolverton, W. L. and Johnson, K. M. (1992). Neurobiology of cocaine abuse. *Trends in Pharmacological Sciences* 13, 193–200.

Wootton, B. (1959). *Social science and social pathology*, pp. 203–26. George Allen and Unwin, London.

Worden, J. W. (1991). *Grief counselling and grief therapy: a handbook for the mental health practitioner* (2nd edn). Tavistock Routledge, London.

World Health Organization (1973). *Report of the International Pilot Study of Schizophrenia*, Vol. 1. World Health Organization, Geneva.

World Health Organization (1978a). *Mental disorders: glossary and guide to their classification in accordance with the ninth revision of the International Classification of Diseases*. World Health Organization, Geneva.

World Health Organization (1978b). *Alma-Ata 1978: primary health care*. World Health Organization, Geneva.

World Health Organization (1979). *Schizophrenia: an initial follow up*. Wiley, Chichester.

World Health Organization (1980). Changing patterns in mental health care. *Euro Reports and Studies 25*. World Health Organization, Copenhagen.

World Health Organization (1981). *Current state of diagnosis and classification in the mental health field*. World Health Organization, Geneva.

World Health Organization (1984). *Mental health care in developing countries: a critical appraisal of research findings*. Technical Report Series 698. World Health Organization, Geneva.

World Health Organization (1988). *World Health Organization Psychiatric Disability Assessment Schedule*. World Health Organization, Geneva.

World Health Organization (1989). *Composite International Diagnostic Interview (CIDI)*. World Health Organization, Geneva.

World Health Organization (1992a). *Glossary: differential definitions of SCAN items and commentary on the SCAN text*. World Health Organization, Geneva.

World Health Organization (1992b). *The ICD–10 classification of mental and behavioural disorders*. World Health Organization, Geneva.

Wray, J., *et al.* (1994). Cognitive function and behavioural status in paediatric heart and heart–lung transplant recipients: the Harefield experience. *British Medical Journal* 309, 837–41.

Wright, J. D. and Pearl, L. (1990). The knowledge and experience of young people regarding drug abuse, 1969–1989. *British Medical Journal* 300, 99–103.

Wright, R. G., Heiman, J. R., Shupe, J., and Olvera, G. (1989). Defining and measuring stabilization of patients during 4 years of intensive community support. *American Journal of Psychiatry* 146, 1293–8.

Wu, J. C. and Bunney, W. E. (1990). The biological basis of an antidepressant response in sleep deprivation and relapse: review and hypothesis. *American Journal of Psychiatry* 147, 14–21.

Wyant, G. M. and MacDonald, W. B. (1980). The role of atropine in electroconvulsive therapy. *Anaesthesia and Intensive Care* 8, 445–50.

Wynne, L. C. (1981). Current concepts about schizophrenics and family relationships. *Journal of Nervous and Mental Disease* 169, 82–9.

Wynne, L. C., Ryckoff, I., Day, J., and Hirsch, S. (1958). Pseudomutuality in the family relations of schizophrenics. *Psychiatry* 21, 205–20.

Yalom, I. D. (1980). *Existential psychotherapy*. Basic Books, New York.

Yalom, I. D. (1985). *The theory and practice of group psychotherapy* (3rd edn). Basic Books, New York.

Yalom, I., Hovis, P. S., Newell, G., and Rand, K. H. (1967). Preparation of patients for group therapy. *Archives of General Psychiatry* 17, 416–27.

Yalom, I. D. (1980). *Existential psychotherapy*. Basic Books, New York.

Yalom, I., Lieberman, M. A., and Miles, M. B. (1973). *Encounter groups: first facts*. Basic Books, New York.

Yanovski, S. Z. (1993). Binge eating disorder: current knowledge and future directions. *Obesity Research* 1, 306–24.

Yap, P. M. (1951). Mental diseases peculiar to certain cultures: a survey of comparative psychiatry. *Journal of Mental Science* 97, 313–27.

Yap, P. M. (1965). Koro — a culture-bond depersonalization syndrome. *British Journal of Psychiatry* 111, 43–50.

Yates, M., Leake, A., Candy, J. M., *et al.* (1990). 5-HT_2 receptor changes in major depression. *Biological Psychiatry* 27, 489–96.

Young, A. W., Reid, I., Wright, S., and Hellawell, D. J. (1993). Face-processing impairments and the Capgras delusion. *British Journal of Psychiatry* 162, 695–8.

Young, L. D. (1992). Psychological factors in rheumatoid arthritis. *Journal of Consulting and Clinical Psychology* 60(4), 619–27.

Young, R. J. and Clarke, B. F. (1985). Pain relief and diabetic neuropathy: the effectiveness of imipramine and related drugs. *Diabetic Medicine* 2, 363–6.

Young, W., Goy, R., and Phoenix, C. (1964). Hormones and sexual behaviour. *Science* 143, 212–18.

Yule, W. (1967). Predicting reading ages on Neale's analysis of reading ability. *British Journal of Educational Psychology* 37, 252–5.

Yule, W. (1994). Post-traumatic stress disorders. In *Child and adolescent psychiatry: modern approaches* (3rd edn) (ed. M. Rutter, E. Taylor, and L. Hersov). Blackwell Scientific Publications, Oxford.

Yule, W. and Carr, J. (1987). *Behaviour modification for people with mental handicaps* (2nd edn). Croom Helm, London.

Yule, W. and Rutter, M. (1985). Reading and other learning difficulties. In *Child and adolescent psychiatry: modern approaches* (2nd edn). (ed. M. Rutter and L. Hersov). Blackwell, Oxford.

Yule, W. and Rutter, M. (1987). *Language development and disorders: clinics in developmental medicine*. Blackwell Scientific Publications, Oxford.

Yule, W. and Williams, R. (1990). Post-traumatic stress reactions in children. *Journal of Traumatic Stress* 3, 279–95.

Zedner, L. (1994). Victims. In *The Oxford handbook of criminology* (ed. M. Maguire, R. Morgan, and R. Reiner), pp. 1207. Clarendon Press, Oxford.

Zeitlin, H. (1986). *The natural history of psychiatric disorder in children*. Maudsley Monograph No. 29. Oxford University Press.

Zimmerman, M., Lish, J. D., Farber, N. J., *et al.* (1994). Screening for depression in medical patients: is the focus too narrow? *General Hospital Psychiatry* 16, 388–96.

Zisook, S. and Shichter, S. R. (1993). Uncomplicated bereavement. *Journal of Clinical Psychiatry* 54(10), 365–72.

Zitrin, C. M., Klein, D. F., and Woerner, M. G. (1978). Behaviour therapy, supportive psychotherapy, imipramine and phobias. *Archives of General Psychiatry* 35, 307–16.

Zitrin, C. M., Klein, D. F., Woerner, M. G., and Ross, D. C. (1983). Treatment of phobias: I. Comparison of imipramine hydrochloride and placebo. *Archives of General Psychiatry* 40, 125–38.

Zobeck, T. S., Grant, B. F., Stinson, F. S., and Bertolucci, D. (1994). Alcohol involvement in fatal traffic crashes in the United States: 1979–1990. *Addiction* 89, 227–31.

Zoccolillo, M., Pickles, A., Quinton, D., and Rutter, M. (1992). The outcome of childhood conduct disorder: implications for defining adult personality disorder and conduct disorder. *Psychological Medicine* 22, 971–86.

Zuger, B. (1984). Early effeminate behaviour in boys: outcome and significance for homosexuality. *Journal of Nervous and Mental Diseases* 172, 90–7.

Author index

Subject index

Page numbers denoting tables refer to tables with/without accompanying text on the same page.

^vs^ denotes differential diagnosis or comparisons.

Some psychiatrists and psychologists have been entered in the subject index. Additional references to these persons may be found in the author index.

Alphabetical order. This index is in letter-by-letter order, whereby hyphens, en-rules and spaces within index headings are ignored in the alphabetization.

Abbreviations used in subentries:

ECT   Electroconvulsive therapy
GABA   Gamma-aminobutyric acid
MAOIs   Monoamine oxidase inhibitors
MDMA   3,4-Methylenedioxymethamphetamine (Ecstasy)
SSRIs   Selective serotonin re-uptake inhibitors